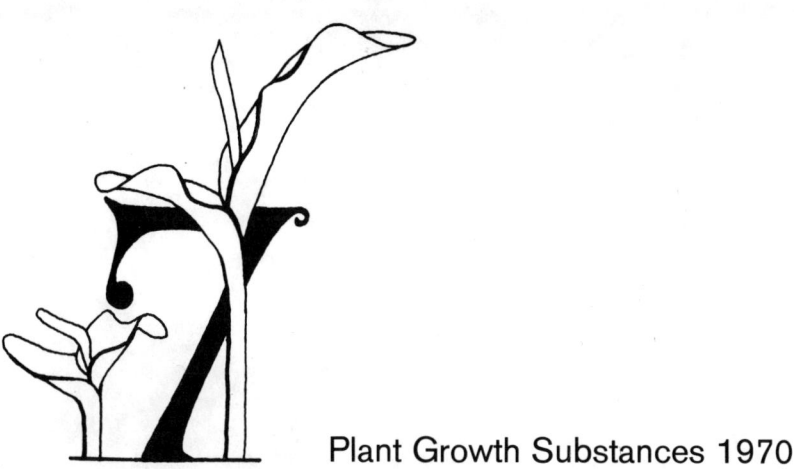

Plant Growth Substances 1970

Plant Growth Substances 1970

Proceedings of the 7th International Conference
on Plant Growth Substances Held in Canberra, Australia
December 7–11, 1970

Edited by Denis J. Carr

With 461 Figures

Springer-Verlag Berlin Heidelberg New York 1972

SEVENTH INTERNATIONAL CONFERENCE ON PLANT GROWTH SUBSTANCES

ADVISORY COMMITTEE

    Dr. D.A. ADAMSON, School of Biol. Sci. Macquarie University, N.S.W.
    Professor D.J. CARR (Organizing Secretary), Res. Schl. Biol. Sci.,
        A.N.U. Canberra.
    Dr. K. GLASZIOU, David North Plant Res. Cen., CSR Co., Brisbane Qld.
    Professor L.G. PALEG, Dept. Plant Physiol., Waite Inst. Adelaide, S.A.
    Professor R.N. ROBERTSON, A.N.U., Canberra.
    Dr. K.S. ROWAN, Botany Dept., Univ. Melbourne, Victoria.
    Professor J.S. TURNER, Botany Dept., Univ. Melbourne, Victoria.

ORGANIZING COMMITTEE

    Dr. E. BACHELARD, Forestry Dept., A.N.U., Canberra.
    Professor D.J. CARR (Chairman)
    Dr. J.V. JACOBSEN, CSIRO Div. Plant Industry, Canberra.
    Dr. R.B. KNOX, Botany Dept. A.N.U., Canberra.
    Dr. D.S. LETHAM, Res. Schl. Biol. Sci., A.N.U., Canberra.
    Dr. D. SPENCER, CSIRO Div. Plant Industry, Canberra.
    Dr. I.F. WARDLAW, CSIRO Div. Plant Industry, Canberra.
    Dr. J.A. ZWAR, CSIRO Div. Plant Industry, Canberra.

Sponsored by the Australian Academy of Science.

ISBN 3-540-05850-8 Springer-Verlag Berlin Heidelberg New York
ISBN 0-387-05850-8 Springer-Verlag New York Heidelberg Berlin

This work is subject to copyright. All rights are reserved, whether the whole or part of the material is concerned, specifically those of translation, reprinting, re-use of illustrations, broadcasting, reproduction by photocoping machine or similar means, and storage in data banks.

Under § 54 of the German Copyright Law, where copies are made for other than private use, a fee is payable to the publisher, the amount of the fee to be determined by agreement with the publisher.

The use of general descriptive names, trade names, trade marks etc. in this publication, even if the former are not especially identified, is not to be taken as a sign that such names, as understood by the Trade Marks and Merchandise Marks Act, may accordingly be used freely by anyone.

© by Springer-Verlag Berlin · Heidelberg 1972. Library of Congress Catalog Card Number 72-80291. Printed in Germany. Offsetprinting: Julius Beltz, Hemsbach/Bergstr. Bookbinding: Konrad Triltsch, Graphischer Betrieb, 87 Würzburg, Germany.

PREFACE

At the 6th International Conference on Plant Growth Substances, held in Carleton University, Ottawa in 1968, it was decided that the 7th should be held in Czechoslovakia, following an invitation by Dr. Kutáček. Historical events intervened and in 1969 another venue was sought. An offer from the Academy of Science in Canberra was accepted by the steering committee. This left rather less time than is desirable to organize an international meeting of this nature and it was with surprise and great relief that the Organizing Committee in Canberra welcomed the arrival of 183 delegates, including a relatively large overseas contingent, to the meeting in December, 1970.

The aim of these Conferences is, of course, to provide a forum for discussion of new work and recent trends, both in the lecture sessions and in conversation. Although many of those who initiated these meetings (e.g. Skoog, Went, Blackman, Bennet-Clark) were absent from the Canberra conference - some have retired - it was good to see present so many of the new generation of research workers in this field. The "generation gap", now so dominant a feature of the world of fashion and of art, is hardly apparent in science. One significant and pleasing fact which emerged from the meetings is the considerable recent progress in understanding of two topics which can be traced right back almost to the beginning of research on plant growth substances - the mechanism of cell extension growth and the biosynthesis of auxins. It was pleasing also to see that those who worked on these topics so long ago could nod assent to much of the recent work which confirms and expands many of the older ideas. While it would be idle to suppose that a conference of this kind could deal with all the many aspects of research on plant growth substances going on all over the world, readers will find in this volume a coverage of some of the more exciting topics currently under investigation. Between the conferences, and especially between the last two, there has been a very large increase in the known number of naturally-occurring plant growth substances, especially in the classes of gibberellins (now approaching 40), inhibitors and cytokinins and it is inconceivable that research on their effects could expand to fit the rate of discovery. Indeed, one of the surprising things at the Conference was that the only report of a totally new effect of growth hormones, a discovery which could have been made 30 years ago, was that of Leonora Reinhold and her colleagues on the role of auxin in thigmotropism. This discovery appeared so anachronistic and its presentation by Dr. Reinhold was so exquisite, that the occasion was reminiscent of the first performance of a newly discovered Mozart symphony!

In their preface to the proceedings of the 6th Conference, Frank Wightman and George Setterfield discussed the growing intensity of interaction between studies of growth hormones and those of molecular biology. It is perhaps too early to see tangible results of this interaction but tactics and initial skirmishes in the struggle to understand the mechanisms of hormone action are recorded in this volume. It is significant and encouraging that so many natural-products chemists have embraced the study of plant hormones and brought with them expertise in automated structural analysis, and separation techniques. The fact that their interest extends, beyond the addition of new active compounds to the already lengthy list, to pathways of biosynthesis and mechanisms of action bodes well for the future. A special evening session of the Conference, which was attended by members of the Australian Society for Spectroscopy and many chemists, was devoted to two papers on applications of gas-liquid chromatography and mass spectroscopy to plant hormone research. It is good to see that in this new area of automated chemical research plant growth substances are well to the fore.

At the general meeting Frank Wightman proposed the formation of an International Plant Growth Substances Association and this having been approved, K.V. Thimann was elected as its first president. Affiliation with the International Association for Plant Physiology has been obtained.

The Canberra conference was held under the auspices of the Australian Academy of Science which provided a substantial part of the subsidy which made it possible, as well as secretarial assistance and venues for the lectures. The assistance of Professors J.S. Turner (Melbourne) and Sir Rutherford N. Robertson (A.N.U.) in making the initial approach to the Academy is gratefully recognized. Australian National University housed the delegates and provided a fitting and pleasant scene for the extra-mural activities as well as various hidden subsidies. In addition to all this support the following companies also made contributions which we gratefully acknowledge as enabling the conference to be held and overseas delegates to be brought to it:-

 Barley Improvement Board - S.A. Dept. of Agriculture

 Boots Pure Drug Co. (Aust.) Pty. Ltd.

 H.V. McKay Charitable Trust

 Shell Chemical (Aust.) Pty. Ltd.

 Colonial Sugar Refining Company

 Arthur Yates & Co. Pty. Ltd.

 The Phosphate Co-Operative Co. of Aust. Ltd.

 Elder Smith Goldsbrough Mort Ltd.

 Union Carbide Australia Ltd.

 Rural Credits Development Fund

 Nicholas Pty. Ltd.

 Commonwealth Development Bank of Australia

 Imperial Chemicals Industries of Australia and New Zealand Ltd.

 Ciba-Geigi Ltd.

As the authors of the manuscripts are well aware, despite a firm contract (not with Springer Verlag), publication was irritatingly delayed by circumstances beyond our control. We have to thank Professor Reinert for his aid in persuading Dr. Konrad Springer to publish the volume. I would like to thank authors for adhering reasonably closely to our fierce instructions and tight deadlines and replying promptly to my editorial queries. For any shortcomings of the volume I am to blame. We are grateful to the editors of Plant Physiology and the Proceedings of the Crop Science Society of Japan for permission to reproduce some diagrams published previously in those journals (acknowledged individually). Thanks are due to Mrs. E. Weil, my secretary, Mr. Peter O'Connor of the Academy of Science, Mrs. Kay Nestler who typed most of the scripts and organized the work of others, and Mr. B. Parr of the photographic unit of R.S.B.S., A.N.U. I would add a special word of thanks to the members of the Organizing Committee for their support especially during the months of despair over publication difficulties and to our eventual publishers.

R.S.B.S., A.N.U.,   D.J. Carr
P.O. Box 475, Canberra City, A.C.T. 2601
Australia

CONTENTS

Section 1

CELL EXTENSION: MODELS, MEMBRANES AND WALL PROPERTIES

Macromolecule synthesis and wall extensibility in relation to the mechanism of auxin-induced cell elongation. *R. Cleland, W.F. Thompson, P.M. Haughton and D.L. Rayle.* — 1

On the biophysical control of growth rate in *Avena* coleoptile sections. *P.B. Green.* — 9

Auxin-induced changes in cell wall properties and growth of *Avena* coleoptiles and green pea epicotyls. *Y. Masuda, R. Yamamoto, and E. Tanimoto.* — 17

Measurements of water potential and hormone transport associated with the growth of cucumber hypocotyls. *D. Cohen and D. Atsmon.* — 23

Promotion of plant and insect hormone action by membrane matrix-matching lipids. *B.B. Stowe.* — 30

Gibberellin and membrane permeability. *A. Wood, L.G. Paleg and R. Sawhney.* — 37

Section 2

RAPID ACTION OF AUXINS

Rapid growth responses in the *Avena* coleoptile: a comparison of the action of hydrogen ions, CO_2 and auxin. *D.L. Rayle and R.E. Cleland.* — 44

Cell elongation and auxin action in lupin hypocotyls. *D. Penny, P. Penny, J. Monro and R.W. Bailey.* — 52

Initial kinetics of auxin-induced cell elongation in coleoptiles. *H. Durand and M.H. Zenk.* — 62

Kinetic studies on the auxin effect and the influence of cycloheximide and blue light. *C.J. Addink and G. Meijer.* — 68

Effects of IAA and cyanide on the growth and respiration of coleoptile sections from *Triticum*. *K.S. Rowan, L.R. Gillbank and A.H. Spring.* — 76

Section 3

BIOSYNTHESIS OF AUXINS

Pathways of auxin biosynthesis in the shoots of higher plants. *Elnora A. Schneider, R.A. Gibson and F. Wightman.* — 82

Enzymic oxidation of indole-3-ethanol. *L.W. Vickery, J.E. Sherwin and W.K. Purves.* — 91

The control of growth by the synthesis of IAA and its conjugation. *R.M. Muir.* — 96

Metabolism of indole-3-acetaldehyde. IV. Electron acceptor studies and physiological significance of the aldehyde oxidase of *Avena* coleoptiles. *R. Rajagopal and P. Larsen.* — 102

Electrophoretic isolation and growth activity of indole-3-acetic acid oxidation products. *W.J. Meudt.* — 110

Conversion of 3-indole acetaldoxime to glucobrassicin and sulphoglucobrassicin by woad (*Isatis tinctoria* L.). *S. Mahadevan and B.B. Stowe.* — 117

Isolation and characterization of indole derivatives in clubroots of Chinese cabbage. *S. Tamura, M. Nomoto, and M. Nagao.* — 127

Section 4

BIOSYNTHESIS OF GIBBERELLINS

Gibberellin biosynthesis and its regulation. *C.A. West and R.R. Fall.* — 133

Recent advances in the metabolism of gibberellins. *G. Sembdner, J. Weiland, G. Schneider, K. Schreiber and I. Focke.* — 143

The biosynthesis of gibberellin precursors in a cell-free system from *Cucurbita Pepo*. L. *J.E. Graebe.* — 151

A polar gibberellin from apricot seed. *B.G. Coombe and M.E. Tate.* — 158

Dwarfing genes in rice and their relation to gibberellin biosynthesis. *Y. Murakami.* — 166

Gibberellins in immature seed of moonflower (*Calonyction aculeatum*). *N. Takahashi, N. Murofushi and T. Yokota.* — 175

Section 5

HORMONES AND NUCLEIC ACIDS

Cytokinin-induced changes in transfer RNA species. *J.H. Cherry and M.B. Anderson.* — 181

Effect of growth substances on rapidly synthesized RNA in sterile tobacco tissue. *D. Klämbt.* — 190

Auxin in sRNA fraction of mung bean hypocotyl. *T. Yamaki and Kō Kobayashi*.   196

ABA- and kinetin-induced changes in cell homogenates, chromatin-bound RNA polymerase and RNA composition. *A.A. Khan*.   207

The temporal separation of transcription and translation and its control in cotton embryogenesis and germination. *J.N. Ihle and L.S. Dure III*.   216

DNA synthesis and hormonal growth response in non-meristematic tissues. *D. Atsmon*.   222

Section 6

HORMONES AND ISOENZYMES

Auxin, macromolecular repressors and the development of isoperoxidases in cultured tobacco pith. *Y. Leshem, A.W. Galston, R. Kaur-Sawhney and L.M. Shih*.   228

Isozymes of cellulase in the abscission zone of *Phaseolus vulgaris*. *L.N. Lewis, F.J. Lew, P.B. Reid and J.E. Barnes*.   234

Section 7

BINDING OF HORMONES TO CELL CONSTITUENTS

Auxin-reactive proteins. *M.A. Venis*.   240

On the significance of cytokinin binding to plant ribosomes. *M.V. Berridge, R.K. Ralph and D.S. Letham*.   248

Binding of indoleacetic acid to isolated pea nuclei. *K.J. Tautvydas and A.W. Galston*.   256

Modification of enzyme activity, conformation and size by indoleacetic acid. *I.V. Sarkissian*.   265

Section 8

INHIBITORS

Biochemical aspects of the action of abscisic acid. *F.T. Addicott*.   272

The biosynthesis and degradation of abscisic acid. *B.V. Milborrow*.   281

The accumulation of abscisic acid in plants during wilting and under other stress conditions. *S.T.C. Wright and R.W.P. Hiron*.   291

Chemistry and biological action of podolactones and other inhibitors of plant growth. *Jenneth M. Sasse, M.N. Galbraith, D.H.S. Horn and D.A. Adamson*.   299

Occurrence of substances in dwarf peas interfering with responses of the same plants to gibberellins. *S. Tamura, S. Ikegami, N. Komoto and M. Noma.* — 306

Plant growth inhibitors in the bulbs of *Lycoris radiata* Herb. *T. Okamoto, Y. Torii and Y. Isogai.* — 311

Morphactin-like activity of benzilate esters in *Arabidopsis thaliana*. *B.T. Brown and L.K. Dalton.* — 318

Chemistry and physiology of rooting inhibitors in adult tissue of *Eucalyptus grandis*. *W. Nicholls, W.D. Crow and D.M. Paton.* — 324

Studies with plant growth inhibitors. *R.L. Wain.* — 330

Section 9

GIBBERELLINS: THE CEREAL ALEURONE LAYER

Control of α-amylase synthesis in isolated barley aleurone layers by gibberellic acid, abscisic acid and ethylene. *J.V. Jacobsen.* — 336

Cytochemical localization of gibberellic acid-induced enzymes in the barley aleurone layer. *J.V. Jacobsen and R.B. Knox.* — 344

Effect of gibberellin A_3 on *in vivo* and *in vitro* induction of α-amylase isozymes. *Y. Momotani and J. Kato.* — 352

Effect of gibberellic acid on ribonucleic acid synthesis in barley aleurone. *J.A. Zwar and J.V. Jacobsen.* — 356

Effect of gibberellic acid on the tRNA methylase activity of barley aleurone cells. *G. Ram Chandra.* — 365

Stages during the induction of α-amylase in barley aleurone layers. I. The timing of sensitivity to Actinomycin D. *P.B. Goodwin and D.J. Carr.* — 371

Stages during the induction of α-amylase in barley aleurone layers. II. Some properties of the pre- and post-transcription stages. *D.J. Carr and P.B. Goodwin.* — 378

The effect of gibberellic acid on the metabolism of soluble nucleotides in aleurone tissue isolated from wheat grain. *G.G. Collins, C.F. Jenner and L.G. Paleg.* — 388

The isolated aleurone layer. *M. Phillips and L.G. Paleg.* — 396

Section 10

GIBBERELLINS: OTHER SYSTEMS

Metabolic changes in internodes of dwarf pea plants treated with gibberellic acid. *A.J. McComb and W.J. Broughton.* — 407

Gibberellin metabolism in the roots of *Phaseolus coccineus* seedlings. *A. Crozier and D.M. Reid.* — 414

Stimulation of the levels of gibberellin-like substances by the growth retardants CCC and AMO 1618. *D.M. Reid and A. Crozier.* 420

DNA analysis of auxin-treated Jerusalem artichoke tuber tissue as a screen for the evaluation of substances influencing cell division. *D. Adamson, R. Hinde and S. Kamisaka.* 428

Promotion by CCC of growth in Jerusalem artichoke tissue. *Heather Adamson and R. Jones.* 435

Gibberellin, a primary determinant in the expression of apical dominance, apical control and geotropic movement of conifer shoots. *R.P. Pharis, Chung-Chi Kuo and J.L. Glenn.* 441

Section 11

CYTOKININS

Active forms of the cytokinins. *J.E. Fox, W.D. Dyson, C. Sood and J. McChesney.* 449

Use of structural analogues in the study of cytokinin action. *Daphne C. Elliott, A.W. Murray, G.T. Saccone and (the late) M.R. Atkinson.* 459

Uptake of cytokinins by *Acer pseudoplatanus* cells: enzymes of the deaminase type as possible regulators of the cytokinin level inside the cell. *C. Terrine, M. Doree, J. Guern and R.H. Hall.* 467

Cytokinins in bleeding sap of the grape vine. *K.G.M. Skene.* 476

Medium and tissue sugar concentrations during cytokinin-controlled growth of tobacco callus tissues. *J.P. Helgeson, C.D. Upper and G.T. Haberlach.* 484

Section 12

ETHYLENE

Studies on the action of ethylene in physiological processes of plant cells. *J.G. Valdovinos, L.C. Ernest and T.E. Jensen.* 493

Functions of naturally-produced ethylene in abscission, dehiscence and seed germination. *Page W. Morgan, D.L. Ketring, E.M. Beyer and J.A. Lipe.* 502

Biosynthesis of ethylene in fruit tissues. *S.F. Yang and A.H. Baur.* 510

The measurement of ethylene from plant tissues and its relation to auxin effect. *R.M. Muir and E.W. Richter.* 518

Auxin and ethylene in adventitious root formation in *Phaseolus aureus* (Roxb.). *M.G. Mullins.* 526

Ethylene and the growth of plant cells: role of peroxidase and hydroxyproline-rich proteins. *Daphne J. Osborne, I. Ridge and J.A. Sargent.* 534

Trauma-induced ethylene production by citrus flowers, fruit and wood.   *W.C. Cooper.* — 543

Thoughts on the role of ethylene in plant growth and development.   *M. Lieberman and A.T. Kunishi.* — 549

Section 13

HORMONES IN RELATION TO SENESCENCE

On the nature of senescence in oat leaves.   *K.V. Thimann, H. Shibaoka and C. Martin.* — 561

Hormonal regulation of leaf senescence in intact plants.   *R.A. Fletcher and N.O. Adedipe.* — 571

Further studies on hormone-regulated senescence in *Rumex* leaf tissue.   *J. Goldthwaite.* — 581

Kinetin treatment and protein synthesis in detached wheat leaves.   *Fong Tung Heng and C.J. Brady.* — 589

Effects of senescence and hormone treatment on the β-1,3-glucan hydrolase in *Nicotiana glutinosa* leaves.   *A.E. Moore and B.A. Stone.* — 598

Ethylene production and biochemical changes in detached leaves of *Nymphoides indica*.   *D.C. Goldney and R.F.M. van Steveninck.* — 604

Increase in ABA-like growth inhibitors and decrease in gibberellin-like substances during ripening and senescence in citrus fruits.   *E.E. Goldschmidt, S.K. Eilati and R. Goren.* — 611

Effects of ABA and kinetin on ultrastructure of senescing wheat leaves.   *Catherine J. Mittelheuser and R.F.M. van Steveninck.* — 618

Section 14

GROWTH AND MORPHOGENESIS

The role of basal and apical factors in the co-ordination of growth in the stems of white clover (*Trifolium repens* L.).   *R.G. Thomas.* — 624

Studies in leaf unrolling in barley.   *D.J. Carr, J.B. Clements and R. Menhenett.* — 633

The effects of growth regulators on RNA metabolism during the unrolling of barley leaf segments.   *R. Poulson and L. Beevers.* — 646

Auxin-induced growth of tuber tissue of Jerusalem artichoke. VII.   Effect of cyclic 3',5'-adenosine monophosphate on the auxin-induced cell expansion growth.   *S. Kamisaka.* — 654

Evidence for the presence of biological activity of a chorionic gonadotropin-like plant growth substance (phytotropin).   *Y. Leshem, A. Shomer-Ilan and R.R. Avtalion.* — 661

Partial and complete growth promoting systems for cultured carrot explants: synergistic and inhibitory interactions. *F.C. Steward and E.F. Bleichert.* 668

Multiple interactions between media, growth factors, and the environment of carrot cultures: effects on growth and morphogenesis. *F.C. Steward and H.W. Israel.* 679

Control of morphogenesis in plant tissue cultures by hormones and nitrogen compounds. *J. Reinert.* 686

Section 15

TRANSPORT AND TROPISMS

Experiments on the mechanism of hormone-directed transport. *J.W. Patrick and P.F. Wareing.* 695

The movement of plant hormones: auxins, gibberellins and cytokinins. *W.P. Jacobs.* 701

Tropic stimuli and the kinetics of basipolar-transport of auxin. *J. Shen-Miller.* 710

The source and transport of growth regulators responsible for the geotropic response of *Zea mays* roots. *M.B. Wilkins, G.S.B. Gibbons and S. Shaw.* 717

Asymmetric "acid"growth" response following gravistimulus. *N. Reinhold and D. Ganot.* 725

The role of auxin in thigmotropism. *N. Reinhold, T. Sachs and L. Vislovska.* 731

Participation of the Golgi apparatus in geotropism. *J. Shen-Miller.* 738

Section 16

HORMONES AND FLOWERING

Hormonal regulation of plant flowering in different photoperiodic groups. *M-Kh. Chailakyan.* 745

The use of aphids in the search for the hormonal factors controlling flowering. *C.F. Cleland.* 753

Hormonal control of flowering in Citrus and some other woody perennials. *E.E. Goldschmidt and S.P. Monselise.* 758

Some growth substances associated with bud failure of peach. *H.D.R. Malcolm.* 767

The flowering process - a new theory. *P. Baxter.* 775

The promotion of ripening in cereal crops by peduncle injection of aqueous solutions of nucleotides. *T. Tomita.* 780

Section 17

APPLICATION OF GLC-MS TO HORMONE STUDIES

A system for the characterisation of plant growth substances based upon the direct coupling of a gas chromatogram, a mass spectrometer and a small computer - recent examples of its application.   *J. MacMillan.*   790

Identification of cytokinins by gas-liquid chromatography and gas liquid chromatography-mass spectrometry.   *C.D. Upper, J.P. Helgeson and C.J. Schmidt.*   798

Index to Authors   808
Index to Organisms   822
Index to Subjects   827

CELL EXTENSION-MODELS, MEMBRANES, AND WALL PROPERTIES

Macromolecule Synthesis and Wall Extensibility in Relation to the Mechanism of Auxin-Induced Cell Elongation

Robert Cleland, William F. Thompson[1], Peter M. Haughton[2] and David L. Rayle[3]
Department of Botany, University of Washington, Seattle, Washington 98105

[1] Present Address: Biology Dept., Harvard University, Cambridge, Mass.
[2] Present Address: Astbury Department of Biophysics, University of Leeds, Leeds, England
[3] Present Address: Botany Department, San Diego State College, San Diego, California

I. INTRODUCTION

Despite extensive investigations, the mechanism of auxin-induced cell elongation remains a mystery. Yet certain facts about the progress are clear. Cell extension requires constant synthesis of RNA and protein (Key, 1969). The extension itself must consist of a continual series of extension events, each of which is composed of a biochemical modification of the wall (wall loosening) followed by a small amount of turgor-driven wall extension (Cleland, 1968).

But many questions remain unanswered. For example, why is RNA synthesis needed for extension? It may be simply to replenish the supply of an unstable but essential constitutive RNA, or it may be that auxin must induce the synthesis of a new species of RNA, the m-RNA for the growth-limiting proteins. It is clear from inhibitor studies (Noodén and Thimann, 1963), that extension is limited by the availability of some growth-limiting proteins (GLP). Are the GLP unstable, or are they used up in extension? Does auxin affect the amount of GLP?

It has been assumed that the wall loosening process can be separated from the actual extension, and can be studied independently (Heyn, 1931; Cleland, 1968). To do this one must be able to measure any potential for wall extension which is built up; i.e. the wall extensibility (WEx). The Instron technique (Olson et al, 1965) has been widely used in recent years as a method to measure WEx. But does this technique actually measure a potential for future wall extension? And can wall loosening actually be separated from wall extension? This research was undertaken in order to provide answers to these questions.

II. MATERIALS AND METHODS

The plant material consisted of 10 or 14 mm sections cut from 25-32 mm coleoptiles of *Avena sativa* L., cv. Victory, starting 3 mm from the top, or 15 mm sections cut from the 3rd internode of 7-day red-light-grown seedlings of *Pisum sativum* L., cv. Alaska. Methods for growing the seedlings are detailed elsewhere (Cleland, 1968; Thompson, 1970).

Extension of coleoptile sections was monitored using the continuous recording technique of Evans and Ray (1969). The sections were always incubated in the chamber for at least 1 hour in buffer in order to reduce the endogenous growth rate, the recording was started, and IAA(5μg/ml) and cycloheximide (CH, 6 μg/ml) were added at the desired times. The following information was then obtained from each record; the induced growth rate, GR_i (the maximum growth rate after auxin minus the endogenous rate); the inhibition time, t_i (the time interval between addition of CH and the midpoint of the growth inhibition); and the amount of induced growth, G_i (the amount of growth induced by auxin after CH addition). See Cleland (1971) for further details on these measurements.

Wall extensibility was measured by means of the Instron technique, using methods already described in detail (Cleland 1967a). The results are expressed as the plastic compliance, DP. The elastic compliance, DE, was also determined, but since in our experience changes in DE are always accompanied by larger changes in DP, we have not reported DE values here.

For hybridization studies, pea stem sections were incubated in K-maleate buffer (2.5 mM, pH 4.7), with or without growth-promoting level of IAA (5×10^{-6}M) and [32p] (50-250 μc/ml). RNA was extracted by the phenol method of Click and Hackett (1966) and was purified by the 2-methoxyethanol-cetyltrimethyl-ammonium bromide (CTAB) technique of Ralph and Bellamy (1964). DNA was prepared from crude chromatin by the method of Marushige and Bonner (1966) followed by purification with CTAB.

Hybridization was carried out by the nitrocellulose membrane filter methods as modified by Church and McCarthy (1968). Conditions used in each hybridization are listed in the figure legends. For further details of these procedures, see Thompson (1970).

III. RESULTS AND DISCUSSION

Effect of auxin on RNA species:

It is now clear that RNA synthesis is needed for auxin-induced growth (Key, 1969), and that auxin can stimulate RNA synthesis, but it is still uncertain whether or not auxin induces the formation of new species of RNA, i.e., acts at the level of gene transcription to induce the m-RNA for the growth-limiting proteins. One of the few techniques capable of answering this question is competitive DNA-RNA hybridization. In this technique reference [32p]-labeled RNA is collected from auxin-treated tissues (+ A), and its ability to hybridize with DNA in the presence of competing unlabeled RNA obtained from auxin-treated or control (-A) tissues is compared. If auxin induces new RNA species, the -A RNA should be a less effective competitor of reference +A RNA than would +A RNA, since it would lack some RNA species which are only found in +A RNA.

This experiment was carried out using DNA and RNA from pea epicotyl sections. It can be seen from Fig. 1 that there is no difference in competitive ability between +A and -A RNA, while both are much better competitors than RNA obtained from *Avena* seedlings or pea leaves. It would appear that auxin has induced no new species of RNA, but it is possible that such new RNA species, even if they existed, might have been missed because they comprise too small a fraction of the reference RNA or turned over too rapidly. To enrich the reference RNA in growth-limiting RNA species, sections were labeled for shorter time periods (down to 2 hours) and 5-fluorouracil was included to reduce labeling of RNA not essential for auxin-induced growth (Key, 1969). Still no differences could be detected by hybridization between +A and -A RNA (Fig. 1).

It must be remembered that only certain species of RNA react in hybridization assays; those produced by a redundant fraction of the DNA (Church and McCarthy, 1968). If auxin causes changes in the RNA complementary to the unique DNA sequences, it will not be detected. Therefore, one cannot unequivocally say that auxin, under these conditions (growth-promoting levels applied to sections) produces no changes in RNA species. But these results, coupled with the fact that auxin can induce cell elongation with a lag of less than one minute (e.g., Rayle et. al, 1970) strongly suggest that the primary mode of action in regard to extension growth is not at the level of gene transcription.

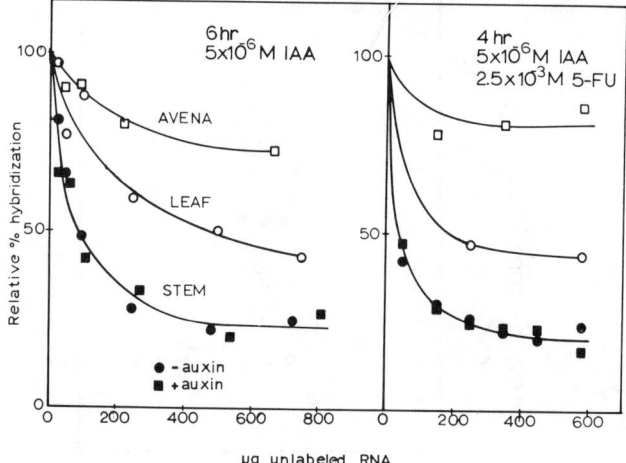

Fig. 1. Ability of various RNA preparations to compete with the hybridization between [^{32}P] -RNA, from auxin-treated pea stem sections and pea DNA. Left: Reference RNA, 5 μg [^{32}P] -RNA, from sections incubated 6 hr + 5x10^{-6} M IAA; competitor RNAs, from pea stem sections incubated 6 hr + 5 x 10^{-6} M IAA, pea leaves and Avena seedlings; DNA, 10 μg; conditions of hybridization, 16 hr at 67°C in 2xSSC. Right: Reference RNA, 10 μg [^{32}P] -RNA from pea stem sections labeled 4 hr in the presence of IAA + 2.5x10^{-3} M 5-FU; competitor RNAs and DNA, same as at left; hybridization conditions, 48 hr at 52°C in 1.33xSSC + 33% formamide.

Auxin and the growth-limiting proteins of the Avena coleoptile:

Inhibition studies have shown that auxin-induced cell elongation depends on the availability of some as yet unidentified growth-limiting proteins (GLP), (Noodén and Thimann, 1963; Key, 1969). When protein synthesis is inhibited, the pool of GLP rapidly disappears and growth ceases. But why do the GLP disappear? Are they unstable, or are they used up in growth process?

To answer this one can make use of the fact that cycloheximide (CH) inhibits protein synthesis in Avena coleoptiles with a lag of less than 5 minutes (Cleland, 1971). Sections are made to grow at various rates (GR_I) by varying the auxin level or by adding mannitol. CH is added, the inhibition time (t_I) and amount of growth subsequent to the CH addition (G_I) are measured and plotted as a function of each other (Fig. 2).

If the GLP are consumed in growth, G_I should be independent of the growth rate, while t_I should vary with GR_I; the points in Fig. 2 should fall along a horizontal line. But if the GLP disappear due to instability, it will be t_I which is independent of the growth rate, and the points should fall along a vertical line. Clearly the GLP are unstable; their average life is only 20-25 minutes.

This makes the GLP one of the least stable proteins yet examined in plants. The reason for the instability is unknown; it could reflect degradation of the GLP, or it could mean that the GLP are rapidly transferred from a functional to a non-functional pool.

Are the GLP constitutive proteins or is their synthesis induced by auxin? This can be tested by adding CH just before or after auxin and determining GR_I, G_I and t_I. If CH is added 5 minutes before the auxin, so that protein synthesis is inhibited at the time auxin is added, there will be no auxin-induced growth if the synthesis of the

GLP is induced by auxin, but a sizable G_I should be obtained if the proteins are constitutive.

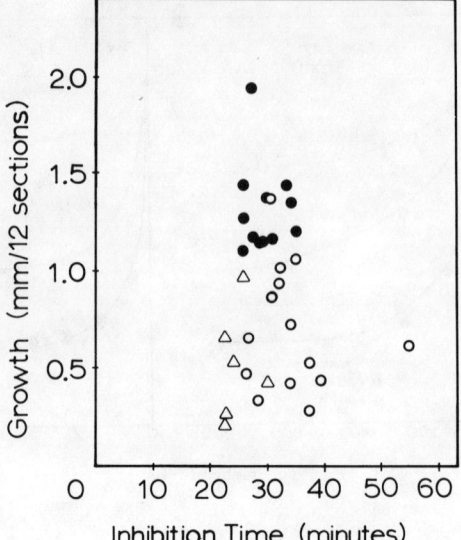

Fig. 2. Demonstration of the instability of the growth limiting proteins of the *Avena* coleoptile. Sections pretreated 1 hr in 5 µg/ml IAA (●), 0.05 µg/ml IAA (△) or 5 µg/ml + 0.03-0.06 M mannitol (○), then 6 µg/ml CH added and the amount of induced growth and inhibition time determined from each growth record.

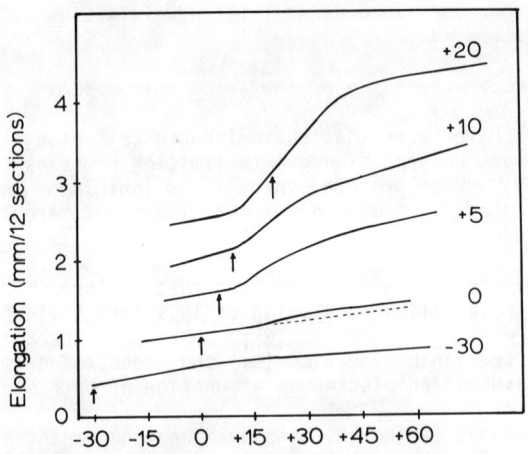

Fig. 3. Typical growth records showing growth response of sections to auxin after treatment just before or after auxin with 6 µg/ml CH. CH added at arrow. Dotted curve, continuation of the control growth rate.

Results from such a study are shown in Figures 3 and 4. Regardless of when CH is added, t_I remains constant, while G_I and GR_I vary. As a result G_I is essentially a measure of the GLP pool size. As expected, no G_I occurs when CH is added 30 minutes before the auxin, since any GLP would have disappeared prior to the auxin treatment. A small but significant G_I is found when CH is added 5 minutes prior to the auxin; apparently a small GLP pool exists in the absence of auxin. The G_I begins to rise when

CH addition is delayed until 2.5 minutes after the auxin, indicating that the increase in the GLP begins 5-7 minutes following addition of auxin and just prior to the start of rapid elongation. Further delay in CH addition results in a progressively larger G_I until a maximum is reached 20 minutes after auxin addition. Clearly, the GLP pool has reached an optimal size about 25 minutes after auxin treatment. It should be noted that G_I and GR_I values are closely related; this suggests that the maximum growth rate is determined by the size of the GLP pool, and that one way in which auxin increases the growth rate is to increase the size of the GLP pool.

The way in which auxin expands the GLP pool remains unknown, but the fact that some GLP are present in non-auxin-treated *Avena* coleoptile and lupin hypocotyl tissues (Penny et al., 1971) suggests that it acts post-translation (Cleland, 1971a).

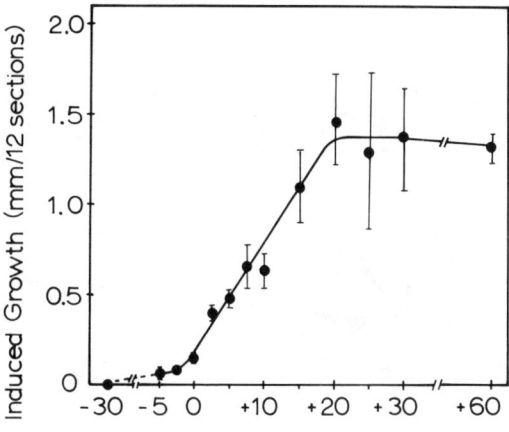

Fig. 4. Relation between time CH added relative to auxin and the amount of induced growth. Sections treated with 5 µg/ml IAA, and CH added at times indicated between 30 minutes before auxin (-30) to 60 minutes after auxin (+60), and amount of induced growth determined. Standard errors shown for all points.

Wall Extensibility and Auxin-action:

Extension analyses have indicated (Cleland, 1971b) that auxin-induced cell elongation involves two processes; a biochemical modification of the cell wall properties (a wall loosening) followed by a turgor-driven physical extension of the wall. Can these two processes be separated and studied individually? It has been generally assumed that the answer is yes; i.e., whenever turgor is reduced, stopping the extension, wall loosenings will continue to accumulate and a potential for future extension will be built up (stored growth). It should be possible to demonstrate this potential, either as an extra extension when turgor is restored, or as a raised plastic extensibility using the Instron technique. Earlier experiments (Cleland and Bonner, 1956) suggested that stored growth could occur, and an auxin-induced increase in extensibility of mannitol-treated sections is well documented (e.g. Cleland, 1967b). However, we recently have had reason to question these results and have now obtained the following data which indicate that wall loosening and wall extensions cannot be separated.

The ability of *Avena* sections to store up a potential for growth during periods of reduced turgor has been examined by adding to sections undergoing rapid elongation enough mannitol to cause growth to cease, and then after 10-100 minutes returning the sections to solutions without mannitol. Stored growth will be indicated if the sections now extend more rapidly than before mannitol treatment. But this does not occur, whether growth is measured for short intervals (data not shown) or long (Fig. 5). Clearly, the sections behave as if the inducing agent was totally inactive during periods of reduced turgor.

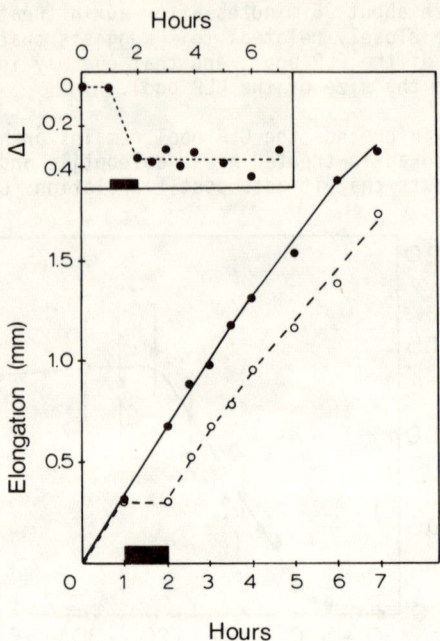

Fig. 5. Demonstration that stored growth does not occur in *Avena* coleoptile sections. Sections incubated in IAA-buffer for 7 hrs (-o-) or in the same solution except for a 1 hr period in which 0.2 mannitol was also present (-o-). Insert: difference in length between control and mannitol-treated sections as a function of time. Solid bars: period of mannitol treatment.

If the auxin-induced increase in plastic extensibility which occurs in mannitol-treated sections were a potential for future extension, when sections are returned to argonated-water (i.e., conditions where extension but not wall loosening can occur), the extensibility (DP) of the auxin-treated sections should fall. On the contrary, it actually increases (Table 1). It is apparent that DP is not a measure of the potential of a section for future extension. Exactly what it does measure will be considered elsewhere.

Table 1. Effect of delayed extension on wall extensibility as measured with the Instron.

Pretreatment	Treatment	Wall Extensibility (DP)
105 min in:	90 min in:	10^{-11} cm^2/dyne
IAA, Mannitol	-	46
IAA, Mannitol	IAA, Mannitol	45
IAA, Mannitol	Water, Argon	63
Mannitol	-	40
Mannitol	Mannitol	38
Mannitol	Water, Argon	47

14 mm coleoptile sections pretreated 105 min in buffered 0.2 M mannitol ± IAA, then treated as indicated prior to killing and extension analysis.

Why can we not separate wall loosening from extension by lowering the turgor?

One possibility is that a reduction in water potential inhibits loosening as well as extension. This seems unlikely, since if an external force is applied to mannitol-treated sections to replace turgor, the sections can respond to auxin and extend (data not shown). Another possibility is that walls extend by chemical creep (Fig. 6). This assumes that the walls are cross-linked by bonds which are continually breaking and re-forming. If the walls are under tension, when a bond breaks a small amount of extension will occur until the broken ends reform with new partners to limit extension. But if the tension is reduced, the bonds are more likely to reform in their original position with the result that the wall's properties are unchanged. The role of auxin would be to facilitate the breakage of these bonds.

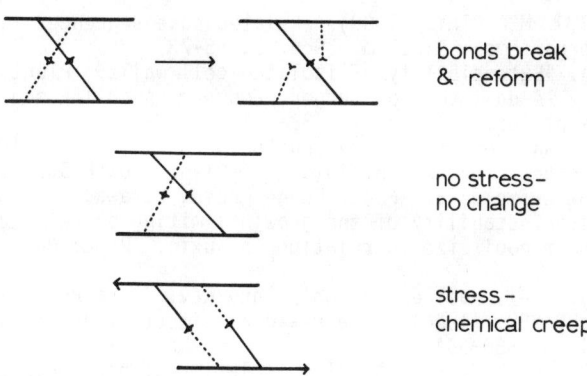

Fig. 6. Mechanisms of chemical creep. Walls are linked by labile crosslinks which break and reform. In the absence of stress, bonds reform in their original positions leaving the wall unchanged. With stress, bond breakage allows extension to occur until bonds reform with new partners, again limiting extension.

Chemical creep is well established as a mechanism of extension in rubbers and other polymers (Tobolsky, 1958). It has been proposed earlier for cell walls (but called chemical stress relaxation) by Ray and Ruesink (1962) but without direct evidence. The present evidence makes it highly likely that it is, in fact, the way in which plant cell walls extend.

IV. SUMMARY

A study has been made of the relation between macromolecule synthesis, wall extensibility and cell elongation, and the following results have been obtained.

The species of RNA from pea stem sections which react in the DNA-RNA hybridization assay are unchanged by auxin, suggesting that the growth-limiting RNA may be a constitutive species.

Auxin does cause a marked increase in the amount of the growth-limiting proteins (GLP) of the *Avena* coleoptile, with the increase starting just before the initiation of rapid elongation, and reaching a maximum about 20 minutes later. These GLP are highly unstable, their average life is only 20-30 minutes.

Wall loosening cannot be separated from wall extension by a reduction in turgor, since this blocks wall loosening as well. The extension which is measured in the Instron technique does not appear to be a potential for growth extension. It appears, now, that wall extension may occur by a form of "chemical creep"; i.e., bonds may break and reform in new configurations, leading to a loosened wall and only if they are under tension.

V. ACKNOWLEDGEMENTS

Supported by Atomic Energy Commission Contract No. AT(45-1)-2217 and a post-doctoral traineeship to DLR from USPHS Developmental Biology Training Grant HD00266.

VI. REFERENCES

CHURCH, R.B., and B.J. MCCARTHY. (1968). Related base sequences in the DNA of simple and complex organisms. Biochem. Gen. 2, 55-73.
CLELAND, R. (1967a). Extensibility of isolated cell walls. Planta 74, 197-204.
CLELAND, R. (1967b). A dual role of turgor pressure in auxin-induced cell elongation in *Avena* coleoptiles. Planta 77, 182-191.
CLELAND, R. (1968). Wall extensibility and the mechanisms of auxin-induced cell elongation, in "Biochemistry and Physiology of Plant Growth Substances" (Eds. F. Wightman and G. Setterfield) 613-624. Runge Press, Ottawa.
CLELAND, R. (1971a). Instability of the growth-limiting proteins of the *Avena* coleoptile and their pool size in relation to auxin. Proc. Nat. Acad. Sci. U.S. (in press).
CLELAND, R. (1971b). Cell wall extension. Ann. Rev. Plant Physiol. 22, (in press).
CLELAND, R. and J. BONNER. (1956). The residual effect of auxin on the cell wall. Plant Physiol. 31, 350-354.
CLICK, R.E. and D.P. HACKETT. (1966). The isolation of ribonucleic acid from plant, bacterial, or animal cells. Biochim. Biophys. Acta 129, 74-84.
EVANS, M.L. and P.M. RAY. (1969). Timing of the auxin response in coleoptiles and its implications regarding auxin action. J. Gen. Physiol. 53, 1-20.
HEYN, A.N.J. (1931). Der Mechanismus der Zellstreckung. Rec. Trav. Bot. Neerl. 23, 113-244.
KEY, J.L. (1969). Hormones and nucleic acid metabolism. Ann. Rev. Plant Physiol. 20, 449-474.
MARUSHIGE,, K. and J. BONNER. (1966). Template properties of liver chromatin. J. Mol. Biol. 15, 160-174.
NOODÉN, L.D. and K.V. THIMANN. (1963). Evidence for a requirement for protein and RNA synthesis for auxin-induced cell enlargement. Proc. Nat. Acad. Sci. U.S. 50, 194-200.
OLSON, A.C., J. BONNER and D.J. MORRÉ. (165). Force extension analysis of *Avena* coleoptile cell walls. Planta 66, 126-134.
PENNY, E.D., J. MONRO and P. PENNY. (1972). Early effects of auxins on macromolecules, in "Plant Growth Substances, 1970" (Ed. D.J. Carr). Springer Verlag (in press).
RALPH, R.K., and A.R. BELLAMY. (1964). Isolation and purification of undergraded ribonuleic acids. Biochim. Biophys. Acta 87, 9-16.
RAY, P.M. and A.W. RUESINK. (1962). Kinetic experiments on the nature of the growth mechanism in oat coleoptile cells. Develop. Biol. 4, 377-397.
RAYLE, D.L., M.L. EVANS and R. HERTEL. (1970). Action of auxin on cell elongation. Proc. Nat. Acad. Sci. U.S. 65, 184-191.
THOMPSON, W.F. (1970). Relationship between RNA species and developmental events in higher plants as studied by DNA-RNA hybridization. Ph. D. Thesis, University of Washington.
TOBOLSKY, A.W. (1958). Stress relaxation studies of the visco-elastic properties of polymers. In "Rheology, Theory and Applications," (F.R. Eirech). Vol. II, 63-81. Academic Press, N.Y.

ON THE BIOPHYSICAL CONTROL OF GROWTH RATE IN *Avena* COLEOPTILE SECTIONS

Paul B. Green

Department of Biological Sciences, Stanford University, Stanford, California, 94305

I. INTRODUCTION

Turgor pressure (P) is regarded as the physical driving force of growth in plants, but it is effective in causing a deformation (elongation) only above a threshold value of turgor which may be regarded as the yielding threshold of the cell wall (Lockhart, 1965). The simplest expression for this relation is:

$$r_s = (P - Y_{min}) m_s \qquad (1)$$

where r_s is steady rate, Y_{min} is the threshold turgor for yielding, and m_s reflects the yielding quality (apparent fluidity) of the cell wall above the threshold. Change in steady rate generally does not reflect a parallel change in P (it sometimes shows an inverse relation to P) so changes in Y_{min} and/or m_s presumably govern rate. The distinction between the extreme possibilities (Y_{min} vs. m_s) is quite clear when r_s rises linearly with ($P - Y_{min}$) as in Fig. 1a, but is less so when the relation is concave upward as in Fig. 1b. Here (Fig. 1b, right) displacement also involves a change in slope at a given P. In terms of equation (1), long-term differences in rate, with and without IAA, appear to reflect changes in m_s (Cleland, 1959) while shorter-term studies indicated a change in Y_{min} (Evans, 1967).

The first question asked in this report is whether the gradual decline in growth rate in coleoptile sections involves primarily a change in m_s or Y_{min}. The second question is whether equation (1) is valid in both the steady-state and the instantaneous sense. That is, will it predict the observed nature of rate transients between steady steady rates? To answer both questions methods were developed which allowed rapid osmotic-elastic equilibration to abrupt shift in osmolarity of the medium, and also permitted accurate measurement of length change in single coleoptile sections.

II. MATERIALS AND METHODS

To accelerate the osmotic-elastic adjustment to turgor shift, a microscope-stage chamber was constructed which bathed the outside of an *Avena* coleoptile at 2 cm/sec and the central cavity at 1 cm/sec. The coleoptile was tied to a modified Leuer-lock syringe tube which screwed into a port in the chamber. Suction drew medium through a single inlet and two outlets (one of which was the coleoptile cavity). Each outlet went through a flow meter so rate could be monitored. A small dichotomous manifold of plastic valves allowed the chamber to be fed by any of eight solutions. The free end of the coleoptile could be observed through a cover-glass which formed the roof of the chamber. A high-power dry objective and an ocular micrometer were used for visual recording (green light illumination) of length change. The osmoticum was NaCl because the low molecular weight of the solute allowed rapid diffusion. To steady the coleoptile against the rapid flow-- and sudden changes in rate during valve switching-- its cavity was partially filled by a long rod. The rod did not prevent flow (but partially accelerated it) nor did it retard growth by friction. Runs were carried out at

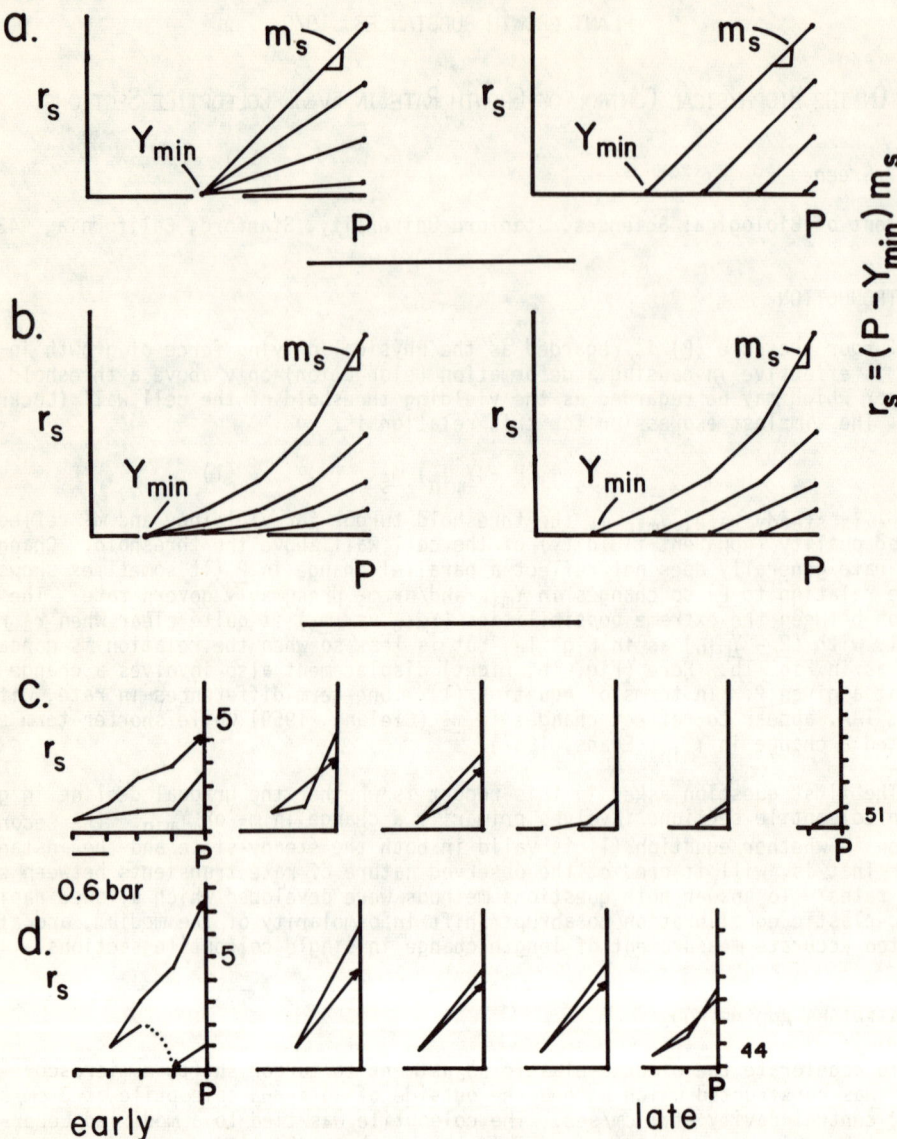

Fig. 1a. Graphical schemes for variation of steady rate, $\underline{r}_s$, at full turgor ($\underline{P}$) as a function of change in slope ($\underline{m}_s$), left graph, or of minimum growth turgor ($\underline{Y}_{min}$), right graph.

1b. The same contrast but now less evident because of the concave upward nature of the rate vs. turgor relation.

1c. Consecutive scans of the rate vs. turgor relation as rate declines during 6 hr. Course of each scan shown by the arrow. If no arrow, no significant hysteresis. Rate units are 2.3 microns/min. Length is about 7 mm; $\underline{r}_s$ of 5 equals about 9%/hr.

1d. Same, with rate initially very slow (high $\underline{Y}_{min}$) in left-hand graph. Dotted line shows increase in rate, fall in $\underline{Y}_{min}$. In 1c and 1d the fall in rate resembles the scheme in 1b, right, where rate is controlled by displacement of the curve and $\underline{Y}_{min}$.

22° on coleoptile sections about 7 mm long. To further speed osmotic equilibration, the epidermis was partially stripped off. Solutions were oxygenated or aerated. Growth proceeded for 6 or more hours, showing a gradual decline from rates (7%/hr) typical for material grown floating on auxin solutions. The time-course of the osmotic-elastic transients was often determined by carrying out the whole range of turgor shifts after growth had been halted by Na azide. The extent of shrinkage was comparable to that seen in the growth records, but sometimes exceeded it. When, after a turgor step-down in growing material, there is a shrinkage followed by complete cessation of length change, it is assumed that the full elastic shrinkage has been seen.

III. RESULTS

Relation between steady rate and turgor pressure

Turgor was decreased, in steps of 0.6 bar, by successive changes of the medium flowing through the chamber. Steady rate was taken over the period 4 to 7 min after the change, or over the first period which showed no apparent change in rate (Fig. 3). Step-downs were given until growth no longer occurred, or until the most concentrated solution had been used. The turgor series was then reversed, with steady rate taken usually 7 or more min after the step-up to allow for the decay of the transiently rapid growth following a step-up. There was noticeable hysteresis in the r_s vs. P relation (Fig. 1c). This bi-directional "scanning" was repeated until growth rate had fallen to a small fraction of the original rate. Because of the hysteresis each such scan is plotted separately. The change in character of the r_s vs. P relation with time is apparent in a succession of graphs (Fig. 1c, 1d). Because the relation is generally concave upward, a displacement of the curve involves a change in slope, at a given turgor.

In general, the data are compatible with a shift of the relation to higher turgor, as rate declines. They resemble the family of curves in Fig. 1b (right). Two detailed observations support this impression. In Fig. 1c, 4th graph, no growth was detected at $(P - 1.2$ bar$)$ while previously (1st graph) growth had been found even at $(P - 2.4$ bar$)$. Also, in the first graph of Fig. 1d, the initial growth rate at full P was very slow, and growth was barely detectable at $(P - 0.6$ bar$)$, indicating Y_{min} was near $(P - 0.6$ bar$)$. Later (dotted line), rate had accelerated and now high rates were found, even at $(P - 1.2$ bar$)$, again indicating a relocation of Y_{min}, this time to lower turgor.

This behavior is compatible with the data of Evans (1967) by which auxin stimulation involved a lateral displacement, toward lower turgor, of a r_s vs. P relation. Auxin appeared to lower P_{min} by several bar. The present results differ from those of Cleland (1959) where P_{min} appeared to be about the same for material with and without auxin. In his study the growth rates of sections in various concentrations of osmoticum were studied in parallel, over several hours. This regime might allow a turgor increase, and hence secondary rate increase, in the -IAA material. In the present study, rate shifts are studied in single sections, exposed to a series of shifts in turgor, allowing little time (15 min) for sections to change their internal solute concentration.

The decline in $(P - Y_{min})$ noted here could involve a fall in P as well as a rise in Y_{min}. The fall in P, if due to sap dilution, would be about 13%, or 1.17 bar, assuming an initial P of 9 bar. The fall in $(P - Y_{min})$ is, however, from $(P - 3)$ to $(P - 0.6)$, or 2.4 bar, hence a rise in Y_{min} appears to be involved. According to Ordin et al. (1956) 2.16 bar of NaCl leads to a 7% increase in P, during 20 hr incubation, so it is possible here that no fall in P occurred, the decrease in $(P - Y_{min})$ being due solely to a rise in Y_{min}.

Is the steady-state relation also valid instantaneously, i.e., during rate transients?

If equation (1), with appropriate modifications to account for non-linearity (Green *et al.*, 1971), is valid in the "instantaneous" sense (through transients) as well as the steady state, it would represent the most detailed expression for the biophysics of growth that could be derived from data taken *in vivo*. In model form rate follows the turgor relation given in Fig. 2A if there were no complications from elasticity, abrupt shifts in P should give abrupt shifts in rate, as in Fig. 2B, dashed line. Length, however, invariably contains an elastic component, present at higher P, which will give transient effects on L independent of growth. This elastic component is the difference between the dashed and solid lines in Fig. 2 and is called ΔL_e. With a fall in P, this length is lost, either as retarded elastic shrinkage (Fig. 2B), or as instantaneous elastic shrinkage accompanying a gradual fall in P, as in Fig. 2C, or both. In the extreme case of pure retarded elasticity, shrinkage (extension) may be slow (a) or fast (b). When this behavior (solid line) is superimposed on the curve for presumed "irreversible" extension, dashed line, one notes a small shrinkage over the period 5 to 8 min, but this is less than ΔL_e due to the increase in irreversible length during the contraction. There is thus a dip in the length curve, but no protracted stoppage. Upon turgor step-up the elastic component of length is regained, rapidly (b) or slowly (a). The extra length gained is equal to the observed contraction-expansion, ΔL_e, in the non-growing state (insert). Note the dotted line connecting ΔL with the "extra length" in the growth trace.

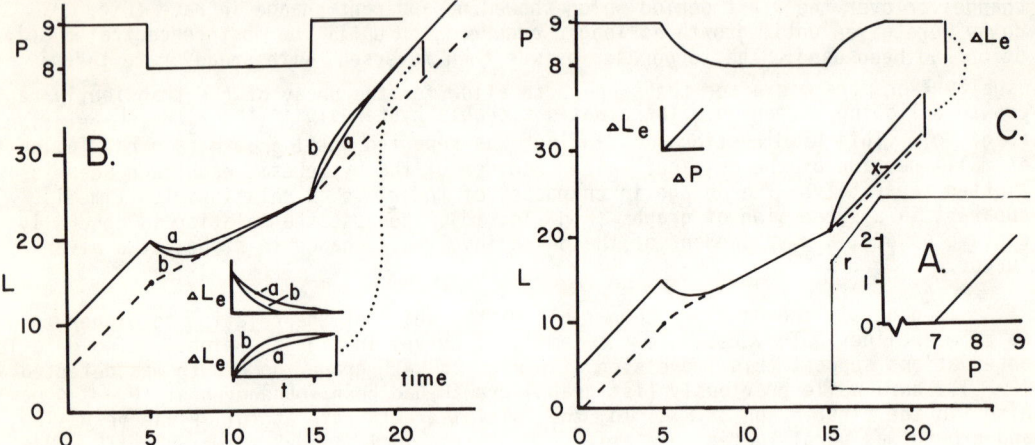

Fig. 2. Schematic responses of length, L, as a function of time, to temporary fall in turgor (P), following the irreversible growth rate diagram in 2A.

2B. Dashed line shows L with no elastic component. Curve a shows a response to abrupt turgor shift where there is a slow retarded elastic shrinkage of magnitude ΔL_e. In curve b the same shrinkage occurs sooner. Note that observed shrinkage in the growth trace is less than that seen in a non-growing system (inserts). Added length due to the rapid extension upon P step-up is identical with ΔL_e (dotted line).

2C. L change through gradual shifts in turgor (P). Elasticity is instantaneous and proportional to change in P (insert). Dashed line, "irreversible extension" is curved (5 to 7 min, 15 to 17 min) so that extra elastic change gives a protracted stoppage or an L increase, after step-up, more than L_e. In growth records (Fig. 3) both are found.

It is more likely that slow elastic contractions are based on essentially instantaneously elastic contraction accompanying a gradual shift in turgor (Ray and Ruesink, 1962) due to diffusion barriers of the tissue. In Fig. 2C the loss in length is taken as proportional to loss in P (insert) and the fall in turgor is not abrupt. With rate following the relation in Fig. 2A, irreversible length (dashed line) shifts only gradually to the new rate (over the periods 5 to 8 min, 15 to 18 min). Superimposing the elastic changes on this curve gives the solid line growth curve. Here there is a similar dip in length, again with no protracted stoppage. The apparent "extra length" following the step-up, would be the vertical distance between the line marked "x" and the growth trace. Due to the bend in the irreversible length curve, this is less than the full ΔL_e that would be observed in the non-growing condition.

When contractions are observed under non-growing conditions, it is not known whether instantaneous or retarded elasticity is involved. Any combination of these should, however, show no protracted stoppage after the step-down, and should show the apparent "extra length" upon turgor step-up, ΔL, as equal to, or less than, the coresponding shrinkage in the non-growing state.

Actual length vs. time records (Fig. 3) show that delay before resumption of growth (Δt) can be protracted, several times longer than the apparent duration of elastic shrinkage, Δt_e. Further, the "extra length" after a step-up, ΔL, is more than the elastic length change (ΔL_e). This indicates that eq. (1) is inadequate to explain the immediate behavior of the tissue after corrections for elasticity.

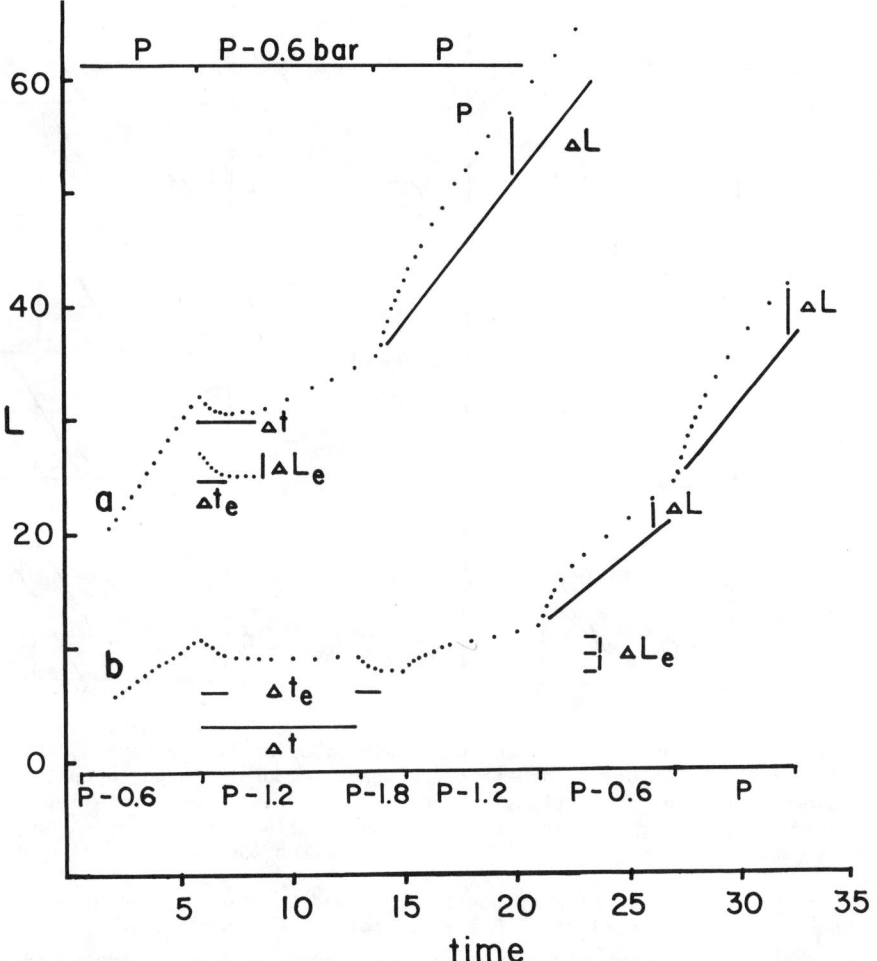

Fig. 3. Growth records, through various reductions in turgor (P).

  a. Note that the lag after turgor drop, Δt, is longer than the course of elastic shrinkage of the identical tissue in azide (insert). Added length after step-up, ΔL_e, is several times greater than the corresponding elastic stretch (ΔL_e).

  b. Similar protracted stoppage, as compared to shrinkage time (seen here as part of growth record at low turgor). Again, ΔL upon turgor step-up is greater than ΔL_e. Time in Min.

IV. INTERPRETATION

The minimal assumption, that the steady-state (eq. 1) and instantaneous relations are the same, is diagrammed in Fig. 4a. Elastic changes are not shown. At all times

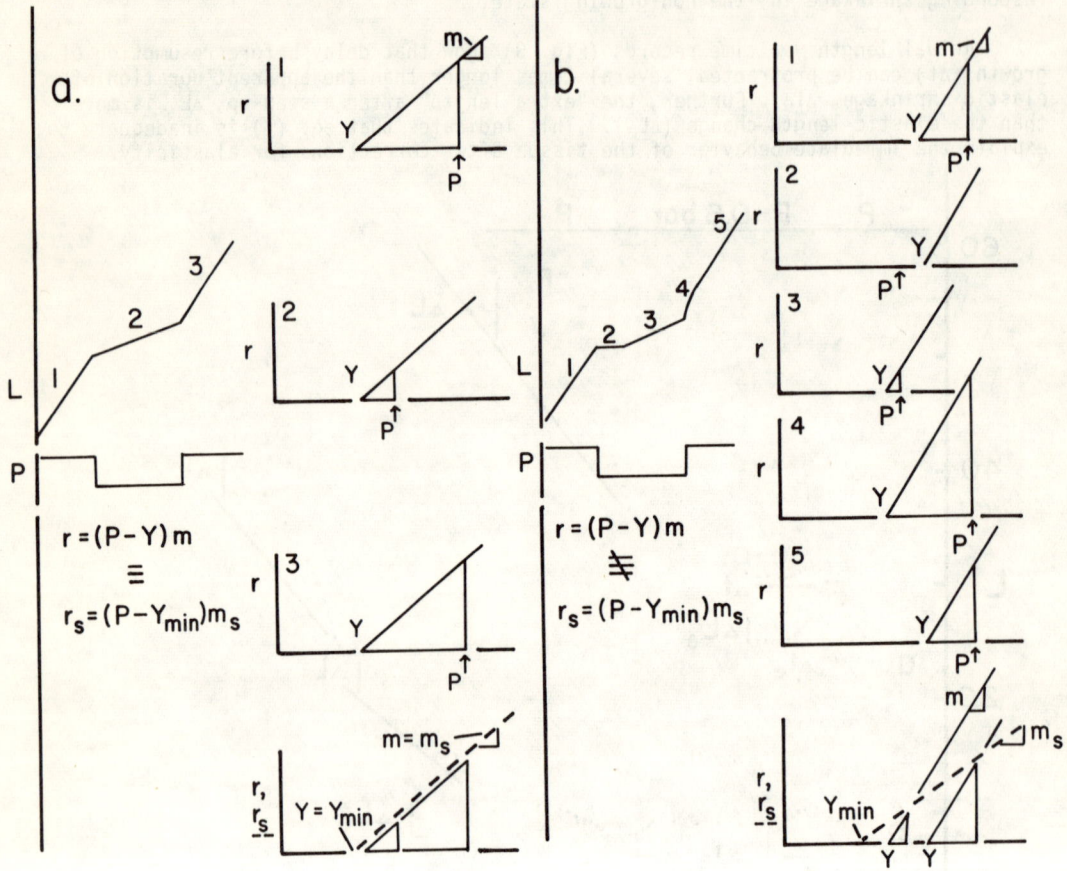

Fig. 4a. Simplest scheme, where instantaneous growth equation is the same as the steady-state one. Only 3 phases of growth are seen through a temporary turgor drop (left). Graphs at right show rate as proportional to $(P - Y)$, decreasing when P falls. Y is constant. Bottom graph shows the various instantaneous relations (triangles on abscissa with common vertex) and the congruent steady state relation (heavy dashed line).

4b. More complex scheme to account for the observed 5 phases of the growth response to temporary turgor drop (left). Phase 2, protracted stoppage, and phase 4 (extra length beyond that expected from elastic properties), are new. Instantaneous relations for the five phases are shown at right. Y is now mobile; m, slope of the r vs. P relation, is now steeper. A fall in P gives no turgor above the yield threshold (Y), phase 2, so there is no growth until Y falls below P, phase 3. Y falls less than P, so rate is reduced. Upon P step-up, phase 4, the driving force, $(P - Y)$ is unusually large and rate is high. Increase in Y reduces $(P - Y)$ to reduce rate to normal, phase 5. Bottom graph shows superimposed instantaneous relations (triangles on abscissa) and the steady-state relation (with different slope, m_s) as the heavy dashed line. This model fits data in Fig. 3.

rate is governed by the difference between the present turgor and a fixed turgor threshold or yield value, $\underline{Y}$. The growth relation is always a triangle, with rate proportion to its height. Since the slope, $\underline{M}$, is constant, rate is proportional to $(\underline{P} - \underline{Y})$. The steady-state relation, shown as the heavy dashed line in the bottom graph, is simply an assembly of such triangles, all with a common vertex at $\underline{Y}_{min}$.

The growth records indicate, however, that the irreversible length changes, as a function of transient fall in $\underline{P}$, are as in Fig. 4b, left. The five phases (especially the new ones, 2 and 4) can be explained by assuming an instantaneously valid relation.

$$\underline{r} = (\underline{P} - \underline{Y})\,\underline{m} \qquad (2)$$

In the first diagram, at full turgor, $(\underline{P} - \underline{Y})$ is smaller than before, but $\underline{m}$ is greater, so the rate is about the same. Upon abrupt fall in $\underline{P}$ (phase 2), growth stops because turgor is now below the yield threshold and there is no effective driving force. With time, however, $\underline{Y}$ falls to lower turgor (phase 3) and growth resumes, but at reduced rate because $\underline{Y}$ does not fall as far as did $\underline{P}$. Return to high turgor gives very rapid growth and added length (phase 4) because $(\underline{P} - \underline{Y})$ is very large. $\underline{Y}$ rises, phase 5, to return $(\underline{P} - \underline{Y})$ and rate to normal.

By this kind of scheme the steady-state relation, dashed line at bottom of Fig. 4B, is derived from the complex behavior of the instantaneous relation in eq. (2). $\underline{Y}$ is now variable, with the lowest attainable value being $\underline{Y}_{min}$. $\underline{Y}$ varies (by falling less than the corresponding fall in $\underline{P}$) so as to give reduced rate at reduced turgor. In this way both the transient and steady-state relations can be accounted for. A similar scheme appears to apply well in *Nitella* (Green et al., 1971) where the differences between instantaneous and steady-state behavior are more pronounced.

By this analysis, the behavior of $\underline{Y}$ becomes central in the control of rate. In *Nitella* its fall appears to be both time and metabolism-dependent; its rise appears to be the physical process of strain-hardening (Green et al., 1971). If this is also true in *Avena*, hormone treatment could involve factors which either raise or lower $\underline{Y}$, since its position is clearly a balance between them.

V. SUMMARY

To study both rapid and long-term relations between growth rate and turgor pressure, coleoptile segments were grown in a microscope stage chamber where medium flowed past the tissue at steady rates greater than 2 cm/sec. The epidermis had been partially stripped off to speed osmotic-elastic equilibration (80% complete in 2 min). The first question was: do changes in steady rate (r_s), at full turgor, reflect a change in slope of an r_s (ordinate) vs. turgor relation or do they reflect a displacement of such a curve, without change in slope, toward lower or higher turgor? Step shifts of 0.6 bar in the medium (NaCl), over a range 0 - 2.4 bar below full turgor, allowed periodic characterization of the relation. It was displaced toward higher turgor as rates declined. This agrees with earlier results of Evans where acceleration of r_s by auxin involved an apparent displacement of such a curve toward lower turgor. The second question was whether the r_s vs. turgor relation was also valid in the instantaneous sense: would it predict, after osmotic and elastic effects were accounted for, the exact nature of the rate transitions between steady rates? Small step-downs in turgor ($\underline{P}$) led to temporary stoppages incompatible with the above relation, small step-ups led to unexpectedly rapid growth transients ("bursts"). That these responses are soon overcome, or damped, to give r_s indicates that r_s is the result of governor-like control. The governor, as in *Nitella*, appears to involve the re-positioning of an instantaneously valid minimum growth turgor ($\underline{Y}$, or apparent yield point) so as to nearly re-establish the pre-existing effective driving force $(\underline{P} - \underline{Y})$, and hence pre-existing rate.

VI. ACKNOWLEDGEMENTS

Supported by N.S.F. grant GB 6055. Written while the author held an N.S.F. Senior Post doctoral Fellowship. The advice of Dr. Peter M. Ray is deeply appreciated.

VII. REFERENCES

CLELAND, R. (1959). Effect of osmotic concentration on auxin action and on irreversible and reversible extension of the *Avena* coleoptile. Physiol. Plantarum 12, 809-825.

EVANS, M.L. (1967). Kinetic studies of the cell elongation phenomenon in *Avena* coleoptile segments. Ph.D. Thesis, University of California, Santa Cruz.

GREEN, P.B., R.O. ERICKSON and J. BUGGY (1971) Metabolic and Physical control of elongation rate - *in vivo* studies in *Nitella*. Plant Physiol. 47, 423-430.

GREEN, P.B., R.O. ERICKSON, and P.A. RICHMOND (1970). On the physical basis of cell morphogenesis. Ann. N.Y. Acad. Sci. 175, 712-731.

LOCKHART, J.A. (1965). Cell extension, in "Plant Biochemistry" (Ed. J. Bonner and J.E. Varner) pp. 826-849. Academic Press, New York, 1965.

RAY, P.M. and A.W. RUESINK (1962). Kinetic experiments on the nature of the growth mechanism in oat coleoptile cells. Dev. Biol. 4, 377-397.

PLANT GROWTH SUBSTANCES, 1970

Auxin-Induced Changes in Cell Wall Properties and Growth of *Avena* Coleoptiles and Green Pea Epicotyls

Yoshio Masuda, Ryoichi Yamamoto and Eiichi Tanimoto

Department of Biology, Faculty of Science, Osaka City University, Sumiyoshi-ku, Osaka, Japan

I. INTRODUCTION

Auxin-induced changes in mechanical properties of the plant cell wall have been confirmed using non-automatic techniques (cf. Heyn and van Overbeek, 1931; Cleland, 1958; Masuda and Wada, 1966). A newly-developed technique using an automatic tensile testing apparatus offers several advantages in operation and in evaluation of the results obtained (Olson et al., 1965; Cleland, 1967). However, mechanical properties of the cell wall measured by a standard tensile technique, as compliance (Cleland, 1967), is not easy to interpret rheologically. We have recently developed a technique using a stress-relaxation process to determine rheological properties of the cell wall. This technique is useful for obtaining better insights into the mechanism(s) of cell growth.

II. MATERIALS AND METHODS

Segments of etiolated *Avena* coleoptile and green pea epicotyl were prepared as reported previously (Yamamoto et al., 1970). Techniques used in stress-relaxation analysis have also been reported elsewhere (Yamamoto et al., 1970).

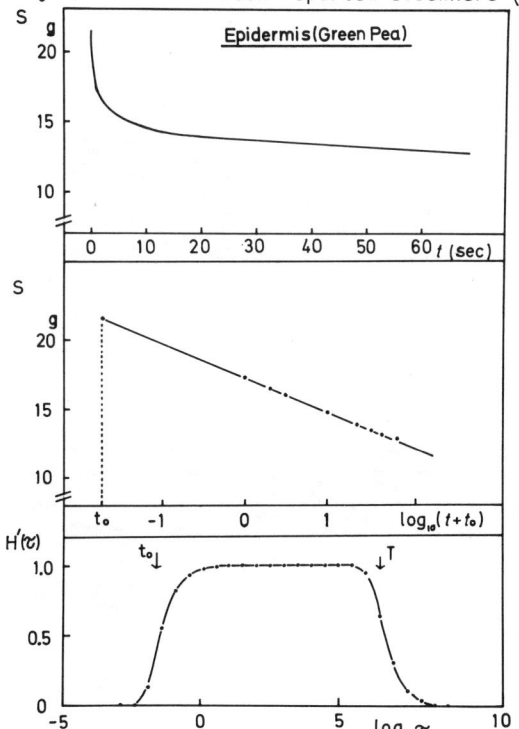

Stress-relaxation analysis

By applying a certain amount of stress by a quick extension of the isolated cell wall, i.e. a methanol-killed, pronase-treated tissue segment, a stress-relaxation curve, as shown in Fig. 1 A, was obtained (Yamamoto et al., 1970).

Fig. 1. Stress-relaxation curves for isolated green pea epicotyl cell wall. A stress of 20 g was applied to methanol-killed pea epicotyl epidermis by extending it at a rate of 20 mm/min. This stress-relaxation was recorded.

A: Ordinate: S (stress, g); abscissa: t (time in seconds)
B: Ordinate: S; abscissa: log $(t + t_o)$
C: H' (τ) spectrum of the isolated epidermal cell wall.

The curve is well represented by a model consisting of at least four Maxwell viscoelastic components. Generally the curve is simulated by a model consisting of i number of Maxwell components:

$$S = \Sigma \gamma \cdot G_i \cdot \exp(-\frac{t}{\tau_i}) \qquad (1)$$

$$(i = 1, 2, ---, n; \tau_1 < \tau_2 < --- < \tau_n)$$

where S = stress, γ = strain, G_i = elastic modulus, τ_i = relaxation time (= η_i/G_i, η: viscosity), t = time. Using a model consisting of four Maxwell viscoelastic components simulating the cell wall of *Avena* coleoptile segments, for example, the parameters could be determined by Procedure X (Tobolsky and Murakami, 1959):

5 min incubation in the presence of 0.25 M mannitol

<u>Without IAA</u>:

$S/S_0 \times 100 = 11.1 \exp(-1.99t) + 10.5 \exp(-0.64t) + 17.3 \exp(0.060t) + 61.1$

<u>With IAA</u>:

$S/S_0 \times 100 = 11.2 \exp(-2.12t) + 10.6 \exp(-0.66t) + 17.2 \exp(-0.059t) + 61.0$

60 min incubation in the presence of 0.25 M mannitol

<u>Without IAA</u>:

$S/S_0 \times 100 = 15.4 \exp(-2.3t) + 7.5 \exp(-0.43t) + 15.0 \exp(-0.055t) + 61.1$

<u>With IAA</u>:

$S/S_0 \times 100 = 11.6 \exp(-2.9t) + 9.6 \exp(-0.43t) + 18.3 \exp(-0.057t) + 60.5$

Thus, the effect of auxin is obvious at the initial portion of the stress-relaxation curve, producing a high value of $1/\tau_1$, i.e. the reciprocal of the stress-relaxation time of the first Maxwell viscoelastic component. Hence the value of $1/\tau_1$ is determined using the following approximation:

$$S = \gamma \cdot G_1 \cdot \exp(-\frac{t}{\tau_1}) + \gamma(G_2 + G_3 + G_4) \qquad (t \ll \tau_2) \qquad (2)$$

Thus,

$$1/\tau_1 = \log_e \left(\frac{S_0 - S_1}{S_1 - S_2}\right) \qquad (3)$$

where S_i denotes the stress at t = i (= 0, 1 and 2 seconds).

III. RESULTS AND DISCUSSION

As an example, the effect of auxin on $1/\tau_1$ of *Avena* coleoptile cell wall is summarized in Table 1.

Table 1. Effect of auxin on $1/\tau_1$ of *Avena* coleoptile cell wall

Treatment	$1/\tau_1$
Initial	1.32 ± 0.02
1 hr with buffer	1.36 ± 0.02
1 hr with IAA	1.49 ± 0.03

Etiolated *Avena* coleoptiles, 30-35 mm long, were decapitated, and 2 hr later a 15 mm segment was excised from each. Segments thus prepared were incubated for 1 hr at 25°C in 0.25 M mannitol solution (solvent: M/400 K-maleate buffer, pH 4.7) in the presence of 5 mg/liter IAA and in its absence, and killed in boiling methanol. $S_0 = 30.0$ g. Mean values (n = 10) with standard errors.

Since auxin causes no consistent effect on G_i, the increase in $1/\tau_1$ caused by auxin may indicate that auxin primarily decreases viscosity of the first Maxwell viscoelastic component.

For the stress-relaxation curve of the plant cell wall, a generalized model consisting of infinite number of Maxwell components was adopted to simulate the mechanical properties of the cell wall (Yamamoto et al., 1970), since a straight line for $S - \log(t + t_0)$ was obtained (Fig. 1 B). Thus,

$$\int_0^\infty G(\tau) \cdot \gamma \cdot \exp(-\frac{t}{\tau}) \cdot d\tau = a - b \cdot \log(t + t_0) \tag{4}$$

where $G(\tau)$ = the τ spectrum of G, a and b = constants, and t_0 is determined by the point for S_0 in the figure (Fig. 1 B). Eq. (4) finally leads to the following equation (Yamamoto et al., 1970):

$$G(\tau) = \frac{b}{\gamma} \cdot \frac{1}{\tau} [\exp(-\frac{t_0}{\tau}) - \exp(-\frac{T}{\tau})] \tag{5}$$

where $T = \exp(\frac{a}{b})$.

From Eq. (5), $H(\tau)$ $(=G(\tau)\tau)$ can be obtained:

$$H(\tau) = \frac{b}{\gamma} [\exp(-\frac{t_0}{\tau}) - \exp(-\frac{T}{\tau})] \tag{6}$$

The $H(\tau)$ spectrum represented by Eq. (6) indicates a box-type distribution (Tobolsky and Murakami, 1959) as illustrated in Fig. 1C, where $H(\tau)$ is converted to $H'(\tau)$ $(= \frac{\gamma}{b} \cdot H(\tau))$. Important parameters determining the shape of the distribution of the $H(\tau)$ spectrum are t_0, T and $\frac{b}{\gamma}$. Values of t_0 and T can be obtained by graphing procedures (including the method of least squares) and $\frac{b}{\gamma}$ is roughly proportional to the stress/strain ratio. Parameters for the *Avena* coleoptile cell wall are indicated in Table II.

Table II. Effect of IAA on the parameters of isolated *Avena* coleoptile cell wall

Treatment	t_0	$T (\times 10^6)$	$1/\tau_1$	Stress/Strain
1 hr with buffer	0.045 ± 0.002	3.6 ± 0.9	1.58 ± 0.02	31.7 ± 0.63
1 hr with IAA	0.027 ± 0.002	10.3 ± 3.0	1.72 ± 0.03	26.7 ± 0.62

Etiolated *Avena* coleoptiles, 30-35 mm long, were decapitated, and 2 hr later a 15 mm segment was excised from each. Segments thus prepared were incubated for 1 hr at 25°C in M/400 M K-maleate buffer solution in the presence and absence of 5 mg/liter IAA in the absence of mannitol, and killed in boiling methanol. S_0 = 30.0 g. Mean values (n = 10) with standard.

A shift in box-type distribution of the τ spectrum to the left (deceased t_0 and T) is said to be caused by a decrease in the mean molecular weight of the constituents and a shift to the right (increased t_0 and T) is caused by an increase in mean molecular weight (Tobolsky and Murakami, 1959). If this is true for the plant cell wall, the auxin must cause a reduction of the mean molecular weights of polysaccharide components in the cell wall. Synthesis of some cell wall components may also be possible since auxin increases T.

Cell elongation due to auxin has been shown conventionally by measuring the change in length of excised tissue segments in auxin solution. Thimann and Schneider (19 pointed out that, in etiolated pea epicotyls and *Avena* coleoptiles, the epidermis and tissue without epidermis have different sensitivities to auxin, with different optimal concentrations of auxin required for growth, thus resulting in differential growth at a certain auxin concentration. In green pea epicotyls (5th internode of 8-day-old seedlings) we found that the growth of both unpeeled and peeled segments has almost the same optimal concentration of 2,4-D (ca. 1 mg/liter). However, as reported by van Overbeek and Went (1937) and Thimann and Schneider (1938) for etiolated pea epicotyl segments, unpeeled segments of green pea epicotyls elongated more conspicuously than peeled segments in response to 1 mg/liter 2,4-D (Table III). In addition, 2,4-D-induced elongation was completely inhibited by 10^{-4} M cycloheximide (CHI). This result may indicate that elongation of epidermal cells is essential for elongation of the whole segment; in other words, auxin-induced elongation of the internode is dependent on the reaction of epidermal cells to auxin which involves protein synthesis.

Table III. Effect of 2,4-D and cycloheximide (CHI) on elongation of peeled and unpeeled segments of green pea epicotyls

Treatment	Length (arbitrary unit) at		
	0 hr	1 hr	3 hr
Unpeeled			
Water	70.0	71.0	71.0 (100)
2,4-D	70.0	72.0	80.0 (114)
2,4-D + CHI	70.0	70.5	71.0 (100)
Peeled			
Water	70.0	75.0	75.0 (100)
2,4-D	70.0	75.0	80.5 (107)

Segments, 10 mm in length, excised from 5th internodes of Alaska pea seedlings grown for 8 days in the light (3000 lux), were treated for 1 and 3 hr with water, 1 mg/liter 2,4-D and 10^{-4} M CHI. Mean values (n = 10).

Stress-relaxation analysis was carried out in order to analyse the role of the epidermis in 2,4-D-induced elongation of green pea epicotyl segments. Segments were incubated for 1 and 3 hr in water, in 1 mg/liter 2,4-D and in 2,4-D + 10^{-4} M CHI. A group of segments was killed in boiling methanol and another group of segments was quickly peeled, and the epidermis and the peeled tissue separately methanol-killed. The three varieties of samples were subjected to stress-relaxation analysis.

Parameters for unpeeled segments are summarized in Table IV. 2,4-D decreased t_o and the stress/strain ratio, and increased T and $1/\tau_1$. CHI reversed the effect of auxin on these parameters. Similar effects of auxin on these parameters were found in the epidermis but not in the peeled segments (Table V), suggesting that promotion of tissue elongation by auxin is primarily due to its effect on the epidermal cell wall.

Table IV. Effect of 2,4-D and cycloheximide (CHI) on the parameters of isolated green pea epicotyl cell wall

Treatment	t_o	T (x10^5)	$1/\tau_1$	Stress/Strain
Initial	0.042±0.006	3.63±1.84	1.55±0.06	39.7±2.0
1 hr				
Water	0.050±0.001	0.91±0.23	1.45±0.06	48.4±1.3
2,4-D	0.026±0.007	18.5±5.0	1.61±0.05	45.1±1.8
2,4-D + CHI	0.056±0.008	0.47±0.23	1.45±0.04	55.0±2.1
3 hr				
Water	0.046±0.008	2.19±1.37	1.45±0.04	58.2±1.6
2,4-D	0.019±0.002	20.0±9.6	1.77±0.05	33.0±1.0
2,4-D + CHI	0.083±0.009	1.36±0.28	1.49±0.05	58.2±1.3

Segments, 15 mm in length, excised from 5th internodes of Alaska pea seedlings grown for 8 days in the light (3000 lux), were treated for 1 and 3 hr with water, 1 mg/liter 2,4-D and 10^{-4} M CHI, and methanol-killed. S_o = 20 g. Mean values (n = 10) with standard errors.

In addition, nucleic acid and protein synthesis seemed to be required for auxin-induced change in properties of the epidermal cell wall, since the effect of 2,4-D on the parameters was inhibited by CHI and also by actinomycin D (data not presented). However, since 2,4-D also promotes elongation of peeled epicotyl segments without affecting the parameters, a factor or factors other than changes in cell wall properties must be taken into consideration when unpeeled segments or intact pea epicotyls grow in response to auxin.

IV. SUMMARY

Using a stress-relaxation technique cell wall properties of etiolated segments of *Avena* coleoptile and internodes of green pea seedlings were simulated with a model consisting of four or an infinite number of Maxwell viscoelastic components. Auxin affected the τ (relaxation time) spectrum of the model, causing changes in parameters which determine the shape of the spectrum, assuming a box-type distribution. Thus, changes in parameters caused by auxin imply auxin-induced changes in mechanical properties of the cell wall, involving partial degradation and possibly some synthesis of cell wall constituents. In pea epicotyl segments auxin-induced changes in parameters were found in the epidermis but not in peeled segments. These effects of auxin on the epidermis were inhibited by cycloheximide, suggesting that protein synthesis is required for the auxin effect on mechanical properties of the epidermal cell wall.

Table V. Effect of 2,4-D and cycloheximide on the parameters of isolated cell wall of epidermis and peeled segments of green pea epicotyls

Treatment	t_o	T (x 10^7)	$1/\tau_1$	Stress/Strain
Epidermis				
0 hr	0.020±0.002	0.55±0.26	1.65±0.02	24.3±1.2
1 hr				
Water	0.016±0.002	1.74±0.45	1.71±0.03	23.3±1.6
2,4-D	0.013±0.001	5.50±1.7	1.94±0.04	24.1±1.0
2,4-D + CHI	0.022±0.002	0.59±0.28	1.56±0.04	24.7±2.9
3 hr				
Water	0.019±0.001	1.62±0.67	1.65±0.03	26.1±1.4
2,4-D	0.012±0.004	5.374±4.4	1.80±0.03	29.2±1.8
2,4-D + CHI	0.018±0.002	0.91±0.57	1.63±0.03	21.3±1.1
Peeled segments				
0 hr	0.091±0.006	0.015±0.020	1.36±0.03	28.9±3.4
1 hr				
Water	0.079±0.004	0.021±0.007	1.43±0.02	29.1±2.2
2,4-D	0.092±0.006	0.007±0.006	1.36±0.03	33.7±4.5
2,4-D + CHI	0.086±0.004	0.017±0.006	1.39±0.03	31.8±4.3
3 hr				
Water	0.088±0.007	0.011±0.001	1.39±0.02	37.7±4.4
2,4-D	0.083±0.007	0.013±0.001	1.42±0.03	34.0±5.8
2,4-D+CHI	0.086±0.006	0.013±0.001	1.37±0.03	25.6±1.8

Segments, 15 mm in length, excised from 5th internodes of Alaska pea seedlings grown for 8 days in the light (3000 lux), were treated for 1 and 3 hr with water, 1 mg/liter 2,4-D and 10^{-4} M CHI. Then epidermis was peeled off and the epidermis and peeled segments were separately methanol-killed. S_o = 20 g. Mean values (n = 10) with standard errors.

V. REFERENCES

CLELAND, R. (1958). A separation of auxin-induced cell wall loosening into its plastic and elastic components. Physiol. Plant., 11, 599-609.

CLELAND, R. (1967). Extensibility of isolated cell walls: Measurement and changes during cell elongation. Planta, 74, 197-209.

HEYN, A.N.J. and J. van OVERBEEK (1931). Weiteres Versuchsmaterial zur plastischen und elastischen Dehnbarkeit der Zellmembran. Kon. Akad. Wet. Amsterdam, 34, 1190-1195.

MASUDA, Y. and S. WADA (1966). Requirement of RNA for the auxin-induced elongation of oat coleoptile. Physiol. Plant., 19, 1055-1063.

OLSON, A.C., J. BONNER and D.J. MORRÉ (1965). Force extension analysis of Avena coleoptile cell wall. Planta, 66, 126-134.

VAN OVERBEEK, J. and F.W. WENT. (1937). Mechanism and quantitative application of the pea test. Bot. Gaz., 99, 22-41.

THIMANN, K.V. and C.L. SCHNEIDER (1938). Differential growth in plant tissues. Am. J. Bot., 25, 627-641.

TOBOLSKY, A.V. and K. MURAKAMI (1959). Existence of a sharply defined maximum relaxation time for monodisperse polystyrene. J. Polymer, Sci., 40, 443-456.

YAMAMOTO, R., K. SHINOZAKI and Y. MASUDA (1970). Stress-relaxation property of plant cell walls with special reference to auxin action. Plant & Cell Physiol., 11, 947-956.

MEASUREMENTS OF WATER POTENTIALS AND HORMONE TRANSPORT ASSOCIATED WITH THE GROWTH OF CUCUMBER HYPOCOTYLS

Daniel Cohen* and Dan Atsmon

Department of Plant Genetics, The Weizmann Institute of Science, Rehovot, Israel

* Present address: Plant Physiology Division, D.S.I.R., Palmerston North, New Zealand

I. INTRODUCTION

Rapid growth responses in non-dividing cells are essentially brought about by water uptake. This increased water uptake (J_v) may result from an increased driving force ($\Delta\psi$) or a change in membrane permeability (L_p), thus:

$$J_v = L_p \Delta\psi$$

A change in L_p may affect the rate at which equilibrium is achieved but not the extent of water uptake needed to reach equilibrium. An increased driving force may arise from a change in turgor pressure (ΔP), a change in osmotic pressure ($\Delta\Pi$) or a change in the reflection coefficient (selective permeability, σ) of the membrane.

Thus the equation for water uptake may be written

$$J_v = L_p (\Delta P - \sigma \Delta\Pi)$$

Since plant membranes can normally be considered effectively semipermeable, σ is close to 1 (Dainty 1969). Therefore a hormonally induced change in J_v should result from a change in either P or Π.

The response of the cucumber seedling to externally applied hormones is of interest for several reasons. Katsumi et al. (1965) have shown that the hypocotyl of the intact seedling elongates in response to either IAA or GA in a very similar way. They also presented evidence that inhibition of the endogenous level or action of one of the hormones interfered with the response to the other hormone. In a continuation of this work, Cleland et al. (1968) have shown that whereas IAA induced an increase in the extensibility of the cell wall, there was little or no increase with GA. They proposed that GA might induce growth by increasing the osmotic potential of the cell sap.

We decided to investigate the water relations of the cucumber hypocotyl at various times after treatment of the intact seedling with either IAA or GA and, in particular, to test the hypothesis that an increase in osmotic potential is the driving force for GA-induced growth.

II. MATERIALS AND METHODS

Cucumber seedlings (c.v. Beit Alpha) were grown as described by Degani and Atsmon (1970).

Hormone solutions were prepared in 10% ethanol with 0.05% Tween 20. In the case of GA_{4+7}, the concentration used was 10ppm and a 10µl drop was placed on the shoot apex. IAA at 100ppm was applied as a 10µl drop to one cotyledon about 1cm from the apex.

To collect cell sap, hypocotyl sections were first frozen in small tubes held in dry ice, thawed at room temperature and then centrifuged at 10,000 x g for 10 min. in small perforated cups suspended from the top of 6ml centrifuge tubes. Using this method, approximately 90% of the fresh weight was recovered in the tube.

$IAA-1-^{14}C$ was purchased as the ammonium salt from The Radiochemical Centre, Amersham (specific activity 57mc/mM). Purity was checked by paper chromatography. In most experiments the specific activity was reduced to 29mc/mM by the addition of unlabelled IAA. Levels of $IAA-1-^{14}C$ in tissue sections were counted following an overnight digestion in Soluene (Packard). Counting efficiency was 80% using this method.

Water potential was measured using a modification of the Schardakov method and also the gravimetric solution method (see Barrs, 1968). Generally ten sections of 15mm (5-20mm from apex) were immersed in a range of sucrose solutions. For the Schardakov method the sections were blotted after 10min and transferred to 1ml of fresh solutions of sucrose in small beakers. Thereafter, the method followed Barrs (1968). For the gravimetric solution method, the change in weight of the sections was followed over two hours. Before each weighing the sections were blotted between moist layers of Wettex (cellulose sponge cloth).

III. RESULTS AND DISCUSSION

The results obtained using the Schardakov method to determine water potential did not show any significance in Ψ between control tissue and tissues treated with IAA 30 min., 1hr or 2hr earlier. A value of approximately 3.6Atm. was found in all cases.

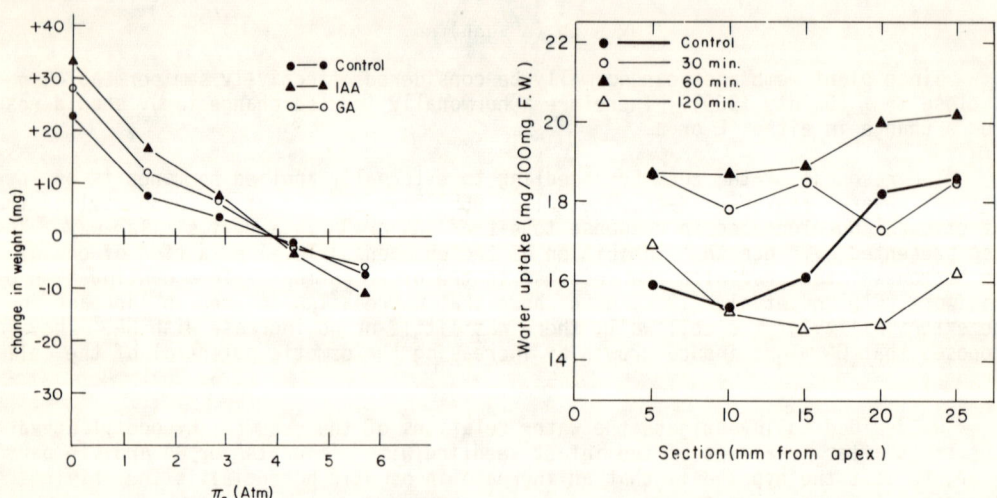

Fig. 1. Measurement of water potential by the gravimetric solution method. Water uptake from sucrose solutions of the indicated osmotic potential, π_e, was measured after 2hr.

Fig. 2. Water uptake by sections of cucumber hypocotyl. Seedlings were treated with a 10µl drop of 100 ppm IAA and, at the times shown, 10 hypocotyls were cut into 5mm sections, weighed and placed in distilled water.

Using the gravimetric solution method, Ψ was found to be 3.8Atm. in the case of sections from either control plants or plants treated 1hr previously with IAA or plants treated 4hr previously with GA_{4+7} (Fig. 1). There appears to be a difference in the shape of the curves, however. This may indicate that although the driving force (Ψ) remains the same, the rate of water entry has been altered by a change in membrane permeability (Lp).

Another method for assessing the water status of a tissue is to express water uptake in terms of initial fresh weight. From observations of seedlings following IAA treatment, the zone of curvature appears to move down the hypocotyl. To ascertain whether there was a change in the water status of the tissue in the zone of curvature, seedlings were treated with IAA and the hypocotyls were cut into 5mm sections after 30, 60 or 120min. The uptake of water was followed over a two hour period. After a rapid initial uptake in the first 15min., the rate declined. The total water uptake over 2 hours is shown in the Figure 2. In comparison with control, IAA increased the water uptake in the upper sections at 30min. and in all sections at 60 min. At 120min. the uptake was actually lower than control in all except the two most apical sections. This experiment has been repeated a number of times and, although water uptake in sections from plants treated 120min. earlier is not always lower than control, it invariably is less than the water uptake in treated sections at 60min. These results suggest that the lower sections have reacted to IAA and have reached a new equilibrium by 120min.

The osmotic potential of cell sap was measured by two methods. Firstly, a direct measurement of freezing point depression using a Knauer micro-osmometer and secondly, a method developed by Shimshe and Livne (1968) which combines the values from separate determinations of the conductivity and refractive index of the sample. Table I shows the results of two independent experiments in which these methods were used. When we consider the cryoscopic data it is clear that π did not increase following GA treatment, in fact, the value appears to decrease. The refractometric data indicate that the concentration of sugars decreased whereas the conductimetric data show that part of this decrease was compensated for by an increase in the concentration of ions. It is apparent from these results that the driving force for GA-induced growth cannot be an increase in π. Rather, the hypocotyl appears to accumulate solutes, particularly ions, in response to elongation. We suggest that the results of Cleland et al. (1968) can be reinterpreted as showing that GA maintains the extensibility of the cell wall during elongation.

Table 1. Measurements of Osmotic Potential (π in Atm.) following hormonal treatments.

Time(hr)	Treatment	A Cryoscopy	B Refractometry	C Conductimetry	B + C
2	C	6.55			
	IAA	6.15			
	GA	5.99			
4	C	6.30	1.30	4.10	5.40
	IAA	5.84	1.25	3.85	5.10
	GA	5.67	1.40	4.10	5.50
8	C	5.84			
	IAA	5.87			
	GA	5.82			
13	C		1.35	4.15	5.50
	IAA		1.30	4.00	5.30
	GA		1.25	4.05	5.30
24	C	6.41	1.05	3.90	4.95
	IAA	5.99	1.05	3.90	4.95
	GA	5.87	1.00	4.10	5.10
48	C	5.90	1.00	4.20	5.20
	IAA	6.01	1.00	4.20	5.20
	GA	5.60	0.95	4.25	5.20
96	C		1.05	4.15	5.20
	IAA		0.95	4.30	5.25
	GA		1.05	4.20	5.25

Auxin transport in cucumber hypocotyls

When cucumber seedlings are treated with 100ppm IAA, the hypocotyl begins to bend after 25-30min. The rapidity of this response raises several important questions. Why does the hypocotyl bend? How much IAA is actually entering the tissue, how fast is it translocated and what is its fate in the tissue?

The rate of IAA movement in cut sections has been measured at 10-15mm/hr by many workers. In this system, if IAA is indeed inducing bending by direct action on hypocotyl tissue, then the IAA must be translocated at least 20mm in less than 30 minutes. It is known that herbicides applied to leaves are sometimes translocated at rapid rates, and that such translocation is in the vascular system. If IAA is being transported in the vascular tissue in the present system then the movement may be non-polar. In order to investigate these points, a series of experiments was undertaken using IAA-1-^{14}C.

First it was necessary to describe the curvature of the hypocotyls quantitatively. The height of the cotyledonary node from the soil surface was measured at the time zero and at intervals after treatment; the loss of height was taken to represent bending. Figure 3 shows that both the extent of bending and the time at which bending was first

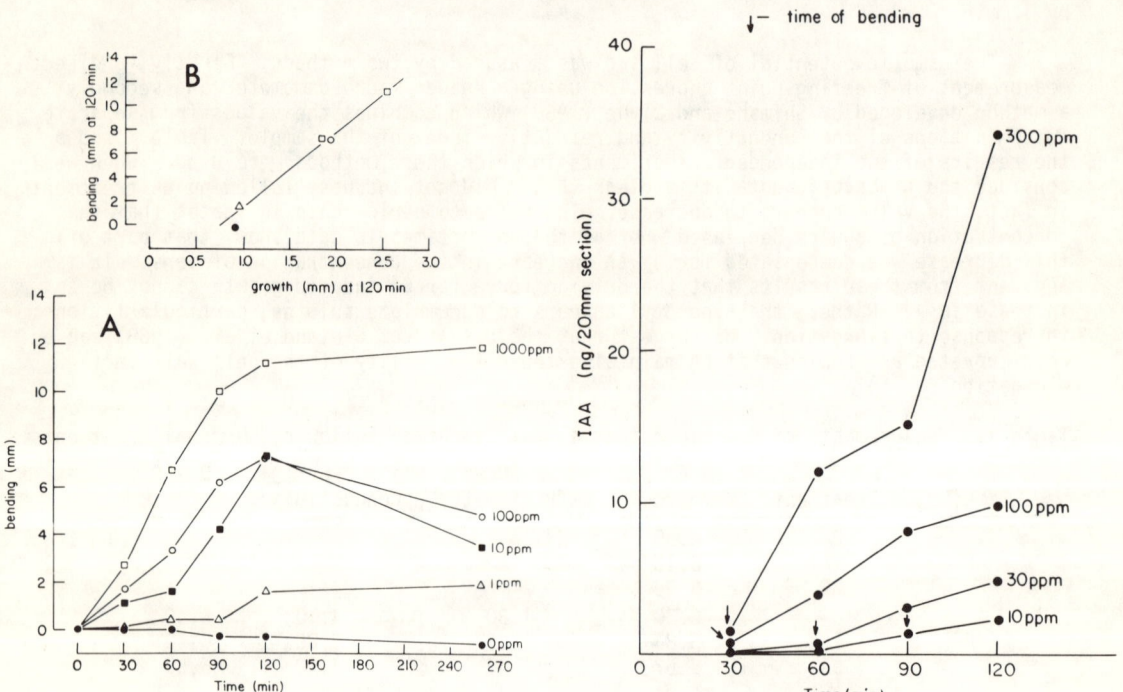

Fig. 3A. Time course and concentration dependence of hypocotyl bending. The height of the cotyledonary node from the soil surface was measured and the loss of height in mm is the unit of bending.

3B. Relationship of hypocotyl bending to elongation of a marked 10mm section of the hypocotyl for various concentrations of IAA. Symbols as in Fig. 3A: ●, Control; △, 1ppm; ■, 10ppm; ○, 100ppm; □, 1000ppm.

Fig. 4. Time course and concentration dependence of IAA transport. The total cpm in sections (5-25mm from apex) were determined. IAA as ng/section has been calculated using counting efficiency and specific activity data. The time at which hypocotyl bending was first noted is indicated by an arrow.

noted were concentration dependent. Furthermore, the extent of bending was directly related to the growth response measured at 2hr.

Figure 4 shows the results of an experiment in which IAA-1-^{14}C was applied at different concentrations. At 30 min. intervals up to 2 hrs the hypocotyls were sectioned and the total counts in these sections measured. Since counting efficiency and the specific activity of each concentration of IAA-1-^{14}C were known, it was possible to convert cpm to ng IAA in the tissue. The time at which bending was first noted is also indicated. It can be seen that the amount of IAA in the section at the onset of bending was approximately 1ng. irrespective of concentration of the applied IAA.

The most obvious explanation for the curvature of the hypocotyl is a difference in the level of IAA on the two sides of the hypocotyl. To test this, hypocotyls were harvested at various times after treatment, cut into 10mm sections and then split lengthwise. Figure 5 shows the distribution of counts in the treated (+) and untreated (-) sides of the seedling. It is clear that there is a rapid movement of IAA down the treated side of the hypocotyl and only at two hours does the level in the untreated side equal that of the treated side when bending was first noted. Since carboxyl-labelled IAA was used in these experiments, it is most likely that the counts measured were in fact IAA, or a conjugation product of IAA.

In order to obtain some information about the fate of IAA in the seedling, cell sap was collected in the usual way from hypocotyl sections of plants treated with IAA-1-^{14}C. Aliquots of the cell sap were applied directly to Whatman No. 1 paper for ascending chromatography. After development in n-butanol:acetic acid:water(5:1:1.2 v/v), the paper was cut into 22 segments and each piece counted directly in a scintillation counter (Fig. 6). The results indicate that the top half of the hypocotyl

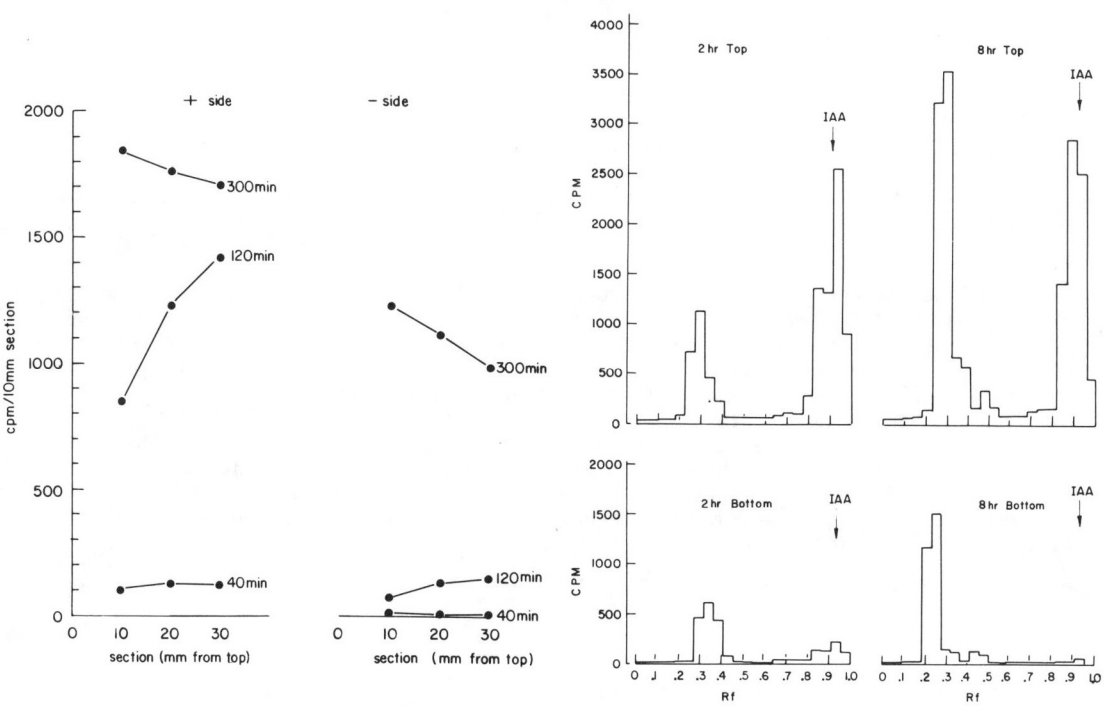

Fig. 5 Distribution of IAA-1-^{14}C in the hypocotyl. Hypocotyls were harvested at various times after treatment, cut into 10mm sections and split lengthwise into treated (+) and untreated (-) halves.

Fig. 6. Chromatography of cell sap from hypocotyl sections. 40µl cell sap was applied to each chromatogram.

contained free IAA and also a second compound at Rf 0.3. The amount of this second
compound increased from 2 to 8hr. In the lower half of the hypocotyl very little free
IAA was found at 2hr and none at 8hr. The compound located at Rf 0.3 was again found
at 2hr and increased at 8hr. The identity of this compound is not known, but its
mobility is less than IAA-Glu or IAA-Asp in this solvent (Zenk 1961). It will be of
interest to characterise this compound and demonstrate whether it is in fact an indole
conjugate.

The residues of the hypocotyl segments after cell sap collection were extracted
using a mixture of methanol, chloroform and water in the ratio 12:5:3 v/v (Bieleski
and Turner, 1967). Aliquots of these extracts were counted. The counts/mg in these
extracts were approximately the same as in the corresponding cell sap sample. There
was, therefore, no evidence for specific binding or compartmentation of IAA. However,
the residue extracts have not yet been chromatographed and the identity of the radioactive compounds in this fraction is not known.

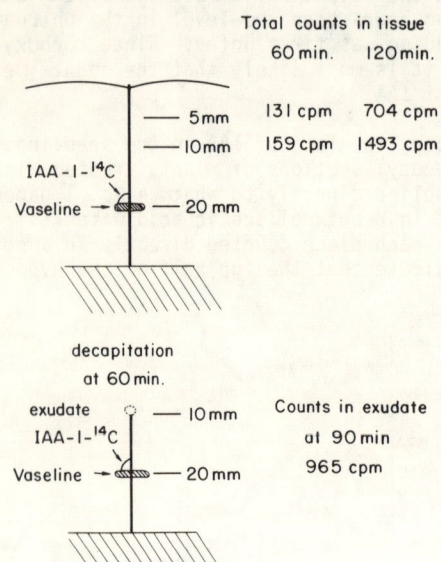

Fig. 7. Demonstration of acropetal movement of IAA-1-^{14}C. For explanation see text.

To find whether externally applied IAA could be transported upwards, a ring of
vaseline was placed 20mm below the cotyledons and a 10μl drop of 100ppm IAA-1-^{14}C was
placed against the hypocotyl. After 60 and 120min. two sections of 5mm were cut from
the top 10mm of the hypocotyl and counted (Fig. 7). Radioactivity did move upwards
in the first 60min. and the level of counts greatly increased by 120min. Plants which
had been cut at 60min. were left *in situ* and at 90min. the drop of fluid which had
exuded from the cut surface was absorbed on filter paper and also counted. This exudate mainly represents the contents of xylem vessels and the high counts measured
suggest that IAA can move acropetally in the xylem. We have evidence, therefore, of
considerable non-polar movement of externally applied IAA in the hypocotyl of an intact cucumber seedling.

IV. SUMMARY

Both exogenous auxins and gibberellins cause significant elongation of the cucumber hypocotyl. Auxins induce both rapid curvature and, as reported by others, cell wall
loosening in the cucumber hypocotyl. Gibberellin does not induce either effect. These
differences in response indicated a possible difference in the mechanism of action of
the two hormones and was investigated further.

The expected changes in water potential following hormone treatment were too small to be measured by a number of techniques including the Schardakov and gravimetric methods. The osmotic potential of cell sap from hypocotyl sections was measured both cryoscopically and by a combined refractometric-conductimetric method. No evidence was found for the hypothesis that an increase in osmotic potential is the driving force for GA-induced growth. This force must arise from the maintenance of the extensibility of the cell wall.

Using IAA-1-^{14}C, it was found that IAA applied to one cotyledon is rapidly transported asymmetrically to and down the hypocotyl, inducing curvature. The concentration of applied IAA controlled the timing and extent of curvature and was linearly related to the radioactive counts measured in the hypocotyl.

V. REFERENCES

BARRS, H.D. (1968). Determination of water deficits in plant tissues, in "Water Deficits and Plant Growth". (Ed. T.T. Kozlowski) Vol. 1. Academic Press, New York.

BIELESKI, R.L. and N.A. TURNER (1967). Separation and estimation of amino acids in crude plant extracts by thin-layer electrophoresis and chromatography. Anal. Biochem. $\underline{17}$, 278-293.

CLELAND, R., M.L. THOMPSON, D.L. RAYLE and W.K. PURVES (1968). Difference in effect of gibberellins and auxins on wall extensibility of cucumber hypocotyls. Nature, Lond. $\underline{219}$, 510-511.

DAINTY, J. (1969). The water relations of plants, in the "The Physiology of Plant Growth and Development". (Ed. M.B. Wilkins) pp 421-452.

DEGANI, Y. and D. ATSMON (1970). Enhancement of non-nuclear DNA synthesis associated with hormone-induced elongation in the cucumber hypocotyl. Expl. Cell Res. $\underline{61}$, 226-229.

KATSUMI, M., B.O. PHINNEY and W.K. PURVES (1965). The roles of gibberellin and auxin in cucumber hypocotyl growth. Physiol. Plant $\underline{18}$, 462-473.

SHIMSHE, D., and A. LIVNE (1967). The estimation of the osmotic potential of plant sap by refractometry and conductimetry: A field method. Ann. Bot. $\underline{31}$, 505-511.

ZENK, M.H. (1961). 1-(Indole-3-acetyl)-β-D-glucose, a new compound in the metabolism of indole-3-acetic acid in plants. Nature, Lond. $\underline{191}$, 493-494.

PLANT GROWTH SUBSTANCES, 1970

Promotion of Plant and Insect Hormone Action by Membrane Matrix Matching Lipids

Bruce B. Stowe

Kline Biology Tower, Yale University, New Haven, Connecticut 06520, U.S.A.

I. INTRODUCTION

At the 1959 meeting of this conference, I reported an unexpected effect of small quantities of diverse lipids on the growth and respiration of pea stem sections (Stowe, 1960). Such lipids often doubled the cell elongation caused by auxin, while having no growth promoting action of their own. Later studies showed that these lipids entered membrane-rich fractions and that the concomitant stimulation of respiration could, in some instances, be elicited by the lipids in the absence of growth (Penny & Stowe, 1965, 1966). Last year, we reported that some analogs of the juvenile hormone show parallel activity in insects and in the pea bioassay, and suggested that in plant and insects such lipids have a sterically similar locus of action in a membrane (Stowe & Hudson, 1969).

The intent of this paper is to examine in more detail the molecular dimensions of lipids biologically active in the pea and insect bioassays. Since other lipid effects on plants have been discussed earlier, they will not be reviewed again here. However, a recent paper by Mitchell *et al.* (1970), who are dealing with a somewhat similar lipid response in an intact bean internode assay, suggests that a new class of hormones, which they name "brassins", may be involved. After a dozen years work with the pea lipid response I remain uncertain that it can be characterized as a hormonal effect. To be sure, very small quantities of these lipids are effective in regulating both growth and respiration of an isolated plant part, but their molecular specificity is remarkably broad for a hormone, and most importantly, the translocation of a specific regulatory substance within the same plant has yet to be shown. This paper will present evidence that juvenile hormones and plant effective lipids have some characteristics in common, but are distinctly different.

II. MATERIALS AND METHODS

The techniques used have already been described (Stowe & Obreiter, 1962). Briefly, the data consist of measurements of elongation in 24hr of 10mm Progress No. 9 pea stem sections cut subapically from 7 day old seedlings grown at 25° and 80% relative humidity while receiving constant very weak red light. Ten sections are incubated in a basal medium of 20ml 50mM KH_2PO_4 at pH 5.5, 50 μM $CoCl_2$, 1 1/2% sucrose and an emulsion stabilizer, usually 0.004% Pluronic F-68 (Wyandotte Chemical Co.). The red light treatment and a readily metabolizable sugar are needed for the lipid response. Auxin, as 1.8μM IAA, and 0.3μM gibberellic acid together provide maximal growth stimulation. Auxin is essential and without it neither lipids nor GA produce significant elongation of the sections.

Conventional gas chromatography was used to check the homogeneity of the lipids tested, which were of the highest purity available. Usually 3% SE-30, or its OV-9 equivalent were used, while some compounds were better resolved on polyethylene glycol succinate columns at appropriate temperatures and loadings. Although some substances contained a few percent of a minor impurity, since the data obtained are consistent with the assumption that the major component is the effective one, no attempt will be made to detail the homogeneity of these many compounds here.

In assessing relative activities of substances tested in a bioassay, it is usually preferable to compare concentrations at some specific value corresponding to less than maximal activity. But for efficiency in screening a large number of compounds, a logarithmic series of concentrations was employed. The lowest concentration showing a 5% significant difference above the controls was used for comparing relative activities. The reciprocal of this micromolar concentration permitted comparison of specific activities on a per molecule basis. The full data and statistical justification are published elsewhere (Stowe & Dotts, 1971).

In calculating lengths of lipid molecules the techniques and assumptions of van den Heuvel (1963) were used, and all cited dimensions are between the limits reached by the van der Waal's radii of the terminal atoms. Scale models constructed from "Framework Molecular Models" (Prentice-Hall, Inc., Englewood Cliffs, N.J.) were helpful in assigning values to some strained structures. Although calculations were made to the nearest 1/10 Å, experience with these methods leads to the belief that a 2 to 4% variation from the maximum length of these non-rigid molecules may exist, quite apart from any modifying effect of the lipid matrix at the active site into which these substances presumably fit.

III. RESULTS AND DISCUSSION

To further test the conclusion of Stowe and Hudson (1969) that molecules effective in the pea bioassay must be at least 20Å in length, three series of alkyl halides were assayed. The results are summarized in Table I. No molecules shorter than 20Å significantly stimulated growth, even when tested over a wide range of concentrations, while all available longer substances were active. The activity of dodecyl iodide indicates that absolute length rather than carbon chain length is crucial, since dodecyl bromide with the smaller van der Waal's radius of bromine is inactive. As length of each molecule increases, so does the specific activity up to the longest substance tested, which in each series was the limit at which a stable emulsion could be maintained.

Table I. Relationship of molecular length of alkyl Halides to growth promotion of pea stem sections

Hydrocarbon Chain	Chloride		Bromide		Iodide	
	Length (Å)	Specific Activity (1/μM)	Length (Å)	Specific Activity (1/μM)	Length (Å)	Specific Activity (1/μM)
Decyl	16.7	nil	17.0	nil	17.3	nil
Undecyl	18.0	--	18.2	nil	18.6	nil
Dodecyl	19.2	nil	19.5	nil	19.8	0.05
Tridecyl	20.5	--	20.7	0.03	21.1	0.1
Tetradecyl	21.7	0.05	22.0	0.05	22.4	--
Pentadecyl	23.0	0.05	23.2	--	23.6	0.1
Hexadecyl	24.2	0.1	24.5	0.1	24.8	0.2
Octadecyl	26.8	--	27.1	--	27.3	0.2

Insect juvenile hormone assays similarly indicate that a minimum length is required for activity, this is well demonstrated in the homologous series tested by Schneiderman *et al*. (1965). Their data also show an optimum length, and a maximum length beyond which activity disappears. Since our inability to emulsify them precluded tests of longer alkyl halides, and since Schneiderman *et al*. (1965) had found ethers and sulfides to be effective mimics of insect juvenile hormone, homologous series of dialkyl ethers and sulfides were tested on peas. The data are summarized in Table II.

Table II. Relationship of molecular length of dialkyl ethers and sulfides to growth promotion of pea stem sections

Hydrocarbon Chain	Ether		Sulfide	
	Length (Å)	Specific Activity (1/μM)	Length (Å)	Specific Activity (1/μM)
Dihexyl	19.2	nil	19.8	nil
Diheptyl	21.7	nil?	22.3	0.03
Dioctyl	24.2	0.05	24.8	0.05
Dinonyl	26.7	...	27.3	0.1
Didecyl	29.2	0.2	29.8	0.2
Diundecyl	31.7	...	32.3	0.1
Didodecyl	34.2	0.1	34.8	0.03
Ditetradecyl	39.2	...	39.8	0.03

The shorter members of these series produced erratic results, a few dishes showing highly significant growth promotion, while overall averages of as many as 8 tests did not show significant differences. This behaviour is not normal to the bioassay, and it may be a result of their relatively high volatility permitting escape from the test solution. In any case, the strict 20Å limit dependence evinced by the alkyl halides is not clearly shown in the ether and sulfide series, the first dependably significant activity being that of diheptyl sulfide of 22Å length. Both series reach and pass a maximum specific activity, at optimum molecular lengths of 29.2Å for didecyl ether and 29.8Å for didecyl sulfide. Longer molecules than those listed here are available, but again stable emulsions are difficult to maintain, and at this stage in the investigation no maximal length is discernible in the scattered data so far obtained.

Table III. Relation of molecular length of alkyl benzenes to growth promotion of pea stem sections

Benzene	Overall Length (Å)	Alkyl Length (Å)	Specific Activity (1/μM)
Decyl	19.3	14.7	nil
Undecyl	20.5	15.9	0.03
Dodecyl	21.8	17.2	0.05
Tridecyl	23.0	18.4	0.05
Pentadecyl	25.5	20.9	0.10
Heptadecyl	28.0	23.5	0.20
Nonadecyl	30.5	25.9	0.33

Another homologous series available were the alkyl benzenes, tests are summarized in Table III. Some tests of the shortest, decyl benzene, reached significant values but the overall average was not significant. It too may be sufficiently volatile to escape from the test solution. Boiling points of all compounds tested which are close to 20Å in length vary over a wide range, and some are solids, so volatility does not provide a simple physical explanation of the minimum length requirement. Nonadecyl benzene of 30.5Å length has one of the highest specific activities yet found, another indication that the optimal molecular length may be near 30Å. The most interesting aspect of the alkyl benzene data is the demonstration that it is the overall length of the molecule, rather than that of the hydrocarbon chain, which is important to activity. Hence even the benzene ring has room to fit into the matrix at the active site.

In this connection, molecules with substituents at *both* ends of the hydrocarbon chain were tested, but all have so far proved negative, even if the length of the overall molecule, or of the hydrocarbon chain internal to it, exceeded 20Å. Thus, dimethyl dodecandioate, dimethyl octadecandioate, tetradecanedinitrile and dimethyl traumatate are biologically ineffective. This implies that one end of the molecule must be free of even slightly polar substituents, to fit the active site.

Modified chains. With length limitations established it should be possible to assess the effect of substituents internal to the chain. For this purpose 8-pentadecanol, 8-pentadecanone and hexadeca-8-yne, all near 24Å in length, were tested. Only hexadeca-8-yne was active. Ethers, sulfides and unsaturated fatty acid esters of similar length were previously shown to be active. Thus double or triple bonds, or a sulfur or oxygen atom within the chain do not strongly modify acceptability at the active site, while carbonyl and hydroxyl groups are inhibitory. Juvabione and other active juvenile hormones are ketones (Schwarz *et al*, 1970).

Although hydrocarbons with terminal alcohol groups are active in the pea test (Stowe 1960), they seem to require chain lengths longer than 20Å, but are difficult to maintain in a stable emulsion. These primary alcohols are the only marked exception to the generality that all simple alkyl hydrocarbons are active down to a 20Å length. This is true of fatty acid esters, alkyl acetylenes, alkyl nitriles, alkyl bromides and alkyl iodides, and probably alkyl chlorides, alkyl benzenes, and dialkyl sulfides and ethers also adhere to this rule.

Branched chains. Despite this near unanimity on the minimum length requirement of a simple unbranched chain, work in progress indicates that introduction of bulky groups on the chain may modify the situation. Natural isoprenoids such as α-tocopherol, vitamin K_1, phytol, were previously reported to be active (Stowe & Obreiter, 1962) and related substances like methyl farnesenate, methyl phytanate, and methyl phytenate can now be added. Tests of a range of ubiquinones show these to have only a moderate change in specific activity with varying chain length. Perhaps these are more readily subject to metabolism and all are convertible to active lengths.

More disturbing is the fact that members of the "iso" and "anteiso" monomethylated fatty acid methyl ester series are in some cases active, in others inexplicably not, and are being tested further. Since the absolute stereochemistry of these synthetic compounds is not defined in most cases, and some are probably RS-mixtures, publication of these results will await better knowledge of their true configuration and composition.

This as yet uncompleted work on methyl-substituted chains, although somewhat at variance with the unbranched chain results, is not necessarily contradictory. If a pit in a membrane is being plumbed with an unencumbered chain having various substituents at one end, it is not surprising that agreement on the minimum depth of the hole might be reached. But when bulky methyl groups enlarge the chain, then steric requirements due to the cross-sectional shape of the hole could well be imposed.

Comparison with insect activity

These studies do make possible an assessment of the analogy between the pea lipid effect and insect juvenile hormones noted by Stowe and Hudson (1969). As Fig. 1 shows, when specific activities of homologous series tested on the pea are compared with Schneiderman *et al's* (1965) *Galleria* moth data, two distinct regions of maximum activity appear. Insect active substances are most active at 21 to 22Å in length while pea stem growth is best promoted by compounds 28 to 30Å. There is an area of overlap between 20 and 24Å where lipids of those lengths might be expected to have activity in both systems.

This result could be chance, resulting from the different homologous series used in the testing. However, when the compounds known to have the highest specific activities in various bioassays are compared, as in Table IV, those substances most active in the plant and on insects continue to fall into two distinct groups, the former near 30Å and the latter near 21Å.

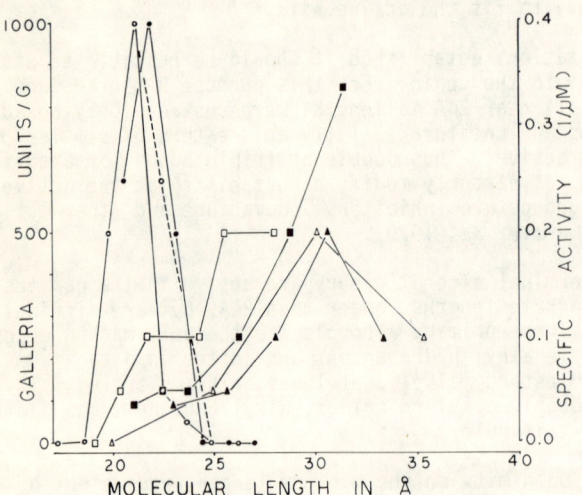

Fig. 1. Comparison of specific activities of homologous series of lipids tested in the *Gall Galleria* wax moth assay and on peas. Referred to left ordinate, *Galleria* data of Schneiderman *et al* (1965) o———o alkyl methyl ethers, o - - - - o dodecyl alkyl ethers, ●———● farnesyl alkyl ethers. Referred to right ordinate, pea bioassay, □———□ alkyl iodides, ■———■ alkyl benzenes, △----△ dialkyl ethers and ▲----▲ dialkyl sulfides.

The molecular requirements of both effects still lead to the assumption that both are matching a lipid matrix, most likely in a membrane, but that matrix or membrane has different dimensions in peas and in insects.

To be speculative, there is still one way to link these effects. We have shown earlier (Penny & Stowe 1965, 1966) that the initial lipid effect in peas is probably a respiratory stimulus, and have noted elsewhere (Stowe & Hudson 1969) that both pea and insect sites of action might be in a membrane controlling respiratory function. In these connections, it is noteworthy that the linear dimensions of 4 classes of lipids known to be involved in respiratory election transport are comparable to our results.

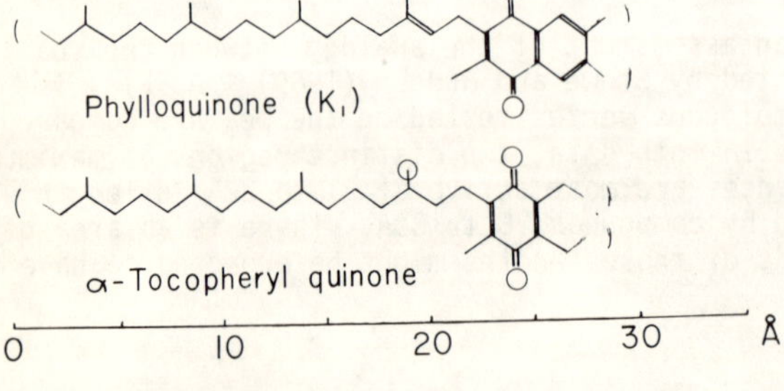

Fig. 2. The linear molecular dimensions of four lipoidal electron transfer co-factors.

As Fig. 2 shows, ubiquinone-20, plastoquinone-20 phylloquinone and α-tocopheryl quinone have isoprenoid chains near 21Å in length while the overall dimensions approach 30Å. Those isomers of ubiquinone and plastoquinone are not known to be common, but phylloquinone and α-tocopherol are of widespread occurrence in plants and animals. It could be then, that insects are modulating a respiratory response via effects on the 21Å isoprenoid moiety of the molecule, while peas require a longer analog of the whole molecule. Obviously, the data herein are only weak and preliminary evidence for such a hypothesis. The concept is advanced in order to emphasise the need to keep comparing these substances in plants and in animals to membrane dimensions and membrane functions.

Table IV.

Comparison of molecular lengths of highly active lipids

Substance tested assay	Length	Specific Activity (1/μM)	Reference
Pea bioassay			
Octadecyl iodide	27.3	0.2	Stowe & Dotts (1971)
α-tocopherol	28.0	0.33	Stowe & Obreiter (1962)
Phylloquinone (K_1)	29.0	0.33	" " "
Didecyl ether	29.2	0.2	Stowe & Dotts (1971)
Didecyl sulfide	29.8	0.2	" " "
Nonadecyl benzene	29.3	0.33	" " "
Galleria bioassay		(G.u./g)	
Dodecyl methyl ether	20.4	1,000	Schneiderman et al (1965)
Farnesyl ethyl ether	21.4	1,000	" " "
Tenebrio bioassay		(T.u/μg)	
Farnesal	17.8	--	Schmialek (1963)
Juvenyl methyl ketone	19.2	33	Schwarz et al (1970)
Homojuvenyl methyl ketone	19.2	1,000	" " "
Farnesyl acetone	20.3	--	Schmialek (1963)
Geranyl methylene dioxyphenyl ether	20.0	3	Zaoral & Sláma (1970)
Methyl homojuvenate (trans,trans,cis)	20.3	5,000	Röller & Dahm (1968)
Farnesyl acetate	21.4	--	Schmialek (1963)
Methyl homojuvenate (trans,trans,trans)	21.5	2,000	Roller & Dahm (1968)
Ethyl homojuvenate (Trans,trans,cis)	21.5	40,000	" " "
Ethyl farnesoate-HCl product	21.5	100	" " "
Ethyl homojuvenate (trans,trans,trans)	22.8	16,000	" " "
Rhodnius bioassay		(1/μg for grade 10 response)	
Farnesyl methyl ether	20.1	0.25	Wigglesworth (1969)
Methyl homojuvenate (trans,trans,cis)	20.3	2.0	" " "
Methyl homojuvenate (trans,trans,trans)	21.5	0.56	" " "
Ethyl homojuvenate (trans,trans,cis)	21.5	2.0	" " "
Ethyl farnesoate + HCl product	21.5	0.56	" " "
Pyrrhocoris bioassay		(1/μg for 50% response)	
Juvabione	17.7	1	Zaoral & Sláma (1970)
Methyl 3,7,11-trimethyl-2-dodecenate	20.3	0.2	" " "
Geranyl methylene dioxyphenyl ether	20.0	0.2	" " "
Geranyl p-hydroxymethyl benzoate ether	22.5	2	" " "

IV. SUMMARY

The growth response to homologous series of lipids by pea stem sections cut from red-light treated seedlings and incubated with auxin and sugar has been correlated with the linear molecular dimension of the lipids. Alkyl halides show no activity until an overall length of 20Å is reached, confirming previous work with fatty acid esters, alkyl nitriles and alkyl acetylenes. Specific activity of these and alkyl benzenes increases with molecular length, and in the case of dialkyl ethers and dialkyl sulfides reaches an optimum just below 30Å. Substituents within the chain have varying effects, carbonyl and hydroxyl groups being inhibitory, while methylation as in ubiquinones and "iso" and "anteiso" fatty acid esters appears to modify the acceptable chain length. Comparison with the dimensions of lipids active as insect juvenile hormones shows their optimum length to be near 21Å, but that lipids of 20 to 24Å length have some activity in both the pea and in insects.

V. ACKNOWLEDGEMENTS

This investigation was in part supported by U.S. Public Health Service research grant GM-06921 from the National Institutes of Health. Mrs. Mary Ann Dotts provided able technical assistance.

VI. REFERENCES

MITCHELL, J.W., N. MANDAVA, J.F. WORLEY, J.R. PLIMMER and M.V. SMITH (1970). Brassins - a new family of plant hormones from rape pollen, Nature, Lond. 225, 1065-1066.

PENNY, D. and B.B. STOWE (1965). The relationship between the growth and respiration induced by lipids in pea stem sections. Pl. Physiol. Lancaster 40, 1140-1145.

PENNY, D. and B.B. STOWE (1966). Relationship of lipid metabolism to the respiration and growth of pea stem sections. Pl. Physiol. Lancaster 41, 360-365.

RÖLLER, H. and K.H. DAHM (1968). The chemistry and biology of juvenile hormone. Recent Prog. Horm. Res. 24, 651-680.

SCHMIALEK, P. (1963). Über Verbindungen mit Juvenilhormonwirkung. Z. Naturf. 18b, 516-519.

SCHNEIDERMAN, H.A., A. KRISHNAKUMARAN, V.G. KULKARNI and L. FRIEDMAN (1965). Juvenile hormone activity of structurally unrelated compounds. J. Insect Physiol. 11, 1641-1649.

SCHWARZ, M., P.E. SONNET and N. WAKABAYASHI (1970). Insect juvenile hormone: Activity of selected terpenoid compounds. Science, N.Y. 167, 191-192.

STOWE, B.B. (1960). Growth promotion in pea stem sections. I. Stimulation of auxin and gibberellin action by alkyl lipids. Pl. Physiol. Lancaster 35, 262-269.

STOWE, B.B. and J.B. OBREITER (1962). Growth promotion in pea stem sections. II. By natural oils and isoprenoid vitamins. Pl. Physiol. Lancaster 37, 158-164.

STOWE, B.B. and V.W. HUDSON (1969) Growth promotion in pea stem sections. III. By alkyl nitriles, alkyl acetylenes and insect juvenile hormones. Pl. Physiol., Lancaster 44, 1051-1057.

STOWE, B.B. and M.A. DOTTS (1971). Probing a membrane matrix regulating hormone action. I. The molecular length of effective lipids. Plant Physiol. 48: 559-565.

VAN DEN HEUVEL, F.A. (1963). Study of biological structure at the molecular level with stereomodel projections. I. The lipids in the myelin sheath of nerve. J. Am. Oil Chem. Soc. 40, 455-471.

WIGGLESWORTH, V.B. (1969). Chemical structure and juvenile hormone activity: Comparative tests on *Rhodnius prolixus*. J. Insect Physiol. 15, 73-94.

ZAORAL, M. and K. SLÁMA (1970). Peptides with juvenile hormone activity. Science, N.Y. 170, 92-93.

PLANT GROWTH SUBSTANCES, 1970

Gibberellin and Membrane Permeability

A. Wood, L.G. Paleg and Ravindar Sawhney*

Department of Plant Physiology, Waite Agricultural Research Institute, The University of Adelaide

* Present address: Department of Biology, Yale University, New Haven, Connecticut

I. INTRODUCTION

Studies of the effects of auxins on the permeability of membranes have a long history. Much of the work was orientated towards an examination of the physical effects of auxins on model systems, e.g., Bungenburg de Jong and Bonner (1935), Veldstra (1947), Brian and Rideal (1952), such as monolayers composed of natural membrane constituents. These have been supplemented by studies with selected tissues, such as bean endocarp which becomes water-logged in the absence of auxins (Sacher, 1959). As early as 1944, Veldstra remarked on the requirement for a lipophilic basal ring structure and a projecting hydrophilic group in active auxins and compared these structures to wetting agents and penetrants. More recently the auxin effect has been shown to be essentially instantaneous (Nissl and Zenk, 1969), lending strong support to the earlier work. It suggests that at least one auxin-induced response does not require the relatively long period which might be associated with direct gene activation by the hormone.

Investigations of the mechanism of gibberellin action have followed a different course derived from results with animal hormone systems. Some animal hormones stimulate adenyl cyclase, located in the target membrane, with the consequent production and release of cyclic 3', 5'-AMP. This nucleotide, in animals, is an important enzyme regulator. The cyclic compound has been identified in barley endosperm and GA_3 is purported to stimulate the incorporation of adenine into AMP (Galsky and Lippincott, 1969; Pollard, 1970), sporadic success having been obtained in stimulating α-amylase production in barley endosperm by the addition of high levels of 3', 5'-AMP. We have failed to confirm this with extensive tests using several sources of nucleotide. Furthermore, Rasmussen (1970) suggests that adenyl cyclase stimulating activity is probably mainly associated with peptide hormones.

An aspect of animal steroid action that has received considerable attention in recent years involves direct effects on various membrane components organized into model systems. Similar systems have given good correlative results when tested with drugs, antibiotics and hormones which are known to cause lysis of various cellular organelles such as mitochondria, lysosomes and erythrocytes (Sessa and Weissman, 1968). We have explored the activity of GA_3 in several model systems composed partly or fully of components of plant origin.

II. MATERIALS AND METHODS

Thirty mM solutions of the lipids (lecithin-PC; phosphatidyl inositol-PI; phosphatidyl serine-PS; phosphatidyl ethanolamine-PE; cholesterol; sitosterol) were prepared in chloroform or petroleum ether and isopropanol (1:1). In some cases dicetyl phosphate or cetyl pyridinium bromide was also used. The lipids were deposited in the required volumes into a 250 ml round bottom flask and evaporated with a nitrogen stream. Ten 1 of 0.15M glucose and a small glass bead to aid dispersion were also added. The flask was again flushed with nitrogen, stoppered tightly and shaken vigorously at a controlled temperature, usually 25°C, for 4 hrs. The resultant milky dispersion was placed, 3.0 ml at a time, on a Sephadex G25 coarse grade column (Vo approx. 16.0 ml) and eluted with 0.15 M KCl. Three and a half ml were collected immediately after the void volume (established by blue dextran), and 0.1 ml of the eluted dispersion was analysed for phosphorus (Bartlett, 1959). One ml aliquots were placed in test tubes which contained GA_3 of appropriate concentration, previously evaporated from a methanolic solution with nitrogen. The tubes were flushed, stoppered and placed for 1 hr or less in a water bath. At the end of the chosen period the solutions were again freed of external glucose by a Sephadex column, (Vo approx. 7.2 ml). Five ml, containing all micelles, was collected after the void volume, and the free glucose was collected in the next 25 ml. One and 2.0 ml respectively of the above solutions were used in the Somogyi (Paleg, 1959) procedure for glucose assay.

III. RESULTS AND DISCUSSION

The effects of several concentrations of GA_3 on the leakage of chromate ions from micelles composed of crude PC, sitosterol, cetyl pyridinium bromide and linoleic acid are shown in Figure 1. This was an early experiment in which micelles were confined to a dialysis sac and the external medium was changed every 15 minutes. Increasing the concentration of GA_3 increases the response up to a maximum effectiveness; the limiting feature is probably the diffusion of the marker through multiple layers of lipid.

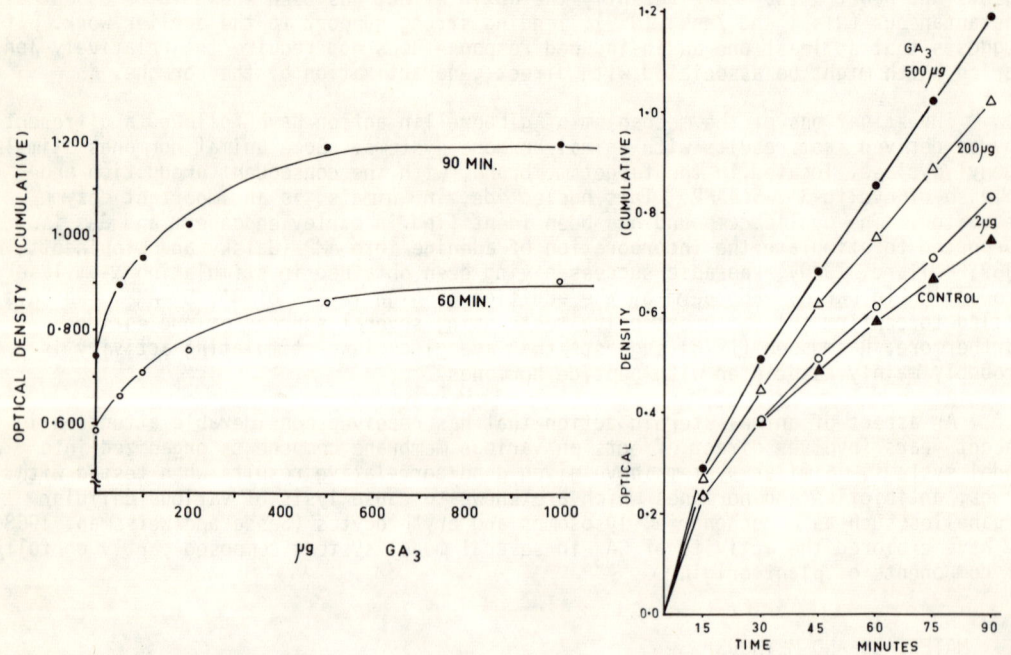

Fig. 1. GA_3-induced chromate leakage. *Conditions:* Micelles of lecithin-cetyl pyridinium bromide-sitosterol-linoleic acid 70 : 10 : 15 : 5 mole %. 0.15M potassium chromate removed on Sephadex G25. 2.0 ml aliquots incubated with GA_3 at 30°C for 15 min intervals against fresh 0.15 M NaCl/KCl. O.D. read at 340 nm.

Fig. 2. GA_3-induced chromate leakage. *Conditions:* as for Fig. 1.

The time course of the response is shown in Fig. 2 for three different concentrations of GA_3. As the volume of micelle solution was initially 2.0 ml, the lowest GA_3 treatment represents a concentration of 1 µg/ml added to approximately 10 mg of lipid. The continued steady leakage rate at 75 minutes suggests that the micelles are not leaking rapidly enough to attain equilibrium with the dialysis medium during each time interval, and that GA_3 is therefore probably genuinely modifying the permeability of the membranes, permitting the leakage of greater amounts of the trapped chromate. With the exception of the first 15 min interval, where there are larger errors of timing involved in the handling and tying off of the dialysis sacs, no GA_3 treated sample has a leakage rate below that of the control and the effect of the hormone appears to be immediate. After 60 mins the 500 µg GA_3 treatment had leaked 54% more chromate than the controls. The steroid-like membrane-active compound diethylstilbestrol, in the same system, attained a maximum rate of 165% that of the control, compared with a maximum rate of 140% for GA_3.

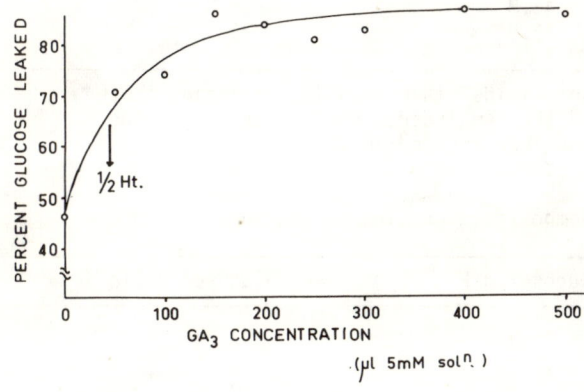

Fig. 3. GA_3-induced glucose leakage. *Conditions:* Micelles of crude lecithin and sitosterol 75 : 25 mole %. Incubation 1 hr at $35°C$ in 0.15 M KCl.

Figure 3 shows the effect of GA_3 using glucose as the trapped marker. The composition of the micelles differed from the previous experiments, and were crude lecithin and sitosterol in the molar proportions of 75:25. A similar curve can be seen, but here the 1 hr reading shows that over 85% of the marker has leaked. The half-effect concentrations of GA_3, at $35°C$ for 1 hr, was approx. 80 µg, acting on 10 mg lipid. Levels of 70 to 100 µg GA_3 have been consistently obtained as that which causes half maximum leakage with different micelle compositions. Varying the temperature at which the incubation with GA_3 is carried out has little influence on the shape of the curve although it influences other parameters. This again suggests that the diffusion of glucose is limiting, since the higher concentrations at the lower temperature ($25°C$) would be expected to approach the limiting leakage percentage at the higher temperature ($35°C$). The results of several experiments suggest that micelles consisting of unpurified lecithin, sterol and dicetyl phosphate give a maximum of 40% additional leakage during 1 hr at $35°C$ with GA_3, compared to untreated controls. If the dicetyl phosphate is omitted, the enhancement falls to 25-30%. The difference is probably largely accounted for by the greater intrinsic leakage of the control micelles. Purified lecithin gave micelles that did not show a GA_3 effect. Again, the leakage levels of the control were so high as to eliminate the response. Table 1 shows the influence of other pure phospholipids on the behaviour of the micelles and the hormone response. Other lipids which occur in membranes of certain plant organelles, such as monogalacto-diglyceride and digalacto-diglyceride, proved too hydrophobic to allow dispersion of mixed lipid films in water. Lipids such as these may be important in controlling permeability characteristics of specific membranes.

The data in Table 1 are arranged in order of control leakiness, and it can be seen that both phosphatidyl serine (PS) and phosphatidyl ethanolamine (PE) contributed to control of leakiness when added to PC and sterol, although no significant GA_3 effect was seen. In other similar experiments, data were also obtained on the amount of total glucose entrapped by the micelle immediately following dispersal (Table 2), and related to the amount of lipid present. This provides information on the surface area covered per unit of lipid and, on this basis, the very marked effect of small additions of phosphatidyl serine becomes apparent.

Table 1. The Effect of Micelle Composition on the GA_3 enhanced Leakage of Glucose.

*Composition	Glucose (% leaked)	GA enhancement
PC+PI	78.8	7.2
PC	68.3	7.1
PC+PE	63.0	1.5
PC+PS	50.8	nil
PC+PS+PE	47.3	nil
part purif. PC	46.6	nil
crude PC	43.8	28.8
crude PC (sitosterol)	42.9	25.3

*20% cholesterol, 20% DCP; phospholipids other than lecithin constitute 10% each and lecithin comprised the remainder. Conditions:- 1 h incubation at 35°. 200μl of 5 mM GA_3 per treated incubate.

Table 2. The Influence of Micelle Composition on Glucose Entrapment

*Composition	μg glucose/μg P	μg glucose/μg lipid
PC+PE+PS	11.7	0.49
PC+PS	10.7	0.45
PC+PI+PS	9.4	0.39
crude PC	8.6	0.30
crude PC	8.3	0.29
PC+PE	7.6	0.32
PC+PI+PE	6.1	0.29
PC+PI	5.6	0.24

*20% cholesterol, 20% dicetyl phosphate, and phospholipid other than PC constitute 10% each. (The remainder composed of PC.) Conditions:- 35°, 1 h. Averaged values of GA treated and untreated incubates.

The figures illustrate that there is a considerable change in leakiness as well as glucose retaining capacity of the micelles brought about by fairly small percentage changes in phospholipids. The contribution of PS to the intrinsic properties of micelles was particularly interesting as it seemed to control leakiness to the greatest extent; in fact to the point where GA effects might be expected to appear. To determine whether the proportion of sterol was also important in relation to PS we compared various percentages of sterol and PC, with and without PS (Fig. 4).

It is clear that unlike PS, which completely eliminated the GA effect, the leakiness control exerted by the sterol allows expression of the hormonal effect. At 30% sitosterol, glucose retention by the micelles is also at a maximum (Fig. 5). Part of the fall at lower sterol concentrations is probably due to the more rapid glucose leakage through PC during preparatory gel filtration, rather than a reduction in amount enclosed.

We can conclude that a membrane of pure plant-source PC is very permeable. There is some evidence, with micelles, (Demal et al., 1968) that unsaturation in the phospholipid fatty acids causes greater leakiness in membranes. Since there is, in general, a greater degree of unsaturation of the fatty acids of plant phospholipid, this feature could be a more active control mechanism in plants than in animals.

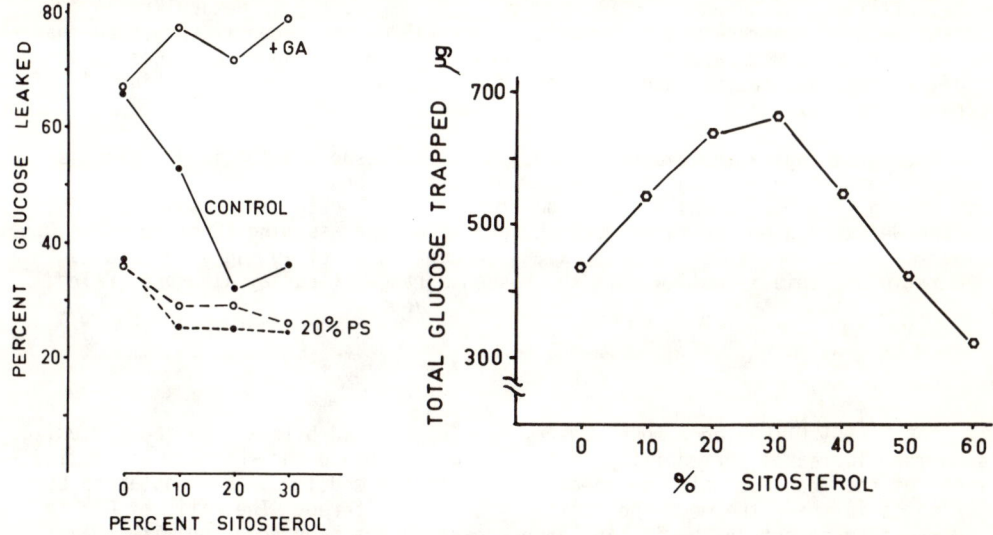

Fig. 4. The effect of sitosterol and phosphatidyl serine on micelle leakage. *Conditions*: Crude lecithin 100 mole%. Incubation 1 hr at 35°C. 200μl of 5 mM GA_3/treatment in 0.15M KCl.

Fig. 5. Glucose trapped versus micelle composition. *Conditions*: Lecithin (crude). 100 mole%. 1 hr incubation at 35°C. Glucose assayed by Somogyi procedure; each point average of four values.

Gibberellin induces leakiness without great structural damage to the bilayers. In experiments in which rates of leakage of different sized molecules were simultaneously compared, the larger molecule, sucrose, was always found to diffuse more slowly than the glucose, as shown in Table 3. The effect of GA was apparent with both molecules and since the percentage of sucrose leakage never approached the final high figure shown by glucose, it is clear that there was no major loss of membrane integrity. The effect of the hormone is superimposed on the intrinsic rate limiting parameters of the system.

Table 3. The Comparison of Simultaneous Sucrose and Glucose Leakage Rates

	% leakage	
	Glucose	Sucrose
Control	68	8.2
GA	82	20.6
Enhancement	14	12.4

Conditions: Micelles pure lecithin - dicetyl phosphate - cholesterol 78:11:11. Incubation 1 h at 35° GA_3 200μl of 5mM aq. solution.

The model system is a gross oversimplification of most cellular membranes. The micelles contain no enzymes, and no intervening layer of protein between lipid and water. The hydration of the phospholipid can greatly influence its properties (Ladbrooke, Williams and Chapman, 1968), and the micellar system is metabolically static whereas the cell membrane may be dynamic. Lucy (1970) has recently suggested that the micellar form of a membrane is a possible configuration during such processes as cell fusion, through the enzymic manipulation of acylation and hence the packing fit of membrane components.

The physiological relevance of model membrane responses to gibberellins is clearly demonstrated when the relative amounts of hormone and lipid are compared. The ratio of GA_3 to lipid at which a leakage stimulation is detectable is 1:500. The aleurone, weighing 3 mg, contains 1.7% phospholipid, and assuming 1% of this, or 0.5µg, is located in a responsive membrane, a GA_3 concentration of 10^{-9}µg also results in a 1:500 ratio and this is well within the response range of barley aleurone (Paleg, 1965).

IV. SUMMARY

The effect of GA_3 on a model system of phospholipid bilayers has been studied. The hormone increases permeability of the lipid to glucose and chromate ions. Indications of the means of building specificity into the model have been obtained by experiments in which the membrane constituents were altered. The ratio of GA_3 to lipid required to obtain the instantaneous response has been compared with that required for an aleurone response and the two were found to be of the same magnitude.

V. ACKNOWLEDGEMENTS

The technical assistance of B. Hancock and G.A.J. Dostal is gratefully acknowledged, as is support from the Australian Research Grants Committee.

VI. REFERENCES

BARTLETT, G.R. (1959). Phosphorus assay in column chromatography. J. Biol. Chem. 234, 466.

BRIAN, R.C. and E.K. RIDEAL (1952). On the action of plant growth regulators. Biochim. biophys. Acta 9, 1.

BUNGENBURG, DE JONG, H.G. and BONNER, J. (1935). Phosphatide auto-complex coacervates as ionic systems and their relation to the protoplasmic membrane. Protoplasma 24, 198.

DEMAL, R.A., S.C. KINSKY, C.B. KINSKY and L.L.M. VAN DEENAN (1968). Effects of temperature and cholesterol on the glucose permeability of liposomes prepared with natural and synthetic lecithins. Biochim. biophys. Acta 150, 666.

GALSKY, A.G. and J.A. LIPPINCOTT (1969). Promotion and inhibition of α-amylase production in barley endosperm by cyclic 3'5'-adenosine monophosphate and adenosine diphosphate. Plant and Cell Physiol. 10, 607.

LADBROOKE, B.D., R.M. WILLIAMS and D. CHAPMAN (1968). Studies on lecithin-cholesterol-water interactions by differential scanning calorimetry and x-ray diffraction. Biochim. biophys. Acta 150, 333.

LUCY, J.A. (1970). A working hypothesis for the fusion of biological membranes. Nature 227, 815.

NISSL, D. and M.H. ZENK (1969). Evidence against induction of protein synthesis during auxin-induced elongation of *Avena* coleoptiles. Planta 89, 323.

PALEG, L.G. (1959). Citric acid interference in the estimation of reducing sugars with alkaline copper reagents. Anal. Chem. 31, 1902.

PALEG, L.G. (1965). Physiological effects of gibberellic acid. Ann. Rev. Plant Physiol. 16, 291.

POLLARD, C.J. (1970). Influence of gibberellic acid on the incorporation of 8-^{14}C adenine into adenosine 3', 5' cyclic phosphate in barley aleurone layers. Biochim. biophys. Acta 201, 511.
RASMUSSEN, H. (1970). Cell communication, calcium ion, and cyclic adenosine monophosphate. Science 170, 404.
SACHER, J.A. (1959). Studies on auxin-membrane permeability relations in fruit and leaf tissues. Plant Physiol. 34, 365.
SESSA, G. and G. WEISSMANN (1968). Review describing properties of phospholipid spherules (liposomes) as a model for biological membranes. J. Lipid Res. 9, 310.
VELDSTRA, H. (1944). Researches on plant growth substances IV. Relation between chemical structure and physiological activity. (I). Enzymologia 11, 97.
VELDSTRA, H. (1947). Considerations on the interaction of ergons and their "substrates". Biochim. Biophys. Acta 1, 364.

RAPID ACTION OF AUXINS

Rapid Growth Responses in the Avena Coleoptile: A Comparison of the Action of Hydrogen Ions, CO_2, and Auxin

David L. Rayle[1] and Robert Cleland

Department of Botany, University of Washington, Seattle, Washington 98105

[1] Present Address: Department of Botany, San Diego State College, San Diego, California

1. INTRODUCTION

The ability of auxin to induce rapid cell elongation is well known, and it is often believed that auxins are the only compounds capable of causing a rapid growth response. Often overlooked is the fact that two other agents, CO_2 and hydrogen ions, can also induce a limited amount of rapid elongation (Harrison, 1965; Bonner, 1934). Until recently these responses received little attention because of the lack of suitable methods for continuously monitoring extension, but the advent of such techniques (Ray and Ruesink, 1962; Evans and Ray, 1969) has changed this situation. Studies on these two responses have recently been initiated (Rayle and Cleland, 1970; Rayle et al, 1970; Evans et al, 1970) with the hope that they might provide us with a clue as to how rapid cell elongation is regulated.

In this paper we have attempted to summarize the information now available concerning the growth responses to CO_2 and low pH, and to compare and contrast them with the growth response to auxin. In particular, we will show that while all three responses are distinctly different, the similarities between the auxin and low pH responses suggest that these two responses possess a common cell wall loosening process. As a consequence, the low pH response may prove to be a useful model for the study of the mechanism of cell elongation.

Finally, we will show that the hydrogen ion response can be demonstrated in frozen-thawed sections. This finding provides us with the tools for simulating cell extension in an essentially *in vitro* system, and therefore allows us to pursue questions about the mechanism of cell extension which were previously impossible to attack for technical reasons.

11. MATERIALS AND METHODS

The plant material consisted of 10 or 14 mm sections cut from 25-32 mm coleoptiles of *Avena sativa*, cv. Victory, beginning 3 mm below the tip. *Avena* seedlings were grown as described earlier (Cleland, 1967). Sections were either used fresh or were subjected to two cycles of rapid freezing with freon followed by thawing in 0.01 M phosphate buffer at room temperature.

Two methods were used to monitor cell extension. Columns of 12 fresh 10 mm sections were placed in the continuously-recording apparatus of Evans and Ray (1969) and their growth was recorded photographically (Method A). Alternatively, single 14mm sections, either fresh or frozen-thawed, were fixed between two sets of clamps held 5 or 7 mm apart, a constant force was applied and the resulting extension was read directly from dials (Method B) or was recorded automatically (Method C). Method B made use of an Instron extensometer while Method C used a specially fabricated apparatus (details of this apparatus will be described elsewhere). In every case the sections were allowed to incubate in the apparatus for at least one hour in the presence of buffer alone in order to reduce the endogenous growth rate.

Extensibility measurements were performed with an Instron TM-S linear extensometer, using methods already described in detail (Cleland, 1967). The capacity for irreversible extension is expressed as percentage of irreversible extension per 100 gram load.

All incubations at pH 7.0 made use of a 0.01 M phosphate buffer, and incubations at pH 3.0 were usually carried out with a 0.01 M citrate buffer although other buffer systems were checked in order to make certain that the observed effects at pH 3 were in response to the pH and not the type of buffer. Previous experiments showed that an optimal growth response occurred at a pH of 3 (Rayle and Cleland, 1970). For CO_2 treatment, the appropriate solutions were saturated with 100% CO_2 before their addition to the growth chamber, and then CO_2 was continuously bubbled into the chamber during the duration of the experiment. Except in the CO_2 experiments, air was continuously bubbled through the growth apparatus whenever fresh sections were employed.

Additional details of these procedures can be found in previous publications (Rayle and Cleland, 1970; Rayle et al, 1970).

111. RESULTS

Treatment of coleoptile sections with pH 3 buffer or CO_2-saturated water leads to rapid elongation after a lag of not more than 1 minute (Fig. 1). The initial growth rate at pH 3, which is similar in magnitude to that induced by optimal IAA, persists for about 30-50 minutes, then gradually declines and terminates after about 2 hours (Fig. 2).

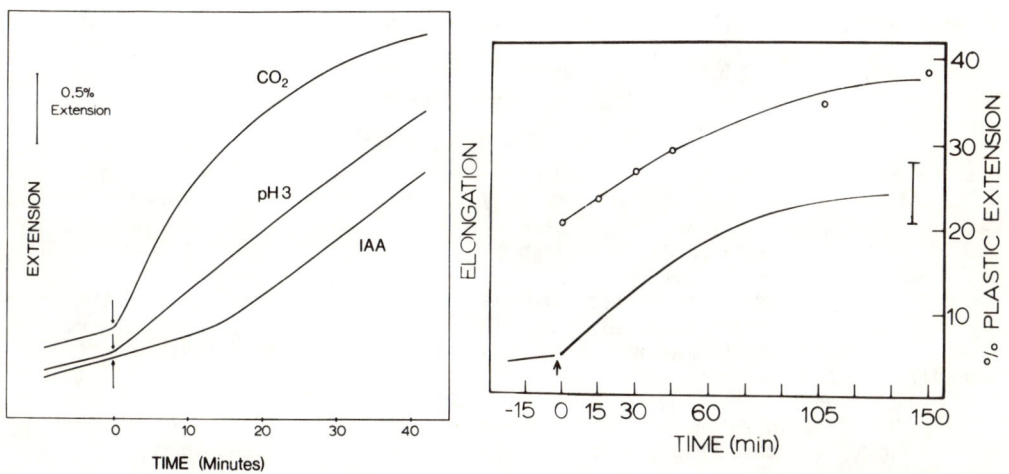

Fig. 1. Comparison of growth promotions induced by IAA (5 µg/ml), CO_2-saturated water and hydrogen ions (pH 3). Sections incubated 1 hr in pH 7 buffer, then buffer replaced by indicated solution, and growth monitored by method C. Note differences between the 3 responses in length of lag and initial growth rate.

Fig. 2. Time course of growth and wall extensibility at pH 3. Upper curve; wall extensibility expressed as percentage plastic extension/100 gram load. Lower curve is elongation, monitored by method A. At arrow, solution changed from pH 7 to pH 3.

The initial rate for the CO_2 response is considerably greater (2-3x) than any rate ever achieved with auxin or low pH, but does not persist; rather, the rate continuously falls off and reaches the control rate after 40-80 minutes. For both responses the inducing agent must continuously be present in order to maintain the elevated growth rate; transfer of sections back to a CO_2 free solution at pH 7 causes a rapid termination of both responses (Fig. 3). Obviously neither hydrogen ions nor CO_2 can act as a trigger for growth.

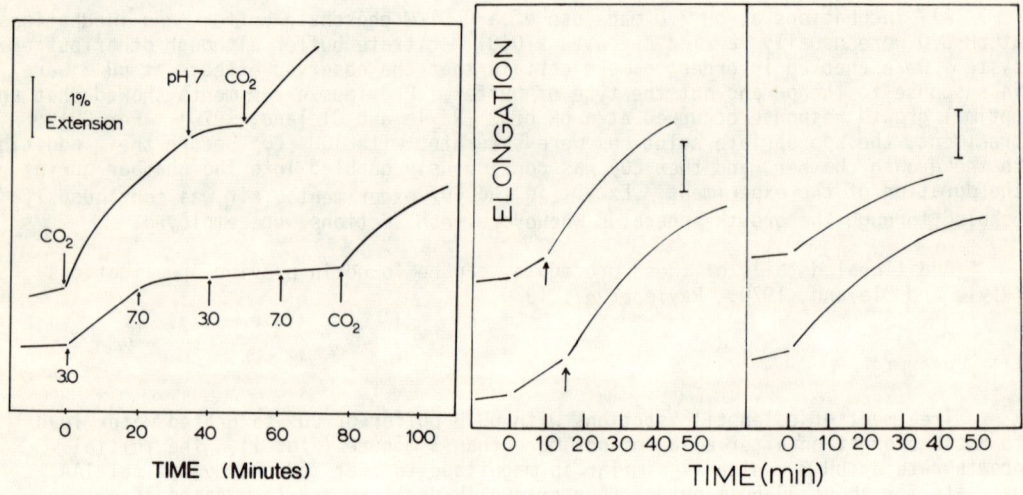

Fig. 3. Reversibility of growth responses to CO_2 and pH 3. Sections pre-treated 1 hr at pH 7, then growth monitored by method A. Upper curve: sections treated sequentially with CO_2-saturated water, pH 7 buffer, and again with CO_2. Lower curve: treated sequentially with pH 3 buffer, pH 7 buffer, pH 3 buffer, pH 7 buffer, then CO_2-saturated water. Note that the inducing agent must be present continuously in order to maintain the rapid growth response.

Fig. 4. Demonstration that response to CO_2-saturated water is not a pH effect. Left side: At time 0, segments changed from pH 7 to pH 3.8. At arrow solution changed to CO_2-saturated water (pH 3.8). Right side: At time 0, sections transferred from pH 7 buffer to CO_2-saturated buffer (final pH 5.9).

Since the pH of CO_2-saturated water is about 3.8, it seemed possible that the CO_2 response is, in fact, a pH response. This now seems unlikely for three reasons. First, the response to CO_2 at pH 3.8 is much greater than to pH 3.8 alone (Fig. 4A). Secondly, a growth response is induced by a CO_2-saturated phosphate buffer (final pH 5.9) even though this hydrogen ion concentration itself produces no growth (Fig. 4B). Finally, sections in which a pH response has been terminated by a pH 7 treatment cannot undergo a second response to hydrogen ions, but can respond to CO_2 (Fig. 3).

The pH response mimics the auxin response in several ways. For example, both produce almost the same maximum growth rate (Fig. 1) and wall extensibility. As with auxin (Cleland et al, 1971), wall extensibility rises rapidly following a lowering of the pH, but does not reach a maximum until 1.5-2 hours later (Fig. 2).

Table 1. Comparison of Q_{10} values for different growth responses. Growth monitored by method B.

Growth response	Q_{10} 15°C-25°C	25°C-35°C
pH 3.0, fresh	4-5	1.1
pH 3.0, frozen-thawed	5	1.2
IAA, 3×10^{-5} M, fresh	4	1.2

The Q_{10} values for the auxin and hydrogen-ion-induced growth rates are also similar (Table 1). Note especially the rather low values for the 25-35°C range. Finally, both auxin (Cleland et al, 1971) and hydrogen ions require that the walls be under normal tension in order to act as growth-inducing agents. This is shown by the fact that if the tension is temporarily reduced with mannitol, the growth rate upon resumption of the normal tension is no higher than the original growth rate; there has been no potential for growth stored up during the period of reduced turgor (Fig. 5).

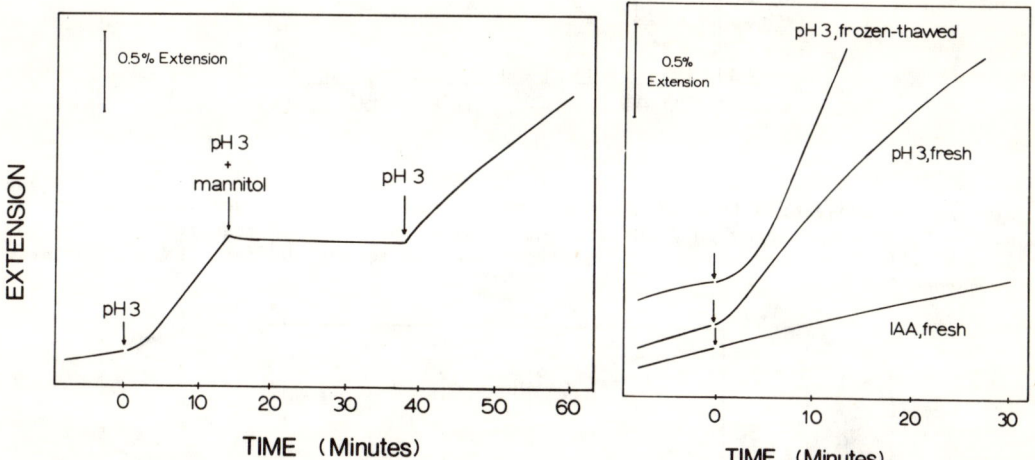

Fig. 5. Inability of hydrogen ions to act as a growth inducer when tension in walls is reduced. Sections first treated with pH 3 buffer, then 0.1 M mannitol added after 15 min to reduce wall tension. Sections returned to pH 3 after 25 minutes. Note absence of stored growth. Growth monitored by method A.

Fig. 6. Difference in auxin and low pH responses in regard to requirement for protein synthesis. Live sections pretreated 30 min (fresh) or 2hr (frozen-thawed) at pH 7 with 6 µg/ml cycloheximide. Fresh sections then given IAA at pH 7 or pH 3 buffer and growth monitored by method C. Other sections frozen-thawed, preincubated for 30 min in pH 7 buffer with applied load of 20 grams, then solution changed to pH 3 buffer and extension monitored by method C.

The pH and auxin responses do differ in one important respect; the growth-limiting proteins are required for the auxin response, but not for the response to low pH. Thus the growth of sections pretreated 30 minutes with 6 µg/ml cycloheximide cannot be enhanced by auxin but can still be stimulated by a pH 3 buffer (Fig. 6).

The effect of hydrogen ions is not restricted to fresh sections; a significant extension response can also be detected with frozen-thawed sections. To show this, frozen-thawed sections were subjected to an applied force of 20 grams (to replace the driving force of turgor pressure), the viscoelastic extension was allowed to occur, and when the extension rate had reached a low level the inducing agents were added. Neither auxin nor CO_2 had any effect, but a pH 3.6 buffer caused a rapid extension response after a lag of 3-10 minutes (Fig. 7). The response required the continued presence of the hydrogen ions (Fig. 7) but not the growth-limiting proteins (Fig. 6), and continued until extensions of greater than 30% have been achieved (Fig. 8).

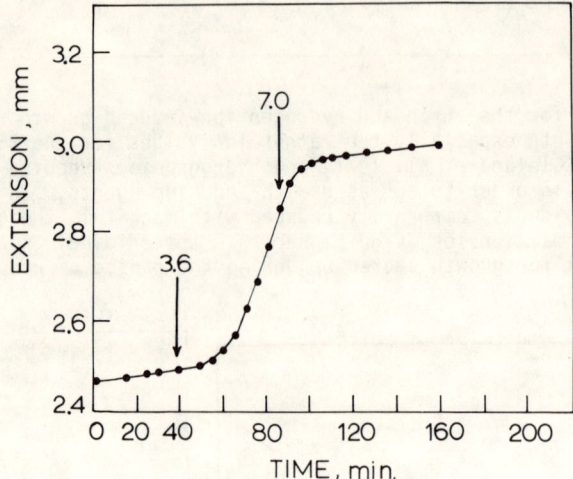

Fig. 7. Response of a frozen-thawed coleoptile section to hydrogen ions and termination of the response by pH 7 buffer. At first arrow solution changed from pH 7 to pH 3.6 buffer, and at second arrow section returned to pH 7 buffer. Extension measured by method B.

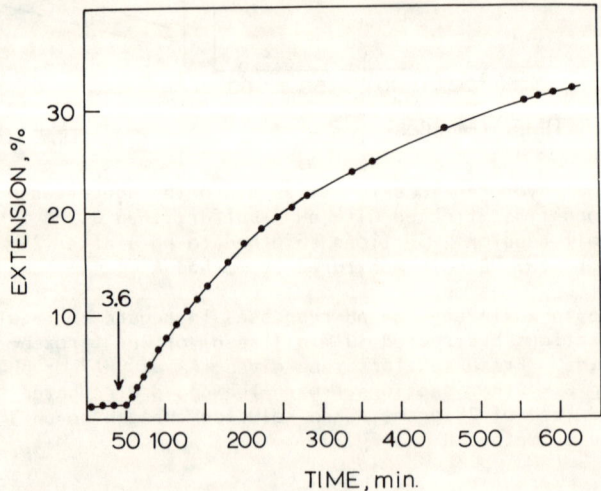

Fig. 8. Time course of response of frozen-thawed section to low pH. At 1st arrow section changed from pH 7 to pH 3.6 buffer. Elongation plotted at % initial length.

IV. DISCUSSION

Rapid extension of *Avena* coleoptile sections can be induced by three agents; auxin, CO_2-saturated water and acidic solutions (pH 2.8-4.0). The three responses have many characteristics in common, but it is shown here that each is distinctly different. For example, both the CO_2 and low pH responses begin after a lag of less than 1 minute, the duration of both responses is short, and neither requires protein synthesis. Evans *et al* (1970) have pointed out that the pH of CO_2-saturated water is about 3.8, and have suggested that the CO_2 response might, in fact, just be a response to the lowered pH. However, it is shown here that a CO_2 response can occur under conditions where the response to hydrogen ions is eliminated; either by use of solutions buffered at pH's above 5 or by use of sections whose pH response had been terminated by a treatment at pH 7. In both cases the initial growth rate induced by CO_2 was lower than that normally obtained with CO_2-saturated water. The data suggest that the response to CO_2 in water is actually a dual one; partly a response to CO_2 itself and partly to the hydrogen ions. Future studies on the CO_2 effect should make use of solutions buffered at pH's which are not themselves growth-promotive.

The auxin and low pH responses are similar in several ways. They have the same maximum growth rate and wall extensibility (Rayle and Cleland, 1970). In both cases wall extensibility begins to rise at the time rapid extension commences, but does not reach a maximum value until 1½ - 2 hours later (Cleland *et al*, 1971). The rather unusual Q_{10} for growth, high in the 15-25°C range and low between 25-35°C, is found for both responses (Ray and Ruesink, 1962). In both cases the action of the inducing agent ceases whenever the tension in the walls is reduced. But the two responses are distinctly different. Of greatest importance is the fact that continued protein synthesis is needed for the auxin response but not for the growth induced by low pH (Evans *et al*, 1970); apparently, the growth-limiting proteins are only needed for the auxin response.

These facts suggest to us that while auxin and hydrogen ions do not act in an identical fashion, the final step leading to growth may be the same for both responses. It is believed (Cleland, 1967; Ray, 1969) that these final steps involve a modification of the cell wall properties (a wall loosening), either by cleavage of some crosslink in the wall or by an alteration in wall synthesis. Presumedly both auxin and hydrogen ions must set in motion the events leading to this wall loosening process.

But what is the nature of the wall loosening process? Some indication of this may have come from our studies using frozen-thawed coleoptile sections. We wish to point out that this essentially *in vitro* system has certain advantages over the normal *in vivo* systems for studying the mechanism of cell extension. For example, because an applied force is used in place of turgor pressure, the response is not sensitive to the molarity of the incubating solutions or to materials which affect membrane permeability. This allows one to use solvents and solutions which could not be used with turgid sections. Another advantage of this system is the elimination of synthetic activity and wall synthesis, with the result that their contribution to the wall loosening process can be assessed.

Of particular interest is the finding that frozen-thawed walls can undergo rapid extension when treated with hydrogen ions, but not auxin, CO_2 or hydroxyl ions (Rayle *et al*, 1970). This *in vitro* pH response resembles the *in vivo* response in regard to its temperature dependence (Q_{10}) and its requirements that the walls be under tension and that the hydrogen ions be present continuously in order for rapid extension to continue. The two responses do differ in two respects; the lag for the *in vitro* response is considerably longer, and the *in vitro* response persists for considerably longer periods of time. We have no explanation for the former difference. The short duration for the *in vivo* pH response may be due to a general toxicity and loss of turgor caused by this level of hydrogen ions.

Although the evidence on hand is certainly not conclusive, it suggests to us that the *in vitro* extension is similar to the *in vivo* extensions which occur in response to low pH and auxins, and that the biochemical mechanisms underlying these reactions will be the same. If so, two conclusions can be drawn from these results. First, it is apparent that cell wall synthesis is not directly involved in the initiation of rapid cell extension. Wall synthesis will certainly be important for long-term cell elongation, but its role may only be to maintain the properties of the wall so further extension can take place. Secondly, it appears possible that the rate of wall extension is controlled by hydrolysis of acid-labile, alkali-stable cell wall bonds. One such bond, the hydroxyproline-arabinose link is known to occur in cell walls (Lamport, 1969) and other such bonds are thought to exist there as well (Lamport, 1970).

We wish to propose that the action of auxin may be to insure the presence in the cell wall of some factor which is capable of causing the rupture of acid-labile cell wall bonds. This in turn would lead to an increase in the extensibility of the wall and thus to cell extension. The way in which auxin exerts its effect on this factor (e.g. stimulates its synthesis or transport), the nature of this factor, and the identity of the acid-labile bonds remain to be determined.

V. SUMMARY

Rapid cell elongation can be induced by three agents; auxin, CO_2 and hydrogen ions. Although the three responses have many common characteristics, each is distinctly different. The characteristics of each are detailed in this paper.

The similarities between the auxin and low pH responses in regard to maximum growth rates, wall extensibility, temperature dependence and requirement for wall tension suggest that the final step in both responses, the loosening of the cell wall, is the same.

A system has been developed which allows one to study cell wall extension *in vitro*. It makes use of frozen-thawed tissues and an applied force to replace turgor. Extension in such a system can be induced by low pH but not by auxin or CO_2. The similarities between this response and the *in vivo* auxin and low pH growth responses suggests a common mechanism of wall extension. We propose that extension is controlled by cleavage of acid-labile, alkali-stable wall crosslinks, and auxin acts to insure the presence in the wall of some factor which can cause this cleavage.

VI. ACKNOWLEDGEMENTS

Supported by Atomic Energy Commission Contract No. AT(45-1)-2217, National Science Foundation Grant GB-5385, and a postdoctoral traineeship to DLR from USPHS Developmental Biology Training Grant HD00266.

VII. LITERATURE CITED

BONNER, J. (1934). The relation of hydrogen ions to the growth rate of the *Avena* coleoptile. Protoplasma 21, 406-423.
CLELAND, R. (1967). Extensibility of isolated cell walls. Planta 74, 197-209.
CLELAND, R., W.F. THOMPSON and P.M. HAUGHTON (1972). Macromolecule synthesis and wall extensibility in relation to the mechanism of auxin-induced cell elongation, in "Plant Growth Substances, 1970" Ed. D.J. Carr.
EVANS, M., and P.M. RAY (1969). Timing of the auxin response in coleoptiles and its implications regarding auxin action. J. Gen. Physiol. 53, 1-20.
EVANS, M.L., P.M. RAY and L. REINHOLD (1970). Stimulation of coleoptile elongation by carbon dioxide. Plant Physiol. 46
HARRISON, A. (1965). Auxanometer experiments on extension growth of *Avena* coleoptiles in different carbon dioxide concentrations. Physiol. Plant. 18, 321-328.
LAMPORT, D.T.A. (1969). The isolation and partial characterizations of hydroxyproline-rich glycopeptides obtained by anzymic degradation of primary cell walls. Biochemistry 8, 1155-1163.

LAMPORT, D.T.A. (1970). Cell wall metabolism. Ann. Rev. Plant Physiol. $\underline{21}$, 235-270.

RAY, P.M. (1969). The action of auxin on cell enlargement in plants, in "Communication in Development" (Ed. A. Lang) 173-205. Academic Press, New York.

RAY, P.M., and A.W. RUESINK. (1962). Kinetic experiments on the nature of the growth mechanism in oat coleoptile cells. Develop. Biol. $\underline{4}$, 377-397.

RAYLE, D.L. and R. CLELAND (1970). Enhancement of wall loosening and elongation by acid solutions. Plant Physiol. $\underline{46}$, 250-253.

RAYLE, D.L., P.M. HAUGHTON and R. CLELAND (1970). An *in vitro* system which simulates cell extension growth. Proc. Nat. Acad. Sci. $\underline{67}$, 1814-1817.

Cell Elongation and Auxin Action in Lupin Hypocotyls

David Penny, Pauline Penny, J. Monro and R.W. Bailey

Department of Botany and Zoology, Massey University, and Applied Biochemistry Division, D.S.I.R., Palmerston North, New Zealand

I. INTRODUCTION

The mechanism of action of auxins on cell expansion has been a major interest in plant physiology for forty years and despite slow but steady progress the molecular mechanisms involved are still not understood. In studying such an intractable subject our approach has been to measure as accurately as possible the growth response to auxin and then to measure selected parameters after auxin addition. If any change is found it is then tested to see whether the parameter changes at the same time as the growth rate and if it can quantitatively and not just qualitatively predict the observed growth rate response to auxin.

II. MATERIALS AND METHODS

Seedlings of bitter blue lupin *(Lupinus angustifolia)* or of Alderman peas (a tall variety) were grown in the light as described in Penny (1969). The method of measuring the elongation of single segments every minute is also described in that paper. On non-moving objects the measuring accuracy is $\pm$ one unit and each unit is about 0.226 μm. The preparation of segments for testing on the Instron Universal Testing Instrument is described in Penny et al. (1971) and is based on the method of Cleland (1967). The extraction of cell wall polysaccharide is based on the procedure used by Gaillard and Bailey (1968).

III. RESULTS AND DISCUSSION

Short Term Kinetics. Short term kinetics of the response of *Avena* coleoptiles to auxin were measured several years ago by Yamaki (1954) and Köhler (1956) who were able to show a latent period before auxin-induced growth occurred. By using light grown lupin hypocotyl and pea epicotyl segments it was shown (Penny 1969) that the latent period occurs not only in monocotyledonous plants but in dicotyledonous light-grown plants. Barkley and Evans (1969) demonstrated the latent period in etiolated pea epicotyl. The latent period is usual in auxin responses - it occurs in a wide variety of higher plants, with light-and dark-grown material, and with coleoptiles, hypocotyls and epicotyls.

When the growth response of lupins or peas is plotted as a rate there is a rapid rise after the latent period to a maximum after which the rate falls to a minimum and then rises to a second maximum (Fig. 2A). It is not clear whether the minimum is of general occurrence. It occurs in our light-grown tissue and is apparent when the graphs of Barkley and Evans (1969) for etiolated peas are replotted as rate curves. Although Köhler's work suggests that it may occur in *Avena* coleoptiles, later work (e.g. Rayle et al., 1970) does not show it. However, this may be a reflection of the different methods and sensitivity of measurement. The data in this paper were expressed as a rate because they were obtained in digital form. The system was more sensitive by an order of magnitude than that of other workers.

We have also found that if one averages the growth/min. of several segments, each of which clearly demonstrated the fluctuations in growth rate, then because of differences in timing between the sections, the minimum was much less apparent. This is illustrated in Fig. 1 which is an average of 7 segments taken from a table in Penny et al. (1971). The triangles represent the average time for the sections to reach their maxima and minimum and the average maxima or minimum rates of the different segments. The variation is that of individual segments in different experiments. It is not known what is the variation between segments where the growth of a file of segments is recorded and so it is not possible to tell to what extent individual differences between segments mask the fine structure of the short term kinetics.

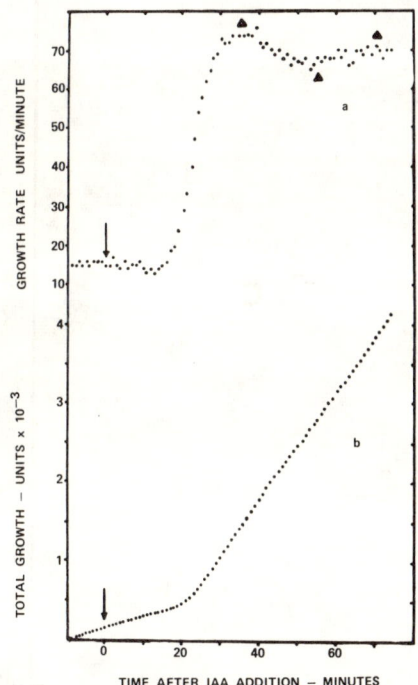

Fig. 1. Effect of averaging the growth rates of 7 segments on the a) growth rate curve and b) on the total growth curve. The Δ's in a) represent the average of the actual maximum, minimum and second maximum of the segments. 3×10^{-5}M IAA added at arrow. 1 unit = 0.23 μm. 22°C.

Protein Synthesis. Since it is unlikely that RNA synthesis is necessary for the initial action of auxin (e.g. Penny et al., 1971), the requirement for protein synthesis was studied. This was begun by examining the effect of cycloheximide (10 μg/ml) in the short term kinetics. Cycloheximide has been used by Evans and Ray (1969) and Barkley and Evans (1970) who showed that it had strongly inhibited auxin-induced growth. Two approaches have been used in our study of the effect of cycloheximide on IAA-induced growth:

1) after various periods of pretreatment before IAA addition

2) on the IAA-induced steady-state rate.

The effect of pretreatment in cycloheximide for 0, 5, 10, 15, 20, 30 min. is illustrated in Fig. 2. The amount of IAA-induced growth was reduced with increasing periods of pretreatment. The IAA-induced growth was measured by subtracting the endogenous growth rate from the auxin-induced growth and summing the area under the curve.

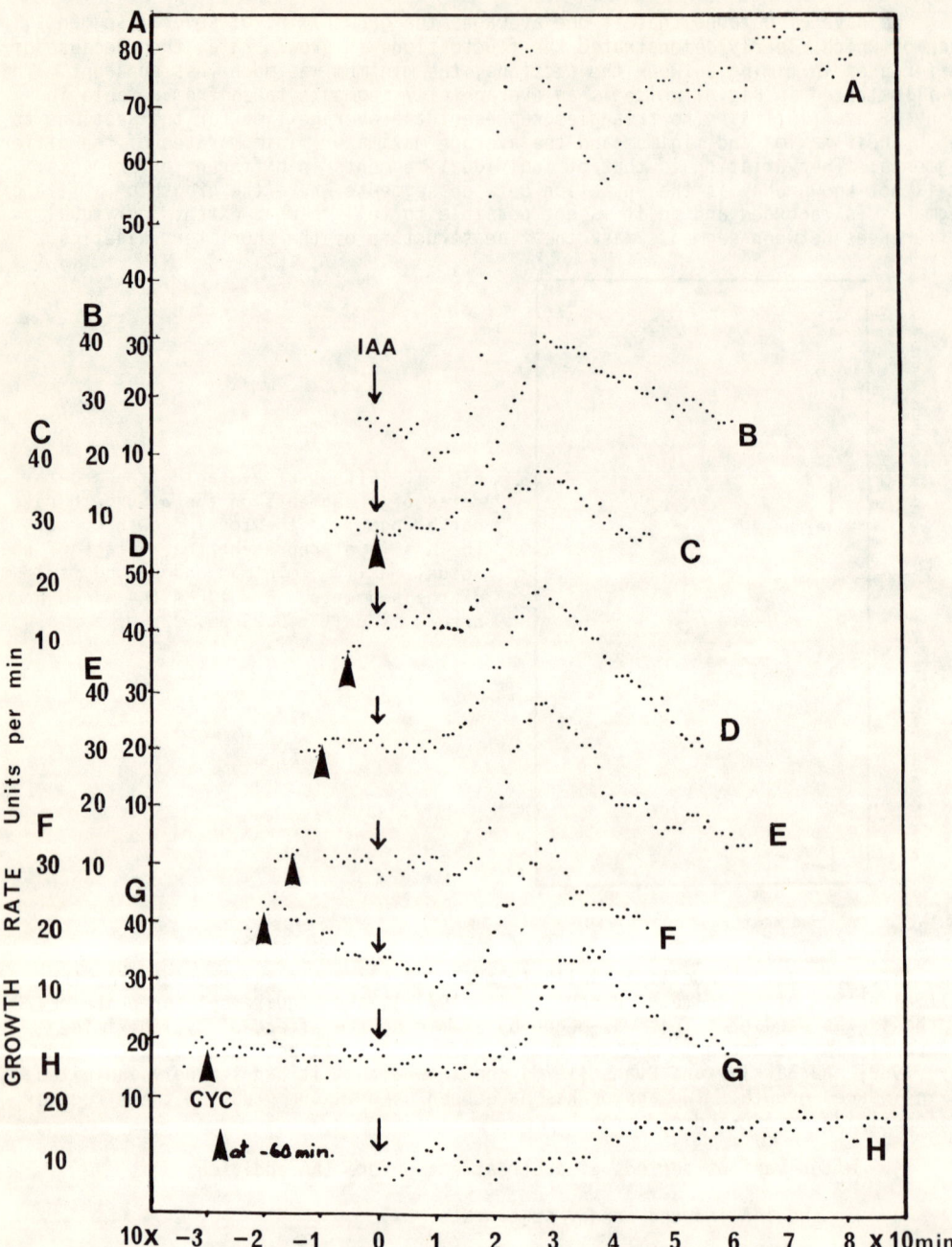

Fig. 2. Effect of various times of pretreatment in cycloheximide on the auxin-induced growth rate of lupin hypocotyl segments. Cycloheximide (10 μg/ml) added at upward pointing arrow. IAA (3×10^{-5}M) added at downward pointing arrow. Tris-maleate buffer. With supplementary light. 1 unit = 0.23 μm. 28°C.

Two main types of explanations of the short-term kinetics are possible. These are either that the two maxima are the result of separate responses to auxins that occur at different times. The alternative is that the response is due to a negative feedback system with under-damping. With either short applications (5 mins.) of auxin (Penny 1969) or pretreating with cycloheximide (Fig. 2) only the first peak is obtained. So far we have been unable to obtain the second peak without the first and it is thus not yet possible to decide which type of explanation is correct. A model based on negative feedback in protein synthesis is developed in more detail later in the paper.

The results indicate that auxin action depends on pre-existing protein(s) that have been synthesized a short time before auxin addition. This protein would probably be equivalent to the "growth limiting protein" (GLP) of Cleland (1972).

When the total IAA-induced growth was plotted against time of pretreatment, an exponential curve was obtained as illustrated by the curved line in Fig. 3. Also illustrated in Fig. 3. is the straight line which was obtained by plotting the logarithm of the IAA-induced growth. This curve could result from first order kinetics where the rate of removal of GLP from the compartment is a constant (k) times the concentration of GLP at any time. The amount of GLP at any time (t) (in minutes) is given by:

$$[GLP] = [GLP]_o \, e^{-kt} \quad\quad (1)$$

where $[GLP]_o$ is the concentration of GLP at time zero (i.e. when protein synthesis is stopped by cycloheximide). The semilog plot of the amount of auxin-induced growth was used to obtain the line of best fit by the method of least squares. The equation of the form y = mx+c is shown in Fig. 3. When converted into the form of (1) the equation becomes:

$$[GLP] = 1194 \, e^{-.060t}$$

This was used to calculate the apparent half-life and mean life of the protein, which were 11.6 and 16.7 min. respectively.

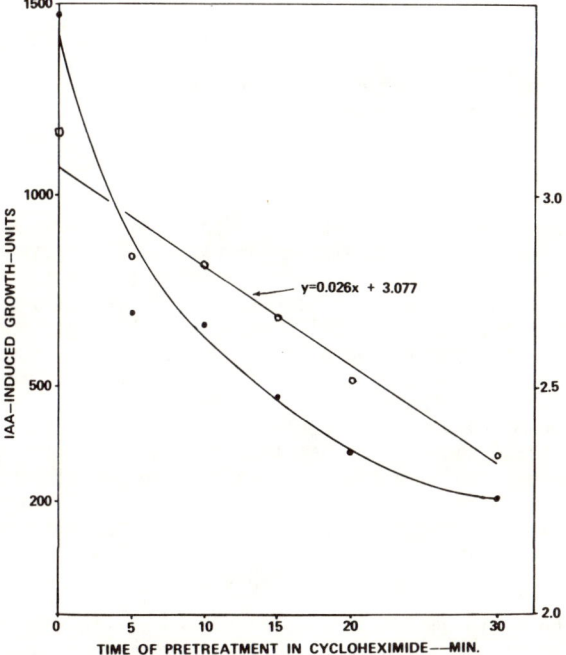

Fig. 3. Effect of various times of pretreatment in cycloheximide in auxin-induced growth. ● and ○ experimental points; ●———● linear plot, o ____ o semilog plot, line of best fit. 10 µg/ml cycloheximide; 3 x 10^{-5}M IAA; 1 unit = 0.23 µm.

This is an unusually short half-life for a protein from a higher organism (Evans and Ray, 1969; Dowben, 1969). It is similar to that estimated by Evans and Hokanson (1969). At present it seems more probable that the half-life obtained was a measurement not of rate of degradation of an enzyme but of the time that the protein is available to react with auxin either because the protein is removed from a compartment or undergoes an irreversible conformational change.

Computer Simulation. The information on the apparent half-life has been used to develop a model to explain the observed kinetics (Fig. 4).

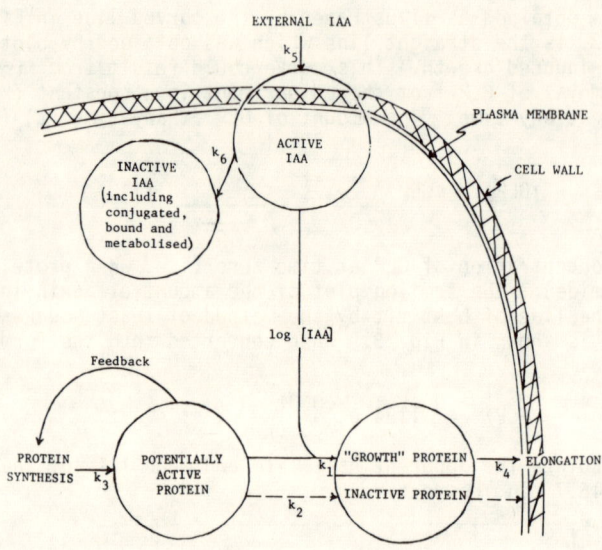

Fig. 4. Schematic representation of a model for the interaction of IAA and protein in explaining the kinetics of cell elongation.

It has been assumed that growth rate is proportional to an internal pool of IAA and not the total IAA taken up by the tissue (Andreae 1967). Evans and Ray (1969) have clearly demonstrated that it is not possible using the standard assumptions of compartmentation analysis to account for both the latent period and the comparatively rapid change in growth rate. It is normally an essential assumption in considering compartments that there is rapid mixing of their contents i.e. once a molecule has entered a compartment it has the same chance of being removed as any other molecule in the compartment. In developing the present model we have assumed that the GLP enters a completely non-mixed pool. Molecules leave the compartment (or series of compartments) in almost the same order that they arrive. The biological equivalent could be a vesicle where there is a time delay after the protein enters the vesicle and before it is released into the cell wall or membrane. A similar approach to unmixed compartments (without the biological interpretation) has been used by Schotz *et al.* (1964). Chrispeels (1970) has reported an apparently similar system of a protein pool and a delay before the protein appears in the wall.

For a model to be fully useful it should be possible to show that the model both quantitatively predicts the observed response and makes predictions on the response under conditions that have not yet been tried. The model has been evaluated on a digital computer using a standard digital analog simulator programme. By selecting likely values for the unknown constants the output in Fig. 5 was obtained.

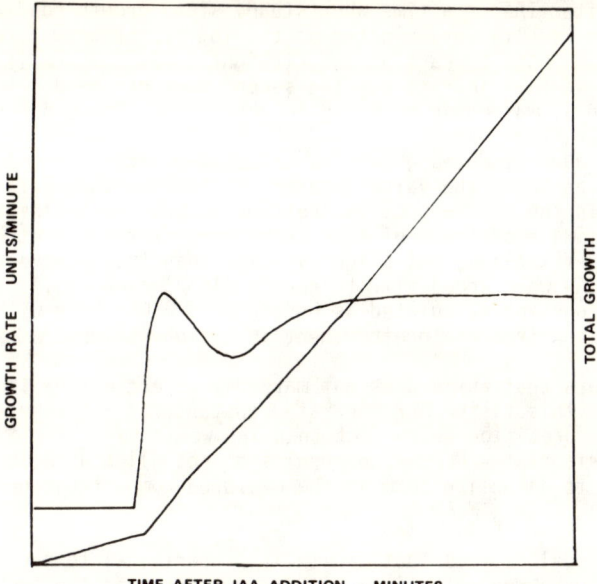

Fig. 5. A computer-drawn output evaluating the model in Fig. 4.

The model was evaluated for several different experiments and successfully predicted several results such as the pretreatment in cycloheximide for varying times before IAA addition. However when the model was applied to steady state growth conditions in IAA it was found that the predicted results did not agree with the experimental results.

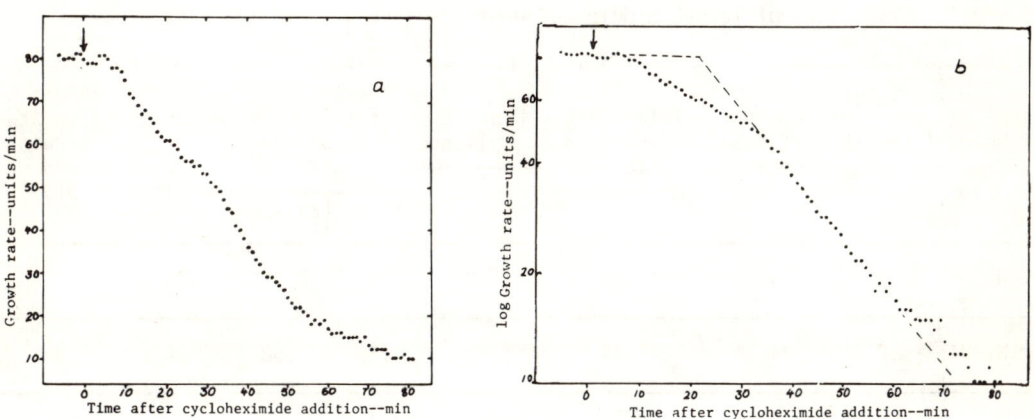

Fig. 6. Effect of cycloheximide on the IAA-induced steady-state growth rate average of 4 experiments a) linear plot; b) log plot of the data in a). 3×10^{-5}M IAA; 10 μg/ml cycloheximide. 1 unit = 0.23 μm. 28°C.

Pretreatment in IAA. Fig. 6 is an average of four experiments where cycloheximide was added 90 minutes after IAA at a time when steady state growth conditions had been attained. When the results were plotted as the log of the growth rate then the decline of the growth rate appears as two straight lines. There is no explanation at present for the first fall in rate but the second part of the decline is most likely the same phenomenon as was being measured in the cycloheximide pretreatment (Fig. 2).

If the total auxin-induced growth is calculated after the end of the latent period (15 minutes at $28°C$) the value obtained is 1421 units. This is an estimate of the amount of GLP in the active pool at the time of protein inhibition. It is slightly larger than the estimate of the pool size calculated from the pretreatment experiments by the line of best fit (1194) but slightly lower than the observed pool size when cycloheximide and IAA were added simultaneously (1472) (see Fig. 3). The numerical value of the rate constant calculated as before is .78 which is within experimental error of that obtained from the pretreatment in cycloheximide.

It thus appears that auxin does not markedly affect either the steady state pool size of GLP in this tissue nor the rate constant for its removal from the compartment. If this interpretation is correct then IAA would have little direct effect on the rate of synthesis of the GLP. The results do not allow a decision about whether IAA alters the GLP to an active form in the measured compartment or at a subsequent step.

The present model assumes that growth is proportional to the amount of GLP and to the log of the auxin concentration. It can not describe the mechanism by which the GLP can control elongation. So far our measurements on the physical properties of the wall (compliance and creep) do not give measurements that correlate with the rate of extension.

Comparison of Elongating and Non-elongating Regions. An attempt has been made to correlate the growth rates of the lupin hypocotyl with the physical properties and chemical structure of its cell wall.

Measurements were made of relative growth rate, extensibility, water potential and carbohydrate composition of three 2 cm regions from a 6 cm hypocotyl. Some results are given in Table 1.

Table 1. Properties of lower, middle and upper regions of 6 cm lupin hypocotyl

Hypocotyl Region	Length (cm)	% of total elongation	Total (DT) Compliance $X10^{-3}$ (m^2/newton)	Elastic Compliance Plastic	% extension for 10 gm increase in load		Molarity of isotonic mannitol	Relative Calculated Strain (based on DT)
					Plastic	Elastic		
Upper	2	76	1.69	1.58	0.57	0.92	.28	1.71
Middle	2	24	1.49	1.73	0.39	0.68	.24	1.34
Lower	2	0	1.16	2.28	0.22	0.50	.23	1.0

The upper 2 cm contributed about 75% to growth while the lower 2 cm had ceased elongation. Results from the Instron analysis are given for both the compliance and for the percentage increase in length for a 10 gram increase in load. The compliance takes into account the difference in the weights of the wall of the different segments and shows that some of the differences in extension between the different three regions could be accounted for by increasing wall thickness but there is still an effect of the different composition of the walls from the three regions.

Measurements of water potential of the three regions using a modification of the Schardakov dye method were made to see if a prediction of growth rate from values of extensibility and turgor pressure could be made. The calculated strain bore no resemblance to actual growth. Elongation is apparently not merely a function of the measured extensibility and turgor pressure in this tissue. A more thorough investigation of its rheology seems necessary.

The composition of the hypocotyl regions was investigated first in terms of total pectic, hemicellulosic and cellulosic sugars and then in terms of alkali-soluble fractions.

In the first case pectic material was extracted by refluxing in 0.5% ammonium oxalate, and total hemicellulose by 1.0 N H_2SO_4 hydrolysis at $100°C$. The final cellulose residue was acid hydrolysed and with the hemicellulose hydrolysate subjected to quantitative paper chromatography and total reducing sugar estimation by the method of Nelson (see Gaillard and Bailey 1968). Pectic uronic acids were measured by the carbazole reaction (Montreuil and Spik (1963)).

The alkaline fractionation was essentially the method used by Gaillard and Bailey (1968) which enables separation of hemicellulose into hemicellulose A and linear and branched hemicellulose B fractions. The tissue was treated with the neutral detergent method of Van Soest (removes pectin and protein) and delignified with acetic acid and Chloramine T (see Gaillard and Bailey (1968)) because of indications in the earlier acid hydrolyses of a carry over of mannose and xylose into the cellulose component. A portion of each alkali soluble fraction was acid hydrolysed and analysed by quantitative paper chromatography.

The proportions of carbohydrates did not appear to differ significantly between the three regions of hypocotyl, except possibly in the case of cellulose, in either the gross estimation or the alkali fractionation. However a considerable loss of galactose (70%) and arabinose (90%) from the alkali-extracted polysaccharides when compared with total acid hydrolysis precludes the conclusion that there *are* no differences, in fact.

Subsequent investigation has shown the loss of galactose and arabinose to be due to the delignification treatment, where a pH of 3.6 is maintained for 6 hours at $100°C$. The extracted sugars were non-dialysable. Hence it appears that there exists in lupin hypocotyl arabinogalactan resistant to usual pectic extraction but released by mild acid treatment.

Initial experiments suggest that pronase treatment of crude cell wall preparations obtained by grinding fresh hypocotyls in buffer does not enhance the extraction of carbohydrate by cold water, 0.5% ammonium oxalate or 10% KOH.

Deoxyglucose. It is not known whether the synthesis of any particular polysaccharide or glycoprotein is necessary for auxin action on cell expansion (Ray 1967). 2 deoxy-D-glucose has been used to inhibit synthesis and secretion of a polysaccharide component of a glycoprotein in yeast protoplasts (Farkaš et al., 1970). The effect of this compound on the long term kinetics of auxin action is shown in Fig. 7.

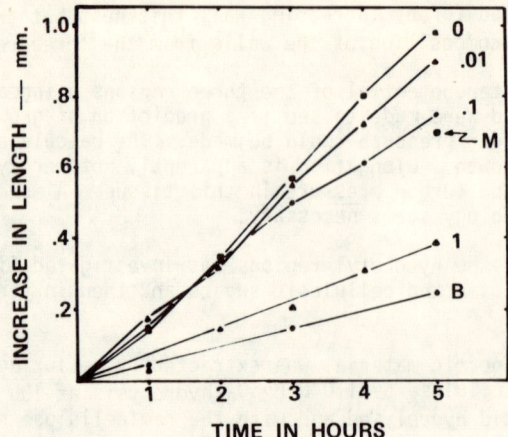

Fig. 7. Effect of different concentrations of 2-deoxyglucose on the IAA-induced growth (in mm) of 5, 10 mm lupin hypocotyl segments. 0, 0.01, 0.1, 1% deoxyglucose, B, buffer control. M, growth of segments treated with mannitol of the same molarity as 1% 2-deoxyglucose.

Comparatively high levels of deoxyglucose were required to give a strong inhibition of growth but this could give some evidence that at least one polysaccharide is necessary for continued auxin induced growth. The inhibition of growth is also shown with short term kinetics but the first peak of the response to auxin still occurs with a sixty minute pretreatment in deoxyglucose. However the specificity of deoxyglucose as an inhibitor in higher plants is still unknown so that no definite conclusion can be drawn. The inhibition is not an osmotic effect because a mannitol solution at the same molarity has a much smaller effect. At present we are studying the specificity of deoxyglucose inhibition.

IV. SUMMARY

The short term kinetics of the response to auxin have been studied with lupin hypocotyls, by a method that gives a high resolution with single segments. The growth rate shows a latent period, first maximum, a minimum and then a second maximum. Possible explanations for the kinetics are given. Hypocotyls were pretreated in cycloheximide, an inhibitor of protein synthesis, before auxin addition. This permitted estimates of pool size and rate constants for the removal of a hypothetical growth limiting protein(s) (GLP). A model to explain the observed kinetics is proposed and simulated with the aid of a computer. By using the measured parameters and selecting reasonable values for the unmeasured constants it is possible to explain the observed kinetics and to predict the growth rate under different conditions. By adding cycloheximide after the auxin-induced growth rate had reached a steady state it was calculated that auxin had not changed the pool size or rate constants of the GLP. Segments from elongating and non-elongating regions of the hypocotyl were studied in an attempt to find physical properties or components of the cell wall that paralleled the growth rates. 2 deoxy-D-glucose which inhibits polysaccharide synthesis in yeast was shown to inhibit elongation of the hypocotyls although at comparatively high concentrations.

V. ACKNOWLEDGEMENTS

This work has been supported in part by grants to Prof. R.G. Thomas and Dr. E.D. Penny from the New Zealand University Grants Committee Research Committee.

VI. REFERENCES

ANDREAE, W.A. (1967). Uptake and metabolism of indoleacetic acid, naphthalene acetic acid, and 2,4-dichlorophenoxy-acetic acid by pea root segments in relation to growth inhibition during and after auxin application. Can.J.Bot. $\underline{45}$: 737-753.

BARKLEY, G.M. and M.L. EVANS. (1970). Timing of the auxin response in etiolated pea stem sections. Pl. Physiol., Lancaster $\underline{45}$: 143-147.

CHRISPEELS, M.J. (1970). Synthesis and Secretion of Hydroxyproline-containing Macromolecules in carrots. Pl.Physiol., Lancaster $\underline{45}$: 223-227.

CLELAND, R. (1967). Extensibility of isolated cell walls: measurement and changes during elongation. Planta $\underline{74}$: 197-209.

CLELAND, R. (1972). Instability of the growth limiting proteins of the *Avena* coleoptile and their pool size in relation to auxin, in "Plant Growth Substances, 1970" Ed. D.J. Carr.

DOWBEN, R.M. (1969). "General Physiology. A molecular approach." Harper and Row. London.

EVANS, M.L. and HOKANSON, R. (1969). Timing of the response of coleoptiles to the application and withdrawal of various auxins. Planta $\underline{85}$: 85-95.

EVANS, M.L. and P.M. RAY. (1969). Timing of the auxin response in coleoptiles and its implication regarding auxin action. J. gen. Physiol. $\underline{53}$: 1-20.

FARKAŠ, V. , A. SVOBODA and Š. BAUER. (1970). Secretion of cell wall glycoproteins by yeast protoplasts. Effect of 2-deoxy-D-glucose and cycloheximide. Biochem. J. $\underline{118}$: 755-758.

GAILLARD, B.D.E. and R.W. BAILEY. (1968). The distribution of galactose and mannose in the cell wall polysaccharides of red clover (*Trifolium pratense*) leaves and stems. Phytochem. $\underline{7}$: 2037-2044.

KÖHLER, D. (1956). Über die Beziehungen zwischen der Länge von Haferkoleoptilen und der Wachstumsgeschwindigkeit ihrer isolierten Ausschnitte. Planta $\underline{47}$: 159-164.

MONTREUIL, J. and G. SPIK. (1963). Microdosage des Glucides. Vol. 1., p. 59. Univ. de Lille, France.

PENNY, P. (1969). The rate of response of excised stem segments to auxins. N.Z. J. Bot. $\underline{7}$: 290-301.

PENNY, D., P. PENNY, D.C. MARSHALL and J.K. HEYES (1971). Early responses of excised stem segments to auxins. J exp. Bot. $\underline{22}$.

RAY, P.M. (1967). Radioautographic study of cell wall deposition in growing plant cells. J. Cell Biol. $\underline{35}$: 659-674.

RAYLE, D.L., M.L. EVANS and R. HERTEL. (1970). Action of auxin on cell elongation. Proc. Natn. Acad. Sci. U.S.A. $\underline{65}$: 184-191.

SCHOTZ, M.C., N. BAKER and M. CHAVEZ. (1964). Effect of carbon tetrachloride ingestion on liver and plasma triglyceride turnover rates. J. Lipid. Res. $\underline{5}$: 569-577.

YAMAKI, Toshio. (1954). Effect of indoleacetic acid upon oxygen uptake, carbon dioxide fixation and elongation of *Avena* coleoptile cylinders in the darkness. Scient. Pap. Coll. gen. Educ. Tokyo $\underline{4}$: 127-154.

PLANT GROWTH SUBSTANCES, 1970

Initial Kinetics of Auxin-induced Cell Elongation in Coleoptiles

H. Durand and M.H. Zenk

Institute of Plant Physiology, The Ruhr-University, Bochum, Germany

I. INTRODUCTION

The timing of the first detectable response of cell elongation of plant organs to auxins has drawn considerable attention in the past two years (Evans and Ray, 1969; Evans and Hokanson, 1969; Uhrström, 1969; Nissl and Zenk, 1969; Rayle et al., 1970; Barkley and Evans, 1970; Evans and Rayle, 1970; de la Fuente and Leopold, 1970). The objective of these studies was to find out if the lag observed in the growth response to exogenous auxin is compatible with the hypothesis that auxin acts by gene activation and induction of messenger-RNA synthesis, which results in the synthesis of new enzymes and ultimately in the modification of the cell wall, allowing cell expansion (for review, see Key, 1969). It was found (Nissl and Zenk, 1969) that the lag period before a steady-state rate of growth is reached is dependent on the hormone concentration, and the maximal initial growth rate is constant over a 10^5-fold concentration range at 21°C. These results, obtained with *Avena* coleoptiles using the microscope technique of Ray and Ruesink (1962), strongly suggest that auxins do not induce or promote protein synthesis during the initial elongation phase of these plant organs, whether at the transcriptional or at the translational level. Using a high-resolution continuous recording optical method, Rayle, Evans and Hertel (1970) have recently studied the early time course of auxin action on the elongation of corn coleoptile segments. Their results, as well as those of Evans and Ray (1969), indicate that (1) the lag period before onset of auxin-stimulated elongation is independent of the IAA concentration; (2) after addition of a low concentration of auxin to 4 mm sections, a very rapid enhancement of elongation occurred (similar and even more rapid responses were found with low concentrations of IAA-methylester); and (3) large step-ups in the auxin level induced a rapid transient decrease in growth rate, lasting 10 to 15 minutes.

The results presented by Rayle et al. (1970) are in considerable conflict with the results obtained by Nissl and Zenk (1969). To clarify this situation, we repeated the work of Rayle et al. (1970) using the same plant material and a high resolution auxanometer developed by de la Fuente and Leopold (1970).

II. MATERIALS AND METHODS

Corn seeds (*Zea mays* L., Golden Bantam 8 row) were obtained from Vaughan's Seed Co., Chicago, Illinois. The seeds were sown and coleoptile segments prepared as described by Evans and Hokanson, (1969) and Hertel et al. (1969). Oats (*Avena sativa* L. c.v. Victory) as well as the other cereals were grown as described by Nissl and Zenk (1969). The experimental conditions were as those described by Rayle et al. (1970); however, fresh growth medium was pumped through the growth chamber at a rate of 25 ml/min. Each experiment was repeated from 4 to 15 times.

III. RESULTS

Since it was found that temperature has a marked effect on the growth response of oat coleoptiles towards IAA (Nissl and Zenk, 1969), its effects have been checked using corn coleoptiles. Figure 1 shows that both the endogenous and the IAA-induced growth rate is stimulated considerably by temperature, showing a distinct optimum at 35°C. The growth rate in the presence of 10^{-5}M IAA follows a logarithmic relationship up to 35°C. The activation energy for this process was calculated to be E = 13,240 cal/mol, which corresponds to a Q_{10} of 2.05 for the temperature interval between 19°C and 32°C. It has been shown previously that in the case of oat coleoptiles the steady-state elongation rate of these organs at 21°C is constant over a concentration range from 10^{-8} to 10^{-3}M IAA (Nissl and Zenk, 1969).

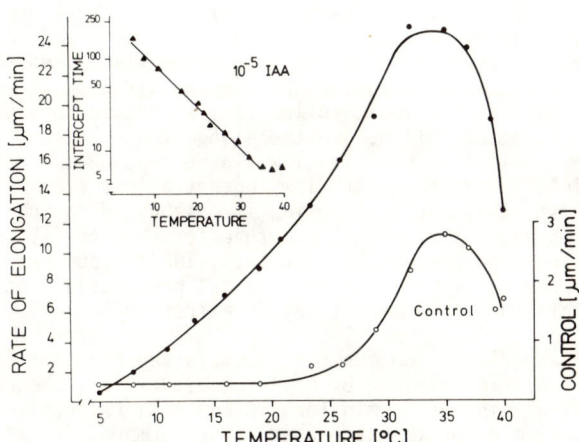

Fig. 1. Influence of temperature on initial steady-state rate of elongation in corn coleoptile sections. 10^{-5}M IAA (●); Control (O). Insert: Plot of log intercept time of IAA-induced coleoptiles *versus* temperature.

In contrast, corn coleoptiles in short term experiments show a marked concentration dependence, and yield, as the initial steady-state growth rate, the familiar (in long term experiments with oats) bell-shaped optimum curve (Fig. 2A). Concentrations above 10^{-4}M IAA at both 21° and 37°C are inhibitory to maize coleoptiles.

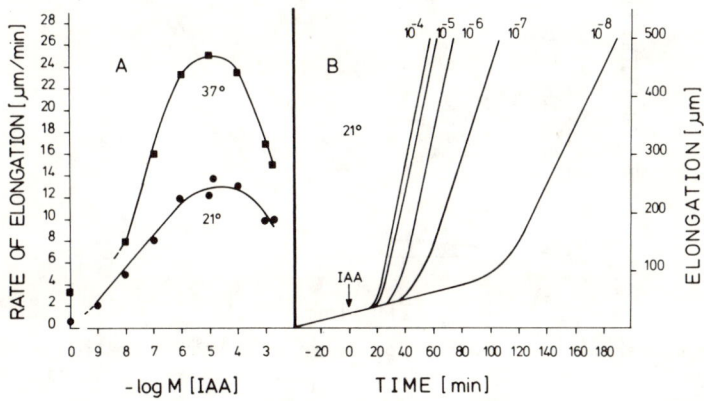

Fig. 2. A. Initial steady-state elongation rates of corn coleoptile sections in different IAA concentrations at 21° and 37°. B. Time course of elongation of corn coleoptiles in different IAA concentrations at 21°. Phosphate buffer replaced by the appropriate concentration of IAA in buffer at the time indicated by the arrow.

Thus it would appear that oat coleoptiles during short term experiments are much more resistant to high IAA concentrations (Housley et al., 1954; Marinos, 1957; Cleland, 1960; Truelsen, 1961) than the apparently more sturdy corn coleoptiles. However, there is no qualitative difference in their response to different IAA concentrations. Figure 2B shows the time course of growth promotion of corn coleoptiles by IAA in the concentration range between 10^{-8}M and 10^{-4}M. The lag periods (here the lag period is defined as the intercept time) in the different auxin solutions differed from as much as 115 min at 10^{-8}M to 20 min at 10^{-4}M IAA. The same held true for the latent period which was reduced from 80 min at 10^{-8}M to 15 min at 10^{-4}M IAA. Thus it has been demonstrated here that the lag phase observed is dependent on the auxin concentration supplied, in contrast to previous observations by Evans and Ray (1969). This result suggests that some portion of the observed lag can be attributed to the time required for the uptake of auxin into the tissue and for the subsequent active transport or physical distribution within the tissue. If this interpretation of our observations is correct, it would follow that a certain threshold amount of auxin has to be reached inside the cell before the elongation reaction starts, in agreement with the data in that this threshold level would be reached much earlier with high concentrations of IAA (10^{-4}M) than with low concentrations (10^{-8}M) (Nissl and Zenk, 1969). Evidence exists that entry of auxins is most rapid through the cut ends and the hormone may first exert its maximal growth effects on the tissue near the ends and then move away from the ends (Housley et al., 1954; Saunders et al., 1965; Nissl and Zenk, 1969). However, it has been pointed out previously that in the case of *Avena* coleoptiles this threshold amount may be extremely low (Nissl and Zenk, 1969).

Repetition of experiments on the fast stimulatory reaction (lag 2 to 3 min) of the methylester of IAA (MIA) claimed by Rayle et al. (1970), showed that at 21°C the intercept times were 31 min and 35 min for MIA and IAA, respectively, and the latent periods, 21 min and 22 min for the same pair. Furthermore, MIA at concentrations of 10^{-5}M showed no inhibition period.

The claim that a short lag can be seen if one applies low IAA concentrations, such as 2×10^{-7}M, to 4 mm sections of Bantam corn coleoptiles while a long lag appears with 8 mm sections (Rayle et al., 1970) has also been tested.

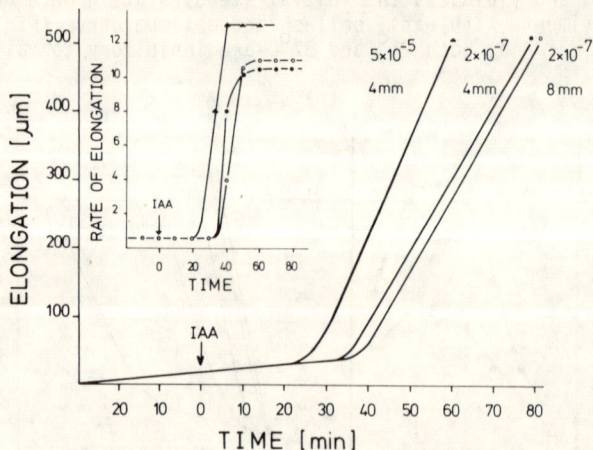

Fig. 3. Time course of the growth response to a low concentration of IAA in 4 and 8 mm corn coleoptiles. Inset: Data from Fig. 3 plotted as growth rate *versus* time.

As shown in Figure 3, the curves for 4 mm and 8 mm sections are almost identical. Intercept times and latent periods had almost identical values (34 min and 36 min, respectively). The most important point is, however, that 4 mm sections supplied with 5×10^{-5}M IAA showed, in contrast to the report of Rayle et al. (1970), a much shorter lag than those supplied with 2×10^{-7}M IAA. This observation is in accord with our results given in Figure 2B. A comparison of the growth responses of 2.5 mm sections, cut out of a 1 cm segment, and 1 cm sections showed that the latent period of the longer ones was about 25 per cent greater. This indicates again that, indeed, part of the lag may be due to the time required for the uptake of IAA, in agreement with de la Fuente and Leopold (1970). In their paper Rayle et al. (1970) state that the ability of low concentrations of MIA and IAA to produce short lag times was not under strict control and occurred only in roughly 50 per cent of the cases. Each of the experiments reported here was repeated from 4 to 15 times. We have never seen this sort of effect.

Finally, we have attempted to reproduce the alleged fast inhibitory action of auxins on cell elongation. It has been reported (Rayle et al., 1970) that when a large step-up in IAA concentration is given to coleoptile sections in buffer the first observable reaction is a rapid but transient reduction of the growth rate below the endogenous one; this effect was observed in three species of plants tested. As shown in Figure 4B, in experiments in which five different species of the family Gramineae were treated with high, nonphysiological concentrations of IAA (10^{-5}M to 10^{-3}M) we never observed a transient inhibition or inhibition lag, regardless of the temperatures and IAA concentrations at which the experiments were conducted.

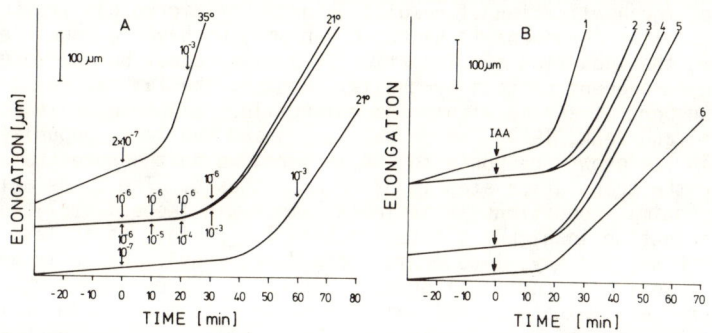

Fig. 4. A. Effect of sequential increases in IAA concentration on the time course of elongation of corn coleoptiles.
B. Effect of different IAA concentrations on the time course of elongation in 5 plant species. Growth medium changed from phosphate buffer to the solution indicated at the arrows.

1. *Secale cereale*, 10^{-5}M IAA; 2. *Hordeum vulgare*, 10^{-5}M IAA; 3. *Triticum vulgare*, 10^{-5}M IAA; 4. *Zea mays*; 2×10^{-4}M IAA; 5. *Zea mays*, 10^{-3}M IAA; 6. *Avena sativa*, 10^{-4}M IAA.

Furthermore, no inhibition was found in pea or sunflower stems. Also, the effect of sequential increases in IAA concentration was re-examined (Fig. 4A), but no inhibition was found. The addition of high auxin concentrations to tissue growing in low concentrations of IAA at different temperatures did not give any indication of a fast transient inhibition (Fig. 4A). In each case the growth rate continued undiminished until the end of the latent period.

Neither Ray and Ruesink (1962), using a microscopic measuring device for auxin-induced cell elongation, nor Nissl and Zenk (1969), using the same method, nor Evans and Ray (1969), using the same optical method as Rayle et al. (1970), have observed a transient inhibition. No such inhibitory phase has ever been observed in the present study. Therefore we have to conclude that the assumption of Rayle et al. (1970) that "any model of primary auxin action has to account ... also for the observed transient inhibitions caused by a large step-up in auxin concentration" is not correct.

IV. DISCUSSION

The results presented in this paper support our original concept (Nissl and Zenk, 1969) that the mode of auxin action in causing cell enlargement does not involve activation of gene transcription and specific protein synthesis through messenger-RNA synthesis (cf. also Evans and Ray, 1969; Nelson et al., 1969); furthermore, the kinetics of the growth response of coleoptiles at high auxin concentrations at elevated temperatures indicate that the primary step in auxin-induced growth is also very unlikely to involve the translational step in protein synthesis in these organs (Nissl and Zenk, 1969). It is well known from work on enzyme induction in bacteria that the biosynthesis of inducible enzymes is preceded by a latent period (time of gene translation) of 3 to 4 min (Branscomb and Steward, 1968) which is followed by an acceleration lag of about 3 min, during which time messenger-RNA accumulates until a steady state of messenger-RNA is achieved which represents the mean life of the messenger-RNA (Kepes, 1967). The time for gene translation and the mean life of messenger-RNA is a constant which cannot be influenced by the concentration of the inducer applied. Since it has now been shown beyond any doubt that the lag phase of the growth response toward auxins in coleoptiles is definitely dependent on the IAA concentration (10^{-4}M and 10^{-8}M IAA at $21°C$: latent period, 15 min and 80 min, respectively; acceleration lag, 10 min and 45 min, respectively), and if we accept the validity of the application of results gained with microorganisms to the discussion of similar phenomena in higher organisms, we have to conclude that, by these criteria, the induction of elongation by auxins cannot be due to gene activation and subsequent protein synthesis. Rather, the lag may be partly due to uptake and transport of auxins within the coleoptile. Since in at least two instances, in *Avena* (Nissl and Zenk, 1969) and in *Secale* (Durand and Zenk, unpublished), the observed lag in the growth response to IAA is zero, we must assume that auxin does not even influence the translation step in protein synthesis. The translation time in *E. coli* under optimal conditions is at least 1 to 2 min (Kepes, 1967). On the other hand, enzyme induction in higher plants exhibits lag periods of at least 1 to 2 hr (literature in Evans and Ray, 1969; Nissl and Zenk, 1969). We fully agree with Evans and Ray (1969) that continual protein synthesis may be important in the mechanism of cell wall expansion and maintenance of cell metabolism during later phases in growth. Some of the clearly established effects of auxin on RNA synthesis (e.g., Masuda and Kamisaka, 1969) may be related to these inductions and certainly also to the induction of L-aspartate-N-acyl-ligase which, however, shows an induction lag of 1 to 2 hr (Zenk, 1962).

The argument that shifts in pH value could cause the rapid responses in elongation is ruled out since the pH has been carefully controlled in these experiments at 6.3. We have to assume, therefore, that auxin action does not involve induction of protein synthesis at either the transcriptional or the translational level; it has to be assumed that auxins act on preformed systems within the cell and thus cause cell enlargement.

V. SUMMARY

Using a high resolution continuous technique (sensing transducer), the timing of the auxin response in corn coleoptile tissue was determined. The growth response toward IAA is maximal at $35°C$, showing an activation energy of 13,240 cal/Mol. Growth rates of corn coleoptiles are maximal at a concentration of 10^{-5}M to 10^{-4}M IAA. Higher auxin concentrations are inhibitory. The latent period and intercept times are dependent on the IAA concentration supplied. Addition of low auxin concentrations (2×10^{-7}M) to short coleoptile sections (4 mm) does not give a rapid enhancement of elongation. Also, no such rapid response was observed using the methylester of IAA as compared to IAA. Following large step-ups in auxin level (10^{-5}M to 10^{-3}M), no rapid transient decrease in growth rate was observed after IAA addition. The inhibitory lag reported in the literature was not found in any of the 5 species of grasses examined.

Analysis of the growth kinetics indicate that in the first growth responses auxin does not induce protein synthesis at either the transcriptional or the translational level but rather acts on a preformed system.

VI. ACKNOWLEDGEMENTS

We would like to thank Dr. A.C. Leopold for making the transducer method available to us prior to publication. We also gratefully acknowledge the stimulating and critical discussion of this manuscript with Dr. H.V. Marsh. The financial support of the "Landesamt für Forschung, Nordrhein-Westfalen" is gratefully acknowledged.

VII. REFERENCES

BARKLEY, G.M. and M.L. EVANS (1970). Timing of the auxin response in etiolated pea stem sections. Plant Physiol., 45, 143-147.
BRANSCOMB, E.W. and R.N. STUART (1968). Induction lag as a function of induction level. Biochem. Biophys. Res.Comm., 32, 731-738.
CLELAND, R. (1960). The effect of high concentrations of auxin on the extension of *Avena* coleoptile sections. Plant Physiol., 35, Supp. 6.
EVANS, M.L. and R. HOKANSON (1969). Timing of the response of coleoptiles to the application and withdrawal of various auxins. Planta, Berl. 85, 85-95.
EVANS, M.L. and P.M. RAY (1969). Timing of the auxin response in coleoptiles and its implications regarding auxin action. J.gen.Physiol., 53, 1-20.
EVANS, M.L. and D.L. RAYLE (1970). The timing of growth promotion and conversion to indole-3-acetic acid for auxin precursors. Plant Physiol., 45, 240-243.
De la FUENTE, R.K. and A.C. LEOPOLD (1970). Time course of auxin stimulations of growth. Plant Physiol., 46, 186-189.
HERTEL, R., M.L. EVANS, A.C. LEOPOLD, and H.M. SELL (1969). The specificity of the auxin transport system. Planta, Berl. 85, 238-249.
HOUSLEY, S., J.A. BENTLEY, and A.S. BICKLE (1954). Studies on plant growth hormones. III. Application of enzyme reaction kinetics to cell elongation in the *Avena* coleoptile. J.exp.Bot., 5, 373-388.
KEPES, A. (1967). Sequential transcription and translation in the lactose operon of *Escherichia coli*. Biochim. Biophys.Acta, 138, 107-123.
KEY, J.L. (1969). Hormones and nucleic acid metabolism. Ann.Rev.Plant Phys., 20, 449-474.
MARINOS, N.G. (1957). Responses of *Avena* coleoptile sections to high concentrations of auxin. Austr.J.Biol.Sci., 10, 147-163.
MASUDA, Y. and S. KAMISAKA (1969). Rapid stimulation of RNA biosynthesis by auxin. Plant and Cell Physiol., 10, 79-86.
NELSON, H., I. ILAN, and L. REINHOLD (1969). The effect of inhibitors of protein and RNA synthesis on auxin-induced growth. Israel J.Bot., 18, 129-134.
NISSL, D.F. and M.H. ZENK (1969). Evidence against induction of protein synthesis during auxin-induced initial elongation. Planta, Berl. 89, 323-341.
RAY, P.M. and A.W. RUESINK (1962). Kinetic experiments on the nature of the growth mechanism in oat coleoptile cells. Develop.Biol., 4, 377-397.
RAYLE, D.L., M.L. EVANS, and R. HERTEL (1970). Action on auxin on cell elongation. Proc.N.A.S., 65, 184-191.
SAUNDERS, P.F., D.V. JENNER, and G.E. BLACKMAN (1965). The uptake of growth substances. V. Variation in the uptake of a series of chlorinated phenoxyacetic acids by stem tissues of *Gossypium hirsutum* and its relationship to differences in auxin activity. J.exp.Bot., 16, 697-713.
TRUELSEN, T.A. (1961). Growth stimulation by high indole-3-acetic acid concentrations. Nature, 191, 1410-1411.
UHRSTRÖM, I. (1969). The time effect of auxin and calcium on growth and elastic modulus in hypocotyls. Physiol. Plant., 22, 271-287.
ZENK, M.H. (1962). Aufnahme und Stoffwechsel von α-Naphthyl-Essigsäure durch Erbsenepicotyle. Planta, Berl. 58, 75-94.

Kinetic Studies on the Auxin Effect and the Influence of Cycloheximide and Blue Light

C.C.J. Addink and G. Meijer

Philips Research Laboratories, Eindhoven-Netherlands

I. INTRODUCTION

Whether indoleacetic acid (IAA) has a promotive or an inhibitory effect on the elongation of comparable zones of the hypocotyl of gherkin seedlings depends not only on the amount of IAA applied but also on the physiological condition of the test object.

IAA-induced promotion can be obtained not only with intact irradiated seedlings but also with isolated sections of dark-grown seedlings - floating on the IAA solution - either in darkness or exposed to light (Engelsma and Meijer, 1965). With intact dark-grown seedlings, however, it was found that the elongation was never promoted but always appeared to be inhibited after a treatment with IAA.

Following an earlier kinetic investigation on the influence of light on the growth rate of etiolated gherkin seedlings (Meijer, 1968) we have now studied the effect of IAA in darkness and in relation to blue-light activity on the growth rate of etiolated hypocotyls. The influence on IAA activity of cycloheximide, an inhibitor of protein synthesis, has also been investigated.

II. MATERIALS AND METHODS

Dark-grown gherkin seedlings were raised and used - three days old - at a temperature of 25°C and a humidity of 85%. The electronic growth recorder has been described before (Meijer, 1968). Changes in length of an object are transduced to a voltage output which is recorded continuously. From this the growth rate can be calculated. In the course of this investigation one of us (A) extended the equipment with a growth-rate recorder. By electronic means the electric output voltage of the growth recorder is differentiated. The voltage output obtained in this way is proportional to the *growth rate*. It is now possible to obtain a continuous, instantaneous record of the growth rate as a function of time. To avoid complications like the opening of the plumular hook, etc. (Galston *et al*, 1964) only the elongation below this hook was recorded. For this purpose a horizontal clip, with a vertical needle, was placed around the hypocotyl at the transition of the cell-dividing and the cell-elongating zones. This place can easily be seen as an abrupt change in diameter of the hypocotyl. The vertical needle touches the base of the central core of the growth detector.

IAA and cycloheximide were dissolved in 95% ethanol and 0.01 ml of this solution was applied to the cotyledons. The solution entered the space between the cotyledons by capillary force. Ethanol alone did not influence the growth rate (E in Fig. 2e). The characteristics of the blue light have been given earlier (Meijer, 1957).

The figures represent the response curves of a single plant but are representative of the specific treatment. Response times mentioned in the text of this paper are averages of at least 10 experiments.

III. RESULTS AND DISCUSSION

IAA effect in darkness

The effect of an application of IAA on the growth rate of the hypocotyl of an etiolated seedling, kept in darkness, is shown in Fig. 1. About 15 min after the application of 10µg of IAA (Fig. 1a) the growth rate increases rapidly and doubles in about another 15 min. After this the growth rate remains more or less constant or decreases slightly for a period of one hr, followed by a more rapid decrease of the growth rate.

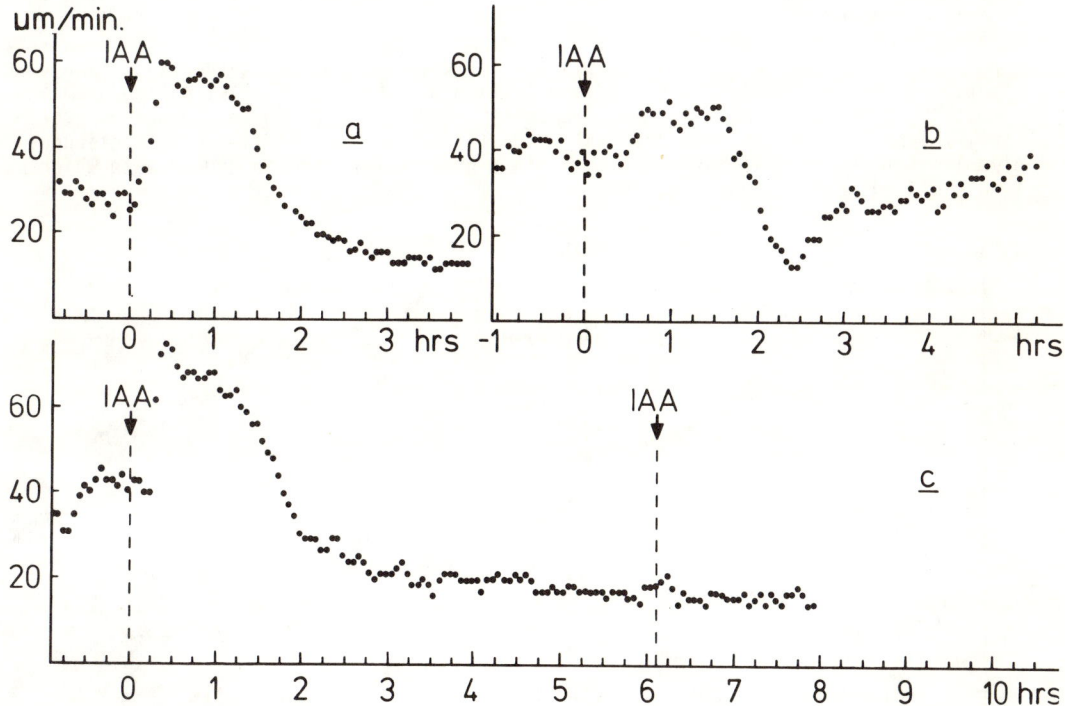

Fig. 1. The effect of IAA on the growth rate in darkness of etiolated gherkin hypocotyls. a: 10µg of IAA; b: 0.1µg of IAA; c: 10µg of IAA followed after 370 min by a second application of 10µg of IAA.

After another 1 to 1½ hr a constant growth rate is obtained, which is about half of the original value. A treatment with 1µg of IAA shows the same result. The effect of an application of 0.1µg of IAA is shown in Fig. 1b; a gradual increase of the growth rate is now followed by only a temporary decrease. The original growth rate is restored, but the time in which this occurs varies from experiment to experiment. A second application of IAA (10µg), after about 6 hours, does not influence the lower, constant growth rate (Fig. 1c).

It may be mentioned here that it is not possible with this method to determine the lag of the response to IAA. The distance between the site of application and the place where the response is measured is about 5 to 10 mm. If the response to IAA were to occur immediately, without a lag, the transport velocity would be about 20-40 mm/hr. In the case of a measurable lag the velocity would be higher. It should be noted that transport velocities of this order of magnitude are not extremely high compared with values reported in literature, which range from 11 mm/hr up to 200-240 mm/hr (Morris *et al*, 1969; Little and Blackman, 1963).

Interaction of IAA and blue light

In the following experiments the interaction of blue light and IAA is investigated. Blue light may influence the IAA effect in at least three different ways. 1. The IAA content can be decreased by photo-oxidation, in which riboflavin is thought to be the pigment involved (Galston, 1950). 2. In addition to this decrease of the IAA concentration, degradation products of IAA have been reported to have an inhibitory effect (Hager and Schmidt, 1968). 3. Another light-controlled decrease of IAA has been found to be of enzymatic (IAA oxidase) nature (Hillman and Galston, 1957). In gherkin seedlings it has been found that light induces the synthesis of p-coumaric acid and of ferulic acid, the one being a co-factor and the other an inhibitor of IAA oxidase (Engelsma and Meijer, 1965).

In these experiments IAA was applied after the onset of the irradiation with blue light. As we have seen before, the application of 10μg of IAA on etiolated seedlings - kept in darkness - results after a temporary stimulated growth in a constant growth rate, which is about half the original value (Fig. 2a). Irradiation with continuous blue light (500 μW/cm^2) causes within 1 min a drastic decrease in growth rate which levels off to nearly zero (Fig. 2b).

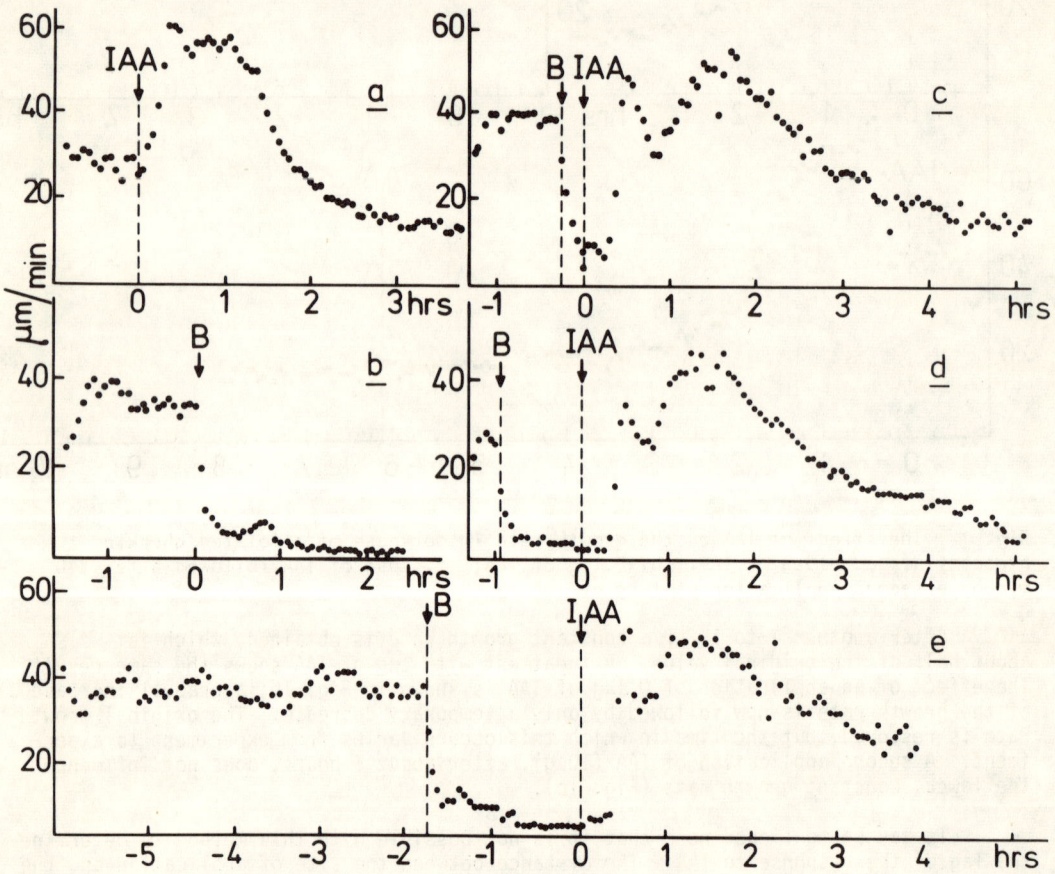

Fig. 2. The interaction of the effects of blue light (500μW/cm^2) and IAA (10μg) on the growth rate of dark-grown gherkin hypocotyls; a: dark control with IAA, b: blue light (B) without IAA, c-e: IAA applied 15, 60 and 115 min respectively after the onset (B) of the irradiation.

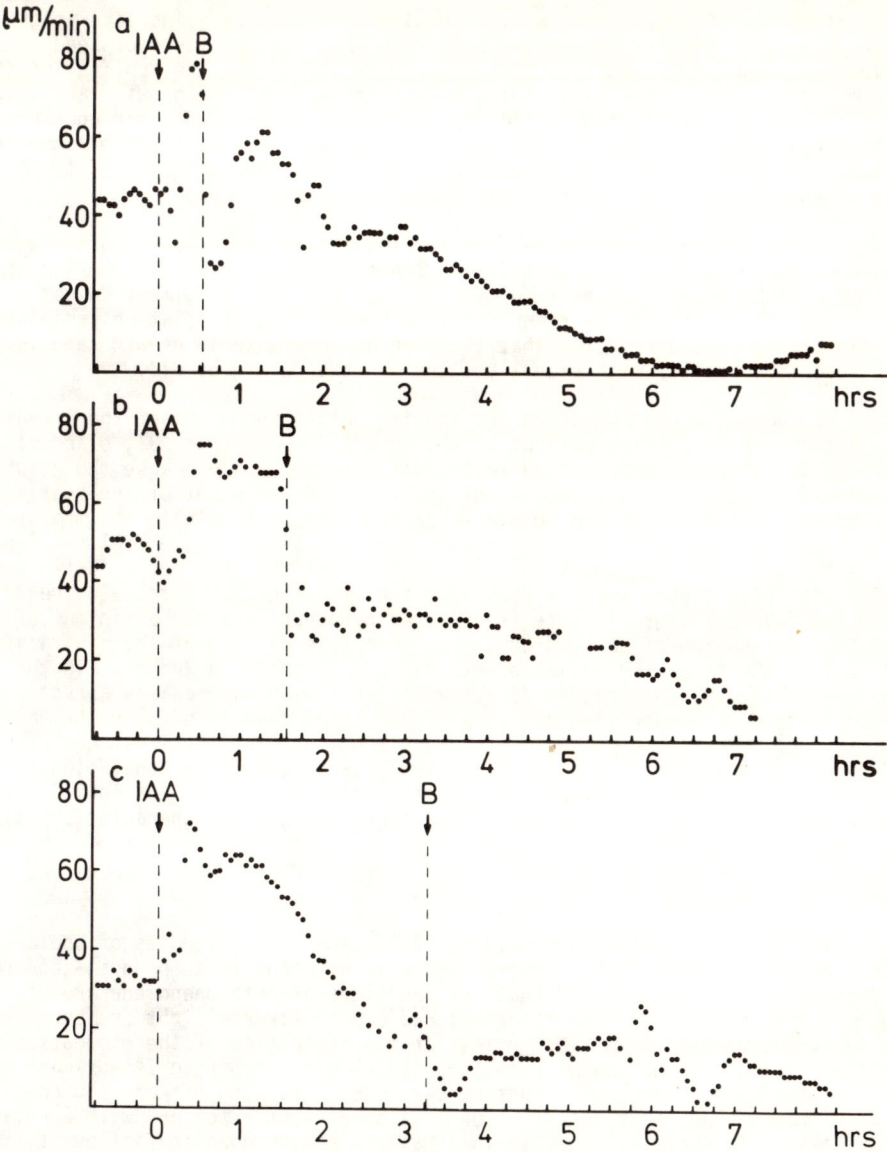

Fig. 3. The interaction of the effects of IAA (10μg) and of blue light (500μW/cm^2) on the growth rate of dark-grown gherkin seedlings. a-c: IAA applied respectively 30, 90 and 200 minutes before the onset (B) of irradiation. Pure ethanol (E) did not influence the growth rate.

The application of 10µg of IAA always results in a temporary increase in growth rate, irrespective of whether dark-grown seedlings (Fig. 2a) are used or seedlings which have been irradiated with blue light 15, 60 or 115 before the application of IAA (Fig. 2c, d and e). As in the dark controls, the increase of the growth rate starts about 15 min after the application of IAA and a maximum is reached after another 15 min. Only in the irradiated seedlings is this maximum followed by a distinct minimum in growth rate. This minimum is always reached 45 min after the application of IAA, irrespective of the moment at which irradiation with blue light begins.

In the experiment in which IAA was applied 115 min after the onset of irradiation (Fig. 2e) one might expect an increase in the IAA-oxidase activity, as the lag of this process is over. Nevertheless the results obtained with this treatment do not differ essentially from those of the experiments with a shorter pre-irradiation period (Fig. 2c and d). Finally it should be noted that the eventual decrease in growth rate is slower in seedlings treated with blue light than in the dark controls.

In the subsequent experiments IAA was applied before the onset of the irradiation with blue light, for successive periods of 30, 90 and 200 min (Fig. 3a, b and c). The curve in Fig. 3c can be regarded as an additional dark control, because the growth rate has already reached the lower constant level before the onset of irradiation. The other two times chosen, 30 and 90 min, correspond respectively to the moment of the maximum growth rate induced by IAA and to the beginning of the rapid decline (Fig. 3b and c). In the two cases last mentioned (30 and 90 min) the onset of a continuous irradiation with blue light causes a fast decrease of the growth rate. In the 30 min experiment the decreasing growth rate is restored temporarily after a minimum of 45 min following the application of IAA, which is the same time as that of the minima after IAA treatment in the foregoing series of experiments (Fig. 2c, d and e). In the 90 min experiment (Fig. 3b) the drastic decrease of the growth rate ceases abruptly and remains constant for a longer time and at a higher level than it does in the dark control (Fig. 3c). Continuous blue light, given 200 min after the treatment with IAA, resulted only in a temporary decrease of the low growth rate. The conclusion of the last two sets of experiments is that IAA cannot restore or prevent the inhibitory effect of blue light. On the other hand, blue light changes the characteristics of the response to IAA.

Blue light and IAA-oxidase activity

In earlier papers (Engelsma and Meijer, 1965) time course studies of photo-inhibited elongation and of light-induced synthesis of phenolic compounds - controlling IAA-oxidase activity - revealed that the lag times of both phenomena are of the same order of length as far as the effect of red and of far-red light are concerned. In blue light, however, a lag of 10 minutes for the inhibition of the elongation has been reported, which does not agree with a possible relationship to IAA-oxidase activity. This discrepancy, however, might be caused by a masking effect - in continuous light - of a "rapid blue-light reaction" on a "slower light reaction" with a relatively long lag time. To ascertain whether this masking effect does in fact exist, the following conditions have to be met. Firstly, it must be possible to induce the synthesis of the phenolic compounds not only in continuous light but also by a certain dose of blue light, and secondly the induction must take place in a restricted period of time. Both conditions can be fulfilled as has been reported by Engelsma (1967 and 1969). A third condition is that the decreased growth rate, caused by the "rapid blue-light reaction", does not continue at this low level but recovers partly or totally after the light treatment is stopped. This condition has been investigated with the growth-rate recorder. In Fig. 4 replicates are presented of the records of growth (a) and of growth rate (b) of an etiolated seedling irradiated for 20 min with blue light (B-D) at an intensity of 400µW/cm^2. It is obvious that the growth rate decreases again 116 min after the onset of the blue period of 20 min.

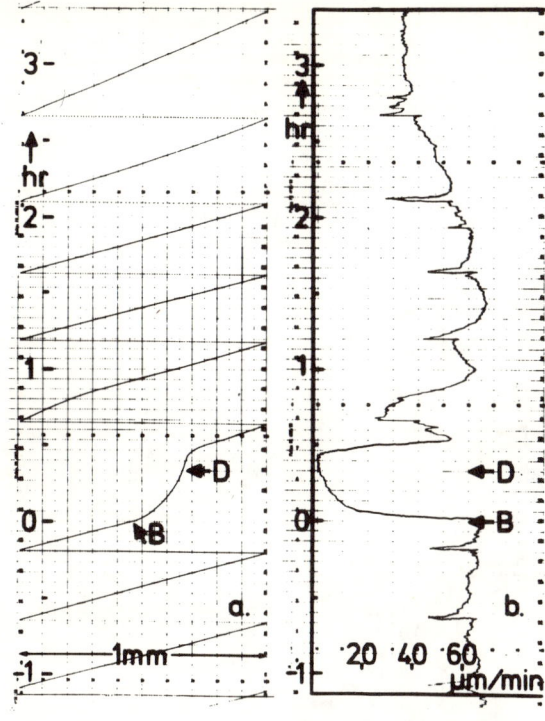

Fig. 4. Continuous recording of the growth (a) and the growth rate (b) of a gherkin hypocotyl in darkness, irradiated 20 minutes with blue light ($400\mu W/cm^2$), B-D. The growth rate in darkness decreases again 110 minutes after the beginning of the irradiation.

This result shows that in continuous blue light a second photoreaction is masked by the one that started about 30 sec after the onset of irradiation. The lag of this second photo-induced inhibition is of the same order of length as the lag found for increasing IAA-oxidase activity, which indicates that in blue light the IAA-oxidase system, too, may be involved in the photo-control of elongation.

IAA and cycloheximide

In the last decade much attention has been paid to the question of whether the primary effect of IAA should be sought at the DNA, RNA or protein synthesis level. Masuda (1959) reported that *Avena*-coleoptile sections pretreated with the enzyme RNAase did not show IAA-induced elongation within 90 min. The influence of IAA on the rate of RNA synthesis and on the incorporation of precursors into RNA and proteins has been studied (Noodén and Thimann 1966; Trewavas, 1968 etc.). Inhibitors of RNA- and of protein synthesis have also been used to study the nature of the activity of IAA (Evans and Ray, 1969; Nelson et al.,1969; etc.). The results of these experiments are often controversial.

It is a prerequisite that the time courses of the response to a given activator and those of the underlying mechanism correspond with each other. Noodén and Thimann (1966) reported that the inhibition of elongation and of incorporation of C^{14}-leucine into protein went closely parallel. Evans and Ray (1969) have reported that cycloheximide - at a concentration partially inhibiting the elongation - did not extend the time of the response to applied IAA. They concluded, however, that the speed with which cycloheximide inhibits elongation suggests that continuous protein synthesis is important in the mechanism. Nissl and Zenk (1969), using another approach, found that the lag for IAA-induced promotion of growth decreased with an increase of the IAA concentration. At 40°C the lag was reduced to zero. The authors concluded that these results do not support the idea that a primary effect of IAA is to be found at the DNA, RNA or protein synthesis level.

In our experiments the IAA-induced change in growth was recorded after a previous application of 10 μg of cycloheximide, an inhibitor of protein synthesis. The calculated growth rates are given in Fig. 5. The effect of IAA, applied 330 min after a pretreatment with cycloheximide, is very similar to the blank. Our results justify the conclusion that IAA activity does not depend on *de novo* protein synthesis.

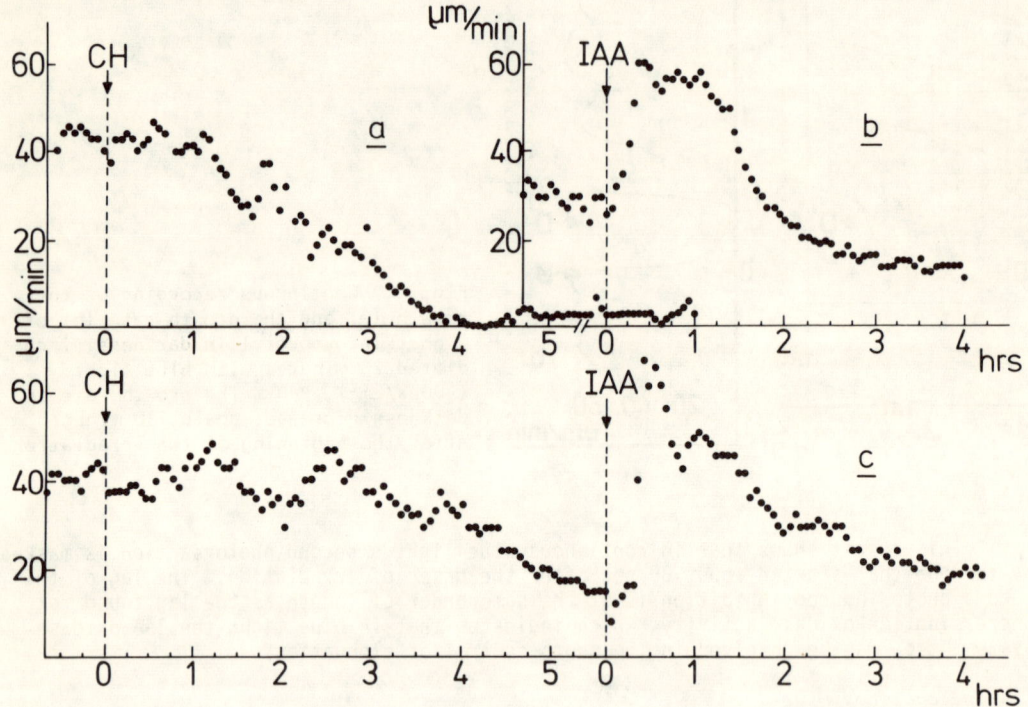

Fig. 5. Effects of cycloheximide (10μg) and of IAA (10μg) on the growth of dark-grown gherkin hypocotyls. a: Cycloheximide (CH), b: IAA, c: Cycloheximide followed by an application of IAA after 330 min.

IV. SUMMARY

Following an earlier kinetic study on the photoinhibition of etiolated gherkin hypocotyls, this paper presents results of experiments on the effect of IAA on the growth rate of etiolated gherkin hypocotyls in darkness and after exposure to blue light. Using an electronic growth recorder (and occasionally a growth-rate recorder) it was found that IAA always induces a temporary increase in growth rate whether in darkness or after irradiation. In the irradiated seedlings, however, the growth maximum was always followed by a distinct minimum, 45 min after the application of IAA but irrespective of the time the irradiation was started. It is also clear that blue light delays the inhibiting period of IAA and that IAA partly prevents the inhibition by blue light. A limited dose of blue light - instead of a continuous irradiation - reveals that in blue light two different reactions occur which inhibit elongation. The rapid reaction with a lag of 30 sec inhibits elongation almost completely and is sustained in continuous light. The second reaction - only found after a short irradiation - shows a lag of light-induced inhibition of growth which is of the same order of magnitude as the lag of the light-induced increase of IAA-oxidase activity. The application of cycloheximide, an inhibitor of protein synthesis, essentially does not influence the response to IAA.

V. REFERENCES

ENGELSMA, G. (1967). Photoinduction of phenylalanine deaminase in gherkin seedlings. I. Effect of blue light. Planta (Berl.) 75, 207-219.
ENGELSMA, G. (1969). The influence of light of different spectral regions on the synthesis of phenolic compounds in gherkin seedlings, in relation to photomorphogenesis VI. Phenol synthesis and photoperiodism. Acta Bot. Neérl. 18, 347-352.
ENGELSMA, G. and G. MEIJER (1965). The influence of light of different spectral regions on the synthesis of phenolic compounds in relation to photomorphogenesis. I en II. Acta Bot. Neérl. 14, 54-72 & 72-92.
EVANS, M.L. and P.M. RAY (1969). Timing of the auxin response in coleoptiles and its implication regarding auxin action. J. Gen. Physiol. 53, 1-20.
GALSTON, A.W. (1950). Riboflavin, light and the growth of plants. Science III, 619-624.
GALSTON, A.W., A.A. TUTTLE and P.J. PENNY (1964). A kinetic study of growth movements and photomorphogenesis in etiolated pea seedlings. Am. Journ. Bot. 51, 853-858.
HAGER, A. und R. SCHMIDT (1968). Auxintransport und Phototropismus. I & II. Planta (Berl). 83, 347-371 & 372-386.
HILLMAN, W.S. and A.W. GALSTON (1957). Inductive control of indoleacetic-acid oxidase activity by red and near infrared light. Plant Physiol. 32, 129-135.
LITTLE, E.C.S. and G.E. BLACKMAN (1963). The movement of growth regulators. III. New Phytol. 62, 173-197.
MASUDA, Y. (1959). Role of cellular RNA in the growth response of *Avena* coleoptile to auxin. Physiol. Plant 12, 324-335.
MEIJER, G. (1957). The influence of light quality on the flowering response of *Salvia occidentalis*. Acta Bot. Neérl. 6, 395-406.
MEIJER, G. (1968). Rapid growth inhibition of gherkin hypocotyls in blue light. Acta Bot. Neérl. 17, 9-14.
MORRIS, D.A., R.E. BRIANT and P.G. THOMPSON (1969). The transport and metabolism of ^{14}C-labelled IAA in intact pea seedlings. Planta 89, 178-179.
NELSON, H., I. ILAN and L. REINHOLD (1969). The effect of inhibitors of protein and RNA synthesis on auxin induced growth. Israel J. Bot. 18, 129-134.
NISSL, D. and M.H. ZENK (1969). Evidence against induction of protein synthesis during auxin-induced initial elongation of *Avena* coleoptiles. Planta 89, 328-341.
NOODÉN, L.D. and K.V. THIMANN (1966). Action of inhibitors of RNA and protein synthesis on cell enlargement. Plant Physiol. 41, 157-164.
TREWAVAS, A. (1968). Effect of IAA on RNA and protein synthesis. Arch. Biochem. 123, 324-335.

EFFECTS OF IAA AND CYANIDE ON THE GROWTH AND
RESPIRATION OF COLEOPTILE SECTIONS FROM *Triticum*

K.S. Rowan, Linden R. Gillbank and Ahava H. Spring

Botany School, University of Melbourne, Parkville, Victoria, 3052, Australia

I. INTRODUCTION

Some workers have suggested that parallel stimulation of growth and respiration of plant tissue by auxin indicates that auxin-induced growth and respiration are closely connected (reviewed by Audus, 1960). However, the manometric techniques used did not show changes in rate of respiration within minutes of treatment and therefore we have re-examined rates of growth and respiration of the *Triticum* coleoptile using the microscopic technique of Ray and Ruesink (1962) and the oxygen electrode. Using these techniques we have also investigated inhibition of respiration and growth by cyanide.

II. MATERIALS AND METHODS

Wheat seeds (*Triticum vulgare,* cultivar"Gabo") were germinated in darkness at $25^{\circ}C$. Coleoptile sections 10 mm long were cut 3 mm below the tip using a Thimann coleoptile microtome, except when single sections were required. All experiments were carried out in 10 mM phosphate buffer pH 4.7 or 5.5. Rate of growth of single sections was measured at $25^{\circ}C$ using a modification of the method of Ray and Ruesink (1962). Growth of batches of 10 sections was measured by threading on nylon wire and incubating in shaken glass dishes at $25^{\circ}C$. The length of the combined sections was measured with a graduated scale. The rate of oxygen uptake of batches of sections (10 to 20) was measured at $25^{\circ}C$ using a Clark oxygen electrode (Yellow Springs Instrument Co.) and potentiometric recorder. Sections were protected from damage by placing a screen of fibre-glass mesh in the "Perspex" reaction chamber between the magnet and the sections.

III. EXPERIMENTAL

Figure 1 shows that the lag period occurring before IAA stimulates the rate of growth of wheat coleoptile sections is between 12 and 15 min at $25^{\circ}C$, similar to that reported by Ray and Ruesink (1962) and Evans and Ray (1969) in *Avena* and *Zea*.

We occasionally see the transient decrease in growth reported by Rayle, Evans and Hertel (1970) during the lag phase. Adding mannitol (150 mM) with the IAA inhibits growth almost entirely during the lag phase but does not prolong it since rate of growth increases rapidly within a minute of withdrawing the mannitol. Pretreating sections with cycloheximide (2.5 mg/l) completely suppressed the response to IAA, in contrast to the partial inhibition reported by Evans and Ray (1969). Table 1 shows that uptake of oxygen did not increase significantly until at least 90 min after treatment, though the rate increased to 24% above the control by 210 min.

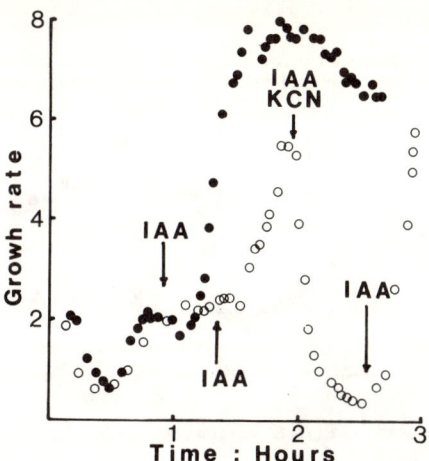

Fig. 1. The inhibition of growth by KCN (100 μM) applied to a single coleoptile 35 min after treating with IAA (10 μM). Growth unit = 0.231 mm/hr.

Adding 100 μM KCN to a section previously treated for about 40 min in IAA inhibits the rate of growth very markedly (Fig. 1); on the other hand, KCN added at this concentration concurrently with IAA (Fig. 2 and 4) or before IAA (Fig. 3) has much less effect (48% in Table 2).

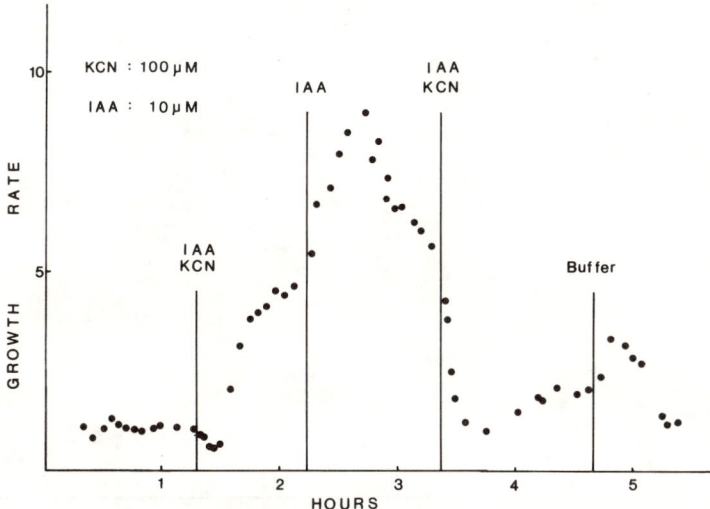

Fig. 2. The inhibition of growth by KCN (100 μM) applied to a single coleoptile concurrently with IAA (10 μM). Growth unit = 0.231 mm/hr.

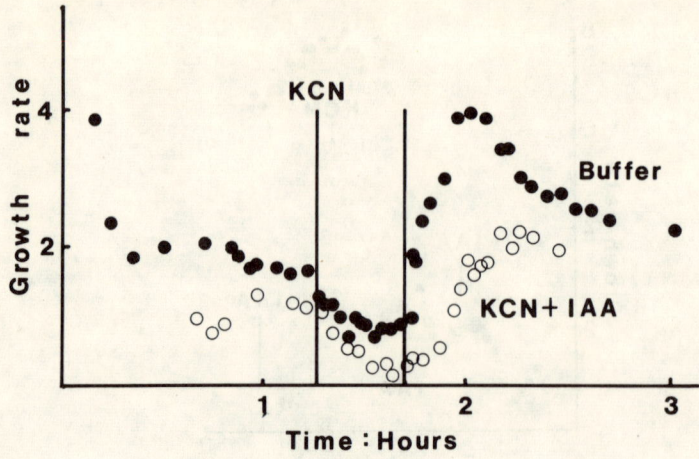

Fig. 3. The inhibition of growth by KCN (100 μM) applied to single coleoptiles followed by withdrawal of KCN (●) and by concurrent treatment with KCN + IAA (10 μM) (O). Growth unit = 0.231 mm/hr.

Table 1. The Effect of IAA on the Rate of Respiration of Coleoptile Sections

Time after Treatment	Rate of Oxygen Uptake of 20 coleoptile sections at 25°C (Arbitrary Units)	
(min)	Control	20 μM IAA
0	4.2	4.2
30	-	4.0
45	4.2	-
60	-	4.2
75	4.1	-
90	-	4.2
120	4.2	-
150	-	4.5
180	4.3	-
210	-	5.2
240	4.2	-

These two types of response to KCN are shown in Fig. 2; in this experiment a period of treatment with IAA alone is inserted between treatments in KCN + IAA. In the first period in KCN, growth is inhibited during the lag phase, but the response to IAA that follows is quite distinct.

Table 2. The Inhibition of Respiration and IAA-induced Growth by Cyanide.

KCN Concn (μM)	Percentage Inhibition 20 min after treating Batches of Coleoptiles with Cyanide	
	IAA-induced Growth Rate (25μM IAA)	Respiration Rate
1	-	6
3.3	-	19
10	0	24
100	48	35
250	65	-
500	70	-
1000	-	51

In contrast to the effect of 500 μM KCN on *Avena* coleoptile sections reported by Evans and Ray (1969), KCN does not extend the length of the lag period at 100 μM in our experiments. In the second period, when KCN is withdrawn, the growth rate increases even more rapidly but falls again after 30 min; a second treatment with KCN then inhibits growth rapidly, as shown in Fig. 1. Figures 3 and 4 show that when KCN treatment applied without IAA to a coleoptile section is withdrawn, the rate of growth increases rapidly without a lag phase.

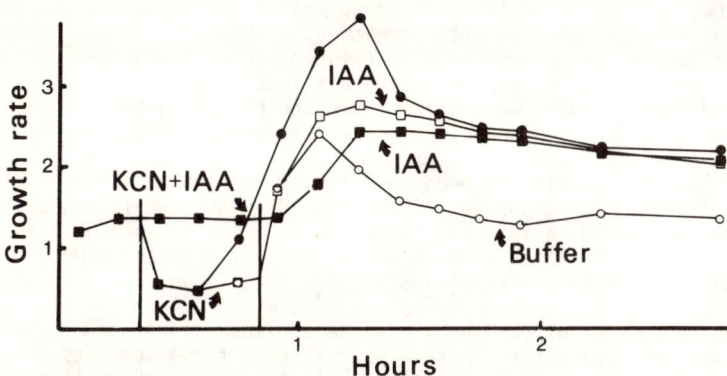

Fig. 4. The effect of concurrent and sequential treatments of batches of coleoptiles with KCN (250μM) and IAA (10μM). Values are means of two batches of 10 sections. Growth unit = mm/10 sections/10 min.

Treatment	20 to 50 min	50 to 180 min
●	KCN + IAA	IAA
□	KCN	IAA
■	Buffer	IAA
○	KCN	Buffer

Figure 4 shows that this after-effect following cyanide treatment and the effect of IAA are approximately additive (i.e. ■ + ○ = □). Since the total length of the batches of sections treated with IAA is the same after 3 hr with or without pre-treatment with cyanide (Table 3) this inhibitor does not effect permanently the growth response of the sections to IAA.

Table 2 shows the differences in response of respiration and IAA-induced growth to cyanide; 10 μM KCN distinctly inhibits respiration but does not affect growth, while higher concentrations of cyanide inhibit growth more than respiration. Figures 1 and 2 also show marked inhibitions of growth in single sections treated with 100 μM KCN.

Table 3. A Comparison of the Growth of Batches of Sections receiving KCN (250μM) or Phosphate Buffer before IAA (10μM) Treatment. Data of Fig. 4.

Time of Treatment		Growth of Batches of 10 Sections (mm) after:	
20 to 50 min	50 to 180 min	2 hr	3 hr
KCN	IAA	18.8))19.4 19.5)	32.2))31.6 31.0)
Buffer	IAA	19.7))20.3 20.8)	30.2))31.0 31.8)

IV. DISCUSSION

Although growth of coleoptile sections increases up to 5-fold within 30 min after treating with IAA at 25°C (Fig. 1), respiration rate does not increase significantly until at least 2 hr after treatment (Table 1). Correlation between respiration and IAA-induced growth in tissue inhibited with cyanide is low also; in particular, 10 μM KCN inhibits oxygen uptake distinctly but does not inhibit growth (Table 2). From these results we conclude that the rate of growth of coleoptile sections treated with IAA is not closely correlated with the rate of respiration.

Lack of close correlation between respiration and growth can be explained by high affinity of growth reactions for ATP, since 10 μM KCN, while inhibiting oxygen uptake, need not seriously inhibit ATP synthesis. In the absence of cyanide, growth does not appear to be correlated with concentrations of ATP, as shown by Spring and Rowan (1966) and Ray and Abdul-Baki (1968).

Comparing inhibition of growth by cyanide applied (a) before or concurrently with IAA treatment and (b) after IAA treatment (Fig. 1 to 4) suggests that cyanide affects growth both directly and indirectly. The following scheme is suggested as a possible explanation of the different intensity of inhibition in (a) and (b):

$$A \rightarrow B \begin{array}{c} \xrightarrow{IAA} C \rightarrow D \xrightarrow{\not{KCN}} E \rightarrow \text{Growth} \\ \xrightarrow{KCN} F \rightarrow \text{Other processes} \end{array}$$

Reaction C to D is activated by IAA; all reactions are assumed irreversible. In (a), cyanide would divert B almost entirely to C by blocking conversion of B to F. C would then accumulate until converted to D when the conversion was activated by IAA. Since D is now relatively concentrated, KCN would only be moderately inhibitory. In (b), IAA would convert C to D, but the level of D would be less than in (a) since C would be lower. Hence cyanide would be more inhibitory. In tissue treated with cyanide alone, C would accumulate but would not be converted readily to D and cyanide would be inhibitory. When cyanide was withdrawn, the accumulated C would give rise to the after-effect (Fig. 3 and 4). Identifying the hypothetical compound C accumulating in cyanide-treated tissue would help in determining the mechanism of auxin action.

V. SUMMARY

Treating sections cut from coleoptiles of *Triticum* with IAA stimulates growth after 15 minutes at $25°C$ but the rate of respiration does not increase for at least 90 minutes. Studies of the inhibition of respiration and IAA-induced growth by cyanide show that the rate of growth is not closely correlated with the rate of respiration.

The inhibition of growth of coleoptile sections by cyanide applied after IAA treatment is more marked than when applied before treatment. We propose that direct and indirect action of cyanide accounts for these differences.

VI. REFERENCES

AUDUS, L.J. (1960). Effect of growth-regulating substances on respiration, in "Encyclopedia of Plant Physiology" (Ed. W. Ruhland) Vol. 12/2, p. 360. Springer, Berlin. 1960.
EVANS, M.L. and P.M. RAY (1969). Timing of the auxin response in coleoptiles and its implications regarding auxin action. J. Gen. Physiol. 53, 1-20.
RAY, P.M. and A.A. ABDUL-BAKI (1968). Regulation of cell wall synthesis in response to auxin, in "Biochemistry and Physiology of Plant Growth Substances" (Ed. F. Wightman and G. Setterfield) p. 647. Runge, Ottawa.
RAY, P.M. and A.W. RUESINK (1962). Kinetic experiments on the nature of the growth mechanism in oat coleoptile cells. Devl. Biol. 4, 377-397.
RAYLE, D.L., M.L. EVANS and R. HERTEL (1970). Action of auxin on cell elongation. Proc. natn. Acad. Sci. U.S.A. 65, 184-191.
SPRING, A. and K.S. ROWAN (1966). Phosphorus metabolism and auxin action. Nature, Lond. 210, 1166-1167.

BIOSYNTHESIS OF AUXINS

Pathways of Auxin Biosynthesis in the Shoots of Higher Plants

Elnora A. Schneider, R.A. Gibson[1] and F. Wightman

Department of Biology, Carleton University, Ottawa, Canada

[1]Present address: Department of Plant Physiology, Waite Agricultural Research Institute, Glen Osmond, South Australia

I. INTRODUCTION

On the basis of previous work, at least five possible pathways have been proposed (Gordon, 1961; Wightman, 1964; Wightman and Cohen, 1968; Kutáček and Kefeli, 1968; Phelps and Sequeira, 1968; Kindl, 1968) for the biosynthesis of the plant growth hormone, 3-indoleacetic acid (IAA) from tryptophan (TPP), namely:

I.   TPP $\longrightarrow$ IPyA $\longrightarrow$ IAAld $\longrightarrow$ IAA
II.   TPP $\longrightarrow$ TNH$_2$ $\longrightarrow$ IAAld $\longrightarrow$ IAA
III. TPP $\longrightarrow$ ILA $\longrightarrow$ TOL $\longrightarrow$ IAAld $\longrightarrow$ IAA
IV.   TPP $\longrightarrow$ GLUBR $\longrightarrow$ IAN $\longrightarrow$ IAA
V.   TPP $\longrightarrow$ IAOX $\longrightarrow$ IAN $\longrightarrow$ IAA

The aim of the present study was to determine which of these pathways is operating in the *in vivo* biosynthesis of IAA in the shoots of a typical dicotyledon, tomato, and a typical monocotyledon, barley.

Three lines of evidence will be presented which throw light on this problem:

1. Identification of the indoles naturally occurring in the shoots of young tomato and barley plants.

2. Identification of the radioactive indoles obtained in *in vivo* metabolism experiments in which ^{14}C-labelled tryptophan, and other possible precursors of IAA, were fed to excised tomato and barley shoots.

3. Demonstration of the presence of several of the enzymes involved in IAA biosynthesis in cell-free systems prepared from tomato and barley shoots.

II. MATERIALS AND METHODS

Plant Material

Barley (Var. Keystone) and tomato (Var. Big Boy) plants were grown in air-conditioned greenhouses at 75°F under natural daylight supplemented to 18 hours with incandescent light. Shoots were excised from barley plants after 14 days growth, whereas tomato plants were allowed to grow for six weeks and were approximately 12" high and at the eight-leaf stage when they were harvested for extraction or for use in metabolism experiments.

Extraction and Identification of Native Indoles

Shoots of tomato and barley were macerated in chilled methanol in a Waring blendor and allowed to extract for 24 hr in the cold. The methanol was evaporated at $45°C$ under reduced pressure and the chlorophyll in the aqueous concentrate was removed by Celite filtration. Indole compounds in the filtrate were then separated into fractions soluble in ether at pH 7, pH 3 and pH 11, in butanol at pH 11 and in a residual aqueous fraction. Each fraction was analysed by paper and thin-layer chromatography and the indoles present were identified by comparison of their Rf values with authentic compounds in several solvent systems, by their colour reactions with DMAC, ninhydrin, Ehrlich and DNPH reagents, by their UV, fluorescence and (in the case of tryptamine) IR spectroscopy, and by bio-assay (Schneider, Gibson and Wightman, 1971). The compounds were measured quantitatively by direct densitometry of paper chromatograms after reaction with DMAC reagent, or of thin-layer chromatograms after reaction with Ehrlich reagent. The following solvent systems were used in this work.

For paper chromatography:

IAW; Isopropanol : ammonia : water, 8:1:1 v/v

BuAW; Butanol : glacial acetic acid : water, 60:15:10 v/v

IBeW; Isopropanol : benzene : water, 55:30:11 v/v, using paper buffered at pH 6.5 and dried before use.

BeAW; Benzene : glacial acetic acid : water, 2:2:1 v/v (upper phase).

8%NaCl; 8%NaCl in water.

For thin-layer chromatography:

BeAE; Benzene : glacial acetic acid : ethyl acetate, 90:5:5 v/v.

EIA; Ethyl Acetate : isopropanol : ammonia, 45:35:20 v/v.

Be; Benzene.

CEF; Chloroform : ethyl acetate : formic acid, 35:55:10 v/v.

In vivo Metabolism Experiments

Solutions of DL-tryptophan-3-^{14}C (total activity, 100 μc) and tryptamine-2-^{14}C (total activity, 100 μc) were fed to excised shoots of tomato (200 g FW) and barley (500 g FW) through the cut ends of the stem. 3-Indolelactic acid-3-^{14}C (total activity, 20 μc) was also fed to tomato, and all compounds were supplied at the rate of 0.1 mg (barley) and 0.25 mg (tomato) of indole compound per gram fresh weight (FW) of shoot. Plants absorbed the feeding solutions within 1-3 hr. and were then supplied with distilled water and allowed to metabolize the indole compound for 24 hr in a growth chamber in constant light. The shoots were then homogenized and the radioactive indole metabolites were extracted into methanol and fractionated into different ether and butanol fractions, as previously described. The fractions were analysed by chromatography and the indoles identified by their Rf value, reaction with DMAC and other reagents, and by bio-assay. Radioactivity was located on chromatograms by autoradiography and then quantitatively measured by liquid scintillation counting.

Enzyme Experiments

Enzymes were assayed in a 37,000 x g supernatant fraction prepared from tomato or barley shoots homogenized in a buffered medium containing 0.1M KH_2PO_4, 0.01M EDTA and 0.01M 2-mercaptoethanol, and adjusted to pH 8.0. The final supernatant fraction was partially purified by passage through 2.5 x 30 cm columns of Sephadex G-25, and was then sterilized by filtration through a Millipore filter with a 0.45 μ pore size. Tryptophan transaminase, aldehyde dehydrogenase and alcohol dehydrogenase were assayed according to the method of Wightman and Cohen (1968). Tryptophan decarboxylase, indolepyruvic acid decarboxylase and indolelactic acid decarboxylase activities were measured by the method reported by Gibson, Schneider and Wightman (1972).

III. RESULTS AND DISCUSSION

Native indoles in barley and tomato shoots

The indole compounds found to occur in normal barley and tomato shoots are given in Table I. As was expected, considerable free pools of tryptophan were found in both species and the concentration of this amino acid was similar to that reported in other species (Forest and Wightman, 1971). The presence of free IAA in both plants was fully established by chromatography, UV analysis and bioassay. The values recorded here for tomato (50 µg/kg) are close to those found by Pegg and Selman (1959) for tomato leaf tissue (77 µg/kg) using bioassay techniques. The amount of IAA found in our examination of barley tissue (3.2 µg/kg var. Keystone) is also similar to that reported previously for barley (2.3 µg/kg var. Atlas) by Shaw and Hawkins (1958). Undoubtedly, the native IAA content of plants will vary considerably with such factors as age and variety, but a certain consistency between different determinations made on the same species is emerging.

TABLE I. Amounts of indole compounds present in shoots (µg/g fresh weight) of 6-week old tomato plants and 2-week old barley seedlings.

COMPOUND	TOMATO var. Big Boy	BARLEY var. Keystone
Tryptophan	12.0	20.0
3-Indoleacetic acid	0.05	0.003
Tryptamine	1.0	1.69
Tryptophol	0.005	-
3-Indolelactic acid	0.004	-
3-Indolealdehyde	0.005	-

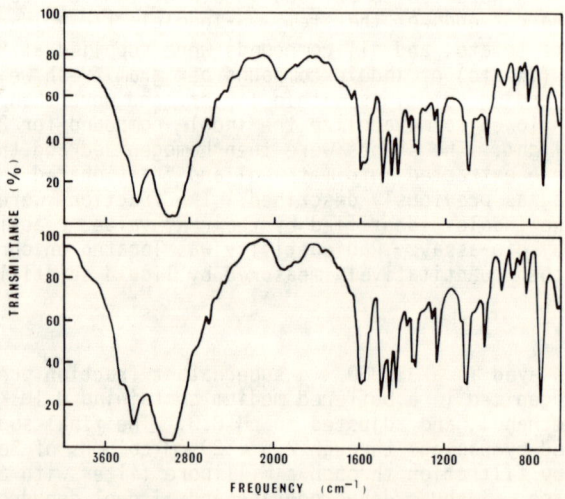

Fig. 1. Infrared absorption spectra of suspected tryptamine isolated from tomato shoots (lower scan) and of authentic tryptamine (upper scan). Both tryptamine samples were in the form of the hydrochloride salt.

With regard to the natural occurrence of compounds that could be considered as intermediates between tryptophan and IAA, only tryptamine (TNH_2) was present in both barley and tomato, which suggests that biosynthesis Pathway II could be operating in these species. Proof of the identity of this amine was obtained by isolating the compound from basic ether fractions, as the hydrochloride salt, and comparing its IR absorption spectrum with that shown by authentic tryptamine-HCl (Fig. 1). Trace amounts of 3-indolelactic acid (ILA) and tryptophol (TOL) were found in tomato only, raising the possibility that Pathway III might also be operating in this species. 3-indolepyruvic acid (IPyA) and 3-indoleacetaldehyde (IAAld), which are the two most likely intermediates between tryptophan and IAA, were not detected in our extracts from either tomato or barley shoots, but this may have been due to the instability of these compounds in methanol, and in solvent fractionation procedures. For this reason, the possibility that Pathway I might be operating in tomato and barley tissues under natural conditions cannot be ruled out from the present observations. 3-indole-acetonitrile (IAN), which is a relatively stable compound, was not detected in any of our extracts, and it appears that this compound and its involvement as an intermediate in IAA biosynthesis (Pathways IV or V) is restricted to Brassicaceae and closely related families.

In vivo metabolism experiments with excised shoots

In these experiments, excised shoots of barley and tomato were fed with solutions of either tryptophan-3-^{14}C, tryptamine-2-^{14}C or indolelactic acid-3-^{14}C. The results presented in Table II show the distribution of radioactivity found in the indole metabolites of these compounds. Radioactivity from both tryptophan-3-^{14}C and tryptamine-2-^{14}C was incorporated into IAA in both species, indicating that both of these compounds can serve as precursors of the growth hormone in these plants. The relative differences in the amount of radioactivity incorporated into IAA in barley shoots, for example, when fed with either tryptophan-3-^{14}C or tryptamine-2-^{14}C is shown in Fig. 2.

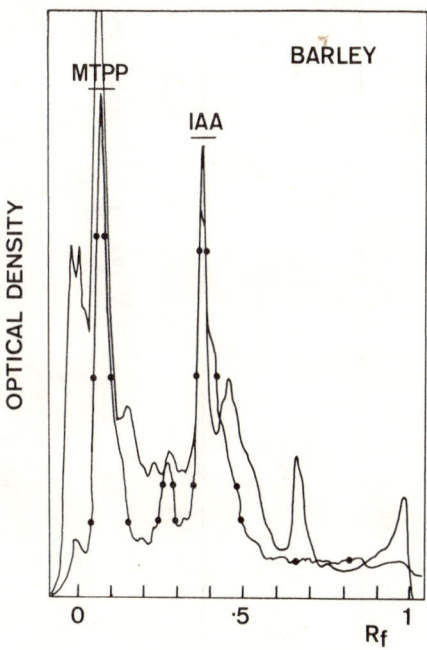 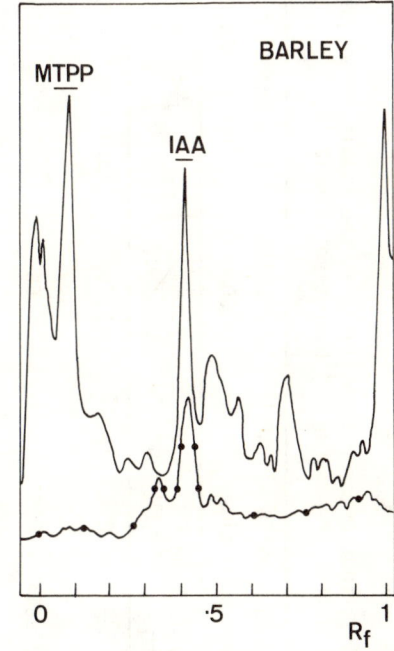

Fig. 2. Densitometer scan (continuous line) and radioactivity scan (line with circles) of thin-layer chromatograms of Acid ether fractions obtained from barley shoots after feeding with either tryptophan-3-^{14}C (LEFT) or tryptamine-2-^{14}C (RIGHT). Solvent system: EIA

TABLE II. Distribution of radioactivity in the indole compounds isolated from tomato and barley shoots 24-hours after feeding with either tryptophan-3-^{14}C or tryptamine-2-^{14}C.

Compounds from:	UNFED 0 time µg/g FW	UNFED 24 hr µg/g FW	FEEDING SOLn dpm/µg	Shoots fed TRYPTOPHAN-3-^{14}C			Shoots fed TRYPTAMINE-2-^{14}C			Shoots fed INDOLELACTIC ACID-3-^{14}C		
				µg/g FW	cpm*/g FW	Dilution**	µg/g FW	cpm*/g FW	Dilution**	µg/g FW	cpm*/g FW	Dilution**
TOMATO SHOOTS												
Tryptophan	12.0	12.0	2,750	313.0	497,000	1.5	14.0	600	52.2	71.0	32,000	1.52
Tryptamine	1.0	1.0	2,600	1.6	710	5.4	160.0	28,000	1.3	0.98	77	6.4
Indolelactic acid	0.005	0.005	682	0.102	127	2.0	0.005	-	-	7.6	4,200	1.24
Tryptophol	0.005	0.005	-	0.014	185	1.9	0.066	154	1.0	0.04	17	1.6
IAA	0.05	0.05	-	0.06	53	2.8	0.056	19	6.6	0.07	23	2.1
Indolealdehyde	0.005	0.005	-	0.02	13	3.7	0.020	10	4.5	-	-	-
BARLEY SHOOTS												
Tryptophan	20.0	76.5	3,100	123.0	71,275	5.4	77.5	-	-	ILA-3-^{14}C was not fed to barley shoots		
Tryptamine	1.69	8.3	3,050	8.3	871	29.3	52.0	52,700	3.0			
IAA	-	0.012	-	0.017	49	1.1	0.014	28	1.5			

* Corrected for counting efficiency of 89%

** Dilution = $\dfrac{\text{Sp. Act. of Feeding Soln., dpm/µg}}{\text{Sp. Act. of Compound, cpm/µg}}$

After feeding tryptophan-3-^{14}C, carrier 3-indolepyruvic acid or 3-indoleacetaldehyde was added to the extracts, and the 2,4-dinitrophenyl-hydrazone of each of these compounds was isolated. Both hydrazones were found to be radioactive, as shown in Figs. 3 and 4. Therefore, all the compounds proposed to participate in the Indolepyruvic acid pathway (Pathway I) were found to be labelled after administration of tryptophan-3-^{14}C.

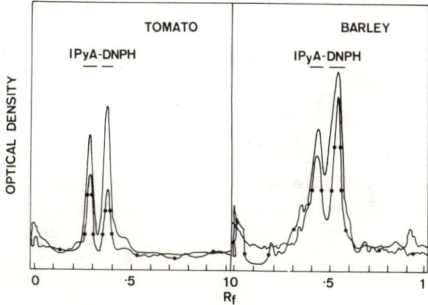

Fig. 3. Densitometer scans of thin-layer chromatograms of 3-indolepyruvic acid-2,4-dinitrophenylhydrazones (IPyA-DNPH) isolated from tomato and barley shoots after feeding with tryptophan-3-^{14}C. Continuous line; Optical density of TL chromatogram at 410 mµ. Circles; Optical density of autoradiograph of TL chromatogram. Solvent system: BeAE. Note that in this solvent system the *cis*- and *trans*- isomers of IPyA-DNPH are separated.

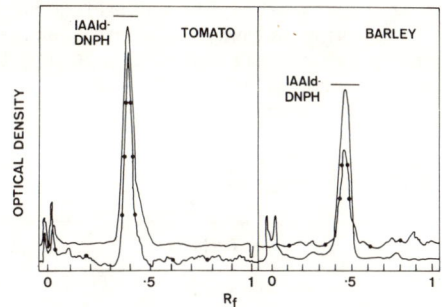

Fig. 4. Densitometer scans of thin-layer chromatograms of 3-indoleacetaldehyde-2,4-dinitrophenylhydrazones (IAAld-DNPH) isolated from tomato and barley shoots after feeding tryptophan-3-^{14}C. Continuous line; Optical density of TL chromatogram at 410 mµ. Circles; Optical density of autoradiograph of TL chromatogram. Solvent system: Be.

Radioactivity from tryptophan-3-^{14}C was also found in tryptamine (Table II). When tryptamine-2-^{14}C was fed to the shoots, radioactivity could be found not only in IAA but also in 3-indoleacetaldehyde, the proposed intermediate between tryptamine and IAA, when this compound was isolated as the 2,4-dinitrophenylhydrazone (Fig. 5). These findings are clearly consistent with the operation of biosynthesis Pathway II in barley and tomato.

When 3-indolelactic acid-3-^{14}C was fed to shoots to tomato, in which this compound occurs naturally, radioactivity was found in IAA. However, since the major metabolite of indolelactic acid-3-^{14}C was radioactive tryptophan (Table II), it would appear that the route between indolelactic acid and IAA may be *via* tryptophan rather than the more direct pathway *via* tryptophol as proposed in earlier work (Wightman, 1964).

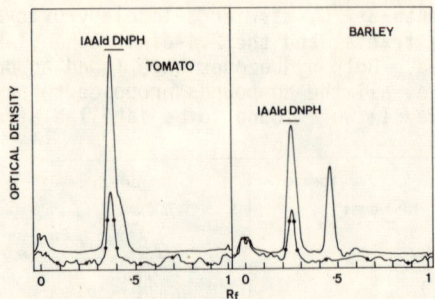

Fig. 5. Densitometer scans of thin-layer chromatograms of 3-indoleacetaldehyde-2,4-dinitrophenylhydrazones (IAAld-DNPH) isolated from tomato and barley shoots after feeding with tryptamine-2-^{14}C. Continuous line; Optical density of TL chromatogram at 410 mµ. Circles; Optical density of autoradiograph of TL chromatogram. Solvent system: Be.

Demonstration of enzymes of IAA biosynthesis in cell-free systems

All the enzymes required for the biosynthesis of IAA by Pathways I, II and III, with the exception of amine oxidase, were assayed in sterile, cell-free systems prepared from tomato shoots. The results are presented in Table III. The values obtained in assays of several of these enzymes in barley cell-free systems are also shown. It can be seen that in tomato preparations, all the enzymes of both the Indolepyruvic acid and the Tryptamine pathways can be demonstrated, with the exception of the amine oxidase catalysing the conversion of tryptamine to 3-indoleacetaldehyde. This enzyme is currently under investigation.

TABLE III. Activities of some enzymes involved in IAA biosynthesis found in cell-free preparations of tomato or barley shoots.

ENZYME ASSAYED	TOMATO		BARLEY	
	µg/mg protein per hr	µg/g FW/hr	µg/mg protein per hr	µg/g FW/hr
L-Tryptophan transaminase (TPP ⟶ IPyA)	48.0	4.65	44.0	8.8
L-Tryptophan decarboxylase (TPP ⟶ TNH$_2$)	2.3	0.23	1.42	0.28
3-Indolepyruvic acid decarboxylase (IPyA ⟶ IAAld)	present			
3-Indolelactic acid decarboxylase (ILA ⟶ TOL)	not detected			
Aldehyde dehydrogenase (IAAld ⟶ IAA)	7.8	0.75	10.0	1.95
Alcohol dehydrogenase (IAAld ⟶ TOL)	17.4	1.7	7.7	1.5

No indolelactic acid decarboxylase could be demonstrated in the tomato preparation, and this finding lends further support to the view that the conversion of 3-indolelactic acid to IAA occurs *via* tryptophan and not *via* tryptophol.

IV. CONCLUSIONS

Figure 6 presents a summary of the evidence obtained in these experiments on the pathways of biosynthesis of IAA in tomato shoot tissue. The natural occurrence of the indole compounds involved, the results obtained from *in vivo* metabolism experiments in which tryptophan-3-^{14}C and tryptamine-2-^{14}C were fed to intact shoots, and the enzyme activities demonstrated in cell-free preparations, all combine to suggest that both the Indolepyruvic acid and Tryptamine pathways can operate in tomato shoots. The results from barley shoots lead to essentially identical conclusions. We have found, however, in other investigations that while tryptamine is widely distributed in many families of plants, it is certainly not ubiquitous and was not found, for example, in pea, bean, squash or cabbage. It is, therefore, unlikely that the Tryptamine pathway is an important mechanism of IAA formation in all higher plants.

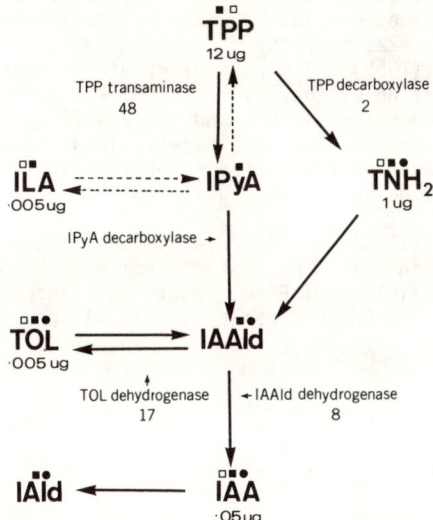

Fig. 6. Summary of evidence for the pathways of IAA biosynthesis in tomato shoots. Numbers below names of indole compounds refer to the concentration of the compound in unfed shoots in µg/g fresh wt. Numbers below names of enzymes represent the activity of the enzyme in µg product/g fresh wt shoot tissue/hr. Black squares indicate the compounds that were labelled when tryptophan-3-^{14}C was fed; open squares indicate the compounds that were labelled when 3 3-indolelactic acid-3-^{14}C was fed; and black circles indicate the compounds that were labelled when tryptamine-2-^{14}C was fed.

Further work in this area must be directed toward the problem of discovering whether both the Indolepyruvic acid and Tryptamine pathways are actually used under natural conditions, or whether one pathway or the other is predominant at different stages of plant development, or in different organs, or perhaps under different physiological conditions.

V. REFERENCES

FOREST, J.C. and F. WIGHTMAN. (1971). Metabolism of amino acids in plants. I. Changes in the soluble amino acid fractions of bushbean seedlings (*Phaseolus vulgaris* L.) and the development of transaminase activity. Can. J. Biochem., 49, 709-720.

GIBSON, R.A., ELNORA A. SCHNEIDER and F. WIGHTMAN. (1972). Biosynthesis and metabolism of Indol-3yl-acetic acid. II. *In vivo* metabolism experiments with ^{14}C-labelled precursors of IAA in tomato and barley shoots. J. exp. Bot., 23, in press.

GORDON, S.A. (1961). The biogenesis of auxin. In: "Encyclopedia of Plant Physiology. Vol. XIV. Growth and Growth Substances". Springer-Verlag, Berlin. pp. 620-646.

KINDL, H. (1968). Oxydasen und Oxygenasen in höheren Pflanzen, I. Über das Vorkommen von Indolyl-(3)-acetaldehydoxim und seine Bildung aus L-Tryptophan. Hoppe-Seyler's Z. Physiol. Chem., 349, 519-520.

KUTÁČEK, M. and V.I. KEFELI. (1968). The present knowledge of indole compounds in plants of the Brassicaceae family. In: "Biochemistry and Physiology of Plant Growth Substances." Eds. Wightman, F. and G. Setterfield. The Runge Press Ltd., Ottawa. pp. 127-152.

PEGG, G.F. and I.W. SELMAN. (1959). An analysis of the growth response of young tomato plants to infection by *Verticillium albo-atrum*. II. The production of growth substances. Ann. Appl. Biol., 47, 222-231.

PHELPS, R. and L. SEQUEIRA. (1968). Auxin biosynthesis in a host-parasite complex. In: "Biochemistry and Physiology of Plant Growth Substances." Eds. Wightman, F. and G. Setterfield. The Runge Press Ltd., Ottawa. pp. 197-212.

SCHNEIDER, ELNORA A., R.A. GIBSON and F. WIGHTMAN. (1971). Biosynthesis and metabolism of Indole-3yl-acetic acid. I. The native indoles of barley and tomato shoots. J. exp. Bot., 22, in press.

SHAW, M. and A.R. HAWKINS. (1958). The physiology of host-parasite relations. V. A preliminary examination of the level of free endogenous indoleacetic acid in rusted and mildewed cereal leaves and their ability to decarboxylate exogenously supplied radioactive indoleacetic acid. Can. J. Bot., 36, 1-16.

WIGHTMAN, F. (1964). Pathways of tryptophan metabolism in tomato plants. In: "Régulateurs Naturels de la Croissance Végétale," Coll. int. C.N.R.S., 123, 193-212.

WIGHTMAN, F. and D. COHEN. (1968). Intermediary steps in the enzymatic conversion of tryptophan to IAA in cell-free systems from mung bean seedlings. In: "Biochemistry and Physiology of Plant Growth Substances". Eds. Wightman, F. and G. Setterfield. The Runge Press Ltd., Ottawa. pp. 273-288.

ENZYMIC OXIDATION OF INDOLE-3-ETHANOL

Larry E. Vickery, John E. Sherwin and William K. Purves

Department of Biological Sciences, University of California, Santa Barbara

I. INTRODUCTION

The current level of interest in pathways of auxin biosynthesis is indicated by the substantial number of papers on this topic in this symposium volume and its predecessor (Wightman and Setterfield, 1968). One topic of concern is the placement of certain intermediates, such as indole-3-ethanol, in the pathway. Here we shall direct our attention to the metabolism of indole-3-ethanol (IEt).

We have already shown that IEt is a naturally occurring component of cucumber seedlings (Rayle and Purves, 1967a), and it also occurs in sunflower, pea, and zucchini squash seedlings (Rajagopal, 1967; Rayle and Purves, 1967a, 1968). IEt-^{14}C is converted to labelled indole-3-acetic acid (IAA-^{14}C) in cucumber seedlings (Rayle and Purves, 1967b), and Sherwin and Purves (1969) have shown that this conversion is not dependent upon the presence of epiphytic microorganisms. The pathway of synthesis of IEt is uncertain. Cucumber seedlings treated with tryptamine-^{14}C produce IEt-^{14}C (Sherwin and Purves, 1969), while the data of Wightman (1964) suggest the production of IEt from indole-3-pyruvic acid or indole-3-lactic acid rather than from tryptamine in tomato seedlings. In any case, it is likely that indole-3-acetaldehyde (IAAld) is the immediate precursor of IEt in higher plants (Libbert et al., 1968; Rayle and Purves, 1967b; Wightman and Cohen, 1968).

We have now extracted, from cucumber seedlings, an enzyme preparation which catalyzes the conversion of IEt to IAAld. The isolation and some properties of this preparation are described in this preliminary paper. The activity of the enzyme appears to be partially inhibited by IAA, suggesting a feedback control of auxin levels in the cucumber seedling. This presents an appealing alternative to the hypothesis that auxin levels are determined by the activity of an IAA oxidase (see, for example, Galston and Hillman, 1961).

II. MATERIALS AND METHODS

Seeds of *Cucumis sativus* L. cv. National Pickling were sown in vermiculite and grown under light and temperature conditions described by Purves et al (1967). After eight to ten days the shoots were harvested and weighed. The seedlings were extracted for enzyme by grinding in a Waring blendor, using 2 litres of 0.05M Na_2HPO_4 (pH 7.5) per kilogram of tissue. The homogenate was squeezed through cheesecloth and centrifuged at 20,000 xg for 30 minutes. The supernatant was decanted and brought to a final volume of approximately 4 litres (per kilogram of tissue) with 0.02M Na_2HPO_4, pH 7.5. A batch extraction procedure was repeated three times with the diluted supernatant fluid, as follows. Approximately 100 ml (wet volume) of Bio-Rex 70 resin was added to each 4 l beaker and stirred for 2 hr. The resin was allowed to settle and the supernatant decanted for another extraction. The resin from the three extractions was successively poured into a column of an appropriate size and washed with 0.02M Na_2HPO_4, pH 7.0, until the eluate was clear. Enzyme was then eluted using 0.5M NaCl in 0.05 M Na_2HPO_4, pH 7.0. The concentrated enzyme solution was frozen, thawed, centrifuged to remove precipitated particles, and dialyzed for 4 hr against 0.05M Na_2HPO_4, pH 7.0. This solution was absorbed onto a 2.5 x 16 cm Bio-Rex 70 column in the same buffer. The column was eluted with a linear salt gradient, the first stage containing 1 litre of column buffer and

the second containing 1 litre of buffer which was 0.5M in NaCl. Fractions exhibiting
activity which was equal to 50 per cent or more of the most active fraction were
combined and used for further study.

A colorimetric assay for reaction product (IAAld) utilizing Salkowski reagent was
used throughout. The H_2SO_4 reagent (Tang and Bonner, 1947) and the $HClO_4$ reagent,
(Gordon and Weber, 1951) gave similar results, but the latter technique was chosen
because of its greater sensitivity. Absorbance measurements were made at 528 nm using
a Beckman model DB spectrophotometer after incubation of the acid-treated reaction
mixture in the dark for 1 hr. Absorption spectra of reaction products were determined
with a Cary model 15 recording spectrophotometer. The use of Salkowski reagent as a
test for IAAld depends upon the presence of IEt. Standard preparations of the reagent
give no stable color with either IEt or IAAld alone, but a quantitative reaction with
IAAld is observed in the presence of an excess of IEt. Protein was determined according to the method of Lowry et al.(1951).

An alternate preparation of the enzyme has also been used. In this method the
crude supernatant from the initial centrifugation step was heated for 30 min at $65^°$,
centrifuged to remove precipitate, dialyzed overnight, and again centrifuged clear.
This solution was chromatographed on a cellulose-phosphate column as with the resin.
Active fractions were assayed by the method of Gordon and Weber (1951) but also
showed the ability to reduce 2,6-dichloroindophenol (DCIP) in the presence of substrate.

III. RESULTS AND DISCUSSION

Salt gradient chromatography

An elution profile for an enzyme preparation derived from 1 kg of tissue and
chromatographed on Bio-Rex 70, a carboxylic cation exchange resin, is illustrated in
Fig. 1. Enzyme activity was determined by incubation of 0.2 ml of 4×10^{-4} MIEt with
0.2 ml of each fraction for 10 min, followed by treatment of the reaction mixture with
Salkowski reagent. By combining all fractions having specific activities at least 50
per cent of that of the peak tube, we obtained a solution 2-3,000 fold purified as
compared with the crude homogenate. This pooled solution was dialyzed and, following
concentration on a small Bio-Rex 70 column, used as a stock enzyme solution for the
experiments described below. Some degree of further purification has been obtained by
taking advantage of the enzyme's relative heat stability, being stable at $70^°$ for up
to 1 hr. Gel chromatography (Sephadex G-150) has also afforded further purification
and suggests that the molecular weight of the enzyme(s) involved is slightly greater
than 100,000 daltons.

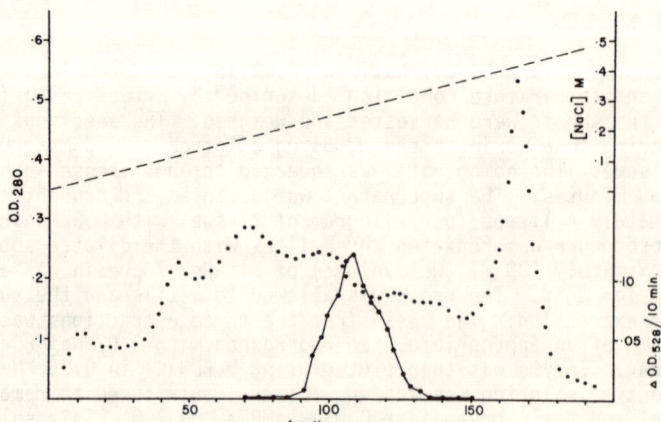

Fig. 1. Cation exchange chromatography of enzyme activity from batch extraction.
Fraction volume = 10 ml. For details, see text.

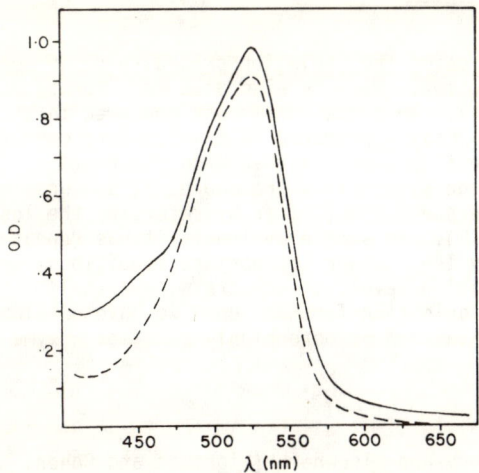

Fig. 2. Absorption spectra of enzyme reaction mixture (solid line) and 2×10^{-4}M IAA (dashed line) following treatment with Salkowski reagent.

Characterization of the reaction product

A solution of enzyme containing 0.3 mg/ml protein was incubated for 1 hr at $22°$ with 0.1M Na_2HPO_4 (pH 7.8) and 3×10^{-4}M IEt. The reaction was stopped by the addition of $HClO_4$ Salkowski reagent and, after 1 hr in the dark, the resultant absorption spectrum was compared with that of 2×10^{-4}M IAA in Salkowski reagent (Fig. 2). The two spectra are quite similar; however, the spectrum for the enzyme reaction mixture shows a prominent shoulder at about 450 nm. Such a shoulder is characteristic of mixtures of IAAld and IEt treated with Salkowski reagent; and, of course, the reaction mixture contained IEt (excess substrate). Results of thin layer chromatography in two solvent systems also indicate the presence of IAAld in the reaction product. Thin layer radiochromatography experiments utilizing IEt-^{14}C are also being carried out to confirm the identity of IAAld as the reaction product.

Time course

The time course of the reaction at two concentrations of enzyme is shown in Fig. 3. For each curve the IEt concentration was 1.5×10^{-4}M in 0.1M Na_2HPO_4, pH 7.8. For the upper plot, the protein concentration was 0.30 mg/ml and, for the lower plot, 0.15 mg/ml.

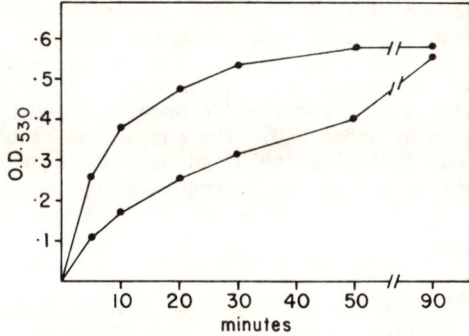

Fig. 3. Time course of the reaction as a function of enzyme concentration.

Inhibition by IAA

The preceding results were all obtained using the isolation procedure described at the outset of the Materials and Methods section. We have also used an alternative isolation procedure (see Materials and Methods). When the enzyme was prepared in the latter fashion, it coupled readily with DCIP. Enzyme prepared in the standard manner also reduced DCIP, but at a slower rate, suggesting either a change in the protein or the loss of one or more cofactors. Using the alternative procedure, it was possible to follow the oxidation of IEt directly in the spectrophotometer by recording the loss of optical density at 600nm (reduction of DCIP). In such experiments it was repeatedly observed that enzyme activity was inhibited by IAA. Under appropriate conditions, the inhibition was as great at 40%, even at saturating levels of substrate. We shall study the characteristics and specificity of inhibition further, once we have developed suitable methods for following the reaction catalyzed by more highly purified enzyme preparations.

IV. DISCUSSION

It seems clear from work in our laboratory and elsewhere (Wightman and Cohen, 1968; Wightman and Gibson, personal communication) that the terminal steps in the formation of IAA in several plants are as follows:

$$\text{Precursors} \longrightarrow \text{Indole-3-acetaldehyde} \longrightarrow \text{IAA}$$
$$\updownarrow$$
$$\text{IEt}$$

If this scheme is correct, and if the conversion of IEt to IAA is inhibited by IAA, this suggests a tempting model for the regulation of auxin levels by feedback inhibition. Clearly, our tentative model requires further study.

At the preceding symposium, Wightman and Cohen (1968) reported the isolation of two enzymes (from mung bean) which together catalyzed the oxidation of IEt to IAA. One, an alcohol dehydrogenase, catalyzed the interconversion of IEt and indoleacetaldehyde; the second, an aldehyde dehydrogenase, catalyzed the oxidation of indoleacetaldehyde to IAA. The relationship between our enzyme preparation and these two enzymes is unclear. However, the mung bean enzymes were strongly promoted by exogenously supplied NAD, while our preparation is unaffected by the addition of NAD.

V. SUMMARY

Previous work in this laboratory has shown that cucumber seedlings contain indole-3-ethanol (IEt), and that IEt is readily metabolized to IAA, accounting for the growth-promoting activity of IEt on cucumber hypocotyls. We now report the isolation of an enzyme preparation, catalyzing the oxidation of IEt to indole-3-acetaldehyde (IAAld), from green cucumber seedlings. The identification of the product as IAAld was based on chromatographic R_f values, adding Salkowski reagent to the reaction mixture. The preparation has been purified in excess of 2,000-fold by a series of steps including salt gradient chromatography on Bio-Rex 70. Preliminary gel filtration studies (Sephadex G-150) indicated a molecular weight somewhat in excess of 100,000. The activity of the preparation was unaffected by added NAD. The enzyme was stable at 70° for at least 1 hr. Enzyme activity was partially inhibited by IAA, suggesting the possibility that auxin levels in cucumber seedlings may be regulated by feedback inhibition of IAA synthesis.

VI. ACKNOWLEDGEMENTS

This work was supported by Grant GB-13219 of the U.S. National Science Foundation, awarded to W.K. Purves. L.E. Vickery was a National Defence Education Title IV Predoctoral Fellow, and J.E. Sherwin a U.S. Public Health Service Predoctoral Fellow.

VII. REFERENCES

GALSTON, A.W. and W.S. HILLMAN (1961). The degradation of auxin, in "Encyclopedia of Plant Physiology" (Ed. W. Ruhland) Vol. 14, pp 647-670. Springer, Berlin, 1961.

GORDON, S.A. and R.P. WEBER (1951). Colorimetric estimation of indoleacetic acid. Pl. Physiol., Wash. $\underline{26}$, 192-195.

LIBBERT, E., S. WICHNER, E. DUERST, W. KAISER, R. KUNERT, A. MANICHI, R. MANTEUFFEL, E. RIECKE and R. SCHRÖDER (1968). Auxin content and auxin synthesis in sterile and non-sterile plants, with special regard to the influence of epiphytic bacteria. in Wightman and Setterfield (1968), pp. 213-230.

LOWRY, O.H., J.N. ROSEBROUGH, A.L. FARR and R.J. RANDALL (1951). Protein measurement with the Folin phenol reagent. J. Biol. Chem. $\underline{193}$, 265-275.

PURVES, W.K., D.L. RAYLE and K.D. JOHNSON (1967). Actions and interactions of growth factors on cucumber hypocotyl segments. Ann. New York Acad. Sci. $\underline{144}$, 169-179.

RAJAGOPAL, R. (1967). Occurrence of indoleacetaldehyde and tryptophol in the extracts of etiolated shoots of *Pisum* and *Helianthus* seedlings. Physiol. Plantarum $\underline{20}$ 655-660.

RAYLE, D.L. and W.K. PURVES (1967a). Isolation and identification of indole-3-ethanol (tryptophol) from cucumber seedlings. Pl. Physiol., Wash. $\underline{42}$, 520-524.

RAYLE, D.L. and W.K. PURVES (1967b). Conversion of indole-3-ethanol to indole-3-acetic acid in cucumber seedling shoots. Pl. Physiol., Wash. $\underline{42}$, 1091-1093.

RAYLE, D.L. and W.K. PURVES (1968). Studies on 3-indoleethanol in higher plants, in Wightman and Setterfield (1968), pp 153-161.

SHERWIN, J.E. and W.K. PURVES (1969). Tryptophan as an auxin precursor in cucumber seedlings. Pl. Physiol., Wash. $\underline{44}$, 1303-1309.

TANG, Y.W. and J. BONNER (1947). The enzymatic inactivation of indoleacetic acid. I. Arch. Bioch. Bioph. $\underline{13}$, 11-25.

WIGHTMAN, F. (1964). Pathways of tryptophan metabolism in tomato plants, in Colloq. Intern. Centre Natl. Rech. Sci. Paris $\underline{123}$, 193-212.

WIGHTMAN, F. and D. COHEN (1968). Intermediary steps in the enzymatic conversion of tryptophan to IAA in cell free systems from mung bean seedlings, in Wightman and Setterfield (1968), pp. 273-288b.

WIGHTMAN, F. and G. SETTERFIELD, Eds. (1968). "Biochemistry and Physiology of Plant Growth Substances". Runge Press, Ottawa.

PLANT GROWTH SUBSTANCES, 1970

The Control of Growth by the Synthesis of IAA and its Conjugation

R.M. Muir

Department of Botany, University of Iowa, Iowa City, Iowa, U.S.A.

I. INTRODUCTION

In many studies of growth the responses of plant tissue to treatments with IAA, gibberellins and phytokinins have been given simplistic interpretations in which the treatment is represented as an aspect of the natural control system. In such interpretations the evidence concerning the internal control of growth by IAA has been minimised or ignored although the measurement of diffusible auxin is the most accurate and realistic indicator of the endogenous control of elongation. The experiments reported here have examined the control of growth by the synthesis of IAA in the plant and the combination of IAA with other substances.

II. MATERIALS AND METHODS

Seedlings of the oat plant (*Avena sativa* L. cv. Garland) were grown in moist sterile sand in small trays, 20 x 3 x 2cm. After 63 hours in the dark at 24° and 85% relative humidity, the coleoptiles were just emerging from the sand. At this time trays with 20 seedlings in each were placed under near-red or under far-red radiation. These conditions were obtained with interference filters (Bausch and Lomb, designated as 660 mu and 730 mu) and a 1500 w incandescent light source 50 cm above the filters. The coleoptiles under near-red received 0.5 uw/cm^2/mu at 660 mu while those under far-red received 0.4 uw/cm^2/mu at 730 mu. After 24 hours the diffusible auxin for 10 coleoptiles and the height of 20 coleoptiles were measured for seedlings under each type of radiation and seedlings that remained in the dark for the period. The diffusible auxin was determined by placing the apical segment 2-mm long on a 2 x 2 x 2 mm agar block (1.5%) which had been equilibrated in 0.05 M phosphate buffer at pH 3.5. The diffusion period was 3 hours under weak green light. The auxin content was then measured by the *Avena* curvature bioassay.

Seeds of the Alaska (normal) and Little Marvel (dwarf) cultivars of the pea (*Pisum sativum* L.) were soaked in water for 18-24 hours with aeration and planted in sterilized sand. The seedlings were then grown either in the dark at 24° and 85% relative humidity or in the greenhouse. Treatment with GA consisted of the application of 0.01 ml of 10^{-3}M solution between the young stipules of the apical bud. Diffusible auxin was determined by placing the tissue on 2 x 2 x 2 mm agarose blocks (1%) for 2 to 3 hours under fluorescent light (250 fc).

The determination of indoleacetyl aspartate in the stem tissue of the peas employed a modification of the procedure of Good *et al,* (1956) which has been described elsewhere (Lantican and Muir, 1969).

III. RESULTS AND DISCUSSION

One of the generalizations that have been developed concerning the phytochrome system and the effects of near-red and far-red radiation on plants is that near-red reduces stem growth while far-red promotes it (Downs *et al*, 1957). Such observations suggest that the synthesis of IAA is reduced by exposure to near-red and increased

Table 1. Effect of near-red and far-red radiation on the synthesis of IAA in the apex of the *Avena* coleoptile.

Condition	Average IAA Equivalents (ng/apex)	Average Coleoptile Height (cm)
24 Hr Dark	.22	2.54
12 Hr Dark ≠ 12 Hr Near-Red	.08	2.44
12 Hr Dark ≠ 12 Hr Far-Red	.19	2.44
4 Hr Near-Red ≠ 20 Hr Dark	.05	1.76
4 Hr Far-Red ≠ 20 Hr Dark	.22	2.20
12 Hr Near-Red ≠ 12 Hr Dark	.09	1.75
12 Hr Far-Red ≠ 12 Hr Dark	.22	2.17

Table 2. Reversibility of the effect of near-red radiation on the synthesis of IAA in the apex of the *Avena* coleoptile.

Condition	Average IAA Equivalents (ng/apex)	Average Coleoptile Height (cm)
24 Hr Dark	.25	2.82
4 Hr Near-Red ≠ 20 Hr Far-Red	.23	2.89
8 Hr Near-Red ≠ 16 Hr Far-Red	.23	2.88
4 Hr Far-Red ≠ 20 Hr Near-Red	.08	2.53
8 Hr Far-Red ≠ 16 Hr Near-Red	.05	2.58

by exposure to far-red. This possibility has been examined by comparing the synthesis of IAA in the apex of the *Avena* coleoptile grown in darkness with the synthesis in seedlings exposed to near-red and seedlings exposed to far-red. Data obtained by Miss Katherine W. Chen in such an experiment are given in Table 1. When the coleoptiles are exposed to the radiation during the 12 hours immediately preceding auxin measurement, near-red reduces IAA synthesis to one-third the synthesis in the dark while far-red has no effect, and no effect on coleoptile growth is found in either case. As little as 4 hours of near-red causes a reduction in IAA synthesis which persists through 20 hours of dark. In this case the growth of the coleoptile is also reduced. Exposure to far-red followed by 12 to 20 hours of dark has no effect on IAA synthesis although some reduction in coleoptile growth occurs. The obligatory experiment is to follow the exposure to near-red with exposure to far-red and the data obtained by Miss Katherine W. Chen are given in Table 2. The results show that far-red does reverse the effect of near-red, IAA synthesis is restored to the level occurring in the dark, and the growth of the coleoptile is the same as the growth in the dark. Briggs (1963) has reported that auxin production is suppressed by exposure of corn coleoptiles to radiation from ruby-red darkroom safelights, but Hopkins and Hillman (1966) have found that exposure of apical segments (6 mm long) of the *Avena* coleoptiles to 5 minutes of radiation from red fluorescent tubes behind red Plexiglas promotes elongation of the segments. The latter finding probably has its explanation in the IAA gradient of control in the coleoptile and therefore is unrelated to the growth of the intact coleoptile.

Table 3. Response of Alaska and Little Marvel pea seedlings to light and GA.

	Stem Height (cm)		
	Grown in Dark 7 days	Grown in Light 7 days	10 Days
Alaska	12.5	5.8	16.7 (treated with GA)
		5.1	15.4 (no GA treatment)
Little Marvel	11.6	3.1	9.9 (treated with GA)
		3.1	5.1 (no GA treatment)

Table 4. Diffusible auxin from apical tissues of Alaska and Little Marvel pea seedlings.

Condition	Average IAA Equivalents (n/g apex)	
	Alaska	Little Marvel
Grown in Dark		
Apex	0.3	0.3
Grown in Light		
Apex	1.3	1.4
Apex - 1 mm Internode	0.5	0.3
Apex - 5 mm Internode	0.3	0.2
Apex -10 mm Internode	0.2	0.0

The analysis of the synthesis of IAA in relation to elongation of the stem of pea seedlings includes the comparison of growth of the dwarf cultivar (Little Marvel) with the growth of the normal cultivar (Alaska) in darkness and in sunlight, and the growth response to treatment with GA 7 days after planting. The measurements of growth under our conditions are given in Table 3 as averages for 20 seedlings. The amounts of diffusible auxin from apical portions of these plants are given in Table 4 as averages of 6 determinations. As is well known, the growth of the dwarf and normal cultivars is nearly the same in darkness and the synthesis of IAA is also the same. In sunlight the growth of Little Marvel is less than a third of its growth in darkness while the growth of Alaska is about one-half its growth in darkness, yet the synthesis of IAA is 4 times greater in sunlight and that in Little Marvel is always a little more than that in Alaska. It is clear that the differences in growth of the pea stem are not the result of differences in the synthesis of IAA in the apex. When the diffusible auxin is measured in the subtending internode the amount is only about one-third as much as that from the apex and in Little Marvel no auxin is found when 10 mm of internode subtends the apex.

The observations of Andreae and Good (1955) on the formation of indoleacetyl aspartate in the epicotyl of the pea suggested that this might be the natural conjugate of the IAA rendering it ineffective as an auxin. A sample of the conjugate was supplied by Dr. Good and it was found to be inactive in the *Avena* curvature bioassay. When tested for its effect on the straight growth of coleoptile segments, some activity is observed at a concentration of 4×10^{-4}M which is probably the result of hydrolysis in the tissue during the 24 hours. Measurements of the indoleacetyl aspartate in the stem tissue of Alaska and Little Marvel seedlings are given in Table 5. No conjugate is found in either of the seedlings grown in darkness but it is present in both when grown in sunlight with more than 3 times as much in Little Marvel as in Alaska. Thus the greater growth in darkness with lower synthesis of IAA is possible because of the lack of conjugate formation. The lesser growth of the Little Marvel in sunlight is due to the greater degree of conjugation in its tissue. The greater promotive effect of GA on the growth of Little Marvel results from the increased synthesis of IAA which apparently saturates the conjugating system and provides free IAA for the promotion of elongation. The formation of the indoleacetyl aspartate synthetase in plants grown in sunlight is probably the result of induction by higher IAA levels in light (Lantican and Muir, 1969).

Table 5. Indoleacetyl aspartate content of Alaska and Little Marvel pea seedlings.

	nmol/g	nmol/plant
Grown in Dark		
Alaska	0	0
Little Marvel	0	0
Grown in Light		
Alaska	6	3
Little Marvel	20	11

The control of growth by IAA in plants of the Cruciferae involves an additional component in the auxin physiology, glucobrassicin (GLBR). The information on this substance was reviewed by Kutáček and Kefeli (1968) at the Conference at Ottawa. It is now generally agreed that ascorbigen is a degradation product of GLBR in the presence of ascorbic acid. Experiments with tryptophan labelled with ^{14}C and ^{15}N indicate the simultaneous biosynthesis of both IAA and GLBR from this substrate in *Brassica* plants with GLBR exceeding IAA by 26:0.25. The appearance of label in indoleacetonitrile (IAN) is interpreted as the result of breakdown of GLBR leading ultimately to IAA.

The diffusible auxin from the stem apex of Savoy cabbage has been shown to consist of IAA, IAN and GLBR (Skytt Andersen and Muir, 1969). While the *Avena* curvature response to IAN is similar to the response to IAA, the response to GLBR requires about 100 times more GLBR than IAA (Skytt Andersen and Muir, 1966). Since the young stem tissue of Savoy cabbage is most responsive to exogenous treatment with IAA, somewhat less to IAN, and not at all to GLBR, it is reasoned that IAA is the substance controlling elongation in the stem, that IAN is an intermediate in the conversion of GLBR to IAA, and that GLBR is a translocation and storage form of the precursor of IAA.

After a period of normal stem elongation in Savoy cabbage the elongation stops and the head forms. During the elongation phase diffusible auxin equivalent to 1-2 ng of IAA is measured for the stem apex or young leaves just below the apex. At the head stage the diffusible auxin from the apex or young leaf is equivalent to 0.2-0.3 ng of IAA. It is assumed that at this time the auxin is largely GLBR which though active in the coleoptile is inactive in the stem of the Savoy cabbage. Indirect evidence for this interpretation is found in the observed accumulation of GLBR with age in the leaves and stem at the head stage.

Treatment of young Savoy cabbage plants with GA results in an increase in the amount of IAA and a decrease in the amount of GLBR in the diffusible auxin followed by an increase in elongation of the stem. When the enzymatic conversion of tryptophan and tryptamine to IAA was examined with preparations from cabbage tissue similar to those used in the investigation of IAA synthesis in the pea (Muir and Lantican, 1968), tryptamine was found to be converted more readily than tryptophan and GA had no effect on the conversion of either substrate. The evidence suggests that the effect of treatment with GA is to promote the conversion of GLBR to IAA.

IV. SUMMARY

Exposure of *Avena* coleoptiles to near-red radiation results in a reduction of IAA synthesis to one-third of the synthesis in coleoptiles in darkness. If given early enough in the growth of the coleoptile, the near-red reduces the growth. Exposure to far-red radiation has no effect on IAA synthesis or coleoptile growth. Exposure to far-red after exposure to near-red reverses the effect of near-red. Dwarf and normal cultivars of the pea have similar growth in darkness and similar levels of synthesis of IAA. Grown in light the dwarf and normal seedlings have the same high levels of IAA synthesis but the IAA in the internode subtending the apex is greatly reduced in the dwarf. Indoleacetyl aspartate which is inactive as an auxin is absent in plants grown in dark but appears in plants grown in light and there is three times as much of it in the dwarf as in the normal. The reduced growth of the dwarf due to greater conjugation of IAA is overcome by treatment with GA which causes greater synthesis of IAA. In Savoy cabbage plants the synthesis of IAA is accompanied by the synthesis of glucobrassicin. Both substances are then translocated through the plant. Treatment with GA results in more IAA and less glucobrassicin and greater growth.

V. REFERENCES

ANDREAE, W.A., and N.E. GOOD (1955). The formation of indoleacetyl aspartic acid in pea seedlings. Pl. Physiol. 30, 380-382.

BRIGGS, W.R. (1963). Red light, auxin relationships, and the phototropic responses of corn and oat coleoptiles. Amer. Jour. Bot. 50, 196-207

DOWNS, R.J., S.B. HENDRICKS and H.A. BORTHWICK (1957). Photoreversible control of elongation of Pinto beans and other plants under normal conditions of growth. Bot. Gaz. 118, 199-208.

GOOD, N.E., W.A. ANDREAE and M.W.H. VAN YSSELSTEIN (1956). Studies on 3-indoleacetic acid metabolism. II. Some products of the metabolism of exogenous indoleacetic acid in plant tissues. Pl. Physiol. 31, 231-235.

HOPKINS, W.G., and W.S. HILLMAN (1966). Relationships between phytochrome state and photosensitive growth of *Avena* coleoptile segments. Pl. Physiol. 41, 593-598.

KUTÁČEK, M. and V.I. KEFELI (1968). The present knowledge of indole compounds in plants of the Brassicaceae family, in "Biochemistry and Physiology of Plant Growth Substances" (Ed. F. Wightman and G. Setterfield) p. 127, Runge Press, Ottawa. 1968.

LANTICAN, B.P. and R.M. MUIR (1969). Auxin physiology of dwarfism in *Pisum sativum* Physiol. Plant. 22, 412-423.

MUIR, R.M. and B.P. LANTICAN (1968). Purification and properties of the enzyme system forming indoleacetic acid, in "Biochemistry and Physiology of Plant Growth Substances" (Ed. F. Wightman and G. Setterfield) p. 259 Runge Press, Ottawa. 1968.

SKYTT ANDERSEN, A. and R.M. MUIR (1966). Auxin activity of glucobrassicin. Physiol. Plant. $\underline{19}$, 1038-1048.

SKYTT ANDERSEN, A. and R.M. MUIR (1969). Gibberellin induced changes in diffusible auxins from Savoy cabbage. Physiol. Plant $\underline{22}$, 354-363.

METABOLISM OF INDOLE-3-ACETALDEHYDE. IV. ELECTRON ACCEPTOR STUDIES AND PHYSIOLOGICAL SIGNIFICANCE OF THE ALDEHYDE OXIDASE OF Avena COLEOPTILES

R. Rajagopal and Poul Larsen

Institute of Plant Physiology, Aarhus University, Aarhus C, Denmark

I. INTRODUCTION

In many higher plants, indole-3-acetaldehyde (IAAld) is believed to be the immediate precursor of the endogenous auxin, IAA. An enzyme system catalysing the oxidation of IAAld to IAA is present in the coleoptiles of Avena. The preparation, partial purification and some characteristics of this enzyme were reported earlier (Rajagopal 1971). This paper records the results of electron acceptor studies in IAAld oxidation and also discusses the physiological significance of the enzyme.

II. MATERIALS AND METHODS

Etiolated oat coleoptiles (Avena sativa cv. Seger I, lot E-2903 from Svalof), without the first leaf, were collected from 25 to 30 mm tall, 5-days old seedlings grown on vermiculite in the same way as for coleoptile section tests. In experiments reported herein, the protein fraction precipitated between 0 and 30 per cent saturation of the cell-free coleoptile supernatant with ammonium sulphate was used as the enzyme source. It was dissolved in KH_2PO_4-Na_2HPO_4 buffer, pH_{25} 7.2 (I = 0.05, β = 2.3) and stored at 2 to 4°C. It contained 0.5 mg/ml streptomycin sulphate (Sigma) to combat microbial contamination.

Assay

Unless otherwise stated, the reaction was carried out in a final volume of 2 ml, in acetic acid-sodium acetate buffer, pH_{25} 4.4 (I = 0.1, β = 30) at 25°C with optimum substrate concentration. Where necessary, 0.5 mg (1.6 μmoles) of phenazine methosulphate (PMS) were added. Enzyme and substrate were separately brought to temperature equilibrium over 10 minutes, prior to mixing. IAA formed was extracted and determined spectrophotometrically according to Larsen (1966).

III. RESULTS AND DISCUSSION

Electron acceptor

The normal, physiological electron acceptor in IAAld oxidation by this enzyme is oxygen. In its absence, the dye PMS is able to accept electrons leading to IAA formation. With PMS as the acceptor, IAAld is oxidized at a rate 2.5 to 3-fold that obtained with atmospheric oxygen. However, on isolation and standing in solution at 2 to 4°C, without any treatment, the enzyme gradually loses part of its capacity to react with oxygen, whereas its activity with PMS is not so affected (Fig. 1).

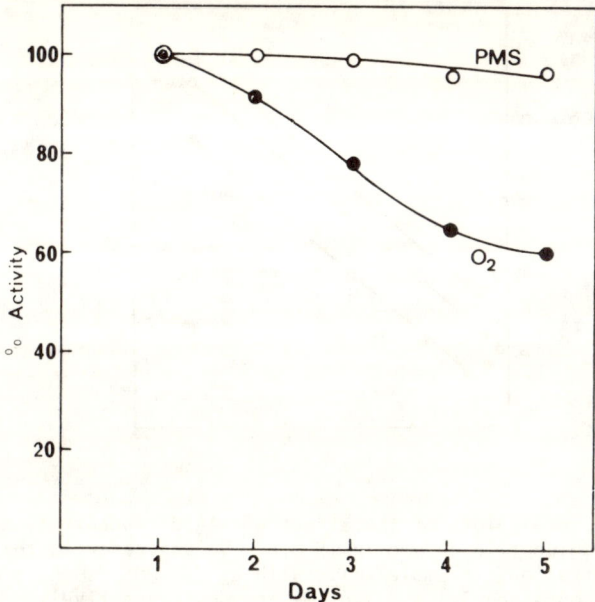

Fig. 1. Activity with increasing age of the *Avena* coleoptile enzyme in oxidizing IAAld to IAA in 1 hr. When O_2 was the electron acceptor, the medium contained 1.6 mg enzyme + 600 nmoles IAAld; with PMS as the acceptor, it contained 1.6 mg enzyme + 1,200 n moles IAAld + 0.5 mg (1.6 μmoles) PMS.

Activity with oxygen dropped by about 30% in 5 days of storage but activity with PMS showed a decline of less than 10%. This dichotomous behaviour poses the question - Are there two enzymes acting on the substrate, or is only one enzyme responsible for transferring electrons to both oxygen and PMS? The best proof for the occurrence and operation of two enzymes is to physically separate them. That we were not able to achieve:

1. Both activities occur in the soluble fraction. A centrifugal force of up to 210,000 x *g* failed to separate them. Both are precipitated together by saturation with ammonium sulphate to a level of 30%.

2. Acidification to about pH 4 precipitates both activities together (Table 1). They share a common isoelectric point, pH 4.05.

3. Filtration through gel types Sephadex G-100 and G-150 and Agarose 1.5, 5 and 15 M failed to separate them.

4. On lyophilisation, the capacity of the enzyme to react with both oxygen and PMS is reduced drastically, *but to an equal extent*, as shown in Table 1.

However, inability to separate the two activities is not convincing proof of their identity. A different approach was made by comparing certain characteristics of IAAld oxidation by this enzyme when oxygen or PMS was the electron acceptor. Among the parameters investigated were the optimal pH value and temperature, but the nature of the electron acceptor had no influence on any of these.

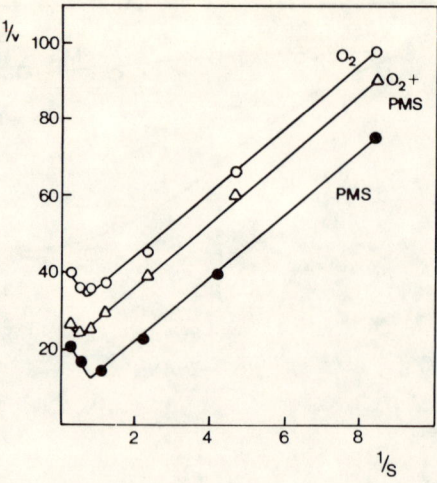

Fig. 2. Double reciprocal plots of the effect of IAAld concentration on its oxidation to IAA by the *Avena* coleoptile enzyme in 15 min. Values refer to the reciprocals of IAAld and IAA concentrations in μmole/2ml. With O_2 and PMS as the electron acceptors, the reaction medium contained 3.3 and 1.65 mg enzyme, respectively. In the presence of both O_2 and PMS, it had 3.3 mg enzyme. PMS added : 0.5 mg (1.6 μmoles). Since PMS is a potent inhibitor of catalase, in experiments with O_2 + PMS, 400 Sigma units of beef liver catalase were added to decompose H_2O_2 formed in the spontaneous reoxidation of reduced PMS.

The effect of IAAld concentration on the reaction velocity with either of the electron acceptors alone or in combination is shown in the form of double reciprocal plots in Fig.2. The following points emerge:

1. There is a striking parallelism in the enzyme activity under the 3 conditions studied. High substrate concentrations inhibit the reaction even in the presence of both oxygen and PMS. Were two enzymes present, with increasing substrate concentrations, the plot would curve downwards, towards the abscissa.

2. Velocity in the presence of both oxygen and PMS is only intermediate between the velocities with either oxygen or PMS alone. A purely additive effect is absent, suggesting the occurrence of only one IAAld oxidizing activity.

Lyophilisation reduces the capacity of the enzyme to utilize either of the electron acceptors to 7 or 8 per cent of the normal (Table 1). The constancy of the activity ratio *i.e.* activity with PMS : activity with O_2, before and after freeze-drying, following prolonged dialysis or acid precipitation is of critical significance. Even infiltration of intact coleoptiles yielded a similar ratio. This evidence unequivocally suggests that one and the same enzyme transfers electrons to both oxygen and PMS.

Inhibitors

Results of some inhibiting treatments, performed with a view to selectively suppress electron transfer to one of the acceptors, are shown in Table 2. Loss of activity after boiling or urea treatment indicates the protein nature of the enzyme. Reversible urea inhibition reveals that the enzyme is made up of subunits. In the presence of urea, they dissociate with loss of activity but they possess the inherent capacity to reunite yielding the catalytically active molecule, once urea is

Table 1. Effect of various treatments of the IAAld oxidizing enzyme of *Avena* coleoptiles on its subsequent activity with oxygen or PMS as the electron acceptor. Activities in absolute terms are comparable only in experiments 2, 3 and 5, carried out with the same enzyme preparation.

No.	Treatment	nmoles IAA/mg protein and hr		Ratio Activity with PMS / Activity with O_2
		O_2	PMS	
1	None - 1.5 g living coleoptiles infiltrated	320	914	2.86
2	None - Fresh enzyme preparation	79	227	2.87
3	Lyophilised	5.5	16	2.9
4	Acid precipitated (5-days old prepn.)	58	174	3.0
5	Dialysed for 18 hr	37	104	2.8

Table 2. Effect of some inhibitors on the activity of the IAAld oxidizing enzyme of *Avena* coleoptiles with oxygen or PMS as the electron acceptor.

Treatment	nmoles IAA/mg protein & hr	
	O_2	PMS
Control	86	235
Boiled enzyme	0	0
Butanol - 2% - preincubated for 3 hours	0	0
Butanone - 4% - added at start of reaction	41	-
Lipase - preincubated for 30 min	88	118
Phospholipase D (Cabbage) 30 min preincubation	45	117
$Na_2S_2O_4$, 10^{-3} M - added at start of reaction	0	185
$SnCl_2.H_2O$, 2×10^{-6} M	0	0
Urea 4M - present throughout	0	0
Urea 4M - dialysed for 2 hr after 30 min preincubation	84	-

removed. Butanol and butanone are commonly employed protein denaturants, especially of lipoproteins. Though heavy metal inhibition of enzymic activity is well known, this is believed to be the first instance of so potent an inhibition by tin. Preincubation with pancreatic lipase for 30 minutes did not affect the activity with oxygen but reduced the activity with PMS by about half. Preincubation for 30 min with phospholipase D from cabbage, reduced electron transfer to both the acceptors by about 50 per cent, suggesting the possible association of lipids with the enzyme. It is noteworthy that most of these compounds suppress electron transfer to oxygen and PMS alike; yet another pointer for the operation of only one enzyme. The only clear-cut exception is dithionite, in the presence of which the enzyme was unable to oxidise IAAld with oxygen, but did so with PMS, though at a reduced level. This prompted a study into the mechanism of dithionite inhibition.

Dithionite can suppress the reaction (a) by directly reducing the enzyme or (b) by abstracting the oxygen from the medium so that the reaction suffers for want of an electron acceptor or (c) by inactivating the enzyme with the H_2O_2 it forms in aqueous solutions. When dithionite is present throughout the incubation period, no IAA is formed (Table 3). But if it is removed by 2 hr dialysis following 20 to 30 min preincubation with the enzyme, activity is almost fully restored. In short-term incubations therefore, dithionite apparently inhibits by sequestrating the oxygen, the natural electron acceptor from the enzyme. Presence of an alternate acceptor like PMS leads to IAA formation under similar circumstances. But contact of the enzyme with dithionite for 2 hours led to about 57 per cent irreversible inactivation, perhaps the result of peroxide formation.

Table 3. Effect of sodium dithionite on the activity of the IAAld oxidizing enzyme of *Avena* coleoptiles with oxygen as the electron acceptor.

Treatment	nmoles IAA/mg protein & hr
Control - aerobic	66
$Na_2S_2O_4$ 10^{-3}M present throughout	0
$Na_2S_2O_4$ 10^{-3}M preincubated for 30 min; then dialysed	64
$Na_2S_2O_4$ 10^{-3}M preincubated for 2 hours; then dialysed	28

Association of lipids with the enzyme, in one form or another, is indicated by the following:

1. Acid insolubility of the enzyme.

2. Sensitivity to butanol and butanone. Contact with butanol, for 2 hours or more, irreversibly inactivates the enzyme. Even shaking the enzyme with ether for 4 min reduces the activity by about 30 per cent.

3. Reduction of the activity with O_2 or PMS by half, following pretreatment with phospholipase D for 30 min. Partial suppression of electron transfer to PMS following 30 min preincubation of the enzyme with lipase. (This is not consistently reproducible. The reasons are obscure.)

4. Considerable loss of the activity following lyophilisation. Freeze-drying is believed to sever the lipid-protein linkages.

5. Desoxycholate addition, besides greatly clarifying the turbid enzyme solution, also enhanced the activity by 10 to 15 per cent. This is commonly used to release membrane or particle-bound enzymes. We suspect that the *Avena* enzyme is probably bound to fragmented membrane particles, perhaps of the endoplasmic reticulum. The high particle weight of about 2 million indicated by gel filtration coupled with the

tendency of the isolated enzyme to aggregate and fall out of solution, lend credence to this view.

Mechanism of IAAld oxidation

Though molecular oxygen is essential for the reaction, the oxygen atom introduced into the aldehyde molecule seemingly emanates from water. This is inferred from electron transfer to PMS under anaerobiosis with concurrent IAA formation. This means that IAAld oxidation takes place by dehydrogenation of the hydrated *gem*-diol form of the aldehyde, Wieland's time-honoured concept. How electrons are transported to oxygen under natural conditions is not clear. In all probability an electron transport chain of intermediate carriers is involved. Cyanide inhibition and the differential activity of the ageing enzyme with oxygen and PMS under certain conditions are suggestive of such a transport chain. Possibly oxygen and PMS accept electrons at different sites in the chain, as envisaged below:

$$R.CH_2.\underset{OH}{\overset{H}{C}}\text{-}OH \longrightarrow 2H^+ + 2e^- \longrightarrow A \xrightarrow{Tin} \!\!\!\!/\!\!\!\!\longrightarrow B \xrightarrow{PMS} C \longrightarrow D \xrightarrow{Na_2S_2O_4} \!\!\!\!/\!\!\!\!\longrightarrow O_2$$

The degeneration or masking of that part of the chain reacting with oxygen may explain the gradual reduction in activity of the ageing enzyme. That part of the chain delivering electrons to PMS may be relatively more stable. Further, the enzyme preparation contained streptomycin, which itself has an aldehyde group. Though there is no indication of streptomycin acting as a substrate for the enzyme, it is not known if it has any adverse effect on the stability of the enzyme.

Table 4-A. Oxidation of infiltrated IAAld to IAA by *Avena* coleoptiles. About 80 coleoptiles, (12 to 15 mm long tips) with a fresh weight of 1.5 g were infiltrated with 10 ml buffer containing 3.5 μmoles IAAld and 5 mg streptomycin sulphate, by removing the air first under suction and then letting in N_2 or air, as the case may be.

Treatment	nmoles IAA formed in 1 hr	
	Expt. 1	2
Aerobic	320	365
Anaerobic	43	57
Anaerobic + PMS - 2 mg	914	943

Physiological significance

1. As demonstrated in an earlier paper (Rajagopal, 1971) there is a gradient in the concentration of this enzyme along the length of the *Avena* coleoptile and the first internode. This may be of significance in regulating the level of IAA along the length of the coleoptile.

2. Evidence we have indicates the occurrence and operation of only one enzyme in *Avena* preparations oxidizing IAAld to IAA. Even in the intact coleoptile, this appears to be the case, as infiltration experiments indicate (Table 4-A). In the absence of oxygen or PMS, IAA formed is significantly less. Even the small amounts formed should perhaps be attributed to the residual oxygen since it is not possible to evacuate all the oxygen from intact coleoptiles under tap suction. Unambiguous results were obtained with cell-free homogenates (Table 4-B). IAA is formed only when oxygen or PMS is present. Ineffectiveness of added NAD and NADP indicate the

absence of phridinoprotein dehydrogenases capable of oxidizing IAAld. We conclude that there is only one enzyme in the *Avena* coleoptile oxidizing IAAld, an oxidase whose only natural, physiological electron acceptor is oxygen. We further believe that this is *the* enzyme leading to the biogenesis of IAA in the *Avena* coleoptile. This is indeed surprising since in the living cell one would expect more than one aldehyde oxidizing enzyme, particularly NAD- or NADP-linked dehydrogenases. Such enzymes probably do occur, but because of their restricted specificity do not act on IAAld, even as the enzyme under discussion fails to act on lower aliphatic aldehydes like acetaldehyde and propionaldehyde.

3. The possible occurrence or association of this enzyme with subcellular membranes was mentioned earlier. If this is the case, its physiological significance is considerably enhanced since one site of auxin action is believed to be on cell membranes. That view gains credence if IAA is formed on or contiguous to sub-cellular membranes. The aggregating tendency of the isolated enzyme preparations reinforces this belief.

4. Our best preparation, under optimal conditions, formed about 2 nmoles of IAA/min and mg of the partially purified protein at 25°C. Assuming a molecular weight of 300,000 and that our preparation contained only 10 per cent active enzyme, a turnover number of 6 molecules of IAA/min is obtained; far from an impressive velocity. But this sluggishness may be of decisive importance and advantage as the hormonal role of IAA is compatible only when it occurs in very low concentrations.

Table 4-B. Oxidation of IAAld to IAA by the cell-free dialysed supernatant of *Avena* coleoptiles. The 2 ml reaction volume contained 860 nmoles IAAld and 1.38 and 0.69 mg enzyme in aerobic and anaerobic experiments, respectively. PMS added = 1.6 μmoles; NAD or NADP = 1 mg.

Treatment	nmoles IAA/mg protein & hr
Aerobic control	48
Anaerobic control	0
Anaerobic + PMS	120
Anaerobic + NAD or NADP	0

IV. SUMMARY

An oxidase in etiolated *Avena* coleoptiles catalyses the oxidation of IAAld to IAA, using either oxygen or phenazine methosulphate (PMS) as the electron acceptor. When stored at 2 to 4°C, it lost about 30% of its capacity to react with oxygen in 5 days but its activity with PMS declined by less than 10% in the same period. This raised the problem of the possible presence of two enzymes. Experiments performed to clarify this issue showed unequivocally the operation of only a single enzyme transferring electrons to oxygen and PMS. $Na_2S_2O_4$ at $10^{-3}M$ abolishes electron transfer to oxygen (but not to PMS) and does so by abstracting the oxygen from the reaction medium. Probably PMS and oxygen accept electrons at two different sites in the electron transport chain. Lipids are likely to be associated with the enzyme. *In vivo*, it probably occurs in association with subcellular membranes. The enzyme is made up of subunits. Tin ($SnCl_2 \cdot H_2O$) at a concentration of $2 \times 10^{-6}M$ totally inhibits electron transfer to oxygen as well as PMS. IAAld oxidation proceeds through the dehydrogenation of the hydrated aldehyde molecule. As *Avena* coleoptiles contain only this one enzyme transforming IAAld to IAA, this is believed to be *the* enzyme concerned in the biogenesis of the endogenous IAA.

V. REFERENCES

LARSEN, P. (1966). Quantitative determination of indole-3-acetaldehyde. II. Physiol. Plant. 19, 780-784.
RAJAGOPAL, R. (1971). Metabolism of indole-3-acetaldehyde. III. Some characteristics of the aldehyde oxidase of *Avena* coleoptiles. Physiol. Plant. 24, 272-281.

We gratefully acknowledge the support received from the Danish Natural Science Research Council.

Electrophoretic Isolation and Growth Activity of Indole-3-Acetic Acid Oxidation Products

Werner J. Meudt

U.S. Department of Agriculture, Crops Research Division, Agricultural Research Service, Plant Industry Station, Beltsville, Maryland 20705, U.S.A.

I. INTRODUCTION

Indole-3-acetic acid (IAA) is oxidized by peroxidase enzymes. The physiological significance ascribed to this reaction is that it controls or regulates the biological activity of IAA. There are two views as to the function of auxin oxidases, one degradative and the other an activation role. The prevailing concept is that the oxidation of IAA represents a detoxification reaction considered necessary for plant to keep the level of IAA at an optimum for maximal physiological response (Galston and Davies 1969). Contrary to the concept of inactivation of IAA, it is proposed that the oxidative transformation of IAA also activates IAA (Meudt, 1965 and 1967). Furthermore, the function of an IAA oxidase, in a system other than for the inactivation of IAA, is suggested by our findings that binding of IAA, first shown by Siegel and Galston (1953), depends on prior oxidation of IAA (Meudt and Galston, 1962). It was suggested that the binding of IAA depended on at least two reactions; (1) enzymatic transformation of IAA by the IAA oxidase system followed by; (2) the binding of a primary oxidation product to RNA, (Galston, *et al* 1964). In 1967, Meudt and Gaines proposed that the binding involved an indolenine derivative conjectured to be an early oxidation product of IAA (Ray and Thimann, 1955 and Hinman and Lang, 1965). The IAA oxidase system thus assumes a new significance. The present paper collates data on the isolation, electrophoretic mobility, and biological activity of 3 oxidation products of IAA.

II. MATERIALS AND METHODS

The oxidation of IAA was catalyzed by electrophoretically purified horseradish peroxidase (HRP) obtained from Worthington Biochemical Corporation*. The enzyme had a specific activity of 2,900 µ/mg. Oxidation reaction mixtures usually contained 1 x 10^{-4}M each of $MnCl_2$ and, 2,4-dichlorophenol (DCP); 1 x 10^{-7}M HRP, 1 x 10^{-3}M pH 6.1 sodium phosphate buffer, and 1 to 4 x 10^{-4}M IAA. In the case of short time reactions, oxidation was stopped with .1M KCN. The progression of the oxidation of IAA was followed either by measuring the residual IAA with Salkowski reagent or by assaying for the accumulation of oxidation intermediates with the modified Ehrlich reagent p-dimethylaminocinnamaldehyde (DMACA), (Meudt and Gaines, 1967).

Electrophoretic separations

Electrophoresis of the various samples was carried out with a Spinco Continuous Flow Electrophoresis Apparatus Model CP, equipped with S and S No. 470 paper curtains*. Samples were injected into the separating area at a flow rate of 0.49 to 2.10 ml/hr and carried through a voltage gradient of 28V/cm. Except where stated, 1 x 10^{-3}M pH 6.1 sodium phosphate buffer was used as the electrolyte. Each sample resolved into

---

* Mention of a trademark name or a proprietary product does not constitute a guarantee or warranty of the product by the USDA, and does not imply its approval to the exclusion of other products that may also be suitable.

32 fractions which were collected at a rate of about 1.2 ml/hr. All electrophoretic operations were carried out at 6°C. After each run, which lasted about 12 hrs. the paper curtains were recovered, dried and sprayed with either Salkowski or DMACA reagent. The biological activity of each fraction was determined by the *Avena* first internode test of Nitsch and Nitsch (1959). Routinely, ten 4 mm *Avena* sections were placed into test tubes which contained 0.2 mml of eluates, 1.6 ml of 2% sucrose and 0.2 ml of 1 x 10^{-3} M MnCl. The tubes were placed into a tube roller apparatus and rotated at a speed of 4 rpm for 1 to 12 hrs. All operations were performed in the dark under green safety light.

III. RESULTS AND DISCUSSION

Under the experimental conditions used, a sample of completely oxidized IAA resolves upon electrophoresis into 4 oxidation products. Three products are negatively charged and one is positively charged. With the zero voltage position at the centre of the curtain (tube 16) two of the three negatively charged products migrated toward the anode at a relative mobility of 1.28 and 2.38 x 10^5 cm^2 sec^{-1} $volt^{-1}$ respectively. The third negatively charged product accumulated near the origin of application and was carried only a short distance down the curtain and towards the anode. The fourth product migrated toward the cathode at a relative mobility of 1.8 x 10^5 cm^2 sec^{-1} $volt^{-1}$. All three products that moved down the curtain yield triple UV absorption bands in the 280 mµ range, characteristic of most indoles. Fig. 1 shows the electrophoretic pattern and the relative absorption at 280 mµ of the three oxidation products. The shaded area represents the color obtained after the curtain was dried and

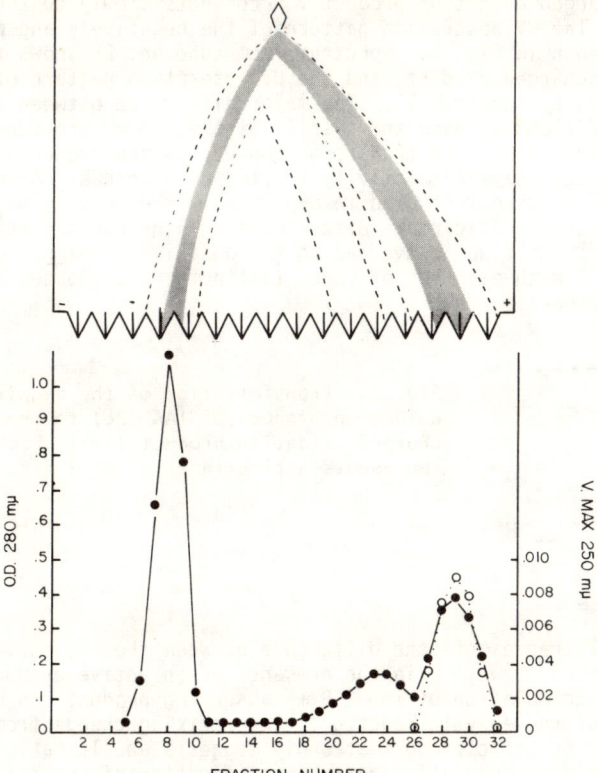

Fig. 1. Electrophoretogram of an IAA oxidation reaction mixture containing initially 4 x 10^{-4}M IAA, 1 x 10^{-4}M DCP and $MnCl_2$, and 1 x 10^{-7} HRP. Above: shaded area represents migration path of oxidation products after curtain was sprayed with DMACA reagent. The dotted line delineates the migration on paths as revealed under a UV lamp. Below: optical densities at 280 mµ.

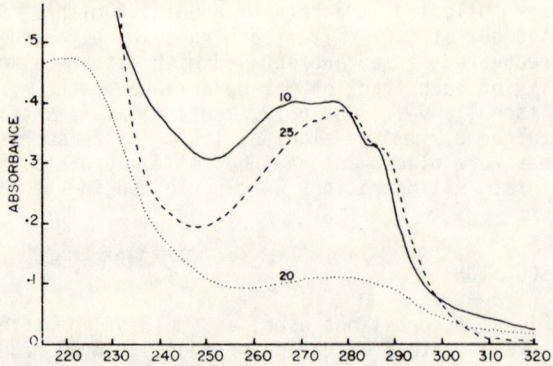

Fig. 2. UV absorption pattern of the negatively charged oxidation products (Spectrum 20 and 25) and of the positively charged oxidation product (Spectrum 10).

sprayed with Salkowski or DMACA reagents. The dotted line delineates the migration path of the products as revealed under a UV lamp.

Under similar electrophoretic conditions, authentic IAA migrates towards the anode with a relative mobility of 2.35×10^5 cm^2 sec^{-1} volt^{-1}, practically coincident with the more negatively charged oxidation product (Fraction 28 in Fig. 1). Unlike IAA, the negatively charged oxidation product "C" converts slowly to the positively charged product ("A"). The UV absorption patterns of the negatively and positively charged products are shown in Fig. 2. Spectrum from tube No. 25 shows the UV absorption of the negatively charged product; and the UV absorption pattern of the positively charged product is shown by spectrum 10. The major difference between spectrum 10 and 25 is the appearance of a 268 mµ band and a shift of the 288 mµ shoulder of spectrum 25 to a 286 mµ absorption peak. The transformation of the two products is spontaneous and takes place even after enzymatic activity is stopped with KCN. Although the kinetics of this reaction was not studied in any detail, the conversion can be observed during a prolonged electrophoretic run. After about a 15 hr run and at 6°C, most of the negatively charged product was converted to the positively charged product. The data are shown in Fig. 3 with each set of tubes (set number) collected at 3 hr intervals, (4.5 ml fraction/tube).

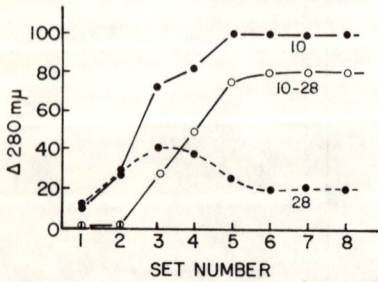

Fig. 3. Transformation of the negatively charged oxidation product of IAA (28) to the positively charged oxidation product (10). Each set number represents a time interval of 3 hrs.

The curve labelled 10-25 represents the difference between the two curves 10 and 25. The conversion is enhanced, however, in the presence of an active enzyme and, as reported in 1967, the accumulation of the 268 mµ absorbing product can be detected 5 min after the start of the enzymic reaction. The striking change from a strongly acidic to a less acidic or basic product strongly suggests an alteration of an acidic group such as by decarboxylation or an esterification of the carboxyl group. At pH 6.0, non-acidic indoles, such as indole, skatol, indole-3-carboxylaldehyde, and ethyl-3-indoleacetate, all migrate with similar affinity for the cathode. But none match exactly the oxidation product when other criteria, such as a UV absorption spectrum, or reactivity with Salkowski or DMACA reagents, are employed. We must also rule out 2-oxindole, a known final oxidation product of oxidized IAA, because oxindoles absorb strongly in the 240-250 mµ range of the UV spectrum; and they do not react with the two colour reagents used (Meudt and Gaines, 1967).

Biological activity

If the concept of oxidative activation of IAA is to be considered, it is important that the same biological criteria, which establishes IAA as an auxin, be applied to and met by the oxidation products of IAA. Previous work from this laboratory (Meudt 1967), has shown that oxidation products have biological activity. Furthermore, it was shown that the stimulation of growth by the oxidation products was greater than that induced by IAA. The present objective is to inquire about the extent the oxidation products of IAA simulate the biological activity of authentic IAA. In this endeavor, 2 criteria were investigated. First, the response to pH - because, as shown by Nitsch and Nitsch (1956) and confirmed in our tests, IAA stimulates growth maximally at pH 5.0. Our second criterium was that the concentration response curve be two-phasic. The compound should be effective at extremely low concentrations (oligodynamic) and should inhibit growth at relatively high concentrations.

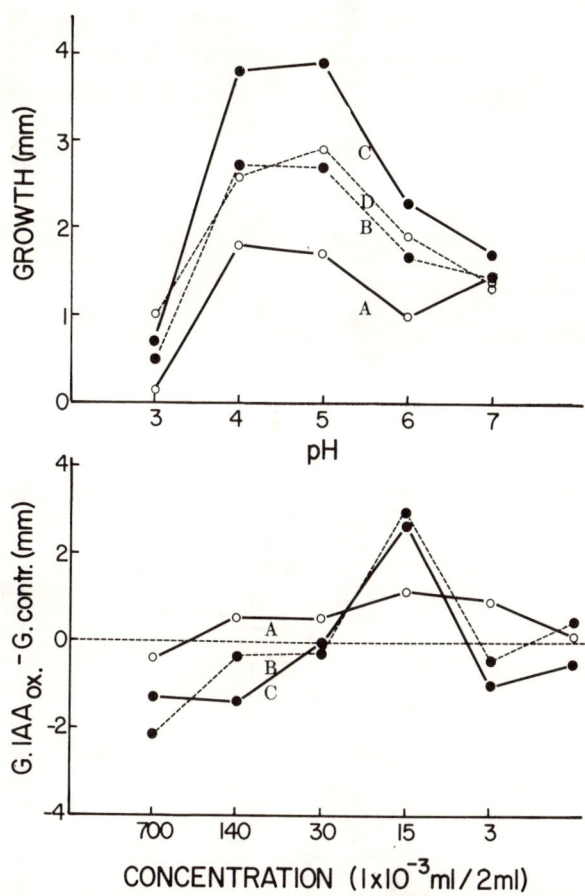

Fig. 4. The effect of pH on the growth of first internode section in 0.001M phosphate-citrate buffer plus 0.15 ml effluent of oxidation product/2ml buffer. A - positively charged product, B and C - negatively charged products. D - 1 x 10^{-5}M IAA.

Fig. 5. The response to various concentrations of oxidation products to a first internode section growth. Points below the zero line (control) show growth inhibition. Fraction A, B, and C are the positively charged and the two negatively charged products, respectively. Reading from left to right, the concentrations are equivalent to approximately 70.0, 14.0, 3.0, 1.5, and 0.3x10^{-7}M.

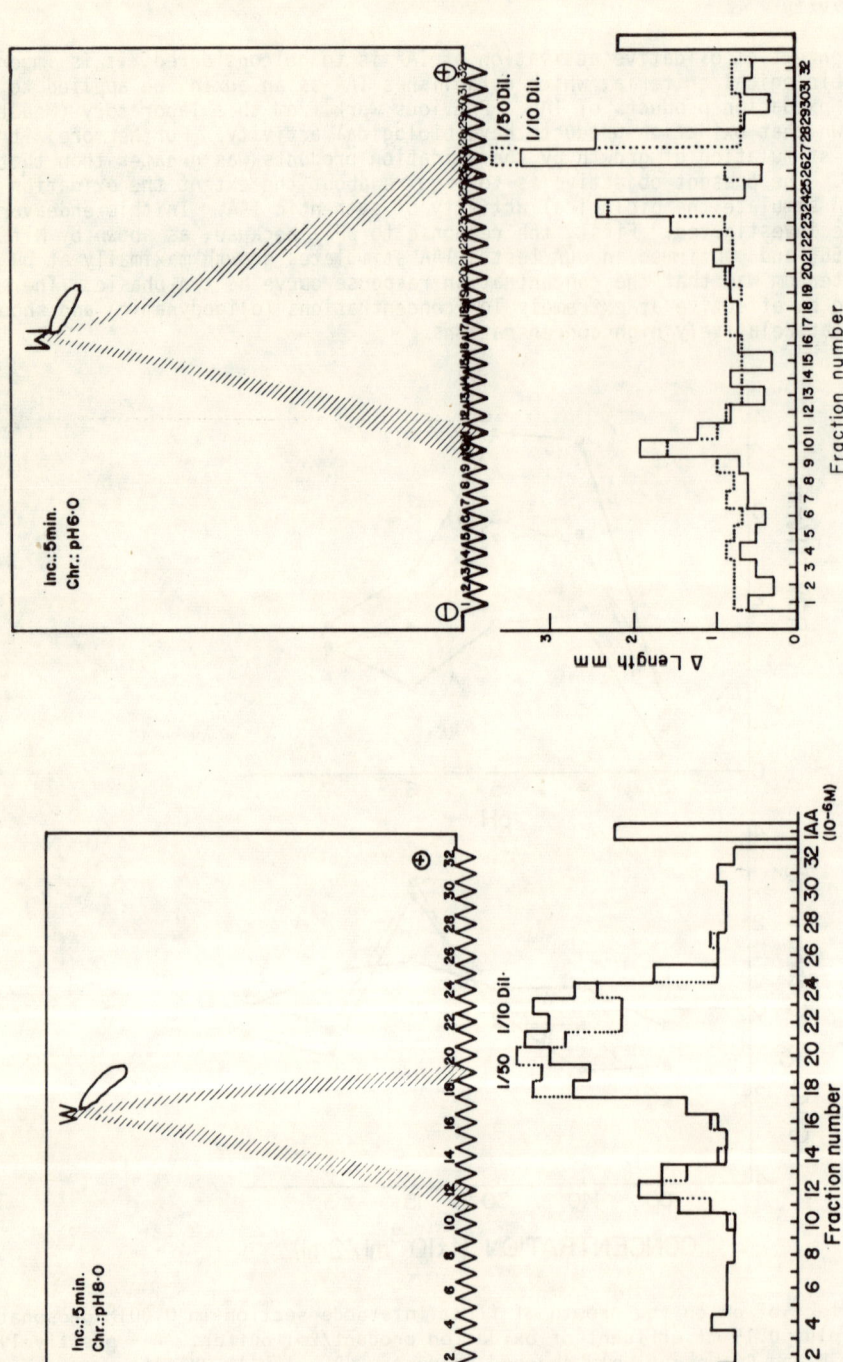

Fig. 6 and 7. Above: Electrophoretic migration of IAA oxidation products in the presence of pH 8.0 buffer (Fig. 6) and pH 6.0 buffer (Fig. 7). The shaded area represents the migration paths of oxidation products after curtains were sprayed with Salkowski reagent. Below: Growth respnse of the various fractions in the first internode test. For the bio-assay, the pH of each fraction was adjusted to 5.0 after the appropriate dilutions were made.

To investigate the effect of pH on the biological activity of the three oxidation products, the first internode sections were incubated in 0.001M sodium phosphate - citrate buffers plus 0.15 ml eluates/s ml buffer equivalent to about 1.5×10^{-7}M. Fig. 4 shows that all three products A, B, and C stimulate growth maximally at pH 4 - 5 and parallel the results obtained with 1×10^{-5}M IAA (Curve D). Curve A, B and C correspond to the positive and the two negative charged oxidation products, respectively. In this test, IAA induced maximum growth at 1×10^{-5}M which is equivalent to the growth obtained with 1.5×10^{-7}M of oxidation products B and C. At concentrations equivalent to 1×10^{-5}M IAA, products B and C strongly inhibited growth. The effect of various concentrations of oxidation products is shown in Fig. 5 and is a characteristic auxin response; for it is known that high auxin concentrations inhibit growth and also show the oligodynamic effect typical of a growth hormone. In this test, 5×10^{-4}M IAA inhibits growth, whereas the negatively charged oxidation products inhibited growth when present at a concentration of 1.4×10^{-6}M or higher. The positive charged product is less active.

It should be noted that product B and C yielded similar biological responses which may indicate that they are also chemically closely related. This is further substantiated from experiments in which we changed the pH of electrolytes to obtain comparable concentrations of B and C. For a good yield of product B the pH of the electrolyte used was 8.0. The pH 6.0 electrolyte was used if product C was required. The data for both runs are given in Figures 6 and 7. The changed equilibrium concentration of the two acidic oxidation products is attributed to an indolenine that reacts in their tautomeric forms to a change of pH. We have shown earlier that indolenines are primary oxidation products of IAA and that they react in the form of their 1,3-tautomers with the aldehyde DMACA to form a quinoidal red-violet compound (Meudt and Gaines, 1967). The results obtained may thus be resolved into the following sequence of reactions:

(1) An initial enzyme reaction which involves the formation of free radicals as first experimentally demonstrated by Yamazaki, et al. (1959). The formation of free radicals was also demonstrated within the first 30 seconds of the oxidation reaction, with bisulfite, a sensitive detector of free radicals (Meudt 1970).

(2) The subsequent reaction involves the formation of acidic products B and C, which are of great interest to us because they are chemically and biologically very active. These negatively charged products react readily with DMACA to form an aldehyde condensation product, and are more oligodynamic than IAA or any of the other oxidation products tested for effectiveness as growth hormones. Considering these data, one cannot refrain from suggesting that the hormonal property of IAA is an attribute of its acidic oxidation products. This transformation might be the amplificatory mechanism necessary for IAAto attain its activity as a growth hormone.

(3) The next reaction in this sequence involves the transformation of product C to the basic product A. Product A, which migrates to the negative pole, is biologically less active than B or C and may represent a decarboxylated or an esterified derivative of product C. Like products B and C, product A reacts with DMACA and Salkowski reagent and its absorption in the UV resembles that of an indole. After prolonged oxidation of IAA (several hours) the oxidation reaction mixture yields a reddish precipitate and oxindole. Both of the latter products are biologically inactive in the *Avena* test and are thus of little physiological significance.

IV. SUMMARY

Three biologically active IAA oxidation products were isolated by continuous flow electrophoresis. Two oxidation products are acidic and migrate to the anode and one is less acidic and migrates to the cathode. Spectrophotometric analysis indicate that they are indolic. The biological activity of the negatively charged products is such that they are more oligodynamic than IAA. Like IAA, all three products promote growth maximally at pH 5.0. It is suggested that the two negatively charged products are tautomeric species. The more negatively charged product transforms to a positive charged product involving most likely a decarboxylation reaction. It is suggested that the oxidative transformation of IAA is essential for IAA to attain its activity as a growth hormone.

V. REFERENCES

GALSTON, A.W. (1969). Hormonal regulation in higher plants. Science, 163: 1288-1297.

GALSTON, A.W., P. JACKSON, R. KAU-LAWBNEY, N.P. KEFFORD and W.J. MEUDT. (1964). Interactions of auxins with macromolecular constituents of pea seedlings, in Régulateurs Naturels de la Croissance Végétale. Coll. Int. C.N.R.S. 123, 251-264.

HINMAN, R.L. and J. LANG (1965). Peroxidase catalyzed oxidation of indole-3-acetic acid. Biochem., 4: 144-158.

MEUDT, W.J. (1965). Studies on the oxidation of indole-3-acetic acid by horse-radish peroxidase. Pl. Physiol., 40: (suppl.), 71-72.

MEUDT, W.J. (1967). Studies on the oxidation of indole-3-acetic acid by peroxidase enzymes. Ann. N.Y. Acad. Sci. 144: 118-128.

MEUDT, W.J. (1970). Oxidation of indole-3-acetic acid by peroxidase enzymes. II. Sulfite and manganous ion interaction with intermediate oxidation products of IAA. Phytochem., (In Press).

MEUDT, W.J., and A.W. GALSTON (1962). Binding of an indole-3-acetic acid metabolite to macromolecules of pea plants. Fed. Proc. 21: 399.

MEUDT, WJ. and T.P. GAINES (1967). Studies on the oxidation of indole-3-acetic acid by peroxidase enzymes. I. Colorimetric determination of indole-3-acetic acid oxidation products. Pl. Physiol., 42: 1395-1399.

NITSCH, J.P. and C. NITSCH (1956). Studies on the growth of coleoptile and first internode sections. A new, sensitive, straight-growth test for auxins. Pl. Physiol., 31: 94-111.

RAY, P.M. and K.V. THIMANN (1955). Steps in the oxidation of indoleacetic acid. Science, 122: 187-188.

SIEGEL, S.M. and A.W. GALSTON (1953). Experimental coupling of indoleacetic acid to pea root protein *in vivo* and *in vitro*. Proc. Nat. Acad. Sci. (U.S.), 39: 1111-1118.

YAMAZAKI, I., H.H. MASON and L. PIETTE (1959). Identification by electron paramagnetic resonance spectroscopy of free radicals generated from substrates by peroxidase. J. Biol. Chem., 235: 2444-2449.

PLANT GROWTH SUBSTANCES, 1970

Conversion of 3-Indoleacetaldoxime to Glucobrassicin and Sulfoglucobrassicin by Woad (*Isatis tinctoria* L.)

S. Mahadevan* and Bruce B. Stowe**

* Department of Biochemistry, Indian Institute of Science, Bangalore-12, India

**Department of Biology, Kline Biology Tower, Yale University, New Haven, Conn., 06520, U.S.A.

I. INTRODUCTION

The high indole content of cruciferous plants has been known for a number of years (Kutáček and Kefeli, 1968 and references therein). The crucifer *Isatis tinctoria* L. or the common woad, however, is remarkable in that it produces at least two classes of indole compounds, the indigo precursor isatan B, and the indole glucosinolates. Three of the latter have been described to date, Glucobrassicin (GB) (Gmelin and Virtanen, 1961), Neoglucobrassicin (NGB) (Gmelin and Virtanen, 1962) and Sulfoglucobrassicin (SGB) (Elliot and Stowe, 1970) and all these three are present in the woad plant. The structures of these compounds are given in Figure 1.

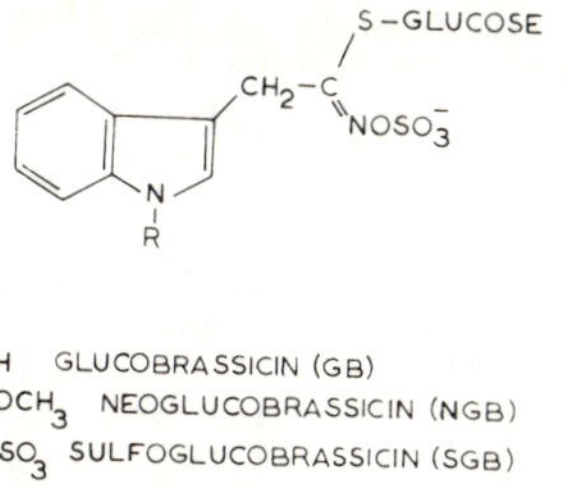

R=H   GLUCOBRASSICIN (GB)
R=OCH$_3$ NEOGLUCOBRASSICIN (NGB)
R=SO$_3^-$ SULFOGLUCOBRASSICIN (SGB)

Fig. 1. Naturally occurring indole glucosinolates.

The indole glucosinolates attracted the attention of auxinologists when GB was shown to be converted to indoleacetic acid precursor indoleacetonitrile (IAN), by the enzyme myrosinase under acid pH conditions (Gmelin and Virtanen, 1961). This demonstration, besides throwing light on the nature of IAN formation in crucifers, also suggested that IAN itself may be an artefact of the extraction procedure. Indeed it is now believed that IAN does not exist *per se* in the cruciferous tissue but is formed from GB during extraction if myrosinase is not properly inactivated (Schraudolf and Bergmann, 1965). This argument was strengthened by the observation that IAN has been conclusively shown to be present only in those plants where GB also occurs. However, reports of the occurrence of small amounts of IAN in cruciferous tissues, even after myrosinase inactivation, have been made (Kutáček and Procházka, 1964).

In 1963, the direct formation of IAN from 3-indoleacetaldoxime (IAOX) was shown in some higher and lower plants (Mahadevan, 1963) and the enzymic nature of this conversion was established (Kumar and Mahadevan, 1963). A significant finding in this connection was that all plants converting IAOX to IAN also converted IAN to IAA, suggesting that these two reactions are consecutive steps of a general pathway. The conversion of an aldoxime to a nitrile, besides spurring investigations on the biosyntheses of cyanogenic glycosides (Conn and Butler, 1969), have helped to focus attention on aldoximes as intermediates in plant metabolism. Independent investigations from two laboratories have established aldoximes as intermediates in the conversion of amino acids to glucosinolates (Underhill, 1967; Tapper and Butler, 1967; Kindl and Underhill, 1968). Recently, Underhill and Wetter (1969) have proposed a pathway for the synthesis of the glucosinolate, glucotropaeolin, from phenylacetaldoxime in *Tropaeolum majus*, based on (i) the conversion of chemically synthesized radioactive phenylacetothiohydroximate and desthioglucotropaeolin to glucotropaeolin, (ii) formation of labelled phenylacetothiohydroximate from labelled phenylacetaldoxime and (iii) conversion of thiohydroximate to desthioglucotropaeolin by an UDPG-mediated enzyme reaction (Matsuo and Underhill, 1969) (Figure 2). Whereas the chemical identity of the divalent sulfur donor compound SX (Figure 2) is not known, DL-cysteine has been found to be a good source of the sulfur atom (Wetter and Chisholm, 1968).

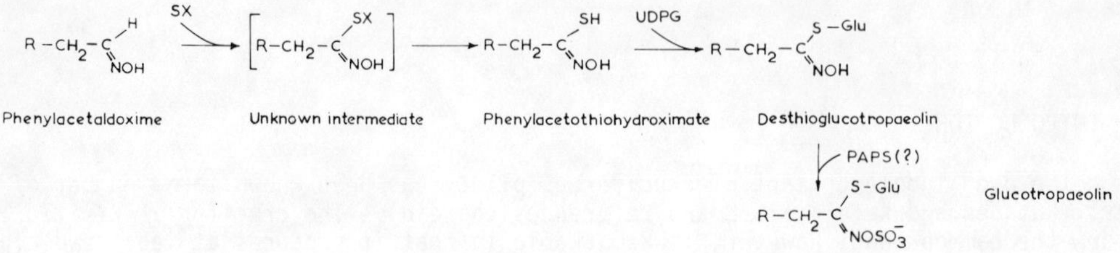

Fig. 2. Pathway proposed for glucotropaeolin biosynthesis (Underhill and Wetter, 1969). R = phenyl.

The demonstration by Kindl (1968) of the occurrence of ^{14}C-IAOX in cabbage plants grown in a $^{14}CO_2$ atmosphere has added IAOX to the list of naturally occurring indoles in crucifers. He has further demonstrated that IAOX is a precursor of GB. Two modes of IAN formation from IAOX has thus become evident, (i) a direct conversion of IAOX to IAN and (ii) an indirect formation of IAN *via* GB. Interestingly, both these processes seem to involve sulfur bound intermediates (Shukla and Mahadevan, 1970).

This paper reports in brief, some observations made on the formation of indole glucosinolates in the woad leaf using IAOX as the source of the indole ring and the side chain and L-cystine as the source of divalent sulfur. A detailed report will appear elsewhere. L-cystine was deliberately chosen as the source of divalent sulfur on the premise that it could donate sulfur in at least three ways, (i) after reduction to L-cysteine by the leaf and subsequent condensation with the aldoxime, (ii) degradation to thiocysteine (Tishel and Mazelis, 1966) and subsequent condensation with the aldoxime and (iii) an unknown way. Possible intermediates of steps (i) and (ii) are given in Figure 3.

(a)   (b)

Fig. 3. Hypothetical condensation products of cysteine and thiocysteine with IAOX. R = Indole.

II. MATERIALS AND METHODS

Woad plants (*Isatis tinctoria*) were grown from seeds, either in vermiculite irrigated automatically with nutrient solution in a constant temperature room at 23°C under 16 hour photoperiods or in Yale University's Marsh Botanic Garden. Fully expanded leaves from plants one to four months old were taken for experimentation.

Paper chromatographic separations were by descending development using Whatman 3 MM paper. Thin layer chromatography was done using Eastman TLC Chromogram 6061 silica gel coated plastic plates. Such TLC plates could be cut into 1/2 to 1 cm by 2 cm pieces and counted in a liquid scintillation counter for radioactivity measurements. Solvent systems used were:

1) 1-Butanol-ethanol-water 4:1:3 (upper phase), v/v/v.

2) 1-Butanol-acetic acid-water 6:1.5:2.5, v/v/v.

3) Chloroform (1% ethanol)-ethanol 10:0.2, v/v.

Indole compounds were routinely detected by means of p-dimethyl-aminocinnamaldehyde reagent (PDAC: 10% solution w/v in a mixture of equal volumes of 37% HCl and ethanol). Zones of radioactivity on paper chromatograms were located by Vanguard paper strip gas-flow, 4π counter. An Ansitron liquid scintillation spectrometer was routinely used for counting fluid samples of TLC pieces in 10 ml of Bray's solution (Bray, 1960).

Cold IAOX was prepared by the procedure of Ahmad $et\ al.$ (1960). Crystalline ^{3}H-IAOX (uniformly labelled, m.p. 137-138°C) was synthesized from uniformly labelled ^{3}H-tryptophan by a combination of the procedures of Gray (1959) and Ahmad $et\ al.$ (1960). IAOX had a specific activity of 7 mC/mM and was over 97% radiochemically pure as determined by thin layer chromatography. ^{3}H-IAOX was dissolved in reagent grade acetonitrile (80 μg/ml solution) and stored in a -20°C freezer in rubber-capped serum vials.

Potassium selenate was prepared from selenious acid by the procedure of Rosenfeld and Beath (1964). ^{35}S-L-Cystine was a Schwarz sample, over 99% pure, with a specific activity of 20.2 mC/mM when received. A 1-mC aliquot (11.94 mg) was dissolved in 10 ml of 0.02 M HCl, and stored in a refrigerator. All other chemicals were reagent grade.

Myrosinase was prepared from white mustard ($Sinapis\ alba$ L.) seeds according to the method of Neuberg and Wagner (1926).

Feeding solutions to leaves

Compounds in solution were allowed to be taken up by transpiration through the freshly cut end of petioles. All experiments were conducted in a constant temperature (25°C), continuous light room. Small samples of ^{3}H-IAOX or ^{35}S-cystine solutions (50-200 μl) were taken in horizontally placed pieces of Tygon tubing of a bore slightly larger than that of the petiole diameter. The freshly cut petiole was quickly inserted into one end of the solution meniscus. Solutions were taken up by the leaf within 5-15 min. 100 μl water was then placed in the Tygon tubing and allowed to infuse into the leaf. The petiole was rinsed with distilled water and the leaf was than placed in a designated solution for metabolism. Radioactivity remaining in the tubing and petiole rinses was measured, which was usually less than 5% of the total used in the experiment.

Extraction and isolation procedures

Leaves were inactivated by quickly chopping and dropping them in 10 volumes of boiling methanol and the boiling continued for one more minute (Gmelin and Virtanen, 1961). The leaf pieces were then extracted several times with warm 80% methanol till they became colorless. The methanolic solution was processed for the separation of glucosinolates as described earlier (Gmelin and Virtanen, 1961; Elliott and Stowe, 1970). Following flash evaporation at 30°C to remove methanol, the aqueous layer was shaken with 2½ volumes of heptane to remove most green coloring matter and placed on an alumina (acid) column and eluted with water to remove neutral compounds. The indole glucosinolates were then eluted with 1% K_2SO_4 solution. The K_2SO_4 eluate was concentrated by flash evaporation at 30°C and desalted by passing through a column of Sephadex G-10 (Stowe and Elliott, 1970). GB and SGB were eluted as overlapping peaks and well separated from K_2SO_4.

III. RESULTS AND DISCUSSION

Conversion of ^{3}H-IAOX to ^{3}H-GB and ^{3}H-SGB

Preliminary experiments indicated that ^{3}H-IAOX was efficiently converted to GB by woad leaves. Results of two such experiments are given in Table 1. The Sephadex G-10 eluates (Materials and Methods) were chromatographed (TLC, solvent 2), and radioactivity measured. In both experiments, only one radioactive peak corresponding to GB (Rf 0.36) was obtained which accounted for 78 and 91% of the radioactivity respectively on TLC plates. The percentage incorporation (35 and 37%) of ^{3}H-IAOX into GB was somewhat higher than that reported for the incorporation of other aldoximes to glucosinolates (Conn and Butler, 1969). No significant incorporation into SGB was observed. However, when a larger amount of IAOX (2.3 x 10^6 cpm) was fed to woad leaves and the glucosinolate fraction chromatographed on paper, two radioactive peaks, one large and corresponding to GB and a smaller one corresponding to SGB were observed. (Fig. 4). The spots were identified by their color reaction with PDAC reagent. No NGB spot was seen however. ^{3}H-IAOX was therefore a precursor for both GB and SGB, indicating that the sulfonation of the indole-nitrogen in SGB formation occurs after the formation of IAOX during the biosynthesis of SGB.

Table 1. Conversion of ^{3}H-IAOX to ^{3}H-GB by Woad Leaves

Expt.	Weight of leaves	Hours of metabolism	Total radioactivity in ^{3}H-IAOX fed	Total radioactivity in ^{3}H-GB formed	% Incorporation
			(cpm)	(cpm)	
1.	1.8gm(2)*	29	115400(8)**	42688	37
2.	5.5gm(4)	26	981036(68)	352208	35

* Number of leaves is given in parenthesis
**µgm of IAOX

Incorporation of ^{35}S from ^{35}S-L-cystine into GB and SGB

^{35}S-L-Cystine (33 gn; 4 x 10^6 cpm) was fed to woad leaves and allowed to metabolize for 28 hours. The leaves were extracted and processed as described earlier. The radioactivity in the indole glucosinolate fraction contained 1.01 x 10^6 cpm, accounting for about 25% of the total radioactivity administered. Chromatography in solvent 1 revealed two radioactive peaks corresponding to GB and SGB (Fig. 5). No NGB spot was observed. Surprisingly, the radioactivity of ^{35}S-SGB was more than that of the ^{35}S-GB in contrast to the relative radioactivities of ^{3}H-SGB and ^{3}H-GB in Fig. 4. Such a pattern suggested that GB may have been an acceptor for ^{35}S atom derived from the oxidation of ^{35}S-cystine in forming ^{35}S-SGB. Time course experiments of incorporation of ^{35}S into GB and SGB (Mahadevan and Stowe, 1972) have indeed shown this to be so.

A new neutral intermediate in the conversion of IAOX to GB. Its tentative identification as desthioglucobrassicin

The water eluate obtained during alumina column chromatography of extracts of woad leaves fed ^{3}H-IAOX (cf. materials and methods) contained some radioactivity. TLC of the residue of the eluate gave several PDAC-reacting spots; those having the Rf of IAOX and IAN contained some radioactivity. In addition a spot more polar than IAOX, IAN or IAA but less polar than GB was found to contain radioactivity indicating the formation of a new compound from IAOX. When both ^{35}S-L-cystine and cold IAOX were administered to woad leaves, a similar sulfur-labelled compound, more polar than IAOX and less polar than GB was again formed. On a premise that this new compound X may be

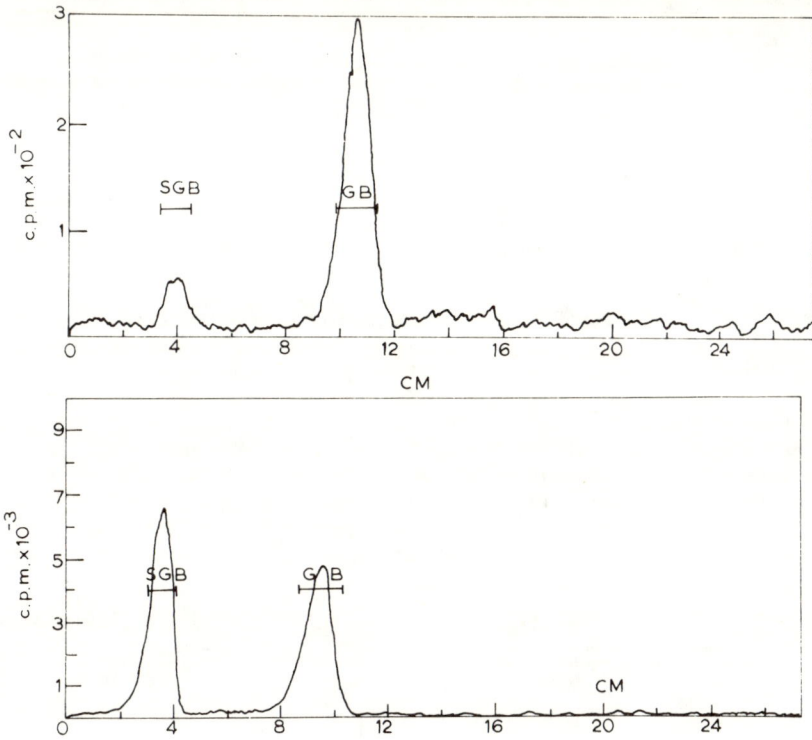

Figs. 4 & 5. Pattern of Radioactivity distribution on the chromatogram of the indole glucosinolate fraction extracted from woad leaves following the administration of ^{3}H-IAOX (Fig. 4) or ^{35}S-L-cystine (Fig. 5).

a non-sulfated precursor of glucobrassicin, an attempt was made to accumulate it by inhibiting the incorporation of sulfate with a suitable inhibitor. Selenate was the inhibitor of choice as it has been known to inhibit biological sulfatation processes (Boyd and Neuman, 1954). Administration of selenate along with ^{35}S-cystine and cold IAOX was indeed found to increase the formation of ^{35}S labelled compound X (Fig. 6). Larger quantities of ^{35}S-X were isolated by bulk administration of IAOX, ^{35}S-cystine and selenate to a large number of woad leaves, followed by extraction, paper chromatography and elution of the zone corresponding to the Rf (0.58) of compound X.

When fresh woad leaves were fed ^{35}S-X and allowed to metabolize for 18 hrs, 55 to 60% of the label was recovered in GB and SGB indicating that ^{35}S-X was converted to the glucosinolates with a greater efficiency than IAOX. The tritium labelled compound X, ^{3}H-X, was prepared by similarly feeding ^{3}H-IAOX, cold cystine and selenate to woad leaves. This compound moved with the same Rf (0.57) as ^{35}S-X and both compounds showed no mobility on electrophoresis. Compound X was therefore a neutral compound containing both an indole ring and sulfur. Further purification of compound X was achieved by chromatography and electrophoresis. The purified compound gave a purple color reaction with PDAC reagent which turned bluish with time. It was ninhydrin negative, indicating that it was not an amino acid derivative as shown in Fig. 3. The stability of compound X during the several purification procedures indicated that it was not the labile indoleacetothiohydroximate (Fig. 7a). These observations pointed to the last possibility that it may be the thioglucoside, desthioglucobrassicin (Fig. 7b).

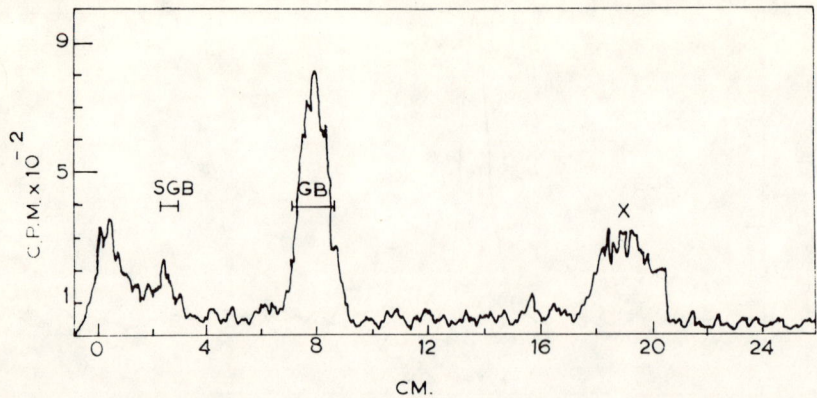

Fig. 6. Pattern of Radioactivity distribution on the chromatogram of the total methanolic extract of woad leaf fed ^{35}S-L-cystine, cold IAOX and potassium selenate. Woad leaves were placed in a solution of IAOX (51 µgm/ml) containing 10^{-4}M K_2SeO_4 for $3\frac{1}{2}$ hr and then administered 100 µl of ^{35}S-cystine (110 µgm; 3.9×10^6 cpm) solution. The leaves were then placed in the original IAOX-selenate solution to metabolize for 570 min and extracted with methanol. Aliquot of the total extract was chromatographed using solvent 1.

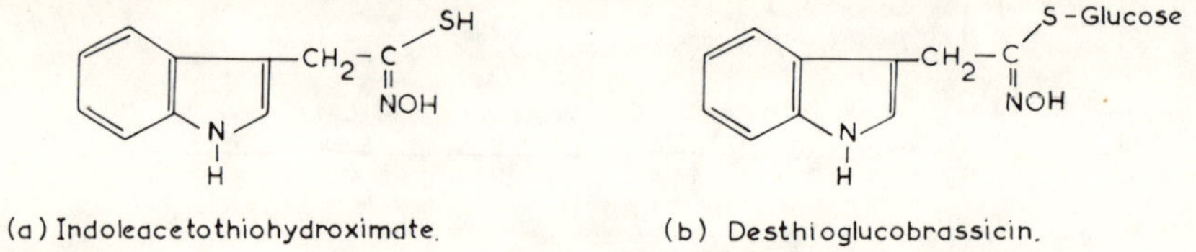

(a) Indoleacetothiohydroximate.   (b) Desthioglucobrassicin.

Fig. 7.

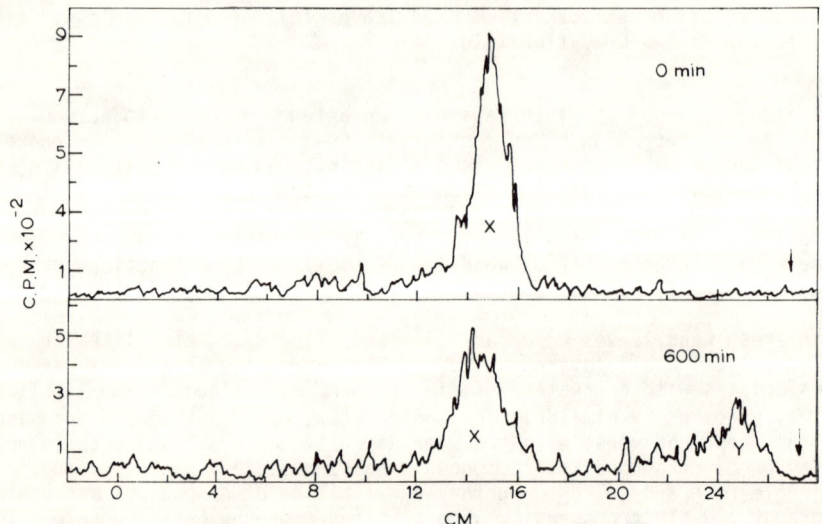

Fig. 8. Radioactivity scans of chromatograms of ^{35}S-X before and after treatment with myrosinase. Reaction mixture consisted of 50 µl of ^{35}S-X and 6 mg of myrosinase in 90 µl of 0.1 M phosphate buffer, pH 7, containing 2×10^{-3}M ascorbate. Incubation was at $37°C$. 20 µl aliquots, removed at 0 and 600 mins were transferred to 50 µl of ethanol, centrifuged to remove proteins and chromatographed in solvent 1.

In order to test whether compound X was indeed the thioglucoside desthiogluco-brassicin, ^{35}S-X and ^3H-X were treated with the thioglucosidase, myrosinase. Myroxinase has been variously reported to act either very feebly on synthetic desthioglucotropaeolin (Ettlinger and Lundeen, 1957) or not at all (Underhill and Wetter, 1969). Preliminary investigations indicated that at pH 7 and in the presence of ascorbate (2×10^{-3}M), myrosinase did act very slowly on ^{35}S-X, yielding an ether-soluble radioactive compound Y. Chromatography of the products of a ten hour reaction of myrosinase on ^{35}S-X (Fig. 8) indicated the formation of a new radioactive peak, Y, which had an Rf of 0.92 in solvent 1. This new radioactive zone gave a purplish color with PDAC reagent thereby showing the presence of an indole ring in the structure of the compound. Treatment of ^3H-X with myrosinase gave ^3H-Y. Compound Y was therefore a sulfur-containing indole compound, less polar than compound X. ^3H-Y and ^{35}S-Y in aqueous ethanolic solution and at slightly acidic pH, gradually decomposed to ^3H-IAN and cold IAN respectively, as demonstrated by chromatography. Since thiohydroximates are known to decompose spontaneously to the nitrile (Ettlinger and Lundeen, 1957), this and other properties of compound Y suggested that it was indoleacetothiohydroximate (Fig. 7a).

The other product of myrosinase action on compound X, if it were desthioglucobrassicin, would be glucose. Glucose was indeed shown to be the other product of myrosinase action on ^3H-X, by means of paper chromatography and color reaction with aniline-phthalate reagent.

The properties of compound X as well as its decomposition product Y indicated compound X to be desthioglucobrassicin. Attempts to synthesize this compound by a procedure for desthioglucosinolate synthesis (Benn, 1963) failed owing to the great reactivity of the indole ring.

The formation of desthioglucobrassicin in woad leaves, though enhanced by selenate inhibition, was not an artefact of inhibition, for small amounts of it were formed even in the absence of selenate. Desthioglucobrassicin is therefore a true intermediate in the conversion of IAOX to GB and SGB. We believe this is the first demonstration of the *in vivo* formation of a desthioglucosinolate. These findings support the pathway proposed by Underhill and Wetter (1969) for glucosinolate biosynthesis.

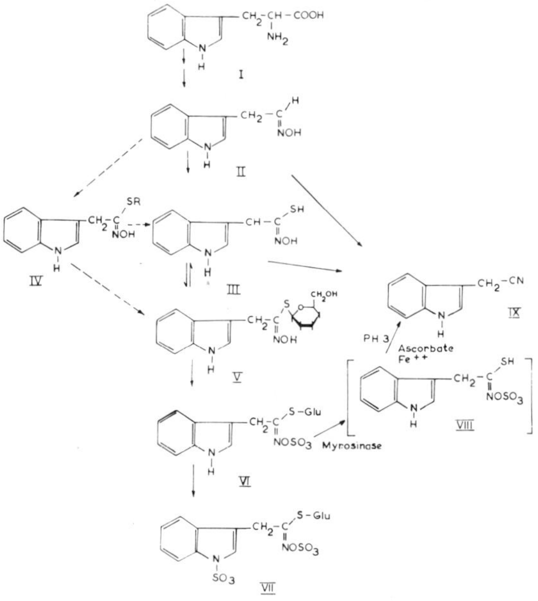

Fig. 9. Pathway for GB and SGB biosynthesis showing branch-off points for IAN formation.

Figure 9 depicts the pathway for GB and SGB biosynthesis from tryptophan (I) via IAOX (II). The conversion of tryptophan to IAOX by cabbage has been shown by Kindl (1968). IAOX is converted to indoleacetothiohydroximate (III) via the hypothetical intermediate IV which may have either of the structures shown in Fig. 3. The thiohydroximate (III) may be converted to desthioglucobrassicin (V) by an UDPG-dependent reaction. A possibility that IV may be directly converted to V by a transglycosylation reaction is not overruled. Desthioglucobrassicin (V) can be then sulfated to yield GB (VI), the step inhibited by selenate treatment. GB is finally converted to SGB (VII).

IV. EPILOGUE: MODES OF BIOSYNTHESIS OF IAN IN CRUCIFERS

Four possible ways for the formation of IAN can be visualized from the precursors of GB and from GB itself, as shown in Fig. 9. 1) A direct conversion of IAOX (II) to IAN; 2) the conversion of indoleacetothiohydroximate (III) to IAN, which is a facile chemical reaction; 3) the conversion of desthioglucobrassicin (V) to IAN via the thiohydroximate, a myrosinase (thioglucosidase) catalyzed reaction; 4) the conversion of GB (VI) to IAN by myrosinase via the sulfated thiohydroximate intermediate VIII. The last reaction, which occurs *in vitro* only below pH 5 may occur in tissues at higher pHs in the presence of ferrous ions and ascorbate, in analogy with the reaction described by Tookey and Wolff (1970).

Enzyme preparations from cabbage leaves have been shown to convert IAOX to IAN (Kumar and Mahadevan, 1963) and this conversion appeared to depend on the physiological condition of the plant. The observation that gibberellic acid treatment leads to increases in the IAN content with a concomitant lowering of the GB content in cabbage plants (Skytt-Andersen and Muir, 1969) suggests that the biosynthesis of these indole compounds depends on the physiological condition of the plant. IAOX could thus be channelled either into the direct synthesis of IAN or towards formation of glucobrassicin.

The ready formation of IAN from GB during extraction due to incomplete inactivation of myrosinase has led to the common belief that IAN is an artefact of isolation. Some IAN has been demonstrated in tissues even when all precautions have been taken to inactivate myrosinase (Kutáček and Procházka, 1964). The presence of a nitrilase in the crucifers which converts IAN to IAA would prevent the accumulation of substantial amounts of IAN in the tissue. In view of the alternate routes available for IAN formation, and the ability of tissues forming IAN being also capable of converting it to IAA, IAN may be regarded as a natural metabolite in the crucifers, its formation being dependent on the physiological status of the plant. Thus the sequential conversion of tryptophan, IAOX, IAN and IAA can be regarded as one of the pathways for the biosynthesis of IAA in cruciferous plants.

V. SUMMARY

Woad leaves were found to incorporate efficiently ^{3}H-indoleacetaldoxime (IAOX) and sulfur of ^{35}S-L-cystine into glucobrassicin (GB) and sulfoglucobrassicin (SGB). A neutral, indolic, sulfur-containing intermediate was detected during this conversion which was efficiently converted to GB and SGB. The formation of this intermediate was enhanced by administration of selenate along with cystine and IAOX. The intermediate was tentatively identified as desthioglucobrassicin. Various possibilities for the biogenesis of indoleacetonitrile (IAN) have been discussed. It is suggested that in crucifers, IAA may be also formed by the following pathway:

Tryptophan ⟶ IAOX ⟶ IAN ⟶ IAA.

VI. ACKNOWLEDGEMENTS

This investigation was supported by U.S. Public Health Service Grant GM-06921 from the National Institutes of Health to B.B. Stowe. We thank Prof. G.R. Wyatt for the generous loan of a paper strip counter.

VII. REFERENCES

AHMAD, A., I. EELNURME, and I.D. SPENSER (1960). Indolyl-3-acetaldoxime. Can. J. Chem. 38, 2523.

BENN, M.H. (1963). A new mustard oil glucoside synthesis. The synthesis of glucotropaeolin. Can. J. Chem. 41, 2836-2838.

BRAY, C.A. (1960) A simple efficient liquid scintillator for counting aqueous solutions in a liquid scintillation counter. Anal. Biochem. 1, 279-285.

BOYD, E.S. and W.F. NEUMAN (1954). Chondroitin sulfate synthesis and respiration in chick embryonic cartilage. Arch. Biochem. Biophys. 51, 475-486.

CONN, E.E. and G.W. BUTLER (1969). The biosynthesis of Cyanogenic Glycosides and other Simple Nitrogen Compounds, in "Perspectives in Phytochemistry". 47-74. Ed. J.B. Harborne and T. Swain, Academic Press, N.Y., London.

ELLIOTT, M.C. and B.B. STOWE (1970). A novel sulphonated natural indole. Phytochemistry, 9, 1629-1632.

ETTLINGER, M.J. and A.J. LUNDEEN (1957). First synthesis of a mustard oil glucoside; The enzymatic Lossen Rearrangement. J. Am. Chem. Soc. 79, 1764-1765.

GMELIN, R. and A.I. VIRTANEN (1961). Glucobrassicin der Prekursor von SCN, 3-indoleacetonitril und Ascorbigen in *Brassica oleracea* species. Annales Acad. Sci. Fennicae, Ser A.II, Chemica, 1-23.

GMELIN, R. and A.I. VIRTANEN (1962). Neoglucobrassicin, ein zweiter SCN-Precursor von Indoltyp in *Brassica*-Arten. Acta Chem. Scand. 16, 1348-1384.

GRAY, R.A. (1959). Preparation and Properties of 3-Indoleacetaldehyde. Arch. Biochem. Biophys. 81, 480-488.

KINDL, H. (1968). Oxydasen und Oxygenasen in höheren Pflanzen, I. Über das Vorkommen von Indolyl-(3)-acetalhydoxim und seine Bildung aus L-Tryptophan. Hoppe Seyler's Z. Physiol. Chem. 349, 519-520.

KINDL, H. and E.W. UNDERHILL (1968). Biosynthesis of mustard oil glucosides: N-Hydroxy-phenylalanine, a precursor of glucotropaeolin and a substrate for the enzymatic and non-enzymatic formation of phenylacetaldehyde oxime. Phytochemistry, 7, 745-756.

KUMAR, S.A. and S. MAHADEVAN (1963). 3-Indoleacetaldoxime Hydro-lyase: A pyridoxal-5'phosphate activated enzyme. Arch. Biochem. Biophys. 103, 516-518.

KUTÁCEK, M. and V.I. KEFELI (1968). Indole compounds of the Brassicaceae family, in "Biochemistry and Physiology of Plant Growth Substances." The Runge Press Ltd. Ottawa, p, 127-152.

KUTÁCEK, M. and Z. PROCHÁZKA (1964). Méthodes de détermination et d'isolement des Composés indoliques chez les Crucifères, in Régulateurs naturels de la Croissence végétale. Colloques de CNRS - Nro. 123, Paris, p. 445-456.

MAHADEVAN, S. (1963). Conversion of 3-Indoleacetaldoxime to 3-Indoleacetonitrile by plants. Arch. Biochem. Biophys. 100, 557-558.

MAHADEVAN, S. and B.B. STOWE, (1972). An intermediate in the synthesis of glucobrassicins from 3-indoleacetaldoxime by woad leaves. Plant Physiol. In press.

MATSUO, M. and E.W. UNDERHILL (1969). A UDP-Glucose: Thiohydroximate glucosyl transferase from *Tropaeolum majus* L. Biophys. Res. Commun. 36, 18-23.

NEUBERG, C. and J. WAGNER (1926). Uber die Verschiedenheit der Sulfatase und Myrosinase. VIII. Mitteilung über Sulfatase. Biochem. Z. 174, 457-463.

ROSENFELD, I. and O.A. BEATH (1964). "Selenium". Academic Press, New York, p. 305.

SCHRAUDOLF, H. and F. BERGMANN (1965). Der Stoffwechsel von Indolderivaten in *Sinapis alba* L. II. Untersuchungen zur Biogenese und Umsetzung von Indol-Glucosinolaten mit Hilfe von Ringmarkiertem C^{14}- Tryptophan und S^{35}- sulfat. Planta (Berl.) 67, 75-95.

SHULKA, P.S. and S. MAHADEVAN (1970). Indoleacetaldoxime Hydro-lyase (4.2.1.29). III. Further studies of the Nature and Mode of Action of the Enzyme. Arch. Biochem. Biophys. 137, 166-174.

SKYTT-ANDERSEN, A. and R.M. MUIR (1969). Gibberellin Induced Changes in Diffusible Auxins from Savoy Cabbage. Physiol. Plantarum 22, 354-363.

TAPPER, B.A. and G.W. BUTLER (1967). Conversion of oximes to mustard oil glucosides (glucosinolates). Arch. Biochem. Biophys. 120, 719-721.

TISHEL, M. and M. MAZELIS (1966). Enzymatic Degradation of L-cystine by Cytoplasmic Particles from Cabbage Leaves. Nature (Lond.) 211, 745-746.

TOOKEY, H.L. and I.A. WOLFF (1970). Effect of organic reducing agents and ferrous ion on thioglucosidase activity of *Crambe abyssinica* seed. Can. J. Biochem. 48, 1024-1028.

UNDERHILL, E.W. (1967). Biosynthesis of Mustard Oil Glucosides: Conversion of phenylacetaldoxime and 3-phenylpropionaldoxime to Glucotropaeolin and Gluconasturtiin. European J. Biochem. 2, 61-63.

UNDERHILL, E.W. and L.R. WETTER (1969). Biosynthesis of Mustard Oil Glucosides: Sodium Phenylacetothiohydroximate and Desulfobenzylglucosinolate, Precursors of Benzylglucosinolate in *Tropaeolum majus* L. Plant Physiol., 44, 584-590.

WETTER, L.R. and M.D. CHISHOLM (1968). Sources of sulfur in the thioglucosides of various higher plants. Can. J. Biochem. 46, 931-935.

PLANT GROWTH SUBSTANCES, 1970

Isolation and Characterization of Indole Derivatives in Clubroots of Chinese Cabbage.

Saburo Tamura, Michio Nomoto and Minoru Nagao

Department of Agricultural Chemistry, the University of Tokyo, Bunkyo-ku, Tokyo, Japan

I. INTRODUCTION

It is well-known that when Chinese cabbage, *Brassica pekinensis* Rupr. is infected with *Plasmodiophora brassicae* Woronin, its root enlarges abnormally to form a club. Up to now, however, the cause of this malformation has not been clarified. Katsura *et al.* (1966, 1969) using paperchromatography demonstrated the presence of two unidentified auxins, one neutral, the other basic, in clubroots of *Brassica rapa* var. *neosuguki* Kitam. infected with the same pathogen. Similarly, Kavanagh *et al.* (1969) conducted a qualitative chromatographic analysis for indole auxins in *Plasmodiophora brassicae*-infected cabbage hypocotyls and detected IAN. To elucidate the causes of abnormal growth of this nature, we have attempted to isolate the indole derivatives contained in Chinese cabbage club roots.

II. MATERIALS AND METHODS

Plant materials

Clubroots (30 kg) of Chinese cabbage, *Brassica pekinensis* Rupr. were collected in 1969 in Ibaragi Prefecture, Japan. At the same time, a small amount of healthy root was harvested to be used for comparison.

Extraction

Roots were covered with methanol, ground with a blendor and filtered. The methanol extracts thus obtained were subjected to further fractionation and purification.

Paper chromatography

Preparative paper chromatography was conducted on Toyo Filter Paper No. 51 with the solvent system isopropanol-ammonium hydroxide-water (10:1:1).

Adsorption chromatography

Column chromatography was carried out on silica gel (Mallinckrodt) with the solvent system hexane-ethyl acetate, increasing stepwise the ratio of the latter or with ethyl acetate alone.

Thin-layer chromatography (TLC)

TLC was preparatively conducted using silica gel GF_{254} (Merck) and the following solvent systems: (1) hexane-ethyl acetate (4:1) ; (2) benzene-isopropyl ether (4:1) ; (3) dichloromethane-methanol (10:1) ; (4) isopropyl ether-acetic acid (9:1).

Detection of indole compounds. After chromatography spots of indole compounds were detected as shadows under ultra-violet light and by coloration with Ehrlich's reagent.

Spectral measurements

UV spectra were measured in ethanol with a Cary-PM spectrophotometer. IR spectra were determined in nujol mulls or film with a JASCO IR-S KCl spectrometer. NMR spectra were measured in $CDCl_3$ with a JNM-4H-100 spectrometer at 100 Mc with tetramethylsilane as the internal standard. Mass spectra were obtained by use of a RMS-4 spectrometer with a direct inlet system. The electron accelerating energy was 70 eV, and the temperature of the ionization source 130 - 160°C.

Synthesis of 4-, 5- and 6-methoxy-IAN

These compounds were synthesized in our laboratory.

Bioassay

The assay was carried out using the *Avena* coleoptile straight-growth test (Nitsch and Nitsch, 1956).

III. RESULTS AND DISCUSSION

As a preliminary trial, methanol extracts from small amounts of healthy and diseased roots (taken separately) were separated into ethyl acetate-soluble, acidic and neutral fractions in the following way. Each methanol extract was evaporated under reduced pressure, and the aqueous residue was extracted at pH 2.5 with ethyl acetate. The extract was shaken with dilute sodium bicarbonate solution and evaporated to give the neutral fraction. The aqueous layer was acidified to pH 2.5 and extracted with ethyl acetate. The organic layer was separated and evaporated, yielding the acidic fraction. One mg aliquots of each of the acidic and neutral fractions obtained from healthy and diseased roots were subjected to preparative paper chromatography. Each paper was cut into ten pieces, which were then bioassayed.

As illustrated in Fig. 1, the acidic fractions from both healthy and diseased plants had hardly any biological activity. On the other hand, a remarkable increase in auxin activity of the neutral fraction was observed, due to infection. Further, there were distinct differences in the patterns of the histograms from the two kinds of plant material and marked activity was detected at Rf 0.6 - 0.9 for diseased roots.

Isolation of active compounds was then carried out using 30 kg of clubroots and the procedure shown in Fig. 2. The ethyl acetate-soluble neutral fraction (6 g) was chromatographed on a silica gel column, which was eluted with hexane-ethyl acetate. The eluates obtained with solvents containing 10 and 15% ethyl acetate revealed biological activity. Preparative TLC of the former eluate by use of solvent system (1) (see "Materials and Methods") gave substance II, while the latter eluate by the successive use of the solvent systems of (1) and (2) afforded substances I, III and IV. Yields of I, II, III and IV were 20, 10, 5 and 1 mg, respectively.

The aqueous layer remaining after the first separation of ethyl acetate-soluble fractions was extracted with *n*-butanol. After evaporation of the solvent, the residue was chromatographed on a silica gel column, which was eluted with ethyl acetate to give substance V in 1 mg yield. Structural elucidation of the compounds thus obtained necessitated the use of physico-chemical methods due to the small size of the samples.

The presence of the indole nucleus in the molecule of I was suggested by UV measurement as well as coloration with Ehrlich's reagent. λ_{max} mμ (log ε) : 218 (4.43), 270 (3.93), 278 (3.92), 289 (3.66). ν_{max} cm^{-1} : 3480 (NH), 2240 (CN). δ : 3.80 (2H, s, CH_2). m/e : 156 (M^+). These data are completely identical with those of an authentic sample of IAN.

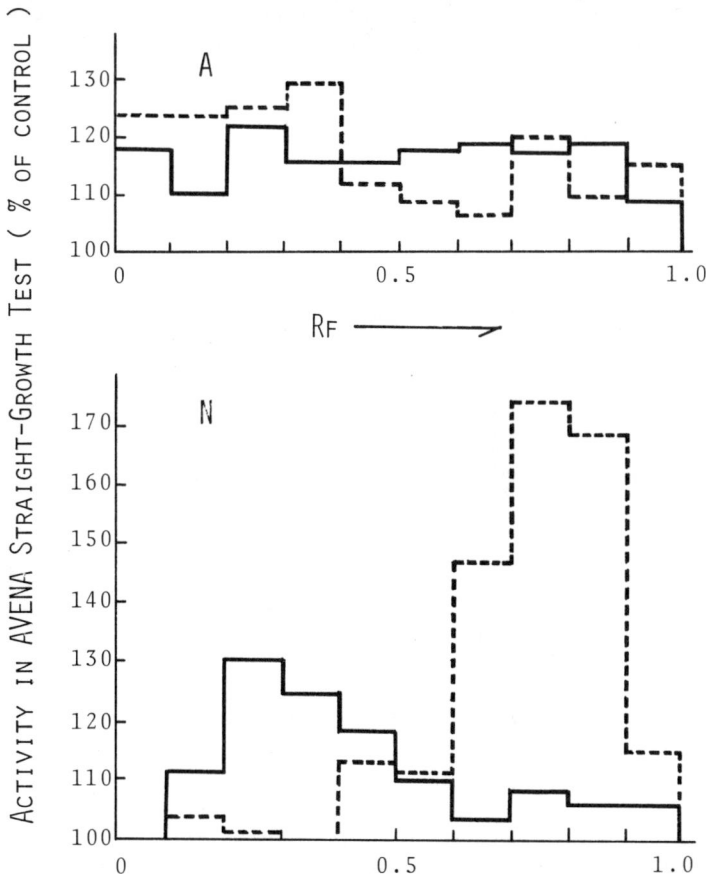

Fig. 1. Biological activity of Acidic(A) and Neutral(N) Fractions Separated from Healthy and Diseased Roots of Chinese Cabbage. ———— Healthy, ------ Diseased.

Also with III, the presence of the indole nucleus was similarly indicated. λ_{max} mμ (log ε) : 220 (4.64), 265 (3.86), 278 (3.73), 289 (3.71). ν_{max} cm^{-1} : 3480 (NH), 2240 (CN). δ : 3.95 (3H, s, OCH$_3$), 4.05 (2H, s, CH$_2$). m/e : 186 (M$^+$), 171 (M-CH$_3$), 155 (M-OCH$_3$). These data suggest that III is a substituted IAN carrying a methoxyl on the benzene ring. When the NMR spectrum of III was compared with those of methyl chloroindoleacetates, each containing a chlorine atom at C-4, 5, 6 or 7, close similarity was observed between III and the 4-chloro compounds. We therefore synthesized 4-methoxy-IAN starting from 2-amino-6-nitrotoluene and confirmed its complete identity with III on the basis of spectral measurements.

II : X = OCH$_3$, Y = H

III : X = H , Y = OCH$_3$

II was shown to be an indole derivative by the color reaction and the UV spectrum. λ_{max} mμ (log ε) : 218 (4.42), 270 (3.70), 286 (3.66), 297 (3.58). ν_{max} cm^{-1} : 2240 (CN). δ : 3.78 (2H, s, CH$_3$), 4.07 (3H, s, OCH$_3$). m/e : 186, 171, 155. In the IR spectrum of II, the band at 3480 cm^{-1} characteristic of the NH group constituting part of the indole nucleus was not observed. The methoxyl group can be assigned to either C-1 or C-3 of the nucleus.

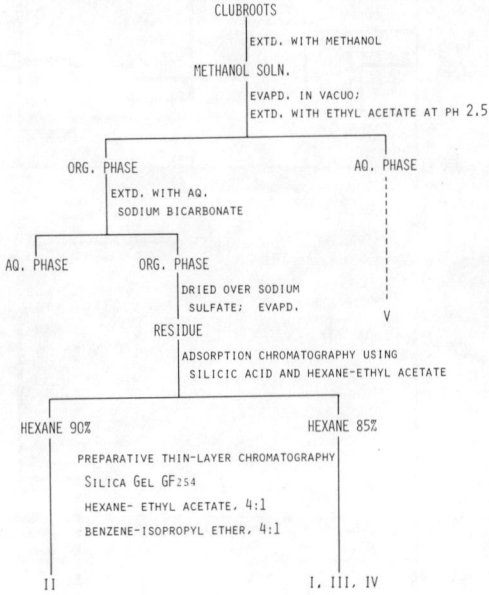

Fig. 2. Isolation Procedure of Indole Derivatives from Clubroots of Chinese Cabbage.

The chemical shifts of methoxyl and methylene (-CH$_2$CN) cited above are too low to be located together at C-3 to form 3-cyanomethyl-3-methoxyindolenine (Cohen et al., 1960). Thus the structure of II has been assigned as 1-methoxy-IAN. This assignment will be supported by the occurrence of neoglucobrassicin or 1-methoxyglucobrassicin as a normal constituent of higher plants (Gmelin, 1964; Schraudolf, 1965). II and III are the first compounds of their kind to have been isolated from higher plants.

The identification of IV as methyl indoleacetate is based on mass spectrometry and behaviour on TLC. Similarly, V was identified as indoleacetamide.

Auxin activities of the new indole derivatives, II and III, were assayed. Since the activity of methoxy-IANs have not hitherto been reported, we attempted synthesis of IAN containing a methoxyl at C-5, 6 or 7 position. Though the 5- and 6-substituted compounds could be obtained, all our attempts to obtain the 7-methoxy derivative ended in failure. Contrary to our expectation, these substituted IANs revealed only slight auxin activity or were almost inactive, as shown in Table 1. Therefore, the significant activity observed in the neutral fraction from diseased roots should be ascribed mainly to IAN itself.

At this stage, the origin of the large quantities of IAN and its methoxy derivatives accumulated in clubroots must be taken into consideration. The possibility that these compounds are metabolites of the pathogen, Plasmodiophora brassicae, must be rejected, since IAN has already been demonstrated to be the principal auxin of the members of the genus Brassica. Moreover, 1-methoxy-IAN isolated here is a constituent of neoglucobrassicin which has been specifically isolated from a limited variety of plants including members of Brassicaceae (Schraudolf, 1965; Kutáček and Kefeli, 1968).

There remains the further question as to whether IAN and its methoxyl derivatives isolated by us might be artefacts produced from glucobrassicin and neoglucobrassicin during the extraction and purification procedure. Examination of the histograms shown in Fig. 1 makes this improbable except for the methoxy compound with slight

Table 1. Activities of IAN and its Methoxyl Derivatives in Avena Coleoptile Straight Growth Test.

Compd.	Concentration (mg/l)			
	0.2	1.0	5	25
N-CH_3O-IAN	1.0	1.1	1.2	1.2
4-CH_3O-IAN	1.0	1.1	1.0	0.9
5-CH_3O-IAN	1.0	1.1	1.2	1.1
6-CH_3O-IAN	1.0	1.2	1.2	1.1
IAN	1.5	1.7	1.6	--
IAA	1.5	1.6	1.6	--

activity. We could not detect activity due to IAN in the neutral fraction from healthy roots, processed in the same way as diseased roots. Further, Schraudolf and Weber (1969) showed that neither IAN nor other indole compounds are formed by enzymatic hydrolysis of glucobrassicin at pH values higher than 5.2.

We must now pursue the possible change in content of glucosinolates in healthy and diseased roots to investigate further the host-parasite interaction responsible for the formation of IANs. Isolation of other growth regulators including cytokinins is in progress. The isolation of 4-methoxy-IAN suggests the possibility of the occurrence of a new glucosinolate based on this nitrile.

IV. SUMMARY

Plasmodiophora brassicae-infected roots of Chinese cabbage, *Brassica pekinensis*, were extracted with methanol and separated into ethyl acetate-soluble neutral and acidic fractions. The acidic fraction from diseased roots as well as that from healthy roots hardly showed any auxin activity. On the other hand, there is a remarkably large activity of the neutral fraction of the infected roots.

From the ethyl acetate-soluble neutral fraction separated from diseased roots, IAN, 1-methoxy-IAN, 4-methoxy-IAN and methyl indoleacetate were isolated. Furthermore, indoleacetamide was obtained from the water-soluble fraction by extraction with *n*-butanol.

V. ACKNOWLEDGEMENTS

We are grateful to Dr. S. Marumo of Nagoya University for providing the NMR data of methyl chloroindoleacetates.

VI. REFERENCES

COHEN, L.A., J.W.DALY, H. KNY and B. WITKOP (1960). Nuclear magnetic resonance spectra of indoles. J. Am. Chem. Soc. 82, 2184-2187.
GMELIN, R. (1964). Occurrence, isolation, and properties of glucobrassicin and neo-glucobrassicin, *in* Coll. Int. C.N.R.S., 123, 159-167.
KAVANAGH, J.A., M.N. REDDY and P.H. WILLIAMS (1969). Growth regulators in clubroot of cabbage. Phytopath. 59, 1035.

KATSURA, K., H. EGAWA, T. TOKI and S. ISHII (1966). On the plant hormones in the healthy and *Plasmodiophora*-infected roots of *Brassica rapa* var. *neosuguki* Kitam. (1). Ann. Phytopath. Soc. Japan 32, 123-129.
KATSURA, K., H. EGAWA, M. MASUKO and A. EUYAMA (1969). On the plant hormones in the healthy and *Plasmodiophora*-infected roots of *Brassica rapa* var. *neosuguki* L. (2). Proc. Kansai Pl. Protection Soc. (Japan) No. 11, 23-27.
KUTÁČEK, M. and VI. KEFELI (1968). The present knowledge of indole compounds in plants of the Brassicaceae family, *in* "Biochemistry and Physiology of Plant Growth Substances" (Ed. F. Wightman and G. Setterfield) p. 127-152. Runge Press, Ottawa. 1968.
NITSCH, J.P. and C. NITSCH (1956). Studies on the growth of coleoptile and first internode sections. A new, sensitive, straight growth test for auxins. Pl. Physiol. 31, 94-111.
SCHRAUDOLF, H. (1965). Zur Verbreitung von Glucobrassicin und Neoglucobrassicin in höheren Pflanzen. Experientia 21, 520-522.
SCHRAUDOLF, H. and H. WEBER (1969). IAN-Bildung aus Glucobrassicin: pH-Abhängigkeit und wachstumsphysiologische Bedeutung. Planta (Berl.) 88, 136-143.

BIOSYNTHESIS OF GIBBERELLINS

Gibberellin Biosynthesis and Its Regulation

Charles A. West and R. Ray Fall*

Division of Biochemistry, Department of Chemistry,
University of California, Los Angeles, California, 90024

*Present address: Department of Biological Chemistry, School of Medicine,
Washington University, St. Louis, Missouri

I. INTRODUCTION

Studies of the mode of action of a family of plant hormones are of great interest to the plant physiologist. But at the same time the mechanisms which govern the production of that group of growth regulatory substances must be examined for possible clues as to how environmental factors interact with the plant and regulate its development. The problem of the biosynthesis of the plant hormones can be considered at several levels, although naturally these different aspects are highly interrelated: (1) the biosynthetic pathway, that is the sequence of intermediate structures leading from metabolites of the central metabolic pathways to the active regulatory agent or agents; (2) the nature of the enzymes involved in these steps; (3) the localization and organization of this biosynthetic machinery within the various tissues, cell-types and organelles of the plant; and (4) the regulation of activity of this biosynthetic sequence by the metabolic and environmental factors to which the plant responds. Information in these areas will lead to a better understanding of the natural regulation of plant development and may suggest ways of modifying a plant's response to its environment.

In the last ten or so years since the structures of some of the gibberellins and their role as natural growth regulators in higher plants were established, we have accumulated some knowledge about the various aspects of their biosynthesis mentioned above. Our knowledge is still quite fragmentary in terms of what we would like to know and is hampered by a lack of clear understanding of where the biosynthetic metabolism ends and functional or catabolic metabolism begins. An overall pattern of the biosynthetic pathway is evident and some of the intermediates and sequences of conversions have been established with reasonable certainty for a few systems. The general characteristics are known for a few of the enzymes, especially those early in the pathway, but detailed examinations of these enzymes are just beginning. Both direct and indirect evidence has been advanced to implicate as sites of gibberellin biosynthesis the endosperm and cotyledons of immature seeds, the scutellum and embryonic axes, shoot apices and young leaves, root tips, and leaf chloroplasts; however, this information comes from relatively few examples and the picture is quite incomplete. We know a number of chemicals and environmental factors such as photoperiod and temperature which appear to influence gibberellin levels and/or the response to applied gibberellins of appropriate test organisms, but only in the case of the synthetic plant growth retardants is there an explanation at the molecular level of how some of these agents may be influencing the biosynthetic pathway.

This paper will attempt to summarize briefly the present state of knowledge of the gibberellin biosynthetic pathway, the general characteristics of some of the enzymes involved, and the regulation of these reactions by plant growth retardants. Reference will be made to some recent observations from our laboratory relevant to these problems. Recent reviews by Cross (1968) and Lang (1970) and a chapter in a monograph by Hanson (1968) discuss the background of much of this material and give additional citations to the original literature.

II. PATHWAY OF BIOSYNTHESIS

The isoprenoid nature of the gibberellins and their general biosynthetic relationship to the diterpenes is clear. The overall scheme envisions (1) mevalonate as the precursor of the acyclic diterpene geranylgeranyl-PP, (2) cyclization of this substance to kaurene, and (3) modification of kaurene by oxidation, ring contraction, and elimination of an angular methyl group to form the C_{19}-gibberellins.

The sequence of steps leading from mevalonate to kauren-7β-ol-19-oic acid, an oxidized metabolite of kaurene, appears to be common to all systems studied to date. The evidence from several laboratories in support of it will not be summarized here. Figure 1 shows the steps from geranylgeranyl-PP to kauren-7β-ol-19-oic acid.

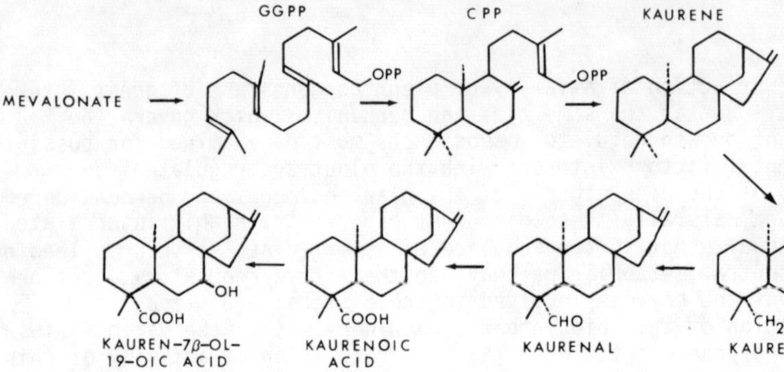

Fig. 1. Sequence of conversions from geranyl-geranyl-PP to kauren-7β-ol-19-oic acid in gibberellin biosynthesis.

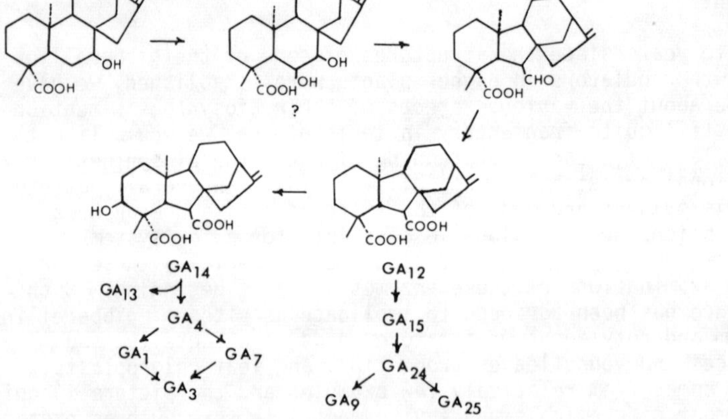

Fig. 2. Possible conversions in gibberellin biosynthesis beyond kauren-7β-ol-19-oic acid in *F. moniliforme*.

The sequence beyond kauren-7β-ol-19-oic acid is much less firmly established. A speculative scheme for these interconversions in *Fusarium moniliforme (Gibberella fujikuroi)* is shown in Figure 2. It must be remembered that this scheme pertains to studies in the fungus and sequences may differ from one to another organism even if the patterns are similar in an overall sense. The contraction of the B-ring, a process indicated by early labeling studies of Birch *et al.* (1958), is thought to proceed prior to the elimination of the C-20 carbon and the formation of the C-19→C-10 lactone characteristic of the C_{19} gibberellins. The widespread occurrence of C_{20}-gibberellins with the C-20 carbon present supports this view. A substrate for the ring-contraction step has not been identified. Kauren-6β,7β-diol-19-oic acid, which seemed a logical choice, is pictured in this role in Figure 2. However, Cross et al. (1970) recently tested this diol in cultures of F. moniliforme and did not find it to be a precursor of GA_3 even though it was metabolized by an oxidative opening of the B-ring to fujenal. Evidence was presented for the natural occurrence of the diol in the culture filtrate and it was further demonstrated that the diol did show gibberellin-like biological activity in four higher plant assay systems. On this basis they speculated that a derivative of the diol such as the 6β-O-pyrophosphate might be the actual substrate for contraction. It seems most likely that the product of ring contraction has an aldehyde substituted in the 6-position which subsequently becomes oxidized to a carboxylic acid. The aldehyde analogue of gibberellin A_{12} was synthesized and shown to be incorporated relatively efficiently into other gibberellins in the fungus (Cross et al., 1968). This aldehyde was also shown to be a normal metabolite in the fungus (Hanson and White, 1969a).

A cell-free system has been obtained from F. moniliforme which may prove useful in establishing the sequence of transformations at this stage of the pathway. A microsome-like particulate fraction prepared from cell-free extracts of the fungal mycelia supplemented with NADPH ($5 \times 10^{-4}\underline{M}$), FAD ($10^{-5}\underline{M}$) and Tricine, pH 8.0 ($0.10\underline{M}$) converts ^{14}C-kaurenoic acid to a series of products which have been separated by thin layer chromatography. Figure 3 suggests a tentative identification of some of these products from thin layer chromatographic comparisons with reference compounds. It should be emphasized that

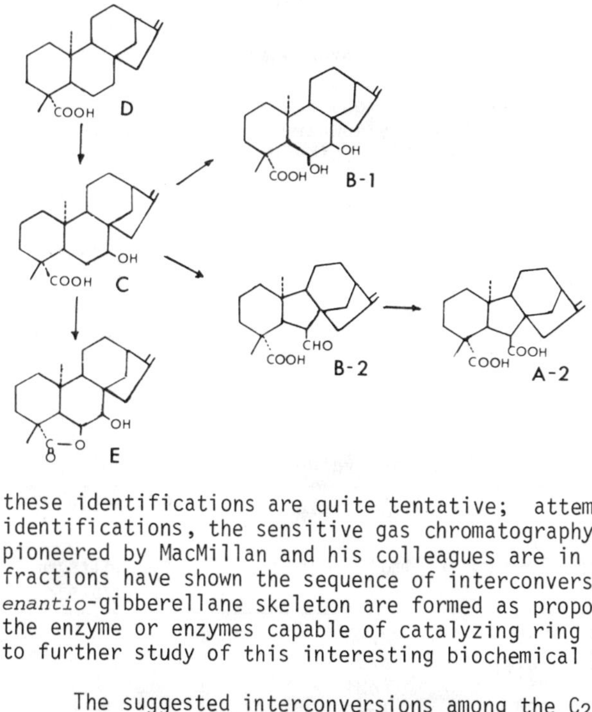

Fig. 3. Possible metabolites formed from kaurenoic acid in cell-free extracts of F. moniliforme.

these identifications are quite tentative; attempts to apply, for more rigorous identifications, the sensitive gas chromatography-mass spectrometry techniques pioneered by MacMillan and his colleagues are in progress. Reincubations of isolated fractions have shown the sequence of interconversions shown. If products with the enantio-gibberellane skeleton are formed as proposed, then this preparation contains the enzyme or enzymes capable of catalyzing ring contraction and should be amenable to further study of this interesting biochemical problem.

The suggested interconversions among the C_{20}- and C_{19}-gibberellins in F. moniliforme shown in Figure 2 stem from labeling studies in the intact fungus carried out by several groups of investigators and from an examination of the structural relationships among these fungal gibberellins. It should be noted that most of the proposed steps are oxidative. The evidence indicates that the loss of the C-20 carbon occurs (1) while it is more reduced than a carboxylic acid and (2) without the loss of H from C-1, C-5, or C-9 (Hanson and White, 1969b). It has been further suggested that lactonization may accompany the elimination of this carbon. The sequence of interconversions at this stage of biosynthesis may well differ in other organisms from that shown here for F. moniliforme. Indeed, there may not be a unique sequence within one organism.

III. BIOSYNTHETIC ENZYMES

The several enzymes responsible for the transformation of mevalonate to prenyl pyrophosphates have not been so extensively studied in higher plants as in animals or yeast, but their general properties appear to be similar. These enzymes appear in the soluble fraction of cell extracts; ATP and divalent metal ions (Mg^{2+} or Mn^{2+}) are the only low molecular weight cofactors required. The prenyl transferase of some plant systems produces predominantly *trans*-geranylgeranyl-PP which is involved in the biosynthesis of diterpenes and carotenes. The factors which govern the pattern of prenyl pyrophosphates produced in a given tissue have not been established. In fact, the relationship between the biosynthetic pathways leading to the different classes of isoprenoid products which occur in plants is very little understood.

The most studied enzymic reactions in the pathway are those involved in the overall cyclization of geranylgeranyl-PP to kaurene catalyzed by kaurene synthetase. Crude or somewhat purified preparations of kaurene synthetase from wild cucumber (*Echinocystis macrocarpa*) endosperm (Upper and West, 1967) and castor bean (*Ricinus communis*) seedlings (Robinson and West, 1970) were found in the soluble fraction and were shown to require a divalent metal ion, preferably Mg^{2+}, for activity. Copalyl-PP appears to be an intermediate in the transformation in both preparations even though it does not normally accumulate in incubation mixtures (Shechter and West, 1969). For convenience, the conversion of geranylgeranyl-PP to copalyl-PP is referred to as activity A and the conversion of copalyl-PP to kaurene as activity B.

We have recently succeeded in purifying kaurene synthetase from cell-free extracts of *F. moniliforme* to a considerable extent and have investigated some properties of the purified enzyme. The purification steps are summarized in Table 1. Details of the procedures involved will be published elsewhere. The washed mycelial mat from 6 1-l. cultures of *F. moniliforme* ACC 917 M419 (kindly supplied by Dr. R.H.B. Galt of the Imperial Chemical Industries Pharmaceuticals Division) grown for 90 hr at room temperature on a standard glucose-ammonium nitrate medium served as the source of enzyme. The frozen cell mass was crushed in a Sagers press and debris and other materials were removed by a preliminary centrifugation at 26,000 x g followed by a protamine sulfate precipitation. Subsequent steps included two ammonium sulfate precipitations under different conditions and a combination of hydroxylapatite and QAE-Sephadex chromatography. It was found to be important for recovery of activity to carry out all steps in the presence of 1-2 mM dithiothreitol and the later steps in the presence of 25% glycerol.

Table 1. Purification of kaurene synthetase from *F. moniliforme*

	Kaurene Synthetase Activity				
	Total units		Specific activity		
Purification Step	A^a (nmoles/min.)	B^b (nmoles/min.)	A (units/mg protein)	B (units/mg protein)	A/B
1. S-26,000	4840	2730	1.44	0.81	1.8
2. Protamine sulfate	4127	2799	2.12	1.44	1.5
3. Ammonium sulfate	4862	2681	21.9	12.1	1.8
4. Hydroxylapatite - QAE-Sephadex	2584	1393	208	112	1.8

[a] Initial rates of kaurene production were measured in incubation mixtures containing 0.5 nmole ^{14}C-geranylgeranyl-PP, 25 µg bovine serum albumin, 1 µmole $MgCl_2$, 0.2 µmole dithiothreitol, 50 µmoles potassium phosphate, pH 7.5, and 0.10 ml of appropriately diluted enzyme source in 0.5 ml total volume. Incubations were carried out for 5 min at 30°C.

[b] Initial rates of kaurene formation were measured under the conditions described in footnote a except that 0.5 nmoles of ^{14}C-copalyl-PP was substituted for ^{14}C-geranylgeranyl-PP.

The ratio of activities A to B remained constant throughout the various stages which resulted in about 140-fold purification and 52% recovery of both activities. Neither these nor any of several other protein fractionation techniques attempted gave any resolution of activity A from activity B. The approximately 1.8 times higher rate of kaurene formation consistently seen from geranylgeranyl-PP in comparison with copalyl-PP was shown to be due to substrate inhibition of activity B by relatively high levels of copalyl-PP.

Table 2. Properties of purified kaurene synthetase from *F. moniliforme*.

	Activity A	Activity B
Molecular weight		
Sucrose gradient centrifugation	$4.3\text{-}4.6 \times 10^5$	$4.3\text{-}4.6 \times 10^5$
Sepharose 4B gel filtration	$4.6\text{-}4.9 \times 10^5$	$4.6\text{-}4.9 \times 10^5$
pH optima	7.5	6.9
Divalent metal ion activation	$Mg^{2+} > Ni^{2+} = Co^{2+}$	$Mg^{2+} > Co^{2+} > Mn^{2+} = Ni^{2+}$
Stimulation by dithiothreitol		
in the present of 10^{-4} M EDTA	none	+
in the absence of EDTA	+	++
% Inhibition by 10^{-3} M sulfhydryl agents in the presence of 10^{-4} M EDTA		
Iodoacetamide	6	35
p-Hydroxymercuribenzoate	94	99
N-Ethylmaleimide	12	98
$HgCl_2$	93	96
$CuSO_4$	95	98
K_m (app) for substrate	0.7 µM	1.0 µM
Inhibition by 10^{-3} M GA_3	None	None

Some of the properties of the purified enzyme are summarized in Table 2. Both activities A and B are associated with a relatively high molecular weight protein of about 4.6×10^5 daltons. Polyacrylamide gel electrophoresis of the purified kaurene synthetase at pH 8.0 showed a major protein band from which both activities could be recovered as well as a few minor bands. Electrophoresis in SDS revealed the presence of multiple protein bands. These properties suggest that the kaurene synthetase is associated with a protein complex. It was considered possible that some other closely related isoprenoid biosynthetic enzymes might also be present in the complex; however, no evidence for prenyl transferase, squalene synthetase, 2,3-oxidosqualene cyclase or kaurene oxidase could be found in the purified enzyme. Therefore, there is no indication of the nature of any other proteins which may be present. The somewhat different properties of the two cyclization activities suggest, as predicted from mechanistic considerations, that they reside on two different catalytic sites in the complex in spite of the fact that it has not been possible to resolve them from one another. Dual isotope studies with mixtures of ^{3}H-copalyl-PP and ^{14}C-geranylgeranyl-PP as substrates have suggested that copalyl-PP produced by activity A serves as a substrate for activity B in preference to the general pool of copalyl-PP present.

Gibberellic acid at the relatively high level of $10^{-3}\underline{M}$ does not inhibit activity A or B significantly. This suggests that gibberellins do not act as feedback inhibitors at this step in this organism.

Purification of the kaurene synthetase from *E. macrocarpa* endosperm is in progress. A comparison of the properties of the fungal and higher plant enzymes should prove of interest.

Most of the reactions leading from kaurene to the C_{19}-gibberellins must be oxidative in nature. Although the detailed characteristics of these oxidative enzymes are not known, each of the four steps leading from kaurene to kauren-7β-ol-19-oic acid has been shown to be catalysed by a microsome-like particulate fraction of *E. macrocarpa* endosperm and to require oxygen and a source of reduced pyridine nucleotide (Murphy and West, 1969). NADPH or an ATP-dependent transhydrogenation between NADH and NADP appear to be the best sources of reduced pyridine nucleotide, with NADH itself functioning somewhat less efficiently. FAD at the optimal concentration of $10^{-5}M$ stimulates these conversions. The characteristics of inhibition by CO and its photoreversal have implicated cytochrome P-450 in these reactions. The characteristics of these reactions indicate their close similarity to the mixed function oxygenases which have been described in mammalian microsomes. The enzymes involved in other oxidative steps have not been characterized even to this degree, although there are indications that the enzymes catalyzing the conversions of kaurenoic acid to incompletely identified metabolites in both *E. macrocarpa* endosperm and *F. moniliforme* extracts share many of these same properties. It is attractive to speculate that the common features of the many oxidative reactions involved might serve as a common focus for regulatory control of the entire process of gibberellin production.

IV. REGULATION OF BIOSYNTHESIS

Kende *et al.* (1963) provided the first direct experimental evidence that the plant growth retardants CCC and Amo 1618 (see Figure 4 for systematic chemical names for these retardants) were potent inhibitors of gibberellin synthesis in cultures of *F. moniliforme*. Subsequent studies by this group indicated a similar interference by these substances with gibberellin biosynthesis in higher plants. Dennis *et al.* (1965) demonstrated that the kaurene synthetase of *E. macrocarpa* was a likely enzymic site of action of Amo 1618 and Phosfon D. This site has been further implicated in other systems including *F. moniliforme* (Shechter and West, 1969; Barnes *et al.*, 1969; Cross and Myers, 1969). Lang (1970) has recently reviewed the development of this subject.

The availability of the purified preparation of *F. moniliforme* kaurene synthetase described above enabled a more thorough and quantitative investigation of its interaction with growth retardants. The data of Figure 4 summarize the percent of inhibition of activity A seen with different levels of several growth retardants and related substances. Highly effective inhibitors included Amo 1618, Carvadan and Phosfon D; Phosfon S, a series of limonene derivatives (Q-64, Q-58 and Q-53) identified as growth retardants by Newhall (1969), and SKF 525A and SKF 3301A, two substances normally classified as a steroid synthesis inhibitors (Holmes and DiTullio, 1960), were of intermediate potency; CCC and a group of related substances were not significantly inhibitory at concentrations below $10^{-3}M$ and B-995 had no inhibitory effect at any level tested. Amo 1618, Phosfon D and SKF 3301A were non-competitive with respect to geranylgeranyl-PP. The K_i for Amo 1618 was estimated to be $2.3 \times 10^{-7}M$.

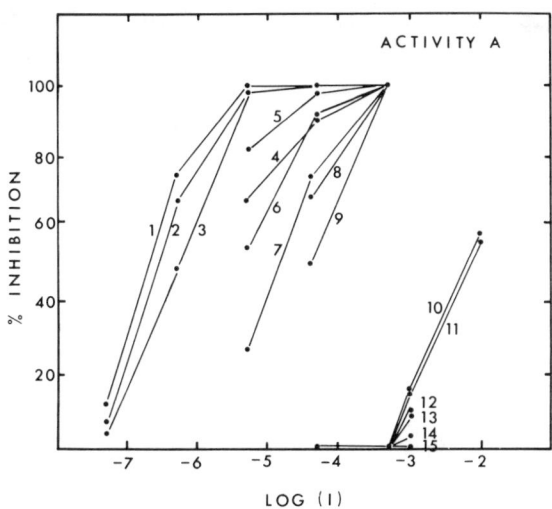

Fig. 4. Percent inhibition of kaurene synthetase activity A as a function of the concentration of various growth retardants and related substances. Incubated for 5 min at 30°C in a total volume of 0.5 ml were 0.5 nmole ^{14}C-geranylgeranyl-PP, 0.25 mg polyvinylpyrollidine 40, 2.5 μmoles $MgCl_2$, 0.10 ml purified enzyme, appropriately diluted, and the indicated concentration of inhibitor. Reaction was stopped with 2 μM p-hydroxymercuribenzoate and the mixture was hydrolyzed for 90 min at 30°C in the presence of 100 μg *Escherichia coli* alkaline phosphatase. Activity A was measured as the sum of kaurene and copalol present. The inhibitors tested were:

1 = Amo 1618 = 2'-isopropyl-4'-(trimethylammonium chloride)-5'-methylphenyl piperidine-1-carboxylate

2 = Carvadan = 3'-isopropyl-4'-(trimethylammonium chloride)-6'methylphenyl piperidine-1-carboxylate

3 = Phosfon D = tributyl-2,4-dichlorobenzylphosphonium chloride

4 = Phosfon S = tributyl-2,4-dichlorobenzylammonium chloride

5 = Q-64 = 2-(N,N-dimethyl-N-octylammonium bromide)-p-menthan-1-ol

6 = Q-58 = 2-(N,N-dimethyl-N-heptylammonium bromide)-p-menthan-1-ol

7 = Q-53 = 2-(N,N-dimethyl-N-3,4-dichlorobenzylammonium chloride)-p-menthan-1-ol

8 = SKF 3301A = N,N-dimethylaminoethyl 2,2-diphenylpentyl ether

9 = SKF 525A = N,N-dimethylaminoethyl 2,2-diphenylpentanoate

10 = CCC = 2-chloroethyltrimethylammonium chloride

11 = BCB = 2-bromethyltrimethylammonium bromide

12 = TMAB = tetramethylammonium bromide

13 = choline

14 = cholamine = 2-aminoethyltrimethylammonium bromide

15 = B-995 = N,N-dimethylsuccinamic acid

The effects of most of these same substances on activity B of the purified enzyme are summarized in Figure 5. In this case it can be seen that Amo 1618, Carvadan, Phosfon D and Phosfon S were not significantly inhibitory below 10^{-3}M. Interestingly, the limonene derivatives and the steroid synthesis inhibitors exhibited inhibitory properties for activity B approaching those seen with these same substances for activity A. Once again CCC and BCB were not significantly active, and B-995 not at all active, at $10^{-3}\underline{M}$ or lower concentrations. A test of the mode of inhibition of activity B by Q-64 and SKF 3301A showed that both were non-competitive with respect to copalyl-PP.

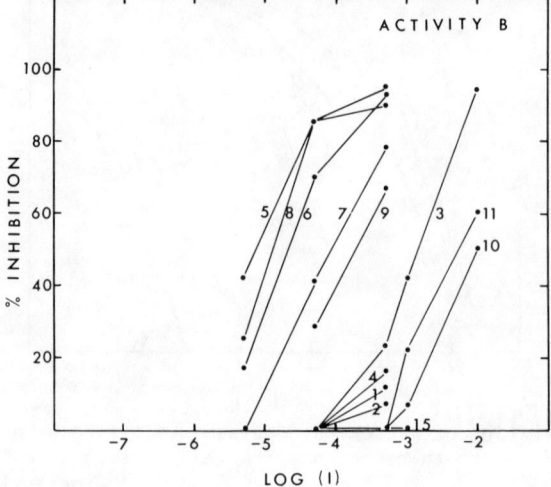

Figure 5. Per cent inhibition of kaurene synthetase activity B as a function of the concentration of various growth retardants and related substances. Incubated for 5 min at 30°C in a total volume of 0.5 ml were 0.5 nmoles ^{14}C-copalyl-PP, 0.25 mg polyvinylpyrollidone 40, 2.5 μmoles $MgCl_2$, 0.10 ml of purified enzyme, appropriately diluted, and the indicated concentration of inhibitor. Activity B was measured as the amount of kaurene formed under these conditions. The inhibitors tested were:

1 = Amo 1618	7 = Q-53
2 = Carvadan	8 = SKF 3301A
3 = Phosfon D	9 = SKF 525A
4 = Phosfon S	10 = CCC
5 = Q-64	11 = BCB
6 = Q-58	15 = B-995

The behaviour of Amo 1618 and Phosfon D (and by analogy Carvadan and Phosfon S) was as might be predicted from earlier studies (Shechter and West, 1969). However, the behaviour of CCC and its analogues is puzzling. Studies with the intact fungus had shown that as little as 6×10^{-7}M CCC in the growth medium reduced gibberellin production by 50% (Ninnemann, et al., 1964). We checked the response of the fungal strain used as the source of kaurene synthetase in the present study and found that $10^{-4}\underline{M}$ CCC in the growth medium inhibited gibberellin production by 97% and $10^{-5}\underline{M}$ CCC inhibited 93%. Barnes et al. (1969) concluded from tracer studies with the intact fungus that relatively low concentrations of CCC (0.13 m$\underline{M}$) blocked gibberellin synthesis between geranylgeraniol and kaurene (hence at kaurene synthetase). Cross and Myers (1969) reached similar conclusions from analysis of the effect of CCC on metabolite production in cultures of the fungus. From these observations we have concluded that either the biosynthetic site inhibited by CCC in vivo is other than kaurene synthetase or the response of kaurene synthetase to CCC has been modified in the course of its removal from the cell. In view of the results of Barnes et al. and Cross and Myers, the latter explanation seems more likely. Alternatively, CCC may be metabolized in vivo to some other inhibitory compound. It is interesting to note that none of the cell-free systems containing kaurene synthetase studied to date has been very responsive to CCC in comparison with Amo 1618 or Phosfon D.

The failure of B-995 to inhibit kaurene synthetase is consistent with its failure to inhibit gibberellin synthesis in any systems examined. This contrasts with its effects on plant development which in many cases are very similar to other growth retardants and are reversible by applied gibberellins (Lang, 1970).

The mode of inhibition of the growth retardants in all cases tested proved to be noncompetitive with respect to substrate. This suggests that they are acting at some site or sites other than the catalytic sites and prompts us to speculate that the growth retardants may be binding to a natural regulatory site and thereby mimicing natural regulators which act at this site.

V. SUMMARY

The present state of knowledge of the biosynthetic pathway leading from mevalonate to the gibberellins is summarized briefly. A cell-free preparation of the fungus *Fusarium moniliforme* which catalyzes the conversion of kaurenoic acid to a number of metabolites is described. Among the products which have been tentatively identified are *enantio*-gibberellane derivatives. It is hoped that this system will be of use in studying the biosynthetic mechanism of ring contraction of kaurane to *enantio*-gibberellane derivatives. Kaurene synthetase, which catalyzes the overall conversion of geranylgeranyl pyrophosphate to kaurene, has been purified from *F. moniliforme* extracts and a number of its properties are reported. It has not been possible to resolve the catalytic activity for the conversion of geranylgeranyl pyrophosphate to copalyl pyrophosphate (activity A) from the activity for the conversion of copalyl pyrophosphate to kaurene (activity B). The interactions of a number of plant growth retardants and related substances with the purified enzyme have been studied. Amo 1618, Phosfon D and Carvadan were the most effective inhibitors of activity A. Several limonene derivatives and steroid synthesis inhibitors were found to inhibit both activities A and B. All of the inhibitions tested were found to be non-competitive with respect to substrate. CCC and related substances and B-995 were found not to be significantly inhibitory for either activity at concentrations below $10^{-3}\underline{M}$.

VI. ACKNOWLEDGEMENTS

The authors are grateful to Mr. Dennis Nakata in our laboratory who performed some of the experiments with kaurenoic acid as substrate with cell-free preparations of *F. moniliforme*.

VII. REFERENCES

BARNES, M.F., E.N. LIGHT, and A. LANG (1969). The action of plant growth retardants on terpenoid biosynthesis. Planta, 88, 172-182.
BIRCH, A.J., R.W. RICKARDS and H. SMITH (1958). The biosynthesis of gibberellic acid. Proc. chem. Soc., 192-193.
CROSS, B.E. (1968). Biosynthesis of the gibberellins. Progr. Phytochem., 1, 195-222.
CROSS, B.E. and P.L. MYERS (1969). The effect of plant growth retardants on the biosynthesis of diterpenes by *Gibberella fujikuroi*. Phytochemistry, 8, 79-83.
CROSS, B.E., K. NORTON and J.C. STEWART (1968). The biosynthesis of the gibberellins, Part III. J. chem. Soc. (C), 1054-1063.
CROSS, B.E., J.C. STEWART and J.L. STODDART (1970). 6β,7β-Dihydroxy-kaurenoic acid: Its biological activity and possible role in the biosynthesis of gibberellic acid. Phytochemistry, 9, 1065-1071.
DENNIS, D.T., C.D. UPPER and C.A. WEST (1965). An enzymic site of inhibition of gibberellin biosynthesis by Amo 1618 and other plant growth retardants. Pl Physiol., 40, 948-952.
HANSON, J.R. (1968). The biosynthesis of the tetracyclic diterpenes, in "The Tetracyclic Diterpenes" Chapter 8, pp 114-121, Pergamon Press, London. 1968.
HANSON, J.R. and A.F. WHITE (1969a). The oxidative modification of the kauranoid ring B during gibberellin biosynthesis. Chem. Commun., 410-411.

HANSON, J.R. and A.F. WHITE (1969b). Studies in terpenoid biosynthesis. Part IV. Biosynthesis of the kaurenolides and gibberellic acid. J. chem. Soc. (C), 981-985.

HOLMES, W.L. and N.W. DiTULLIO (1960). Inhibitors of cholesterol biosynthesis which act at or beyond the mevalonic acid stage. Am. J. clin. Nutr. $\underline{10}$, 310-322.

KENDE, H., H. NINNEMANN and A. LANG (1963). Inhibition of gibberellic acid biosynthesis in *Fusarium moniliforme* by Amo 1618 and CCC, Naturwissenschaften, $\underline{50}$, 599-600.

LANG, A. (1970). Gibberellins: Structure and Metabolism. A. Rev. Pl. Physiol. $\underline{21}$, 537-570.

MURPHY, P.J. and C.A. WEST (1969). The role of mixed function oxidases in kaurene metabolism in *Echinocystis macrocarpa* Greene endosperm. Archs. Biochem. Biophys., $\underline{133}$, 395-407.

NEWHALL, W.F. (1969). Correlation of pseudocholinesterase inhibition and plant growth retardation by quaternary ammonium derivatives of (+)-limonene. Nature,Lond. $\underline{223}$, 965-966.

NINNEMANN, H., J.A.D. ZEEVAART, H. KENDE and A. LANG (1964). The plant growth retardant CCC as inhibitor of gibberellin biosynthesis in *Fusarium moniliforme*. Planta, $\underline{61}$, 229-235.

ROBINSON, D.R.and C.A. WEST (1970). Biosynthesis of cyclic diterpenes in extracts from seedlings of *Ricinus communis* L. II. Conversion of geranylgeranyl pyrophosphate into diterpene hydrocarbons and partial purification of cyclization enzymes. Biochemistry, $\underline{9}$, 80-89.

SHECHTER, I. and C.A. WEST (1969). Biosynthesis of gibberellins. IV. Biosynthesis of cyclic diterpenes from *trans*-geranylgeranyl pyrophosphate. J. biol. Chem., $\underline{244}$, 3200-3209.

UPPER, C.D. and C.A. WEST (1967). Biosynthesis of gibberellins. II. Enzymic cyclization of geranylgeranyl pyrophosphate to kaurene. J. biol. Chem., $\underline{242}$, 3285-3292.

PLANT GROWTH SUBSTANCES, 1970

Recent Advances in the Metabolism of Gibberellins

G. Sembdner, J. Weiland, G. Schneider, K. Schreiber and I. Focke

Institute of Plant Biochemistry, Halle/Saale, of the German Academy of Sciences at Berlin, and Institute of Cereals Research, Bernburg-Hadmersleben, of the German Academy of Agricultural Sciences at Berlin, German Democratic Republic

I. INTRODUCTION

Metabolism of gibberellins (GAs) includes biosynthesis as well as biochemical transformations, which possibly take place in connection with hormone action, translocation, deactivation, and degradation, respectively. In some of these processes reversible or irreversible formation of conjugated GAs (e.g. glucosides) seem to be important. Both higher plants and microorganisms are involved in GA metabolism. Some of these topics have been studied in our laboratory.

II. GA BIOSYNTHESIS

In biosynthetic studies some types of inhibitors are of scientific and practical value, as for instance 2-chloroethyl-trimethylammonium chloride (CCC) and related compounds. It has been shown that CCC inhibits the GA pathway between the geranylgeraniol and the tetracyclic diterpene stage both in the fungus *Gibberella fujikuroi* (Saw.) Woll. (Shechter and West, 1969) and in higher plants (Robinson and West, 1970; see also Lang, 1970). For some CCC analogs a correlation has been demonstrated between growth retarding activity and inhibition of GA biosynthesis (Harada and Lang, 1965). Despite these results there are still unanswered questions concerning the role of CCC in growth retardation. Therefore, we studied the inhibition of GA biosynthesis in *Fusarium moniliforme* Sheld. by the following hydrazonium salts, analogous to CCC which had been synthesised by König (1968) in order to get non-toxic growth retardants: 2-chloroethyl-dimethyl-hydrazonium chloride (CMH), allyldimethyl-hydrazonium chloride (AMH), and isopropyl-dimethyl-hydrazonium bromide (IMH).

Using concentrations up to 10^{-2} M the growth of fungus mycelium was not significantly affected by the inhibitors. Fig. 1 shows that the hydrazonium salts have much lower activity than CCC in inhibiting GA_3 production. The same results were obtained by measuring the GA_4/A_7 contents (Sembdner et al., 1970a).

These data are unexpected if we consider that CCC and its hydrazonium analogs are of the same potency in growth retardation in wheat (Jung, 1967; Schilling, 1970). Therefore, it may be concluded the growth retardation by CMH, AMH, and IMH may not be correlated to inhibition of GA biosynthesis, or GA biosynthesis in higher plants may be much more sensitive to the hydrazonium salts than GA biosynthesis in the fungus.

Abscisic acid (ABA) has been shown to antagonize GA biosynthesis or GA metabolism in higher plants (Wareing et al., 1968). Therefore it would be interesting to have detailed information about ABA effects on GA biosynthesis in the fungus *Fusarium moniliforme*. The results obtained (Sembdner et al., 1970a) are summarized in Fig. 2. These data suggest that ABA may inhibit GA biosynthesis only *via* inhibition of fungal growth and not by blocking any step in the GA pathway. Similar results have been published recently by Smith and Sadri (1970).

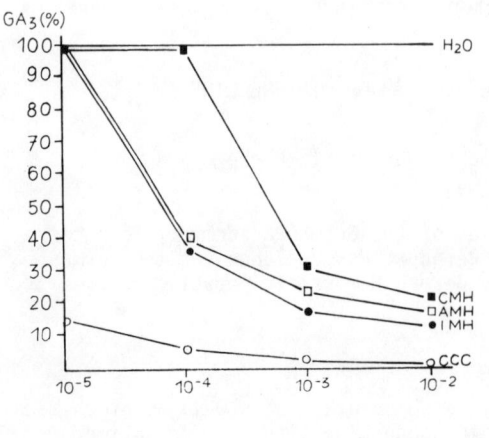

Fig. 1. Inhibition of GA biosynthesis in *Fusarium moniliforme* by hydrazonium analogs of CCC.

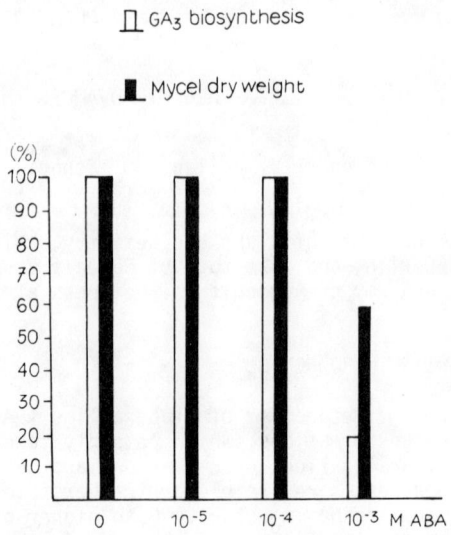

Fig. 2. Effect of ABA on growth and GA biosynthesis of *Fusarium moniliforme*.

III. GA GLYCOSYLATION

In plants GAs are not always to be found in "free" form, but they may be conjugated with other low molecular compounds or bound to macromolecules or attached to cell structures. As to the conjugated GAs, till now 5 native GA glucosides have been isolated and structurally elucidated: $O(2)$-β-D-glucosyl GA_8 from *Phaseolus coccineus* (Schreiber et al., 1967, 1970), *Althaea rosea* (Harada and Yokota, 1970), and *Pharbitis nil*; from this plant also $O(3)$-β-D-glucosyl GA_3 (Tamura et al., 1968; Yokota et al., 1969a) as well as the $O(2)$-β-D-glucosides of GA_{26}, GA_{27} (Yokota et al. 1969b), and GA_{29} (Yokota et al., 1970).

In *Phaseolus coccineus* L.cv. 'Preisgewinner' the GA_8 glucoside arises from non-conjugated GAs in maturing fruits as could be shown by feeding (^{14}C)-GA_3 and (^3H)-GA_6, respectively (Sembdner et al., 1968). During the ripening process the content of GA_8 glucoside increases and that of non-conjugated GAs decreases. Finally, in mature dry seeds only the glucoside is detectable. The question, whether GA_8 glucoside is metabolized during seed germination has been studied in the following manner: After application of (^{14}C)-GA_3 to maturing fruits of bean plants growing in the field, ripe seeds were harvested and studied with regard to the occurring radioactive GAs, one part of the seeds before and the other part after germination (daily 16 h 10000 lux, 25°C, and 8 h dark, 23°C, until root length reached 2-3 cm).

The ethyl acetate extract (mainly free GAs) of the seeds showed a low radioactivity before and a high one after germination. The butanol fraction (GA glucosides) showed reversed proportions. According to the results of thin-layer chromatography (TLC), followed by dwarf pea bio-assay the high radioactivity of the butanol fraction from ungerminated seeds is mainly to be attributed to GA_8 glucoside. In germinated seeds this glucoside had disappeared. They possessed only low radioactivity in the butanol fraction, which chiefly belongs to a less polar GA-like substance with a RF value near to GA_3. The ethyl acetate fraction of germinated seeds showed high radioactivity, the location of which on TL plates was between GA_3 and GA_6 (Sembdner et al., 1970b).

These results suggest that the GA_8 glucoside in bean seeds is reconvertible and, therefore, possesses a depot function.

The occurrence of a polar depot GA of yet unknown structure has been demonstrated in peas by Barendse et al. (1968). But the possible depot function of *Pharbitis* GA glucosides needs further investigation (Barendse et al., 1968; Reinhard and Sacher, 1967).

Further studies have been done on the glycosylation of exogenously applied GAs. In experiments with ripening bean fruits it had been shown (Sembdner et al., 1968), that exogenously applied (^{14}C)-GA_3 in part was converted to a GA_3 hexopyranoside. In order to establish its structure and to get material for detailed biological studies we repeated these investigations on a large scale by applying 4 g of unlabelled GA_3 to 33 kg of detached, maturing bean fruits (Weiland et al., 1970). After 2 days incubation the fruits were extracted with 100 l 80% methanol. The extract was evaporated *in vacuo* to an aqueous residue (7 l), which after acidifying (pH 2.8) was extracted with 14 l n-butanol (H_2O-saturated). The residue (240 g) from the butanol extract after repeated chromatography on silica gel gave a crude GA_3 glycoside fraction (22 g), one half of which was purified by TLC (n-propanol : acetic acid : ethyl acetate = 7 : 2 : 2) yielding 487 mg GA_3 glucoside ($[\alpha]_D^{24.5}$ + 30°, c = 0.45, methanol). The second part (11 g) of the crude glycoside fraction was acetylated to 744 mg GA_3 glucoside tetra-acetate (mass spectrum: M^- = 676 m/e). Hydrolysis of the glycoside with N H_2SO_4 (2 h, 100°C) yields 1 mole glucose/1 mole glucoside, identified and determined quantitatively by TLC. Though hydrolysis with β-glucosidase was very incomplete the β-configuration of the glucosidic linkage was proved by the molecular rotation difference between the glucoside ($[M]$ D(glucoside) = + 152.6°) and its aglycone ($[M]$ D(GA_3) = + 258.1°), according to the rule of Klyne (1950), as well as by the n.m.r.spectrum of the GA_3 glucoside pentaacetate (δ 4.585 ppm, d, $J_{1'-2'}$ = 7 Hz, glucose-1-H). Infrared, mass spectrometry (M^+ = 718 m/e) and n.m.r.spectroscopy (in $CDCl_3$, using hexamethyldisiloxane as an interior standard) of this pentaacetate established all structural characteristics still to question. Typical signals for the GA_3 skeleton were shown at δ = 2.69 ppm (d, J_{6-5} = 10 Hz, 6-H), 3.175 ppm (d, J_{5-6} = 10 Hz, 5-H), 5.835 ppm (dd, J_{2-1} = 9 Hz, J_{2-3} = 4 Hz, 2-H), and 6.21 ppm (d, J_{1-2} = 9 Hz, 1-H).

Thus, the structure of the GA glucoside isolated from beans after application of GA_3 has been shown to be GA_3 O(3)-β-D-glucopyranoside.

Glycosylation of exogenously applied GA was also observed in cucumber seedlings and in dwarf maize plants (Sembdner, 1969). After application of (^{14}C)-GA_3 to cotyledons of *Cucumis sativus* L. cv. 'Eva' in all seedling organs, and outside the roots in the nutrient solution, too, a radioactive polar fraction was found besides other GA metabolites. Hydrolysis with β-glucosidase of the polar compound yielded a radioactive substance soluble in ethyl acetate, which proved to be identical to GA_3 by TLC.

In *Zea mays* L. (dwarf mutant d_1), supplied with tritiated GA_4, glycosylation took place to a higher extent in light-grown plants than in dark-grown plants. The glycosidic fraction was partially soluble in ethyl acetate at pH 2.8 and yielded by enzymatic hydrolysis 2 radioactive compounds with R_F values similar to GA_4 and GA_1/A_3 in TLC. Therefore, not only exogenously GA_4 seemed to have been glycosylated but also a GA_1-like metabolite, which could be found in the seedling in the non-conjugated form, too.

The natural occurrence of GA glucosides in plants as well as the rapid glycosylation of exogenously applied GA, suggest that these glucosidic forms play a special role within the GA metabolism of plants. In order to get further information about their possible functions the biological activities of the GA_8 and GA_3 glucosides have been studied in different bioassays. Results on the potency of GA_8 glucoside have been published already by Sembdner et al. (1968) and Crozier et al. (1970). In the dwarf pea bioassay under red light conditions GA_8 glucoside is of low but significant activity of the same order as GA_8. The same results were obtained with dark grown peas which were retarded by AMO-1618. In order to study the question, whether GA_8 glucoside is hydrolysed in the plant, we supplied pea plants with condurit B epoxide, a specific inhibitor of β-glucosidase (Legler, 1966). Representative results of repeated experiments are shown in Tab. 1. From these data it may be concluded, that enzymatic hydrolysis of GA_8 glucoside in fact takes place and, therefore, measurable activity in dwarf pea bioassay of the glucoside must depend on its hydrolysis.

With regard to the potency of GA_3 glucoside only few data are yet available (Tamura et al., 1968). To get more detailed information we compared the activities of free and glucosylated GA_3 in different bioassays (Sembdner et al., 1970b).

Table 1. Effect of condurit B epoxide on the biological activity of GA_8 glucoside in the dwarf pea bioassay (cv. 'Meteor', shoot length (mm), 5d, red light, $25°C$).

gibberellin	μg / plant			
	0	0.01	1(1.5)+	10(15)+
A_3	46	105	193	198
A_8	46		63	107(a)
A_8 gluc.	46		66	98(a)
A_8 + condurit B epoxide ++			73	116(a)
A_8 gluc. + " ++			54	60(b)

+ 1,5 and 15 μg GA_8 gluc. corresponding to 1 and 10 μg GA_8, respectively.

++ 5×10^{-3}M, 50 μl/plant 5 h before, 1 h before, 24 h after and 48 h after GA application.

    a - a : not significant (> 5%)

    a - b : significant (≤ 1%)

The results, summarized in Table 2, show that GA_3 glucoside has a very low potency in most of the bioassays studied. It is highly active in the dwarf pea and dwarf rice seedlings only when applied via the roots. Leaf application gives no effect even in the dwarf pea bioassay, where the GA_8 glucoside is active. Therefore, penetration will not be limiting. Rather, it may be assumed, that plant glucosidases cannot hydrolyse GA_3 glucoside, perhaps due to stereochemical differences between O(3)- and O(2)-glucosides. These data coincide with the almost negative results on enzymatic hydrolysis in vitro using a highly active β-glucosidase. During root application, giving a high activity of GA_3 glucoside in dwarf pea and dwarf rice, hydrolysis may take place at the root surface before penetration or in the surrounding medium. Under these conditions microbial effects have not been excluded.

The studies on metabolism and biological activity of GA glucosides suggest, that glycosylation of GA leads to a deactivation, which is reversible at least with O(2)-glucosides, e g. GA_8 O(2)-β-D-glucoside. Endogenous GA glucosides, therefore, are to be assumed as deactivated stages of GA metabolism in plant, and glucosylation of exogenous GAs seems necessary for momentary deactivation, if they are applied in a superoptimal dosis.

Table 2. Biological activity of GA_3 glucoside

bioassay	relative activity (%)[+]	
	GA_3	GA_3 glucoside
Lactuca sativa L. cv. 'Grand Rapids'	100	0.1
Zea mays L. d_1 (light)	100	0.5
" " d_5 (light)	100	0.1
Pisum sativum L cv. 'Meteor' (red light)		
leaf application	100	0.05
root "	100	$\geq$ 50
Oryza sativa L. 'Tan-ginbozu'		
root application	100	40
Triticum aestivum L. cv. 'Derwisch'		
α-amylase assay	100	0.2

[+] ascertained from the graphical functions relating "log. dose " to "effect" by comparing parallel segments of the GA_3 curve and the GA_3 glucoside curve within regions of linearity.

IV. MICROBIAL TRANSFORMATION OF GA

Many years ago Brian et al. (1954) mentioned that biological activity of GA_3, kept in the soil, rapidly disappeared, probably due to microbial degradation. Riviere et al. (1966) found few strains of *Pseudomonas fluorescens* and *Xanthomonas* sp. being able to destroy GA_3. Microbial transformations of GAs are of interest not only for preparing special derivatives but also with regard to natural degradation in the soil of GA secreted by plant roots.

Investigations on these topics done in our laboratory have involved many defined strains of fungi and bacteria as well as mixed isolates from different soils, which have been studied both for GA production and for metabolizing exogenous GA_3 or GA derivatives. One part of these investigations will now be described (cf. Schneider et al. 1970). Among fungi which do not produce GA we found some strains able to transform GA_3 O(3,13)-diacetate to a more polar substance, which could not be identified by TLC. For detailed studies on this microbial transformation *Phoma medicaginis* var. *pinodella* (L.K.Jones) Boerema has been used. After growing the fungus for 3 days on a liquid medium, containing 5% glucose, 2% peptone, 0.5% yeast extract and 0.2% casamino acids, the diacetate was added (4 mg/100 ml medium) and incubated for 3 days. Extraction of altogether 800 ml culture filtrate with ethyl acetate (pH 2.8) and purification of the extract by repeated chromatography on silica gel yielded 5.2 mg of the crystallized metabolite (mp. 124-125°C). In thin-layer electrophoresis (Schneider et al., 1965) the substance proved to be a monocarbonic acid (U_F = 0.52 μ.sec^{-1}.cm.Volt^{-1}). Infrared spectroscopy (chlf.) gave evidence for the following groups: -COOH (1720 cm^{-1}), O-acetyl (1745), γ-lactone (1782),

= CH_2 (1665, 895), sec. OH (3625). The existence of a secondary hydroxy group also is required because the substance could be acetylated very easily. Mass spectrometry (M^+ = 388 m/e) gives the molecular formula $C_{21}H_{24}O_6$. Therefore, the metabolite cannot possess an additional OH-group, but seems to be deacetylated at position C-3. From these data 3 possible structures (I, II, III) may be formulated (Fig. 3). Substance II has been synthesised; it is not identical with the metabolite (infra-red spectrum, TLC), and structure III could be excluded on the basis of chemical studies, too. Therefore, the structure of the metabolite has been established to be GA_3 O(13)-acetate (I), which has been confirmed by re-acetylation of I yielding the original diacetate.

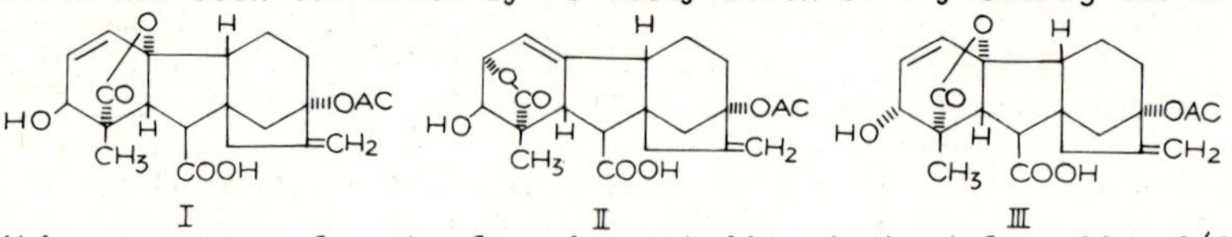

Fig. 3. Possible structures for the fungal metabolite derived from GA_3 O(3,13)-diacetate.

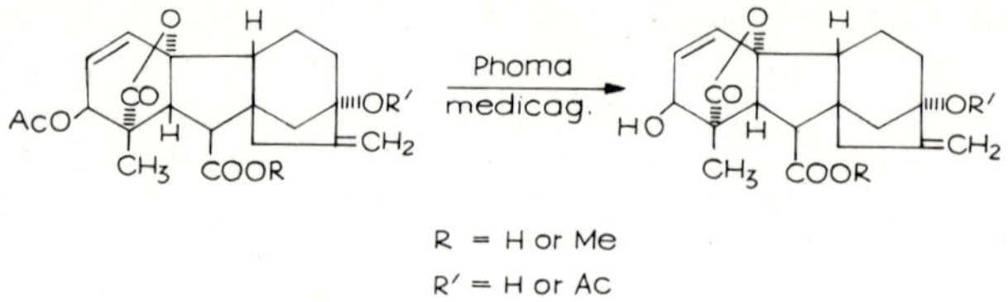

R = H or Me
R' = H or Ac

Fig. 4. Scheme of deacetylation of GA_3 acetates by the fungus *Phoma medicaginis* var. *pinodella*.

The microbial deacetylation takes place in the same manner with the methyl ester of GA_3 O(3,13)-diacetate, the GA_3 O(3)-acetate, and its methyl ester, respectively (Fig. 4). From the fact that neither the susceptible 19,10-lactone bridge nor other ester bindings are affected by the fungus, the existence of a specific enzyme may be supposed.

V. SUMMARY

The report deals with investigations on different aspects of GA metabolism.

1. Some hydrazonium analogues of CCC with the same potency in growth retardation of wheat are much less active in inhibiting GA biosynthesis in *Fusarium moniliforme* than CCC.

ABA inhibits GA biosynthesis in *F. moniliforme* only at high concentrations (10^{-3}M), which inhibit growth of the fungus, too.

2. GA_8 O(2)-β-D-glucoside, arising from non-conjugated GAs during fruit ripening in *Phaseolus coccineus* proved to be a depot GA in mature seeds, because it is reconvertable during seed germination. Glycosylation takes place also with exogenous GAs, as could be shown by application of (^{14}C)-GA_3 to bean fruits and cucumber seedlings and of (^{3}H)-GA_4 to dwarf maize plants. Glycosylated GA_3 in bean fruits has been isolated on a large scale. Its structure has been shown by chemical and physical methods to be the GA_3 O(3)-β-D-glucoside.

Glucosylation of GA_3 leads to deactivation. The biological activity of GA_8 glucoside depends on enzymatic hydrolysis. GA_3 glucoside is of high potency in dwarf pea and dwarf rice bioassay only when applied *via* the roots, at the surface of which microbial effects are not excluded.

3. Microbial deacetylation of GA_3 O(3,13)-diacetate to GA_3 O(13)-acetate has been studied in detail using the fungus *Phoma medicaginis* var. *pinodella*.

VI. REFERENCES

BARENDSE, G.W.M., H. KENDE, and A. LANG (1968). Fate of radioactive gibberellin A_1 in maturing and germinating seeds of peas and Japanese Morning Glory. Plant Physiol. 43, 815-822.

BRIAN, P.W., G.W. ELSON, H.G. HEMMING, and M. RADLEY (1954). The plant-growth-promoting properties of gibberellic acid, a metabolic product of the fungus *Gibberella fujikuroi*. J.Sci.Food Agric. 5, 602-612.

CROZIER, A., C.C. KUO, R.C. DURLEY, and R.P. PHARIS (1970). The biological activities of 26 gibberellins in nine plant bioassays. Canad.J.Bot. 48, 867-877.

HARADA, H. and A. LANG (1965). Effect of some (2-chloro-ethyl) trimethylammonium chloride analogs and other growth retardants on gibberellin biosynthesis in *Fusarium moniliforme*. Plant Physiol. 40, 176-183.

HARADA, H. and T. YOKOTA (1970). Isolation of gibberellin A_8-glucoside from shoot apices of *Althaea rosea*. Planta, 92, 100-104.

JUNG, J. (1967). Wachstumsregulierende Wirkung von Hydrazoniumverbindungen bei Weizen. Z. Acker- und Pflanzenbau 125, 124-129.

KLYNE, W. (1950). The configuration of the anomeric carbon atoms in some cardiac glycosides. Biochem.J. 47, XLI - XLII.

KÖNIG, K.H. (1968). N,N-Dimethylhydraziniumsalze als neue Wachstumsregulatoren. Naturwissenschaften 55, 217-219.

LANG, A. (1970). Gibberellins: Structure and metabolism. Annu. Rev. Plant Physiol. 21, 537-570.

LEGLER, G. (1966). Untersuchungen zum Wirkungsmechanismus glykosidspaltender Enzyme, I. Darstellung und Eigenschaften spezifischer Inaktivatoren. Hoppe-Seyler's Z. physiol. Chem. 345, 197-214.

REINHARD, E und R. SACHER (1967). Versuche zum enzymatischen Abbau der "gebundenen" Gibberelline von *Pharbitis purpurea*. Experientia, Basel 23, 415-416.

RIVIÈRE, J., P. LABOUREUR, et M. SÉCHET (1966). Dégradation microbienne de l'acide indole-3-acétique et de la gibbérelline A_3 dans le sol. Ann. Physiol. vég. 8, 209-221.

ROBINSON, D.R. and C.A. WEST (1970). Biosynthesis of cyclic diterpenes in extracts from seedlings of *Ricinus communis* L. II. Conversion of geranylgeranyl pyrophosphate into diterpene hydrocarbons and partial purification of the cyclization enzymes. Biochemistry 9, 80-89.

SCHILLING, G. (1970). Unpublished results.

SCHNEIDER, G., G. SEMBDNER, I. FOCKE und K. SCHREIBER (1970). Mikrobielle Umwandlung von Gibberellinen - Entacetylierung von 0(3,13)-Diacetylgibberellin-säure. Phytochemistry, in preparation.

SCHNEIDER, G., G. SEMBDNER und K. SCHREIBER (1965). Gibberelline VI. Mitt. Die Dunnschichtelektrophorese von Gibberellinen. J. Chromatogr., Amsterdam 19, 358-363.

SCHREIBER, K., J. WEILAND und G. SEMBDNER (1967). Isolierung und Struktur eines Gibberellinglucosids. - Tetrahedron Letters, London, 4285-4288.

SCHREIBER, K., J. WEILAND und G. SEMBDNER (1970). Isolierung von Gibberellin-A_8-0(3)-β-D-glucopyranosid aus Fruchten von *Phaseolus coccineus*. Phytochemistry 9, 189-198.

SEMBDNER, G. (1969). Untersuchungen uber Vorkommen, Stoffwechsel und biologische Wirksamkeit von Gibberellinen. Habilitationsschrift, Halle/Saale.

SEMBDNER, G., I. FOCKE und K. SCHREIBER (1970a). Wirkung von Abscisinsaure und einigen CCC-Analoga auf die Gibberellinbildung von *Fusarium moniliforme* Sheld. Biochem. Physiol. Pflanzen, in preparation.

SEMBDNER, G., J. WEILAND, O. AURICH, and K. SCHREIBER (1968). Isolation, structure, and metabolism of a gibberellin glucoside, in "Plant Growth Regulators, p. 70-86. S.C.I. Monograph No. 31, London. 1968.

SEMBDNER, G., J. WEILAND und K. SCHREIBER (1970b). Gibberellin Glucoside. - Metabolismus, Funktion und biologische Aktivitat von 0(3)-Gibberellin-A_3-glucosid und 0(2)-Gibberellin-A_8-glucosid. Phytochemistry, in preparation.

SHECHTER, I. and C.A. WEST (1969). Biosynthesis of gibberellins. IV. Biosynthesis of cyclic diterpenes from transgeranylgeranylpyrophosphate. J. biol. Chemistry 244, 3200-3209.
SMITH, O.E. and H.A. SADRI (1970). Effect of abscisic acid on gibberellin biosynthesis in *Fusarium moniliforme*. Plant Cell Physiol. 11, 345-348.
TAMURA, S., N. TAKAHASHI, T. YOKOTA, and N. MUROFUSHI (1968). Isolation of water soluble gibberellins from immature seeds of *Pharbitis nil*. Planta 78, 208-212.
WAREING, P.F., J. GOOD, and J. MANUEL (1968). Some possible physiological roles of abscisic acid, in "Biochemistry and Physiology of Plant Growth Substances" (Ed. F. Wightman and G. Setterfield) p. 1561-1579. The Runge Press, Ottawa. 1968.
WEILAND, J., G. SEMBDNER, O. AURICH und K. SCHREIBER (1970). Gibberellin-Glucoside, - Glucosylierung von GA_3 und GA_8 in *Phaseolus coccineus*. Phytochemistry, in preparation.
YOKOTA, T., N. TAKAHASHI, N. MUROFUSHI, and S. TAMURA (1969a). Structures of new gibberellin-glucosides in immature seeds of *Pharbitis nil*. - Tetrahedron Letters, London, 2081-2084.
YOKOTA, T., N. TAKAHASHI, N. MUROFUSHI, and S. TAMURA (1969b). Isolation of gibberellins A_{26} and A_{27} and their glucosides from immature seeds of *Pharbitis nil*. Planta 87, 180-184.
YOKOTA, T., N. MUROFUSHI, and N. TAKAHASHI (1970). Structure of new gibberellin glucoside in immature seeds of *Pharbitis nil*. Tetrahedron Letters, London, 1489-1491.

PLANT GROWTH SUBSTANCES, 1970

THE BIOSYNTHESIS OF GIBBERELLIN PRECURSORS IN A CELL-FREE SYSTEM FROM *Cucurbita Pepo* L.

Jan E. Graebe

Pflanzenphysiologisches Institut der Universität, Untere Karspüle 2,
Göttingen, Germany

I. INTRODUCTION

Although many steps in the gibberellin biosynthesis have been elucidated, others still are unknown (see reviews by Cross, 1968 and Lang, 1970). Using the fungus *Gibberella fujikuroi* and a cell-free system from *Echinocystis macrocarpa* it has been shown that gibberellins are biosynthesized along the common isoprenoid pathway from mevalonic acid over phosphorylated intermediates to the diterpenoid cyclic hydrocarbon (-)-kaur-16-ene (kaurene), which is further transformed into gibberellins by oxidations and a rearrangement of the ring structure. The first intermediates in the transformation of kaurene are known to be (-)-kaur-16-en-19-ol (kaurenol), (-)-kaur-16-en-19-al (kaurenal), (-)-kaur-16-en-19-oic acid (kaurenoic acid), and 7β-hydroxy-(-)-kaur-16-en-19-oic acid (7β-hydroxykaurenoic acid) (Fig. 1).

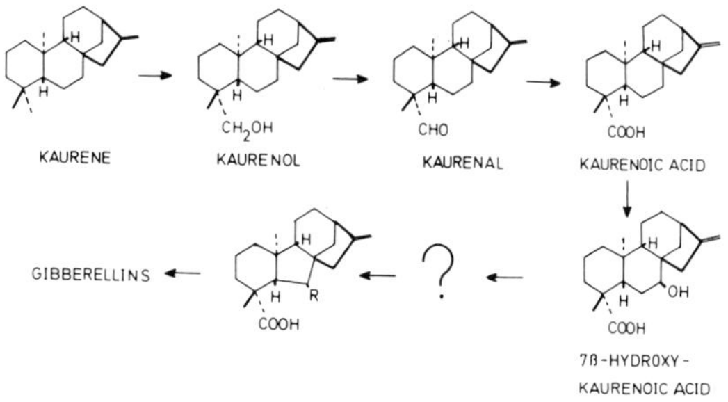

Fig. 1. Steps in the conversion of kaurene to gibberellins.

The latter is the last product to have been identified in the *Echinocystis* system (Murphy and West, 1969). After its formation, a contraction of ring B, a loss of the C-20 methyl group, and the formation of the characteristic lactone group must take place. The sequence of these events is largely unknown although several compounds have been obtained from *G. fujikuroi* that most likely are intermediates in the pathway following the ring contraction.

The purpose of the present work is to obtain a cell-free system in which the knowledge of the pathway can be extended and which produces labelled gibberellin precursors in sufficient amounts for further experimentation with other biological objects. An enzyme system of this kind has been found in immature seeds of *Cucurbita pepo*.

II. MATERIALS AND METHODS

Fruits of *Cucurbita pepo* L. were harvested when they were fully grown but not yet ripe. The endosperm was obtained by cutting off the tip of each seed and squeezing. It was gently homogenized in a glass homogenizer ground to a suitable clearance. The homogenate (2 - 20 ml) was centrifuged at 20,000 xg for 15 min, dialyzed 3 h against 3 one-litre changes of potassium phosphate (0.05 M, pH 7.6) with $MgCl_2$ (2.5 mM), and centrifuged again. The opalescent supernatant fluid was divided into aliquots and frozen in liquid nitrogen. The activity remained unchanged for at least two years.

Incubation mixtures (0.2 ml) contained $2\text{-}^{14}C$-DL-mevalonate (195,000 count/min, 5.85 mc/mmole), potassium phosphate (0.05 M, pH 7.6), $MgCl_2$ (5 mM), $MnCl_2$ (1 mM), ATP (5 mM), phosphoenol pyruvate (10 mM), NADPH (0.5 mM), and 0.15 ml endosperm preparation. The incubation in air at 30° lasted 2 h.

The reaction was stopped by addition of N HCl (0.05 ml), acetone (1 ml), and ethyl acetate (1 ml). After separation of the phases with water (0.8 ml), the organic phase was removed, and the water phase extracted twice more with ethyl acetate (0.5 ml). The pooled organic phases were washed with water (2 x 0.5 ml) and evaporated under nitrogen. The extracts were subjected to multiple development TLC on silica gel G: Benzene to 15 cm; petroleum ether/benzene (100/0.5) to 18 cm; benzene/ethyl acetate to kaurenal (approximately 12 cm); $CHCl_3$/ethyl acetate/acetic acid (85/15/1) two times to kaurenol (approximately 8 cm); same solvent system (70/30/1) to kaurenoic acid (approximately 5.5 cm).

III. RESULTS

When preparations of *Cucurbita* endosperm were incubated with $2\text{-}^{14}C$-mevalonate, Mg^{++}, Mn^{++}, ATP and regenerator (see "Methods"), the main radioactive product was kaurene. In addition, diterpenoid alcohols, mainly geranylgeraniol, and an unidentified product were obtained (Fig. 2A). Optimisation of the incubation conditions led to a very high activity. As an example, a 10-ml incubation containing 7.5 ml of endosperm and 13.5×10^6 count/min of racemic $2\text{-}^{14}C$-mevalonate yielded 3.4×10^6 count/min of kaurene and 0.9×10^6 count/min of other petroleum ether-soluble material. This represents a 25 per cent conversion of the total label to kaurene or 50 per cent of the active isomer.

When the incubations were performed in the presence of NADPH, the label from mevalonate became incorporated into more polar material. The main products were acid in character and more polar than kaurenoic acid (Fig. 2B). Only very little material with the chromatographic properties of kaurene was obtained in such incubations. Therefore the question arose whether the unidentified acids were formed from kaurene or whether they represented a diversion of the pathway before kaurene, thus preventing the latter from being formed.

Direct evidence that the unidentified acids lie after kaurene in the biosynthetic sequence was obtained by incorporation of ^{14}C-kaurene into the acid material. For this purpose radioactive kaurene was generated in the absence of pyridine nucleotides, isolated, and fed back to the complete system (omitting $2\text{-}^{14}C$-mevalonate). As shown in the first line of table 1, kaurene was converted to the acids very efficiently, 66 per cent of the label turning up in them. Since the acids obtained from kaurene showed the same chromatographic properties as those obtained from mevalonate, it was concluded that they were the same.

In working with dialyzed extracts, an absolute requirement for ATP in the conversion of kaurene to the acids was discovered. Table 1 shows the effects of ATP and phosphoenol pyruvate (PEP) on the system with ^{14}C-kaurene as a precursor. In the absence of ATP no acids were formed, kaurenol and what is presently believed to be kaurenal accumulated, and kaurene was much less utilized than in the complete system. The interpretation of these results, assuming that the sequence in Fig. 1 operates in the system, is that each step after kaurenal has a requirement for ATP. Table 1 also shows that in the presence of ATP but absence of PEP, the more polar acids (ref. Fig. 2B) did not form to any greater extent, while the less polar acids, kaurenoic acid, and the material believed to be kaurenal accumulated. Kaurenol did not accumulate as much as in the absence of ATP and kaurene was better utilized.

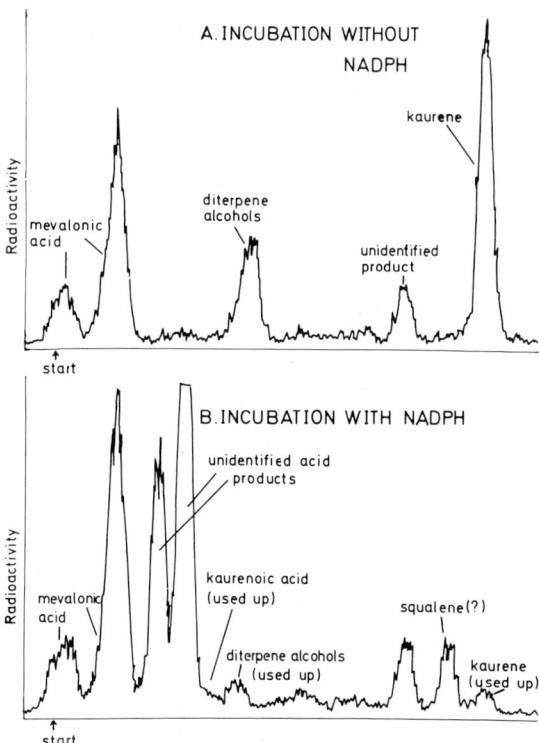

Fig. 2. Multiple development TLC of products obtained from 2-^{14}C-mevalonate with *Cucurbita* endosperm preparations. Incubations with and without NADPH as described under "Methods".

Table 1. Requirement for ATP in the conversion of ^{14}C-kaurene to acids

Incubation mixture[1]	Kaurene recovered (cpm)	Incorporation into				
		Kaurenol (cpm)	Kaurenal? (cpm)	Kaurenoic acid (cpm)	Acids[2] (cpm)	Acids[3] (cpm)
Complete	19,176	297	1,818	4,035	62,445	25,455
ATP omitted	65,734	21,731	14,960	1,261	2,651	578
PEP omitted	40,761	5,794	29,015	22,167	29,444	3,498
ATP and PEP omitted	88,441	20,270	12,983	2,029	2,946	433

[1] With 133,000 cpm kaurene. [2] Less polar. [3] More polar.

The known function of PEP is to regenerate ATP from ADP in the presence of pyruvate kinase, which apparently is present in the crude preparation. Apparently ATP becomes limiting in the absence of PEP, which indicates a very high requirement for ATP in the reactions, since it was present at an initial concentration of 5 mM. The effect of omitting both ATP and PEP was the same as the effect of omitting ATP only. This shows that PEP did not have an effect of its own, but only in conjunction with ATP, which confirms its role as regenerator. Since kaurenoic acid accumulates when

ATP becomes limiting but the unidentified acids when it is not limiting, it is believed that the latter lie after kaurenoic acid in the biosynthetic sequence. For the same reason, it is believed that the more polar acids are formed after the less polar ones.

Further evidence that the unidentified acids are formed *via* kaurenoic acid was obtained by the addition of unlabelled kaurenoic acid to the incubation mixture before incubation. In such mixtures there was labelling of kaurenoic acid, kaurenol, the product migrating like kaurenal, and kaurene but much less of the unidentified acids (Fig. 3). Apparently the added unlabelled kaurenoic acid both diluted the newly formed radioactive kaurenoic acid and backed up the reactions so that precursors accumulated. A direct incorporation of kaurenoic acid has not yet been attempted.

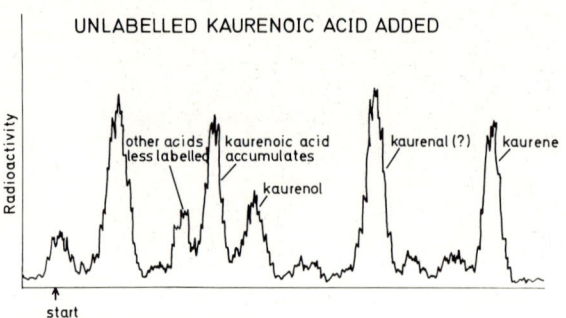

Fig. 3. Trapping of kaurenoic acid formed from $2-^{14}C$-mevalonate by addition of unlabelled kaurenoic acid to the incubation.

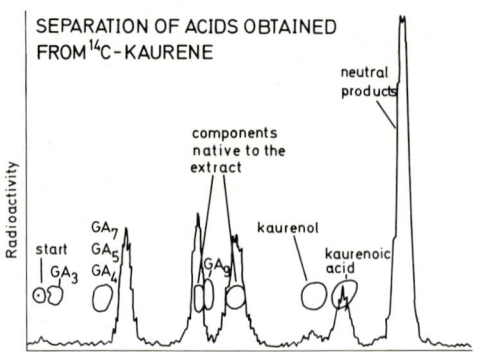

Fig. 4. Chromatographic separation of acids obtained from ^{14}C-kaurene. TLC solvent system: Chloroform/ethyl acetate/acetic acid (70/30/1).

A further separation of the unidentified acids obtained from ^{14}C-kaurene is shown in Fig. 4. There are at least three main components, which are all more polar than kaurenoic acid and therefore probably represent higher stages of oxidation. One component is somewhat less, one is slightly more polar than GA_9. The dialyzed endosperm preparations contain materials that have identical chromatographic properties with these two labelled fractions. Sprayed with anis aldehyde reagent, the less polar fraction becomes salmon pink, similar to kaurenoic acid, whereas the more polar component becomes dark blue, similar to the color of GA_9 or GA_7. The third labelled fraction, which is the most polar, does not correspond to any material present in sufficiently high concentration in the dialyzed preparations to give color with the anis aldehyde reagent. Several chromatograms have indicated the presence of at least one smaller component migrating approximately like GA_9, and some extracts incorporated label into material which was somewhat more polar than GA_3.

Extracts from different fruits differed in activity. Almost all were efficient in making kaurene, but only some got beyond this stage, and relatively few showed full activity with respect to their ability to incorporate label into the unidentified acids. An attempt was made to correlate this activity with the developmental stages of the seeds. During the maturation of a *Cucurbita* seed, the cotyledons of the embryo start growing only after the seed itself has reached its full length (Fig. 5). The seeds of a single fruit are all at the same stage. The length of the cotyledons expressed as per cent of the length of the seed lumen as measured on a sample therefore

Fig. 5. Developmental stages of *Cucurbita* seeds proceeding from upper left to lower right. Longitudinal cuts. Line = 1 cm.

provides a reasonable index of the developmental stage of the seeds. The results from testing the activity of seeds from 29 fruits showed that although the correlation was not perfect, most active preparations originated from seeds having cotyledons 13 - 46 per cent of the seed lumen (table 2). High or even medium activities were seldom observed in preparations from seeds at developmental stages outside of this range. Unfortunately there are no known means of determining the stage of development without opening the fruit.

Table 2. Activity ratings with respect to acid production of preparations obtained from seeds at different development stages

Cotyledon development (per cent of seed length)	Number of preparations with		
	High activity	Medium activity	Low activity
0 - 12	1	1	9
13 - 46	4	4	2
47 -100	0	1	7

IV. DISCUSSION

The cell-free system from *Cucurbita* shows a high incorporation of $2-^{14}C$-mevalonic acid or ^{14}C-kaurene into the known gibberellin precursors kaurenol, kaurenoic acid, and a substance preliminarily identified as kaurenal. The main products under certain conditions are acids that are believed to follow kaurenoic acid in the biosynthetic sequence. The experiments that would prove conclusively whether these acids are formed from kaurenoic acid and whether they are precursors of gibberellins have not yet been performed. If they are indeed precursors of gibberellins, their number and position on thin-layer chromatograms indicate that the system may be active past the stage of known precursors and perhaps also past the stage of ring contraction (Fig. 1).

The requirement for ATP in the formation of acids from kaurene was unexpected insofar as both kaurenoic acid and 7β-hydroxykaurenoic acid has been obtained without added ATP in the *Echinocystis* system (Dennis and West, 1967; Murphy and West, 1969). Although a stimulation of conversion of kaurene and kaurenal was noted by Murphy and West (1969), they obtained high conversion without added ATP even when working with buffer-washed microsomes. The activity can therefore hardly be due to endogenous ATP in their system. This difference between the systems from *Echinocystis* and *Cucurbita* remains unexplained.

In contrast to the formation of the first acid intermediates, an ATP-requirement might be predicted for the ring contraction, since a pyrophosphorylated derivative would be a likely starting material for this reaction. This was also pointed out by Cross, Stewart, and Stoddart (1970), who found that 6β,7β-dihydroxykaurenoic acid did not get incorporated into gibberellic acid in *G. fujikuroi* although it showed gibberellin activity in a number of biological assays. They concluded that 6β,7β-dihydroxykaurenoic acid, once it had left the enzyme surface, could not undergo ring contraction via a phosphorylated intermediate. An alternative explanation would be a requirement for phosphorylation at a lower oxidation level and therefore earlier in the pathway.

The appreciation of the ATP-requirement has led to enough accumulation of the unknown acids to perform an analysis, which is pending. The activity of the system is high enough to produce gibberellin precursors for further experimentation and sufficiently characterized to enable any one of the precursors to be obtained as the main component.

V. SUMMARY

A cell-free system from immature seeds of *Cucurbita pepo* L. incorporates the label from 2-^{14}C-mevalonate into several substances, among which the known gibberellin precursors kaurene, kaurenol, and kaurenoic acid have been identified. In addition a substance believed to be kaurenal and at least three unidentified acids are produced in the system. These acids are also obtained from ^{14}C-kaurene and some evidence is presented that they form via kaurenoic acid. It has not yet been established whether they play a role as gibberellin precursors. The degree of conversion is very large and the system has been characterized so that different products can be obtained by varying the incubation mixture. Thus kaurene is obtained as the main product from mevalonate in the absence of pyridine nucleotides. Kaurenol and kaurenal (tentative identification) are obtained if ^{14}C-kaurene is incubated with NADPH but without ATP. Kaurenal, kaurenoic acid, and one of the fractions of unidentified acids are obtained from kaurene with NADPH and limiting amounts of ATP, whereas the unidentified acids are main products in the presence of NADPH, ATP, and regenerator. An absolute requirement for ATP in the production of kaurenoic acid and the unidentified acids was found.

VI. ACKNOWLEDGEMENTS

This work was supported by the Deutsche Forschungsgemeinschaft.

VII. REFERENCES

CROSS, B.E. (1968). Biosynthesis of the gibberellins. Progr. Phytochem. 1, 195-222.
CROSS, B.E., J.C. STEWART, and J.L. STODDART (1970). 6β,7β-dihydroxykaurenoic acid: Its biological activity and possible role in the biosynthesis of gibberellic acid. Phytochem. 9, 1065-1071.
DENNIS, DT. and C.A. WEST (1967), Biosynthesis of gibberellins III. The conversion of (-)-kaurene to (-)-kauren-19-oic acid in endosperm of *Echinocystis macrocarpa* Greene. J. Biol. Chem. 242, 3293-3300.

LANG, A. (1970). Gibberellins: Structure and metabolism. Ann. Rev. Plant Physiol. 21, 537-570.

MURPHY, P.J. and C.A. WEST (1969). The role of mixed function oxidases in kaurene metabolism in *Echinocystis macrocarpa* Greene endosperm. Arch. Biochem. Biophys. 133, 395-407.

A Polar Gibberellin from Apricot Seed

B.G. Coombe and M.E. Tate

Departments of Plant Physiology and Agricultural Biochemistry,
Waite Agricultural Research Institute, The University of Adelaide, Glen Osmond,
South Australia, 5064

I. INTRODUCTION

Polar gibberellins were unrecognised during the early work on gibberellins, but in the last decade several workers have described substances which cannot be extracted from acidified aqueous solutions with ethyl acetate, but which can be with n-butanol. An example is the gibberellin-like substance found in apricot fruit (Jackson and Coombe, 1966). Until recently, these compounds have all proved to be glucosides and have been classed as conjugated gibberellins (Lang, 1970).

In this paper we describe the purification of a free (unconjugated) polar gibberellin from seed of apricot, *Prunus armeniaca* L., and its identity with the recently-described GA_{32} found in peach seed (Yamaguchi *et al.*, 1970).

II. METHODS

Bioassay

The presence of gibberellin activity was detected using the barley endosperm bioassay. In the absence of suitable alternatives the general procedure was to purify to maximal biological potency.

The bioassay was used as published (Coombe, Cohen and Paleg, 1967) except that the endosperm was incubated for 42 hours and the response was measured by the increase in optical refraction of the ambient solution. The incubates were diluted 10-fold with water and the solution transferred to a Waters Differential Refractometer and read digitally. Filter cups were inserted only when solid material such as silica gel was present.

These changes improved the bioassay's precision and speed. GA_3 equivalents were determined by assaying GA_3 and unknowns at three levels of dilution beginning with aliquots of 1/1000th. to 1/10,000th. Concurrent tests without endosperm were made to assess background refraction although these were necessary only during the early stages of purification.

Extraction and purification

Apricot seed (15 kg fresh wt., harvested 27 days after anthesis and stored frozen) was macerated in the cold with 10 l water containing $0.5MNa_2SO_4$, $1 mMNa_2S_2O_5$ and sufficient H_2SO_4 to make the pH 2.5. The mixture was further blended with 12 l n-butanol, centrifuged and the butanol layer removed. The aqueous layer was re-extracted with a further two 9 l of butanol and the combined butanol extracts partitioned with 10 l water adjusted to pH 10 with NaOH.

After acidification by exchanging with Amberlite IR120-H the aqueous solution was adsorbed on to a 300 g DEAE Sephadex column and the activity was eluted with a water (2 l)- 2N acetic acid (2 l) linear gradient.

The active fractions were rotary film evaporated, streaked on to silica gel H TLC plates and developed with solvent mixture C, (Fig. 1.). Strips were scraped, eluted with MeOH and the activity measured. Active zones (at Rf 0.1) were further chromatographed by the same method except that the plates were developed three times. Finally, the active zones (at Rf 0.3) from these plates were applied to a single 20 cm plate and developed in solvent mixture A, Fig. 1. The active zone, at Rf 0.5, was eluted, applied to acid-washed Whatman 3MM paper and developed in a descending direction with iso-propanol: NH_4OH : H_2O : 10 : 1 : 1. The most active eluate when evaporated yielded semicrystalline apricot gibberellin (AG) as the ammonium salt. It weighed 230 µg and its activity was equivalent to 600 µg GA_3 i.e. it was 2½ times as potent as GA_3.

The reduction in weight and the increase in potency achieved at each step are summarised in Table 1.

Table 1. Changes in dry weight and biological potency during the purification of polar gibberellin from 15 kg. fresh weight of apricot seed, 1968-69 season

Purification stage	Dry Wt. mg.	GA_3 equivs. mg.	Recovery %	Potency mg. GA_3 equivs. per mg. dry wt.
Starting material	2×10^6	-	-	-
n-BuOH partition	45,000	15	100	0.0003
DEAE Sephadex	135	5	33	0.037
T.L.C.	4.2	1.2	8	0.28
Paper chromatography	0.23	0.6	4	2.6

III. PROPERTIES AND IDENTIFICATION OF APRICOT GIBBERELLIN

Melting point, solubility and spot tests

The NH_4 salt did not melt but gradually decomposed between 250 and 330°C, the maximum temperature tested. It was very soluble in water and methanol, moderately soluble in n-BuOH, sparingly soluble in acetone and relatively insoluble in ethyl acetate and less polar solvents. One microgram portions showed no reaction to silver nitrate, ninhydrin, phosphomolybdate, iodine vapour, anthrone, α-naphthol and antimony trichloride reagents. The compound did react with 0.5% (w/v) $KMnO_4$ spray giving a brown colour after 30 min. Unlike known gibberellins and gibberellin-glucosides it did not fluoresce on silica gel plates after spraying with ethanolic-H_2SO_4 and heating; in amounts greater than 5 µg, AG reacted to this test by giving a light tan absorbance under UV light, barely visible under white light.

Thin layer chromatography

Thin layer plates, 5 x 20 cm, coated with 0.25 mm silica gel H, were spotted with 0.1 and 15 µg AG and with a standard mixture containing 10 µg of glucose, GA_3, GA_{4-7}, and in one instance, GA_8. These were developed in four solvent mixtures described in Fig. 1, dried, and the 0.1 µg AG strip divided into 15 or 20 sections which were scraped into vials and bioassayed. The rest of the gel was sprayed with ethanolic-H_2SO_4 and heated to 110° for 10 min.

The results (Fig. 1) show that the biological activity was confined to a single zone which corresponded in position to a UV absorbant spot on the 15 µg AG strip. They also show the mobility of AG relative to glucose and known gibberellins. This is especially clear in Fig. 1C: from a comparison of mobility and the number of hydroxyls in GA_{4-7}, GA_3, and GA_8 (1, 2 and 3 respectively) we concluded that AG was likely to have more than three hydroxyls.

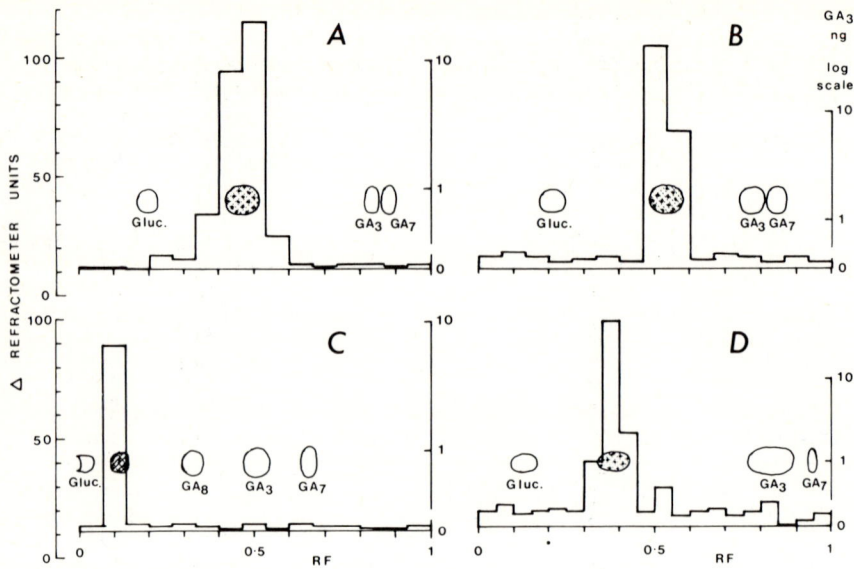

Fig. 1. Thin layer (silica gel H) chromatography of gibberellins developed in four solvent systems: A. CHCl3: Ethyl acetate: iso-propanol: HAc :: 4 : 6 : 9 : 1. B. n-BuOH : HAc : H2O :: 25 : 6 : 25 (upper phase). C. CHCl3 : EtOH : HCOOH :: 85 : 15 : 1. D. CHCl3 : MeOH : HAc : H2O :: 45 : 15 : 3 : 2. Each figure is a composite of the bioassay results from the 0.1 µg strip and of the spots visible under UV light after spraying with EtOH-H2SO4 and heating. The cross-hatched areas represent AG.

Electrophoresis

AG, together with GA and fructose markers, was applied to acid-washed 3MM paper and run in a flat-bed electrophoresis unit using three different buffer systems: formic acid at pH 2.0, ammonium bicarbonate at pH 8.1 and ammonium borate at pH 9.3. The biological activity ran with GA3 marker in the first two buffers (Fig. 2). In borate, however, the activity moved beyond GA3 to a position opposite GA8. The simplest explanation for this data is that AG is a monocarboxylic acid which has a vicinal glycol capable of forming a borate complex.

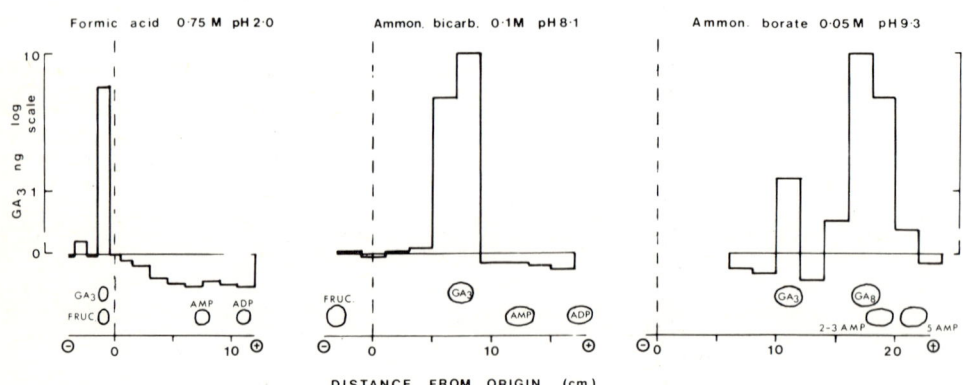

Fig. 2. Paper electrophoresis of gibberellins on a flat bed apparatus at 2000 V for 1 hr using the three buffers indicated. The AG strip was cut into appropriate lengths and bioassayed; the histograms are drawn so that the responses to 0 and 10 ng GA3 are equal. Marker spots were located under UV light and after spraying with KMnO4 solution.

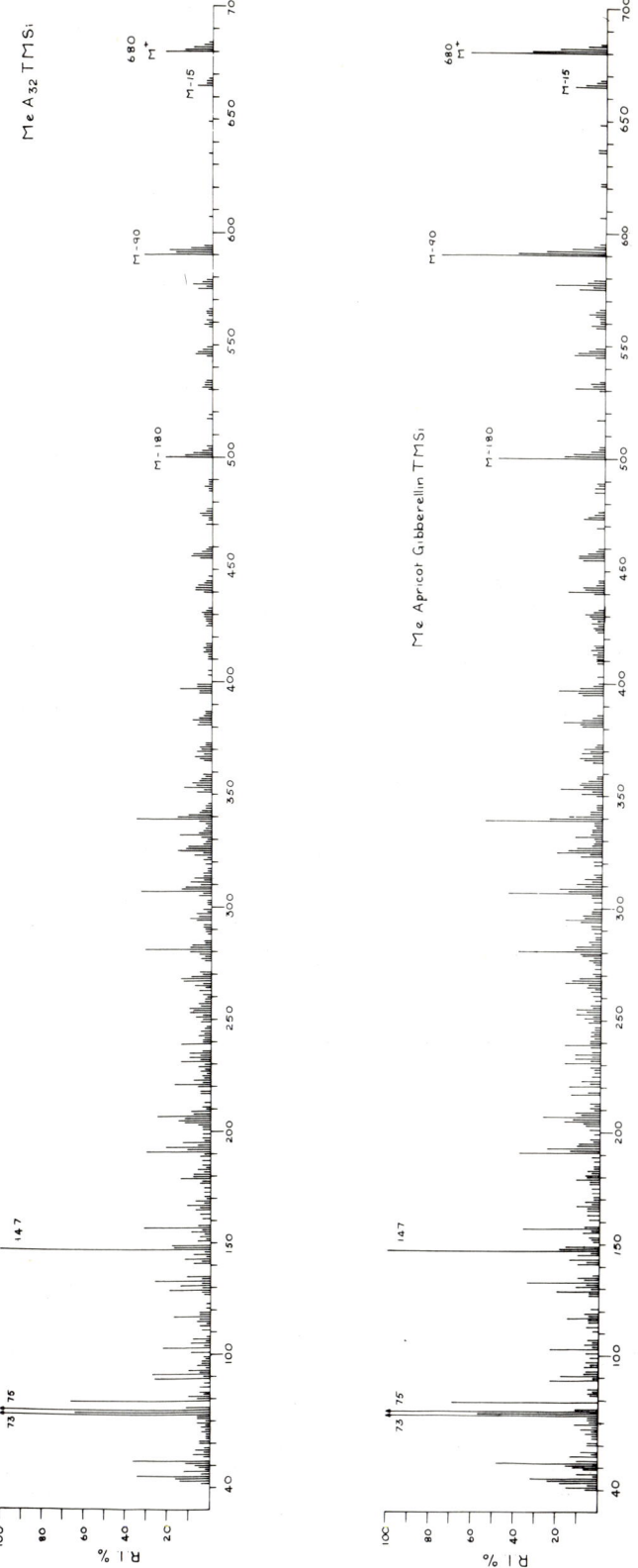

Fig. 5. Line diagrams of the mass spectra of the methyl ester-TMS ether of GA_{32} and apricot gibberellin. Fig. 6. GA_{32}.

Infra-red spectroscopy

A MeOH solution of the 230 μg NH₄ AG was evaporated onto the surface of a pressed KCl disc, and the disc was again pressed. A similar disc was made using a slightly larger amount of GA_3. Segments of both IR spectra are compared in Fig. 3.

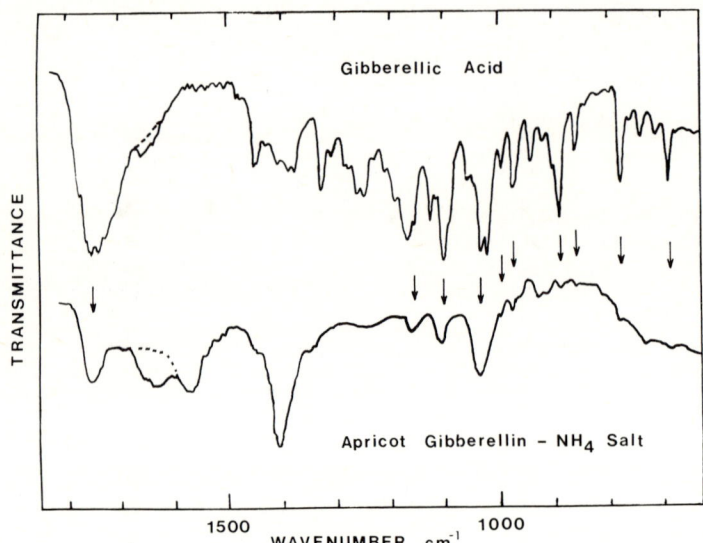

Fig. 3. Infra-red spectra of NH₄AG and GA_3 applied to the surface of pressed KCl discs. The broken line represents the change after moistening the disc with D_2O.

The spectrum of AG is weak but it clearly demonstrates similarities to that of GA_3 in the fingerprint region. The large signals in the AG spectrum at 1575 and 1400 cm^{-1} are attributable to the ammonium ion.

The physical and chemical properties of the compound are obviously consistent with a non-conjugated polyhydroxy gibberellin nucleus. TMS derivatives enable separation by GLC with subsequent recovery of biological activity in fractions which are collected, since the TMS ethers and esters are readily hydrolysed (Davis, Heinz and Addicott, 1968).

The physical and chemical properties of the compound are obviously consistent with a non-conjugated polyhydroxy gibberellin nucleus.

Gas liquid Chromatography

We tested the GLC of the trimethylsilyl (TMS) ester/ether of AG to prepare for GLC - mass spectrometry and to provide an alternative to bioassay for locating the compound. TMS derivatives enable separation by GLC with subsequent recovery of biological activity in fractions which are collected, since the TMS ethers and esters are readily hydrolysed (Davis, Heinz and Addicott, 1968).

NH₄ AG was converted to the free acid using Amberlite IR-120 H resin and, after thorough drying, was heated with bis-(trimethyl silyl) acetamide (BSA) at 60° for 2 days. The derivative was injected on to the GLC glass column using excess BSA as the solvent. Columns with OV17 liquid phase were more discriminating than QF_1 or SE30.

The results of three consecutive GLC traces are illustrated in Fig. 4. TMS derivatives of GA_4, GA_7 and GA_3 were chromatographed first and a portion of that trace is inserted in this diagram. TMS-AG was then chromatographed and the full trace is shown. From this were selected the limits of appropriate fractions which were collected from a repeat injection of TMS-AG. Fractions of the column effluent were collected in tubes packed with glass wool and moistened with aqueous ethanol.

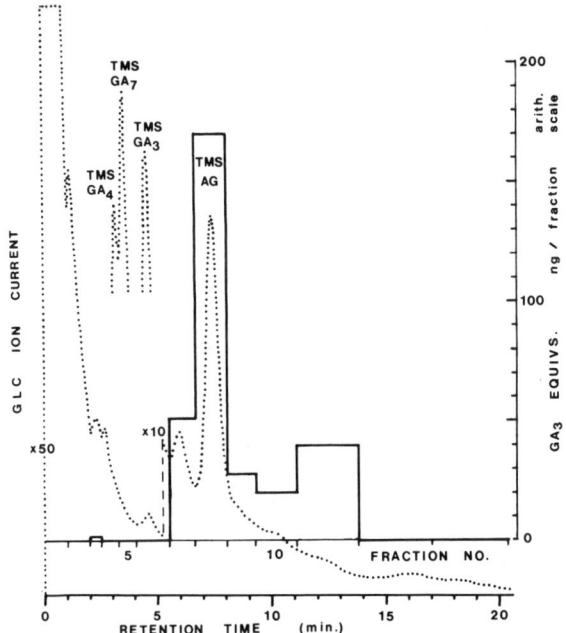

Fig. 4. Gas liquid chromatography traces (dotted lines) of trimethyl silyl ether/esters of an apricot gibberellin extract run on a 2m x 4mm glass column packed with 1% OV 17 on Gaschrom Q (100-120 mesh) at 230°. A trace of GA standards is shown as an insert to demonstrate relative retention times. The solid line histograms represent biological activity expressed as GA_3 equivalents and plotted arithmetically.

After elution with 2 ml of solvent, aliquots of 200, 20 and 2 µl were bioassayed. The GA_3 equivalents so obtained are plotted in Fig. 4 on an arithmetic scale.

The results show that the trimethylsilyl ester/ether of AG was successfully gas chromatographed and the biological activity of collected fractions was established. The activity shown after the main peak could have been caused by TMS-AG less 1 or more TMS units. The active eluate was chromatographed on TLC and bioassayed and found to consist of several non-polar compounds; after hydrolysis in water for 10 days, however, activity was found at the Rf of AG.

It was then possible to supplement bioassay data with a reasonably sensitive physical measure. Similar GLC traces were obtained with the methyl ester/trimethylsilyl ether of AG, though with shorter retention time. Since methylation reduces biological activity 300-fold, fractions were not collected for bioassay.

Mass spectrometry

Our attempts to obtain satisfactory mass spectra by direct insertion into the mass spectrometer of TMS derivatives were unsuccessful, probably due to loss of some TMS units during transfer. Recently Dr. MacMillan of the University of Bristol has informed us that a sample of apricot gibberellin we had sent him has GLC retention times and mass spectra which are identical with those of a sample of GA_{32} supplied by Prof. Takahashi, University of Tokyo. This is illustrated in the mass spectra of the methyl esters/TMS ethers of GA_{32} and AG obtained by combined gas chromatography - mass spectrometry (Fig. 5, kindly supplied by Dr. MacMillan). Similar identity is shown by the TMS ester/ethers. Thus apricot gibberellin is GA_{32}.

The structure of GA_{32}, (Fig. 6) isolated from immature seeds of peach (*Prunus persica* L.) has been elucidated by elegant nuclear magnetic resonance and mass spectrometric studies published by Yamaguchi *et al.* (1970).

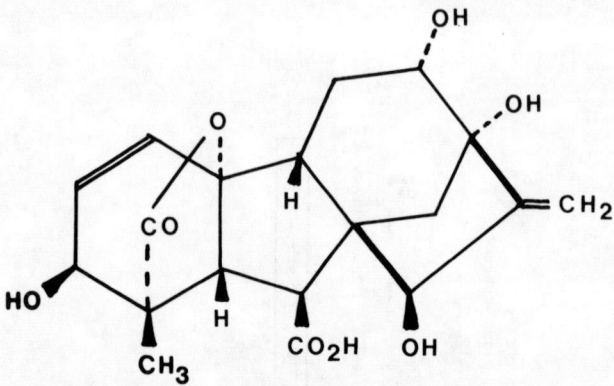

Fig. 6. GA_{32}

Biological Activity

Apricot gibberellin (GA_{32}) is very active biologically. In the barley endosperm bioassay it is the most potent gibberellin we have tested being three times as active as GA_3 and of the same order of activity as GA_1. It is also active in seven other bioassay systems: oat mesocotyl, oat leaf, dwarf corn leaf sheath, lettuce hypocotyl, *Rumex* leaf senescence, and ovaries of seedless grape and unfertilised apricot, (Coombe, unpublished data).

IV. SUMMARY

A polar gibberellin was extracted from apricot seed and purified by butanol partition with subsequent chromatography on DEAE Sephadex, silica gel plates and paper. The location of biological activity was followed by the barley endosperm bioassay.

A partly crystalline substance, as the ammonium salt, was obtained. Its non-reactivity to various reagents and the similarity of its I.R. spectrum to GA_3 suggested it was an unconjugated gibberellin. Chromatographic behaviour indicated that the compound probably possessed more than three hydroxyls while electrophoretic properties suggested the presence of one carboxyl and a vicinal glycol which would complex with borate.

G.L.C.-mass spectrometry as the methyl ester-trimethylsilyl ether has established its identity with GA_{32}. This gibberellin is active in eight biological systems and is the most potent compound we have tested in the barley endosperm bioassay, being three times as potent as GA_3.

V. ACKNOWLEDGEMENTS

This work was supported in part by an Australian Research Grant to B.G.C. The authors are grateful to Mrs. P. Phillips and Mr. P. Hoskyns for assistance, to Dr. R.F. Seamark and his staff at the Queen Elizabeth Hospital for the use of their GLC equipment, to Dr. J. MacMillan and Mr. P. Gaskin, University of Bristol, for the GLC-mass spectra and for preparing Fig. 5, and to Dr. L.C. Luckwill, Long Ashton Research Station, in whose laboratory this investigation began.

VI. REFERENCES

COOMBE, B.G., D. COHEN and L.G. PALEG (1967). Barley endosperm bioassay for gibberellins. II. Application of the method. Plant Physiol. <u>42</u>, 113-119.

DAVIS, L.A., D.E. HEINZ and F.T. ADDICOTT (1968). Gas-liquid chromatography of trimethylsilyl derivatives of abscisic acid and other plant hormones. Plant Physiol. 43; 1389-1394.
JACKSON, D.I. and B.G. COOMBE (1966). Gibberellin-like substances in the developing apricot fruit. Science 154; 277-278.
LANG, A. (1970). Gibberellins: structure and metabolism. Ann. Rev. Plant Physiol. 21; 537-570.
YAMAGUCHI, I., T. YOKOTA, N. MUROFUSHI, Y. OGAWA and N. TAKAHASHI (1970). Isolation and structure of a new gibberellin from immature seeds of *Prunus persica*. Agr. Biol. Chem. 34; 1439-1441.

Dwarfing Genes in Rice and their Relation to Gibberellin Biosynthesis

Yutaka Murakami

National Institute of Agricultural Sciences, Nishigahara, Kita-ku, Tokyo, Japan

I. INTRODUCTION

The dwarf mutants of maize have been used to study the mechanism of action of dwarfing genes in the control of plant growth (reviewed by Pelton, 1964). Phinney (1960) has suggested that each dwarfing gene of maize controls a different step in a biochemical pathway leading to the production of GA necessary for normal growth. The present investigation was undertaken to confirm Phinney's suggestion by using dwarf mutants of rice available in Japan.

II. MATERIALS AND METHODS

Plant material

Rice plants were grown in the paddy field. Young bamboo shoots were obtained at a local market. Samples of apple (var. Star King Delicious) were collected from orchard trees. Immature seeds (3 g) and sarcocarps (320 g) were taken from 15 fruitlets, approximately 3 cm in diameter, 150 flowers (32 g) were gathered when the petals were starting to unfold, and 50 shoots (65 g), approximately 10 cm long, were cut from branches.

Extraction of GA

After sampling, each material was ground in a blendor with 5 times its fresh weight of 70 per cent aqueous acetone, and this blended mixture was allowed to stand overnight at room temperature before filtration. After the acetone was evaporated, the resulting aqueous solution was adjusted to pH 2.5 with phosphoric acid and extracted three times with ethyl acetate. The combined ethyl acetate phase was extracted three times with 1 M potassium phosphate buffer of pH 7.0. The buffer phase was then adjusted to pH 2.5 with phosphoric acid and again extracted three times with ethyl acetate. The acidic ethyl acetate was dried over anhydrous sodium sulfate and evaporated to dryness.

Each evaporated extract was taken up in a small volume of 70 per cent acetone and applied as a band to the starting line of a 20 x 20 cm silica gel (G) thin-layer plate with 0.5 mm thickness. The plate was developed in di-isopropyl ether:acetic acid (95:5, v/v) by the ascending method, unless specifically stated otherwise. The plate was dried when the solvent front reached 10 cm from the starting line and was divided into 10 equal zones between the starting line and the solvent front (the first zone was further subdivided into two zones). Each zone was scraped off into a small beaker (10 ml) and eluted three times with 2 ml of 50 per cent acetone. The eluates were taken up into a small glass bottle (1.8 x 3.5 cm), evaporated to dryness using a hair dryer, and dissolved again in 0.1 ml of 50 per cent acetone for bioassay.

Reference chromatograms were sprayed with 5 per cent sulfuric acid in ethanol and heated for 10 min at $120°C$. Known GAs were detected under a UV lamp as fluorescent spots.

Bioassay

The bioassay was carried out by the method of microdrop application (Murakami, 1968a). Two dwarf varieties of rice, Tan-ginbozu and Waito-C, were mainly used. Seeds were disinfected by soaking for 6 hr in a 0.1 per cent solution of "Uspulun" tablet. Germinated seeds with 2 mm coleoptiles were planted in a group of five on the surface of 0.9 per cent water-agar in a cylindrical bottle 2.8 cm diameter and 6 cm deep. The bottles were placed in a tall Petri dish of about 18 cm diameter, 12 cm deep, and incubated for 45 hr in a cabinet under continuous fluorescent light (about 5,000 lux) at 32°C. When the second leaf emerged from the first leaf, eluates were applied as a single 1 μl droplet to the surface of each coleoptile with a micropipette. The treated seedlings were then grown for a further 3 days under the same conditions as before the treatment. The length of the second leaf sheath was then measured with a ruler. The results are shown in the form of histograms each indicating the average of 5 test plants. Assay responses were not regarded as positive unless they were 10 per cent greater than the controls.

III. RESULTS

A rice mutant deficient in GA

An attempt was made to find a rice mutant deficient in GA (Suge and Murakami, 1968). GAs were estimated in shoots using Tan-ginbozu seedlings as the assay plant. Among normal and 20 dwarf varieties surveyed for endogenous GA, two dwarf varieties, Tan-ginbozu and Sankei-10 were found to contain no detectable amounts of GA. The Sankei-10 is a recessive progeny of a cross between the dwarf Tan-ginbozu and the normal Ayanishiki. This indicates that the lack of native GA in both dwarfs is under the control of a recessive gene. The lack of GA in the Tan-ginbozu dwarf was also observed when other dwarf or normal rice was used as the bioassay plant.

Genetic control of GA production

As stated above, Tan-ginbozu is a simple recessive dwarf. Another dwarf variety, Kotake-tamanishiki, is also a simple recessive dwarf (Morinaga and Fukushima, 1943). These two dwarf varieties have almost the same height at about the fifth leaf stage. As part of the investigation on the relationship between the endogenous GA and the dwarf habit of growth, crosses were made between these two dwarfs: "Tan-ginbozu".x "Kotake-tamanishiki", "Kotake-tamanishiki" x "Tan-ginbozu". The direction of the cross did not affect the F_1 or F_2 progeny. The F_1 progeny was normal in height. The F_2 progeny segregated into three distinct classes, normal, single dwarf and double dwarf, in agreement with the ratio of 9:6:1 (Table 1). Thus height in these two dwarf varieties is controlled essentially by the two different genes. Their causal dwarf genes are tentatively called dx for Tan-ginbozu and dy for Kotake-tamanishiki.

Table I. Segregation of height in F_2 plants from the cross between Tan-ginbozu (dx) and Kotake-tamanishiki (dy) dwarf varieties of rice.

F_2 progeny		Height* (cm)	Observed No.	Expected no. from 9:6:1
Normal	dx^+dy^+	30.0	328	302.1
Single dwarf	$dx^+dy, dxdy^+$	22.0	180	201.4
Double dwarf	$dxdy$	10.5	29	33.6

* Height was the mean length of 20 plants from ground level to the tip of the longest leaf at the age of 5th leaf stage.

F_2 plants were grown in the paddy field for 2-3 months prior to the estimation of endogenous GA. The results of assays of extracts from representative plants are given in Fig. 1. The normal plant (dx^+dy^+) produced GA, but not the double dwarf plant ($dxdy$). With respect to single dwarf plants (dx^+dy, $dxdy^+$), there were two kinds of dwarf which differed in their GA-producing phenotypes. Among 16 single dwarf plants surveyed for endogenous GA, 9 plants were GA-producers and 7 plants were non-producers. Thus it was confirmed that there is a 1:1 segregation for GA production in the single dwarf type. These facts indicate that the F_2 progeny segregates in a 3:1 ratio, GA production *versus* no production, and the gene dx controls GA production. As shown in Fig. 1, almost no difference was detected on histograms between the normal (dx^+dy^+) and the single dwarf (dx^+dy).

Responses to kaurenes and GAs

Recent work has shown that mevalonate, (-)-kaurene, (-)-kaurenol, (-)-kaurenoic acid and steviol are intermediates in GA biosynthesis of the fungus, *Fusarium moniliforme* (see, for example Cross, 1968). GA-like activity of kaurene and its derivatives has been demonstrated in dwarf-5 mutants of maize (Ruddat *et al.*, 1963; Katsumi *et al.*, 1964; Phinney *et al.*, 1964; Brian *et al.*, 1967). Biological activity of these compounds was then tested in the rice seedling assay (Murakami, 1968b and 1968c).

The results are reproduced in Table II. For Tan-ginbozu seedlings kaurene and kaurenol showed a slight activity at the dosages of 4 and 2 µg per plant, respectively. Kaurenoic acid and steviol showed clear activity at dosages above 0.2 µg per plant. The amount of growth response to 0.2 µg kaurenoic acid and steviol was approximately the same for that induced by 0.5 mµg GA_3. Normal and two dwarf varieties, Kotake-tamanishiki and Waito-C, having the same dwarfing gene, dy, showed no response to these kaurene compounds at the dosages tested. Mevalonic acid was inactive even in Tan-ginbozu seedlings.

Table II. Effect of mevalonic acid, (-)-kaurene, (-)-kaurenol, (-)-kaurenoic acid and steviol on the elongation of the second leaf sheath of rice seedlings (control = 100). (From Murakami, 1968b, 1968c).

Variety	Mevalonic acid (µg)		Kaurene (µg)		Kaurenol (µg)		Kaurenoic acid (µg)		Steviol (µg)		GA_3 (mµg)
	0.4	4	0.4	4	0.2	2	0.2	2	0.2	2	0.5
Tan-ginbozu (dwarf)	100	95	109	124	110	129	148	160	139	170	145
Kotake-tamanishiki (dwarf)			98	95	101	103	97	102	101	101	132
Waito-C (dwarf)			100	98	101	105	102	102	100	102	164
Norin 25 (normal)			96	100	97	100	98	97	98	100	116

Each compound was dissolved in ethanol and applied to rice seedlings. Assay responses were not regarded as positive unless they were greater by 10 per cent than control.

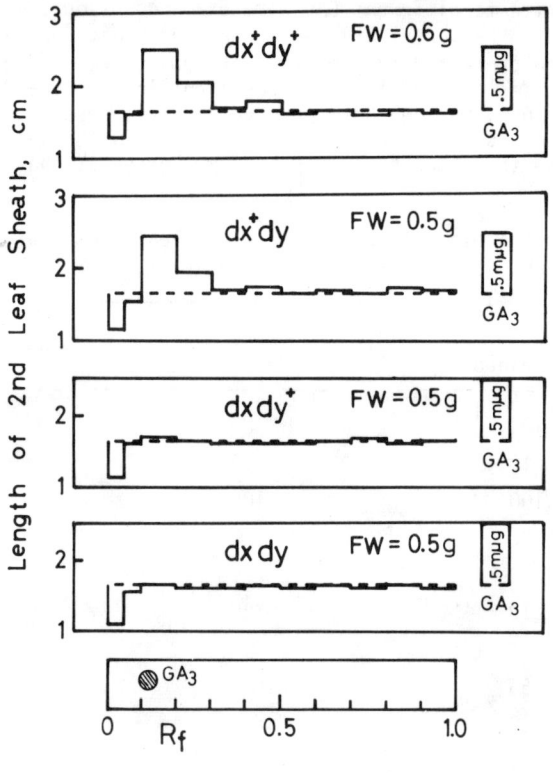

Fig. 1. Histograms showing GA-activities of eluates from chromatograms of shoot extracts of F_2 plants of the cross between Tan-ginbozu (dx) and Kotake-tamanishiki (dy). Eluates were bioassayed with Tan-ginbozu seedlings. TLC solvent, di-isopropylether:acetic acid (95:5, v/v). FW= Fresh weight equivalent of each extract applied to one rice seedling.

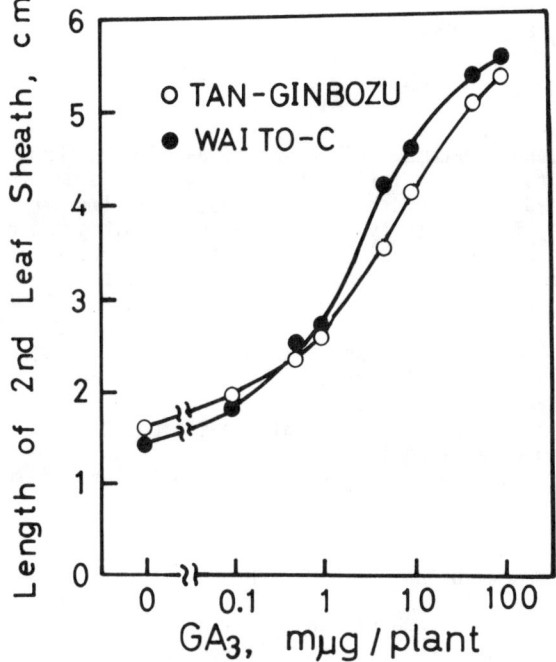

Fig. 2. Responses of the second leaf sheath of two dwarf rice seedlings to GA_3.

Responses of dwarf varieties of rice to different GAs were examined using the microdrop method (Murakami, 1968a, 1969). Responses of the second leaf sheath of Tan-ginbozu and Waito-C seedlings to GA_3 are presented in Fig. 2. Both dwarfs responded almost equally to GA_3. Relative activities of GAs at 10 mµg per plant in both dwarf seedlings are summarized in Table III. GA_5, GA_9 and GA_{20} were very mutant-specific. They were highly active in the Tan-ginbozu seedling but had low activity in the Waito-C seedling. It should be noted that these three GAs have no β-hydroxyl group at the C-2 position in the A ring of gibbane.

Table III. Relative activity of GAs on the elongation of the second leaf sheath of rice seedlings at the concentration of 10 mµg per plant. (From Murakami, 1969).

GA	Dwarf variety of rice	
	Tan-ginbozu	Waito-C
A_1	100	80
A_2	10	10
A_3	100	100
A_4	30	20
A_5	30	5
A_7	10	20
A_8	1	0.5
A_9	50	1
A_{18}	8	3
A_{20}	90	1

GA_7 is a mixture of GA_7 and isoGA_7

Endogenous GA of rice

Endogenous GA of rice shoots (var. Kotake-tamanishiki) was compared with that of bamboo shoots, since both plants belong to the same family, Gramineae, and the main GA of bamboo shoots has been established as GA_{19} (Murofushi et al., 1966). Preliminary experiments showed that the distribution of GA-activity on thin-layer chromatograms of extracts of bamboo shoots was similar to that of rice. To compare the properties of GA-activity from both plants, eluates from the active zone of rice and bamboo extracts were re-chromatographed and bioassayed with Tan-ginbozu and Waito-C seedlings. The results are summarized in Fig. 3. A growth-promoting zone which almost coincided with the Rf of an authentic sample of GA_3 was found in the upper histograms (assayed with Tan-ginbozu seedlings) of rice and bamboo. This activity was not thought to be due to GA_3, since the lower histograms (assayed with Waito-C seedlings) had no clear growth-promoting zone. The bamboo GA, GA_{19}, has no OH group at the C-2 position of gibbane. As already described, the Waito-C seedling is only slightly responsive to GAs having no OH group at the C-2 position of gibbane. Consequently the eluates from bamboo shoots did not affect the growth of Waito-C seedlings at all.

Because of the apparent similarity of GA extracted from bamboo and rice shoots, further examinations were made of the extracts using different chromatographic solvent systems. The results are summarized in Table IV. As can be seen, activity in the Tan-ginbozu seedling assay was obtained from the same Rf regions of chromatograms of both rice and bamboo. The active zones of the chromatograms differed from the known position of GA_3, when solvent systems containing ammonia were used. The high activity in the Tan-ginbozu assay, the non-activity in the Waito-C assay, and the Rf evidence all suggest the presence of GA_{19} in rice shoots.

Table IV. Comparison of GA-activity in extracts from rice and bamboo shoots following thin-layer chromatography in five different solvent systems.

Solvent system	Rf of GA$_3$	Plant material	Zone of growth promotion* Rf
1	0.2	Rice	0.1 - 0.2
		Bamboo	0.1 - 0.2
2	0.9	Rice	0.8 - 1.0
		Bamboo	0.8 - 1.0
3	0.75	Rice	0.7 - 0.9
		Bamboo	0.7 - 0.9
4	0.65	Rice	0.05 - 0.2
		Bamboo	0.05 - 0.2
5	0.4	Rice	0.05 - 0.1
		Bamboo	0.05 - 0.1

* Bioassay was carried out using Tan-ginbozu seedlings. Solvent systems:
  1) Chloroform:ethyl acetate:acetic acid (60:40:5, v/v).
  2) Isopropanol:acetic acid (95:5, v/v).
  3) Isopropanol:acetic acid (4:1, v/v).
  4) Isopropanol:ammonia (28 %):water (10:1:1, v/v).
  5) n-Butanol:1.5 N ammonia (3:1, v/v).

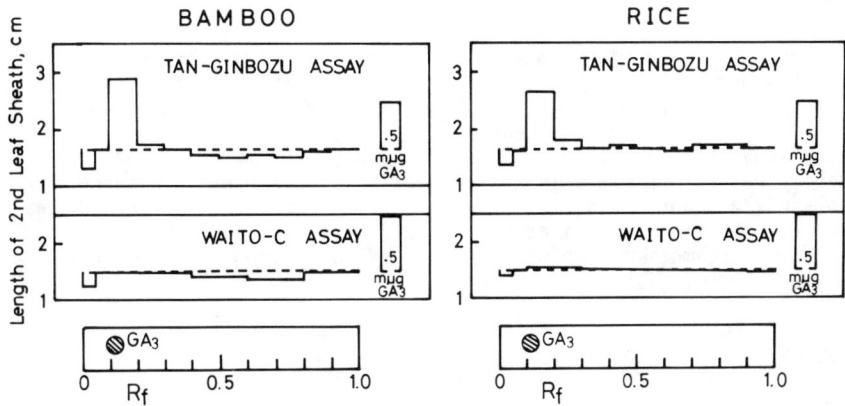

Fig. 3. Comparative rice seedling assays of thin-layer chromatograms loaded with extracts of bamboo and rice shoots. About 50 g fresh weight equivalent of each extract was chromatographed on silica gel (G) in the solvent system, di-isopropylether acetic acid (95:5, v/v). GA activities were determined by Tan-ginbozu and Waito-C seedlings. These are mentioned as Tan-ginbozu assay and Waito-C assay, respectively.

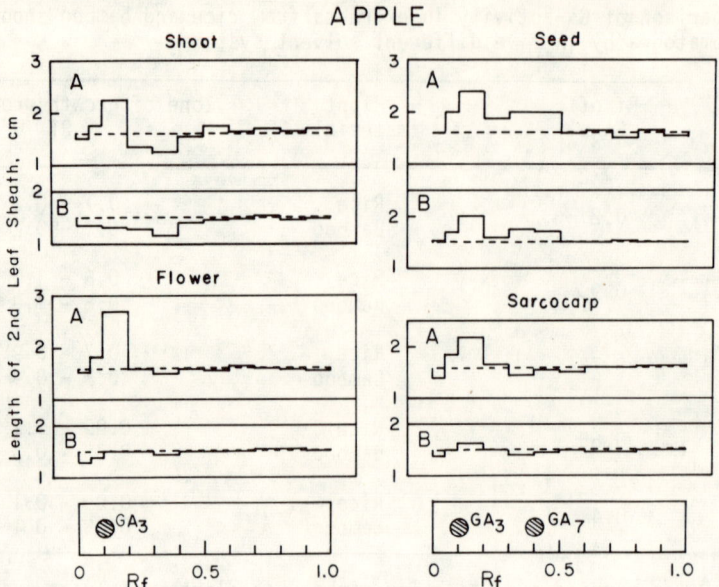

Fig. 4. Histograms showing GA-activities of eluates from chromatograms of extracts of different tissues of apple.

A, Tan-ginbozu assay; B, Waito-C assay
TLC solvent, di-isopropylether:acetic acid (95:5, v/v).

Endogenous GA of other plants

The Tan-ginbozu and Waito-C varieties of rice are especially useful single gene dwarfs for detecting GA in plant extracts because of the specificity of the two dwarf seedlings in response to GAs with similar structures. Fig. 4. shows the results of experiments on apple (var. Star King Delicious). Extracts from immature seeds, already been known to contain GA_4 and GA_7 (Dennis *et al.*, 1966; Luckwill *et al.*, 1969), gave almost the same pattern of histograms when assayed by the two dwarf varieties of rice. But GA-activity was not detected in extracts of shoots, flowers and sarcocarps, unless Tan-ginbozu seedlings were used as the assay plant. Extracts from many other plant shoots also gave the same results as that of apple shoots (Murakami, 1970). These results indicate that most of the GAs found in extracts from shoots do not possess the C-2 β-hydroxyl group of gibbane.

IV. DISCUSSION

Shoot extracts of the Tan-ginbozu dwarf are inactive when tested on the GA_3-responding dwarf varieties of rice (Suge and Murakami, 1968). It was also found that the lack of endogenous GA in this dwarf variety is under the control of a recessive gene, *dx* (Table I, Fig. 1). Phinney (1960) has reported that extracts of the d-5 mutants of maize have no GA-like substances. The d-5 mutants of maize show growth response to kaurene, kaurenol, kaurenoic acid and steviol (Ruddat *et al.*, 1963; Katsumi *et al.*, 1964; Phinney *et al.*, 1964; Brian *et al.*, 1967). Phinney and Katsumi (1967) have suggested that the gene d-5 of maize might block a chemical reaction between mevalonate and kaurene and that the biological activity of the kaurene compounds may be a result of their conversion to GAs by the d-5 mutants of maize. A similar explanation for the dwarfism of the Tan-ginbozu variety of rice is supported by the finding that this dwarf differed from others in lacking endogenous GA and showing growth response to added kaurene compounds (Table II).

Kotake-tamanishiki and Waito-C seedlings which have the dy gene respond as well to GA_1 and GA_3 as the Tan-ginbozu dwarf (Fig. 2). But they respond weakly to GAs such as GA_9 and GA_{20} which have no β-hydroxyl group at the C-2 position in the A ring of gibbane (Table III). For example, GA_{20} is only one per cent as active as GA_3 on the Waito-C seedling, but it is as active as GA_3 on the Tan-ginbozu seedling. Kotake-tamanishiki and Waito-C seedlings are also non-responsive to kaurene and its derivatives (Table II). These characters of the dy mutants of rice are similar to those of d-1 mutants of maize.

Thin-layer chromatographic studies indicate that GA_{19} is the main GA present in shoots of the dy mutants of rice (Fig. 3, Table IV). Murofushi et al. (1966) have suggested that GA_{19} is a key intermediate in biological transformation from C_{20} to C_{19} GAs. Thus, it seems to be possible that the gene dy might interfere with a step between GA_{19} and much oxidized GAs such as GA_1.

Waito-C seedlings did not respond to the extracts of shoots of many plants including apple (Fig. IV), while Tan-ginbozu seedlings were well responsive to the same extracts (Murakami, 1970). This means that GAs such as GA_1 and GA_3 are absent from shoots or present only in trace amounts in them. Such GAs are known to be more highly active in many bioassay systems than other GAs. The formation of the highly active GAs might be a limiting factor for the elongation of shoots.

<u>Note added in proof</u>. Recently I tested GA_{19} in the dwarf rice seedling. GA_{19} was almost as active as GA_3 in the Tan-ginbozu seedling, but it was only 0.1 per cent as active as GA_3 in the Waito-C seedling.

VI. SUMMARY

Dwarf varieties of rice, the dwarfness of which is interpretable in terms of a single gene mechanism, were examined for GA production. One dwarf variety, Tan-ginbozu (dx), had no GA in shoots. Other dwarfs and normals produced GA. Crosses were made between two different dwarfs bearing the genes dx and dy (Kotake-tamanishiki). The F_1 progeny produced GA and the F_2 segregated in a 3:1 ratio, GA production versus no production. Thus one gene is present which controls GA production in these dwarfs. The dx mutant showed elongating response to (-)-kaurene, (-)-kaurenol, (-)-kaurenoic acid and steviol, but the dy mutant did not respond to these compounds. The gene dx might block a chemical reaction between mevalonate and kaurene in the pathway of GA biosynthesis because of non-response of the dx mutant to mevalonate. The dx mutant, Tan-ginbozu, and the dy mutant, Waito-C, differed in the ratio of elongating response toward GAs. Tan-ginbozu seedlings responded well, but Waito-C seedlings much less to GAs which do not possess the β-OH group at the C-2 position in the A ring of gibbane. However, both mutants responded equally well to GA_1 and GA_3. The chromatographic and biological behaviour of the active zone of extracts from rice shoots was identical to that of bamboo GA (GA_{19}). The gene dy might block a chemical reaction between GA_{19} and GA_1 because of the high response of the dy mutant to GA_1. The two dwarf mutants can be used together as a 'multiple plant assay' to distinguish among GAs with similar structures. The results for extracts from various tissues of apple are presented as an example.

VII. LITERATURE CITED

BRIAN, P.W., J.F. GROVE and T.P.C. MULHOLLAND (1967). Relationships between structure and growth-promoting activity of the gibberellins and some allied compounds, in four test systems. Phytochem. 6, 1475-1499.
CROSS, B.E. (1968). Biosynthesis of the gibberellins, in "Progress in Phytochemistry" (Ed. L. Reinhold and Y. Liwschitz) Vol. 1, p. 195. Interscience Publishers, London. 1968.
DENNIS, F.G. and J.P. NITSCH (1966). Identification of gibberellins A_4 and A_7 in immature apple seeds. Nature 211, 255-256.
KATSUMI, M., B.O. PHINNEY, P.R. JEFFERIES and C.A. HENRICK (1964). Growth response of the d-5 and an-1 mutants of maize to some kaurene derivatives. Science 144, 849-850.

LUCKWILL, L.C., P. WEAVER and J. MACMILLAN (1969). Gibberellins and other growth hormones in apple seeds. J. Hort. Sci. 44, 413-424.

MORINAGA, T. and E. FUKUSHIMA (1943). Heritable characters in rice. I. Abnormal mutant characters and their mode of inheritance. Gakugei Zasshi 10, 301-339.

MURAKAMI, Y. (1968a). A new rice seedling bioassay for gibberellins, "Microdrop Method", and its use for testing extracts of rice and morning glory. Bot.Mag. Tokyo 81, 33-43.

MURAKAMI, Y. (1968b). Gibberellin-like activity of (-)-kaurene, (-)-kauren-19-ol and (-)-kauren-19-oic acid in leaf sheath elongation of 'Tan-ginbozu' dwarf of *Oryza sativa*. Bot. Mag. Tokyo 81, 100-102.

MURAKAMI, Y. (1968c). Gibberellin-like activity of steviol in leaf sheath elongation of the 'Tan-ginbozu' dwarf of *Oryza sativa*. Bot. Mag. Tokyo 81, 464-466.

MURAKAMI, Y. (1969). A new rice seedling test for gibberellins, 'Microdrop Method'. Shokubutsu No Kagaku Chosetsu (Chem. Regul. Plant) 4, 78-83.

MURAKAMI, Y. (1970). A survey of gibberellins in shoots of angiosperms by rice seedling test. Bot. Mag. Tokyo 83, 312-324.

MUROFUSHI, N., S. IRIUCHIJIMA, N. TAKAHASHI, S. TAMURA, J. KATO, Y. WADA, E. WATANABE and T. AOYAMA (1966). Isolation and structure of a novel C_{20} gibberellin in bamboo shoots. Agr. Biol. Chem. 30, 917-924.

PELTON, J.S. (1964). Genetic and morphogenetic studies on angiosperm single-gene dwarfs. Bot.Rev. 30, 479-512.

PHINNEY, B.O. (1960). Dwarfing genes in *Zea mays* and their relation to the gibberellins, in "Plant Growth Regulation" (Ed. R.M. Klein) p. 489. Iowa State Univ. Press, Ames. 1960.

PHINNEY, B.O., P.R.JEFFERIES, M. KATSUMI and C.A. HENRICK (1964). The biological activity of kaurene and related compounds. Plant Physiol. 39, (Suppl.) xxvii.

PHINNEY, B.O. and M. KATSUMI (1967). Genetic control of gibberellin production. Shokubutsu No Kagaku Chosetsu (Chem. Regul. Plant) 2, 79-84.

RUDDAT, M., A. LANG and E. MOSETTIG (1963). Gibberellin activity of steviol, a plant terpenoid. Naturwissenschaft. 50, 23-24.

SUGE, H. and Y. MURAKAMI (1968). Occurrence of a rice mutant deficient in gibberellin-like substances. Plant and Cell Physiol. 9, 411-414.

PLANT GROWTH SUBSTANCES, 1970

Gibberellins in Immature Seed of Moonflower

(Calonyction aculeatum)

Nobutaka Takahashi, Noboru Murofushi and Takao Yokota

Department of Agricultural Chemistry, The University of Tokyo, Bunkyo-ku, Tokyo, Japan

I. INTRODUCTION

Up to now, thirty-two gibberellins of established structures have been isolated from fungus and higher plants.

Murakami (1959) has reported that immature seed of Convolvulaceae is a rich source of gibberellin. From seed of morning-glory (*Pharbitis nil*), which is a member of Convolvulaceae we have isolated GA_3, GA_5, GA_8, GA_{20}, GA_{26} and GA_{27} (Murofushi, 1968; Yokota, 1969; Takahashi, 1969) together with glucosides of GA_3, GA_8, GA_{26}, GA_{27} and GA_{29}, gibberellenic acid and isoGA$_3$ (Tamura, 1967; Yokota, 1969, 1970).

More recently, we have examined gibberellins in immature seed of another member of Convolvulaceae, moonflower (*Calonyction aculeatum*). We wish to report here the results of this investigation.

II. MATERIALS AND METHODS

Materials

Eleven and 23 kg of seeds of moonflower were harvested about 20 days after flowering in 1968 and 1969, respectively.

Extraction

Seed was extracted with methanol. Ethyl acetate-soluble acidic fraction obtained in the usual way was subjected to further purification.

Counter-current distribution

The ethyl acetate-soluble acidic fraction was subjected to twenty transfers countercurrent distribution in the same way as already described (Tamura, 1968). After each fraction was tested by bioassay and TLC, fractions containing no gibberellin-like substances were discarded and those containing similar components were combined. Thus six fractions (F-I~VI) were obtained finally.

Adsorption chromatography

The F-I~VI were separately purified by silicic acid adsorption chromatography in the same way as already described (Tamura, 1968). All gibberellin-like substances were eluted from the column with benzene containing 35-60% ethyl acetate.

Thin layer chromatography (TLC)

TLC was conducted on plates of silica gel GF_{254} (Merck) and the following solvent systems: (1) benzene-acetone-acetic acid (13:6:1); (2) isopropyl ether-acetic acid (19:1). Spots of gibberellin-like substances were detected by fluorescence under ultraviolet light, after treatment with 70% sulfuric acid followed by heating at 120°C for 10 min.

Bioassay

Dwarf maize (*Zea mays*, dwarf mutants d_1 and d_5) and dwarf rice (*Oryza sativa*, dwarf mutant Tan-ginbozu) seedling tests (Tamura, 1968) were used as guidance in the purification process and for the evaluation of biological activity of purified gibberellins.

III. RESULTS AND DISCUSSION

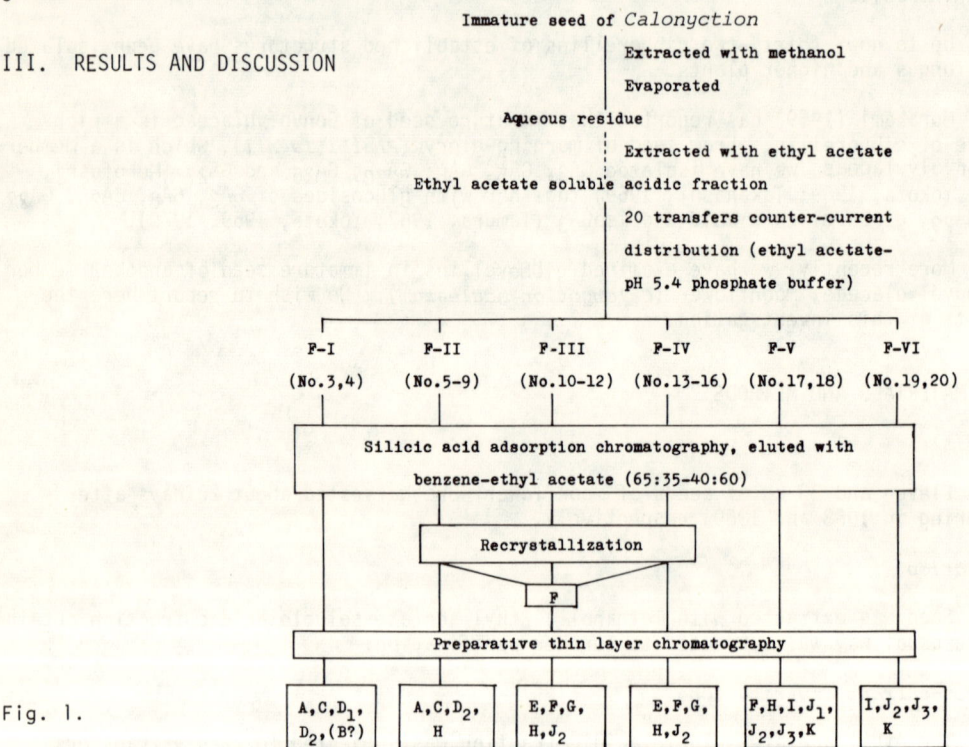

Fig. 1.

An outline of the isolation procedure of gibberellins from seed of moonflower is shown in Fig. 1. One crystalline gibberellin was obtained by recrystallization of some fractions from silicic acid adsorption columns of the F-II-IV. It was identified as GA_{17}, which has been isolated from immature seed of *Phaseolus multiflorus* by Pryce and MacMillan (1967) as a non-crystalline powder. Other gibberellins were obtained in a pure state by repeated preparative TLC. In some cases, non-crystalline fractions were methylated with diazomethane and further purified through TLC. Thus fourteen gibberellins were isolated and their properties are summarized in Table 1.

Five of them, namely, compounds A, C, E, F and I, were identified as GA_8, GA_{29}, GA_{19}, GA_{17} and GA_{27}, respectively, by direct comparison. Although GA_{29} has been isolated from *Pharbitis nil* as its glucoside (Yokota, 1970), this is the first example of its isolation in the free form. Others were considered to be new gibberellins from their characteristic mass spectra and molecular compositions. The structures of four of these, compounds G, H, J_1 and K were decided as I, II, III and IV, illustrated in Fig. II. Hence they were named GA_{30}, GA_{33}, GA_{34} and GA_{31}, respectively. We wish now to describe briefly the structural determination of these new gibberellins.

Table 1. Properties of gibberellin-like substances isolated from immature seed of moonflower.

	R_{GA_3}*	M.p. (°C)	Molecular formula	Identification
A	0.54	145-148(hydrate)	$C_{19}H_{24}O_7$	GA_8
B			$C_{19}H_{22}O_7$	New GA**
C	0.83	Non-crystalline	$C_{19}H_{24}O_6$	GA_{29}
D_1	0.90	167-168	$C_{19}H_{20}O_8$	New GA
D_2	0.90	196-198	$C_{19}H_{22}O_7$	New GA
E	1.10	236-237	$C_{20}H_{26}O_6$	GA_{19}
F	1.13	170-171	$C_{20}H_{26}O_7$	GA_{17}
G	1.13	188-191	$C_{19}H_{22}O_6$	New GA (GA_{30})
H	1.23	219-221	$C_{19}H_{22}O_7$	New GA (GA_{33})
I	1.24	163-165(solvate)	$C_{20}H_{26}O_6$	GA_{27}
J_1	1.48	218-219	$C_{19}H_{24}O_6$	New GA (GA_{34})
J_2	1.72	Non-crystalline	$C_{20}H_{24}O_7$	New GA
J_3	1.93	203-206	$C_{19}H_{22}O_6$	New GA
K	1.95	227-231	$C_{19}H_{22}O_5$	New GA (GA_{31})

* Movement relative to GA_3 in solvent system benzene-acetone-acetic acid (13:6:1).
** Isolated as a methyl ester from seed of 1968 only.

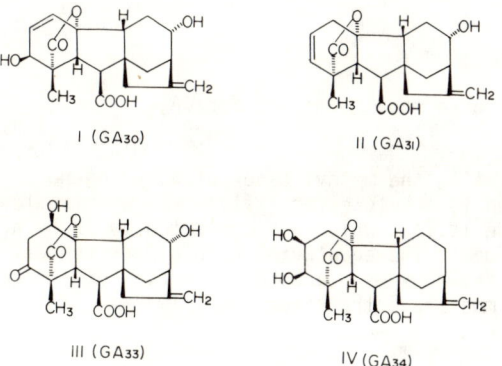

Fig. 11. Structures of new gibberellins, GA_{30}, GA_{31}, GA_{33} and GA_{34}.

GA$_{30}$ has the molecular formula C$_{19}$H$_{22}$O$_6$, deduced from the high resolution mass spectrum of its monomethyl ester. The presence of a carboxyl, two hydroxyl, an exomethylene, and a γ-lactone groups was deduced on the basis of spectral properties. The NMR spectrum of GA$_{30}$ methyl ester in the region of τ 3.5 to 5.5 is quite similar to that of GA$_3$, suggesting the presence of the same ring A system in GA$_{30}$ and in GA$_3$. The location of the other hydroxyl group at C-12 was decided on the following basis. An oxidation product of GA$_{30}$, obtained by Jones reagent, shows the λ_{max} at 286 mμ (ε 1000), which is characteristic of β,γ-unsaturated ketones in an appropriate conformation as in GA$_{26}$. Thus the structure I was assigned to GA$_{30}$.

GA$_{31}$ has the molecular formula C$_{19}$H$_{22}$O$_5$ and it contains a carboxyl, a hydroxyl, an exomethylene and a γ-lactone groups. The presence of the same ring A system as in GA$_5$ was indicated by the similar NMR pattern to that of GA$_5$ in the region of τ4.2. The presence of the α-orientated hydroxyl group at C-12 was shown by a 1H triplet similar to that of GA$_{30}$ and by the λ_{max} 287 mμ (ε 300) of its oxidation product. Thus the structure II was assigned to GA$_{31}$.

GA$_{33}$ has the molecular formula C$_{19}$H$_{22}$O$_7$ and it contains a carboxyl, two hydroxyl, a carbonyl, an exomethylene and a γ-lactone groups. The absence of a hydroxyl group at C-13 was indicated by no down-field shift of one of the exomethylene signals in the NMR spectrum of GA$_{33}$ in the change of solvent from deuteriochloroform to deuteriopyridine. The presence of a 1H triplet at τ5.88 is indicative of the C-12 α-hydroxyl group. This was further confirmed by the λ_{max} at 284 mμ (ε 400) of the oxidation product. The presence of the system -CO-CH$_2$-CH(OH)- was demonstrated by analysis of the NMR spectrum and from a double resonance experiment. Thus only two structures, III and V are possible for GA$_{33}$. Of these, III was selected on the following basis.

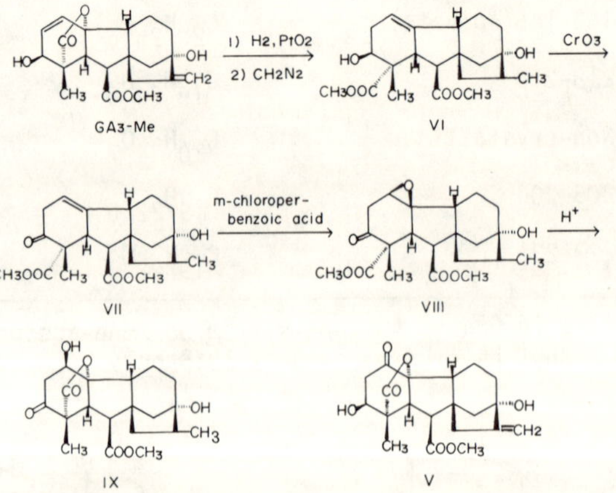

Fig. III. Preparation of a model compound for GA$_{33}$ from GA$_3$.

As shown in Fig. III, the methyl ester of a hydrogenolysis product of GA$_3$ methyl ester (VI) was oxidized to a keto ester (VII). VII was treated with *m*-chloroperbenzoic acid to give an epoxide (VIII), which was converted to IX by acid treatment. IX shows a 1H doublet at τ5.5 due to the equatorial carbinyl proton at C-1. This signal is quite similar to that shown by GA$_{33}$ and the CD curves of IX and GA$_{33}$ are almost identical. Thus GA$_{33}$ must have the structure III.

GA$_{34}$ has the molecular formula C$_{19}$H$_{24}$O$_6$ and contains two hydroxyl groups in addition to the normal C$_{19}$ gibberellin functionality. The locations of the hydroxyl groups at C-2 and C-3 were indicated by the consumption of periodate, the formation of acetonide and the similar NMR pattern of the carbinyl proton region to those of GA$_8$, GA$_{26}$ and GA$_{27}$. Thus the structure IV was assigned to GA$_{34}$.

The structural studies of the other new gibberellins of the immature seed of morning-glory are still in progress.

The structural features of the gibberellins isolated from seed of moonflower can be summarized in the following way.

1) The first occurrence of C-12 hydroxylated gibberellins such as GA$_{30}$, GA$_{31}$ and GA$_{33}$.

2) The first occurrence of a gibberellin, GA$_{33}$, containing a C-3 carbonyl group and a C-1 β-hydroxyl group.

3) The occurrence of a large number of 2,3-dihydroxy-gibberellins, such as GA$_8$, GA$_{27}$ and GA$_{34}$.

4) The occurrence of gibberellins such as GA$_8$, GA$_{27}$ and GA$_{29}$ which are also among the gibberellins isolated from seed of morning-glory.

5) A very high content (15 mg/kg of fresh seed) of GA$_{17}$.

Table II. Effects of new gibberellins on elongation of the second leaf sheath of rice seedlings (dwarf mutant, Tan-ginbozu, microdrop method*).

mμg/plant	GA$_{29}$	GA$_{30}$	GA$_{31}$	GA$_{33}$	GA$_{34}$
			% of control		
0.1	104	92	108	96	100
1	100	108	100	96	117
10	96	133	113	92	113
100	104	208	146	100	121

mμg/plant	GA$_3$	GA$_5$	GA$_8$
		% of control	
0.1	125	125	104
1	142	129	104
10	183	142	104
100	242	167	100

* Murakami, Y. (1968). Bot. Mag. Tokyo, 81, 33-43.

It has been suggested by Pryce and MacMillan (1967) that C-13 hydroxylation may occur in an earlier stage of gibberellin biosynthesis in green plants than in fungi. But the occurrence of many gibberellins lacking the C-13 hydroxyl group in Convolvulaceae suggests the presence of a gibberellin biosynthetic pathway in higher plants, which does not require C-13 hydroxylation.

The growth-promoting activity of the new gibberellins on rice seedlings is shown in Table II. GA_{30} and GA_{31} are fairly active, their activity being half to one tenth that of GA_3 and GA_5, respectively, while GA_{29}, GA_{33} and GA_{34} show only slight activity, comparable with that of GA_8.

IV. SUMMARY

Fourteen gibberellin-like substances were isolated from immature seed of moonflower (*Calonyction aculeatum*) and five of them were identified as GA_8, GA_{17}, GA_{19}, GA_{27} and GA_{29}. The others are confirmed to be new gibberellins. The structures of four new gibberellins, GA_{30}, GA_{31}, GA_{33} and GA_{34}, were established as *ent*-3α, 10, 12β-trihydroxy-20-norgibberella-1,16-diene-7,9-dioic acid 19, 10-lactone, *ent*-10, 12β-dihydroxy-20-norgibberella-2,16-diene-7,19-dioic acid 19,10-lactone, *ent*-1α,10,12β-trihydroxy-3-oxo-20-norgibberell-16-ene-7,9-dioic acid 19,10-lactone and *ent*-2α,3α,10-trihydroxy-20-norgibberell-16-ene-7,19-dioic acid 19,10-lactone, respectively. GA_{30} and GA_{31} are fairly active in the rice seedling test but GA_{33} and GA_{34} show only slight activity.

V. REFERENCES

MURAKAMI, Y. (1959). The occurrence of gibberellins in mature dry seed. Bot. Mag. Tokyo, 72, 438-442.

MUROFUSHI, N., N. TAKAHASHI, T. YOKOTA and S. TAMURA (1968). Gibberellins in immature seeds of *Pharbitis nil*. Part I. Isolation and structure of a novel gibberellin, gibberellin A_{20}. Agr. Biol. Chem. 32, 1239-1245.

PRYCE, R.J. and J. MACMILLAN (1967). A new gibberellin in the seed of *Phaseolus multiflorus* by combined gas chromatography-mass spectrometry. Tetrahedron Lett., No. 49, 5009-5011.

TAKAHASHI, N., T. YOKOTA, N. MUROFUSHI and S. TAMURA (1969). Structures of gibberellin A_{26} and A_{27} in immature seeds of *Pharbitis nil*. Tetrahedron Lett., No. 25, 2077-2080.

TAMURA, S., N. TAKAHASHI, T. YOKOTA, N. MUROFUSHI and Y. OGAWA (1967). Isolation of water-soluble gibberellins from immature seeds of *Pharbitis nil*. Planta (Berl.) 78, 208-212.

TAMURA, S., N. TAKAHASHI, N. MUROFUSHI, T. YOKOTA and J. KATO (1968). Isolation of new gibberellins from higher plants and their biological activity, in "Biochemistry and Physiology of Plant Growth Substances" (Ed. F. Wightman and G. Setterfield) p.85-99. The Runge Press Ltd., Ottawa, Canada. 1968.

YOKOTA, T., N. TAKAHASHI, N. MUROFUSHI and S. TAMURA (1969). Isolation of gibberellins A_{26} and A_{27} and their glucosides from immature seed of *Pharbitis nil*, Planta (Berl.) 87, 180-184.

YOKOTA, T., N. TAKAHASHI, N. MUROFUSHI and S. TAMURA (1969). Structures of new gibberellin glucosides in immature seed of *Pharbitis nil*. Tetrahedron Lett., No. 25, 2081-2084.

YOKOTA, T., N. MUROFUSHI and N. TAKAHASHI (1970). Structure of new gibberellin glucoside in immature seed of *Pharbitis nil*. Tetrahedron Lett., No. 18, 1489-1491.

HORMONES AND NUCLEIC ACIDS

Cytokinin-Induced Changes in
Transfer RNA Species

Joe H. Cherry and Marianne B. Anderson

Horticulture Department, Purdue University, Lafayette, Indiana, U.S.A.

I. INTRODUCTION

Ribonucleic acids, particularly transfer RNAs, exhibit cytokinin bioassay activity. Isolated RNA of rat and sheep liver, and cultured tobacco cells is active in promoting growth of carrot phloem callus while the DNA of these respective tissues is inactive (Bellamy, 1966). Hydrolysates of sRNA, but not ribosomal RNA, from roots and shoots of germinating sweet corn possesses strong cytokinin activity in the carrot root tissue culture assay (Letham and Ralph, 1967). Highly purified tRNA of yeast, rat liver, *E. coli* (Skoog et al., 1966), chick embryo, calf and human liver (Robbins et al., 1967) stimulates cell division and differentiation. Again, rRNA from similar tissues exhibited no cytokinin activity. RNA rich in seryl-, isoleucyl-, and tyrosyl-tRNAs are very high in cytokinin activity in contrast to arginyl$_2$-, glycyl-, phenylalanyl-, valyl-, and alanyl-tRNAs which had little growth promoting activity (Skoog et al., 1966). Thus, tRNA is the only known nucleic acid fraction active in cytokinesis and more specifically only certain tRNA species are functional. It is calculated that one molecule of isopentenyladenine (IPA) per 20 molecules of tRNA could account for the cytokinin activity of the tRNA fraction (Skoog et al., 1966).

N^6-(γ,γ-dimethylallyl) adenosine (IPA) is the first naturally occurring cytokinin to be found as an integral part of nucleic acid (Hall et al., 1967). IPA is isolated from Baker's yeast, calf liver (Hall et al., 1967), *Zea mays* (Hall, 1967), *Corynebacterium fascians* (Matsubara et al., 1968), spinach leaves and pea seedlings (Hall et al., 1967). In yeast, IPA comprises 0.1 mole % of the nucleotides, hence statistically affirming the idea that IPA occurs only in certain tRNA molecules (Hall et al., 1967). The tRNA of *E. coli* lacks IPA but possesses instead a methythionyl-IPA, 6-(3-methyl-2-butenylamino)-2-methythio-9-β-D-ribofuranosylpurine (Burrows et al., 1968).

The native cytokinin, IPA, has been identified in the primary structure of 3 tRNAs; seryl-tRNA (Zachau et al., 1967), tyrosyl-tRNA (Madison et al., 1967; Gefter and Russell, 1969), and cysteinyl-tRNA (Hecht et al., 1969). The sequence of alanyl-tRNA, the first tRNA for which the primary structure was known, does not contain IPA or other cytokinin nucleosides (Holley et al., 1965).

According to Fox (1966) a small portion of labeled 6-benzyladenine is selectively incorporated into sRNA. Fifty percent of the radioactivity chromatographs similarly to 6-BA. However, *in vitro* biosynthesis of IPA and tRNA shows that Δ^2-isopentenyl groups of IPA residues of tRNA are derived from mevalonic acid or Δ^3-isopentenyl pyrophosphate (Fittler et al., 1968). Radioactive mevalonic acid is utilized by yeast, rat liver, and *Lactobacillus acidophilus*, and recovered in tRNA, but not rRNA. The enzyme system for incorporation is apparently specific for certain adenosine residues in selected tRNA molecules. In yeast, 2-^{14}C acetate is also incorporated into IPA-tRNA.

II. MATERIALS AND METHODS

Soybean seeds (*Glycine max* L. cv. Wayne) were sown in moist Vermiculite and placed in a dark, humid chamber at $29°$. After 4 days the seedlings (tray $2\frac{1}{4}$ x 12 x 17 inches containing about 1000 seedlings) were sprayed with 10 ml of water or a solution of cytokinin as indicated. At various times after treatment the cotyledons and hypocotyls (including meristematic and mature tissue) were removed and aminoacyl tRNA synthetase and tRNA preparations were isolated by the methods previously described (Anderson and Cherry, 1969). Acylation of tRNA with ^{3}H-leucine or ^{14}C-leucine employing synthetase preparations from cotyledon or hypocotyl tissues was the same as previously described. The leucyl-tRNA isoacceptor species were fractionated on Freon reverse phase columns (RPC) using linear gradients of NaCl as previously used. Acylation of tRNA with other amino acids and the conditions for separation on Freon column are described in the appropriate legend.

III. RESULTS

Cytokinin-Induced Swelling

Soybean seedlings were selected to determine whether or not cytokinins affected tRNAs because, a) tRNAs and aminoacyl tRNA synthetases are easily isolated from the tissue and b) soybean hypocotyls are responsive to treatment with cytokinin. As noted in Table 1, 6-BA induced marked swelling of the hypocotyl. However, it was noted that the seedlings sprayed with 6-BA did not elongate to the same extent as control seedlings. The cotyledons of 6-BA treated plants were slightly swollen and bulged away from each other in comparison to control cotyledons.

Table 1. 6-Benzyladenine-induced swelling of the soybean hypocotyl.

Treatment	Hypocotyl fresh wt. (mg/4 cm section)
Control	122
1×10^{-4}M 6-benzyladenine	186
2×10^{-4}M 6-benzyladenine	237

Soybean seedlings, $3\frac{1}{2}$ days after planting, were sprayed with 6-benzyladenine solutions and kept in the humidity chamber for 23 hrs. Subsequently, 4 cm sections of the hypocotyl (immediately below the cotyledons) were cut and weighed.

Changes in Leucyl-tRNA Isoacceptor Species

Charging tRNA preparations from control and 6-BA-treated hypocotyls with leucine by enzyme preparations from either control or treated tissues revealed that the enzyme from 6-BA treated tissue was less active than that from the control (Table 2). The leucine acceptor activities of the tRNA from control and 6-BA hypocotyls were the same. Experiments presented in Figure 1 show that 6-BA treatment of soybean tissue leads to major changes in the relative amounts of three leucyl-tRNA species. Twenty-four hours following 6-BA treatment soybean hypocotyls contained much more leucyl-tRNA$_{5\ \&\ 6}$ while the amount of leucyl-tRNA$_1$ was less (Figure 1A).

Although the leucyl-tRNA profiles of homologous 6-BA and control systems were identical on the Freon column (Figure 1B), large differences in the amount of leucyl-tRNA$_{5\ \&\ 6}$ were noted only if the cotyledon enzyme was used in the acylation reaction (Figure 1A). It is of interest that 6-BA increased the relative amount of leucyl-tRNA$_{5\ \&\ 6}$ species in cotyledon tissue even though these species comprise 25% of the

Table 2. Comparison of leucyl-tRNA synthetase activities from control and 6-benzyladenine-treated soybean hypocotyls with sRNA from both tissues.

Source of Synthetase	Source of sRNA	cpm ^{3}H-leucine incorporated x 10^{-3}M
6-BA	6-BA	14.8
Control	6-BA	28.4
6-BA	Control	19.2
Control	Control	28.4

A one ml reaction mixture containing 0.4 mg enzyme and 0.5 mg tRNA was incubated for 25 min at 37°C. One-tenth ml aliquots were spotted on filter disks, processed and the radioactivity determined. The incorporation of leucine is based on the one ml reaction.

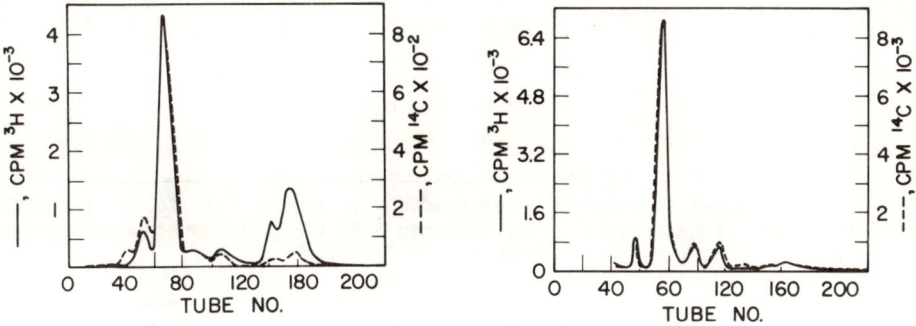

Fig. 1. A. Leucyl-tRNA profiles of tRNA isolated from benzyladenine-treated and control hypocotyls and then acylated with cotyledon enzyme. The dashed line represents control hypocotyl tRNA (2.4 mg) acylated with cotyledon protein (1.44 mg) and 10 μc ^{14}C-L-leucine. The solid line represents tRNA from 6-BA treated hypocotyls (2.8 mg) acylated by cotyledon protein (1.44 mg) with 0.1 mc ^{3}H-L-leucine.

B. A comparison of leucyl-tRNA from homologous control and 6-benzyladenine hypocotyl systems on the Freon column. The solid line represents tRNA (2.8 mg) from 6-BA treated hypocotyls and acylated with enzyme from the same tissue (1.62 mg) in the presence of 0.1 mc ^{3}H-L-leucine. The solid line is control hypocotyl tRNA (4.5 mg) charged with control hypocotyl enzyme (1.74 mg) in the presence of 10 μc ^{14}C-L-leucine.

leucyl-tRNAs in that tissue (Table 3). As noted in Figure 1 the cytokinin-induced changes in leucyl-tRNA species are of a quantitative nature. No new peaks or shifting of peaks were noted by cytokinin treatment.

Table 3. The effect of 6-benzyladenine on the acylation of leucyl-tRNAs of the cotyledons.

Source of cotyledon tRNA	Relative amount of leucyl-tRNA acylation of each peak (% of total)					
	1	2	3	4	5	6
Control	5.1	59.6	6.3	4.3	9.4	15.3
6-BA	4.8	48.8	5.9	4.6	12.9	22.8

6-BA was sprayed onto plants at a concentration of 3×10^{-4}M. The acylated tRNA was fractionated on a Freon column.

The effects of various concentrations of 6-BA on the leucyl-tRNA species of soybean hypocotyls are presented in Table 4. In general, 6-BA concentrations of 1×10^{-5}M and higher enhanced changes in leucyl-tRNA acceptor activity while lower concentrations, 10^{-7} and 10^{-8}M, had little effect.

Table 4. The effect of pretreatment of the hypocotyl with various 6-benzyladenine concentrations on the acylation of leucyl-tRNAs.

Concentration of Benzyladenine	Relative amount of leucine acylation of each peak (% of total)					
	Species of tRNA					
	1	2	3	4	5	6
Control	23.8	61.1	7.7	4.3	1.1	1.6
1×10^{-8}M	12.6	67.4	7.7	7.0	1.6	3.6
1×10^{-7}M	9.6	64.2	7.4	7.5	3.5	7.8
1×10^{-5}M	6.0	52.8	5.8	5.7	29.7	
6×10^{-5}M	8.0	58.8	6.2	6.8	3.7	16.5
3×10^{-4}M	6.8	60.5	3.7	3.8	6.5	18.7

Aminoacyl-tRNA synthetase isolated from the cotyledons was used to acylate each sample of control or 6-BA treated tRNA. The acylated tRNA was fractionated on a Freon column and the relative percentage of each species calculated.

A time-course study (Table 5) of 6-BA treatment showed that hormone application began to increase the level of leucyl-tRNA$_{5\ \&\ 6}$ within 3 hours and the changes in leucyl-tRNAs continue to occur during the entire 24 hours following 6-BA treatment. Again, there was a direct correlation between a decrease in leucyl-tRNA$_1$ and an increase in leucyl-tRNA$_{5\ \&\ 6}$.

Table 5. Changes in acylation of leucyl-tRNA species with time after 6-benzyladenine treatment.

Time after benzyladenine treatment (hrs)	Relative amount of leucyl-tRNA acylation of each peak (% of total)					
	Species of tRNA					
	1	2	3	4	5	6
0	24.2	62.1	6.7	4.3	1.1	1.6
3	12.0	59.5	12.3	8.2	7.9	
6	15.8	61.2	7.5	8.1	2.9	4.6
12	14.7	56.1	8.0	8.7	6.5	6.1
24	8.4	56.7	4.7	4.8	6.2	19.3

Transfer RNA was isolated from hypocotyls treated with 3×10^{-4}M 6-BA and acylated with cotyledon enzyme. The acylated tRNA was fractionated on a Freon column.

The Effect of Other Hormones on Leucyl-tRNA

Three other hormones, the naturally occurring cytokinin (6-dimethylallyl amino purine, DMAAP), an auxin (2,4-dichlorophenoxyacetic acid, 2,4-D) and an ethylene-producing chemical (2-chloroethanephosphoric acid) were tested to determine whether the changes in leucyl-tRNAs were specific to 6-BA. All three of these compounds caused a very slight swelling of the hypocotyl; however, none of them substantially stimulated changes in leucyl-tRNA$_{5+6}$ of the hypocotyl (Table 6). In other experiments (data not included) zeatin treatment of soybean hypocotyls resulted in changes in leucyl-tRNA species similar to those induced by 6-BA.

Table 6. The effect of various chemicals on leucyl-tRNA species of the hypocotyl.

Hypocotyl Treatment	Relative amount of leucyl-tRNA acylation of each peak (% of total)					
	1	2	3	4	5	6
Control	21.7	62.5	7.4	4.4	1.2	2.2
2,4-D (100 µg/ml)	13.5	64.9	12.4	4.6	1.3	2.9
2,4-chloroethane-phosphoric acid (250 µg/ml)	10.0	68.9	8.7	7.2	5.1	1.0
DMAAP (1×10^{-5}M)	11.2	65.1	6.4	8.5	2.7	6.1
6-BA (3×10^{-4}M)	7.6	58.6	4.2	4.3	6.2	19.0

The various chemicals were sprayed (10 ml) on 4-day-old soybeans at the concentration indicated. After 24 hours the tissue was harvested and the sRNA extracted. The enzyme preparation from cotyledons was used to charge the tRNA samples. The acylated tRNA was fractionated on a Freon column.

The Effect of 6-Benzyladenine on Other Amino Acid tRNA Species

To determine whether 6-BA treatment of hypocotyls results in changes in the amounts of other amino acid tRNA isoacceptor species, possible differences in valyl-, tyrosyl-, seryl- and phenylalanyl-tRNAs were tested using the Freon column. As illustrated there were no apparent differences between the valyl- and phenylalanyl-tRNAs of control and 6-BA-treated hypocotyls (Figure 2).

However, as noted in Figure 3, the seryl-tRNA profiles of acylated tRNAs from control and 6-BA-treated hypocotyls are different. The tRNA from 6-BA treated hypocotyls contains an additional seryl-tRNA peak. This third peak elutes after the 2 major seryl-tRNA peaks which were identical in both tRNA preparation. Data from preliminary experiments indicate that 6-BA treatment also results in changes in tyrosyl-tRNA isoacceptor species. However, because of inconsistent elution patterns for tyrosyl-tRNA, little can be concluded from the data. From our present data it is noteworthy that 6-BA treatment appears to result in changes in tRNA species (leucine, serine and tyrosine) which contain the cytokinin, IPA.

IV. DISCUSSION

Cytokinins are found in transfer RNAs of animals, microorganisms and plants. Furthermore, in those cases where the nucleotide sequence is known, the native cytokinin, IPA (or a modified form), is found adjacent to the anticodon. At the moment, it appears that all IPA-containing species of tRNA recognize codon triplets which

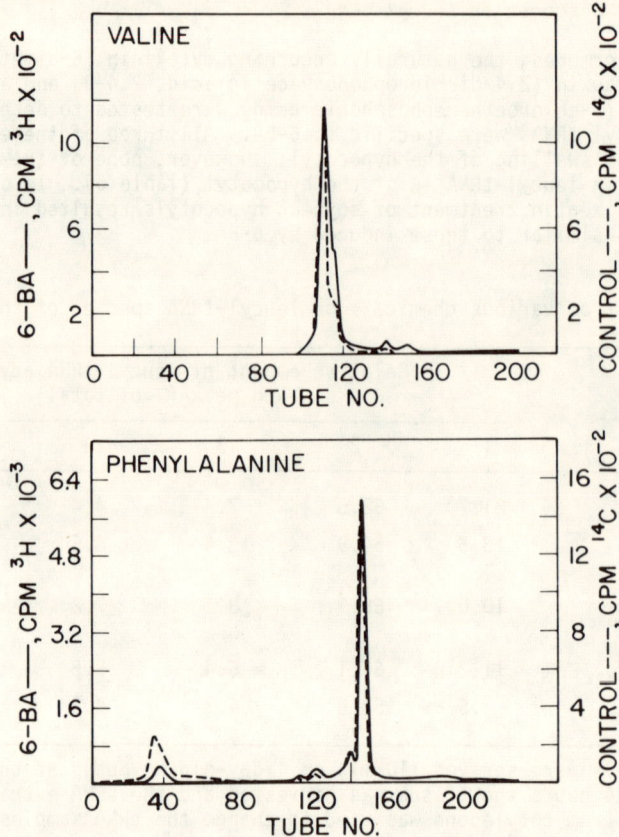

Fig. 2. A. Fractionation of valyl-tRNA of control and 6-benzyladenine treated hypocotyls on a Freon column. The solid line represents tRNA from 3×10^{-4}M 6-BA treated hypocotyls (2.5 mg) acylated by cotyledon enzyme (2.0 mg) with 0.5 mc ^{3}H-L-valine. The dashed line is control tRNA (2.46 mg) acylated by cotyledon enzyme (2.0 mg) with 10 μc ^{14}C-L-leucine. The gradient was run from 0.2 M to 1.0 M NaCl.

B. Fractionation of phenylalanyl-tRNA of control and 6-benzyladenine treated hypocotyls on a Freon column. The solid line represents tRNA from 3×10^{-4}M 6-BA treated hypocotyls (2.5 mg) acylated by cotyledon enzyme (2.0 mg) with 0.3 mc ^{3}H-L-phenylalanine. The dashed line is control tRNA (2.46 mg) charged by cotyledon enzyme (2.0 mg) with 0.1 mc ^{14}C-L-phenylalanine. The gradient was run from 0.2 M to 1.0 M NaCl.

begin with U. Thus, the anticodon loop of the cytokinin-containing tRNAs could be of at least 16 codon sequences of the following type: A⎯ ⎯IPA. By eliminating the nonsense, terminator and phenylalanine codons it seems that IPA-containing tRNAs would be confined to less than 10 codons.

It can be assumed that the IPA adjacent to the anticodon of these tRNA species plays some role in mRNA-tRNA-ribosome recognition and that these tRNA species are required for the translation of some mRNAs into specific proteins. Based on these assumptions it is possible to speculate on the mechanism of action of cytokinins in plants.

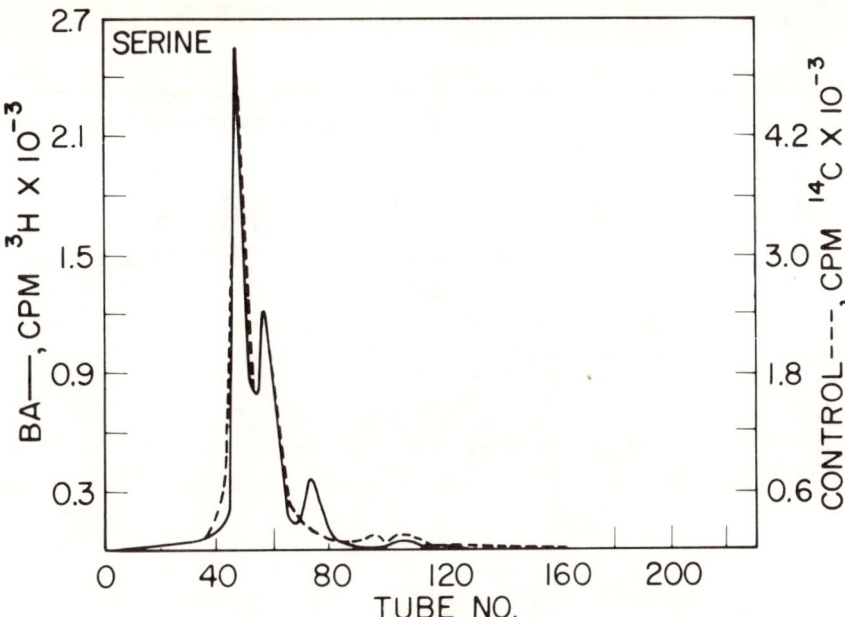

Fig. 3. Fractionation of seryl-tRNA of control and benzyladenine treated hypocotyls on a Freon column. The solid line represents tRNA from 1×10^{-5}M 6-BA treated hypocotyls acylated by cotyledon enzyme with 0.5 mc ^{3}H-L-serine. The dashed line denotes control tRNA acylated by cotyledon enzyme with 25 µc ^{14}C-L-serine. In both reactions 2.4 mg tRNA and 2.0 mg enzyme were incubated for 30 min at 25°C. The gradient was run from 0.5 to 1.0 M NaCl.

From the observation that cytokinins appear not be incorporated into tRNA as precursor units, but rather the isopentenyl group comes from mevalonic acid, it seems likely that cytokinins have no effect on the synthesis of IPA-containing tRNA. Furthermore, since a number of chemicals including various isomers of IPA, kinetin (6-furfurylaminopurine), 6-benzyladenine and even diphenylurea have cytokinin biological activities, it appears that these cytokinins do not participate in tRNA synthesis. If this is true, then what is the mechanism of action of cytokinins?

A model presented in Figure 4 is based on the speculation that specific nucleases break the primary structure of the IPA-containing tRNA. The IPA would provide an unique attachment for an enzyme to bind and subsequently break the phosphodiester bonds. The action of this enzyme would destroy the function of the tRNA and yield free cytokinin, IPA. Cytokinins, including 6-benzyladenine and diphenylurea, added exogenously to plants or tissue cultures would defer senescence by essentially protecting the cytokinin-containing tRNA species. This protection would be mediated by the exogenously added cytokinin binding to the nuclease and competitively inhibiting its action. Therefore, tissue treated with cytokinin would retain its IPA-containing tRNA species and continue to synthesize essential proteins. This tissue would continue to undergo cell division and regeneration.

Preliminary unpublished work of Cherry and Osborne provide suggestive evidence of a tRNA-specific nuclease. It is of interest that Chen and Hall (1969) have suggested that the pathway of IPA synthesis could be via tRNA biosynthesis. It is to be emphasized that this hypothetical model is presented as a working base to further investigate the mechanism of action of the cytokinins at the molecular level.

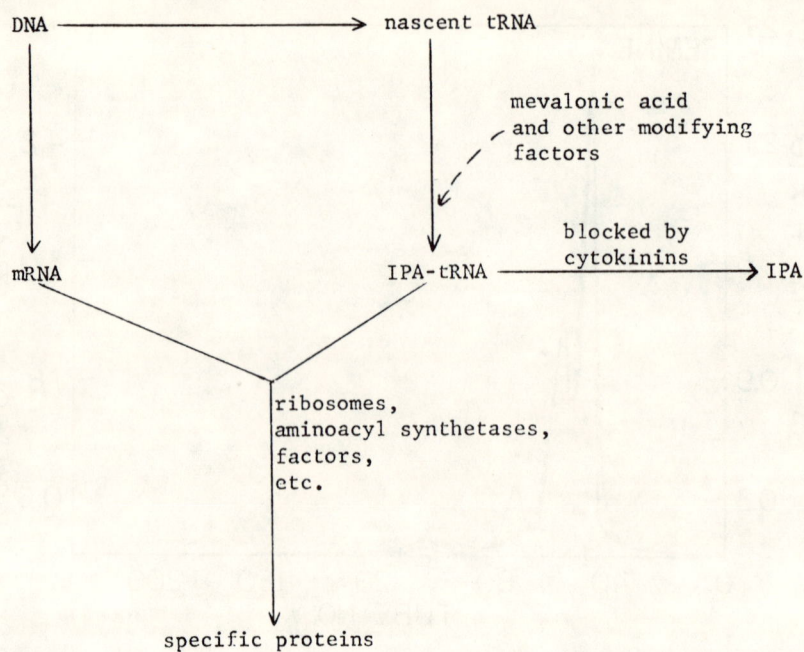

Fig. 4. A hypothetical model of the role of cytokinins in tRNA metabolism. The model implies that the cytokinins act mainly as a sparing agent to protect isopentenyladenine (IPA)-containing tRNA species. The cytokinin would provide this protection by inhibiting a specific nuclease.

V. SUMMARY

Preparations of soluble RNA and aminoacyl-sRNA synthetase were isolated and purified from soybean hypocotyl and cotyledon tissues. The amount and rate of aminoacylation by the soybean systems are comparable to those from bacterial and animal systems. The charged leucyl-transfer RNA (leucyl-tRNA) of soybean cotyledons can be fractionated into six radioactive peaks on a Freon column. Four leucyl-tRNA peaks are observed with the homologous hypocotyl system.

The results show that 6-benzyladenine (6-BA) increased the amount of leucyl-tRNA$_{5\ \&\ 6}$ while decreasing the amount of leucyl-tRNA$_1$ of the soybean hypocotyl. 6-Benzyladenine while lowering the amount of total aminoacyl-tRNA synthetase activity, did not effect the activity of leucyl-tRNA$_{5\ \&\ 6}$ synthetase. Therefore, it was necessary to use the enzyme of cotyledons to detect the changes in tRNA isoacceptor species. The effect of 6-BA on tRNA was only noted for leucyl-tRNA species; no changes in valyl- or phenylalanyl-tRNAs were obvious. The fractionation of seryl- and tyrosyl-tRNA on Freon columns indicated changes in isoacceptor species as a result of 6-BA treatment. However, because changes in one or more of the species were not consistently observed from experiment to experiment the data must be considered with caution.

VI. ACKNOWLEDGEMENTS

The research was supported by a grant (GB-7415) from the National Science Foundation. This report is journal paper number 4264 of the Purdue Agriculture Experiment Station.

VII. LIST OF REFERENCES

ANDERSON, M.B. and J.H. CHERRY (1969). Differences in leucyl-transfer RNAs and synthetase in soybean seedlings. Proc. Natl. Acad. Sci. U.S. 62, 202-209.
BELLAMY, A.R. (1966). Cytokinins in ribonucleic acids. Nature 211, 1093-1095.
BURROWS, W.J., D.J. ARMSTRONG, F. SKOOG, S.M. HECHT, J.T.A. BOYLE, N.J. LEONARD and J. OCCOLOWITZ. Cytokinin from soluble RNA of *Escherichia coli*: 6-(3-methyl-2-butenylamino)-methythio-9-D-ribofuranosylpurine. Science 161, 691-693.
CHEN, C.M. and R.H. HALL (1969). Biosynthesis of N^6-(Δ^2-isopentenyl) adenosine in transfer ribonucleic acid of cultured tobacco pith tissue. Phytochem. 8, 1687-1695.
FITTLER, F., L.K. KLINE and R.H. HALL (1968). Biosynthesis of N^6-(2-isopentenyl)$_2$ adenosine. The precursor relationship of acetate and mevalonate to the Δ^2-isopentenyl group of the transfer ribonucleic acid of microorganisms. Biochem. 7, 940-944.
FOX, J.E. (1966). Incorporation of a kinin, N, 6-benzyl-adenine into soluble RNA. Plant Physiol., 41, 75-82.
GEFTER, M.L. and R.L. RUSSEL (1969). Role of modifications in tyrosine transfer RNA: A modified base affecting ribosomal binding. J. Mol. Biol., 39, 145-147.
HALL, R.H. (1967). An N^6-(alkyl) adenosine in the sRNA of *Zea mays*. Ann. N.Y. Acad. Sci. 144, 258-259.
HALL, R.H., L. CSONKA, H. DAVID and B. McLENNON (1967). Cytokinins in the soluble RNA of plant tissues. Science 156, 69-71.
HOLLEY, R.W., J. APGAR, G.A. EVERETT, J.T. MADISON, N. MARQUISUS, S.N. MERRILL, J.R. PENSWICK, A. ZAMIR (1965). Structure of ribonucleic acid. Science 147, 1462-1465.
LETHAM, D.S. and R.K. RALPH (1967). A cytokinin in soluble RNA from a higher plant. Life Sci. 6, 387-394.
MADISON, J.T., G.A. EVERETT and HUEI-KUEN KONG (1967). Oligonucleotides from yeast tyrosine transfer ribonucleic acid. J. Biol. Chem. 242, 1318-1323.
MATSUBARA, S., D.J. ARMSTRONG and F. SKOOG (1968). Cytokinins in tRNA of *Corynebacterium fascians*. Plant Physiol. 43, 451-453.
ROBBINS, M.J., R.H. HALL and R. THEDFORD (1967). N^6-(Δ^2-isopentenyl) adenosine. A component of the transfer ribonucleic acid of yeast and mammalian tissues, methods of isolation and characterization. Biochem. 6, 1837-1848.
SKOOG, F., D.J. ARMSTRONG, J.D. CHERAYIL, A.C. HAMPEL and R.M. BOCK (1966). Cytokinin activity: Localization in transfer RNA preparations. Science 154, 1354-1356.
ZACHAU, H.G., D. DUTTING and H. FELDMAN (1966). Nucleotidsequence zweier serinspezifischer Transfer-Ribonucleinsauren. Angew. Chem. 78, 392.

PLANT GROWTH SUBSTANCES, 1970

Effect of Growth Substances on Rapidly Synthesized RNA in Sterile Tobacco Tissue

Dieter Klämbt

Botanisches Institut der Universität Bonn, Germany

Dedicated to Professor K. Mothes on the occasion of his 70th birthday.

I. INTRODUCTION

It has long been assumed that plant growth substances act through the stimulation of RNA and protein synthesis (Key 1969). But all recent papers confirming such a hypothesis report a more or less unspecific increase of all classes of RNAs. Masuda et al. (1967) demonstrated a growth independent enhancement of RNA-synthesis in *Avena* coleoptiles. If these results are correct, as expected, we should think about an influence of growth substances on the RNA-polymerase itself.

Growth substances cause - as their name indicates - cell multiplication and/or cell elongation. It is obvious that such growth stimulation implies a general increase of all or nearly all cell structures and populations of molecules. That means that after a certain time of incubation with the growth substance, an increase of all RNA and protein fractions should be found. This general enhancement is not necessarily effected directly by growth substances. In looking for the site of primary action of growth substances we have to shorten the time of incubation and to work with sterile plants or plant parts (Burdett and Wareing 1968). There is another difficulty to overcome. If growth substances have a regulating capacity one has to assume that changes may occur primarily on a qualitative rather than a quantitative scale. Methods have therefore to be developed to test for such qualitative differences.

II. MATERIAL AND METHODS

Material

Sterile tobacco callus tissues - Wisconsin 38 - were used throughout the experiments. About 2.0 to 3.0 g of second subculture "T_{II}", grown on 0.02 mg/l kinetin and 0.2 mg/l naphthalene-acetic acid for 2 to 3 weeks were incubated for 60 min and 90 min in 3 ml of the basic medium (Linsmaier and Skoog 1965) without PO_4^{3-} and agar without or with auxin (4.0 mg/l naphthalene-acetic acid) and cytokinin (0.4 mg/l kinetin). Each incubation sample contained 0.5 or 2 mC carrier-free $^{32}PO_4^{3-}$ diluted 1:1000 by $^{31}PO_4^{3-}$ and sterilized by autoclaving at $123°C$ for 20 min. During treatment samples were shaken at $27°C$ and diffuse dim light.

RNA-preparation

At the end of the incubation the tissues were collected on fiberglass filters and immediately frozen under liquid nitrogen. In a mortar precooled with liquid nitrogen the tissues were ground under nitrogen and transferred to 50 ml flasks each containing: 5 ml of 0.01 M Tris-HCl, 0.06 M KCl, 0.01 M $MgCl_2$, pH 7.6, and 5.0 mg bentonite (Serva 14515), and 5 ml water saturated phenol containing 2% sodium dodecyl sulfate and 0.1% 8-hydroxyquinoline. The flasks were vigorously shaken for 90 min at $27°C$. After cooling to $4°C$ the extracts were centrifuged at 4000 x g, 20 min at $4°C$. Deproteination with phenol was usually repeated once more. To the upper aqueous phases 2.5 parts of 2% potassium acetate ethanol were added for RNA precipitation. After aging for at least 4 hr at $-20°C$ the precipitates were collected by centrifugation and lyophylised.

RNA-fractionation

Neither sucrose gradient centrifugation nor chromatography on MAK columns is a convenient method of separating rapidly synthesized RNA from rRNA. Therefore column chromatography on benzoylated-naphthoylated DEAE-(BND) cellulose was used (Sedat *et al.* 1967, Kiger and Sinsheimer 1969). This ion-exchange cellulose fractionates bacterial nucleic acids in respect to their primary and secondary structure. Since the properties of the mRNA of bacterial and higher plant cells ought to be similar, we tried to adapt this method. The results will show that although the method is useful its development is still in progress.

The BND-cellulose (Serva 45025) was washed with 1.0 M NaCl in 0.01 M Tris-HCl, 0.001 M EDTA, pH 8.1, to remove UV-absorbing material and finally equilibrated with 0.3 M NaCl in the same Tris-EDTA-buffer.

The lyophylised RNA-samples were suspended and dissolved in 0.3 M NaCl in the same buffer at $0°-4°C$ and applied to the columns. Each loaded column (1 x 4 cm) was extensively washed with 0.3 M NaCl before starting elution with the first linear gradient of 15:15 ml, 0.3 M NaCl to 1.0 M NaCl in Tris-EDTA-buffer. The gradient elution was followed by 5.0 or 10.0 ml of 1.0 M NaCl and finally by a second linear gradient of 10:10 ml, 0.0 to 2.0% caffein, 1.0 M NaCl in Tris-EDTA-buffer. For complete elution of all radioactivity the second gradient was followed by 0.1 N NaOH, 2% caffein, 1.0 M NaCl, Tris-EDTA-buffer.

For further differentiation and characterization sucrose gradient centrifugation (20-5%) was used after extensive dialysis against distilled water and lyophylisation of the various fractions. The linear sucrose gradient was made up in 0.01 M sodium acetate, 0.1 M NaCl, and 0.001 M $MgCl_2$, pH 5.1, and the lyophylised samples were applied in 1.0 ml of the same buffer.

After centrifugation at 24,000 rpm for 19 hr in the SW 25.1 Rotor of a Spinco L50 at $8°C$, 8 drops equal to about 0.5 ml were collected, after puncturing the centrifuge tube. All measurements of radioactivity were made by use of Cherenkov radiation of the whole fractions in a Packard Tri-Carb Scintillation Spectrometer 3320.

UV absorption was measured with or without dilution in a Zeiss Spectrophotometer PMQ II.

III. RESULTS AND DISCUSSION

The figures 1 and 2 show typical profiles of radioactive nucleic acids extracted from tobacco tissue after short-time incubation.

At first we were looking for early visible differences on such profiles due to the different treatments of the tissues. But no reproducible quantitative changes were observed. Therefore we further analysed the main peak, because the data of Sedat *et al.* (1967) and Kiger and Sinsheimer (1969) predict that the caffein peak should contain most parts of RNA in primary structure.

Since the profiles of the 60 min samples from the BND-cellulose column do not show reproducible quantitative differences, samples of the different caffein peaks are adjusted and corrected to comparable total radioactivities. Therefore differences in the profiles are now due to qualitative differences between the various samples.

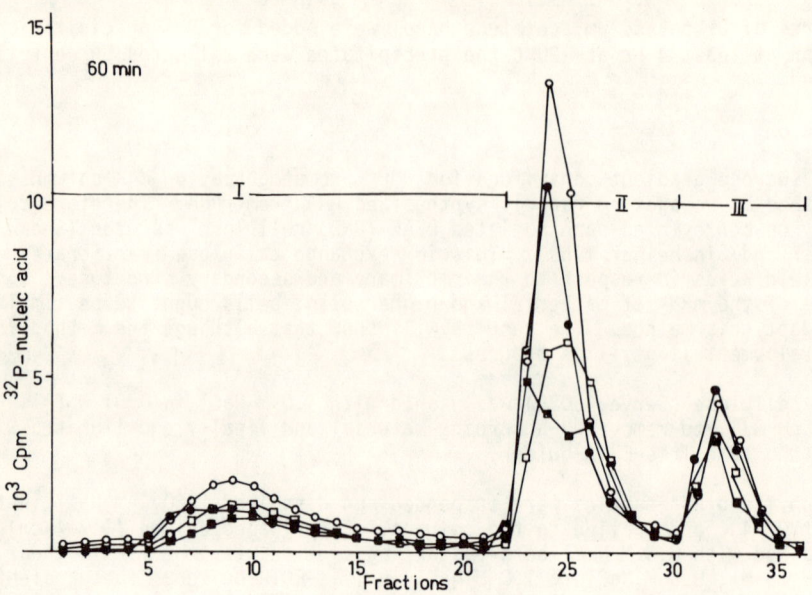

Fig. 1. RNA elution profiles on BND-cellulose in terms of ^{32}P-activity from samples incubated for 60 min.

I: NaCl gradient: 0.3 to 1.0 M NaCl + 5.0 ml 1.0 M NaCl in Tris-EDTA-buffer.

II: Caffein gradient: 0.0 to 2.0% caffein in 1.0 M NaCl, Tris-EDTA.

III: 0.1 N NaCl in 2.0% caffein, 1.0 M NaCl, Tris-EDTA.

Symbols: 0 ——— 0 sample of tissue incubated without growth substances; □ ——— □ sample of tissue incubated with auxin; ■ ——— ■ sample of tissue incubated with cytokinin; ● ——— ● sample of tissue incubated with auxin and cytokinin.

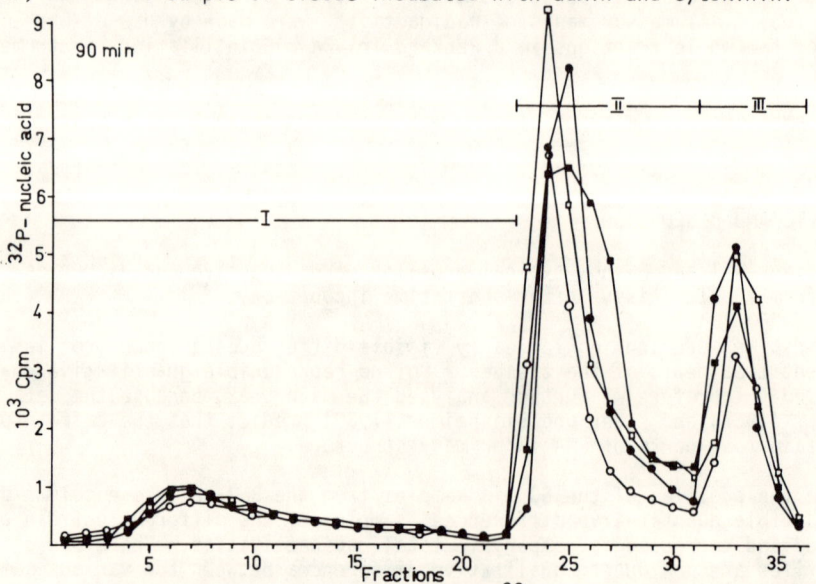

Fig. 2. RNA elution profiles on BND-cellulose. ^{32}P-activity from samples incubated for 90 min. Explanations and symbols as in fig. 1.

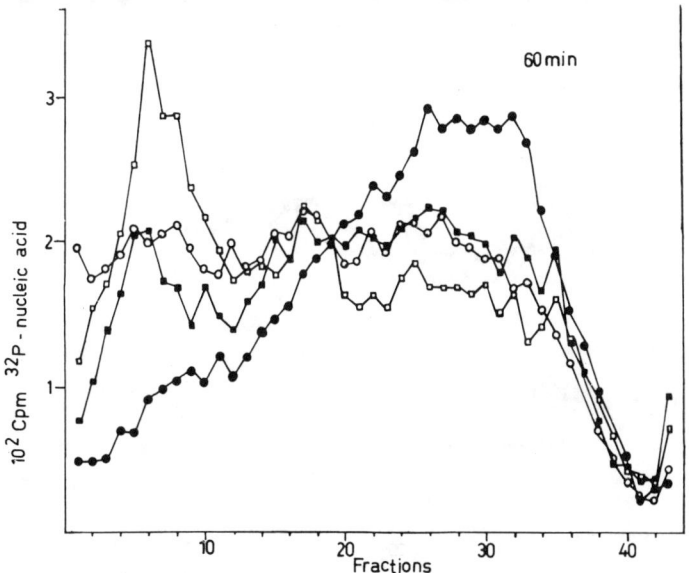

Fig. 3. RNA sedimentation profiles on sucrose gradients in terms of ^{32}P-activity from the caffein peaks of fig. 1. Symbols as in fig. 1.

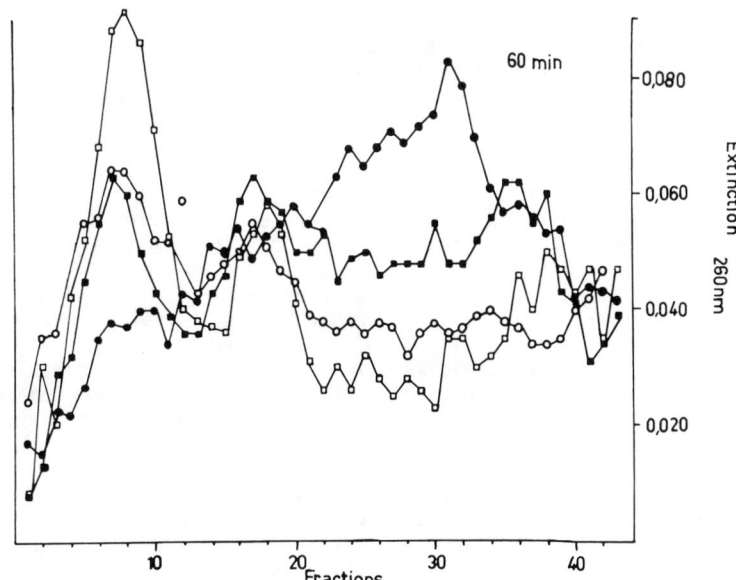

Fig. 4. The same sedimentation profile as in fig. 3 but read by UV-absorption at 260 nm. Symbols as in fig. 1.

Figure 3 and 4 clearly demonstrate differences of the profiles of the ^{32}P radio-activity and the UV-absorption of the RNA-samples of differently treated tissues after 60 min of incubation. The differences obtained were even more pronounced for similar RNA-samples after 90 min of tissue incubation (fig. 5).

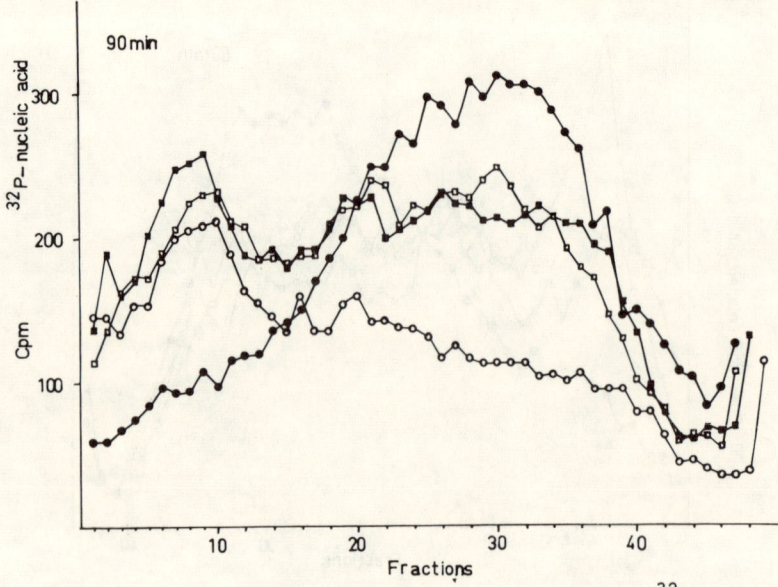

Fig. 5. RNA sedimentation profiles on sucrose gradients read by ^{32}P-activity from the caffein peaks of fig. 2. Symbols as in fig. 1.

After 60 min of incubation the main differences exist between the RNA of tissues treated in the complete medium with both auxin and cytokinin, and the RNA of tissues untreated by either growth substance. The other experimental RNA profiles are similar to those of the controls. Nevertheless after 90 min of incubation the RNA of samples treated with auxin or cytokinin begin to show intermediate profiles.

Differences in the distribution of RNA along the sucrose gradient must be very cautiously interpreted. We observed peaks for the UV-absorption and the ^{32}P-activity, at the region of the heavy and light rRNA. It is possible that rRNA or at least part of it is eluted from the column by the caffein gradient. In any case, these peaks are missing in the auxin + cytokinin samples although it is highly improbable that the complete medium for tissue growth would reduce the rRNA. Another fact is the more or less similar specific activity found in the large molecular weight RNA region for auxin or cytokinin treated samples and in regions of lighter molecular weight RNA for the (auxin + cytokinin) -treated samples. We expected some remarkable differences in specific activities between rRNA and rapidly labelled RNA. This point is open for discussion.

It seems to us that under short-time incubation, RNA synthesis is not regulated by a limiting enzyme situation but by limitation at the level of the genes, restricting transcription. Therefore each early regulation at the level of the genes has first to effect qualitative changes of the total informational RNA population. We assume that our results demonstrate such differences. If this conclusion is valid, auxin and cytokinin could be held to act through inductor properties.

IV. SUMMARY

In experiments on RNA synthesis in sterile tobacco tissues it could be demonstrated that auxin and cytokinin together and, to some extent, either substance by itself, cause qualitative differences in rapidly synthesized RNA. The method used was a combination of column chromatography on benzoylated-naphthoylated DEAE-cellulose and sucrose gradient centrifugation. Although adaptation of the method is still in progress, the results described are tentatively discussed as indicating possible inductor functions of auxin and cytokinin.

V. ACKNOWLEDGEMENTS

This work is supported by a grant from the Landesamt für Forschung, Düsseldorf, Germany.

VI. REFERENCES

BURDETT, A.N. and P.F. WAREING (1968). The effects of kinetin and contaminating bacteria on the incorporation of ^{32}P-orthophosphate into various fractions of nucleic acid extracted from radish leaves. Planta (Berl.) 81, 88-96.
KEY, J.L. (1969). Hormones and nucleic acid metabolism. in Ann. Rev. Plant Physiol. 20, 449-474.
KIGER, J.A. Jr. and R.L. SINSHEIMER (1969). Vegetative lambda DNA. IV. Fractionation of replicating lambda DNA on benzoylated-naphthoylated DEAE-cellulose. J. Mol. Biol. 40, 467-490.
LINSMAIER, E.M. and P. SKOOG (1965). Organic growth factor requirements of tobacco tissue cultures. Physiol. Plant. 18, 100-127.
MASUDA, Y., E.TANIMOTO, and S. WADA (1967). Auxin-stimulated RNA synthesis in oat coleoptile cells. Physiol. Plant. 20, 713-719.
SEDAT, J.W., R.B. KELLY, and R.L. SINSHEIMER (1967). Fractionation of nucleic acid on benzoylated-naphthoylated DEAE-cellulose. J. Mol. Biol. 26, 537-540.

Auxin in sRNA Fraction of Mung Bean Hypocotyl

Toshio Yamaki and Kō Kobayashi

Department of Pure and Applied Sciences and Biological Institute, College of General Education, University of Tokyo, Komaba, Meguro-ku, Tokyo, Japan

I. INTRODUCTION

Auxin regulates many aspects of the growth of plants. At the cellular level, it regulates cell expansion and cell division, at the subcellular level, the synthesis of many kinds of enzymes, e.g. acid phosphatase (Palmer, 1967), cellulase (MacLachlan, et al, 1967), hemicellulase, β-1,3-glucanase (Masuda, 1967), peroxidase (Glasziou et al, 1967, Galston et al, 1967). This indicates the multiplicity of auxin action even in enzyme synthesis. This multiplicity will be observed in nucleic acid synthesis, when we consider the correspondence of an enzyme to a messenger RNA. Actually, auxin promotes the synthesis of mRNA (Key and Ingle, 1967), and there must be many kinds of mRNAs whose synthesis are regulated by auxin. What is the main point of action of auxin? O'Brien et al. (1967) and others tried to find a repressor protein(s) for a gene which is derepressed by auxin. This is a possible way to elucidate the multiple action of auxin monistically.

When we consider the localization of auxin in a cell, both free and bound auxin are in high concentrations in the ribosomal fraction or the soluble fraction of cell homogenate (Nakamura et al, 1963). And Galston et al. (1964), Key and Ingle (1967) observed the incorporation of radioactivity from ^{14}C-labelled auxin into sRNA fraction, and in case of 2,4 D-^{14}C, a larger amount of radioactivity was found in sRNA than in rRNA and DNA fractions. Gray and Lane (1967) found 5-carboxymethyluridine in tRNA of wheat embryo and yeast, Murao et al. (1970) found uridine-5-oxyacetic acid as a minor constituent of E. coli valine tRNA.

These findings suggest the possibility that an auxin(s) regulates enzyme synthesis as a minor constituent of a tRNA. In the present experiment we tried to find the incorporation of IAA-2-^{14}C in tRNA or tRNA-like compounds and investigate the presence of native auxin in sRNA.

II. MATERIALS AND METHODS

Materials and IAA treatments

Cotyledons, apical buds, roots were removed from etiolated mung bean (*Phaseolus aureus* Roxb. cv. Black) hypocotyls and the hypocotyls were placed vertically in water 24 hrs. in the dark at 32°C. This treatment increases their sensitivity to auxin (Kuraishi et al, 1968). Five mm sections were excised from these preincubated hypocotyls, 2.5 mm below the tip. About 1000 sections were floated on 25 ml of 10^{-4}M IAA solution containing 1μCi/ml IAA-^{14}C. After 3 or 4 hrs of incubation at 32°C in the dark, they were thoroughly washed with deionized water. For fractionation on a MAK column (Sueoka et al, 1962), nucleic acids were extracted as indicated below; for the benzoylated DEAE-cellulose (BD-cellulose) column (Murao et al, 1969), the acids were extracted according to Zubay's method. The latter column was used mainly for the fractionation of native sRNA binding IAA. In this case, hypocotyls were used for extraction without preincubation.

EXTRACTION of NUCLEIC ACID from ETIOLATED MUNG BEAN HYPOCOTYL

etiolated mung bean hypocotyl sections
|
floated on 25 ml IAA-^{14}C (10^{-4}M) solution for 4 hr. at 32°C
|
washed with distilled water
|
homogenized in 1% SLS-Tris buffer and 90% phenol
|
centrifuged at 1300 x g for 15 min.
|
├─ phenol layer + plant residue ───────────── aqueous layer
|
added 5 ml 1% SLS-Tris buffer
|
stirred in ice box for 10 min.
|
centrifuged at 1300 x g for 15 min.
|
├─ phenol layer ──── aqueous layer
   plant residue |
                    added 2 vol. of ethanol
                           |
                    kept in deep freezer, overnight
                           |
                    centrifuged at 3500 x g for 10 min.
                           |
                    loaded on MAK column
                           |
                    washed with 250 ml Tris buffer
                           |
                    fractionated with NaCl solution

RNA hydrolysis and detection of nucleotides

The mixture of the radioactive sRNA fraction obtained from MAK column, yeast RNA, 2 volumes of ethanol and 0.15 M potassium acetate (final conc.) was kept at -10°C at least 1 hr. The precipitate was hydrolysed with 0.3 N KOH at 37°C for 18 hrs, neutralized and centrifuged. The supernatant was loaded on Dowex 1 (x2) of formic type and the four nucleotides were eluted with 0 - 4 M formic acid; the remaining bound material was eluted with 0.5 N KOH.

Detection of IAA-^{14}C

0.5 N KOH hydrolyzate of RNA or the above mentioned 0.5 N KOH eluate was acidified to pH 3.0 - 3.5 with 1.0 M tartaric acid and extracted with ethyl ether. The ether layer was dried and chromatographed, using isopropanol - 28% ammonia - water (8/1/1, v/v) and butanol - acetic acid - water (4/1/1, v/v). 0.5 µg of authentic IAA was always cochromatographed.

Measurement of radioactivity

A low background gasflow counter (Aloka) and a liquid scintillation counter (Packard, model 574) were used for the measurement of radioactivity. The scintillater contained 3 g of 2,5-diphenyloxazole and 100 mg of 1,4 bis 2-(4-methyl-5-phenyloxazolyl) benzene in 1 liter of toluene.

III. RESULTS

Incorporation of radioactivity into nucleic acid

The nucleic acid obtained from hypocotyl sections treated with IAA-^{14}C was loaded on a MAK column, washed with 200 ml of 0.05 M Tris-HCl buffer of pH 7.5 and fractionated with a linear gradient of NaCl. The typical pattern of nucleic acid on MAK column was obtained (Fig. 1).

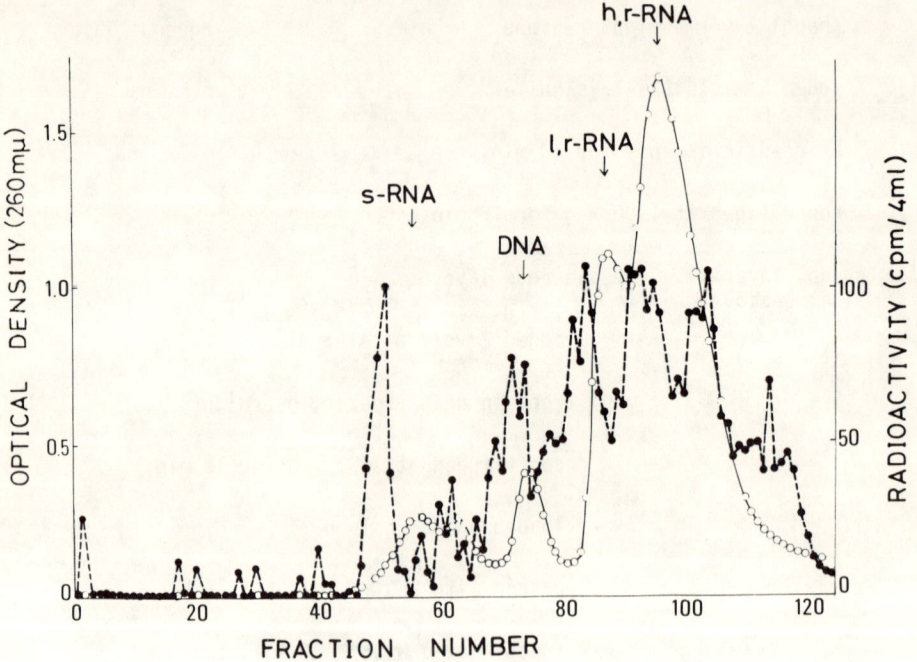

Fig. 1. Chromatogram of nucleic acids on MAK column. The nucleic acids were extracted from sensitized hypocotyls treated with 10^{-4}M IAA containing 1μCi/ml IAA-2-^{14}C and fractionated with a linear gradient of NaCl(0-1M) in Tris-HCl buffer of pH 7.5. O———O :optical density; ●- - -● : radioactivity.

In this column, peaks of radioactivity were found in fractions 46-55(sRNA), 60-80(DNA), 81-88, 89-98, 99-108 (rRNAs). The radioactivity in these fractions did not derive from contamination by IAA-^{14}C because radioactivity was not detected on Millipore filters when a mixture of the nucleic acid and IAA-^{14}C (20,000 cpm) was loaded on a MAK column and fractionated using the method described above. But almost the original amount of free IAA-^{14}C was recovered in the Tris-HCl buffer washing of the column.

Isolation of IAA-^{14}C from radioactive RNA fractions

The ether-soluble acidic substance in the alkali hydrolysates of radioactive sRNA or other radioactive RNA fractions was paper chromatographed with 2 solvent systems. A radioactive substance having the same Rf values as those of authentic IAA was detected when sRNA was examined; however, very little amount of IAA-^{14}C was recovered from other RNA fractions. The alkali hydrolysate of rRNA and other fractions were loaded on Dowex 1 (x2) column and were eluted stepwise with a linear gradient of formic acid and 0.5 N KOH.

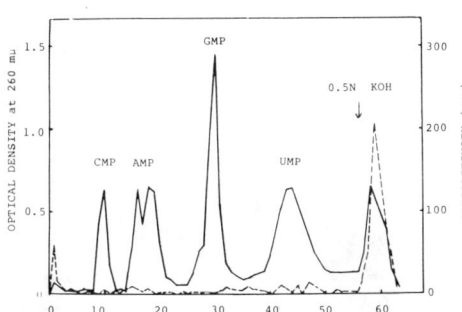 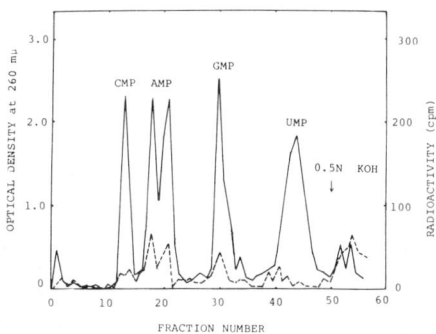

Fig. 2. Chromatogram of nucleotides on Dowex 1 column. Upper figure: nucleotides obtained from sRNA fraction of MAK column (Fig. 2). Lower figure: nucleotides obtained from other RNA fractions of Fig. 2. Solid line: optical density; dashed line: radioactivity.

As shown in Fig. 2, the hydrolysate from rRNA had radioactivity in adenilic and guanilic acids, and also a small peak in the fraction eluted with 0.5 N KOH. On the other hand, the hydrolysate from sRNA had no peak of radioactivity in either of the four nucleotides, but a large peak in the fraction eluated with KOH. The ether-soluble acidic substance in fractions eluted by 0.5 N KOH from hydrolysed sRNA and sRNA were paper chromatographed. Most of the radioactivity obtained from the sRNA column was recovered as IAA-^{14}C, but from the rRNA column the amount of IAA-^{14}C was very little. These results show that the radioactivity in the sRNA fraction is attributable to IAA-^{14}C, but this is not the case with rRNA.

Rechromatography of the radioactive sRNA fraction on a DEAE-cellulose column

A mixture of the radioactive sRNA fractions obtained from a MAK column and 'cold' nucleic acid from mung bean hypocotyl was loaded on the DEAE-cellulose column. Nucleic acid was eluted with a linear gradient of NaCl containing 0.01 M Tris-HCl buffer (pH 7.6) and 0.005 M Mg-acetate. The remaining material on the column was eluted with KOH (Fig. 3). The radioactivity remains with the sRNA fraction even in this fractionation.

The radioactive sRNA fractions from the DEAE-cellulose column were combined and hydrolyzed with 0.5 N KOH. From this hydrolysate almost all the radioactivity was recovered as IAA-^{14}C on paper chromatograms. These results show that the compound(s) binding IAA-^{14}C behaves like sRNA not only on a MAK column but also on the DEAE-cellulose column.

Rechromatography of the radioactive sRNA fraction on a BD-cellulose column

A mixture of the radioactive sRNA fractions obtained from the MAK column and cold *E. coli* sRNA was chromatographed on a BD-cellulose column with a linear gradient of NaCl (0 - 1.0 M) in 0.02 M Na-acetate (Gillam *et al*, 1967) and next with a linear gradient of NaCl (L.0 - 2.0 M) in ethanol having a linear gradient of 0 to 20% (v/v). The result shows that the radioactivity is always located in a determined fraction(s) of sRNA on the column (Fig. 4). The presence of bound IAA-^{14}C in these radioactive fractions was also detected by paper chromatography of the ether soluble acidic substances in the alkali hydrolysate of the fractions.

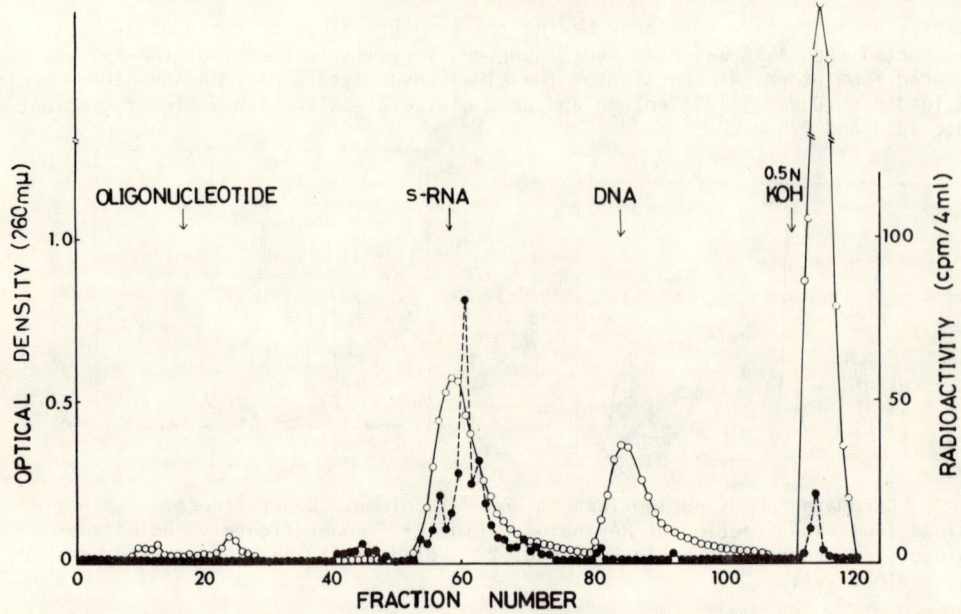

Fig. 3. Chromatogram of the radioactive sRNA on DEAE-cellulose column. The details are described in the text. o———o : optical density   ●- - -● : radioactivity.

pH 9 treatment of the radioactive sRNA fraction

The aqueous solution of the radioactive sRNA fraction obtained from MAK column was adjusted to pH 9 with Tris-buffer and incubated 90 min at 37°C. After addition of 2 volumes of ethanol, it was kept 120 min at -20°C and filtered using a Millipore filter (Table 1). The result indicates only 26% loss of the radioactivity from sRNA by this treatment.

Table 1. *pH 9 treatment of sRNA fraction*. The suspension of radioactive sRNA fraction is adjusted to pH 7 with Tris-buffer and incubated 90 min. at 37°C. After addition of 2 volumes of ethanol, the suspension is kept 120 min. at -20°C and filtered on a Millipore filter. Radioactivity on the filter is counted.

	cpm	%
Initial	342	100
Hydrolized sample	254	74.3

Dialysis of the radioactive sRNA fraction

The aqueous solution of radioactive sRNA fraction obtained from MAK column was dialysed with deionized water, 10^{-3}M IAA and 7M urea in a cold room overnight. The dialyzed solution was added to 2 volumes of ethanol and was kept overnight at -10°C. RNA was collected on a Millipore filter. The filter was dried and the radioactivity was counted (Table 2).

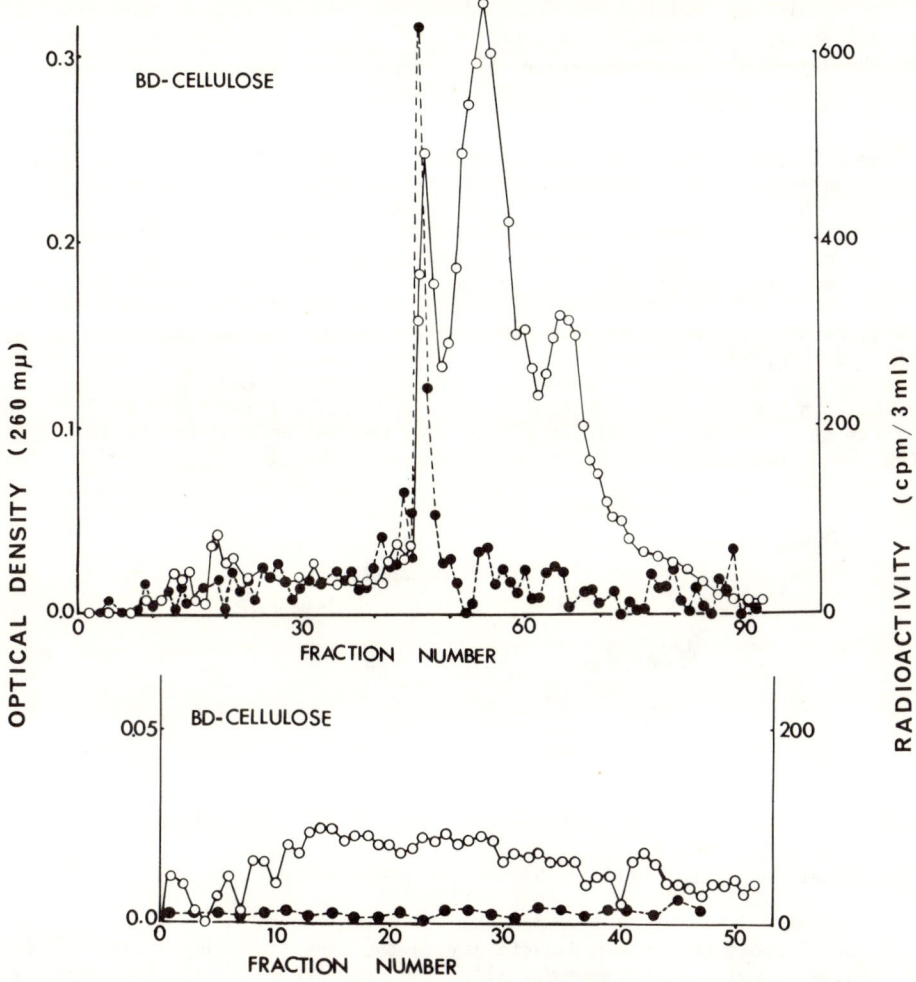

Fig. 4. Chromatogram of the radioactive sRNA on BD-cellulose column. Upper figure: sRNA fractionated with a linear gradient of NaCl (0-1.0M) in 0.02M Na-acetate. Lower figure: sRNA fractionated with a linear gradient of NaCl (1-2M) in ethanol having a linear gradient of 0 to 20%. The details are described in the text. o——o : optical density •---• : radioactivity.

Table II. *Dialysis of radioactive s-RNA fraction*. Aqueous solution of radioactive s-RNA fraction is dialyzed with deionized water, 10^{-6} M IAA and 7 M urea.

Treatment	Counts (on millipore filter) (cpm/10 ml)
Initial	141.8
Dialyzed with	
water	133.1
10^{-6} M IAA	148.5
7 M Urea	148.2

The results in Tables 1 and 2 show that the radioactivity of this fraction is caused neither by the absorption of IAA-^{14}C, nor by hydrogen bonding of IAA-^{14}C, but by the production of a compound(s), which binds IAA-^{14}C by covalent bonds.

Enzyme digestion of the radioactive sRNA fraction

The radioactive sRNA fraction obtained from a MAK column was dialyzed against deionized water in a cold room overnight and used for enzyme digestion.

Digestion with RNase: One hundred units of RNase T_1, 1 μg/mg of bovin pancreatic RNase and 0.1 ml of 2x10^{-2} M EDTA were added to the aqueous solution of the sRNA fraction at pH 7.4 and incubated at 37°C. An aliquot of ten ml of the reaction mixture was pipetted out, mixed with 10% acetone and 20% TCA, and left in an ice box for 1 hr. This mixture was filtered through a membrane filter and the radioactivity on the filter was counted (Fig. 5).

Digestion with pronase: The reaction mixture (pH 7.4) was incubated at 40°C, and the radioactivity in the macromolecular compound(s) was measured using the same method as for the RNase treatment (Fig. 5).

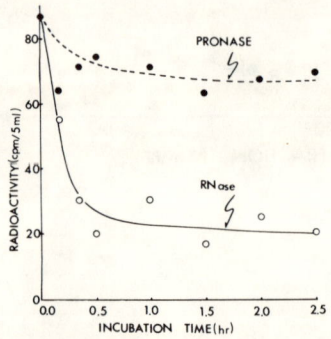

Fig. 5. Digestion of the radioactive sRNA by pronase and RNase.

Figure 5 shows that the radioactivity in the sRNA fraction decreases 78% by RNase treatment, while by pronase digestion the decrease is only 23%. These facts show that the compound(s) which binds IAA-^{14}C is not protein but RNA or oligonucleotides.

Digestion with T_1 RNase: The radioactive sRNA fraction of MAK column was concentrated with DEAE-cellulose column, desalted and lyophilized. The lyophilized material was dissolved in 100 μl of water. Fourty μl of this solution (5 OD) was mixed with 40 μl of *E. coli* tRNA (20 OD) and 20 μl of 1 mg/ml T_1 RNase, and incubated 6 hrs at 37°C, pH 7.0. The hydrolysed material was chromatographed on a DEAE-Sephadex A-25 column with a linear gradient of NaCl in 7 M urea (pH 7.5). The results (Fig. 6) show that the largest amount of radioactivity is located in fractions corresponding to nona- or deca-nucleotides.

Native auxin in sRNA fraction

Soluble RNA was prepared from 2 Kg of etiolated hypocotyls by Zubay's method (1962) and treated with 0.5 N KOH, 18 hrs, at 37°C. The hydrolyzate was acidified to pH 3.5, and extracted with ether. The ether extract was evaporated and the residue was dissolved in 0.25 ml of water. Ten agar blocks 2x2x2 mm were immersed in one half of this solution and another 10 agar blocks were immersed in 1/10 diluted solution. These agar blocks were used for the *Avena* curvature test. The result indicates the presence of auxin in the unfractionated sRNA fraction. When the amount of auxin is expressed as IAA equivalents, auxin content is less than one molecule of native IAA to 5,000 molecules of tRNA.

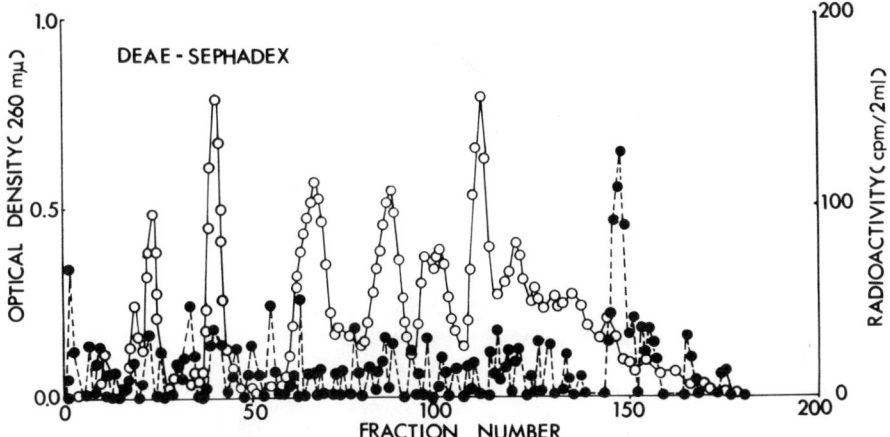

Fig. 6. Chromatogram of T1RNase hydrolysate of the radioactive sRNA on DEAE-Sephadex column. o———o : optical density   •– – –• : radioactivity.

The sRNA preparation was chromatographed on a BD-cellulose column and, according to the optical density of each fraction, the fractions were divided into 7 groups, as indicated in Fig. 7. Each group was hydrolysed with KOH and auxin was determined by the *Avena* curvature test. Auxin was obtained only from one group of sRNA fractions, where IAA-^{14}C was always found to be located in the previous experiments (Fig. 7).

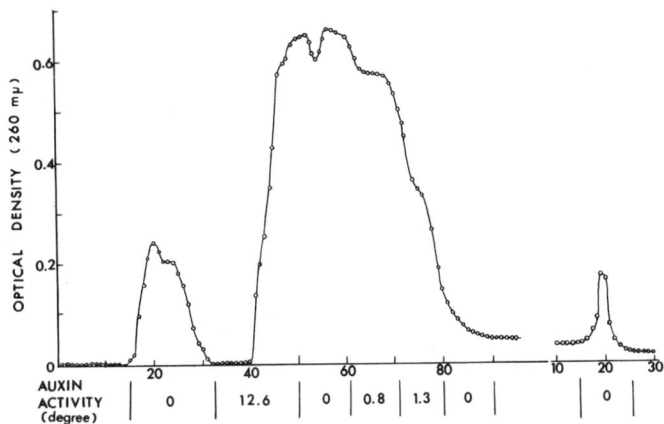

Fig. 7. Chromatogram of sRNA of mung bean hypocotyl extracted by Zubay's method on BD-cellulose column. Auxin activities found in the different groups of fractions are shown under the abscissa. Fractions 0-108: fractionated with a linear gradient of NaCl (0-1M) in 0.02M Na-acetate. Fractions 10-30 on the right side of the figure: fractionated with a linear gradient of NaCl (1-2M) in ethanol having a linear gradient of 0 to 20%. Fractions 10-32 are rich in protein; fractions 33-108 are RNA fractions and radioactivity locates in fractions 33-50.

This auxin was chromatographed on paper and the localization of auxin was determined using the *Avena* curvature test. Active spots were observed at Rf 0.05, 0.25-0.55 and 0.75 on a chromatogram developed by isopropanol-ammonia-water. The Rf of the main spot (0.25-0.55) seemed to correspond to the Rf of IAA. The results of these experiments show the possible presence of native IAA in a certain fraction of sRNA, where fed IAA-^{14}C is always located.

Amount of radioactive sRNA

The sections obtained from pre-incubated hypocotyls were floated on 10^{-4}M IAA containing IAA-^{14}C, at 32°C. Each 300 sections were taken up from the reaction solution 0.5, 1, 2, 3, 4 and 6 hrs after the start of the incubation. Nucleic acids were extracted by the method outlined in Fig. 1 and were fractionated on a MAK column. From measurements of the optical density and the radioactivity of each fraction, its specific radioactivity (cpm/OD) was calculated. The result is plotted against the growth rate in Fig. 8. It shows that specific radioactivity increases until 2 hrs of incubation and remains constant afterwards. If the amount of sRNA were constant during the whole course of the experiment or were only slightly increased, this curve would indicate a change in the amount of specific sRNA binding IAA-^{14}C.

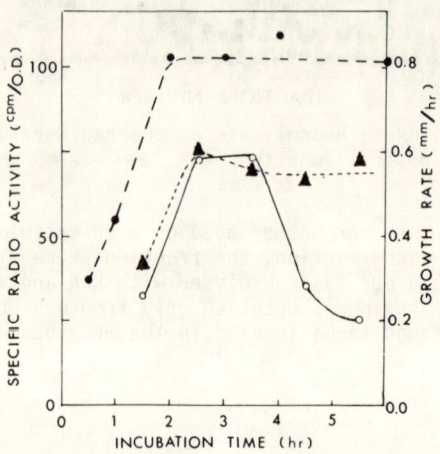

Fig. 8. Time course of the production of radioactive sRNA and changes in growth rate of hypocotyl sections.

●- - -● : specific radioactivity

▲- - -▲ : growth rate in presence of 10^{-4}M IAA

○———○ : growth rate in presence of 10^{-5}M IAA

The change in the amount of sRNA which binds IAA is parallel to the change of growth rate, which could mean that the amount of sRNA which binds IAA determines the growth rate.

IV. DISCUSSION

In the present experiment, incorporation of radioactivity from IAA-2-^{14}C into a nucleic acid fraction was observed; and only sRNA fractions on MAK, DEAE-cellulose and BD-cellulose columns released IAA-^{14}C by alkali hydrolysis and other nucleic acids released radioactive nucleotides or other compounds metabolized from IAA-2-^{14}C.

The compound(s) which binds IAA-^{14}C is not protein but tRNA(s) or oligonucleotide(s), because RNase releases the radioactivity from the radioactive sRNA fraction and pronase does not. The type of binding of IAA-^{14}C is due to covalent rather than hydrogen bonding because of the dialysis of the radioactive sRNA does not release IAA-^{14}C, while T$_1$ RNase hydrolysis produces a radioactive compound which behaves like a nona-nucleotide on a DEAE-Sephadex column. If the substance is tRNA or a tRNA like compound, the position where IAA is bound seems to be not the terminal -C-C-A, but near the anti-codon of tRNA, because pH 9 treatment of the sRNA does not release radioactivity but radioactive substance released by TRNase behaves on a DEAE-Sephadex column like nona-nucleotides having only one terminal molecule of guanilic acid. A nucleotide of this size and composition mostly occupies the position near the anticodon of tRNA.

Native auxin can be isolated from a specific part of the sRNA fraction of mung bean hypocotyl, where incorporated IAA-^{14}C always becomes located, and on the basis of paper chromatography, this auxin is IAA or at least contains IAA. These results also indicate that at least a part of the behaviour of native IAA in a cell or tissue can be traced qualitatively by the feeding of IAA-2-^{14}C as a tracer.

The amount of the tRNA or tRNA like substance which binds IAA is directly correlated with the growth rate of the tissue as indicated above. This and the results previously mentioned are consistent with the view that the mechanism of auxin action involves the production of a tRNA or a tRNA-like compound which binds auxin, and that this compound or compounds regulates protein (enzyme) synthesis at a translational level.

V. SUMMARY

Although the bulk of the IAA-^{14}C-2 supplied to auxin sensitized tissue is degraded, a small portion remains intact and is to be found mostly accompanying sRNA and partly rRNA or mRNA on MAK columns.

sRNA isolated from the MAK column releases ^{14}C-IAA only by alkaline hydrolysis or RNase digestion, but not with ethanol, ethyl ether or pH 9 treatments nor by dialysis with 7M urea. When radioactive sRNA isolated from the MAK column is further fractionated on a column of DEAE-cellulose or BD-cellulose, the radioactivity is always localized solely in a specific part of the sRNA fraction and this fraction releases IAA-^{14}C by alkaline hydrolysis.

The result of these experiments suggest the existence of some kinds of sRNA capable of binding IAA. The genesis of this type of sRNA is apparently a biological phenomenon, because it is produced only when IAA-^{14}C is supplied to living tissues. The formation of the sRNA which binds IAA-^{14}C is observed within 30 minutes after the supply of IAA-^{14}C and the sRNA becomes saturated with IAA at about 2 hrs after incubation. The amount of this sRNA is directly correlated with, and may regulate, the growth rate. Native auxin is isolated from specific parts of sRNA on BD-cellulose column where fed IAA-^{14}C is always located.

VI. ACKNOWLEDGEMENTS

We are indebted to Dr. S. Nishimura of the National Cancer Center Research Institute of Japan, for his helpful suggestions and donation of *E. coli* tRNA. We also express our deep thanks to Dr. S. Nitta of Tokyo University for his invaluable advice.

VII. REFERENCES

GALSTON, A.W., S. LAVEE and B. SIEGEL (1968). The induction and repression of peroxidase isozymes by 3-indoleacetic acid, in "Biochemistry and Physiology of Plant Growth Substances" (Ed. F. Wightman and G. Setterfield) p. 445-472. Runge Press, Ottawa. 1968.
GILLAM, Ian, S. MILLWARD, Daphne BLEW, Margalet VON TIGERSTROM, E. WIMMER and G.M. TENER (1967). The separation of soluble ribonucleic acids on benzoylated diethyl amino ethyl cellulose. Biochemistry, 6, 3043-3056.
GRAY, M.W. and B.G. LANE (1967). 5-carboxymethylaridine, a new nucleotide derived from yeast and wheat embryo transfer ribonucleates. Biochemistry 7, 3441-3454.
KEY, J.L. and J. INGLE (1968). RNA metabolism in response to auxin, in "Biochemistry and Physiology of Plant Growth Substances" (Ed. F. Wightman and G. Setterfield) p. 711-722. Runge Press, Ottawa. 1968.

KURAISHI, S., K. KASAMO and T. YAMAKI (1968). The relationship between growth and proline incorporation after auxin treatment of mung bean hypocotyls. Physiologia Plantarum 21, 842-850.

NAKAMURA, T., H. ISHII and T. YAMAKI (1962). Intercellular localization of native auxin in *Avena* coleoptile. Plant and Cell Physiol. 3, 149-156.

MacLACHLAN, G.A., E. DAVIES and D.F. FAN (1968). Induction of cellulase by 3-indoleacetic acid, in "Biochemistry and Physiology of Plant Growth Substances" (Ed. F. Wightman and G. Setterfield) p. 443-453. Runge Press, Ottawa. 1968.

MURAO, K., M. SANEYOSHI, F. HARADA and S. MISHIMURA (1970). Uridin-5-oxy acetic acid: A new minor constituent from *E. coli* valine transfer RNA I. B.B.R.C. 38, 657-660.

O'BRIEN, T.J., B.C. JARVIS, J.H. CHERRY and J.B. HANSON (1968). The effect of 2,4-D on RNA synthesis by soybean hypocotyl chromatin, in "Biochemistry and Physiology of Plant Growth Substances" (Ed. F. Wightman and G. Setterfield) p. 747-760. Runge Press, Ottawa. 1968.

TANIMOTO, E. and Y. MASUDA (1968). Effect of auxin on cell wall degrading enzymes. Physiol. Plant. 21, 820-826.

ZUBAY, G. (1962). The isolation and fractionation of soluble ribonucleic acid. J. Mol. Biol. 4, 347-356.

PLANT GROWTH SUBSTANCES, 1970

ABA- AND KINETIN-INDUCED CHANGES IN CELL HOMOGENATES, CHROMATIN-BOUND RNA POLYMERASE AND RNA COMPOSITION

A.A. Khan

Cornell University, New York State Agricultural Experiment Station, Geneva, New York, U.S.A.

I. INTRODUCTION

Recent studies show that abscisic acid (ABA), a potent inhibitor of seed germination, is found in a number of seeds. The ABA-induced inhibition is overcome in many instances by the cytokinins (Khan, 1967, 1968a, 1969a; Aspinall et al, 1967; Sankhla and Sankhla, 1968). A treatment with cytokinins also results in the release of dormancy in seeds (Frankland, 1961; Khan, 1966). During dormancy release in seeds and buds by cytokinins, by cold treatment or by other means, a decrease in the level of ABA or other inhibitors has been observed (Hemberg, 1965, 1970; Sondheimer and Galston, 1968). These and other results have led to the postulate that cytokinins and inhibitors along with gibberellins participate in the expression of seed dormancy and germination (Khan, 1968a, 1969a,b).

We have recently examined the quantitative and qualitative effects of ABA and cytokinins on RNA metabolism in pear embryos (Khan and Heit, 1969c; Khan and Anojulu, 1970a). Here we report further studies on the ability of ABA and kinetin to induce profound changes in cell homogenates of pear embryos, chromatin preparations from dormant and non-dormant pear embryos, and RNA composition of rapidly labeled RNA species of excised lentil roots.

II. MATERIALS AND METHODS

Preparation of dialyzed and undialyzed homogenates

Pear (*Pyrus communis*, var Bartlett) embryos from unchilled (soaked for 1 day at room temperature to facilitate excision) and chilled (at 5^0C on moist blotters) seeds were excised. Excised embryos were homogenized with cold 0.01 M tris-HCl, pH 7.6, with or without hormones (ABA, 2×10^{-5} M; kinetin, 5×10^{-5}M; 100 embryos/15 ml solution). The extract was centrifuged at 30,000 x g. The supernatant was recentrifuged. A 5 ml aliquot of the clear supernatant, henceforth referred to as a cell homogenate, was dialyzed against 4 liters of tris-HCl, pH 7.6 for 20 hours at 5 ± 1^0C with one change. Following dialysis, volume was recorded.

32p labeling and RNA extraction

Forty pear embryos were incubated in 7 ml of cell homogenates containing the same embryo equivalents (33.3 embryos) and carrier-free $H_3{}^{32}PO_4$ for 1 hour at 25^0 in a metabolic shaker. After washing RNA was extracted from the embryos by the phenol-bentonite-sodium lauryl sulphate method and purified as described before (Khan and Anojulu, 1970a).

Sucrose density gradient sedimentation

Purified RNA preparations were layered on 5-20 per cent sucrose gradients in tris-HCl, pH 7.6 containing 0.01 M $MgCl_2$ and centrifuged at 41,000 rev./min. for 6 hours in a Spinco Model SW41Ti rotor in L2-65B ultracentrifuge. Fractions (15 drops) were collected by bottom puncture. 1.6 ml of water was added to each fraction and absorbance at 260 mµ and radioactivity determined.

Nucleotide composition analysis

Fractions corresponding to the 18s region (tubes 17-24, see Fig. 1) were pooled. 1 mg yeast RNA was added and RNA precipitated, hydrolyzed and mononucleotides separated and counted as described before (Khan and Anojulu, 1970a).

Preparation of embryos and chromatin extraction

Fifty excised, chilled, pear embryos were grown for various lengths of time on 0.75 per cent agar (Bactoagar) containing various hormones in Petri plates. Chromatin was prepared from the embryos essentially by the method of Huang and Bonner (1962) as described by Jarvis *et al* (1968). The purified chromatin was suspended in 1 ml of 0.05 M tris-HCl, pH 8.0 containing 0.01 M mercaptoethanol. Aliquots of this preparation (4-8 µg DNA) were used in the assay of chromatin-bound RNA polymerase.

Chromatin-bound RNA polymerase system

The chromatin assay system consisted of the following (all in µ moles): ATP, 0.2; CTP, 0.2; GTP, 0.2; $MgCl_2$, 1.0; $MnCl_2$, 0.25; mercaptoethanol, 3.0; Tris-HCl, pH 8.0, 20.0; ^{3}H-UTP, 0.01 (2.02 mc/µ mole) and chromatin in a final volume of 4.0 ml unless specified otherwise. The reaction was carried for 20 min at 37^oC and stopped by the addition of 5 ml of cold trichloracetic acid (TCA). After 20 minutes in ice, TCA-precipitable material was collected on a membrane filter (Schleicher and Schuell type B-6). The filters were dried and counted in the Nuclear Chicago Unilux II liquid scintillation system.

DNA, RNA and Protein determination

An aliquot of purified chromatin was hydrolyzed in 0.5 M $HClO_4$ at 70^oC for 40 minutes and DNA was determined by diphenylamine test according to Burton (1956). Protein was determined in an aliquot of alkaline hydrolyzate (0.5 N KOH, 18 hr, 37^oC) according to Lowry *et al* (1951). RNA was determined spectrophotometrically in the perchloric acid soluble supernatant.

III. RESULTS AND DISCUSSION

Effect of ABA and kinetin on cell homogenates

Kinetics of ^{32}P incorporation (details to be published elsewhere) into RNA of excised pear embryos showed that in 1 hour the labeling was largely in the 18s region of the absorption profile (Fig. 1).

A nucleotide composition analysis of the radioactive peak (tubes 17-24) showed that this peak might be a mixture of ribosomal RNA (rRNA) and DNA-like RNA. The extent of ^{32}P incorporation by pear embryos in this region of the RNA profile in presence of variously treated pear homogenates was examined.

The effect of dialyzed and undialyzed pear homogenates on ^{32}P incorporation by chilled pear embryos is shown in Figs. 2A and 2B.

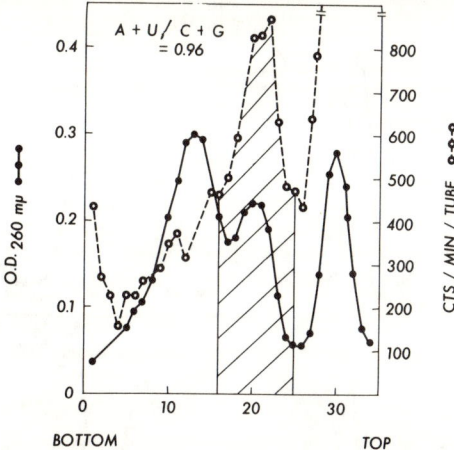

Fig. 1. Incorporation of ^{32}P (60 μc) into RNA of excised pear embryos (from seeds chilled for 47 days at 5°C) in 1 hr. Contents of tubes 17-24 were pooled and composition of RNA determined. See "Materials and Methods" Section.

Incorporation of label in the 18s as well as other regions was completely inhibited by the undialyzed pear homogenate. Dialysis of the homogenate failed to overcome the inhibition. Pear homogenates prepared from unchilled embryos similarly inhibited ^{32}P labeling (data not shown). The effect of ABA treated and kinetin treated homogenates on ^{32}P labeling in the 18s regions is shown in Figs. 2C to 2H. In these homogenates, following dialysis, there was a marked reduction in their capacity to inhibit labeling in the 18s and other regions of the RNA profile (2D, 2F, 2H). The hormones had little effect, however, when undialyzed homogenates were used (Figs. 2C, 2E, 2G). These results clearly show that pear homogenates contain some factor(s) whose presence in the growth medium restricts rapid labeling of RNA in growing pear embryos. ABA and kinetin appear to profoundly alter the physical properties of the cell homogenate by causing release of a dialyzable factor needed for entry of the RNA precursor to the embryo and/or for rapid labeling of the RNA species. Exactly how the hormones bring about the change in the homogenate is not known. This may involve a binding of the hormone with some factor bound to the macromolecule by weak secondary bonds. Hormone-macromolecule complexes appear to be necessary for hormone regulated RNA transcription in plant cells (Pearson and Wareing, 1969; Matthysse and Abrams, 1970).

Requirements of pear embryo chromatin RNA polymerase

The ratio of DNA:RNA:protein in purified chromatin was 2.1:1.3:1.0. ^{3}H-UTP incorporation into acid-insoluble material was inhibited to varying degrees by absence of any one of the three nucleoside triphosphates (Table 1). The absence of all three gave most inhibition. The reaction appeared to be absolutely dependent upon the presence of divalent cations (Mg^{++}). Actinomycin D, an accepted inhibitor of DNA-directed RNA synthesis, inhibited the system at high concentration. Ribonuclease and deoxyribonuclease inhibited the system nearly completely. Deoxyribonuclease given prior to the commencement of the reaction was most active indicating that DNA is the template which directs ^{3}H-UTP incorporation into RNA. This system (Table 1) thus appears to be similar to other plant chromatin or nuclear systems which support RNA synthesis (Pearson and Wareing, 1969; Varner and Johri, 1968; Jarvis et al., 1968; O'Brien et al., 1968).

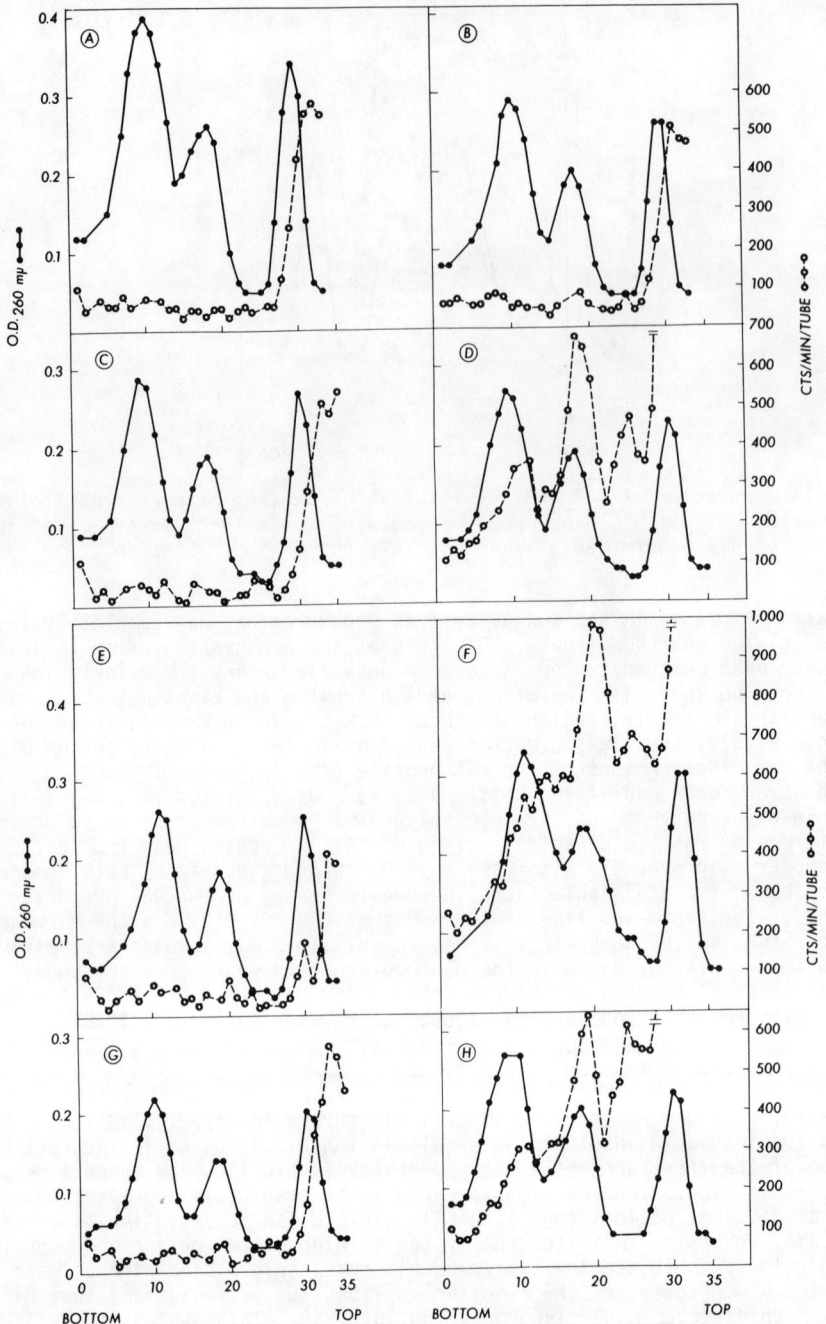

Fig. 2. Incorporation of ^{32}P (1 hr, 40 μc) into RNA of excised pear embryos (from seeds chilled for 38 days at $5°C$) in presence of ABA-treated and kinetin-treated pear homogenates (dialyzed and undialyzed). Cell homogenates in each case represent 33.3 embryo equivalents. A, C, E and G : undialyzed water, kinetin, ABA and (ABA + kinetin)-treated cell homogenates respectively. B, D, F, and H : dialyzed cell homogenates respectively.

Table 1. Requirements of chromatin-bound RNA polymerase. Incorporation of ^{3}H-UTP per 100 μg DNA was approximately 68 μμ moles in the complete system. Chromatin isolated from chilled (42 days at 5°C) embryos.

System	Percent of control	System	Percent of control
Complete	100	Actinomycin D 2.5 μg	75
- ATP	50	25 μg	30
- CTP	44	RNase 25 μg	29
- GTP	41	*25 μg	14
- ATP, CTP, GTP	31	DNase 25 μg	9
- Mg^{++}	5	*25 μg	43

Reaction was carried out in 0.5 ml. Chromatin was treated in ice-bath with actinomycin D for 10 min at 0°C and with ribonuclease and deoxyribonuclease for 30 min at 26°C prior to adding the nucleotides. For details see "Materials and Methods" section.

*The reaction mixture was cooled following completion of ^{3}H-UTP incorporation, enzymes added and the mixture reincubated for 10 min at 37°C. TCA-precipitable counts were determined as described in "Materials and Methods" section.

RNA polymerase activity in chromatin from chilled and unchilled embryos

The chromatin from chilled (non-dormant) embryos had far greater capacity to support ^{3}H-UTP incorporation into acid-insoluble material than the chromatin from unchilled (dormant) embryos (Table 2). This is in close agreement with the finding that there is a progressive increase in growth as well as in the incorporation of ^{32}P into RNA species of dormant pear embryos with increasing periods of cold treatment (Khan et al, 1968b). The chromatin from chilled embryos rapidly acquired (24 hr growth) a high capacity for ^{3}H-UTP incorporation into acid-insoluble material presumably due to increased production of chromatin-bound RNA polymerase. This increased capacity to incorporate label decreased appreciably after 72 hours of growth. This effect was similar to the chromatin of gibberellin (GA$_3$)-treated dormant hazel embryos whose capacity to incorporate ^{3}H-UTP decreased after 66 hours of growth (Jarvis et al, 1968). On the other hand, in the chromatin from unchilled embryos, the capacity to incorporate label continued to increase steadily during growth up to 72 hours. These results suggest that the chromatin of unchilled embryos is in a state of partial "repression" from which it is slowly released. Chilling treatment causes rapid "derepression" of the genome.

Effect of ABA and kinetin on chromatin-bound RNA polymerase activity

The rapid loss in RNA polymerase activity of the chromatin from chilled embryos is checked by growing these embryos in medium containing ABA (Table 2). The activity of chromatin from ABA-treated embryos appears to resemble that of chromatin from dormant (unchilled) embryos. Thus ABA action may be related to slowing down or arresting gene derepression.

The effect of various hormones on the activity of chromatin-bound RNA polymerase in embryos grown for 22 hours are shown in Table 3. Kinetin was most active in stimulating polymerase activity in chromatin from unchilled embryos. Although GA$_3$ was also active in increasing the chromatin activity, it was not as active as kinetin. ABA reversed the increase in activity caused by kinetin, but had little effect on GA$_3$-induced activity. In chromatin from chilled embryos, kinetin as well as GA$_3$ reduced the polymerase activity. The reason for this decrease is not known. It may be due to further "acceleration" of gene derepression. However, ABA as before overcame the

Table 2. The incorporation of ^{3}H-UTP into TCA-precipitable material by chromatins from chilled (45 days at 5°C) and unchilled embryos grown for various lengths of time. ABA, 20 μM.

Embryos grown (hr)		cpm/100 μg DNA	
		Unchilled	Chilled
Water	0	6,640	28,680
	24	17,888	55,095
	72	64,058	8,526
ABA	24	-	12,345
	72	-	26,568

Table 3. Effect of ABA, kinetin and gibberellin (GA$_3$) on the activity of chromatin-bound RNA polymerase from chilled (50 days at 5°C) and unchilled pear embryos grown for 22 hours. Concentrations: ABA, 20 μM; kinetin, 50 μM; GA$_3$, 50 μM.

Treatment	cpm/100 μg DNA	
	Unchilled	Chilled
Water	5,737	55,020
ABA	6,648	13,929
Kinetin	34,832	26,763
GA$_3$	11,062	24,464
ABA + kinetin	10,017	52,791
ABA + GA$_3$	12,761	25,887

decrease in chromatin activity caused by kinetin but not that caused by GA$_3$. The effect of hormones on chromatin-bound RNA polymerase closely parallels the effect of these hormones on the growth of chilled and unchilled embryos (Khan et al, 1969c). A cytokinin-inhibitor antagonism at the level of chromatin-bound RNA polymerase suggests that such interaction may play a part in the control of dormancy and germination in vivo.

Effect of ABA and kinetin on nucleotide composition of rapidly labeled RNA

ABA profoundly altered the nucleotide composition of rapidly labeled RNA species by excised pear embryos (Khan and Anojulu, 1970a) and by excised lentil roots (Khan et al, 1970b). In the case of lentil roots, ABA induced profound changes in RNA composition and increased ^{32}P labeling (Table 4). The increase in A+U:C+G ratio was largely due to an increase in the UMP content (Khan et al., 1970b). The effects of ABA on RNA composition as well as ^{32}P labeling were reversed by kinetin. In the case of pear embryos, ABA also induced changes in composition of DNA-like RNA (for example tbRNA, A+U:C+G = 1.30 to 1.38). This effect, however was not reversed by kinetin (Khan and Anojulu, 1970a). Furthermore, in contrast to lentil roots, ^{32}P incorporation into RNA species of pear embryos was inhibited by ABA (Khan and Heit, 1969c). These results indicate that the quantitative effects of ABA on ^{32}P labeling of RNA probably vary in different tissues or organs while its qualitative effects on RNA composition are independent of the quantitative changes.

Table 4. ABA-kinetin interaction on labeling of ^{32}P into 25S and 18S ribosomal RNAs of excised lentil roots and on nucleotide composition of rapidly labeled RNA species. Numbers in parentheses are specific activities (cpm/O.D.). ABA, 20 μM; kinetin, 10 μM.

Treatments	Radioactivity cpm x 10^{-3}	Nucleotide ratios	
		A+U:C+G	U:G
Water			
25S	696 (245)	1.18	1.03
18S	307 (244)	1.46	1.36
ABA			
25S	1202 (302)	1.47	1.48
18S	460 (297)	2.05	3.09
Kinetin			
25S	668 (198)	1.25	1.38
18S	290 (203)	1.37	1.59
Kinetin + ABA			
25S	863 (315)	1.23	1.20
18S	314 (243)	1.37	1.42

Data taken from Khan et al.(1970b).

IV. CONCLUSION

The data presented here show profound qualitative changes by ABA and kinetin in pear embryos and lentil roots. There is a close parallel between the chromatin-bound RNA polymerase activity and growth responses of pear embryos with or without the presence of hormones. Kinetin and GA_3 increase RNA polymerase activity in dormant embryos as well as releasing dormancy. ABA inhibits growth of pear embryos and appears to slow down the rapid decrease in chromatin activity. ABA and kinetin cause physical changes in macromolecules and alter the composition of RNA. These results suggest that the actions of these hormones are closely related to mechanisms controlling read-out pattern of the genome and DNA-to-RNA transcription. It is noteworthy that the interaction of kinetin and ABA in the control of chromatin-bound RNA polymerase activity resembles the effects of these hormones on amylase activity (Khan, 1969a) and chromatin-bound nucleases (Srivastava, 1968).

V. SUMMARY

ABA and kinetin altered the physical properties of pear homogenates. These hormones appear to release a dialyzable factor associated with the macromolecules in the cell homogenates. This factor seems to be essential for inhibiting rapid ^{32}P incorporation into RNA of excised pear embryos. Kinetin, and to a lesser degree gibberellin, increased the activity of the chromatin-bound RNA polymerase in dormant (unchilled) pear embryos. ABA countered the kinetin-induced increase in chromatin activity in dormant embryos and the kinetin-induced decrease in chromatin activity in chilled embryos. Furthermore, ABA appeared to slow down the rapid loss of RNA polymerase activity in chromatin from chilled embryos with time of growth.

It is suggested that more than one hormone may regulate germination and dormancy by increasing or decreasing RNA polymerase activity or DNA-template availability or both.

VI. ACKNOWLEDGEMENTS

This research was supported by a grant from the American Seed Research Foundation. The capable technical assistance of Miss Catherine Roe and Miss Laurel Andersen is gratefully acknowledged. Abscisic acid was a generous gift from Dr. Bauernfeind of Hoffmann-La Roche, Inc., Nutley, New Jersey.

VII. REFERENCES

ASPINALL, D., L.G. PALEG and F.T. ADDICOTT (1967). Abscisin II and some hormone-regulated plant responses. Aust. J. Biol. Sci. 20, 869-882.

BURTON, K. (1956). A study of the conditions and mechanism of diphenylamine reaction for the colorimetric estimation of DNA. Biochem. J. 62, 315-323.

FRANKLAND, B. (1961). Effect of gibberellic acid, kinetin and other substances on seed dormancy. Nature 192, 678-679.

HEMBERG, T. (1965). The significance of inhibitors and other chemical factors of plant origin in the induction and breaking of rest periods. Handb. Pflanzenphysiol.15(2), 669-698.

HEMBERG, T. (1970). The action of some cytokinins on the rest period and the content of acid growth-inhibiting substances in potato. Physiol. Plant. 23, 850-858.

HUANG, R.C. and J. BONNER (1962). Histone, a suppressor of chromosmal RNA synthesis. Proc. Nat. Acad. Sci., Wash. 48, 1216-1222.

JARVIS, B.C, B. FRANKLAND and J.H. CHERRY (1968). Increased nucleic-acid synthesis in relation to the breaking of dormancy of hazel seed by gibberellic acid. Planta, Berl. 83, 257-266.

KHAN, A.A. (1966). Breaking of dormancy in *Xanthium* seeds by kinetin mediated by light and DNA-dependent RNA synthesis. Physiol. Plant. 19, 869-874.

KHAN, A.A.(1967). Antagonism between cytokinins and germination inhibitors. Nature 216, 166-167.

KHAN, A.A. (1968a). Inhibition of gibberellic acid-induced germination by abscisic acid and reversal by cytokinins. Plant Physiol. 43, 1463-1465.

KHAN, A.A. (1969a). Cytokinin-inhibitor antagonism in the hormonal control of α-amylase synthesis and growth in barley seed. Physiol. Plant. 22, 94-103.

KHAN, A.A. and C.C. ANOJULU (1970a). Abscisic acid-induced changes in nucleotide composition of rapidly labelled RNA species of pear embryos. Biochem. Biophys. Res. Commun. 38, 1069-1075.

KHAN, A.A. and C.E. HEIT (1969c). Selective effect of hormones on nucleic acid metabolism during germination of pear embryos. Biochem. J. 113, 703-712.

KHAN, A.A. and E.C. WATERS (1969b). On the hormonal control of post-harvest dormancy and germination in barley seeds. Life Sci. 8, 729-736.

KHAN, A.A., L. ANDERSEN and T. GASPAR (1970b). Abscisic acid-induced changes in nucleotide composition of rapidly labeled ribonucleic acid species of lentil root. Reversal by kinetin. Plant Physiol. 46, 494-495.

KHAN, A.A., C.E. HEIT and P.C. LIPPOLD (1968b). Increase in nucleic acid-synthesizing capacity during cold treatment of dormant pear embryos. Biochem. Biophys. Res. Commun. 33, 391-396.

LOWRY, O.H., N.J. ROSEBROUGH, A.L. FARR and R.J. RANDALL (1951). Protein measurement with the Folin phenol reagent. J. Biol. Chem. 193, 265-275.

MATTHYSSE, A.G. and M. ABRAMS (1970). A factor mediating interaction of kinins with the genetic material. Biochim. Biophys. Acta 199, 511-518.

O'BRIEN, T.J., B.C. JARVIS, J.H. CHERRY and J.B. HANSON (1968). The effect of 2,4-D on RNA synthesis by soybean hypocotyl chromatin, in "Biochemistry and Physiology of Plant Growth Substances" (Eds. F. Wightman and G. Setterfield), p. 747. Runge Press, Ottawa, 1968.

PEARSON, A.P. and P.F. WAREING (1969). Effect of abscisic acid on activity of chromatin. Nature 221, 672-673.

SANKHLA, N. and D. SANKHLA (1968). Reversal of (±)-abscisin II induced inhibition of lettuce seed germination and seedling growth by kinetin. Physiol. Plant. 21, 190-195.

SONDHEIMER, E., D. TZOU and E.C. GALSON (1968). Abscisic acid level and seed dormancy. Plant Physiol. 43, 1443-1447.
SRIVASTAVA, B.I.S. (1968). Acceleration of senescence and of the increase of chromatin-associated nucleases in excised barley leaves by abscisin II and its reversal by kinetin. Biochim. Biophys. Acta 169, 534-536.
VARNER, J.E. and M.M. JOHRI (1968). Hormonal control of enzyme synthesis, in "Biochemistry and Physiology of Plant Growth Substances" (Eds. F. Wightman and G.Setterfield), p. 793. Runge Press, Ottawa, 1968.

PLANT GROWTH SUBSTANCES, 1970

The Temporal Separation of Transcription and Translation and its Control in Cotton Embryogenesis and Germination

James N. Ihle and L.S. Dure, III
Department of Biochemistry, The University of Georgia, Athens, Georgia 30601, U.S.A.

I. INTRODUCTION

The separation in time of mRNA synthesis from its translation into protein was shown to be a developmental regulatory mechanism in animal embryogenesis as early as 1963 (Gross and Cousineau). This phenomenon, in which the transcription of genes into mRNA occurs in one developmental stage, the mRNA "stored" in some as yet undiscovered form, and subsequently translated at a later stage, has now been observed in a large number of studies of animal and fungal development studies, (reviewed by Gross, 1968; Davidson, 1968; Brown and Dawid, 1969).

In higher plants the first indication that a temporal separation of transcription and translation may be a developmental regulatory scheme was the observation of Waters and Dure (1965, 1966) that an inhibition of RNA synthesis by actinomycin D failed to inhibit much of the protein synthesis that takes place in the cotyledons of cotton seeds during the first 3 days of germination. Subsequently, the use of mRNA that apparently exists in the dry seed by its cotyledons in early germination has been suggested for peanuts (Cherry, 1968) and black-eyed peas (Chakravorty, 1968).

We have continued the investigation of Waters and Dure, utilizing the embryogenesis and germination of cotton seeds, with the following aims: (a) to establish that certain specific enzymes are synthesized *de novo* during early germination from existing mRNA, (b) to determine when in embryogenesis the mRNA for these specific enzymes is synthesized and (c) to determine how the translation of this body of mRNA for these specific enzymes is prevented until germination begins. As indicatory enzymes we choose a protease specific for hydrolysing BAEE (benzoyl-arginine ethyl ester) and isocitratase, since their necessity in germination and their absence in embryogenesis appeared to be a reasonable initial assumption. Preliminary reports of some of our findings with this system have appeared (Ihle and Dure, 1969, 1970).

II. METHODS AND MATERIALS

The methods used to germinate mature and precocious cotton embryos and to assay the protease have been described previously (Ihle and Dure, 1969, 1970). Procedures for purifying the protease and for determining its physical and chemical properties will be published elsewhere. Isocitratase activity was measured by the method of Dixon and Kornberg (1959), and an enzyme unit defined as an increase in A_{324} of 0.01/min/ml.

Actinomycin D was obtained from Calbiochem, cycloheximide from Pierce Chemical Co., and D,L- abscisic acid was generously provided by the Shell Development Company.

III. RESULTS AND DISCUSSION

The appearance of the protease activity in cotton cotyledons during germination and the effects of transcription and translation inhibitors on its appearance are shown in Fig. 1. This figure shows that the normal appearance of activity by 24 hours and its increase to a maximum value by 96 hours is completely inhibited by the continuous presence of cycloheximide (1 mg/ml) from imbibition on, but is essentially unaffected by the continuous presence of actinomycin D (20 μg/ml). (The other plantlet tissues are severely affected by this prolonged exposure to actinomycin D, and their necrotic state presumably causes the small actinomycin D effect on activity observed by the third day). Completely analogous results to those in Fig. 1 were obtained for isocitratase activity.

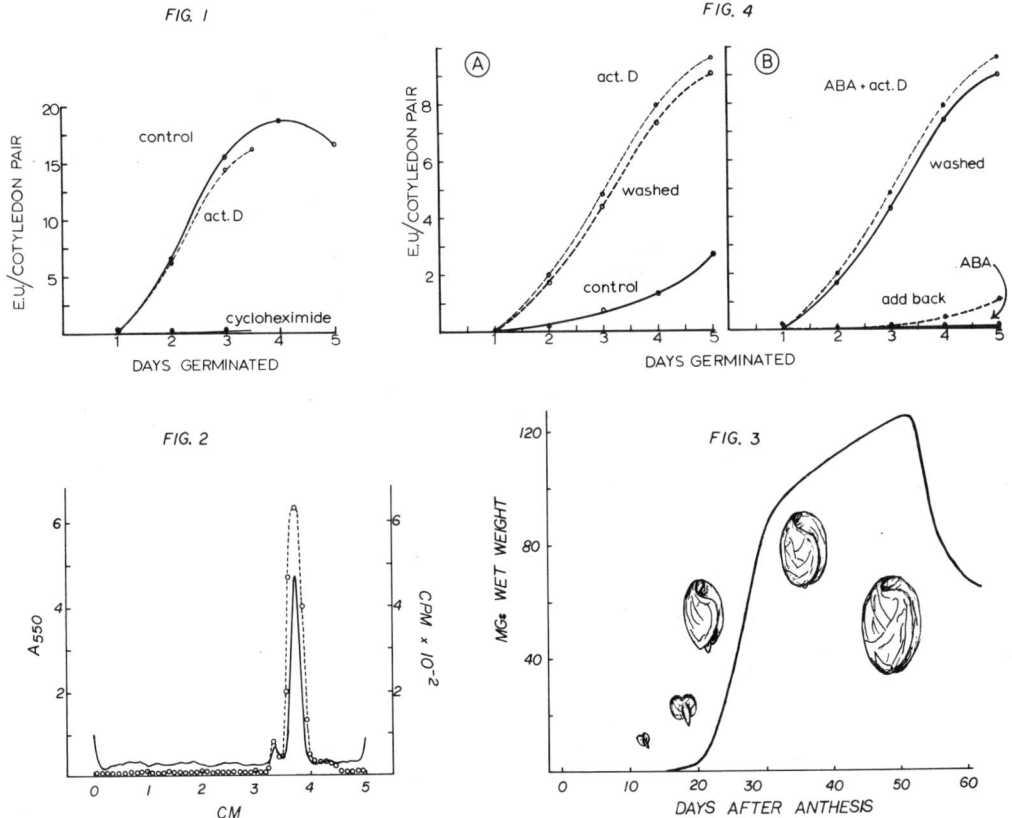

Fig. 1. The development of a specific protease activity in germinating cotton cotyledons, and the effects of cycloheximide and actinomycin D (act D) on this development.

Fig. 2. Profiles of absorbancy at 550 nm (solid line) and radioactivity (dashed line) of a polyacrylamide gel electropherogram of the purified protease.

Fig. 3. Wet weight of cotton embryos during embryogenesis.

Fig. 4A & B. The development of the protease activity in the cotyledons during the precocious germination of embryos larger than 85 mgs, and the effects of various treatments on this development. act D represents actinomycin D.

These results suggest that the appearance of the protease and isocitratase activities is the result of the *de novo* translation of pre-existing mRNA. To substantiate this possibility for the protease, seeds were germinated in the presence of ^{14}C amino acids, the protease extracted and purified, and the final preparation electrophoresed on polyacrylamide gel. Fig. 2 shows the A_{550} profile of the stained gel and the radioactivity profile of gel sections. The single major peak of radioactivity which corresponds to the major single band of protein indicates that ^{14}C amino acids were incorporated into the protease during germination. Furthermore, the specific activity (CPM/mg protein) of the purified protease was 18 fold greater than that of the supernatant protein of the initial extract. These results reinforce the indication that the protease and isocitratase are synthesised *de novo* during germination.

Since the above experiments suggest that the mRNA for the two enzymes is present in the dry seed, the transcription of the mRNA must occur at some point in embryogenesis The time course of embryogenesis in cotton has been plotted with respect to wet weight of the embryo in Fig. 3. After a 20 day lag period, the wet weight of the embryo increases rapidly until it reaches 85 mg by 30 days. After 30 days the embryo growth is somewhat slower and the embryo reaches 125 mg wet weight by 50 days, after which desiccation of the embryo and sclerification (followed by death) of the ovule wall commences.

Since cotton embryos will germinate precociously when dissected from the ovule, we are able to determine the time in embryogenesis at which the transcription of the protease and isocitratase mRNA's appears to occur by precociously germinating embryos of successively younger ages until an embryo age was reached at which the cotyledons failed to develop protease and isocitratase activity in the presence of actinomycin D (4 µg/ml). Table 1 shows that embryos 85 mg and larger will develop protease and isocitratase activity in their cotyledons in the presence of actinomycin D, whereas embryos smaller than 85 mg will not. These results indicate that when embryos reach approximately 60% of their final wet weight and approximately 20 days before embryogenesis is complete, mRNA for the protease and isocitratase and presumably a number of other enzymes necessary for germination is transcribed in the cotyledons. Curiously, these experiments show a stimulation of both enzyme activities in all embryos between 85 and 125 mg that are precociously germinated in the presence of actinomycin D.

Table 1. Enzyme activities during precocious germination.

		Enzyme Units/Cotyledon Pair			
Embryo Wet Weight	Days Germinated	+Actinomycin D. Protease	Isocitratase	-Actinomycin D. Protease	Isocitratase
110-125	3	12.48	----	9.33	----
100-110	4	13.44	20.03	5.46	9.12
85-100	4	7.11	21.32	1.33	6.46
60-85	5	0	0	2.21	5.25

Since these mRNAs are apparently present in the cotyledons during the last 20 days of embryogenesis, their translation during this time must be inhibited in some manner. Data on the responses of embryos weighing 85-125 mgs to various treatments during precocious germination presented in Fig. 4, suggest how this translation inhibition may be accomplished. Fig. 4A shows that protease activity develops slowly during precocious germination relative to normal germination, but is greatly stimulated by actinomycin D (4 µg/ml) as was also apparent in Table 1. Furthermore, Fig. 4A shows that simply washing embryos in distilled water prior to placing them in germination dishes causes a stimulation in the development of protease activity equivalent to that caused by actinomycin D. Completely analogous results were obtained for isocitratase activity. The fact that washing embryos results in a more rapid development of enzyme activity suggests that a compound that somehow inhibits translation may be located on the cotyledon surface (presumably originating in the ovule wall and being absorbed by the cotyledons). The fact that enzyme activity does develop in time in unwashed cotyledons suggests that the compound can be metabolized by cotyledon cells. The fact that actinomycin D can effect the same stimulation as does washing, suggests that the inhibition somehow requires RNA synthesis.

To test this possibility washed embryos larger than 85 mg were precociously germinated in the presence of an aqueous extract of ovule walls. The inhibitory effect of this extract (called the "add back" fraction) on the development of protease activity in the cotyledons of these washed embryos is shown in Fig. 4B. Again, completely analogous results were obtained for isocitratase activity. These results prompted us to investigate the effect of abscisic acid (ABA) on protease and isocitratase appearance during precocious germination, since ABA was first isolated from cotton bolls (Ohkuma et al, 1963) and its inhibitory action on plant development is well documented (reviewed by Addicott and Lyon, 1969). Fig. 4B shows that 10^{-6}M ABA completely inhibits the appearance of the protease during precocious germination, and also shows that if actinomycin D is supplied to the embryos along with ABA, the ABA inhibition is overcome, and protease activity develops as rapidly as it does in washed embryo cotyledons. Again, the development of isocitratase activity is affected in the same manner. Because of the similarity of action of ABA and the ovule extract and of the demonstrated presence of ABA in embryonic cotton seed of this age (Addicott and Lyon, 1969), we feel that one of the functions of ABA in cotton embryogenesis is to inhibit the translation of mRNA destined to be utilized during germination.

The mode of action of ABA here is not known, but the actinomycin D data suggest that ABA may act on DNA to induce via RNA and protein the synthesis of an inhibitor that prevents the translation of a specific body of mRNA. Or ABA may act in concert with a product of RNA synthesis which turns over rapidly to effect this translation inhibition. Oddly, ABA has no effect on the appearance of the protease or isocitratase in the cotyledons during the germination of embryos that have begun to desiccate or of mature seeds. This transition from sensitivity in immature embryos to insensitivity to ABA in mature embryos is not understood as yet.

Since it appears from these data that the transcription of the mRNA for the protease and isocitratase takes place thirty days after anthesis, we proceeded to explore the induction of this transcription. Table II shows that although embryos smaller than 85 mg do not develop either enzyme activity when precociously germinated in the presence of actinomycin D, both enzyme activities do develop in the presence of actinomycin D, if the embryos are allowed to precociously germinate for 24 hours before being exposed to the inhibitor. Thus, presumably, the requisite mRNA for these two enzymes (and presumably a large number of other "germination" enzymes) is synthesised in cotyledons within 24 hours after removal from the boll. Table II also shows that the 24 hour period is sufficient time for embryos of different ages to produce this body of mRNA and that the presence of ABA does not influence this induction of transcription. Subsequently we observed that this premature transcription could be induced in cotyledons in 24 hours by simply removing the bolls from the plant.

Table II. Induction of Transcription

	Enzyme Units/Cotyledon Pair	
70 mg Embryos + Abscisic Acid	Protease	Isocitratase
+ Actinomycin D	0	0
+ Actinomycin D after 24 hours	1.10	6.16
60 mg Embryos - Abscisic Acid		
+ Actinomycin D	0	0
+ Actinomycin D after 10 hours	0	0
+ Actinomycin D after 24 hours	1.00	2.86
No additions	1.10	2.15

The results of these experiments with embryos smaller than 85 mg imply that the transcription of the mRNA for the "germination enzymes" is induced by severing of the vascular connection between the ovule and the maternal plant. This implication is strengthened by the observation that in vivo the funiculus connecting the ovule to the placenta breaks when the embryos reach approximately 85 mg. This corresponds to the time in embryogenesis when the production of mRNA occurs normally as indicated by the experiments with actinomycin D. Curiously, the severing of the vascular connection also appears to be related to the synthesis of ABA in the ovule tissue, since the inhibitory action of ovule extracts on the development of the protease and isocitratase activity during precocious germination is observed only with ovule extracts obtained from ovules that contained embryos larger than 85 mg. Finally, the breaking of the funiculus also appears to stop DNA synthesis and cell division in the cotyledons (unpublished observations).

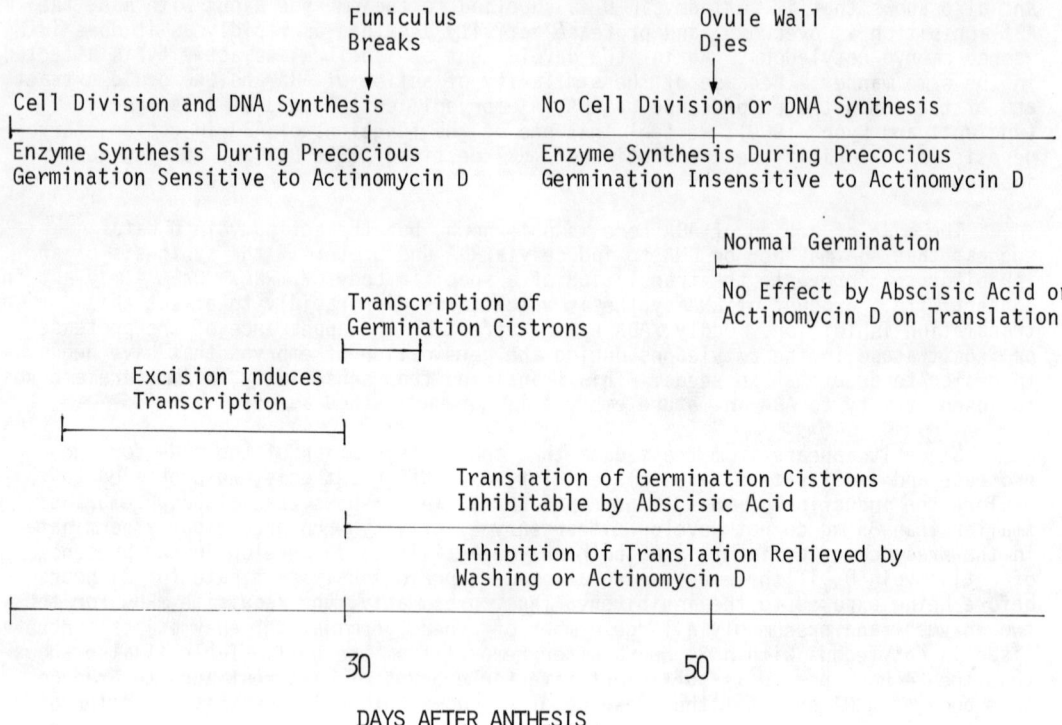

Fig. 5. Postulated sequence of regulatory events occurring in cotton embryogenesis.

The sequence of some of the events controlling cotton embryogenesis that is suggested by our experiments to date is presented diagrammatically in Fig. 5.

IV. SUMMARY

The appearance of the enzyme activities of a specific protease and isocitratase in the cotyledons have been studied during germination of cotton seeds and embryos. Both enzymes appear to be synthesised *de novo* during germination from mRNA which is transcribed at a point representing 60% completion to embryogenesis. The translation of these mRNAs during the last 40% of embryogenesis is prohibited by the presence of abscisic acid diffusing into the embryo from the ovule wall. The mode of action of abscisic acid in maintaining translation inhibition appears to involve RNA synthesis. Transcription of the requisite mRNAs for the germinating enzymes which normally takes place at 60% completion of embryogenesis may be induced prematurely by simply dissecting the ovules from the plant suggesting that the breaking of the connection between the ovule and the mother plant may be required *in vivo* for induction of transcription. This hypothesis is further substantiated by the observation that in *in vivo* the funiculus connecting the ovule to the placenta is normally severed at 60% completion of embryogenesis. The breaking of the funiculus may also be involved in the induction of the synthesis of ABA by the ovule wall and the cessation of cell divisions in the cotyledons of the embryo.

V. ACKNOWLEDGEMENTS

This work was supported in part by funds from the U.S. N.F.S. and the U.S. A.E.C. J.N. Ihle was a U.S. Public Health Service Pre-doctoral Fellow. L.S. Dure was the recipient of a U.S. Public Health Service Career Development Award.

VI. REFERENCES

ADDICOTT, F.T. and J.L. LYON (1969). Physiology of abscisic acid and related substances. Annu. Rev. of Plant Physiol., 20, 139-164.
BROWN, D.D. and I.B. DAWID (1969). Developmental Genetics. Annu. Rev. Gen., 3, 127-154.
CHAKRAVORTY, A.K. (1968). Ribosomal RNA synthesis in the germinating Black Eye Pea II. Biochem. Biophys. Acta. Amsterdam, 179, 83-96.
CHERRY, J.H. (1967). Nucleic acid metabolism in ageing cotyledons. Symp. Soc. Exp. Biol., Cambridge, 21, 247-268.
DAVIDSON, E.H. (1968). Gene Action in Early Development. Academic Press, New York.
DIXON, G.H. and H.L. KORNBERG (1959). Assay methods for key enzymes of the glyoxylate cycle. Biochem. J., London, 72, 3P.
GROSS, P.R. and G.H. COUSINEAU (1963). Effects of Actinomycin D on macromolecule synthesis and early development in Sea Urchin eggs. Biochem. Biophys. Res. Comm., New York, 10, 321-325.
GROSS, P.R. (1968). Biochemistry of Differentiation. Annu. Rev. Biochem., 37, 631-660.
IHLE, J.N. and L.S. DURE, III (1969). Synthesis of a protease in germinating cotton cotyledons catalysed by mRNA synthesized during embryogenesis. Biochem. Biophys. Res. Comm., New York, 36, 705-710.
IHLE, J.N. and L.S. DURE, III (1970). Hormonal regulation of translation inhibition requiring RNA synthesis. Biochem. Biophys. Res. Comm., New York, 38, 995-1001.
OHKUMA, K., J.L. LYON, F.T. ADDICOTT and O.E. SMITH (1963). Abscisin II, an abscission-accelerating substance from young cotton fruit, Science, New York. 142, 1592-1593.
WATERS, L.C. and L.S. DURE, III (1965). Long-lived messenger RNA: Evidence from cotton seed germination, Science, New York, 147, 410-412.
WATERS, L.C. and L.S. DURE, III (1966). Ribonucleic Acid Synthesis in Germinating Cotton Seeds. J.Mol. Biol., Cambridge, 19, 1-27.

PLANT GROWTH SUBSTANCES, 1970

DNA Synthesis and Hormonal Growth Response
in Non-meristematic Tissues*

D. Atsmon

Plant Genetics, The Weizmann Institute of Science, Rehovoth, Israel

I. INTRODUCTION

In view of the regulatory functions of hormones, much work has been done on the molecular aspects of hormonal responses. It has been demonstrated in numerous experiments that RNA and protein syntheses are essential for an optimal hormonal response. This situation is true for all hormones and plant tissues tested, except for the short term, immediate response to auxins - cell wall loosening - which takes place before any major metabolic changes can occur. Most of the evidence comes from inhibition of the hormonal response by specific inhibitors of RNA and protein synthesis. In some isolated cases information is more specific. Masuda (1965) has demonstrated the need for a GA-induced RNA in order to get a full IAA response; Matthysse and Phillips (1969) have shown that IAA interacts with the target cells' chromatin, via a specific protein; Johri and Varner (1968) measured a significant increase in RNA synthesis by isolated nuclei which were treated by GA.

Most of the thoughts on molecular aspects of hormonal response were concentrated around transcription (DNA-dependent RNA synthesis) and translation ("new" protein synthesis). Less attention has been paid to the involvement of DNA in hormonal responses. One possible approach is to test for specific and direct interaction between the hormone and plant DNA; this approach has yielded some very interesting information, according to Kessler and Snir (1969 and personal communication). Their results seem to indicate that the physico-chemical properties of the DNA change as a result of *in vitro* interaction with gibberellin. Moreover, it seems that such an interaction causes a change in base composition of the newly synthesized DNA.

II. THE INVOLVEMENT OF DNA SYNTHESIS AND CELL DIVISION IN HORMONE-INDUCED CELL ELONGATION.

The metabolic approach includes the reports on inhibition of the hormonal response by inhibitors of DNA synthesis. Most of the experiments along this line have been done with tissues which expand (or elongate only) in response to hormonal treatment. These systems have the obvious disadvantage of not being truly conditioned by the hormone treatment - the exogenous hormone only enhances otherwise existing processes which are, in all probability, normally regulated by endogenous levels of the same hormone.

---

* Part I of this work was carried out in the MSU/AEC Plant Research Laboratory, E. Lansing, Mich. in collaboration with A. Lang; part II was carried out at The Weizmann Institute of Science in collaboration with Y. Degani, A.H. Halevy and A. Kadouri.

The results obtained are different in different experimental system (see table 1), and it is of interest to review them briefly. GA- or dark-induced elongation in mustard, cucumber and lentil hypo- or epicotyls is inhibited by 5-fluorodeoxyuridine (FUdR) an inhibitor of DNA synthesis (Bopp, 1967; Nitsan and Lang, 1965, 1966; Degani and Atsmon, 1970a; Atsmon and Lang, unpublished). It should be noted that although these organs elongate mostly by cell elongation, they are all adjacent to apical meristems which merge gradually into the elongating zone; they also show cambial activity.

Table 1. Hormone-induced elongation in several plant organs and its interaction with FUdR (10^{-5}M) and colchicine (0.001%)

Plant organ	Hormone and concentration	Final length (in percent. of corresponding control)	
		FUdR	colchicine
lentil epicotyl	GA_3; 5×10^{-5}M	40	41
mustard hypocotyl	GA_3; 1.5×10^{-4}M	65	41
cucumber hypocotyl	GA_4; 10 mg/l	30	53
cucumber hypocotyl	IAA; 100 mg/l	15	-
Avena coleoptile sections	IAA; 10 mg/l	95	91
gamma wheat (coleoptiles)	GA_3; 10^{-4}M	94	45
gamma wheat (first leaf)	GA_3; 10^{-4}	97	38

In several other cases, where the hormonal response is shown by a tissue that is absolutely non-dividing and, therefore, probably also not synthesizing DNA, the response is not inhibited by inhibitors of DNA synthesis. Thus, Rose and Adamson (1969) as well as Atsmon and Lang (unpublished) have found that the GA response in wheat gamma plantlets is not inhibited by FUdR; these plantlets develop from seeds which got a high dose of ionizing radiation and, therefore, are devoid of any meristematic activity or DNA synthesis (Haber and Foard, 1964). Haber et al (1969) reported that the GA effect on the germination of gamma-irradiated lettuce seeds was also independent of DNA synthesis. The GA-induced α-amylase production by barley aleurone cells is also independent of DNA synthesis and is not inhibited by FUdR. The most striking example along this line was provided by Holm and Key (1969). They found that GA-induced elongation in soybean hypocotyl was inhibited by FUdR only in the apical section, but not in the subapical one. In both sections it is an elongation response; in the apical one it is associated with cell division and DNA synthesis, while in the subapical it is not. Thus, it seems that in systems which normally do not synthesize DNA the hormonal response, if any, is independent of DNA synthesis. This general correlation led us to test the effect of colchicine (table 1) on the hormonal response, since this drug inhibits meristematic activity but not DNA synthesis. Colchicine was a potent inhibitor of the hormonal response in all systems where FUdR was an effective inhibitor. The auxin-induced elongation of Avena coleoptile sections, as well as the GA-induced α-amylase production by barley aleurone were not inhibited by colchicine, nor by FUdR. The GA-response of wheat gamma plantlets was inhibited by colchicine but not by FUdR. A possible explanation is that colchicine disrupts the microtubules, which are possibly associated with the stretching of the cell wall. The otherwise complete parallelism between the effectiveness of FUdR and that of colchicine in the various systems may suggest that they act through one common mechanism - the inhibition of meristematic activity which may, for some yet unknown reason, be a pre-requisite for the hormone-induced elongation in the adjoining region.

This requirement for meristematic activity may explain also the fact that decapitated organs respond much less to GA treatment, as compared to the intact organ. The data in Table 2 show very clearly that decapitation *after* treatment with GA is relatively more inhibitory than decapitation *before* the treatment.

Table 2. Effect of decapitation (24 hrs after beginning of experiment) on elongation and GA response in lentil epitocyls. H - Hoagland's solution

Treatment No.	Medium in first 24 hrs	Medium in second 24 hrs	± apex	Length at 48 hrs (mm)	% inhibition by decapitation
1	H	H	+	24.0	
2	H	H	-	20.0	17
3	H	H + GA	+	31.7	
4	H	H + GA	-	26.5	16
5	H + GA	H	+	38.3	
6	H + GA	H	-	28.3	26
7	H + GA	H + GA	+	39.6	
8	H + GA	H + GA	-	26.6	33

III. HORMONE-ENHANCED DNA SYNTHESIS IN ELONGATING CELLS: LOCALIZATION AND CHARACTERIZATION

Most of the tissues which show hormone-induced elongation are characterized also by enhanced DNA synthesis, associated with the hormonal response. Since most of the tissues of the elongating organs are non-meristematic, some obvious questions arise in connection with this newly synthesized, hormone-induced DNA: a) Is it mostly nuclear or extra-nuclear? b) If it is nuclear - does it represent passage of cells from G_1 to G_2 resulting in the doubling of DNA content per nucleus which is one of the stages in the normal cell cycle, or, does it rather result in polyteny - i.e. multiple amounts of DNA per nucleus? c) Is this newly synthesized DNA qualitatively identical to the 'old' DNA, characteristic of the organism, or does it show changes, for instance, in density, molecular weight, base composition, etc.? d) Is it stable or, rather, a so-called 'metabolic DNA'?

Before attempting to answer these questions it is of importance to verify that the enhanced DNA synthesis cannot be accounted for simply by hormone-enhanced cell divisions in an otherwise non-meristematic tissue. Thus, it was found by Atsmon *et al* (unpublished; see table 3) that in a GA-treated lentil epicotyl there is enough meristematic activity, especially around and within the vascular bundles, to account for the rise in DNA synthesis.

Table 3. Gibberellin effect on number of nuclei in longitudinal sections on lentil epicotyls.

Treatment	Number of nuclei
initial state	1200
control	1380
GA_3 ($10^{-5}M$)	2400

Degani and Atsmon (1970a) have examined the hormone-induced DNA synthesis in the cucumber hypocotyl. This tissue has several advantages: a) It responds to both IAA and GA in the intact seedling; b) at the time of treatment and thereafter the analyzed hypocotyl section is normally devoid of any cell divisions which may complicate the interpretation of the data. Hormonal treatments, and especially GA, caused some increase in the number of nuclei within the experimental period (table 4), but it was small in relation to the DNA synthesis recorded. In general, the kinetics of growth rate paralleled very much that of ^{3}H-thymidine incorporation into DNA in the various treatments - the peak of growth rate and of rate of DNA synthesis was 12 hours after treatment for IAA, 36 hours for GA; at 60 hours both growth and DNA synthesis dropped back to control level in the IAA-treated seedlings, while it stayed much higher in the GA-treated ones. Preliminary results (Kadouri and Atsmon, unpublished) indicate that inhibition of the hormone-induced elongation, by mannitol or by decapitation (see Katsumi et al, 1965), brings about a parallel inhibition of ^{3}H-thymidine incorporation. It seems therefore, that in this system there is a close correlation between elongation and DNA synthesis, but this does not mean necessarily a causal relationship between the two processes.

Table 4. Hormonal effects on number of nuclei in cross sections of vascular bundles in cucumber hypocotyl.

Treatment	Number of nuclei
initial state	308
48 hrs control	327
48 hrs GA	461
48 hrs IAA	429

Autoradiography of transverse and longitudinal sections of tissue which has incorporated ^{3}H-thymidine showed that most of the incorporation into DNA took place in chloroplasts, while the proportion of labelled nuclei was very low. When such labelled material was separated by sucrose gradient into chloroplast and nuclear fractions, it was found that the specific labelling of the chloroplast DNA was about 10-20 times as high as that of the nuclear DNA (tables 5 and 6, and fig. 1). In summary, DNA synthesis in the elongating cucumber hypocotyl parallels growth and is mostly extra-nuclear; this is true for normal, accelerated (by hormones) or inhibited (by mannitol or decapitation) rates of growth (Degani and Atsmon, 1970b).

Table 5. Amounts of DNA in nuclear and non-nuclear fractions of cucumber hypocotyl.

	µg DNA/hypocotyl unit
nuclear	0.52
non-nuclear	0.03

Table 6. Hormonal effects on ^{3}H-thymidine radioactivity in DNA of nuclear and non-nuclear fractions of cucumber hypocotyls (12 hrs treatments).

	cpm/50 hypocotyl sections	
Treatment	Nuclear fraction	Non-nuclear fraction
initial state	1126	884
control	2307	2053
GA	4338	3892
IAA	5288	3926

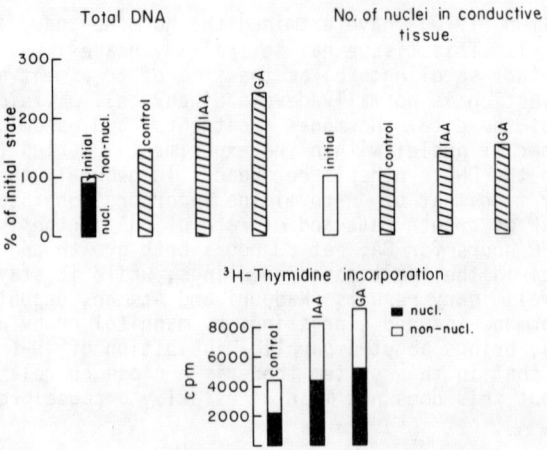

Fig. 1. Summary of evidence showing that major part of hormone-induced DNA synthesis in cucumber hypocotyl is extra-nuclear (details in text).

Is this enhanced, extra-nuclear DNA synthesis a pre-requisite for the hormone-induced elongation or just another, parallel manifestation of the increased growth? At present the second possibility seems more likely. Kern and Lang (personal communication) as well as recent findings in our laboratory indicate that chloramphenicol, a selective poison for chloroplasts, brings down the level of DNA synthesis in GA-treated material to the control level. Yet, the GA-induced growth is not inhibited by this treatment, indicating that the enhanced DNA synthesis is not required for it.

Some preliminary experiments (Kadouri and Atsmon, unpublished) indicate that much of the newly synthesized DNA shows rapid turnover and might be some kind of 'metabolic' DNA.

Further work will attempt to answer questions about the quality of this new, hormone-induced DNA, whether it is different from the already existing DNA or identical with it.

IV. SUMMARY

Research on the molecular aspects of hormone-response concentrates mainly on the possible mechanisms of regulation associated with the transcription and translation levels; i.e. RNA and protein syntheses. Less work has been done on DNA synthesis in relation to hormonal response. In general, inhibitors of DNA synthesis (like 5-fluoro-deoxyuridine, FUdR) inhibit also the hormonal response, mainly GA responses. This is true for tissues in which the response is associated with DNA synthesis; in others, like the barley aleurone cells, gamma plantlets and *Avena* coleoptiles, the hormonal response is independent of DNA synthesis. Experiments with colchicine have shown that if the responding tissue is synthesizing DNA and is adjacent, *in situ*, to a meristematic zone, this chemical, which does not inhibit DNA synthesis but inhibits meristematic activity, is a potent inhibitor of the hormonal response. It is, therefore, possible that the FUdR inhibition of hormonal-induced elongation is due to the inhibition of meristematic activity within and around the elongating region, rather than the inhibition of DNA synthesis. The cause for this requirement of dividing cells for full elongation of *other* cells is yet unclear.

Hormonal growth response is mostly associated with enhanced DNA synthesis. In many cases cell division cannot account for this newly synthesized DNA. It was found that in the hormone-responding cucumber hypocotyl the enhanced DNA synthesis is mostly extra-nuclear and is concentrated in the chloroplasts. The kinetics of this enhanced synthesis parallels that of growth. The newly synthesized DNA seems to be of the 'metabolic' type.

V. REFERENCES

BOPP, M. (1967). Hemmung des Streckungswachstums etiolierter Sprossachsen durch FUdR. Z. Pflanzenphysiol. 57: 173-187.

DEGANI, Y. and D. ATSMON (1970a). DNA synthesis and hormone-induced elongation in the cucumber hypocotyl. Nature 228: 554-555.

DEGANI, Y. and D. ATSMON (1970b). Enhancement of non-nuclear DNA synthesis associated with hormone-induced elongation in the cucumber hypocotyl. Exptl. Cell Res. 61: 226-229.

HABER, A. and D.E. FOARD (1964). Further studies of gamma-irradiated wheat and their relevance to use of mitotic inhibition for developmental studies. Am. J. Bot. 51: 151-159.

HABER, A., D.E. FOARD and Stella W. PERDUE (1969). Action of gibberellic and abscisic acids on lettuce seed germination without action on nuclear DNA synthesis. Plant Physiol. 44: 463-467.

HOLM, RE. and J.L. KEY (1969). Hormonal regulation of cell elongation in the hypocotyl of rootless soybean: an evaluation of the role of DNA synthesis. Plant Physiol. 44: 1295-1302.

JOHRI, M.M. and J.E. VARNER (1968). Enhancement of RNA synthesis in isolated pea nuclei by gibberellic acid. Proc. Nat. Acad. Sci. 59: 269-276.

KATSUMI, M., W.K. PURVES, B.O. PHINNEY and J. KATO (1965). The role of cotyledons in gibberellin and auxin induced elongation of the cucumber hypocotyl. Physiol. Plant. 18: 550-556.

KESSLER, B. and I. SNIR (1969). Interactions *in vitro* between gibberellins and DNA. Biochim. Biophys. Acta 195: 207-218.

MASUDA, Y. (1965). RNA in relation to the effect of auxin, kinetin and gibberellic acid on the tuber tissue of Jerusalem artichoke. Physiol. Plant 18: 15-23.

MATTHYSSE, Ann G. and C. PHILLIPS (1969). A protein intermediary in the interaction of a hormone with the genome. Proc. Nat. Acad. Sci. 63: 897-903.

NITSAN, J. and A. LANG (1965). Inhibition of cell division and cell elongation in higher plants by inhibitors of DNA synthesis. Develop. Biol. 12: 358-376.

NITSAN, J. and A. LANG (1966). DNA synthesis in the elongating non-dividing cells of lentil epicotyl and its promotion by gibberellin. Plant Physiol. 41: 965-970.

ROSE, R.J. and D. ADAMSON (1969). A sequential response to growth substances in coleoptiles from gamma-irradiated wheat. Planta 88: 274-281.

HORMONES AND ISOENZYMES

Auxin, Macromolecular Repressors and the Development of Isoperoxidases in Cultured Tobacco Pith

Y. Leshem[1], A.W. Galston, R. Kaur-Sawhney and L.M. Shih

Department of Biology, Kline Biology Tower, Yale University, New Haven, Connecticut, U.S.A.

[1]Permanent address: Department of Biology, Bar Ilan University, Ramat Gan, Israel

I. INTRODUCTION

The inductive and repressive actions of the plant growth hormone, indole-3-acetic acid (IAA) on a peroxidase-based enzyme system which may control endogenous levels of IAA have been described in detail (Galston, Lavee & Siegel, 1968). In this paper we show that the repressive effect of IAA may be duplicated by a macromolecular fraction extracted from the tissue. Such a fraction may be under the direct control of IAA, and could represent the mechanism through which the hormone affects the enzyme activity.

The peroxidases of many plant species can be resolved into a series of isozymes by electrophoresis on starch-gel or polyacrylamide gel (Galston & McCune, 1961; McCune, 1961; Shannon, Kay & Lew, 1966; Siegel, 1966; Lavee & Galston, 1968a, 1968b). The number and relative activities of such isozymes vary according to position on the plant, hormone status, environmental factors during growth and age of the plant (Birecka & Galston, 1970). In freshly excised tobacco pith, the peroxidase isozyme pattern after electrophoresis at pH 9.0 is essentially anodic, showing two major bands, A_2 and A_3. After incubation in sterile culture medium for 24 hours the pith develops five new isoperoxidases: three of these are cathodic, C_1, C_2, and C_3; and two anodic, A_0 and A_1 (Lavee & Galston, 1968a). The appearance of these new peroxidase isozymes is inhibited by actinomycin D and thus probably represents *de novo* synthesis of protein (Galston, Lavee & Siegel, 1968). The appearance of several of the new isozymes is prevented by IAA, which acts later on, after 48-72 hours, to induce another rapidly moving cathodic isoperoxidase. Both IAA and actinomycin D have only a slight inhibitory effect on the activity of the two parental isozymes in 24 hour culture; this decline presumably represents their turnover during the culture period.

The appearance in culture of new isoperoxidases that are absent in the parent tissue lends itself to several possible interpretations. One possibility is that high endogenous levels of IAA in the parent tissue are lowered upon removal of tissue to culture, and that this decline permits the synthesis of the previously repressed isozymes. Another possibility is that a macromolecular repressor in fresh tissue prevents expression of genetic loci which is involved in the formation of new peroxidases, and that this repressor declines following excision. We have examined these two possibilities and wish now to present evidence indicating an inhibitory effect of exogenously applied parent tissue RNA and protein on the development of new peroxidase isozymes during aseptic tissue culture.

II. MATERIALS AND METHODS

Upper internodes of preflowering tobacco plants *cv* Wisconsin 38 were used as the material for isozyme studies. Methods of sampling and aseptic culture were essentially as outlined by Galston, Lavee and Siegel (1968). In these experiments pith cylinders were grown in culture medium or in moist air. After 24 hr. culture period cylinders were removed, gently washed three times with distilled water, blotted with filter paper and ground with solid CO_2. After thawing, the expressed sap was centrifuged for 15 min. at 14,000 x *g* and two to three replicates of 40 μl each used for vertical starch-gel electrophoresis. The gel plate for starch gel electrophoresis was made up in a modified Buchler apparatus having the origin in the centre of the plate. It contained 18.4 ml borate buffer pH 9.0 and 381.6 ml distilled H_2O in which 40 g of hydrolyzed starch had been solubilized by heating. The hot sol was degassed by vacuum and allowed to set for at least 1 hr. at $4°C$ before use. After insertion of the 40 µl samples into the origin slots with a microsyringe, the origin opening and slots were sealed off with warm Vaseline and the plates subsequently placed vertically in the bridge containing NaOH-borate buffer pH 8.3 as the mobile phase, and run at 300V (10 V/cm) for $3\frac{1}{4}$ hrs. at $4°C$. After completion of the separation, the gels were developed with guaiacol - H_2O_2 reagent for 10 min. and finally fixed with Smithies' solution (1962), containing 2 parts glycerine, 2 parts glacial acetic acid, 5 parts H_2O and 5 parts methyl alcohol. Gels were scanned for optical density at 470 nm by placing longitudinal gel strips, each representing a 40 µl slot, in a Gilford Model 2410 spectrophotometer equipped with a linear transport scanner attachment.

RNA was extracted from *ca.* 120 g tobacco pith tissue pooled from four different preflowering and flowering plants. The leaves were removed and the stem surface sterilized with 70% ethanol. The pith was extracted with a No.4 cork borer from internodes 10-23. Half of the extracted pith was frozen immediately and the remaining half cultured aseptically for 24 or 48 hrs. in moist chambers at $24°C$. For extraction we used the procedure of Loening and Ingle (1967), which utilizes a phenol reagent containing 0.1% by wt. of hydroxyquinoline and 10% *m* - cresol, together with an equal volume of 10 mM Tris buffer of pH 7.4 containing 1% sodium lauryl sulphate and 12 mg/ml bentonite. Two additional cycles of solution in buffer and cold ethanol precipitation (Kaur-Sawhney, Bara & Galston, 1967) were included. The absorption pattern of the final RNA precipitate redissolved in 10 mM tris pH 7.4 was checked with a Model 350 Perkin-Elmer Spectrophotometer. It showed a sharp peak at 257 nm with a 260:280 nm ratio of 2.8:1. For application to tissues the RNA precipitate thus obtained was dissolved in the above buffer and successive 0.05 ml aliquots applied by microsyringe injection into the interior of the pith cylinders, followed by vacuum infiltration for three min. in the excess solution which did not enter tissue upon injection. Controls were similarly treated with buffer. In some instances, infiltration alone was used, without prior injection.

A complete inhibition of the formation of all new isozymes was caused by 0.33 mg/ml RNA extracted from parent pith; hence this concentration was used for most experiments. Hydrolyses of the extracted material were carried out with KOH and enzymes: KOH was used at a concentration of 0.3 N at $30°C$ for 18 hrs. Ribonuclease (Worthington Biochemical Co.) was used at 10 µg/mg RNA at $37°C$ for 1 hr. in acetate buffer pH 5.0. Trypsin (Worthington) and pronase (Cal. Biochem.) were used at 30 or 60 µg/mg RNA at $37°C$ for 1 hr. in Tris buffer pH 8.1 and 8.5 respectively. Treatments containing ribonuclease and trypsin, and ribonuclease and pronase were also used. The RNase solution was heated for 5 min. in a boiling water bath to denature DNase and protease prior to its use. Pronase solution was placed at $37°C$ for 1 hr. to inactivate RNase and DNase prior to its use. RNase, trypsin or pronase when infiltrated alone showed no effect on the development of new isozymes.

III. RESULTS AND DISCUSSIONS

The RNA extract was applied to pith culture in several basic experiments: The first involved extract from 120 g parent pith; the second involved dilutions of the original RNA extract prior to application. Typical results are presented in Figures 1 and 2. It is evident from Figure 1 that the application of RNA from parent tissue having the anodic isozymic pattern inhibits formation of the induced anodic and cathodic isozymes which normally appear during culture. Figure 2 shows the effect of dilutions of an RNA extract (BM + 10 RNA). This concentration of RNA is slightly less than that required for complete inhibition of the formation of new isozymes. Further dilution of the extract decreases the inhibition as indicated by treatments BM + 3 RNA (3-fold dilution) and BM + 1 RNA (10-fold dilution).

RNA extracted from pith cultured for 24 or 48 hours in a moist chamber is less effective in producing this inhibition than RNA extracted from fresh pith (Figure 3,

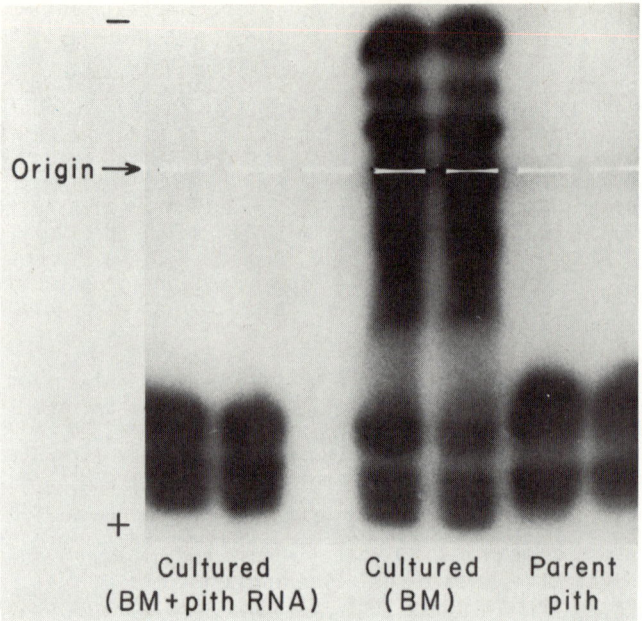

Fig. 1. Peroxidase isozyme patterns obtained by starch-gel electrophoresis of extracts of tobacco pith from intact plants (right); pith cultured in basal nutrient medium (BM) for 24 hours (center), and in basal medium to which RNA from parent pith has been applied (left). (Arrow indicates origin, + the anodic, and - the cathodic sides of the zymogram).

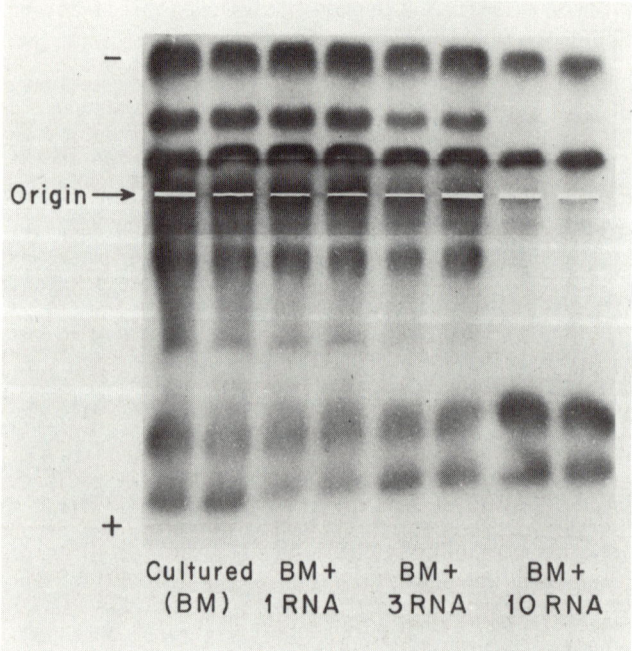

Fig. 2. Effect of concentration of total RNA extract upon peroxidase isozyme development in tobacco pith in 24 hour cultures as determined by starch-gel electrophoresis, BM + 10 RNA denotes highest concentration of RNA used; BM + 3 RNA, diluted three times; BM +1 RNA, diluted 10 times; cultured (BM), no RNA.

Table 1. Effect of duration of culture of donor pith on the ability of the extracted "RNA" to repress new isoperoxidase formation in receptor pith.

Duration of culture of pith used for RNA extraction	Percentage inhibition of new isoperoxidases at RNA concentrations		
	0.03 mg/ml	0.10 mg/ml	0.30 mg/ml
0 hrs	29	40	100
24 hrs	0	22	100
48 hrs	-	-	0

Table 2. Effect of treatment with KOH and hydrolytic enzymes on the repressor activity of fresh and cultured donor pith "RNA".

Source of RNA	Treatment	Percent inhibition of new isozymes
Fresh pith	None	100
Pith cultured 24 hours	None	83
Pith cultured 48 hours	None	-21 (promotion)
Yeast RNA	None	-114 (promotion)
Fresh pith	None	100
Fresh pith	0.3N KOH	51
Fresh pith	RNAse	49
Fresh pith	Trypsin	18
Fresh pith	Trypsin + RNAse	-72 (promotion)
Fresh pith	Pronase	33
Fresh pith	Pronase + RNAse	-50 (promotion)

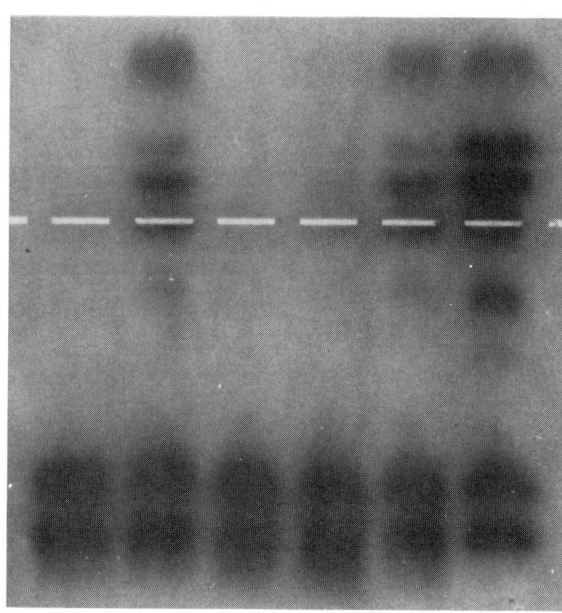

Fig. 3. Peroxidase isozyme patterns obtained by starch-gel electrophoresis of extracts from tobacco pith of intact plants. The pith was treated as follows. Left to right: (1) fresh pith, no treatment; (2) infiltrated with Tris buffer pH 7.4; (3) infiltrated with RNA from fresh pith of intact plants; (4) infiltrated with RNA from pith cultured for 24 hrs.; (5) infiltrated with RNA from pith cultured for 48 hrs. and (6) infiltrated with yeast RNA. Numbers 2-6 were incubated for 24 hrs. in moist chambers.

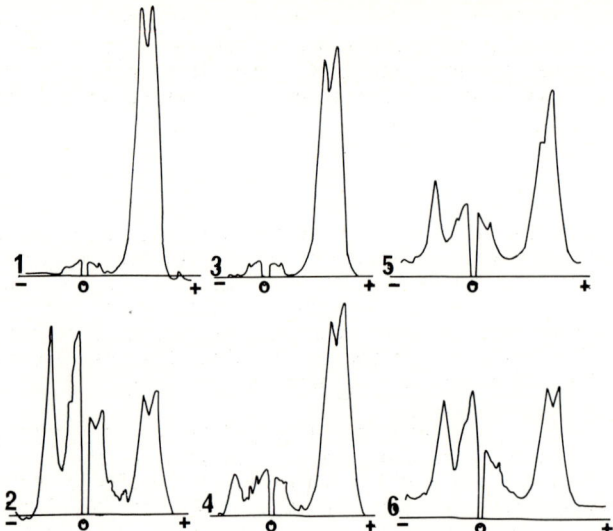

Fig. 4. Scan diagram of peroxidase isozymes obtained by starch-gel electrophoresis of extracts from tobacco pith. The pith was treated as follows: (left to right) (1) fresh pith, no treatment, (2) infiltrated with Tris buffer pH 7.4, (3) infiltrated with RNA from pith of intact plants, (4) infiltrated with RNA from pith cultured for 24 hours, (5) infiltrated with RNase hydrolysate from pith of intact plants and (6) infiltrated with RNase hydrolysate from pith cultured for 24 hours. Numbers 2 - 6 were incubated for 24 hours in moist chambers.

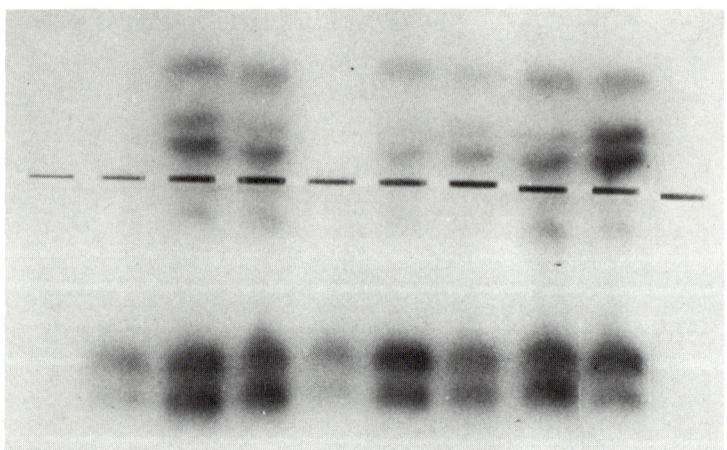

Fig. 5. Peroxidase isozyme patterns obtained by starch-gel electrophoresis of extracts from tobacco pith. The pith was treated as follows: (left to right) (1) fresh pith, no treatment, (2) infiltrated with Tris buffer, pH 7.4; (3) infiltrated with buffer + RNase + trypsin; (4) infiltrated with RNA; (5) infiltrated with KOH hydrolysate; (6) infiltrated with RNase hydrolysate; (7) infiltrated with trypsin hydrolysate; and (8) infiltrated with RNase plus trypsin hydrolysate. RNA for the last 5 treatments was from fresh pith of intact plants. Numbers 2-8 were incubated for 24 hrs. in moist chambers.

Table 1). Thus it appears that a repressive macromolecular fraction extractable from pith diminishes in quantity during *in vitro* culture, but is stabilized or its synthesis promoted by exogenous RNA.

Hydrolysis of the extracted RNA with 0.3 N KOH or with ribonuclease diminished its effectiveness (Figures 4 & 5, Table 2), while yeast RNA promoted the formation of the new isoperoxidases (Figure 3). These observations would suggest that the effective molecule is a specific active type of RNA. However, pronase and trypsin were also found to be effective in destroying the repressor activity of the extracted macromolecule (Figure 5, Table 2). We thus envision that a protein repressor of isoperoxidase formation as well as an RNA coding for its formation may be present in the active extract.

IV. SUMMARY

Previous work has shown that freshly excised tobacco pith tissue contains two isoperoxidases migrating toward the anode at pH 9.0. Within 24 hours of aseptic culture on basal medium or in moist air, such tissue develops a considerably greater peroxidase activity, localized in *ca*. five new isoperoxidases, three cathodic and two anodic. The appearance of the new isoperoxidases and to a certain extent the augmentation of the old ones is inhibited by the plant hormone indole-3-acetic acid (IAA) as well as by inhibitors of RNA and protein synthesis.

Phenol RNA extracts from fresh pith injected or vacuum infiltrated into fresh pith explants prevent the appearance of the new isozymes during the usual 24 hour incubation. Hydrolysis with 0.3 N KOH, crystalline ribonuclease, pronase or trypsin significantly diminishes this inhibitory activity. Similar phenol extracts from pith cultured in moist air for 24 hours have diminished inhibitory activity; after 48 hours of culture, the inhibitory activity disappears completely. RNA from other sources, such as yeast, has no inhibitory action. We conclude that isoperoxidase formation in tobacco pith is under the control of macromolecular repressors, whose activity in turn may be determined by IAA.

V. ACKNOWLEDGEMENTS

This project was supported by USPHS grant to the second author.

VI. REFERENCES

BIRECKA, H. and A.W. GALSTON (1970). Peroxidase ontogeny in a dwarf pea stem as affected by gibberellin and decapitation. Jour. Exp. Bot., 21, 735-745.
GALSTON, A.W., S. LAVEE and B.Z. SIEGEL (1968). The induction and repression of peroxidase isozymes by 3-indoleacetic acid, in "Biochemistry and Physiology of Plant Growth Substances" (Ed. F. Wightman and G. Setterfield), p.455, Runge Press, Ottawa, Canada.
GALSTON, A.W. and D.C. McCUNE (1961). An analysis of gibberellin-auxin interaction and its possible metabolic basis, in "Plant Growth Regulation", p. 611, Iowa State College Press, Iowa, U.S.A.
KAUR-SAWHNEY, R., M. BARA and A.W. GALSTON (1967). Analysis of labelling patterns in soluble RNA preparations from green pea stem sections supplied with ^{14}C-carboxyl-labelled indoleacetic acid. Ann. New York Acad. Sci., 144, 63-67.
LAVEE, S. and A.W. GALSTON (1968a). Structural, physiological, and biochemical gradients in tobacco pith tissue. Pl. Physiol., 43, 1760-1768.
LAVEE, S. and A.W. GALSTON (1968b). Hormonal control of peroxidase activity in cultured *Pelargonium* pith. Amer. J. Bot., 55, 890-893.
LOENING, U.E. and J. INGLE (1967). Diversity of RNA components in green plant tissues. Nature, 215, 363-367.
McCUNE, D.C. (1961). Multiple peroxidases in corn. Ann. New York Acad. Sci., 94, 723-730.
SHANNON, L.M., E. KAY, and J.Y. LEW (1966). Peroxidase isozymes from horseradish roots. 1. Isolation and physical properties. J. Biol. Chem., 241, 2166-2172.
SIEGEL, B.Z. (1966). The molecular heterogeneity of plant peroxidases. Ph. D. Dissertation, Yale University.
SMITHIES, O. (1962). Molecular size and starch-gel electrophoresis. Arch. Biochem. Biophys. Suppl. 1, 125-131.

ISOZYMES OF CELLULASE IN THE ABSCISSION ZONE OF *Phaseolus vulgaris*

L.N. Lewis, F.T. Lew, P.D. Reid and J.E. Barnes[1]

Department of Plant Sciences, University of California, Riverside, California, 92502

[1]Present address: Department of Cell Biology, The John Hopkins University, Baltimore, Maryland

I. INTRODUCTION

Cellulase activity in the abscission zone of Red Kidney bean was previously shown to be partially soluble in a low salt buffer and partially soluble in a high salt buffer (Lewis and Varner, 1970). This suggested the presence of more than one molecular form of the enzyme. The work reported here was done to elucidate the multiple nature of cellulase in the abscission zone.

II. MATERIALS AND METHODS

Abscission zones were taken from the primary leaves of *Phaseolus vulgaris* L., cultivar Red Kidney. The plants were grown for about 12 days in the greenhouse at $27 \pm 2°C$ in a peat-sponge rock mix and watered daily with a 0.5 strength Hoagland's solution.

Seedlings were debladed, cotyledons removed and the plant was cut off 2-4 cm below the cotyledonary node leaving the main axis plus the petiole and distal abscission zone of each primary leaf as a unit. Groups of these seedling explants were placed in cups of H_2O, exposed to 50 ppm ethylene, and their abscission zones harvested about 48 hours after deblading. This abscission zone represented about 1 cm of petiole and the entire distal abscission zone. Pull force necessary to separate the abscission zone (Lewis and Varner, 1970) averaged less than 10 grams at the time of harvest.

Cellulase was extracted by grinding with a Waring blender first in 0.02 M phosphate pH 6.1 and then in 1 M NaCl plus 0.02 M phosphate pH 6.1. Extracts were filtered through nylon cloth and centrifuged for 10 min at 10,000 xg. The supernatant solutions were then centrifuged for 60 min at 100,000 xg, dialyzed against 0.1 M NaCl in 0.02 M phosphate, pH 6.1, and the dialysate used for the enzyme characterization studies.

Cellulase activity was assayed viscometrically and drainage times were converted to relative activity units (B) as previously described (Lewis and Varner, 1970).

Isoelectric focusing was carried out according to the procedure described by Haglund (1967). Samples were focused for 36-40 hr at 300 volts.

Molecular weights of the isozymes were estimated by gel filtration (Ackers, 1964; Ting, 1968) through calibrated 1.5 x 75 cm Bio Gel P-150 columns. The gel was equilibrated with 0.1 M NaCl in 0.02 M phosphate buffer, pH 6.1. Two ml of enzyme preparation containing abscission zone cellulase and at least one marker enzyme were applied to the column, 2 ml fractions of the eluate were collected. The Bio Gel column was calibrated with the following marker enzymes: equine heart cytochrome C (assumed mol wt = 12,500; California Corporation for Biochemical Research), and rabbit muscle lactic dehydrogenase (assumed mol. wt = 135,000; Sigma Co.). The enzyme activities of the dehydrogenases were measured spectrophotometrically by following the oxidation of NADH at 3400 Å. The substrate for lactic dehydrogenase was pyruvate. Cytochrome C was reduced by dithionite and measured spectrophotometrically at 5400 Å and 5500 Å.

III. RESULTS

The bulk of the cellulase activity in the abscission zone of Red Kidney bean exists in two different molecular forms as shown by isoelectric focusing separation (Fig. 1). One form has a pI (isoelectric point) of 4.2 to 4.6 (Fig. 2) and will henceforth be referred to as cellulase 4.5. The other form has a pI of 9.2 to 9.6 (Fig. 3) and will be designated as cellulase 9.5. The peaks at 5.0, 6.5 and 7.5 were observed consistently; however, they represent a very small per cent of the total activity. The properties of these other cellulases have not been studied at this time.

Cellulase 4.5 was found primarily in the 0.02 M phosphate buffer, and the major portion of cellulase 9.5 was extracted with the 1 M NaCl fortified phosphate buffer.

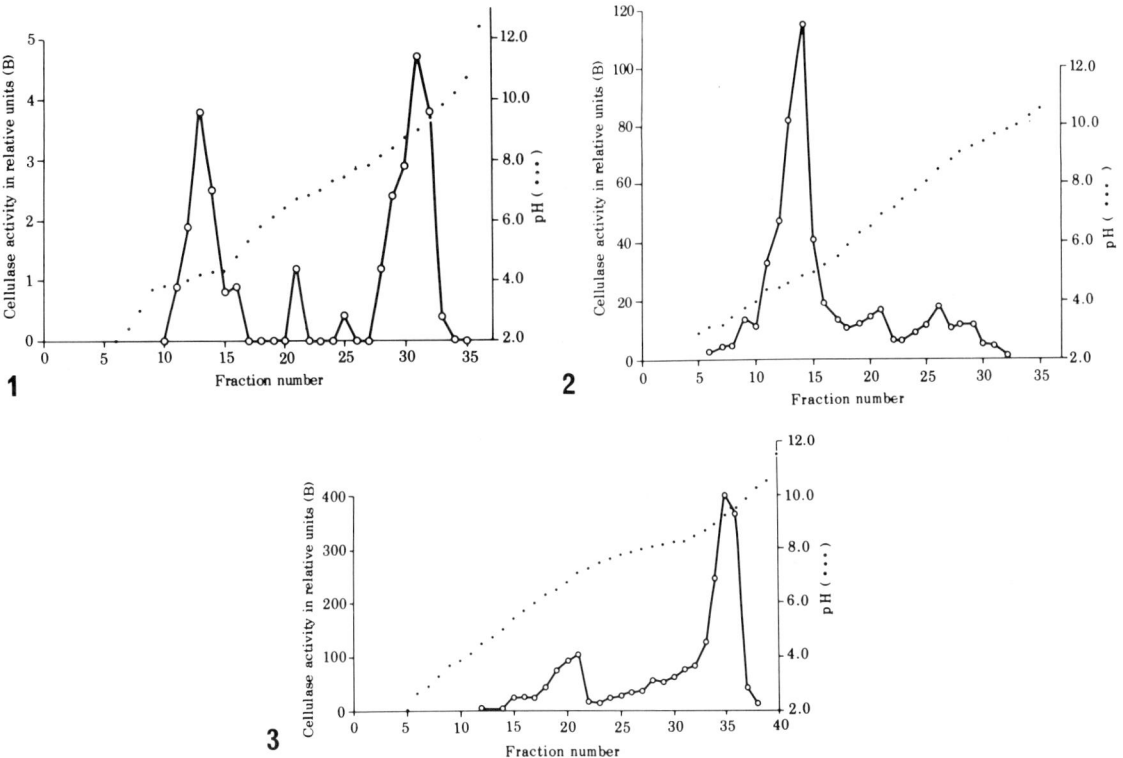

Fig. 1, 2, 3. Isoelectric characterization of abscission zone cellulase 4.5 and 9.5 by isoelectric focusing in an ampholine buffer, pH 3-10. Fig. 1 shows the fractionation of a mixture of the two cellulases. Fig. 2 is primarily cellulase 4.5. Fig. 3 is cellulase 9.5 following its separation from cellulase 4.5 on CM Sephadex C-50.

Further comparisons of the enzymatic properties of cellulase 4.5 and 9.5 are reported in the following sections.

Heat stability

Cellulase 4.5 and 9.5 seemed about equally stable when dialyzed against buffers from pH 4.2 to 9.0 (Fig. 4 & 5). Exposing these variously buffered solutions of the cellulase 4.5 to 50°C showed that the enzyme was about equally stable from pH 6.0 to 7.8. At pH 6.0 the enzyme lost about one half its activity in 30 min and it lost over 80 per cent of its activity in 8 hr. At the pH extremes of 4.2 or 9.0, 80 per cent of the activity was lost in 30 min.

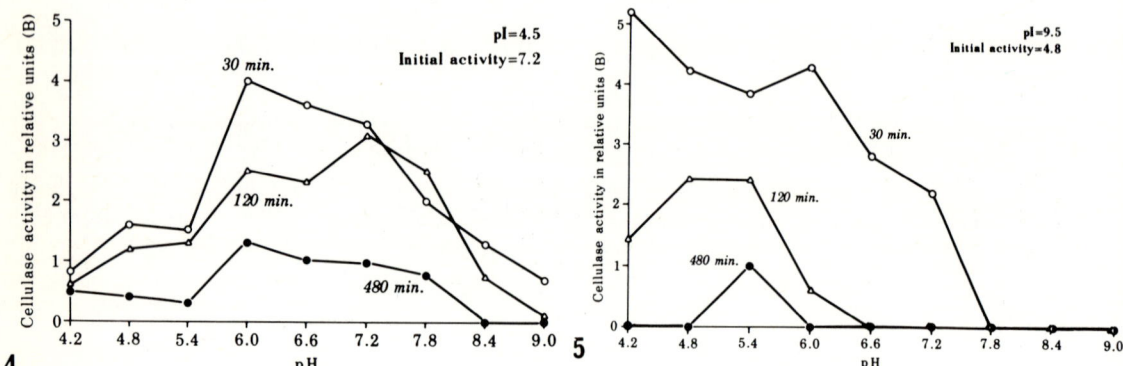

Fig. 4 and 5. Heat stability of abscission zone cellulase 4.5 (top) and 9.5 (bottom) at various pH values. Buffers for pH 4.2, 4.8, and 5.4 were 0.1 M NaCl and 0.02 M acetate; for pH 6.0, 6.6, 0.1 M NaCl and 0.02 M phosphate; and for pH 7.2, 7.8, 8.4 and 9.0, 0.1 M NaCl and 0.02 M Tris-HCl. Temperature used was 50°C. Enzyme activity is expressed in relative units of cellulase activity (B).

Cellulase 9.5 had a narrower pH range for optimum stability with peak stability at pH 5.4 This enzyme lost less than 10 per cent of its activity during a 30 minute exposure to 50°C at pH values from 4.2 to 6.0. Above pH 7.8, all activity was lost in 8 hr and it lost 80 per cent of its activity in 30 min. At the pH extremes of 4.2 or 9.0, 80 per cent of the activity was lost in 30 min.

Optimum reaction pH

Both isozymes showed the optimal hydrolytic activity between pH 5.0 and 5.5 as measured by increase in reducing power (Fig. 6 & 7). Cellulase 9.5 retained some activity from pH 6.0 to 8.0, but only about 30% or less of its optimal activity.

Sulfhydryl sensitivity

Cellulase 9.5 does not seem to require sulfhydryl groups for its enzymatic activity (Table 1). Neither PCMB (p-chloromercuribenzoate) nor NEM (n-ethylmaleimide) reduced the activity of the enzyme. Amino acid analysis of cellulase 9.5 showed no sulfur-containing amino acids.

Table 1. Effect of sulfhydryl inhibitors on cellulase activity. Shown as per cent of original activity.

	p-chloro-mercuribenzoate	n-ethyl-maleimide
cellulase 9.5	100%	100%
cellulase 4.5	50%	100%

Cellulase 4.5 was insensitive to NEM but PCMB reduced enzyme activity about 50 per cent. PCMB is more effective on sulfhydryl groups in the interior of the molecule than NEM (Frankel-Conrat, 1957). Since PCMB did not completely inactivate the cellulase 4.5 it may be that the sulfhydryl groups alter the structure slightly to maximize activity but that their presence is not essential for the hydrolytic action of the enzyme. It is also possible that cellulase 4.5 represents more than one form of cellulase, some of which are insensitive to PCMB.

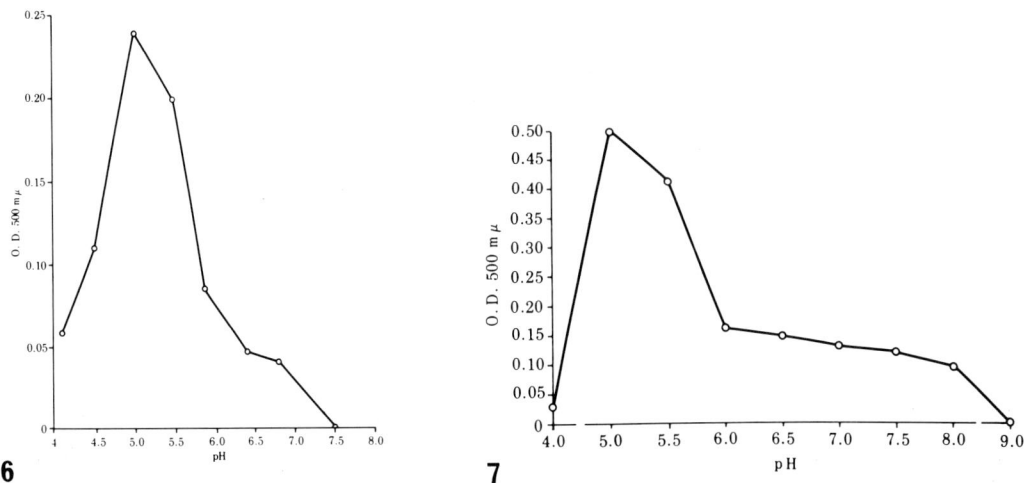

Fig. 6 and 7. pH optimum for the hydrolytic activity of abscission zone cellulase 4.5 (Fig.6) and 9.5 (Fig.7). Buffers for pH 5.5 and below were 0.2 M Acetate, for 6.0 to 7.0, 0.2 M phosphate, and for 7.5 to 9.0, 0.2 M Tris-HCl. Enzyme activity was based on the increase in reducing power of the CM-Cellulose, type 7HP, according to the Nelsen-Somogyi method (Nelsen 1944) for reducing sugar.

Molecular weight

Cellulase 9.5 was a smaller enzyme with a molecular weight of about 22,000 to 26,000 (Fig. 8). Cellulase 4.5 is a larger protein with a molecular weight of 72,000 to 78,000.

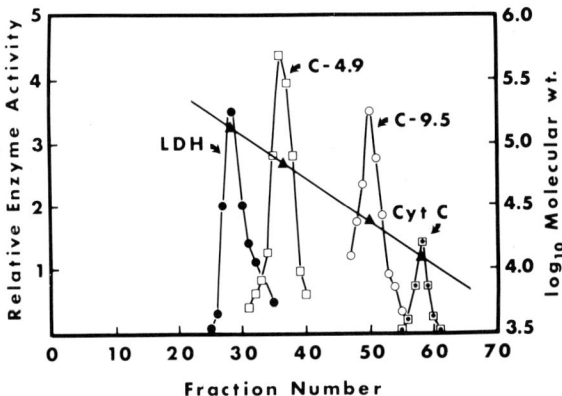

Fig. 8. Gel filtration of a preparation of abscission zone cellulase 4.5 and 9.5. Column: Bio Gel P-150, 1.5 x 75 cm; elution buffer = 0.1 M NaCl and 0.02 M phosphate pH 6.1. Elution volumes of cellulase 4.5 (□) corresponded to 72,000-78,000 mol wt and the cellulase 9.5 (O) to 22,000-26,000, LDH (●)=rabbit muscle lactic dehydrogenase, mol wt= 135,000; cyto C (▣)=equine heart cytochrome C, mol wt=12,500.

Changes during abscission

Fractionation by isoelectric focusing and gel filtration failed to detect any cellulase 9.5 at the time the seedlings were debladed; i.e. zero time (Fig. 9). The cellulase activity at zero time was due to cellulase 4.5 plus some uncharacterized activity in the pI 5.0 to 7.0 range. Forty-eight hours after deblading, when abscission was almost completed due to the ethylene treatment, cellulase 9.5 was the main form of cellulase recovered. It accounted for most of the total cellulase in abscising tissue and there was about 5 times more cellulase in abscising tissue as in zero time tissue as reported previously by several workers (Abeles, 1969; Horton and Osborne, 1967; Lewis and Varner, 1970).

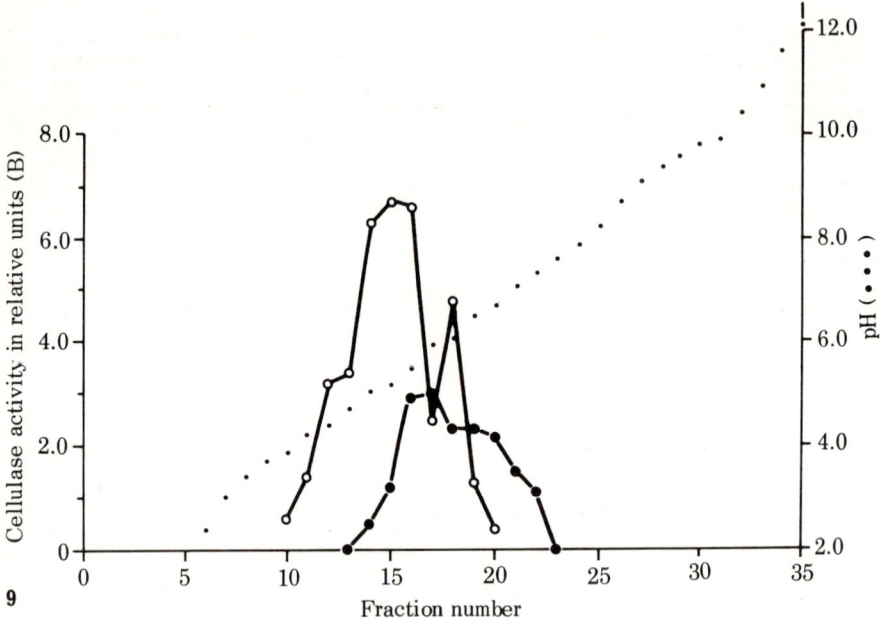

Fig. 9. Isoelectric characterization of abscission zone cellulase at zone time by isoelectric focusing in an ampholine buffer, pH 3-10. Combined low salt and high salt extract = (O), high salt extract only = (●).

IV. DISCUSSION

Two cellulase isozymes from the abscission zone of bean were separated by isoelectric focusing in ampholine buffer. At the time abscission was initiated by deblading the seedlings, there was no detectable amount of cellulase 9.5. Cellulase 9.5, the isozyme primarily soluble in the high salt buffer, is synthesized *de novo* during abscission (Lewis and Varner, 1970) and its accumulation is closely related to the abscission process. Its role in abscission or the role of any other isozyme of cellulase is still not known.

Comparisons of the enzymatic properties of these two isozymes of cellulase showed that they differ widely. Cellulase 4.5 has an acidic isoelectric point, is a much larger molecule than cellulase 9.5, is partially sensitive to PCMB and is more stable from pH 6.0 to 7.2 than at more acidic or basic pH values. Cellulase 9.5 on the contrary has a basic isoelectric point, is a smaller molecule, is insensitive to sulfhydryl inhibitors, is most stable at pH 5.4 and unstable at pH 7.2 or higher. Both enzymes are optimally active at pH 5.0.

Although isozymes of fungal cellulase have been described in the literature (Bjorndal and Eriksson, 1968), this is the first report of isozymes of cellulase in higher plants.

Although it is not possible at this time to ascribe any detailed physiological roles to these isozymes, it seems reasonable to speculate that cellulase 4.5 and 9.5 have substantially different roles within the cell. It also seems reasonable to speculate that cellulase 9.5 and abscission are closely interrelated. In this case it appears that the correlation between the developmental process of abscission and the increase of the activity of cellulase results from an increase in activity of one molecular form of the enzyme, namely, cellulase 9.5. Further, this cellulase isozyme which is responsible for the observed increase in activity apparently is not present in the tissue prior to the beginning of the abscission process.

V. SUMMARY

Cellulase activity in the abscission zone of *Phaseolus vulgaris* variety Red Kidney exists in several isozymes. Two of the isozymes which represent the majority of the cellulase activity have been characterized. Cellulase 9.5 accumulated in the abscission zone during abscission but was not detectable prior to abscission. Cellulase 4.5 does not change appreciably during ethylene-induced abscission.

These two cellulases differ markedly in their physical properties.

VI. ACKNOWLEDGEMENTS

This work was supported by the National Science Foundation through grant #GB-17850.

VII. REFERENCES

ABELES, F. (1969). Abscission: Role of cellulase. Pl. Physiol. 44, 447-452.
ACKERS, G.K. (1964). Molecular exclusion and restricted diffusion processes in molecular sieve chromatography. Biochemistry 3, 723.
BJORNDAL, A. and K.E. ERIKSSON. (1968). Extracellular enzyme system utilized by the rot fungus *Stereum sanguinolentum* for the breakdown of cellulose III: characterization of two purified cellulase fractions. Arch. Biochem. Biophys. 124 149-153.
FRANKEL-CONRAT, H.(1957). Methods for investigating the essential groups for enzyme activity, in "Methods in Enzymology" (Ed. S.P. Colowick and N.O. Kaplan) vol. IV. P. 268. Academic Press, New York. 1957.
HAGLUND, H. (1967). Isoelectric focusing in natural pH gradients-A technique of growing importance for fractionation and characterization of proteins. Science Tools 14, 17-23.
HORTON, R.F., and D.J. OSBORNE. (1967). Severence, abscission, and cellulase activity in *Phaseolus vulgaris*. Nature 214, 1086-1088.
LEWIS, L.N. and J.E. VARNER. (1970). Synthesis of cellulase during abscission of *Phaseolus vulgaris* leaf explants. Pl. Physiol. 46, 194-199.
NELSEN, N. (1944). A photometric adaptation of the Somogyi method for the determination of glucose. J. Biol. Chem. 153, 375.
TING, I.P. (1968). Malic dehydrogenases in corn root tips. Arch. Biochem. Biophys. 126, 1-7.

BINDING OF HORMONES TO CELLULAR CONSTITUENTS

Auxin-Reactive Proteins

M.A. Venis

Canada Department of Agriculture, Research Institute, University Sub Post Office, London 72, Ontario, Canada

I. INTRODUCTION

The idea that expression of auxin activity may involve an interaction with specific proteins or other macromolecules has a long history, but activity in this area has been fairly sporadic (e.g. Siegel and Galston, 1953; Zenk, 1964; Venis, 1967). More recently, interest in this field has been stimulated by preliminary reports that in the presence of protein mediators, 2,4-D (Matthyse and Phillips, 1969) and cytokinins (Matthyse and Abrams, 1970) are able to enhance RNA synthesis supported by isolated pea chromatin or DNA. Similarly, Johri and Varner (1968) have obtained circumstantial evidence that some factor required for stimulation of RNA synthesis in dwarf pea nuclei by GA is rapidly lost during nuclear isolation.

The application of conventional protein purification techniques to this problem has yielded few positive results, despite the successful isolation of numerous animal hormone-binding proteins by standard procedures (e.g. Chader and Westphal, 1968; Giorgio and Tabachnik, 1968; Soloff and Szego, 1969). A potentially attractive alternative lies in the technique of affinity chromatography, i.e. the design of a chemically-modified adsorbent that will specifically and reversibly bind only the desired macromolecule from a complex mixture such as a crude tissue extract. Although such a technique was first used by Campbell *et al.* (1951) for antibody isolation, the principle has until recently been used relatively sparingly and with limited success. However, Cuatrecasas *et al*, (1968) defined clearly the conditions necessary for satisfactory application of the method. Using the cyanogen bromide activation method of Porath *et al*, (1967) they prepared specific adsorbents by coupling various substrate analog derivatives to agarose gel and obtained highly effective purifications of the appropriate enzymes. Since then the technique has found increasing use for single step enzyme isolation and has also been used successfully in separating thyroxine-binding globulin from human serum (Pensky and Marshall, 1969). The present report describes initial attempts to apply the principle of affinity chromatography to the isolation of auxin-reactive macromolecules.

II. MATERIALS AND METHODS

Synthesis of auxin derivatives

Indole-3-acetyl-ε-L-lysine (IAA-lysine) and 2,4-dichlorophenoxyacetyl-ε-L-lysine (2,4-D-lysine, Fig. 1) were synthesized by the methods of Hutzinger and Kosuge (1968 and personal communication). N-(6-aminocaproyl)-Amiben (Fig. 1) was prepared by general methods of peptide synthesis (Schröder and Lübke, 1965). Briefly, the carbobenzoxy derivative of 6-aminocaproic acid was reacted with Amiben methyl ester by the carbodiimide method. The protecting groups were removed by catalytic hydrogenation and by saponification respectively. Amiben (2,5-dichloro-3-aminobenzoic acid) was kindly supplied by Amchem Products Inc. and was recrystallized from water before use.

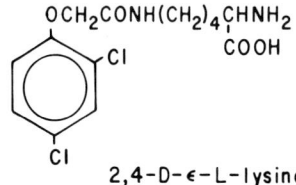

Fig. 1. The structures of auxin derivatives used to prepare affinity adsorbents.

Affinity adsorbents were prepared by the general method of Cuatrecasas et al., (1968). To 15 ml of Sepharose 4B and 10 ml water, 1.5 g of cyanogen bromide in 25 ml water was added. The pH was adjusted immediately to 11 with 4N NaOH and kept there for the duration of the reaction (8-10 min). The Sepharose was then washed on a Buchner funnel with about 200 ml of cold 0.1 M $NaHCO_3$, suspended in 15 ml of cold 0.1 M $NaHCO_3$ and 50 μ moles of the auxin derivative added in 1 ml of 0.4 N NaOH and 0.5 ml water. The mixture (final pH 9.2) was placed on a slowly rotating turntable at 2° for 24 hr, then washed with water. The amounts coupled to the Sepharose (determined spectroscopically from the amounts not recovered in the final washings) were as follows (in μ moles): IAA-lysine 10; 2,4-D-lysine 15; aminocaproyl-Amiben 8. The affinity columns prepared from these materials are referred to in the text as the IAA-lysine, 2,4-D-lysine and Amiben-caproic columns respectively.

Extracts were prepared from 72 hr shoots of hybrid corn (*Zea mays*, var. Golden Cross Bantam) and from third and second internodes of pea (*Pisum sativum*, var. Alaska), both grown in the dark at 25°. With corn, the initial 4000 x g supernatant from chromatin preparation (see below) was used, after recentrifuging at 30,000 x g for 10 min. Pea segments were homogenized in a Waring blendor for 1 min in an equal volume of TM (0.02 M tris-HCl, pH 7.6, 0.01 M mercaptoethanol), strained through Miracloth and centrifuged at 30,000 x g for 10 min. This supernatant was used directly, or sometimes after recentrifugation at 270,000 x g for 30 min (Type 65 Spinco rotor, 65,000 r.p.m.). Some extracts were adjusted to 0.1 M NaCl by the addition of one-nineteenth volume of 2 M NaCl before application to the column. All extracts were prepared at $0-4^\circ$.

Affinity chromatography. Typically, a supernatant from 15-25 g corn shoots or 50 g pea segments was pumped through a 3-5 cm x 1.2 cm column of adsorbent at ca. 200 ml/hr. The column was jacketed and maintained at 1° by circulated coolant. The effluent was monitored at 280 nm using an Isco UA 2 flow-cell and analyzer, with a Sargent 10" recorder. After passage of the extract, the column was washed with starting buffer at the same flow rate until the absorbance trace returned to the original baseline. Elution was then effected with a variety of eluants (see Results) at ca. 150 ml/hr and 2 ml fractions collected when UV-absorbing material began to emerge. When 0.002 N KOH, pH 10.2 was used as eluant, the fractions were collected into tubes containing 0.02 ml of 1M tris-HCl, pH 7.5, 1 M mercaptoethanol, the final mixture having a pH of 8.1. All eluates were dialyzed against water before being tested in assays for RNA synthesis. They were assayed for protein content by the method of Lowry *et al.* (1951) and for DNA by the diphenylamine method of Giles and Myers (1965). Amounts in Tables refer to μg protein.

After cleaning off residual protein with 0.5 per cent sodium lauryl sulfate, the columns could be re-used repeatedly. The ultimate limitation was generally an eventual deterioration in flow properties. However, the IAA-lysine column appeared to be somewhat less stable than the others, showing significant loss of capacity during a couple of months of repeated use.

Chromatin was prepared from corn shoots and from 4-6 day pea buds by the methods of Fambrough and Bonner (1966) except that: a) all solutions contained 0.01 M mercaptoethanol; b) centrifugation of crude chromatin through 1.7 M sucrose was for 105 min at 22,000 r.p.m. in the SW 27 Spinco rotor.

RNA polymerase was prepared from *E. coli* K-12 (frozen cells from General Biochemicals) by the method of Burgess (1969), to include sigma factor.

RNA synthesis by chromatin or purified DNA was carried out in a final volume of 0.4 ml containing the following, in μ moles: tris-HCl (pH 8.0), 16; $MgCl_2$ 1.6; $MnCl_2$ 0.4; mercaptoethanol 4; CTP, UTP, GTP 0.16 each; ATP-8-^{14}C, 0.03 (0.25 μc); polymerase 1-5 units and DNA (0.1-1 μg) or chromatin (equivalent to 3-5 μg DNA). After 15 min at $37°$, the reaction was stopped by the addition of 5 ml of cold 10 per cent TCA (trichloroacetic acid) containing 0.04 M sodium pyrophosphate. After 10 min the precipitates were transferred to glass fiber discs (Reeve Angel) prewashed with 5 per cent TCA, and washed with 20 ml of cold 5 per cent TCA and 10 ml of cold ethanol. The filters were dried under infra-red lamps and counted, in 2 ml of scintillation fluid, in a liquid scintillation spectrometer at ca. 80 per cent efficiency.

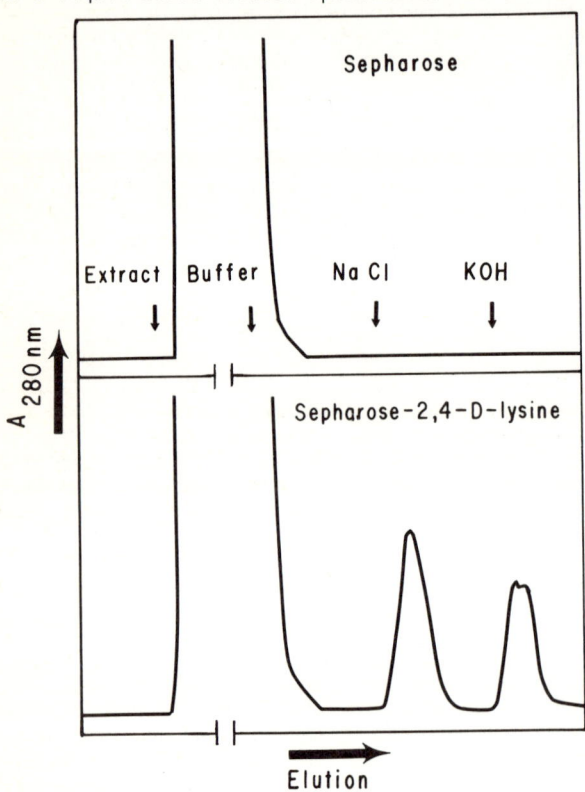

Fig. 2. Elution behaviour of unsubstituted Sepharose and of Sepharose-2,4-D-lysine.

III. RESULTS

Elution characteristics of affinity columns. Crude cell-free extracts applied to columns of unsubstituted Sepharose are eluted completely with starting buffer (Fig. 2). Columns to which auxin derivatives have been covalently attached however, retain small quantities of protein which may be eluted by a variety of means. The IAA-lysine and 2,4-D-lysine columns can be eluted with dilute acid (1N acetic) or alkali (0.002N KOH, pH 10.2) or sequentially with 1M NaCl and 0.002N KOH (Fig. 2). Acid or alkali are inferior eluants for the Amiben-caproic column, while 1M NaCl is highly efficient. The columns can also be partially eluted with high concentrations (ca 0.05M) of 2,4-D.

Despite the differing elution characteristics of the columns, no marked qualitative differences have been observed in the patterns obtained on disc electrophoresis of the various fractions. These patterns show clearly however, that the columns are not acting as indifferent adsorbents; only four to five bands are discernible compared with at least thirty bands in a crude extract.

Effects on chromatin-dependent RNA synthesis. Extensive preliminary experiments with the different columns and various eluants failed to yield any fractions with consistent effects on the activity of isolated chromatin. When using a supernatant from corn shoots however, it was observed that after 1M NaCl elution of the amiben-caproic column, a very small absorbing peak sometimes emerged when the column was washed with water. It was found that if the supernatant and starting buffer were adjusted to 0.1M NaCl, then this water peak, though still small, was much enhanced, suggesting that a certain minimal ionic strength may be necessary for this fraction to bind to the column. Furthermore, this fraction proved to stimulate RNA synthesis primed by corn chromatin (and supported by *E. coli* polymerase), while the protein eluted by 1M NaCl prior to the water fraction was inactive (Table 1).

Table 1. Activity of corn fractions on RNA synthesis primed by isolated corn chromatin $\pm$ 0.05 mg/l 2,4-D

Fraction added	ATP incorporation, p moles	
	No 2,4-D	+ 2,4-D
Control	165	196
NaCl, 2 µg	155	160
Water, 1 µg	206	212

This activity was not dependent on the addition of 2,4-D, although 2,4-D alone also produced a slight stimulation. Both these effects were only observable using freshly-isolated chromatin; within 90 min. of its isolation, corn chromatin has lost its ability to respond both to 2,4-D and to the active fraction. That it is the chromatin and not the active factor that has deteriorated was shown clearly by the fact that the corn fraction was also active 24 hours later on freshly-isolated pea chromatin (30 per cent stimulation). Further, this pea chromatin also lost within 90 min. its ability to respond to the corn factor.

When a similar elution strategy was applied with peas, (i.e adjusting extract and starting buffer to 0.1M NaCl, then eluting with 1M NaCl followed by water), little or no material emerged during the water wash, but the 1M NaCl fractions were slightly active, promoting RNA synthesis by freshly-prepared pea chromatin by 15 per cent in the presence of exogenous polymerase.

Effects on DNA-dependent RNA synthesis. Since the chromatin preparations remained responsive to the active fractions for such a short period, it was decided to examine their activity further using purified DNA templates. From Table 2 it is evident that the pea and corn fractions (prepared as described above) dramatically enhance RNA synthesis primed by the homologous DNA. The corn factor is also active, though to a somewhat lesser extent, on pea and calf thymus DNA, while the factor from peas is without effect on calf thymus DNA. No incorporation is obtained with either factor in the absence of added polymerase, i.e. the factors do not have RNA polymerase or polynucleotide phosphorylase activity. Heating at 100° for 15 min. results in ca. 50 per cent loss of activity.

Table 2. Activity of corn and pea factors on DNA-dependent RNA synthesis

Factor added	Source of DNA		
	Corn	Pea	Calf thymus
	ATP incorporated, p moles		
Control	212	297	321
Corn, 1 µg	558	499	575
Control		220	252
Pea, 2 µg		417	252

The effect of different concentrations of pea factor is shown in Fig. 3; apparently there is no effect below a certain threshold concentration of factor, resulting in a sigmoidal shape curve. This may possibly indicate a cooperative interaction of some sort.

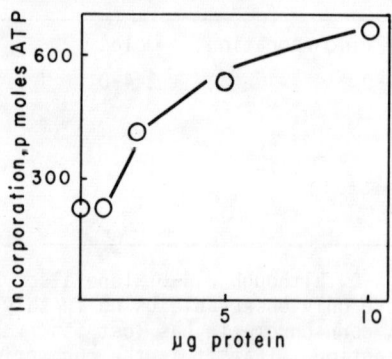

Fig. 3. The effect of different concentrations of pea factor on RNA synthesis primed by pea DNA.

It should be noted that different regions of the active pea eluate (1M NaCl peak) may contain varying amounts (up to 9 per cent) of DNA and hence possess varying degrees of inherent template activity. The results presented here have either been obtained with fractions lacking template activity, or else have been corrected by amounts corresponding to the incorporation obtained in the absence of added DNA. Generally, the template activity did not exceed 25 per cent of the total stimulatory activity, but occasionally fractions with template activity only were obtained. This variability suggests that the DNA is fortuitously associated with protein fractions possessing true stimulatory activity, rather than being an integral part of a stimulatory DNA-protein complex. Whether or not the retention of this small fraction of DNA on an auxin-based affinity adsorbent is of any significance is not known.

Molecular sieve chromatography of active pea fraction. Passage through a column of Sephadex G200 resolved the pea factor (1M NaCl eluate) into two active fractions (Fig. 4), one of molecular weight 200,000 or more, and the other 100,000-120,000. The lighter fraction possessed stimulatory activity only, while the heavier fraction, as anticipated from the relative 260:280 nm absorbance values, had in addition some template activity. As noted above, the DNA is probably not integrally associated with the stimulatory protein and hence it is possible that the two active fractions bear a monomer-dimer relationship, though no evidence is available on this point.

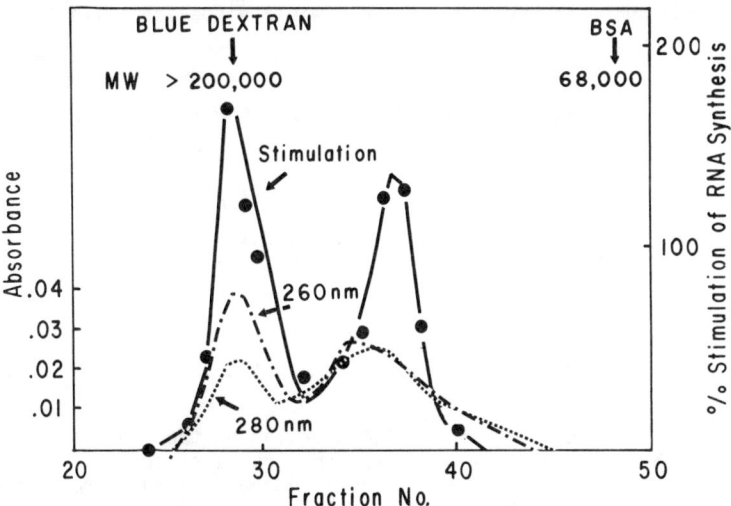

Fig. 4. Molecular sieve chromatography of active pea factor. 1 ml of factor was chromatographed on a column of Sephadex G 200, 85 x 1.5 cm. Fractions of 1.55 ml were collected and 0.1 ml aliquots tested in reactions for pea DNA-dependent RNA synthesis. The activities in the higher M.W. peak have been corrected for template activity.

2,4D-dependent active fraction from peas. The fractions described so far were derived from 30,000 x g supernatants passed through the Amiben-caproic column. When the initial pea extract was centrifuged at 270,000 x g for 30 min. before passage through the column, the amount of protein subsequently eluted with 1M NaCl was greatly reduced and it exhibited different behaviour in assays for RNA synthesis. The fraction was itself devoid of template activity and was without effect on DNA-primed RNA synthesis when added alone. However, in the presence of 2,4-D the factor enhanced RNA synthesis by over 100 per cent (Table 3). The extreme lability of this fraction (activity disappears in a few hours) has so far precluded any further study of its properties.

IV. DISCUSSION

These results indicate that affinity chromatography holds great promise for the isolation of auxin-reactive proteins. All the data described were obtained using the Amiben-caproic column, but preliminary experiments indicate that with a similar elution strategy, active fractions can also be obtained from a 2,4-D-lysine column.

Centrifugation at speeds which sediment ribosomes altered the quantity and nature of the fraction(s) retained by the Amiben-caproic column (Table 3). However, ribosomes *per se* do not appear to contribute to the activity of the 1M NaCl eluate derived from 30,000 x g pea supernatants, since centrifugation of the active fraction at 270,000 x g for 30 min. does not result in any loss of activity. Rather, it is likely that a protein fraction which adsorbs (specifically or otherwise) to the ribosomes is stripped off during passage of a 30,000 x g supernatant through the affinity column. Presumably this fraction either displaces or masks the 2,4-D-dependent activity that is manifested when a 270,000 x g supernatant is used.

Although the fractions derived from 30,000 x g supernatants did not require added 2,4-D for activity, it is surely of significance that they are retained on auxin-based adsorbents. It may be that mere passage through the affinity column is sufficient to transform the factor to an active configuration in which further contact with auxin is not required. Alternatively, as suggested in a preliminary report by O'Brien et al. (1969), the *in vivo* situation may be that auxin transports an already active regulatory protein to the nucleus. In other words, this would visualize auxin acting as a protein carrier, rather than a protein behaving as an auxin carrier.

Table 3. Activity of a fraction derived from a 270,000 x g pea supernatant on RNA synthesis primed by pea DNA ± 0.05 mg/l 2,4-D

	ATP incorporated, p moles	
	No 2,4-D	+ 2,4-D
Control	269	271
+Factor, 2 µg	250	560

Attempts to detect binding of ^{14}C-2,4-D or IAA by the isolated fractions, using equilibrium dialysis, have not met with any conspicuous success. It is of course possible that the structural modifications introduced into the hormone molecules result in affinity adsorbents that do not in fact select for auxin-reactive proteins. However, there are other, less depressing possibilities. For example, as already mentioned, passage through the column may alter the configuration of the factor to one in which auxin is not bound, either by virtue of contact with the adsorbent or through destruction of the binding site under the conditions of elution. Alternatively, the binding constant may be so great (say 10^8 or more) that binding is not detectable at the specific activities of commercially available labelled auxins. Equally, binding of a weak and transient nature may not be detectable by equilibrium dialysis. In this connection it may be noted that the auxin-reactive factor described by Matthyse and Phillips (1969) does not appear to bind 2,4-D *in vitro* (Matthyse, personal communication).

Time course experiments suggest that the factors enhance the initial reaction rate, rather than prolonging the reaction beyond its normal 15-20 min. term, i.e. that they are not acting by permitting chain release and re-initiation at original growth points. However, it is not yet known exactly how the factors described operate in promoting DNA-dependent RNA synthesis, i.e. whether they increase the rate of polynucleotide chain growth, prevent premature chain termination, or permit initiation at template regions which are not read in their absence.

V. SUMMARY

The possibility of using specific affinity adsorbents to isolate proteins which may mediate auxin action has been investigated. When crude supernatants from corn or pea shoots are passed through columns of agarose to which derivatives of IAA, 2,4-D or Amiben have been attached, certain protein fractions are retained, whereas unmodified agarose adsorbs no protein. This protein can be eluted under appropriate conditions and is found to stimulate slightly chromatin-dependent RNA synthesis (supported by exogenous polymerase); however, chromatin only remains responsive to the factors for a very short period after its isolation. A more dramatic and more easily studied response is obtained using RNA synthesis conducted by purified DNA, where the factors may enhance the reaction by nearly 200 per cent. Under slightly different conditions another factor can be prepared which is inactive alone but which in the presence of 2,4-D promotes DNA-dependent RNA synthesis by over 100 per cent.

VI. ACKNOWLEDGEMENTS

I thank Henry Bork for excellent technical assistance.

VII. REFERENCES

BURGESS, R.R. (1969). A new method for the large scale purification of *Escherichia coli* deoxyribonucleic acid-dependent ribonucleic acid polymerase. J. Biol. Chem. 244, 6160-6167.

CAMPBELL, D.H, E. LUESCHER and L.S. LERMAN (1951). Immunologic adsorbants I. Isolation of antibody by means of a cellulose-protein antigen. Proc. Natl. Acad. Sci. U.S. 37, 575-579.

CHADER, G.J. and U. WESTPHAL (1968). Steroid-protein interactions XVIII. Isolation and observations on the corticosteroid-binding globulin of the rat. Biochemistry, 7, 4272-4282.

CUATRECASAS, P., M. WILCHEK and C.B. ANFINSEN (1968). Selective enzyme purification by affinity chromatography. Proc. Natl. Acad. Sci. U.S. 61, 636-643.

FAMBROUGH, D.M. and J. BONNER (1966). On the similarity of plant and animal histones. Biochemistry, 5, 2563-2570.

GILES, K.W. and A. MYERS (1965). An improved diphenylamine method for the estimation of deoxyribonucleic acid. Nature, 206, 93.

GIORGIO, N.A. and M. TABACHNIK (1968). Thyroxine-protein interactions. V. Isolation and characterization of a thyroxine-binding globulin from human plasma. J. Biol. Chem. 243, 2247-2259.

HUTZINGER, O. and T. KOSUGE (1968). Microbial synthesis and degradation of indole-3-acetic acid. III. The isolation and characterization of indole-3-acetyl-ε-L-lysine. Biochemistry 7, 601-605.

JOHRI, M.M. and J.E. VARNER (1968). Enhancement of RNA synthesis in isolated pea nuclei by gibberellic acid. Proc. Natl. Acad. Sci. U.S. 59, 269-276.

LOWRY, O.H., N.J. ROSEBROUGH, A.L. FARR and R.J. RANDALL (1951). Protein measurement with the Folin phenol reagent. J. Biol. Chem. 193, 265-290.

MATTHYSE, A.G. and M. ABRAMS (1970). A factor mediating interaction of kinins with the genetic material. Biochem. Biophys. Acta 199, 511-518.

MATTHYSE, A.G. and C. PHILLIPS (1969). A protein intermediary in the interaction of a hormone with the genome. Proc. Natl. Acad. Sci. U.S. 63, 897-903.

O'BRIEN, TJ., J.W. HARDIN, R.J. SPERCA and J.H. CHERRY (1969). Factors which may mediate hormone-induced chromatin-bound RNA polymerase activity in soybean hypocotyl. XI Interntl. Bot. Congress Abstracts, p. 161.

PENSKY, J. and J.S. MARSHALL (1969). Studies on thyroxine-binding globulin. II. Separation from human serum by affinity chromatography. Arch. Biochem. Biophys. 135, 304-310.

PORATH, J., R. AXÉN and S. ERNBACK (1967). Chemical coupling of proteins to agarose. Nature 215, 1491-1492.

SCHRÖDER, E. and K. LÜBKE (1965). "The Peptides. Vol. 1. Methods of Peptide Synthesis". Academic Press, New York.

SIEGEL, S.M. and A.W. GALSTON (1953). Experimental coupling of IAA to pea root protein *in vivo* and *in vitro*. Proc. Natl. Acad. Sci. U.S. 39, 1111-1118.

SOLOFF, M.S. and C.M. SZEGO (1969). Purification of estradiol receptor from rat uterus and blockade of its estrogen-binding function by specific antibody. Biochem. Biophys. Res. Comm. 34, 141-147.

VENIS, M.A. (1968). Auxin-histone interaction, in "Biochemistry and Physiology of Plant Growth Substances" (Eds. F. Wightman and G. Setterfield) p. 761-775. Runge Press, Ottawa.

ZENK, M.H. (1964). Isolation, biosynthesis and function of indole-acetic acid conjugates, in Coll. Int. C.N.R.S. Paris 123, 241-249.

On the Significance of Cytokinin Binding to Plant Ribosomes

M.V. Berridge and R.K. Ralph

Department of Cell Biology, University of Auckland, Auckland, New Zealand

and D.S. Letham

Research School of Biological Sciences, Australian National University, Canberra, A.C.T., Australia

I. INTRODUCTION

Considerable experimental evidence supports the view that phytohormones affect transcription (see Letham, 1969). However certain other observations suggest that cytokinins may also modify translational processes. For example cytokinins maintain the level of ribosomes in excised leaf tissue (Shaw and Manocha, 1965; Srivastava and Arglebe, 1968; Berridge and Ralph, 1969). Although the conclusion that cytokinins promote protein synthesis in excised leaf tissue undergoing senescence (Osborne, 1962) no longer seems valid (see Tavares and Kende, 1970; Kuraishi, 1968), cytokinins markedly increase the specific activity of protein in purified nuclei, mitochondria and plastids incubated with labelled amino acids *in vitro* (Datta and Sen, 1965; Bhattacharyya and Roy, 1969; Davies and Cocking, 1967). The rapidity of the response to cytokinin in plastids and mitochondria (lag period only a few minutes) suggests that the hormone may promote protein synthesis in a direct way, possibly by interaction with the ribosome.

Cytokinins are known to occur adjacent to the 3' end of the anticodon in several tRNA species from *E. coli* and yeast (for review see Skoog and Armstrong, 1970). The presence of cytokinins in tRNA has been shown to be functionally significant (Fittler and Hall, 1966; Gefter and Russell, 1969). These modified bases adjacent to anticodons may play some regulatory role in protein synthesis possibly by interacting with a site on the ribosome. The occurrence of cytokinins in tRNA hydrolysates from higher plants (Hall *et al.*, 1966; Letham and Ralph, 1967; Burrows *et al.*, 1970) raises the possibility that cytokinins as free bases may interact with a site on plant ribosomes and thus modify ribosomal processes.

In a previous study we reported an equilibrium-type binding of cytokinins to purified plant ribosomes (Berridge, Ralph and Letham, 1970). We present here a summary of this work, and of our attempts to demonstrate an effect of cytokinins on *in vitro* protein synthesis in plant systems.

II. MATERIALS AND METHODS

1. Determination of cytokinin binding to ribosomes

Ribosomes were prepared from Chinese cabbage leaves and characterized as previously described (Berridge, Ralph and Letham, 1970). Ribosomes were sedimented from, and resuspended in Nirenberg buffer containing the radioactive compound under investigation. The ribosomes were applied to a column of Sephadex G200 (0.25 cm x 20 cm) pre-equilibrated with the same buffer which also served as eluent. Samples from each fraction were used to determine A_{260} and radioactivity.

2. In vitro protein synthesis

(a) Protein synthesis by nuclei-chloroplast preparations

Young Chinese cabbage leaves (6 gm) were deribbed and plunged into iced water. The leaves were blotted dry and chopped with sharp blades in 15 ml Spencer medium (Spencer and Whitfield, 1967) containing 0.4M sucrose. The extract was filtered through 4 layers of muslin and 2 layers of Miracloth and the filtrate centrifuged at 900 xg for 10 min at $2°C$. The pellet was resuspended in cold Spencer medium and 0.4 ml aliquots used for each time point. Each assay mixture (0.57 ml total) contained tris HCl buffer pH 7.6 (5 μmoles); $MgCl_2$ (2.5 μmoles); KCl (15 μmoles); mercaptoethanol (1 μmole); $(NH_4)_2SO_4$ (5 μmoles); ATP (0.2 μmoles); GTP (0.1 μmole); phosphoenolpyruvate (1.5 μmoles); pyruvate kinase (5 μg); an amino acid mixture containing 0.04 μmoles of each amino acid minus phenylalanine and valine; ^{14}C-phenylalanine (0.45 μC, 225 mC/mmole); ^{14}C-valine (0.45 μC, 149 mC/mmole). The assay tubes were incubated at $25°C$. At time intervals individual assay mixtures were removed and treated with 5% trichloroacetic acid at $80°C$ for 15 min. Following millipore filtration radioactivity in protein was assessed in a gas flow counter.

(b) Protein synthesis by isolated plant ribosomes

Ribosomes and a dialyzed supernatant fraction (termed S30D) were prepared from young Chinese cabbage leaves or etiolated dwarf pea shoots by grinding the tissue in PS buffer (50 mM tris HCl pH 7.6; 60 mM NH_4Cl, 8 mM $MgCl_2$, 0.5 mM spermidine, 0.5 mM dithiothreitol, 10% glycerol) at $2°C$. The homogenate was strained through muslin and centrifuged at 30,000 xg for 30 mins. The supernatant was again centrifuged at 30,000 xg for 30 min. A sample of the resulting supernatant was dialysed against PS buffer for 3 hr yielding S30D. The remaining supernatant was centrifuged at 250,000 xg for 2 hr. The resulting pellet of ribosomes was rinsed and resuspended in PS buffer. Washed ribosomes were prepared by re-sedimenting them once from PS buffer.

Chinese cabbage tRNA was prepared by DEAE-cellulose fractionation of total leaf nucleic acid which had been extracted with phenol. The tRNA was further purified by the isopropanol method of Zubay (1962) and deacylated in $2\underline{M}$ tris HCl pH 8.0 at $37°C$ for 2 hr.

Incorporation of amino acids was measured in 0.1 ml reaction mixtures containing tris HCl pH 7.6 (5 μmoles); $Mg(OAc)_2$ (0.8 μmoles); NH_4Cl (6 μmoles); glutathione (0.8 μmoles); ATP (0.13 μmoles); phosphoenolpyruvate (0.5 μmoles); spermidine (87 nmoles); dithiothreitol (4 nmoles); an amino acid mixture containing 4 nmoles of each amino acid minus phenylalanine or serine; ^{14}C-amino acid (0.2 μC); S30D, ribosomes, and tRNA were added as described in the text. Samples were incubated for the desired time and the reactions terminated by the addition of $0.2\underline{M}$ KOH (0.2 ml). Incubation at $30°C$ for 10 min resulted in the release of amino acids bound to tRNA. The samples were then precipitated with TCA, membrane filtered and radioactivity assessed in a liquid scintillation spectrometer.

III. RESULTS

1. Cytokinin binding to plant ribosomes

When ribosomes were suspended in buffer containing 8-^{14}C-kinetin (16.5 mC/mmole; 23 μ$\underline{M}$) or ^3H-6-benzylaminopurine (27 mC/mmole; 23 μ$\underline{M}$) and a sample passed through a column of Sephadex G-200 that had been previously equilibrated with non-radioactive buffer, the radioactivity separated from the ribosomes (Fig. 1A). However when ribosomes suspended in radioactive buffer were passed through Sephadex G-200 that had been previously equilibrated with the same radioactive buffer, a peak of radioactivity eluted together with the ribosomes (Fig. 1B). In separate experiments at $4°C$, one molecule of kinetin and 1.34 molecules of 6-benzylaminopurine were bound per ribosome (assuming a ribosome molecular weight of 2.7×10^6 daltons and an A_{260} of 16 for 1 mg/ml of ribosomes). At $20°C$ 0.5 molecules of kinetin were bound per ribosome indicating that binding is temperature dependent. Equilibrium binding of kinetin and 6-benzylaminopurine to ribosomes was confirmed by sucrose density gradient centrifugation and by equilibrium dialysis. Using 8-^{14}C-kinetin or ^{3}H-6-benzylaminopurine at a range of specific activities we demonstrated both by gel filtration and by a modified procedure of Vasquez (Berridge, Ralph and Letham, 1970) that binding increased with cytokinin concentration over the concentration range 0-20 µg/ml.

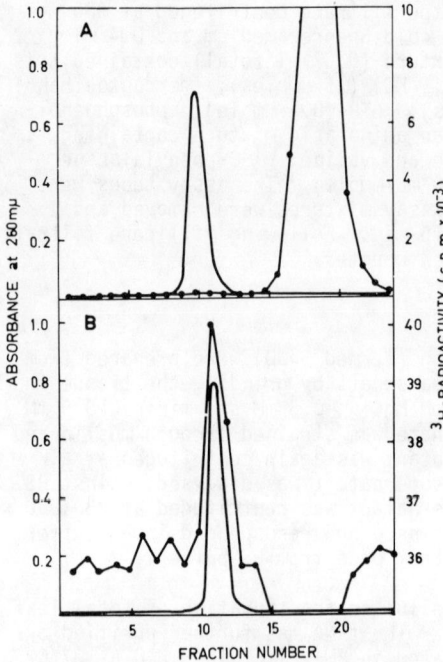

Fig. 1. Elution profile from Sephadex G200 of ribosomes suspended in buffer containing ^{3}H-6-benzylaminopurine (23 µM, 27 mC/mmole). (a) Column and eluent did not contain ^{3}H-6-benzylaminopurine. (b) Column previously equilibrated with buffer containing ^{3}H-6-benzylaminopurine and eluted with this buffer.
●——● ^{3}H-radioactivity in 75 µl sample; ——— A_{260} (100 x dilution)

Washing ribosomes with 0.5M ammonium chloride or 1% Triton X-100 did not significantly affect binding as determined by gel filtration. Ribosomes extracted with buffer containing 1% Triton X-100 contained a lower proportion of 83S ribosomes (51%) than ribosomes prepared in buffer without detergent (75-90% 83S ribosomes). With the former preparations kinetin binding was reduced to 0.52 molecules of kinetin bound per ribosome suggesting that binding may be mainly to 83S ribosomes. At low magnesium concentrations increased binding of kinetin and 6-benzylaminopurine was observed.

2. *Specificity and significance of cytokinin binding*

To establish the functional significance of cytokinin binding to ribosomes the biological activities and binding affinities of a series of substituted adenines were determined (Table 1). The biological activities of the various compounds in both leaf disc expansion and chlorophyll retention assays were positively and significantly correlated with their capacity to bind to ribosomes. 6-Morphilinopurine (at 23 µM) exhibited greatest divergence from the general correlative trend. In subsequent leaf disc expansion bioassays 6-morphilinopurine exhibited lower activity and the disc expansion data correlated better with ribosomal binding.

By sucrose density gradient centrifugation it was shown that kinetin did not bind to purified turnip yellow mosaic virus.

3. *Effects of kinetin on in vitro protein synthesis*

The observation that cytokinins bind to purified Chinese cabbage ribosomes suggested that protein synthesis might be directly affected by cytokinins. To test this possibility *in vitro* amino acid incorporating systems from Chinese cabbage and etiolated pea shoots were developed.

Table 1. Comparison of the activities of purine derivatives in causing expansion and retarding senescence of Chinese cabbage leaf-discs, and correlation of these activities with degree of ribosome binding

Test compound	Disc expansion* (% increase in fresh weight over control)		Index of cytokinin-induced chlorophyll retention†	No. of molecules bound/ribosome
	23µM	2.3µM	2µM	
Adenosine	1.7	0.5	4	0.34
Adenine	5.8	0.0	0	<0.10
3-Benzyladenine	8.9	6.8	0	0.51
9-Benzyladenine	9.3	7.1	3	0.53
6-Morpholinopurine	17.0	4.6	7	<0.10
Kinetin	23.8	18.7	36	1.00±0.05
6-Benzylaminopurine	29.0	23.4	52	1.35±0.05

* see Berridge and Ralph (1970)

† see Letham (1967)

(a) Protein synthesis in chloroplast-nuclei preparations

The kinetics of incorporation of ^{14}C-amino acid into protein in chloroplast-nuclei preparations from Chinese cabbage are shown in Fig. 2A. Kinetin (5-15 µg/ml) had no effect on incorporation over a period of 90 min. Chloramphenicol (20 µg/ml) inhibited incorporation by 50% while actidione (20 µg/ml) had no effect suggesting that much of the protein synthesis occurred on 68S chloroplast ribosomes. Omission of ATP from the reaction mixture reduced incorporation by 90% but did not make the system sensitive to kinetin.

(b) In vitro protein synthesis by Chinese cabbage ribosomes

To test the effects of cytokinins on protein synthesis on isolated ribosomes, *in vitro* amino acid incorporating systems from Chinese cabbage were developed. The lower set of curves in Fig. 2B shows that kinetin (10 µg/ml) did not affect the incorporation of ^{14}C-phenylalanine into protein directed by endogenous messenger RNA. The upper set of curves demonstrates that kinetin (25 µg/ml) did not appreciably affect ^{14}C-phenylalanine incorporation in the presence of polyuridylic acid (0.2 mg/ml). Further investigation with kinetin and zeatin at concentrations between 10-50 µg/ml failed to reveal significant effects of these cytokinins on endogenous or poly-U-directed protein synthesis. In these systems incorporation was inhibited 70-80% by puromycin (0.1 mg/ml). Chloramphenicol (1 mg/ml) reduced incorporation by 40-70% while actidione (1 mg/ml) caused less than 14% inhibition.

(c) In vitro protein synthesis by etiolated pea ribosomes

Attempts to isolate 83S ribosomes active for *in vitro* protein synthesis from Chinese cabbage were unsuccessful. Since protein synthesis in *in vitro* systems from etiolated peas occurs mainly on 80S ribosomes (Davis and MacLachlan, 1969), we used pea ribosomes to determine the effect of cytokinins on protein synthesis on 80S

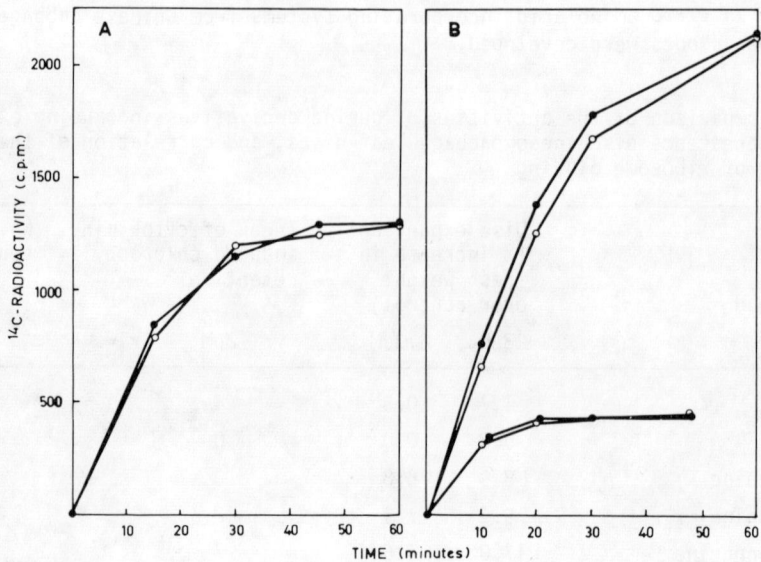

Fig. 2. Effect of kinetin on kinetics of protein synthesis *in vitro*. A. ^{14}C-amino acid incorporation in chloroplast-nuclei preparations (see Methods). ● — ●, Control; 0 — 0, kinetin (13 µg/ml). B. ^{14}C-phenylalanine incorporation on Chinese cabbage ribosomes. 0.1 ml reaction mixtures (see Methods) contained 0.2 µC ^{14}C-phenylalanine (225 mC/mmole) and 20 µg Chinese cabbage tRNA. The reaction mixtures represented by the upper set of curves also contained 0.36 A_{260} S30D; 0.32 mg ribosomes and 40 µg polyuridylic acid per assay, while the lower set of curves contained 2.0 A_{260} S30D, and 0.3 mg ribosomes per assay. Incubation was at 20°C. ● — ●, control; 0 — 0, kinetin 10 µg/ml in lower set of curves, 25 µg/ml in upper set of curves.

ribosomes. Table 2 shows that kinetin (10-150 µg/ml) and zeatin (50 µg/ml) did not affect the incorporation of ^{14}C-serine into protein in an *in vitro* system from etiolated pea shoots. Kinetic studies confirmed these results. Puromycin and actidione inhibited incorporation 80-90% while chloramphenicol had no effect on incorporation.

Table 2. Effects of cytokinins on *in vitro* protein synthesis on 80S ribosomes from etiolated pea plants.

Legend: 0.1 ml reaction mixtures (see Methods) contained 0.2 µC ^{14}C-serine (130 mC/mmole); 20 µg Chinese cabbage tRNA; 10 µl S30D; 0.75 mg washed ribosomes. Assay tubes were incubated at 20°C for 20 min prior to alkaline hydrolysis of aminoacyl tRNA and estimation of acid-precipitable radioactivity. Time zero binding blanks were subtracted from each value.

Reaction Mix	^{14}C-radioactivity (c.p.m.)
Complete	7334
+ 100 µg puromycin	972
+ 100 µg actidione	1941
+ 1 µg kinetin	7131
+ 5 µg kinetin	7175
+ 10 µg kinetin	7621
+ 15 µg kinetin	7348
+ 5 µg zeatin	7391

IV. DISCUSSION

Many biologically active small molecules and drugs exert their effects through reversible interactions with receptor molecules (see Rose, 1969). The rapid effects of cytokinins on *in vitro* protein synthesis in several systems (Davis and Cocking, 1967; Battacharyya and Roy, 1969) suggest that they may interact directly with a component of the translational complex. In attempts to demonstrate such an interaction we discovered that the cytokinins kinetin and 6-benzylaminopurine bind reversibly to Chinese cabbage leaf ribosomes, 1.00 and 1.35 molecules being bound per ribosomes at 4^0C and 23 μM (5 μg/ml). The fact that 23 μM is the minimum concentration of kinetin or 6-benzylaminopurine required to produce maximal expansion of floated leaf discs suggests that binding may be functionally significant. This idea is further supported by the observation that the capacity of ribosomes to bind cytokinins increases with the biological activity of the cytokinin (Table 1). Binding to ribosomes increased with 6-benzylaminopurine concentrations (0-20 μg/ml). At 20 μg/ml 4 molecules of 6-benzylaminopurine were bound per ribosome. No limit to binding was demonstrated because of physical limitations of the assay system. The extent of kinetin binding was reduced with increasing temperature and with increasing Mg^{++} concentration, but was not affected by washing the ribosomes with 0.5M NH_4Cl or 1% Triton X-100, suggesting that the observed binding was not to protein associated with the ribosomes, but easily removed from them.

The reversible binding observed may stabilise ribosomes. It might also affect protein synthesis by regulating initiation, peptide elongation or termination, or by altering codon specificity. To investigate the effects of cytokinins on protein synthesis we have developed *in vitro* protein-synthesizing systems from Chinese cabbage and etiolated peas.

Kinetin appeared to have no effect on *in vitro* protein synthesis in chloroplast-nuclei preparations from Chinese cabbage or in reconstituted systems from Chinese cabbage consisting of ribosomes, tRNA and supernatant factors. Both of these systems were sensitive to chloramphenicol but not actidione suggesting that most of the amino acid incorporation observed was associated with 68S ribosomes. Boardman et al. (1967) have previously shown that in tobacco leaf extracts, 83S ribosomes are relatively inactive. Because we could not obtain active 83S ribosomes from Chinese cabbage we tested the effects of cytokinins on an 80S-ribosome, amino-acid incorporating system from etiolated pea shoots. We could not demonstrate an effect of kinetin or zeatin in this system. It is probable that amino-acid incorporation in the above systems does not involve normal initiation processes since we could detect no stimulation of incorporation with added natural m-RNA. In our systems (8mM Mg^{++}) translation of poly-U probably does not require special initiation factors, since it has been shown that in reticulocyte and bacterial systems poly-U translation at 10mM Mg^{++} occurs in the absence of initiation factors (Millar and Schweet, 1968; Lipmann, 1969).

If cytokinin binding to plant ribosomes is of functional significance to *in vitro* protein synthesis its effect appears to be on some process other than the rate or extent of peptide elongation.

In preliminary experiments we have studied the interaction of cytokinins with isolated nuclei prepared from pea buds. By equilibrium dialysis both 6-benzylamino-purine and kinetin were found to bind to the nuclei (1.28 and 0.73 mμmoles/mg nuclear DNA); zeatin however did not bind significantly. These observations are surprising in view of the recent report by Matthysse and Abrams (1970) that both kinetin and zeatin promote RNA synthesis *in vitro* using pea bud chromatin and *E. coli* RNA polymerase. These investigators have also reported cytokinin-induced RNA synthesis in purified nuclei of soybean callus tissue and of pea buds. We have not been able to confirm these experiments with nuclei. When ATP, GTP, UTP and CTP were all present in the *in vitro* system, active RNA synthesis could be demonstrated in the nuclei, but no reproducible stimulation of synthesis was observed on addition of cytokinin. Hence we have observed binding of certain cytokinins to purified nuclei but have no evidence that this binding is of functional significance.

V. SUMMARY

The synthetic cytokinins kinetin and 6-benzylaminopurine bind reversibly to purified Chinese cabbage leaf ribosomes. At 23 μM and 4°C one molecule of kinetin and 1.34 molecules of 6-benzylaminopurine are bound per ribosome. Pretreatment of ribosomes with 0.5M NH_4Cl or Triton X-100 did not reduce the extent of binding of kinetin. Binding was temperature, magnesium and concentration dependent and appeared to be to the 83S ribosome species. Adenine and adenine derivatives that were inactive as cytokinins showed much less affinity for ribosomes. A positive correlation between the extent of binding and the biological effect of various cytokinin analogues was demonstrated.

In attempts to determine the biological significance of cytokinin binding to ribosomes we tested the effects of cytokinins on *in vitro* protein synthesis in (i) chloroplast-nuclei preparations from Chinese cabbage; (ii) systems comprising isolated ribosomes and supernatant factors from Chinese cabbage and from etiolated pea shoots. We could detect no effects of cytokinins in these systems. The results are discussed in terms of possible mechanisms of cytokinin control of growth processes.

Binding of certain cytokinins to purified nuclei prepared from pea buds has also been demonstrated.

VI. ACKNOWLEDGEMENTS

This research was supported in part by a grant from the N.Z. Cancer Society (Auckland Division) to M.V.B.

VII. REFERENCES

BHATTACHARYYA, J. and ROY, S.C. (1969). Growth promoters and the synthesis of protein in plant mitochondria. Part I. Effect of kinetin on the incorporation of amino acids into mitochondrial protein. Biochem. Biophys. Res. Comm. 35, 606-610.

BERRIDGE, M.V. and RALPH, R.K. (1969). Some effects of kinetin on floated Chinese cabbage leaf discs. Biochim. et Biophys. Acta 182, 266-269.

BERRIDGE, M.V, RALPH, R.K. and LETHAM, D.S. (1970). The binding of kinetin to plant ribosomes. Biochem. J. 119, 75-84.

BOARDMAN, N.K., FRANCKI, R.I.B. and WILDMAN, S.G. (1966). Protein synthesis by cell-free extracts of tobacco leaves. III. Comparison of physical properties of protein synthesizing activities of 70S chloroplast and 80S cytoplasmic ribosomes. J. Mol. Biol. 17, 470-489.

BURROWS, W.J., ARMSTRONG, D.J., KAMINEK, M., SKOOG, F., BOCK, R.M., HECHT, S.M., DAMMANN, L.G., LEONARD, N.J., OCCOLOWITZ, J. (1970). Isolation and identification of four cytokinins from wheat germ transfer ribonucleic acid. Biochemistry 9, 1867-1872.

DATTA, A. and SEN, S.P. (1965). The mechanism of action of plant growth substances: Growth substance stimulation of amino-acid incorporation into nuclear protein. Biochim. et Biophys. Acta 107, 352-357.

DAVIES, J.W. and COCKING, E.C. (1967). Protein synthesis in tomato-fruit locule tissue: Incorporation of amino-acids into protein by aseptic cell-free systems. Biochem. J. 104, 23-33.

DAVIS, E. and MACLACHLAN, G.A. (1969). Generation of cellulase activity during protein synthesis by pea microsomes *in vitro*. Archiv. Biochem. Biophys. 129, 581-587.

FITTLER, F. and HALL, R.H. (1966). Selective modification of yeast seryl-tRNA and its effect on the acceptance and binding functions. Biochem. Biophys. Res. Comm. 25, 441-446.

GEFTER, M.L. and RUSSELL, R.L. (1969). Role of modifications in tyrosine transfer RNA: A modified base effecting ribosome binding J. Mol. Biol. 39, 145-157.

HALL, R.H., ROBINS, M.H., STASIVK, L. and THEDFORD, R. (1966). Isolation of N^6-(γ,γ-dimethylallyl)adenosine from soluble ribonucleic acid. J. Am. Chem. Soc. 88, 2614-2615.

KURAISHI, S. (1968). The effect of kinetin on protein level of *Brassica* leaf discs. Physiol. Plantarum 21, 78-83.

LETHAM, D.S. (1967). Regulators of cell division in plant tissues: V. A comparison of the activities of zeatin and other cytokinins in five bioassays. Planta 74, 228-242.

LETHAM, D.S. (1969). Cytokinins and their relations to other phytohormones. Bioscience 19, 309-316.

LETHAM, D.S. and RALPH, R.K. (1967). A cytokinin in soluble RNA from a higher plant. Life Sci. 6, 387-394.

LIPMANN, F. (1969). Polypeptide chain elongation in protein biosynthesis. Science 164, 1024-1031.

MATTHYSE, A.G. and ABRAMS, M. (1970). A factor mediating interaction of kinins with the genetic material. Biochim. Biophys. Acta 199, 511-518.

MILLER, R.L. and SCHWEET, R. (1968). Isolation of a protein fraction from reticulocyte ribosomes required for *de novo* synthesis of hemoglobin. Archiv. Biochem. Biophys. 125, 632-646.

OSBORNE, D.J. (1962). Effect of kinetin on protein and nucleic acid metabolism in *Xanthium* leaves during senescence. Pl. Physiol. 37, 595-602.

ROSE, M.S. (1969). Reversible binding of toxic compounds to macromolecules. British Medical Bulletin 25, 227-230.

SHAW, M. and MANOCHA, M.S. (1965). Fine structure in detached and senescing wheat leaves. Canad. J. Bot. 43, 747-755.

SKOOG, F. and ARMSTRONG, D.J. (1970). Cytokinins. Ann. Rev. Plant Physiol. 21, 359-384.

SPENCER, D. and WHITFIELD, P.R. (1967). Ribonucleic acid-synthesizing activity of Spinach chloroplasts and nuclei. Archiv. Biochem. Biophys. 121, 336-345.

SRIVASTAVA, B.I.S. and ARGLEBE, C. (1968). Effect of kinetin on ribosomes of excised barley leaves. Physiol. Plantarum 21, 851-857.

TAVARES, J. and KENDE, H. (1970). The effect of 6-benzylaminopurine on protein metabolism in senescing corn leaves. Phytochem. 9, 1763-1770.

ZUBAY, G. (1962). The isolation and fractionation of soluble RNA. J. Mol. Biol. 4, 347-356.

BINDING OF INDOLEACETIC ACID TO ISOLATED PEA NUCLEI

Kestutis J. Tautvydas[2], and Arthur W. Galston

Department of Biology, Yale University, New Haven, Connecticut, 06520

[2] Present address: Department of Biology,
Marquette University,
Milwaukee, Wisconsin, 53220

I. INTRODUCTION

Although there is no doubt that auxins can enhance RNA synthesis in many plants (Key, 1969), it is not known what part of this increase is caused by auxin activity in cell nuclei nor how it is accomplished. The many known examples of the binding of small effector molecules to macromolecular acceptors (Monod et al., 1965; Gilbert and Müller-Hill, 1966; Changeux et al., 1968) as well as the binding of steroid hormones to nuclear proteins (Fanestil & Edelman, 1966; Maurer & Chalkley, 1967; Talwar et al., 1968; Bruchovsky & Wilson, 1968) suggest that auxin activity in plant cell nuclei might involve binding to specific nuclear receptors. The possibility of IAA binding in isolated nuclei was suggested by the findings of Roychoudhury et al. (1965), Maheshwari et al. (1966), and Cherry (1967) that IAA and other auxins enhanced RNA synthesis in isolated nuclei.

We report here that nuclei isolated from light- and dark-grown peas retain their ability to bind IAA to a small number of binding sites. Various parameters of this binding are presented.

II. MATERIALS AND METHODS

The method of growing plants and isolating nuclei has been published previously (Tautvydas, 1971). Briefly, the isolation procedure consisted of the following steps. Apical buds of 7 day-old peas were infiltrated with a solution containing gum arabic, sucrose, buffer, and salts and then steeped in the same solution for 14 hours at $10°C$. Next, the tissues were homogenized in a Sorvall Omni-mix and filtered through a series of nylon screens. The filtrate was immediately layered on a discontinuous gradient of gum arabic and centrifuged to pellet the nuclei. A second centrifugation of the resuspended pellets gave a pure preparation of structurally intact nuclei. The yield of nuclei was about 32% of the total number present in apical buds.

Procedure for IAA binding. Aliquots of a suspension of isolated nuclei, each containing 5.0×10^7 nuclei, were centrifuged in 10 x 75 mm culture tubes for 3-4 minutes at 1000 xg in an International Clinical Centrifuge equipped with a swing-out bucket rotor (No. 221). The pelleted nuclei were resuspended in 0.15 ml of an incubation solution (IS) containing sucrose, 0.15 M; N-2-hydroxyethylpiperazineN-2-ethanesulfonic acid (HEPES) buffer, 20 mM; potassium acetate, 60 mM; magnesium acetate, 4 mM; 2-mercaptoethanol, 5 mM; 2x glass redistilled water. The pH was adjusted to pH 6.9 with KOH. After 15 minutes of equilibration at $3°C$, the culture tubes were transferred to a Dubnoff water bath-shaker (American Instrument Co.) with a water temperature of $20°C$. Then, 0.15 ml of IS containing ^{14}C-1-IAA (New England Nuclear Corp., 16.9 mc/mmole) or ^{14}C-2-IAA (Amersham-Earle Co., 49 mc/mmole) was pipetted into each culture tube. After the desired reaction time, the binding was stopped by dilution through a rapid delivery of 2.5 ml of an ice-cold rinse solution containing sucrose, 0.15 M; magnesium acetate, 4 mM; morpholinoethenesulfonate (MES) buffer, 5 mM; non-radioactive IAA, 0.1 M; 2x distilled water; pH 5.5 with KOH. The nuclear suspensions were centrifuged 4 minutes to pellet the nuclei, and the pellets were next rinsed 6x more by resuspension in 1 ml of the rinse solution followed by centrifugation for 2 minutes. By the sixth rinse, the radioactivity in the supernatant liquids had reached a plateau and more than 98% of the initial radioactivity had been removed. The rinsed pellets were transferred quantitatively with Bray's scintillation fluid (Bray, 1960) to vials and their radio-activity was determined with an Ansitron scintillation counter (Picker Nuclear Co.).

III. RESULTS

Preliminary experiments *in vivo* revealed that incubation of apical buds of dark-grown pea seedlings in solutions of ^{14}C-1-IAA at $10°C$ for 14 hours resulted in maximum uptake of radioactivity. About 1.2% of the radioactivity in the apices was retained by the cell nuclei after isolation. Preliminary experiments with isolated nuclei incubated at 10^{-5}M ^{14}C-1-IAA or ^{14}C-2-IAA showed that, as *in vivo*, about 1% of the radioactivity in the incubation solution was taken up by the nuclei. Maximum uptake was achieved in 2 minutes or less. Addition of nonradioactive IAA to the solutions used to stop the binding reaction and to rinse the nuclei decreased the total radioactivity retained by the nuclei by about one-half due to exchange with ^{14}C-IAA adsorbed to nuclear surfaces, but did not affect the bound IAA characterized by the kinetics shown later. Subsequently, nonradioactive IAA was routinely included in the rinse solution to remove most of the exchangeable ^{14}C-IAA. Less than 22% of the remaining radioactivity could be removed after 24 hours of incubating radioactive nuclei in solutions containing 0.1 mM non-radioactive IAA at $3°C$. However, essentially all of the radioactivity could be removed readily by treatment of the radioactive nuclei with ice-cold 95% ethanol, during which treatment the nuclei remained structurally intact. It thus appeared that the binding of ^{14}C-IAA was not covalent, or, if covalent, involved receptor molecules soluble in 95% ethanol. This type of reaction to aqueous and non-aqueous solutions is characteristic of hormone binding to nuclear proteins (Talwar et al., 1968; Bruchowsky and Wilson, 1968; Wira & Munck, 1970) as well as the binding of actinomycin-D to DNA and of dyes such as neutral red to proteins, but it is not true of all types of effector binding (Gilbert and Müller-Hill, 1966).

Since nuclei contain a variety of molecules to which IAA might bind or adsorb, it was necessary to distinguish possible sites having a high affinity from sites having a low affinity for IAA. The former would be good candidates for physiologically significant binding sites while the latter would represent adsorption sites. Binding sites having a high affinity for IAA would necessarily have to be few relative to the number of sites available for adsorption. Thus, if such high affinity binding sites were present in isolated pea nuclei, one would expect to find, at non-saturating concentrations of IAA, regions in the IAA binding isotherm which would deviate from the linearity shown by adsorption alone. These deviations could be in the form of a sigmoidal twist or of a brief plateauing in an otherwise linear binding isotherm.

Initial experiments with ^{14}C-1-IAA and ^{14}C-2-IAA indicated that sigmoidal or plateau regions did exist in the IAA binding isotherm (Figure 1). Of the two plateau regions observed, we decided to focus our attention on the one occurring at about 4×10^{-6} M IAA (plateau #1) since physiologically significant binding sites are more likely to occur at relatively low concentrations of effector molecules. However, the presence of a relatively high level of adsorbed radioactivity, which often varied from experiment to experiment, made it impossible to study the binding of IAA to high affinity sites merely by measuring total radioactivity bound to the nuclei. Since the binding of IAA to isolated pea nuclei showed a measurable time-course, we reasoned that the difference between the radioactivity bound at time 0 and the amount bound at the end of the reaction could represent a binding of IAA to high affinity binding sites.

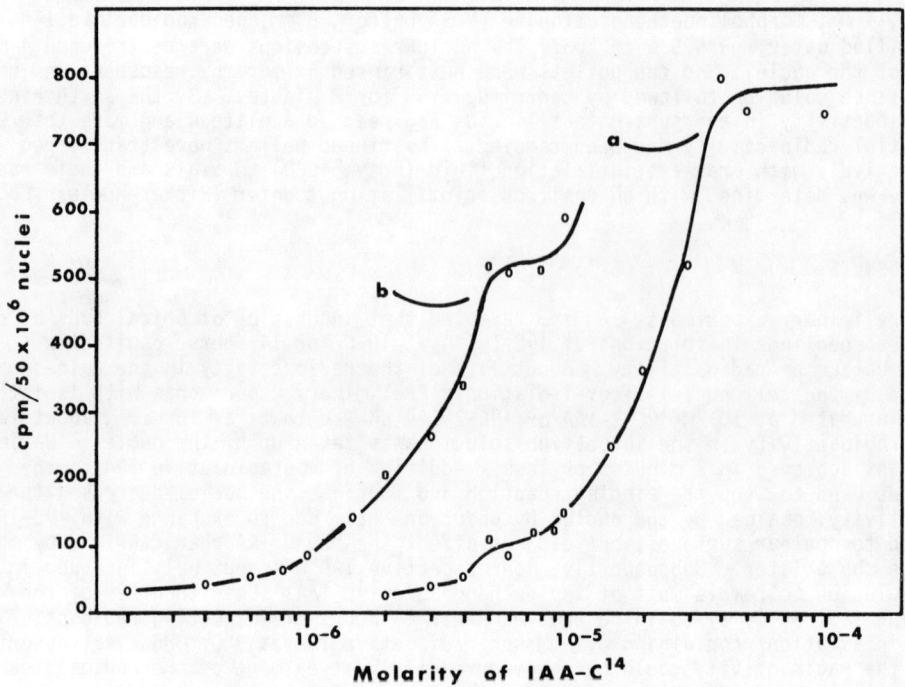

Fig. 1. Binding of ^{14}C-IAA as a function of the concentration of IAA. The data represent total radioactivity recovered. (a) ^{14}C-1-IAA (b) ^{14}C-2-IAA (see Materials and Methods).

The (a) curve is the average of 5 experiments, three of which were used to define the binding from 10^{-5}M to 10^{-4} M, and the other two were used to define the binding curve from 2×10^{-6} M to 8×10^{-6}M. In these experiments, each treatment was duplicated. The variation between duplicates within treatments ranged from a minimum of 5% to a maximum of 20%.

The (b) curve is the average of 2 experiments, but in each experiment the treatment were not duplicated. The treatments 2x, 4x, and 6×10^{-7} M are from only one experiment. The variation here is the same as in (a).

Going from the lowest to the highest concentrations of IAA, the first saturation plateau reached is plateau #1 and the second is plateau #2.

Figure 2 shows the results of time-course experiments performed at several concentrations of IAA in the range of plateau #1. The binding reaction appeared to start without a lag period and was complete within 50-60 sec. after the addition of ^{14}C-1-IAA or ^{14}C-2-IAA to the nuclei. When the difference in the radioactivity bound to the nuclei at 0 sec. and 60 sec. of reaction was determined from each of the kinetic curves and was plotted as a function of the concentration of IAA in the incubation mixture, the binding isotherm shown in fig. 3 was obtained. It appears from this sigmoidal curve that saturation of the plateau #1 (pl-1) binding sites occurs at about 4×10^{-6}M, the same as first indicated in figure 1. Using 180 dpm/5.0×10^7 nuclei as the mean saturation value (fig. 3), we calculated a maximim of 2×10^4 molecules of IAA or derivative bound per nucleus to the pl-1 binding sites.

We have not proven that the radioactivity bound at the pl-1 binding sites was still IAA. Some indirect evidence suggests, however, that the binding isotherm in figure 3 does represent the binding of IAA. The maximum binding obtained with ^{14}C-2-IAA was 180 dpm ±18 dpm/5.0×10^7 nuclei and the maximum binding obtained with ^{14}C-1-IAA was 63 dpm ±8 dpm/5.0×10^7 nuclei. Converting to moles of IAA bound, we get

1.655 pmoles/5.0 × 10^7 nuclei with ^{14}C-2-IAA and 1.699 pmoles/5.0 × 10^7 nuclei with ^{14}C-1-IAA. We regard these values as very close together.

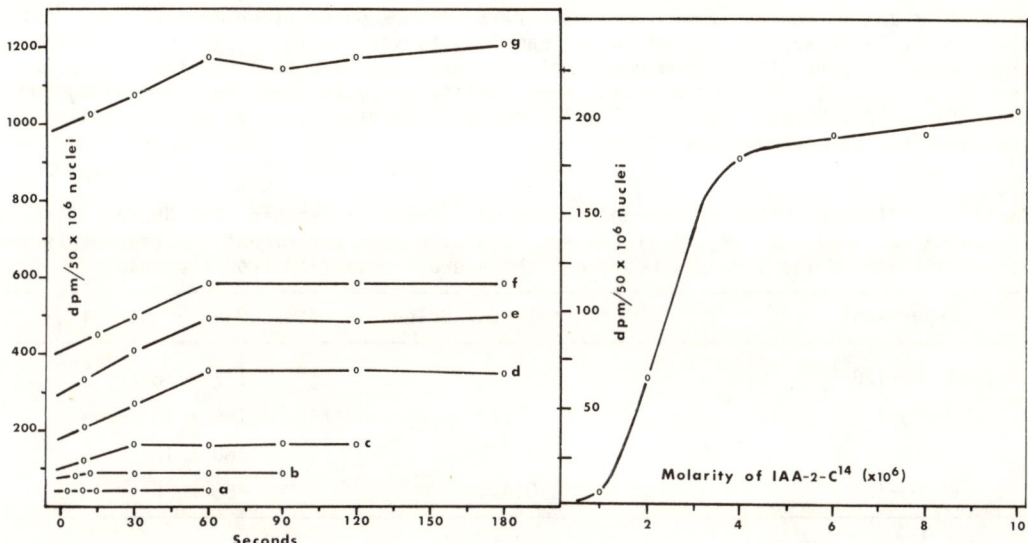

Fig. 2. (Left) Family of time-course curves for the various concentrations of ^{14}C-2-IAA indicated below:

a) 5 × 10^{-7} M
b) 1 × 10^{-6} M
c) 2 × 10^{-6} M
d) 4 × 10^{-6} M
e) 6 × 10^{-6} M
f) 8 × 10^{-6} M
g) 1 × 10^{-5} M

Standard procedures were used (see Materials and Methods) for these experiments. Each curve is the result of one separate experiment. In each experiment each treatment was duplicated. The deviation from the mean is a minimum of 1% and a maximum of 10%. Experiments using 4 × 10^{-6} M and 8 × 10^{-6} M ^{14}C-2-IAA have been repeated many times. Although there was variation in total radioactivity recovered, the kinetics were the same and the difference between the extrapolated 0 time value and maximum quantity bound was the same (within <10%) from experiment to experiment.

Fig. 3. (Right) Binding isotherm of ^{14}C-2-IAA at the pl-1 binding sites. The data for this curve were obtained by subtracting the extrapolated 0 time value from the maximum value for each time-course curve in Figure 2. These differences were then plotted as a function of the concentrations of ^{14}C-2-IAA. The variation for the plateau values is about ±10 dpm (5.5%).

Further support for the argument that IAA binds to the pl-1 binding sites comes from time-course experiments with 10^{-5}M IAA, as shown in Fig. 2. In this experiment the specific radioactivity of ^{14}C-2-IAA was reduced by half with non-radioactive IAA. When the dpm values were corrected for the dilution of specific radioactivity, the value obtained for the pl-1 binding sites, 205 dpm ±20 dpm/5.0× 10^7 nuclei, fit the expected plateau value. This result is possible only if the binding were due to IAA or if the IAA were converted to a radioactive derivative during the binding. If the binding was that of a radioactive contaminant already present in the ^{14}C-IAA solutions, diluting the ^{14}C-IAA with nonradioactive IAA would have reduced the concentration of the contaminant to half its concentration at 10^{-5}M. Then, since the amount bound at 5 × 10^{-6}M ^{14}C-2-IAA is about 185 dpm/5.0 × 10^7 nuclei, and since the specific radioactivity of the supposed contaminant was not reduced by adding IAA, correcting for the dilution would have given a binding value of 370 dpm/5.0 x 10^7 nuclei, which is much higher than the plateau value for pl-1 binding sites.

Binding of IAA to Nuclei from Light-Grown Peas. Time-course experiments using ^{14}C-2-IAA and ^{14}C-1-IAA were performed in the same way as with nuclei isolated from dark-grown peas. The results of two representative experiments with ^{14}C-2-IAA are shown in Fig. 4. The pl-1 binding kinetics were qualitatively (60 sec. to completion) and quantitatively (180 dpm ±18 dpm/5.0 x 10^7 nuclei) the same as those with nuclei isolated from dark-grown peas.

Table 1. Effect of Chloroplasts on Binding of ^{14}C-IAA to Isolated Pea Nuclei. In (A) ^{14}C-2-IAA was used. In (B) ^{14}C-1-IAA was used. In each experiment the treatments were duplicated. The ± values represent the average deviation from the mean.

Experiment	Chloroplasts/nucleus	dpm/5.0 x 10^7 nuclei
A. 68-128	3.770	117 ± 18
69-4	0.274	184 ± 18
69-7	0.772	180 ± 18
B. 69-13	0.430	63 ± 6

Since the nuclear preparations obtained from light-grown peas were contaminated with chloroplasts, it was possible to check if the binding of IAA to pl-1 binding sites observed with nuclei may also occur with other organelles such as chloroplasts. As shown in Table 1, varying the chloroplast concentration in the reaction mixture did not affect the pl-1 binding. These results indicate that at least *in vitro*, pea chloroplasts do not contain pl-1 binding sites of the kind found in pea nuclei.

Other Characteristics of IAA Binding. If the pl-1 binding sites are in protein molecules, it is possible that sulfhydryl-binding agents such as N-ethylmaleamide (NEM) or iodoacetamide (IAM) could inhibit the binding of IAA. In preliminary experiments, it was found that the total radioactivity recovered from nuclei pretreated for 20 minutes at 0°C with 3 mM NEM or 3 mM IAM was reduced by 19% and 26% respectively. To test whether the binding of IAA to the pl-1 binding sites was inhibited by such treatments, a time-course experiment was performed after pretreatment of isolated pea nuclei with IAM. The results shown in Fig. 5 indicate that pl-1 binding was inhibited by 60% while the total radioactivity bound was reduced by 27% as in previous experiments. The 12% reduction in the radioactivity by IAM at 0 sec. of reaction may represent an inhibition of binding at the high affinity binding sites suggested by plateau #2 in figure 1.

The kinetics of IAA binding to pl-1 binding sites could not be demonstrated in nuclei isolated from mature leaves, or in nuclei derived from diseased or unusually weak plants which could not respond to IAA treatment *in vivo*. In all such cases, the radioactivity bound at 0 sec. of the binding kinetics was the same as at 3 minutes and comparable to the 0 sec. radioactivity bound to nuclei which showed pl-1 binding kinetics. Whether this indicates a complete absence of the pl-1 binding sites from tissues which do not respond *in vivo* to IAA treatment or merely a greater instability of such sites in mature or necrotic tissues is not clear.

IV. DISCUSSION

A direct effect of IAA on RNA synthesis was implicated by the discovery that IAA promotes RNA synthesis in isolated plant cell nuclei (see Introduction). This discovery has led to the present finding that IAA binds to high affinity binding sites in

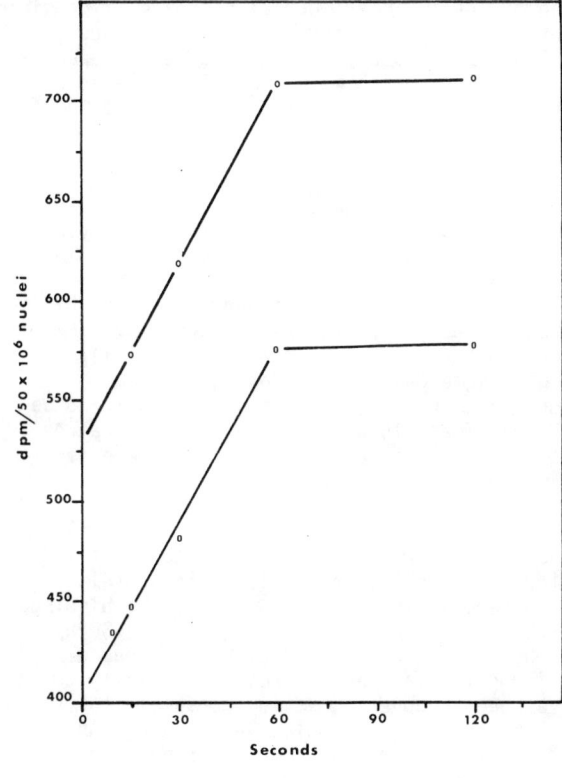

Fig. 4. (Top) Time-course of ^{14}C-2-IAA binding to nuclei isolated from light-grown peas. The concentration of IAA was 8×10^{-6} M. The two curves are results of two separate experiments performed 3 months apart using separately prepared solutions. In each experiment each treatment was duplicated. The results are averages of these duplicated treatment. The deviation from the mean ranged from a minimum of 2% to a maximum of 10%. The difference between the background levels in the two experiments was probably due to a higher concentration of breakdown products in an old, frozen solution of ^{14}C-2-IAA used in the experiment represented by the upper curve. Freshly prepared solutions of ^{14}C-2-IAA always have much lower background levels.

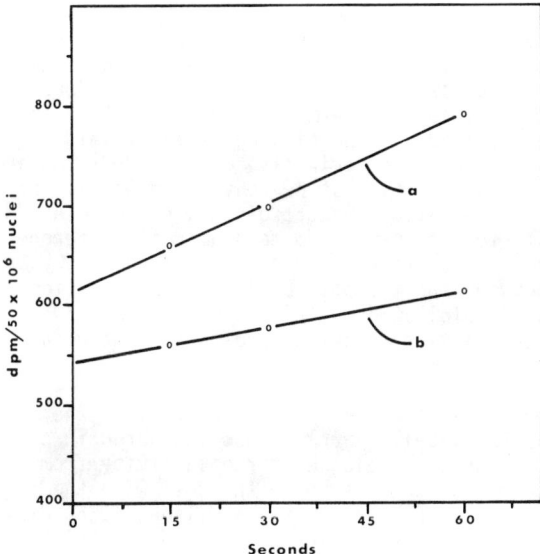

Fig. 5. (Bottom) Effect of iodoacetamide (IAM) on the binding of ^{14}C-2-IAA (8×10^{-6} M). (a) is the time-course of the binding in the absence of IAM. (b) is the time-course of the binding in the presence of IAM. These are results from one experiment in which each treatment was duplicated. The data are the averages of two duplicated treatments. The maximum deviation from the mean is 10%. In (b) the IAM was added to the nuclei at a concentration of 3 mM 20 minutes before adding ^{14}C-2-IAA. The IAM was present during the binding reaction.

isolated pea nuclei. Studying the binding of IAA in intact nuclei has made it possible to avoid the equilibrium dialysis procedure normally used for such work and thus to explore the time-course of IAA binding. Equilibrium dialysis is suitable for quantitative study of binding to purified macromolecules, which will bind ligands even when removed from their original site in the cell but it is first necessary to find these macromolecules.

Physical adsorption or non-specific binding of a solute from aqueous solutions is a linear function of its concentration at non-saturating concentrations of solute (Kipling, 1965). Thus, any deviation of the IAA binding isotherm from linearity at non-saturating concentrations of IAA would be a strong indication that some binding was occurring at sites other than those available for physical adsorption. Such binding was indicated in our work by the sigmoidal twist in an otherwise linear binding isotherm when total radioactivity recovered in nuclei treated with increasing concentrations of IAA was plotted as a function of IAA concentration (Figure 1). This deviation from linearity was more strikingly sigmoidal in the binding isotherm obtained for the pl-1 binding sites (figure 3). The pl-1 binding sites appear to be saturated at about 4×10^{-6} M IAA, which is at least 15 x lower than the saturating concentrations of IAA. This suggests that IAA binds to a small number of receptor sites having a high affinity for IAA and not merely to adsorption sites on nuclear surfaces.

The sigmoidal shape of the slow binding isotherm could be due to cooperative interactions of two or more similar binding sites on the same allosteric protein, as has been shown for the binding of oxygen to hemoglobin (Wyman, 1963) and the binding of CTP and succinate to aspartate transcarbamylase (Changeux et al., 1968). Such a curve could also result from a heterogeneous population of molecules which has a statistically normal distribution of binding sites varying in their sensitivity to IAA. Although such a situation has not yet been uncovered, it is impossible to distinguish between two alternatives without isolating the molecules binding IAA, determining their chemical and structural identity, and analyzing the binding of the purified molecules via the Scatchard (1949) method.

In the concentration range of plateau #1 (figure 1), the time-course for the completion of pl-1 binding was 50-60 seconds. Since little is known about the kinetics of hormone binding, the meaning of the time-course is not clear. Although radioactive IAA was the starting compound, it was not possible to determine whether the time course for the pl-1 binding represented the time required for the binding of IAA itself or the time required to form a strongly-bound radioactive derivative of IAA *after* IAA had entered the slow binding site. The latter possibility must be considered in view of the finding of Kefford et al. (1963) that an oxidation product of IAA, but not IAA itself, binds to purified RNA. Also, testosterone is reduced to dihydrotestosterone, a more active form of the androgen, during binding to nuclear proteins (Bruchowsky & Wilson, 1968). Attempts to distinguish between the two possibilities by extraction of radioactive compounds from the labelled nuclei followed by thin-layer and paper chromatography were inconclusive due to incomplete recovery of radioactive material from the gummy residue which formed during the extraction.

The binding of ^{14}C-1-IAA to isolated nuclei appeared to be strong since the radioactivity was removed only very slowly when labelled nuclei were incubated in a solution of non-radioactive IAA at $4^{\circ}C$. Whether the release of radioactivity from the nuclei might be faster at higher temperatures, as appears to be the case with the binding of some animal hormones (Talwar et al., 1968; Bruchowsky & Wilson, 1968; Wira & Munck, 1970), was not investigated extensively. Incubation of labelled nuclei (^{14}C-2-IAA) at $20^{\circ}C$ in non-radioactive IAA for one hour did diminish pl-1 binding by about 30% but did not reduce the extrapolated 0 time radioactivity. The loss of radioactivity from the labelled nuclei could have been due to a loss of the binding sites themselves and not necessarily to a release of IAA or its derivative.

Although the release of radioactivity from nuclei labelled with ^{14}C-1-IAA or ^{14}C-2-IAA was slow in aqueous solutions, treatment with organic solvents such as ethanol, ethyl acetate or ether completely removed the radioactivity within minutes. These findings suggest that the pl-1 binding sites have a high affinity for IAA or its derivative, but that the binding is probably not covalent. If it is covalent, then the IAA would have to be bound to small molecules soluble in organic solvents.

That the pl-1 binding of IAA to isolated nuclei is a physiologically normal reaction and not an artefact of the *in vitro* system is indicated by the following findings: (a) no detectable pl-1 binding occurs to chloroplasts, (b) there is no pl-1 binding to nuclei isolated from mature leaves of light-grown peas which do not respond to exogenous IAA, (c) the pl-1 binding of IAA is absent from nuclei isolated from apical buds of weak or necrotic seedlings. Furthermore, there is a correlation between the concentrations of IAA used in the binding studies and those used in the studies on the growth of pea stem sections as affected by IAA. The pl-1 binding sites are saturated at about 4×10^{-6} M, and maximum stimulation of elongation and increase in fresh weight of etiolated pea stem sections is achieved with 10^{-6} to 10^{-5} M IAA solutions containing 2% sucrose (Galston & Kaur, 1961). Concentrations of IAA above 10^{-5} M inhibit elongation and have no further effect on the increase in fresh weight. Although this is merely a correlation and the concentrations of IAA used in the solutions bathing the stem sections do not necessarily reflect the concentrations of IAA inside the tissue, the binding of IAA at the pl-1 binding sites could have something to do with its promotion of growth in excised stem sections.

Although we were not able to demonstrate IAA promotion of RNA synthesis in our system, there is agreement between the pl-1 binding of IAA found in this research and the promotion of RNA synthesis found in coconut and pea nuclei by Roychoudhury *et al.* (1965). They found that auxins promoted RNA synthesis slightly at a low concentration of 2×10^{-6} M, but that maximum promotion was obtained at a concentration of 10^{-5} M. Since the pl-1 binding sites begin to be filled at a concentration of 2×10^{-6} M and are saturated at concentrations of 4×10^{-6} M to 10^{-5} M, it is possible that the pl-1 binding of IAA is involved in the regulation of RNA synthesis. In the same context it would be interesting to know if the pl-1 binding sites are on the proteins obtained from pea nuclei by Matthyse and Phillips (1969), which appear to be necessary for auxin promotion of RNA synthesis on isolated chromatin. On the other hand, since nuclei are capable of other reactions besides RNA synthesis, it is possible that the pl-1 binding of IAA or its derivative found in this research serves another function.

V. SUMMARY

IAA, supplied as either ^{14}C-1-IAA or ^{14}C-2-IAA bound readily to isolated pea nuclei. In addition to undefined binding to various organelles, a high affinity binding, requiring 60 seconds to reach completion at a given concentration of IAA, was found in nuclei isolated from apices of light or dark-grown peas. When plotted against concentration of IAA, it yielded a sigmoidal curve saturating at ca. 4×10^{-6}M. No detectable high affinity binding was found with chloroplasts or with nuclei isolated from mature pea leaves. The binding of IAA did not appear to be covalent and may not be related to RNA synthesis.

VI. ACKNOWLEDGEMENTS

K.J. Tautvydas was aided by a Title IV NDEA fellowship.

A.W. Galston was aided by a grant from the National Institutes of Health.

VII. REFERENCES

BRAY, G.A. (1960). A simple efficient liquid scintillator for counting aqueous solution in liquid scintillation counter. Analyt. Biochem. 1, 279-285.

BRUCHOVSKY, N. and J.D. WILSON (1968). The intranuclear binding of testosterone and 5 α-androstan-17β-ol-3-one by rat prostate. J. Biol. Chem. 243, 5953-5960.

CHANGEUX, J.P., J.C. GERHART, and H.K. SCHACHMAN (1969). Allosteric interactions in aspartate transcarbamylase. I. Binding of specific ligands to the native enzyme and its isolated subunits. Biochemistry 7, 531-538.

CHERRY, J.H. (1967). Nucleic acid biosynthesis in seed germination: influences of auxin and growth-regulating substances. Ann. N.Y. Acad. Sci. 144, 154-168.

FANESTIL, D.D. and I.S. EDELMAN (1966). Characteristics of the renal nuclear receptors for aldosterone. Proc. Natl. Acad. Sci. U.S.A. 56, 872-879.

GALSTON, A.W. and R. KAUR (1961). Comparative studies on the growth and light sensitivity of green and etiolated pea stem sections. In: "Light and Life," Chapter 10. McElroy and Glass, Eds., Johns Hopkins Press (1961).

GILBERT, W. and B. MÜLLER-HILL (1966). Isolation of the lac repressor. Proc. Natl. Acad. Sci. U.S.A. 56, 1891-1898.

JOHRI, M.M. and J.E. VARNER (1969). Enhancement of RNA synthesis in isolated pea nuclei by gibberellic acid. Proc. Natl. Acad. Sci. U.S.A. 59, 269-276.

KEFFORD, N.P., KAUR-SAWHNEY, R. and A.W. GALSTON (1963). Formation of a complex between a derivative of the plant hormone indoleacetic acid and ribonucleic acid from pea seedlings. Acta Chem. Scand. 17, S313-S318.

KEY, J.L. (1969). Hormones and nucleic acid metabolism. Ann. Rev. Plant Physiol. 20, 449-474.

KIPLING, J.J. (1965). Adsorption from solutions of non-electrolytes. Academic Press, Publs. (1965).

MAHESHWARI, S.C., S. GUHA and S. GUPTA (1966). The effect of indoleacetic acid on the incorporation of P^{32}-orthophosphate and C^{14}-adenine into plant nuclei in vitro. Biochim. Biophys. Acta 117, 470-472.

MATTHYSSE, A.G. and C. PHILLIPS (1969). A protein intermediary in the interaction of a hormone with the genome. Proc. Natl. Acad. Sci., U.S.A., 63, 897-903.

MAURER, H.R. and G.R. CHALKLEY (1967). Some properties of a nuclear binding site of estradiol. J. Mol. Biol. 27, 431-441.

MONOD, J., J. WYMAN and J. CHANGEUX (1965). On the nature of Allosteric transitions: a plausible model. J. Mol. Biol. 12, 88-118.

ROYCHOUDHURY, R., A. DATTA and S.P. SEN (1965). The mechanism of action of plant growth substances: the role of nuclear RNA in growth substance action. Biochim. Biophys. Acta 107, 346-351.

SCATCHARD, G. (1949). The attractions of proteins for small molecules and ions. Ann. N.Y. Acad. Sci. 51, 660-672.

TALWAR, G.P., M.L. SOPORI, D.K. BISWAS, and S.J. SEGAL (1968). Nature and characteristics of the binding of oestradiol-17β to a uterine macromolecular fraction. Biochem. J. 107, 765-774.

TAUTVYDAS, K.J. (1971). Mass isolation of pea nuclei. Plant Physiology, 47, 499-503.

WIRA, C. and A. MUNCK (1970). Specific glucocorticoid receptors in thymus cells. Localization in the nucleus and extraction of the cortisol-receptor complex. J. Biol. Chem., 245, 3436-3438.

WYMAN, J. (1963). Allosteric effects in hemoglobin. Cold Spring Harbor Symp. Quant. Biol., 28, 483-490.

PLANT GROWTH SUBSTANCES, 1970

Modification of Enzyme Activity, Conformation and Size by Indoleacetic Acid.

Igor V. Sarkissian

Institute of Life Science, Texas A&M University, College Station, Texas, 77843

I. INTRODUCTION

Indoleacetic acid (IAA) is a plant hormone whose molecular mode of action is not understood. While its effects on plant growth are well known, it is not known how they are brought about. This is surprising since IAA has been known for many years and much work has been devoted to this hormone. In seeking to understand the molecular action of IAA it is obvious that we must isolate the *act* of the hormone to the molecular, sub-tissue level.

Considerable interest is focused at present on the nature of allosteric regulatory enzymes and the changes produced in them by their effectors. According to the model proposed by Monod, Wyman, and Changeux (1965), an allosteric enzyme, which is itself a specific association of subunits, can exist reversibly in at least 2 states differing in protein conformation. The different states induced by low molecular weight, non-substrate compounds, possibly hormones, provide a mechanism for control of metabolism by such compounds.

The proposition that hormones are allosteric effectors now permits a means for comprehension of regulatory control which would have been considered inconceivable a few years ago (Koshland and Neet, 1968). It has been suggested that hormones have the capability of triggering transitions in a variety of different proteins (Monod *et al.*, 1963) and that hormones act not as co-enzymes or intermediates, but as materials that favorably alter a conformation of the active sites of the appropriate enzymes (Koshland, 1959).

Citrate synthase (E.C. 4.1.3.7.) is one enzyme which can be regulated *in vivo* or *in vitro* by indoleacetic acid (Sarkissian and Schmalstieg, 1969a, 1969b). The importance of citrate synthase and its regulation as related to growth cannot be overemphasized (Atkinson, 1968). Furthermore, in keeping with the discussion above and with the notion that the initial molecular *act* of IAA is the same irrespective of the protein affected by IAA, knowledge of the manner whereby IAA regulates one enzyme will contribute significantly to a better understanding of regulation of metabolism and growth by IAA. In this presentation I shall discuss some of the physicochemical changes induced by indoleacetic acid in purified citrate synthase.

II. MATERIALS AND METHODS

Citrate synthase was purified about 400-fold from root tips of 7-day old seedlings of bean (*Phaseolus vulgaris*, var. Burpee's Stringless Greenpod) or from buds of cauliflower (*Brassica oleracea*, var. *botrytis*) obtained from local markets. The enzyme was extracted from mitochondria which were isolated by differential centrifugation of tissue homogenates as previously described (Mulliken and Sarkissian, 1970). The mitochondria were then suspended in 0.02 M phosphate buffer, pH 7.4 (40 ml buffer/ kg of original tissue) and sonicated for 2 minutes.

The solution of sonicated mitochondria was fractionated by ammonium sulfate followed by de-salting on Sephadex G-50 and chromatography on DEAE cellulose (Sarkissian and Schmalstieg, 1969b). All steps during extraction were done between 1 and $4°$. Other details are given in figures or tables.

III. RESULTS AND DISCUSSION

Indoleacetic acid induces three molecular changes in citrate synthase. The changes, although related, will be considered separately for ease of presentation. They are, (1)-Activation of the isolated citrate synthase, (2)-Alteration of the conformation of the enzyme, and (3)-Apparent increase in size of the enzyme.

Activation and Conformational Changes of Citrate Synthase Induced by IAA

Citrate synthase can be activated by IAA under appropriate conditions. While isolation and purification of the enzyme under reducing conditions (about 10^{-4} M dithiothreitol or cysteine) may be practised, it is absolutely essential that activation of enzyme be studied at high ionic strength, 0.1 to 0.2 M (Table 1) and below the apparent K_m for the condition of assay. The competitive interaction of IAA and acetyl CoA (AcSCoA) concentration yielded with bean citrate synthase a 3-fold decrease in the K_m for acetyl CoA in the presence of 1×10^{-11}M IAA (Sarkissian, to be published). The nature of the competitive effect is not yet known but it is clear that at saturating substrate concentrations it is impossible to observe activation (Zenk and Nissl, 1968), (Fig. 1). Activation also may not be detected at low ionic strength (Brock and Fletcher, 1969). The optimum pH for activation is 7.5 (Table 2). At pH 8.0, the conventional pH for assay for citrate synthase (Lowenstein, 1969), activation drops off markedly.

Table 1. Activation of cauliflower citrate synthase by IAA as a function of molarity of the buffer

Phosphate Buffer M	Percent Activation by 10^{-7}M IAA
0.02	0
0.05	20.2
0.10	23.6
0.20	23.0
0.50	0

Substrate concentrations were: AcSCoA, 10 uM and OAA, 5 uM. Other conditions were as in legend to Fig. 1.

Table 2. Activation of cauliflower citrate synthase by IAA as a function of pH

pH	Percent Activation by 10^{-7}M IAA
7.0	0
7.5	23.6
8.0	9.5

Other reaction conditions were as in Table 1.

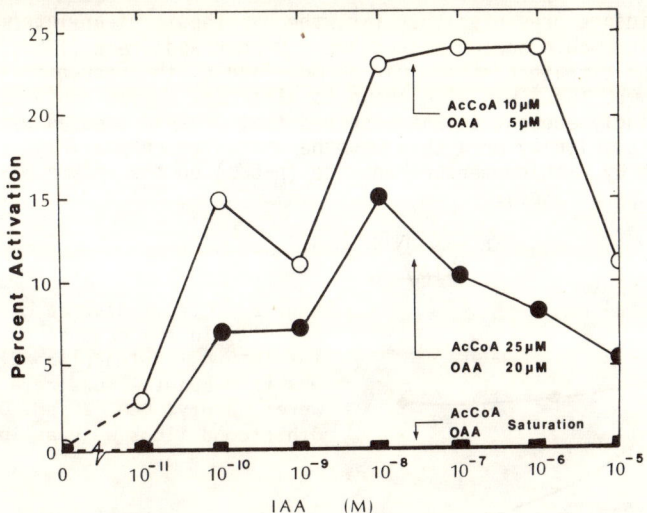

Fig. 1. Activation of cauliflower citrate synthase by IAA as a function of substrate concentration. The reaction mixture contained 100 mM phosphate buffer, 10 μg of enzyme, substrates and IAA as shown. Enzyme, buffer and IAA were mixed first; the reaction was started by addition of substrates. Saturation under present ionic strength was 250 μm acetyl CoA and 600 μM OAA. Total volume was 1.0 ml, pH 7.4. Activity at 25°C was measured at 233 nm with the recorder set at high sensitivity (0.25 O.D. full scale).

Given these conditions, one is still not assured of consistent activation *in vitro*. This is not surprising since endogenous IAA activates the enzyme and desensitizes it to further activation by IAA (Sarkissian and Schmalstieg, 1969b). Furthermore, the substrates oxalacetate (OAA) and acetyl CoA "protect" the enzyme from activation or retard the effect of IAA (Table 3). The question now arises as to what type of action does IAA have on the native enzyme. Dialysis of bean citrate synthase treated with IAA either *in situ* or *in vitro* did not result in diminution of stimulated activity (Sarkissian and Schmalstieg, 1969b). This, coupled with the fact that the maximum activation occurs at micromolar or lower concentrations of IAA, strongly suggests that the action of IAA is catalytic in nature. It has been shown that the hormone binds with the plant enzyme but not reversibly (Sarkissian, 1970 and unpublished observations).

Table 3. Protection from activation by IAA of bean citrate synthase by its substrates

Preparation*	Reaction Started By	Percent Enhancement
Enzyme** + IAA	AcSCoA + OAA ***	67
Enzyme** + IAA + AcSCoA	OAA	20
Enzyme** + IAA + OAA	AcSCoA	15

Activity was measured as described in the legend to Figure 1. IAA was 10^{-11} M.

* The combinations were pre-incubated 5 min. at 25°C.

** 12 ug protein

*** AcSCoA - 10 uM; OAA - 5 uM.

From the evidence presented thus far, the hypothesis of an intermolecular protein aggregation reaction would appear attractive. Broder and Srere (1963) have shown that the conformation of citrate synthase may be modified by the reduction state of the SH groups of the enzyme. Urea was also shown to alter the enzyme conformation (Srere, 1965a). Interestingly enough, it was observed that protein treated with IAA was more resistant to denaturation by urea than was the untreated enzyme (Fig. 2). Similarly, the rate of attack by p-chloromercuribenzoate (p-CMB) on the enzyme was retarded by IAA (Sarkissian, unpublished).

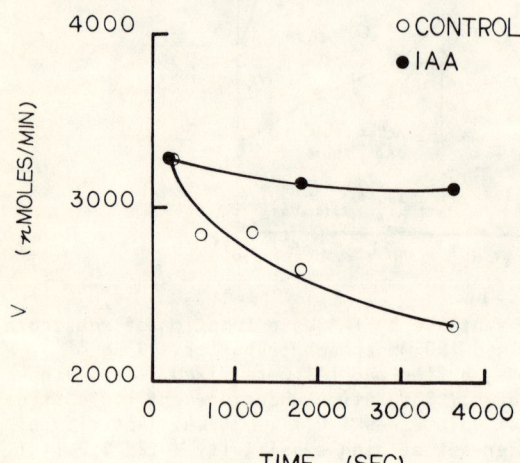

Fig. 2. Effect of urea on citrate synthase activity in the absence or presence of 1×10^{-8} M indoleacetic acid. The enzyme was from beans; substrate concentrations were: acetyl CoA, 20 µM, OAA, 20 µM. Other conditions were as in Fig. 1.

In Table 4 are shown the effects of p-CMB and mercuric acetate on the activity of citrate synthase. Basal enzyme activity was not affected by p-CMB. However, the stimulatory effect was completely abolished by the mercurial. Treatment with mercuric acetate led to marked inhibition. These results agree with those of Srere (1965b) who has shown that SH groups of citrate synthase react easily with low concentrations of Hg^{++} and less so with p-CMB.

Two conclusions may be drawn from this information. First, it can be inferred that the conformation change or aggregation involves a change in the region of the external SH groups of citrate synthase. Coupled with the observation that p-CMB eliminated the effect of IAA without significantly altering basal enzyme activity, this tends to support the conclusion that the site of action of IAA is distinct from the catalytic site (Sarkissian, 1968).

Further attention must be drawn to the importance of the ionic environment of citrate synthase. This is prompted by the observation that ionic strength affects the Michaelis constant for both substrates (Poulsen and Sarkissian, 1971). This is probably related to the conformation and thus to the allosteric properties of the enzyme. The finding that conformation of citrate synthase is changed by ionic strength with a resulting change in the degree of interaction with substrates further supports the concept that IAA may control the activity of the enzyme depending on the conformational state of the protein.

Changes in Structure of Citrate Synthase Induced by IAA

Chromatography of cauliflower or bean citrate synthase on Sephadex G-200 yielded one peak. The molecular size of this enzyme was consistent with the reported size of about 100,000 daltons for a mammalian citrate synthase (Wu and Yang, 1970). However, incubation of plant citrate synthase for 5 minutes in 10^{-5}M or less IAA followed by filtration through Sephadex yields a fraction with citrate synthase activity which is eluted earlier than the untreated fraction (Fig. 3). This has been interpreted as increase in molecular size of citrate synthase treated with IAA; the apparent increase in size is accompanied by increased enzyme activity and by binding of the hormone to the enzyme (Sarkissian, 1970).

Table 4. Effects of p-mercuribenzoate and mercuric acetate on activity of bean citrate synthase.

Preparation	Activity nmoles/min	Percent Activity
Enzyme	1650	100
", 1.54×10^{-11}M IAA	2470	150
", pCMB[a] (1×10^{-4}M)	1700	103
", pCMB, IAA[b]	1440	87
", IAA, pCMB[c]	1950	118
", $Hg(CH_3COO)_2$[a] (1.5×10^{-5}M)	1440	87
", Hg++, IAA[b]	1600	90
", IAA, Hg++[c]	2060	125

a Incubated 5 min. with mercurial before initiation of reaction by addition of substrates.
b Incubated 5 min. with mercurial, then 5 min. with IAA before initiation of reaction.
c Incubated 5 min with IAA, then 5 min. with mercurial before initiation of reaction.

Activity was measured as described in the legend to Fig. 1.

The reaction mixture was composed as follows: 0.10 M phosphate buffer (pH 7.4), 5 uM oxalacetate (freshly made to pH 7.4), 10 uM acetyl coenzyme A, about 10 ug citrate synthase, other addenda as shown. The total volume was 1.0 ml. Other conditions were as in Fig. 1.

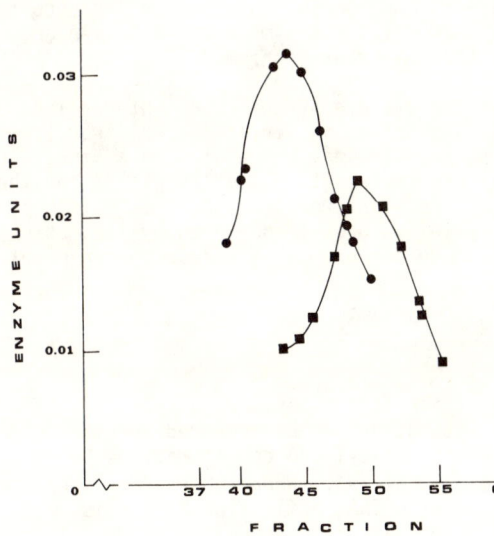

Fig. 3. Comparative elution patterns of bean citrate synthase on Sephadex G-200. ———■——— control enzyme eluted with 0.05 M tris, pH 7.4; ———●——— citrate synthase incubated 10 min. in 1×10^{-9}M IAA at $4°$ and eluted with 1×10^{-11} M IAA in 0.05 M tris, pH 7.4. Flow rate was 0.5 ml/min, each fraction was 3 ml. One unit of citrate synthase is that amount of enzyme which will utilize one μmole of acetyl-coenzyme A in 1 minute at $25°$. 0.4 enzyme units were applied to each column (from Sarkissian, 1970).

Ultracentrifugation of cauliflower citrate synthase in a 5 - 20% sucrose gradient revealed that native citrate synthase has an s_{20w} of 5.99 S. It is of interest that this value compares favorably with the values for pigheart citrate synthase, 6.0 (Wu and Yang, 1970) and 6.2 (Singh *et al.*, 1970). Furthermore, it places the native cauliflower citrate synthase in the class of "small" citrate synthases (Weitzman and Dunmore, 1969). While the pattern of elution from Sephadex G-200 columns may suggest increase in molecular size, ultra-centrifugation of citrate synthase treated with IAA yielded an s_{20w} of 5.99 S, identical to the value of the native enzyme (Table 5). The most plausible conclusion is that IAA modifies the conformation rather than the molecular size of citrate synthase. A change in either size or conformation could be reflected in modified elution behavior from Sephadex G-200 (Sarkissian, 1970). However, a change in conformation cannot be readily discerned in the ultracentrifuge. Thus, if anything, the suggestion that IAA modifies the conformation of citrate synthase is supported by the ultracentrifuge data.

Table 5. Ultracentrifugation of cauliflower citrate synthase.

PROTEIN	S_{20w}
Native citrate synthase, 1.2 mg	5.99
1×10^{-5} M IAA citrate synthase, 1.2 mg	5.99

Centrifugation was for 16 hrs. at 45,000 rpm in 5-20% sucrose gradient in 0.02 M potassium phosphate, pH 7.4 and in the absence or presence of IAA. Beef heart lactic dehydrogenase (S_{20w} 7.32) was used as a marker protein.

IV. CONCLUSION

Citrate synthase is affected by indoleacetic acid. The related effects are (1) -activation and (2) - change in conformation of the enzyme. High ionic strength and non-saturating acetyl CoA concentrations, those below the apparent K_m for the conditions of assay are necessary for activation. Change in conformation is supported by the altered elution pattern from Sephadex G-200 of the enzyme exposed to IAA and by the resistance to denaturation by urea of the IAA-treated enzyme.

As to the physiological significance of the citrate synthase-IAA interaction, I suggest that regulation resulting from the interaction is very different from feedback regulation. The two types of regulation, however, may be related in a "throttle" and "governor" relationship with citrate synthase representing the "throttle" and the feedback regulation involved in homeostasis of the plant representing the "governor". In this scheme, the "throttle" of the Krebs cycle could be thrown open by IAA, but in turn it would be held in check by feedback regulation until a stress on the plant demanded a response.

V. SUMMARY

Isolated and highly purified citrate synthase from two higher plants, cauliflower and bean, can be activated by IAA *in vitro*. Activation can be measured spectrophotometrically in the presence of high ionic strength, acetyl CoA concentrations below the apparent K_m for conditions of the assay and pH of 7.4. When the enzyme is exposed to either of the substrates prior to incubation with IAA, activation by IAA is considerably reduced. Activation apparently occurs at the external SH groups of citrate synthase distinct from the active site.

Change in conformation of citrate synthase is suggested by dependence of activation on ionic strength and substrate (acetyl coenzyme A) concentration. It is supported by earlier elution from Sephadex G-200 of the enzyme treated with IAA. In the ultracentrifuge, however, the treated and the untreated enzyme showed an s_{20w} of 5.99 S, confirming that the earlier elution from Sephadex G-200 was due to change in conformation of the enzyme rather than increase in molecular size.

VI. ACKNOWLEDGEMENTS

The support of National Science Foundation (Grant GB-7618) and Robert A. Welch Foundation is gratefully acknowledged.

VII. REFERENCES

ATKINSON, D.E. (1968). Citrate and the citrate cycle in the regulation of animal metabolism, *in* "The Metabolic Roles of Citrate" (Ed. T.W. Goodwin) p. 23-40, Academic Press, New York.

BROCK, B.L.W. and R.A. FLETCHER (1969). Activation of citrate synthase by indoleacetic acid. Nature 224: 184-185.

BRODER, I. and P.A. SRERE (1963). Starch-gel electrophoresis of condensing enzyme from pig heart. Biochim. Biophys. Acta 67: 626-632.

KOSHLAND, D.E. (1959). Enzyme flexibility and enzyme action. J. Cellular and Comp. Phys. 54: 245-258.

KOSHLAND, D.E. and K.E. NEET (1968). The catalytic and regulatory properties of enzymes. Ann. Rev. Biochem 37: 359-410.

LOWENSTEIN, J.M. (Editor) (1969). Citrate synthase of mammals, yeast, *E. coli* and lemon fruit. Methods in Enzymol. 13: 3-26, Academic Press, New York.

MONOD, J.J., J. WYMAN and J.P. CHANGEUX (1965). On the nature of allosteric transitions: a plausible model. J. Mol. Biol. 12: 88-118.

MONOD, J., J.P. CHANGEUX and F. JACOB (1963). Allosteric proteins and cellular control systems. J. Mol. Biol. 6: 306-329.

MULLIKEN, J.A. and I.V. SARKISSIAN (1970). Uncoupler-insensitive respiratory control of α-ketoglutarate oxidation by plant mitochondria. Biochem. Biophys. Res. Comm. 39: 609-615.

POULSEN, L.L. and I.V. SARKISSIAN (1971). On the regulation of the regulatory enzyme, citrate synthase, by salt. Life Science 10: 91-97.

SARKISSIAN, I.V. (1968). Nature of molecular action of 3-indoleacetic acid, *in* "Biochem. & Physiology of Plant Growth Substances" (Eds. F. Wightman and G. Setterfield) p. 473-485, Runge Press, Ottawa.

SARKISSIAN, I.V. (1970). Hormonal modifications of plant citrate synthase *in vitro*. Biochem. Biophys. Res. Comm. 40: 1385-1390.

SARKISSIAN, I.V. and F.C. SCHMALSTIEG (1969a). Confirmation of the effect of indole-3-acetic acid on citrate synthase. Naturwissenschaften 56: 284.

SARKISSIAN, I.V. and F.C. SCHMALSTIEG (1969b). Citrate synthase of bean seedlings - comparison of activity following *in vitro* and *in vivo* treatments of enzyme. Life Science 8: 933-938.

SINGH, M., G.C. BROOKS and P.A. SRERE (1970). Subunit structure and chemical characteristics of pig heart citrate synthase. J. Biol. Chem. 245: 4635-4640.

SRERE, P.A. (1965a). Conformation changes in citrate condensing enzyme. Arch. Biochem. Biophys. 110: 200-204.

SRERE, P.A. (1965b). The sulfhydryl groups of citrate condensing enzyme. Biochem. Biophys. Res. Comm. 18: 87-91.

WEITZMAN, P.D.J. and P. DUNMORE (1969). Citrate synthase: allosteric regulation and molecular size. Biochim. Biophys. Acta 171: 198-200.

WU, J.Y. and J.T. YANG (1970). Physicochemical characterization of citrate synthase and its subunits. J. Biol. Chem. 245: 212-218.

ZENK, M.H. and D. NISSL (1968). Evidence against an allosteric effect of indole-3-acetic acid on citrate synthase. Naturwissenschaften 55: 84.

INHIBITORS

Biochemical Aspects of the Action of Abscisic Acid

F.T. Addicott

University of California, Davis, California, U.S.A. 95616

I. INTRODUCTION

Abscisic acid (ABA) is a plant hormone which has important functions in the control of abscission, dormancy and germination. Further, applications of ABA can inhibit growth and promote senescence in a large variety of experimental materials (see Addicott & Lyon, 1969). The broad spectrum of responses to ABA and its wide distribution in plants suggest that it may be one of the more important hormonal factors in the control of plant behaviour, and has stimulated numerous investigations of the chemistry and biochemistry of the physiological action of ABA. This paper reviews present knowledge of the biochemical responses to ABA.

Biochemistry encompasses the synthesis and activities of enzymes; this review will be concerned broadly with the experimental evidence of the influence of ABA on the enzymes of higher plants. Some of the now classical investigations of the kinetics of auxin-induced growth suggested that certain growth inhibiting substances could compete with auxin for a site on an enzyme. It is not surprising therefore that some of the first experiments with ABA included investigation of its kinetic interactions with the other hormones. A discussion of these investigations will form the first section of this review. The second section will consider those experiments in which changes in enzyme activity or synthesis have been induced by ABA, including both inhibition and promotion of synthesis (or activity). The third section will consider the experiments which have attempted to pinpoint the steps in nucleic acid or protein synthesis which may be sensitive to ABA.

II. INTERACTIONS OF ABA AND OTHER HORMONES

During the past decade it has become abundantly clear that each of the major processes of plant growth and development can be influenced by more than one hormone and there has emerged a body of evidence which suggests that each process is controlled by an interplay of hormonal factors. Further, when the plant is considered as a whole, the interplay of hormones has a homeostatic effect, tending to maintain the integrity of the plant as a functional organism. In respect to any given process, some hormonal factors tend to promote the process and some tend to inhibit it. Thus when ABA became available, its interactions with the other hormones were examined in some detail.

The early experiments demonstrated that it was a powerful inhibitor of auxin-induced growth (e.g., Addicott et al., 1964; Wareing et al., 1964) and in retrospect: Hemberg (1946, 1961); Stewart & Caplin, 1952; Bennet-Clark & Kefford (1953); Carns et al., (1954); van Steveninck (1959), and others of that period who were investigating the role of "inhibitor β" and other auxin antagonists, appear to have been in fact investigating ABA (Milborrow, 1967). A kinetic study of the interaction of IAA and ABA in the control of coleoptile growth gave clear evidence of non-competitive inhibition (figure 1; Rothwell & Wain, 1964), indicating that the two substances act at different biochemical sites.

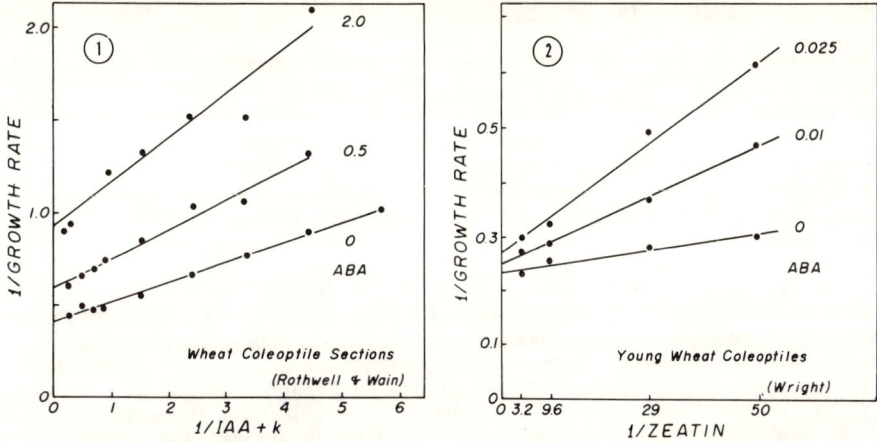

Fig. 1. Interaction of ABA and IAA in the growth of wheat coleoptile sections (from data of Rothwell & Wain, 1964). These Lineweaver-Burk plots indicate that the inhibition by ABA is non-competitive.

Fig. 2. Interaction of ABA and zeatin (a cytokinin) in the growth of wheat coleoptiles (from data of Wright, 1968). These Lineweaver-Burk plots indicate that the inhibition by ABA is non-competitive.

For the interaction of GA and ABA in coleoptile growth the results of Thomas et al., (1965) and Wright (1968) indicate a similar, non-competitive inhibition of GA-induced growth by ABA. A recent investigation by Jacobsen (1972) has disclosed that ethylene (ETH) also can to some extent counteract the ABA inhibition of α-amylase synthesis.

Interacting with cytokinin (CK) in coleoptile growth, ABA also showed a non-competitive inhibition (figure 2; Wright, 1968). However the antagonism of ABA to the CK-promoted germination of lettuce seed was somewhat different (Wareing et al., 1968b; Wareing & Ryback, 1970). The data of that experiment indicated that ABA and CK interacted competitively at the lower concentrations of the two substances, but interacted non-competitively at higher concentrations. Such results suggest the possibility that each substance acts at two biochemical sites, one site being the same for the two substances, the other sites being different.

Surprisingly, in a few instances ABA has promoted growth. Applied to emasculated flower buds of *Rosa sherardii*, ABA promoted parthenocarpic development (Jackson & Blundell, 1966). Interacting with a mixture of GA_4 + GA_7, ABA induced a synergistic promotion of the growth of cucumber hypocotyls (Aspinall et al., 1967).

The kinetics of the interactions of ABA with the other hormones in the process of abscission has received only limited attention. The results of Smith et al., (1968) indicate that the opposing influences of ABA and IAA are biochemically non-competitive. The same investigators found that the abscission-promotive effects of ABA and GA were not additive when the two substances were combined. Somewhat similar results were obtained by Böttger (1970) with explants from the nodes of mature and senescent leaves of *Coleus*. However, with explants from young nodes he found that the combination of ABA and GA gave remarkable acceleration of abscission over that which could be induced by either substance alone.

The above evidence indicates that for the most part ABA has a strongly inhibitory, but biochemically non-competitive interaction with the growth-promoting hormones. However, there is a small body of provocative evidence that in certain circumstances ABA can promote growth and/or interact additively with the other hormones.

III. ABA AND ENZYME BIOCHEMISTRY

The biochemical influences of a hormonal substance ultimately become manifest in the activity of enzymes. For most enzymatic responses to plant hormones, it is not yet clear whether the influence of the hormone is a direct one, bearing immediately on the synthesis or activity of the enzyme, or whether it is an indirect manifestation of some more general influence on growth processes. Regardless of whether the influence is direct or indirect, the enzymatic responses to hormones have a significant role in the physiology of plants and the responses to ABA must be included in our considerations.

The inhibitory effects of ABA on germination and growth are correlated with inhibitory effects on enzymes. The most noteworthy instance is in the germination of barley where ABA prevents the development of α-amylase, protease, and ribonuclease (Thomas et al., 1965; Chrispeels & Varner, 1967; Filner et al., 1969). Alpha-amylase is synthesized *de novo* in the aleurone layer of the endosperm, and ABA is a strong inhibitor of this synthesis (figure 3; see Filner et al., 1969). There is also a report that ABA can inhibit the activity of purified preparations of α-amylase but not β-amylase (Hemberg, 1967). Further, ABA has been implicated in the suppression of protease synthesis in embryos of cotton (Ihle & Dure, 1970). Inhibition of the development of fatty acid synthetase by ABA has been demonstrated in germinating castor beans (figure 4; Glew, 1968). In excised axes of germinating beans, ABA promotes the development of phenylalanine ammonia-lyase (figure 5; Walton & Sondheimer, 1968).

For the enzyme invertase, ABA at low concentrations inhibited the development of invertase activity in carrot root slices (Saunders & Poulson, 1968). With commercial yeast invertase these investigators found a relatively small inhibition by ABA. In contrast, ABA promoted the development of invertase in sugar beet discs (Cherry, 1968) and in sugar cane tissue (Gayler & Glasziou, 1969).

Reduced activity of chromatin-dependent RNA polymerase has followed addition of ABA at an early stage during the homogenisation of radish hypocotyls (Pearson & Wareing, 1969), and a lowered level of soluble RNA polymerase followed treatment of etiolated barley leaf segments with ABA (Poulson & Beevers, 1970; see also Khan & Anojulu, 1970).

Correlated with promotion of senescence and abscission ABA treatments have increased the activity of chromatin-associated ribonuclease in excised barley leaves (Srivastava, 1968). Similar enhancements of ribonuclease activity by ABA occur in sugar cane tissue (J.C. Waldron, unpublished, cited by Glasziou, 1969) in *Rhoeo* leaves (de Leo & Sacher, 1970a) and in bean endocarp (de Leo & Sacher, 1971). In excised abscission zones ABA accelerates the synthesis of cellulase even beyond the rate inducible by ethylene (figure 6; Cracker & Abeles, 1969; and see Lewis & Varner, 1970). Also during abscission there is a rapid increase in pectinase (Morré, 1968) a decrease in pectin methylesterase (Osborne, 1958), as well as increases in peroxidase, dehydrogenases, esterase and phosphatases (Sutcliffe et al., 1969). Such changes can be presumed to be promotable by ABA as a consequence of its promotion of abscission.

Careful investigations disclosed that ABA has essentially no effect on the development of nitrate reductase activity in barley aleurone (Ferrari & Varner, 1969) or in *Lemna* (G.R. Stewart & H. Smith, unpublished); the enzyme is substrate-induced in both instances (cf., Filner, et al., 1969). Also Gayler & Glasziou (1969) found that ABA did not affect the synthesis of peroxidase in sugar cane internodal tissue.

While the information on the influence of ABA in enzyme synthesis and activity is obviously still fragmentary, it correlates well with the knowledge of ABA's physiological influences. That is, where ABA is retarding or inhibiting a process such as growth or germination, its effect is to retard or inhibit the development of enzymes required for growth or germination. Conversely, when ABA is promoting a process such as senescence or abscission, it promotes the development of the special enzymes required for those processes. At the same time it may inhibit enzyme development and related events involved with the delay of senescence or abscission (as discussed in the next section).

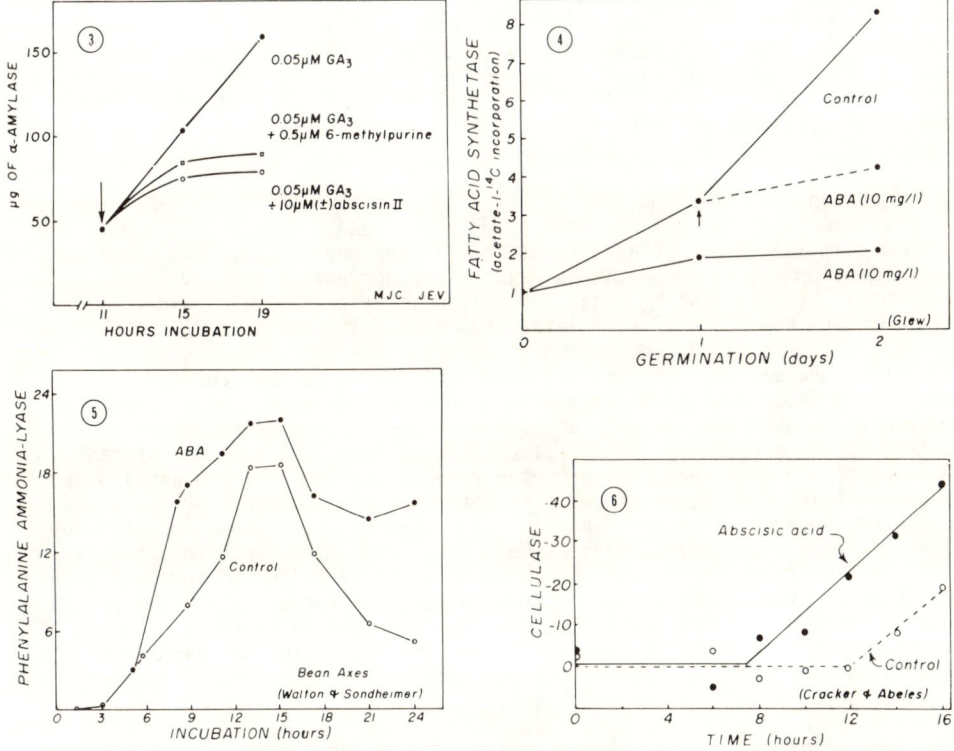

Fig. 3. Inhibition by ABA {(±) abscisin II} of GA-induced synthesis of α-amylase in barley aleurone (from data for Chrispeels & Varner, 1967).

Fig. 4. Inhibition by ABA of development of activity of fatty acid synthetase in germinating castor beans (from data of Glew, 1968).

Fig. 5. Promotion by ABA of the development of phenylalanine ammonia-lyase activity in bean seed axes (from data of Walton & Sondheimer, 1968).

Fig. 6. Promotion by ABA of the development of cellulase activity in the presence of saturating levels of ethylene (control) (from data of Cracker & Abeles, 1969).

IV. ABA AND NUCLEIC ACID METABOLISM

The influences of the plant hormones are essentially catalytic in nature; clues to their mode of action are being sought in the nucleic acid biochemistry which underlies and controls the synthesis of enzymes, the organic catalysts of living things (see: Audus, 1968; Filner et al., 1969; Galston & Davies, 1969). After the broad spectrum of physiological responses to ABA became generally appreciated (see papers in Wightman & Setterfield, 1968), ABA was included in many of the current investigations of the influences of the other hormones on nucleic acid metabolism.

Since in many instances the physiological influences of ABA are antagonistic to the other hormones, it is not surprising that ABA tends to inhibit the nucleic acid changes that are promoted by IAA, GA and CK. Inhibitions of nucleic acid synthesis and/or related protein synthesis by ABA have been found in: (a) germinating seeds and sprouting buds (Chrispeels & Varner, 1967; Glew, 1968; Madison & Rappaport 1968; Villiers, 1968; Khan & Heit, 1969; Khan & Anojulu, 1970; Walton et al., 1970); (b) growing tissues, organs and whole plants (Esashi & Leopold, 1968; van Overbeek et al., 1968; Pearson & Wareing, 1969, 1970; Pilet, 1970; Poulson & Beevers, 1970; H.Smith & G.R.Stewart, unpublished); (c) and mature tissues such as leaf discs and stem sections (Osborne, 1967, 1968; Wareing et al., 1968a,b; Oritani et al., 1969; Mullins & Osborne, 1970). Such inhibitions of nucleic acid metabolism have the net effect of inhibiting or retarding the enzymatically controlled reactions that support the processes of germination and growth.

As for the processes of senescence and abscission, both of which are promoted by ABA, the evidence is still meager, but it indicates that ABA can *promote* changes in nucleic acid metabolism that are associated with these processes. Srivastava (1968) found that as ABA accelerated senescence in detached barley leaves, it also accelerated development of chromatin-associated ribonuclease and deoxyribonuclease. In the synthesis of invertase by sugar cane tissue, Gayler's (1969) evidence indicated that ABA specifically stimulates the translation steps of RNA metabolism. The results of Osborne (1968) are more difficult to interpret. With both leaf rings and abscission zone explants she usually found ABA decreased incorporation of leucine into protein and incorporation of adenine into RNA, at the same time she noted that total uptake of the radioactive substances was slightly enhanced by ABA in all the tissues tested. More recently Khan et al., (1970) showed that in lentil roots, where ABA inhibits growth, ABA also promoted P incorporation into 25S and 18S species of RNA, and modified the ratios of bases in both species. They appear to have observed changes not detected earlier. In bean explants Cracker and Abeles (1969) found ABA induced a rapid development of cellulase, even more rapid than that from saturating levels of ethylene. The increased activity of cellulase in abscission has been shown to result from *de novo* synthesis (Lewis & Varner, 1970) and should thus be a reflection of increased activity of the appropriate 'cellulase-RNAs'. Thus the present evidence suggests that in senescence and abscission ABA can induce the nucleic acid changes necessary to promote the synthesis of the special enzymes which develop in association with these processes.

The results of many of the investigations of ABA and nucleic acid metabolism have given some indication of the step in the molecular biological pathway at which an effect was manifest. As research techniques have increased in precision, the accuracy of the indications have also increased. To date some of the most careful attention has been given to the nucleic acid metabolism involved with GA-induced hydrolase syntheses in aleurone during germination, syntheses strongly inhibited by ABA. The earlier work of Chrispeels & Varner (1967) indicated that ABA might act by inhibiting the synthesis of enzyme-specific RNA molecules. Khan & Anojulu (1970) investigated the nucleotide composition of six rapidly labelled RNA fractions in germinating embryos of pear. Abscisic acid increased the UMP content and decreased the CMP content of the RNA fractions. Also ABA increased the U/G ratio in all fractions, but especially in the "DNA-RNA" fraction. Walton et al., (1970) working with germinating bean axes found ABA inhibited P incorporation into ribosomal RNA and transfer RNA; they suggested that one possible effect of ABA may be at the level of translation. The results of Ihle & Dure (1970) indicate that in embryos of cotton ABA inhibits translation of the protease-RNA. The most recent results from Varner's laboratory (W.H. Evins, unpublished) show that GA rapidly induces polysome formation in barley aleurone and that the induction can be prevented by ABA. Poulson and Beevers (1970) found a similar effect of GA and ABA on the polysomes of barley leaves, and also a general inhibition by ABA of the incorporation of P into a series of RNA fractions. The latter type of general inhibition of RNA synthesis was also found by Wareing et al., (1968b) in radish leaf discs and by H. Smith and G.R. Stewart (unpublished) in *Lemna*. In radish leaf discs ABA reduced the levels of polyribosomes and monoribosomes (Wareing et al., 1968b) but in wheat leaves no effect of ABA on the percentage of polysomes was found (Pearson & Wareing, 1970).

Effects at an earlier step in the molecular biological pathway, transcription, have been detected by Pearson & Wareing (1969) who found in growing radish hypocotyls that ABA could inhibit chromatin dependent RNA polymerase.

Inhibition of DNA synthesis (replication) by ABA has also been observed. The results of the first report (van Overbeek et al., 1968) were difficult to interpret as the effect was observed in growing cultures of *Lemna* and could have been the result of a non-specific inhibition of growth rather than a direct effect on DNA replication. However, the inhibition of DNA synthesis in *Lemna* by ABA has been confirmed in experiments which demonstrated the inhibition of thymidine incorporation (H.Smith & G.R.Stewart, unpublished). Walton et al.,(1970) also observed ABA inhibition of DNA synthesis in growing bean axes; however in their experiments the inhibition was not related to ABA's inhibition of growth. In a related investigation, Haber et al., (1969) showed that ABA inhibits lettuce seed germination in circumstances which do not involve DNA synthesis.

Thus a substantial body of evidence has developed indicating the action of ABA on nucleic acid metabolism as being largely that of inhibiting or modifying the synthesis of one or more major fractions of RNA. Inhibition of DNA synthesis by ABA also occurs, but this inhibition does not yet closely correlate with ABA's physiological effects. Further there is evidence of a dual effect of ABA in some physiological processes such as senescence and abscission, simultaneously inhibiting some aspects of nucleic acid metabolism and promoting others.

The results of several physiological investigations suggest the possibility of still other biochemical effects of ABA. For example, in their work with the growth factors of carrot tissue cultures Bleichert & Steward (1970) found that ABA inhibited the cell enlargement moiety of growth rather than cell division. This suggested that ABA interacts with "growth factor system I" which involves inositol, rather than with "system II", the IAA system (see Steward, 1970). Further, the literature reviewed in this paper has been concerned almost exclusively with factors influencing enzyme synthesis; future investigators might note the observation of Filner et al., (1969) that the control of enzyme degradation is also a very important aspect of the biochemistry of enzymes.

V. SUMMARY

Abscisic acid is a plant hormone that promotes senescence and abscission, initiates and prolongs dormancy, and inhibits growth. The kinetics of the interactions of ABA with the other hormones indicates that for the most part their antagonistic interactions are of a biochemically non-competitive nature, i.e., ABA did not act as the same biochemical site as the other hormones. Inhibition of the synthesis of α-amylase and of protease by ABA has been demonstrated by rigorous methods; inhibition of the development of activity of a number of other enzymes of germination and growth has also been shown. Promotion of the synthesis of cellulase by ABA has been demonstrated as well as the development of phenylalanine ammonia-lyase, invertase and ribonuclease; it is probable that at least the development of other enzymes of senescence and abscission is promoted by ABA. The effects of ABA on enzyme biochemistry are underlain by more fundamental effects on nucleic acid metabolism. The synthesis and composition of several RNA fractions are especially sensitive to ABA and would appear to account for the altered enzyme patterns induced by ABA. The physiological significance of the limited evidence that ABA inhibits DNA synthesis remains obscure. There are some exceptions to the above generalizations; these may ultimately prove as important as the rule.

VI. ACKNOWLEDGEMENTS

The assistance of Alice B. Addicott in the preparation of figures and bibliography, and in proofreading is gratefully acknowledged.

VII. LIST OF REFERENCES

ADDICOTT, F.T. and J.L. LYON (1969). Physiology of abscisic acid and related substances Ann. Rev. Pl. Physiol. 20, 139-64.
ADDICOTT, F.T. H.R. CARNS, J.L. LYON, O.E. SMITH and J.L. McMEANS (1964). On the physiology of abscisins, in "Régulateurs Naturels de la Croissance Végétale" (Ed. J.P. Nitsch) p. 687-703. Coll. Int. C.N.R.S. 123, Paris 1964.
ASPINALL, D., L.G. PALEG and F.T. ADDICOTT (1967). Abscisin II and some hormone-regulated plant responses. Aust. J. biol. Sci. 20, 869-82.
AUDUS, L.J. (1968). Plant growth substances - past, present and future. Adv. Sci. 25, 1-12.
BENNET-CLARK, T.A. and N.P. KEFFORD (1953). Chromatography of the growth substances in plant extracts. Nature 171, 645-7.
BLEICHERT, E.F. and F.C. STEWARD (1970). The inhibitory effects of abscisic acid on the growth and metabolism of cultured carrot tissue. Pl. Physiol., Wash. 46, (Suppl.), 33.
BÖTTGER, M. (1970). Die hormonale Regulation des Blattfalls bei *Coleus rehneltianus* Berger. I. Die Wechselwirkung von Indol-3-essigsaure, Gibberellin-und Abscisinsaure auf Explantate. Planta. Berl. 93, 190-204.
CARNS, H.R., J. HACSKAYLO and J.L. EMBRY (1954). Relation of an indole-3-acetic acid inhibitor to cotton boll development. (Abstr.) A.I.B.S. meetings. Gainsville, Florida.
CHERRY, J.H. (1968). Regulation of invertase in washed sugar beet tissue, in "Biochemistry and Physiology of Plant Growth Substances" (Eds. F. Wightman and G. Setterfield) p. 417-31. Runge Press, Ottawa. 1968.
CHRISPEELS, M.J. and J.E. VARNER (1967). Hormonal control of enzyme synthesis: on the mode of action of gibberellic acid and abscisin in aleurone layers of barley. Pl. Physiol., Wash. 42, 1008-16.
CRACKER, L.E. and F.B. ABELES (1969). Abscission: role of abscisic acid. Pl. Physiol., Wash. 44, 1144-9.
ESASHI, Y and A.C. LEOPOLD (1968). Regulation of tuber development in *Begonia evansiana* by cytokinin, in "Biochemistry and Physiology of Plant Growth Substances" (Eds. F. Wightman and G. Setterfield) p. 923-41. Runge Press, Ottawa. 1968.
FERRARI, T.E. and J.E. VARNER (1969). Substrate induction of nitrate reductase in barley aleurone layers. Pl. Physiol., Wash. 44, 85-8.
FILNER, P., J.L. WRAY and J.E. VARNER (1969). Enzyme induction in higher plants. Science 165, 358-67.
GALSTON, A.W. and P.J. DAVIES (1969). Hormonal regulation in higher plants. Science 163, 1288-97.
GAYLER, K.R. (1969). Mechanisms of regulation in plant development. Ph. D. thesis, Univ. Queensland, Brisbane.
GAYLER, K.R. and K.T. GLASZIOU (1969). Plant enzyme synthesis: hormonal regulation of invertase and peroxidase synthesis in sugar cane. Planta. Berl. 84, 185-94.
GLASZIOU, K.T. (1969). Control of enzyme formation and inactivation in plants. Ann. Rev. Pl. Physiol. 20, 63-88.
GLEW, R.H. (1968). Developmental aspects of lipid metabolism in the developing and germinating castor bean seed. Ph. D. dissertation, Univ. Calif., Davis.
HABER, A.H., D.E. FOARD and S.W. PERDUE (1969). Actions of gibberellic and abscisic acids on lettuce seed germination without actions on nuclear DNA synthesis. Pl. Physiol., Wash. 44, 463-7.
HEMBERG, T. (1946). Wachstumhemmende und wachstumfördernde Stoffe bei der Kartoffel. Arkiv. f. Bot. Stockholm 33 B, Häfte 2, no. 2, 1-4.
HEMBERG, T. (1961). Biogenous inhibitors. Encycl. Pl. Physiol. 14, 1162-84.
HEMBERG, T (1967). Abscisin II as an inhibitor of α-amylase. Acta Chem. Scand. 21, 1665-6.
IHLE, J.N. and L. DURE III (1970). Hormonal regulation of translation inhibition requiring RNA synthesis. Biochem. biophys. Res. Comm. 38, 995-1001.
JACKSON, G.A.D. and J.B. BLUNDELL (1966). Effect of dormin on fruit-set in *Rosa*. Nature, Lond. 212, 1470-1.

JACOBSEN, J.V. (1972). Control of α-amylase synthesis in isolated barley aleurone layers by gibberellic acid, abscisic acid and ethylene in 'Plant Growth Substances 1970' Ed. D.J. Carr.

KHAN, A.A., L. ANDERSON and T. GASPAR (1970). Abscisic acid-induced changes in nucleotide composition of rapidly labeled ribonucleic acid species of lentil root. Pl. Physiol., Wash. 46, 494-5.

KHAN, A.A. anc C.C. ANOJULU (1970). Abscisic acid induced changes in nucleotide composition of rapidly labelled RNA species of pear embryos. Biochem. biophys. Res. Comm. 38, 1069-75.

KHAN, A.A. and C.E. HEIT (1969). Selective effect of hormones on nucleic acid metabolism during germination of pear embryos. Biochem. J. 113, 707-12.

DE LEO, P. and J.A. SACHER (1970a). Control of ribonuclease and acid phosphatase by auxin and abscisic acid during senescence of *Rhoeo* leaf sections. Pl. Physiol., Wash. 46, 806-11.

DE LEO, P and J.A. SACHER (1971). Effect of abscisic acid and auxin on ribonuclease during aging of bean endocarp tissue sections. Plant & Cell Physiol. 12, 791-6.

LEWIS, L.N. and J.E. VARNER (1970). Synthesis of cellulase during abscission of *Phaseolus vulgaris* leaf explants. Pl. Physiol., Wash. 46, 194-9.

MADISON, M. and L. RAPPAPORT (1968). Regulation of bud rest in tubers of potato, *Solanum tuberosum*, L.V. Action of abscisic acid and inhibitors of nucleic acid and protein syntheses. Plant Cell Physiol. 9, 147-53.

MILBORROW, B.V. (1967). The identification of (+)-abscisin II [(+)-dormin] in plants and measurement of its concentration. Planta, Berl. 76, 93-113.

MORRÉ, D.J. (1968). Cell wall dissolution and enzyme secretion during leaf abscission. Pl. Physiol., Wash. 43, 1545-59.

MULLINS, M.G. and D.J. OSBORNE (1970). Effect of abscisic acid on growth correlation in *Vitis vinifera* L. Aust. J. biol. Sci. 23, 479-83.

ORITANI, T., R. YOSHIDA and T. ORITANI (1969). Studies on the nitrogen metabolism of crop plants. VI. Interactive effects of abscisic acid and kinetin on the changes of the amount of chlorophyll and nucleic acid in the rice leaf sections during senescence. Proc. Crop Sci. Soc. Japan 38, 587-92.

OSBORNE, D.J. (1958). Changes in the distribution of pectin methylesterase across leaf abscission zones of *Phaseolus vulgaris*. J. exp. Bot. 9, 446-57.

OSBORNE, D.J. (1967). Hormonal regulation of leaf senescence, in "Aspects of the Biology of Ageing". p. 305-21. Academic Press, New York, 1967.

OSBORNE, D.J. (1968). Hormonal mechanisms regulating senescence and abscission, in "Biochemistry and Physiology of Plant Growth Substances" (Eds. F. Wightman and G. Setterfield). p. 815-40. Runge Press, Ottawa. 1968.

VAN OVERBEEK, J., J.E. LOEFFLER, and M.I.R. MASON (1968). Mode of action of abscisic acid, in "Biochemistry and Physiology of Plant Growth Substances" (Eds. F. Wightman and G. Setterfield) p. 1593-1607. Runge Press, Ottawa. 1968.

PEARSON, J.A. and P.F. WAREING (1969). Effect of abscisic acid on activity of chromatin. Nature, Lond. 221, 672-3.

PEARSON, J.A. and P.F. WAREING (1970). Polysomal changes in developing wheat leaves. Planta, Berl. 93, 309-13.

PILET, P.-É. (1970). The effect of auxin and abscisic acid on the catabolism of RNA. J. exp. Bot. 21, 446-51.

POULSON, R. and L. BEEVERS (1970). Hormonal regulation of nucleic acid metabolism-during photomorphogenesis. Pl. Physiol., Wash. 46, (Suppl.) 20.

ROTHWELL, K and WAIN, R.L. (1964). Studies on a growth inhibitor in yellow lupin (*Lupinus luteus* L.), in "Régulateurs Naturels de la Croissance Végétale" (Ed. J.P. Nitsch) p. 363-75. Coll. Int. C.N.R.S. 123, Paris 1964.

SAUNDERS, P.F. and R.H. POULSON (1968). Biochemical studies on the possible mode of action of abscisic acid: an apparent allosteric inhibition of invertase activity, in "Biochemistry and Physiology of Plant Growth Substances" (Eds. F. Wightman and G. Setterfield) p. 1581-91. Runge Press, Ottawa. 1968.

SMITH, O.E., J.L. LYON, F.T. ADDICOTT and R.E.JOHNSON (1968). Abscission physiology of abscisic acid, in "Biochemistry and Physiology of Plant Growth Substances" (Eds. F. Wightman and G. Setterfield) p. 1547-60. Runge Press, Ottawa. 1968.

SRIVASTAVA, B.I.S. (1968). Acceleration of senescence and of the increase of chromatin-associated nucleases in excised barley leaves by abscisin II and its reversal by kinetin. Biochim. biophys. Acta 169, 534-6.

VAN STEVENINCK, R.F.M. (1959). Factors affecting the abscission of reproductive organs in yellow lupins (*Lupinus luteus* L.) III. Endogenous growth substances in virus-infected and healthy plants and their effect on abscission. J. exp. Bot. 10, 367-76.

STEWARD, F.C. (1970). From cultured cells to whole plants: The induction and control of their growth and morphogenesis. Proc. Roy. Soc. London, B, 175, 1-30.

STEWARD, F.C. and S.M. CAPLIN (1952). Investigations on growth and metabolism of plant cells. III. Evidence for growth inhibitors in certain mature tissues. Ann. Bot. 16, 477-89.

SUTCLIFFE, J.F., P.D. ARCH, P.A. LEGGETT, B.J. PHILLIPS and R. SEXTON (1969). Enzymic changes occurring during the development of the abscission zone of *Coleus blumei*. p. 213 in "Abstracts, XI International Bot. Congr." Seattle. 1969.

THOMAS, T.H., P.F. WAREING and P.M. ROBINSON (1965). Action of the sycamore 'dormin' as a gibberellin antagonist. Nature, Lond. 205, 1271-6.

VILLIERS, T.A. (1968). An autoradiographic study of the effect of the plant hormone abscisic acid on nucleic acid and protein metabolism. Planta, Berl. 82, 342-54.

WALTON, D.C. and E. SONDHEIMER (1968). Effect of abscisin II on phenylalanine ammonia-lyase activity in excised bean axes. Pl. Physiol., Wash. 43, 467-9.

WALTON, D.C., G.S. SOOFI and E. SONDHEIMER (1970). The effects of abscisic acid on growth and nucleic acid synthesis in excised embryonic bean axes. Pl. Physiol., Wash. 45, 37-40.

WAREING, P.F., C.F. EAGLES and P.M. ROBINSON (1964). Natural inhibitors as dormancy agents, in "Régulateurs Naturels de la Croissance Végétale" (Ed. J.P. Nitsch) p. 376-86. Coll. Int. C.N.R.S. 123, Paris 1964.

WAREING, P.F., J. GOOD and J. MANUEL (1968a). Some possible physiological roles of abscisic acid, in "Biochemistry and Physiology of Plant Growth Substances" (Eds. F. Wightman and G. Setterfield) p. 1561-79. Runge Press, Ottawa. 1968.

WAREING, P.F., J. GOOD, H. POTTER and A. PEARSON (1968b). Preliminary studies on the mode of action of abscisic acid, in "Plant Growth Regulators" p. 191-207, Monograph No. 31. Soc. Chem. Ind., London. 1968.

WAREING, P.F. and G. RYBACK (1970). Abscisic acid: a newly discovered growth-regulating substance in plants. Endeavour 29, 84-8.

WIGHTMAN, F. and G. SETTERFIELD (Eds.) (1968). Biochemistry and Physiology of Plant Growth Substances. Runge Press, Ottawa. 1642 pp.

WRIGHT, S.T.C. (1968). Multiple and sequential roles of plant growth regulators, in "Biochemistry and Physiology of Plant Growth Substances" (Eds. F. Wightman and G. Setterfield) p. 521-42. Runge Press, Ottawa. 1968.

PLANT GROWTH SUBSTANCES, 1970

THE BIOSYNTHESIS AND DEGRADATION OF ABSCISIC ACID

B.V. Milborrow

Shell Research Limited, Milstead Laboratory of Chemical Enzymology, Broad Oak Road, Sittingbourne, Kent

I. INTRODUCTION

Many of the questions we have been asking about the role played by abscisic acid (ABA) in plants are at present unanswerable because we do not know enough of the metabolism of the hormone or the factors which influence its transport. As usual, the more we have found out the less sure we are of earlier results.

We have been investigating the biosynthesis and degradation of ABA because it is necessary to know something of both processes before the mechanisms which affect the concentrations in plants can be understood.

II. MATERIALS AND METHODS

Tomato and wheat plants used were grown in a glasshouse with supplementary illumination in winter; imported avocado fruits were bought locally. Isolation techniques and chromatographic procedures were as described previously (Milborrow, 1970; Milborrow and Noddle, 1970).

III. RESULTS AND DISCUSSION

Degradation of Abscisic Acid

At the last conference $(\pm)$-$[2-^{14}C]$ABA was reported to be converted to three products: "Metabolites A, B and C".

"Metabolite B" has been identified (Milborrow, 1970) as abscisyl-β-D-glucopyranoside (I) which had been characterized by Koshimizu et al (1968). We do not know whether this bound ABA is re-used or further metabolised. The glucose ester appears to be stable in the plant but it is rapidly hydrolysed by expressed plant juices. Consequently, the form in which ABA is translocated is now uncertain; although petiole segments absorb $[2-^{14}C]$ ABA from agar and pass $[2-^{14}C]$ ABA out into receiver blocks of agar it is possible that the glucose ester is transported and is hydrolysed by enzymes released from the cut surfaces.

I   (+)-abscisyl-β-D-glucopyranoside

The acetylation of the glucose ester by acetic anhydride in pyridine, followed by
chromatography, gave methyl abscisate by an acid-catalysed trans-esterification
reaction in the methanol in which the material was loaded onto a chromatogram. Later
the supposed Metabolite A was identified as methyl abscisate and was probably formed
by a similar reaction in the acidic, methanolic plant extract and hence is probably
an artefact.

Metabolite C was isolated in crystalline form and showed an intense ORD spectrum
with a positive Cotton effect, closely similar to that of (+)-ABA, but with extrema
at 298nm (+) and 254nm (-). It formed an acetyl derivative, as determined by a
change in Rf. Mass spectrometry of the methyl ester showed that it contained 16 mass
units more (294) than methyl ABA (278) and therefore probably contained an extra oxygen
atom. A major fragment ion at m/e 125 indicated that the ABA side chain was intact
but the methylated material was found to have rearranged to a less polar product.
This rearranged material had the same physical properties, MS, ORD, UV, IR and NMR,
as phaseic acid which had been isolated from bean seeds *(Phaseolus vulgaris)* by
MacMillan and Pryce (1968) and given in structure II

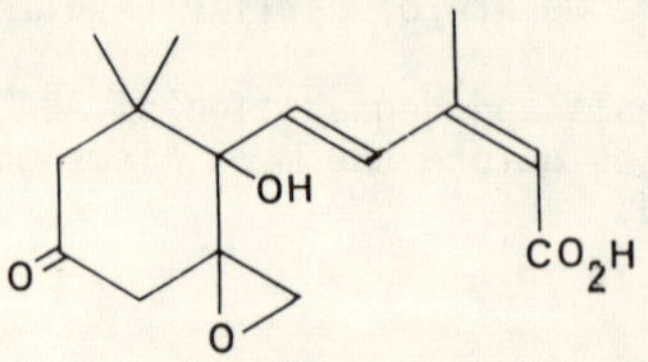

II 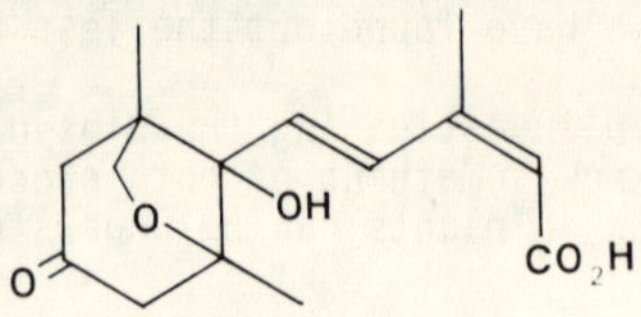   III

It was difficult to understand how Metabolite C, a hydroxylated derivative of ABA,
could rearrange to an epoxide so an alternative interpretation of the structure of
phaseic acid was sought. Professor Cornforth suggested structure III. This could
arise by the hydroxylation of one of the 6' *gem* methyl groups of ABA to give Metabolite
C (IV)

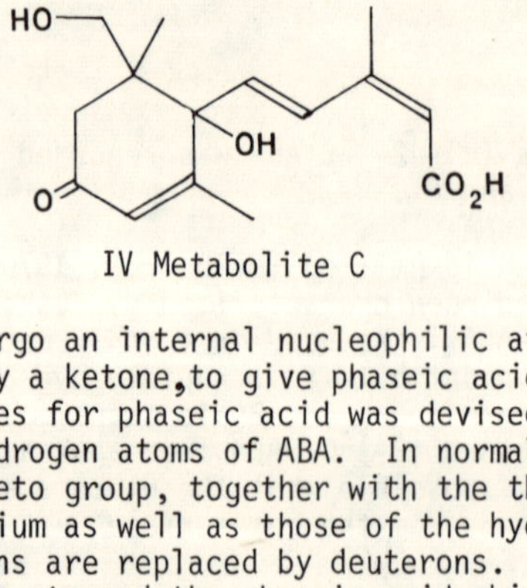

IV Metabolite C

Metabolite C could then undergo an internal nucleophilic attack, of a hydroxyl group
on a double bond activated by a ketone, to give phaseic acid. The crucial test between
the two alternative structures for phaseic acid was devised by making use of a base-
catalysed exchange of six hydrogen atoms of ABA. In normal alkali the three protons
of ABA adjacent to the 4'-keto group, together with the three protons of the 2'-methyl
group, exchange with the medium as well as those of the hydroxyl and carboxyl groups.
In N NaOD in D_2O these protons are replaced by deuterons. Deuterated abscisic acid
(V) was supplied to tomato shoots and the phaseic acid which had been formed from it
was isolated 48 hours later.

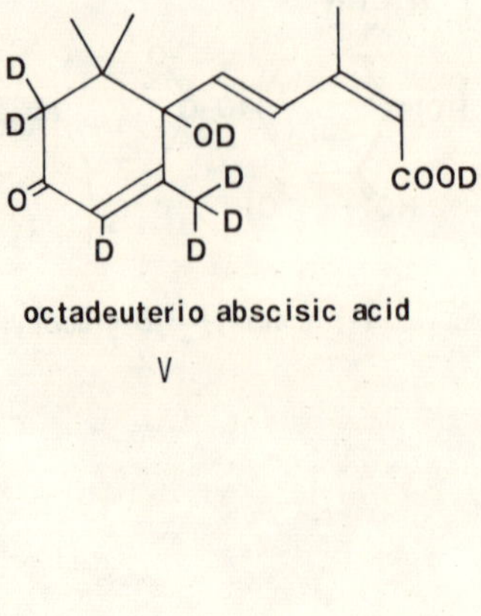

octadeuterio abscisic acid
V

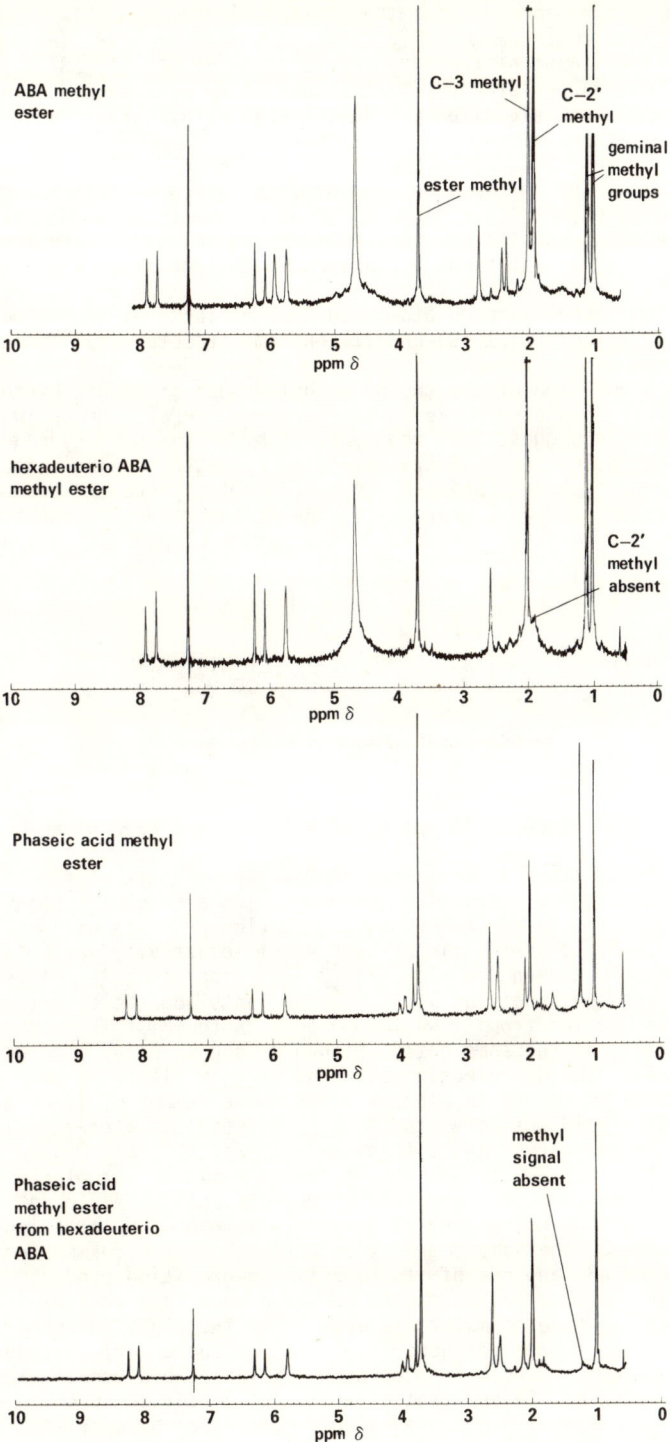

Fig. 1. 100 mHz n.m.r. spectra of methyl esters of abscisic acid (top), hexadeuterio-abscisic acid, phaseic acid and trideuteriomethyl phaseic acid, (bottom) in deuteriochloro form. The signal of the C-2' methyl group of ABA (1.928) is missing from the spectrum of the deuteriated ABA. After oxidation by tomato plants and rearrangement to phaseic acid the signal of the deuteriated methyl group is absent (1.238) from the spectrum of phaseic acid.

If the N.M.R. signals of the apparent *gem* dimethyl group of phaseic acid arose from what had originally been the *gem* dimethyl group of ABA then the NMR signal of these methyl protons would be unaltered. If, however, phaseic acid were formed, as we believed, by an internal attack of a 6'-hydroxymethyl group on C-2', the signal of what had been the C-2' methyl of ABA in the NMR spectrum of phaseic acid should be absent because the three hydrogen atoms had been replaced by deuterium. This can be seen in Figure 1 (Milborrow, 1969) and it confirms structure III for phaseic acid.

We do not know which of the two 6'-methyl groups of ABA is hydroxylated but if the tetrahydrofuran bridge were *cis* to the tertiary hydroxyl group in phaseic acid then molecular models suggest that the hydroxyl hydrogen could hydrogen-bond to the ether oxygen. No indication of intra-molecular hydrogen-bonding can be seen in the infra-red spectrum of phaseic acid at about 35000mm^{-1}, therefore the bridge is probably *trans* to the tertiary hydroxyl group. (+)-ABA is the major, if not the sole, source of phaseic acid therefore the absolute configuration of phaseic is tentatively assigned as VI.

tentative configuration of phaseic acid

VI

Five subsequent attempts to reisolate Metabolite C have given phaseic acid and so its metabolism has not been investigated. The role of phaseic acid also remains obscure, it may be a naturally-occurring degradation product of ABA or it may be an artefact formed during the extraction procedure. An attempt to choose between these two hypotheses was made by comparing the degradation products of [^{14}C]ABA and [^{14}C]-phaseic acid. [2-^{14}C]phaseic acid of high specific activity was obtained from young tomato shoots which had been fed with (±)-[2-^{14}C] abscisic acid. This phaseic acid was then fed to excised shoots of young tomato plants and the distribution of [^{14}C] label in products derived from it was compared with the distribution of labelled products from plants of the same batch which had been supplied with an equal quantity of (±)-[2-^{14}C] ABA. The distribution patterns of [^{14}C]labelled products were different. ABA was converted to glucose ester and a number of other acids, whereas most of the phaseic acid was converted to a hydrolysable, water-soluble material, which may be a glucose ester, and a trace only of another acid was detected on an autoradiogram. This metabolite was absent from the extract of plants supplied with ABA. It appears from this that the fate of phaseic acid is different from that of ABA (Table 1.). This is not a very satisfactory method of determining whether phaseic acid is a precursor of the same degradation products as are formed from ABA because equal quantities of ABA and one of its potential degradation products would probably not occur *in vivo*. Also great care must be taken when interpreting differences between the metabolic fate of materials applied to the plant in comparison with that of the same compounds formed endogenously. For instance, early attempts to obtain incorporation of mevalonic acid into carotenoids failed because of slow penetration and the glucose ester of ABA and that of *trans*-ABA appear to be quite stable in the plant when formed *in vivo* but are rapidly hydrolysed by expressed leaf juices. The differences between the fates of [2-^{14}C]ABA and [2-^{14}C]phaseic acid, therefore, may represent differences in uptake and compartmentalisation within the cells, rather than differences between routes of katabolism. However, other derivatives of ABA are metabolised within a short time of application and the stability of phaseic acid is unlikely to arise from a failure to penetrate cells because a large part of it was converted to a conjugate. At the moment the weight of the evidence suggests that phaseic acid is an artefact.

Table 1. R_fs of labelled, ether-soluble, acidic metabolites formed from $(\pm)$-$[2\text{-}^{14}C]$ ABA and $[2\text{-}^{14}C]$ phaseic acid during 2 and 4 days incubation by excised tomato shoots. The acid fractions were chromatographed in toluene, ethyl acetate, acetic acid (50:20:2, v/v) on Merck F_{254} silica gel TLC plates and the autoradiograms were exposed for three weeks.

Marker Compounds and R_fs		Abscisic Acid	
		Relative Intensities	
		2 Days Incubation	4 Days Incubation
2-*trans*-ABA	0.5		
ABA	0.43	(++++)	(+++)
Phaseic acid	0.26		
GA$_3$	0.12		
R_fs of metabolites			
Phaseic acid	0.26	+++	++
γ	0.19	+	++
β	0.13	+	+
α	0.09	+	+

		Phaseic Acid	
		2 Days Incubation	4 Days Incubation
2-*trans*-ABA	0.51		
ABA	0.43		
Phaseic Acid	0.27	(++)	(++)
GA$_3$	0.12		
Metabolite	0.38	+	+

The two enantiomers of racemic $[2\text{-}^{14}C]$ ABA are treated differently by the enzymes of tomato shoots and after a few hours there is a preponderance of the unnatural (-) enantiomer in the free acid fraction. The missing (+)-ABA was sought in the glucose ester but after hydrolysis and isolation this bound ABA was also found to contain excess (-). The fate of the three components of ABA, endogenous (+), $[2\text{-}^{14}C]$ (+) and $[2\text{-}^{14}C]$ (-) was followed in an experiment whose analysis required the assumption that all the (-)-ABA is exogenous. This assumption is justified because all isolates of natural ABA have been (+) and the specific rotations of a number of samples of crystalline material have been virtually identical. The ABA from a plant which had been supplied with a known quantity of racemic $[2\text{-}^{14}C]$ ABA of known specific activity is isolated and purified until the profile of its UV absorption spectrum is identical with that of authentic ABA. The total quantity of ABA in a methanolic solution can then be calculated from the optical density; the excess (-) in the same solution can be determined by an optical rotatory dispersion measurement (ORD) and finally the amount of $[2\text{-}^{14}C]$ ABA in the same solution can be determined from its radioactivity. From these data it is possible to calculate the amounts of ABA from the different sources.

Total ABA (by UV) − (−)-ABA (by ORD) = Racemate
Racemate is, by definition half (+), half (−).
∴ Total (−)-[2-^{14}C] ABA is given by :−
(−)-ABA (by ORD) + half racemate.
(+)-[2-^{14}C] ABA is given by :−
Total [2-^{14}C] ABA − total (−)-[2-^{14}C] ABA
Remaining endogenous (+)-ABA is given by :−
Total ABA (by UV) − [2-^{14}C]ABA.

The same technique has been applied to the ABA released from the glucose ester by alkaline hydrolysis. The data show an excess of (−)-[2-^{14}C] ABA and most of the missing (+) can be accounted for by the [^{14}C] Metabolite C or phaseic acid (Table 2). Apart from its intrinsic interest this technique has been used to follow the stereochemistry of formation of ABA from synthetic precursors *in vivo*.

Table 2. Measurement of (+)-ABA, (+)-[2-^{14}C] ABA and (−)-[2-^{14}C] ABA as free acids and in the hydrolysate of their abscisyl-β-D-glucopyranosides.
1.5mg (±)-ABA (27.6 μCi/Mm) was supplied to 250g tomato shoots and isolated 15 h later. 37% of the radioactivity was recovered in the final extracts as ABA, glucose ester and "Metabolite C". Autoradiography of chromatograms of the original extracts showed that these were the only radioactive compounds detectable.

		Free ABA μg	ABA from hydrolysis of abscisyl-β-D-glucopyranoside, μg
(a)	Total ABA by UV absorption	199	305
(b)	Excess (−)-ABA over racemate	56.2	203
	Racemate (a−b)	142.8	102
(c)	Total (−)-ABA	127.6	254
(d)	Radioactive ABA	195	269
	(+)-ABA (a−b)	4	36
	(+)-[2-^{14}C] ABA (d−c)	67.4	15
	"Metabolite C"	93	0

Biosynthesis

The structure of abscisic acid immediately suggests that it is a terpenoid and that its carbon skeleton is formed from three isoprene residues by the now well-known reactions. Noddle and Robinson (1969), in Milstead Laboratory first demonstrated that labelled mevalonic acid is incorporated into ABA. Later, Robinson and Ryback (1969) used stereospecifically tritiated mevalonic acids to show that the 2-*cis*-double bond of ABA is probably formed in the *trans* configuration. They did this in an attempt to decide between the two main routes which have been proposed for the biosynthesis of ABA − the "direct synthesis pathway" and the "carotenoid pathway". ABA is envisaged, by the former route, to be elaborated from a fifteen carbon monocyclic precursor which could possibly be formed with a 2-*cis* double bond. The latter route is postulated to occur via an oxidative cleavage of a 40 carbon carotenoid which gives rise to one, or perhaps, two fifteen carbon fragments with the carbon skeleton of ABA and a 2-*trans* double bond. The retention of two 4R tritium atoms from [(4R)-4-^{3}H]MVA per three MVA residues in ABA and the absence of tritium derived from [(4S)-4-^{4}H]MVA indicates that the 2-*cis* double bond of ABA is formed in the *trans* configuration. This result failed to differentiate between the two alternative routes of ABA biosynthesis.

We have made two further attempts to investigate the two pathways. In the first of these, Dr. Robinson, to whom I am grateful for allowing me to cite his unpublished results, prepared [^{14}C] labelled phytoene, an uncyclized 40 carbon carotenoid which is believed to be a precursor of the more familiar 2-ring carotenoids. This compound was supplied to avocado pears together with [2-^{3}H] mevalonic acid and the ABA was isolated after 24 hours incubation. The [^{14}C] label was found to have been incorporated into carotenoids but was absent from ABA. The ABA, however, was heavily labelled with tritium. This result indicates that ABA is synthesised by a route which does not involve a carotenoid. Furthermore, avocado and tomato fruit can synthesise ABA from MVA in darkness so a photolytic step is not obligatory.

The second experiment was carried out in cooperation with Professor P. Jefferies who is spending part of his sabbatical leave in Milstead. We considered that 2-*trans*-cyclofarnesol (VII) could be one of the early intermediates along the "direct synthesis pathway" and Professor Jefferies synthesised the material. It was then supplied as a "cold trap" to avocado fruit tissue which was synthesising [^{3}H] labelled ABA from [2-^{3}H] mevalonic acid and after 24 hours incubation the cyclofarnesol was reisolated and found to contain tritium. In contrast to this result only a negligible amount of tritium was present in cyclofarnesol when it was added as a "cold scavenger". In this part of the experiment the other half of the same avocado was incubated with [2-^{3}H] mevalonic acid but the cyclofarnesol was added after the fruit had been homogenised in methanol. The 2-*trans*-cyclofarnesol from the "cold trap" lost the tritium during successive recrystallisations so it is probably not a natural intermediate although it may resemble an intermediate closely.

The result of the experiment with [^{14}C] labelled phytoene is as expected from the "direct synthesis pathway" hypothesis and argues strongly against the "carotenoid pathway " being the normal route of ABA biosynthesis."

Conversion of Analogues to Abscisic Acid

A number of analogues of ABA have been synthesised and detailed structure/activity relationships have been proposed on the basis of their bioassay results but no account was taken of penetration rates or metabolism of the compounds. A number of analogues of ABA have also been prepared at Woodstock Agricultural Research Centre by Dr. M. Anderson (1969) and one of these (VIII) attracted our attention because its biological activity was almost equal to that of ABA.

VII 2-*trans*-cyclofarnesol   1,2′-epoxy-β-ionylideneacetic acid   VIII

Dr. Ryback and Mr. Mallaby synthesised [2-^{14}C] epoxide and its metabolism has been investigated by Mr. R.C. Noddle. It is converted to ABA in avocado and tomato fruit and also in wheat leaves (Milborrow and Noddle, 1970) so its apparent biological activity can be attributed to the ABA formed from it.

We used wheat leaves because Wright and Hiron (1969) had found that their ABA concentration increased 40-fold on wilting. We were able to show that nine times as much of the [^{3}H] mevalonic acid fed just prior to wilting was incorporated into ABA by wilting leaves in comparison with the other half of the batch which was kept wet and turgid. Thus most, if not all, of the increase can be attributed to biosynthesis of ABA rather than to its release from a conjugate. The technique also provided a means for testing possible biosynthetic intermediates. When the epoxide was supplied to wheat and then half the plants were wilted there was more incorporation of epoxide into ABA by the wilted plants than by the turgid ones. This suggests that the incorporation of the epoxide, or a derivative of it, is regulated by a control mechanism which regulates normal ABA biosynthesis. A "cold trap" experiment carried out by Dr. Robinson failed to demonstrate incorporation of [^{3}H] mevalonic acid into epoxide although [^{3}H]ABA was formed. The epoxide, therefore, is probably not a natural precursor of ABA.

We have found one metabolite of the epoxide; [^{14}C]labelled material derived from [2-^{14}C]epoxide co-chromatographed with ABA in butanol, propanol, ammonia, water (2:6:1:2, v/v; Rf 0.6) and in toluene, ethyl acetate, acetic acid (50:20:2, v/v; Rf 0.3) and, after methylation, with methyl ABA in hexane, ethyl acetate (2:1, v/v; Rf 0.65). After treatment of the eluted ABA zone of the chromatogram with aqueous, methanolic sodium borohydride at 0°, traces of labelled material again chromatographed adjacent to methyl ABA whereas the bulk of the radioactivity present in ABA co-chromatographed with the 1', 4'-*cis* and 1', 4'-*trans* diol esters of ABA. Unlabelled (±)-ABA was added to the eluate containing the unreacted material and treated with borohydride as before but the labelled material was unaffected and unlabelled diol esters were separated by TLC. Miss Garmston isolated a derivative of the material after treatment with acetic anhydride in pyridine. An increase in Rf (to 0.8) after acetylation showed that the metabolite had probably contained a hydroxyl group. Chemical and mass spectrometric evidence suggests structure IX for the original metabolite.

Drs. Taylor and Burden generously provided us with a sample of the O-acetyl methyl ester of IX which they had prepared from violaxanthin (1970). The acetyl methyl ester of the metabolite co-chromatographed with synthetic material. It may be an intermediate between the epoxide (VIII) and ABA and a natural precursor of ABA.

IX

[^{18}O]labelling has also demonstrated that the 1', 2'-oxygen of the epoxide VIII becomes the oxygen of the tertiary hydroxyl group of ABA. Wilting wheat incorporated the epoxide into ABA so that 96% of free ABA was derived from the synthetic analogue as calculated from measurements of the optically active ABA and the [^{14}C] content, and 97% by measurement of ^{18}O contents. The optical activity of the ABA in conjunction with the [^{14}C] and [^{18}O] data enabled us to calculate that only one enantiomer of the epoxide was utilised. Therefore a stereoselective reaction occurs between epoxide and ABA.

Finally the metabolism of the (±)-[2-^{14}C]1', 4'-*cis*- and *trans*-diols (X, XI) was investigated (Noddle, in preparation).

1;4'-*cis*-diol of (-)-abscisic acid

1;4'-*trans*-diol of (-)-abscisic acid

X   XI

We have failed to find any indication that these compounds occur naturally but their biological activity can be attributed to their oxidation to ABA. The conversion is not stereospecific because both *cis* and *trans* epimers of the diols of (+)- and (-)-ABA are oxidised, nor was there any difference between their rates of conversion to ABA in turgid and wilted wheat. Measurements of the optical activity of the ABA extracted, together with its [2-^{14}C]content, showed that the wilted wheat contained very little (+)-ABA formed from unlabelled precursors, less than 8% of the amount biosynthesised in wilted controls without diols (Table 3). It appears, therefore, that the presence of ABA, formed by an adventitious route from an unnatural precursor, "switches off" the wilt-induced biosynthesis. Thus ABA (and possibly its diols) can operate a feed-back mechanism which limits its own biosynthesis. It is also significant that most of the free ABA is (-), it appears, therefore, that the (-) enantiomer is effective at the site of regulation of ABA biosynthesis as well as at the site where it exerts its growth-inhibiting effect.

Table 3. Conversion of (±)-1',4'-*cis*-[2-^{14}C] and (±)-1',4'-*trans*-[2-^{14}C] diols of ABA (21.9 μCi/mM) to ABA by turgid and wilting wheat shoots. 0.96mg of each diol in 50ml 0.01M potassium phosphate buffer, pH 7.0, was absorbed by a 400g batch of wheat shoots during 16 hours at 7°. The batches were divided, 200g of each were wilted, the other two 200g subsamples were supplied with water and remained turgid until all were extracted 4 hours later. Controls were supplied with phosphate buffer only.

Compound Supplied	Treatment	ABA by ORD measurement μg/Kg (+)	ABA by ORD measurement μg/Kg (-)	ABA by [^{14}C] measurement, μg/Kg	Endogenous (+)-ABA present μg/Kg
(±)-1',4'-*cis*-diol	Water	-	180	205	<25
(±)-1',4'-*cis*-diol	Wilt	-	200	193	(0)
(±)-1',4'-*trans*-diol	Water	-	86	163	<77
(±)-1',4'-*trans*-diol	Wilt	-	165	178	<13
Control	Water	1.5	-	-	1.5
Control	Wilt	168	-	-	168

IV. SUMMARY

When racemic [2-^{14}C]labelled abscisic acid is supplied to excised tomato shoots the (-) enantiomer is converted to abscisyl-β-D-glucopyranoside (Metabolite B) faster than the (+). The (+)-ABA is hydroxylated on one of the C-6' *gem* methyl groups to form Metabolite C and this material rearranges to phaseic acid for which a new structure has been proposed.

1',2'-Epoxy-β-ionylideneacetic acid is rapidly converted to ABA in plants and hence its biological activity can be attributed to the ABA formed from it. The oxygen of the epoxy group was shown by [^{18}O]labelling to become the tertiary hydroxyl of ABA.

The 1',4'-*cis*- and 1',4'-*trans*-diols of ABA are also converted to ABA, but equally by wilted and turgid wheat. High concentrations of ABA formed adventitiously from the diols strongly inhibited biosynthesis of ABA. ABA, therefore, (and/or the diols) regulates its biosynthesis by a feed-back mechanism.

The incorporation of [^{14}C]phytoene into carotenoids by avocado fruit which were simultaneously synthesising [^{3}H]labelled but [^{14}C]unlabelled ABA from [2-^{3}H]mevalonic acid suggests that carotenoids are not intermediates in ABA biosynthesis in this tissue.

V. REFERENCES

ANDERSON, M. British Patent No. 1164564.

KOSHIMIZU, K., M. INUI, H. FUKUI, T. MITSUI (1968). Isolation of (+)-abscisyl-β-D-glucopyranoside from immature fruit of *Lupinus luteus*. Agric.biol.Chem. $\underline{32}$, 789-791.

MACMILLAN, J. and R.J. PRYCE (1968). Phaseic acid, a putative relative of abscisic acid, from seed of *Phaseolus multiflorus*. Chem. Comm. 124-126.

MILBORROW, B.V. (1969). Identification of "Metabolite C" from abscisic acid and a new structure for phaseic acid. Chem. Comm., 966-967.

MILBORROW, B.V. (1970). The metabolism of abscisic acid. J. exp. Bot. $\underline{21}$, 17-29.

MILBORROW, B.V. and R.C. NODDLE (1970). Conversion of 5-(1,2-epoxy-2,6,6-trimethyl-cyclohexyl)-3-methylpenta-*cis*-2-*trans*-4-dienoic acid into abscisic acid in plants. Biochem.J., 119, 727-734.

NODDLE, R.C. and D.R. ROBINSON (1969). Biosynthesis of abscisic acid: Incorporation of radioactivity from [2-^{14}C]mevalonic acid by intact fruit. Biochem.J., $\underline{112}$, 547-548.

ROBINSON, D.R. and G. RYBACK (1969). Incorporation of tritium from $[(4R)-4-^3H]$ mevalonate into abscisic acid. Biochem.J., $\underline{113}$, 895-897.

TAYLOR, H.F. and R.S. BURDEN (1970). Xanthoxin, a new naturally occurring plant growth inhibitor. Nature, Lond., 227, 302-304.

WRIGHT, S.T.C. and R.W.P. HIRON (1969). (+)-Abscisic acid, the growth inhibitor induced in detached wheat leaves by a period of wilting. Nature, Lond., $\underline{224}$, 719-720.

The Accumulation of Abscisic Acid in Plants During Wilting and Under Other Stress Conditions

S.T.C. Wright and R.W.P. Hiron

ARC Plant Growth Substance and Systemic Fungicide Unit, Wye College, University of London, Nr. Ashford, Kent, U.K.

I. INTRODUCTION

When excised wheat leaves are wilted and maintained in a wilted condition for a period of three to four hours there is a large increase in the "inhibitor β" content of the leaves. The amount of "inhibitor β" accumulating in the leaves was shown to be dependent on the length of the period in the wilted condition, the temperature of the wilted leaves and the degree of leaf-water deficit (Wright, 1969). Because of the inhibitor's acidic nature and its R_F in several solvent systems, ABA was suspected as being the substance concerned in the "inhibitor β" complex. This was later confirmed by optical rotatory dispersion (Wright and Hiron, 1969).

Although other growth inhibiting substances have been reported to increase during wilting (Pustovoitova, 1967) our results would suggest that ABA is the principal growth inhibitor involved. In addition to wheat leaves we have obtained evidence (ORD measurements and bioassay) of similar increases in the excised leaves of other species including cotton, pea and dwarf bean.

In this paper the study has been extended to the wilting of whole plants.

II. MATERIALS AND METHODS

The degree of water stress

The degree of water stress in plants was estimated by leaf-water deficit (saturation deficit). This is the difference between the water content of the leaf at sampling time and the water content at full turgor, expressed as a percentage of the latter.

The weight of a leaf sample taken from wilted plants was adjusted to allow for the estimated water deficit.

Extraction, purification and bioassay procedure for endogenous ABA

Immediately following treatment the leaf samples were cut into 5 mm strips and extracted for 20 h in cold ether (2^0C) (100 ml/5 g fresh wt.). The ether extract was shaken with two 30 ml volumes of 2% sodium bicarbonate solution. The bulked bicarbonate was acidified to pH 3 and shaken with two volumes of 200 ml ether. These extracts containing the acidic substances were taken to dryness. Each acid fraction was purified on a fluorescent silica gel coated TLC plate developed in a mixture of n-butanol, n-propanol, 0.88 ammonia and water (2:6:1:2 v/v) and the zone corresponding to ABA marker spots (that is, quenched spots under u/v) were scraped off and eluted with ethanol. The eluate was re-applied to another TLC plate which was developed six times in a mixture of chloroform, benzene, and acetic acid (100:100:1.5 v/v). The zone in the ABA region was eluted again. The eluates were then applied to a paper chromatogram (Whatman No. 1) which was developed in an ascending manner in a mixture of isopropanol,

0.88 ammonia and water (10:1:1 v/v) allowed to run 20.5 cm from the origin. After drying overnight the chromatograms were divided into 21 horizontal sections, but only those sections corresponding to the ABA zone (i.e. R_F 0.4 - 0.7) were bioassayed. Each section was placed in a plastic dish containing 1 ml distilled water and bioassayed in the wheat coleoptile straight growth test (Wright, 1969). The growth inhibitory activity per segment was converted into µg equivalents of ABA using a dosage response curve of ABA (Fig. 1).

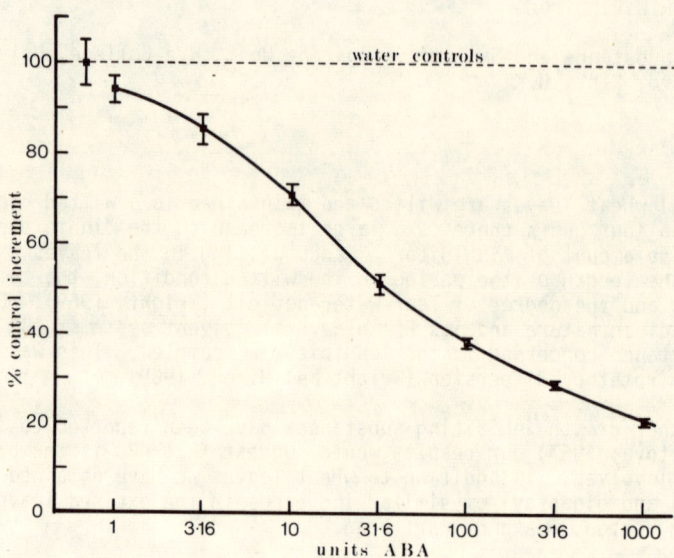

Fig. 1. Dosage response curve for ABA in the wheat coleoptile straight growth test. The vertical lines indicate the 2 S.E. limits of the ABA growth increments (1 unit = 0.00528 µg/ml ABA).

III. RESULTS AND DISCUSSION

ABA content of wheat seedlings wilted in carbowax solutions

Wheat seedlings (cv. Hybrid 46) were grown with their roots in water in a room providing 14 h photoperiods (800 lumens/ft^2 from fluorescent tubes giving "natural daylight") at 22.5°C (night temperature 20°C) and a relative humidity of 75%. The seedlings were supported on rafts kept afloat by polystyrene beads (Taylor and Knight, 1968). After 10 days some of the rafts were carefully lifted off the water, the roots of the seedlings dried with filter paper, and the rafts refloated on carbowax solutions (275 g/l) for 4 h (Carbowax, M 20, Gurr Ltd., London). The leaves were seen to wilt visibly within 30 min. and at the end of the 4 h treatment had a leaf-water deficit of 9%. This degree of water stress had been found to be optimal for ABA accumulation in excised wheat leaves and therefore was applied again to these plants by use of predetermined concentrations of carbowax.

Samples (10 g) of leaves were taken from the wilted plants, and from plants kept continuously on water (i.e. non-wilted controls). In order to study a period of recovery from wilting, some of the rafts were transferred from the carbowax back to water, after first washing the seedlings' roots thoroughly. These plants were left for a further 3 h in which time they regained their full turgor. Duplicate samples were extracted of all treatments.

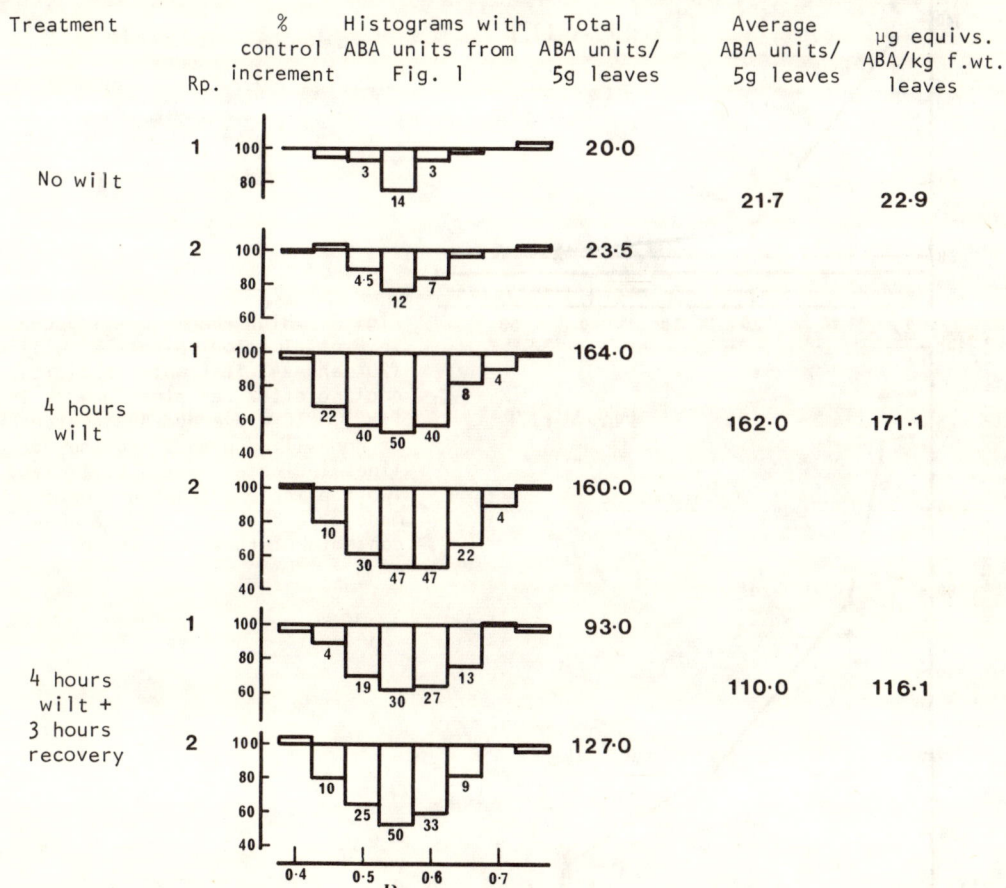

Fig. 2. The ABA content of leaves sampled from wheat seedlings wilting in carbowax solutions, and after a recovery period in water. The figure illustrates how the growth inhibitor as shown in the chromatogram histograms is converted into μg equivalents ABA using Fig. 1 (R_p = replicate extractions).

Histograms of the growth inhibiting activity of the extracts are shown in Fig. 2. The ABA content of the wilted leaves increased 8-fold. After a recovery period of 3 h, during which time the plants regained their full turgor, 63% of the accumulated ABA was still present. A 9% leaf-water deficit is typical of a temporary or transient wilt and wilts of this kind are common on a hot sunny mid-summer day in temperate regions.

ABA content of young Brussels sprout plants during a recovery period from permanent wilting

Leaf-water deficits of 20% and 44% were induced in 4-week old greenhouse-grown Brussels sprout plants (cv. Prize Taker) by withholding water for 64 and 72 h respectively. The wilted plants were then thoroughly watered and duplicate 20 g samples taken at intervals, over several days, to study the after-effects of wilting on ABA content. Other plants which had not been wilted were sampled concurrently as controls.

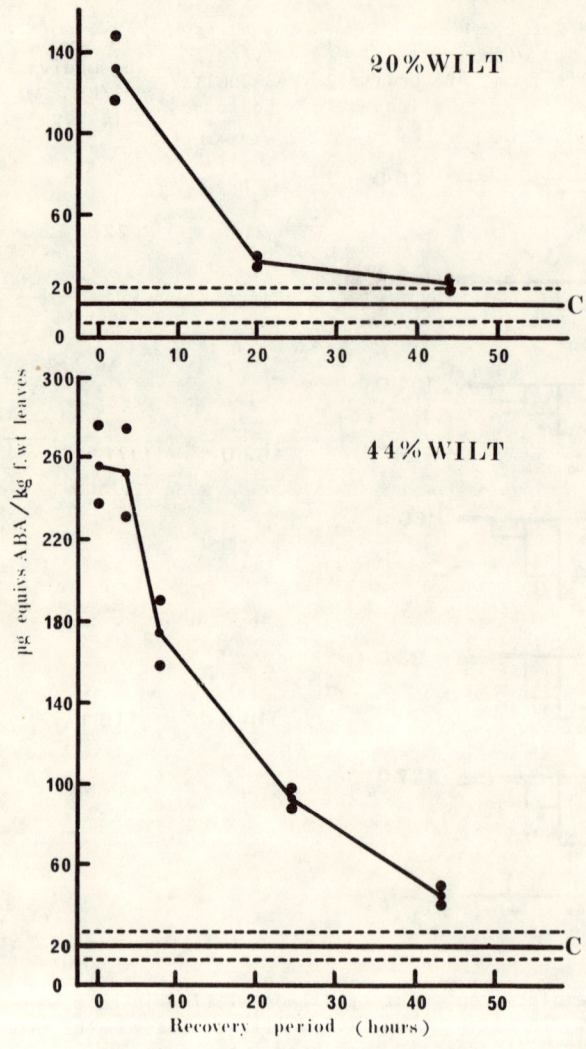

Fig. 3. High ABA levels induced in Brussels sprout plants by wilting (20% and 44% leaf-water deficits) decline after re-watering (0 h in fig.). (C = average ABA content of non-wilted plants and the dotted lines indicate the 2 S.E. limits). Two replicate extractions were made at each stage and the ABA contents of these are shown.

At the end of the wilting treatment (0 h in Fig. 3) there was a large increase in the ABA content of the wilted plants representing an 8-fold and a 13-fold increase over control levels for the 20% and 44% wilts respectively. It was not until full turgor had been attained, usually a period of 2 - 4 h from the time of re-watering, that the ABA levels began to fall rapidly (see 44% wilt curve). The ABA content remained significantly above the control levels for a long period after re-watering. The ABA content of the plants recovering from the 44% leaf-water deficit was still twice that of the controls after nearly two days.

ABA content of dwarf bean plants exposed to a continuous stream of warm air

Dwarf bean plants (*Phaseolus vulgaris* L. cv. Canadian Wonder) were raised in 9 cm pots containing John Innes potting compost until the first two primary leaves were approximately half expanded (9 days in a growth room, as above). The plants were then grouped together and exposed to continuous streams of warm air from electric hair driers spaced 40 cm away from the leaves. The temperature of the air in the vicinity of the leaves was $33^{\circ}C$. Within 20 min the leaves began to curl and wilt severely and the leaves remained in this condition for a further 20 min. Then slowly they began to regain turgor and by 60 - 90 min were fully turgid despite the continuous passage of warm air.

Duplicate samples of leaves (10 g) were taken at intervals during this experiment. One sample was extracted in ether immediately and the other sample allowed to incubate for a further 4 h at $23^{\circ}C$ in a closed container, in order to allow the ABA to accumulate and to see whether the ABA inducing system had been "turned on or off". Silicone rubber imprints of the leaves were made from each sample to study the stomatal apertures. The endogenous ABA contents are shown in the histograms Fig. 4.

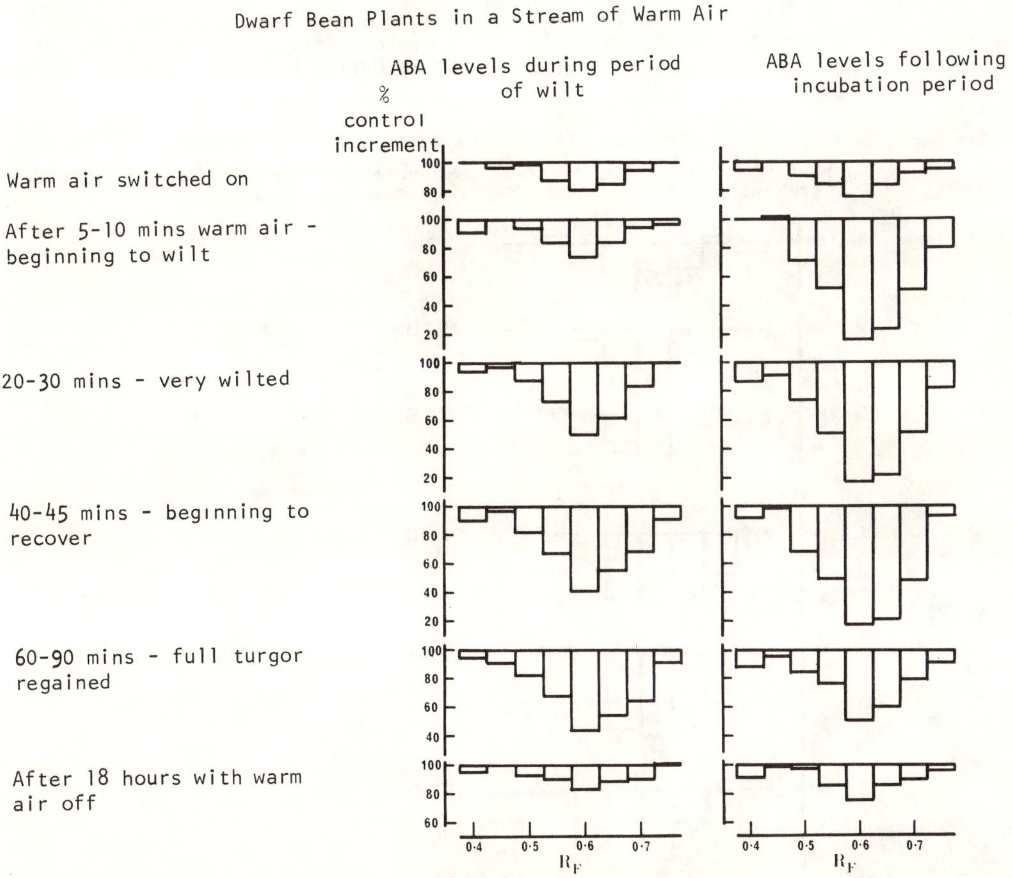

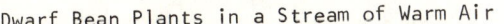

Dwarf Bean Plants in a Stream of Warm Air

Fig. 4. ABA contents of dwarf bean plants, wilting and recovering in a continuous stream of warm air.

The initial low levels of ABA were accompanied by stomata which were partially open. Soon after the warm air was switched on the stomata opened fully. After 20-30 min. when the leaves had developed a severe water-deficit there was an appreciable increase in the level of ABA and this was further increased by incubation. At this stage the stomata were closing. After 40 - 45 min. when the leaves were beginning to regain turgor the stomata were tightly closed and the ABA levels high, even in the sample not incubated. At 60 - 90 min. when full turgor had been regained the ABA levels were still high and a period of incubation made little significant difference. At this stage the warm air was switched off and the stomata were still closed. Next day a further leaf sample was taken, (warm air switched off 18 h) the ABA content was back to normal and the stomata displayed their original apertures.

The effect of waterlogging on ABA levels in dwarf bean plants

Dwarf beans (cv. Canadian Wonder) were sown 2.5 cm deep in seed trays of vermiculite. The plants were grown under the stated conditions in a growth room for nine days. Half the seed trays were then flooded with water for a further 5 days. At the time of flooding the primary leaves were approximately half expanded. Triplicate 10 g leaf samples were taken of both flooded and normally grown plants. The leaves of the waterlogged plants were a deeper green and had not expanded as much as the control leaves. The histograms showing the ABA activity of the leaf extracts are shown in Fig. 5. The ABA levels of the waterlogged leaves were five times greater than those of the control plants.

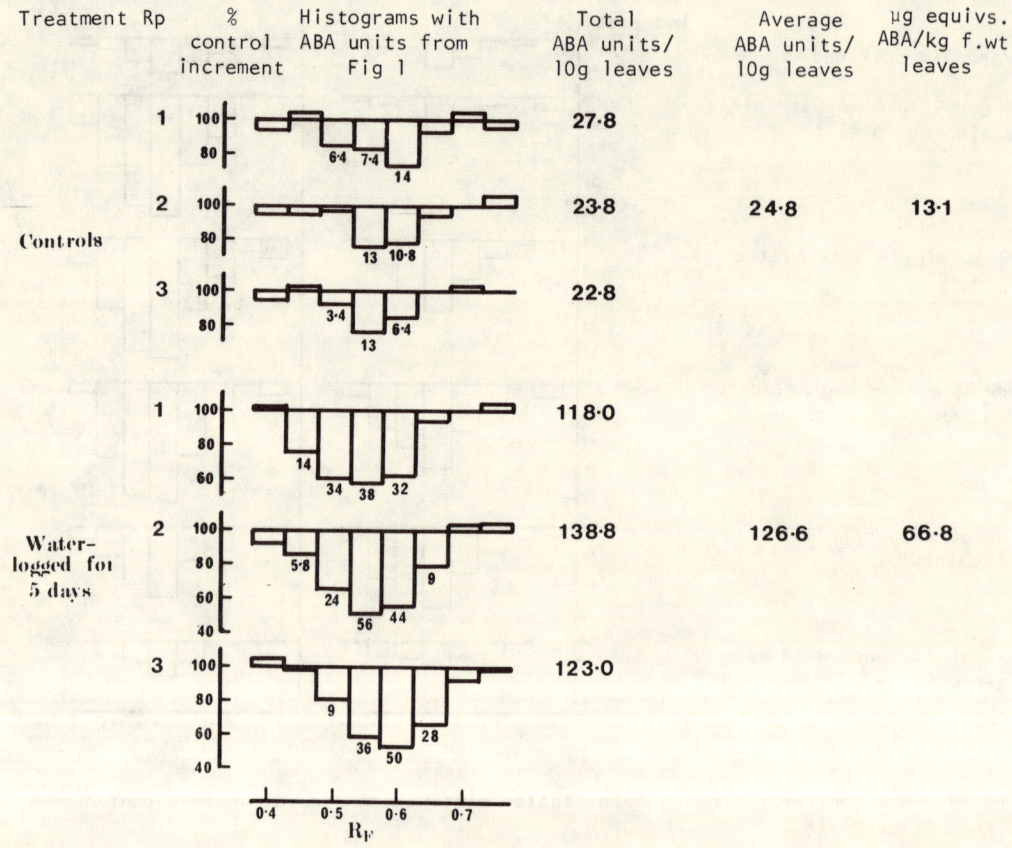

Fig. 5. The effect of 5 days waterlogging on the ABA levels in dwarf bean plants.

Flooding of plants soon leads to an inadequate aeration of the roots. Consequently, there is a decreased permeability of the root tissues to water, an associated decreased absorption and a resulting leaf-water deficit (Kramer, 1969). Thus the increase in ABA levels of the waterlogged plants could have resulted from a leaf-water deficit in the same way as in the more obvious wilting conditions studied in the previous experiments. We also have evidence that this type of wilting (i.e. incipient wilting) can lead to an increase in the content of ABA in tomato plants. Plants grown in a growth room having a relative humidity of 68% contained twice as much ABA as plants grown at a relative humidity of 94% (unpublished results).

IV. GENERAL DISCUSSION

The induction of an accumulation of ABA by a period of wilting, first shown in detached leaves (Wright, 1969; Wright and Hiron, 1969), has now been confirmed in whole plants. Leaf-water deficits of varying degrees of severity have been induced in plants by exposing them to a variety of stress conditions, and all have resulted in an accumulation of ABA within the plants. As in detached leaves the increase in ABA was dependent on the magnitude of the wilt and its duration.

The uptake of ABA solutions by excised cuttings or leaves has been reported to lead to a reduced transpiration rate (e.g. Little and Eidt, 1968) and stomatal closure (e.g. Mittelheuser and Van Steveninck, 1969) and we have established a correlation between increase in endogenous ABA content and stomatal closure in dwarf bean plants. Taken together these observations provide strong grounds for believing that under stress conditions the closure of stomata in plants is controlled by endogenous ABA levels. Furthermore, as demonstrated in the Brussels sprout seedlings experiment the ABA which accumulates during wilting disappears very slowly. The ABA content of plants which had experienced a 44% leaf-water deficit was still twice that of the controls two days after re-watering. Because of this lag period we should expect the opening of stomata to be inhibited for several days after a serious water stress and, indeed, other workers have reported such a phenomenon (e.g. Fischer, Hsiao, and Hagan, 1970). The slow dispersal of ABA after recovery from wilting would also be beneficial to a plant after transplanting, allowing it time to re-establish its rooting system (unpublished results).

The alleviation of stress may be one of the major functions of ABA in plants. As suggested by the dwarf bean warm air experiment, the accumulation of ABA helped the plants to reduce their transpiration rate sufficiently to recover from wilting, even though the stress-inducing factor was still present (i.e. stream of warm air).

Jones and Mansfield (1970) noted that the application of ABA levels sufficient to cause stomatal closure for several days had no observable effect on growth rate. Thus the exciting possibility arises that ABA could be added to overhead irrigation water. Moreover, since cytokinins (e.g. Livné and Vaadia, 1965) have been reported to stimulate stomatal opening, a computerised system controlled by a unit monitoring the total crop environment could achieve a fine control of transpiration rate in a crop by adding cytokinin or ABA to the irrigation water. This could lead to a dramatic reduction in the water needed per acre.

V. SUMMARY

Leaf-water deficits were induced in plants by i) immersing the roots in carbowax, ii) withholding water, iii) subjecting plants to a stream of warm air and iv) water-logging. All these treatments stimulated many-fold increases in ABA. The greater the degree of wilt (i.e. without desiccation) and the longer the duration, the greater the ABA accumulation.

Dwarf bean plants exposed to a *continuous* stream of warm air, wilted and then recovered within 90 min. ABA increased during the wilting and, it is proposed, this was responsible for causing the stomata to close, or to be held closed, enabling the plants to regain full turgor.

Plants wilted by withholding water soon recovered full turgor upon re-watering, but the high ABA levels did not return to normal for several days. This lag period probably explains reports in the literature of stomata remaining closed and insensitive to light for long periods after recovery from wilting.

An increase in ABA was observed in waterlogged plants. This increase probably resulted from an incipient wilting of the leaves caused by a reduced efficiency in water uptake. An increase in ABA could stimulate stomata closure and thereby help to reduce the water stress.

The alleviation of water-stress may be one of the major functions of ABA in plants.

VI. ACKNOWLEDGEMENTS

We should like to express our thanks to Professor R.L. Wain, F.R.S., for his advice and encouragement in these investigations.

VII. REFERENCES

FISCHER, R.A., T.C. HSIAO and R.M. HAGAN (1970). After-effect of Water Stress on Stomatal Opening Potential. J. Exp. Bot., 21, 371-404.
JONES, R.J., and T.A. MANSFIELD (1970). Suppression of Stomatal Opening in Leaves treated with Abscisic Acid. J. Exp. Bot., 21, 714-719.
KRAMER, P.J. (1969). Plant and Soil Water Relationships: A Modern Synthesis. McGraw-Hill, New York, pp. 201-207.
LITTLE, C.H.A. and D.C. EIDT (1968). Effect of Abscisic Acid on Budbreak and Transpiration in Woody Species. Nature, 220, 498-499.
LIVNE, A. and Y. VAADIA (1965). Stimulation of Transpiration Rate in Barley Leaves by Kinetin and Gibberellic Acid. Physiol. Plant., 18, 658-664.
MITTELHEUSER, C.J. and R.F.M. VAN STEVENINCK (1969). Stomatal Closure and Inhibition of Transpiration induced by (RS)-Abscisic Acid. Nature, 221, 281-282.
PUSTOVOITOVA, T.N. (1967). Formation of Growth Inhibitors in Wilting Apricot Leaves. Fiziologiya Rastenii, 14, 90-97.
TAYLOR, H.F. and B.E.A. KNIGHT (1968). Wheat Seedling Method. Agriculture Handbook No. 336. U.S. Dept. Agric. pp. 69-72.
WRIGHT, S.T.C. (1969). An Increase in the "Inhibitor β" Content of Detached Wheat Leaves Following a Period of Wilting. Planta, 86, 10-20.
WRIGHT, S.T.C. and R.W.P. HIRON (1969). (+)-Abscisic Acid, the Growth Inhibitor induced in Detached Wheat Leaves by a Period of Wilting. Nature, 224, 719-720.

CHEMISTRY AND BIOLOGICAL ACTION OF PODOLACTONES AND OTHER
INHIBITORS OF PLANT GROWTH

Jenneth M. Sasse[1], M.N. Galbraith, D.H.S. Horn[2], and D.A. Adamson[3],

[1] School of Biological Sciences, Macquarie University, North Ryde, N.S.W. Australia

[2] C.S.I.R.O., Division of Applied Chemistry, P.O. Box 4331, Melbourne, Australia

[3] School of Biological Sciences, Macquarie University, North Ryde, N.S.W. Australia

I. INTRODUCTION

The responses to auxin and gibberellin of a system of excised stem segments from etiolated dwarf peas have been described previously (Adamson, Low and Adamson, 1968). In this system, differences in the effects of these two hormones could be distinguished on excised hook and tip segments.

This system appeared to be one that might be developed as a primary screen for the discovery of new plant growth regulators that affect the expansion growth of plant cells. It was hoped to distinguish between the effects of the various types of plant growth regulators so that compounds that had interesting structures could be tested quickly and easily for plant growth regulatory activity in a single assay system.

The responses of the system to plant growth regulators other than auxin and gibberellin were studied, and it was found that many of their effects were distinctive.

II. MATERIALS AND METHODS

The methods of growth and harvest of the dwarf pea plants var. "Greenfeast" have been described previously (Adamson, Low and Adamson, 1968). It was essential to support the hook segments at the surface of the treatment solution; any variation of this parameter of the method affected the growth of controls. For routine screening purposes, hook segments were grown in continuous undefined white fluorescent light of dim but undefined intensity, while tip segments were grown in the same light, but at greater intensity. Compounds were tested preliminarily as "saturated" solutions in 7.5 ml of 2% w/v sucrose by placing a small amount (ca. 0.1 mg) of material directly into the 5 cm dish in which the segments were grown. This method can give solutions above 10 ppm, depending on the solubility of the compound, and is very economical on material, a point which is important when working with new, rare, or heat- or light-sensitive compounds. Signs of promotive or inhibitory activity were followed up by tests with three replicates of each treatment solution, at known concentrations (usually a range from below 10^{-6}M to 10^{-5}M was covered). Ten segments were allocated randomly to each treatment solution and initial measurements of the total fresh weight, length and width for each group of ten were taken. Hooks were grown for 24-30 hours and tips for 72 hours, then final measurements taken. Growth was expressed as increase/initial value as a percentage. Where possible, the results were submitted to analyses of variance and control ranges or significant differences between means calculated at the 5% level. The basal swellings or "knobs" were scored in arbitrary units where the initial appearance of the tip was scored as "0", control as "2", 2,4-D at 1 ppm as "4" and gibberellin at 1 ppm as "1", as illustrated previously (Adamson, Low and Adamson, 1968).

III. RESULTS AND DISCUSSION

The hook and tip sections showed a distinctive response to cytokinin (Table 1). Lateral enlargement of the hook segments was promoted by zeatin (Calbiochem. Lot 800848), but not elongation. Pronounced enlargement of the swellings at the base of the tips was induced, and these swellings were different in appearance from those produced by auxin, which are more bulbous. Examination of hand sections showed that cytokinin-treated tissue contained cortical cells that were more lignified than those of controls, while the auxin-treated tips had swellings in which the corresponding cells were less lignified than those of controls. There was no obvious change in cell number in either case. Both hook and tip segments were sensitive to kinetin as well as to zeatin, but the response to concentration was not as linear.

Table 1. Effect of Zeatin on Excised Pea Stem Segments

HOOKS	Increase/Initial Value (%)		
	Fresh Wt.	Length	Diameter
Control	66 ± 5*	28 ± 7*	14 ± 7*
10^{-7}M	77	29	14
10^{-6}M	81	25	18
10^{-5}M	91	19	38
TIPS			Knobs #
Control	40 ± 10*	15 ± 7*	2
10^{-7}M	56	16	3
10^{-6}M	60	22	4
10^{-5}M	76	16	5

Arbitrary units

Ethylene (Commonwealth Industrial Gases, Pty. Ltd. Sydney) while inhibiting the increases in fresh weight and length of both segments, affected tip tissue preferentially. The formation of the basal swelling of the tip was also inhibited. Tip tissue treated with the gas resembles that cultured in the dark (Adamson, Low and Adamson, 1968) except that ethylene does not prevent the greening of the tissue on exposure to light. No effect on diameter was observed after treatment of the hook tissue with ethylene (Table 2) or with Ethrel (Amchem. Inc.) over a wide range of concentrations.

Table 2. Effect of Ethylene on Excised Pea Stem Segments

HOOKS	Increase/Initial Value (%)		
	Fresh Wt.	Length	Diameter
Control	65 ± 6*	35 ± 5*	20 ± 5*
0.14 ppm	61	29	20
1.2 ppm	44	19	19
L.S.D.*	11	9	9
TIPS			Knobs (arb. unit)
Control	39 ± 4*	20 ± 5*	2
0.14 ppm	26	12	1
1.2 ppm	16	9	0
L.S.D.*	7	9	

The mixture of racemates of abscisic acid (Reynolds Tobacco Co. Lot 371) inhibited the growth of both types of segment (Table 3) including knob formation, but the sections became green, so that they resembled control segments, but were smaller. Separation of the mixture into the racemates of the geometrical isomers showed that the *trans, trans* racemate was virtually inactive in this system.

Table 3. Effect of ABA[a] on Excised Pea Stem Segments

HOOKS	Increase/Initial Value (%)		
	Fresh Wt.	Length	Diameter
Control[b]	65 ± 10*	31 ± 5*	11 ± 6*
10^{-6}M[b]	62	32	8
10^{-5}M[b]	48	25	10
10^{-4}M[b]	33	20	3
Control	58 ± 3*	24 ± 3*	
2×10^{-5}M[c]	31	16	
2×10^{-5}M[d]	41	18	
L.S.D.*	6	4	
TIPS			Knobs (arb. unit)
Control	40 ± 7*	17 ± 4*	2
10^{-6}M	41	18	2
10^{-5}M	34	14	1
10^{-4}M	19	8	0
Control	33 ± 4*	12 ± 3*	1
2×10^{-5}M[c]	24	9	0
2×10^{-5}M[d]	29	11	0-1
L.S.D.*	7	6	

a. 50:50 mixture of synthetic racemates
b. Initial measurements in light, growth in dark
c. (±) - *cis, trans* racemate
d. (±) - *trans, trans* racemate

The effects of some of the main plant growth regulators on the system are summarized below (Table 4) and it can be seen that, provided appropriate concentrations are used, the use of both types of segment allows one to distinguish most kinds of regulators by their different effects. The characteristic patterns of growth are distinctive over a wide range of concentrations, but some relative inhibitions occur with high concentrations of auxin and gibberellin.

One compound that was examined because its structure was of interest was nagilactone C, a norditerpenoid dilactone originally isolated from *Podocarpus nagi* (Hayashi, Takahashi, Ono and Sakan, 1968) but found since in other *Podocarpus* species. It is a strong inhibitor in our screening system (Galbraith, Horn, Sasse and Adamson, 1970).

This compound was unusual in affecting the expansion growth of the hook segments more than that of the tips, preventing the formation of the swelling at the base of the tip, and inhibiting the greening of the stem tissue on exposure to light.

Table 4. Effects of Growth Regulators at $10^{-6} - 10^{-5}$M on Pea Stem Segments as compared with Controls

Regulator	HOOKS			TIPS		
	Fr. Wt.	Length	Diam.	Fr. Wt.	Length	Knob
Auxin	+	+	+	+	-	+
Gibberellin[a]	+	+	o	+	+	-
Cytokinin	+	o	+	+	o	+
Ethylene	-	-	o	-	-	-
ABA	-	-	-	-	-	-
TIBA[b]	o	o	o	o	o	-
Morphactin[c]	o	o	o	-	-	-

+ = increased, o = no change, - = decreased

a. Gibberellic acid. (K & K Laboratories, Inc. Plainview, N.Y. Lot. 83235. (mainly GA_3)).
b. 2,3,5-Triiodobenzoic acid (C.S.I.R.O., Div. of Plant Ind.)
c. E. Merck A.G., Darmstadt.

Following the discovery of the novel inhibitory activity of nagilactone C, a search was made for other such diterpenoid dilactones in *Podocarpus* extracts. So far, 16 such lactones, from both Australia and Japan, have been screened, and most have been found to be active. The effects of one of the most active, podolactone DHR21A, are shown below (Table 5).

Table 5. Effect of Podolactone DHR 21A on Excised Pea Stem Segments

HOOKS	Increase/Initial Value (%)		
	Fresh Wt.	Length	
Control	51 ± 3*	19 ± 3*	
10^{-7}M	50	19	
10^{-6}M	24	10	
10^{-5}M	5	2	
L.S.D.*	6	6	
TIPS			Knobs (arb. unit)
Control	46 ± 5*	18 ± 2*	2
10^{-7}M	42	20	1-2
10^{-6}M	29	10	0
10^{-5}M	28	7	0
L.S.D.*	10	4	

The effects of this compound exemplified those of the whole group. Hook tissue was kept pale, expansion was strongly inhibited and straightening did not occur at the higher concentrations. With tip tissue, knob formation was prevented, and toxic effects, such as bleaching and infection were observed at the higher concentrations, while at the lower concentrations, leaf spread appeared to be promoted.

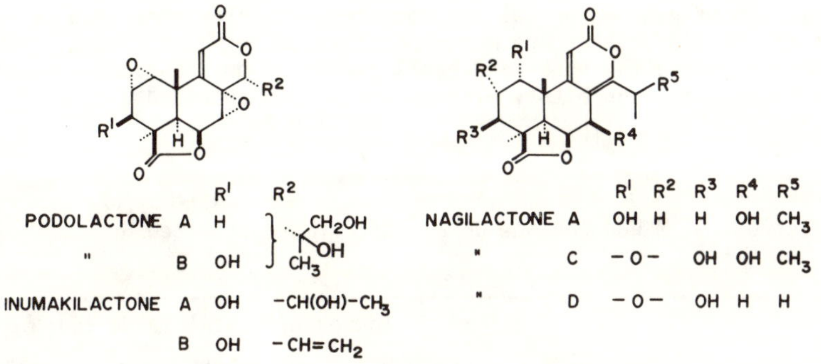

	R^1	R^2			R^1	R^2	R^3	R^4	R^5
PODOLACTONE A	H	CH_2OH	NAGILACTONE A		OH	H	H	OH	CH_3
" B	OH	CH_3 /OH	"	C	—O—		OH	OH	CH_3
INUMAKILACTONE A	OH	$-CH(OH)-CH_3$	"	D	—O—		OH	H	H
B	OH	$-CH=CH_2$							

Fig. 1. Structures of norditerpenoid dilactones isolated from *Podocarpus* species.

The detailed structures of seven of these compounds (Figure 1) have been elucidated (Galbraith, *et al*, 1970; Hayashi, *et al*, 1968; Itô, *et al*, 1968; Itô, *et al*, 1971) and six show a wide range of activity (Table 6).

Table 6. Effects of Inhibitors on Pea Stem Segments

COMPOUND	Increase/Initial Fresh Weight (%)			
	HOOKS		TIPS	
Control	100 ± 12*		100 ± 14*	
	$10^{-6}M$	$10^{-5}M$	$10^{-6}M$	$10^{-5}M$
Nagilactone A		79		72
Podolactone B	82	60	100	70
Nagilactone C	89	53	80	70
Podolactone A	88	37	81	68
Inumakilactone A	66	34	73	46
Inumakilactone B	51	18	86	48
Lycoricidinol	47	37	58	46
(±) - *cis, trans*- Abscisic Acid[a]		54		43

a) While the type of inhibition caused by ABA in this system is distinct from that due to the other inhibitors listed, it has been included as a convenient standard inhibitor for comparison.

Lycoricidinol, a plant growth inhibitor active in the *Avena* coleoptile, rice seedling and tobacco callus tests (Okamoto, Torii and Isogai, 1968) has the characteristic podolactone-type inhibitory effect in our system (Table 6). Inspection of models reveals that its structure resembles those of the podolactone group. From these results it is suggested that the most important molecular requirements for activity may be the tricyclic ring system, unsaturation and oxygen functions in ring C, while the arrangement of the unsaturation and oxygen function in the 7-, 8-, and 14-positions also seems important.

In our system of excised apical and hook segments from etiolated dwarf peas, those regions of the tissue that are most sensitive to the promotion of growth by auxin or cytokinin, viz., the hooks and basal swellings on the tips, are the ones that are most sensitive to the inhibitory effects of the new compounds. The effects of treatment of these tissues with mixtures of promotive hormone and inhibitor are shown below (Table 7).

Table 7. Effect of 10^{-6}M Podolactone DHR21A on Hormone-promoted Growth of Pea Stem Segments

HOOKS	Increase/Initial Value (%)		
	Fresh Wt.	Length	Diameter
Untreated	49	28	6
10^{-6}M GA	78	56	6
10^{-6}M IAA	72	32	20
10^{-5}M Zeatin	75	26	19
10^{-6}M DHR21A	26	16	0
IAA + DHR21A	27	18	5
GA + DHR21A	36	24	0
Zea. + DHR21A	37	16	10
L.S.D.*	12	11	11
TIPS			Knob (arb. unit)
Untreated	42	20	2
10^{-6}M GA	71	39	1
10^{-6}M 2,4-D	44	12	4
10^{-5}M Zeatin	62	17	5
10^{-6}M DHR21A	42	21	0
2,4-D + DHR21A	26	19	0
GA + DHR21A	54	32	0
Zea. + DHR21A	28	6	1
L.S.D.*	14	11	

There is strong inhibition of the auxin and zeatin-induced growth of the hooks and basal swellings of the tips by podolactone DHR21A, while the growth induced by gibberellin is less affected. However, in the latter case, the elongation of the hook tissue is relatively more sensitive to the inhibitor than the tip tissue, illustrating, as does the tissue treated with inhibitor alone, the sequential response to hormones proposed earlier (Adamson, Low and Adamson, 1968) for pea stem cells as they mature.

It is possible that inhibitors such as the new group we have described may have a natural role in the control of plant growth, particularly in its prevention or by causing its cessation. The multiplicity of oxygenated structures may represent secondary metabolism of more highly active precursors that are more unsaturated. Or they might be formed directly under the influence of light and oxygen, serving as mediators in the interaction of the plant with the environment. However, while such ideas are only speculative, we think that we have demonstrated that our approach to the discovery of new plant growth regulators by testing compounds that have been selected from the rich supply of identified plant products is a fruitful and promising one.

IV. SUMMARY

The responses of excised apical and hook sections from etiolated dwarf peas to auxin, gibberellin, zeatin, abscisic acid and ethylene are sufficiently distinctive to suggest the use of such sections as a primary screen for new plant growth regulators. Only very small quantities of the chemicals are required and results are available in 1-3 days. The utility of this system of pea stem segments is illustrated by the discovery of a new group of naturally occurring inhibitors, the podolactones. These compounds are norditerpenoid dilactones which, in contrast to many other inhibitors, particularly affect the expansion growth of hook sections at concentrations at or below 10^{-5}M. Other properties of these compounds include inhibition of knob formation on the tips, inhibition of greening in response to light, and strong inhibition of auxin- and zeatin-induced growth. A study of the detailed structures of 6 of the new group of inhibitors permits some correlation of structure and activity.

V. ACKNOWLEDGEMENTS

We wish to thank Professors S. Itô, T. Sakan, T. Okamoto and Merck A.G. and Amchem. Inc. for gifts of compounds, and Dr. B. McGlasson for facilities for work with ethylene.

VI. REFERENCES

ADAMSON, D., VERONICA, H.K. LOW and HEATHER ADAMSON (1968). Transitions between different phases of growth in cells from etiolated pea stems, Jerusalem artichoke tubers and wheat coleoptiles. in "Biochemistry and Physiology of Plant Growth Substances". Ed. F. Wightman and G. Setterfield, The Runge Press, Ottawa, 1968.

GALBRAITH, M.N., D.H.S. HORN, JENNETH M. SASSE and D. ADAMSON (1970). The Structures of Podolactones A and B, Inhibitors of Expansion and Division of Plant Cells. Chem. Commun., 1970, 170-171.

HAYASHI, Y., S. TAKAHASHI, H. ONO and T. SAKAN (1968). Structures of nagilactones A, B, C and D, novel nor-and bisnorditerpenoids. Tetrahedron Lett., No. 17, 2071-2076.

ITÔ, S., M. KODAMA, M. SUNAGAWA, T. TAKAHASHI, H. IMAMURA and O. HONDA (1968). Structure of inumakilactone A, a bisnorditerpenoid. Tetrahedron Lett., 2065-70.

ITÔ, S., M. SUNAGAWA, M. KODAMA, H. HONMA, and T. TAKAHASHI (1971). Structures of inumakilactones B and C. Chem. Commun., 91

OKAMOTO, T., Y. TORII and Y. ISOGAI (1968). Lycoricidinol and Lycoricidine, New Plant-growth Regulators in the Bulbs of *Lycoris radiata* HERB. Chem. Pharm. Bull. 16, 1860.

OCCURRENCE OF SUBSTANCES IN DWARF PEAS INTERFERING
WITH RESPONSES OF THE SAME PLANTS TO GIBBERELLIN

Saburo Tamura, Susumu Ikegami, Nobuo Komoto and Masana Noma

Department of Agricultural Chemistry, The University of Tokyo, Bunkyo-ku, Tokyo
Japan

I. INTRODUCTION

Köhler and Lang (1963) reported that immature limabean seeds contain substances which reduce the GA response of dwarf pea seedlings but have no effect on the growth of the latter in the absence of GA. Further, they demonstrated the occurrence in pea seedlings and hemp plants of similar substances which are soluble in chloroform and ethyl acetate at pH 5 and 7.

Following this study, we attempted to isolate the substances in dwarf pea seedlings interfering with the response of the same plant to GA. Though we have not accomplished our purpose, here we wish to describe the preliminary experimental results so far obtained.

II. MATERIALS AND METHODS

Plant materials

Seedlings of dwarf pea (*Pisum sativum* L., var. Progress No. 9) to be used for the extraction of inhibitors were cultivated in the following way. Seeds were soaked overnight in tap water, planted in vermiculite and grown for 7-9 days at 20 - 23°C under red light, simulated daylight and darkness. The red light source was two FL-40-RF fluorescent tubes spaced 1 m from test plant level. The daylight fluorescent lighting was supplied from two FL-40 D tubes. Plant materials thus obtained were extracted and the extracts fractionated into the ethyl acetate-soluble, neutral and acidic extracts and water-soluble extract.

Bioassay

The bioassay for each extract was conducted according to the procedure described by Köhler and Lang (1963) with slight modification. Dwarf pea seeds were soaked, planted in vermiculite on the next day and grown for 4 days at 20 - 23°C in darkness. Seedlings with heights of 20 - 25 mm were selected and transferred to water culture. After being kept overnight under red light, the terminal bud of each seedling was treated with 10 μl of the test solution containing a definite amount of the extract, 0.015 μg of GA_3, 50% of acetone and 0.05% of Tween 20. For the assay of the water-soluble extract, acetone was omitted. The plants were then grown at 20 - 23°C under red light, and their stem lengths were measured 5 days later. Inhibitory activity was expressed by the following equation.

$$\text{Inhibition (\%)} = \frac{\text{Stem length (GA)} - \text{Stem length (GA+Extract)}}{\text{Stem length (GA)} - \text{Stem length (No } GA_3)} \times 100$$

III. RESULTS AND DISCUSSION

Fractionation of inhibitory substances

As a preliminary trial, dwarf pea seedlings grown for 7 days under red light were ground with a blendor and repeatedly extracted with methanol. After filtration, the extract was evaporated under reduced pressure. The aqueous residue was extracted at pH 3 with ethyl acetate, which was subsequently shaken with dilute sodium bicarbonate solution. The organic layer was washed with water, dried over anhydrous sodium sulphate and evaporated to dryness to give the neutral fraction. The aqueous layer was adjusted to pH 3 and extracted with ethyl acetate. The extract was dried and evaporated to afford the acidic fraction.

As illustrated in Fig. 1, the both fractions showed inhibitory activities to dwarf pea seedlings in the presence of GA_3 and caused some injury on the test plants at higher doses. Contrary to expectation, hardly any difference was observed in the inhibitory substances of either fraction whether the plant materials were cultivated under red light, daylight or darkness. Interestingly, an anti-GA effect was given by the water-soluble fraction, whose separation is described later.

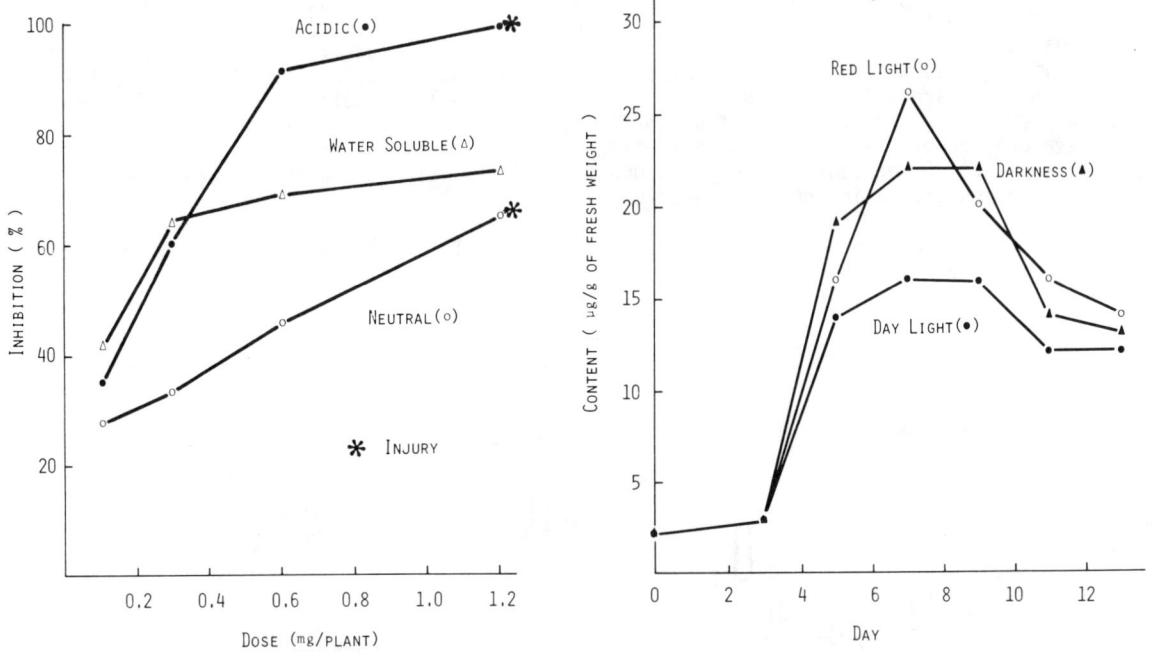

Fig. 1. Inhibitory Effects of Neutral○, Acidic● and Water Soluble△ Fractions from Dwarf Pea Seedlings on the Growth Response of the Same Plant to GA_3 (0.015 μg).

Fig. 2. Time-Course Changes in the Content of Pisatin in Dwarf Pea Seedlings Grown in Red Light○, Day Light● and Darkness▲.

Inhibitors in the neutral fraction

The main principle responsible for the inhibitory effect of the neutral fraction was ultimately identified as pisatin. Since we did not expect the formation of this compound in healthy plants at the beginning of isolation, we adopted rather tedious procedures for isolation and purification of the neutral fraction and finally separated 15 mg of crystals melting at 67 - 70°C from 20 kg of seedlings (Cruickshank and Perrin, 1961). The UV and IR spectra of the crystals thus obtained coincided with the data presented by Perrin and Bottomley (1962).

Pisatin has been considered until now to be a phytoalexin which is formed as a response to inoculation of pea pod endocarp tissues with fungi. In our experiment, however, this compound has been isolated as a normal constituent of seedlings. We then examined the time-course change in pisatin content at earlier stages of seedling growth. Quantitative analysis of pisatin extracted from tissues was carried out by the measurement of UV absorption at 358 mμ due to anhydropisatin which is easily and quantitatively formed by dehydration of the former with dilute sulfuric acid. As indicated in Fig. 2, the pisatin content reached a maximum 7 days after germination (16 - 28 μg per g of fresh weight) and rapidly decreased thereafter. Though red light seems somewhat favorable for pisatin formation, the content decreased to the same level as that under daylight or darkness after 13 days cultivation.

Anti-gibberellin activity of pisatin was not very strong since 40 μg of this substance was required for 30% inhibition of the elongation of pea seedlings in the presence of 0.015 μg of GA_3. In the *Avena* straight-growth test, pisatin at a concentration of 50 mg/l produced 60% inhibition of the elongation caused by IAA at 1 mg/l.

Inhibitors in the acidic fraction

According to the procedure cited above, the acidic fraction was separated from 6 kg of dwarf pea seedlings grown for 9 days under red light. The crude fraction was applied to a column (3 x 50 cm) packed with a 1:2 mixture of charcoal (Wako) and celite (Wako) which had been treated with 6N hydrochloric acid, washed successively with water and methanol, and dried. The column was eluted with 400 ml each of water-acetone increasing the content of the latter by steps of 10% (10 - 100%). After evaporation of the solvent, each fraction was applied to the bioassay. Fractions No. 3 and 8 at 50 μg/plant strongly antagonised the GA_3 effect. Further, both showed growth inhibition in the absence of GA_3.

Each fraction was then subjected to TLC on silica gel (Mallinckrodt) using ABA as a marker. The chromatogram developed with n-butanol-n-propanol-ammonium

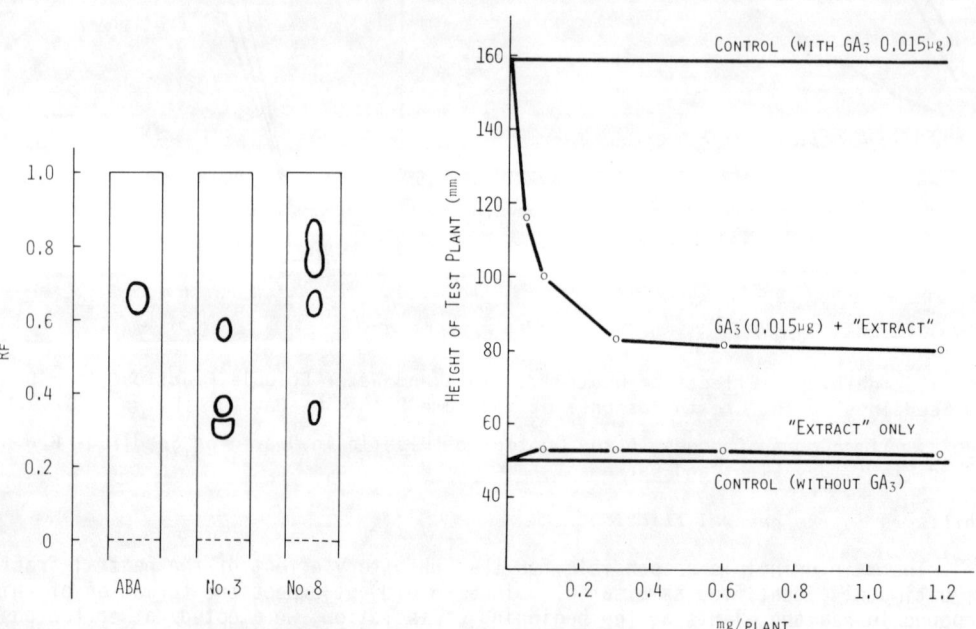

Fig. 3. TLC of the Acidic Fraction from Dwarf Pea Seedlings after Charcoal-Celite (1:2) Column Chromatography Using Water-Acetone (10 - 100%) TLC Solvent System : n-Butonol-n-Propanol-NH_4OH-H_2O (6:2:1:2)

Fig. 4. Growth Response of Dwarf Pea Seedlings to "Water-Soluble Extract" from the Same Plant

hydroxide-water (6:2:1:2) is shown in Fig. 3. A spot corresponding to that of ABA is observed in the fraction No. 8. Isogai *et al.* (1967) reported the isolation of ABA from fresh seeds of garden pea (*Pisum sativum* L.). In our bioassay, ABA applied at 15 μg/plant did not interfere with the response of dwarf peas to GA_3. Therefore, we must assume the presence of an inhibitor or inhibitors other than ABA in the acidic fraction.

Inhibitors in the water-soluble fraction

Dwarf pea seedlings grown for 9 days under red light were covered with a large volume of acetone (five times by weight) and ground with a blendor. The mixture was filtered and repeatedly washed with acetone. The acetone powder was collected on a filter paper, dried in a vacuum desiccator and subsequently extracted with 0.1 M phosphate buffer (pH 6.9). The extract was centrifuged for 50 min at 3,800 x g. The supernatant thus obtained was dialyzed against water for 24 hr in a cellophane tube, and the solution in the tube was lyophilized to yield the "water-soluble extract", which was subjected to bioassay.

As shown in Fig. 4, the "extract" at doses of 50 - 1200 μg/plant counteracted the effect of 0.015 μg of GA_3 applied simultaneously, whereas the "extract" itself, without GA_3, did not inhibit the growth of seedlings indicating that the active principle in the "extract" is incapable of interfering with the action of endogenous GA in the test plants. This might be explained if the active principle forms a complex with GA_3 exogenously applied, reducing the uptake of the latter into plant tissues.

Subsequently, 300 mg of the "water-soluble extract" dissolved in 2 ml of 0.1 M pyridine acetate buffer (pH 6.0) was applied to a Sephadex G-50 column (1.0 x 130 cm) which had been preequilibrated with the same buffer. The column was eluted with the buffer, and 10 ml fractions were collected, lyophilized and bioassayed.

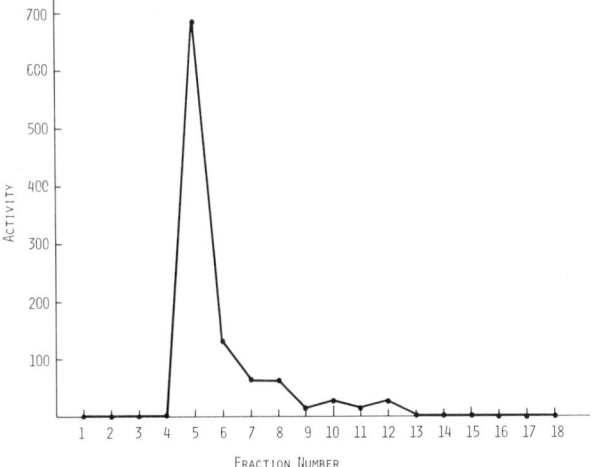

Fig. 5. Biological Activity of Fraction from 300 mg of "Water-Soluble Extract". A Sephadex G-50 Column was used to obtain the fractions. Activity is Expressed in Relative Terms by the Number of Plants whose Growth is Inhibited by 50%.

As illustrated in Fig. 5. almost all the activity was excluded from the gel and concentrated in the fifth fraction. This suggests that the active principle in the "extract" is a substance(s) of high-molecular weight. On heat treatment in aqueous solution or dry, the "extract" lost its biological effect. For example, the aqueous solution lost the greater part of its activity, when kept in boiling water for 5 min. Physico-chemical properties of the material as well as its action mechanism are under investigation.

In our preliminary experiments no significant difference in inhibitory effect was recognized between the water-soluble fractions separated from light - and dark - grown plants. Furthermore, similar activity was detected in the water-soluble fraction from pea seedlings of a tall variety.

IV. SUMMARY

Seedlings of dwarf pea (*Pisum sativum* L., var. Progress No. 9) grown for 9 days under red light were extracted with methanol and separated into ethyl acetate-soluble, neutral and acidic fractions. Purification of the neutral fraction afforded pisatin as a growth inhibitor of dwarf peas. The acidic fraction revealed the occurrence of a substance(s) other than ABA counteracting the effect of GA_3 when simultaneously applied to dwarf peas.

Furthermore, the water-soluble fraction contained high-molecular substances which interfere with the GA_3 response of dwarf peas while showing no effect on the growth of the latter in the absence of GA_3.

V. REFERENCES

CRUICKSHANK, I.A.M. and D.R. PERRIN (1961). Studies on phytoalexins. III. The isolation, assay, and general properties of a phytoalexin from *Pisum sativum* L. Aust. J. Biol. Sci. $\underline{14}$, 336-348.
ISOGAI, Y., T. OKAMOTO and Y. KOMODA (1967). Isolation of a plant growth inhibitory substance from garden pea (*Pisum sativum* L.) and its identification with (+)-abscisin II. Chem. Pharm. Bull. (Japan) $\underline{15}$, 1256-1257.
KÖHLER, D. and A. LANG (1963). Evidence for substances in higher plants interfering with response of dwarf peas to gibberellin. Pl. Physiol. $\underline{38}$, 555 - 560.
PERRIN, D.R. and W. BOTTOMLEY (1962). Studies on phytoalexins. V. The structure of pisatin from *Pisum sativum* L. J. Am. Chem. Soc. $\underline{84}$, 1919-1922.

PLANT-GROWTH INHIBITORS

IN THE BULBS OF *Lycoris radiata* HERB.

T. Okamoto, Y. Torii and Y. Isogai*

Faculty of Pharmaceutical Sciences, University of Tokyo, Hongo, Bunkyo-ku, Tokyo

* Biological Institute, College of General Education, University of Tokyo, Komaba, Meguroku, Tokyo

I. INTRODUCTION

Two growth inhibitors, lycoricidinol and lycoricidine, were isolated[**] from the methanol extract of the bulbs of *Lycoris radiata* Herb. The isolation procedures, structural determination, and some biological activities of the compounds will be reported in the present paper.

II. METHODS OF BIOASSAY USED FOR THE DETECTION OF THE INHIBITORS

Avena straight growth test

Dehusked oat seeds (*Avena sativa* var. Victory) were soaked in water for 2 hr and grown for 72 hr at $25°$ in darkness, except for occasional exposure to red light. Sections of 5 mm in length were cut from coleoptiles 3 mm below the tip, and 15 sections per dish were floated on 3 ml of aqueous solution containing the test substance. After a 24-hr growth in the dark, the length was measured and compared with that of control.

Rice seedling growth test

This test was made by the following procedures. Seeds of rice (*Oryza sativa* var. Norin No. 29) were soaked in ethanol for 10 min and then in a saturated solution of bleaching power for 1 hr. The sterilized seeds were transferred to a large petri dish containing sterilized water 1 cm in depth. The petri dish was kept at $30°$ for 48 hr under white fluorescent and incandescent lamps (about 5000 lux). The germinating seeds were employed in the bioassay.

a) Fifteen of the seedlings with coleoptiles 3-5 mm long were placed in a test tube (3 x 10 cm) containing 1.35 ml of aqueous solution of the test compound. After covering the tubes with polyethylene film, they were kept under the lighting and temperature conditions mentioned above for 5 days. The second leaf sheath of each seedling was measured and the measurements compared with those of controls which grew in water.

b) 20 ml of aqueous test solution was solidified with agar (2%) in a 100-ml dish. Twenty germinating seeds with coleoptiles about 1 mm long were planted on the agar. After covering the dishes with polyethylene film, they were kept under the lighting and temperature conditions mentioned above for 5 days, and measurements taken as in method (a).

[**] Okamoto, Torii and Isogai (1968).

III. ISOLATION OF LYCORICIDINOL AND LYCORICIDINE

Bulbs of *Lycoris radiata* Herb. were collected in Chiba Prefecture in September. After washing, the bulbs, with short stems (0-7 cm in length), were added to methanol, macerated in a Waring blendor, and left for a few days at room temperature. After filtration, the solvent was evaporated at low temperature *in vacuo* to give a crude extract. The raw material was separated into acidic, basic, neutral and water-soluble fractions by treatment as shown in Fig. 1. The acidic and water-soluble fractions showed inhibitory activity in the *Avena* straight growth test and the rice seedling test.

The acidic fraction was separated with silica-gel column chromatography using methanol-chloroform (saturated with water) as a solvent. The fractions eluted with solvent containing 10% methanol showed growth inhibition in the bioassay. These fractions were combined, treated with methanol, and the insoluble residue recrystallized from water to give lycoricidinol (I). The methanol-solubles were further separated by silica-gel chromatography and lycoricidine (II) was obtained.

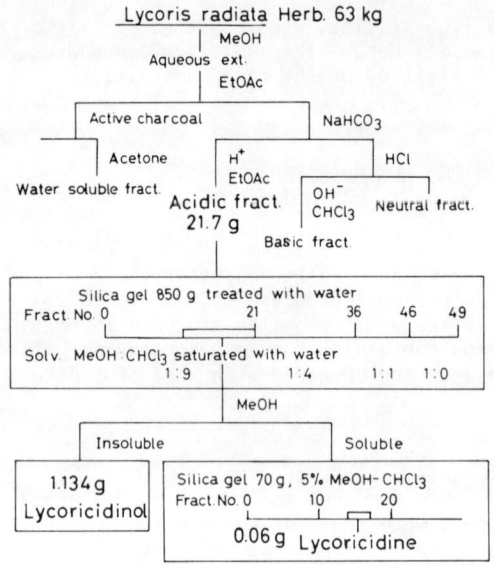

Fig. 1. Isolation of lycoricidinol and lycoricidine.

Lycoricidine was also isolated from the water-soluble fraction of the crude extract by the following procedure: the material adsorbed on active charcoal was eluted with acetone and the eluate was shaken with a butanol-water mixture after evaporation of the acetone. The residue was separated by silica-gel chromatography to give lycoricidine.

IV. CHEMICAL STRUCTURE OF THE ACTIVE SUBSTANCES

Lycoricidinol, $C_{14}H_{13}O_7N$, has no sharp melting point, begins to color about $200°$, and decomposes slowly above $216°$. Ultraviolet (UV) and infrared (IR), spectra are shown in Fig. 2. As the functional groups, a methylenedioxy, an amide, a chelated phenolic group, and a trisubstituted double bond were confirmed by the analyses of UV, IR and nuclear magnetic resonance (NMR) spectra.

Lycoricidine decomposes at $214.5-215.5°$. A methylenedioxy and an amide group, and a trisubstituted double bond were observed but no phenolic group was found in the spectra.

When treated with acetic anhydride in pyridine, lycoricidinol formed a triacetate and a tetracetate, and the triacetate had a free phenolic group. Acetylation of lycoricidine gave only a triacetate. The NMR spectra of tetraacetyllycoricidinol and

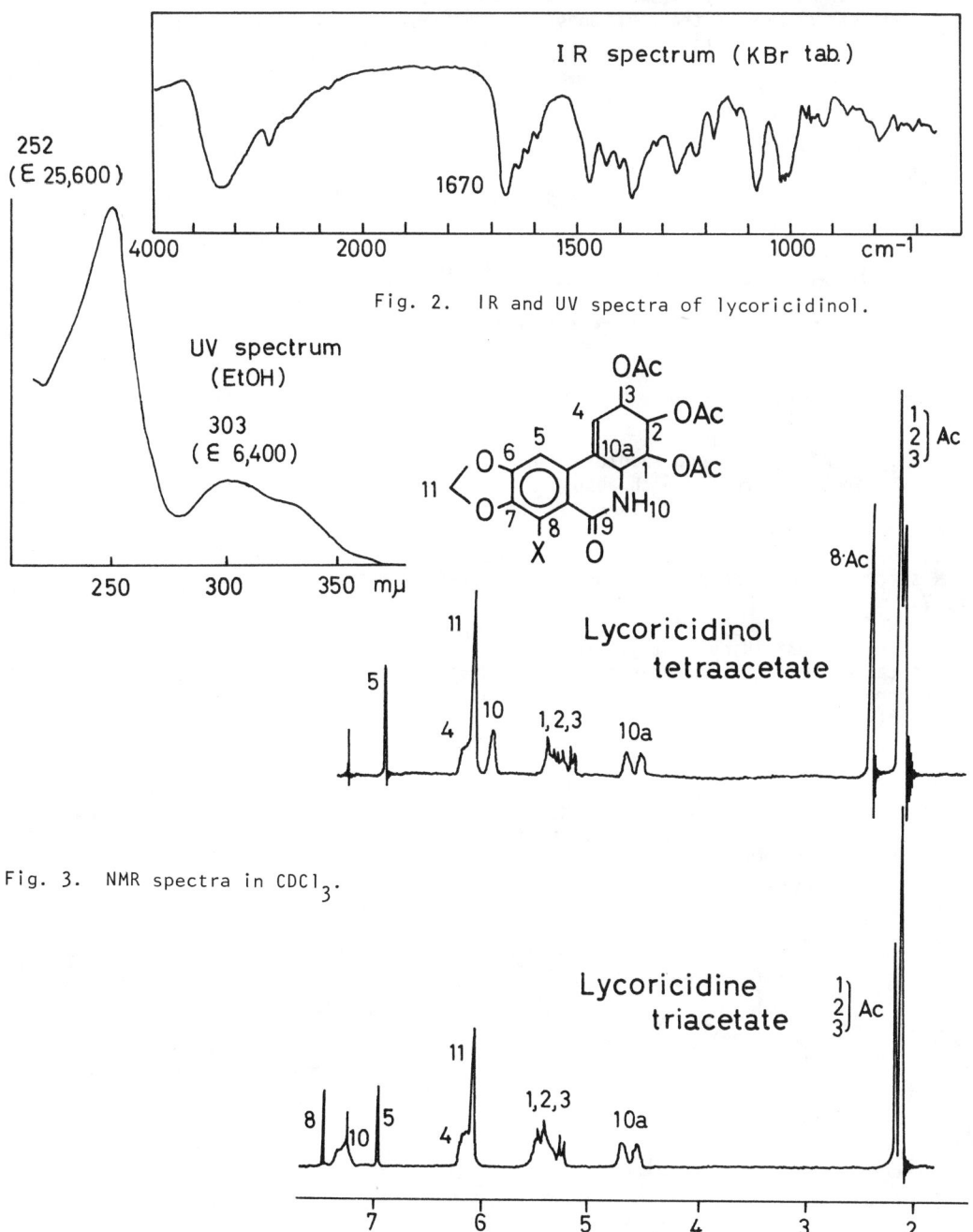

Fig. 2. IR and UV spectra of lycoricidinol.

Fig. 3. NMR spectra in CDCl$_3$.

triacetyllycoricidine are shown in Fig. 3. Both acetates give very similar NMR charts, lycoricidinoltetraacetate shows a signal attributable to a phenol acetate group in place of the aromatic proton signal of lycoricidinetriacetate. Lycoricidinol and lycoricidine require two moles of periodic acid for oxidation and, therefore, the three alcoholic hydroxyl groups are vicinal to each other.

When lycoricidinol and lycoricidine are treated with methanolic hydrochloric acid, arolycoricidinol and arolycoricidine are obtained respectively, eliminating two molecules of water.

The structures of the compounds were confirmed to be III and IV by the sequences of syntheses shown in Fig. 4.

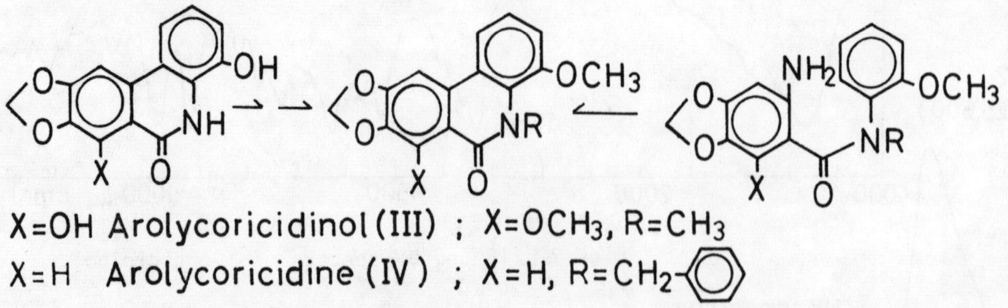

X=OH Arolycoricidinol(III) ; X=OCH₃, R=CH₃
X=H Arolycoricidine (IV) ; X=H, R=CH₂–⬡

Fig. 4. Syntheses of arolycoricidinol and arolycoricidine derivatives.

The conformation and absolute configuration were determined as follows. When lycoricidinol was treated with absolute acetone in the presence of anhydrous copper sulfate, an acetonide was obtained as the sole product. This gave a monoacetate when treated with acetic anhydride in pyridine. From the fact that the proton at C-4 was coupled with the proton adjacent to the acetyloxy group and from the analyses of the NMR spectra using decoupling techniques, stereochemistry in the C ring was established as V.

When lycoricidinol methyl ether was treated with benzoyl chloride, a tribenzoate was obtained. In the circular dichroism curve, a negative Cotton effect was observed. Therefore, the C ring has (VI) configuration when the dibenzoate chirality rule* is adopted for this compound. Finally, we propose I and II for the structures of lycoricidinol and lycoricidine.   (*Nakanishi and Harada (1969).

Identity of lycoricidinol with narciclasine, also that of lycoricidine with margetine was recently reported by Mondon,1970 and Piozzi,1970. Narciclasine and margetine were isolated from   daffodil bulbs and their chemical structures were studied by Piozzi,1968.

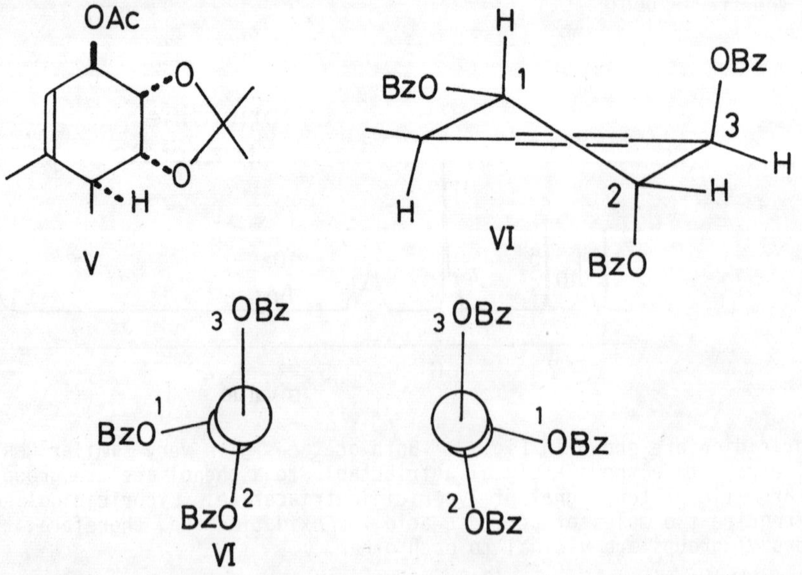

Fig. 5. Conformation and absolute configuration of lycoricidinol.

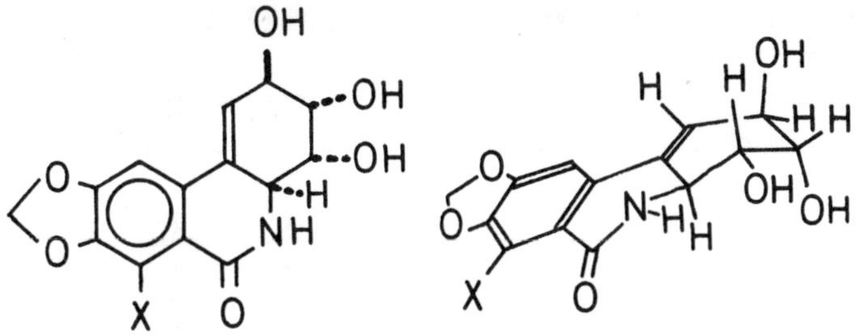

X = OH Lycoricidinol (I)
X = H   Lycoricidine (II)

Fig. 6. Structures of lycoricidinol and lycoricidine.

V. BIOLOGICAL ACTIVITY OF LYCORICIDINOL AND LYCORICIDINE

Avena straight growth test

Lycoricidinol and lycoricidine inhibit the elongation of *Avena* coleoptile sections and the inhibition is also observed in the presence of IAA, as shown in Table 1. The activity of lycoricidinol is a little stronger than that of lycoricidine. Even at 100 ppm, however, the inhibition by either compound is not total, so their activities are not overwhelming.

Rice seedling growth test

The results of the tests by method (b) are shown in Table 2. By method (a), it is rather difficult to measure the length of the leaf sheath, because both compounds inhibit strongly the root growth of rice and this causes the young plants to fall over, with subsequent irregular growth of the stems. Method (b) gave good results. Lycoricidinol and lycoricidine both inhibit the elongation of rice stems and show antagonism to gibberellin A_3.

Other biological activities

The compounds inhibited cell division induced by kinetin in tobacco pith callus, as shown in Table 3. Further, considerable antitumor activities of these inhibitors were observed in the Ehrlich carcinoma test.

VI. CONCLUSION

There is growing evidence that various growth inhibitors take part in the control of the growth of higher plants as well as growth promotors. In the case of *Lycoris*, it seems to be reasonable to assume that, besides these growth inhibitors, the plant

Table 1. Avena straight growth test* using *Avena sativa* var. Victory.

Concentration ppm	Lycoricidinol	Lycoricidinol + IAA (1 ppm)	Lycoricidine	Lycoricidine + IAA (1 ppm)
100	6.0++	6.2++	6.1++	6.1++
10	6.2++	6.6++	7.8	6.9
1	7.3	8.3	9.3	8.1
0.1	8.2	9.0	9.4	8.3
IAA (1 ppm)		9.0		
Control		8.0		

* Average length (mm) of 15 coleoptile sections.
++ Coleoptile sections suffered toxicity and were degenerated.

Table 2. Rice seedling test[1] using *Oryza sativa* var. Norin 29.

Concentration ppm	Lycoricidinol	Lycoricidinol + GA_3 (1 ppm)	Lycoricidine	Lycoricidine + GA_3 (1 ppm)
100	$1.2^{2,4)}$	$1.6^{2,4)}$	$2.9^{2,4)}$	$1.4^{2,4)}$
10	$2.2^{2)}$	$3.1^{2)}$	$5.2^{3)}$	$2.5^{2)}$
1	$2.8^{2)}$	$5.4^{3)}$	7.4	$3.1^{3)}$
0.1	3.2	7.7	9.9	3.8
GA_3 (1 ppm)		9.8		
Control		3.8		

1) Average length (cm) of 15 second leaf sheaths of rice seedlings on agar.
2) Root-growth was completely inhibited.
3) Few roots, growing to 1~2 mm, were observed.
4) Whole length of rice seedling.

Table 3. Tissue culture[1] using pith callus of *Nicotiana tabacum* var. Bright Yellow

Concentration ppm	Lycoricidinol		Lycoricidine	
	Fresh	Dry	Fresh	Dry
1	79	4.5	139	5.1
0.1	511	23.3	1710	64.3
0.01	5182	200	5715	223.4
Control [2]	Fresh, 5344		Dry, 210	

1) Average weight (mg) of 12 calluses.
2) Murashige and Skoog's culture medium + 2 ppm IAA + 0.01 ppm kinetin.

contains some growth promotors and we have tried to detect them. But, except for a small quantity of auxin, no appreciable amounts of growth promotors were detected by the bioassays used here. We have recently detected a growth promoting agent in the methanol extract of *Lycoris* using a new type of rice seedling test and we are now engaging in isolating it.

We hope to elucidate the mechanisms which regulate the rapid growth of the scapes of *Lycoris radiata* through the study of these growth regulators.

VII. SUMMARY

Two plant growth inhibitors, lycoricidinol ($C_{14}H_{13}O_7N$) and lycoricidine ($C_{14}H_{13}O_6N$), were isolated from methanol extracts of the bulbs of *Lycoris radiata* Herb. The chemical structure and the stereochemistry, including absolute configuration of the compounds, were determined.

Lycoricidinol and lycoricidine showed inhibitory activities in the Avena straight growth test, the rice seedling test, and a tobacco tissue culture test. Further, considerable carcinostatic activity was observed in a test using Ehrlich carcinoma.

VIII. REFERENCES

1. MONDON, A. and KROHN, K. (1970). Struktur und Synthese des Narciprimins. Tetrahedron Letters, 2123-2126 (1970).
2. NAKANISHI, K. and HARADA, N. (1969). A method for determining the chiralities of optically active glycols. J. Am. Chem. Soc., 91, 3989-3991.
3. OKAMOTO, T., TORII, Y. and ISOGAI, Y. (1968). Chem. Pharm. Bull. 16, 1860-1864.
4. PIOZZI, F., FUGANTI, C., MONDELLI, R. and CERIOTTI, G. (1968). Narciclasine and narciprimine. Tetrahedron, 24, 1119-1131.
5. SAVONA, G., PIOZZI, F. and MARINO, M.L. (1970). Structure and synthesis of permethynarciprimine. Chem. Comm., 1006 (1970).

MORPHACTIN-LIKE ACTIVITY OF BENZILATE ESTERS
IN *Arabidopsis thaliana*

B.T. Brown* and L.K. Dalton+

* C.S.I.R.O., Division of Plant Industry, Canberra, Australia

+ C.S.I.R.O., Division of Applied Chemistry, Melbourne, Australia

I. INTRODUCTION

Fluorenols (I. R=OH), commonly referred to as morphactins, have been examined intensively for their plant growth regulating properties (Schneider, 1970). They may inhibit bolting and stimulate axillary bud and primary root development. In the case of *A. thaliana* changes in the morphology of the plant have been described in some detail (Sankhla, 1968).

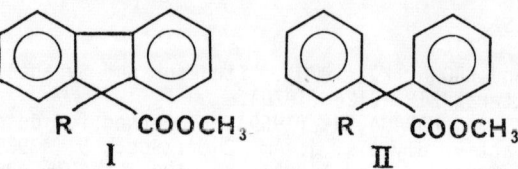

Fig. 1. Structural formulae of fluorene carboxylate and benzilate

This interest in the fluorenols led us to consider the structurally related benzilates (II. R=OH) although Mohr (1965) has suggested that effective activity resides in the skeletal fluorene ring system. Since we completed this work Buchenauer (1970) has reported that 9-xanthenecarboxylic acid exhibits the same morphological effects in the tomato plant as the fluorenols. The only examinations reported in the literature for plant activity of benzilates are those related to the detection of possible phytotoxic action of the miticide, ethyl 4,4'-dichlorobenzilate (Beye, 1961; Eder, 1963).

Substitution of a hydroxyl group at the 9-position in fluorenecarboxylic acid is reported by Mohr (1965) to substantially increase its plant activity although in earlier discussion on aromatic carboxylic acids it had usually been accepted that when a carbon atom is interposed between the carboxyl group and the ring it should have attached a hydrogen atom for the compound to have significant activity. In support of this Heacock (1958) showed that 9-fluorenecarboxylic acid loses its activity in the pea curvature test when a methyl group is substituted at the 9-position. However in this case the comparison has been made with a non-polar substituent.

In the present study changes in the growth pattern of *A. thaliana* have been used to determine the effect of structural differences between fluorenols and benzilates. The effect of the hydroxyl group has also been examined by inclusion of a diphenylacetate and a 9-fluorenecarboxylate.

II. MATERIALS AND METHODS

Growth of Arabidopsis thaliana

A growth nutrient was prepared containing KH_2PO_4 (0.002 M), KNO_3 (0.006 M), $Ca(NO_3)_2$ (0.004 M) $MgSO_4$ (0.002M) and micro elements in distilled water and adjusted to pH 6.0. The compound under test was dissolved in a small quantity of acetone and then diluted with growth nutrient to produce the appropriate concentration. The solution was then heated, 0.75% agar added, boiled for 2 minutes and approximately 7 ml poured into each of 10 test tubes, 18 x 150 mm. Seeds of *Arabidopsis thaliana* were then carefully placed on the agar surface, the tube plugged with cotton wool, and maintained at 65% relative humidity at $25^0 \pm 1^0C$ under continuous artificial illumination (16,000 lux.). The plug of cotton wool was removed after 12 days.

After 21 days growth the plants were carefully examined for morphological changes induced by the test compound. The dry weight of the plants was then determined by heating at 90^0C for 18 hours, and compared to the dry weight of plants grown on nutrient without added test substances. Under these growth conditions morphological changes in individual plants in the same treatment were reproducible.

III. RESULTS AND DISCUSSION

Morphological Effects of Benzilates

At sub-toxic concentrations of methyl benzilate in the nutrient agar distinct and characteristic changes in the growth habit of *A. thaliana* plants were observed. After 21 days growth plants had developed a shortened flower stem, much thickened at the base and tapering to a flower head which was very much underdeveloped when compared with control plants. The root structure was modified and at higher concentrations consisted of a single primary root, thickened and elongated. The rosette leaves were thin, about half the width of the leaves of untreated plants, spatula-like and curled under at the tip. Both the leaves and the lower portion of the stem were more pigmented than those of the control.

Nutrient containing fluorenols resulted in plants with the same kind of growth changes as those grown on benzilates and corresponding to the changes described by Sankhla (1968). The changes in growth habit appear characteristic of both classes of compounds and were different from changes induced by a wide range of growth regulators tested on *A. thaliana* under similar conditions.

Fluorenols and benzilates also reduced plant dry weight.

Activity of Substituted Methyl Benzilates

The effect of symmetrical substitution in the methyl benzilate molecule on the concentration for toxicity and for detectable change in growth habit is shown in Table I.

Halogens enhanced plant activity over the unsubstituted benzilates more when substituted in the 3-position than when substituted in the 4-positions. Chlorine had greatest effect in the 3-positions and fluorine the least effect. Substitution of methoxyl groups in the 3- or 4-positions caused little or no change in activity while the polar nitro- and non-polar isopropyl- groups in the 4-positions actually reduced activity of the molecule.

A striking effect of substitution is that either chlorine or methoxyl groups substituted in the 2-positions removed or very substantially reduced the activity of the benzilate molecule. Possible explanations for this effect are discussed later.

Table 1. Activity of Substituted Methyl Benzilates on *A. thaliana*

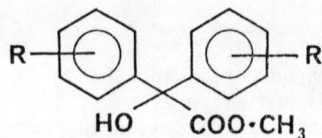

Compound R	Minimum Toxic Concentration p.p.m.	Minimum concentration for change in growth habit p.p.m.
H	16	2
2-OCH$_3$	>8	*
3-OCH$_3$	8	2
4-OCH$_3$	16	2
2-Cl	>8	*
3-Cl	1	0.2
4-Cl	8	1
3-Br	2	0.3
3-F	4	0.5
4-NO$_2$	>8	*
4-CH(CH$_3$)$_2$	>8	*

* No detectable change in growth habit at 8 p.p.m.

Comparison of Benzilates and Fluorenols

The results of a comparative study of two benzilates and two fluorenols are given in Table 2.

Fluorenols are reported to be active in plants as growth regulators over a wider than usual concentration for synthetic compounds and the present results show this to be the case also with activity in *A. thaliana*. On the other hand for both unsubstituted and substituted benzilates the range of concentration over which changes in growth habit are observed is very much narrower than with the fluorenols.

The situation is different for toxic concentrations and the corresponding fluorenols and benzilates are quite close to each other although in the case of the chlorinated compounds the comparison has been made between a mono-chlorinated fluorenol and a dichlorinated benzilate.

Inhibition of shoot elongation by fluorenols is particularly marked in rosette plants at the stage of rapid stem elongation (Schneider 1970). A limited number of foliar applications of chlorofluorenol and 3,3'-di-chlorobenzilate have been made to 11 day old *A. thaliana* at the rate of 1, 0.1 and 0.01 µg per plant. Both compounds induced similar changes in plant growth. Stem bolting was inhibited, axillary bud development was promoted and the plants produced a typical underdeveloped flower head.

Table 2. Comparison of Effects of Fluorenols and Benzilates on A. thaliana

Compound	Minimum Toxic Concentration p.p.m. (A)	Minimum Concentration for change in Growth Habit p.p.m. (B)	Ratio A/B
fluorenol HO COO-CH₃	8	0.05	160
chlorofluorenol HO COO-CH₃ Cl	2	0.01	200
benzilate HO COO-CH₃	16	2	8
dichlorobenzilate Cl HO COO-CH₃ Cl	1	0.2	5

Substitution on the Benzylic Carbon

In both classes of compounds replacement of the benzylic hydroxyl with a hydrogen increased the minimum concentration for change in growth habit with little change in minimum toxic concentration (Table 3). Whereas the loss of the hydroxyl from the fluorenol did not alter the nature of the effect on the plant, the changes in the growth habit induced by benzilates seem to be absent for the diphenylacetate.

Molecular structure and Plant Activity

Of the two classes of compounds only the fluorene molecule is planar and so it possesses the greater potential for charge transfer. The benzene rings in the benzilate molecule may rotate but this is much restricted and they are unable to adopt a planar conformation although the electrostatic charges of the substituted chlorine atoms may anchor the molecule at the site of action with a minimum of twist between the rings. The loss of activity when the 2-positions are substituted may be explained in the same terms since the benzene rings must now adopt planes almost at right angles to each other in order to accommodate the substituents.

Another difference between the structures of the two molecules relates to the freedom of the carboxyl group to rotate. Models show that while in the fluorene molecule this group may rotate freely, in the benzilate molecule its rotation is restricted. In the molecular structure of 9-xanthenecarboxylic acid the carboxyl group is restricted in its freedom to rotate as compared to 9-fluorenecarboxylate and from the brief report of its activity it would seem to be less active than 9-carboxyfluorene although showing growth effects of the same nature.

Table 3. Effect of Hydroxyl Group on Benzylic Carbon Atom

Compound	Minimum Toxic Concentration p.p.m.	Minimum concentration for change in Growth habit p.p.m.
fluorene with HO and COO·CH₃ at C9	8	.05
fluorene with H and COO·CH₃ at C9	16	2
diphenyl with HO and COO·CH₃ on benzylic carbon	16	2
diphenyl with H and COO·CH₃ on benzylic carbon	16	*

* No change in growth habit at 8 p.p.m.

   The most significant difference between the performance of the chlorobenzilate and chlorofluorenol in *A. thaliana* lies in the range of concentration over which the compounds are active. Important in any consideration of the two classes of compounds is the finding that their minimum toxic concentrations are nearly the same. This difference may relate to the different equilibria maintained by the two classes of compounds at the site of action as regulated by their charge transfer potential and the ability of the carboxyl group to orientate itself.

IV. SUMMARY

   The growth inhibition and effects on morphogenesis in *Arabidopsis thaliana* shown by fluorenols (9-hydroxy-9-fluorenecarboxylates) are shown also by benzilate esters but over a narrower concentration range. Addition of compounds of either of these two classes to the growth media brings about similar morphological changes in roots, leaves and flowers, while foliar application inhibits bolting and dwarfs the plant.

   It is suggested that the differences in plant growth activity between the benzilates and fluorenols relate to differences in both planarity of the ring system and freedom to rotate of the carboxyl group.

V. REFERENCES

BEYE, F. (1961). The effects of insecticides on the growth of the roots of cress. Z. Pflanzenkrank. u. Planzenschutz. 68, 6-17.

BUCHENAUER, H. and GROSSMANN, F. (1970). Morphoregulatory effects of xanthene derivatives. Planta, 92, 86-88.

EDER, F. (1963). Phytotoxicity of insecticides. Phytopath. Z. 47, 129.

HEACOCK, R.A., WAIN, R.L. and WIGHTMAN, F. (1958). Studies on plant growth regulating substances. Ann. appl. biol. 46, 352-65.

MOHR. G., ERDMAN, D. and SCHNEIDER, G. (1965). Derivatives of fluorene as novel and highly active plant morpho-regulators. Symp. New Herbicides, 2nd, Paris, 135-40.

SANKHLA, D. and SANKHLA, N. (1968). Growth of *Arabidopsis thaliana,* En-2, in response to added morphactin. Arabidopsis Information Service 5, 19-20.

SCHNEIDER, G. (1970). Morphactins: Physiology and performance. Ann. Rev.Pl. Physiol. 21, 499-536.

PLANT GROWTH SUBSTANCES, 1970

Chemistry and Physiology of Rooting Inhibitors
in Adult Tissue of Eucalyptus grandis

W. Nicholls*, W.D. Crow† and D.M. Paton*

* Botany Department, School of General Studies, Australian National University, Canberra, A.C.T.

† Chemistry Department, School of General Studies, Australian National University, Canberra, A.C.T.

I. INTRODUCTION

We have proposed (Paton et al 1970) that endogenous rooting inhibitors may have an overriding influence on adventitious root formation in adult tissues of E. grandis. This suggestion is supported by several observations indicating that a relationship exists between decreased rooting ability of stem cuttings and increased levels of a rooting inhibitor in the tissue forming the base of the cutting.

Reliable bioassays were available for monitoring the extraction and isolation of the active material and this paper reports on some chemical and physiological properties of three closely related rooting inhibitors isolated from adult tissue of E. grandis.

II. EXTRACTION OF THE INHIBITORS

The procedure for isolation of the inhibitors is summarised in Fig. 1. The final purification step involved column chromatography on cellulose (Whatman Standard Grade) as the support medium. The cellulose was thoroughly impregnated with the stationary phase of ethylene glycol and water (4:1) before packing the column. Elution of the column with cyclohexane (mobile phase) led to the successful separation of two active compounds. A third active compound was eluted with diethyl ether. The three active compounds were designated in order of their elution from the column, as G1, G2 and G3; the "G" being derived from "grandis" to distinguish these inhibitors from other inhibitors that may eventually be isolated from other Eucalyptus species. Throughout the isolation procedure, the inhibitors were monitored using either the cress seed germination or the mung bean rooting bioassays (cf. Paton et al 1970).

The highest recorded yield of the combined inhibitors was 0.056% fresh weight or 560 ppm.

III. CHEMISTRY OF THE INHIBITORS

The chemical identity of the three compounds was investigated by examining their spectral and other chemical and physical properties.

The structure of inhibitor G1 (Fig. 2) was determined by X-ray crystallography (Sterns, 1971, in press). The spectral properties of G1 are in complete agreement with this structure.

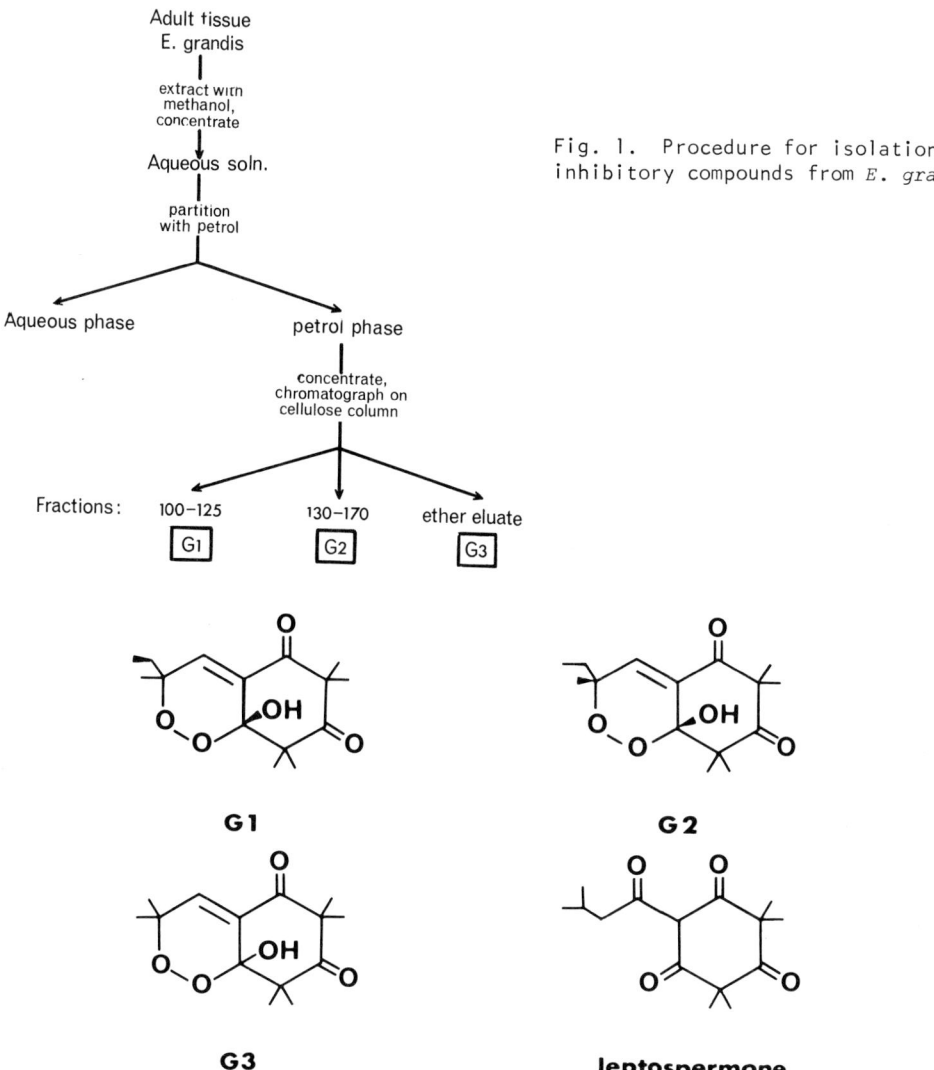

Fig. 1. Procedure for isolation of inhibitory compounds from *E. grandis*.

Fig. 2. Proposed structures for G1, G2 and G3; leptospermone is included for comparison.

G1 recrystallised readily from cyclohexane to yield colourless crystals, m.p. 98.5°C. High resolution mass spectrometry gave an accurate molecular weight of 282.146758 that corresponded to the empirical formula $C_{15}H_{22}O_5$. Mass spectrometry also showed two initial, successive losses of 16 mass units (corresponding to oxygen atoms), as well as concerted O_2 loss, indicating the presence of a peroxide linkage. Ultra-violet spectroscopy of G1 in 90% ethanol solution gave λmax 242 mu (ε, 7440). This is in good agreement with the expected λ_{max} (Williams & Fleming 1966), for an α,β-unsaturated ketone with an exocyclic double bond (cf. Fig. 2). Prominent bands in the infrared (IR) spectrum occurred at 3380 cm^{-1}, corresponding to absorption of a hydroxyl (OH) group; 1715 cm^{-1}, carbonyl (CO) group; and 1690 cm^{-1} and 1635 cm^{-1}, α,β-unsaturated carbonyl group. Nuclear magnetic resonance (NMR) spectrometry of G1 showed the presence of one hydroxyl proton, one olefinic proton, two methylene protons (CH_2 adjacent to an optically active centre) of an ethylene group and a complicated systems of methyl groups.

Inhibitor G2 also recrystallised readily from cyclohexane, yielding colourless crystals, m.p. 127-8°C. The mass spectrum of G2 is very similar to that of G1, the two spectra differing in relative peak heights. The empirical formula of G2 as indicated by high resolution mass spectrometry is also $C_{15}H_{22}O_5$. The broad similarities in spectral properties of G1 and G2 also extend to the IR, UV and NMR spectra. Comparing spectral properties of G1 and G2, the structure for G2, as shown in Fig. 2, is proposed.

On this basis, inhibitors G1 and G2 have the same structural formulae, but differ in the arrangement of their groups in space, i.e. they are diastereomers. In this case, it would be expected that G1 and G2 would interconvert by random, reversible hemi-acetal formation. Under slightly alkaline conditions (pH 8) interconversion of G1 and G2 was indeed shown to occur, and this is strong supporting evidence for the proposed structure of G2 (Fig. 2). Both G1 and G2 are racemic mixtures of two enantiomers, and this explains their observed lack of optical activity.

Inhibitor G3 recrystallised readily from benzene, yielding colourless crystals, m.p. 169-171°C. Although the empirical formula of G3, as indicated by high resolution mass spectrometry, is $C_{14}H_{20}O_5$, the cracking pattern of the molecule during mass spectrometry is similar to that of G1, indicating basic similarities in structure, including the peroxide linkage. Other similarities between G3 and G1, together with the spectra indicating the similarities, are the presence of: one hydroxyl group (IR, NMR), an α,β-unsaturated ketone (IR, UV), an additional carbonyl group (IR), and various methyl but no ethyl, groups (NMR). G3 is also racemic. Careful comparison of the spectral properties of G3 with those of G1, led to the proposal of the structure of G3 (Fig. 2).

IV. PHYSIOLOGY OF THE INHIBITORS

The physiological properties of the three compounds were investigated in several large scale bioassays. These bioassays involved examination of the effect of G1, G2 and G3, over a wide range of concentrations (10^{-3}-10^{-8}M), on adventitous root formation in cuttings of mung bean, *E. deglupta* and *E. grandis*.

The mung bean bioassay used was a modification of that described by Hess (1964). We have reported on the techniques used in the *E. deglupta* bioassay (Paton et al., 1970).

The *E. grandis* bioassay was of importance because of its specificity for the detection of rooting inhibitors isolated from *E. grandis* tissue. For this bioassay, cuttings were prepared from a selected uniform batch of 17-19 cm tall, 11 weeks old, *E. grandis* seedlings. The seedlings were severed between the second and third leaf pair, and the lower nodes of the stem cutting defoliated. The cuttings were placed singly in test tubes (15 x 1.4 cm) containing loosely packed perlite that had been moistened by the test solution. The cuttings were protected by a plastic cover, but no mist spray was used. Daily additions of water to the perlite provided adequte moisture in the rooting medium, while allowing sufficient aeration for rooting. This procedure avoided the difficulties experienced in preliminary trials using water as the striking medium. Measurements of the time at which roots were visible, having penetrated the perlite to the test tube wall, were recorded. The results of the mung bean and *E. grandis* bioassays are summarised in Fig. 3. and Table 1, respectively.

In all bioassays the compounds G1, G2 and G3 completely inhibited root initiation at concentrations of 10^{-4}M whereas 10^{-5}M solutions had little if any effect on adventitious root formation. Results of the bioassays gave no clear indication as to which if any, of the inhibitors is the most active. Studies using differently substituted molecules may prove useful in indicating the portions of the molecule responsible for bioassay activity. There was no firm evidence of any promoting effects of the inhibitors at low concentrations.

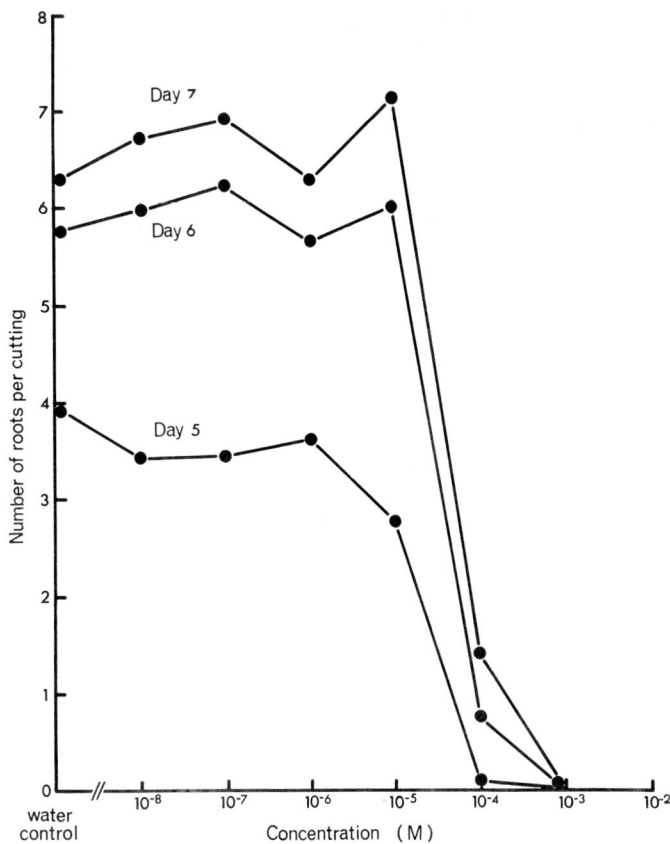

Fig. 3. Mung bean rooting bioassay of inhibitory compounds in *E. grandis*. The activity of compounds G1, G2 and G3 was not significantly different and the mean values given for number of roots per cutting are grouped for the total number of cuttings (84) at each concentration. The number of roots was determined 5, 6 and 7 days after commencement of bioassay.

Table 1. Effect of inhibitors G1, G2 and G3 on root initiation in cuttings of *E. grandis*.

Concentration (M)	Inhibitor			
	G1	G2	G3	Control : 10
10^{-3}	-	-	-	
10^{-4}	6	7	7	
10^{-5}	10	10	10	

Values in table are no. of cuttings (out of 10) rooted.

V. DISCUSSION

These three closely related compounds have several distinctive features which distinguish them from other identified plant growth regulating substances as well as substances previously identified from extracts of *Eucalyptus* species.

The chemical structures of the inhibitors are quite different from those of other identified plant substances. The most closely related known compounds appear to be the β-triketones e.g. leptospermone (Fig. 2), isolated from the steam volatile oils of *Eucalyptus* as well as of other genera of the family Myrtaceae (Hellyer 1968). However, G1, G2 and G3 appear to be the first reported compounds of this type with either a fused bicyclic structure or a peroxide function.

It is perhaps surprising that these unique compounds occurring at such high concentrations, have not been detected in previous studies in *Eucalyptus*. One possible reason is the use of either sodium carbonate, sodium bicarbonate, or both in methods commonly used to extract *Eucalyptus* tissue. Such a technique could lead to the breakdown of the inhibitors. Probably the main reasons for the successful isolation of the three *E. grandis* inhibitors were the availability of a convenient and reliable bioassay combined with the chemical approach.

It is proposed that G1, G2 and G3 are responsible for the increased rooting inhibitor concentration in successively higher leaf pairs of *E. grandis*. We suggest therefore that these inhibitors are involved in the associated decrease in rooting ability observed with ontogenetic ageing in seedlings of this species (Paton et al 1970).

An interesting feature of these inhibitors is the relatively high minimal concentrations (ca 10^{-4}M) required for activity in the bioassays used. This suggests that the physiological basis for activity of the inhibitors may differ from that typical of hormonal regulation where activity is often detected at much lower concentrations. Decomposition of the inhibitors in water complicated attempts to determine their real minimum level of activity, but apparently the low activity of the inhibitors may be reconciled with their relatively high concentration in the plant ($\sim 2 \times 10^{-3}$M). This concentration completely inhibits root development in all our bioassays. The natural concentration of the inhibitors in the plant appears to be sufficient to prevent adventitious root formation in tissue of adult *E. grandis*.

Although the natural concentration was shown to be fully active in rooting bioassays, it is doubtful whether the natural function of the inhibitors in the plant concerns inhibition of root formation in adult stem tissue. We should expect at least one of the natural roles of the inhibitors to be compatible with a gradual increase in concentration at successively higher leaf pairs of a *Eucalyptus* seedling, and also the apparent absence or presence at a much lower concentration, of such inhibitors in *E. deglupta* (Paton et al.,1970).

Among several interesting possibilities for the natural function of the inhibitors are activity as a fungicide, insecticide or an inhibitor of seed germination in the litter beneath *E. grandis* trees.

VI. SUMMARY

Isolation of three closely related inhibitors from adult tissue of *E. grandis* involved non-alkaline extraction and monitoring with appropriate bioassays. The chemical structure of one of the inhibitors (G1) was unequivocally determined by X-ray diffraction.

G1 is a fused bicyclic compound with a peroxide linkage. The empirical formula is $C_{15}H_{22}O_5$. The most closely related known compounds appear to be the β-triketones. A similar general structure for G2 and G3 is proposed, based on the similar spectral properties of all three compounds. The compounds differ in their substitution pattern at one C atom. Each compound is a racemic mixture of two enantiomers.

Each compound inhibited rooting of stem cuttings of *E. grandis* seedlings, *E. deglupta* seedlings and mung bean seedlings at concentrations of 10^{-4}M. Minimal concentrations for root inhibition are difficult to estimate because of the breakdown of the inhibitors in water to inactive compounds. However, the natural concentration of inhibitor in the plant (ca 2×10^{-3}M) appears sufficient to prevent adventitious root formation.

It is proposed that the three inhibitors are responsible for the increased rooting inhibitor concentration in successively higher leaf pairs, and, therefore, the apparently associated decrease in rooting ability, observed in seedlings of *E. grandis*.

VII. REFERENCES

HELLYER, R.O. (1968). The occurrence of β-triketones in the steam volatile oils of some Myrtaceous Australian plants. Aust. J. Chem. 21, 2825-8.

HESS, C.E. (1964). A physiological analysis of root initiation in easy and difficult-to-root cuttings. Proc. 16th Int. Hort. Congr. 1962, 375-381.

PATON, D.M., R.R. WILLING, W. NICHOLLS and L.D. PRYOR. (1970). Rooting of stem cuttings of *Eucalyptus* : a rooting inhibitor in adult tissue. Aust. J.Bot. 18, 175-183.

STERNS, M. (1971) Crystal and molecular structure of a root inhibitor from *Eucalyptus grandis*, 4-ethyl-1-hydroxy - 4,8,8,10,10, - pentamethyl - 7, 9 - dioxo - 2,3 - dioxabicyclo [4.4.0] decene - 5. J. Cryst. Mol. Structure (in press).

Studies with Plant Growth Inhibitors

R.L. Wain

Wye College, University of London, England

I. INTRODUCTION

In this paper I propose to deal with three aspects of our research at Wye all of which, in their different ways, are concerned with plant growth inhibition.

II. 3,5-DICHLOROPHENOXYACETIC ACID (3,5-D) AS A GROWTH RETARDANT

Over the past 25 years, many studies have been made on the relationships between chemical structure and plant growth-regulating activity in phenoxyalkane carboxylic acids (for detailed review see Wain and Fawcett (1969)). In one aspect of this work the carboxyl group and an α-hydrogen atom, together with an unsaturated ring system were shown to be necessary for activity in the phenoxyacetic acids. Substituents in the aromatic ring are also of importance in determining the order of activity shown; in particular, 3,5-dichloro-substitution leads to complete inactivity. All this and further research on stereochemical requirements for growth-regulating activity led to a theory on mode of action which envisaged the active molecule donating three essential groupings to specifically placed receptor groups at site of action (Smith and Wain, 1951). This theory provided a logical approach to competitive antagonism and such antagonism against phenoxy acids was clearly demonstrated in segment tests using a number of structurally related molecules, including 3,5-D, which were themselves inactive (Wain and Wightman, 1957). No inhibition of natural plant growth, however, was shown to occur following treatment with phenoxy acids.

Although 3,5-D is inactive, like 2,4-D and other active auxins it is readily translocated in plants. The possibility that such a foreign molecule, present within the tissues, might alter the chemical composition of the plant was recently examined by Smith and Mohsin (1970). Surprisingly, when tomato plants were sprayed with low concentrations of 3,5-D, a striking reduction in growth occurred. Thus for example, a solution of 10^{-4}M applied to young tomato plants led to a 70 per cent reduction in height after 3 weeks (Fig. 1). The growth of the main axis and apical dominance were markedly reduced and the leaves of the treated plants were darker in colour and somewhat smaller than those of control leaves. This growth retardation effect of 3,5-D on tomato plants could not be demonstrated in peas, beans, wheat, barley or any other of the plant species examined. This unique effect on tomatoes, however, has been shown to operate in presence of added indolylacetic acid (IAA) or gibberellic acid (GA$_3$) so there is no obvious interaction between these growth hormones and 3,5-D within the plant (Firn, 1971). A further series of Firn's experiments, which included studies with ^{14}C-labelled 3,5-D, failed to reveal the physiological mechanism by which 3,5-D exerts its growth retarding effect. Furthermore, no evidence of metabolic breakdown of the compound was obtained, though some conversion to a glucose ester has been shown to occur (Firn, 1971).

III. NEW SYNTHETIC GROWTH RETARDANTS

A number of synthetic compounds have been prepared which inhibit the growth of certain plant species. They are series of phosphonium, ammonium, and sulphonium salts, all of which are characterised by having a free positive pole within the molecule

$$ArCH_2 \overset{+}{P}R_3 X' \qquad ArCH_2 \overset{+}{N}R_3 X' \qquad ArCH_2 \overset{+}{S}R_2 X'$$

$$\text{phosphonium} \qquad \text{ammonium} \qquad \text{sulphonium}$$

(where Ar is an aryl grouping, R is alkyl and X' halogen ion). In the phosphonium series, where Ar is 2,4-dichlorophenyl and R is n-butyl we have the familiar Phosphon D, first described by Preston and Link (1958).

Fig. 1. Tomato plant 3 weeks after spraying with 3,5-dichlorophenoxyacetic acid (3,5-D) at 10^{-4}M. Control plant *on left*.

Some 58 compounds were examined for their activity in retarding the growth of wheat, pea and bean seedlings (Knight, Taylor and Wain, 1969). Of these, quaternary ammonium derivatives were found to be the most active and the following chemical structure/growth retarding activity relationships were found to operate:

1. When R is methyl, the compounds are inactive. Activity increases through ethyl and propyl to butyl and then decreases.

2. In the benzyl series with R = butyl, the 3- and 4- chloro derivatives show very high activity. The 2,3-, 2,4-, 2,5- and 3,4-dichloro- derivatives are all active but the 2,6- isomer is inactive and the 3,5- isomer has very low activity.

3. When Ar is α-naphthyl and R = n-butyl, the compound is highly active.

4. When an oxygen bridge is introduced between the methylene grouping and the ring system, activity is markedly reduced.

The two most active compounds arising from this work are 4-chlorobenzyl- and 1-naphthyl- tributylammonium bromides, both of which are finding practical use for retarding the growth of ornamental plants, e.g. chrysanthemums. They are also very effective with dwarf bean and soya bean (Fig. 2) and experiments are now proceeding to investigate possible uses in agriculture.

Fig. 2. Soya bean plants 25 days after spraying with (*on right*) 4-chlorobenzyl-tri-n-butylammonium bromide (B) at 1,000 ppm. and (*middle*) 1-naphthyl-tri-n-butylammonium bromide (N) at 1,000 ppm. Control plant *on left*.

The mode of action of these compounds is obscure. They do not appear to be anti-gibberellins and do not reduce the capacity of the fungus *Gibberella fujikuroi* to synthesise gibberellin. The molecular requirements for activity and the specificity which operates in the size of the three alkyl groupings would seem to indicate that they act within plant membranes.

Plant membranes are complicated in structure. It has been suggested that they contain a double film of orientated phosphatide ions as well as other materials which are lipophilic. Alternatively one membrane is envisaged as a lipoid barrier made up of steroids, glycerides, free fatty acids and phosphatides bounded on each side by a thin layer of protein (Davson and Danielli, 1952). Both of these conceptions provide for ionic groupings within the membrane and these charged centres must play an important part in the movement of organic anions and other ions through the membrane. There is also evidence that the protoplasmic membrane contains hydrophilic as well as strongly lipophilic groupings and the balance between lipophilic and hydrophilic character of a biologically active molecule is also a factor in relation to its ease of penetration.

From these considerations it is clear that a specifically shaped quaternary ammonium salt of the type we have been discussing might well become incorporated within the matrix of plant membranes. With its positive charge and lipophilic character such a compound might then modify membrane properties thereby influencing the working of the cell and processes associated with growth.

IV. XANTHOXIN, A NEW PLANT GROWTH HORMONE INHIBITOR

At the Vth International Plant Growth Substance Conference held in Paris in 1963 progress reports were presented on growth hormone inhibitors present in immature cotton fruits (Addicott et al., 1964), sycamore leaves (Wareing et al., 1964) and yellow lupin pods (Rothwell and Wain, 1964). In all cases the inhibitor has since been shown to be abscisic acid (ABA). The molecule of this inhibitor bears structural resemblances to that of Vitamin A and the fact that this vitamin can be produced in the animal liver from certain carotenoids led to speculations on whether carotenoid pigments might be precursors of ABA in plants. Furthermore, plants in the dark grow taller than those in the light and etiolated plants contain less hormone inhibitor than those grown in the light (Wright, 1954). Such considerations indicated that light might be involved in any conversion of carotenoid to inhibitor within the plant. To examine these possibilities, nettle leaf carotenoids deposited from an acetone solution on a filter paper disc were held in the light for 1 hr and seeds of cress were then sown on the moistened paper. None of the seeds germinated although germination of the seeds placed on similarly treated papers which had not been held in the light was complete (Taylor and Smith, 1967). Thus, simple exposure of the mixed carotenoids to light had led to the formation of a potent seed germination inhibitor. The inhibitor however was readily shown to be a neutral substance so it was not abscisic acid. A wide range of pure carotenoids was then examined using this procedure and it was found that the inhibitor precursors were certain xanthophylls, not all of which were equally effective; the epoxides violaxanthin and neoxanthin were found to produce the greatest inhibition on exposure to light (Taylor, 1968). The structural relationships between ABA and violaxanthin are shown below.

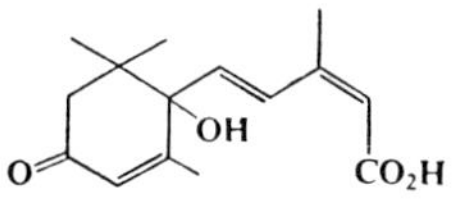

Abscisic acid

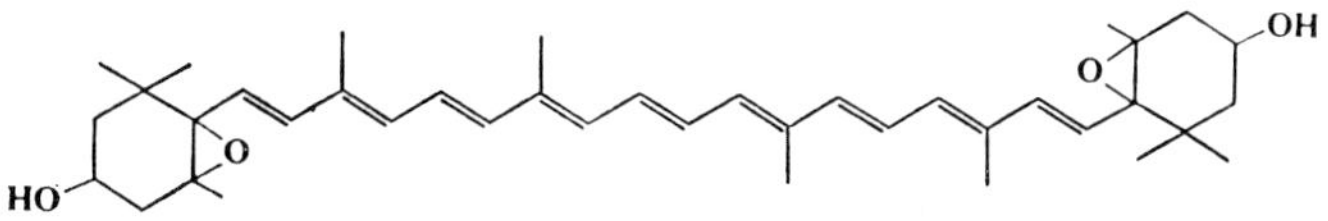

Violaxanthin

Large quantities of violoxanthin were extracted from orange peel, photo-oxidised, and three products were isolated. These were a butenone (I), an unsaturated lactone (II) (loleolide) - known to occur in perennial rye grass and other species, and a substituted pentadienal (III) (Taylor and Burden 1970a).

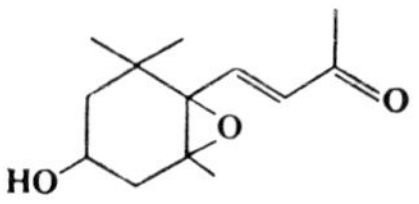

(I) 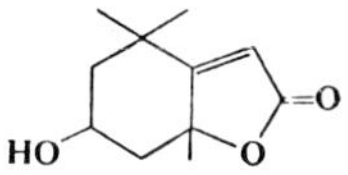   (II)

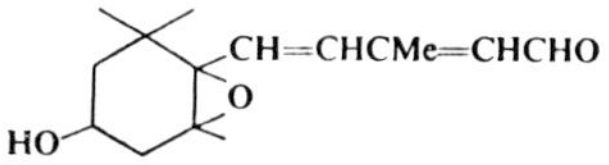

CH=CHCMe=CHCHO

(III)

While each of these showed some growth inhibitory activity, only the dienal was fully comparable with abscisic acid in the tests employed. To increase the yield of this inhibitor a neutral zinc permanganate oxidation was used in place of photolysis and by this means it was obtained as a syrup which ran as a single substance in thin layer chromatography (Burden and Taylor, 1970). In view of its formation by the oxidation of certain xanthophylls this new inhibitor was given the name *xanthoxin*.

Although xanthoxin appeared homogeneous, Burden found that it was readily acetylated with acetic anhydride in presence of pyridine to produce material which could be resolved by gas-liquid chromatography into two acetyl derivatives having almost identical mass spectra. Reduction of xanthoxin with sodium borohydride also gave two isomeric products. From this and other evidence it was shown that two geometric isomers of xanthoxin existed and these were assigned the 2-*cis*4-*trans* and 2-*trans* 4-*trans* configurations (Burden and Taylor, 1970).

cis,trans-   *trans,trans-*
xanthoxin

In biological tests, *cis, trans* xanthoxin possessed much greater inhibitory activity than the *trans, trans*-isomer (Taylor and Burden 1970b). Both isomers of xanthoxin have been found to occur in extracts of dwarf bean and wheat seedlings (Taylor and Burden, 1970c) and, more recently, in the shoots of many higher plants so it is clear that xanthoxin is a widely occurring plant growth hormone inhibitor.

The possibility exists however that the physiological activity of xanthoxin might not arise from the molecule acting *per se* but only after conversion to some other molecule such as abscisic acid. Indeed, Burden has recently developed a mild chemical oxidation procedure whereby the mixed xanthoxin insomers are converted to *cis, trans* abscisic acid and *trans, trans* abscisic acid respectively. These products showed optical rotatory dispersion and circular dichroism properties which are in excellent agreement with those of natural (+)-ABA and (+)-*trans* ABA. All these findings suggest a possible biosynthetic pathway to (+)-ABA by way of xanthophyll epoxides. Whether such a route operates within the living plant has yet to be established but evidence is now being obtained by Taylor which indicates that *cis, trans*-xanthoxin can be converted to (+)-ABA by tomato shoots. Much research however remains to be done, especially on xanthoxin metabolism and the environmental factors which affect its concentration in plant tissues. Such investigations will establish the physiological role of xanthoxin in the hormonal complex which controls plant growth.

V. SUMMARY

The chemistry of a number of synthetic phosphonium, ammonium and sulphonium compounds and their properties as growth retardants and the growth retardation exerted on tomato plants by 3,5-dichlorophenoxyacetic acid (3,5-D) are described. The mode of action of these compounds is discussed.

An account is given of the discovery of xanthoxin, a new naturally-occurring plant growth retardant which is formed from certain xanthophylls on exposure to light. The isolation and identification of the inhibitor and its stereochemical configuration are described, together with some of its biological properties and the possible physiological role played by xanthoxin in the hormonal complex which controls the growth of plants is discussed.

VI. REFERENCES

ADDICOTT, F.T., H.R. CARNS, J.L. LYON, O.E. SMITH and J.L. McMEANS (1964). On the physiology of abscisins. Coll. Int. Cent. Nat. Res. Sci. 123, 687.
BURDEN, R.S. and H.F. TAYLOR (1970). The structure and chemical transformations of xanthoxin. Tetrahedron Letters 47, 4071.
DAVSON, H. and J.F. DANIELLI (1952). "Permeability of Natural Membranes." Cambridge Univ. Press.
FIRN, R.D. (1971). Ph.D. Thesis. University of London.

KNIGHT, B.E.A., H.F. TAYLOR and R.L. WAIN (1969). Studies on plant growth-regulating substances. XXIX. The plant-growth retarding properties of certain ammonium, phosphonium and sulphonium halides. Ann. appl. Biol., 63, 211.

PRESTON, W.H. and C.B. LINK (1958). Use of 2,4-dichlorobenzyl tributylphosphonium chloride to dwarf plants. Plant Phys. (suppl.) 33, xlix.

ROTHWELL, K. and R.L. WAIN (1964). Studies on a growth inhibitor in yellow lupin (*Lupinus luteus* L.) Coll. Int. Cent. Nat. Res. Sci. 123, 363.

SMITH, M.S. and M. MOHSIN (1970). 3,5-dichlorophenoxy-acetic acid as a growth retardant. Ann. appl. Biol., 66, 233.

SMITH, M.S. and R.L. WAIN (1951). The plant growth regulating activity of *dextro*- and *laevo*- d-(2-naphthoxy)propionic acid. Proc. Roy. Soc., B139, 118.

TAYLOR, H.F. (1968). Carotenoids as possible precursors of abscisic acid in plants. Plant Growth Regulators. Soc. Chem Ind. Lond., Monograph No. 31, 22.

TAYLOR, H.F. and R.S. BURDEN (1970a). Identification of plant growth inhibitors produced by photolysis of violoxanthin. Phytochemistry, 9, 2217.

TAYLOR, H.F. and R.S. BURDEN (1970b). Xanthoxin, a new naturally occurring plant growth inhibitor. Nature, 227, 302.

TAYLOR, H.F. and T.A. SMITH (1967). Production of plant growth inhibitors from xanthophylls: a possible source of dormin. Nature, 215, 1513.

WAIN, R.L. and C.H. FAWCETT (1969). "Chemical Plant Growth Regulation. Plant Physiology" Vol. VA, ed. F.C. Steward, Academic Press, New York and London.

WAIN, R.L. and F. WIGHTMAN (1957). Studies on plant growth-regulating substances. XI. Auxin antagonism in relation to a theory on mode of action of aryl- and aryloxy- alkanecarboxylic acids. Ann. appl. Biol., 45, 140.

WAREING, P.F., C.F. EAGLES and P.M. ROBINSON (1964), Natural inhibitors as dormancy agents. Coll. Int. Cent. Nat. Res. Sci., 123, 377.

WRIGHT, S.T.C. (1954). A chromatographic study of auxins in relation to fruit morphogenesis and fruit drop in black currant (*Ribes nigrum*). Ph. D. thesis, University of Bristol.

GIBBERELLINS : THE CEREAL ALEURONE LAYER

Control of α-Amylase Synthesis in Isolated Barley Aleurone Layers by Gibberellic acid, Abscisic acid and Ethylene

John V. Jacobsen

Division of Plant Industry, C.S.I.R.O., Box 109, Canberra City, A.C.T., Australia

I. INTRODUCTION

Abscisic acid (ABA) antagonizes gibberellic acid (GA)-induced α-amylase synthesis in barley aleurone tissue (Chrispeels and Varner, 1966; Thomas et al., 1965). The system would seem to offer a good opportunity to study ABA action but studies so far have left its relation to GA in doubt (Chrispeels and Varner, 1966; Drury, 1969; Thomas et al., 1965).

One study (Chrispeels and Varner, 1966) has used Lineweaver-Burk analysis of data from amylase assays obtained after a certain time of incubation. In order to do this, variations in the time course of amylase synthesis due to the presence of different concentrations of GA and ABA must meet certain requirements. This report shows that these requirements are not met and that another method must be used to decide whether or not GA and ABA are competitive.

Nevertheless, it has been demonstrated that one cannot completely overcome the effects of ABA with additional GA (Chrispeels and Varner, 1966; Thomas et al., 1965) even at high GA excess, which leads to the conclusion that GA and ABA cannot be fully competitive. However, this report describes an effect of ethylene on ABA-inhibited amylase synthesis which may be the 'missing link' and shows that together GA and ethylene can completely annul the effect of ABA.

II. MATERIALS AND METHODS

Seeds of Himalaya barley (*Hordeum vulgare* L.) were used as the experimental material. The methods of seed imbibition and aleurone layer isolation were those of Chrispeels and Varner (1967).

Incubation Conditions

Isolated aleurone layers were incubated in 150 ml Erlenmeyer flasks containing 4.0 ml of medium or in 25 ml flasks containing 2.0 ml of medium. The medium contained 10^{-3}M acetate buffer of pH 4.8, 10^{-2}M calcium chloride and varying amounts of GA and ABA[1]. The flasks containing buffer, calcium chloride and GA were autoclaved and then cold sterilized ABA was added. One drop of a chloramphenicol solution (0.5 mg/ml) was added for each 2.0 ml of incubation medium to control growth of micro-organisms and then 10 aleurone layers were put into each flask.

Large flasks to which ethylene was to be added were sealed with rubber stoppers fitted with a glass tube into which was inserted a rubber septum. The rubber stoppers were sterilized with ethanol before being put into the flasks. The small flasks were

---

[1] RS - abscisic acid (2-cis, 4 trans-isomer)

fitted with rubber injection caps. Ethylene or dilutions of ethylene in air were injected through the septa with a hypodermic syringe. The gas samples injected were no greater than 150 µl. Where ethylene was not required, the cotton wool plugs used during autoclaving were left in the flasks.

Amylase production by aleurone layers was unaffected by sealing the incubation vessels with rubber stoppers so that any increase of carbon dioxide and decrease in oxygen levels in the flasks during incubation was of no apparent importance. The same applied to 25 ml Erlenmeyer flasks containing 20 aleurone layers.

The ethylene present in stoppered flasks was the total of added ethylene and ethylene already present. The ethylene already present was the amount of laboratory air at the time of sealing the flasks and this varied from 0.01 to 0.02 µl/liter of air. Hence, where 0.01 µl/liter of ethylene was added, the true level in the flask was between 0.01 and 0.03 µl/liter above controls which were in equilibrium with the gassing hood (outside) air. These considerations were important in experiments involving 0.01 µl/liter of added ethylene (figure 1) but were of no consequence where higher levels were used.

The flasks were incubated in a thermostated shaking water bath at $25°C$ usually for 24 hr. The bath was fitted with a gassing hood which was flushed continuously with air drawn from outside of the laboratory. If incubation was performed in laboratory air, the effect of ABA on amylase production varied from day to day presumably because of the fluctuating level of ethylene in the air. Outside air contained less ethylene (0.005 - 0.01 µl/liter) and the level was relatively stable. Because of the need for the gassing hood, all incubations occurred in darkness.

Tissue Extraction

After incubation, the medium was decanted and the aleurone layers washed twice in 4 ml of acetate buffer. They were then ground in a mortar and pestle with the aid of acid washed sand. Amylase extracted in this way was taken to be enzyme not yet released from the aleurone cells.

Amylase Assay

Amylase in both the medium (released amylase) and tissue extracts (unreleased amylase) was assayed as described by Chrispeels and Varner (1967).

Ethylene Determination

Ethylene was determined by gas chromatography as described by McGlasson (1969).

III. RESULTS

If 10^{-7}M GA was used to promote amylase synthesis in isolated aleurone layers, 1.5×10^{-8}M ABA was only slightly inhibitory while 2×10^{-7}M ABA was about 90% inhibitory (figure 1). This range of ABA concentrations has been used to demonstrate the effect of ethylene. Responses to ABA and ethylene were seen mainly in the medium amylase. At low levels of ABA, ethylene reversed the inhibition, but at higher levels, the reversal was only partial. Two aspects of these data are noteworthy. Firstly, the amount of ethylene necessary to cause some reversal was very small, 0.01 µl/liter being sufficient to have a marked effect. Secondly, there was a dose response to ethylene and the amount necessary to saturate the reversal increased with increasing ABA concentration. Complete reversal of the effect of 1.5×10^{-8}M ABA was accomplished by 0.01 µl/liter of ethylene while 0.1 and 1 µl/liter caused maximum (but not complete) reversal of 4×10^{-8}M ABA and 2×10^{-7}M ABA respectively. The two higher levels of ABA and the ethylene treatments superimposed on them had no effect on the level of extract amylase (figure 1), but at the lowest level of ABA, ethylene caused significant, although very small, reduction.

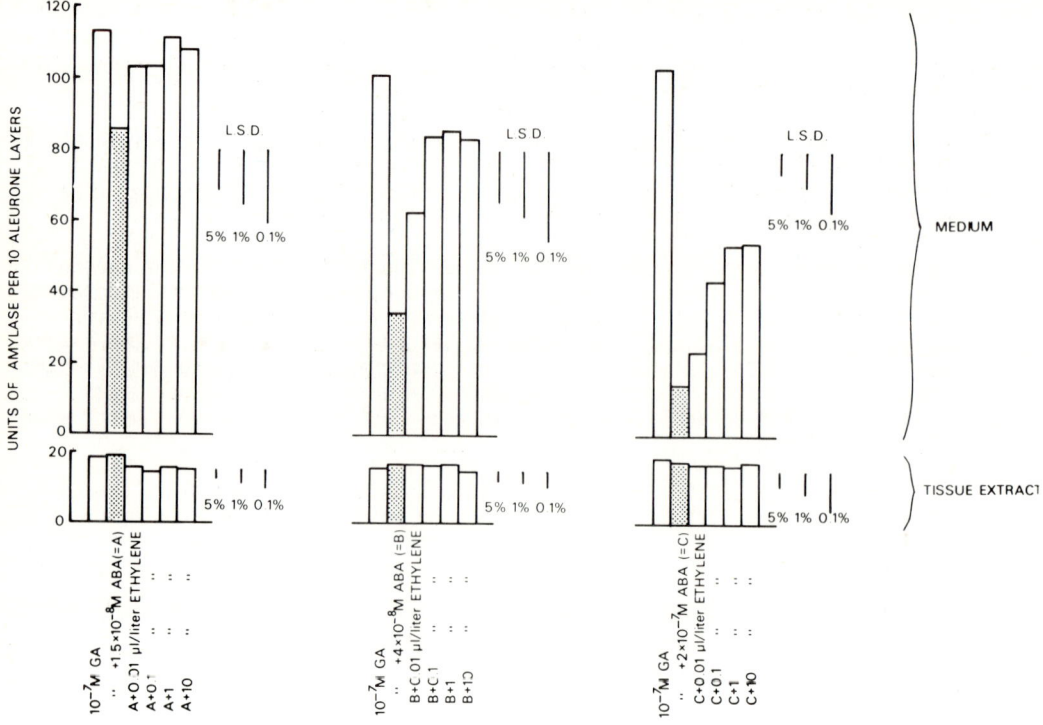

Fig. 1. The reversal of ABA inhibition of amylase synthesis by ethylene. Aleurone layers were incubated with 10^{-7}M GA and three different concentrations of ABA. The effects of 0.01 - 10 μl/liter of ethylene on the inhibition caused by each ABA concentration are shown. The amylase appearing in both the incubation medium and in the tissue extract after 24 hr incubation was measured.

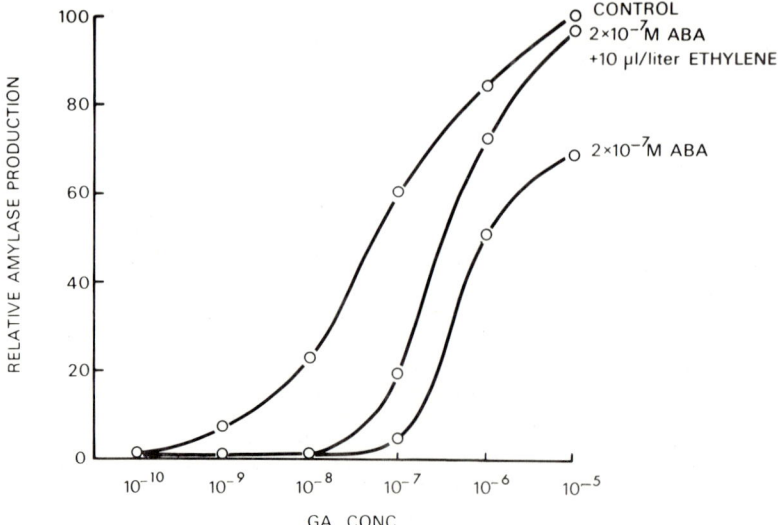

Fig. 2. The reversal of ABA inhibition by increasing amounts of GA in the presence and absence of ethylene. All incubations occurred for 24 hr. The values given are for released amylase only.

The effects of ethylene alone and in the presence of GA were small and probably do not complicate the results involving ABA but nevertheless they were consistent (Table 1). Ethylene alone gave rise to some released and unreleased amylase but the amounts were small in comparison with the effect of GA, and clearly ethylene cannot substitute for GA. In the presence of GA, ethylene caused a small increase in released amylase but had no effect on unreleased amylase. Because this contrasted with previously published data (Jones, 1968), the experiment was repeated on younger seed (1969 harvest) but the effect was the same. These results are similar in principle to those of figure 1 and perhaps they reflect the presence of small amounts of ABA in aleurone tissue although there is no evidence for this.

Table 1. The effect of Ethylene on Amylase Synthesis in the Presence and Absence of GA.

Treatment	Units of amylase/10 aleurone layers		
	Medium	Extract	Total
1964 seed			
Control	4.5	7.3	11.8
Ethylene (10 µl/liter)	8.0	11.4	19.4
GA (10^{-7}M)	61.8	17.7	79.5
GA (10^{-7}M) + ethylene (10 µl/liter)	66.8	17.7	84.5
1969 seed			
Control	1.7	4.3	6.0
Ethylene (10 µl/liter)	4.1	8.7	12.8
GA (10^{-7}M)	73.8	34.7	108.5
GA (10^{-7}M) + ethylene (10 µl/liter)	89.1	32.0	121.1

Gas analysis showed that the ethylene contained 0.015% of ethane, which was the major contaminant, but the effect ascribed to ethylene could not be obtained with pure ethane, methane or propylene at concentrations up to 100 µl/liter. Neither kinetin nor auxin gave the effect.

Figure 2 shows that increasing concentrations of GA partly overcame the effect of ABA. The control and 2×10^{-7}M ABA curves converge to some extent demonstrating some degree of interaction between GA and ABA. Since both GA and ethylene could partially overcome the ABA inhibition, it seemed possible that, combined, they could completely overcome the ABA inhibition. To test this, a situation was required where the effect of a certain concentration of ABA was reversed as much as possible by GA. Such a situation is shown in figure 2 where, in the presence of 2×10^{-7}M ABA, increasing concentrations of GA progressively reversed the inhibition of amylase production until at 10^{-6}M and 10^{-5}M GA, the lines converge no more and the inhibitions of amylase synthesis are essentially the same. Hence at 10^{-5}M GA, the reversal is probably maximal. At this point, addition of 10 µl/liter of ethylene overcame the remaining inhibition. At lower concentrations of GA, the effect of ethylene was progressively diminished until at 10^{-8}M GA, where in the presence of ABA, there was almost no amylase synthesized, ethylene had no effect even at levels of 1000 µl/liter. The data show that ethylene and GA together can completely counteract the inhibition caused by ABA and that for ethylene to produce any response at all, ABA must be partially counteracted by GA.

Figure 1 shows that total amylase synthesis was increased by ethylene under conditions where GA was present at sub-saturating levels and table 2 shows that even when GA is present at saturating levels and ethylene can completely restore amylase release to control levels, the increased amylase release represents an increase in total synthesis and not simply a movement of enzyme from inside to outside of the tissue. Therefore both GA and ethylene are able to restore amylase synthesis in the presence of ABA.

Table 2. The Effect of Ethylene on Amylase Synthesis in a Situation Where the GA Reversible Component of ABA Inhibition was Saturated.

Treatment	Units of amylase/10 aleurone layers		
	Medium	Extract	Total
10^{-5}M GA	135	14	149
10^{-5}M GA, 2×10^{-7}M ABA	79	16	95
10^{-5}M GA, 2×10^{-7}M ABA, 10 µl/liter ethylene	137	11	148

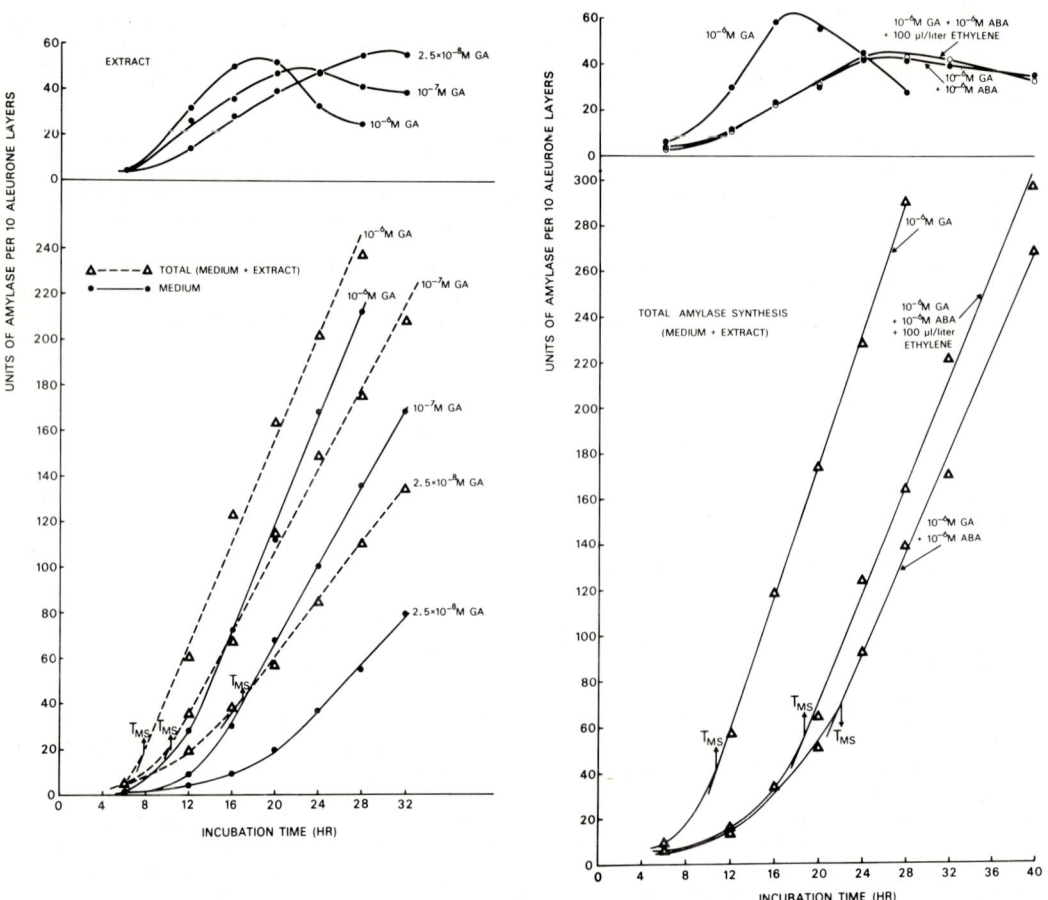

Fig. 3. The time course of amylase synthesis in the presence of different concentrations of GA. T_{MS} is the time taken to reach the steady rate of synthesis R_{MAX}.

Fig. 4. The effects of ABA and ABA + ethylene on the time course of amylase synthesis. T_{MS} is the same as in figure 3.

It has been shown (Chrispeels and Varner, 1967) that the production of amylase begins some hours after the addition of GA (the lag phase) and that the rate of synthesis increases until a constant rate (R_{MAX}) is attained (the linear phase). Use of Lineweaver-Burk kinetics to examine the interactions described above requires that the effects of different GA, ABA and ethylene concentrations be on the slopes of the linear phases of the time course curves and that all such curves must project to meet on the time axis. Time course studies were made to examine this.

The effects of different concentrations of GA are shown in figure 3. As GA concentration increases from $3 \times 10^{-8}M$ to $10^{-7}M$ and from $10^{-7}M$ to $10^{-6}M$, the linear phases of the curves of amylase (both total and released) have increasing slopes demonstrating increasing ultimate rates of amylase synthesis (R_{MAX}) but the time necessary to reach the linear phase (T_{MS}) becomes less. Maximum amylase in the extract occurred earlier as GA concentration was increased and occurred about 11 hours after T_{MS} for each GA concentration. Hence it appears that the overall effect of increasing the GA concentration is to speed up all phases of amylase synthesis and because T_{MS} becomes less, the linear parts of the curves do not meet on the time axis if projected back.

The effects of ABA and ethylene on the time course of amylase synthesis are shown in figure 4. ABA increased T_{MS}, reduced R_{MAX} and increased the time taken for extract amylase to reach its maximum as though the effective concentration of GA had been reduced. Addition of ethylene to ABA-treated aleurone caused an increase in R_{MAX} and although a reduction of T_{MS} appeared to occur, effects of ethylene on T_{MS} and the time taken to reach peak amylase must remain doubtful until better data are obtained.

These results show that although all three compounds regulate R_{MAX}, the fact that T_{MS} changes, at least for GA and ABA, caused projections of the linear parts of the curves to intersect below the time axis. Therefore the use of amylase assays at some time after GA addition to construct a Lineweaver-Burk plot appears to be invalid.

It is apparent that some other measurement of the effect of ABA on amylase synthesis is necessary to establish the relationship between GA and ABA and between ABA and ethylene and R_{MAX} would seem to be the right parameter to measure.

IV. DISCUSSION

This is the first report of ethylene interfering with ABA action and in this system it can be a very potent regulator of amylase synthesis. As little as 0.01 μl/liter of added ethylene was sufficient to cause a profound change in ABA-inhibited amylase synthesis. Since this amount of ethylene was superimposed on atmospheric ethylene, the effect of small quantities of the gas in an ethylene-free system could be much greater than that described here.

The effect of ethylene is to control total amylase synthesis, not simply to promote the transport of amylase from inside to outside the aleurone cells, and it does this at least by raising the maximum rate of amylase synthesis. How it does this is not known but the results given here allow several possibilities to be eliminated. Ethylene does not substitute for GA in inducing amylase synthesis nor does it affect the response to GA. Hence ethylene acts neither by causing the tissue to make GA nor by making GA more effective. If ethylene caused ABA breakdown, presumably this would be progressive with time and the effect on the time course would be to transform the linear phase into a curve of increasing slope. This does not occur. Because ethylene can fully restore amylase synthesis in the presence of a saturating GA concentration the two possibilities I consider to be most likely are that ABA and ethylene are competitive or that ABA inhibits ethylene synthesis by the tissue and ethylene may be required for amylase induction.

Because ethylene can complete the reversal of the ABA inhibition when additional GA is ineffective, the simplest way to explain the results is that ABA has two points of action, one an ethylene block and the other a GA block. Both control amylase synthesis. Ethylene has no effect on amylase synthesis if there is a strong GA block by ABA, and although it has not been shown that a strong ethylene block by ABA renders GA ineffective, the data are consistent with it. It could be that atmospheric ethylene is sufficient to allow the GA effect to be partially expressed and that added ethylene alleviates the ethylene block sufficiently for the full effect of GA to be seen.

The ultimate aim of this work is to establish whether or not GA and ABA are competitive but this has not yet been accomplished. The results indicate that previous studies have been inconclusive for two reasons. Firstly, using the Lineweaver-Burk plot as it has been used appears to be invalid and secondly, the amount of interaction one can show between GA and ABA could depend on the amount of ambient ethylene. If it was high, GA could completely reverse ABA inhibition but if it was low, very little interaction would be demonstrated.

The occurrence of an ethylene-ABA antagonism in barley aleurone leads one to think that perhaps the mechanism may explain some of the effects of ethylene on other plant tissues. For example, ethylene has been shown to break dormancy in seeds of peanuts (Toole *et al.* 1964; Ketring and Morgan, 1969), clover (Esashi and Leopold, 1969), wheat (Balls and Hale, 1940) and other plants (Vacha and Harvey, 1927) and the presence of inhibitors, perhaps ABA, has been linked with dormancy in many seeds (for reviews see Addicott and Lyon, 1960; Wareing 1965). GA also very commonly breaks seed dormancy (Wareing, 1965) and the relation between GA and ABA in aleurone possibly explains this. Very similar arguments can be proposed for bud dormancy. ABA occurs in buds (for review see Addicott and Lyon, 1969) and probably accounts for dormancy in a number of species. GA and ethylene commonly overcome bud dormancy (for review see Vegis, 1965).

V. SUMMARY

Gibberellic acid (GA)-induced α-amylase synthesis in barley aleurone layers was inhibited by abscisic acid (ABA) and the inhibition was partly reversed by additional GA and by ethylene. Together, GA and ethylene could completely reverse the ABA effect. The data are consistent with ABA having two points of action, each exercising control over amylase synthesis.

Because GA can completely annul ABA action in saturating ethylene, it is possible that GA and ABA are competitive but studies of the effects of GA, ABA and ethylene on the time course of amylase synthesis show that care must be exercised in the use of kinetic data to determine whether or not this is so.

The possible relevance of the ethylene effect to other plant systems is discussed.

VI. ACKNOWLEDGEMENTS

The author is indebted to Mrs. Helen Chadim for her skilful technical assistance, to Dr. W.B. McGlasson who helped with and provided gas chromatography equipment for ethylene analyses and who provided samples of methane, ethane and propylene, and to Dr. C.F. Konzak who provided the Himalaya barley seed. I am also grateful to F. Hoffmann-La-Roche and Co. Ltd. (Switzerland) for its generous gift of abscisic acid.

VII. REFERENCES

ADDICOTT, F.T. and J.L. LYON (1969). Physiology of abscisic acid and related substances. Ann. Rev. Pl. Physiol. 20, 139-164.
BALLS, A.K. and W.S. HALE (1940). The effect of ethylene on freshly harvested wheat. Cereal Chem. 17, 490-494.

CHRISPEELS, M.J. and J.E. VARNER (1966). Inhibition of gibberellic acid induced formation of α-amylase by abscisin II. Nature 212. 1066-1067.

CHRISPEELS, M.J. and J.E. VARNER (1967). Gibberellic acid-enhanced synthesis and release of α-amylase and ribonuclease by isolated barley aleurone layers. Pl. Physiol. 42, 398-406.

DRURY, R.E. (1969). Interaction of plant hormones. Science 164, 564-565.

ESASHI, Y. and A.C. LEOPOLD (1969). Dormancy regulation in subterranean clover seeds by ethylene. Pl. Physiol. 44, 1470-1472.

JONES, R.L. (1968). Ethylene enhanced release of α-amylase from barley aleurone cells. Pl. Physiol. 43, 442-444.

KETRING, D.L. and P.W. MORGAN (1969). Ethylene as a component of the emanations from germinating peanut seeds and its effect on dormant Virginia-type seeds. Pl. Physiol. 44, 326-330.

McGLASSON, W.B. (1969). Ethylene production by slices of green banana fruit and potato tuber tissue during the development of induced respiration. Aust. J. Biol. Sci. 22, 489-491.

THOMAS, T.H., P.F. WAREING, and P.M. ROBINSON (1965). Action of the Sycamore 'Dormin' as a gibberellin antagonist. Nature 205, 1270-1272.

TOOLE, V.K., W.K. BAILEY and E.H. TOOLE (1964). Factors influencing dormancy of peanut seeds. Pl. Physiol. 39, 822-832.

VACHA, G.A. and R.B. HARVEY (1927). The use of ethylene, propylene and similar compounds in breaking the rest period of tubers, bulbs, cuttings, and seeds. Pl. Physiol. 2, 187-194.

VEGIS, A. (1965). Dormancy in organs and tissues. Encycl. Pl. Physiol. XV/2. pg 534-668.

WAREING, P.F. (1965). Endogenous inhibitors in seed germination and dormancy. Encycl. Pl. Physiol. XV/2. pg. 909-924.

CYTOCHEMICAL LOCALIZATION OF GIBBERELLIC ACID – INDUCED ENZYMES IN THE
BARLEY ALEURONE LAYER

John V. Jacobsen and R.B. Knox

Division of Plant Industry, C.S.I.R.O., P.O. Box 109, Canberra City, A.C.T., Australia,
and
Department of Botany, Australian National University, P.O. Box 4, Canberra, A.C.T.
Australia

I. INTRODUCTION

The induction of α-amylase in barley aleurone tissue is now well documented and the system has been used by a number of workers to study GA action. It has been established that GA induces the *de novo* synthesis of α-amylase (Filner and Varner, 1967) and protease (Jacobsen and Varner, 1967) and probably other hydrolases which are released by the cells. However relatively little attention has been given to the cytology of enzyme release.

Vesicles apparently derived from endoplasmic reticulum have been described by Van der Eb and Nieuwdorp (1967) and by Jones (1969a), who have proposed that they are associated with enzyme release but no evidence associating these vesicles with GA-induced enzymes has yet been presented.

The aim of the present study has been to use a variety of high resolution cytochemical methods to localize three GA-induced enzymes in aleurone cells at the light microscope level to prepare the way for fine structure studies of the way in which enzyme release occurs.

II. MATERIALS AND METHODS

Tissue preparation

Aleurone layers were obtained from barley seed (*Hordeum vulgare*) cultivar Himalaya. The seeds were imbibed for 72 hr and the aleurone isolated as described by Chrispeels and Varner (1967). For treatment with GA, isolated layers were incubated for 16 or 24 hr in a medium containing 10^{-3}M acetate buffer pH 4.8, 10^{-2}M calcium chloride and 10^{-6}M GA.

Where required, tissue was fixed in 2.5% glutaraldehyde in 0.05M cacodylate buffer at pH 7.2 or in 4% paraformaldehyde in the same buffer, in both cases for 6 hr at ice water temperature. The layers were then washed for 15-24 hr in cold buffer.

The layers were then set in the gelatin-glycerol medium of Knox (1970) for freeze-sectioning. The blocks were rapidly cooled to $-15°$C or $-24°$C and 2 μm sections were cut. The sections were mounted on slides which had been previously coated with 1% gelatin and air dried. This ensured retention of protoplasts in the procedures described below.

Time courses of enzyme production

Aleurone layers were incubated for various periods of time and enzyme was assayed in both the media and tissue extracts as described by Jacobsen et al. (1970). Amylase was assayed using the method of Chrispeels and Varner (1967), acid phosphatase with p-nitrophenyl-phosphate as substrate using standard techniques and peroxidase using the method of Filner and Varner (1967). Esterase was assayed using α-naphthyl acetate as substrate and the α-naphthol was coupled to Fast Blue B. The coupled product was extracted with ethyl acetate and the optical density at 500 nm was measured.

Enzyme localization:

Amylase was localized using two different methods.

(1) Substrate film technique. Starch films were prepared by coating slides with a 6% solution of Connaught starch and allowing them to air dry. Sections of fresh tissue were placed directly on the film and left for 2-4 min. The section was washed from the film under running tap water and the film was then stained in an iodine-potassium iodide solution or alternatively both film and section were stained in iodine. Areas of amylase activity were clear while unaffected areas of the film were blue. Controls were prepared by placing sections of fresh tissue on films which had been pre-stained with iodine-potassium iodide. No activity (clear areas) was detectable after 2-4 min incubation.

(2) Immunofluorescence technique. α-amylase from barley was highly purified as described by Jacobsen et al. (1970) and two rabbits were each immunized at weekly intervals with 3 injections each containing 1-2 mg α-amylase protein. The first injection was given with complete Freund adjuvant and the others with incomplete adjuvant. A high titre anti-serum was obtained and its specificity was confirmed by immunodiffusion and by immunoelectrophoresis against purified amylase antigen and a mixture of GA-induced barley proteins. These data will be presented elsewhere. Normal (pre-immunization) rabbit serum was inactive when tested against amylase antigen. The immunofluorescence procedures used were similar to those described by Knox et al. (1970) as modified by Knox (1971).

Acid phosphatase was localized by the method of Barka and Anderson (1962) using α-naphthyl acid phosphate as the substrate in a simultaneous coupling reaction with hexazonium pararosanilin. The procedures were essentially those described by Knox and Heslop-Harrison (1970). Controls were run by omitting the substrate.

Peroxidase was localized by the method of Straus (1964). Control sections incubated in the absence of hydrogen peroxide showed no staining.

Electron microscopy:

Tissue was fixed in 4% formaldehyde in 0.025 M phosphate buffer at room temperature for 2 hr. It was then washed in the same buffer and post-fixed in osmium tetroxide in the buffer for 2 hr at room temperature. The tissue was washed again in buffer, dehydrated in ethanol and embedded in Epon. Sections were post-stained with uranium and lead.

III. RESULTS

The time courses of enzyme release for four GA-induced enzymes are shown in Figure 1.

Firstly, they show that cytochemical studies of GA-induced enzymes should be done on tissue incubated for 16 hr or more because at this time, enzyme induction is well under way. Secondly, the data show some correspondence in release of the enzymes. The curves for peroxidase and amylase coincide and release commences 10-11 hr after the beginning of GA treatment. The initially high esterase occurred in the control

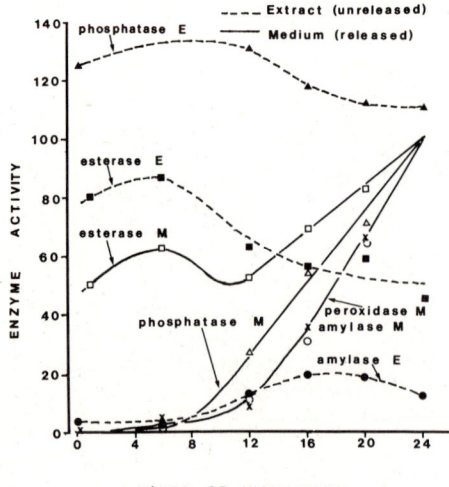

Fig. 1. Time courses of production of amylase, peroxidase, acid phosphatase and esterase in response to gibberellic acid. Except for peroxidase, enzyme activity appearing both in the medium (M) and in an extract of the tissue (E) are shown. The amounts of each enzyme in the medium after 24 hr incubation have been assigned a value of 100 and all other amounts in both media and extracts are relative to this.

(no GA) medium as well and probably represents enzyme solubilized from the surface of the aleurone tissue. The trend of esterase in the medium over the first 11 hr of incubation was also the same for control and GA-treated tissue, but the increase in medium enzyme of GA-treated tissue beginning at about 11 hr did not occur in control and therefore represents GA-induced enzyme. It has been shown previously (Jacobsen and Varner, 1967) that amylase and protease are released simultaneously so there is some evidence that induction and release of amylase, peroxidase esterase and protease is a unified process. However the beginning of release of GA-induced acid phosphatase appears to occur about 2 hr before the other enzymes. Thirdly, figure 1 shows that there was some amylase and relatively large amounts of esterase and acid phosphatase in imbibed tissue before incubation. The tissue also contained peroxidase but the data are not given owing to difficulties in obtaining a quantitative estimate of the enzyme in a tissue extract. In the cytochemical studies it was necessary therefore to compare GA-treated tissue with imbibed tissue or with tissue which had been incubated in the absence of GA.

Amylase localization:

In both fresh imbibed tissue and in that incubated for 16 hr in buffer only, both substrate film and immunofluorescence techniques showed similar patterns (Figs. 2A and 3A). Most of the activity was detected in the peripheral cytoplasm, usually in a thin band adjacent to the plasmalemma especially towards the endosperm side of the cells. The cytoplasmic localization was confirmed using tissue which had been plasmolysed in 1M sucrose prior to freeze-sectioning, where the cytoplasm (and amylase activity) had withdrawn from the cell walls. Some activity was detected around the aleurone grains, apparently at the surface of the grains and especially in cells adjacent to the starchy endosperm (labelled E). In tissue which had been centrifuged as described by Jones (1969b) prior to sectioning, the aleurone grains and amylase sedimented together so that the enzyme is firmly attached to the aleurone grains.

After 16 hr treatment of the tissue with GA, the amylase pattern had changed considerably (Figs. 2B and 3B). The aleurone grains had enlarged and while some still showed intense peripheral staining, most of the enzyme was diffuse throughout the cytoplasm though still most active at the periphery. As for imbibed tissue, the peripheral cytoplasmic amylase tended to be concentrated along the endosperm sides of the cells.

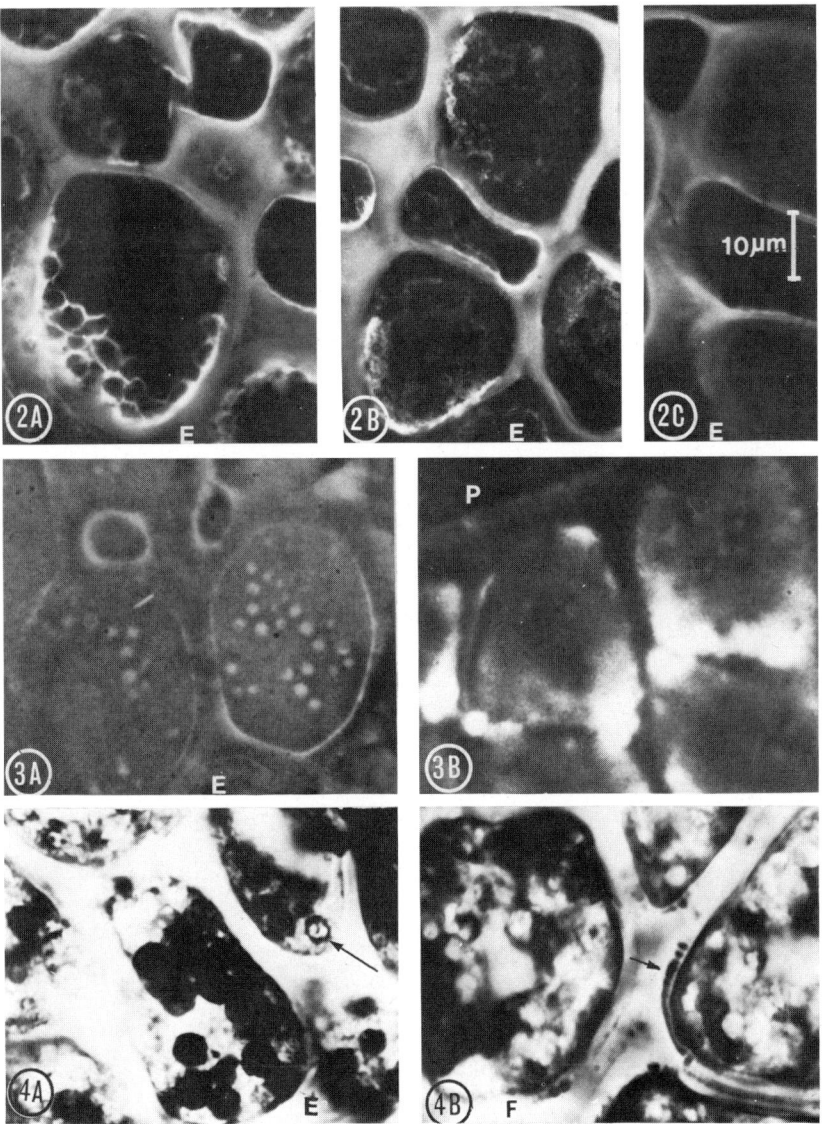

Fig. 2. Immunofluorescence localization of α-amylase. A shows tissue incubated in buffer for 16 hr before sectioning and reacting with anti-amylase serum. Fluorescence appears white in the photograph; B shows tissue incubated for 16 hr in GA, reacted with anti-amylase serum; C control, tissue as A but normal serum used instead of anti-amylase serum. No detectable fluorescence. E indicates starchy endosperm side of tissue. All Figs. at same magnification, indicated on Fig. 2C.

Fig. 3. Starch substrate-film method for amylase localization. A shows imbibed tissue, B shows tissue GA-treated for 16 hr. P indicates pericarp. Magnification - as in Fig. 2C.

Fig. 4. Peroxidase localization. A shows imbibed tissue; arrow shows activity at periphery of a sectioned aleurone grain. B shows tissue treated with GA for 24 hr; arrow shows pockets of activity in cell walls.

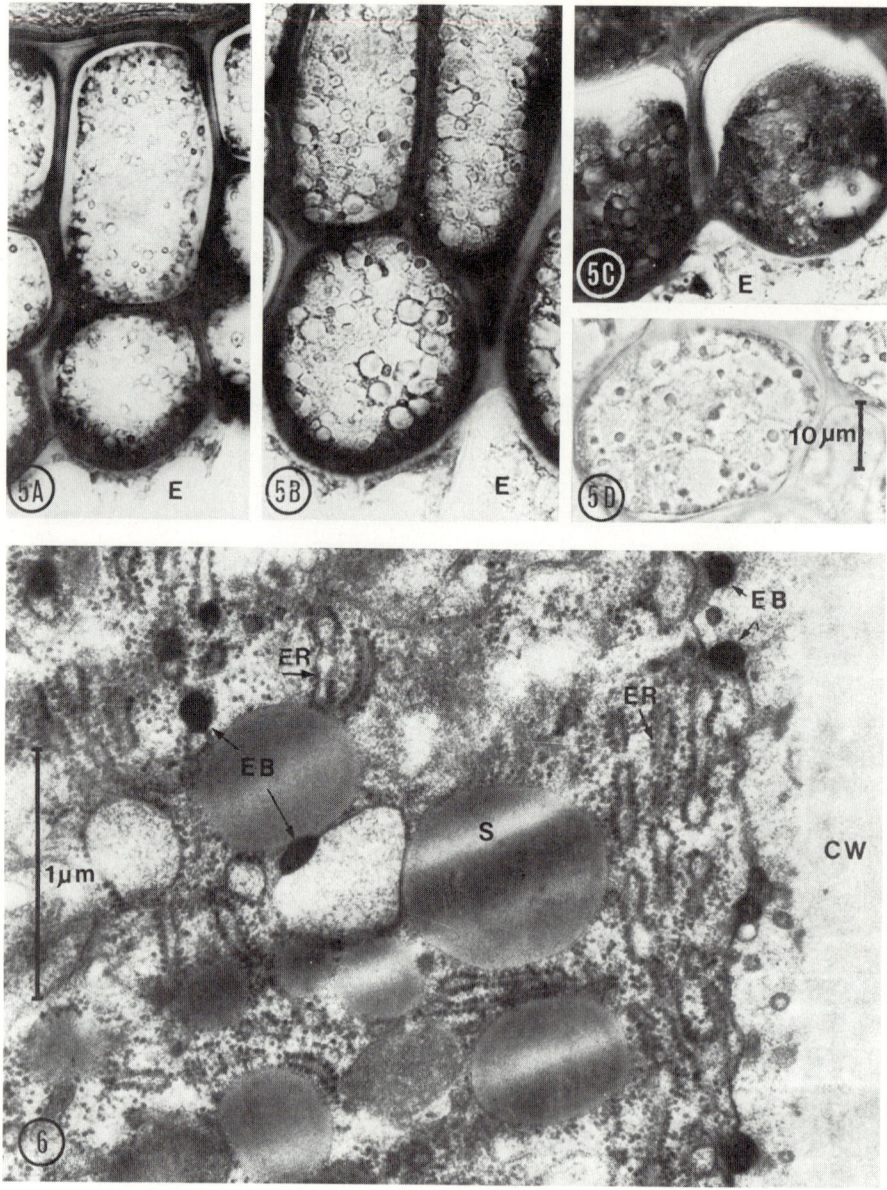

Fig. 5. Acid phosphatase localization. A shows unfixed imbibed tissue; B shows glutaraldehyde-fixed imbibed tissue; C shows glutaraldehyde-fixed tissue incubated with GA for 16 hr; D shows control, glutaraldehyde-fixed imbibed tissue, substrate omitted.

Fig. 6. Electron micrograph from 16 hr GA-treated tissue showing periphery of cytoplasm and cell wall (CW) with stratified ribosomal endoplasmic reticulum (ER), electron dense bodies (EB) and spherosomes (S).

Peroxidase localization:

In fresh imbibed tissue, most of the peroxidase was associated with aleurone grains (Fig. 4A) and where aleurone grains had been sectioned (see arrow) it could be seen that the enzyme was confined to the periphery of the grains. There was some peroxidase in the cytoplasm especially adjacent to the plasmalemma.

GA-treatment of aleurone for 24 hours caused a much more diffuse peroxidase distribution (Fig. 4B). The undilated aleurone grains had lost their dense staining and most of the enzyme was distributed throughout the cytoplasm but concentrated at the periphery. In the cell walls, in addition to some diffuse activity, there were small pockets of enzyme (Fig. 4B) distributed tangentially around the cell and these tended to occur on the endosperm sides of the cells.

Acid phosphatase localization:

In fresh imbibed tissue (Fig. 5A) and in glutaraldehyde-fixed and washed layers (Fig. 5B) acid phosphatase activity was greatest in the peripheral cytoplasm adjacent to the plasmalemma, though in glutaraldehyde-fixed cells some activity was observed around the aleurone grains, and in fresh tissue dense reaction product occurred in the cell wall. It is interesting that this cell wall activity was absent in tissue imbibed for only 6 hr, so the enzyme has accumulated in the cell walls during imbibition. Enzyme activity was greatest in the corners of the cells and at the endosperm ends of the cells. In GA-treated fresh tissue, there was little evidence of the dense cell wall deposits of reaction product, and as in glutaraldehyde-fixed layers (Fig. 5C) most activity was diffuse throughout the cytoplasm, and concentrated around aleurone grains remaining after 16 hours incubation. In both imbibed and GA-treated clutaraldehyde-fixed tissue, the activity is detectable in the light microscope as small particles about the limits of resolution (c. 0.2 μm).

Ultrastructure:

GA-treated tissue has been compared with untreated tissue for structural features which could account for the enzyme localizations seen in the light microscope. In addition to abundant rough endoplasmic reticulum, structures which we consider to be of interest are the small electron dense bodies shown in Figure 6. They are approximately 0.1-0.2 μm in diameter and appear to originate in the cisternae of the rough endoplasmic reticulum. The bodies also occur outside of the endoplasmic reticulum and therefore could provide a vehicle for enzyme release. In tissue treated with GA for 24 hr, we have seen strongly osmiophilic material in pits in the cell wall similar to those shown in Figure 4B. The pits are probably zones of cell wall hydrolysis and osmiophilic substance accumulates in them, much as peroxidase did.

IV. DISCUSSION

The high resolution of the cytochemical techniques used in this study have made it possible to observe and analyze the marked changes in the protoplasm of the aleurone tissue during incubation in GA. The initial low level of enzyme activity detectable in imbibed tissue is readily accounted for by the intense activity detected in the peripheral cytoplasm and around aleurone grains. In 16 and 24 hour GA-treated tissue, however, activity of all 3 enzymes investigated was diffuse throughout the protoplast, but again concentrated towards the periphery of the cytoplasm especially at the endosperm end of the cell. Since release must occur through the plasmalemma, the finding of such high levels of activity associated with the cytoplasm adjacent to it, suggests that this region is the site of production and release of the GA-induced enzymes. Similar conclusions have been reached by Knox and Heslop-Harrison (1970) in cytochemical and ultrastructural studies of the incorporation of various hydrolytic enzymes into the cellulose layer of the pollen grain wall. All the cytochemical techniques used are capable of ready adaption for ultrastructural studies which are in progress. Light microscope observations have, of course, only limited value and are only the first step towards determining how enzyme release occurs in response to GA.

The abundance of stratified ribosomal ER in GA-treated tissue (see Fig. 6) suggest that the peripheral protoplast is amply endowed with the machinery of protein synthesis. The finding of the electron dense bodies closely associated with both the ER cisternae and the plasmalemma is suggestive of release in progress, but it remains to be seen whether these will prove to be the sites of enzyme activity. Their apparent absence in imbibed tissue needs to be explained.

A novel feature of this study has been the use of immunofluorescence for localization of a plant enzyme. This was made possible by the availability of quantities of highly purified α-amylase antigen. Despite the high autofluorescence of the cell walls, the intensity of the specific fluorescein-isothiocyanate-labelled precipitins is clearly detectable even with exposure to antiserum for periods as short as 40 min at room temperature. Should the electron dense bodies (Fig. 6) prove to be enzyme-containing vesicles, then of course, this would be difficult to demonstrate by visual electron-histochemical techniques involving increased osmiophilia at site of reaction product. However we have begun to use ferritin-coupled antibodies to demonstrate the presence of α-amylase. This should overcome the difficulty since the immunological reaction with antigen is a surface phenomenon which would result in a change in the image of these bodies in the electron microscope.

The association of amylase, peroxidase and acid phosphatase with aleurone grains in imbibed tissue is difficult to interpret because we do not know if the enzymes are constitutive and have functions in or near the aleurone grains or whether they are GA-induced enzymes destined to be released. Aleurone tissue does contain some GA (Chrispeels and Jones, unpublished data) and a small amount of enzyme synthesis probably occurs during imbibition. Electrophoretically, the amylase and peroxidase isoenzymes in imbibed tissue are similar to those which are released after GA treatment although there appear to be differences for acid phosphatase and esterase (Jacobsen et al. 1970; Jacobsen and Scandalios, unpublished data). In general, it appears that aleurone grains play some role in the early stages of GA-induced enzyme synthesis. What this role may be is unknown but Paleg and Hyde (1964) have suggested that extensions of the aleurone grain membranes might serve as sites of enzyme synthesis.

There is clearly a preference for enzyme to move towards the endosperm sides of the cells, (c.f. Taiz and Jones, 1970). Why the directional release occurs is unknown but it is perhaps significant that the GA-induced enzymes are more basic than most cytoplasmic proteins. Electrophoretic studies have shown that the peroxidase is still positively charged at pH 8 and while amylase, esterase and acid phosphatase are negatively charged at this pH, their mobility towards the anode is less than for many proteins (Jacobsen and Scandalios, unpublished data). The relatively high isoelectric points of the GA-induced enzymes could be associated with directional flow in the cytoplasm.

The association of aleurone grains and hydrolases has been reported before. Acid phosphatase, protease, RNA-ase, β-amylase, α-glucosidase and esterase have been found in aleurone grains of various plant species (Poux, 1963 and 1965; Yatsu and Jacks, 1968; Matile, 1968). Presumably these enzymes hydrolyze the reserve materials in the aleurone grains.

V. SUMMARY

The time courses of release of GA-induced α-amylase, esterase and peroxidase from isolated barley aleurone layers were similar although release of acid phosphatase appeared to commence earlier. High resolution cytochemical techniques suitable for light microscopy were used to localize the enzymes in untreated and GA-treated tissue. α-Amylase was localized by a substrate film method, and by immunofluorescence techniques using specific rabbit antiserum prepared from highly-purified amylase antigen. Phosphatase and peroxidase were localized using simultaneous coupling methods. In imbibed tissue and in layers incubated in buffer only, most enzyme activity was detectable in the peripheral cytoplasm, though for acid phosphatase intense deposition of reaction product occurred in the cell walls in fresh-frozen sections. After 16-24 hr incubation in GA, enzyme activity was diffuse throughout the cytoplasm, though usually most intense at the periphery. The presence of stratified ribosomal endoplasmic reticulum and associated electron dense vesicles in the peripheral cytoplasm suggested that this is the region of synthesis and release of the enzymes.

VI. ACKNOWLEDGEMENTS

The authors are grateful to Dr. Ian D. Marshall for his help in the preparation of anti-amylase serum and to Mrs. H. Chadim and Mrs. K. Marshall for their technical assistance.

VII. REFERENCES

BARKA, T. and P.J. ANDERSON (1962). Histochemical methods for acid phosphatase using hexazonium pararosanilin as coupler. J. Histochem. Cytochem. 10, 741-753.

CHRISPEELS, M.J. and J.E. VARNER (1967). Gibberellic acid-enhanced synthesis and release of α-amylase and ribonuclease by isolated barley aleurone layers. Pl. Physiol. 42, 398-406.

EB, A.A. Van der. and P.J. NIEUWDORP (1967). Electron microscopic structure of the aleurone cells of barley during germination. Acta Botan. Néerl. 15, 690-699.

FILNER, P. and J.E. VARNER (1967). A test for *de novo* synthesis of enzymes : Density labelling with $H_2^{18}O$ of barley α-amylase induced by gibberellic acid. Proc. Nat. Acad. Sci. U.S.A. 58, 1520-1526.

JACOBSEN, J.V. and J.E. VARNER (1967). Gibberellic acid-induced synthesis of protease by isolated aleurone layers of barley. Pl. Physiol. 42, 1596-1600.

JACOBSEN, J.V., J.S. SCANDALIOS and J.E. VARNER (1970). Multiple forms of amylase induced by gibberellic acid in isolated barley aleurone layers. Pl. Physiol. 45, 367-371.

JONES, R.L. (1969a). Gibberellic acid and the fine structure of barley aleurone cells. II. Changes during the synthesis and secretion of α-amylase. Planta (Berl.) 88, 73-86.

JONES, R.L. (1969b). The effect of ultracentrifugation on fine structure and α-amylase production in barley aleurone cells. Pl. Physiol. 44, 1428-1438.

KNOX, R.B. (1970). Freeze-sectioning of plant tissues. Stain Techn. 45, 265-272.

KNOX, R.B. and J. HESLOP-HARRISON (1970). Pollen wall proteins: localization and enzyme activity. J.Cell Sci. 6, 1-27.

KNOX, R.B., J. HESLOP-HARRISON and C. REED (1970). Localization of antigens associated with the pollen grain wall by immunofluorescence. Nature (Lond.) 225, 1066-1068.

KNOX, R.B. (1971). Pollen wall proteins : Localization, enzymatic and antigenic activity during development in *Gladiolus* (Iridaceae) J. Cell Sci. 9, 209-237.

MATILE, Ph, (1968). Aleurone vacuoles as lysosomes. Z. Pflanzenphysiol. 58, 365-368.

PALEG, L.G. and B. HYDE (1964). Physiological effects of gibberellic acid. VII. Electron microscopy of barley aleurone cells. Plant Physiol. 39, 673-680.

POUX, N. (1963). Localization des phosphates et de la phosphatase acide dans des cellules des embryons de blé (*Triticum vulgare* Vill.) lors de germination. J. Microscopie. 2, 557.

POUX, N. (1965). Localization de l'activité phosphatasique acide et des phosphates dans les grains d'aleurone. 1. Grains d'aleurone renfermant a la fois globoides et crystalloides. J. Microscopie 4, 771-782.

STRAUS, W. (1964). Factors affecting the cytochemical reaction of peroxidase with benzidine and the stability of the blue reaction product. J. Histochem. Cytochem. 2, 462-469.

TAIZ, L. and R.L. JONES (1970). Gibberellic acid, β-1,3-glucanase and the cell walls of barley aleurone layers. Planta (Berl.) 92, 73-84.

YATSU, L.Y. and T.J. JACKS (1968). Association of lysosomal activity with aleurone grains in plant seed. Arch. Biochem. Biophys. 124, 466-471.

PLANT GROWTH SUBSTANCES, 1970

Effect of Gibberellin A_3 on *in vivo* and *in vitro* Induction of α-Amylase Isozymes

Yoshihide Momotani and Jiro Kato

Department of Biology, University of Osaka Prefecture, Sakai, Osaka, Japan

I. INTRODUCTION

Since Galston and McCune (1961) reported that GA altered the isoperoxidase pattern of a dwarf genotype of maize, Galston and his co-workers have shown repressive and inductive effects of auxin on isoperoxidases of green pea stems and tobacco pith cells, respectively (Ockerse *et al.*, 1966; Galston *et al.*, 1968). In previous papers (Momotani and Kato, 1966; 1967), we have reported on the isozymes of the α-amylase induced by GA and helminthosporol (H-ol) in the embryo-less endosperm of barley and discussed their correspondence with those obtained by Frydenberg and Nielsen (1965). More recently, we obtained 18 isozymes of α-amylase from the same preparation by improved gel electrofocusing as developed by Awdeh *et al.* (1968). This paper presents the experimental results together with those of experiments in which synthesis of α-amylase was achieved *in vitro*.

II. MATERIAL AND METHODS

Seeds of *Hordeum distichon* L., "*Shigachusei*" of the 1967 and 1969 harvests were used in all experiments. Sterile, embryo-less endosperms, prepared as described in a previous paper (Momotani and Kato, 1966), were incubated with GA solution with and without addition of ABA for 1-5 days at $25°$. Endosperm halves without embryo were treated with re-distilled water and served as controls. After incubation, ten endosperm halves of each set were homogenized in 10 ml of 10^{-2} M Tris-HCl buffer pH 8.5, containing 10^{-3} M NaCl, with a glass homogenizer. The resulting homogenate was centrifuged at 5,000 X g for 30 min, the supernatant being used as an enzyme solution.

Isozymes of α-amylase were separated by isoelectric focusing in acrylamide gel as follows: 2 ml of 8% (g/v) ampholine solution (LKB-Production), pH 5-8, 4 ml of 30% (g/v) acrylamide monomer solution containing 1% (g/v) of *bis*-acrylamide, 2 ml of 0.004% (g/v) riboflavin and 8 ml of air-free distilled water were combined. The combined solution was poured to a height of 60 mm into a glass tube, 5 mm inside diameter and 80 mm in height, the bottom of which was sealed with a rubber stopper. A layer of distilled water 2 mm in height was layered on the solution. These glass tubes were put under a fluorescent light for 15 min to polymerise the acrylamide to form a gel. Residual distilled water on the gel was removed with an injection tube after gelation. The test solution consisted of 1 ml of 8% (g/v) ampholine solution, 2 ml of 30% (g/v) acrylamide solution, 1 ml of 0.004% (g/v) riboflavin, a trace of dimethyl-aminopropionitrile and 0.1 ml of the enzyme solution, which had been incubated at $70°$ for 15 min at pH 8.5 to eliminate β-amylase activity. This was layered on the gel to a height of 10 mm. Following gelation by light the tubes were inserted into a disc electrophoresis apparatus after removal of the rubber stoppers. Five hundred ml of 4×10^{-1} M acetic acid containing 2×10^{-3} M Ca-acetate and the same volume of 4×10^{-1} M ethylenediamine were poured into the anode and cathode electrode vessels, respectively. A constant voltage of 200 volts was applied to the acrylamide gel for 3 hr at $0°$. The current density was 10mA/cm^2 at the beginning but it decreased rapidly to a constant value of

0.5 mA/cm^2. Under these conditions a pH gradient from 5.0 to 7.5 was obtained in the gel. After electrolysis, the gel cylinders were washed with 10^{-1} M acetate buffer, pH 5.1, for about 1 min and then placed in good contact with an acrylamide gel plate containing 0.4% (g/v) of starch, 0.1 mm in thickness, 7 x 9cm, pH 5.3, for 30 min at 40°. After incubation, the gel plate containing starch was immersed for a few minutes in 3 x 10^{-3} M iodine solution which contained a small amount of potassium iodide and 3 x 10^{-2} N HCl. The relative enzyme activity was expressed as a decrease of optical density at 610 mμ.

The methods used in the experiments on *in vitro* synthesis of α-amylase were described in detail in a previous paper (Momotani and Kato, 1971   ). In this paper GA always refers to gibberellic acid (GA_3).

III. RESULTS AND DISCUSSION

Effect of GA concentration on α-amylase zymogram

By gel electrofocusing, 18 bands with α-amylase activity appeared in the embryoless endosperm treated with 10^{-10} - 10^{-7} M GA. The 18 bands were numbered in the order of their isoelectric points from the acidic side to the alkaline side. Zymograms obtained by scanning with a densitometer were not different at 10^{-8} and 10^{-7} M GA. However, the zymograms obtained from the α-amylase induced by lower concentrations of GA were considerably different from those of higher concentrations of GA, as shown in Figure 1. In preparations obtained after 3 days of 10^{-10} or 10^{-9} M GA treatment, the amounts of isozymes 1 - 5 were higher than those of isozymes 6 - 18. On the other hand, in preparations obtained after 3 days of 10^{-8} or 10^{-7} M GA treatment, the amounts of isozymes 6 - 18 were higher than those of isozymes 1 - 5. The time required to reach the maximum relative enzyme activity was different for different isozymes. For example, as shown in Figure 2, activity of isozyme 2 reached a maximum on the third day while that of isozyme 4 was reached on the second day (Fig. 2).

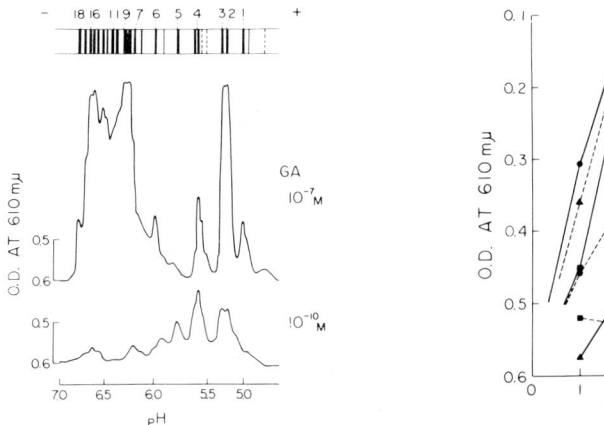

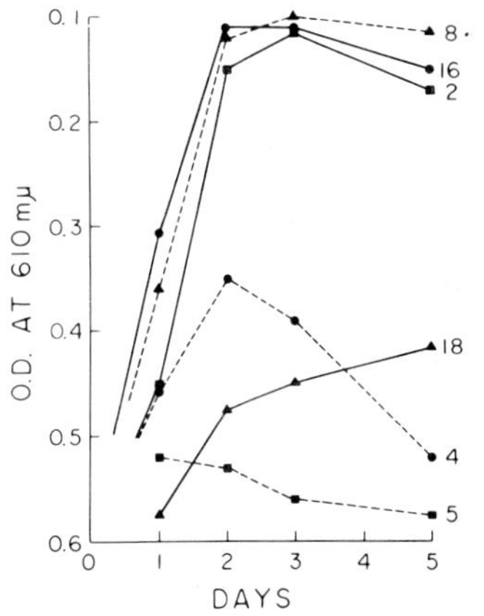

Fig. 1. Zymograms of α-amylase induced by 10^{-7} M (upper) and 10^{-10} M GA (lower) *in vivo*.

Fig. 2. Changes during incubation in the relative enzyme activity of isozymes of α-amylase induced by 10^{-7} M GA *in vivo*.

With addition of 10^{-6} M IAA to 10^{-7} M GA, the total α-amylase activity increased compared with that of GA alone but the activity of isozyme 4 clearly decreased as compared with that produced by GA alone.

Effects of ABA and EPAMe on zymogram of α-amylase induced by GA

The α-amylase induction by 10^{-7} M GA was strongly inhibited by 10^{-5} M ABA but not by 10^{-8} M or less ABA. About fifty percent inhibition was obtained by 10^{-6} M ABA. These results agree with those of Chrispeels and Varner (1967). The zymogram of α-amylase induced by 10^{-7} M GA with the addition of 10^{-5} ABA resembled that induced by distilled water. The induction of individual isozymes, especially of isozymes 6 - 18, was noticeably suppressed by the addition of 10^{-6} M ABA when compared with the zymogram induced by GA alone (Figure 3). There was no difference between the zymogram of α-amylase induced by 10^{-7} M GA in the presence or absence of 10^{-8} M or less ABA.

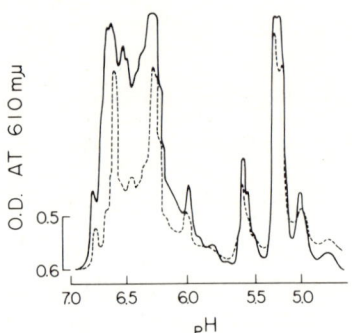

Fig. 3. Effect of 10^{-6} M ABA on the zymogram of α-amylase induced by 10^{-7} M GA *in vivo*. GA alone (———); GA + ABA (-----).

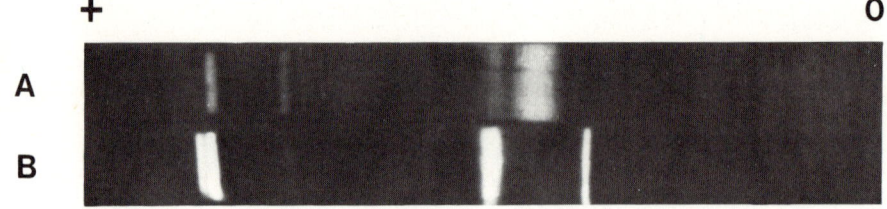

Fig. 4. Comparison of α-amylase zymogram synthesized *in vitro* (A) and *in vivo* (B).

5-(1,2-Epoxy-2,6,6-trimethyl-1-cyclohexyl)-3-methyl-*cis,trans*-2,4-pentadienoic methylester (EPAMe), a compound related to ABA and synthesized by Tamura and Nagao (1969), slightly inhibited α-amylase induction by 10^{-7} M GA at 10^{-5} M. The zymogram resembled that induced by 10^{-7} M GA and 10^{-6} M ABA.

α-Amylase synthesis by a cell free system

A washed microsomal preparation from aleurone layers treated with 10^{-8} M GA (MPG) or 10^{-5} M H-ol (MPH) was found capable of synthesizing α-amylase when incubated with an embryo supernatant as a source of bio-synthetic enzymes, the Duffus (1967) incubation medium and an amino acid solution containing 20 kinds of amino acids. Without the embryo supernatant, no α-amylase synthesis was observed. The microsomal preparation isolated from aleurone layers not treated with GA or H-ol lacked this synthetic ability. The differences in the amount of hydrolyzed starch determined by the blue-value method and the reducing sugar method, indicated that the protein synthesized by the cell free system is likely to be α-amylase, as pointed out by Paleg (1960).

In order to confirm that MPG is able to synthesize proteins, its ability to incorporate L-(U-^{14}C)-threonine and - glutamic acid into proteins was tested. Both labelled amino acids were incorporated into α-amylase. The ratio of incorporated threonine to incorporated glutamic acid approximately equals that of these amino acids in barley seed proteins. The incorporation of labelled glutamic acid was inhibited by the addition of p-fluorophenylalanine or by the pretreatment of MPG with RNAase.

α-Amylase synthesized by the cell free system was separated into isozymes by gel electrofocusing as shown in Figure 4. The zymogram of α-amylase synthesized *in vitro* was clearly different from that of α-amylase induced by GA in the intact aleurone layer (Fig. 4).

Whether or how the isozymes synthesized *in vitro* correspond to those synthesized *in vivo* remains to be investigated.

IV. SUMMARY

The α-amylase induced by GA in the embryo-less endosperm of barley seeds was separated into 18 isozymes by gel electrofocusing. During incubation the time taken for each isozyme to reach its maximum relative activity was different. ABA and EPAMe, a compound related to ABA, inhibited α-amylase induction by GA.

A washed microsomal preparation isolated from aleurone layers treated with GA or H-ol is able to synthesize α-amylase when incubated with an embryo supernatant, the Duffus incubation medium and solution containing 20 kinds of amino acids. The zymogram of α-amylase synthesized by the cell free system was clearly different from that of α-amylase synthesized *in vivo* following GA treatment.

V. REFERENCES

AWDEH, Z.L., A.R. WILLAMSON and B.A. ASKONAS (1968). Isoelectric focusing in polyacrylamide gel and its application to immuno-globulins. Nature (Lond.), 219, 66-67.
CHRISPEELS, M.J. and J.E. VARNER (1967). Gibberellic acid-enhanced synthesis and release of α-amylase and ribonuclease by isolated barley aleurone layers. Pl. Physiol., Wash. 42, 398-406.
DUFFUS, J.H. (1967). A cell-free system for the study of α-amylase synthesis in barley aleurone layers. Biochem. J. 103, 215-217.
FRYDENBERG, O. and G. NIELSEN (1965). Amylase isozymes in germinating barley seeds. Hereditas 54, 123-139.
GALSTON, A.W. and D.C. McCUNE (1961). Analysis of gibberellin-auxin interaction and its possible metabolic basis, in "Plant Growth Regulation" (Ed. R.M. Klein) p. 611. The Iowa State Univ. Press, Iowa. 1961.
GALSTON, A.W., S. LAVEE and B.Z. SIEGEL (1968). The induction and repression of peroxidase isozymes by 3-indoleacetic acid, in "Biochemistry and Physiology of Plant Growth Substances" (Ed. F. Wightman and G. Setterfield) p. 455. The Runge Press, Ottawa. 1968.
MOMOTANI, Y. and J. KATO (1966). Isozymes of α-amylase induced by gibberellic acid in embryo-less grains of barley. Pl. Physiol., Wash. 41, 1395-1396.
MOMOTANI, Y. and J. KATO (1967). Hormonal regulation of the induction of α-amylase isozymes in the embryo-less endosperm of barley. Plant and Cell Physiol. (Tokyo) 8, 439-445.
MOMOTANI, Y. and J. KATO. (1971). α-Amylase synthesis by a cell free system and plant growth substances. *ibid.*, (1971) 12, 405-410.
OCKERSE, R., B.Z. SIEGEL and A.W. GALSTON (1966). Hormone-induced repression of a peroxidase isozyme in plant tissue. Science 151, 452-453.
PALEG, L.G. (1960). Physiological effects of gibberellic acid. II. On starch hydrolyzing enzymes of barley endosperm. Pl. Physiol., Wash. 35, 902-906.
TAMURA, S. and M. NAGAO (1969). Synthesis of novel plant growth inhibitors structurally related to abscisic acid. Agr. Biol. Chem. (Tokyo) 33, 296-298.

Effects of Gibberellic acid on Ribonucleic acid Synthesis in Barley Aleurone

J.A. Zwar and J.V. Jacobsen

C.S.I.R.O., Division of Plant Industry, Box 109 City, Canberra, A.C.T. 2601

I. INTRODUCTION

The induction of α-amylase synthesis in endosperm halves of barley seeds by gibberellic acid (GA) was discovered by Paleg (1960) and by Yomo (1960). Later Paleg (1964) and Varner (1964) showed that the site of enzyme production was the aleurone layer and that the GA effect could be obtained in aleurone layers isolated from the starchy endosperm. Filner and Varner (1967) found that α-amylase was synthesized *de novo* and the inhibitor studies of Varner et al.(1965) suggested the dependence of enzyme synthesis on RNA synthesis. The incorporation studies of Chandra and Varner (1965) supported this.

This work was undertaken to provide more direct evidence for the association of RNA synthesis and enzyme synthesis.

II. MATERIALS AND METHODS

Barley (*Hordeum vulgare* L.) seed, cultivar Himalaya, was used as the experimental material. Seed imbibition, isolation of the aleurone layers and the amylase assays were carried out as described by Chrispeels and Varner (1967). All amylase measurements are given in units; one unit is the amount of amylase necessary to give an OD_{620} change of 1 per minute. In some experiments (indicated in the text) half seeds with embryos removed were used. The incubation medium contained $10^{-2}M$ calcium chloride, $10^{-3}M$ sodium acetate buffer pH4.8 and 1 drop of chloramphenicol solution (0.5mg/ml) in a final volume of 2 ml. GA at $10^{-6}M$ and labelled adenosine and uridine were added when appropriate. In some experiments inhibitors were also present in the incubation medium.

To each of 2 30ml Erlenmeyer flasks (20 layers/flask) 20 μc of ^3H adenosine (2000 mCi/mM Radiochemical Centre) and 20 μc of ^3H uridine (500-3000 mCi/mM) were added, and to 1 flask (20 layers) 5 μc of ^{14}C adenosine (450 mCi/mM) and 5 μc of ^{14}C uridine (405 mCi/mM). The flasks were incubated at $25°C$ for the required time in a shaking water bath in the dark and then the layers were removed and washed with 3 x 3 ml aliquots of $10^{-3}M$ uridine and adenosine. The 40 layers incubated in ^3H were then mixed with the 20 incubated in ^{14}C and RNA was prepared from them. This treatment was designated (= GA) (minus, minus GA) as neither the ^3H nor the ^{14}C labelled layers had been exposed to GA.

Another set of labelled layers was prepared, identical to the first in every respect save that the ^3H labelled flasks contained $10^{-6}M$ GA. After mixing the layers this treatment was called (± GA) as the ^3H labelled layers had been exposed to GA but the ^{14}C had not.

Nucleic acids were prepared by a modification of the method of Ingle and Burns (1968) or by a modification of the method of Solymosy et al.(1968). In the first case the layers were immersed in phenol and buffer, chopped for 1 min in a "Virtis" homogenizer run at high speed and then stirred for 4 min at low speed. The method of Ingle and Burns was then followed to the ethanol precipitation step. In the second case the layers were ground for 1 min on the "Virtis" homogenizer in 20ml of 0.05M Tris-HCl buffer, pH7.6 containing 1 per cent sodium dodecylsulfate and 0.005M $MgCl_2$ and 0.6 ml diethyl pyrocarbonate (DEP) ("Baycovin", Bayer Ltd., Leverkusen, Germany). The homogenate was incubated at $37°C$ for 5 min then centrifuged at 12,000xg for 10 min. Sodium chloride (2g) was added to the supernatant and dissolved and the fluid was incubated again for 5 min at $37°C$ and centrifuged at 12,000xg for 10 min. The supernatant was poured into 50ml of cold absolute ethanol and kept overnight at $-10°C$.

Total uptake of uridine and adenosine was measured by taking a 50 µl aliquot from the initial grindate, counting it for ^3H and ^{14}C and calculating the ratio of ^3H DPM to ^{14}C DPM. The (= GA) gave the standard ratio, comparison of this with the (± GA) ratio showed whether GA had affected uptake.

The ethanol was removed after overnight precipitation and the RNA was dissolved in 3ml of 0.5N sodium acetate pH5.8 equilibrated with DEP. After short term incubations, difficulty was experienced in redissolving the RNA prepared by the phenol method and the DEP method was preferred in these cases. The RNA was precipitated with 0.4ml of 1 per cent cetyltrimethyl-ammonium bromide, washed according to the method of Ralph and Bellamy (1963) and dissolved in electrophoresis buffer.

Thirty to 40 ug of RNA were subjected to electrophoresis on 7mm diameter acrylamide gels according to the method of Loening (1969) but modified by the addition of 0.5 per cent agarose (Dingman and Peacock 1968). The gels were scanned at 265 mµ in a Joyce-Loebl Chromoscan. The gels were then cut into 1mm slices; each slice was transferred to a scintillation vial, 0.5ml of NCS solubilizer was added and the vials were heated for 2 hr at $60°C$. Eight ml of scintillation fluid (6g 2,5-diphenyloxazole/l toluene) were added and after overnight standing in darkness the vials were counted in the ^3H and ^{14}C channels of a Beckman CPM 100 counter equipped with an external standard channel. The DPM values for both ^3H and ^{14}C were calculated from efficiency and "spillover" curves obtained with the aid of internal standards.

Nucleic acids were estimated by the method of Guinn (1966). Total nucleic acid was estimated from the OD_{260} of the 0.5N perchloric acid digest, an aliquot from this digest was taken for the estimation of DNA by the diphenylamine method (Burton 1956). RNA was obtained by difference.

III. RESULTS

Time course of α-amylase production in isolated aleurone layers and half seeds

Table 1 shows total α-amylase production for isolated layers and for half seeds at 0, 4, 8 and 16 hours which were the times used in the labelling experiments. In the first 4 hours little synthesis took place in response to GA, at 8 hours amylase synthesis had begun and a large increase took place between 8 and 16 hours.

Table 1. Units* of α-amylase produced by 10 isolated aleurone layers or half seeds with and without GA $10^{-6}M$

Time of incubation (hr)	Isolated layers -GA	+GA	Half seeds -GA	+GA
0		4.8		
4	3.6	5.0	6.8	8.3
8	7.7	15.2	6.5	10.4
16	4.8	104.0	8.9	50.8

* One unit is the amount of amylase necessary to give an OD_{620} change of 1 per minute.

Effects of GA on label incorporation into RNA of aleurone layers

Figure 1 shows the optical density (E_{265}) traces of the fractionated RNA and the $^3H : ^{14}C$ ratios obtained when isolated layers were incubated for 16 hours with GA $10^{-6}M$ and labelled uridine and adenosine, and the RNA had been prepared by the phenol method. The nucleic acid (E_{265}) peaks from left to right are due to DNA, 25S RNA, 18S RNA and 4S RNA. The plot of the ratios shows that when GA was absent from both the ^3H and ^{14}C incubations (=) there were, as expected, no significant peaks in the ratio line and the deviations from a straight line were a measure of the variability of the procedure. When GA was present in the ^3H and absent from the ^{14}C (±) the ratio was markedly changed. It was higher over the entire scan, indicating that GA stimulated incorporation of uridine and adenosine into all species of RNA. However this stimulation was minimal for the ribosomal and transfer species and maximal in the range 5S to 10S where about 3 times as much uridine and adenosine was incorporated in the presence of GA. This effect could not be accounted for by increased uptake of the labelled compounds since, in the presence of GA, uptake was only slightly increased.

Figure 2 shows that the same effect occurred when half seeds were incubated in the presence of GA and the layers were isolated at the end of the incubation. As in the first case more uridine and adenosine was incorporated into all species of RNA in the presence of GA and there was a three-fold increase in the 5S-10S region. Therefore the stimulation of incorporation occurred also in layers attached to the endosperm.

Figures 3 and 4 show the results obtained when the incubations were for 4 and 8 hours respectively. In these cases the initial RNA preparation was by the DEP method. At 4 hours there was no overall increase in the ratio when GA was present and no suggestion of a localized increase in any particular region of the scan. At 8 hours there was an increase in the overall ratio and some indication of a greater increase in the 5S - 10S region. Thus the ratio rose to 12 in this region in the (±) treatment whilst in the (=) it had a value of about 8. The increase in the ratio over the remaining portion of the scan was somewhat less and was minimal over the light ribosomal peak.

Effects of Actinomycin D on enzyme synthesis and on RNA synthesis

Actinomycin D is known to suppress α-amylase synthesis (Chrispeels and Varner 1967) and is a well known inhibitor of RNA synthesis. Its effect on enzyme production at a concentration of 100 μg/ml is shown in Table 2.

Figure 5 shows the results of an experiment in which Actinomycin D at 100 μg/ml was included in a 16hr incubation with the treatment receiving GA. The OD trace shows that the 18S RNA peak was greatly reduced in the presence of Actinomycin D. A reduction in the ratio was observed for all species of RNA and was greatest in the 5S - 10S region.

Effects of GA on the nucleic acid levels of aleurone layers

Table 3 shows the levels of DNA and of RNA from isolated layers after 0, 4, 8 and 16 hr incubation with or without GA. It is clear that there is a big rise in the levels of both DNA and RNA in the first 4 hours and that the rise continues to 8 hr though at a much lower rate. Between 8 and 16 hr there is a fall in the levels. GA does not affect the changes in the levels of either DNA or RNA.

Table 4 shows the levels of DNA and of RNA from the layers isolated from half seeds after incubation for 0, 4 and 16 hr. In this case the rise seen with isolated layers does not occur. The levels in the presence of GA are lower at 16 hr.

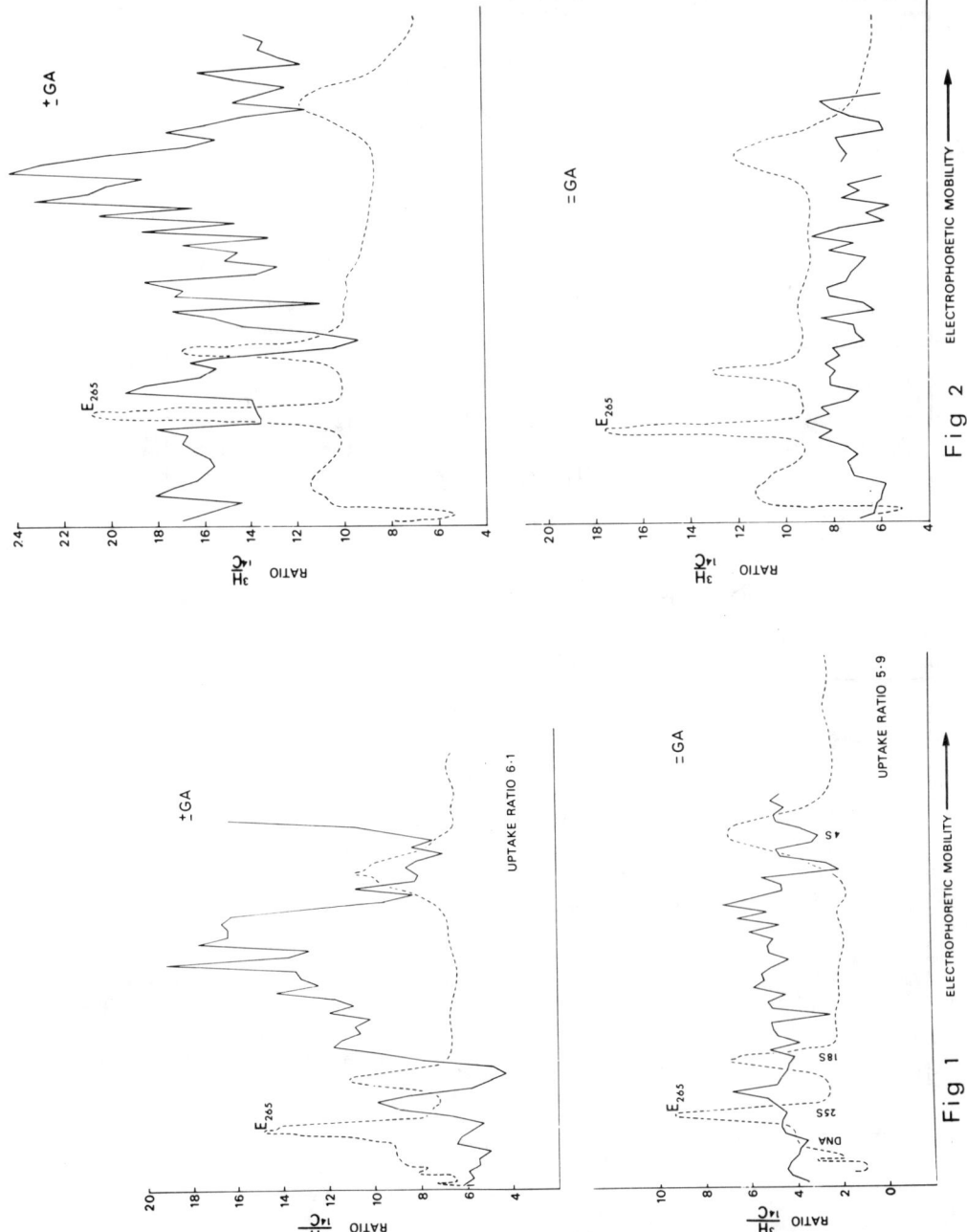

Fig. 1. OD (E_{265}) traces of 16 hr RNA from isolated aleurone layers fractionated on acrylamide gels and $^3H : ^{14}C$ DPM ratios of labelled RNA contained in the gel slices. In (= GA) neither the ^3H nor the ^{14}C labelled layers received GA, in (± GA) the ^3H labelled layers received GA. For further explanation see text.

Fig. 2. OD (E_{265}) traces of the RNA acrylamide gel fractionation of 16 hr RNA from aleurone layers attached to half-seeds and $^3H : ^{14}C$ DPM ratios of labelled RNA contained in the gel slices. In (= GA) neither the ^3H nor the ^{14}C labelled layers received GA, in (± GA) the ^3H labelled layers received GA.

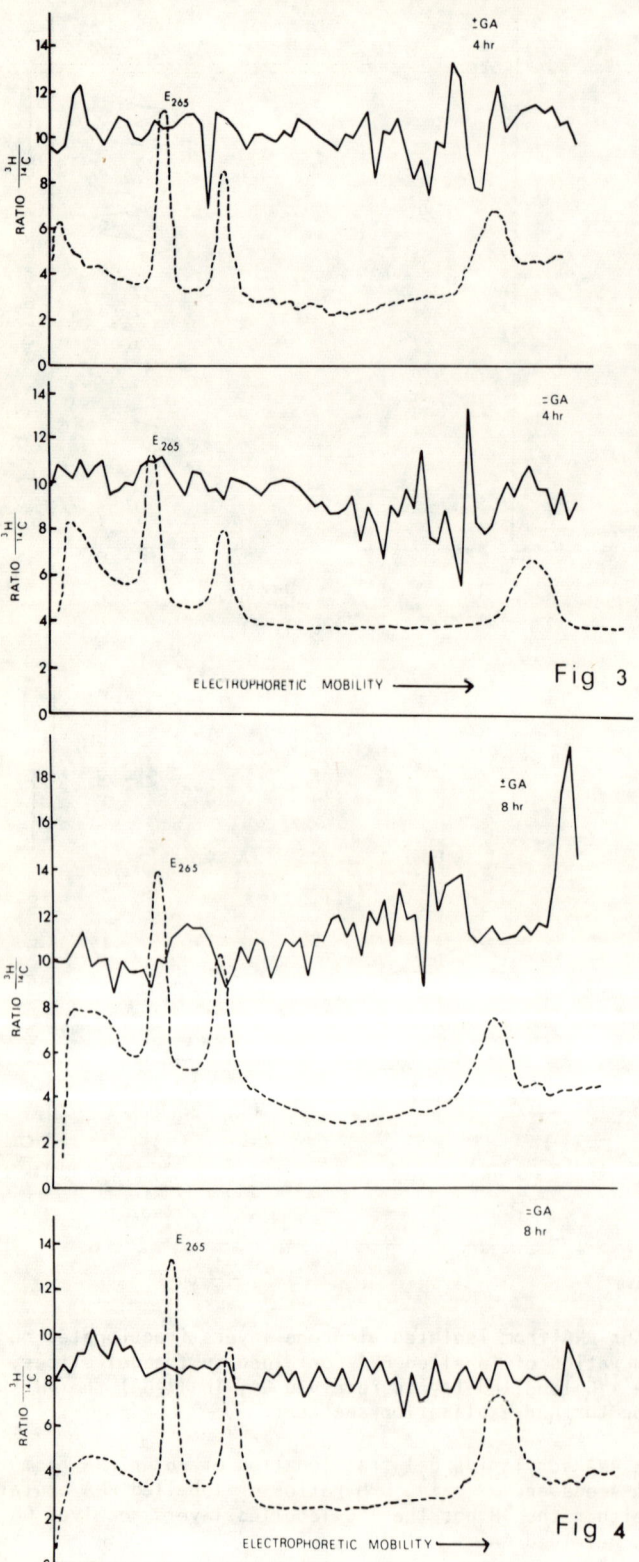

Fig. 3. OD (E_{265}) traces of 4 hr RNA from isolated aleurone layers fractionated on acrylamide gels and ^3H : ^{14}C DPM ratios of labelled RNA contained in the gel slices. In (= GA) neither the ^3H nor the ^{14}C labelled layers received GA, in (± GA) the ^3H labelled layers received GA.

Fig. 4. OD (E_{265}) traces of 8 hr RNA from isolated aleurone layers fractionated on acrylamide gels and ^3H : ^{14}C DPM ratios of labelled RNA contained in the gel slices. In (= GA) neither the ^3H nor the ^{14}C labelled layers received GA, in (± GA) the ^3H labelled layers received GA.

Fig. 3. OD (E_{265}) traces of 4 hr RNA from isolated aleurone layers fractionated on acrylamide gels and $^3H : ^{14}C$ DPM ratios of labelled RNA contained in the gel slices. In (=GA) neither the ^3H nor the ^{14}C labelled layers received GA, in (± GA) the ^3H labelled layers received GA.

Fig. 4. OD (E_{265}) traces of 8 hr RNA from isolated aleurone layers fractionated on acrylamide gels and $^3H : ^{14}C$ DPM ratios of labelled RNA contained in the gel slices. In (= GA) neither the ^3H nor the ^{14}C labelled layers received GA, in (± GA) the ^3H labelled layers received GA.

Table 2. Units* of α-amylase produced by 10 isolated aleurone layers in the absence and presence of Actinomycin D after 16 hours incubation.

	Medium	Extract	Total
No GA	3	9	12
10^{-6}M GA	132	99	231
10^{-6}M GA, + Act D 100 ug/ml	45	64	109

* One unit is the amount of amylase necessary to give an OD_{620} change of 1 per minute.

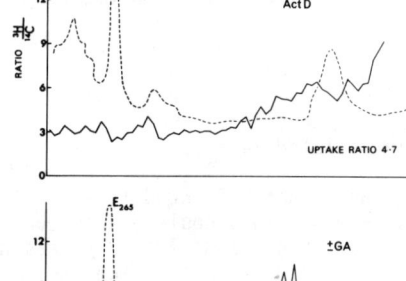

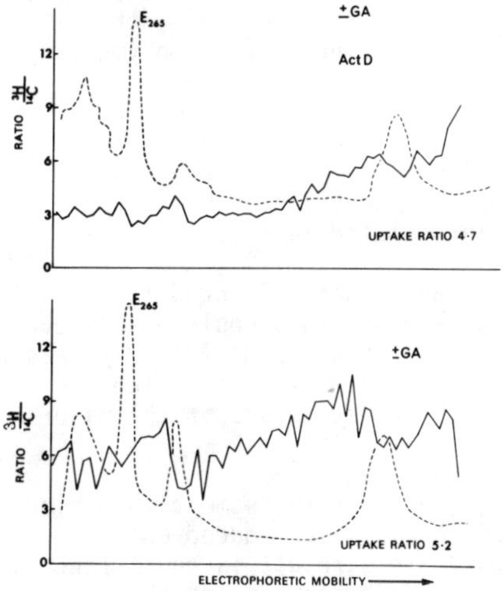

Fig. 5. OD (E_{265}) traces of 16 hr RNA from isolated aleurone layers fractionated on acrylamide gels and $^3H : ^{14}C$ DPM ratios of labelled RNA contained in the gel slices. In (± GA) the ^3H labelled layers received GA, in (± GA Act D) the ^3H labelled layers received GA and actinomycin D.

Table 3. Nucleic acid levels of isolated aleurone layers.

Time of incubation (hr)		DNA(ug/layer)	RNA(ug/layer)
0		0.47	10.1
4	-GA	1.61	23.9
	+GA	1.64	24.9
8	-GA	2.02	28.1
	+GA	1.81	25.2
16	-GA	1.45	22.0
	+GA	1.51	21.7

Table 4. Nucleic acid levels of aleurone layers from incubated half seeds

Time of incubation (hr)		DNA(ug/layer)	RNA(ug/layer)
0		1.80	6.4
4	-GA	1.48	5.9
	+GA	1.83	6.3
16	-GA	1.63	6.6
	+GA	1.14	4.6

IV. DISCUSSION

These results establish that aleurone layers incorporate uridine and adenosine into RNA more rapidly in the presence of added GA and that this increase is minimal for the major species of RNA and maximal for the species sedimenting between 5S and 10S. Further work is necessary to establish whether the increased synthesis represents the formation of new species of RNA or whether it denotes more rapid syntheses of species being made in the absence of GA.

The increase in overall isotope ratio is evident after 8 hours incubation although not after 4 hours. Between 8 and 16 hours the stimulation of incorporation into the 5S - 10S species is greater than in the initial 8 hours. Table 1 shows that most of the enzyme is synthesized in the second 8 hours of incubation hence the increase in the 5S - 10S species is contemporaneous with the increase in enzyme synthesis.

Table 3 shows that isolated layers accumulate a large amount of nucleic acid whilst those still attached to the endosperm do not (Table 4). Thus the GA-induced enzyme production and 5S - 10S accumulation of RNA which take place in both, are independent of the accumulation of RNA which occurs in isolated layers.

The increase in incorporation in the presence of GA is not reflected in an increase in the total nucleic acids of the GA-treated layers. This could be caused by a compensating increase in breakdown in the presence of GA but a more likely reason is that the RNA species whose syntheses are most promoted are minor ones which contribute little to the total amount of nucleic acids. The syntheses of the major species are promoted to only a minor extent.

Actinomycin D inhibits the production of α-amylase. Over the same time period the GA promotion of incorporation into all species of RNA is also reduced and incorporation into the 5S - 10S region is reduced most. This result is consistent with the existence of a causal connection between α-amylase synthesis and 5S - 10S RNA accumulation.

It is tempting to think of the increased incorporation in the 5S - 10S region as representing messenger RNA molecules for α-amylase, ribonuclease and the other hydrolytic enzymes which are synthesized in the presence of GA. The molecular weight of ribonuclease is about 14,000 which would require RNA of molecular weight about 130,000 for its specification. The sedimentation constant corresponding to this molecular weight is approximately 8S. However the sedimentation constant of RNA corresponding to the molecular weight of α-amylase is considerably higher.

These results have established that GA stimulates the incorporation of label into all RNA species and particularly of those which sediment between 5S and 10S. There are some reasons for believing that the 5S - 10S stimulation is connected with enzyme synthesis but more correlative data on the point is required.

V. SUMMARY

The effect of GA on the incorporation of uridine and adenosine into RNA of barley aleurone layers was investigated using a double labelling method combined with acrylamide gel electrophoresis. It was found that after 16 hours incubation GA stimulated the incorporation of uridine and adenosine into all species of RNA but least into ribosomal and transfer RNA and most into those species sedimenting between 5S and 10S. This result was obtained both with layers exposed to GA and label after isolation and with those which received this treatment before isolation. A similar but less marked pattern was found after 8 hours incubation of isolated layers but the effect was not observed after 4 hours incubation.

Synthesis of α-amylase took place contemporaneously with the change in incorporation into RNA. Actinomycin D treatment reduced both the enzyme synthesis and the stimulation of incorporation.

The nucleic acid levels of isolated and attached layers were measured over the 16 hour incubation period. The isolated, but not the attached ones showed a marked rise in both RNA and DNA. In both, however, GA induced enzyme synthesis and the same changes in incorporation into RNA.

VI. ACKNOWLEDGEMENTS

The authors acknowledge the able technical assistance of Mrs. H.E. Chadim, Miss B. Haiderer and Mr. M. McCuaig.

VII. REFERENCES

BURTON, K. (1956). A study of the conditions and mechanism of the diphenylamine reaction for the colorimetric estimation of deoxyribonucleic acid. Biochem. J. 62, 315-323.

CHANDRA, G. RAM and J.E. VARNER (1965). Gibberellic acid controlled metabolism of RNA in aleurone cells of barley. Biochem. Biophys. Acta 108, 583-592.

CHRISPEELS, M.J. and J.E. VARNER (1967). Gibberellic acid-enhanced synthesis and release of α-amylase and ribonuclease by isolated barley aleurone layers. Pl. Physiol. 42, 398-406.

DINGMAN, C.W. and A.C. PEACOCK (1968). Analytical studies on nuclear ribonucleic acid using polyacrylamide gel electrophoresis. Biochemistry 1, 659-668.

FILNER, P. and J.E. VARNER (1967). A test for de novo synthesis of enzymes: density labelling with $H_2{}^{18}O$ of barley α-amylase induced by gibberellic acid. Proc. Nat.Acad.Sci.U.S.A. 58, 1520-1526.

GUINN, G. (1966). Extraction of nucleic acids from lyophilized plant material. Pl. Physiol. 41, 689-695.

INGLE, J. and R.G. BURNS (1968). The loss of ribosomal ribonucleic acid during the preparation of nucleic acid from certain plant tissues by the detergent-phenol method. Biochem. J. 110, 605-606.

LOENING, U.E. (1969). The fractionation of high molecular weight RNA, in "Chromatographic acid electrophoretic techniques" (Ed. I. Smith) Vol.2, p.437, William Heinemann, London. 1969.

PALEG, L. (1960). Physiological effects of gibberellic acid. I. On carbohydrate metabolism and amylase activity on barley endosperm. Pl. Physiol. 35, 293-299.

PALEG, L. (1964). Cellular localisation of the gibberellic-induced response of barley endosperm, in Coll. Int. C.N.R. 123: 303-317.

RALPH, R.K. and A.R. BELLAMY (1963). Isolation and purification of undegraded ribonucleic acids. Biochim. Biophys. Acta 87, 9-16.

SOLYMOSY, F., I. FEDORCSAK, A. GULYAS, G.L. FARKAS, and L. EHRENBERG (1968). A new method based on the use of diethyl pyrocarbonate as a nuclease inhibitor for the extraction of undegraded nucleic acid from plant tissues. European J. Biochem. 5, 520-527.

VARNER, J.E. (1964). Gibberellic acid controlled synthesis of α-amylase in barley endosperm. Pl. Physiol. 39, 413-415.
VARNER, J.E., G. RAM CHANDRA, and M.J. CHRISPEELS (1965). Gibberellic acid-controlled synthesis of α-amylase in barley endosperm. J. Cell and Comp. Physiol. 66, Suppl. 1, 55-68.
YOMO, H. (1960). Studies on the amylase activating substance. V. Purification of the amylase activating substance in barley malt. Hakko Kyokaishi, 18, 603-606 (cited in) Chem.Abs. 55, 26145 (1965).

Effect of Gibberellic Acid on the t RNA Methylase Activity of Barley Aleurone Cells

G.R. Chandra

Research Chemist Plant Science Research Division, Agricultural Research Service, U.S. Department of Agriculture, Beltsville, Maryland 20705

I. INTRODUCTION

In barley endosperm, gibberellic acid (GA_3) treatment causes enhanced synthesis of RNA (Chandra and Varner, 1965; Chandra and Duynstee, 1968). In addition, there appears to be a need for a continuous synthesis of RNA (Chrispeels and Varner, 1967 a,b) to support the hormone-evoked synthesis of alpha-amylase (Varner and Chandra, 1964; Filner and Varner, 1967). Whether or not special kinds of RNA's are synthesized in response to the hormone is still an open question. Recently, we (Chandra and Duynstee, 1971) reported that the *in vivo* methylation of the purine residues of the t- and h-r RNA is enhanced in the tissue treated with GA_3. L-Ethionine inhibited the incorporation of (Me-^{14}C) methionine into RNA, altered the incorporation of (5-^{3}H) uridine into RNA's and decreased the specific activity of (UL-^{14}C) tyrosine labeled alpha-amylase. These results suggest that the methylation of nucleic acids is important for the hormone-evoked synthesis of alpha-amylase. In this communication, I present evidence that enzyme preparations from hormone-treated tissue have enhanced capacity to transfer methyl groups of (Me-^{14}C) S-adenosyl-L-methionine into heterologous t RNA.

II. MATERIALS AND METHODS

Embryo-free endosperm halves of barley seeds (*Hordeum vulgare,* var. Himalaya) were washed with sodium hypochlorite solution (1%) and aseptically incubated with and without gibberellic acid (GA_3) as detailed in Fig. 1. After incubation, the aleurone cell layers were detached, washed with sterile water, and frozen in dry ice for enzyme extraction.

Extraction and assay of methylase

Frozen aleurone tissue (ca.1g) was homogenized with 1 g of glass powder and 10 ml of buffer containing 0.05 M Tris-HCl, pH 8.0, 0.02 M magnesium acetate, 0.02 M potassium chloride, 0.005 M dithiothreitol. The homogenate was centrifuged at 12,000 x g for 15 min. Bentonite (100 µg/ml) was added to the supernatant and centrifuged at 100,000 x g for 1 hr. Four volumes of saturated ammonium sulfate solution (at room temperature) was mixed with the resulting high-speed supernatant and stored in ice for 1 hr. The precipitate was recovered by centrifugation (12,000 x g for 15 min), dissolved in 1.5 ml of Tris buffer, the solution clarified by centrifugation, and used as a source of t RNA methylase. The incorporation of (Me-^{14}C) groups of S-adenosyl-L-methionine into *E. coli* K12 t RNA was used as a measure of methylase activity (see Fig. 1 for details). The protein content of the enzyme preparation was determined (Lowry 1951) in suitable aliquots with bovine serum albumin as standard.

(Me-^{14}C) S-adenosyl-L-methionine (SA 46.5 mC/m mole) was purchased from ICN Corporation. *Escherichia coli,* strain K12 t RNA was obtained from General Biochemical Corporation.

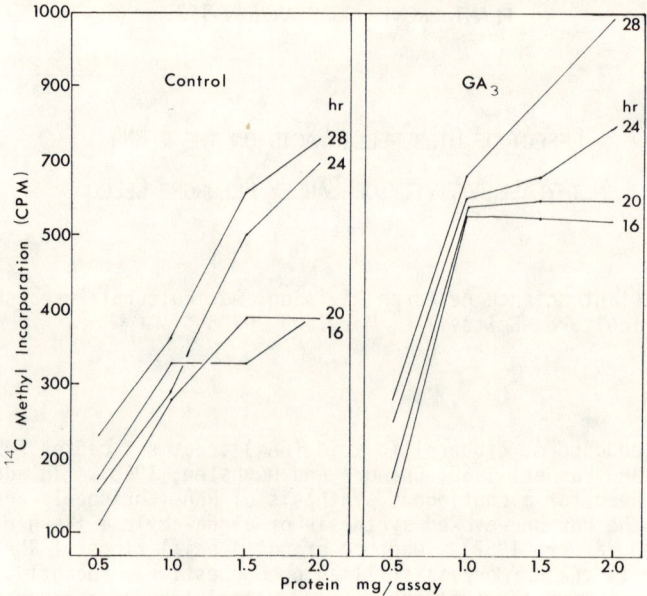

Fig. 1.

Methylation of *E. coli* K12 t RNA by enzyme preparations from aleurone cells treated *in vivo* with and without gibberellic acid. Ten embryo-free endosperm halves of barley seeds were aseptically incubated (25° C) with 2 ml of a reaction mixture containing sodium acetate (10^{-3} M, pH 5.0), calcium chloride (10^{-2} M), ± gibberellic acid (GA$_3$, 10^{-7} M), penicillin (50 µg) and streptomycin (50 µg). The flasks were incubated for 16, 20, 24 and 28 hr. respectively. Enzymes were extracted as detailed in the text. The reaction mixture for methylase assay contained: Tris HCl, pH 8.0 (50 µ moles), magnesium acetate (20 µ moles), potassium chloride (10 µ moles), dithiothreitol (5 µ moles), *E. coli* K12 t RNA (100 µg), (Me-^{14}C), S-adenosyl-L-methionine (1 µC, SA 46.5 µC/m mole) and enzyme in a total volume of 0.5 ml. Control tubes received no t RNA. The tubes were incubated at 25°C for 30 min. The reaction was terminated by adding 5 ml of 10% trichloroacetic acid containing 100 µg methionine. The tubes were stored in ice for 1-2 hr. The precipitate was washed, by centrifugation (12,000 x g), three times with excess of cold 10% trichloroacetic acid and twice with cold ethanol. The washed precipitate was dissolved in 1 ml of ammonium hydroxide (0.1N) and quantitatively transferred to a scintillation vial using excess of ethanol as the solvent. The vials were dried (80° C for 24 hr) and the radioactivity was measured, at 50% counting efficiency, in 10 ml of Bray's solution (Bray, 1960) by standard liquid scintillation procedures.

III. RESULTS

Enzyme preparations from aleurone cells incubated with and without GA$_3$ for 16, 20, 24 and 28 hr., respectively, were assayed for methylase activity. The enzyme-dependent incorporation of (Me-^{14}C) groups into t RNA was expressed as saturation capacity (Sharme and Borek, 1970); i.e., increase of enzyme concentration in the assay produced no further incorporation of methyl groups. Each point on the activity curves (Fig. 1) represents the average value of duplicate (1-2% variation) determinations. The amount of radioactivity incorporated into t RNA varied considerably (±50%) with different supplies of the radiochemical. The results of a typical experiment using the same batch of (Me-^{14}C) S-adenosyl-L-methionine are summarized in Fig. 1. When compared to the controls, the enzyme preparations from the GA$_3$-treated tissue showed increased incorporation of (Me-^{14}C) groups into t RNA (Fig. 1). For instance, the amount of radioactivity incorporated into t RNA by 1 mg of GA$_3$ enzymes was almost 100% greater than that of the control enzymes. In the 16- and 20-hour samples, wherein the methylase assay showed saturation, the GA$_3$ enzymes exhibited 48-63% enhanced capacity to transfer (Me-^{14}C) groups to t RNA (Fig. 1).

Under the conditions of methylase assay, the enzyme preparations from tissue incubated for 24 and 28 hr showed no saturation capacity (Fig. 1). In fact, the incorporation of (Me-^{14}C) groups into t RNA was nonlinear with respect to the enzyme concentration. In the methylase assay, one is dealing with the site and type of methylation on a methyl deficient heterologous t RNA. The enzyme preparation is expected to contain different amounts of methylases, each specific for specific bases (C, A, U, G) of the polynucleotide. The methylase assay measures the bulk incorporation of (Me-^{14}C) groups on the available sites in the t RNA. The assay does not necessarily give the best measure of the activity of each individual methylase. Hence, the nonlinear incorporation of (Me-^{14}C) into t RNA could be partly due to differences, both quantitative and qualitative, in the methylase complements of the enzyme preparations.

Analysis of the pattern of methylation showed that the enzyme preparations from aleurone tissue (± GA$_3$, 24 hr) methylated all of the four bases (C, A, U, G) of the t RNA (Fig. 2). The elution profile of the methylated components of t RNA, methylated *in vitro* by enzymes from control tissue was identical to that of the GA$_3$-enzymes (Fig. 2). Earlier results on the thin-layer chromatographic analysis of RNA, methylated *in vivo*, had indicated that in the column separations ca.90% of the radioactivity associated with the UV absorbancy profile represents the methylated components of the respective base. The quantitative results of the chromatographic separation, enabling a comparison of the *in vitro* methylation of t RNA by ±GA$_3$-enzymes to be made, are summarized in Table 1.

Table 1. Comparison of the pattern of *in vitro* methylation of *E. coli* K12 t RNA by enzymes of aleurone cells treated *in vivo* with and without gibberellic acid.

RNA bases	Control			Gibberellic acid		
	Total OD	Total CPM	% CPM	Total OD	Total CPM	% CPM
X	0.84	830	7.54	0.84	1240	8.77
C	10.14	4890	44.46	9.57	5615	39.72
A	7.23	1640	14.91	7.62	1800	12.73
Y	0.40	450	4.09	0.40	1060	7.50
U	6.07	1700	15.45	5.38	1678	11.87
G	11.04	1480	13.45	11.67	2740	19.38
$\frac{A+G}{C+U}$			0.473			0.628

The results of the column separation were integrated (see x' axis Fig. 2) to give the total absorbancy and radioactivity values for the respective nucleoside monophosphates. The total absorbancy units recovered from the column separations have not been corrected for the solvent blank value. Experimental details as in the legend for Fig. 2. Fractions designated as X and Y have not been identified.

The total absorbancy units of the nucleotides recovered from the two column separations was comparable. However, the total amount of the radioactivity associated with the individual nucleotides was different in the t RNA's methylated by the enzymes from the hormone-treated and those from untreated tissue (Table 1). The total (Me-^{14}C) activity associated with the respective nucleotide was calculated from column fractions pooled on the basis of the absorbancy profile (see x' axis, Fig. 2). Calculations of the percentage distribution of radioactivity show that a large proportion of (Me-^{14}C) groups is associated with the C residues of t RNA. Comparisons, based on the ratio of labeling in C and U residues to A and G residues, indicate that the degree of methylation of the purine residues of t RNA is greater with the GA$_3$-enzymes. The results suggest that the enhanced methylase activity of the enzyme preparation from tissue treated with GA$_3$ is probably due to increased methylation of purine residues.

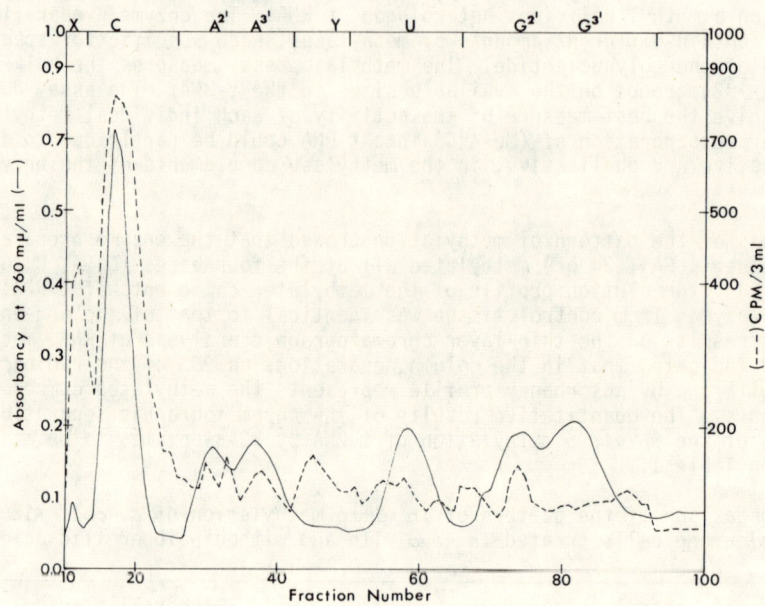

Fig. 2.

Column chromatographic separation of ribonucleotides of *E. coli* K12 t RNA methylated *in vitro* with enzymes from gibberellic acid treated aleurone cells.

Enzyme preparations were made from aleurone cells incubated with and without GA$_3$ for 24 hr at 25° C as detailed in the text. The reaction mixture contained: Tris-HCl, pH 8.0 (100 μ moles), magnesium acetate (40 μ moles), potassium chloride (40 μ moles), dithiothreitol (10 μ moles), *E. coli* K12 t RNA (500 μg), (Me-^{14}C) S-adenosyl-L-methionine (10 μC, SA 55.4 μC/m mole) and enzyme (5 mg protein) in a total volume of 1.25 ml. The tubes were incubated at 25° C for 30 min and the reaction terminated with 10 ml of 10% trichloroacetic acid containing 1 mg of methionine. The precipitate was washed by centrifugation three times with excess of cold 10% trichloroacetic acid and twice with cold ethanol. The pellet was air dried, dissolved in 1 ml of potassium hydroxide (0.3 N), and incubated at 37°C for 16 hr. The resulting alkaline hydrolysate was directly chromatographed.

The column packing consisted of 15 g of anion exchanger (BioRad AG1, 8X, 200-400 mesh, formate form) packed in a 50 x 1 cm column. The nucleotides were eluted with a linear gradient of ammonium formate (0.025 M-0.50 M, pH 2.5 200 ml each) solution. Three and one-half ml fractions were collected at a flow rate of 1 ml/min. After recording the absorbancy at 260 mμ, 3 ml aliquots from each fraction were taken to dryness (85° C for 48 hr) in a scintillation vial and radioactivity was measured, in 10 ml of Bray's solution at 50% counting efficiency. Fraction C, A, U, and G refer to the nucleoside cytidine, adenosine, uridine and guanosine 2', 3' monophosphate, respectively. Fractions X and Y have not been identified. The absorbancy values and the elution profile of the nucleotides of t RNA methylated *in vitro* by enzymes of the control aleurone cells were identical to those of GA$_3$-treated tissue. Differences in the pattern of methylation of t RNA between enzymes of control and GA$_3$-treated tissue are summarized in Table 1.

IV. GENERAL DISCUSSION

The methylation of polynucleotides is accomplished by specific methylase with S-adenosyl methionine as the donor compound (Fleissner and Borek, 1962; Srinivasan and Borek, 1963). Although many studies (Peterkofsky et al., 1966; Shugart et al., 1968, Capra and Peterkofsky, 1968; Gefter and Russell, 1969) suggest the involvement of methylated bases of t RNA in protein synthesis, the cellular mechanisms for the control of methylases is not known. In rat liver, cortico-steroids were shown (Foo Pan et al., 1968) to induce methionine adenosyl transferase activity. Estrogen was found (Hacker, 1969) to enhance (65 fold) the t RNA methylase activity of chick oviduct. Recently, Sharma and Borek (1970) showed that estradiol, when administered to ovariectomized pig will restore the methylase capacity and the elution profile of t RNA^{Ser} to that of the normal uterus. This investigation led to the suggestion that, in the uterus, t RNA methylases might be under estrogenic regulation.

While considerable information on the regulation of nucleic acid metabolism is documented (Key, 1970) information on the hormonal regulation of nucleic acid methylation is lacking for higher plants. Recently, we (Chandra and Duynstee, 1971) showed that the in vivo methylation of the purine residues of t and h-r RNA is enhanced in aleurone tissue treated with GA_3. The data presented in this paper show that enzyme preparations from the hormone-treated tissue have an enhanced capacity to transfer (Me-^{14}C) groups of S-adenosyl methionine into heterologous t RNA. Part of the enhancement in methylation is probably due to an increase in the degree of methylation of the purine residues of t RNA. This conclusion, drawn from in vitro assays, agrees with our earlier finding that the in vivo methylation of the purine residues of t RNA is enhanced in the tissue treated with GA_3.

In the barley endosperm, ethionine treatment inhibits the methylation of nucleic acids, alters the synthesis of polynucleotide and reduces the hormone-evoked synthesis of alpha-amylase. Stone et al. (1970) also found that ethionine will inhibit the development of invertase, methylation of RNA and leucine charging of t RNA in sugar beet root tissue. These studies with ethionine suggest that nucleic acid methylation is important for protein synthesis.

Evidence from bacterial and mammalian systems suggests that alterations in t RNA methylases will allow structural and functional modification of RNA. Such modifications need not be restricted to t RNA. In hormone treated tissues methylation of h. r RNA is also enhanced. The mechanism by which the hormone might directly regulate the activity of methylase(s) is not known. Evidence for indirect regulation of methylase activity by methylase inhibitor(s) is available (Sharma and Borek, 1970). In any event, cellular regulation of methylase(s) activity might allow specific structural alterations in the transcribed substances (RNA's) and consequently translational modulation of protein synthesis. The influence of GA_3 on methylase(s) could, therefore, be an important regulatory mechanism for the control of the hormone-evoked, RNA-dependent, enzyme synthesis in aleurone cells.

V. SUMMARY

The methylase activity of the enzyme preparations from aleurone tissue incubated with and without GA_3 for 16 to 28 hr was measured. When compared with preparations from controls, enzyme preparations from tissue treated with GA_3 had enhanced capacity to transfer (Me-^{14}C) groups of S-adenosyl-L-methionine into E. coli K12 t RNA. Chromatographic analysis suggests that the enhanced methylase capacity of the enzyme from hormone-treated tissue is probably due to increased methylation of the purine residues of the heterologous t RNA.

VI. LIST OF REFERENCES

BRAY, G.A. (1960). A simple efficient liquid scintillator for counting aqueous solutions in a liquid scintillation counter. Anal. Biochem., 1, 279-285.

CAPRA, J.D. and A. PETERKOFSKY (1968). Effect of *in vitro* methylation on the chromatographic and binding properties of methyl-deficient Leucine transfer RNA. J. Mol. Biol., 33, 591-607.

CHANDRA, G.R. and J.E. VARNER (1965). Gibberellic acid-controlled metabolism of RNA in aleurone cells of barley. Biochim. Biophys. Acta, 108, 583-592.

CHANDRA, G.R. and E.E. DUYNSTEE (1968). Hormonal regulation of nucleic acid metabolism in aleurone cells, in "Biochemistry and Physiology of Plant Growth Substances" (Ed. F. Wightman and G. Setterfield) pp. 723-745. The Runge Press Ltd., Ottawa. 1968.

CHANDRA, G.R. and E.E. DUYNSTEE (1970). Methylation of ribonucleic acid and hormone induced alpha-amylase synthesis in the aleurone cells. Biochim. Biophys. Acta 232, 514-523.

CHRISPEELS, M.J. and J.E. VARNER (1967a). Gibberellic acid-enhanced synthesis and release of α-amylase and ribonuclease by isolated barley aleurone layers. Pl. Physiol. 42, 398-406.

CHRISPEELS, M.J. and J.E. VARNER (1967b). Hormonal control of enzyme synthesis: On the mode of action of gibberellic acid and abscisin in aleurone layers of barley. Pl. Physiol., 42, 1008-1016.

FLEISSNER, E and E. BOREK (1962). A new enzyme of RNA synthesis: RNA methylase. Proc. Natl. Acad. Sci., U.S.A. 48, 1199-1203.

FILNER, P. and J.E. VARNER (1967). A simple and unequivocal test for *de novo* synthesis of enzymes: density labeling with H_2O^{18} of barley α-amylase induced by gibberellic acid. Proc. Nat. Acad. Sci., U.S.A., 58, 1520-1526.

GEFTER, M.L. and R.L. RUSSELL (1969). Role of modifications in tyrosine transfer RNA: A modified base affecting ribosome binding. J. Mol. Biol. 39, 145-157.

HACKER, B. (1969). Estrogen-induced transfer RNA methylase activity in chick oviduct. Biochim. Biophys. Acta, 186, 214-216.

KEY, J.L. (1969). Hormones and nucleic acid metabolism. Ann. Rev. Pl. Physiol., 20, 449-474.

LOWRY, O.H., N.J. ROSEBROUGH, A.L. FARR, and R.J. RANDALL (1951). Protein measurement with the Folin phenol reagent. J. Biol. Chem., 193, 265-275.

PAN, FOO (1968). Induction of methionine adenosyltransferase in rat liver by corticosteroids. Proc. Soc. Exp. Biol. and Med., 128, 611-616.

PETERKOFSKY, A., C. JESENSKY, and J.I. CAPRA (1966). The role of methylated bases in the biological activity of *E. coli* leucine t RNA. Cold Spring Harbor Symp. Quant. Biol., 31, 515-524.

SHARMA, O.K. and E. BOREK (1970). Hormonal effect on transfer ribonucleic acid methylases and on Serine transfer ribonucleic acid. Biochem., 9, 2507-2513.

SHUGART, L.S., B.H. CHASTAIN, E.D. NOVELLI, and M.P. STULBERG (1968). Restoration of aminoacylation activity of undermethylated transfer RNA by *in vivo* methylation. Biochem. Biophys. Res. Commun., 31, 404-409.

SRINIVASAN, P.R. and E. BOREK (1963). The species variation of RNA methylases. Proc. Nat. Acad. Sci., U.S.A., 49, 529-533.

SRINIVASAN, P.R. and E. BOREK (1964). Enzymatic alteration of nucleic acid structure. Science, 145, 548-553.

STONE, B.P., C.D. WHITTY, and J.H. CHERRY (1970). Effect of ethionine on invertase development and methylation of ribonucleic acid. Pl. Physiol., 45, 636-638.

VARNER, J.E. and G.R. CHANDRA (1964). Hormonal control of enzyme synthesis in barley endosperm. Proc. Nat. Acad. Sci. 52, 100-106.

Stages During the Induction of α-Amylase by Gibberellic Acid in Barley Aleurone Layers.

I. The Timing of Sensitivity to Actinomycin D.

P.B. Goodwin and D.J. Carr

Research School of Biological Sciences, Australian National University, Canberra, A.C.T., Australia

I. INTRODUCTION

Gene activation in response to external or internal stimuli, is most simply studied by examination of the induction of new enzymes which are dependent on RNA synthesis. The induced synthesis of such enzymes in plants occurs after a lag period. In order to define the events during a typical lag period, we have investigated the time of transcription in the induction of amylase synthesis by gibberellic acid. When applied to barley aleurone layers at the same time as gibberellic acid, Actinomycin D (AM) causes 65% inhibition of induced enzyme synthesis (Chrispeels and Varner, 1967b), and when applied to embryo-less half seeds at imbibition it completely suppresses the response to gibberellic acid (Varner and Chandra, 1964). AM is thought to act primarily by binding to DNA and inhibiting RNA transcription (Wells and Larson, 1970). It suppresses the incorporation of labelled uridine into the trichloracetic acid-insoluble fraction from barley aleurone layers. The response to gibberellic acid is also suppressed by the base analogues, 6 - methylpurine and 8-azauracil (Chrispeels and Varner, 1967b). These results have been interpreted as suggesting that gibberellic acid induces transcription of enzyme-specific RNA molecules required for the synthesis of the enzyme.

The lag period between gibberellin application and the beginning of rapid, induced enzyme synthesis in the experiments of Varner and Chandra (1964) and Chrispeels and Varner (1967a,b) was 7 to 8 hours. AM was much less inhibitory when applied at the end of the lag period than when applied at the beginning or middle, although it was just as effective in inhibiting nucleoside incorporation at the end of the lag period. That is to say, by the end of the lag period, enzyme synthesis is apparently no longer dependent on RNA synthesis. The following experiments were conducted in an attempt to define precisely when, during the lag period, the RNA synthesis related to α-amylase synthesis takes place.

II. METHODS

The procedures are given in detail in Goodwin and Carr, 1970, 1971a. Embryo-less barley half seeds, *Hordeum vulgare* c.v. Himalaya, 1965 harvest, were preincubated in water at $20°C$, or, when stated, at $22.5°C$, for 3 days, and then the aleurone layers were removed and incubated in a medium containing, per ml; 100 μmoles of calcium chloride, 1 μmole of acetate buffer pH 5.05, 100 mμ moles of ferric chloride and 1 mμ mole of gibberellic acid. Since AM is known to inhibit amylase secretion (Chrispeels and Varner, 1967a), total amylase activity was measured after grinding the tissues in the medium, and centrifuging down the debris. Aleurone layers were incubated at $30°C$ for 8 hours and then the amylase activity was measured following the method of Chrispeels and Varner (1967a). Most of the activity measured is due to α-amylase synthetised *de novo*, with a small contribution from β-amylase (Jacobsen, Scandalios and Varner, 1970). The time course of amylase production under these conditions is as follows: there is a lag period of about 4 hours, and then a period of rapid enzyme synthesis (Goodwin and Carr, 1971a; this volume p. 378).

III. RESULTS

1. Optimum concentration of AM for inhibition of the induction process

Actinomycin D was added to the aleurone layers with or without gibberellic acid, at concentrations between 0 and 200 µg/ml, and 8 hours later the total amylase activity was measured. AM had no effect on the background activity determined in the absence of gibberellic acid (Fig. 1), but concentrations of 150 and 200 µg/ml caused effectively 100% inhibition of the response to gibberellic acid. In all further experiments AM was applied at 150 µg/ml, and the results are expressed as percentage inhibition on a scale fixed between zero inhibition (plus hormone, minus AM) and 100% inhibition (minus hormone, plus AM).

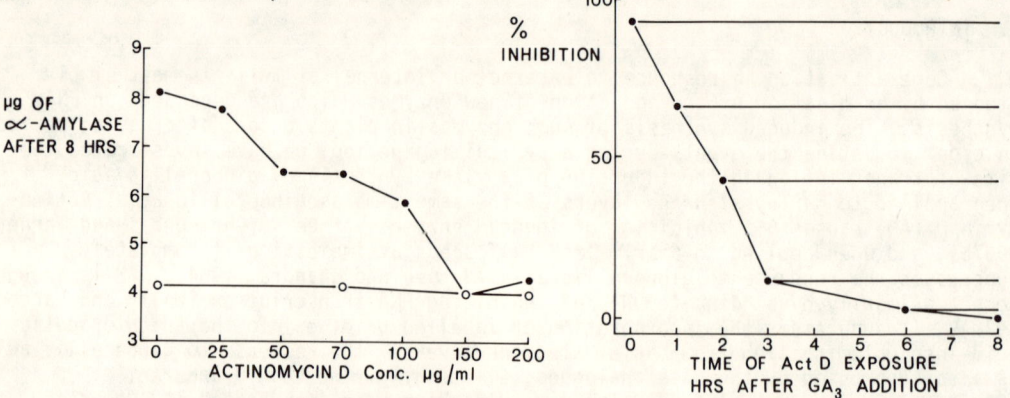

Fig. 1. Relationship between AM concentration and amylase activity after 8 hours, in the presence (solid circles) or absence (open circles) of gibberellic acid. µg α-amylase equivalents per aleurone layer.

Fig. 2. Time of addition of AM, and effect of inhibition of amylase activity measured at 8 hours.

2. Effect of time of application of AM

The tissue was treated with 150 µg/ml AM at 0, 1, 2, 3, 6 and 8 hours after gibberellic acid, and the amylase activity measured at 8 hours. AM had no effect on enzyme activity (8 hr. addition, Fig. 2). There was a rapid fall in the effectiveness of the inhibitor with delay in application, and when supplied later than 2 hours after gibberellic acid it is essentially non-inhibitory.

This fall in inhibition with delay in application could be due to a number of factors. Penetration might be so slow as to require 4 or more hours for an effective internal concentration to be reached, with the inhibitor acting immediately before enzyme synthesis, or it could take this length of time to exhaust the pool of RNA species whose synthesis was being inhibited, or the tissue might simply become impermeable to AM after 3 hours. Alternately, the process it inhibits could be completed within the first 4 hours.

3. Effect of 2-hour pulses

To elucidate the reason for the fall in sensitivity to AM with time the tissues were given 2-hour pulses of AM, at overlapping periods from 3 hours before gibberellin application to 6 hours after, and the amylase activity measured 8 hours after gibberellin addition. Exposures to AM were terminated by rinsing the aleurone layers 10 times over a period of 1 hour with 1 ml of medium lacking the inhibitor. Controls were given the same rinsing treatments at the same time as the AM - treated tissue.

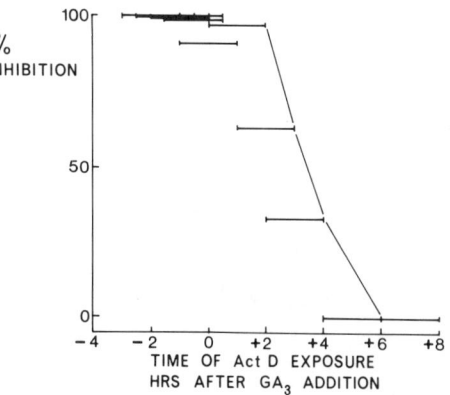

Fig. 3. Effect of time of exposure to AM (horizontal bars) on amylase activity (ordinates, inhibition scale)- see text.

A 2 hour exposure to AM, starting at the same time as gibberellic acid addition, was sufficient to cause 100% inhibition (Fig. 3). Hence penetration takes less than 2 hours, so that the fall in inhibition with time in Fig. 2 is not due to slow penetration and consequent delay to the action of the inhibitor. Secondly, the inhibitor is fully effective even if the exposure to it terminates 1 hour before gibberellic acid addition (-3 to -1 hour pulse). That is to say, rinsing does not lower internal concentrations of the inhibitor to ineffective levels. Thirdly, the fall in inhibition with delay in application is very similar whether or not the external AM is rinsed off after 2 hours. Hence, the fall in inhibition with delay in exposure in Fig. 2 is not due to a shortening of the total time of exposure to AM.

In order to pinpoint more precisely the time at which the tissue is sensitive to AM, it is necessary to know how long the tissue has to be exposed to external AM before the compound becomes inhibitory.

4. *Duration of exposure*

Aleurone layers were exposed to AM between 0 and 1, 1 and 2, 0 and 1.5, and 0 and 2 hours after gibberellic acid application, and at 8 hours the amylase activity was measured. Exposure to AM for 1 hour caused about 65% inhibition, (Fig. 4) and for 2 hours 100% inhibition.

From the results it may be suggested that AM causes, on average, about 35% inhibition of the induction over the first hour after application, about 80% over the second hour, and 100% inhibition later than 2 hours. Since exposure to AM beginning at the same time as GA addition causes 100% inhibition, the process sensitive to AM must occur more than 2 hours after gibberellin application. Application of AM starting 1 hour after GA causes about 66% inhibition (Figs. 2 and 3). This agrees reasonably well with the average inhibition proposed to occur in the second hour after application (80%) i.e. between 2 and 3 hours after GA addition. Application of AM starting at 2 hours causes 33%-42% inhibition, in agreement with the suggested average inhibition over the 1st hour after AM application, that is, once again, 2 to 3 hours after GA addition. Hence, the simplest explanation of the results is that the time of sensitivity of the induction of inhibition by AM is between 2 and 3 hours after gibberellic acid addition. That is to say, if AM acts *only* on RNA synthesis, this is the time of production of enzyme-specific RNA.

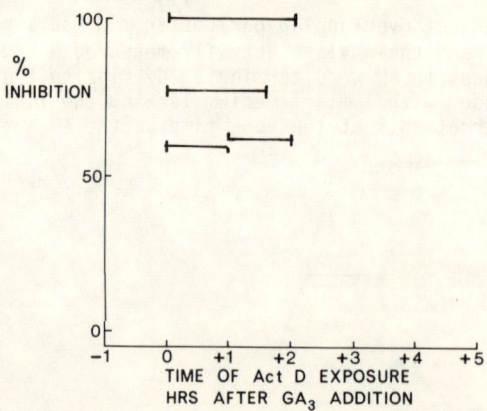

Fig. 4. Effect of duration of exposure to AM (horizontal bars) on amylase activity (ordinates, inhibition scale).

5. *Responses with preincubation at 22.5°C*

It was observed in other work that preincubation at 22.5°C, instead of the standard 20°C, gave aleurone layers which separated from the endosperm easily, and which were very plastic. It was thought that such layers might be more permeable to AM. This was found to be the case, 1 hour exposure to AM causing 80 to 90% inhibition. Hence, in order to test the hypothesis based on the results obtained above, aleurone layers from seeds preincubated at 22.5°C were given pulse exposures to AM.

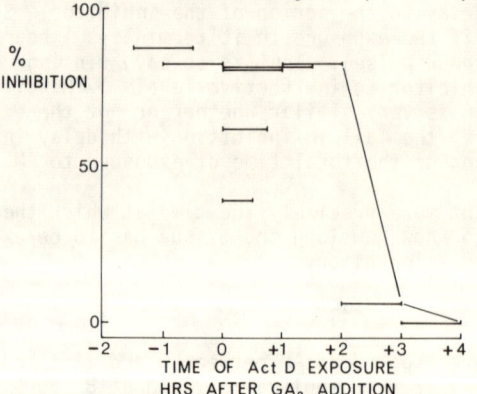

Fig. 5. Time and duration of exposure to AM (horizontal bars) on amylase activity (ordinates, inhibition scale). Layers preincubated at 22.5°C.

Application of AM between - 1.5 and - 0.5 hours followed by thorough rinsing was as inhibitory as exposure between 0 and + 1 hours (Fig. 5), that is to say, once again, rinsing does not remove the inhibitor. Secondly, the inhibition by AM increased with increasing duration of exposure, up to a 1 hour exposure. Hence, the 80% inhibition was developed essentially from the end of 1 hour of exposure. Thirdly, there is a very marked fall in inhibition when AM is applied later than + 1 hour. These results show again that an effective concentration of AM between 2 and 3 hours after GA addition is both necessary and sufficient to inhibit amylase induction. The experiments described in sections 2, 3, 4 and 5 have been repeated a number of times. The data all point to the critical period being between 2 and 3 hours, or in some experiments between 1½ and 2½ hours after GA addition.

There is an alternative explanation for the decrease in sensitivity to AM with delay in application, namely that between 2 and 3 hours after GA addition there is a rapid fall in the permeability of the tissue to AM, or a sudden increase in its ability to degrade AM.

6. *Permeability of aleurone layers to AM*

To check on the possibility that there is a fall in permeability, aleurone layers preincubated at 22.5°C, were exposed to AM in the usual way between 0 and 0.5, 0 and 0.75, 0 and 1, 2 and 3, and 4 and 5 hours after gibberellic acid addition, and at the end of the rinsing period the tissues were blotted dry, weighed, and ground in methanol to extract the AM. The absorbance at 442 mµ was measured, and the AM concentration calculated from the extinction coefficient.

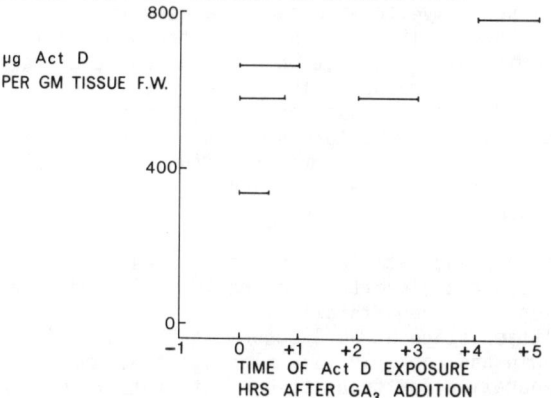

Fig. 6. Effect of time and duration of exposure to AM on tissue concentration.

The tissues accumulate AM from the external solution (Fig. 6). The amount accumulated increases with duration of exposure to AM. There is no appreciable fall in permeability to the inhibitor with time, in fact in some experiments, there is a suggestion of an increase in permeability later during the incubation.

To test for degradation of tissue AM the extracts were run on thin layer chromatography, using 0.25 mm layers of Kieselgel GF254 (Merck), and the solvents: water saturated ethyl methyl ketone (1); isopropanol/ammonia/water 7:2:1 (2); and n-butanol/acetic acid/water 45:113:188 (3). All chromatograms were run 10 cms. The extracted AM following each exposure ran at the same Rf as pure AM. The Rf's with solvents (1), (2) and (3) were 0.75, 0.86 and 0.69 respectively. Hence, it was concluded that the tissue AM was undegraded.

It is thus established that the fall in inhibition with delay in supply of AM is not due to decreased permeability of the tissue or to increased degradation, but to the process of amylase induction becoming insensitive to the inhibitor.

7. *Effect of respiration*

Since, at high concentrations, AM may inhibit protein synthesis at stages after RNA synthesis (Collier, 1966; Weisner, Acs, Reich and Shafiq, 1965), and at low concentrations it preferentially inhibits ribosomal RNA synthesis (Penman, Vesco and Penman, 1968), inhibition by AM is not necessarily due to effects on m-RNA synthesis, or even on RNA synthesis. As a check for non-specific effects of AM the respiration rate of aleurone layers was measured with an oxygen electrode between 2 and 2.5 hours after gibberellic acid addition, either with or without previous exposure to AM (0 to + 1 hours). This exposure is sufficient to strongly inhibit the induction response. The respiration rates were: +AM: 11.2 µl O_2/gm FW/min; - AM: 11.7 µl O_2/gm FW/min. The difference is not significant.

IV. DISCUSSION

Actinomycin D inhibits the synthesis of ribonuclease (Chrispeels and Varner, 1967a) and protease (Jacobsen and Varner, 1967) as well as amylase, by barley aleurone layers in response to gibberellic acid. The inhibitor also prevents labelled uridine incorporation (Chrispeels and Varner, 1967b). AM has been shown to inhibit the synthesis of a unique 28S II RNA, found in aleurone layers only after treatment with gibberellic acid, and tentatively regarded as messenger RNA (Chandra and Duynstee, 1968). Thus AM in this tissue is an inhibitor of transcription. The inhibition of post-transcription stages (e.g. Rich and Goldberg, 1964; Collier, 1966) seems unlikely since AM can be applied immediately before or during translation without affecting enzyme synthesis, and it does not inhibit respiration.

Since the process sensitive to AM appears largely to occur between 2 and 3 hours (or in some experiments between 1½ and 2½ hours) after gibberellic acid addition, this is probably the period of transcription. It follows that the lag period can be broken up into 3 phases: a pretranscription phase, 0 to 2 hours; the transcription period, 2 to 3 hours; and a post-transcription, pretranslation period, 3 to 4 hours after gibberellic acid addition. The length of the post-transcription stage, a little less than 1 hour, agrees well with the time of diffusion of labelled RNA from the nucleus to the cytoplasm (Prescott, 1964). There is a requirement for calcium during this period (Goodwin and Carr, 1971b; this volume p. 385.).

Previous studies have indicated a number of processes which occur during the pretranslation stage. Immediately after gibberellic acid addition the system becomes vulnerable to inhibition by o-phenanthroline, irreversible by iron (Goodwin and Carr, 1971b). The pretranslation stage at $30^{\circ}C$ occurs at the same time as the temperature-sensitive period (Goodwin and Carr, 1971a), although the temperature-sensitive reaction need not itself be preparation for translation. Attempts to shorten the lag period by treatment with cyclic AMP or dibutyryl cyclic AMP have been unsuccessful (Goodwin and Carr, unpublished), suggesting that the pretranslation stage does not involve synthesis of this secondary messenger (Butcher, Robison, Hardman and Sutherland, 1968; Pastan and Perlman, 1970). Pretranslation lags have been also shown following cortisone or hydrocortisone administration in liver (Kenny, Greenman, Wicks and Albritton, 1965; Feigelson and Feigelson, 1966), and following cell aggregation in *Dictyostelium discoideum* (Sussman, 1966). It is evident that hormone action involves more than the rapid derepression of the amylase operon, leading to translation. Elucidation of the events of the pre- and post-transcription phases will throw light not only on the mechanism of hormone action, but also on that of gene activation and gene expression in eucaryotes.

V. SUMMARY

Actinomycin D (AM) is able to inhibit the induction of amylase synthesis in barley aleurone layers in response to gibberellic acid, provided it is present in the tissue at an effective concentration between 2 and 3 hours after gibberellic acid addition. It does not inhibit respiration at this time. The inhibitor is ineffective if supplied later than 3 hours after gibberellic acid addition, even though still penetrating into the tissue. It is suggested that the process sensitive to AM occurs between 2 and 3 hours after gibberellic acid addition, and that the sensitive process is RNA synthesis.

VI. REFERENCES

BUTCHER, R.W., G.A. ROBISON, J.G. HARDMAN and E.W. SUTHERLAND (1968), The role of cyclic AMP in hormone actions. Adv. Enz. Reg. 6, 357-389.
CHANDRA, G.R. and E.E. DUYNSTEE (1968). Hormonal regulation of nucleic acid metabolism in aleurone cells. *in* Wightman and Setterfield, (1968), 723-745.
CHRISPEELS, M.J. and J.E. VARNER (1967a). Gibberellic acid-enhanced synthesis and release of α-amylase and ribonuclease by isolated barley aleurone layers. Pl. Physiol. Lancaster, 42, 398-406.

CHRISPEELS, M.J. and J.E. VARNER (1967b). Hormonal control of enzyme synthesis: on the mode of action of gibberellic acid and abscisin in aleurone layers of barley. Pl. Physiol. Lancaster. $\underline{42}$, 1008-1016.

COLLIER, J.R.(1966). The transcription of genetic information in the spiralian embryo. Current Topics in Dev. Biol. $\underline{1}$, 39-59.

FEIGELSON, M. and P. FEIGELSON (1965). Metabolic effects of glucocorticoids as related to enzyme induction. Adv. Enz. Reg. $\underline{3}$, 11-27.

GOODWIN, P.B. and D.J. CARR (1970). Ferric ion requirement for amylase synthesis by barley aleurone layers in response to gibberellic acid. Cytobios. $\underline{2}$, 165-174.

GOODWIN, P.B. and D.J. CARR (1971a). The induction of amylase synthesis by gibberellic acid. I. Response to temperature. J. exp. Bot. (in press).

GOODWIN, P.B. and D.J. CARR (1971b). The induction of amylase synthesis by gibberellic acid. II. Timing of the requirement for iron and calcium ions. J. exp. Bot. (in press).

KENNY, F.T., D.L. GREENMAN, W.D. WICKS and W.L. ALBRITTON (1965). RNA synthesis and enzyme induction by hydrocortisone. Adv. Enz. Reg. $\underline{3}$, 1-10.

JACOBSEN, J.V., J.G. SCANDALIOS and J.E. VARNER (1970). Multiple forms of amylase induced by gibberellic acid in isolated barley aleurone layers. Pl. Physiol. Lancaster, $\underline{45}$, 367-371.

JACOBSEN, J.V. and J.E. VARNER (1967). Gibberellic acid-induced synthesis of protease by isolated aleurone layers of barley. Pl. Physiol. Lancaster. $\underline{42}$, 1596-1600.

PASTAN, I. and R. PERLMAN (1970). Cyclic adenosine monophosphate in bacteria. Science, $\underline{169}$, 339-344.

PENMAN, S., C. VESCO and M. PENMAN (1968). Localisation and kinetics of formation of nuclear heterodisperse RNA, cytoplasmic heterodisperse RNA and polyribosome-associated messenger RNA in HeLa cells. J. Mol. Biol. $\underline{34}$, 49-69.

PRESCOTT, D.M. (1964). Cellular sites of RNA synthesis. Prog. Nuc. Acid Res. $\underline{3}$, 33-57.

REICH, E. and I.E. GOLDBERG (1964). Actinomycin and nucleic acid formation. Prog. Nuc. Acid. Res. $\underline{3}$, 183-234.

SUSSMAN, M. (1966). Some genetic and biochemical aspects of the regulatory program for slime mold development. Current Topics in Dev. Biol. $\underline{1}$, 61-83.

VARNER, J.E. and G.R. CHANDRA (1964). Hormonal control of enzyme synthesis in barley endosperm. Proc. natn. Acad. Sci. U.S.A. $\underline{52}$, 100-106.

WELLS, R.D. and J.E. LARSON (1970). Studies on the binding of Actinomycin D to DNA and DNA model polymers. J. Mol. Biol. $\underline{49}$, 319-342.

WEISNER, R., G. ACS, E. REICH and A. SHAFIQ (1965). Degradation of ribonucleic acid in mouse fibroblasts treated with actinomycin. J. Cell. Biol. $\underline{27}$, 47-52.

WIGHTMAN, F. and G. SETTERFIELD (1968). Biochemistry and Physiology of Plant Growth Substances. The Runge Press, Ottawa.

Stages during the Induction of α-Amylase by Gibberellic Acid in Barley Aleurone Layers.

II. Some properties of the Pre-and Post-transcription Stages

D.J. Carr and P.B. Goodwin

Research School of Biological Sciences, Australian National University, Canberra, A.C.T. Australia

I. INTRODUCTION

Gibberellic acid applied to barley aleurone layers induces the synthesis and/or secretion of a considerable number of enzymes, such as α-amylase, protease and ribonuclease (Paleg, 1960; Yomo, 1960; Chrispeels and Varner, 1967a; Filner and Varner, 1967; Jacobsen and Varner, 1967). These changes begin only after a lag period, which varies between 4 and 20 hours, depending on the enzyme and the experimental conditions (Yung and Mann, 1967; Pollard and Singh, 1968; Pollard, 1969). There is, during the lag phase, a complex sequence of events which should, in principle, be capable of analysis and it has already been shown (Goodwin and Carr, 1972, this volume) that a process sensitive to inhibition by actinomycin D occurs between 2 and 3 hours after the addition of gibberellic acid to the aleurone layers; since we believe, on a number of grounds, that this process is RNA synthesis, probably specific to the induction of enzyme synthesis by gibberellic acid, we have referred to this period as the transcription phase.

In this paper we will deal with some of the physiological characteristics of the period which precede transcription as well as of the rest of the lag phase.

II. MATERIALS AND METHODS

The methods used are those outlined already by Dr. Goodwin in the paper cited. Briefly, barley grains, c.v. Himalaya, harvest 1965, are sterilised and pre-incubated in sterile water at 20°C for 3 days. The aleurone layers are then removed under conditions as far as possible aseptic and incubated in the standard medium (see Goodwin and Carr, 1970). Amylase activity is measured in the standard manner 8 hours after adding the gibberellic acid.

III. RESULTS AND DISCUSSION

1. Temperature and the time course of appearance of amylase activity

As far as we are aware, no data have been reported on the effects of incubation temperature *per se*, for the isolated aleurone layer. Reducing sugar formed by half seeds is maximal at 30°, according to Paleg (1961) and MacLeod and Miller (1962) although the optimum for α-amylase activity *in vitro* is 45-50° (Greenwood and MacGregor, 1965). In the absence of gibberellic acid (Fig. 1, a-d) there is a small increase in amylase activity in both the tissue and the medium. This background activity is little affected by temperature of incubation up to 30°; at 35° (Fig. 1,e) there is a significant increase in the background activity. Using the same system, Jacobsen, Scandalios and Varner (1970) have shown that most of the background starch-hydrolysing activity is attributable to α-amylase isozyme 2, while the activity induced by gibberellic acid is mainly in isozymes 3 and 4. There is a small contribution by β-amylase to both the background and the induced activity.

The time course of appearance of amylase activity in the tissue shows a slow rise during incubation with gibberellic acid, followed by a sharp rise at the end of

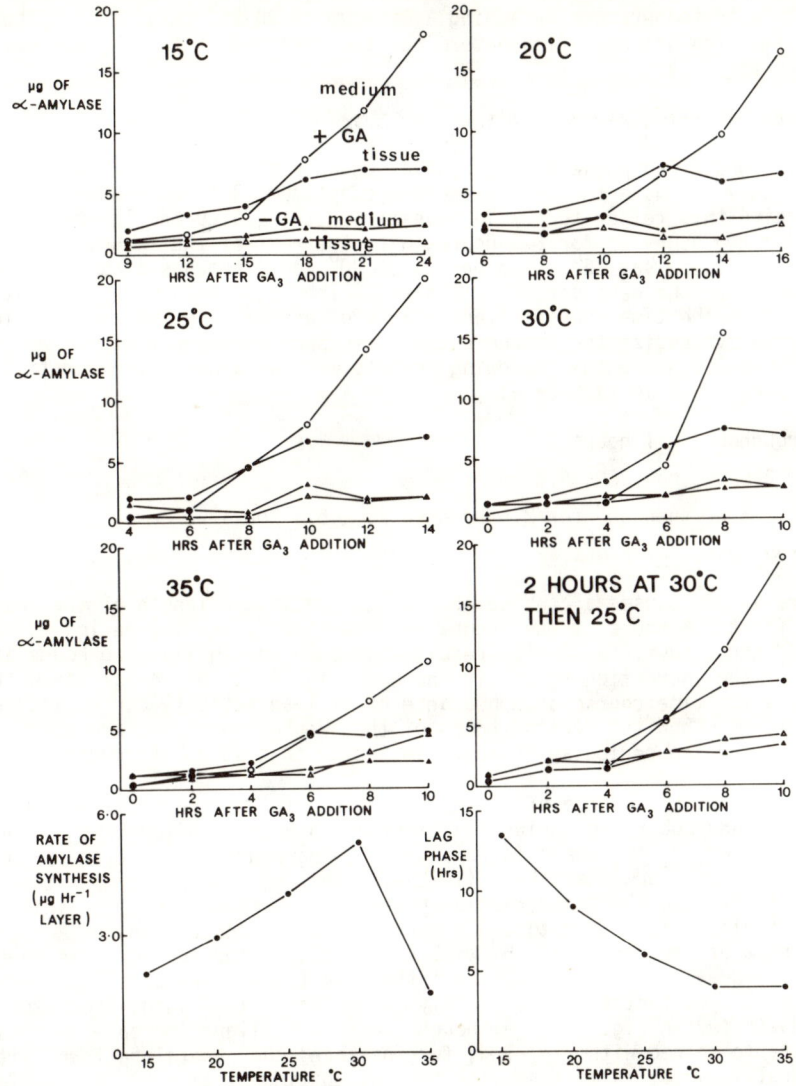

Fig. 1. a-h. a-c, Time course of amylase production at different incubation temperatures. (μg of α-amylase equivalents per aleurone layer) in the presence (circles) or absence (triangles) of gibberellic acid, measured in the tissue (solid symbols) or medium (open symbols). (a) 15°, (b) 20°, (c) 25°, (d) 30°, (e) 35°C. f, Time course of amylase production after stepdown incubation. Details as for 1 a-e, and in the text. g and h, Effect of incubation temperature on the rate of synthesis (g) and duration of the lag phase (h).

the lag period. Enzyme is then accumulated in the tissue and it is also secreted into the medium. Eventually the content of enzyme in the tissue ceases to rise while that in the medium continues to do so (Fig. 1, a-e). Incubation temperature has little effect on the tissue enzyme plateau; it is as though there is some storage compartment, or perhaps the secretary apparatus itself, which becomes fully loaded at the plateau and which is not appreciably enlarged by temperature in the range we have investigated. The optimum temperature for the rate of amylase production appears to be 30° (Fig. 1 g) and the Q_{10} between 15 and 30° is about 1.9. This rate is obtained from the slope of the graphs showing development of activity in the medium. Further, and perhaps more important, the duration of the lag period, defined as above, falls steadily with temperature, reaching a minimum of 30-35° (Fig. 1 h). The Q_{10} for the rate-limiting process which determines the duration of the lag process is about 2.2 over the interval 15-30°.

2. Temperature sensitive periods of the lag phase

Since the whole lag phase consists of a sequence of steps, it is of interest to learn whether they all have the same temperature optimum (30°) or whether only some of the steps require this temperature. To investigate this, aleurone layers were incubated at 30° for two hours or for 2.5 hours, with the rest of the incubation period at some other temperature, say 25°. The period at 30° could be chosen as either the initial two hours or the next period of 2 hours. The lag phases were calculated by extrapolation from the time course of appearance of activity in the medium, minus the time required to synthesize the tissue enzyme. In such an experiment, in which the period at 30° was 2.5 hours the following results were obtained - the duration of the lag phase is underlined in each case:

30° throughout, 4.1 hours

30° for 2.5 hr then 25°, 4.0 hours

25° for 2.5 hr then 30° for 2.5 hr then 25°, 5.4 hours

25° throughout, 7.1 hours

Thus 2.5 hours at 30° causes a considerable reduction in the length of the lag period and when the first 2.5 hours of the incubation are at 30°, the lag period is reduced to that at 30° throughout. A similar result with only the initial two hours at 30° and the rest of the incubation at 25° is shown in Fig. 1 f, for comparison with Fig. 1 d, representing the time course of appearance of amylase activity during incubation at 30°. Similarly, when the first 2 hours of the incubation period are at 30° and the rest at 15° the lag period again approximates to that at 30° throughout.

We therefore have evidence of a temperature sensitive event limiting the length of the lag phase and occurring during its first 2-2.5 hours. After that period has elapsed, the rest of the lag phase is relatively temperature-insensitive. From the evidence presented in Goodwin and Carr (1972 - in this volume) the temperature sensitive events evidently precede the transcription phase and represent processes initiated by the addition of gibberellic acid to the aleurone layers. Incubation for 2 hours at 30° with the rest of the period at 25° will be termed, in the rest of this paper, the *temperature step-down incubation*. The fact that the temperature stepdown incubation is effective on enzyme synthesis, not solely on secretion is evident from experiments like that illustrated in Fig. 1 f. Although there is a slight increase in tissue enzyme activity (compared with Fig. 1 d, for instance) the general pattern, with or without gibberellic acid, is very similar to that obtained from a 30° incubation, except for the lower rate of enzyme synthesis characteristic of 25° incubation.

3. pH optimum

Fig. 2 shows the results of experiments in which the pH of the medium was adjusted over the range 4.45 to 5.35, either at constant temperature of incubation (30 or 25°) or with the stepdown incubation. At 30° throughout, there is a weak optimum at 5.0-5.2, at 25° inconsistent evidence of an optimum pH. Stepdown incubation conditions induce a highly reproducible and strong response to pH with a remarkably sharp optimum at pH 5.0-5.05. The effectiveness of the temperature stepdown incubation is lost unless the pH of the medium is kept within this very narrow range; at pH 5.4 or 4.85 stepdown incubation is hardly more effective in shortening the lag phase than incubation at the lower temperature throughout. Hence pH 5.00 to 5.05 could be regarded as the pH at which the temperature-sensitive reaction can be completed within 2 hours.

4. Requirement for iron

Ferric chloride is routinely added to the medium in our experiments, at 100 μM. The reason for this is that we have found that there is a requirement for iron in the medium during incubation, after pre-incubation of half-seeds in water, prior to stripping off the aleurone layers (Goodwin and Carr, 1970). Figure 3 shows that the concentration we use greatly increases the amylase activity in the medium and, if anything, tends to depress the background activity.

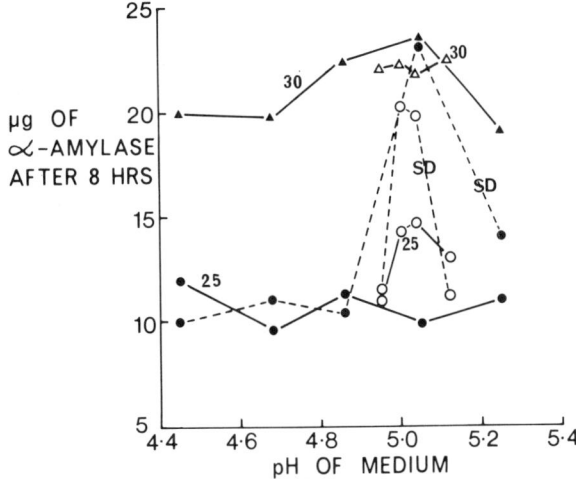

Fig. 2.

Effect of pH of incubation medium on amylase production (μg of α-amylase equivalents per aleurone layer) in the medium after 8 hours with aleurone layers incubated at 30° (triangles, continuous lines), 25°C (circles, continuous lines) or stepdown (SD) incubation conditions (circles, dotted lines). The data are drawn from a number of experiments.

It can also be shown that iron acts during the incubation period, i.e., while gibberellic acid is present, since the response to iron added then is greater than that obtained by supplying iron only during pre-incubation. The effect of iron can also be shown to be on total amylase synthesis rather than on the ratio of retained to released enzyme (Table 1). Iron appears to affect the rate of synthesis of amylase rather than the duration of the lag phase (Fig. 4.). In the absence of exogenous iron, the optimum pH for enzyme synthesis is somewhat sharply 5.05. Adding iron to the medium promotes the synthesis of amylase at pH values either side of this optimum.

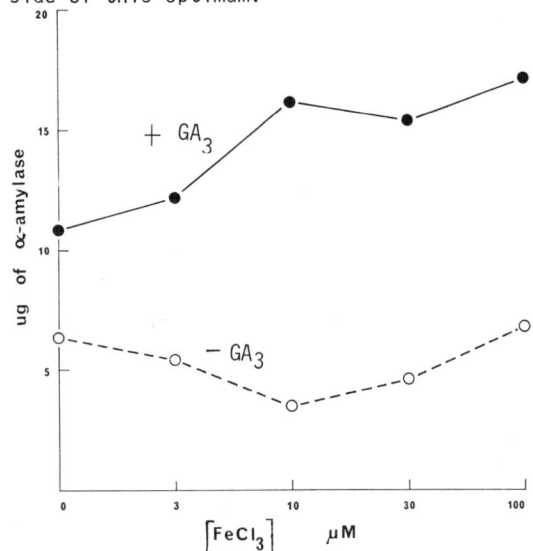

Fig. 3.

Effect of $FeCl_3$ concentration in the medium on α-amylase formation in the presence (solid line) or absence (dotted line) of 1.0 μM gibberellic acid. μg of α-amylase per aleurone layer in the medium after 8 hr.

If we now examine the temperature step-down incubation, it is found to have a requirement for iron (as ferric chloride) in the medium at a concentration of at least 30 μM. In the absence of added iron, the step-down incubation is no more effective than incubation at the lower temperature throughout (Fig. 5). This is so, even when the pH is kept at 5.05, as was the case in the experiment illustrated in Fig. 5.

Table 1. Effect of 10 μM FeCl$_3$ on α-amylase synthesis and secretion. α-amylase in μg per aleurone layer

		-FeCl$_3$: Secreted	Tissue	Total	S/T*	+FeCl$_3$: Secreted	Tissue	Total	S/T*
EXPERIMENT 1									
6 hr	-GA3	5.20	5.40	10.60	(0.49)	4.30	4.20	8.50	(0.505)
	+GA3	4.90	7.88	12.78	(0.38)	3.61	7.55	11.16	(0.32)
8 hr	-GA3	6.00	4.80	10.80	(0.54)	4.52	3.36	7.88	(0.57)
	+GA3	12.00	7.65	19.65	(0.61)	14.30	8.63	22.93	(0.62)
EXPERIMENT 2									
8 hr	-GA3	2.93	6.18	9.11	(0.32)	3.00	4.92	7.92	(0.38)
	+GA3	15.15	8.85	24.00	(0.63)	19.20	11.30	30.50	(0.63)

* S = secreted; T = total.

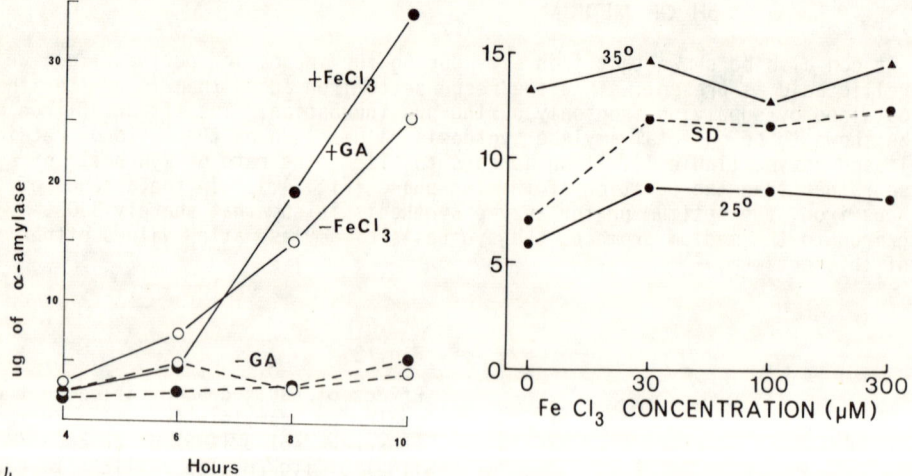

Fig. 4.

Time course of α-amylase release into media with (solid line) or without (dotted line) 1.0 μM gibberellic acid and containing (solid circles) or lacking (open circles) 10 μM ferric chloride. μg of α-amylase per aleurone layer.

Fig. 5.

Effect of ferric chloride concentration in the incubation medium on amylase activity in the medium after 8 hours. Incubation at 35° throughout (triangles) 25°C throughout (circles, continuous lines) or stepdown incubation conditions (circles, dotted lines). Ordinates, as in Fig. 4.

The need for iron for the full effect of the step-down incubation has been used to explore the timing of the requirement for iron. Aleurone layers were incubated in media lacking ferric chloride and at either 30 or 25° throughout or under stepdown incubation conditions. Ferric chloride was added (100 μM) 0,2,4 or 6 hours after gibberellic acid. In presenting the data (Fig. 6) the results from the incubation at 30° with iron present throughout are treated as showing zero inhibition, those at 25° as showing 100% inhibition. The data from the step-down treatments, with iron added at different times, are shown against this inhibition scale. Iron evidently must be present from the beginning of stepdown incubation to be effective.

Addition of iron after 2 or after 4 hours brought the amylase activity resulting from stepdown incubation to below the 25° control level.

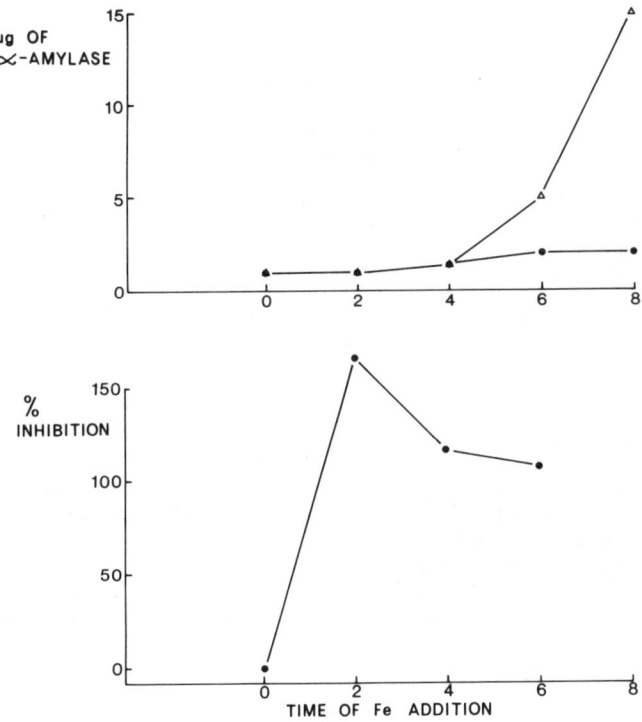

Fig. 6.

Effect of time of addition of ferric chloride, 100μM, on the stepdown incubation. 100% of greater inhibition indicates that amylase activities were below those in the 25°C control. For a time scale, the upper figures shows the time course of production of amylase by barley aleurone layers in the presence (triangles) or absence (circles) of gibberellic acid. μg of α-amylase equivalents per aleurone layer.

5. *Effects of complexing agents in the medium*

The chelating agents, 8-hydroxyquinoline sulphate (8-HQS) and O-phenanthroline (O-PH) and to a lesser extent, bathophenanthroline sulphate, added to the incubation medium, strongly inhibit the response to gibberellic acid. Concentrations of 8-HQS and O-PH above 100 μM virtually eliminate the response (Fig. 7), which can be restored by adding ferric chloride to a concentration of 300 μM to the medium, together with the chelator (Table 2). This information has also been used to explore the timing of the requirement for iron in the lag phase.

Table 2. Effect of ferric chloride on the inhibition of α-amylase synthesis by chelators. μg of α-amylase/aleurone layer produced after 8 hr incubation in combinations of 1.0 μM gibberellic acid (GA), 100 μM 8-hydroxyquinoline sulphate (HQS) or 1:10 O-phenanthroline(O-PH), and 300 μM ferric chloride (Fe).

Chelating agent	− Gibberellic acid		+ Gibberellic acid	
	−Fe	+Fe	−Fe	+Fe
No chelator	1.16	2.98	9.40	18.60
HQS	3.20	2.99	3.12	7.78
O-PH	1.69	2.40	2.88	12.00

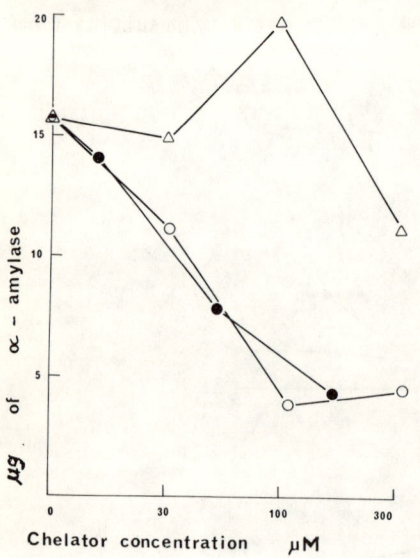

Fig. 7.

Effect of increasing concentrations of chelators in the medium on the formation of α-amylase. Triangles, open circles and closed circles refer to bathophenanthroline sulphate, 1:10 phenanthroline and 8-hydroxyquinoline sulphate respectively. μg of α-amylase per aleurone layer in the medium after 8 hr.

Aleurone layers were incubated in media lacking iron. O-PH was added at 100 μM to the medium 3, 2 or 1 hours before adding GA_3 and at one hour intervals afterwards. Two hours after adding the chelator it was "antidoted" by adding 200 μM ferric chloride. In one case, the chelator and the gibberellic acid were added together and the ferric chloride one hour later. In all cases, amylase activity of the medium was measured 8 hours after adding the gibberellic acid and compared with controls given gibberellic acid and no O-PH (zero inhibition) or O-PH but no gibberellic acid (100% inhibition).

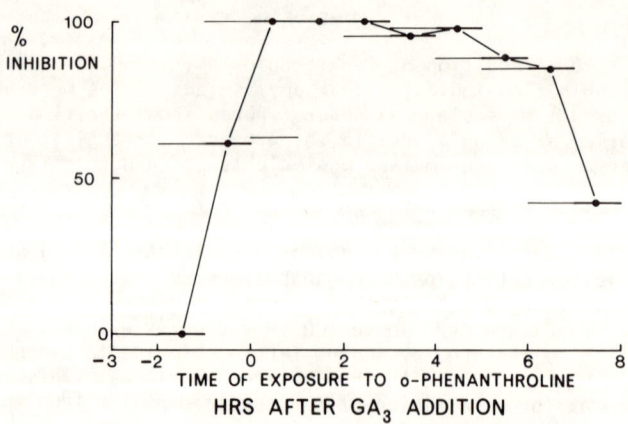

Fig. 8.

Effect of time of exposure to 100 μM o-phenanthroline on the inhibition of amylase production. Solid horizontal lines: time of exposure to chelator. Ordinates, % inhibition on a scale between 0° (GA_3 no O-PH) and 100% (no GA_3, plus O-PH) inhibition.

The results are shown in Fig. 8. (1) When O-PH is added 3 hours before GA_3 and $FeCl_3$ 2 hours later, there is no inhibition, i.e. ferric chloride reverses any inhibition due to the chelator before GA_3 is added. (2) Once GA_3 is present, exposure to O-PH causes 100% inhibition, and it cannot be reversed by ferric chloride. (3) Exposure to O-PH for one hour causes only 63% inhibition, for two hours, 100%: the shorter exposure leaves the tissue incompletely inhibited. The second hour of exposure to O-PH is shown in Figure 8 as a thickening of the line. (4) Exposure to O-PH between -2 hours and 0 hours (the time at which GA_3 is supplied) causes about 60%

inhibition. That is, in the presumably very short period between penetration (including the processes leading to effective action) of GA_3 and the penetration of the antidoting ferric ions, both applied at the same time, the inhibition due to O-PH becomes appreciably irreversible. (5) Sensitivity to inhibition by O-PH is present all through the lag and enzyme synthesis phases, the falling-off towards the end reflecting, quite probably, the amylase synthesized before O-PH becomes effective when added late. These results are to be compared with those shown in Fig. 6.

6. *Requirement for calcium*

Chrispeels and Varner (1967) have shown that calcium is required in the medium for maximal rates of enzyme production and have argued that it is needed to stabilise amylase. However, calcium deficiency in the medium has been shown to inhibit the production of isozymes 3 and 4, and has little or no effect on isozymes 1 and 2 (Jacobsen, Scandalios and Varner, 1970). Moreover the concentration of calcium chloride required for enzyme production - 10 to 20 mM - is well above the concentration needed to stabilise the enzyme (1 to 2mM, Briggs, 1964; Greenwood and MacGregor, 1965). It is therefore not excluded that exogenous calcium may be needed for enzyme induction or synthesis and we have examined the timing of the requirement for calcium.

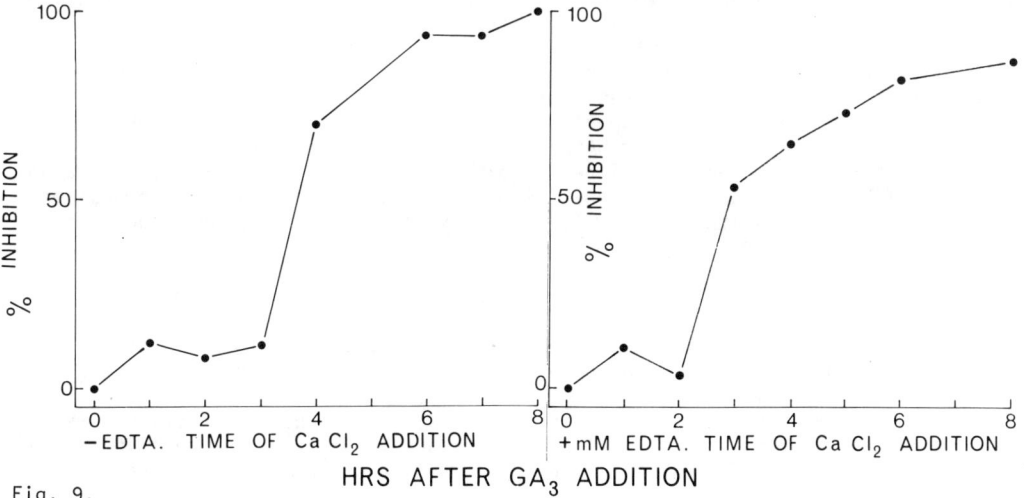

Fig. 9.

Inhibitory effect of delay in addition of calcium chloride 100mM. Inhibition is scaled in terms of incubation throughout with (0%) or without (100%) calcium.

Fig. 10.

Inhibitory effect of delay in addition of calcium chloride 100mM to aleurone layers incubated in a medium containing 1.0mM EDTA. Inhibition is scaled in terms of incubation throughout with calcium and with (0%) or without (100%) gibberellic acid.

Aleurone layers were incubated in a medium lacking calcium chloride or with this salt added at 0, 1,2 --- 8 hours after addition of gibberellic acid. Were calcium ions required only for enzyme stabilisation, the time course of its requirement should follow closely the pattern of onset of enzyme synthesis. Typical data are shown in Figure 9. While there is indeed no requirement for calcium in the early part of the lag phase, it must be added well before its end for maximum effectiveness. If it is added at 4 or 6 hours, before most of the enzyme is synthesized, it is too late to prevent a very considerable or even complete inhibition.

To explore further the time of onset of the requirement for calcium during the lag phase, experiments have been carried out using a medium containing 1.0 mM ethylene diamine tetraacetic acid (EDTA) to chelate extracellular calcium. In these experiments the control, setting the level of 100% inhibition, lacked gibberellic acid. Figure 10 shows that omitting calcium in the presence of gibberellic acid causes at most 86% inhibition of the response, in agreement with the results of Chrispeels and Varner (1967). The time course of the requirement for calcium obtained by this method is in good agreement with that found in the absence of the chelator (Fig. 9). In the presence of EDTA however, the time of sensitivity to calcium deficiency is brought forward to about 3 hours after addition of gibberellic acid.

IV. CONCLUSIONS

The value of the Q_{10} for the temperature-sensitive process which affects the duration of the lag phase suggests a chemical, rather than a physical reaction. The temperature-sensitive process resembles the rate of enzyme synthesis in optimum pH, promotion by iron (Goodwin and Carr, 1970) and insensitivity to high temperature shocks (Goodwin and Carr, unpublished). This process occupies a period of time before the onset of the process sensitive to Actinomycin D (Goodwin and Carr, 1971 - this volume) and well may well be preparatory to it. There is a requirement for iron from the beginning of the incubation period but not, apparently, for calcium. Indeed the earliest event uncovered so far in these investigations is the development of irreversible inhibition in the presence of O-Phenanthroline. The metal chelated is probably iron (Goodwin and Carr, unpublished). Presumably there is an iron-binding factor, present even before GA addition. It is stable then, with or without the metal, but in the presence of gibberellic acid, it becomes unstable if the metal is removed. The factor must be present for enzyme synthesis because the removal of iron, even when enzyme synthesis is well under way, stops further production. One possibility is that this is a gibberellin-binding protein.

The later stages of the process of enzyme induction have a requirement for calcium. It would be necessary to explore the timing of penetration of calcium to obtain a more precise timing of the *internal* requirement for the metal but it seems likely that it penetrates rapidly and that the external supply of calcium is needed for events late in the lag phase, as well as to stabilise the amylase isozymes. Since calcium deficiency causes inhibition well before there has been appreciable synthesis of the enzymes it is possible that its lack results in a dismantling of the machinery for synthesis of isozymes 3 and 4, production of which is inhibited.

V. SUMMARY

The lag phase between application of gibberellic acid to barley aleurone layers and the beginning of amylase synthesis is at a minimum at 30°C. This temperature is also the optimum for enzyme synthesis. Incubation temperature has little effect on the background activity found in the absence of gibberellic acid or on the concentration of enzyme retained in the tissue in the presence of the hormone. When aleurone layers spend 2 hours of the lag period at 30° and the rest of the incubation is at 25 or 15° the lag period is considerably reduced. Short-term incubation at 30° is most effective in reducing the lag period at 25° when the higher temperature is applied during the first two hours after gibberellic acid addition. Under these conditions (temperature stepdown incubation) the lag period is reduced to that resulting from incubation at 30° throughout. The production of amylase after stepdown incubation has a pH optimum of 5.00 -5.05 and is stimulated by the presence of iron. There are evidently two sets of processes during the lag phase, one set highly temperature sensitive, the other relatively temperature insensitive.

Amylase synthesis in response to gibberellic acid is inhibited by O-phenanthroline, a powerful ferrous ion chelator. The inhibition is reversible by ferric chloride if it is supplied before gibberellic acid.

The supply of GA to the aleurone layers results in an immediate development of irreversible inhibition, if the chelator is added. The irreversible inhibition is developed following exposure of the tissue to O-phenanthroline at any time subsequent to gibberellin addition, i.e., during both lag and enzyme synthesis phases.

Iron is required from the beginning of the lag phase for the temperature-sensitive processes as well as, possibly, for others. A requirement for calcium is shown for the later part of the lag period. The inhibition of enzyme production by calcium deficiency occurs earlier and is much greater than would be expected if it is required only to stabilize isozymes of α-amylase.

VI. REFERENCES

BRIGGS, D.E. (1964). 'Origin and distribution of α-amylase in malt'. J.Inst.Brew. 70, 14-24.
CHRISPEELS, M.J. and VARNER, J.E. (1967). 'Gibberellic acid-enhanced synthesis and release of α-amylase and ribonuclease by isolated barley aleurone layers'. Plant Physiol. 42, 398-406.
FILNER, P. and VARNER, J.E. (1967). 'A simple and unequivocal test for de novo synthesis of enzymes: density labelling of barley α-amylase with $H_2^{18}O$. Proc. Natl. Acad. Sci. U.S. 58, 1520-6.
GOODWIN, P.B. and CARR, D.J. (1970). 'Ferric iron requirement for amylase synthesis by barley aleurone layers in response to gibberellic acid'. Cytobios, 2, 165-174
GOODWIN, P.B. and CARR, D.J. (1972). 'Stages during the induction of amylase by gibberellic acid in barley aleurone layers. I. The timing of sensitivity to Actinomycin D.' in: "Plant Growth Substances, 1970" Ed. D.J. Carr, p. 371-377.
GREENWOOD, C.T. and MACGREGOR, A.W. (1965). 'The isolation of α-amylase from barley and malted barley, and a study of the properties and action patterns of the enzyme'. J. inst. Brew. 71, 405-417.
JACOBSEN, J.V., SCANDALIOS, J.G. and VARNER, J.E. (1970). 'Multiple forms of amylase induced by gibberellic acid in isolated barley aleurone layers'. Pl. Physiol. Lancaster. 45, 367-371.
JACOBSEN, J.V. and VARNER, J.E. (1967). 'Gibberellic acid-induced synthesis of protease by isolated aleurone layers of barley'. Plant Physiol. 42, 1596-1600.
MACLEOD, A.M. and MILLER, A.S. (1962). 'Effects of gibberellic acid on barley endosperm'. J. Inst. Brew. 68, 322-32.
PALEG, L.G. (1960). 'Physiological effects of gibberellic acid: I. On carbohydrate metabolism and amylase activity of barley endosperm'. Plant Physiol. 35, 293-9.
POLLARD, C.J. and SINGH, B.N. (1968). 'Early effects of gibberellic acid on barley aleurone layers'. Biochem. Biophys. Res. Commun. 33, 321-6.
POLLARD C.J. (1969). 'A survey of the sequence of some effects of gibberellic acid in the metabolism of cereal grains'. Plant Physiol. 44, 1227-32.
YOMO, H. (1960). 'Studies on the amylase activity substance. IV. On the amylase activity action of gibberellin'. Hakko Kyokaishe. 18, 600-2.
YUNG, H.K. and MANN, J.D. (1967). 'Inhibition of early steps in the gibberellin-activated synthesis of α-amylase'. Pl. Physiol. Lancaster. 42, 195-200.

PLANT GROWTH SUBSTANCES, 1970

The Effects of Gibberellic Acid on the Metabolism of Soluble Nucleotides in Aleurone Tissue Isolated from Wheat Grain

G.G. Collins, C.F. Jenner and L.G. Paleg

Waite Agricultural Research Institute, University of Adelaide, Glen Osmond, South Australia, 5064

I. INTRODUCTION

During the germination of cereals, gibberellic acid (GA) produced by the embryo moves to the aleurone layer where it stimulates the *de novo* synthesis of a number of hydrolytic enzymes (Paleg, Coombe and Buttrose, 1962; Briggs, 1963; Varner, Chandra and Chrispeels, 1965; Filner and Varner, 1967).

Treatment with GA has been reported to affect the nucleic acid metabolism of several tissues (Roychoudhury and Sen, 1965; Giles and Myers, 1966; Kamisaka, Masuda and Yanagishima, 1967; Johri and Varner, 1968) including the aleurone tissue of barley (Varner, *et al.*, 1965; Chandra and Duynstee, 1968). However, reports on the inhibitory effect of Actinomycin D on the GA-induced synthesis of enzymes in barley aleurone are conflicting (Paleg, 1964; Varner and Chandra, 1964; Pollard and Singh, 1968), and there is clearly a need for more information on the mode of action of GA in this tissue.

In some tissues at least it is apparent that protein synthesis can proceed without being preceded or accompanied by detectable changes in nucleic acid metabolism (Waters and Dure, 1966; Chen, Sarid and Katchalski, 1968; Chen and Osborne, 1970). These findings suggest that the messenger RNA necessary to support such protein synthesis is already present in the tissue, albeit in an inactive form. However, no evidence has yet been reported indicating that a mechanism of this kind operates in cereal aleurone tissue.

Soluble nucleotides participate both as precursors of nucleic acid synthesis and as co-factors in many biochemical reactions. It was considered that an investigation of these compounds in cereal aleurone treated with GA could determine the extent to which RNA metabolism is altered by the hormone, and might also reveal facets of the mode of action of the hormone not detected by other investigations.

Although barley has been used for most of the work reported, it is difficult to obtain large amounts of tissue from this species because the aleurone layer is peeled from the endosperm by hand. However, a method has been developed for wheat (Phillips and Paleg, 1971) yielding up to 20 g of aleurone tissue in less than an hour. Increases in the activity of several hydrolytic enzymes occur after treatment of this material with GA and the response appears to be identical to that of barley aleurone. Accordingly, wheat aleurone tissue was used in this investigation.

II. MATERIALS AND METHODS

Authentic specimens of nucleotides were purchased from P-L Biochemicals Inc., Milwaukee, Wisc., U.S.A.; gibberellic acid from Merk and Co. Inc., New Jersey, U.S.A. (97.4% GA_3), and ^{32}P (as carrier-free orthophosphate) from the Australian Atomic Energy Commission, Sydney, N.S.W.

The following solvents were re-distilled before use: triethylamine, iso-butyric acid, ethanol, diethyl ether, and n-propanol. Water was distilled twice and then de-ionised. Solutions and equipment used for removing aleurone layers were autoclaved before use. Solutions of GA were sterilized by filtering through "Millipore" filters, porosity of 0.22 micron (Millipore Filter Corporation, Bedford, Mass., U.S.A.).

Wheat seeds (cv. Olympic) were bisected laterally and the halves containing the embryo discarded. The distal halves were sterilized and the aleurone layers were separated from the endosperm by the mechanical method described by Phillips and Paleg (1972). The isolated layers were washed, excess water was removed by blotting on sheets of filter paper, and portions of tissue (about 4 g wet weight) were weighed into 250 ml conical flasks. Water (30 ml) was added and the flasks were shaken gently on a water bath at $30°C$. After 6 hr, 7.5 ml of the ambient solution were removed and replaced with 7.5 ml of water or GA (400 μg/ml) and the flasks shaken further for periods of up to 120 min. Fifteen min before the end of each period, the tissue was drained and re-suspended in 30 ml of fresh media made 0.05 mMolar with respect to KH_2PO_4 and containing 50μc ^{32}P.

In one experiment (see Discussion) the isolated tissue was weighed into 9 cm petri dishes with 10 ml of water. After standing for 6 hr, 0.5 ml of water or GA (2 mg/ml) were added, the suspension was mixed and left for 15 min. Then 50 μc ^{32}P (carrier free) were added with mixing and the tissue incubated without shaking for a further 15 min.

Soluble nucleotides were extracted by a method developed from techniques reported by Cole and Ross (1966), Isherwood and Barrett (1967) and Jenner (1968). Briefly, the tissue was homogenized in cold 5% (w/v) trichloroacetic acid containing 0.15% (w/v) 8-hydroxyquinolene. After 15 min an equal volume of chloroform - iso-amyl alcohol (24:10, v/v) was added and the mixture was shaken and centrifuged at 4000 r.p.m. for 5 min. The upper aqueous layer was extracted twice with twice its volume of ether, and passed through a column of cation-exchange cellulose (Whatman P 11) in the H^+ form. The effluent was neutralised and passed through a column of DEAE cellulose (Whatman DE 11 in the HCO_3^- form). The anions were eluted with triethylammonium bicarbonate (0.5M at pH 7.4), and the eluate was taken to dryness *in vacuo* at temperatures below $30°C$. The extract was applied as a thin band across a full sheet of chromatography paper (Whatman 3 mm, washed in 0.1 M oxalic acid and 2N acetic acid) and the paper developed in n-propanol - ammonia - water - 0.2 M EDTA (72.5:0.1:27.3:0.1, by vol.) adjusted to pH 7 with acetic acid. When dry, the areas of u.v. absorbing material were eluted with water, taken to dryness again, and applied as a band 8 cm long near one corner of another sheet of paper (Whatman No. 1) and separated in two dimensions with iso-butyric acid - aq. ammonia (Sp.gr. 0.88) - water (57:4:39, v/v/v) followed by 95% (v/v) aq. ethanol - 1M ammonium acetate pH 3.8 (and 0.01 M in EDTA) (7:3, v/v).

Compounds absorbing u.v. light were eluted from the paper with 0.01 N HCl and their optical density was measured at 260 nm. Portions of the eluates were dried on copper planchets, and the radioactivity measured in a gas-flow counter.

Nucleotides were identified by their absorption spectra at pH 2 (P-L Biochemicals Circular OR-10, 1956), by hydrolysis in 98% formic acid (Wyatt, 1951) and in N HCl (Amos and Korn, 1958) and by their chromatographic properties in 6 different solvents.

More than 80% of known amounts of authentic nucleotides which were added to the tissue before extraction were recovered.

Table 1. **Amounts of soluble nucleotides extracted from wheat aleurone tissue.**

Distal halves of wheat grain were soaked in water for 24 hr, and the aleurone layers separated from the endosperm by rolling. Samples (about 4 g fresh weight) were extracted with trichloroacetic acid and the nucleotides separated by paper chromatography.

Nucleotide	n moles/ g dry weight	Nucleotide	n moles/ g dry weight
ATP	218	UTP	72
ADP	20	GTP	34
UDP-glucose	204	CTP	34

III. RESULTS

In common with reports for a number of tissues (Cole and Ross, 1966; Jenner, 1968; Bieleski 1968, 1969) derivatives of adenosine and uridine together comprise the bulk of the nucleotides extracted. ATP, UDP-glucose and UTP are the most abundant compounds with smaller amounts of CTP, GTP, NAD and ADP (Table 1). No AMP, ADP-glucose, UMP, CMP or GMP was detected. The weight of starch adhering to the aleurone amounted to about 15% of the dry weight, but, on investigation, this residual starch was found to contain negligible quantities of nucleotides.

In isolated aleurone tissue cultured in water there are appreciable changes with time in the amounts of some of the nucleotides extracted. For example, in tissue isolated from half seeds soaked for 24 hr, the levels of ATP and UDP-glucose fall by about 20% and 40% respectively after 2 hr incubation in water (Table 2). Analyses of the ambient solution failed to detect material resembling purine or pyrimidine bases, so the possibility that these nucleotides were simply lost from the tissue can be ruled out.

Table 2. **Changes with time in the levels of soluble nucleotides extracted from isolated wheat aleurone tissue cultured in water.**

Distal halves of wheat grain were soaked in water for 24, 26 or 30 hr and the aleurone layers isolated. The tissue was extracted immediately or incubated in water for a further 2, 4 or 6 hr before extraction. Nucleotide levels are expressed as a percentage of the values recorded at the time of isolation.

Nucleotide	Time of soaking prior to isolation of aleurone tissue hr	Time of incubation of aleurone tissue per cent			
		0hr	2hr	4hr	6hr
ATP	24	100	80	82	96
	26	100	96	98	103
	30	100	91	92	91
UTP	24	100	135	135	160
	26	100	158	146	158
	30	100	126	117	90
UDP-glucose	24	100	57	70	81
	26	100	80	88	101
	30	100	73	78	82

Moreover, after incubation for longer periods the levels appear to rise closer to values recorded immediately after isolation of the tissue. On the other hand, the amounts of UTP extracted increase after isolation and remain higher than the initial value. Although soaking the half seeds for longer periods than 24 hr before preparing the tissue appears to diminish the subsequent changes in ATP, significant changes still occur in the amounts of UTP and UDP-glucose. Even after 30 hr of soaking, the trend in UDP-glucose observed after 24 hr is still evident, although the fluctuation is not so conspicuous.

From these and other data it seems clear that at least part of the fluctuations in the levels of nucleotides is caused by the method used to isolate the tissue. The fact that the quantities of some of the nucleotides recorded after 6 hr of incubation in water are similar to the initial values, and that thereafter the levels remain more or less constant for periods up to 24 hr suggests that the effects resulting from isolation persist for only a few hours. For this reason all tissue isolated in this way and used to investigate the effects of GA was first pre-incubated for 6 hr in water before the hormone was applied.

Wheat aleurone tissue used here responds to treatment with GA by producing α-amylase. The relationship between the concentration of GA and the production of α-amylase is similar to that observed by other workers using wheat (Phillips and Paleg, 1972), barley (Varner and Johri, 1967), and rice (Ogawa, 1966). With barley (Chrispeels and Varner, 1967) the production of α-amylase is first detected 6 to 8 hr after the application of the hormone. Here (Fig. 1) in tissue pre-incubated in water the enzyme is detected soon after 6 hr of treatment with the hormone, and the rate of production increases rapidly over the next 4 hr. By way of contrast, in tissue treated with GA immediately after isolation, α-amylase is not detected until later, between 8 and 10 hr, and the enzyme accumulates much more slowly.

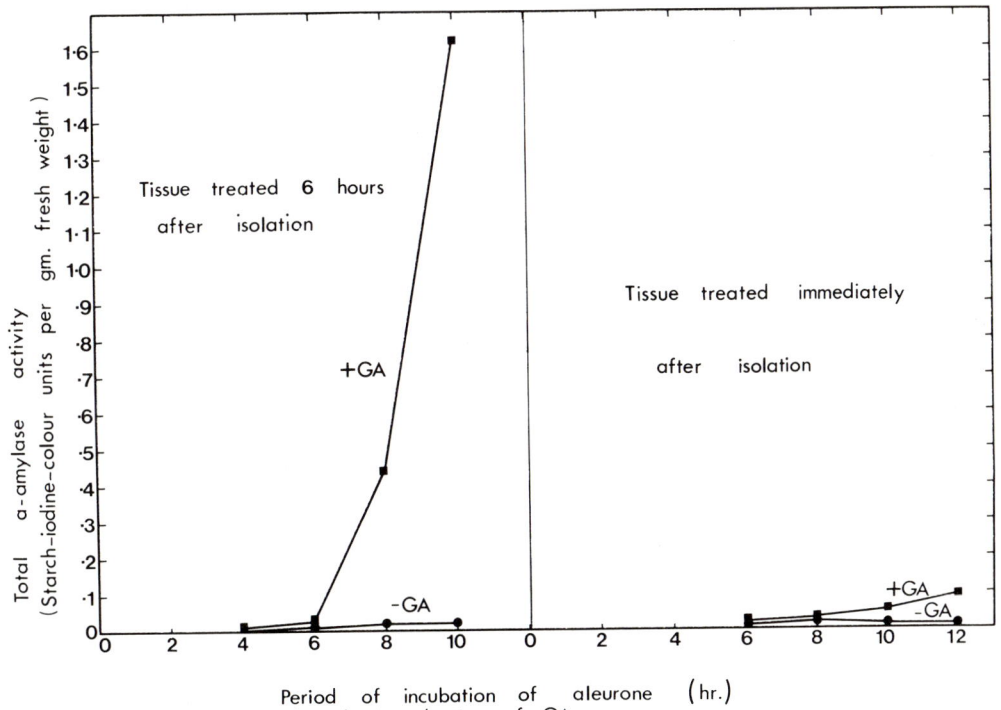

Fig. 1. *The production of α-amylase by aleurone tissue treated with GA.* Aleurone layers were cultured in water or GA (100 μg/ml) immediately after isolation, or after a further 6 hr pre-incubation in water. The ambient solution and the tissue were assayed for α-amylase (Filner and Varner, 1967) and the results calculated by the method of Briggs (1967).

Aleurone tissue was incubated in water or solutions of GA (100 μg/ml) for periods varying from 0.25 to 24 hr, and the quantities of nucleotides extracted from the tissue were measured. When the results of several experiments were pooled and analysed collectively, no significant effects of GA were found on the overall level of nucleotides. However, analyses of individual nucleotides showed small but significant effects on some of them. For example, after 1 hr the amounts of ADP, UTP and UDP-glucose were depressed by 14%, 6% and 23% respectively, but after a further 2 hr none of these effects were evident. In another experiment the level of NAD was increased by about 18% after 0.25 to 2 hr treatment with the hormone, and during the same period the level of CTP was depressed by about 7%.

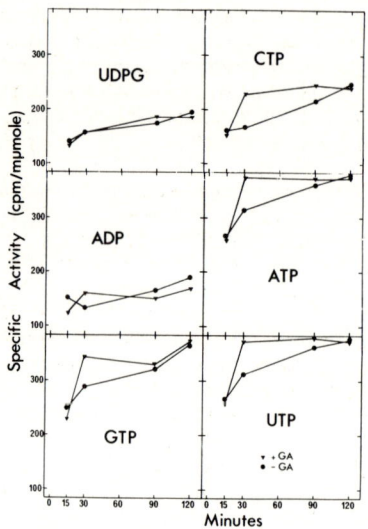

Fig. 2. *The effect of GA on the incorporation of ^{32}P into soluble nucleotides.* Isolated aleurone tissue was pre-incubated for 6 hr in water and then transferred to water or GA (100 μg/ml) for periods of up to 120 min. ^{32}P was added 15 min prior to the end of each period.

Although GA has no large or consistent effects on the levels of nucleotides in the tissue, it does influence the amount of ^{32}P incorporated into the nucleotides (Fig. 2). The hormone slightly but significantly reduces the incorporation of ^{32}P into all six compounds after treatment for 0.25 hr. Thereafter the hormone has no effect on the radioactivity of ADP and UDP-glucose. In the absence of the hormone, the specific activity of all four triphosphates increases progressively with time, and in three both the levels of radioactivity and the rate of increase are comparable. For the fourth, CTP, the initial value for specific activity is lower than for the others, but the rate of increase is similar. For all four triphosphates the greater incorporation at 0.5 hr in the presence of GA is significant. In quantitative terms the responses of ATP, GTP and UTP are similar (about 25%) but for CTP it is greater (38%). Moreover, at 1.5 hr the value for CTP in the presence of the hormone is 27% greater than in its absence, but the effect of GA on the other three triphosphates is no longer evident. After 2 hr GA has no effect on any of the triphosphates, and in another experiment of 3 hr duration GA reduced equally the incorporation of ^{32}P into all four compounds by about 20%.

IV. DISCUSSION

The method used here to separate aleurone layers from the starchy endosperm of wheat grain appears to initiate changes in the nucleotide metabolism of the tissue.

An increase in ribonuclease activity is known to follow the removal of barley aleurone from the endosperm (Chrispeels and Varner, 1967), and to follow wounding in other tissues (Bagi and Farkas, 1967). Thus, it may well be that the transient changes in the levels of ATP and UDP-glucose are a manifestation of similar responses in wheat aleurone. However, the finding that the tissue responds to GA by producing α-amylase supports the contention that the tissue is responding normally to the hormone.

The lack of any marked or consistent effects of GA on the amounts and composition of the soluble nucleotides indicates that the levels of these compounds are under rigid metabolic control. Nevertheless, as GA stimulates the incorporation of ^{32}P into ATP (Fig. 2), although it has no effect on the radioactivity of ADP, it is quite clear that the hormone enhances the rate of turnover of ATP, and perhaps the other triphosphates as well, in reactions involving the terminal phosphate moiety. This is the first report of the effects of GA on the metabolism of soluble nucleotides. Compared with its effect on the production of enzymes this response to the hormone is relatively rapid and is apparently transient: the response is clearly evident after only 0.5 hr of treatment, and yet is not detectable after a further 1.5 hr.

A very similar pattern of response was reported by Chandra and Duynstee (1968) who investigated the effect of GA on the incorporation of ^{32}P into RNA in aleurone layers of barley. They concluded that the primary effect of GA is probably on RNA metabolism. However, the increases caused by GA here of the specific activities of the four nucleotide triphosphates are very nearly equal, although the levels of UTP and ATP are 2 and 6 times as high respectively, as those of GTP and CTP. If nucleic acid synthesis alone were stimulated by GA then the increments in specific activity would be expected to be inversely proportional to the amount of each triphosphate present (assuming equimolar incorporation into RNA). As this is not the case, the response cannot be attributed solely to the synthesis of RNA. Moreover, in one experiment where the conditions of incubation differed from those imposed in Fig. 2 (see Materials and Methods), GA slightly, but not significantly, reduced the incorporation of ^{32}P into ATP, UTP and GTP. Values for specific activity after 0.5 hr of treatment with GA are as follows: ATP 440 (416), UTP 484 (451) and GTP 384 (377) counts/min/nmole (values in parentheses are for treatment with the hormone). The tissue used in this experiment responded by producing α-amylase in an apparently normal way. If it can be assumed that any metabolically significant changes in RNA synthesis would bring about detectable changes in the rates of turnover of the precursors of RNA, then the finding that GA has no effects on the rates of turnover of three of the precursors, while at the same time causing the production of α-amylase, throws some doubt on the concept that GA might exert a primary effect on RNA synthesis.

Cogent evidence that GA is acting on other areas of metabolism is adduced from the unique effects of the hormone on CTP. From Fig. 2 it is clear that the response of CTP is greater and detectable over a longer period than that of the other triphosphates. And in the experiment reported above, although GA had no effect on three of the triphosphates, the hormone significantly increased the incorporation of ^{32}P into CTP from 267 to 303 counts/min/nmole. There is no evidence that cytidine-rich nucleic acids are synthesised in plants as a result of treatment with hormones. In fact Johri and Varner (1968) report a decrease in the number of uracil-cytosine pairs in RNA extracted from dwarf pea seedlings treated with GA. Thus it seems reasonable to conclude that GA is acting on an aspect of metabolism which is dependent on CTP, or derivatives of it, and which is not related to nucleic acid metabolism.

The work of Kennedy and Weiss (1956) clearly establishes that cytidine nucleotides participate in phospholipid synthesis and the more recent investigations of Carter (1968) and Sumida and Mudd (1968) confirm this relationship. It has also been established that phospholipids are an essential part of the structure of membranes (Robertson, 1966; Rothfield and Finkelstein, 1968) and so it may be inferred that cytidine nucleotides participate in the biosynthesis of membranes. Some reports (Tata, 1967, 1970; Kerkof and Tata, 1969) indicate that both the incorporation of radioactive precursors into, and the rate of formation of, endoplasmic reticulum is increased in various animal tissues after the administration of hormones. Furthermore, Tata's results suggest that the induction of enzymes by hormones is closely linked with the increase in the rate of membrane formation. Accordingly, it is suggested that the increase in specific activity of CTP which occurs in wheat aleurone tissue after treatment with GA is a manifestation of an effect of the hormone on membrane synthesis.

V. SUMMARY

The effect of gibberellic acid (GA) on the soluble nucleotides of wheat aleurone layers was investigated. The hormone had little influence on the amounts of most of the nucleotides extracted from the tissue. However, the quantity of ^{32}P incorporated into some of the nucleotides was affected by GA. In particular, CTP responded uniquely to the presence of the hormone, and the implication of this finding in relation to the action of the hormone is discussed.

VI. ACKNOWLEDGEMENTS

One of us (G.G. Collins) held a Senior Postgraduate Scholarship granted by the Australian Wool Board. The support of the Commonwealth of Australia Wheat Industries Council is also gratefully acknowledged.

VII. REFERENCES

AMOS, H. and M. KORN (1958). 5-methyl cytosine in the RNA of *Escherichia coli*. Biochim. biophys. Acta. 29, 444-445.
BAGI, G. and G.L. FARKAS (1967). Of the nature of increase in ribonuclease activity in mechanically damaged tobacco leaf tissues. Phytochem. 6, 161-169.
BIELESKI, R.L. (1968). Effect of phosphorus deficiency on levels of phosphorus compounds in *Spirodela*. Pl. Physiol., Wash. 43, 1309-1316.
BIELESKI, R.L.(1969). Phosphorus compounds in translocating phloem. Pl. Physiol., Wash. 44, 497-502.
BRIGGS, D.E. (1963). Biochemistry of barley germination. Action of gibberellic acid on barley endosperm. J. Inst. Brew. 69, 13-19.
BRIGGS, D.E. (1967). Modified assay for α-amylase in germinating barley. J. Inst. Brew. 73, 361-370.
CARTER, J.R. Jnr. (1968). Cytidine triphosphate : phosphatidic acid cytidyltransferase in *Escherichia coli*. J. Lipid Res. 9, 748-754.
CHANDRA, G.R. and E.E. DUYNSTEE (1968). Hormonal regulation of nucleic acid metabolism in aleurone cells, *in* Wightman and Setterfield (1968). 723-745.
CHEN, D. and D.J. OSBORNE (1970). Hormones in the translational control of early germination in wheat embryos. Nature, Lond. 226, 1157-1160.
CHEN, D., S. SARID and E. KATCHALSKI (1968). Studies on the nature of messenger RNA in germinating wheat embryos. Proc. Natl. Acad. Sci. U.S.A. 60, 902-909.
CHRISPEELS, M.J. and J.E. VARNER (1967). Gibberellic acid-enhanced synthesis and release of α-amylase and ribonuclease by isolated barley aleurone layers. Pl. Physiol., Wash. 42, 398-406.
COLE, C.V. and C. ROSS (1966). Extraction, separation, and quantitative estimation of soluble nucleotides and sugar phosphates in plant tissues. Analyt. Biochem. 17, 526-539.
FILNER, P. and J.E. VARNER (1967). A test for *de novo* synthesis of enzymes: density labelling with H_2O^{18} of barley α-amylase induced by gibberellic acid. Proc. Natl. Acad. Sci. U.S.A. 58, 1520-1526.
GILES, K.W. and A. MYERS (1966). The effects of gibberellic acid and light on RNA, DNA and growth of the three basal internodes of dwarf and tall peas. Phytochem. 5, 193-196.
ISHERWOOD, F.A. and F.C. BARRETT (1967). Analysis of phosphate esters in plant material. Extraction and purification. Biochem. J. 104, 922-933.
JENNER, C.F. (1968). The composition of soluble nucleotides in the developing wheat grain. Pl. Physiol., Wash. 43, 41-49.

JOHRI, M.M. and J.E. VARNER (1968). Enhancement of RNA synthesis in isolated pea
    nuclei by gibberellic acid. Proc. Natl. Acad. Sci. U.S.A. 59, 269-276.
KAMISAKA, S., Y. MASUDA and N. YANAGISHIMA (1967). Gibberellin-induced yeast sporul-
    ation in relation to RNA and protein metabolism. Physiologia Pl. 20, 98-105.
KENNEDY, E.P. and S.B. WEISS (1956). The function of cytidine coenzymes in the
    biosynthesis of phospholipids. J. biol. Chem. 222, 193-214.
KERKOF, P.R. and J.R. TATA (1969). The subcellular distribution of ^{32}P-labelled
    phospholipids, ^{32}P-labelled ribonucleic acid and ^{125}I-labelled iodoprotein in
    pig thyroid slices. Effect *in vitro* of thyrotrophic hormone and dibutyryl-3',
    5'- (cyclic)-adenosine monophosphate. Biochem. J. 112, 729-739.
OGAWA, Y. (1966). Effects of various factors on the increase of α-amylase activity
    in rice endosperm induced by gibberellin A_3. Pl. Cell Physiol., Tokyo. 7,
    509-517.
PALEG, L.G. (1964). Cellular localisation of the gibberellin-induced response of
    barley endosperm, in: Coll. Int. C.N.R.S. (Paris). 123, 303-317.
PALEG, L.G., B.G. COOMBE and M.S. BUTTROSE (1962). Physiological effects of
    gibberellic acid, V. Endosperm responses of barley, wheat and oats. Pl.
    Physiol., Wash. 37, 798-803.
PHILLIPS, M.L. and L.G. PALEG (1972). The isolated aleurone layer. *in* "Plant
    Growth Substances, 1970". Ed. D.J. Carr. Springer Verlag. 396-406.
POLLARD, C.J. and B.N. SINGH (1968). Early effects of gibberellic acid on barley
    aleurone layers. Biochem. biophys. Res. Commun. 33, 321-326.
ROBERTSON, J.D. (1966). Granulo-fibrillar and globular substructure in unit membranes.
    Ann. N.Y. Acad. Sci. 137, 421-440.
ROTHFIELD, L. and A. FINKELSTEIN (1968). Membrane biochemistry. Ann. Rev. Biochem.
    37, 463-496.
ROYCHOUDHURY, R. and S.P. SEN (1965). The effect of gibberellic acid on nucleic acid
    metabolism in coconut milk nuclei. Pl. Cell Physiol., Tokyo. 6, 761-765.
SUMIDA, S. and J.B. MUDD (1968). Biosynthesis of cytidine diphosphate diglyceride by
    cauliflower mitochondria. Pl. Physiol., Wash. 43, 1162-1164.
TATA, J.R. (1967). The formation and distribution of ribosomes during hormone-induced
    growth and development. Biochem. J. 104, 1-16.
TATA, J.R. (1970). Co-ordination between membrane phospholipid synthesis and acce-
    lerated biosynthesis of cytoplasmic ribonucleic acid and protein. Biochem. J.
    116, 617-630.
VARNER, J.E. and G.R. CHANDRA (1964). Hormonal control of enzyme synthesis in barley
    endosperm. Proc. Natl. Acad. Sci. U.S.A. 52, 100-106.
VARNER, J.E., G.R. CHANDRA and M.J. CHRISPEELS (1965). Gibberellic acid-controlled
    synthesis of α-amylase in barley endosperm. J. cell. comp. Physiol. 66, suppl.
    1., 55-68.
VARNER, J.E. and M.M. JOHRI (1967). Hormonal control of enzyme synthesis, *in*
    Wightman and Setterfield, 1968. 793-814.
WATERS, L.C. and L.S. DURE (1966). Ribonucleic acid synthesis in germinating cotton
    seeds. J. molec. Biol. 19, 1-27.
WIGHTMAN, F. and G. SETTERFIELD (1968). Biochemistry and Physiology of Plant Growth
    Substances, Runge Press, Ottawa.
WYATT, G.R. (1951). The purine and pyrimidine composition of the deoxypentose nucleic
    acids. Biochem. J. 48, 584-590.

THE ISOLATED ALEURONE LAYER

M. Phillips* and L.G. Paleg

Department of Plant Physiology, Waite Agricultural Research Institute, The University of Adelaide

* Present address: Department of Biochemistry, University of Toronto, Toronto 5, Canada

I. INTRODUCTION

Ten years ago the first reports appeared of the effects of gibberellic acid (GA) on cereal endosperm (Paleg, 1960; Yomo, 1960). Since that time much work has been published on both the role of GA during cereal seed germination, and the mechanism and modes of action of the hormone on the cereal aleurone layer. In the past few years, however, it has become increasingly clear that definitive work on these problems was being hampered by the lack of a suitable technique for securing large amounts of relatively "clean" aleurone (i.e., free from starchy endosperm) rapidly, and in a responsive state.

The aleurone layer, in theory, is an admirably adapted tissue for hormone work. In practise, however, it has intrinsic problems and is far from ideal. On the one hand it is homogeneous, non-meristematic, non-photosynthetic, and responsive to a specific hormone in a specific way. On the other hand, it has the drawbacks of very thick cell walls, and of being extremely difficult and tedious to isolate in a viable form. In spite of the thick walls, however, penetration into, and release from the aleurone cells seems to be readily accomplished. Consequently, our attention has been directed towards ways of rapidly obtaining large quantities of viable, responsive aleurone, relatively free of contaminating endosperm or micro-organisms.

II. DEVELOPMENT OF THE METHOD

Three main techniques have been commonly used to secure aleurone tissue from barley endosperm: 1) hand-peeling half-seed which had been preincubated in water (Paleg, 1964; Varner and Ram Chandra, 1964), 2) digesting the starchy endosperm with a mixture of hydrolytic enzymes from *Trichoderma viride* (Yomo and Iinuma, 1964), and 3) grinding soaked half-seed by hand in a glass homogenizer (MacLeod, *et al.*, 1964). All three techniques have intrinsic difficulties, and none really result in the easy isolation of substancial amounts of tissue. In addition, the techniques were developed for use with barley, the aleurone layer of which is three cells thick.

We chose to use wheat rather than barley, for two reasons: 1) we hoped there would be less variability between cells since wheat aleurone is only one cell thick, and 2) the proteins of wheat are very much more water soluble than those of barley, and we hoped this would aid in washing the endosperm away from the aleurone layer. There was already sufficient information to indicate that wheat responded to gibberellin in a manner analogous to that of barley (Paleg, *et al.*, 1962; Rowsell and Goad, 1963). The difference between wheat and barley was evident in preliminary experiments in which only 16 hrs were necessary to soften wheat half-seed for hand-peeling whereas 72 hrs were required for barley. Graham, *et al.* (1963) used a technique for squeezing the milky endosperm of developing wheat grains out of the seed coats. Their report was the starting point for the following work.

Roller mill

The motive force of the technique is a roller-mill, consisting of an outer screw-top glass jar (diameter 12 cm) and an inner polythene bottle filled with sand (diameter 7 cm, weight 1169 g), mounted on a set of motor-driven rollers. A variable-speed motor was used initially so that the speed of rotation could be altered as desired. The method depends upon a) softening the starchy endosperm to such a degree that it can be squeezed out of the tough outer layers of the seed when it is crushed between the two bottles, and b) rinsing away the remaining starchy endosperm. Three variables were found to affect the degree to which the aleurone layers could be freed of starchy endosperm:

- a) time of imbibition of the half-seed;
- b) addition of suitable amounts of water at appropriate time intervals during the rolling process;
- c) speed of rotation of the mill.

The following are some observations which led to the final standardized method:

a) Time of imbibition Half-seed, imbibed at 30°C for various periods of time after the initial 2 hr sterilization period (see later), were placed in the roller-mill and rolled for 30 to 45 mins with the addition of 10 ml aliquots of water at suitable time intervals. At the end of the rolling procedure the contents of the mill were poured onto a sieve and the starchy endosperm rinsed away with water. The aleurone layers with their attached testa-pericarp remained on the sieve. They were stained with IKI (0.2% iodine in 2% potassium iodide) and the amount of starch adhering to them observed. Half-seed soaked for 22 to 24 hrs yielded the cleanest tissue. After shorter times of imbibition (12 and 16.5 hrs), the starchy endosperm immediately adjacent to the aleurone layer was not softened enough and could not be removed. After longer times (26 hrs), the endosperm was too soft and formed a glutinous dough which adhered tenaciously to the aleurone layer. Twenty four hrs after the initial two hr sterilization period was selected as the standard imbibition time.

b) Addition of water during rolling If too much water was added initially to the roller-mill, the half-seed simply floated behind the inner polythene bottle and were not crushed between it and the glass jar. If too little water was added initially, the half-seed were crushed between the two bottles and the endosperm became sticky and doughy, adhered to the glass jar and was difficult to rinse away. The final procedure adopted is indicated below.

c) Speed of rotation of the roller mill The same difficulties were encountered for too high or too low a speed of rotation as for the addition of too much or too little water. Furthermore, the tissue was damaged if the speed was too high. The optimum speed was about 50 rpm and the variable-speed motor was replaced by an appropriate constant-speed motor.

Standardized procedure

The final standardized procedure is as follows:

1) Cut seed in half with razor blade and discard embryo-half.

2) Transfer 5 g of half-seed to a 100 ml glass-stoppered Erlenmeyer flask. Surface-sterilize half-seed for 2 hrs at 30°C in 10 ml of 5% calcium hypochlorite and 1 drop of detergent.

3) Rinse half-seed 10 times with 10 ml of water using an automatic syringe. Flame mouth of flask and tip of syringe after each operation.

4) Finally, add 10 ml of water, flame mouth of flask, and pour the mixture

of water and half-seed into a 9 cm petri dish enclosed within a 15 cm petri dish.

5) Incubate the half-seed in petri dishes for 24 hrs at 30°C.

6) Transfer the inner petri dish containing the half-seed to a sterile cabinet containing previously sterilized equipment. (All subsequent procedures are carried out inside the sterile cabinet.)

7) Rinse half-seed 3 times with water.

8) Transfer up to 3 lots of half-seed to roller-mill.

9) Roll at 50 rpm adding 10 ml water initially and every 5 mins for a total of 30 mins.

10) Rinse away starchy endosperm through a sieve (mesh 2 mm^2) with 1.5 litres of water.

11) Replace aleurone tissue in roller-mill.

12) Roll for a further 5 mins adding 10 ml water every 1.5 mins.

13) Rinse tissue on sieve with an additional 1 litre of water.

14) Remove excess moisture by placing the aleurone tissue between layers of nylon gauze sandwiched between a single layer of filter paper and several outer layers of absorbent tissue in a 15 cm petri dish. Blot for 5 mins.

15) Weigh out required amounts of tissue and place in incubation flasks or petri dishes.

16) Add appropriate test solutions with sterilized pipettes or syringes.

17) Incubate at 30°C for desired times.

Maintenance of sterility

It is obvious from the procedure outlined above that the tissue could become contaminated with micro-organisms at almost every step especially if all procedures are carried out in the open laboratory. It was necessary, therefore, to sterilize all solutions and equipment and to carry out most of the procedures inside a sterile cabinet. If any part of the technique was carried out in the laboratory, the bench was wiped, and the area over the bench was sprayed with 70% ethanol to settle dust and spores before opening flasks and/or petri dishes.

A few hours before the isolation began, all equipment to be used for rolling was autoclaved and placed in the cabinet which was then sterilized by fumigating with   propylene glycol. Hands and arms were also thoroughly scrubbed with a detergent and 70% ethanol. The only piece of equipment that could not be autoclaved was the polythene bottle of sand and this was stored in 70% ethanol and wiped dry with sterilized tissue paper before placing in the roller mill.

As the initial step in each experiment, every 5 g quantity of half-seed was soaked for 2 hrs at 30°C in 10 ml of calcium hypochlorite (5 g calcium hypochlorite shaken in 100 ml of water for 10 mins and filtered) to which was added 1 drop of detergent (Teepol, Shell Chemicals, Australia Pty. Ltd.).

Microbial contamination was determined at various times throughout the procedure in several different experiments by plating aliquots of solutions and pieces of tissue or half-seed onto nutrient agar. One ml aliquots of solutions, or pieces of tissue or half-seed were plated on nutrient agar (1.0% agar, 0.1% Difco yeast extract, 0.1%

peptone in tap water) and incubated at 30°C for 48 hrs. Bacterial and fungal colonies were counted, and the results of a typical experiment are indicated in Table 1. The bacterial and fungal counts were extremely low indicating that the procedure was adequate for the maintenance of reasonable sterility.

Table 1.

Microbial Contamination at Different Stages of Isolation Procedure

		Replicate or treatment	No. of colonies on nutrient agar	
			Bacterial	Fungal
1.	Solutions after	1	7	0
	24 hr. imbibition	2	2	5
	period	3	5	0
2.	Half-seed after	1	0	2
	24 hr. imbibition	2	0	0
	period (3/petri)	3	1	0
3.	Aleurone layers	H_2O	0	0
	after 24 hr. incubation	"	0	0
	(6 tissue pieces from 6	"	0	0
	different incubations/petri)	GA*	0	0
		"	0	0
		"	0	0
4.	Ambient solutions after	H_2O	63	0
	9 hrs. incubation	"	13	0
		GA*	15	0
		" **	635	0
5.	Ambient solutions after	H_2O	18	0
	36 hrs. incubation	GA*	29	0

* GA solution filtered through millipore filter

** "   "   prepared with sterilized but unfiltered water

Reproducibility

Two criteria have been used to determine the reproducibility of the isolation procedure. Firstly, the fresh weight to dry weight ratios were used to reflect any difference in the tissue from sample to sample. Secondly, the amount of contamination by starchy endosperm was estimated by measuring the amount of starch present in the isolated tissue.

Samples of tissue were taken from 3 separate isolations carried out on different days. The fresh weight of the samples was determined and the tissue was then frozen on dry ice and lyophilized to a constant weight. The fresh weight to dry weight ratio (Table 2) was almost constant from one batch of tissue to another, as would be expected for tissue prepared by a reproducible method.

Table 2. Relationship Between Fresh and Dry Weight of Isolated Aleurone

Sample	Fresh Weight	Dry Weight	Ratio FW/DW
1	10.5g	2.6g	4.0
2	8.0	2.0	4.0
3	13.0	3.2	4.1

Samples of tissue from several separate isolations were lyophilized and the starch content determined according to the method of Pucher *et al.* (1948) and modified by Jenner (1967) using methods of McCready and Hassid (1943) and McCready *et al.* (1950). Samples (about 200 mg) of the dried material as well as samples of standard potato starch (about 50 mg) were heated with water for 15 mins in a boiling water bath, cooled, and the starch extracted with perchloric acid. After clarification of these extracts by centrifugation, starch was precipitated from the supernatants with iodine. The iodine complex was decomposed with alcoholic sodium hydroxide and the starch dissolved by heating in 0.5 N NaOH. The starch content was determined with anthrone and is indicated in Table 3. Starch varied from about 2.0 to 2.5% of the fresh weight, indicating that the isolation method yields a tissue which is reproducible insofar as contamination from the starchy endosperm is concerned.

Response of Isolated Aleurone from Ten Varieties to GA

Aleurone tissue was isolated from 10 wheat varieties and 1 g samples of each were incubated with 5 ml of either water or GA (100 µg/ml) for 22 hrs at 30°C. The α-amylase activity of both the ambient solutions and the extracts of the tissues was determined and the results (Table 4) were calculated from slopes of the % initial OD versus time curves for each sample.

The ideal variety would produce little or no α-amylase when incubated with water but substantial amounts when incubated with GA. Olympic and Javelin were the most suitable in these two respects. Although Crete showed a much greater response to GA, it produced considerable amounts of enzyme in the water controls. It is evident, however, that the different responses noted previously with barley varieties (Paleg *et al.*, 1962) also occur with the isolated aleurone of different wheat varieties.

Table 3. Starch Analysis of Isolated Aleurone

Sample	Starch content (mg/250mg D.W.)	% Dry Weight	% Fresh Weight
1	19.9	7.96	1.99
2	23.7	9.49	2.37
3	21.8	8.72	2.18
4	20.5	8.20	2.05
5	24.8	9.92	2.48
6	22.6	9.05	2.26
Ave.	22.7	9.08	2.27

Table 4. Production of α-amylase by Isolated Aleurone of Ten Wheat Varieties

α-amylase
(Δ% initial O.D. at 600 nm/hr/g F.W.)

Varieties	Water controls[+]			GA_3[+]		
	Ambient	Tissue	Total	Ambient	Tissue	Total
Olympic*	0	3	3	955	364	1319
Crete*	185	44	229	3859	1839	5698
Mentana*	0	8	8	847	442	1289
Javelin	0	0	0	843	243	1086
Heron	0	4	4	53	100	153
Mengavi	0	4	4	28	70	98
Gamenya	0	42	42	27	28	55
Persia	21	34	55	1736	1468	3204
Warrigo	1212	115	1327	2032	646	2678
Gabo	1290	192	1482	1611	627	2238

+ Tissue incubation medium - H_2O or 100 µg/ml GA_3; temperature - 30°C; time - 22 hrs.

* Average of 3 determinations; others, 1 determination.

Effect of Increasing GA Concentrations

Since Olympic was more readily available than Javelin, it was used in the subsequent work. The response of isolated aleurone to different concentrations of GA is indicated in Table 5. The response is sensitive, occurs over a wide GA concentration range, and levels off at about 100 µg/ml.

III. ELECTRON MICROSCOPY OF ISOLATED ALEURONE

Not only is the physiological response to GA of the isolated wheat aleurone layer similar to that of barley, but, not surprisingly, wheat aleurone is structurally similar to barley. The micrograph in Plate 1 is of control (H_2O-treated) aleurone isolated with the technique described above. Aleurone grains with the two types of inclusions, surrounded by spherosomes are clearly visible. In addition, ER, plastids, mitochondria, etc., as well as the very thick cell wall with plasmodesmata, are all present.

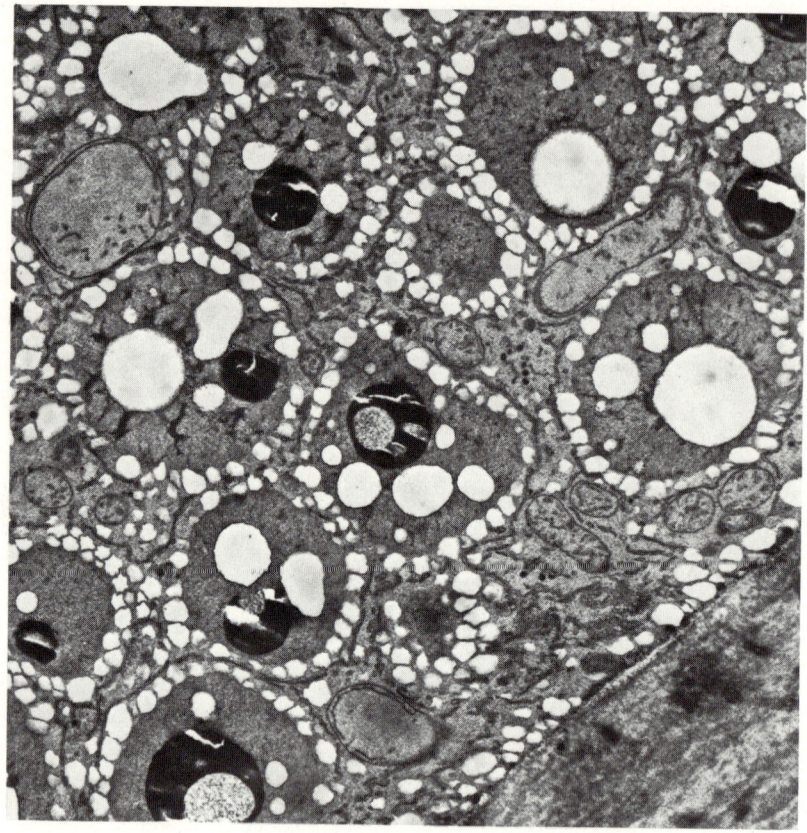

Plate 1.

Section of cell in isolated wheat aleurone tissue. Tissue incubated 18 hrs. in water, fixed in 5% glutaraldehyde in 0.1 M phosphate buffer (pH 7.1), stained in KMnO4 and embedded in araldite. Magnification x12,000. Cell wall in lower right corner.

Table 5. GA_3-induced α-amylase Production by Isolated Aleurone of Olympic Wheat

GA_3 (μg/ml)	α-amylase (Δ% initial O.D. at 600nm/hr/g F.W.)		
	Ambient	Tissue	Total
0	0	0	0
0.01	0	20	20
0.1	191	230	421
0.2	1657	536	2193
1.0	1660	270	1930
2.0	2366	618	2984
20.0	2503	984	3487
100.0	3531	1647	5178
200.0	4278	1169	5447

Plate 2a is a micrograph of a GA-treated cell after 24 hrs. The most noteworthy features of this cell are the two indications of polarity in the GA-induced response. The portion of the cell adjacent to the starchy endosperm (the inner part) has the relatively thinner cell walls and in that area dissolution of aleurone grains is essentially complete. The aleurone grains located in the outer half of the cell (the portion furthest away from the starchy endosperm) are still easily recognizable, and apparently have not yet been subjected to the same intensity of lytic effects.

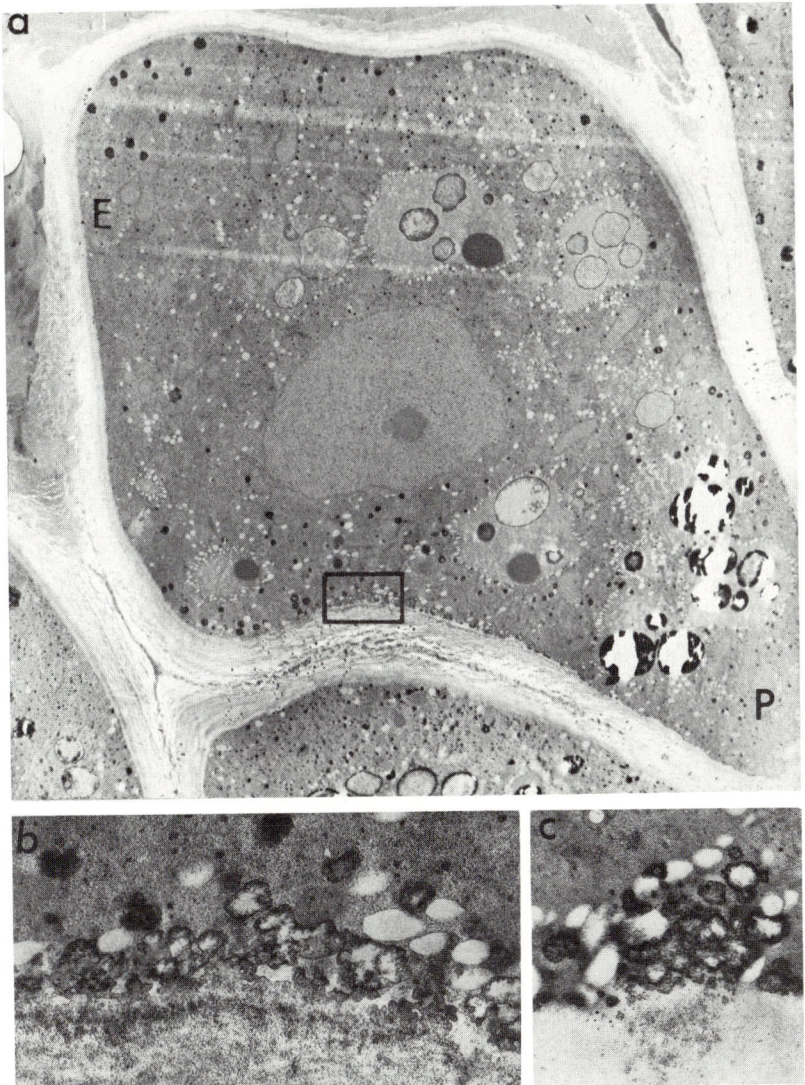

Plate 2.

Cells from isolated wheat aleurone tissue incubated 24 hrs. in 100 µg/ml GA_3.

a) Section in which aleurone grains in part of cell (E) adjacent to starchy endosperm have undergone dissolution before those in part of cell closer to pericarp (P). Magnification x2,700.

b) Outlined area in 2a) showing vesiculation at edge of cell. Magnification x27,000.

c) Edge of another cell showing vesicles and their apparent release of contents into the cell wall. Magnification x27,000.

The second, less obvious though probably related manifestation of polarity, in the hormonally-induced response of aleurone, is indicated by the enclosure on 2a, shown at higher magnification in 2b. Vesiculation at the cell surface is evident, and is also visible at the uppermost edge of the opposite side of the cell. The vesiculation has already occurred below (i.e., closer to the endosperm), but not yet above this point in the cell. The origin of the vesicles is obscure at present, but the contents are apparently released into the cell wall. Plate 2C is a more definite indication of the release of vesicular contents into the cell wall obtained from yet another cell.

IV. CONCLUSIONS

Various workers in the past have commented (Varner, et al., 1965; Duffus, 1967) on the difficulties in securing appreciable amounts of aleurone tissue in a homogeneous reproducible state. The method for isolating wheat aleurone, developed in this work, overcomes these difficulties.

The isolation of barley aleurone tissue required pre-incubation of the half-seed in water for 2 to 3 days, whereas the wheat system requires only 26 hrs (including the 2 hr sterilization period in calcium hypochlorite). Hand-peeling the barley half-seed was time-consuming, the tissue obtained was heterogeneous because of variations in time and handling of individual layers, and only relatively small quantities could be obtained. The wheat system, on the other hand, requires only 45 mins for the isolation of 10 to 20 g of tissue. Larger or smaller amounts can be obtained with slight modifications to the procedure. The nature of the method allows for very strict control over treatment of the tissue from one experiment to another and produces a reproducible population of uniformly treated aleurone layers. Although there is still some contamination from the starchy endosperm, this is also reproducible and is probably not appreciably different from that of the hand-peeled barley aleurone tissue.

The procedure outlined above is relatively rapid, maintains reasonable sterility, and yields fairly large amounts of a viable, responsive and reproducible tissue. The effect of various GA concentrations, the ultrastructure, the time course of enzyme formation and release (Collins, et al., 1972), the influence of Ca^{++}, Actinomycin D, pH (unpublished information), are all similar to, if not identical with similar parameters reported for barley. Furthermore, the tissue isolation technique has made feasible studies of the influence of gibberellin hormones on aspects of metabolism (e.g., Collins, et al., 1972) that were previously impractical.

The presence of polarity in the response of aleurone cells to GA, and the identification of vesiculation at the cell surface as one aspect of GA-induced metabolism, raise interesting questions about the nature of the hormone effect. For example, if the locus of action of GA is in the nucleus, one might expect the response to emanate outward from the nucleus in all directions equally. On the other hand, perhaps the site of the hormonal mechanism of action is not in the nucleus, and the reason for the polarity in aleurone cell response is that GA penetrates the thinner walls adjacent to the starchy endosperm more easily and induces its initial response at or close to its point of entry into the cell. Certainly the data does not yet allow conclusions on these points to be drawn.

The presence of vesicles at the surface of the cell, the contents of which are released into the cell wall, is very suggestive of the lysosomal situation which occurs in animal tissues (de Duve and Wattiaux, 1966). Membrane-enclosed vesicles containing different types of hydrolytic enzymes have been visualized with electron microscopy and isolated biochemically. MacLeod and Millar (1962) suggested that the aleurone response might involve the release of pre-formed lysosomal enzymes, but after the demonstration by Filner and Varner (1967) that α-amylase was synthesized *de novo,* the suggestion received little support. If the vesicles, in fact, do contain α-amylase (and preliminary, unpublished results have demonstrated that more than 50% of the tissue α-amylase can be sedimented), it is likely that the GA-induced response involves the induction of *de novo* lysosomal hydrolytic enzyme synthesis, "packaging" and eventual release from the aleurone cell. The evidence obtained by Collins, *et al*. (1972) of an early specific hormonal stimulation of CTP turnover is probably an indication of the formation of membranes associated with either the synthesis or the packaging of the lysosomal enzymes. Finally, the fact that GA causes drastic and immediate changes in the permeability of model membranes (Wood, *et al*., 1972) strongly suggests that the initial hormonal triggering action is a biophysical modification of pre-existing membrane components, resulting in the *de novo* synthesis of lysosomal enzymes from m-RNA molecules that are already present in the mature aleurone cell.

V. SUMMARY

A procedure is described for isolating aleurone tissue from wheat grain. The method is relatively rapid, maintains reasonable sterility, and reproducibly yields fairly large amounts of viable tissue in a responsive state. The tissue consists of the aleurone layer, the testa-pericarp and some cells of the starchy endosperm, and results indicating lysosome activity obtained with electron microscopy are discussed.

VI. ACKNOWLEDGEMENTS

The award of a Commonwealth Scholarship and a University of Adelaide Research Grant Scholarship to one of us (M.L. Phillips) is gratefully acknowledged, as is financial assistance from the Barley Improvement Trust Fund.

VII. REFERENCES

COLLINS, G.G., C.F. JENNER and L.G. PALEG (1972). The effect of gibberellic acid on the metabolism of soluble nucleotides in aleurone tissue isolated from wheat grain. *in* "Plant Growth Substances 1970", Ed. D.J. Carr, Springer Verlag.

DE DUVE, C. and R. WATTIAUX (1966). Functions of lysosomes. Ann. Rev. Physiol. 28, 435-92.

DUFFUS, J.H. (1967). A cell-free system for the study of α-amylase synthesis in barley aleurone layers. Biochem. J. 103, 215-7.

FILNER, P and J.E. VARNER (1967). A test for *de novo* synthesis of enzymes: density labelling with H_2O^{18} of barley α-amylase induced by gibberellic acid. Proc. Natl. Acad. Sci. U.S. 58, 1520-6.

GRAHAM, J.S.D., R.K. MORTON and J.K. RAISON (1963). Isolation and characterization of protein bodies from developing wheat endosperm. Aust. J. Biol. Sci. 16, 375-83.

JENNER, C.F. (1967). Synthesis of starch in detached ears of wheat. Aust. J. Biol. Sci. 21, 597-608.

MACLEOD, ANNA M. and A.S. MILLAR (1962). Effects of gibberellic acid on barley endosperm. J. Inst. Brew. 68, 322-32.

MACLEOD, ANNA M., J.H. DUFFUS and C.S. JOHNSTON (1964). Development of hydrolytic enzymes in germinating grain. J. Inst. Brew. 70, 521-8.

MCCREADY, R.M., J. GUGGOLZ, V. SILVIERA and H.S. OWENS (1950). Determination of starch and amylose in vegetables. Application to peas. Anal. Chem. 22, 1156-8.

MCCREADY, R.M. and W.Z. HASSID (1943). The separation and quantitative estimation of amylose and amylopectin in potato starch. J. Amer. Chem. Soc. 65, 1154-7.

PALEG, L.G. (1960). Physiological effects of gibberellic acid. I. On carbohydrate metabolism and amylase activity of barley endosperm. Pl. Physiol. Wash. 35, 293-9.

PALEG, L.G. (1964). Cellular localization of the gibberellin-induced response of barley endosperm, in, "Régulateurs Naturels de la Croissance Végétale" (Ed. J.P. Nitsch). Coll. Int. C.N.R.S. 123, 303-317 Paris.

PALEG, L.G., B.G. COOMBE and M.S. BUTTROSE (1962). Physiological effects of gibberellic acid. V. Endosperm responses of barley, wheat and oats. Pl. Physiol. Wash. 37, 798-803.

PUCHER, G.W. C.S. LEAVENWORTH and H.G. VICKERY (1948). Determination of starch in plant tissues. Anal. Chem. 20, 850-3.

ROWSELL, E.V. and L.J. GOAD (1963). The release of hydrolytic enzymes from isolated wheat aleurone layers activated by gibberellic acid. Biochem. J. 90, 12P.
VARNER, J. and G. RAM CHANDRA (1964). Hormonal control of enzyme synthesis in barley endosperm. Proc. Natl. Acad. Sci. U.S. 52, 100-6.
VARNER, J., G. RAM CHANDRA and M.J. CHRISPEELS (1965). Gibberellic acid-controlled synthesis of α-amylase in barley endosperm. J. Cell and Comp. Physiol. 66, Supplement 1, 55-68.
WOOD, A., L.G. PALEG and RAVINDAR SAWHNEY (1972). Gibberellin and membrane permeability. *in* "Plant Growth Substances 1970", Ed. D.J. Carr, Springer Verlag.
YOMO, H. (1960). Studies on the amylase activating substances. II. On the amylase activating substance in the embryo culture medium and the barley malt extract. Hakko Kyokai Shi. 18, 494-9.
YOMO, H. and H. IINUMA (1964). The enzymes of the aleurone layer of barley endosperm. Amer. Soc. Brew. Chem. Proc. Ann. Meetings.

GIBBERELLINS: OTHER SYSTEMS

Metabolic Changes in Internodes of Dwarf Pea Plants treated with Gibberellic Acid

A.J. McComb and W.J. Broughton

Botany Department, and Department of Soil Science and Plant Nutrition, University of Western Australia, Nedlands, Western Australia 6009

1. INTRODUCTION

This paper summarises our work on the metabolic changes which take place in expanding pea internodes after intact plants have been treated with gibberellic acid (GA_3). The investigations are concerned primarily with the way in which the metabolic machinery of the internode is altered to maintain an increased growth response.

11. MATERIALS AND METHODS

Plant material

In most experiments, plants of *Pisum sativum* L. cultivar 'Meteor' (Sutton and Sons, Reading, U.K.) were raised in the glasshouse for about 14 days, when the fifth internodes were expanding. Some of the plants were then treated with 1 or 5 µg of GA_3, applied at node 3 in a 5 µl droplet of 50% ethanol. Expanding fifth internodes were removed from one group of plants at the time of treatment ('zero time'), and subsequently from groups of treated and control plants.

Electrophoresis

Plant material (10-20 internodes) was ground in cold Tris-glycine buffer (0.025M Tris, pH 8.3) containing 5% w/v sucrose, made up to 2 ml, centrifuged, and samples of 50 µl used in electrophoresis. This was carried out at 150V in the cold on 3 mm slabs of 6% acrylamide, using the apparatus of E.C. Corporation (Philadelphia, U.S.A.). Electrode chambers contained Trisglycine (0.025M Tris, pH 8.3), the gels, Tris-HCl (0.1M Tris, pH 8.9). Gels were incubated at $22°$, and after staining scanned with a Joyce Loebl Chromoscan (Gateshead, U.K.).

For detection of amylase, gels containing 0.1% starch were stained with iodine after incubation for 1.25 hr in 0.1M phosphate buffer, pH6.3 (Mills and Crowden, 1968). For starch phosphorylase, gels were stained with iodine after overnight incubation in 0.1M citrate buffer (pH 5.5) containing 0.5% w/v glucose-1-phosphate and 10^{-3} M cysteine.

Fractionation of RNA

Extracts were made from 20 internodes at various times after injection of 10 µl of solution containing 10 µc $NaH_2\ ^{32}PO_4$ (with or without 7.5×10^{-4}M GA_3), using the method of Click and Hackett (1966) with diethyl pyrocarbonate added as a nuclease inhibitor. The phenol-purified extract was fractionated on 2.5 x 45 cm columns of 'Sepharose 4 B' (Pharmacia, Uppsala, Sweden) using the procedure of Oberg and Philipson (1967), and absorbance and radioactivity of each 5ml fraction measured.

Recording of growth rates

The apical region of a plant was connected with cotton to the short arm of a lever. The long arm of the lever was connected to the short arm of a second lever, which magnified the signal from the first, giving a total magnification of 28, and traced a record of growth on a smoked drum rotating at 0.076 rev/hr. Before use each plant was watered thoroughly, and sprayed with 20% v/v ethanol in water containing 0.1% Tween-80 (Honeywill - Atlas, London, U.K.). Some 2 - 3 hours later GA3 (10^{-3}M) was sprayed on the plant in this solvent.

III. RESULTS AND DISCUSSION

After application to a lower leaf, GA_3 is translocated to the expanding internodes and beyond, into the apical region and young leaves (McComb, 1964). The application results in a marked increase in the rate but not the duration of internode expansion (Brian et al., 1958). The increase in length is accompanied by increased dry weight, and during expansion there is a direct relation between dry weight and volume of water in the internode. The crude cell-wall fraction of the dry weight also increases greatly, and there is a direct relation between internode volume and amount of wall (McComb, 1966). The increase in dry weight not accounted for by cell wall is in part due to an increase in protein content (Broughton and McComb, 1967). Internode elongation and the effect of GA_3 are apparently dependent on protein synthesis, as growth is blocked by cycloheximide (Broughton, 1969).

The cells of control internodes are elongating at the time of treatment, and there is a further increase in cell length in internodes treated with GA_3. Subsequently, the number of cells increases in the controls, and there is an additional increase in the GA_3-treated internodes (Fig.1). The level of DNA is closely related to the number of cells in treated and controls (Fig.1). FUDR inhibits the growth of control and GA_3-treated internodes, but its effects, which are less immediate than those of cycloheximide, become obvious when cell numbers are normally increasing (Broughton 1969). The earlier effect of GA_3 is on cell elongation in this system.

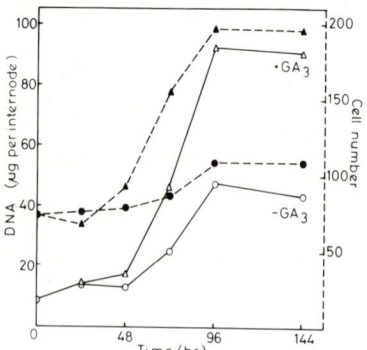 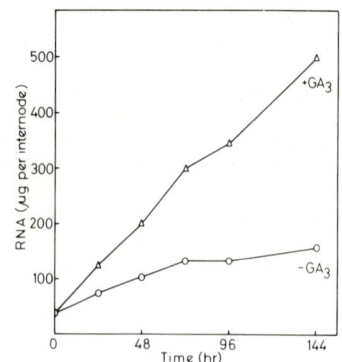

Fig.1. Effect of gibberellic acid on cell number and DNA content of dwarf pea internodes. (Redrawn from Broughton, 1969)

Fig.2. Effect of gibberellic acid on RNA content of dwarf pea internodes. (Data from Broughton, 1969)

The treated internodes show a marked increase in RNA content, and the onset of this effect precedes the increase in DNA (Fig.2). Actinomycin D blocks growth in control and treated tissues soon after application (Broughton, 1969). An increase in RNA occurs on a per DNA basis, suggesting that a given amount of DNA supports a higher rate of RNA synthesis after GA_3 treatment (Broughton, 1968). McComb, McComb and Duda (1970) found that RNA synthesis by chromatin from pea stems was not increased when GA_3 was included in the reaction mixture. However, chromatin isolated from internodes after GA_3 treatment supported synthesis of more RNA than did chromatin from the controls, and this was attributed to an increase in RNA polymerase. When RNA polymerase from E. coli was added to the chromatin of control and treated internodes, no evidence was obtained for an effect of GA_3 on the amount of DNA available for transcription. The increase in polymerase was readily detected 12 hr after treatment, and there was evidence for a small increase at 6 hr.

The RNA formed by this *in vitro* system has not been fractionated, but preliminary investigations by Broughton and Dr. R.C. Jennings have been carried out on the RNA formed by internodes of intact plants. Labelled phosphate was injected into internodes at the time of GA3 treatment, and extracts made as described above. When separated on beaded agarose, most of the label present in the RNA corresponded with a low molecular weight fraction (Fig. 3). An increase in label in this fraction could be readily detected 6 hr after treatment.

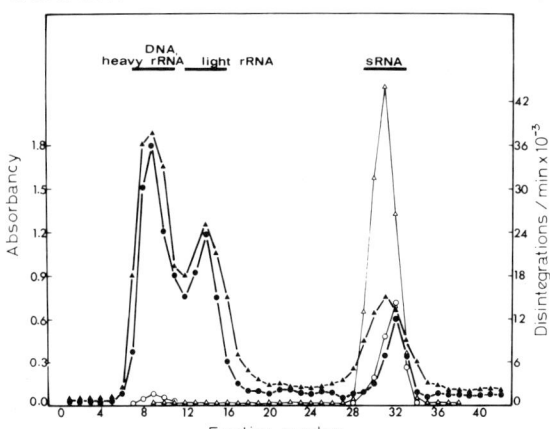

Fig 3. Separation of nucleic acids on beaded agarose.
Extracts were made 12 hr after treatment with GA$_3$. Triangles represent internodes from GA$_3$-treated plants, circles the controls. Closed symbols represent absorbancy ($A^{1 cm}_{260 nm}$), open symbols radioactivity.

In considering protein metabolism, we may well ask if, in response to GA$_3$, there is an increase in the complete complement of enzymes or, alternatively, if there is an alteration in the pattern of enzyme development. We have therefore looked at a number of enzyme systems, chosen partly because of their possible significance in controlling dry weight deposition in the expanding internode, and partly because of their altered activity in other systems after hormone treatment. In view of the general increase in protein, it is not surprising to find that there are increases in the activities of these enzymes, when results are expressed on a per internode basis (Table 1). For some, e.g. β-fructofuranosidase (Fig. 4), the increase is closely correlated with growth, and it is tempting to suppose that the activity of the enzyme is causally related to internode expansion.

Differences between the rates of increase of different enzymes suggest that there is, indeed, an alteration in the pattern of enzyme development, and this is seen more clearly when results are expressed per unit protein, i.e. as specific activities. In Fig. 5 the specific activities of enzymes from treated internodes are expressed as percentages of corresponding specific activities for the controls. Data for other enzymes are included in Table 1. There are clearly major changes in patterns of enzyme development as a result of GA$_3$ treatment. When amylases were examined on acrylamide gels, it was found that extracts from treated and control internodes show the same 6 bands (Fig. 6a). There is some evidence for alterations in the relative activities of different bands, but it is clear that the effect of GA$_3$ on amylolytic activity cannot be attributed to the appearance of a unique enzyme. Similarly, no new bands of starch phosphorylase activity were detected after GA$_3$ treatment (Fig. 6b).

We may usefully concentrate on three interrelated points, which arise in connection with these experiments.

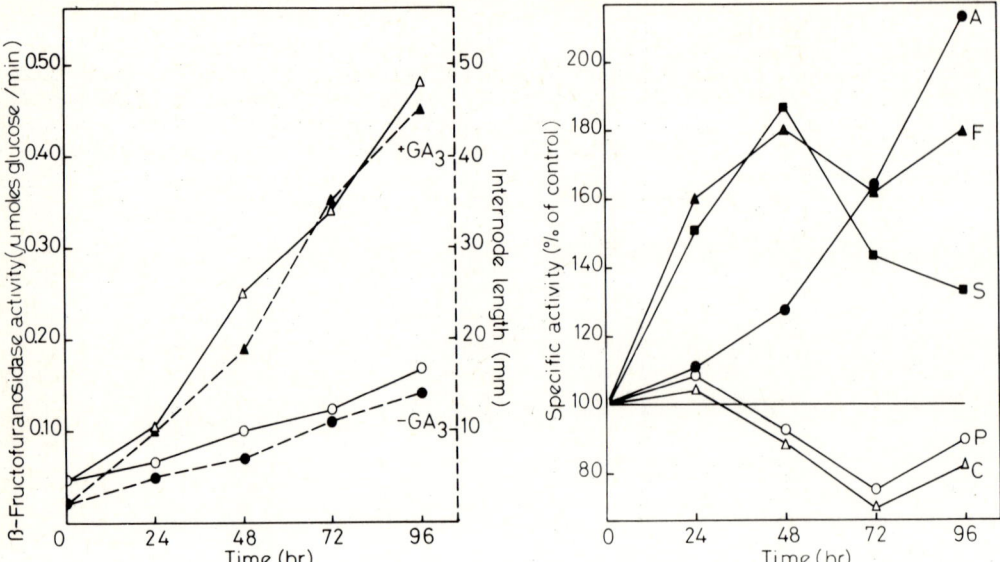

Fig. 4. Effect of gibberellic acid on β-fructofuranosidase activity in dwarf pea internodes. The activity is expressed in units per internode, and internode length is included. (Redrawn from Broughton and McComb, 1971).

Fig. 5. Changes in the specific activities of enzymes in dwarf pea internodes treated with gibberellic acid. Activities were calculated per unit soluble protein, and the specific activities of treated plants expressed as a percentage of specific activities of the corresponding controls. A, amylases; F, β-fructofuranosidase; S, starch phosphorylase; P, pectinesterase; C, cellulase, (Redrawn from Broughton and McComb, 1971).

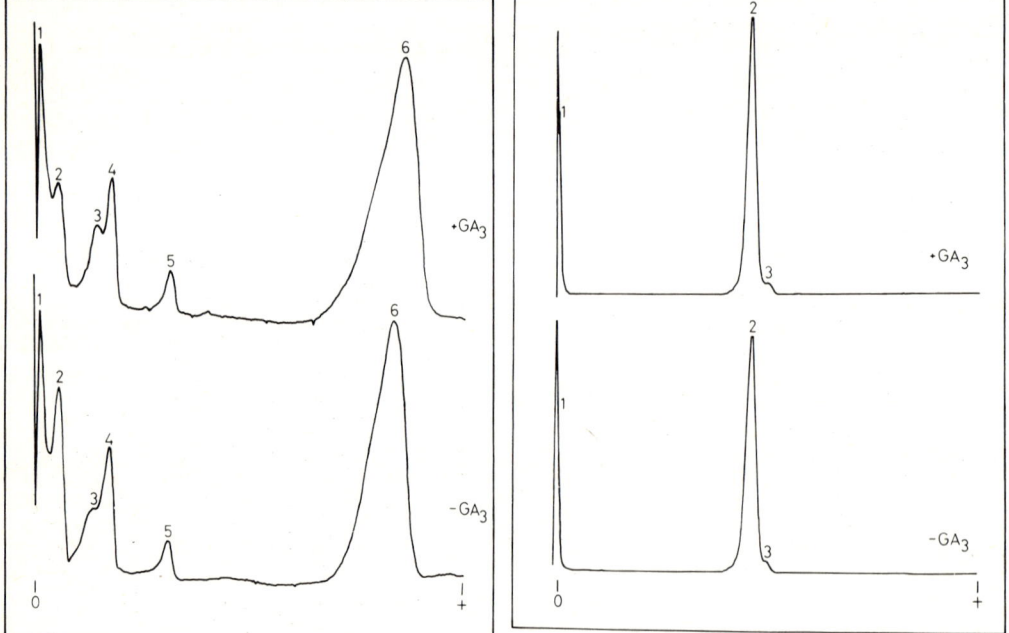

Fig. 6. The effect of gibberellic acid on enzymes of pea internodes, as detected on acrylamide gels. Left, amylase; Right, starch phosphorylase. Extracts were made from internodes 72 hr after GA_3 treatment (above), or from corresponding controls (below). 0 indicates the end of the well, and + the position of Bromphenol blue, some 60 mm from the well.

Table 1.

The activities of enzymes extracted from internodes of control plants, and plants treated with gibberellic acid

(Data from Broughton and McComb, 1971 and Broughton, Hellmuth and Yeung, 1970)

	ACTIVITY PER INTERNODE		ACTIVITY PER UNIT PROTEIN	
	Control (units/internode)	Treated (as % of control)	Control (units/mg protein)	Treated (as % of control)
Amylases	0.15	440	0.54	160
Starch phosphorylase (EC 2.4.1.1)	0.006	210	0.021	140
β-Fructofuranosidase (EC 3.2.1.26)	0.13	260	0.46	160
Cellulase (EC 3.2.1.4)	0.39*	130	1.6*	70
Pectinesterase (EC 3.1.1.11)	0.54	130	2.0	80
Ribulosediphosphate carboxylase (EC 4.1.1.39)	0.009	310	0.031	110
Phosphophyruvate carboxylase (EC 4.1.1.31)	0.002	270	0.006	100
Phosphopyruvate carboxylase (EC 4.1.1.32)	-	-	0.0006	70
Malate dehydrogenase (EC 1.1.1.37)	0.29	290	1.1	100

* relative activity only

Firstly, as each enzyme reflects genetic information, we may conclude that in a general sense GA_3 treatment alters gene expression during development. The problem is in understanding how direct the control may be. Our data offer no strong support for the suggestion that the changes in activity described here are due to the expression of parts of the genome which were previously masked. We do not detect an increase in template by direct assay, and marked increases in enzyme activity are not accompanied by the appearance of unique enzymes.

Secondly, as seen in Table 1, increases in activity occur for systems which might be concerned with providing more carbohydrate to the internodes for general metabolism including wall synthesis. The suggestion that the enzymes function to provide more carbohydrate does not seem unreasonable in view of the increases in dry weight and cell wall weight which are observed to keep pace with internode expansion. Further, increases in the growth rate of dwarf pea plants have been observed to occur after injection of hexose at rates which approximate to the requirements for maintaining wall synthesis in GA_3-treated plants (Broughton and McComb, 1971).

Thirdly, we are faced with the question of understanding how the processes we have considered are interrelated in time. The simplest overall sequence to suggest is that GA_3 treatment brings about increased RNA synthesis, which leads to increased protein and an alteration in the balance between different enzymes. Finally, the altered enzyme pattern leads to increased carbohydrate availability for cell development. If we are to propose that such a simple sequence is responsible for initiating as well as maintaining increased growth, we must suggest that the changes which we observe occur sufficiently soon after GA_3 treatment to account for the earliest detectable growth response.

The timing of the growth response has been studied by linking plants with a system of levers to a smoked, rotating drum. In previous work GA_3 was applied at some distance below the elongating tissue, and the time elapsing between treatment and effect attributed in part to the time taken for GA_3 to be translocated from the site of application to the site of action (McComb, 1962, 1964). In the present work, GA_3 was sprayed directly onto the expanding internodes.

A selection of results is presented in Fig. 7, where it will be seen that the time elapsing between treatment and effect is substantially eliminated by this technique. Thus, the reactions which initiate the growth effect must be quite rapid. Until suitable enzyme assays have been carried out we cannot decide finally whether control of enzymes via RNA metabolism could initiate more rapid growth. However, we must bear in mind the possibility that the changes we have been considering, important as they no doubt are in maintaining increased growth, may occur subsequent to the initiation of the growth response.

IV. SUMMARY

Gibberellic acid brings about a marked increase in the RNA content of expanding pea internodes and, subsequently, in DNA and cell number. Much of the increased RNA is in a soluble fraction. Chromatin RNA polymerase activity is enhanced without an increase in DNA template. Total protein increases, and there is an alteration in the pattern of enzyme development. Levels of enzymes which may provide carbohydrate are increased, though new forms of the enzymes were not detected. These metabolic changes may maintain rather than initiate increased growth, as the time elapsing between treatment and growth response is short.

V. ACKNOWLEDGEMENTS

We are indebted to the University of Western Australia, the Australian Research Grants Committee, and the Plant Research Laboratories of Michigan State University.

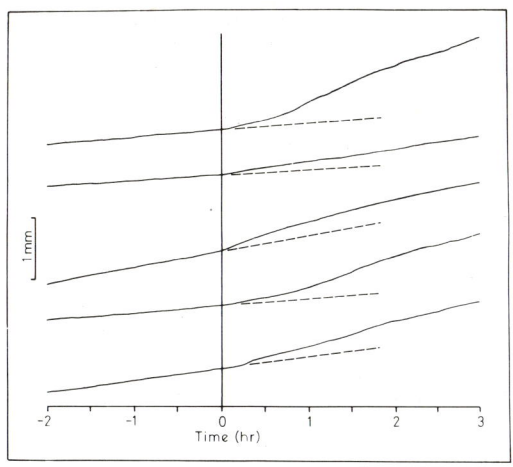

Fig. 7. Increases in the height of pea seedlings, traced from records taken on smoked paper. Each plant was sprayed at 0 hr with GA_3 (10^{-3}M).

VI. REFERENCES

BRIAN, P.W., H.E. HEMMING and E. LOWE (1958). Effects of gibberellic acid on rate of expansion and maturation of pea internodes. Ann. Bot. N.S. 22, 539-542.

BROUGHTON, W.J. (1968). Influence of gibberellic acid on nucleic acid synthesis in dwarf pea internodes. Biochem. biophys. Acta, 155, 308-310.

BROUGHTON, W.J. (1969). Relations between DNA, RNA and protein synthesis, and the cellular basis of the growth response in gibberellic acid-treated pea internodes. Ann. Bot. N.S. 33, 227-243.

BROUGHTON, W.J. and A.J. McCOMB, (1967). The relation between cell-wall and protein synthesis in dwarf pea plants treated with gibberellic acid. Ann. Bot. N.S. 31, 359-366.

BROUGHTON, W.J. and A.J. McCOMB (1971). Changes in the pattern of enzyme development in gibberellin-treated pea internodes. Ann. Bot. N.S. 35, 213-228.

BROUGHTON, W.J., E.O. HELLMUTH and D. YEUNG (1970). Role of glucose in development of the gibberellin response in peas. Biochem. biophys. Acta, 222, 491-500.

CLICK, R.E. and D.P. HACKETT (1966). The isolation of ribonucleic acid from plant, bacterial or animal cells. Biochem. biophys. Acta 129, 74-84.

McCOMB, A.J. (1962). An effect of gibberellic acid on circumnutation. New Phytol. 61, 128-131.

McCOMB, A.J. (1964). The stability and movement of gibberellic acid in pea seedlings. Ann. Bot. N.S. 28, 669-687.

McCOMB, A.J. (1966). The stimulation by gibberellic acid of cell wall synthesis in the dwarf pea plant. Ann. Bot. N.S. 30, 155-163.

McCOMB, A.J., J.A. McCOMB and C.D. DUDA (1970). Increased RNA polymerase activity associated with chromatin from internodes of dwarf pea plants treated with gibberellic acid. Pl. Physiol., Lancaster 46, 221-223.

MILLS, A.K. and R.K. CROWDEN (1968). Distribution of soluble proteins and enzymes during early development of *Pisum sativum*. Aust. J. biol. Sci. 21, 1131-1141.

OBERG, B. and L. PHILIPSON (1967). Gel filtration of nucleic acids on sphere-condensed agarose. Archs. Biochem. Biophys. 119, 504-509.

GIBBERELLIN METABOLISM IN THE ROOTS OF *Phaseolus coccineus* SEEDLINGS

A. Crozier and D.M. Reid

Botany Department University of Canterbury Christchurch 1 New Zealand and Biology Department University of Calgary, Calgary 44, Alberta Canada

1. INTRODUCTION

Roots have been implicated in the GA metabolism of a number of plants (Jones and Phillips, 1966, 1967, Sitton *et al*, 1967, Reid and Carr, 1967, Reid and Crozier, 1971). However, it does not necessarily follow that they are the site of all steps in the GA biosynthetic pathway (Crozier and Reid, 1971). Root apices have been shown to be the focal point of GA metabolism in the roots of *Helianthus* seedlings (Jones and Phillips, 1966). This paper reports the effects of root apex removal on the GA content of *Phaseolus* seedlings. The rationale behind this experiment was that an examination of the GA content of leaves, stems and cotyledons several days after root surgery could possibly yield information on the precise role of roots in GA biosynthesis in *Phaseolus* seedlings.

11. MATERIALS AND METHODS

Seedlings of *Phaseolus coccineus* var. Prizewinner were grown both with and without root apices using the methods described by Crozier and Reid (1971). After 8 days' growth, an equal number of control plants and plants without root apices were separated into leaves and apical buds, stems and petioles, cotyledons and roots or subapical root remnants.

Diffusion experiments were carried out using the techniques of Jones and Phillips (1964). Shoots and roots were excised from seedlings 1 cm from their junction with the cotyledons. The cut surfaces of the shoots and roots were placed in agar. Agar blocks were also placed on the stem and root sections that remained attached to the cotyledons. After 24 hours, the agar containing the shoot, root, cotyledon ⟶ shoot and cotyledon ⟶ root diffusates was collected and frozen. In a further experiment, shoots were diffused into agar for 4 days. The agar was changed daily so the 0-1, 1-2, 2-3 and 3-4 day diffusates could be collected.

The various tissues and agar diffusates were extracted with methanol and the ethyl acetate-soluble fraction purified by Sephadex G-10 column chromatography prior to either silicic acid partition column chromatography (Crozier and Reid, 1971) or thin layer chromatography (Reid and Crozier, 1971). The chromatographed extracts were tested for biological activity on the barley aleurone α-amylase bioassay (Jones and Varner, 1967) and/or the 'Tan-ginbozu' dwarf rice bioassay (Murakami, 1968).

111. RESULTS

The effect of root apex amputation on the GA content of *Phaseolus* seedlings, as determined by the rice bioassay, is illustrated in figure 1.

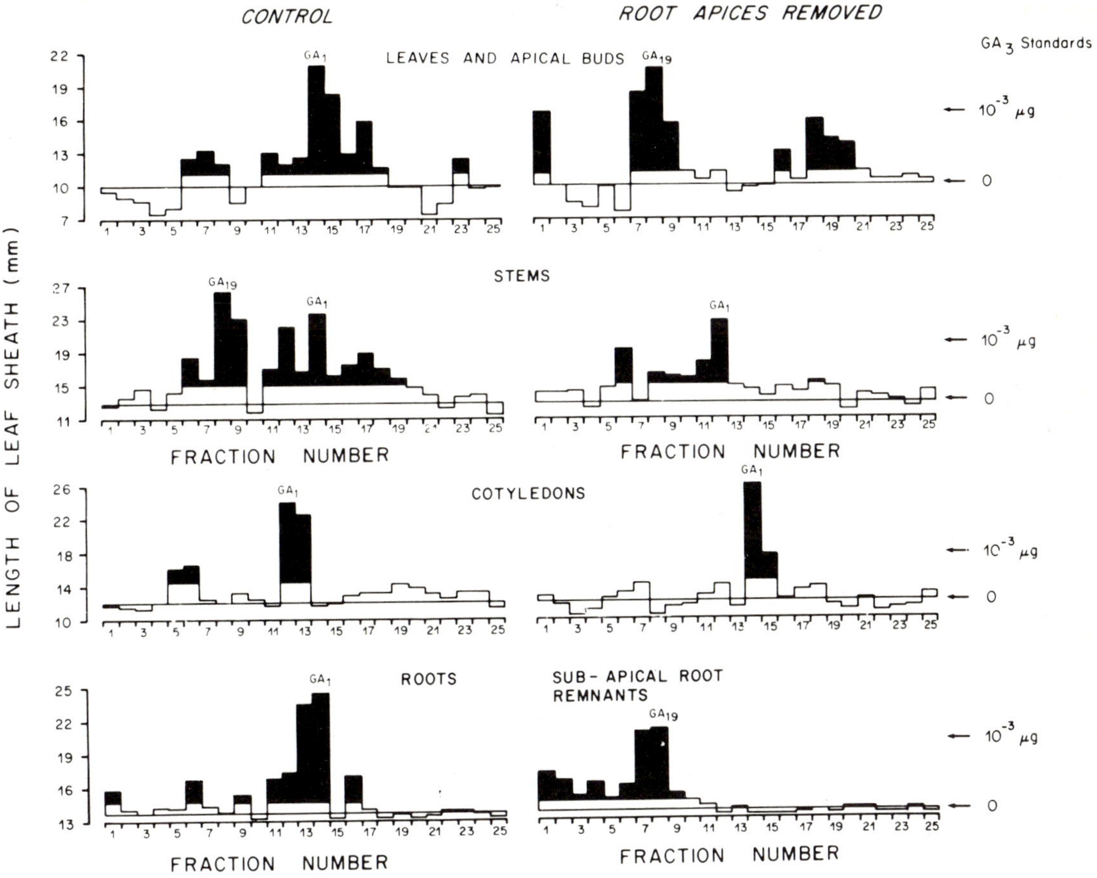

415

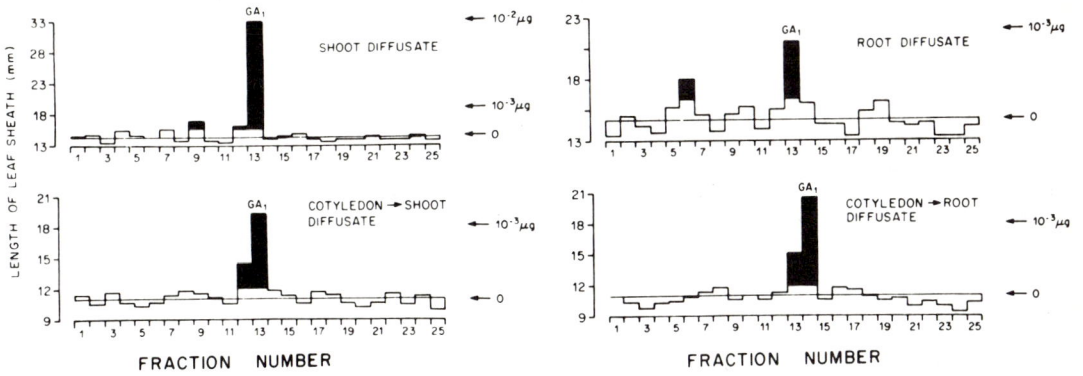

Fig. 1. "Tan-ginbozu" dwarf rice bioassays of silicic acid partition column chromatograms of 1/50th of the purified acid, ethyl acetate-soluble fractions from 100 *Phaseolus coccineus* seedlings with and without root apices. Shaded areas of the histograms represent activity that is significant at the 5 percent level.

Fig. 2. "Tan-ginbozu" dwarf rice bioassays of silicic acid partition column chromatograms of the 1/40th of the purified acid, ethyl acetate-soluble fractions from shoot, root, cotyledon ⟶ shoot and cotyledon ⟶ root diffusates from 200 *Phaseolus coccineus* seedlings. Shaded areas of the histograms represent activity that is significant at the 5 percent level.

There was a total of at least 7 GA-like compounds in the various extracts, with major activity being located in fractions 7-9 and 12-15. These two GA's have also been detected in large scale extracts from dark-grown *Phaseolus* seedlings (Crozier et al, 1971). Based on biological activities and gas-liquid chromatography data, the fraction 7-9 GA appears to be GA_{19} and the fraction 12-15 is most probably GA_1. The exact location of GA_1 in the extracts shown in figure 1 varied slightly from chromatogram to chromatogram. However, it was easily distinguishable from GA-like compounds that were eluted in fractions 11-12 and 17 in the leaf and stem extracts of control plants as it was the only GA to exhibit significant activity in the barley aleurone bioassay.

Gibberellin A_1 was the predominant GA in the leaves, cotyledons and roots of control plants. Stem tissue from control seedlings contained GA_{19} in addition to a small amount of GA_1. Root apex removal resulted in the disappearance of GA_1 and an accumulation of GA_{19} in the leaves and subapical root remnants. Gibberellin A_{19} disappeared from the stems despite its accumulation elsewhere. Gibberellin A_1 in the cotyledons appeared unaffected by root apex amputation.

Gibberellin A_{19} was tested on the rice bioassay at 50-, 100-, 250- and 1000-fold dilutions and the results indicated that its accumulation in the leaves and apical buds was 20 times greater than in the root remnants. Tests using a similar dilution series showed that the leaves and apical buds of each seedling contained approximately 10^{-1} µg of GA_3 equivalents of GA_1.

The results of the rice bioassays on the shoot, root, cotyledon ⟶ shoot and cotyledon ⟶ root diffusates are presented in figure 2. Despite the presence of several GA's in the tissues, only GA_1 could be detected in the diffusates. The shoot diffusates contained the largest amount of GA_1, however, the level was less than 10 percent of that found in shoot tissues. The results of the 4 day shoot diffusion experiment are shown in figure 3. Although there was a small day-to-day variation, there was no significant fall in the level of GA_1 moving into the agar over the 4 day diffusion period. Once again, the quantities of GA_1 in the diffusates were well below the levels found in the shoot tissue of control plants.

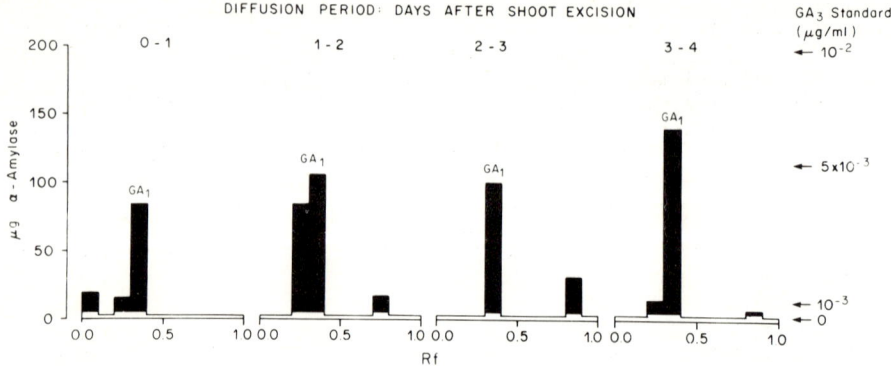

Fig. 3. Barley aleurone α-amylase bioassays of thin layer chromatograms of 1/20th of the purified and ethyl acetate-soluble fraction from shoot diffusates from 200 *Phaseolus coccineus* seedlings collected 0-1, 1-2, 2-3 and 3-4 days after shoot excision. Shaded areas of the histograms represent activity that is significant at the 5 percent level.

IV. DISCUSSION

Root apex removal resulted in the disappearance of GA_1 in the leaves and apical buds, and root remnants. This implies that GA_1 is produced in the root and exported to the shoot. Concurrent with the disappearance of GA_1 in these tissues is the accumulation of GA_{19}. The most likely explanation for this change in GA content is that roots are a site for the conversion of GA_{19} to GA_1. Removal of the root apex removes the site of the conversion process and, as a result, GA_1 disappears and the level of the precursor, GA_{19}, increases. Possible intermediates in the conversion of GA_{19} to GA_1 would appear to be either GA_{20} or GA_{23} (Figure 4). It may be of relevance that GA_{20} has been found in immature *Phaseolus coccineus* seed (MacMillan and Pryce, 1968).

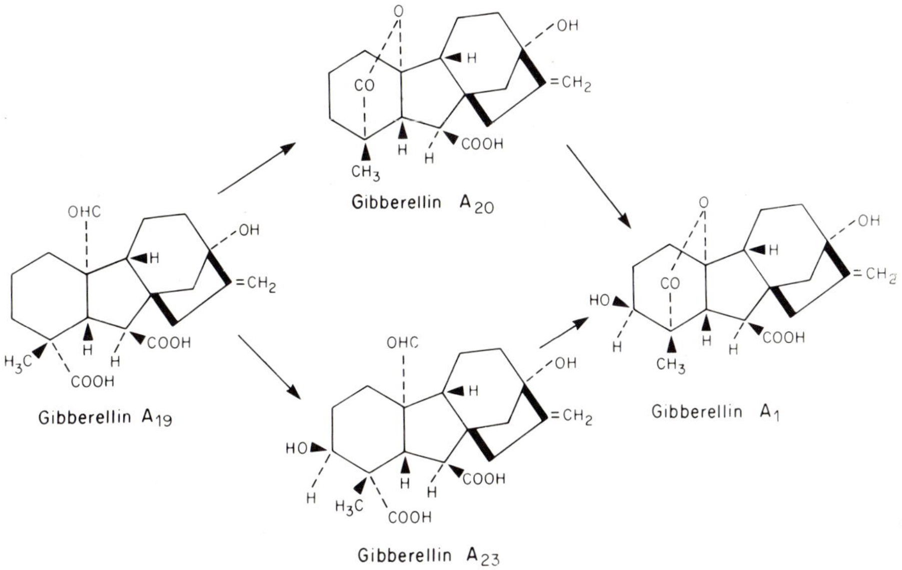

Fig. 4. Possible pathways for the conversion of GA_{19} to GA_1.

While GA_1 from the leaves most probably originates in the roots, it is somewhat more difficult to determine the site of synthesis of GA_{19}. The most likely candidate is the shoot as it is improbable that the subapical root remnants would be able to synthesize GA_{19}, yet unable to convert it to GA_1. If, however, this were the case and the root remnants did synthesize GA_{19}, it would necessitate intercellular compartmentalization of the various stages of GA biosynthesis in the root system. This would restrict the synthesis of GA_{19} to the subapical sections of the root, and its conversion to GA_1 to the root apices. We know of no evidence to support the existence of such a system. The fact that the accumulation of GA_{19} in the leaves is 20 times greater than it is in the subapical root remnants makes the probable site for the synthesis of GA_{19} the shoot. This contention is further supported by the report that *Helianthus* subapical root tissue does not appear to be able to synthesize GA (Jones and Phillips, 1966).

Although root apex removal results in the disappearance of GA_1 in the roots and leaves, the level of GA_1 in the cotyledons appears unchanged. It may be that the synthesis of the cotyledonary GA_1 is not related to GA synthesis in the root. Alternatively, GA_1 in the cotyledons could be produced in the roots but metabolized at a slower rate than GA_1 in the leaves.

The tissues contain several GA-like compounds in addition to GA_1 and GA_{19} and the levels of a number of these GA's are also affected by root apex removal. A not unreasonable general assumption is that the GA's which disappear after root apex removal represent steps in the GA synthesis pathway after GA_1, while those that accumulate are precursors of GA_{19}.

Based on the above arguements, we feel that the data supports the existence of a shoot ⟶ root ⟶ shoot recycling scheme in GA biosynthesis of *Phaseolus* seedlings. Gibberellin A_{19} may be synthesized in the shoot from a precursor such as (-)-kaurenoic acid. It is then transported to the roots where it is converted to GA_1 via either GA_{20} or GA_{23}. The GA_1 is exported to the shoot where it constitutes the main GA in the leaves and apical buds.

A major point in the proposed recycling scheme is the transport of GA's within the seedling, with GA_{19} moving from the shoot to the root and GA_1 moving in the reverse direction. The diffusion data in figures 2 and 3 indicate that although the tissues contain several GA-like compounds, only GA_1 diffuses into agar in significant quantities. The proposed *in vivo* movement of GA_1 and GA_{19} in the *Phaseolus* seedling may require a selective and active transport system. Under these circumstances, either separation of the shoot from the root or root apex removal could stop or severely reduce normal GA transport. The diffusion data may then not give a true indication of GA transport in the intact seedling but merely reflect the fact that GA_1 is not bound to the plant tissue as strongly as GA_{19} and the other GA-like compounds. If the GA content of stems is at least partially representative of GA's being transported within the seedling, then an intact root system may well be necessary for movement of GA_{19} from the shoot to the root as GA_{19} is present in the stems of control seedlings but absent in stems of seedlings that have undergone root apex removal (Figure 1). Perhaps data more relevant to GA transport would be obtained by studying the movement of labelled GA_1 and GA_{19} in intact seedlings and comparing their uptake by excised shoot and root tissues.

V. SUMMARY

Root apices were removed from light-grown *Phaseolus coccineus* seedlings and the endogenous GA's examined. Gibberellin A_1 was the main GA in control plants. Root apex removal resulted in the disappearance of GA_1 and the accumulation of GA_{19} in the leaves and apical buds and in the subapical root remnants. Gibberellin A_1 in the cotyledons was unaffected by root apex removal.

Gibberellin A_1 was the only GA to be detected in significant quantities in 24 hour shoot, root, cotyledon ⟶ shoot and cotyledon ⟶ root diffusates. There was no reduction in the amounts of GA_1 found in 24 hour shoot diffuses over a 4 day diffusion period. The level of GA_1 in 24 hour shoot diffusates was less than 10% of that found in shoot tissue.

VI. ACKNOWLEDGEMENTS

A. Crozier is grateful to Dr. R.P. Pharis, University of Calgary for financial support from NRC Grant A-2585. D.M. Reid was supported by NRC Grant A-5727.

VII. REFERENCES

CROZIER, A., H. AOKI and R.P. PHARIS. (1969). Efficiency of counter-current distribution, Sephadex G-10, and silicic acid partition chromatography in the purification and separation of gibberellin-like substances from plant tissue. J. Expt. Bot. 20, 786-795.

CROZIER, A., D.H. BOWEN, J. MacMILLAN, D.M. REID and B.H. MOST (1971). Characterization of gibberellins from dark-grown *Phaseolus coccineus* L. seedlings by gas-liquid chromatography and combined gas chromatography-mass spectrometry. Planta (Berl.) 37, 142-154.

CROZIER, A., C.C. KUO, R. C. DURLEY and R.P. PHARIS (1970). The biological activities of 26 gibberellins in 9 plant bioassays. Can. J. Botany. 48, 867-877.

CROZIER, A. and D.M. REID (1971). Do roots synthesize gibberellins? Can. J. Botany 43, 967-975.

JONES, R.L. and I.D.J. PHILLIPS (1964). Agar-diffusion technique for estimating gibberellin production by plant organs. Nature, $\underline{204}$, 497-499.

JONES, R.L. and I.D.J. PHILLIPS (1966). Organs of gibberellin synthesis in light-grown sunflower plants. Plant Physiol. $\underline{41}$, 1381-1386.

JONES, R.L. and I.D.J. PHILLIPS (1967). Effect of CCC on the gibberellin content of excised sunflower organs. Planta (Berl.) $\underline{72}$, 53-59.

JONES, R.L. and J.E. VARNER (1967). The bioassay of gibberellins. Planta (Berl.) $\underline{72}$, 155-161.

MacMILLAN, J. and R.J. PRYCE (1968). Further investigations of gibberellins in *Phaseolus multiflorus* by combined gas-chromatography-mass spectrometry - the occurrence of gibberellin A_{20} (*Pharbitis* gibberellin) and the structure of compound *b*. Tetrahedron Letters, 1537-1542.

MURAKAMI, Y. (1968). A new rice seedling bioassay for gibberellins, "Microdrop Method", and its use for testing extracts of rice and morning glory. Bot. Mag. (Tokyo). $\underline{81}$, 33-43.

REID, D.M. and D.J. CARR (1967). Effects of a dwarfing compound, CCC, on the production and export of gibberellin-like substances by root systems. Planta (Berl.) $\underline{73}$, 1-11.

REID, D.M. and A. CROZIER (1971). The effects of waterlogging on the gibberellin content and growth of tomato plants. J. Expt. Bot. $\underline{22}$, 39-48.

SITTON, D., A. RICHMOND and Y. VAADIA (1967). On the synthesis of gibberellins in roots. Phytochem. $\underline{6}$, 1101-1105.

Stimulation of the Levels of Gibberellin-like Substances by the Growth Retardants, CCC and AMO 1618

D.M. Reid and A. Crozier

Biology Department University of Calgary, Calgary 44, Alberta, Canada and Department of Botany University of Canterbury, Christchurch 1, New Zealand

1. INTRODUCTION

It has often been assumed that, when a plant is treated with a growth retardant such as CCC ((2-chloroethyl) trimethyl ammonium chloride) or AMO (2-isopropyl-4-dimethylamino-5-methylphenyl-1-piperidine carboxylate methyl chloride), the subsequent reduction in growth results directly from an inhibition of endogenous GA synthesis (Zeevaart and Osborne, 1965, Bristow, 1966, Cross, 1968, Dale and Felippe, 1968, Felippe and Dale, 1968, Ockerse, 1970). We have found that in pea seedlings CCC will reduce growth, yet at the same time (depending upon the concentration of retardant) either have little effect on the quantity of extractable GA or, actually increase GA levels (Reid and Crozier, 1970). This paper reports further investigations into the effects of CCC and AMO on growth and endogenous GA content of three plant species.

11. METHODS AND MATERIALS

Three species of plants were used : *Pisum sativum* L. cv. Alaska, *Hordeum vulgare* cv. Parkland, and *Phaseolus coccineus* cv. Prizewinner. All plants were supplied with either CCC or AMO via the roots which were immersed in aqueous solutions of the retardants. Details of these methods can be found in Crozier and Audus (1968) and in Reid and Crozier (1970). The only departure from these methods was that the barley was supplied with the retardant on the third day of its growth and the plants were measured and harvested on the 7th day.

Methods of harvesting, extraction, chromatography, bioassay and application of GA are described in Reid and Crozier (1970).

111. RESULTS

The effect of CCC on the levels of extractable GA in peas is shown in Figure 1. The rice assay showed that 1 mg/l CCC stimulated GA levels 150-fold, while the barley assay (of a different extract) showed a 7-fold rise. Plants treated with 1000 mg/l CCC contained approximately the same quantity of GA as was found in the untreated seedlings. In the rice assay of the extracts from the plants grown in 1 mg/l CCC, three peaks of GA activity were detected. Only two peaks were present in the controls. Choline chloride (results not shown here), an analogue of CCC, had no effect on GA levels or stem growth at either 1 or 1000 mg/l. As expected, CCC inhibited, and GA_3 stimulated, stem growth (Figure 1). It was not possible to completely overcome the effects of the CCC by the application of up to 30 ug of GA_3 per plant.

The effects of AMO on peas are shown in Figure 1. As with CCC, treatment with a small quantity of AMO (1 mg/l) stimulated the quantity of GA found in the extracts. There is evidence that more than one GA type is affected. Unlike CCC, AMO at a concentration of 200 mg/l reduced the amount of GA to a level below that found in the untreated plants. A lettuce bioassay of the same extracts showed a very similar result. The effects of AMO on growth are similar to those of CCC. AMO inhibits stem elongation

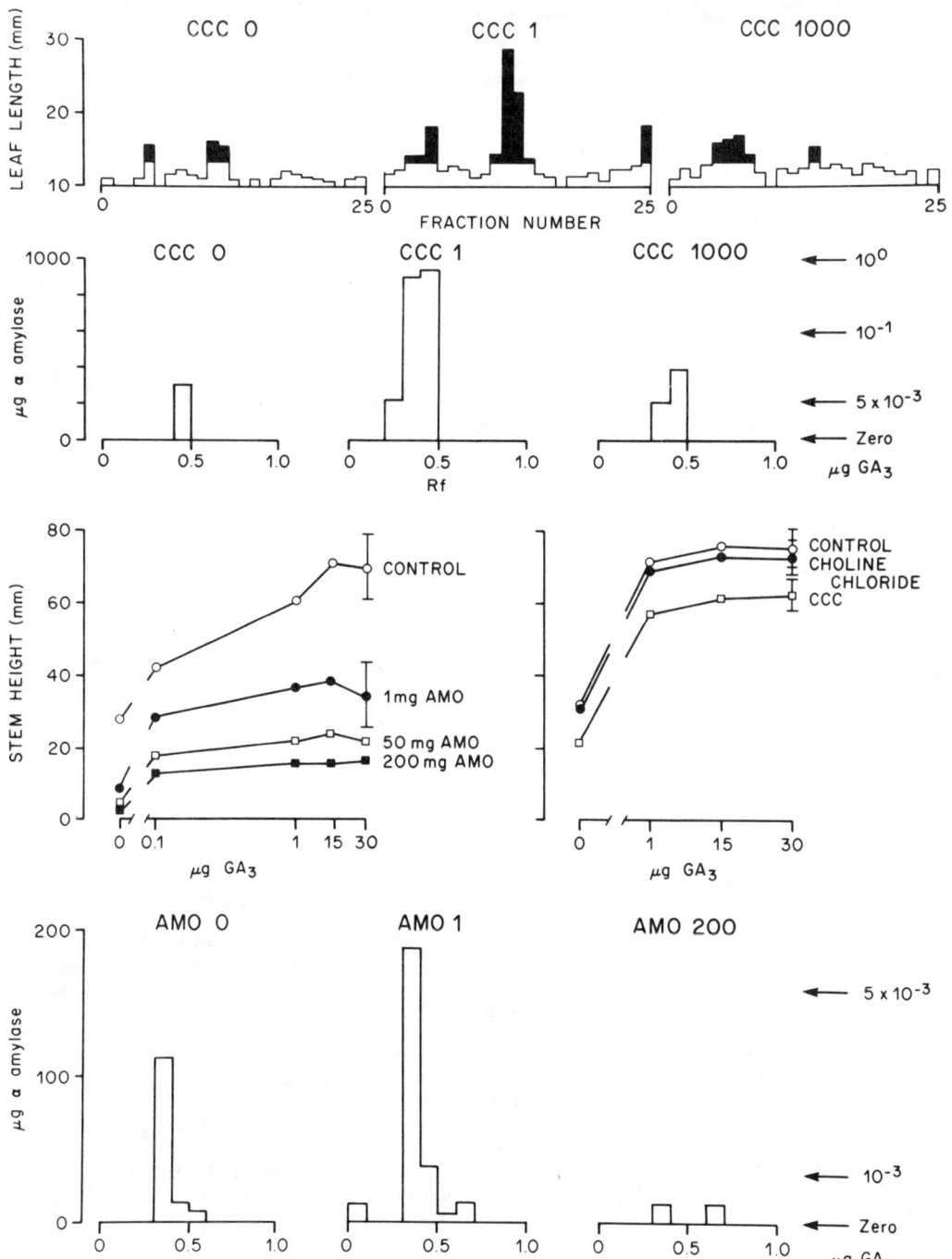

Fig.1. Top: the effects of CCC on GA levels in pea seedlings. GA estimated by a dwarf rice assay of the fractions from a silicic acid partition column. Shaded area represents statistically significant (p = 0.05) difference from the controls. GA also estimated by a barley aleurone assay following TLC. Middle: the effects of GA_3 on the growth of pea plants grown in CCC, choline chloride and AMO. Vertical lines represent 95% confidence limits. Bottom: the effects of AMO on the GA levels in pea seedlings. GA estimated by the barley aleurone assay following TLC.

and plants supplied with 30 μg/GA$_3$ have much shorter stems than control plants treated with smaller quantities of GA$_3$.

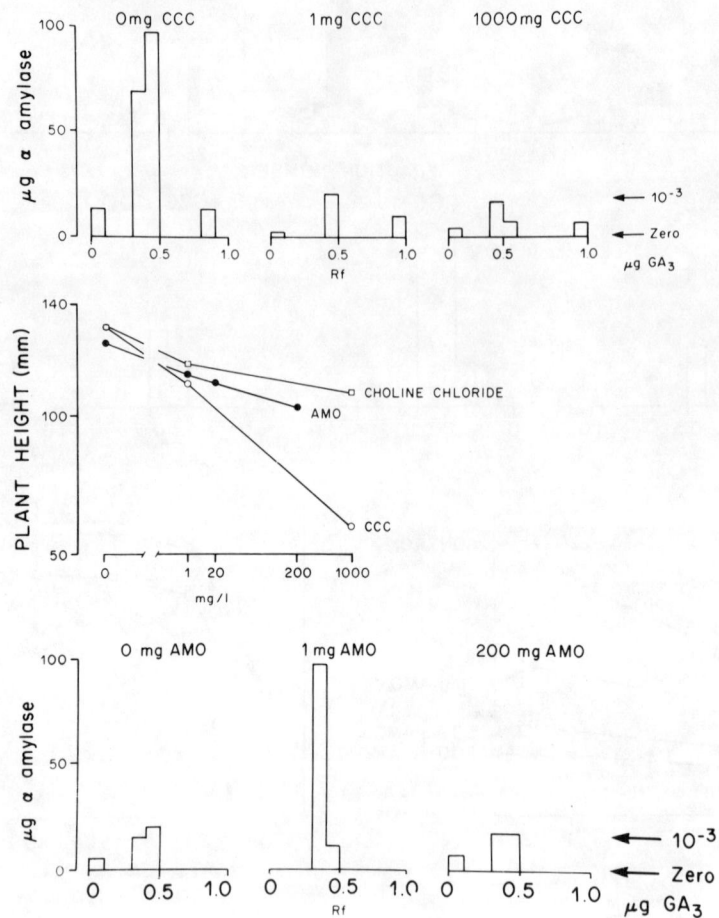

Fig. 2. Top: the effects of CCC on GA levels in barley seedlings. GA estimated by barley aleurone bioassay of TLC. Middle: the effects of CCC and AMO on the growth of barley seedlings. Bottom: the effects of AMO on the GA levels in barley plants. GA estimated by the barley aleurone bioassay following TLC.

The response of the barley seedlings is shown in Figure 2. One and 1000 mg/l CCC significantly reduce the quantity of GA. On the other hand, plants grown in 1/mg/l AMO appear to contain more GA than do controls. There is no evidence that there is any reduction (below the control levels) in GA content of those plants treated with 200 mg/l AMO. CCC is (at the concentrations used) more effective in inhibiting stem growth than is AMO. Choline chloride had a slight inhibitory effect on growth and GA content of barley. We have not as yet tested the combined effects of AMO (or CCC) and GA$_3$ on the growth of barley seedlings.

The effects of AMO on the growth and GA content of bean plants are shown in Figure 3. Plants treated with 50 mg/l AMO are 1/4 the size of the control seedlings, yet this quantity of retardants does not bring about any very drastic reduction in the extractable GA. Not until the beans are grown in 200 mg/l AMO is there any marked fall in the quantity of GA in the extracts. Application of GA$_3$ to these plants results in a situation rather similar to that found with pea plants. Application of GA$_3$ will stimulate growth. However, plants which are grown in 50 or 200 mg/l AMO and, at the

Table 1. Summary of results. The effects of CCC and AMO on stem height and the GA content of pea, barley and bean seedlings.

Tissue	Treatment (mg/l)		GA_3 equivalents (m µg)	Stem height mm ± 95% confidence limits
Peas	CCC	0	5 (α-amylase assay)	60 ± 8
		1	750	62 ± 7
		1000	8	35 ± 4
Peas	CCC	0	10 (Rice assay)	55
		1	70	64
		1000	13	57
Peas	AMO	0	3	50
		1	9	30
		200	>1	23
Barley	CCC	0	5	132
		1	1	112
		1000	>1	69
Barley	AMO	0	1	126
		1	6	115
		200	1	103
Bean	AMO	0	11	109 ± 10
		1	7	- -
		10	7	39 ± 6
		50	8	25 ± 6
		200	3	19 ± 5

same time, treated with 100 µg GA$_3$ were much smaller than control plants treated with 1 µg GA$_3$. Indeed, they were not much taller than those plants not treated with GA$_3$. The results are summarized in Table 1.

We have not as yet tested the response of bean seedlings to CCC.

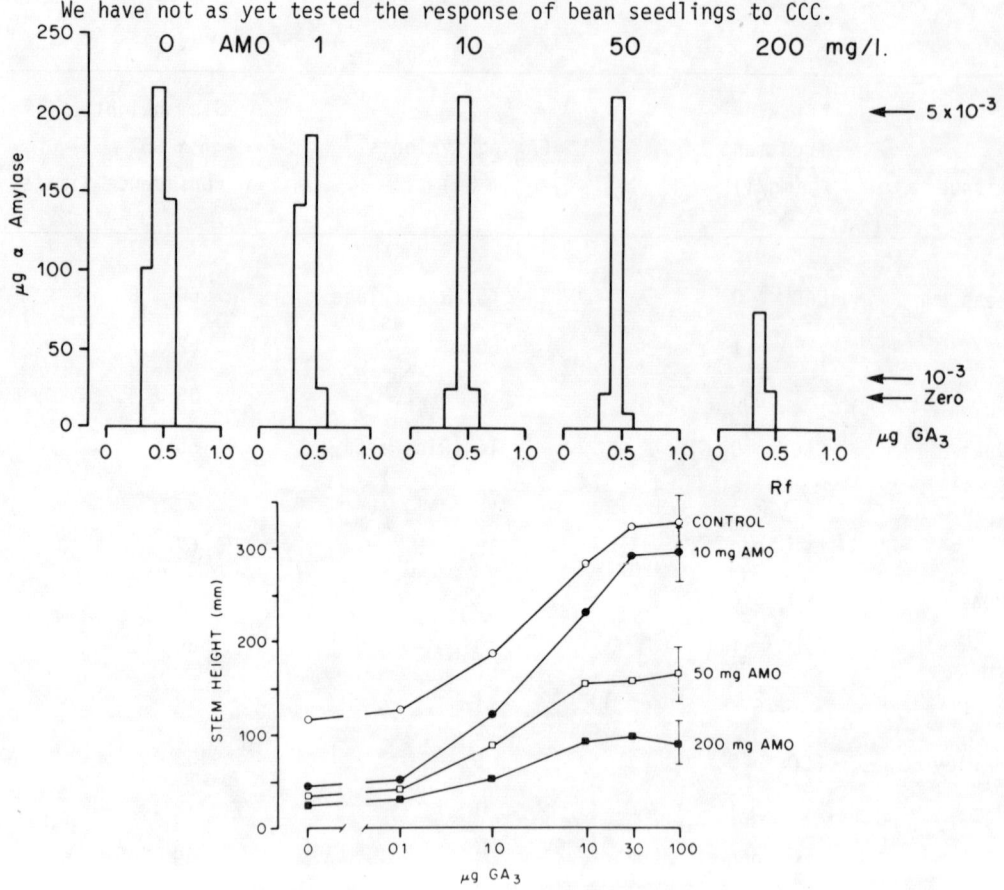

Fig.3. Top: the effects of AMO on the GA levels in bean seedlings. GA estimated by the barley aleurone bioassay following TLC. Bottom: the effects of GA$_3$ on the stem growth of bean seedlings grown in various concentrations of AMO. Vertical lines represent 95% confidence limits.

IV. DISCUSSION

From the above data, it would appear that in pea seedlings grown in CCC or AMO, and barley grown in AMO, there is no correlation between stem height and the levels of GA in plant extracts (see Table 1). As shown, CCC and AMO can at 1 mg/l stimulate GA levels, yet no parallel stimulation of stem growth occurs. On the other hand, at a high dose rate of CCC the growth of pea seedlings is strongly inhibited, while the plants contain levels of extractable GA equal to that found in the control plants. AMO will also stimulate GA levels in barley plants without any corresponding increase in growth. In fact, these plants were smaller than controls. Van Bragt (1969) and Halevy and Shilo (1970) have also shown that CCC treatment can lead to increases in GA levels. Both CCC and AMO also appear to reduce the ability of pea and bean seedlings to respond to application of GA$_3$.

These data appear to contradict the conclusions of workers who have demonstrated that CCC and/or AMO can inhibit kaurene biosynthesis (Barnes et al, 1969) and GA biosynthesis (Zeevaart and Osborne, 1965) in the fungus, *Fusarium moniliforme*.

Others have found that in cell-free homogenates of various plants, including peas, the retardants block the synthesis of kaurene (Anderson and Moore, 1967; Robinson and West, 1970) and GA (Dennis et al, 1965). Direct comparison of our results (Reid and Crozier, 1970) with those just mentioned is difficult. We used intact higher plants and they used cell free homogenates and fungal cultures. An intact higher plant shows a higher degree of complexity and organization than either a cell free homogenate or a fungus. The inability of CCC to reduce GA levels in our system, or its ability to stimulate them, could well be a function of this complexity and organization. However, these are a few cases where high levels of growth retardants will very effectively reduce GA levels in higher plants (Zeevaart, 1966).

We have shown that CCC-or AMO-treated plants do not respond as expected to high levels of exogenous GA_3. Also, plants containing abnormally high levels of endogenous GA do not grow faster than a normal plant. Thus, the retardants, in addition to any effects of GA synthesis *per se,* may inhibit GA action. This inhibition need not be a specific and direct one, but could result from an inhibitory effect of the retardants on some aspect of metabolism not directly related to GA's (Baldev et al, 1965, Cleland, 1965, Harada and Lang, 1965, Berry and Smith, 1970). The above arguments, and those in many other papers, are of course based upon two assumptions that are not yet proven. Firstly, that GA's that are extracted from macerated plant tissue do not necessarily give a true picture of what GA is actually in the tissue. Secondly, it is not yet known with any certainty to what extent these GA's are involved in stem elongation.

A number of hypotheses can be proposed to explain how CCC or AMO might bring about an increase in GA levels. The increase in GA might be due to a retardant-induced inhibition of GA action which might lead to a reduction in GA utilization and a build-up of endogenous GA. On the other hand, it is possible that there is more than one pathway involved in GA biosynthesis (Mertz and Henson, 1967, West et al, 1968), and that some pathways are more susceptible to CCC or AMO inhibition than others. The retardants might block one pathway and the excess of GA precursors that build up could move to an alternate pathway, resulting in the production of a different spectrum of GA's. This could explain why 1 mg CCC/l induces the appearance of GA activity in Fraction 25 of the silicic acid column (Figure 1). Earlier work has shown that CCC treatment will affect the spectrum of endogenous GA's (Reid and Carr, 1967, Carr and Reid, 1968, Bristow and Simmonds, 1968). This hypothesis takes into account changes in the type of GA's but not an overall increase. To account for this, a highly speculative extension of the hypothesis is proposed in Figure 4.

GGPP (geranylgeranyl pyrophosphate) gives rise to kaurene plus another terpenoid (T). T can inhibit a reaction prior to the formation of GGPP by end product inhibition. This inhibition would control the formation of itself and GA. Low levels of CCC (or AMO) might block only the synthesis of T, and not be able to inhibit the synthesis of kaurene.

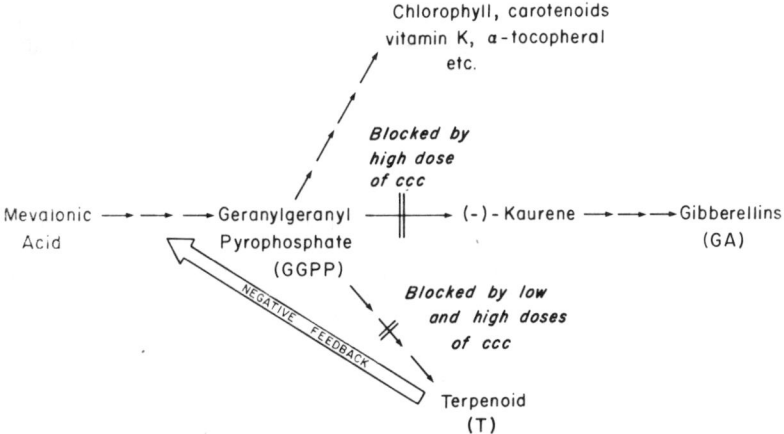

Fig.4. A hypothetical scheme showing a possible mechanism by which CCC or AMO could stimulate GA levels.

A lowering of the levels of T would allow an overproduction of GA. However high doses of CCC (or AMO) might block the synthesis of both T and kaurene from GGPP. Thus low levels of CCC stimulate and higher levels inhibit GA biosynthesis. This scheme postulates that T should inhibit GGPP synthesis and, since GGPP is a precursor of GA and other substances essential for normal growth (e.g. the phytol chain of chlorophyll, carotenoids, vitamin K and α-tocopherol), a reduction in GGPP could result in a reduction of growth. It is of interest that some of the diterpenes, which are synthesized from GGPP together with kaurene, can cause inhibition of stem growth (West *et al*, 1968). It is implicit in the hypothesis that CCC (or AMO) should inhibit the synthesis of these terpenoids. Again West *et al* (1968) have shown that CCC can, in some cases, accomplish this inhibition, further suggesting that these compounds may be acting in a manner similar to T.

This hypothesis can also explain the small increases in chlorophyll found in leaves after treatment with low concentrations of CCC (Kessler *et al*, 1967, Beevers and Guernsey, 1967, Beevers, 1968). CCC inhibits the formation of T and, as a result, removes its inhibiting effects on GGPP synthesis. With the subsequent increase in GGPP levels, one can expect stimulation of the synthesis of terpenoid derivatives such as the phytol chain of chlorophyll as well as an increase in GA levels.

Another explanation for CCC-induced increase of GA levels may be that at low dosages CCC only partially inhibits the action and/or synthesis of enzymes involved in the cyclization of GGPP to (-)-kaurene. The seedling, by some unknown means, responds and overcomes the effects of CCC. In doing so, it overcompensates and as a result the rate of GA biosynthesis increases and the GA levels rise.

V.   SUMMARY

Extracts from pea plants treated with 1 mg/l CCC contain higher quantities of GA than do extracts from untreated plants. 1000 mg/l CCC had no effect on GA levels (as compared to controls). Treatment with 1 mg/l AMO also stimulated GA levels, however, 200 mg/l AMO reduced the amount of GA to below that found in untreated plants. CCC at all concentrations lowered the quantity of GA in extracts from young barley plants, but AMO at 1 mg/l stimulated the level of GA. In bean seedlings, the amount of GA appeared to be somewhat reduced after AMO treatment.

CCC and AMO, at all concentrations used, inhibited stem elongation even though some of these treatments brought about significant increases in the quantity of GA found in the extracts. In some cases, the retardants reduced the ability of pea and bean seedlings to respond to an exogenous application of GA_3.

VI. ACKNOWLEDGEMENTS

This work was supported by NRC (Canada) Grant A-5727 to D.M. Reid. A.Crozier was supported by a post doctoral fellowship from NRC (Canada) Grant A-2585 to R.P. Pharis.

VII. REFERENCES

ANDERSON, J. D. and T. C. MOORE. (1967) Biosynthesis of (-)-kaurene in cell-free extracts of immature pea seeds. Plant Physiol. 42, 1527-1534.
BALDEV, B., A. LANG and A. D. AGATEP. (1965) Gibberellin production in pea seeds developing in excised pods: Effect of growth retardants. Science 147, 155-157.
BARNES, M. F., E. N. LIGHT and A. LANG. (1969) The action of plant growth retardants on terpenoid biosynthesis. Inhibition of gibberellic acid production in *Fusarium moniliforme* by CCC and AMO-1618; action of these retardants on sterol biosynthesis. Planta (Berl.) 88, 172-182.

BEEVERS, L. and F. S. GUERNSEY. (1967) Interaction of growth regulators in senescence of nasturtium leaf disks. Nature, 214, 941-942.
BERRY, D. R. and H. SMITH. (1970) The inhibition by high concentrations of (2-chloroethyl)-trimethyl ammonium chloride (CCC) of chlorophyll and protein synthesis in excised barley leaf sections. Planta (Berl.) 31, 80-86.
BRISTOW, J. M. (1966) The effects of gibberellic acid and cycocel on the growth of cultured leaf tissue. Can. J. Bot. 44, 513-518.
BRISTOW, J. M. and J. A. SIMMONDS. (1968) The effect of CCC on the growth and levels of endogenous gibberellins in *Helianthus* crown gall tissue. In: "Biochemistry and Physiology of Plant Growth Substances", F. Wightman and G. Setterfield, eds., pp. 911-919. Runge Press, Ottawa.
CARR, D. J. and D. M. REID. (1968) The physiological significance of the synthesis of hormones in roots and of their export to the shoot system. In: "Biochemistry and Physiology of Plant Growth Substances", F. Wightman and G. Setterfield, eds., pp. 1169-1185. Runge Press, Ottawa.
CLELAND, R. (1965) Evidence on the site of action of growth retardants. Plant and Cell Physiol. 6, 7-15.
CROSS, B. E. (1968) Biosynthesis of gibberellins. In: "Progress in Phytochemistry", L. Reinhold and Y. Liwschitz, eds., 1, 195-222. Interscience, London.
CROZIER, A. and L. J. AUDUS. (1968) Biological and chromatographic properties of two gibberellin-like compounds from etiolated *Phaseolus multiflorus* seedlings. Phytochem. 7, 1923-1931.
DALE, J. E. and G. M. FELIPPE. (1968) The gibberellin content and early seedling growth of plants of *Phaseolus vulgaris* treated with the growth retardant CCC. Planta (Berl.) 80, 288-298.
DENNIS, D. T., C. D. UPPER and C. A. WEST. (1965) An enzymic site of inhibition of gibberellin synthesis by AMO-1618 and other plant growth retardants. Plant Physiol. 40, 948-952.
FELIPPE, G. M. and J. E. DALE (1968) Effects of a growth retardant, CCC, on leaf growth in *Phaseolus vulgaris*. Planta (Berl.) 80, 328-343.
HALEVY, A. H. and R. SHILO. (1970) Promotion of growth and flowering and increase in content of endogenous gibberellins in *Gladiolus* plants treated with the growth retardant CCC. Physiol. Plant., 23, 820-827.
HARADA, H. and A. LANG. (1965) Effect of some (2-chloroethyl) trimethyl-ammonium chloride analogs and other growth retardants on gibberellin biosynthesis in *Fusarium moniliforme*. Plant Physiol. 40, 176-183.
KESSLER, B., S. SPIEGEL and Z. ZOLOTOV. (1967) Control of leaf senescence by growth retardants. Nature. 213, 311-312.
MERTZ, D. and W. HENSON. (1967) The effect of the plant growth retardants AMO-1618 and CCC on gibberellin production in *Fusarium moniliforme*. Light stimulated biosynthesis of gibberellins. Physiol. Plant., 20, 187-199.
OCKERSE, R. (1970) The dependence of auxin-induced pea stem growth on gibberellin. Botan. Gaz., 131, 95-97.
REID, D. M. and D. J. CARR. (1967) Effects of a dwarfing compound, CCC, on the production and export of gibberellin-like substances by root systems. Planta (Berl.) 73, 1-11.
REID, D. M. and A. CROZIER (1970) CCC-induced increase in gibberellin levels in pea seedlings. Planta (Berl.) 94, 95-106.
ROBINSON, D. R. and C. A. WEST. (1970) Biosynthesis of cyclic diterpenes in extracts from seedlings of *Ricinus communis* L. II. Conversion of geranyl geranyl pyrophosphate into diterpene hydrocarbons and partial purification of the cyclization enzymes. Biochem. 9, 80-89.
VAN BRAGT, J. (1969) The effect of CCC on growth and gibberellin content of tomato plants. Neth. J. Agric. Sci. 17, 183-188.
WEST, C.A., M. OSTER, D. ROBINSON, F. LEW and P. MURPHY. (1968) Biosynthesis of gibberellin precursors and related diterpenes. In: "Biochemistry and Physiology of Plant Growth Substances". F. Wightman and G. Setterfield, eds., pp. 313-332. Runge Press, Ottawa.
ZEEVAART, J. A. D. and H. D. OSBORNE. (1965) Comparative effects of some AMO-1618 analogs on gibberellin production in *Fusarium moniliforme* and on growth in higher plants. Planta (Berl.) 66, 320-330.
ZEEVAART, J. A. D. (1966) Reduction in the gibberellin content of *Pharbitis* seeds by CCC and after effects in the progeny. Plant Physiol. 51, 856-862.

PLANT GROWTH SUBSTANCES, 1970

DNA ANALYSIS OF AUXIN-TREATED JERUSALEM ARTICHOKE TUBER TISSUE AS A SCREEN FOR THE EVALUATION OF SUBSTANCES INFLUENCING CELL DIVISION

D. Adamson, R. Hinde and S. Kamisaka*

School of Biological Sciences, Macquarie University, North Ryde, N.S.W. Australia

* Permanent address: Department of Biology, Osaka City University, Sumiyoshi-ku, Osaka, Japan

I. INTRODUCTION

The action of plant growth substances on induction, maintenance and inhibition of cell division is a relatively neglected field, as illustrated by the small number of papers on the topic presented at this Conference. This paper describes a simple system for detecting substances which promote or inhibit the division of plant cells, using tissue slices from Jerusalem artichoke tubers.

The short-term culture method for studying cell division used by Adamson (1962, 1968) is suitable for studying the induction of mitosis, the first few cycles of division, and the effect of inhibitors. Useful characteristics of the system include:- induction of an high mitotic activity within 30 to 40 hours by auxin; no requirement for exogenous nutrients; dependence of division upon exogenous auxin in tissue from mature tubers; uniformity of initial DNA content per cell (2c).

One disadvantage relates to tissue from dormant tubers in which cell division is mostly confined to the periphery of the slices. However tissue from non-dormant tubers, which show slow root and shoot growth after 4 months or more of cold storage, produces a much more uniform distribution of cell division and cell expansion throughout the thickness of the slices.

The uniformity of initial DNA content per cell makes artichoke tuber tissue ideal for estimating mitotic activity by chemical analysis of DNA using whole tissue slices. Brief microscopic examination, parallel to the DNA analysis, should be used to check for abnormalities of mitosis or cytokinesis. The application of these two techniques to the auxin-induced cell division of artichoke tissue provides a simple system for studying the induction, the promotion and the inhibition of cell division by substances other than auxin.

II. MATERIALS AND METHODS

The methods of storage and preparation of the tissue were similar to those used by Adamson (1962). In experiments using potentially inhibitory substances, the tubers were sliced and the tissue was transferred to the treatment solutions at the storage temperature (2 to $3^{\circ}C$). Treatment was carried out in 9 cm petri dishes containing 20 to 25 ml of solution. Smaller dishes may also be used. Submilligram quantities of substances may be tested at concentrations of $10^{-4}M$ or less. The tissue was supported on Whatman No. 1 filter paper held at the surface of the treatment solution by bent glass rods. The lids of the dishes were kept slightly ajar for ventilation. The initial fresh weight of each sample was obtained and used as the basis for expressing DNA values and cell expansion.

For microscopic examination, parallel samples to those for DNA estimation were fixed in acetic alcohol, stored in 70% (v/v) ethyl alcohol, sectioned free-hand and stained in aceto-orcein.

For DNA estimation, between 1 and 2 g of tissue was weighed, blotted free of excess moisture and frozen by immersion in liquid air in a pre-cooled mortar. The frozen tissue was ground to a powder in cold 0.05 M formic acid in methanol to produce a slurry. DNA extraction was by a method of Schmit-Tannhauser as modified by Holdgate and Goodwin (1965). DNA was analysed by the method of Burton (1956). Extraction and analysis were carried out in duplicate.

III. RESULTS AND DISCUSSION

In tissue treated with auxin and cytokinin the DNA content rose to double the initial value in about 3 days (Figure 1). No significant change in DNA content occurred in the absence of auxin. These results are consistent with the microscopic observations of mitotic activity obtained in Setterfield (1963), Mitchell (1967), Yeoman et al (1967), and Adamson (1968) using artichoke tissue.

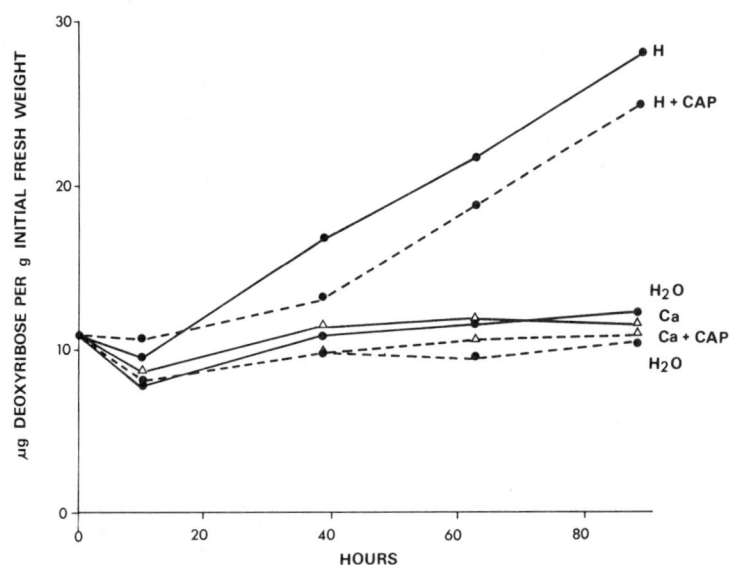

Fig. 1.

The time course of DNA change in freshly-cut tissue treated with growth substances (H), or calcium chloride (5ml/1), or water, in the presence and absence of chloramphenicol (CAP) at 50 mg/1. The growth substance mixture consisted of 2,4-dichlorophenoxyacetic acid (2,4-D) 1mg/1, kinetin 1mg/1, and calcium chloride 5me Ca^{++}/1.

In the experiment shown in Figure 1 no precautions were taken to exclude bacteria. Chloramphenicol (CAP) was used to suppress bacterial growth and hence the contribution of bacterial DNA to the total DNA. CAP at 50 mg/1 is within the concentration range of 10 to 250 mg/1 which was shown by Edelman and Hall (1965) to prevent bacterial growth in artichoke tissue without changing invertase development by the tissue slices. CAP appeared to increase the lag in DNA synthesis but did not change the rate of DNA increase during a period of 3 days. The lengthened lag is consistent with an interference by CAP with cellular metabolism. The results do not suggest that bacterial DNA was making a significant contribution to the total DNA content of the tissue.

The maximum likely contribution of bacterial DNA was calculated by estimating the number of bacteria (Edelman and Hall, 1965) associated with tissue slices grown without CAP and without any precautions to reduce bacterial numbers. After 24 and 48 hours bacterial numbers were of the order of 2×10^6 and 5×10^7 respectively per gram fresh weight of tissue slices. These numbers of bacteria would contribute about 1/300 amd 1/30 respectively to the total DNA, so that the contribution of bacterial DNA was less than the combined sampling and analytical errors of the DNA measurement.

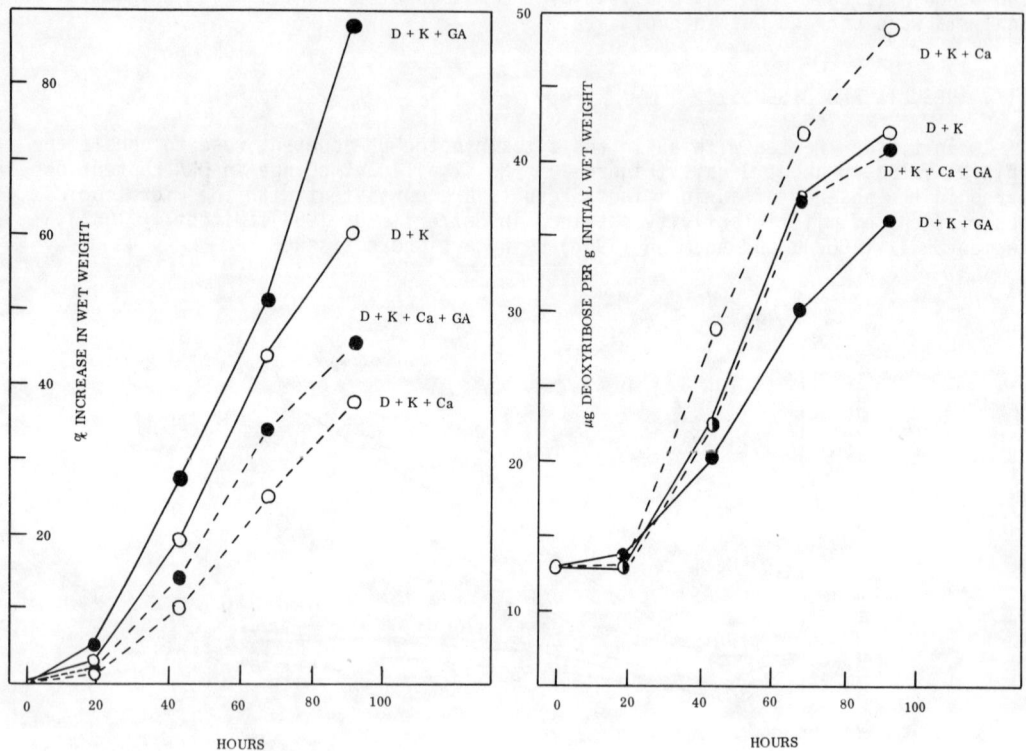

Fig. 2.

The time-course of cell expansion of freshly-cut tissue treated with various combinations of 2,4-D 1mg/1(D), kinetin 1mg/1(K), GA 10mg/1(G) and calcium chloride 5me Ca^{++}/l.

Fig. 3.

The time-course of increase in DNA in the experiment shown in Figure 2.

Figures 2 and 3 show the time-course of expansion and DNA synthesis in tissue treated in auxin plus kinetin with and without GA and $CaCl_2$. Setterfield (1963) found that GA stimulated the auxin-induced expansion and reduced the mitotic frequency of freshly-cut tissue. The results in Figure 3 confirm Setterfield's mitotic observations. GA reduced the amount of DNA synthesis in freshly-cut tissue treated in auxin and cytokinin. The presence of calcium did not alter this result.

In Figures 2 and 3 note that the treatment which caused the highest expansion rate produced the lowest DNA synthesis. This result is consistent with the observation of an inverse relationship between high mitotic frequency and high expansion rate (Adamson, 1962, 1968).

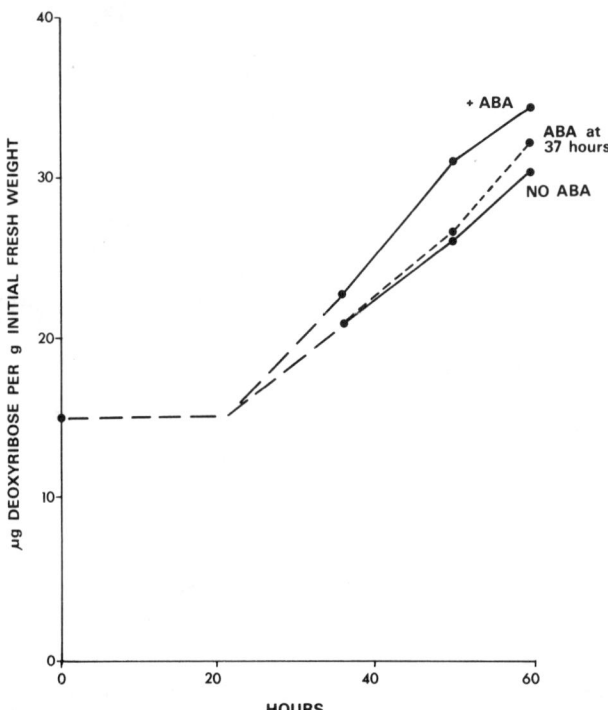

Fig. 4.

The time-course of increase in DNA in freshly-cut tissue treated in 1mg/l 2,4-D plus 1mg/l kinetin plus 10me Ca^{++}/l calcium chloride, in the presence and absence of 10^{-5}M ABA (mixed racemate, Reynolds Tobacco Co.). Each value is the mean of duplicate analyses of triplicate tissue samples. The degree of confidence (95% level) of each value is in the range 0.9 to 1.8.

ABA is an effective inhibitor of auxin-induced cell expansion in freshly-cut tissue slices (Table 1) and in slices which have been soaked in water for 24 hours prior to auxin treatment (Table 2). However ABA caused no inhibition of DNA synthesis in either case. A significant stimulation of DNA synthesis by ABA was shown in Figure 4 after 49 and 60 hours when ABA was present from the beginning of auxin treatment. Addition of ABA at 37 hours had no effect on subsequent DNA values. These results were obtained using non-dormant tubers which had been stored at 2 to 3°C for 5 months. The effect of ABA on tissue growth needs to be checked at different stages of tuber growth and storage.

The use of artichoke tuber tissue as a screen for inhibitors of cell division is illustrated in Table 3. The 3 substances showed a strong delaying and inhibitory action on mitosis similar to that shown by Podolactone A (Galbraith et al, 1970). The action of these 3 substances on mitosis and on cell expansion should be compared with their effect, and that of related substances, on the expansion of pea stem tissue reported by Sasse et al.(1971). This group of related diterpenoids from various Podocarpus species, together with the alkaloid lycoricidinol, appear to be powerful inhibitors of cell division and cell expansion.

The results in Table 3 illustrate the limitations of mitotic frequency as a measure of mitotic inhibition. After 40 hr treatment mitosis was severly inhibited by the 3 substances. If measurements had been carried out only after 40 hr treatment, the substances would have appeared to be stimulators of mitosis, especially as no mitotic abnormalities were observed. Analyses which sum the cumulative mitotic activity, such as DNA estimation, avoid the sampling difficulties associated with mitotic frequency estimation. Checking for mitotic abnormalities by microscopic observation remains a necessary supplement to DNA analysis.

Table 1. Effect of ABA on expansion and DNA content of freshly-cut tissue.

Treatment	Increase in fresh weight after 43 hr(%)	Deoxyribose per g initial fresh weight after 64 hr (μg)
Water	9	16
Kinetin	23	15
ABA	11	14
Kinetin + ABA	17	15
2,4-D	51	21
2,4-D + ABA	29	23
2,4-D + Kinetin	47	25
2,4-D + Kinetin + ABA	29	30

Kinetin 1mg/1; 2,4-D 1mg/1; ABA 10^{-5}M (mixed racemate, Reynolds Tobacco Co.).

Table 2. Effect of ABA on expansion and DNA content of tissue soaked in water for 24 hr prior to treatment.

Treatment	Increase in fresh weight after 30 hr (%)	Deoxyribose per g initial fresh weight after 64 hr (μg)
Water	4	14
Kinetin	7	13
ABA	4	15
Kinetin + ABA	7	16
2,4-D	45	20
2,4-D + ABA	28	24
2,4-D + Kinetin	48	23
2,4-D + Kinetin + ABA	34	24

Kinetin 1mg/1; 2,4-D 1mg/1; ABA 10^{-5}M (mixed racemate Reynolds Tobacco Co.).

Table 3. Effect of 3 inhibitory substances on mitotic frequency and expansion of freshly-cut tissue.

	2,4-D + Kinetin + CaCl$_2$ mitotic frequency (%)		2,4-D Increase in fresh weight (%)
	after 40 hr	after 49 hr	after 72 hr
No inhibitor	18	6	44
Nagilactone A 10^{-5}M	7	10	29
Inumakilactone A 10^{-5} M	0	13	32
Lycoricidinol 10^{-7} M	0	10	37
Lycoricidinol 10^{-6} M	0	0	19

2,4-D 1mg/1; kinetin 1mg/1; CaCl$_2$ 10me Ca^{++}/1.

IV. SUMMARY

Jerusalem artichoke tuber tissue treated with auxin so as to induce cell division is a suitable material for studying the induction and inhibition of division. Because of the 2c condition of the cells, DNA analysis can be used to estimate the cumulative amount of cell division. In combination with microscopic checking for mitotic abnormalities, the system is a simple and rapid screen for substances which influence cell division.

Using this system, the inhibition of cell division and the stimulation of cell expansion by GA was confirmed with freshly-cut tissue from non-dormant cold-stored tubers. With similar tissue, ABA inhibited cell expansion but not DNA synthesis. Low concentrations of nagilactone A, inumakilactone A and lycoricidinol had strong inhibitory effects on division and expansion, similar to the known effects of Podolactone A.

V. ACKNOWLEDGEMENTS

We thank Dr. Ito, Dr. Okamoto and Dr. Sakan for samples of inhibitors, details of which are described by Sasse *et al.* (1972).

VI. REFERENCES

ADAMSON, D. (1962). Expansion and division in auxin-treated plant cells. Can. J. Bot. 40, 719-44.
ADAMSON, D., V.H.K. LOW and H. ADAMSON (1968). Transitions between different phases of growth in cells from etiolated pea stems, Jerusalem artichoke tubers and wheat coleoptiles, in "Biochemistry and Physiology of Plant Growth Substances" (Ed. F. Wightman and G. Setterfield) p. 505. Runge Press, Ottawa.
BURTON, K. (1956). A study of the conditions and mechanism of the diphenylamine reaction for the colorimetric estimation of deoxyribonucleic acid. Biochem. J. 62, 315-23.
EDELMAN, J and M.A. HALL (1965). Enzyme formation in higher plant tissues. Development of invertase and ascorbate-oxidase activities in mature storage tissue of *Helianthus tuberosus* L. Biochem. J. 95, 403-10.

GALBRAITH, M.N., D.H.S. HORN, J.M. SASSE and D. ADAMSON (1970). The structures of podolactones A and B, inhibitors of expansion and division of plant cells. Chem. Communications 1970, 170-1.

HOLDGATE, D.P. and T.W. GOODWIN (1965). Quantitative extraction and estimation of plant nucleic acids. Phytochemistry $\underline{4}$, 831-43.

MITCHELL, J.P. (1967). DNA synthesis during the early division cycles of Jerusalem artichoke callus cultures. Ann. Bot. $\underline{31}$, 427-35.

SASSE, J.M., M.N. GALBRAITH, D.H.S. HORN and D. ADAMSON (1972). Chemistry and Biological action of podolactones and other inhibitors of plant growth. in "Plant Growth Substances, 1970". (Ed. D.J. Carr).Springer Verlag.

SETTERFIELD. G. (1963). Growth regulation in excised slices of Jerusalem artichoke tuber tissue. Symp. Soc. Exp. Biol. 98-126.

YEOMAN, M.M., P.K. EVANS and G.G. NAIK (1967). Growth and differentiation of plant tissue cultures. II. Synchronous cell divisions in developing callus cultures. Ann. Bot. $\underline{31}$, 323-32.

PLANT GROWTH SUBSTANCES, 1970

Promotion by CCC of Growth in Jerusalem Artichoke Tissue

Heather Adamson and R. Jones

School of Biological Sciences, Macquarie University, North Ryde, Australia

I. INTRODUCTION

The response of Jerusalem artichoke tuber tissue to added growth substances varies depending on the length of time the tubers have been stored at low temperature after harvest (Adamson 1962). The present work is part of a study of the response of tissue from non-dormant tubers to growth substances. Immature tubers which were still capable of growth on the plants, and tubers which were sprouting at low temperature after lengthy cold storage were studied.

This paper deals with cell expansion and cell division. Kamisaka and Masuda (1968) showed that 2-(chloroethyl)- trimethyl-ammonium chloride (CCC) suppressed the increase in endogenous GA-like substances which normally occurs in Jerusalem artichoke tissue after cutting. With tissue from dormant tubers they found that the reduction in growth in the presence of CCC was reversed by exogenous GA. Setterfield (1963) had previously shown that GA increased cell expansion and reduced cell division in auxin-treated tissue. We were interested to study the effect of CCC on non-dormant tissue and to find out whether the reduced expansion caused by CCC was associated with increased cell division as the previous work with CCC and GA would imply.

In non-dormant tissue CCC caused a marked stimulation of both cell expansion and division. Non-dormant tissue from immature tubers also showed considerable cell expansion and cell division in the absence of exogenous auxin, in contrast to the well-known dependence of cell division and cell expansion of stored artichoke tissue on exogenous auxin.

II. MATERIALS AND METHODS

The methods used have been described in detail elsewhere (Adamson 1962). Briefly, freshly-harvested or cold-stored (2-3°C) artichoke tubers were peeled and cut transversely into slices 1mm thick. After rinsing in water, the slices were transferred to treatment solutions in Petri dishes where they were supported on moist filter paper spread over a bent glass rod, or they were pretreated in 1mm deep liquid in plastic trays before being transferred to appropriate treatment solutions in Petri dishes. Petri dish lids were kept slightly ajar for ventilation and pretreatment solutions were changed about four times during the course of pretreatment.

Kinetin, GA and 2,4-D were always used at a concentration of 1ppm. CCC was used at 400ppm or 10ppm.

Expansion was measured as an increase in the fresh weight of tissue slices, expressed as a percentage of the fresh weight of the tissue sample at the time of its transfer to treatment solution.

Cell division was observed in hand-cut sections of tissue stained in aceto-orcein. Orientation of the sections was radial and longitudinal with respect to the long axis of the original tuber. The number of nuclei per field was estimated by counting the number of nuclei in sharp focus within a 400X field of view using a constant vertical traverse of the fine focus knob.

Table 1. Maximum rate of expansion of artichoke tissue slices from newly formed tubers, in a range of growth substance solutions, after cutting (no pretreatment) and after 33 hours pretreatment in water, Kinetin, GA, CCC, kinetin +CCC and Kinetin + GA. (percentage increase in fresh weight/hr) 2,4-D, GA and kinetin each 1ppm. CCC 400 ppm.

Pretreatment Treatment	None	H_2O	CCC	GA	Kin	Kin +CCC	Kin + GA
H_2O	0.18	0.08	0.21	0.09	0.02	0.40	0.04
Kin	0.15	0.07	0.15	0.19	0.02	0.70	0.03
GA	0.18	0.07	0.20	0.06			
Kin + GA	0.12	0.11	0.30	0.08			
2,4-D	0.75	0.60	0.55	0.90	0.48	0.70	0.70
2,4-D + GA	1.00	0.75	1.00	0.80			
2,4-D + Kin	0.90	0.64	0.45	0.80	0.48	0.95	0.70
2,4-D + Kin + GA	1.00	0.80	0.65	0.95			

III. RESULTS AND DISCUSSION

Table 1 summarises the results of an experiment carried out in late summer 1969, using freshly harvested tubers taken from the plants at the time of flowering. It shows the maximum rate of expansion of tissue treated in a variety of growth substance solutions, following pretreatment in different combinations of CCC, GA and kinetin. Figure 1 shows progress data for some of the more interesting treatments shown in Table 1.

Freshly harvested tissue was quite responsive to added growth substances when transferred to treatment solutions immediately after cutting. The same tissue pretreated in water for 33 hours was slightly less responsive to added growth substances (Table 1), whilst kinetin pretreatment markedly reduced the expansion of tissue in water, kinetin, 2,4-D and 2,4-D+kinetin (Table 1, Fig 1b). Including CCC with kinetin during pretreatment not only overcame the inhibition by kinetin but greatly stimulated the subsequent expansion of the tissue, particularly in water and kinetin (Table 1, Fig. 1c).

Table 1 shows that tissue pretreated in CCC+kinetin expanded in water twenty times as rapidly as tissue pretreated only in kinetin. The effect was even more marked when tissue pretreated in CCC+kinetin was transferred to kinetin. In this case tissue expanded in kinetin as rapidly as in 2,4-D and at a rate comparable to that of tissue pretreated in water and transferred to 2,4-D in the usual manner.

Although the promotive effect of CCC was most marked when kinetin was present during the pretreatment period, pretreatment in CCC alone more than doubled the subsequent rate of expansion of tissue in water, kinetin, GA and GA+kinetin. Added auxin was not required for this response and indeed the promotive effect of CCC was largely obscured by the addition of auxin to the treatment solution.

The parallel between growth induced by 2,4-D+kinetin following pretreatment in water (Fig. 1a) and growth in water or kinetin following pretreatment in CCC+kinetin (Fig. 1c) is quite striking. It is tempting to postulate that CCC promoted growth by increasing the effective concentration of endogenous auxin in the tissue. If this were the case, it would account for the fact that the promotive effect of CCC was most obvious in the absence of applied auxin. This fact also argues against the possibility of CCC stimulating growth by increasing the levels of gibberellins or cytokinins in the tissue since it is usually necessary to add auxin to Jerusalem artichoke tissue in order to demonstrate the promotive effects of these compounds. The fact that GA stimulated growth in the present experiment (Table 1) also argues against the possibility of CCC promoting growth by reducing the gibberellin concentration of the tissue.

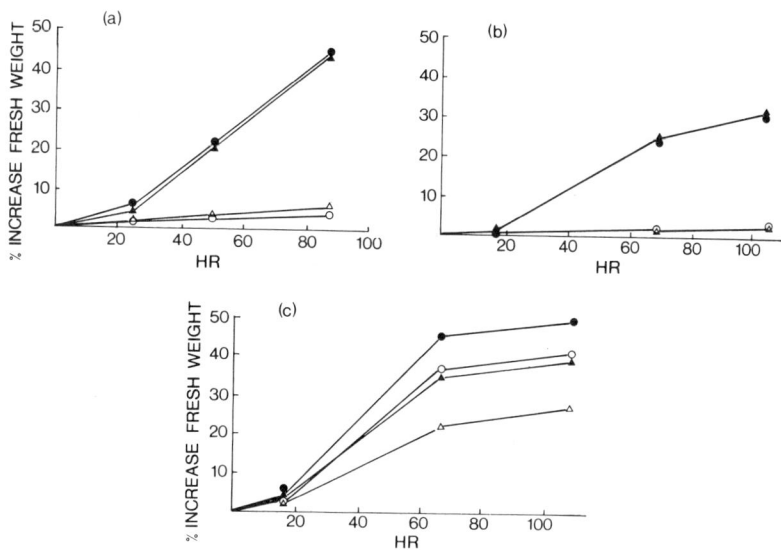

Fig. 1.

Time-course of expansion of artichoke tissue slices from newly formed tubers (1969) in water ∆, kinetin O, 2,4-D▲, 2,4-D + kinetin ●, after 33 hours pretreatment in (a) water, (b) kinetin, (c) kinetin + 400ppm CCC.

In the experiment summarised in Table 1, growth promotion by CCC was accompanied by increased cell division. Figure 2 indicates that the increase in cell number induced by CCC pretreatment was not dependent on the presence of auxin in the treatment solution. When 2,4-D was included with kinetin in the treatment solution, the number of nuclei per field after 70 hours was approximately the same regardless of pretreatment and was about equal to the maximum number of nuclei per field shown in Fig. 2. In other words, the amount of division induced by auxin was no more than that induced by pretreatment in CCC+kinetin followed by treatment in kinetin alone.

It was observed that cell division occurred when freshly-cut tissue from non-dormant tubers was placed in moist filter paper without added growth substances, but washing in water, kinetin or GA virtually abolished the capacity of the tissue to divide without added growth substances. Washing in CCC did not have this effect. Tissue washed for 33 hours in CCC and then transferred to water divided almost as well as fresh tissue in the absence of added growth substances. In view of the importance of auxin in inducing cell division in Jerusalem artichoke tissue this observation lends weight to the possibility that CCC promoted growth by maintaining the level of endogenous auxin during the washing period.

The ability of CCC to promote the growth of tissue from newly formed tubers was confirmed in 1970. The tubers used were a little older than those used the previous year and were harvested before the onset of cold weather from plants with senescent leaves and well past flowering. Growth promotion by CCC was less than in 1969 although still marked, and CCC was effective at lower concentrations. Pretreatment in 400ppm CCC did not affect growth. Figure 3 shows the percentage increase in fresh weight of Jerusalem artichoke tissue treated in water or 2,4-D following 28 hours pretreatment in water or 10ppm CCC. Whereas 2,4-D obscured the effect of CCC in promoting the growth of tissue harvested at flowering in 1969, its addition was essential to show the growth promoting effect of CCC the following year. This probably reflects differences in the stage of development of the two lots of tubers. It does not reflect differences in the duration of pretreatment.

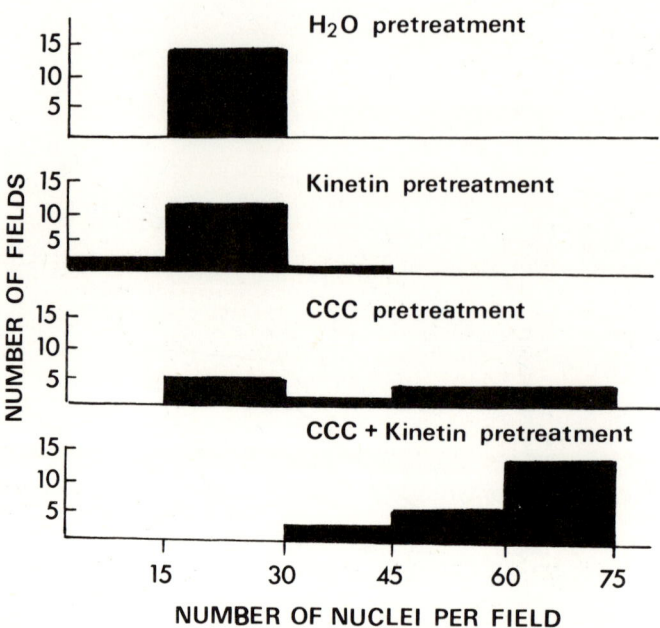

Fig. 2.

Frequency of nuclei in artichoke tissue slices treated for 70 hours in kinetin, following 33 hours pretreatment in water, kinetin, CCC(400ppm) and CCC + kinetin.

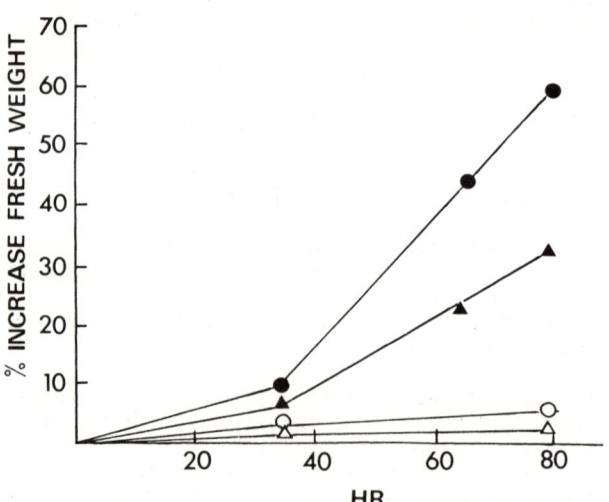

Fig. 3.

Time course of expansion of artichoke tissue slices from newly formed tubers (1970) in water (open symbols) and 2,4-D (closed symbols) following 28 hours pretreatment in water (triangles) and CCC(10ppm. circles).

The growth of auxin-treated tissue slices cut from tubers which were sprouting slowly at 2 to 3°C after 9 months cold storage, was also promoted by CCC (Fig. 4). Comparing Figures 1 and 4, it is clear that tissue stored for 9 months was more responsive to kinetin and GA than newly formed tubers. The expansion of the cold-stored tissue in kinetin and GA was inhibited by CCC.

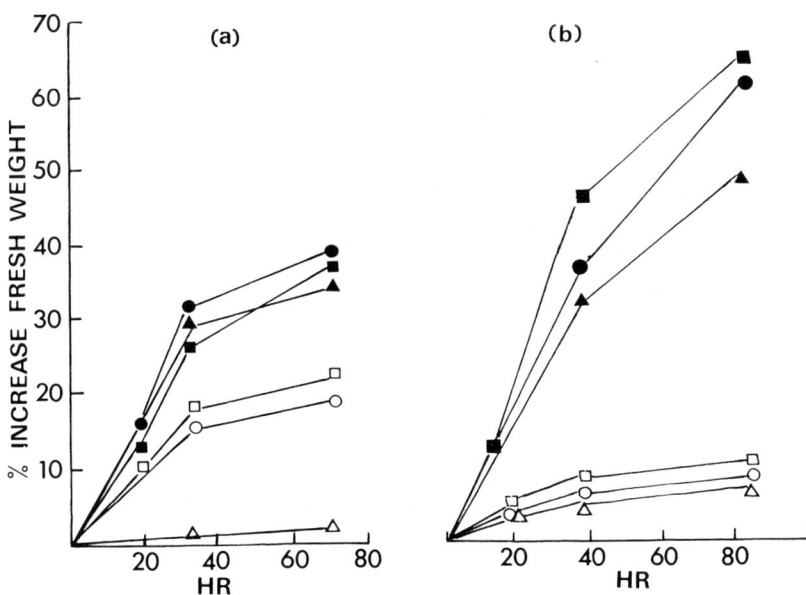

Fig. 4.

Time course of expansion of artichoke tissue slices from tubers sprouting in cold storage, in water △, kinetin ○, GA □, 2,4-D ▲, 2,4-D + kinetin ●, 2,4-D + GA ■, following 28 hours pretreatment in (a) water and (b) CCC (400ppm).

Growth of tissue from dormant artichoke tubers was not enhanced by CCC pretreatment. In general, our findings with dormant tissue confirm the observation of Kamisaka and Masuda (1968) that pretreatment in CCC inhibited subsequent auxin-induced expansion.

Thus CCC when applied to Jerusalem artichoke tissue can act as both a growth retardant and a growth promoter. It is possible that its action is much the same in both cases, the difference in response being due to differences in the relative concentrations and perhaps even nature of the growth substances present in the tissue at different stages of development. More information is needed on this. Certainly there is not enough information available to be able to attribute the promotive effect of CCC in artichoke tissue to effects on the metabolism of the gibberellins or any other group of growth substances.

Tissue which has been soaked, or pretreated, in CCC appears to resemble freshly cut tissue in its ability to grow. Tissue soaked in the absence of CCC usually differs markedly in growth capacity from freshly cut tissue (Adamson 1962). It is as though CCC minimises or slows down the changes which occur during the so-called ageing process so that the tissue is close to the freshly cut condition. With tissue from non-dormant tubers this action of CCC is difficult to account for on terms of an inhibition of GA biosynthesis.

IV. SUMMARY

CCC promotes the growth of tissue from non-dormant artichoke tubers by increasing both cell expansion and cell division. In tissue from growing tubers, exogenous auxin was not necessary for the CCC effect and tissue pretreated in CCC for 1 to 2 days expanded as rapidly in water and kinetin as tissue pretreated in water and transferred to auxin. Slightly older tubers which had ceased growing and tubers which were sprouting slowly after a long period of cold storage however did require added auxin for the promotive effect of CCC to occur.

V. REFERENCES

ADAMSON, D. (1962). Expansion and division in auxin-treated plant cells. Can. J. Bot., 40, 719-44.

KAMISAKA, S. and Y. MASUDA (1968). Auxin-induced growth of tuber tissue of Jerusalem artichoke IV. Significance of gibberellin biosynthesis and basic proteins in chromatin in aging process. Plant and Cell Physiol., 9, 61-67.

SETTERFIELD, G. (1963). Growth regulation in excised slices of Jerusalem artichoke tuber tissue. Symp. Soc. Exptl. Biol., 17, 98-126.

GIBBERELLIN, A PRIMARY DETERMINANT IN THE EXPRESSION OF APICAL DOMINANCE, APICAL CONTROL AND GEOTROPIC MOVEMENT OF CONIFER SHOOTS

Richard P. Pharis, Chung-chi Kuo, James L. Glenn
Department of Biology, University of Calgary, Calgary, Alberta, Canada

I. INTRODUCTION

Early studies with GA on correlative inhibition of lateral buds or shoots were somewhat conflicting, depending on species, and whether or not the plant had been decapitated. In general though, application of GA to intact plants increases apical dominance. It results in lateral bud or shoot growth only when correlative inhibition is diminished by some other means (reviewed by Phillips, 1969). Where GA maintains or enhances apical dominance, an attractive hypothesis is that it acts with auxin, and experiments by Jacobs and Case (1965) and Scott et al (1967) with pea, by H.F. Taylor, (unpublished, cited by Catalano and Hill, 1969) with tomato, and with pine (Tomaszewski 1970) demonstrate an interaction between GA and auxin. The role that GA plays in this interaction remains unclear, although results of Jacobs and Case (1965) support the argument for an effect of GA on "auxin-saving" rather than auxin transport. It was also noted that GA treatment of intact shoots leads to increased auxin levels (reviewed by Phillips, 1969). A GA-kinetin interaction has been observed in tomato, where both compounds enhanced the growth of lateral buds subsequent to their inhibition by IAA applied to the stump of the decapitated main stem (Catalano and Hill, 1969). Evidence that endogenous GA is necessary for maintaining apical dominance of intact plants, or allowing lateral buds or shoots to replace the missing terminal in decapitated plants has been put forward by Pharis et al (1965) and Ruddat and Pharis (1966) using several growth retardants, one of which (AMO-1618; Rainbow Color and Chemical, Sepulveda, Calif.) is known to inhibit endogenous GA synthesis in higher plants (see review by Lang, 1970). Application of B-995 (Naugatuck Chemical Naugatuck, Conn.) a growth retardant whose effects can also be reversed by addition of GA showed decreased apical dominance in pear (Brooks, 1964) and redwood (Ruddat and Pharis, 1966).

The orientation of branches and leaves is not only a geotropic phenomenon, but also involves apical dominance (Palmer, 1955; Booth, 1959). The normal plagiotropic habit of lateral branches, rhizomes, or stolons can be modified in intact plants through the use of auxins, gibberellins, and cytokinins (Booth, 1959 and 1963; Kumar, 1966; Pharis and Morf, 1967; Montaldier, 1969; Halevy et al, 1969) growth retardants (Pharis et al 1965; Halevy et al, 1969), and by a number of enironmental factors such as light intensity, photoperiod, temperature, and mineral nutrition (see reviews by Champagnat, 1965 and Phillips, 1969).

Interrelated to all of the above may be "hormone-directed transport" (Phillips, 1969) where application (or synthesis?) of the hormone "directs" the flow of other hormones and/or organic and inorganic compounds into the target organ. Seth and Wareing (1967) showed a significant increase in the transport of ^{32}P when GA and/or kinetin were applied simultaneously with IAA to the cut stumps of bean, and Harris et al (1969) showed that ^{14}C administered as CO_2 during photosynthesis was diverted preferentially to the carnation flower when GA_3 was applied topically. Hence, GA influences the movement of organic substances within plants, although auxin may have to be present either naturally, or as exogenously applied IAA. Movement of GA itself within plants is known to occur in both xylem and phloem, and while most reports have shown a lack of polarity (see review by Lang, 1970), Jacobs and Kaldewey (1970) noted polar movement of GA_3 in excised *Coleus* petioles, and ABA, a sesquiterpene-derived acid, moves with basipetal polarity in excised petiole and internode segments of *Coleus* (Dörffling and Böttger, 1968). However, a polarity in movement of GA was not noted for corn coleoptile sections (Hertel, *et al*, 1969).

In the present paper we will discuss work, much of it of a preliminary nature, which indicates that at least for certain conifers, GA is a primary determinant in the control of apical dominance and apical control (i.e. tree form, Brown *et al*, 1967), which would include orientation of lateral branches and the terminal shoot.

II. METHODS AND MATERIALS

Plant material

Plants were grown from seed, or year-old transplants were obtained and grown in a greenhouse under 18 hours of natural and supplemental incandescent light (supplemental light 0400 to 0900 hr, 1600 to 2200 hr; approximately 22ºC from 0500 to 1900 hr and 18ºC from 1900 to 0500 hr). Where special photoperiods were used, plants were placed in growth chambers under 8,000 μW cm^{-2} (approximately 3,000 f.c.) where the temperature during the light period was 23ºC, and during the dark 18ºC.

GA application

GA_3 was applied by spraying to drip-off in an aqueous solution of 0.05% Tween 20 at concentrations ranging from 10 to 500 mg/l, soil drench at weekly intervals (100ml/plant of GA solution ranging from 10 to 500 mg/l), or topical application in ethanol by injection or surface application to woody tissue. Specific concentrations and exact methodology are given in Figure or Table legends.

Growth retardant application

AMO-1618 and B-995 were applied in volumes of 100 ml/plant thrice weekly for 2 months at 1,000, 2,000 or 4,000 mg/l prior to GA treatment, and continued at the same rate for 3 weeks after GA injection. Amounts of GA injected were sufficient to induce flowering, but insufficient to overcome growth retardation.

^3H-GA_1 application

One μl of GA_1-3, 4-H^3 (whose radioactive composition was approximately 35% A_1, 50% tetrahydro derivatives, and 15% unknown compound) with a specific activity of 753.4 mc/mM in a mixture of ethanol: benzene was taken to dryness and re-dissolved in 10 μl of 80% ethanol. Portions of the treated plants were oven-dried at 105ºC and oxidized in a Packard Tri-Carb Tritium Oxidizer. Radioactivity was counted in a mixture of dioxane: toluene: naphthalene: PPO and dimethyl POPOP: 720ml: 180ml: 100g: 5g and 0.3g in a Packard Tri-Carb Liquid Scintillation Spectrometer.

III. RESULTS AND DISCUSSION

Apical dominance, apical control, and geotropic responses

We have noted over the past 5 years a number of examples of an influence on apical dominance in a wide variety of conifers by GA's, compounds known to inhibit GA biosynthesis or antagonize a GA-mediated effect, or environmental factors postulated to affect endogenous GA levels. The response usually entails differential cambial growth of a lateral branch or terminal shoot, although one species, *Sequoia sempervirens*, normally expresses its apical dominance through the control of lateral bud growth, and through diminished elongation of lateral shoots. In the families Cupressaceae, Taxodiaceae, and Pinaceae, species within the genera *Cupressus*, *Thuja* (see Figure 2, Pharis and Morf, 1967), *Juniperus, Pinus, Picea, Tsuga,* and *Sequoia gigantea* react to GA application by a decrease in their apical dominance, whereas *Sequoia sempervirens*

(Figure 1) reacts with an increase in apical dominance. Decrease is defined here as the negatively geotropic movement of lateral branches in the presence of the terminal apex to a point where lateral branches appear to actively compete with the terminal shoot for dominance. An increase in apical dominance as noted for *Sequoia sempervirens* entails complete inhibition of lateral bud growth, and diminished elongation of lateral shoots.

The application of AMO-1618 and B-995 to intact plants of *Cupressus arizonica* results in a positively geotropic growth of lateral branches (Pharis et al, 1965), and for *Sequoia sempervirens,* a release of lateral buds. The response for *Sequoia* qualifies as a decrease in apical dominance, but whether the downward growth of lateral branches in *Cupressus* should be classified as an enhancement of apical dominance or a geotropic response is debatable. Both effects can be overcome by simultaneous or subsequent application of GA.

It should be noted that both the negative and positive geotropic growth of lateral branches exhibited by these conifers involves differential cambial growth, that plagiotropic growth of the lateral shoot involves essentially equal cambial growth, and that cambial growth continues, even under the influence of growth retardants.

When either *Cupressus* or *Sequoia* is decapitated, replacement of the missing terminal occurs naturally over the course of some weeks or months. A study of the kinetics of this movement in *Cupressus* shows that GA speeds up the replacement process over and above control plants by 2 to 3 weeks. Application of B-995 or AMO-1618 retards it completely, with the growth retardant effect being overcome by simultaneous or subsequent application of GA (Pharis et al,1965 and Ruddat and Pharis, 1966).

Effects similar to those observed with the use of growth retardants were noted on *C. arizonica* with low nitrogen nutrition (Kuo and Pharis, unpublished results), but since this long term experiment is still in progress, the possibility that exogenous GA might reverse the positive geotropic growth has not yet been tested. As well, we have noted that photoperiod influences the angle of lateral branches of intact *C. arizonica* plants, with longer photoperiods producing a negative geotropic response, intermediate or short photoperiods a plagiotropic growth response, and very short photoperiods (4 hours) producing a positive geotropism, where even the terminal shoot grows straight down (Figure 2). A reversal of the positive geotropism under the 4-hour photoperiod is obtained by treatment with GA (Fig. 2).

It appears then, that endogenous GA most likely plays a major role in the geotropic growth response, expression of apical dominance, and apical control in conifers. Analysis of our tissues for endogenous GA levels has been delayed until purification techniques are devised which would allow us to quantitate accurately the GA levels within conifers. However, the circumstantial evidence seems overwhelming, especially if one assumes that (1) at least one growth retardant used by us acts to inhibit endogenous GA synthesis, (2) the negative geotropic response of a large number of conifer species to additional GA indicates that endogenous GA is a limiting factor, or is in a rather delicate balance with other hormones influencing the response, and (3) photoperiod influences endogenous GA level, the shorter the day, the less GA present. There is good evidence for the first (see review by Lang, 1970) and third (Cleland and Zeevaart, 1970 plus references cited therein). However, the mechanism by which GA influences differential cambial growth or inhibits lateral bud growth remains speculative. Our data indicate that clear distinctions between the expression of apical dominance, apical control (or tree form), and shoot geotropisms do not exist for most conifers.

Hormone-directed transport

It has been found in *Pseudotsuga* and *Pinus* species that topical application of GA, alone, or together with NAA, to the unfertilized female strobilus (personal communication, S. Krugman, Pacific S.W. Forest and Range Expt. Sta., Berkeley, Calif.)

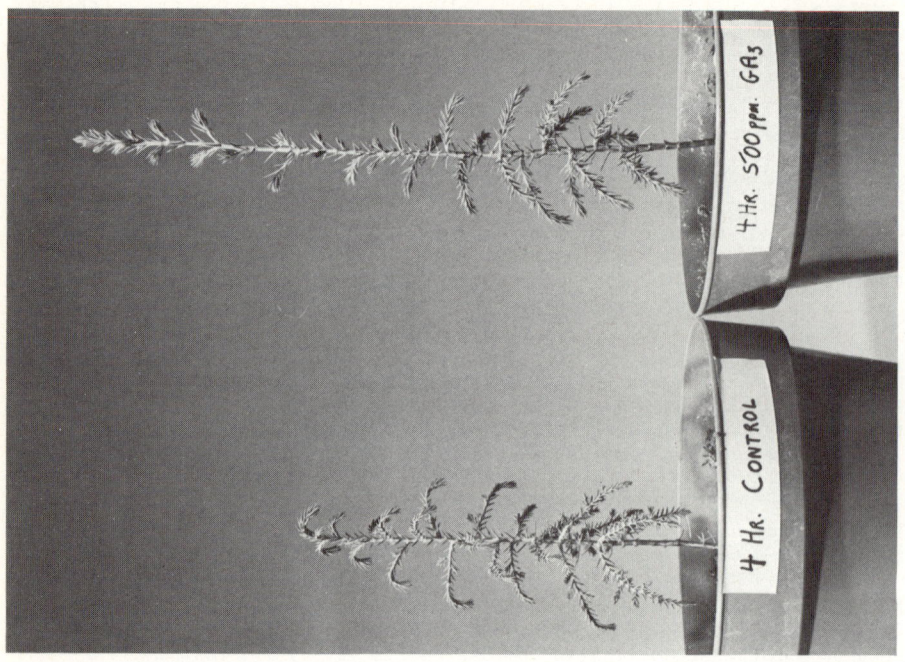

Fig. 2. Effect of a 4 hour photoperiod on the geotropic response of lateral and terminal shoots of *Cupressus arizonica*, and the reversal of this response by soil drench application of a total of 550 mg GA_3 over an 11 week period.

Fig. 1. Effect of topical application of a total of 125 μg GA_3 injected in 80% ethanol at weekly intervals to *Sequoia sempervirens* on the expression of apical dominance. Control (ethanol alone) seedling on right.

will allow parthenocarpy and parthenogenesis, and we have found a similar situation in *Cupressus* and *Thuja* (R. Pharis and W. Morf, unpublished results). Parthenocarpy is known to occur in certain dicotyledonous fruit trees in response to GA application (Dennis, 1970, and references cited therin).

Table 1. Effects of two growth retardants on the distribution of male strobili[1] on *Cupressus arizonica* seedlings induced to flower by GA_3[2] applied topically to 6 branches in the middle of the seedling.

Treatment	Induced Strobili Quotient[3]		
	Top Branches	Middle Branches	Bottom Branches
B-995			
1000 mg/1	7/7	7/7	6/7
2000 mg/1	9/9	9/9	8/9
4000 mg/1	7/7	7/7	5/7
AMO-1618			
1000 mg/1	5/5	5/5	4/5
2000 mg/1	2/2	2/2	2/2
4000 mg/1	-	-	-
Control	8/8	8/8	1/8
(w/o retardant)			

[1] Counts made 50 days from time of injection
[2] Injected in amounts ranging from 0.02 to 10 µg/meristem
[3] Number of trees showing a flowering response in upper, middle or lower branches over number of trees in experiment.

We have obtained indirect evidence through the use of topical application (by injection, or surface application in 95% ethanol of GA to plants treated with AMO-1618 and B995) that *C. arizonica* may use endogenous GA to direct transport of exogenously applied GA_3 (Table 1). Plants not treated with retardants may have directed the GA (or some substance resulting from the injection) in an upward direction *as evidenced by the flowering response,* whereas plants treated with retardants showed little or no polarity of GA movement (again as evidenced by the flowering response). Further evidence for a polarity of movement upward in control plants is shown in Figure 3 where ^3H-GA_1 was injected and the radioactivity measured after 24 and 72 hours. Here, we should note that in other experiments injections of GA_1 show a diminished polarity of movement upwards compared to injection of GA_3, as evidenced by the flowering response.

We realize that we have not yet verified that the movement of radioactivity (Figure 3) was indeed in the form of GA_1, and will continue this work in the future through the use of co-chromatography of unlabeled GA's with radioactive extracts on a gas chromatograph-radio-chromatogram scanner. Technical problems with purity of the labeled GA_1, and instrumentation have delayed these experiments until now. We are also aware that the observation of a flowering response is not synonymous with transport of the injected GA_3.

It is now apparent that GA can move within plants in a polar manner (see Introduction). Since our technique of topical application on, or into the woody portion of the branch undoubtedly results in the major amount of the GA going into the xylem initially, it could be argued that the injected GA is merely being carried upwards in the xylem transpiration stream, and that retardant-treated plants are not normal in their transpiration patterns. While we have not examined transpiration patterns of retardant-treated conifers, work by Halevy (1967) and Plaut et al (1964) shows no consistent effect of growth retardants on transpiration in several crop plants. In fact, bean seedlings treated with B-995 actually transpired more than controls.

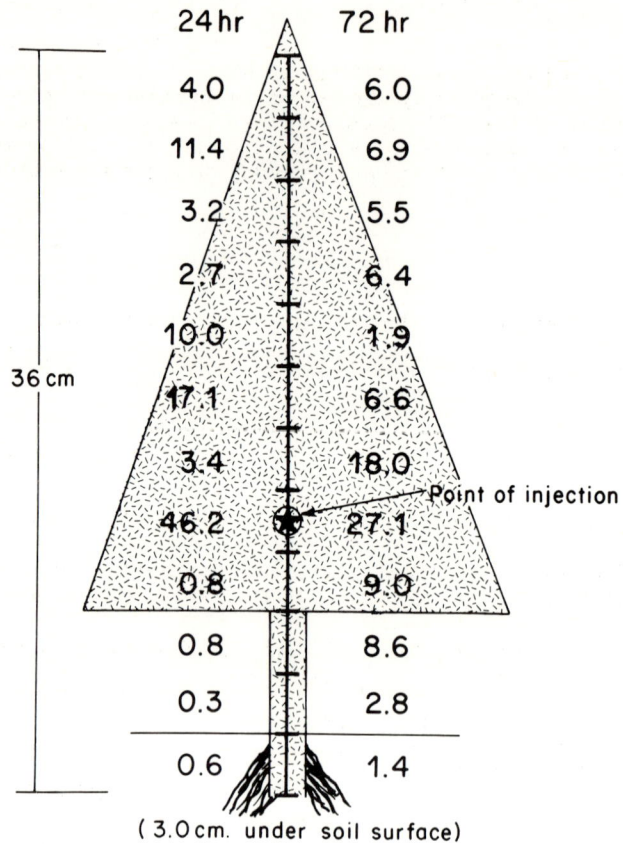

Fig. 3.

Distribution of radioactivity in seedlings of *Cupressus arizonica* 24 and 72 hours after topical application of 1 µl ethanol containing 0.1 µCi of ^{3}H-GA$_1$ per plant (0.046 µg GA$_1$). Plants maintained at 25°C under a 16 hour photoperiod before and after GA application. Each section represents 3.0 cm and radioactivity is expressed as % of total cpm. Total cpm was 43,989 for the 24 hour plant, and 47,871 for the 72 hour plant.

Hence, we would speculate that retardant-treated plants may be unable to mobilize applied GA in a polar manner due to their lack of endogenous GA, and that endogenous GA may act in this conifer as a natural mobilizing hormone.

IV. SUMMARY

Responses to the exogenous application of GA growth retardants, and manipulation of photoperiod lead us to conclude that GA is a primary determinant in the expression of apical dominance, apical control (tree form), and geotropic responses in shoots of a number of species of conifers. In general, positively geotropic growth of the woody shoot occurs under situations assumed to reduce GA levels, negatively geotropic growth in response to added GA. The lateral bud inhibition response is less simple, and generalizations cannot be made at this time. Clear distinctions between the expression of apical dominance, apical control, and shoot geotropism do not appear to exist for most conifers.

Preliminary data indicating that endogenous GA in *Cupressus arizonica* may be responsible for mobilizing exogenously applied GA in an upward direction is presented. This assumption is based on (1) distribution of the flowering response after topical injection of GA_3 to control and retardant-treated plants, and (2) movement of radioactivity after topical injection of $^3H-GA_1$ in seedlings not treated with growth retardants.

V. ACKNOWLEDGEMENTS

We are indebted to Mr. W. Morf for his able technical assistance. Support for this research came from tne National Research Council of Canada Grant No. A-2585, a Canada Department of Forestry Extramural Grant, and a Weyerhaeuser Foundation Grant to R. Pharis.

VI. REFERENCES

BOOTH, A. (1959). Some factors concerned in the growth of stolons in potato. J. Linn. Soc. 56, 166-169.

BOOTH, A. (1963). The role of growth substances in the development of stolons, in "The Growth of the Potato". (Ed. J.D. Ivins and F.L. Milthorpe) pp. 99-113. Butterworths, London.

BROOKS, H.J. (1964). Responses of pear seedlings to N-dimethylamino-succinamic acid, a growth retardant. Nature (Lond.) 203:1303.

BROWN, C.L., R.G. McALPINE, and P.P. KORMANIK (1967). Apical dominance in woody plants: A reappraisal. Am. J. Bot. 54:153-162.

CATALANO, M. and T.A. HILL (1969). Interaction between gibberellic acid and kinetin in overcoming apical dominance, natural and induced by IAA, in tomato (*Lycopersicum esculentum* Mill. Cultivar Potentate). Nature (Lond.) 222:985-986.

CHAMPAGNAT, P. (1965). Physiologie de la croissance et l'inhibition des bourgeons: Dominance apicale et phénomènes analogues. In "Encyc. Pl. Physiol.", (Ed. W. Ruhland) XV/I, pp. 1106-1164. Springer-Verlag, Berlin.

CLELAND, C.F. and J.A.D. ZEEVAART (1970). Gibberellins in relation to flowering and stem elongation in the long day plant *Silene armeria*. Pl. Physiol., (Wash.) 46:392-400.

DÖRFFLING, K and M. BÖTTGER (1968). Transport von Abszisinsäure in Explantaten von Blattstiel- und Internodialsegmenten von *Coleus rheneltianus*. Planta (Berl.) 80:299-308.

DENNIS, F.G. Jr. (1970). Effect of gibberellins and naphthalenacetic acid on fruit development in seedless apple clones. J. Amer. Soc. Hort. Sci. 95:125-128.

HALEVY, A.H. (1967). Effect of growth retardants on drought resistance and longevity of various plants. Proc. 17th Int. Hort. Congr. 3:277-283.

HALEVY, A.H., A. ASHRI and Y. BEN-TAL (1969). Peanuts: Gibberellin antagonists and genetically controlled differences in growth habit. Sci. 164:1337-1338.

HARRIS, G.D., B. JEFFCOAT, and J.G. GARROD (1969). Control of flower growth and development by gibberellic acid. Nature (Lond.) 223:1071.

HERTEL, R., M.C. EVANS, A.C. LEOPOLD, and H.M. SELL (1969). The specificity of the auxin transport system. Planta (Berl.) 85:238-249.

JACOBS, W.P. and D.B. CASE, (1965). Auxin transport, gibberellin, and apical dominance. Sci. 148:1729-1731.
JACOBS, W.P. and H. KALDEWEY (1970). Polar movement of gibberellic acid through young *Coleus* petioles. Pl. Physiol. (Wash.) 45:539-541.
KUMAR, D. (1966). The physiology of stolon development, tuberization and dormancy in the potato. Ph.D. Thesis, Univ. of Wales.
LANG, A. (1970). Gibberellins: Structure and Metabolism. Ann. Rev. Pl. Physiol 21:537-570.
MONTALDI, E.R. (1969). Gibberellin-sugar interaction regulating the growth habit of Bermuda grass, *Cynodon dactylon* (L.) . Experientia 25:91-92.
PALMER, J.H. (1955). An investigation into the behavior of the rhizome of *Agropyron repens* Beauv., with special reference to the effect of light. Ph.D. Thesis, Univ. of Sheffield.
PHARIS, R.P., M. RUDDAT, C. PHILLIPS, and E. HEFTMAN. (1965). Gibberellin, growth retardants, and apical dominance in Arizona cypress. Naturwiss. 52:88-89.
PHARIS, R.P. and W. MORF (1967). Experiments on the precocious flowering of western red cedar and four species of *Cupressus* with gibberellins A_3 and A_4/A_7 mixture. Can. J. Bot. 45:1519-1524.
PHILLIPS, I.D.J. (1969). Apical Dominance. in "The Physiology of Plant Growth and Development" (Edited by M.B. Wilkins) pp. 165-204. McGraw-Hill, London.
PLAUT, Z., A.J. HALEVY and E. SHMUELI (1964). The effect of growth-retarding chemicals on growth and transpiration of bean plants grown under various irrigation regimes. Israel J. Agric. Res. 14:153-158.
RUDDAT, M. and R.P. PHARIS (1966). Participation of gibberellin in the control of apical dominance in soybean and redwood. Planta (Berl.) 71:222-228.
SCOTT, T.K., D.B. CASE, and W.P. JACOBS (1967). Auxin-gibberellin interaction in apical dominance. Pl. Physiol. (Wash.) 42:1329-1333.
SETH, A.K. and P.F. WAREING (1967). Hormone-directed transport of metabolites and its possible role in plant senescence. J. Exp. Bot. 18:65-77.
TOMASZEWSKI, M. (1970). Auxin-Gibberellin interactions in apical dominance. Bull. Acad. Polon. Sci., Ser. Sci. Biol. 18:361-366.

CYTOKININS

ACTIVE FORMS OF THE CYTOKININS

J. Eugene Fox, W.D. Dyson[1], Chander Sood[2] and J. McChesney

Department of Botany University of Kansas, Lawrence, Kansas. 66044 U.S.A.

1. Present Address: Department of Biochemistry, McMaster University, Hamilton, Ontario, Canada

2. Present Address: Department of Biology, University of Windsor, Ontario, Canada

I. INTRODUCTION

In recent years there has been a steady stream of reports linking the biological activity of the cytokinins with transfer RNA (see the review by Key, 1969). Not only are molecules with cytokinin activity found in the transfer RNA of plants, animals and microorganisms, but also where the structure has been investigated they are found adjacent to the anti-codon. The possibilities for an exceedingly delicate control of protein synthesis offered by such a site have convinced many that the biological activities of the cytokinins were very near to being explained. In recent months, however, Chen and Hall (1969) published data which indicate that the situation may be more complex than previously suspected. They discovered that even cytokinin-dependent tissues, i.e. those which will require an exogenous cytokinin in the medium in order to grow, already possess as part of the structure of their transfer RNA, molecules with potent cytokinin activity; furthermore the cytokinin provided in the medium is not the metabolic precursor of the cytokinin in the transfer RNA. This observation argues that the role of the exogenous cytokinin is distinct in at least some important way from that of the cytokinin *in situ* in transfer RNA.

In an attempt to discover the relationship between the exogenous and endogenous cytokinins we have been following the metabolism of the former in cytokinin-requiring tissues. In addition we have investigated the short term metabolism of certain non-puring cytokinins in order to ascertain whether or not these have metabolites in common with the purine cytokinins. Also, we have studied the effects of substituents at the 9 position on cytokinin metabolism. The major finding to be reported here is that a purine cytokinin is converted to a stable, long-lived metabolite which may be an active form of the cytokinin.

II. METHODS

Tissue Culture Studies

Culture methods, media, and the origin of the tissues used in this study have been previously described (Fox, 1964). For the studies with urea cytokinins an especially adapted strain of tobacco tissue was used whose isolation and properties have been described (Dyson *et al.* 1971). For short term metabolic studies, tissues were starved for 5 days on a basal medium containing auxin but no cytokinin and then transferred to fresh medium where they were exposed to the labeled compound. Cytokinins were applied directly to the tissue with a sterile pipette in a solution sterilized by filtration at $27°$. The solution applied as a single drop was immediately taken up by the tissue; no other cytokinin was provided in the medium, i.e. the tissue was pulse labeled.

Radioactive Substrates

The synthesis and characterization of each of the following compounds used in this study has been described elsewhere (Fox, *et al.* 1971, Dyson *et al.* 1971) 6-benzylaminopurine (BA), 6-benzyl-9-B-D-ribofuranosylaminopurine (BA riboside), BA-methylene-14_C, BA-8-14_C, BA-methylene-3_H, diphenylurea (DPU), DPU-benzyl (U.L.)-14_C, DPU-carbonyl-14_C, 6-benzyl-9-methylaminopurine (MBA), MBA-methyl-14_C, MBA-methyl-14_C-methylene-3_H. Benzyladenosine-5'-monophosphate was a gift of Professor Jean Guern, Laboratoire de Physiologie Végétale, Sorbonne, Paris.

III. RESULTS AND DISCUSSION

Cytokinin Metabolism in Soybean

Soybean tissue incubated on 6-benzylaminopurine-methylene-14_C produces at least 4 major metabolites containing radioactivity within 12 hours (Fig. 1). Fraction A (Fig. 1) is almost certainly unmetabolized benzyladenine. It co-chromatographs on paper with authentic BA in several systems, and the two have identical UV spectra. In addition, fraction A was co-crystallized with BA and the crystalline material sublimed. Essentially all of the radioactivity remained in the sublimed BA in those instances where fraction B was clearly separated from A. In many instances, however, B was imperfectly separated from A in this solvent system, appearing as a shoulder on the latter or masked altogether by A. The two could be clearly distinguished by paper chromatography in water where fraction B ran substantially faster (average Rf value 0.55) than A (average Rf 0.40).

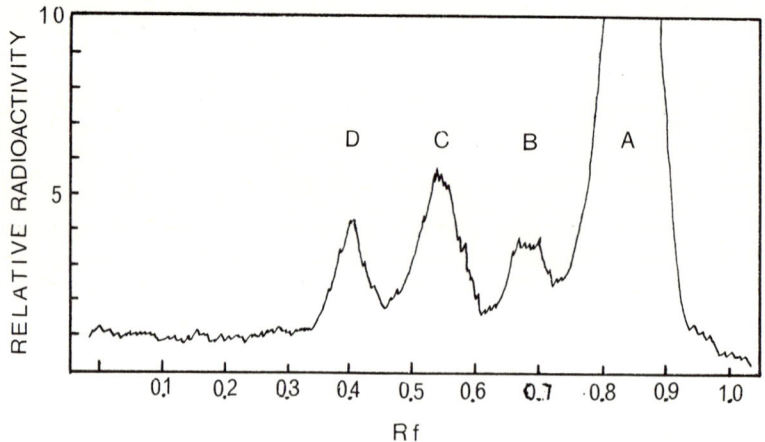

Fig. 1.

70% ethanol soluble, radioactive metabolites of BA from soybean tissue incubated 12 hours on 0.5 mg/1 BA-methylene-14_C, sp. ac. 1.6 mC/mm. The tissue is washed on a Buchner funnel with a small amount of distilled water to remove any unincorporated precursor and then ground in a mortar and pestle with glass beads, (0.2 mm diam.) in sufficient 95% ethanol at 0° to achieve a final concentration of 70% ethanol. The debris is pelleted at 25,000 x G for 10 min. in a refrigerated centrifuge and the supernatant is collected. The pellet is similarly extracted three additional times, and the combined supernatants reduced *in vacuo* at 27°. The extract is streaked onto Whatman 3 mm filter paper and the chromatogram developed in n-butanol, acetic acid and H_2O (4:1:2). The developed chromatogram was analyzed for radioactivity with a Nuclear Chicago Actigraph III paper scanner.

Fraction B is very likely the ribonucleoside of BA. If fraction B is mixed with a sample of authentic benzyladenosine and chromatographed on paper, radioactivity from fraction B is conincident with the ultraviolet absorbing spot due to benzyladenosine in three solvent systems n-butanol, acetic acid, H$_2$O (4:1:2 v/v); n-butanol NH$_4$OH, H$_2$O (96:1:13 v/v); and H$_2$O. In addition treatment of fraction B with 1 N HCL at 100° for one hour converts this material to a second substance which is indistinguishable from BA and which co-crystallizes and co-sublimes with BA. To further characterize this substance, fraction B was adjusted to pH 4.5 with acetic acid, and a 1.0 ml aliquot was reacted at 27° for 30 minutes with 0.3 ml of 0.1 M NaIO$_4$. The pH was adjusted to 13 with KOH and the reaction allowed to proceed an additional 3 hours. After this period the pH was adjusted to 8.0 with acetic acid and the product was co-crystallized and co-sublimed with BA. Again all the radioactivity was associated with BA.

The material at Rf 0.4 (fraction D, Fig. 1) was suspected of being N,6-benzyladenosine-5'-monophosphate on the basis of comparative Rf values of the authentic compound. Not only does authentic BAMP have the same Rf value as D in n-butanol acetic acid and water (3:1:1 v/v) (BAW), but the two can also be co-chromatographed in water. To further characterize fraction D, it was treated with alkaline phosphatase; the product has chromatographic properties identical with benzyladenosine. This product was then degraded with acidic periodate followed by base hydrolysis as described previously. The resulting radioactive material was identified as BA on the basis of its co-crystallization and sublimation with authentic BA.

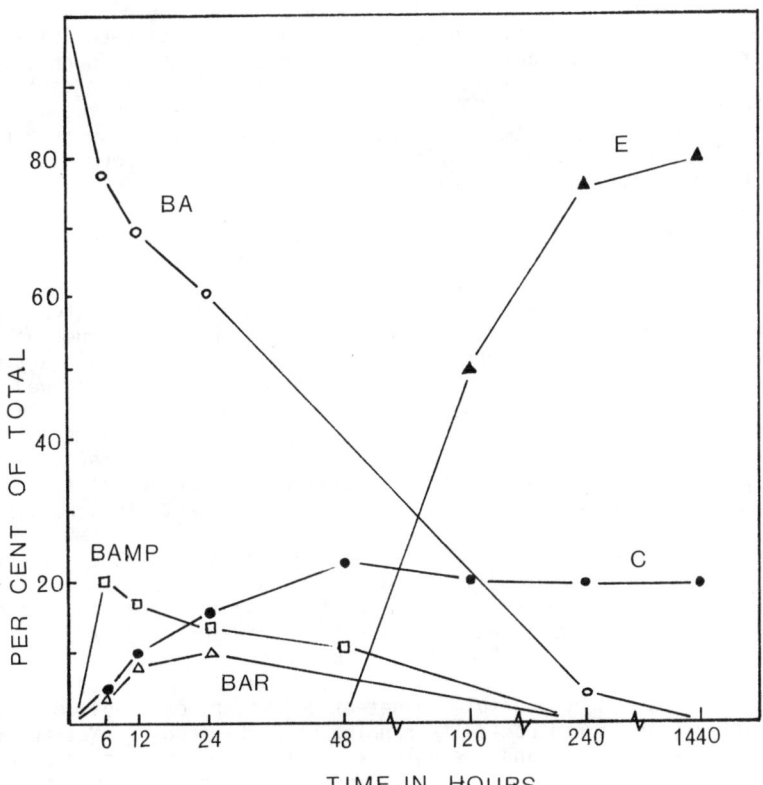

Fig. 2.

Time course of BA metabolism. This figure is a summary of several experiments in which the tissue was incubated on variously labeled BA for varying periods of time. Tissue extracted, the extracts chromatographed, and radioactivity estimated as in figure 1. BAR is benzyladenosine, see text for details of fractions C and E.

Fraction C (Rf 0.5, Fig. 1) is unique among the metabolites of BA which still contain the intact benzyladenine moiety in that it is a very long lived substance. This behavior was revealed in a series of studies where the tissue was exposed to BA labeled in various ways for periods ranging from 2 hours to 60 days. In general the data (summarized in Fig. 2) show that the ribonucleoside and the ribonucleotide appear very early in significant amounts. Thus BAMP accounts for 20% of the total radioactivity with 2 hours of incubation. Within a short time, however, BAMP begins to disappear and this change is paralleled very closely by the formation of fraction C suggesting a metabolic relationship between the two. Within 48 hours BAMP has fallen to about 11% of the total radioactivity while the new metabolite which was undetectable at 2 hours, now accounts for some 22% of the whole. Once having reached this level, the substance is remarkably stable while the free BA, its ribonucleoside and ribonucleotide rapidly disappear, so that by 12 days it is no longer possible to detect them even though the tissue is vigorously growing at this stage. The new fraction continues to persist so that even at the end of 60 days it still accounts for about 20% of the total radioactivity.

Preliminary evidence indicates that fraction C has substantial cytokinin activity itself in the soybean assay. Furthermore this substance appears as a major, radioactive metabolite whether the BA is initially supplied with ^{14}C in the methylene carbon of the side chain, or in the purine 8 carbon, or with ^3H attached to the methylene carbon of the side chain. Thus this metabolite very likely contains an intact benzyladenine moiety. Although not enough of the substance has been collected for a full characterization, it is known that it is retained by both anion and cation exchange resins under conditions which also retain BAMP. That it is somewhat less polar than BAMP, however, is shown by its migration on paper chromatograms. In addition, the substance is adsorbed by charcoal. Preliminary work with ^{33}P indicates that the substance contains phosphorous although interestingly enough it is not produced by cell free systems which readily yield BAMP.

The significance of fraction C as a growth substance is unclear. The fact that benzyladenine itself as well as its riboside and ribotide exists only transiently in the cell argues that they are not themselves growth active forms. It is possible of course that BA initiates a series of reactions which culminate in a growth response long after the inciting molecules have disappeared. If that were true, however, one wonders why there persists a substance (fraction C) with cytokinin activity itself which contains the intact BA moiety, and which is accompanied by continued growth on the part of the tissue. Is it perhaps that we are dealing with a protected form of the cytokinin capable of resisting the degradation which destroys some 75% of the BA within a few days? Or is it that fraction C is an active form of the cytokinin? The fact that it persists in the tissue over a relatively long period of time and that it is the only BA containing metabolite to do so strongly suggests this possibility.

Apparently the remainder of the BA is degraded in the tissue by side chain cleavage and this can be conveniently followed when the cytokinin is supplied as BA-methylene-^{14}C. In this instance a new metabolite appears after about 5 days whose properties indicate that it no longer contains the purine moiety. When the incubation is extended for 60 days, fraction C continues to appear (fig. 3).

However it is no longer possible to detect BA, its ribonucleoside or ribonucleotide. The large peak of radioactivity running near the front (fraction E, fig. 3) was rechromatographed in water and a single radioactive spot was noted at Rf 0.59. Thus this material is not BA (Rf 0.4 in this system). Although it has the same Rf value as benzyladenosine in H_2O, it clearly differs from the latter, as well as from BAMP, by its Rf in BAW chromatography. Furthermore it could not be co-crystallized with either. Perhaps the most revealing characteristic of fraction E, however, is that it cannot be detected as a radioactive containing moiety when the tissue is incubated with BA-8-^{14}C or with BA-methylene-^3H. It appears therefore as though fraction E represents a side chain metabolite of BA presumably oxidized to something like benzoic acid. This substance is quite acidic and may be conveniently separated from neutral and basic metabolites on a cation exchange resin. Although we initially suspected the fraction E was benzoic acid, co-chromatography of the two reveals they are distinct (fig. 3) and that there exist at least 3 minor acidic components. Although the major peak has chromatographic properties similar to that of phthalic acid, its UV spectrum plus the fact that it fails to vaporize at 90° indicates that it cannot be that compound. Apparently the metabolism of BA takes place in such a way that adenine is one of the products (fig. 4). We have previously shown (Fox, 1966) that the purine moiety of BA is scavenged for a wide variety of purposes by the tissue. In figure 5 is shown a summary of the way in which we believe soybean tissues metabolize benzyladenine.

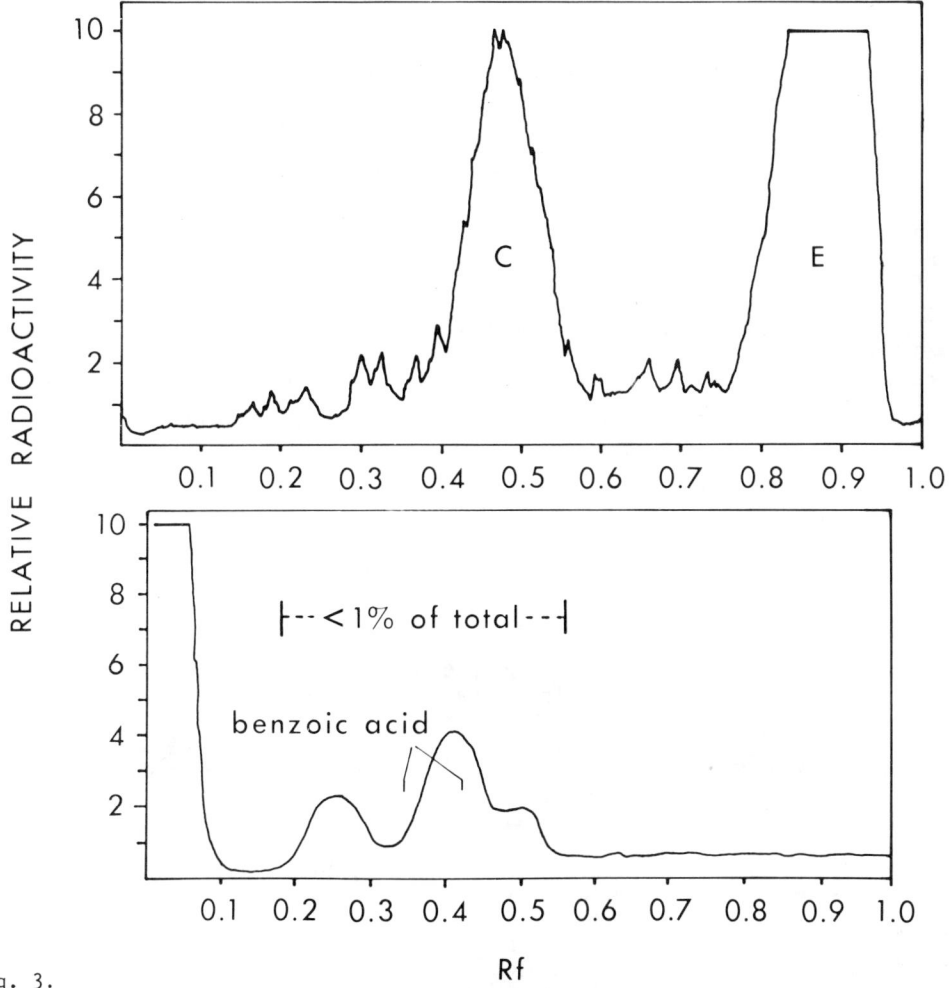

Fig. 3.

(top) 60 day metabolites of BA-methylene -^{14}C. Soybean tissue was incubated on 1 mg/l BA-methylene-^{14}C sp. ac. 1.6 mC/mm for 60 days, and then extracted as in Fig. 1. The extract was chromatographed in n-butanol acetic acid, H_2O (4:1:2 v/v) and assayed for radioactivity with a strip scanner as previously described. (bottom) Paper chromatography of fraction E (above) in n-butanol, NH_4OH, H_2O (96:1:13 v/v).

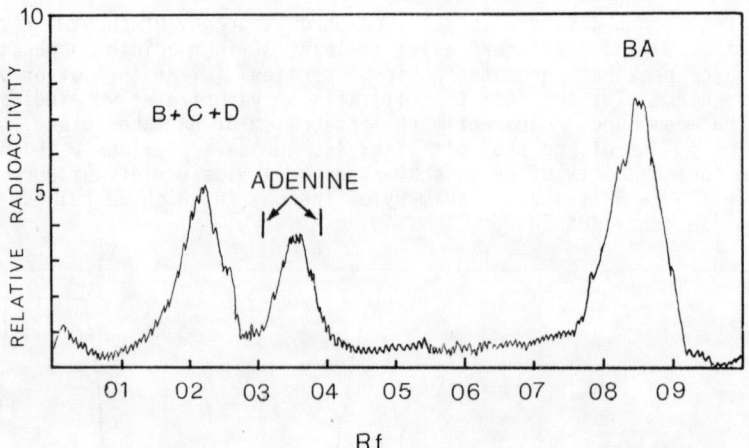

Fig. 4.

5 day metabolites of BA-8-^{14}C. Tissue extracted as in Fig. 1 and chromatographed on paper in n-butanol, NH4OH, H$_2$O (85:5:14 v/v). Radioactivity estimated with a strip scanner. Control spots of adenine and benzyladenine run on the same chromatogram as shown. All of the radioactive material at Rf 0.8 co-crystallized with BA. Likewise all of the material at Rf 0.35 co-crystallized with adenine. 5% of the radioactivity at Rf 0.2 co-crystallized with BA-riboside. The remainder is presumed to be BAMP (D) and fraction C as well as other metabolites. Radioactive material near the baseline is at least partially inosine.

The Metabolism of Non-Purine Cytokinins

If the transformation of purine cytokinins into active metabolites by way of the ribonucleotide is a necessary concomitant to their biological activity, it is not immediately clear how the non-purine cytokinins fit into that scheme. Miller (1961) pointed out several years ago, however, that the substituted ureas might well be used by the cell as precursors of a kinetin-like compound in view of the well-known catabolic relationship between the purine ring and urea. We recently observed that tobacco tissues (particularly an adapted strain) grow very well with diphenylurea as the sole cytokinin source while soybean tissues exhibit no growth at any concentration of DPU tested. In order to gain insight into the marked difference in biological activity of DPU in soybean as compared with tobacco tissue, a study of the metabolic fate of DPU in these tissues was made. The tissue, either soybean or DPU adapted tobacco, was transferred from stock cultures onto a medium complete except for a cytokinin. After a 5 day cytokinin starvation period, the tissue was exposed under sterile conditions to ^{14}C labeled DPU for varying periods.

The results (fig. 6) indicate that soybean is unable to metabolize DPU at all. After 81 hours incubation, essentially all of the radioactivity was coincident with DPU in 3 solvent systems. By contrast, metabolites of DPU could be detected very quickly in tobacco tissue and by 81 hours at least 4 new substances had appeared which account for more than 20% of the total radioactivity. These studies suggest that DPU is not active as such but must be metabolically transformed into active substances. Tissues which can utilize DPU as the sole cytokinin source might well be those capable of such conversions.

Metabolic Studies with 9-Substituted Cytokinins

If the ribonucleotide is a necessary intermediate in the formation of an active substance, one should be able to block otherwise active compounds with suitable groups at the 9-position. To test this we have prepared and assayed a large number of 9-substituted BA derivatives including the methyl, ethyl, propyl, hydroxypropyl, and cyclohexyl analogs. In every case the biological activity of these was considerably less than the free base, and the cyclohexyl analog had only barely detectable activity. These results contradict other work such as that of Kende and Tavares (1968) who described studies with 6-benzylamino-9-methyl purine-benzyl-7-^{14}C. Although radioactivity was incorporated into the RNA of soybean callus tissue when the free base was employed, methylation at the 9 position seemingly prevented incorporation while in no way reducing biological activity of the cytokinin. Letham (1969) and Young and Letham (1969) reported that the 9 cyclohexyl analog of isopentenyladenine possesses considerable cytokinin activity; Letham considers that, "It is extremely unlikely that this compound with a group not subject to hydrolytic cleavage on the 9 position could be converted to a ribotide and incoporated into RNA as a nucleotide." Likewise Shaw et al. (1968) reported that 9-methylzeatin gave biological activity of the same order as zeatin and concluded that, "the results imply that the mechanism of cytokinin activity in substituted adenines does not require prior formation of nucleotide derivatives."

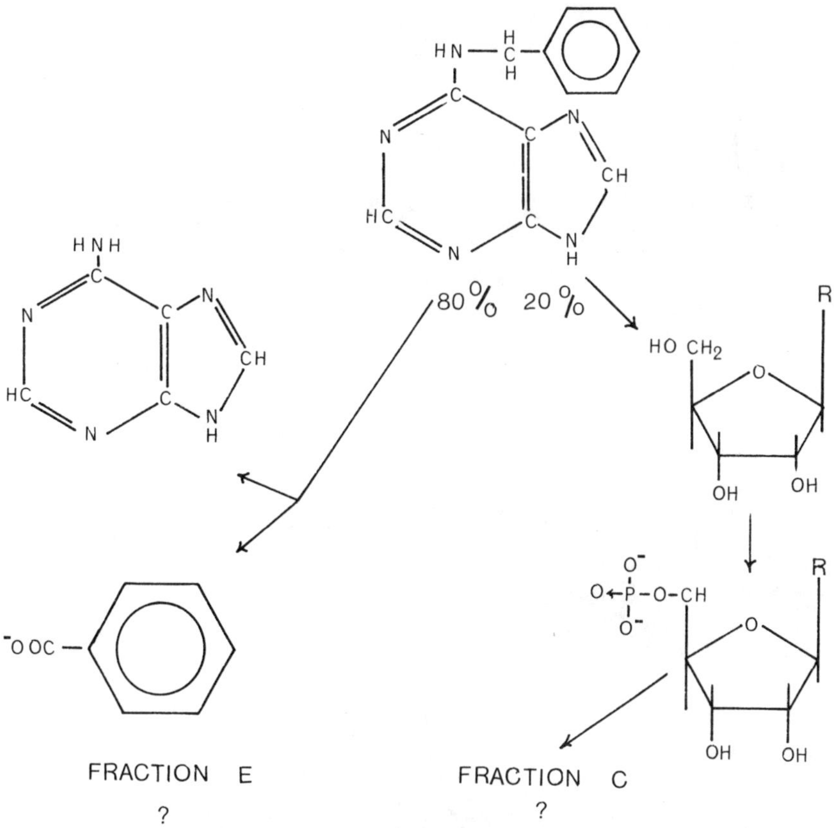

Fig. 5.

Summary of the metabolic fate of BA in soybean tissue. Although fraction E is depicted as benzoic acid, it is not that compound but something similar (see text).

Clearly, however, such conclusions are premature unless it can be demonstrated that the substituent at the 9 position is not removed by the plant tissue and thus constitutes a truly effective block to ribonucleotide synthesis. In order to test this point, the cytokinin derivative 6-benzylamino-9-methyl purine has been synthesized and labeled with ^{14}C in the 9-methyl carbon or doubly labeled with ^{3}H in the methylene moiety of the side chain. Although the substance is chemically stable, cytokinin-requiring tissues begin removing the 9-substituent in as little as 10 minutes (Table I).

Both soybean and tobacco tissue on either liquid or solid media metabolize MBA in such a way that at least some portion of the methyl moiety is released as CO_2. Furthermore the amount of $^{14}CO_2$ released is a function of time. Only a tiny fraction of the counts supplied to the tissue as MBA-methyl-^{14}C are represented by $^{14}CO_2$. Extracts of the tissue chromatographed in a variety of systems have made it clear, however, that MBA persists for only a very short time and is quickly metabolized by the tissue. The most instructive experiments were done with tobacco tissue incubated with MBA doubly labeled with ^{14}C and ^{3}H. The data (Table II) show quite clearly that a major metabolite is benzyladenine itself which after 6 hours of incubation may account for as much as 10% of the total counts recovered as tritium. Thus the biological activity of the 9-substituted cytokinins could be accounted for by their conversion to the free base and then subsequent metabolism to an active form.

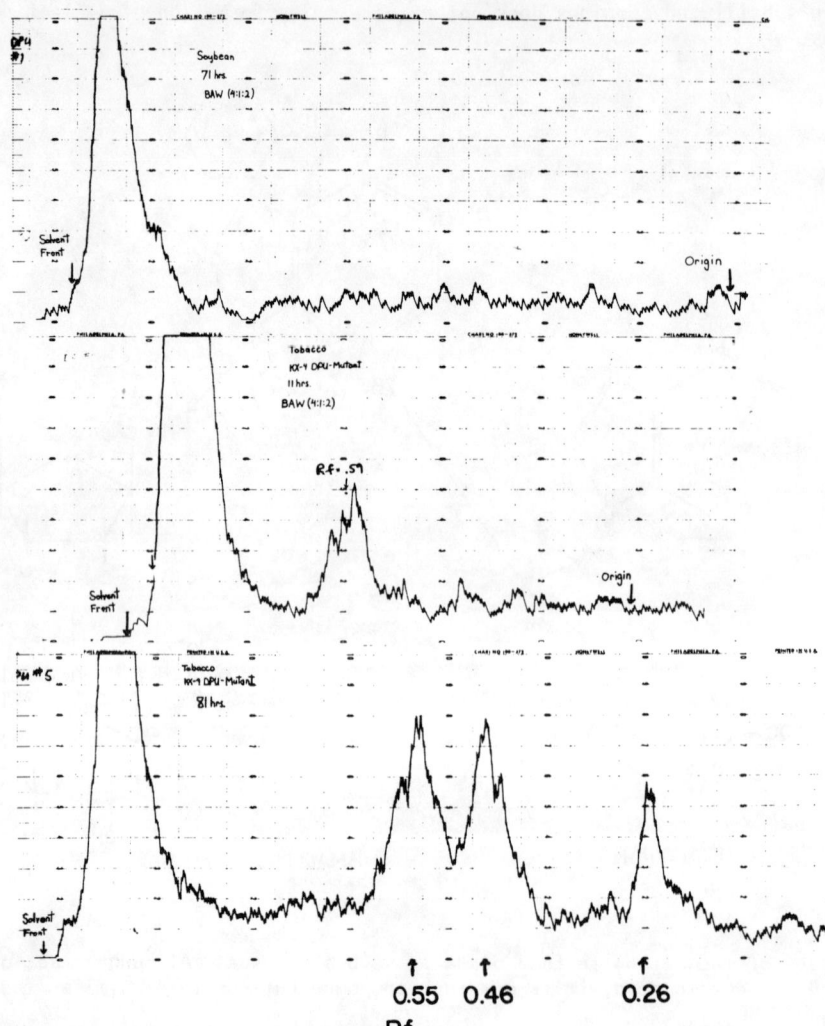

Fig. 6.

Radioactive metabolites of DPU in soybean and tobacco tissues. Cytokinin starved tissue (see text) was incubated with DPU-benzene-^{14}C, sp. ac. 37 mC/mm at the rate of 750,000 cpm/10 grams fresh weight of tissue. The tissue was extracted after the incubation period as in Fig. 1, chromatographed and radioactivity estimated as in Fig. 1. (top) Soybean extract from tissue incubated 71 hrs. (middle) Tobacco extract from tissue incubated 11 hrs. (bottom) Tobacco extract from tissue incubated 81 hrs. See text for further details.

Table 1. $^{14}CO_2$ Evolution from Tobacco and Soybean Tissues Incubated on a Culture Medium Containing MBA-methyl-^{14}C.

Tissue	Incubation Period	CPM - background
Tobacco	10 min.	283
Control	10 min.	19
Tobacco	60 min.	1680
Control	60 min.	2
Tobacco	360 min.	3550
Control	360 min.	8
Tobacco	48 hrs.	4469
Soybean	48 hrs.	352
Control	48 hrs.	6
(Solid Medium)		
Tobacco	48 hrs.	4633
Soybean	48 hrs.	792
Control	48 hrs.	7

Tissues were in the log phase of growth but starved for 72 hours prior to use on a basal medium complete except for the cytokinin. At the end of the incubation period on 1 mg/l of MBA-methyl ^{14}C sp. ac. 43 mC/mm the gases present in the flask were passed through a column containing finely divided filter paper soaked in a concentrated solution of barium hydroxide in order to trap CO_2. The filter paper was air-dried and suspended in a standard fluid for liquid scintillation spectrometry. In order to rule out the possibility of microbial contamination during the manipulations required by the experiment, a control flask complete except for the tissue was subjected to the same operations of each experiment.

Table II. Chromatographic Distribution[1] of Radioactive Material Found in the Ethanol Extracts of Tobacco Tissue Incubated 4 Days on 1 mg/l MBA-methyl-^{14}C-methylene-^{3}H sp. ac. ^{14}C is 10 mC/mm and ^{3}H 39 mC/mm.

Fraction Number	Rf	^{14}C DPM	^{3}H DPM
1	baseline - 0.125	525	3200
2	0.126 - 0.250	300	700
3	0.251 - 0.375	110	180
4	0.376 - 0.500	50	120
5	0.501 - 0.625	40	125
6	0.626 - 0.750	10	---
7	0.751 - 0.875	80	160
8	0.876 - front	5	---

1. Chromatography performed on 1mm layers of aluminum oxide G (Merck) activated at 60° for 12 hours. The solvent was chloroform, ethanol (97:3 v/v). Rf of MBA control spot was 0.81 and BA 0.13 although most of the spot was in fraction 1. Each fraction was eluted in 70% ethanol at 100° and the eluate counted for both ^{14}C and ^{3}H by liquid scintillation spectrometry. Although some ^{14}C-containing materials appeared at the Rf of BA, when this fraction was co-crystallized and co-sublimed with authentic BA, the product contained essentially no ^{14}C but a substantial amount of tritium.

IV. SUMMARY

Approximately 20-25% of the cytokinin, benzyladenine, taken up by soybean tissues in culture is converted to a stable, long lived derivative which contains BA as part of its structure. This derivative may be metabolically related to BAMP. In very short term *in vivo* incubations of two hours or less, we recover only BA, benzyladenosine and BAMP. Benzyladenosine never accounts for more than 10% of the total radioactivity while BAMP builds up to about 20% of the total within 2-4 hours. After this period it begins to disappear and a new, unidentified substance arises at a rate which roughly parallels the loss of BAMP. After about 48 hours this substance, which has substantial cytokinin activity, accounts for some 20-25% of the total radioactivity and persists at this level for at least 60 days. In the meantime the remainder of the BA, as well as benzyladenosine and BAMP disappear completely. Evidence is also presented which suggests that the urea cytokinins are not active as such but only after metabolic transformation into other substances. In addition, it is shown that substituents at the 9 position probably do not block ribonucleotide synthesis in the purine cytokinins since they are metabolically unstable.

V. REFERENCES

CHEN, C., and R. HALL, (1969) Biosynthesis of N^6 (Δ^2-isopentenyl) adenosine in the transfer ribonucleic acid of cultured tobacco pith tissue. Phytochemistry. $\underline{8}$, 1687-1695.

DYSON, W.,J. FOX, and J MCCHESNEY. (1972) Short term metabolism of urea and purine cytokinins. Plant Physiology (in press).

FOX, J.E. (1964). Indoleacetic acid-kinetic antagonism in certain tissue culture systems. Plant Cell Physiol. $\underline{5}$, 251-254.

FOX, J.E. (1966) Incorporation of a kinin, N, 6-Benzyladenine into soluble RNA. Plant Physiol. $\underline{41}$, 75-82.

FOX, J.E. C. SOOD, B. BUCKWALTER, and J. MCCHESNEY. (1971). The metabolism and biological activity of a 9-substituted cytokinin. Plant Physiology $\underline{47}$, 275-281

KENDE, H and J.E. TAVARES. (1968). On the significance of cytokinin icnorporation into RNA. Plant Physiol. $\underline{43}$, 1244-1248.

KEY, J. (1969). Hormones and nucleic acid metabolism. Ann. Rev. Physiol. $\underline{20}$, 449-474.

LETHAM, D.S. (1969). Cytokinins and their relation to other phytohormones. Bioscience $\underline{19}$, 309-316.

MILLER, C. (1961). Kinetin and related compounds in plant growth. Ann. Rev. Plant Physiol. $\underline{12}$, 395-408.

SHAW, G., B.M. SMALLWOOD, and F.C. STEWARD. (1968). Synthesis and cytokinin activity of the 3-, 7- and 9-methyl derivatives of zeatin. Experientia $\underline{24}$, 1089-1090.

YOUNG, H and D. LETHAM. (1969). 6- substituted amino purines: synthesis by a new method and cytokinin activity. Phytochem. $\underline{8}$, 1199-1203.

Use of Structural Analogues in the Study of Cytokinin Action

Daphne C. Elliott, A.W. Murray, G.T. Saccone and M.R. Atkinson

School of Biological Sciences, Flinders University of South Australia, Bedford Park, S.A.

I. INTRODUCTION

The occurrence of cytokinins in certain tRNAs from a wide variety of organisms (reviewed by Skoog and Armstrong, 1970) has led to much speculation that these plant hormones may exert their physiological effects through regulation of protein biosynthesis, by modifying the synthesis and function of specific tRNAs (Anderson and Cherry, 1969; Gefter and Russell, 1969). For this to be the mode of action of cytokinins it would need to be shown both that exogenous cytokinins are the source of the cytokinin residues in tRNA and also that such incorporation results in a biochemical control mechanism for growth and morphogenesis.

The experiments reported here are designed to throw some light on whether incorporation into tRNA is related to the observed physiological effects of cytokinins on growth of soybean callus tissue and induction of betacyanin synthesis in *Amaranthus* seedlings.

Prior experimental evidence has been presented both for (Fox and Chen, 1967) and against (Kende and Tavares, 1968) incorporation into tRNA as a prerequisite for cytokinin action (see discussion in Skoog and Armstrong, 1970, and Galston and Davies, 1969). Arguments against incorporation have relied partly on the demonstration that the isopentenyl side-chain is derived from mevalonic acid (Chen and Hall, 1969) and that the mechanism involved is transfer from Δ^2-isopentenyl pyrophosphate to specific adenosine residues of preformed tRNA (Kline, Fittler and Hall, 1969). Further evidence comes from the high cytokinin activity of certain 9-substituted benzyladenines, 6-benzylamino-9-tetrahydropyranylpurine (van Overbeek, 1962) and 6-benzylamino-9-cyclohexylpurine (Young and Letham, 1969), although there has been no evidence that these 9-substituents are stable in biological systems. Kende and Tavares (1968) have shown that 9-methyl-6-benzyladenine has a high cytokinin activity in the soybean callus test, and that this cytokinin is not incorporated into tRNA. This result has been criticised (Skoog and Armstrong, 1970) since it was shown that the ^{14}C-methyl group in the 9-position was metabolised within a few minutes by soybean and tobacco tissues (Fox *et al.*, 1969). However, if, in fact, dealkylation had taken place in Kende and Tavares' experiment, it is difficult to see why no incorporation of the resulting benzyladenine (BA) into tRNA was observed, as had been shown by Fox and Chen (1967).

In attempts to resolve this question we have used 9-butyl-6-benzyladenine (9-BuBA), since studies with animal tissues have shown that whereas an ethyl group in the 9-position of 6-mercaptopurine and azathioprine is subject to dealkylation, a butyl group is not (Hansen *et al.*, 1963; Chalmers *et al.*, 1969).

II. MATERIALS AND METHODS

Bioassay systems

The procedure of Miller (1965) was used for testing the effect of cytokinins on growth of soybean callus tissue. For the induction of betacyanin synthesis in

Amaranthus tricolor seedlings the method of Bigot (1968) was followed except that the seeds were germinated at 35°. In both assays cytokinins were added prior to autoclaving. Autoclaving resulted in no dealkylation of 9-BuBA to BA within the limits of detection by chromatography; 2.5% conversion would have been detectable.

Enzyme assays

Adenine phosphoribosyltransferase was prepared and assayed as described by Nicholls and Murray (1968). In experiments to see if kinetin would form a mononucleotide in this system, adenine was replaced by kinetin-8-^{14}C (specific activity 780 cpm per nanomole). The kinetin-8-^{14}C as obtained from the Radiochemical Centre, Amersham, contained an impurity (25% of the labelled material) which ran in the same position as adenine in two chromatographic solvents. In *n*-butanol-acetic acid-water (20 : 3 : 7, v/v/v), R_f adenine is 0.54, R_f kinetin, 0.85. In methanol-water (50 : 50, v/v), R_f of adenine is 0.38, R_f kinetin, 0.51. The kinetin-8-^{14}C was separated from this impurity on methanol-washed Whatman 3 MM paper developed with 50% (v/v) methanol, and eluted from the paper with MeOH.

In phosphoribosyltransferase assays, incubations were carried out at 25°C for 15 min and the reactions stopped with 10 μl 10 N HCl. Aliquots (0.1 ml) were chromatographed on 3 MM paper with *n*-butanol-acetic acid-water (20 : 3 : 7, v/v/v) as solvent. In the case of adenine phosphoribosyltransferase an internal marker of 10 μl of 2×10^{-2} M AMP was run and radioactivity associated with the nucleotide was measured directly in a Mark I Nuclear Chicago liquid scintillation counter. In experiments with kinetin-8-^{14}C the chromatographic strips were cut into 0.5" segments and each segment counted.

Rabbit muscle (Boehringer and Soehne) and *E. coli* adenylate kinases were assayed as described by Murray and Atkinson (1968). A preparation from *E. coli* containing a mixture of ribo- and deoxyribonucleotide kinases was prepared as described by Hurlbert and Furlong (1967).

Synthesis of 9-butyl-6-benzyladenine and 6-benzyladenosine monophosphate

9-Butyl-6-benzyladenine was synthesised by the following method (Chalmers, 1970). 1.1 g 6-chloro-purine were dissolved in 40 ml dry dimethylsulphoxide, and 2.0 g anhydrous potassium carbonate and 1.2 g redistilled 1-iodobutane were added. This solution was stirred for 16 hr and then added to 500 ml water. The aqueous solution was extracted three times with ether. The ether extract was dried with anhydrous sodium sulphate and ether evaporated to give a yellow oil (80% yield). This oil was dissolved in 20 ml absolute ethanol and to this was added 1.3 g benzylamine (2 moles of benzylamine per mole 9-butyl-6-chloropurine, assuming all the oil to be the 9-butyl derivative). This solution was refluxed for 4 hr at 100°C. The resulting solution was allowed to stand overnight in an open beaker at RT. It was then rotary evaporated to dryness. The residue was triturated with water leaving an off-white residue of 9-butyl-6-benzyladenine. Yield, 85%. The product was recrystallised twice from methanol-water (1 : 1) and dried over phosphorous pentoxide for 5 hr at 70°/1 mm resulting in white needle-like crystals, m.p. 108-110° (decom.). The crystals gave the following analysis : C, 68.24; H, 6.69; N, 24.8; calculated for $C_{16}H_{19}N_5$: C, 68.5; H, 6.85; N, 25.1.

The compound had λ_{max} 272 nm (mε, 22.0) at pH 7.4 (0.05 M phosphate buffer, K^+); λ_{min} 238 nm (mε, 3.5). In 0.1 N KOH λ_{max} 272 nm (mε, 23.2), λ_{min} 236 nm (mε, 6.1). In 0.1 N HCl λ_{max} 269-270 nm (mε, 21.0), λ_{min} 233 nm (mε, 3.5).

The nuclear magnetic resonance spectrum gave a singlet at 1.8 τ (C_2-H proton on the purine ring), a two proton doublet at 5.1 τ (methylene protons next to the benzene ring, split by the adjacent NH group), a two proton triplet at 5.95 τ (methylene group of the butyl group adjacent to position 9 of the purine ring and split to a triplet by the adjoining methylene in the butyl group), a six proton multiplet at 2.5-3.0 τ (comprising the 5 aromatic benzene hydrogen atoms and the C_8-H proton of the purine ring), and a seven proton multiplet at 7.8-9.3 τ (the remaining 7 atoms of the butyl group, β to the 9 position of the purine ring). The addition of D$_2$O resulted in partial exchange with the hydrogen atom of the NH group and a conversion of the two proton doublet at 5.1 τ to a two proton singlet.

6-Benzyladenosine monophosphate was synthesised in about 58% yield by the reaction of 6-benzyladenosine with phosphoryl-chloride in triethylphosphate essentially as described for N^6-methoxyadenosine by Donaldson, Atkinson and Murray (1969). The product contained ribose and organic phosphate in the ratio 1.0 : 1.0, and contained no contaminants detectable under ultraviolet light after chromatography in n-butanol-acetic acid-water (20:3:7, v/v/v; R$_f$ 0.32) or electrophoresis in 0.05 M tris (citrate, pH 4.5; same ionic mobility as AMP). For calculation of concentrations the value mε$_{268}$ (H$_2$O, pH 6.0) 20.85, as published for 6-benzyladenosine (Kissman and Weiss, 1956) was used. The nucleotide analogue was a poor substrate for snake venom 5'-nucleotidase but was slowly converted into 6-benzyladenosine (detected by electrophoresis at pH 4.5) confirming that the phosphate was present in the 5'-position.

III. RESULTS

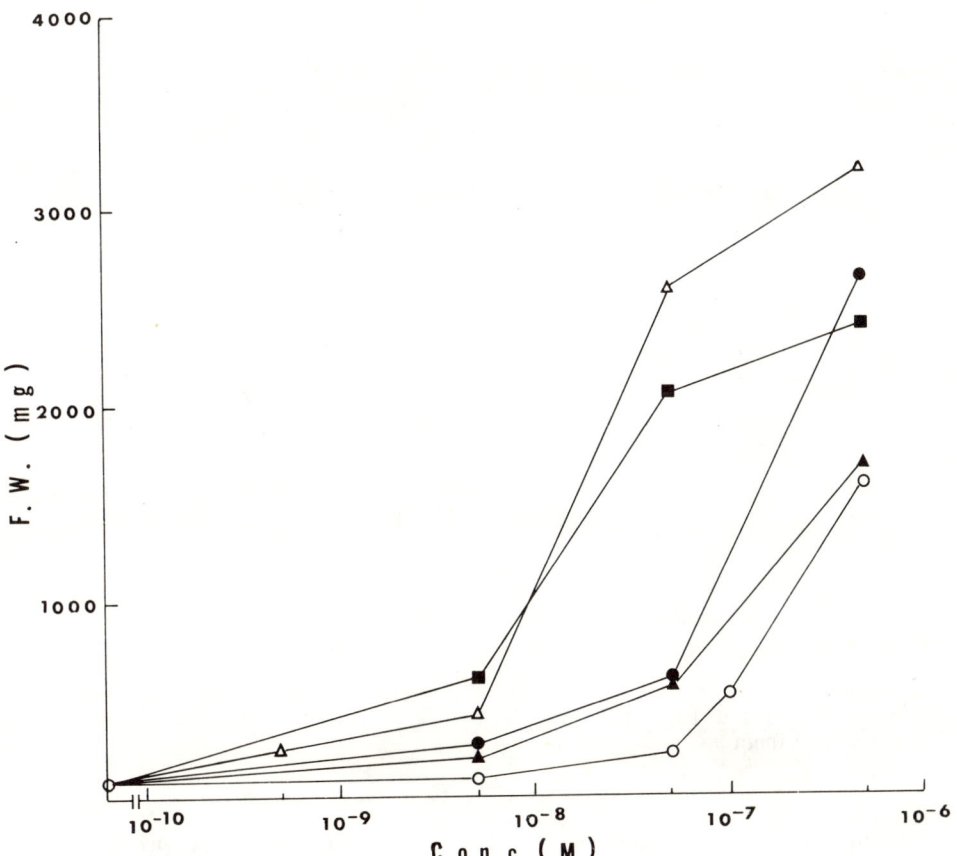

Fig. 1. Cytokinin activity in soybean callus tissue bioassay. O, kinetin; ▲, 9-methyl-benzyladenine; ●, benzyladenine; ■, benzyladenosine; △, 9-butyl-benzyladenine.

Activity of cytokinins in soybean callus tissue assay

As shown in Figure 1, 9-BuBA was a better promoter of callus tissue growth at low concentrations than BA. At concentrations above 5×10^{-7} M BA was a better promoter of growth than the 9-butyl analogue (not shown on Figure 1).

Activity of cytokinins as inducers of betacyanin synthesis

As with the callus growth assay, 9-BuBA stimulated betacyanin formation at low concentrations more than did BA (Figure 2). The values reported in Figure 2 are for the development of pigment after a 72 hr incubation. Time courses of pigment production were the same for both cytokinins. In both cases significant pigment production was observed between 8 and 12 hours.

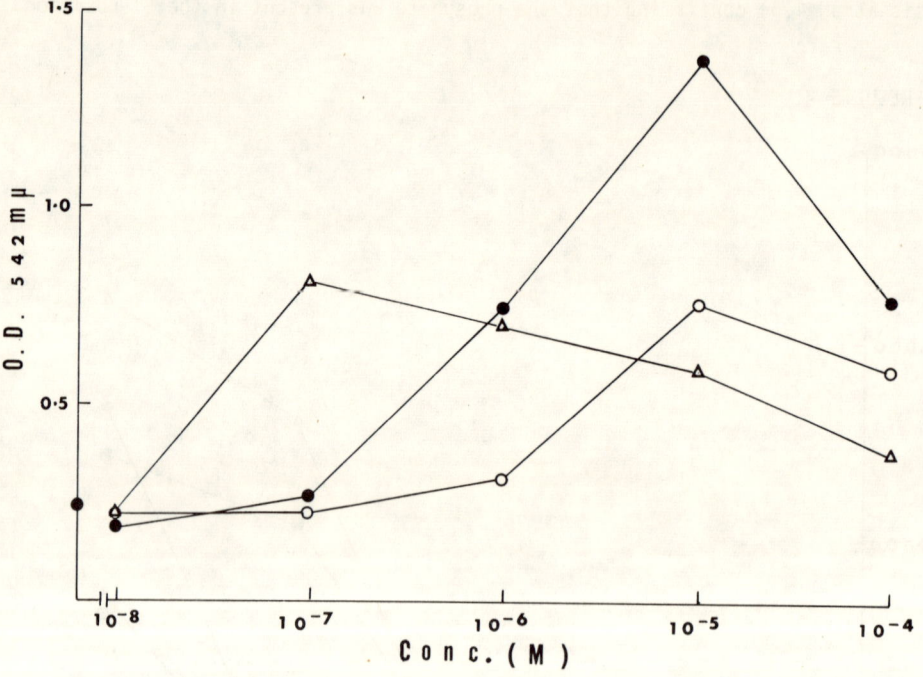

Fig. 2. Cytokinins as inducers of betacyanin synthesis in *Amaranthus* seedlings. ○, kinetin; ●, benzyladenine; △, 9-butyl-benzyladenine.

Seedlings grown on 9-BuBA had a markedly different appearance from those grown on BA. The seed coats did not fall off in the 72 hr assay period and the seedlings were longer and thinner.

Effect of adenine on biological activity of cytokinins

If cytokinins are incorporated into regulatory tRNA species by any of the mechanisms known for adenine incorporation (see Discussion) it might be expected that excess adenine would compete, and so decrease the cytokinin effects. The results in Table 1 show that adenine even in large excess (x 100) does not markedly decrease either the induction of betacyanin synthesis or the promotion of growth of soybean callus by BA. Preliminary results have indicated that excess adenine also does not inhibit the cytokinin activity of 9-BuBA (tested at 5×10^{-7} M).

Table 1. Effect of Adenine on the Biological Activity of Benzyladenine

Adenine conc.	Soybean callus growth (mg F.W.)		Betacyanin induction (O.D. 542 nm)	
	—	5×10^{-7} MBA	—	10^{-5} MBA
0	24	3404	.11	.83
5×10^{-7} M	19	3010	.09	-
5×10^{-6} M	22	2866	.09	-
5×10^{-5} M	343	3208	-	-
10^{-5} M	-	-	.12	.66
10^{-4} M	-	-	.11	.74
10^{-3} M	-	-	.19	.77

Effect of cytokinins on adenine phosphoribosyltransferase

As an extension to the above experiment the cytokinins were tested as inhibitors of soybean callus adenine phosphoribosyltransferase. If cytokinins will act as substrates for this enzyme one would expect reduction in the amount of AMP formed in the presence of unlabelled cytokinin. On the contrary in many experiments a stimulation of this enzyme was shown with kinetin, zeatin and BA. The stimulation varied between 8% for 10^{-4} M kinetin and 30% for 10^{-5} M kinetin. 10^{-4} M zeatin gave 50% stimulation and benzyladenine varied between no stimulation at 10^{-6} M to 20% at 10^{-4} M.

Search for a kinetin phosphoribosyltransferase

Although the experiments reported above make it unlikely that cytokinins are substrates for adenine phosphoribosyltransferase, the possibility remains that there is a specific enzyme for them or perhaps one with less specificity than adenine phosphoribosyltransferase, such as the hypoxanthine phosphoribosyltransferase. Kapoor and Waygood (1965) have shown that benzimidazole, which, like kinetin, delays senescence in detached leaves, is the substrate for an enzyme, benzimidazole nucleotide: pyrophosphate phosphoribosyltransferase, from wheat embryos. The enzyme was assayed indirectly by apparent inhibition of the analogous orotidine 5'-phosphate pyrophosphorylase. Benzimidazole nucleotide was identified as the product. Kinetin showed similar competitive behaviour, but no reaction product was identified. However, total scans of chromatograms from assays with soybean callus extract carried out with kinetin-8-^{14}C in the presence or absence of 5-phosphoribosyl pyrophosphate gave no indication of nucleotide formation.

Studies with 6-benzyladenosine monophosphate and adenylate kinases

With rabbit muscle adenylate kinase no activity was detectable when 0.27 mM 6-benzyladenosine monophosphate replaced AMP in the assay mixture. An activity of about 0.3% of that with AMP would have been detected. The nucleotide analogue was an inhibitor of the reaction of AMP with the muscle kinase. At an AMP concentration of 0.036 mM, 50% inhibition was obtained with 0.096 mM 6-benzyladenosine monophosphate.

No activity was measurable when the analogue was tested as a substrate for *E. coli* kinases. In this case the relatively crude preparation gave a small rate in the absence of added monophosphate, presumably due to ATPase action and this rate was not increased on addition of 6-benzyladenosine monophosphate. Although unlikely, the continued rate on addition of nucleotide could be a summation of a reduced ATPase activity plus a slow kinase activity. If addition of the analogue resulted in complete inhibition of the ATPase the maximum rate of kinase activity would be 4% of that with AMP.

IV. DISCUSSION

While the possibility of dealkylation of 9-BuBA cannot be ruled out there is circumstantial evidence that this does not occur. If the better biological response to 9-BuBA at low concentrations (Figures 1,2) were due to better penetration or transport, followed by conversion to BA one would expect (a) the time course of the betacyanin induction by 9-BuBA to be faster than by BA and (b) the maximum activity more nearly approaching that of BA. The different appearance of the seedlings grown on 9-BuBA is significant and has also been noted in bioassays using *Spirodela oligorhiza* (Elliott, 1969). Both BA and 9-methyl-6-benzyladenine produce large epinastic fronds in this plant, while those plants grown on 9-BuBA more closely resemble controls on water or on N-6-isopentenyladenine. One would expect dealkylation of 9-BuBA to produce epinastic fronds.

If 9-BuBA is dealkylated there still remains the question of whether cytokinins can be converted to the corresponding nucleoside triphosphates and then incorporated into tRNA. Nucleotide biosynthesis from purines can occur by two pathways (Murray, Elliott and Atkinson, 1970; Murray, 1971):

(A) Purine + ribose-1-P $\rightleftharpoons$ nucleoside + P_i;
   nucleoside phosphorylase

   Nucleoside + ATP $\rightarrow$ nucleotide + ADP
   nucleoside kinase

(B) Purine + 5-PRPP $\rightarrow$ nucleotide + P - P
   purine phosphoribosyltransferase

A very active adenine phosphoribosyltransferase has been demonstrated in soybean callus extracts by Nicholls and Murray (1968). We have now shown (Table 1) that it is unlikely that cytokinins are converted to mononucleotides by this enzyme, or by adenosine phosphorylase. Nor have we been able to demonstrate a phosphoribosyl pyrophosphate-dependent nucleotide formation from kinetin by any other phosphoribosyltransferase.

However the possibility remains that cytokinins are converted to mononucleotides by pathway A using a specific cytokinin nucleoside phosphorylase and kinase, since McCalla, Morré and Osborne (1962) and Guern (1968) have shown the formation of benzyladenosine from labelled benzyladenine, and have also demonstrated traces of a compound which gave rise to benzyladenosine on treatment with a phosphomonoesterase, presumably 6-benzyladenosine monophosphate.

Given that cytokinin mononucleotides are formed by some such pathway a further difficulty in visualising incorporation into tRNA is the failure to demonstrate a kinase that will phosphorylate N^6-benzyl AMP to N^6-benzyl ADP. Thus one of the major problems in postulating a role of cytokinins after incorporation into tRNA is the present inability to demonstrate the appropriate enzyme pathways for this incorporation. So far as we are aware there has been no unequivocal demonstration of the incorporation of intact cytokinin molecules into tRNA and certainly no demonstration that the presence of cytokinins in tRNA can lead to an effect equatable with the observed physiological effects of cytokinins. We feel that the question of incorporation or no incorporation is quite open and that there is a strong possibility that the effects of cytokinins are mediated at the free base, nucleoside or nucleotide level.

V. SUMMARY

At very low concentrations, 9-butyl-6-benzyladenine, which cannot be converted into nucleoside derivatives without dealkylation, is more effective than 6-benzyladenine in promoting growth of soybean callus tissue and in the induction of betacyanin synthesis in *Amaranthus* seedlings.

Excess adenine (100 x) does not inhibit either promotion of soybean callus growth or induction of betacyanin synthesis by either benzyladenine or 9-butyl-6-benzyladenine, so this would seem to exclude incorporation into t-RNA by any mechanisms which are known to apply to adenine incorporation.

6-Benzyladenosine monophosphate is not a substrate for the animal and bacterial mononucleotide kinases we have tested.

These experiments do not exclude the possibility of specific enzyme systems for incorporation of cytokinins into t-RNA, nor direct N^6-side-chain transfer to adenosine residues in t-RNA.

VI. ACKNOWLEDGEMENTS

The authors are indebted to A.H. Chalmers for the preparation of the analytical sample of 9-butyl-benzyl-adenine, and for details of the nuclear magnetic resonance spectrum. We should also like to thank Dr. K.G.M. Skene for the original soybean callus line. The financial support of The Reserve Bank of Australia is gratefully acknowledged.

VII. REFERENCES

ANDERSON, M.B. and J.H. CHERRY (1969). Differences in leucyl-transfer RNAs and synthetase in soybean seedlings. Proc. Nat. Acad. Sci. U.S. 62, 202-209.

BIGOT, C. (1968). Action d'adénines substituées sur la synthèse des bétacyanines dans la plantule d'*Amaranthus caudatus* L. Possibilité d'un test biologique de dosage des cytokinines. C.R. Acad. Sci. Paris, D. 266, 349-352.

CHALMERS, A.H., P.R. KNIGHT and M.R. ATKINSON (1969). 6-Thiopurines as substrates and inhibitors of purine oxidases: a pathway for conversion of azathioprine into 6-thiouric acid without release of 6-mercaptopurine. Aust. J. exp. Biol. med. Sci. 47, 263-273.

CHALMERS, A.H. (1970).Ph.D. Thesis. Flinders University of South Australia.

CHEN, C.M. and R.H. HALL (1969). Biosynthesis of N^6-(Δ^2-isopentenyl) adenosine in the transfer ribonucleic acid of cultured tobacco pith tissue. Phytochemistry 8, 1687-1695.

DONALDSON, G., M.R. ATKINSON and A.W. MURRAY (1969). Synthesis and regulatory properties of an adenosine 5'-phosphate analogue, N^6-methoxyadenosine 5'-phosphate. Biochim. Biophys. Acta 184, 655-657.

ELLIOTT, D.C. (1969). Unpublished results.

FOX, J.E. and C.M. CHEN (1967). Characterisation of labelled ribonucleic acid from tissue grown on ^{14}C-containing cytokinins. J. Biol. Chem. 242, 4490-4494.

FOX, J.E., C. SOOD and J.D. McCHESNEY (1969). Metabolism of 9-substituted cytokinins. Plant Physiol., Wash. 44, S-2.

GALSTON, A.W. and P.J. DAVIES (1969). Hormonal regulation in higher plants. Science 163, 1288-1297.

GEFTER, M.L. and R.L. RUSSELL (1969). Role of modifications in tyrosine transfer RNA: a modified base affecting ribosome binding. J. Mol. Biol. 39, 145-157.

GUERN, J., M. DOREE and P. SADORGE (1968). Metabolism, transport and biological activity of some cytokinins, in "Biochemistry and Physiology of Plant Growth Substances", p. 1155. Runge Press, Ottawa.

HANSEN, H.J., W.G. GILES and S.B. NADLER (1963). Metabolism of 9-ethyl-6-mercaptopurine-^{35}S and 9-butyl-6-mercaptopurine-^{35}S in humans. Proc. Soc. Exptl. Biol. Med. 113, 163-165.

HURLBERT, R.B. and N.B. FURLONG (1967). Biosynthetic preparation of ^{32}P-labelled nucleoside 5'-phosphates and derivatives. Meth. Enzymol. XII A, 193-202.

KAPOOR, M. and E.R. WAYGOOD (1965). Metabolism of benzimidazole in wheat. 1. Formation of benzimidazole nucleotide. Can. J. Biochem. 43, 153-163.

KENDE, H. and J.E. TAVARES (1968). On the significance of cytokinin incorporation into RNA. Plant Physiol., Wash. 43, 1244-1248.

KISSMAN, H.M. and M.J. WEISS (1956). Kinetin riboside and related nucleosides. J. Org. Chem. 21, 1053-1055.

KLINE, L.K., F. FITTLER and R.H. HALL (1969). N^6-(Δ^2-isopentenyl)adenosine biosynthesis in transfer RNA in vitro. Biochemistry 8, 4361-4371.

McCALLA, D.R., D.J. MORRÉ and D.J. OSBORNE (1962). The metabolism of a kinin, benzyladenine. Biochim. Biophys. Acta 55, 522-528.

MILLER, C.O. (1965). Evidence for the natural occurrence of zeatin and derivatives: compounds from maize which promote cell division. Proc. Nat. Acad. Sci. 54, 1052-1058.

MURRAY, A.W. and M.R. ATKINSON (1968). Adenosine 5'-phosphorothioate. A nucleotide analog that is a substrate, competitive inhibitor, or regulator of some enzymes that interact with adenosine 5'-phosphate. Biochemistry 7, 4023-4029.

MURRAY, A.W., D.C. ELLIOTT and M.R. ATKINSON (1970). Nucleotide biosynthesis from preformed purines in mammalian cells: regulatory mechanisms and biological significance. Prog. in Nucleic Acid Res. and Molec. Biol. 10, 87-119.

MURRAY, A.W. (1971). The biological significance of purine salvage. Ann. Rev. Biochem. In press.

NICHOLLS, P.B. and A.W. MURRAY (1968). Adenine phosphoribosyltransferase in plant tissues: some effects of kinetin on enzymic activity. Plant Physiol., Wash. 43, 645-648.

SKOOG, F. and D.J. ARMSTRONG (1970). Cytokinins. Ann. Rev. Plant Physiol. 21, 359-384.

van OVERBEEK, J. (1962). Proc. Plant Sci. Symp., p. 37. Camden, New Jersey.

YOUNG, H. and D.S. LETHAM (1969). 6-[Substituted amino] purines: synthesis by a new method and cytokinin activity. Phytochemistry 8, 1199-1203.

UPTAKE OF CYTOKININS BY *Acer pseudoplatanus* CELLS : ENZYMES OF THE ADENOSINE DEAMINASE TYPE AS POSSIBLE REGULATORS OF THE CYTOKININ LEVEL INSIDE THE CELL

C. Terrine, M. Doree, J. Guern[1] and R.H. Hall[2]

(1) Laboratoire de Physiologie végétale, 1 rue Victor Cousin, PARIS, 5e, FRANCE

(2) Department of Biochemistry, McMaster University, Hamilton, Ontario, Canada

I. INTRODUCTION

The biological activity of an exogenously supplied cytokinin is dependent upon several factors amongst which those controlling the intracellular concentration of biologically active molecules are expected to be very important. Those which control *absorption* from the medium and the *degradation* of the absorbed molecules are two of these factors.

As it is a well known fact that the biological activity of cytokinins is dependent upon their chemical structure (Strong, 1958; Kuraishi, 1969; Letham, 1967a), two questions are raised: 1) how does absorption and degradation of cytokinins depend upon chemical structure; 2) what are the relationships between these two factors and biological activity?

It has been shown (Fox, 1966; Srivastava, 1967b) that cytokinins-8-^{14}C supplied to plant tissues are extensively degraded. The finding of labelled adenine and guanine residues in the nucleic acids of tissues incubated with 6-furfurylaminopurine-8-^{14}C (Srivastava, 1967b) or 6-benzylaminopurine-8-^{14}C (Fox, 1966) implies that these cytokinins undergo removal of the substituent group. Several enzymes are known to be able to catalyse the oxidative removal of an $-NH_2$ or $-NHR$ group from the adenine nucleus; adenosine aminohydrolase (E.C.3.5.4.4.) acting on nucleosides as substrates, is the best known. The substrate specificity of this enzyme is rather broad and it converts a number of N^6-substituted analogues of adenosine (including several cytokinin nucleosides) to inosine (Terrine et al., 1969; Hall et al., 1971).

The enzyme has been used in the present study as an *in vitro* model to compare the rates of degradation of various cytokinin nucleosides and to obtain quantitative data on the influence of the chemical structure of the N^6-substituent group on the rate of breakdown of cytokinins.

Turning to the problem of cytokinin absorption it is astonishing that only a very small number of studies have been devoted to the problem of purine absorption by higher plant cells. The main kinetic characteristics of adenine absorption by *Acer pseudoplatanus* cells have been described recently (Doree et al., 1970 a, b, c, d.). Leguay et al. (1970) have also reported that kinetin is absorbed by these cells at a low rate compared to that of adenine.

Using the same methods we have compared the rates of absorption of various cytokinins with the purpose of obtaining quantitative data on the relation between chemical structure and rate of absorption of cytokinins by *Acer pseudoplatanus* cells.

II. MATERIALS AND METHODS

Different substrates of adenosine aminohydrolase were obtained from various sources (see Terrine *et al.*, 1969; Hall *et al.*, 1971).

Adenosine aminohydrolase from calf intestinal mucosa (spec. act. 200 UI/mg) was purchased from Boehringer and the isolation procedure of the enzyme from chicken bone marrow has already been described (Hall *et al.*, 1971).

The techniques used to study the action of adenosine aminohydrolase from calf intestinal mucosa and chicken bone marrow enzyme have been described (Terrine *et al.*, 1969; Hall *et al.*, 1971).

The 6-furfurylaminopurine-8-^{14}C (spec. act. 16.5 mCi/mMole) was purchased from the Radiochemical Center, Amersham. The 6-propylaminopurine-8-^{14}C and 6-benzylaminopurine-8-^{14}C were synthesized from 6-chloropurine-8-^{14}C (spec. act. 11.2 mCi/mMole - Radiochemical Center) and the corresponding amines according to the procedure of Bullock *et al.* (1956). These substances were purified from any contamination of traces of 6-chloropurine or adenine by paper chromatography.

Acer pseudoplatanus cells were cultivated in liquid medium according to the procedures already reported (Doree *et al.*, 1970 b, c). Cells from the logarithmic growth phase were used to study the kinetics of cytokinin absorption. The labelled cytokinins are added at zero time to a cell suspension (800,000 cells/ml) previously equilibrated at constant temperature (25°C) on a reciprocating shaker. From time to time, 2 ml aliquots are removed and injected into 5 ml cold water contained in a special Millipore filter assembly (Doree *et al.*, 1970c). The mixture is then filtered through a Millipore membrane (porosity : 3 μ), injection and filtration being completed within 20 seconds. The cells are then removed from the Millipore membrane with a small spatula and either processed directly for the measurement, by scintillation spectrometry, of total radioactivity absorbed, or extracted with cold 0.5 M perchloric acid for biochemical analysis.

III. RESULTS

In vitro degradation of cytokinin nucleosides catalyzed by adenosine aminohydrolase

Adenosine aminohydrolase from calf intestinal mucosa catalyses the complete conversion to inosine of various cytokinin nucleosides such as N^6-furfuryladenosine (Terrine *et al.*, 1969) and N^6-Δ^2-isopentenyladenosine (Hall *et al.*, 1971). From kinetic studies of this degradation it is possible to calculate K_m and V_{max} values for different N^6-substituted adenosines used as substrates by adenosine aminohydrolase (table 1). With the exception of N^6 methyladenosine, the influence of the size of the N^6 substituent group on the K_m values is not very great. On the contrary, the V_{max} of degradation seems to be much more affected by the structure of the N^6 substituent group : as the size of this group increases, the rate of degradation is strongly reduced, as shown by the curves (fig. 1) calculated from the kinetic parameters of table 1. Table 1 shows that there is a reciprocal relation between biological activity and degradation rate : the most active compounds are degraded most slowly.

Table 1. Kinetic parameters of various N^6-substituted adenosines as substrates of calf intestinal mucosa enzyme.

N^6-substituted adenosines	Km	Vm
6-methyl	$1.1 \cdot 10^{-5}$M	$3.2 \cdot 10^{-6}$M
6-ethyl	$1.2 \cdot 10^{-4}$M	$2.8 \cdot 10^{-6}$M
6-propyl	$2.2 \cdot 10^{-4}$M	$1.8 \cdot 10^{-6}$M
6-butyl	$2.8 \cdot 10^{-4}$M	$2.0 \cdot 10^{-6}$M
6-furfuryl	$3.2 \cdot 10^{-4}$M	$0.8 \cdot 10^{-6}$M
6-benzyl	$2.9 \cdot 10^{-4}$M	$0.09 \cdot 10^{-6}$M

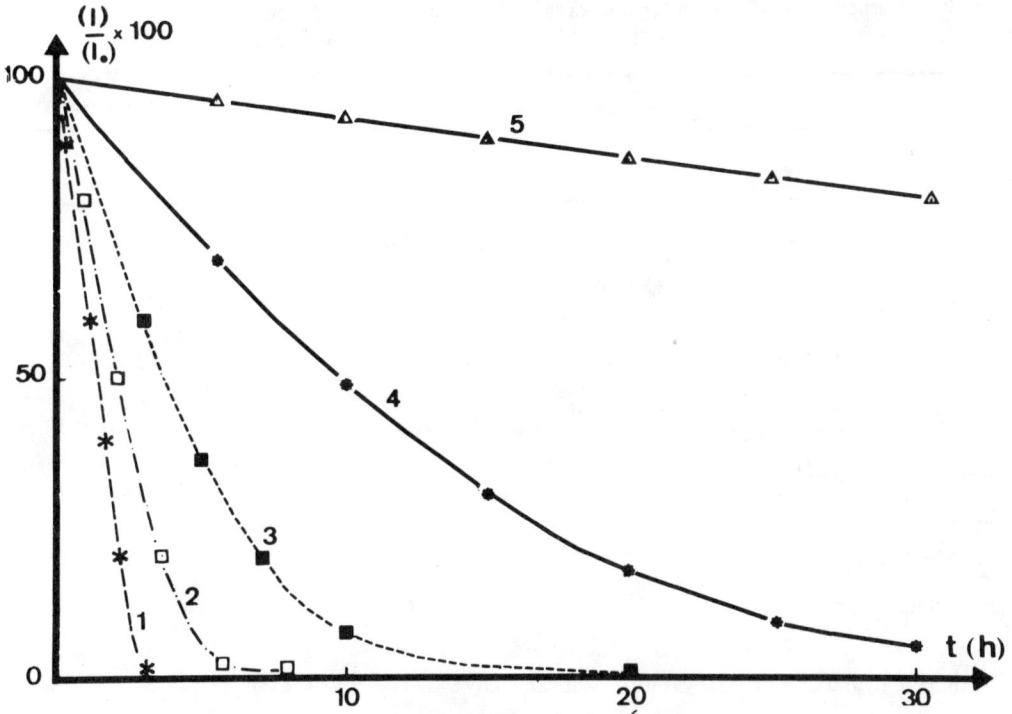

Fig. 1. Time course of the degradation of various N^6 substituted adenosines catalyzed by calf intestinal mucosa adenosine aminohydrolase. The concentration (I) for each substrate is calculated from the relation :

$$t = \frac{[(I_o) - (I)] + 2.3 \, K_m \log \frac{(I_o)}{(I)}}{V_m}$$

I_o is arbitrarily chosen equal to 5.10^{-4} M. K_m and V_{max} values for each substrate are those of table 1. The substrates studied are the 6-methyl (1), 6-ethyl (2), 6-butyl (3), 6-furfuryl (4) and 6-benzyladenosines (5).

With respect to the interest attached to natural cytokinins such as isopentenyladenine, zeatin and their derivatives, a special study of the degradation of these compounds was made (Hall et al., 1971). Two enzyme systems were used, as described in "Methods". There are several differences in the substrate specificity of these two enzymes. The main difference is that the rate of degradation of adenosine is 40 times higher than that of N^6-Δ^2-isopentenyladenosine for the chicken bone marrow enzyme; for the calf intestinal mucosa enzyme, these rates of reaction differ by a factor of 3000. Degradation velocities of various nucleosides related to N^6-Δ^2-isopentenyladenosine are tabulated in table 2. The results show that the sensitivities of the two enzymes to the structural modifications of the isopentenyl side chain are quite similar. The hydroxylation of the side chain results in an important reduction of the degradation rate. It is interesting to note that if one assumes that enzymes of the adenosine deaminase type catalyse the *in vivo* degradation of cytokinins in plant tissues and as ribosyl zeatin occurs in the free state in several such tissues, it is likely that the hydroxylation of N^6- (Δ^2-isopentenyl) adenosine would protect it from degradation.

Table 2. Comparison of Degradation Rates of N^6-substituted adenosines catalyzed by chicken bone marrow and calf intestinal mucosa adenosine aminohydrolases

N^6-substituted adenosines	Relative velocities	
	Bone Marrow enzyme	Calf Intestinal Mucosa enzyme
$-CH_2-CH=C\begin{smallmatrix}CH_3\\CH_3\end{smallmatrix}$ $N^6-\Delta^2$ - isopentenyl	1.0	1.0
$-CH_2-CH_2-C{=}\begin{smallmatrix}CH_2\\CH_3\end{smallmatrix}$ $N^6-\Delta^3$ - isopentenyl	0.8	0.7
$-CH_2-CH_2-C\begin{smallmatrix}CH_3\\CH_3\end{smallmatrix}$ N^6- isopentyl	1.0	0.4
$-CH_2-CH_2=C\begin{smallmatrix}CH_3\\CH_2OH\end{smallmatrix}$ N^6- 4 hydroxy - 3 methylbut - 2 *trans* enyl	0.0	0.1
$-CH_2-CH_2-\underset{OH}{C}\begin{smallmatrix}CH_3\\CH_3\end{smallmatrix}$ N^6- 3 hydroxy - 3 methylbutyl	0.0	0.1
$-CH_2-\underset{OH}{CH}-\underset{OH}{C}\begin{smallmatrix}CH_3\\CH_3\end{smallmatrix}$ N^6- 2,3 dihydroxyisopentyl	—	0.0

Absorption of some cytokinins by Acer pseudoplatanus cells

Three N^6-substituted adenines, namely N^6-propyl, N^6-furfuryl and N^6-benzyladenine-8-^{14}C with different sizes of the N^6-substituent group and different biological activities were studied.

Fig. 2 shows the time course of absorption of these cytokinins by *Acer pseudoplatanus* cells. It is possible to recognise two phases in the absorption curves : a first transient one with a decreasing rate of absorption and a second one during which the rate of absorption is constant over a long period of time. This constant rate of absorption is easily measured and it is possible to compare the rates for different cytokinins at a concentration of $5.10^{-7}M$. The decreasing order of absorption rates is propyladenine > furfuryladenine > benzyladenine (fig. 2).

The relation between rates of absorption and cytokinin concentration in the medium was studied for each of the three adenine analogues. The dual reciprocal plots give linear relations which allow the calculation of maximum velocities (*Vmax*) of absorption and apparent affinity constants (*Km*) of the absorption system for the different cytokinins used.

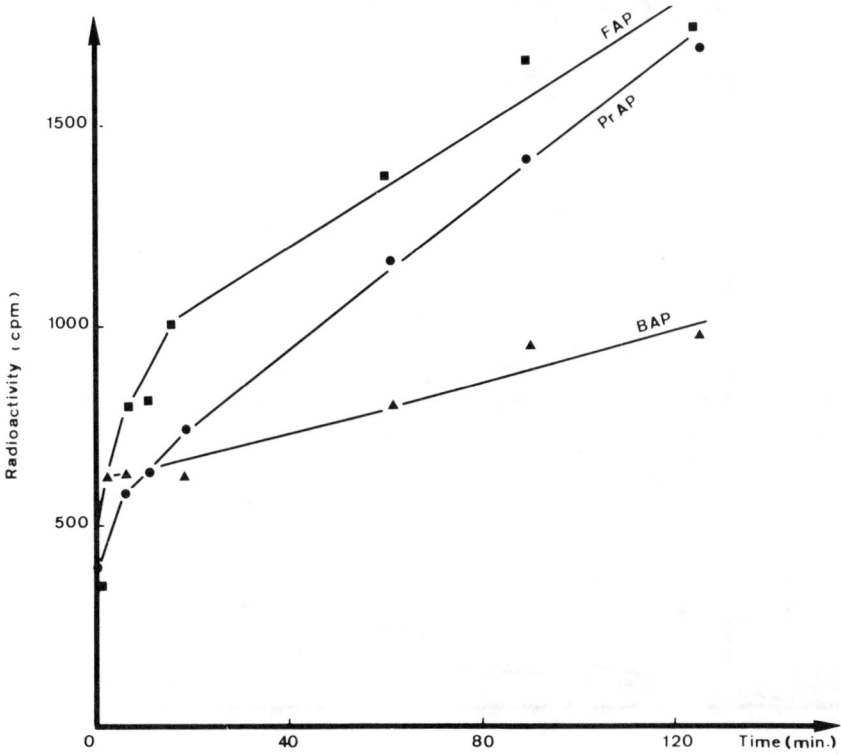

Fig. 2. Kinetics of absorption of various cytokinins by *Acer pseudoplatanus* cells. The radioactivity absorbed by the cells from an aliquot fraction (2 ml) of the suspension (cell density : 800 000 cells/ml) is plotted against time. The concentration of 6-benzylaminopurine (BAP), 6-furfurylaminopurine (FAP) and 6-propylaminopurine (PrAP) is $5.10^{-7}M$.

The results tabulated in table 3 show that the chemical structure of the N^6-substituent group exerts a profound influence not only on the maximum absorption rates of cytokinins but also on the apparent affinity constants for the transport system.

Table 3. Absorption of some cytokinins by *Acer pseudoplatanus* cells - kinetic parameters.

Cytokinins	Km	Vmax (pMoles/min./10^6 cells)
6-propyladenine	$1.2.10^{-6}M$	2.2
6-furfuryladenine	$1.8.10^{-6}M$	1.7
6-benzyladenine	$6.3.10^{-6}M$	0.9
adenine	$2.10^{-7}M$	45

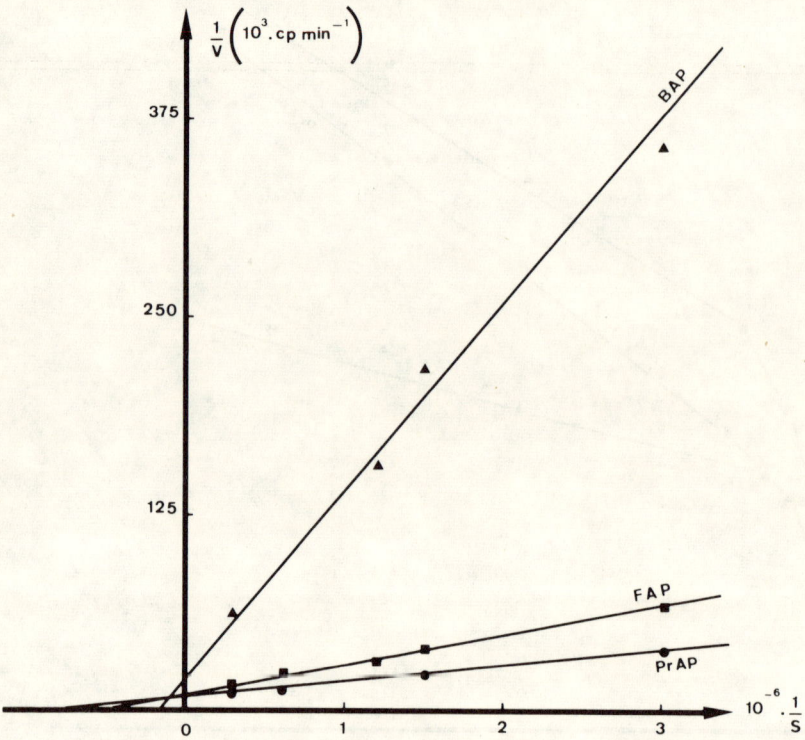

Fig. 3. Double reciprocal plots of the relationship of rate of absorption to concentration for various cytokinins. For experimental procedures, see "Materials and methods". (S = molar concentration)

IV. DISCUSSION

The absorption of cytokinins is a complex phenomenon as 6-furfurylaminopurine (fig. 4) and the other cytokinins used are subjected to an intensive metabolisation linked to their absorption. It can be seen from fig. 4 that free kinetin is not largely accumulated by the cells. The main metabolic fraction is that corresponding to nucleotides; this fraction contains respectively 50 per cent and 75 per cent of the total radioactivity absorbed after 2 hrs and 5 hrs of incubation. So, the absorption measured is the complex result of different events such as crossing the wall, plasmalemma and cytoplasm of the cell and the formation of different metabolites. These metabolic transformations of the absorbed cytokinins prevent the cell from accumulating free cytokinin molecules and as they provide a mean of maintaining a continuous cytokinin absorption, the intensity of these transformations may limit the rate of absorption. This situation is, with respect to several points, comparable to that of adenine absorption by the same cells. We have shown that adenine is absorbed without any intracellular accumulation but is intensively transformed to adenylic nucleotides. The activity of the enzyme system catalyzing the first step of this transformation (i.e. AMP synthesis) seems to be the limiting factor of the absorption rate for low concentrations of adenine in the medium (Doree *et al.*, 1970 c, d,).

If the *order of degradation rates* in the *in vitro* system is compared to the *order of absorption rates* by plant cells, it can be seen that the replacement of an N^6 propyl group by an N^6 furfuryl or N^6 benzyl group results in a decrease in both rates. The question arises, are these two processes related or not?

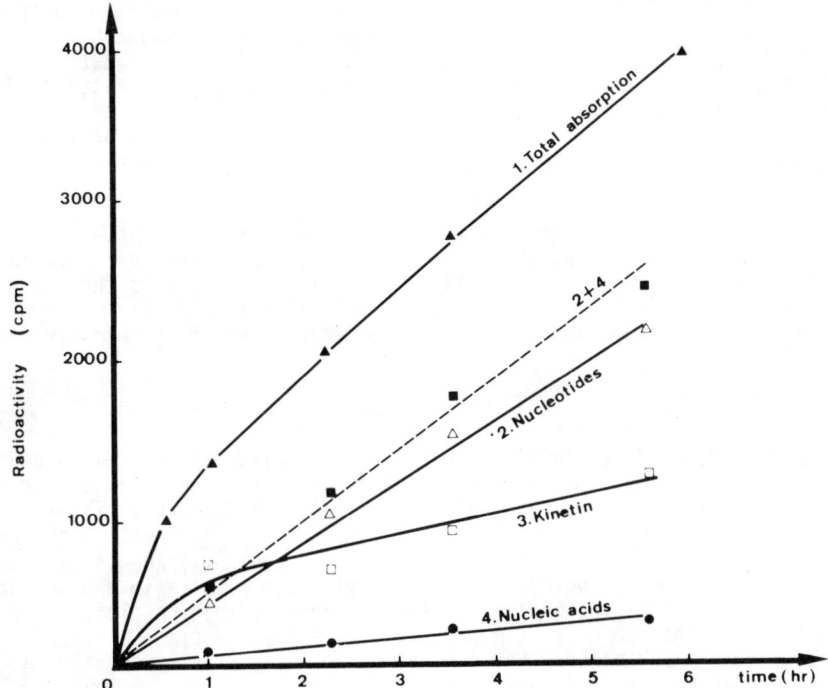

Fig. 4. Kinetics of absorption and metabolism of kinetin by *Acer pseudoplatanus* cells. Total radioactivity absorbed by the cells from an aliquot (2 ml) of the suspension (cell density : 800 000 cells/ml - kinetin concentration : $5.10^{-7}M$) is plotted against time (curve 1). The radioactivity in nucleotides (curve 2) is measured after an electrophoretic separation (Tris-citrate buffer, 0.02 M - pH 3.8 - 3 000 volts) of acid-soluble metabolites extracted by cold perchloric acid 0.5 M, $0°C$ (curve 2). The amounts of free kinetin (with a small amount of nucleoside) are measured after thin-layer chromatography (PEI - cellulose; solvent : Isopropanol:water, 2:1 v/v) of the perchloric acid extracts (curve 3). Nucleic acids are hydrolysed with hot perchloric acid (0.5 N - $70°C$ - 30 min.) after extraction from the cells, 3 times with cold acid (curve 4).

If degradation and absorption are independent processes and if the *in vitro* system has nothing to do with the absorption and metabolisation of cytokinins by plant cells, the fact that rates of both processes are reduced to the same extent by structural modifications of the molecule could be explained if mammalian adenosine aminohydrolase, and certain proteins involved in cytokinin absorption have rather similar substrate specificity so that they are affected to the same extent by the replacement of the N^6 propyl group of the substrate by a benzyl group, for example.

On the other hand, if one supposes that the two processes are linked, the question arises how closely are they linked? This question raises in turn two other questions: are the absorbed cytokinin molecules degraded *in vivo* and if so, is it by an enzyme of the adenosine deaminase type? Analysis of the resultant nucleotide fraction shows that a part of it corresponds to adenylic, guanilic and inosinic nucleotides. These results, and especially the finding of guanilic and inosinic acids, are in good agreement with data reported by McCalla *et al*. (1962) and the incorporation of guanine and adenine residues into nucleic acids. It can be concluded that the appearance of degradation products following removal of the side chain of cytokinins or cytokinin derivatives (nucleosides or nucleotides) is linked to the absorption of cytokinins.

As far as it is possible to conclude at present, judging by the relative proportions of the nucleotides arising from degradation, it appears that IMP might be the first degradation product or closely related to the first one. As the degradation products represent the major metabolic products and if it is assumed that the rate of the reaction catalysed by the degradative enzyme might be the limiting factor of the process of absorption, then it appears that the substrate specificity of the overall process of absorption might be closely related to the substrate specificity of the degradative enzyme.

As for the significance of the absorption and degradation processes to the biological activity of cytokinins, if it is assumed that the intensity of degradation (of cytokinin molecules or biologically active metabolites) represents the limiting factor of cytokinin absorption it would also control the intracellular steady state level of cytokinin molecules or biologically active metabolites of cytokinin.

V. SUMMARY

An *in vitro* model of cytokinin degradation was studied with purified adenosine aminohydrolases using various cytokinin nucleosides as substrates. The substrate specificity of chicken bone marrow and calf intestinal mucosa enzymes is markedly influenced by the structure of the N^6 substituent group. The degradation of various N^6 substituted adenosines, closely related to N^6-(Δ^2-isopentenyl) adenosine, was studied in detail : hydroxylation of the isopentenyl group results in a great decrease of the rate of degradation. The biological significance of this result is discussed with respect to the existence of zeatin and its derivatives in plant tissues. A reciprocal relation was established between the biological activity and the rate of degradation of the adenosine analogues studied.

The kinetic parameters of absorption of N^6-propyl, N^6-furfuryl and N^6-benzyl-adenine by *Acer pseudoplatanus* cells were measured. The cytokinin molecules absorbed are intensively degraded. As the degradative process might represent the limiting factor of the overall rate of absorption, the hypothesis of the regulation of the cytokinin pool inside the cell (cytokinin molecules or biologically active cytokinin metabolic products) by the degradative enzyme is discussed.

VI. REFERENCES

BULLOCK, M.N., J.J. HAND and E.L.R. STOKSTAD (1956). Synthesis of 6-substituted purines. J. Am. Chem. Soc. 78, 3693-3696.

DOREE, M., J.J. LEGUAY and C. TERRINE (1970a). Absorption de l'adénine par des suspensions cellulaires d'*Acer pseudoplatanus*. C.R. Acad. Sci., 270, 2292-2295.

DOREE, M., J.J. LEGUAY, P. SADORGE, C. TERRINE and J. GUERN (1970b). Utilisation d'adénine exogène par les tissus végétaux. II - Rôle de l'adénine pyrophosphoribosyltransférase. Physiol. vég., 8 (3), 515-528.

DOREE, M., J.J. LEGUAY, C. TERRINE, F. TRAPY and J. GUERN (1970c). Adaptation à l'adénine des cellules d'*Acer pseudoplatanus*. Modalités d'utilisation de l'adénine exogène, in II^d International Congress on Plant Tissues Culture, Strasbourg 1970, CNRS ed., Paris, 1971, 345-365.

DOREE, M., J.J. LEGUAY and C. TERRINE (1970d). Existence d'un double système de transport responsable de l'absorption d'adénine par des suspensions cellulaires d'*Acer pseudoplatanus*. C.R. Acad. Sci., 271, 1876-1879.

FOX, J.E. (1966). Incorporation of a kinin N^6-benzyladenine into soluble RNA. Plant Physiol., 41, 75-82.

HALL, R.H., S.N. ALAM, B.D. McLENNAN, C. TERRINE and J. GUERN (1971). N^6-(Δ^2-isopentenyl)adenosine, its conversion to inosine catalyzed by enzymes from Chicken bone marrow and Calf intestinal mucosa. Can. J. Biochem., 49 (6), 623-630.

KURAISHI, S. (1959). Effect of kinetin analogs on leaf growth. Sci. Papers coll. Gen. Educ., Univ. Tokyo, 9, 67-104.

LEGUAY, J.J., M. DOREE and C. TERRINE (1970). Absorption de kinétine par des suspensions cellulaires d'*Acer pseudoplatanus*. Inhibition de l'absorption d'adénine par la kinétine. C.R. Acad. Sci., 270, 3059-3062.
LETHAM, D.S. (1967). Chemistry and physiology of kinetin-like compounds. Ann. Rev. Plant Physiol., 18, 349-364.
McCALLA, D.R., D.J. MORRE and D.J. OSBORNE (1962). The metabolism of a kinin, benzyladenine. Biochim. Biophys. Acta, 55, 522-528.
SRIVASTAVA, S.B.I. (1967a). Effect of kinetin on biochemical changes in excised barley leaves and in tobacco pith tissue culture. Ann. New York Acad. Sci., 144, 260-278.
SRIVASTAVA, S.B.I. (1967b). Mechanism of action of kinetin in the retardation of senescence in excised leaves. in "Biochemistry and Physiology of Plant Growth Substances". Proceedings of the 6th Intern. Conf. on Plant Growth Substances. Runge Press. Ottawa. 1967.
STRONG, F.M. (1958). Kinetin and kinins. in "Topics in Microbial Chemistry". (Ed. Wiley and Sons) Chap. 3, 98-157. New York, 1958.
TERRINE, C., P. SADORGE, M. GAWER and J. GUERN (1969). Etude de la dégradation par voie enzymatique des analogues N^6-substitués de l'adénosine. I. Action de l'adénosine aminohydrolase *in vitro*. Physiol. vég., 7 (4), 425-435.

PLANT GROWTH SUBSTANCES, 1970

Cytokinins in Bleeding Sap of the Grape Vine

K.G.M. Skene

CSIRO Division of Horticultural Research, GPO, Box 350, Adelaide, South Australia 5001

I. INTRODUCTION

Following the conclusive demonstration that bleeding sap, or xylem exudate, contains cytokinins (Loeffler and van Overbeek, 1964; Kende, 1964,1965), a good deal of evidence has accumulated suggesting that cytokinins of root origin exert a regulatory control over the metabolism of the aerial parts of the plant (see Kende and Sitton, 1967; Letham, 1967). As part of a programme on the physiological significance of root cytokinins in grape vines, the writer has been studying cytokinin levels in their bleeding sap. Grape vines provide large quantities of sap (Skene, 1967) containing high levels of cytokinins (Skene, 1970).

The present paper describes some of the properties of cytokinins detected in the xylem transport system of grape vines, and discusses changes in cytokinin content in relation to experimental modifications to the root environment. In addition, a preliminary account is given of quantitative differences in the cytokinin levels of bleeding sap from *Vitis vinifera* and two grape vine rootstocks, Salt Creek and 1613. In past seasons it has been observed that these rootstocks modify the performance of *V. vinifera* scions grafted onto them; it is possible that the influence of rootstocks on scion development in fruit trees (Rogers and Beakbane, 1957) may in part be mediated by hormones produced by the rootstocks.

II. MATERIALS AND METHODS

For most of the experiments described, methods of growing plants and collection of bleeding sap followed the usual procedures (Skene, 1970). In brief, plants of *Vitis vinifera* were grown in aerated Hoagland's nutrient culture solution or perlite-sand mixtures. Bleeding sap was collected twice daily through polythene tubes attached to the bases of the stems after removal of the shoots.

In the rootstock studies, three types of fieldgrown vines were considered:(i) Sultana (*Vitis vinifera*, H_5 clone) growing on its own roots; (ii) Sultana grafted onto the rootstock Salt Creek (*V. Champini*); (iii) Sultana grafted onto the rootstock 1613 (a hybrid derived from the species *V. Longii*, *V. riparia*, *V. lubrusca* and *V. vinifera*). These will be referred to as H_5, H_5/SC and H_5/1613 respectively. Sap was collected from six vines of each type during the natural bleeding period in spring by inserting a rubber stopper and plastic tube into a hole bored in the trunk. In the case of the rootstocks, the hole was about 20 cm above the graft union. In order to establish any possible differences between the rootstocks and Sultana with the greatest possible precision, sap from each plant was processed and assayed separately.

Vacuum-extracted sap was collected from one-year-old canes of *V. vinifera* by the method of Bollard (1953). Canes were taken from dormant vines during winter and stored in plastic bags at 1^oC for about ten months. Sap was collected either directly from these cold-stored canes or from similar stored canes which had subsequently been induced to root in a glasshouse and support growth of a newly emerged shoot.

Sap from all sources was immediately filtered, frozen and freeze-dried. The main method of cytokinin extraction and purification involved extraction of freeze-dried sap in 80% (v/v) ethanol. After removal of the ethanol, cytokinin activity was estimated with soybean callus (Miller, 1965) either directly on serial dilutions of the supernatant, or after chromatography on Whatman No. 3 paper. Assay flasks contained four pieces of callus and 30 ml medium. Each fraction was assayed in triplicate. Chromatographic solvents included n-butanol/acetic acid/water (4:1:1, v/v), sec. butanol/acetic acid/water (70:2:28, v/v) and 0.03 M boric acid adjusted to pH 8.4 with sodium hydroxide. Sometimes additional purification of the extracts involved partitioning into water-saturated n-butanol from aqueous solutions at pH 7.0, especially before chromatography in the borate system.

Properties of the cytokinins in *V. vinifera* sap were further investigated by precipitation from alkaline ethanolic solutions with barium acetate. Cytokinin activity in the supernatant, and activity recoverable from the barium precipitate with excess sodium sulphate at pH 2.0 was assayed after chromatography in sec. butanol/acetic acid/water. Solubility in n-butanol and chromatographic behaviour of the cytokinin precipitated by barium acetate was also determined after incubation overnight with calf intestine alkaline phosphatase (Calbiochem, B. grade, 1 mg/ml) at 32°C. The incubate, at pH 8.3, contained 0.01 M magnesium chloride.

Procedures for investigating the effects of root temperature and treatment with CCC ((2-chloroethyl) trimethylammonium chloride) on cytokinin levels of vine roots have been described elsewhere (Skene and Kerridge, 1967; Skene, 1970).

III. RESULTS AND DISCUSSION

Properties of the cytokinins in bleeding sap

Chromatogrammed extracts of bleeding sap from grape vines grown in nutrient culture solutions contained two regions of cytokinin activity (Fig. 1A), one of low mobility (Peak 1) and another of high mobility (Peak 2).

Mild acid hydrolysis of Peak 1 converted it to a substance having an Rf similar to that of Peak 2; Peak 2 passed into n-butanol from aqueous solutions, and was held by cationic exchangers, from which it could be recovered by elution with ammonium hydroxide. Treatment of 70% (v/v) ethanolic solutions of sap at pH 9.0 with barium acetate also separated the cytokinins of Peaks 1 and 2. The supernatant above the barium precipitate that formed during standing contained Peak 2 (Fig. 1C), its Rf coinciding with that of synthetic zeatin. Peak 1 could be recovered from an aqueous suspension of the barium precipitate by treatment with excess sodium sulphate at pH 2.0 (Fig. 1D). Further treatment with alkaline phosphatase of the activity recoverable from the barium precipitate converted it to a substance soluble in n-butanol with the chromatographic properties of Peak 2 (Fig. 1E). Chromatography of an n-butanol extract of Peak 2 in 0.03 M boric acid adjusted to pH 8.4 with sodium hydroxide resolved the original peak into two regions of activity (Fig. 1B), one of which co-chromatogrammed with zeatin; the other moved to an Rf to which zeatin nucleoside would be expected to migrate (Miller, 1965). Thus Peak 2 appears to consist of a free cytokinin and its nucleoside. The behaviour of Peak 1, especially after barium acetate treatment and hydrolysis with alkaline phosphatase suggests that it is a nucleotide of the Peak 2 cytokinin.

Zeatin or related compounds and their nucleosides and nucleotides have now been isolated from tissues of a wide range of plants (see Letham and Williams, 1969), and it is likely that zeatin is a constituent of the bleeding sap of sunflowers (Kende and Sitton, 1967). Nitsch (1968) has isolated a cytokinin from bleeding sap of grape vines that appears to be a nucleotide of zeatin, and he has suggested that this may be the water-soluble form in which cytokinins circulate. Considering the evidence for the widespread occurrence of zeatin and its derivatives, it is probable that the three cytokinins under consideration in vine sap are the nucleoside and nucleotide of zeatin or a closely related cytokinin, together with the free form of the same cytokinin.

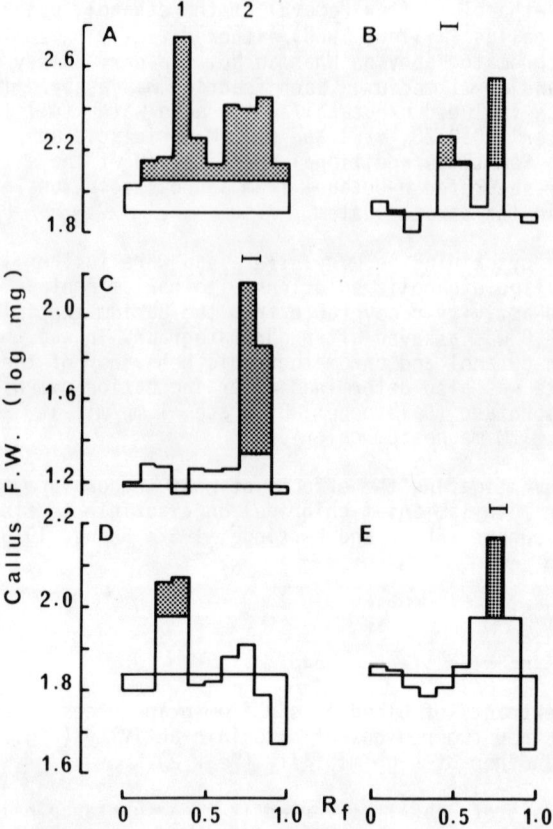

Fig. 1. Response of soybean callus to chromatogrammed extracts of bleeding sap. (A) Unfractionated alcohol extracts developed in n-butanol/acetic acid/water (4:1:1), showing Peaks 1 and 2; (B) n-Butanol fraction developed in 0.03 M borate at pH 8.4; (C, D and E) Fractions separated from alcohol extract with barium acetate and chromatogrammed in sec. butanol/acetic acid/water (70:2:28). C is from the supernatant, D from the barium precipitate and E from the barium precipitate after alkaline phosphatase treatment. Significant responses (P = 0.05) to the extracts are indicated by shading of the histograms. Horizontal lines above the figures show the mobility of synthetic zeatin.

Cytokinins in vacuum-extracted sap

Sap extracted under reduced pressure from cold-stored vine canes or from canes supporting active shoot and root growth caused soybean callus to respond in the manner shown in Figure 2, when assayed at several dilutions. The dilution-response curves for both groups of sap were parallel, with the sap from cold-stored canes containing considerably more activity than that of growing canes. The latter group contained about 35 µg/l kinetin equivalents based on the mid-point of the dilution-response curve, whereas sap from cold-stored canes contained about 130 µg/l kinetin equivalents. These values are only valid for the particular dilution at which they were obtained, as the extract curves are not parallel to the standard response curve.

The responses of soybean callus to chromatogrammed extracts of vacuum-extracted sap are given in Figure 3. Both types of sap contained two regions of cytokinin activity, which at least in the sap from growing plants coincided in Rf with Peaks 1 and 2 of bleeding sap. At present it is not known whether the difference in Rf between the slow running peaks of each group is of any significance. Activity levels, in terms of µg/l kinetin equivalents are shown above each peak of activity in the figure. Total activity is of the same order as determined by dilution-response curves.

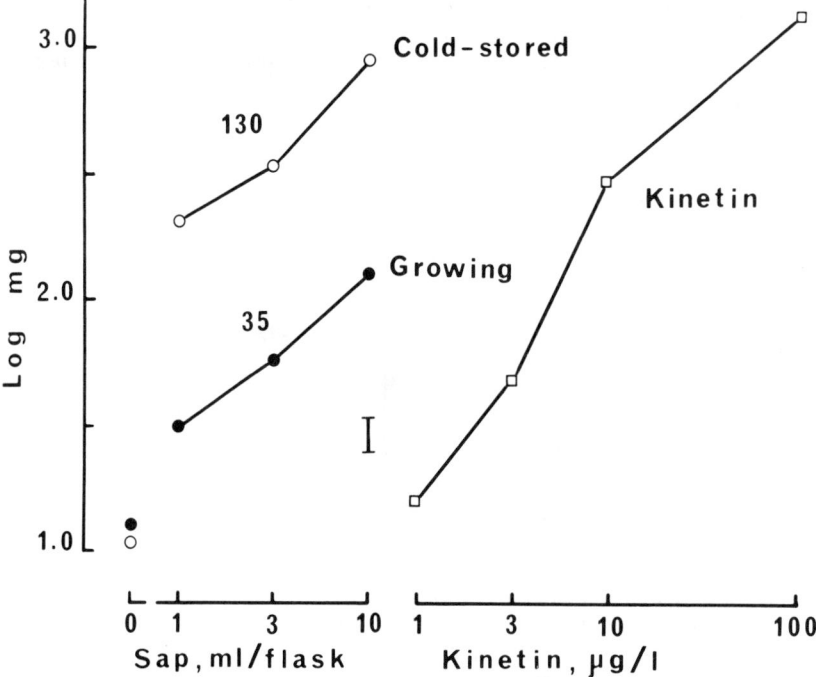

Fig. 2. Response of soybean callus to serial dilutions of sap extracted under reduced pressure from cold-stored and growing canes. Reponse curve to kinetin is shown on the right. Vertical bar is the LSD (P = 0.05) between treatments for each dilution. The numbers adjacent to the curves represent cytokinin concentrations in the sap (μg/l kinetin equivalents) derived from comparing the responses to 3 ml of sap per assay flask with the kinetin standard curve.

There appear to be quantitative and qualitative similarities between the cytokinins of bleeding sap and sap extracted under reduced pressure from growing vines. Jones and Lacey (1968) reached the same conclusions for the gibberellins of apple. The high cytokinin levels in cold-stored canes originally collected in winter were unexpected. Luckwill and Whyte (1968) found marked seasonal changes in the cytokinin content of sap extracted from apple stems under reduced pressure, but no cytokinin activity could be detected during winter. Activity in cold-stored canes may in fact represent cytokinins released or synthesised during storage. However, this aspect of the work, including the estimation of cytokinin levels in dormant canes immediately after collection, is still under investigation.

Environmental effects on cytokinins in bleeding sap

The amounts and types of cytokinins in bleeding sap of grape vines can be influenced by modifications to the root environment. This will be illustrated by examples of both physical and chemical modification.

Temperatures to which the roots of Sultana vines grown in nutrient culture solutions were exposed affected the appearance of Peak 1, the presumed cytokinin nucleotide (Skene and Kerridge, 1967). At $20^{o}C$ Peaks 1 and 2 both were present, whereas at $30^{o}C$

Peak 1 was absent. It is possible that the stimulatory effect of a 30° root temperature on fruit set in Sultana (Woodham and Alexander, 1966) is due to an altered cytokinin production by the roots; perhaps the absence of the cytokinin nucleotide indicates an increased rate of cytokinin turnover by the roots.

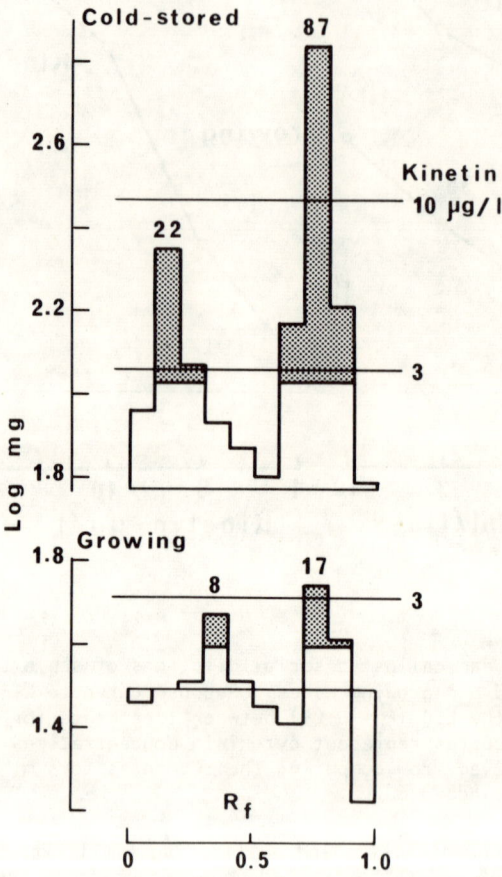

Fig. 3. Response of soybean callus to extracts of sap obtained under reduced pressure from cold-stored and growing canes, and chromatogrammed in sec. butanol/ acetic acid/ water (70:2:28). Cytokinin concentrations (μg/l kinetin equivalents) derived from these responses are shown above each peak of activity on the figure.

Peak 1 is also commonly absent or at a low level in sap from vines grown in solid media (e.g., Skene, 1968). Temperature differences between solid media and culture solutions are not the cause of these cytokinin changes, and it seems that other undefined properties of the root medium can influence the appearance of cytokinins in bleeding sap.

Chemical modification of the root environment with the growth retardant CCC increased the concentration of cytokinins in the bleeding sap of vines and also the amount passing daily to the top of the plant (Skene, 1970). The effect was exerted on both Peaks 1 and 2, but diminished towards the end of the sap collection period, and was likely to be due to CCC directly influencing cytokinin production by the roots (Skene, 1970).

The relative amounts of Peaks 1 and 2 cytokinin activity in sap from vines grown in culture solutions changed with time during bleeding (Table 1). Under the conditions of this set of experiments (Skene, 1970), sap collected within a few hours of shoot removal contained only Peak 2. By the 22nd to 29th hours, Peaks 1 and 2 were both present, with Peak 1 predominating, which is the situation usually observed for nutrient culture solutions. By the 70th hour, Peak 2 had disappeared, or was present in trace quantities. In other words, the free cytokinin and/or its nucleoside were the first detectable cytokinins in sap after shoot removal. As bleeding continued, the free cytokinin, its nucleoside and nucleotide were all present (see also, Fig. 1), and finally just the nucleotide. These results suggest a biosynthetic relationship between the three cytokinins, but do not really assist in deciding in which direction synthesis normally proceeds. Nor should one overlook the possibility of interconversions between the cytokinins during processing of the sap.

Table 1

Changes in cytokinin activity in sap during bleeding*

(Adapted from data in Skene (1970).)

Time after shoot removal (hr)	Kinetin Equivalents (μg/l)			
	Control		Plus CCC	
	Peak 1	Peak 2	Peak 1	Peak 2
5	0	28	0	80
22	22	12	74	46
29	20	11	47	21
70	32	0	31	8

*Extracts of bleeding sap from vines grown in culture solutions were chromatogrammed in n-butanol/acetic acid/water (4:1:1)

Grape vine rootstocks

Bleeding sap collected from six individual fieldgrown plants of H_5 Sultana, H_5/SC and H_5/1613 was freeze-dried, extracted in 80% (v/v) ethanol and assayed at several dilutions with soybean callus. The mean responses for each group of plants are given in Figure 4; kinetin and zeatin standard response curves are included. There was no difference between the responses to H_5 and H_5/SC sap. However, the lower response to H_5/1613 sap was highly significant, indicating lower cytokinin activity in sap of this rootstock. No difference in the slopes of the three curves could be demonstrated statistically, and all three approximate more closely to the slope of the zeatin response curve than to that of kinetin. In terms of zeatin equivalents, H_5 and H_5/SC sap contained 6 μg/l, H_5/1613 3 μg/l.

For the season considered, the amount of vegetative growth for each group of plants was the same. However, the mean yield of fruit for the six vines of each group did show differences in the same order as the differences in cytokinin concentration of the sap. H_5 yielded 45.6, H_5/SC 45.9 and H_5/1613 30.9 Kg fruit/vine, which in part could be accounted for by different numbers of berries in each bunch, viz., H_5 327, H_5/SC 281, H_5/1613 252 ($LSD_{.05}$ = 50).

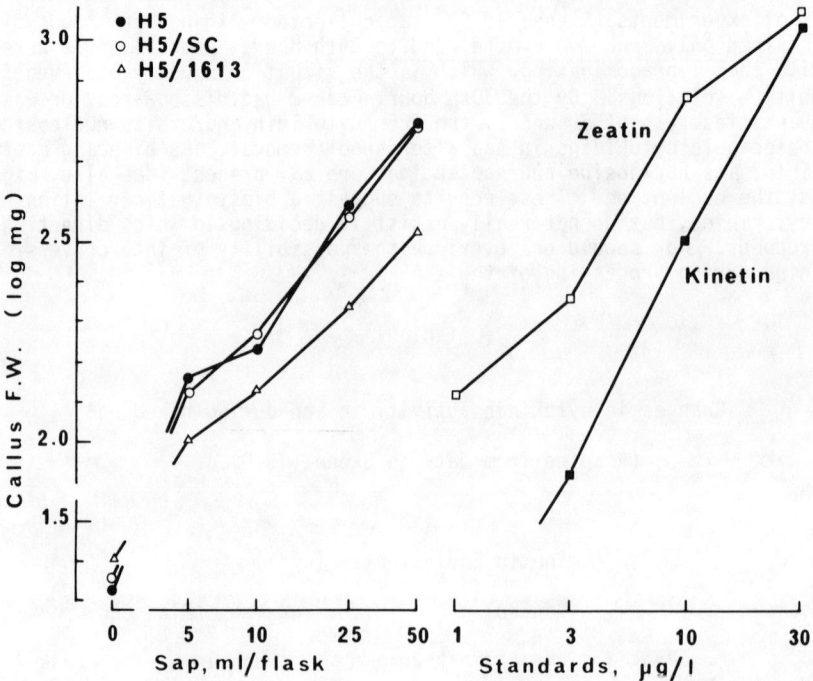

Fig. 4. Response of soybean callus to serial dilutions of bleeding sap from H_5, H_5/SC and H_5/1613 grape vines. Response curves to zeatin and kinetin standards are shown on the right. Each point is the mean of six determinations assayed in triplicate. (Rootstock effect significant at $P = 0.001$; interaction between rootstock and dilution NS.)

The number of berries per bunch is determined firstly by the number of flowers initiated, and secondly by the proportion of flowers developing into berries. Information is not available to distinguish which of these factors was responsible for the observed differences in berry number. However, both processes have been shown to be affected by cytokinins (M.G. Mullins, personal communication; Weaver, van Overbeek and Pool, 1965), and it may well be possible that the lower set for H_5 Sultana grafted onto 1613 roots was due to the reduced cytokinin levels passing from the roots to the aerial parts of the plant. Although an explanation of rootstock-scion relationships merely in terms of hormone production by roots is an over-simplification of the situation (see Jones and Lacey, 1968), the results reported for grape vines do strongly suggest that hormones of root origin play some part in the phenomenon.

IV. SUMMARY

Three cytokinins have been detected in bleeding sap of grape vines. Their properties, taking into account the results of other workers, are suggestive of a cytokinin resembling, or related to zeatin, and its nucleoside and nucleotide. Xylem sap extracted under reduced pressure from canes (one-year-old woody stems) taken from growing plants contains cytokinins which appear similar to those of bleeding sap. The total concentration of activity also is similar; however, activity in sap from isolated canes which had been stored at 1°C is considerably higher. Modifications to the root environment result in quantitative and qualitative changes in cytokinin activity of bleeding sap; these changes are discussed in relation to the root as a biosynthetic site for hormones affecting top growth of the plant. On the same theme, quantitative differences in the cytokinin concentration of bleeding sap from *Vitis vinifera* and two of its rootstocks, Salt Creek and 1613, are described.

V.   REFERENCES

BOLLARD, E.G. (1953). The use of tracheal sap in the study of apple-tree nutrition. J. exp. Bot. 4, 363-8.
JONES, O.P. and H.J. LACEY (1968). Gibberellin-like substances in the transpiration stream of apple and pear trees. J. exp. Bot. 19, 526-31.
KENDE, H. (1964). Preservation of chlorophyll in leaf sections by substances obtained from root exudate. Science 145, 1066-7.
KENDE, H. (1965). Kinetin-like factors in the root exudate of sunflowers. Proc. natn. Acad. Sci. 53, 1302-7.
KENDE, H. and D. SITTON (1967). The physiological significance of kinetin- and gibberellin-like root hormones. Ann. N.Y. Acad. Sci. 144, 235-43.
LETHAM, D.S. (1967). Chemistry and physiology of kinetin-like compounds. A. Rev. Pl. Physiol. 18, 349-64.
LETHAM, D.S. and M.W. WILLIAMS (1969). Regulators of cell division in plant tissues. VIII. The cytokinins of the apple fruit. Physiol. Plant. 22, 925-36.
LOEFFLER, J.E. and J. VAN OVERBEEK (1964). Kinin activity in coconut milk, in "Régulateurs Naturels de la Croissance Végétale" (Ed. J.P. Nitsch) p. 77-82. C.N.R.S., Paris. 1964.
LUCKWILL, L.C. and P. WHYTE (1968). Hormones in the xylem sap of apple trees, in "Plant Growth Regulators" p. 87-101. S.C.I. Monograph No. 31, London. 1968.
MILLER, C.O. (1965). Evidence for the natural occurrence of zeatin and derivatives: compounds from maize which promote cell division. Proc. natn. Acad. Sci. 54, 1052-8.
NITSCH, J.P. (1968) Natural cytokinins, in "Plant Growth Regulators" p.111-23. S.C.I. Monograph No. 31, London. 1968.
ROGERS, W.S. and A.B. BEAKBANE (1957). Stock and scion relations. A. Rev. Pl. Physiol. 8, 217-36.
SKENE, K.G.M. (1967). Gibberellin-like substances in root exudate of *Vitis vinifera*. Planta 74, 250-62.
SKENE, K.G.M. (1968). Increases in the levels of cytokinins in bleeding sap of *Vitis vinifera* L. after CCC treatment. Science 159, 1477-8.
SKENE, K.G.M. (1970). The relationship between the effects of CCC on root growth and cytokinin levels in the bleeding sap of *Vitis vinifera* L. J. exp. Bot. 21 418-31.
SKENE, K.G.M. and G.H. KERRIDGE (1967). Effect of root temperature on cytokinin activity in root exudate of *Vitis vinifera* L. Pl. Physiol. 42, 1131-9.
WEAVER, R.J., J. VAN OVERBEEK and R.M. POOL (1965). Induction of fruit set in *Vitis vinifera* L. by a kinin. Nature 206, 952-3.
WOODHAM, R.C. and D. McE. ALEXANDER (1966). The effect of root temperature on development of small fruiting sultana vines. Vitis 5, 345-50.

Medium and Tissue Sugar Concentrations during Cytokinin-controlled Growth of Tobacco Callus Tissues

John P. Helgeson, C. D. Upper and G. T. Haberlach

Pioneering Research Laboratory, Crops Research Division, Agricultural Research Service, United States Department of Agriculture, Department of Plant Pathology, University of Wisconsin, Madison, Wisconsin, 53706

I. INTRODUCTION

In the course of our work to define the tobacco tissue culture system for molecular studies on growth and development, we have shown that logarithmic growth rates are dependent on the cytokinin concentrations in the medium (Helgeson et al., 1969), and that these rates can be modified to some extent by the addition of gibberellic acid to the medium (Helgeson and Upper, 1970). Maximal yields of tobacco callus tissues are dependent on the amount of carbohydrate in the medium and not influenced by a wide range of hormonal regimes (Upper et al., 1970). We now report the results of our study on the relationships between hormonally programmed growth and sugar concentrations in the medium and the tissue from initial planting of the tissue until senescence.

II. METHODS

Stock pith callus tissues of *Nicotiana tabacum* var. Wisconsin No. 38 were grown continuously on Linsmaier and Skoog's (1965) medium containing 88mM sucrose, 11.5 µM IAA and 0.1 µM 6-(3-methyl-2-butenylamino)- purine (2iP). Unless otherwise noted, the same medium with these IAA and 2iP concentrations was used for experiments. The concentration and type of carbon source used is stated for each experiment. Fresh weights, where given, are for tissues actually homogenized for sugar determinations. Dry weights were determined with tissues planted in the same experiment and harvested at the same time that the other tissues were homogenized.

Tissues were prepared for sugar determinations by weighing and immediately homogenizing in an equal volume (w/v) of 0.2 M Na_2CO_3. The pH of the homogenate was adjusted to 4.5 with glacial acetic acid and the acidified homogenate was centrifuged (25,000×g for 30 min at 4°C). The resulting supernatant was diluted as required for determinations of reducing sugar and total hexose. Provided an adequate volume of Na_2CO_3 was used, there was no hydrolysis of the sucrose in the supernatant, even when it was left at room temperature for two days. Tests of the pellet indicated that less than 3% of the reducing sugar and total hexose remained in this fraction after centrifugation. All media were prepared for analysis as previously described (Upper et al., 1970).

Reducing sugar and total hexose were measured as previously described (Upper et al., 1970). Total hexose is defined as that amount of reducing sugar found after exhaustive treatment with invertase. This value was determined from a sucrose standard curve and is expressed as hexose concentration. Reducing sugar content was determined prior to invertase treatment from a glucose standard curve and expressed as hexose concentration. Sucrose concentrations are expressed in terms of sucrose molarity and were calculated by dividing the difference between total hexose and reducing sugar by two. Total sugar concentrations were calculated by summing sucrose and reducing sugar concentrations. Except where specifically noted, all medium concentrations are corrected for the actual volume of medium remaining at the time the sample was taken.

Tissue sugar concentrations were based on the assumption that 1 gram of tissue fresh weight equals 1 milliliter.

III. RESULTS AND DISCUSSION

Typical results obtained with callus tissues planted on medium initially containing 88 mM sucrose, 11.5 µM IAA and 0.1 µM 2iP are shown in Figure 1. After an adjustment period of 3 days, tissue total hexose was nearly constant during logarithmic growth. During this time the total hexose concentration of the medium also remained relatively constant, even though the volume of the medium decreased from 50 ml at time of planting to 34 ml at 31 days. This indicated that both water and sugars disappeared from the medium at comparable rates. Approximately at the end of log phase, the total hexose concentration of the medium decreased rapidly. This decrease was followed about three days later by a rapid decrease in tissue total hexose. After the initial adjustment period and during early log growth, tissue reducing sugar declined gradually from about 30 to about 15 mM. During this time, the concentration of reducing sugar in the medium gradually increased. Both medium and tissue reducing sugar concentrations then remained relatively constant during the rest of log phase. At the end of log phase the concentration of reducing sugar in the tissue first increased markedly and then declined. (Although only one datum point is shown on the peak in this experiment, this increase in reducing sugar concentration has been confirmed in many other experiments, including one in which daily points were taken during this period). The reducing sugar concentration of the medium also increased at this time, reaching a maximum 4 to 6 days after that of the reducing sugar in the tissue. After the peak, only reducing sugar remained in the medium. When all sugar, both in the medium and in tissue, was nearly gone, no increase in tissue weight occurred.

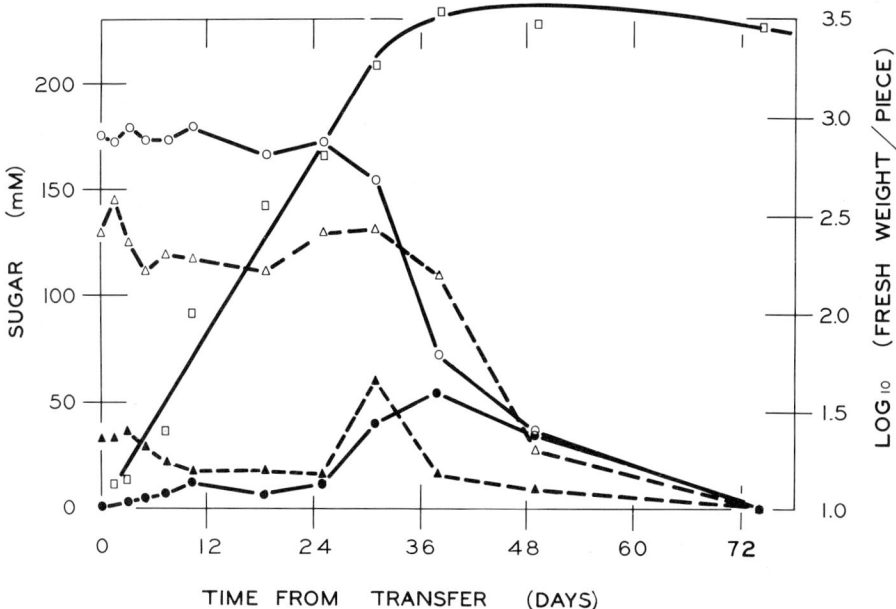

Fig. 1. Tissue and medium sugar concentrations from planting to senescence. Measured quantities were tissue fresh weight (mg/piece, □), tissue total hexose (△), tissue reducing sugar (▲), medium total hexose (O), and medium reducing sugar (●). The medium initially contained 88 mM sucrose, 11.5 µM IAA and 0.1 µM 2iP.

Early fluctuations of reducing sugar were examined in detail (see Fig. 2). Tissues that initially contained about 30-33 mM reducing sugar were cut and planted on new medium. Within an hour the concentration of reducing sugar had dropped to about 15 mM and then, by the 35th hour, had regained a concentration comparable to that in the tissue prior to cutting and transfer. In another experiment the reducing sugar level of this "wounded" tissue dropped to 5 mM before regaining its original concentration. Other tissues were cut, planted, and allowed to stand on new medium. After 69 hours, they were moved gently to other new media. No drastic depletion of reducing sugar was noted in these tissues (see Fig. 2). Thus we concluded that cutting the tissue, and not merely transferring it to new medium, caused this marked, transitory decrease in internal reducing sugar.

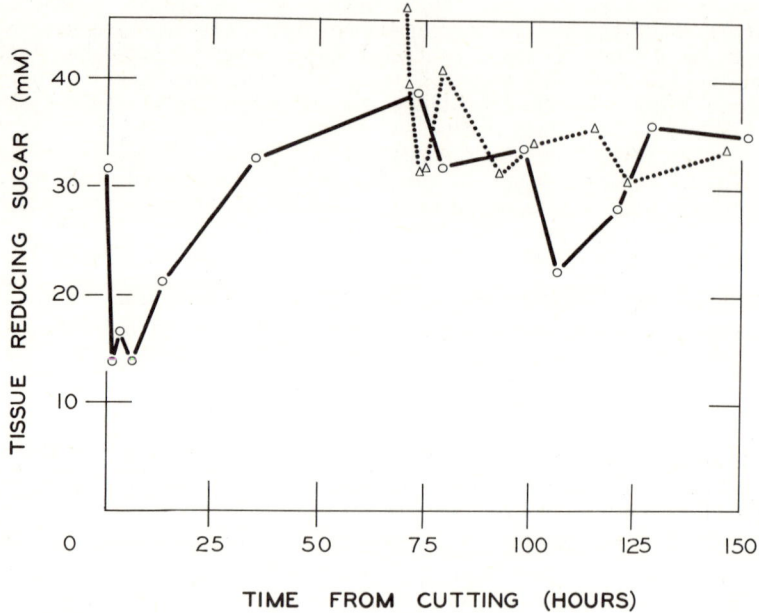

Fig. 2. Tissue reducing sugar changes from planting to 150 hours. Values for tissue cut and planted on new media indicated by circles; values for tissues cut, planted and then transferred after 69 hours indicated by triangles. All media initially contained 88 mM sucrose, 11.5 µM IAA and 0.1 µM 2iP.

In general, depletion of total hexose from the medium appeared similar if the tissues were grown on either sucrose or glucose as carbon source (see Figure 3). In this experiment, tissues from the same stock cultures were planted on media containing either 25 mM sucrose or 50 mM glucose. Tissues grew at nearly the same rate and left log phase at approximately the same time. Final dry weight yields of tissue on the two carbon sources were identical.

Although depletion of total hexose from the medium and growth rates appeared grossly similar on the two media, differences in internal sugars were considerable. Tissues growing on media containing 50 mM glucose, rapidly attained a total hexose concentration close to that value (Figure 4). Then, throughout log phase, the total hexose concentrations of the medium and the tissue were nearly identical. After the end of log phase, the total hexose of the tissue was depleted at a slower rate than the total hexose of the medium. Initially, reducing sugar levels in the tissue were about half those of the medium and gradually declined during growth. When medium sugar was nearly depleted, only reducing sugar was left in the tissue; and this component apparently was the last used during tissue maintenance.

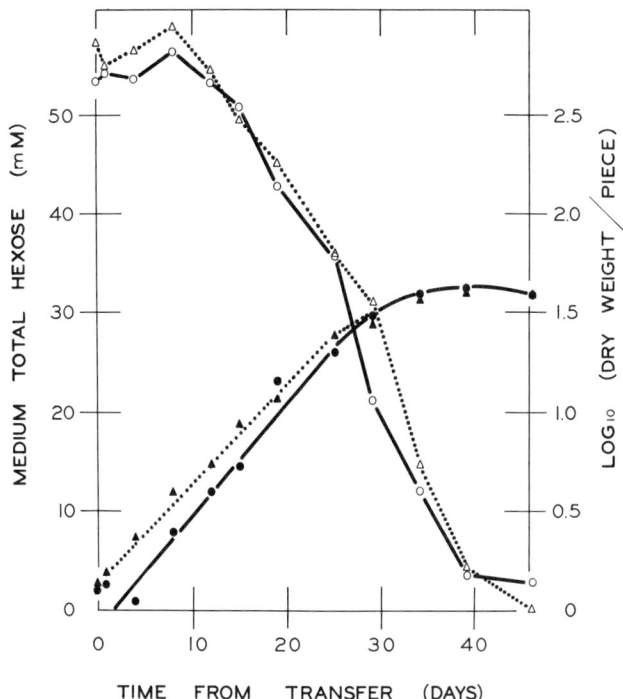

Fig. 3. A comparison of tissue growth (mg/piece, closed symbols) and medium total hexose concentrations (open symbols) for tissues planted on media initially containing 50 mM glucose (circles) or 25 mM sucrose (triangles). Both media initially contained 11.5 µM IAA and 0.1 µM 2iP.

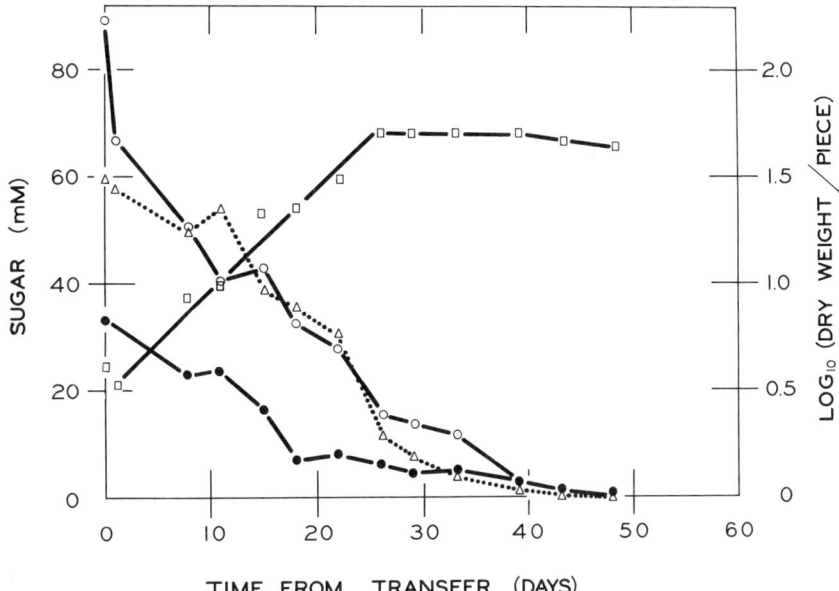

Fig. 4. Tissue and medium sugar concentrations during growth on media initially containing 50 mM glucose. Quantities measured were dry weight (mg/piece, □), tissue total hexose (○), tissue reducing sugar (●), and medium total hexose (all reducing sugar, △). The medium initially contained 11.5 µM IAA and 0.1 µM 2iP.

No difference was found between total hexose and reducing sugar in the medium at any time during growth on media initially containing only glucose. Thus, while the tissues synthesize something that is converted to reducing sugar by invertase (presumably sucrose), this material is not released by the tissues.

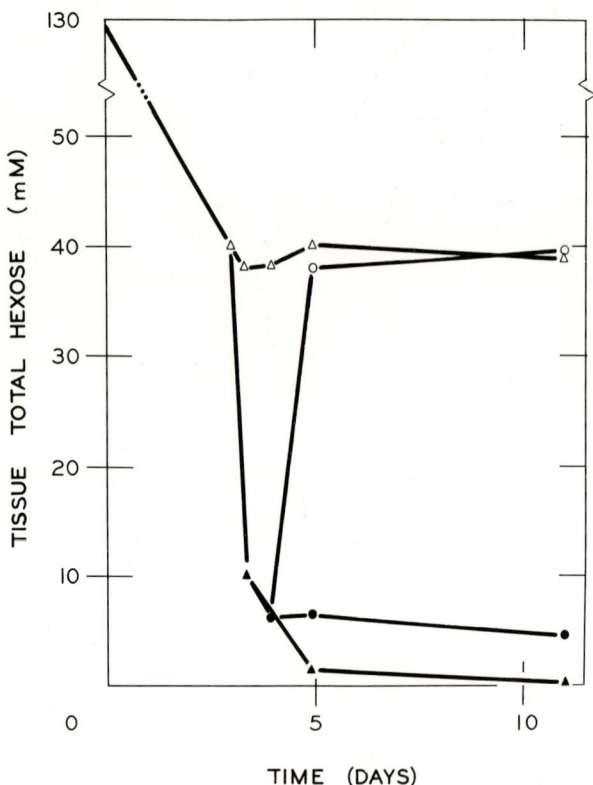

Fig. 5. Total hexose concentrations of tissues transferred to media containing various concentrations of glucose. Tissue transfer regimes were planted on medium containing 48 mM glucose (△), transferred to 0.42 mM glucose after 3 days (▲), transferred to 0.42 mM glucose and then to 6.25 mM glucose after 7 hours (●), transferred to 0.42 mM glucose, then to 6.25 mM glucose after 7 hours, then transferred to 44 mM glucose 16 hours later (○). All media contained 11.5 µM IAA and 0.1 µM 2iP.

The rapid adjustment of tissue total hexose concentration to approach that of medium was further examined in another experiment (Figure 5). Tissue containing 130 mM total hexose was planted on a medium containing 48 mM glucose. After 3 days, the tissue contained 40 mM total hexose, and for the next 8 days this level did not change by more than 2 mM. After the initial 3-day equilibration period, some of the tissues were put through a series of transfers, while others were left on the initial medium as controls. Initially, tissues were transferred to 0.42 mM glucose. Within 7 hours the total hexose of these tissues had declined to 10 mM. Forty hours later, total hexose of these tissues had further declined to less than 2 mM. After the 7-hour period on 0.42 mM glucose, some of the tissues were transferred to medium containing 6.25 mM glucose. After 16 hours, these tissues contained 6.5 mM total hexose and the total hexose concentration remained at approximately this level for the next 6 days. Finally, some tissues which had undergone transfers both to 0.42 and 6.25 mM glucose were transferred back to a medium containing 44 mM glucose. Within 24 hours the total hexose had risen back to the level of tissue which had not been transferred from the original 48 mM glucose. Six days later the two sets of tissues still had identical total hexose levels. The average fresh weight of tissues which had gone through all transfers was nearly equal to that of tissues which had remained on 48 mM glucose (482 vs 508 mg fresh wt/piece at 13 days). Thus, we concluded that even though tissues are remarkably sensitive to sugar concentrations in the medium, and internal sugars can be controlled readily, there was little effect on growth of tissues.

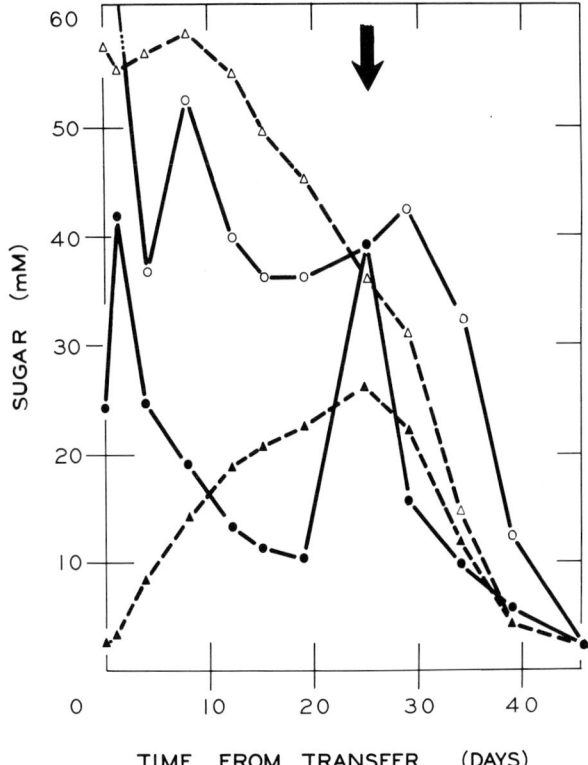

Fig. 6. Tissue and medium sugar concentrations during growth on media initially containing 25 mM sucrose, 11.5 µM IAA and 0.1 µM 2iP. Quantities measured were tissue reducing sugar (●), tissue total hexose (○), medium total hexose (△), and medium reducing sugar (▲). The arrow indicates the time when tissues left log phase (see Fig. 3 for growth data).

Figures 1 and 2 show results for tissues grown on a medium which contained 88 mM sucrose at the time of planting. When the sucrose concentration was 25 mM, the results shown in Figure 6 were obtained. As with tissues grown on 50 mM glucose, the total hexose concentration of the tissues decreased rapidly. This time, however, the total hexose concentration of the tissue dropped to well below that of the medium and remained essentially constant for the duration of log phase. During this time over half the medium total hexose was used. The reducing sugar concentration of tissues growing on 25 mM sucrose dropped slowly during log phase and increased suddenly when tissues were about to leave log phase. An increase in the reducing sugar concentration of the medium was again noted during log phase (see Figure 7). Sucrose concentrations dropped steadily until the tissue left log phase on day 25 (see arrow, Figure 7). When the reducing sugar and sucrose concentrations of the medium were summed it was noted that these values (total sugar concentrations) were essentially constant until a rapid drop occurred at day 29.

At this time very little sucrose (4.4mM) was still available in the medium. It was during the period of relatively constant medium total sugar concentration that the tissue total hexose remained relatively constant (compare Figures 6 and 7).

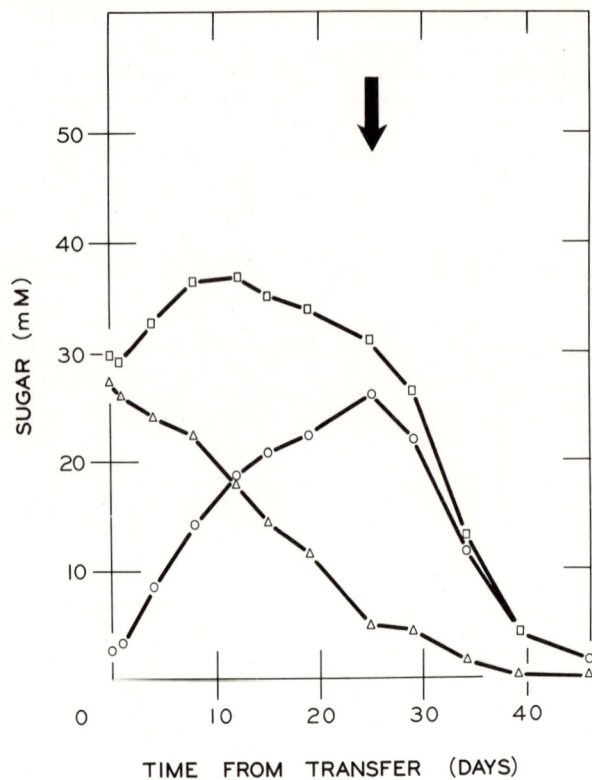

Fig. 7. Medium sugar concentration changes during growth on media initially containing 25 mM sucrose, 11.5 µM IAA and 0.1 µM 2iP. Quantities indicated are measured reducing sugar (O), calculated sucrose concentration (Δ), and calculated total sugar concentration (reducing sugar plus sucrose, □). The arrow indicates the time when tissues left log phase (see Fig. 3 for growth data).

The net result of the relatively constant total sugar concentration is that the tissues actively maintain a relatively constant environment during log phase (ca. 3 to 21 days) in spite of the utilization of about half the medium sugar and water. This effect has been noted over a considerable range of total sugar concentrations (30 to 130 mM). The percent dry weight during this time also remains relatively constant (Helgeson and Upper, 1970). Therefore, it would appear that sucrose would be the carbon source of choice if relatively constant conditions are desired during log phase, and a continuously changing carbon source can be tolerated.

Results presented thus far were obtained with tissues actively growing under nearly optimal conditions on media containing 11.5 µM IAA and 0.1 µM 2iP. Our preliminary results indicated that the addition of GA_3 to the medium (a treatment that increases growth rate) does not appreciably alter sugar balances. However, deletion of IAA from the medium results in drastic increases in reducing sugars (see Figure 8). In this experiment, all tissues were first transferred to medium containing 11.5 µM IAA and 0.1 µM 2iP. Then after 3 days they were transferred to media with (a) 11.5 µM IAA, 0.1 µM 2iP. (b) no 2iP, 11.5 µM IAA (c) no IAA, 0.1 µM 2iP (d) neither IAA nor 2iP. Tissues supplied with 11.5 µM IAA maintained reducing sugar levels similar to those discussed above, with or without 2iP in the medium. However, if IAA was omitted, the reducing sugar concentration of tissue rose rapidly until the 15th day at which time essentially all the tissue sugar was present as reducing sugar. The medium without IAA (with or without 2iP) showed linear increases in reducing sugar from about day 1 until essentially all of the sucrose in the medium had been hydrolyzed. Whether this increase in reducing sugar is due to a release of invertase to the medium, active synthesis of more enzyme (or merely complete loss of control of carbohydrate metabolism) has not been determined. Tissues deprived of IAA for over a day appear to suffer considerable damage when subsequently transferred to complete medium.

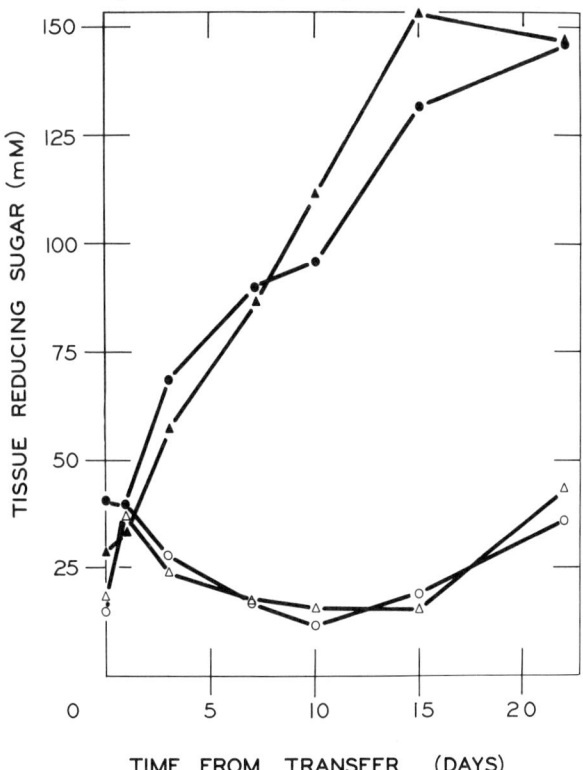

Fig. 8. Reducing sugar concentrations of tissues planted on media differing in growth substance regimes. The regimes tested were 11.5 µM IAA and 0.1 µM 2iP (O), 11.5 µM IAA and no 2iP (Δ), no IAA and 0.1 µM 2iP (●), and neither IAA nor 2iP (▲). All media initially contained 88 mM sucrose.

IV. SUMMARY

Tissue growth, tissue sugar concentrations and medium sugar concentrations were measured from initial planting to senescence of tobacco callus tissues growing on Linsmaier and Skoog's medium containing 11.5 µM IAA, 0.1 µM 6-(3-methyl-2-butenylamino) purine (2iP) and various concentrations of sucrose or glucose. Cutting of tissues caused a "wound" effect which was seen as a sharp drop in reducing sugar immediately after tissue planting. Within two days this effect was no longer apparent and sugar levels in "wounded" and nonwounded tissues were comparable. Tissues rapidly attained total hexose concentrations equal to or less than medium total hexose concentrations. Tissues growing on media containing glucose as a carbon source maintained total hexose concentrations which continually changed to maintain similar levels to those of the media on which they were growing. Tissues growing on sucrose media maintained relatively constant total hexose concentrations which were somewhat below that of the medium. Tissue reducing sugar concentrations initially were about 30 mM after recovery from wounding and the reducing sugar levels gradually decreased during log phase. At the end of log phase both medium and tissue reducing sugar concentrations of sucrose-grown tissues increased markedly and then declined. Omission of IAA from the medium caused a marked rise of reducing sugar, often resulting in complete hydrolysis of both medium and tissue sucrose. This effect occurred regardless of the presence or absence of cytokinin.

V.   REFERENCES

HELGESON, J. P., S. M. KRUEGER and C. D. UPPER (1969). Control of logarithmic growth rates of tobacco callus tissue by cytokinins. Pl. Physiol. 44, 193-198.
HELGESON, J. P. and C. D. UPPER (1970). Modification of logarithmic growth rates of tobacco callus tissue by gibberellic acid. Pl. Physiol. 46, 113-118.
LINSMAIER, E. M. and F. SKOOG (1965). Organic growth factor requirements of tobacco tissue cultures. Physiol. Plant. 18, 100-127.
UPPER, C.D., J. P. HELGESON, and G. T. HABERLACH (1970). Limitation of tobacco callus tissue growth by carbon source availability. Pl. Physiol. 46, 118-122.

ETHYLENE

STUDIES ON THE ACTION OF ETHYLENE IN PHYSIOLOGICAL PROCESSES OF PLANT CELLS[1]

Jack G. Valdovinos, Leland C. Ernest[2], and Thomas E. Jensen

Department of Biological Sciences Herbert H. Lehman College of the City University of New York Bedford Park Blvd. West Bronx, N.Y., U.S.A. 10468

1. Supported in part by National Science Foundation (U.S.A) Grant No. GB-8260, and a grant from Research Corporation.
2. Present Address: Department of Biological Sciences, Eastern Nazarene College, Wollaston, Mass.

I. INTRODUCTION

1. Effect of ethylene on auxin metabolism

The effects of ethylene in lowering levels of auxin diffusing through plant tissues are well known (Laan, 1934; Michener, 1938; Guttenberg and Steinmetz, 1947; Morgan and Gausman, 1966; Burg and Burg, 1967; Valdovinos, Ernest, and Henry, 1967). The manner in which the gas influences levels of diffusible auxin in plant cells is not entirely clear. In a brief report (Valdovinos, Ernest, and Henry, 1967) we have already presented some evidence correlating low levels of diffusible auxin in ethylene treated tissue with decreased auxin formation by enzyme bries from the same tissue. The studies have been expanded in an attempt to correlate several facets of auxin metabolism, auxin biosynthesis, conjugation, and destruction - with levels of extractable and diffusible auxin in the leaf and stem tissue.

11. Effect of ethylene on the ultrastructure of abscission cells

In earlier studies of natural abscission in pedicels of tobacco and tomato flowers, changes were observed in abscission cells at the ultrastructural level (Jensen and Valdovinos, 1967 and 1968; Valdovinos and Jensen, 1968). The increasing amount of biochemical and light microscope data indicates a role for ethylene in the regulation of abscission (Abeles and Rubenstein, 1964; Holm and Abeles, 1967; Horton and Osborne, 1967; Abeles, 1968 and 1969; Dela Fuente and Leopold, 1968 and 1969; Webster, 1968; Jackson and Osborne, 1970). Studies at the fine structural level could give additional insight as to the action of ethylene in abscission. Accordingly, an investigation of the effects of ethylene on abscission at the ultra-structural level has been conducted, the results of which are presented here.

II. MATERIALS AND METHODS

1. Effect of ethylene on auxin metabolism

Plant treatment. Coleus (Coleus blumei Benth 'Scarlet Rainbow') plants derived from the University of Iowa clone (Muir and Valdovinos, 1970) were grown in the greenhouse and in environmental chambers (Sherer Model CEL 25 - 7HL). Plants grown in the growth chambers were exposed to a 16 hr photoperiod (light intensity 1600 ftc), 50 ± 5 percent relative humidity, and to a temperature of 24°C during the light period and 20°C at night. Ethylene treatment consisted of either placing the plants in polyethylene bags and injecting ethylene into the bags or providing a continuous flow of air (30 liter/min) through water to the chambers.

In the case of the plants placed in plastic bags, temperatures inside the bags were maintained at 24°C during the light phase. The plants were exposed to ethylene for

18 hr except in the studies of auxin conjugation where ethylene was provided for an additional 4.5 hr during the IAA-1-^{14}C transport period.

Auxin transport. Following procedures described earlier (Valdovinos and Ernest, 1967), auxin transport was measured through the uppermost internode of *Coleus* stems. Stem segments 5 mm in length were removed from the treated and untreated plants, coated with a ring of vaseline to prevent surface transport, and placed on glass slides. One percent agarose cylinders of a volume of 23 µl containing 5 mg/liter IAA-1-^{14}C (32,000 cpm average per block) served as auxin donors on the physiological apical end and similar cylinders without IAA served as receivers. After five hr in a moist chamber kept in the dark, the donor and receiver cylinders and the stem sections (macerated) were placed in counting vials separately or in groups of five in counting vials and frozen. The radioactivity was assayed with a Nuclear-Chicago liquid scintillation counting system following procedures described in an earlier publication (Valdovinos and Perley, 1966).

Auxin Destruction. Cell free enzyme preparations of apical tissue and untreated *Coleus* plants were prepared using a method similar to that previously described (Valdovinos and Ernest, 1966; Valdovinos, Ernest, and Henry, 1967). Two ml of the supernatant enzyme brei was then incubated for 3 hr at 27 C in sealed flasks with 0.1µCi IAA-1-^{14}C (2 µM). Chloramphenicol was added as a safeguard against bacterial contamination in a final concentration of 50 µg/ml. The carbon dioxide released during the reaction was collected in 0.4 ml of 10% KOH in a well suspended over the mixture. At the end of the incubation period trichloroacetic acid (0.5 ml of 5 percent) was added with a hypodermic syringe (Valdovinos and Ernest, 1966). After twenty min the KOH containing the $^{14}CO_2$ was transferred to counting vials and assayed as described above.

Auxin Extraction. Following the 18 hr pretreatment period with ethylene, 10 plants were removed from each growth chamber and the apical portion of each stem including medium to large leaves (2nd node from apical bud) was collected. The apical bud and leaf blades were excised from the stem and were quickly weighed and homogenized for 3 minutes in a blendor in cold methanol. The cell debris was removed by centrifugation and re-extracted with methanol. The combined methanol extracts were evaporated *in vacuo*. A procedure reported earlier (Perley and Stowe, 1966) with slight modifications, was used to extract acidic indoles. The pH of the aqueous residue was first adjusted to 2.5 and the residue was extracted in a separatory funnel 3 times with methylene chloride. The combined organic extract was extracted 3 times with 0.15 M K_2HPO_4 (pH 8.5). The combined aqueous fraction was then adjusted to pH 2.5 and was extracted with methylene chloride. The final extract was reduced to a small volume and separated by thin layer chromatography with cellulose as the matrix and isopropanol:1.1M ammonia: water (10:1:1) as the solvent. The area of the chromatogram corresponding to authentic IAA (Rf. 0.4-0.6) was scraped off and equilibrated with 25 agarose blocks in 0.2 ml of water. The auxin content was then measured by the *Avena* curvature bioassay. Auxin quantity was computed from the line (determined by the method of least squares) relating log IAA concentration to degrees curvature.

Auxin Formation. The effect of the ethylene pretreatment on auxin formation from L-tryptophan by apical bud tissue was studied following procedures reported earlier (Valdovinos, Ernest, and Henry, 1967). The enzyme breis were prepared as described in the above section on auxin destruction. Chloramphenicol at a final concentration of 50 µg/ml was included in the reaction mixtures. Agarose blocks (block size 2x2x2 mm) were present in the reaction mixture during the 45 min incubation period. These blocks were removed at the end of the incubation period and assayed for auxin using the *Avena* curvature bioassay.

IAA Conjugation. Eighteen hours after beginning the ethylene treatment, the plants were treated with auxin using methods similar to those of Morris, Briant and Thomson (1969) and Morris (1970). 0.025 ml containing 0.5 µg IAA-1-^{14}C (0.04 µCi) in 0.15 percent tween-20 was applied to the apex of each of 10 *Coleus* plants. After 4.5 hours (in the light under continuous exposure to ethylene) the apical portions of the stems were collected, the leaves removed and the remaining 1 to 2 cm of stem rinsed to remove residual IAA. The stem tips were extracted overnight in a freezer with 85 percent ethanol, ground in a blendor, then re-extracted twice with cold 85 percent ethanol. The combined extracts were evaporated to a small volume and separated on thin layer chromatography plates of cellulose in isopropanol: concentrated NH_4OH: water (8:1:1). Portions of the plate were scraped into counting vials and the radioactivity counted as described above.

11. Effect of ethylene on the ultrastructure of abscission cells

The experimental plants (*Nicotiana tabacum* L. cv. Little Turk.) were grown in the greenhouse to the flowering stage. Terminal 25 to 30 cm portions of the plants were then removed and placed in water (basal ends submerged) in an environmental chamber (Sherer Model CEL 25-7 HL). The experimental plant material was exposed to a light intensity of 1600 ft c, a temperature of $27 \pm 1°C$ and a relative humidity of 50 ± 5 percent. A continuous flow of air (60 liter/min) with or without ethylene (5 μl/liter) was provided to the chamber. The air supplied by an air compressor was first passed through an ethylene scrubber. The scrubber consisted of a canister containing Purafil (pellets one-eighth inch in diameter composed of activated Al_2O_3 impregnated with $KMnO_4$), available from Marbon Division, Borg-Warner Corporation, Washington, West Virginia. The canister was constructed of degreased galvanized pipe (10 by 40 cm) with circles of screen wire in drilled pipe caps fitted with laboratory gas outlets. Teflon tape rather than oil was applied to the pipe threads to allow easy disassembly for refilling and to prevent contamination of the Purafil.

Segments of tissue containing the abscission zone were removed from the pedicels of the flowers at hourly intervals starting 1 hr after beginning the ethylene treatment. The tissue was prepared for viewing under the electron microscope as reported earlier (Jensen and Valdovinos, 1967).

During the course of the experiment, the break strength of the abscission zone was determined using a method previously reported (Dela Fuente and Leopold, 1968). The pedicel was pressed against the pan of a top-loading Mettler balance (precision ± 0.01 g). For each determination, an average break strength for 10 pedicels was calculated and plotted versus time.

III. RESULTS

1. Effect of ethylene on auxin metabolism

The effect of ethylene pretreatment on the capacity of *Coleus* stem tissue to transport IAA-1-^{14}C is shown in table I. The amount of labeled auxin delivered to the receiver blocks during the five hour diffusion period is decreased by nearly 50 percent. There appear to be small increases in the uptake of auxin from the donor blocks by the stem segments, as well as in the amount of radioactivity present in the tissue at the end of the transport period.

Table I. Effect of ethylene on basipetal transport of IAA-1-^{14}C. Donor blocks containing an average of 32,600 cpm were applied to the apical end of segments 5 mm in length removed from the uppermost internode of the stem.

Treatment	cpm per receiver	uptake, cpm per section	Radioactivity, cpm per section
Control	442	13,213	5,590
Ethylene, 100 μl/l	228	15,243	6,262

Table II. Effect of ethylene on decarboxylation of IAA-1-^{14}C by apical bud and leaf tissue. The amount of radioactivity provided in the medium was 140,000 cpm.

Treatment	$^{14}CO_2$, cpm obs.	$^{14}CO_2$, cpm/g fresh wt/hr
Control	8,485	7,071
Ethylene, 100 µl/l	8,450	7,041

Table III. Effect of ethylene on levels of extractable auxin in apical bud and leaf tissue. Plants used in the experiments conducted in August were exposed to a continuous flow of air with or without ethylene (25 µl/liter and 100 µl/liter, respectively). Plants used in the remaining experiments were placed in polyethylene bags, treated with ethylene 100 µl/liter) and were placed in the growth chamber.

Month	degrees curvature with SE		Equivalent IAA ng/g fresh wt		t values
	Control	Ethylene	Control	Ethylene	
August	19.3 ± 1.5	9.5 ± 1.3	10.1	4.2	4.8
August	14.9 ± 0.9	10.2 ± 1.1	6.9	4.4	3.3
January	18.3 ± 1.2	3.2 ± 0.6	12.1	1.7	11.1
January	13.9 ± 1.2	8.0 ± 1.2	11.1	5.3	3.4
February	18.7 ± 1.5	15.8 ± 1.2	11.0	9.9	1.5
February	11.9 ± 0.9	7.2 ± 1.0	8.4	4.4	3.4
March	8.4 ± 0.7	5.2 ± 0.8	3.8	2.8	2.8
May	7.1 ± 1.0	2.5 ± 0.7	3.1	1.9	3.7
Averages	14.1 deg.	7.7 deg.	8.3 ng	4.3 ng	

An investigation of the effect of ethylene on decarboxylation of IAA-1-^{14}C by leaf and apical bud tissue shows that the gas has no effect on the breakdown of the auxin (table II). Earlier reports on the breakdown of auxin by *Coleus* tissue are in agreement with this observation (Valdovinos, Ernest, and Henry, 1967).

The results of investigations of the effects of ethylene on levels of extractable auxin from *Coleus* plants are shown in table III. An IBM computer, Model 1130, was used to compute and plot best fit straight lines for the curvature obtained in the *Avena* curvature test from different concentrations of authentic IAA. The auxin activity was determined from the computed straight line. In all experiments except one, results of t tests show that there is a significantly lower level of extractable auxin in ethylene treated plants. The significant difference is at a confidence level of 99 percent or higher.

Table IV. Effect of ethylene on auxin formation and levels of extractable auxin in apical bud and leaf tissue. The data presented are from the same experiment.

Degrees curvature with SE		Equivalent IAA ng/ g fresh wt/hr		t values
Control	Ethylene, 25 µl/l	Control	Ethylene	
Auxin formed from L-Tryptophan				
10.5 ± 0.9	5.5 ± 0.9	6.1	3.7	3.8[1]
Extractable auxin				
18.9 ± 0.9	9.7 ± 0.8	11.1	5.1	7.5[1]

[1] significant at the 99% probability level.

Table IV shows the results of a representative experiment in which the capacity of apical bud tissue to form auxin is compared with levels of extractable auxin present in the tissue. The rate of auxin formation by enzyme breis from ethylene treated plants is roughly one-half that occurring in incubation mixtures containing enzyme from untreated plants. The amount of auxin extracted from tissue harvested from treated plants is also about one-half the amount of extractable auxin present in untreated plants.

The results of studies of the effect of ethylene pretreatment on auxin conjugation show that there is an increased level of conjugation. Thin layer chromatograms of the extracts from stem tissue subtending the apex showed an increase from 340 cpm to 770 cpm in a radioactive substance with an Rf (0.1) equivalent to that of authentic indoleacetyl aspartic acid. The amount of free IAA-1-^{14}C recovered from the thin layer plates was 290 cpm for the control plants and 285 cpm for the ethylene treated plants.

11. Effect of ethylene on the ultrastructure of abscission cells

The break strengths of the abscission zone associated with increasing periods of ethylene treatments were determined. At 2 hr, pedicels bend rather than break. Break strength decreases exponentially with time, beginning at 2.5 hr after commencement of ethylene treatment. The average break strength decreases from 40g at 2.5 hr to about 5 g at 5 hr after treatment.

Examination of the abscission cells at the fine structural level reveals the presence of rough endoplasmic reticulum (ER) 2 hr after exposure of the tissue to ethylene. The amount of rough ER present in the tissue during the first 2 hr of ethylene treatment approximates that observed in abscission cells of naturally abscising tobacco flower pedicels (Jensen and Valdovinos, 1968). Rough ER becomes increasingly abundant in abscission cells exposed to ethylene for 3 and 5 hr (Fig. 1). At this stage, changes are also observed in the vesicles of the Golgi (Fig. 2). Predominantly more electron dense vesicles are produced in the abscission cells of plants treated with ethylene.

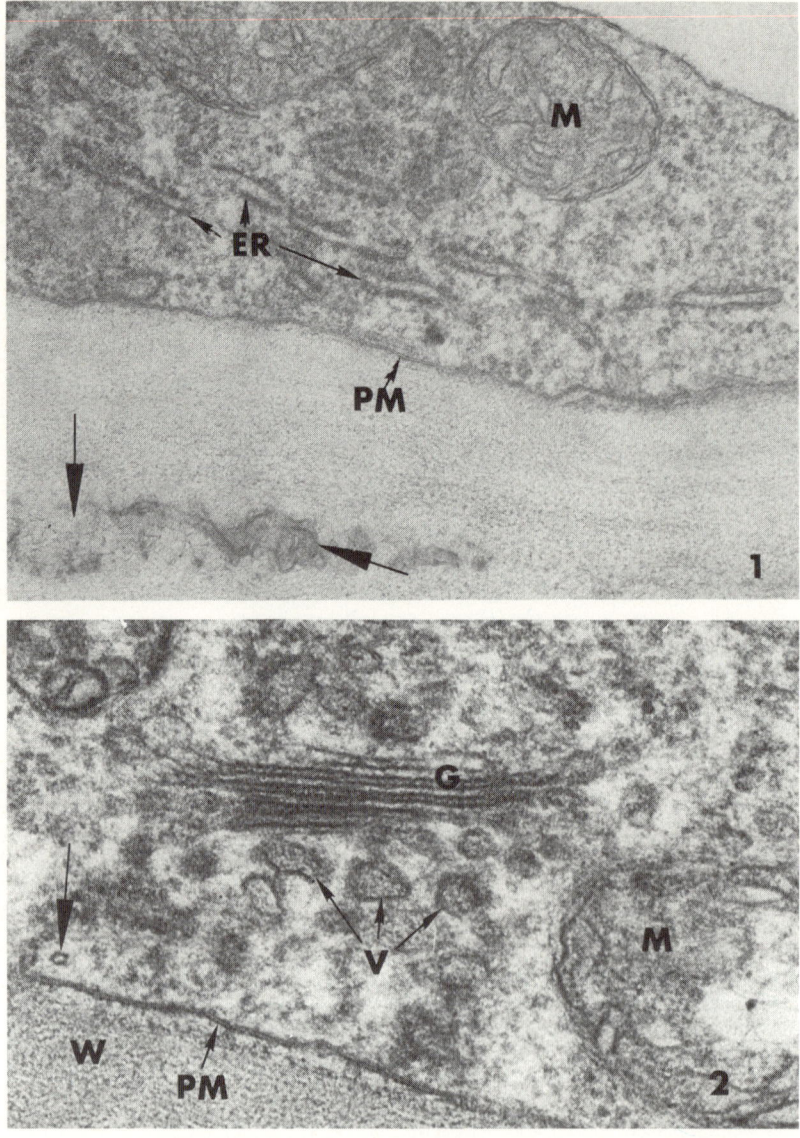

Fig. 1. Portion of an abscission cell of tobacco flower pedicel treated with ethylene for 5 hrs. Note the rough endoplasmic reticulum in the cell (ER). Also visible in the cell are mitochondria (M), the plasma membrane (PM) and an area of wall breakdown (arrows). X 64,000.

Fig. 2. Portion of an abscission cell of tobacco flower pedicel treated with ethylene for 5 hrs. Note the vesicles (V) which are being produced by the Golgi (G). Also visible in the section is a mitochondrion (M), a microtubule (arrow), the plasma membrane (PM) and part of the cell wall (W). X 132,000.

IV. DISCUSSION

1. Effect of ethylene on auxin metabolism

The effect of ethylene in inhibiting transport of IAA in the stem tissues of *Coleus* is similar to that reported for many other plant species. The effect of ethylene on transport cannot be explained in terms of increased auxin destruction, because as seen in this report and in a previous one (Valdovinos, Ernest, and Henry, 1967), ethylene does not enhance decarboxylation of the hormone. If ethylene is acting at sites for auxin transport within the tissue, an accumulation of auxin should occur at regions of synthesis, i.e. in the leaf tissues of the upper region of the plant. This does not happen in *Coleus*.

In experiments where auxin formation and the levels of extractable auxins were examined, there was less auxin formation and less extractable auxin in ethylene-treated plants. Preliminary studies (Valdovinos, Ernest, and Henry, 1967) showed a similar correlation between lowered auxin formation and lower levels of diffusible auxin in ethylene-treated *Coleus* plants.

In studies of the conjugation of IAA-1-^{14}C, it was found that ethylene pretreatment of *Coleus* plants increased the level of a substance characterized as IAAsp (indoleacetylaspartic acid). This substance is a known conjugate of auxin in plants (Andreae and Good, 1955; Good, Andreae, and Ysselstein, 1956). Re-chromatogramming by thin layer chromatography of the substance in tertiary butanol: acetic acid; water (3:1:1) also showed the material to have the same Rf as IAAsp. The level of radioactivity chromatogramming as IAA-1-^{14}C was not significantly higher in ethylene-treated plants than in untreated plants. The effect that increased conjugation of auxin may have on diffusible auxin levels would appear to be one of lowering levels of auxin moving through the tissue (Lantican and Muir, 1969). Beyer and Morgan (1970) have recently reported that an increased conjugation of auxin also occurs in stem sections of cotton plants which were pretreated with ethylene. However, they attribute this to a higher level of auxin present in the tissue, presumably brought about by an inhibition of auxin transport by the pretreatment. Beyer and Morgan (1970) also showed that when NAA-1-^{14}C was applied to stem sections, no increased conjugation of auxin was observed, although the capacity of the tissues to transport auxin was greatly reduced.

Ethylene pretreatment of *Coleus* plants lowers the capacity of the plant to form auxin. The most reasonable explanation for the reduced level of extractable and diffusible auxin in ethylene-treated *Coleus* plants is that less auxin is being formed in the tissue. Further studies are under way in an attempt to elucidate the action of ethylene in its effect on auxin formation by plant tissues.

11. Effect of ethylene on the ultrastructure of abscission cells

Rough endoplasmic reticulum (ER) has been observed in abscission cells prior to natural abscission (Jensen and Valdovinos, 1968). The experiments reported here show rough ER to appear in abscission cells 2 hr after exposure to ethylene. A loss of break strength in the abscission zone began at 2.5 hr. There are substantial biochemical and histochemical data which show increased RNA and protein synthesis in ethylene-treated abscission tissue of bean explants (Holm and Abeles, 1967; Abeles, 1968 and 1969; Webster, 1968). It is tempting to suggest that the site for the new protein synthesis is rough ER such as has been observed in these studies. It might be further speculated that the ribosomes associated with this ER are a result of new RNA synthesis.

The changes observed in the vesicles of the Golgi bodies may be related to the increased activity of the ER and the subsequent breakdown of the cell wall. Whaley and Mollenhauer (1963) found Golgi bodies to be associated with cell wall metabolism. In earlier studies where changes were observed in the fine structure of abscission cells undergoing natural abscission, no differences could be observed in the Golgi bodies of abscising and nonabscising cells (Jensen and Valdovinos, 1967 and 1968).

Other investigators (Bornman, 1967; Morré, 1968) also failed to note any changes in the activity of the Golgi bodies in abscission cells of bean, *Coleus,* and cotton leaf petioles. The short time interval for the occurrence of abscission in ethylene treated tobacco pedicels may allow for detection of changes in the activity of Golgi bodies.

The nature of the activity of Golgi bodies and the function of rough ER in abscission cells is unknown. However, the synthesis of wall-degrading enzymes and their secretion may be involved. Horton and Osborne (1967) and Abeles (1969) have reported that ethylene increases the synthesis of cellulase enzymes by abscission tissue. Halperin (1969) has reported increased acid phosphatase activity at one face of dictyosomes in cultured cells of *Daucus carota*. These cells, which were in a senescent stage, also contained acid phosphatase activity in dictyosome-derived vesicles. Halperin also observed acid phosphatase to be localized in the cell walls of the carrot tissue. This report is of particular interest because wall disintegration in the senescent carrot cells appears essentially the same as in tobacco pedicel abscission cells undergoing cell separation (Valdovinos and Jensen, 1968). Investigations are under way to determine the nature of the activity exhibited by Golgi bodies during abscission and the function of the rough ER in abscission cells.

V. SUMMARY

An investigation of the effects of ethylene pretreatment on several facets of auxin metabolism in *Coleus* revealed a number of changes presumably induced by the gas hormone. Transport of IAA-1-^{14}C in excised segments of the uppermost internode was inhibited by about 50 percent. Decarboxylation of IAA-1-^{14}C by enzyme breis was not affected by the pretreatment. Levels of extractable native auxin in upper leaf and apical bud tissue of the pretreated plants were approximately one-half those present in untreated plants. The rate of formation of auxin from tryptophan by enzyme breis from pretreated plants was approximately one-half that occurring in incubation mixtures containing enzyme from untreated plants. The conjugation of IAA-1-^{14}C in a form characterized chromatographically as indoleacetylaspartic acid was increased twofold in the upper stem region of plants pretreated with ethylene.

The effect of ethylene on abscission of flower pedicels of tobacco plants has been investigated. Two hours after exposure to ethylene, pedicels bend rather than break in response to applied force. After 2.5 hr exposure to the gas, pedicels break at the abscission zone under an applied force of 40 g. The break strength of the abscission zone decreases exponentially with time to 5 g at 5 hr after beginning of the ethylene treatment. An examination of the tissue at the fine structural level 2 hr after exposure to ethylene reveals the accumulation of rough endoplasmic reticulum (ER) in the abscission cells. Rough ER becomes increasingly abundant by 5 hr exposure of the tissue to ethylene. Changes are also observed in the vesicles of the Golgi. Predominantly more electron dense vesicles are produced in the abscission cells of plants treated with ethylene.

VI. REFERENCES

ABELES, F.B. and B. RUBENSTEIN. 1964. Regulation of ethylene evolution and leaf abscission by auxin. Plant Physiol. 39: 963-969.
ABELES, F.B. 1968. Role of RNA and protein synthesis in abscission. Plant Physiol. 43: 1577-1586.
ABELES, F.B. 1969. Abscission: Role of Cellulase. Plant Physiol. 44: 447-452.
ANDREAE, W.A. and N.E. GOOD. 1955. The formation of indoleacetylaspartic acid in pea seedling. Plant Physiol. 30: 380-382.
BEYER E.M. Jr. and P.W. MORGAN. 1970. Effect of ethylene on the uptake, distribution, and metabolism of indoleacetic Acid-1-^{14}C and -2-^{14}C and naphthaleneacetic acid 1-^{14}C. Plant Physiol. 46: 157-162.
BORNMAN, C.H. 1967. Some ultrastructural aspects of abscission in *Coleus* and *Gossypium*. S. Afr. J. Sci. 63: 325-331.

BURG, S.P. and E.A. BURG. 1967. Inhibition of polar auxin transport by ethylene. Plant Physiol. 42: 1224-1228

DELA FUENTE, R.K. and A.C. LEOPOLD. 1968. Senescence processes in leaf abscission. Plant Physiol. 43: 1496-1502.

DELA FUENTE, R.K. and A.C. LEOPOLD. 1969. Kinetics of abscission in the bean petiole explant. Plant Physiol. 44: 251-254.

GOOD, N.E., W.A. ANDREAE, and M.W.H. VAN YSSELSTEIN. 1956. Studies on 3-indoleacetic acid metabolism. II. Some products of the metabolism of exogenous indoleacetic acid in plant tissues. Plant Physiol. 31: 231-235.

GUTTENBERG, H. VON and E. STEINMETZ. 1947. The effects of ethylene on growth hormone and growth. Pharmazie 2: 17-21.

HALPERIN, W. 1969. Ultrastructural localization of acid phosphatase in cultured cells of *Daucus carota*. Planta 88: 91-102.

HOLM, R.E. and F.B. ABELES. 1967. Abscission: The role of the RNA synthesis. Plant Physiol. 42: 1094-1102.

HORTON, R.F. and D.J. OSBORNE. 1967. Senescence, abscission, and cellulase activity in *Phaseolus vulgaris*. Nature. 214: 1086-1089.

JACKSON, M.B. and D.J. OSBORNE. 1970. Ethylene, the natural regulator of leaf abscission. Nature 225: 1019-1022.

JENSEN, T.E. and J.G. VALDOVINOS. 1967. Fine structure of abscission zones. I. Abscission zones of the pedicels of tobacco and tomato flowers at anthesis. Planta 77: 298-318.

JENSEN, T.E. and J.G. VALDOVINOS. 1968. Fine structure of abscission zones. III. Cytoplasmic changes in abscising pedicels of tobacco and tomato flowers. Planta. 83: 303-313.

LAAN, P.A. VAN DER. 1934. Effect of ethylene on growth hormone formation in *Avena* and *Vicia*. Rec. Trav. Botan. Néerl. 31: 691-742.

LANTICAN, B.P. and R.M. MUIR (1969). Auxin physiology of dwarfism in *Pisum sativum*. Physiol. Plant. 22, 412-423.

MICHENER, H.D. 1938. The action of ethylene on plant growth. Amer. J. Bot. 25: 711-720.

MORGAN, P.W. and H. W. GAUSMAN. 1966. Effects of ethylene on auxin transport. Plant Physiol. 41: 45-52.

MORRÉ, J.D. 1968. Cell wall dissolution and enzyme secretion during leaf abscission. Plant Physiol. 43: 1545-1559.

MORRIS, D.A. R.E. BRIANT and P.G. THOMSON. 1969. The transport and metabolism of 14C-labelled indoleacetic acid in intact pea seedlings. Planta. 89: 178-197.

MORRIS, D.A. 1970. Light and the transport and metabolism of indoleacetic acid in normal and albino dwarf pea seedlings. Planta. 91: 1-7.

MUIR, R.M. and J.G. VALDOVINOS. 1970. Gibberellin and auxin relationships in abscission. Amer. J. Bot. 57: 288-291.

PERLEY, J.E. and B.B. STOWE. 1966. The production of tryptamine from tryptophan by *Bacillus cereus* (KVT). Biochem. J. 100: 169-174.

VALDOVINOS, J.G. and L.C. ERNEST. 1966. Gibberellin enhanced CO_2 release from tryptophan-1-C-14 in plant apical tissue. Plant Physiol. 41: 1551-1552.

VALDOVINOS, J.G. and J.E. PERLEY. 1966. Metabolism of tryptophan in petioles of *Coleus*. Plant Physiol 41: 1632-1636.

VALDOVINOS, J.G. and L.C. ERNEST. 1967. Effect of protein synthesis inhibitors, auxin, and gibberellic acid on abscission. Plant Physiol. 20: 1027-1038.

VALDOVINOS, J.G., and C. ERNEST, E. W. HENRY. 1967. Effect of ethylene and gibberellic acid on auxin synthesis in plant tissues. Plant Physiol. 42: 1803-1806.

VALDOVINOS, J.G. and T.E. JENSEN. 1968. Fine structure of abscission zones. II. Cell wall changes in abscising pedicels of tobacco and tomato flowers. Planta. 83: 295-302.

WEBSTER, B.D. (1968). Anatomical aspects of abscission. Plant Physiol. 43, 1512-1544.

WHALEY, W.G. and H.H. MOLLENHAUER. 1963. The Golgi apparatus and cell plate formation- a postulate. Planta. 77: 298-318.

FUNCTIONS OF NATURALLY PRODUCED ETHYLENE IN ABSCISSION,
DEHISCENCE AND SEED GERMINATION

Page W. Morgan, D.L. Ketring[1], Elmo M. Beyer, Jr.[2] and John A. Lipe

Department of Plant Sciences Texas A&M University College Station, Texas 77843, U.S.A.

[1] Agricultural Research Service, United States Department of Agriculture.
[2] Present Address: Central Research Department, E. I. DuPont and Company, Wilmington, Delaware, U.S.A.

I. INTRODUCTION

It is well established that plants produce ethylene as a natural emanation; however, whether ethylene has a functional role in plant growth has often been questioned. Studies of auxin-induced ethylene synthesis and ethylene-mediated reduction of auxin transport capacity (Hall and Morgan, 1963; Morgan, Beyer and Gausman, 1968) have contributed to the recognition of auxin-ethylene interactions as an important area of plant hormone physiology. During recent months these interests have led to studies of some related aspects of ethylene physiology in two plant systems, the role of native ethylene in abscission phenomena in the cotton plant and its involvement in peanut seed germination.

II. MATERIALS AND METHODS

Cotton plants (*Gossypium hirsutum* L., cv. Stoneville 213) were grown in a greenhouse or in a controlled environment room as previously described (Beyer, 1969 and Beyer and Morgan, 1969a). All gases were determined by standard gas chromatographic techniques (Ketring and Morgan, 1969a). For abscission determinations, plants enclosed in bell jars were treated with ethylene and abscission in response to a 3 g force was observed at 12 hr intervals. After each observation, containers were re-sealed and fumigated to treatment levels duplicated ±5% as verified by analysis. Ethylene in control containers never exceeded 0.015 ppm (0.015 µl ethylene/liter air).

For auxin transport inhibition studies, plants were treated with ethylene or room air in 160 liter plexiglass chambers. Carbon dioxide was never higher than 0.20% and O_2 was never below 20%. After treatment, transport of NAA-1-^{14}C was determined in petiole sections as previously described (Beyer and Morgan, 1969a). To evaluate natural transport capacity and ethylene synthesis, 25 green, yellow-green, or yellow (attached, turgid) cotyledons were detached and used. Ethylene production of 15 petioles between 4 and 10 hr after detachment and IAA-1-^{14}C transport for the 10 other petioles was determined. In correlative experiments groups of 20 cotyledons were harvested for the color classes and their internal ethylene concentrations were determined by the vacuum extraction technique (Beyer and Morgan, 1970).

Ethylene production by detached cotton fruits harvested at weekly intervals after anthesis was determined (Lipe and Morgan, 1970). Dehiscence of detached fruits was observed with ethylene or room air in 60 liter plexiglass containers. Dehiscence was observed daily and containers were ventilated and refumigated.

Ethylene and carbon dioxide determinations and germination techniques for peanut seeds (*Arachis hypogaea* L.) have been described (Ketring and Morgan, 1969ab). In addition, single Spanish-type seeds were germinated in individual chambers (about 100 ml volume) provided with flowing air at 10 ± 1 ml/min.

III. RESULTS AND DISCUSSION

Ethylene-mediated auxin transport reduction and abscission

In cotton, auxin transport inhibition by ethylene occurs *in vivo* (Beyer and Morgan, 1969b) and the timing and magnitude of the response are adequate to implicate it in defoliation of plants by exogenous ethylene (Beyer and Morgan, 1969a, 1969b). The next consideration was whether naturally produced ethylene regulates auxin transport *in vivo* and thereby regulates abscission of intact leaves (Beyer, 1969). Although ethylene has long been known as a defoliant, not until the current year was direct evidence published implicating native ethylene in abscission (Jackson and Osborne, 1970). We found that both auxin transport inhibition and abscission acceleration occur when plants were treated with levels of ethylene around 1 ppm or higher, but addition of 0.1 to 0.08 ppm was without effect on either process in 24-hr (Figure 1). We then determined whether transport capacity and ethylene production change as leaves naturally senesce and abscise. Auxin transport capacity declined over 7 fold between petioles of cotton cotyledons of equal age which differed in visible symptoms of senescence (green versus yellow) (Table 1). In the same populations, ethylene production rates of about 2.5 mμl/g/hr were observed and subsequently, we detected internal levels of about 0.7 ppm. These trends for ethylene levels to differ significantly in the cotton cotyledonary petiole with the degree of visible senescence (yellowing) have been verified in several experiments.

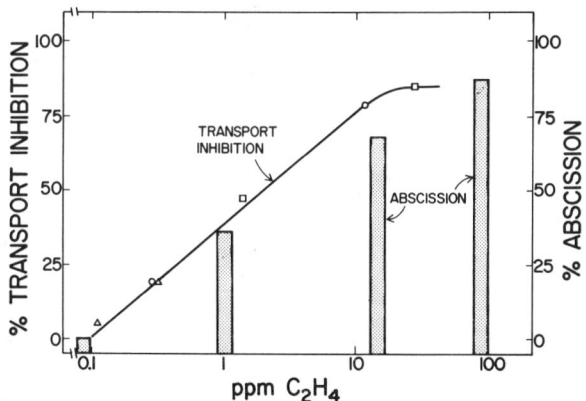

Fig. 1. Effect of 24-hour exposure to ethylene on percent abscission of cotton cotyledons and on percent reduction of auxin transport capacity in cotyledonary petiole segments. Abscission data are averages of 3 experiments using 6 plants, 25 to 29 days old, per treatment. Auxin transport data are averages of 3 experiments, separate from the abscission experiments, which included 6 to 8 plants, 25 days old, per treatment.

These data bring together in one natural physiological system the events concerning ethylene and its initial role in abscission. The levels of ethylene observed appear to be in the physiological range because: (a) naturally rising ethylene production and declining auxin transport capacity correlate with abscission, (b) exogenous ethylene at 1 ppm caused reduced auxin transport and increased abscission (Figure 1) and cotyledons naturally producing ethylene at 2.5 mμl/g/hr were shown to contain over 0.7 ppm ethylene (Table 1), (c) other work suggests that production of 2 to 5 mμl/g/hr approximates from 1 ppm to a few ppm in tissue (Burg, 1968), (d) it has been shown that ethylene production rates and internal levels are from 2 to 6 times higher in cotton leaf petioles than in leaf blades (McAfee, 1970) yet the data in Table 1 are for entire cotyledons, and (e) the yellowing and abscission of cotton cotyledons, under conditions employed here, requires around 11 days thus providing adequate time for levels of ethylene slightly above a threshold level to reduce the supply of auxin reaching the abscission zone.

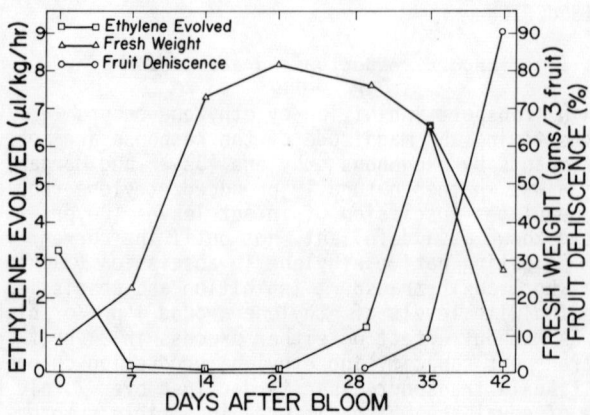

Fig. 2. Time course of ethylene production, weight and dehiscence of cotton fruits following anthesis.

Table 1. Basipetal auxin transport capacity, ethylene production rate and internal ethylene concentration of cotton cotyledons of three physiological ages.

Cotyledon Description	% ^{14}C Auxin Transported	Ethylene mμl/gm/hr	μl/liter air
Green	23[1]	0.95[1]	0.24[2]
Yellow-Green	9	1.53	0.33
Yellow	3	2.49	0.73

[1] Data averages of 4 experiments, 2 employing groups of greenhouse plants 51 and 54 days old and 2 employing groups of growth room plants 36 and 37 days old.
[2] Data averages of 2 experiments separate from others in this table.

Since cotton fruits exhibit a specialized type of abscission, dehiscence of the carpel walls, we investigated the possible involvement of ethylene in this process (Figure 2). After anthesis ethylene production rates are well below 1 mμl/g/hr until a week before initiation of dehiscence when production begins to rise. By the time dehiscence has been initiated, ethylene production rates are above 6 mμl/g/hr and essentially complete dehiscence occurs a week later. The rise in ethylene production and initiation of dehiscence precedes any significant decline in fresh weight. Similar data have been obtained from pecan fruits (Lipe and Morgan, 1970). Dehiscence of detached, mature cotton fruits was accelerated by ethylene concentrations ranging from 10 to 0.1 ppm. The degree of stimulation was concentration dependent, and detached, control fruits dehisced before controls left on plants. Young and mature fruits dehisced to an equal degree in 10 ppm ethylene, but when exposed to 1.0 ppm, young fruits dehisced more slowly than mature ones. Thus, dehiscence of cotton fruits show the events becoming familiar in ethylene-mediated plant responses, a hastening by exogenous ethylene, a rise in ethylene production prior to an observed response (Figure 2) and production at a rate that brings the internal ethylene concentration to a physiologically active level.

These findings suggest that ethylene functions in the regulation of natural cotton leaf abscission and fruit dehiscence. They also provide an overview of natural abscission into which the aging-ethylene hypothesis (Abeles, 1968) may be fitted. Thus, ethylene appears to play a dual role in abscission by first lowering the auxin supply to the separation layer and then inducing synthesis of hydrolytic enzymes (Abeles, 1968). The need for a reduction in auxin supply to the separation layer as a part of natural abscission is also indicated by the fact that application of auxin will prevent abscission when plants are treated with ethylene (Hall, 1952).

Jackson and Osborne (1970) have recently observed a progressive increase in ethylene production by tissue surrounding the abscission zone of bean explants and correlations between ethylene production and abscission in two woody species. Their data and the present findings support a general role of endogenous ethylene in abscission control.

Germination of non-dormant Spanish-type peanut seeds

Non-dormant seeds have been reported to respond to ethylene treatment by improved germination and growth (Balls and Hale, 1940; Haber, 1926; Ruge, 1947). However, ethylene as a natural volatile from seeds was not conclusively shown until Meheriuk and Spencer (1964) studied germination of oat seeds. Figure 3A shows that non-dormant Spanish-type peanut seeds produced ethylene as a natural volatile during germination (Ketring and Morgan, 1969a). These freshly harvested, non-dormant peanut seeds, emerged uniformly, the hypocotyl-radicle attained a length of 1 to 5 mm, and the seeds exhibited an ethylene maximum at 24 hr. Maximum carbon dioxide production occurred at 48 hr of germination, 24 hr after the ethylene maximum. The major site of ethylene production by peanut seeds was in the growing embryonic axis (Ketring and Morgan, 1969a). The data suggests that a large amount of ethylene production is associated with the initiating growth phases of the embryonic axis and that ethylene may have a function in these growth phases.

Germination of single peanut seeds shows that ethylene production begins to rise prior to any visible signs of growth, it peaks at emergence of the hypocotyl-radicle (Figure 3B, E_1). A second ethylene maximum occurs when the radicle emerges from the hypocotyl (Figure 3B, E_2). The peak of ethylene production from large numbers of seeds shown in Figure 3A is then an average value of the ethylene production occurring from both maxima shown in Figure 3B where single seeds were followed during the first 24 hr of germination. That the stage of growth is the critical aspect and not the time of germination is indicated by the following: (a) in lots of seed of lower vigor, emergence (E_1 and E_2) may be delayed, yet ethylene maxima occur at the times of emergence; (b) among the relatively uniform seeds of high vigor used for experiments depicted in Figure 3, radical emergences (E_2) occurred from 22 to 26 hr of germination and ethylene maxima occurred at the times of emergence. A seed that completes emergence at 22 hr may achieve a hypocotyl-radicle length of 5.5 to 6.0 mm by 26 hr of germination, ethylene production declining after the second maximum (Figure 3B). Carbon dioxide production continued to rise during these initial hours of germination (Figure 3B) and had not reached a maximum which agrees with the CO_2 pattern for large numbers of seeds (Figure 3A).

Germination of dormant Virginia-type peanut seeds

The first successful means of breaking dormancy of Virginia-type peanut seeds was heat-treatment (Bailey et al., 1958). When dormancy was broken in this manner, a rise in ethylene production accompanied their germination (Figure 3A, Ketring and Morgan, 1969a). Toole et al. (1964) found that 100 ppm of ethylene gas would induce dormant Virginia-type peanuts to germinate. We confirmed these results and defined the ethylene levels necessary to break dormancy (Ketring and Morgan, 1969a and b).

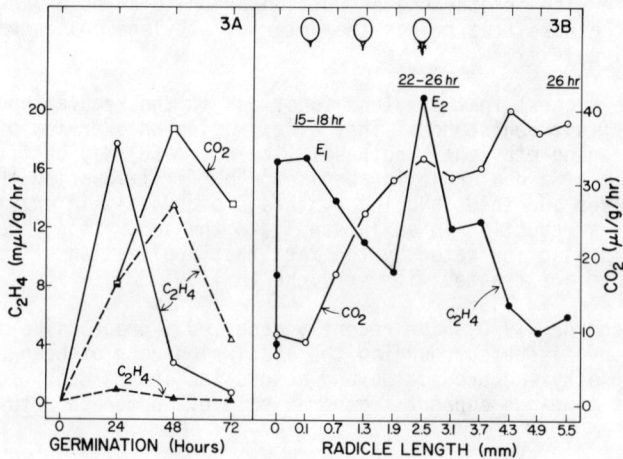

Fig. 3A. The time course of ethylene and carbon dioxide production by germinating peanut seeds. Solid lines, Spanish-type and dashed lines Virginia-type seeds. CO_2 is shown only for Spanish-type seeds. Open triangles, Virginia-type seeds heat-treated to break dormancy; solid triangles, untreated. Data calculated from either 100 seeds (fresh weight) per sample, Spanish-type, or 20 seeds (dry weight) per sample, Virginia-type seeds. One-half open symbols indicate a common point.

Fig. 3B. Ethylene and carbon dioxide production during early phases of non-dormant, Spanish-type peanut seed germination. E_1, emergence of the hypocotyl-radicle; E_2, emergence of the radicle. Hours indicated are hours of germination. Each point represents the mean C_2H_4 and CO_2 produced over a 0.5 mm increment of growth beginning at 0.1 mm; 0.1 to 0.6, 0.7 to 1.2 mm, and so on, for several individual seeds.

Table 2. The germination and ethylene production of dormant Virginia-type peanut seeds as the inherent dormancy declines during storage in sealed containers at 3C.

Weeks After Harvest	Percent Germination at 96 hr		Ethylene mul/50 seeds/hr	
	Apical	Basal	Apical	Basal
6	4	4	-	-
8	8	5	8.5	6.0
12	15	5	-	-
14	32	8	22.9	13.2

Following several weeks of afterripening in sealed containers at $3°$, there is a natural increase of germination (decline of dormancy) and a corresponding ability of the seeds to produce ethylene during germination (Table 2). Apical seeds in the pod, those most distal from the attachment to the plant, are less dormant than the basal seeds. The less dormant apical seeds afterripened more rapidly than the more dormant basal seeds and consequently, germinated to a greater extent and showed a greater increase in ability to produce ethylene.

Ethylene gas applied at 7 ppm for 48 hr effectively induced germination of the dormant seeds, inducing a maximum germination above the control at 48 hr of germination (Figure 4). The decline following this maximum was due to continued germination of the control seeds. The fact that ethylene could induce the more dormant basal seeds to germinate in a manner similar to the less dormant apical seeds indicates it has a natural role in breaking dormancy of these seeds (Figure 4). There was also an enhanced production of ethylene by the ethylene treated seeds (Figure 4) that maximized at 48 hr while CO_2 production continued to rise during the 96 hr test. Ethylene has recently been implicated as a regulator of dormancy in other seeds also (Esashi and Leopold, 1969; Stewart and Freebairn, 1969).

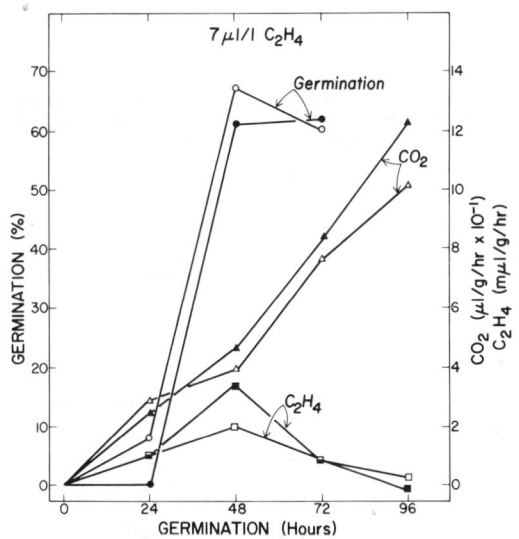

Fig. 4. The effect of 7 μl/l ethylene applied for 48 hr on germination, CO_2 and ethylene production of dormant Virginia-type seeds, 1969 crop. Solid symbols basal seeds; open symbols, apical seeds. One-half solid symbol, common point. Control apical seeds germinated 41 ± 7% and basal seeds 23 ± 9% at 96 hr. Total germination with ethylene treatments was 99 ± 1% and 84 ± 8% at 96 hr for apical and basal seeds, respectively. Ethylene and CO_2 production rates are on a fresh weight basis and expressed as the level above the controls.

Observations from these data that support the concept that ethylene is a natural regulator of abscission and peanut seed germination are: (1) ethylene production by non-dormant seeds and senescing leaves begins to rise prior to any visible signs of growth or abscission and maximizes at emergence of the hypocotyl-radicle and radicle and prior to abscission, (2) treatments that enhance the natural low levels of ethylene production by dormant seeds will stimulate germination, (3) ethylene gas alone is sufficient to stimulate its own synthesis and thus hasten abscission of leaves and fruits and stimulate the germination of dormant peanut seeds, and (4) the ability to produce increased amounts of ethylene occurs during the natural abscission and afterripening processes. Thus, although we are dealing with two quite different plant responses, the ethylene relationships of these processes are quite similar and indicate a function for naturally produced ethylene in abscission and peanut seed germination.

IV. SUMMARY

Ethylene reduces auxin transport capacity and induces abscission of cotton cotyledons at levels between 0.08 and 1 ppm. In plants where cotyledons are allowed to senesce naturally, auxin transport capacity declines and ethylene production increases to maximum rates of 2.5 mμl/g/hr and internal levels of 0.7 ppm, prior to abscission. The time course of ethylene production and cotton fruit dehiscence, a type of abscission phenomenon, revealed that ethylene production increased before dehiscence. A peak level of above 6 mμl/g/hr occurred as dehiscence began. Dehiscence of mature fruits was promoted markedly by 10 or 1 ppm ethylene and slightly by 0.1 ppm. The time course of ethylene production by germinating non-dormant Spanish-type peanut seeds show that ethylene release precedes emergence of the hypocotyl-radicle and that peaks of ethylene occur as the hypocotyl-radicle emerges from the seed and the radicle begins to grow from the hypocotyl. Dormant Virginia-type peanut seeds are stimulated to germinate by application of ethylene or heating, which promotes ethylene production. There is a natural decline in dormancy and increase in ethylene production capacity as Virginia-type seeds after-ripen. The findings suggest that ethylene functions in the regulation of abscission, dehiscence and seed germination. There are several similarities in the ethylene physiology of the three processes.

V. ACKNOWLEDGMENTS

This work was supported in part by National Science Foundation Grant GB-5640 and a grant from the Cotton Producers Institute. Data on dehiscence are from J. A. Lipe's Ph.D. dissertation research project.

VI. REFERENCES

ABELES, F. B. (1968). Role of RNA and protein synthesis in abscission. Plant Physiol. 43:1577-1586.

BAILEY, W. K., E. H. TOOLE, V. K. TOOLE and M. J. DROWNE. (1958). Influence of temperature on the afterripening of freshly-harvested peanuts. Proc. Amer. Soc. Hort. Sci. 71:422-424.

BALLS, A.K. and W.S. HALE. (1940). The effect of ethylene on freshly harvested wheat. Cereal Chem. 17: 490-494.

BEYER, E.M. Jr. (1969). A study of the effect of ethylene on auxin transport in cotton and its relationship to abscission. Ph.D. thesis. Texas A&M University, College Station, Texas, U.S.A., 204pp (also: Plant Physiol. 48: 208-212, 1971).

BEYER, E.M. Jr. and P. W. MORGAN. (1969a). Time sequence of the effect of ethylene on transport, uptake and decarboxylation of auxin. Plant and Cell Physiol. 10: 787-799.

BEYER, E.M., Jr. and P.W. MORGAN. (1969b). Ethylene modification of an auxin pulse in cotton stem sections. Plant Physiol. 44: 1690-1694.

BEYER, E.M., Jr. and P.W. MORGAN. (1970). A method for determining the concentration of ethylene in the gas phase of vegetative plant tissue. Plant Physiol. 46: 352-354.

BURG, S.P. (1968). Ethylene, plant senescence and abscission. Plant Physiol. 43: 1503-1511.

ESASHI, Y. and A. C. LEOPOLD. (1969). Dormancy regulation in subterranean clover seeds by ethylene. Plant Physiol. 44: 1470-1472.

HABER, E.S. (1926). A preliminary report on the stimulation of growth of bulbs and seeds with ethylene. Am. Soc. Hort. Sci. 23: 201-203.

HALL, W. C. and P. W. MORGAN (1964). Auxin-ethylene interrelationships. In: "Régulateurs Naturels de la Croissance Végétale," J.P. Nitsch, ed., Coll. Int. C.N.R.S. 123: 727-745.

HALL, W.C. (1952). Evidence on the auxin-ethylene balance hypothesis of foliar abscission. Botan. Gaz. 113: 310-322.

JACKSON, M.B. and D.J. OSBORNE. (1970). Ethylene, the natural regulator of leaf abscission. Nature 225: 1019-1022.

KETRING, D.L. and P.W. MORGAN. (1969a). Ethylene as a component of the emanations from germinating peanut seeds and its effect on dormant Virginia-type seeds. Plant Physiol. 44: 326-330.
KETRING, D.L. and P.W. MORGAN. (1969b). Physiology of oil seeds. I. Regulation of dormancy in Virginia-type peanut seeds. Plant Physiol. 45: 268-273.
KETRING, D.L. and P.W. MORGAN (1971). Physiology of oil seeds. II. Dormancy release in Virginia-type peanut seeds by plant growth regulators. Plant Physiol. 47: 488-492.
LIPE, J.A. and P.W. MORGAN (1970). Ethylene: Involvement in shuck dehiscence in pecan fruits (*Carya illinoensis* (Wang.) K. Koch). Hort.Science 5: 266-267.
McAFEE, J.A. (1970). A study of the distribution and production rates of ethylene within the cotton plant. M.S. thesis, Texas A&M University, College Station, Texas, U.S.A., 75 pp (also Plant & Cell Physiol. 12 (in press) 1971).
MEHERIUK, J. and M. SPENCER. (1964). Ethylene production during germination of oat seeds and *Penicillium digitatum* spores. Can. J. Botany 42: 337-40.
MORGAN, P.W., E. BEYER, Jr. and H.W. GAUSMAN. (1968). Ethylene effects of auxin physiology. In:"Biochemistry and Physiology of Plant Growth Substances." F. Wightman and G. Setterfield. eds. Runge Press, Ottawa, Canada. Pp. 1255-1273.
RUGE, U. (1947). Untersuchungen über Keimungsfördernde Wuchstoff. Planta, 35: 297-318.
STEWART, E.R. and FREEBAIRN, H.T. (1969). Ethylene, seed germination and epinasty. Plant Physiol. 44: 955-958.
TOOLE, V.K., W.K. BAILEY and E.H. TOOLE. (1964). Factors influencing dormancy of peanut seeds. Plant Physiol. 39: 822-32.

BIOSYNTHESIS OF ETHYLENE IN FRUIT TISSUES[1]

S.F. Yang and A.H. Baur

Department of Vegetable Crops, University of California, Davis, Calif.

[1] The work reported here was supported by a research grant from the National Science Foundation (GB-20336)

I. INTRODUCTION

Although methionine has been established as a precursor of ethylene (Baur and Yang, 1969a, 1969b; Burg and Clagett, 1967; Lieberman et al. 1966; Mapson et al. 1970), whether it is the major precursor *in vivo* or whether there are other precursors is still an open question. Several potential precursors of ethylene, such as ethanol, ethane, propionic acid and fumaric acid (which may directly or indirectly yield ethylene through dehydration, dehydrogenation or decarboxylation), and some representative intermediates of recognized metabolic pathways (including glycolysis, the Krebs cycle and the pentose phosphate cycle) were found to be inactive as ethylene precursors (Jansen, 1965). Other precursors of ethylene which have been suggested include β-alanine (Stinson and Spencer, 1969), ethionine (Shimokawa and Kasai, 1967), β-hydroxypropionic acid (Varner, 1961), propionaldehyde (Lieberman and Kunishi, 1967) and linolenic acid (Lieberman and Mapson, 1964). We have shown that neither propionaldehyde nor linolenic acid was incorporated into ethylene by apple tissue (Baur and Yang, 1969a, 1969b). This paper presents evidence showing that β-alanine, β-hydroxypropionic and ethionine also do not serve as precursors of ethylene. Additional evidence which supports the view that methionine is the biological precursor of ethylene will be presented.

II. MATERIALS AND METHODS

Materials

β-Hydroxypropionic acid-3-^{14}C was prepared from β-alanine-3-^{14}C with nitrous acid. Radioactive L-methionine methyl ester, N-acetyl-L-methionine, L-methionine sulfoxide, α-keto-γ-methylthiobutyric acid (KMB), methional, and S-methylmethionine were prepared from L-methionine-U-^{14}C with diazomethane, acetic anhydride in pyridine, hydrogen peroxide, L-amino oxidase, ninhydrin, and methyl iodide, respectively. β-Methylthiopropylamine-^{3}H was prepared from L-methionine-^{3}H with acetophenone (Baur and Yang, 1969b). Purity of these compounds was confirmed by paper chromatography. The radioactive ^{3}H residing in the ethylene moiety of commercially available L-methionine-^{3}H and L-ethionine-^{3}H was determined after converting the ethylene moiety to ethylene with FMN and light (Yang, *et al*. 1967; Yang, 1970).

Methods

Plugs (1 x 2 cm, 1.3 g) of apple tissue were cut with a cork borer and razor blade and the radioactive substrates in 2% KCl solution were introduced into the plug by a vacuum infiltration technique as described elsewhere (Baur and Yang, 1969b). Samples of the gas phase were analyzed for total and radioactive ethylene and CO_2 by gas chromatography and gas radiochromatography.

III. RESULTS AND DISCUSSION

β-Alanine and β-hydroxypropionic acid pathway

The possibility that β-alanine might serve as an ethylene precursor was proposed by Thompson and Spencer (1967) based on their observation that an enzyme preparation from bean cotyledon catalyzed a slight conversion (less than 0.002% in 21-hour incubation) of β-alanine to ethylene. From the cofactor requirements, they proposed the following scheme for the conversion of β-alanine to ethylene:

β-alanine ⟶ malonic semialdehyde ⟶ β-hydroxypropionic acid ⟶ acrylic acid ⟶ ethylene

We fed to apple tissue β-alanine-3-^{14}C and β-hydroxypropionic acid-3-^{14}C which are the intermediates of the scheme proposed above and have found that none of them were converted to ethylene appreciably, while methionine-U-^{14}C was efficiently converted to ethylene. Earlier investigators (Jansen, 1965) have also reported that malonic acid and propionic acid, which are closely related to malonic semialdehyde and acrylic acid in metabolism, are not efficient precursors of ethylene. If acrylic acid is the immediate precursor of ethylene, then the ethylene anion, which is an intermediate of the decarboxylation, should pick up a proton from water to yield ethylene :

$$CH_2 = CH - COO^- + H^+ \longrightarrow CH_2 = CH_2 + CO_2$$

Additional incorporation of protons into these intermediates may occur as a result of proton exchange between water and these intermediates. Thus, at least one proton will be incorporated into an ethylene molecule. To examine this possibility, we administered tritiated water to apple plugs and determined the specific activity of ethylene produced by the tissue. Although ethylene was found to be labeled, the specific activity of ethylene was only 1/8 of that of the water within the tissue (Yang and Baur, 1969). These results tend to rule out the possibility that ethylene is formed directly from acrylic acid through simple decarboxylation.

β-Hydroxypropionic acid may undergo a concerted dehydration and decarboxylation, yielding ethylene in a manner analogous to the formation of isopentenyl pyrophosphate from mevalonic-5-pyrophosphate as suggested by Varner (1961).

$$HO - CH_2 - CH_2 - C\begin{smallmatrix}O^-\\O\end{smallmatrix} \longrightarrow OH^- + CH_2 = CH_2 + CO_2$$

However, since β-hydroxypropionate was not converted to ethylene by apple tissue as mentioned above, this pathway should also be ruled out.

Ethionine pathway

That ethionine might serve as a precursor of ethylene was suggested by Shimokawa and Kasai (1967) based on their finding that the ethyl moiety of ethionine was converted to ethylene in the presence of FMN and light. Although we confirmed that in the presence of FMN and light the ethyl moiety did convert to ethylene, the ethylene formed represented less than one percent of the total ethylene formed. The bulk of the ethylene was derived from carbons 3 and 4 (Yang 1970). Furthermore, the ethyl moiety of ethionine yields both ethylene and ethane in about equal amounts. When L-ethionine-ethyl-1-^{14}C was administered to apple tissues, no conversion to ethylene was observed. Therefore, we concluded that the ethyl moiety of ethionine is not the source of ethylene *in vivo*.

Methionine pathway

Methionine as a biological precursor of ethylene. Evidence from tracer studies in a number of plant tissues have shown that methionine serves as a precursor of ethylene *in vivo* (Baur and Yang, 1969a,b; Burg and Clagett, 1967; Lieberman *et al.* 1966; Mapson *et al.* 1970). The enzymic conversion of methionine analogues to ethylene, catalyzed by peroxidase, has been elucidated recently (Ku *et al.* 1969; Mapson and Mead, 1968; Mapson *et al.* 1969; Yang, 1967; Yang, 1968; Yang, 1969); KMB and methional (β-methylthiopropionaldehyde) but not methionine, are the active substrates. A chemical mechanism accounting for these enzymatic reactions has been described (Yang,

1967; Yang, 1969). Since then, we have shown that methional is not converted to ethylene in fruit tissues (Baur and Yang, 1969b). Although KMB was readily converted to ethylene in apple tissues, the efficiency of its conversion was found to be less than that of methionine (Baur and Yang, 1969; Baur et al. 1971). These data do not support the view that KMB or methional are intermediates in the formation of ethylene from methionine. Details of the pathway and the intermediates from methionine to ethylene remain to be elucidated. That methionine is indeed a biological precursor of ethylene is further supported by the following observations:

(1) Fruit tissues at different stages of ripening convert methionine into ethylene in parallel with the tissues' ability to produce ethylene endogenously (Baur et al. 1971; Mapson et al. 1970).

(2) Conversion of methionine into ethylene occurs in pea stems which have been pretreated with IAA and which produce significant amounts of ethylene; controls produce little ethylene and do not convert methionine to ethylene (Burg and Clagett, 1967).

(3) Fruit tissues convert the L-form rather than the D-form of methionine to ethylene, indicating that the step is a stereo-specific enzymic reaction (Baur and Yang, 1969a).

Substrate specificity for methionine. In order to study the substrate specificity, several radioactive derivatives of methionine were administered to apple tissues and their efficiency to produce ethylene relative to that of methionine was determined (Table 1). For the calculation of the rate of conversion, it was assumed that only the ethylene moiety of the substrate in question was converted to ethylene. The results indicate that the structural requirements for methionine are very specific. When the carboxyl group is methylated (methionine methyl ester) or deleted (β-methylthiopropylamine), the substrate loses its activity as an ethylene precursor. When the amino group is acetylated (N-acetylmethionine), essentially all of its activity as an ethylene precursor is lost. Unlike methionine methylester, which is very poorly converted to CO_2, N-acetylmethionine is very actively converted to CO_2. When the sulfide function is oxidized (methionine sulfoxide), the activity as an ethylene precursor is reduced to 14%. Since in plant tissues methionine sulfoxide is known to be reduced back to methionine, the active precursor could well be the methionine formed from the administered methionine sulfoxide. When the methyl group is substituted by an ethyl group (ethionine), activity as an ethylene precursor is completely lost. When the CH_3S- group is deleted from methionine (α-aminobutyric acid), all activity as a precursor is lost. Thus, all the functional groups of methionine including the carboxyl, amino, ethylene and the methylmercapto, are essential. Other analogs of methionine which are active as ethylene precursors are KMB, S-methylmethionine and homoserine. Since they are converted into ethylene less efficiently than methionine and are closely related to methionine in metabolism, it is likely that they are converted to methionine prior to their conversion into ethylene.

Degradation products of methionine during its conversion to ethylene. Methionine is known to be converted to ethylene efficiently in a model system consisting of FMN and light (Yang, et al., 1967). The degradation products may be represented by the following equation:

$$\overset{5}{CH_3}-S-\overset{4}{CH_2}-\overset{3}{CH_2}-\overset{2}{\underset{\underset{NH_2}{\|}}{CH}} - \overset{1}{COOH} \longrightarrow \overset{5}{\underset{CH_3-S}{\overset{CH_3-S}{|}}} + \overset{4\ \ \ \ 3}{CH_2=CH_2} + \overset{2}{HCOOH} + \overset{1}{CO_2} + NH_3$$

(also CH_3SH)

In apple tissue, it is known that C-1 is converted to CO_2, C-2 to formic acid, and carbons 3 and 4 to ethylene; sulfur appears to be retained in the tissue, since no volatile radioactivity was detected with ^{35}S-labeled methionine (Lieberman et al., 1966; Burg and Clagett, 1967; Siebert and Clagett, 1969). Thus, except for the methylmercapto group, the products of the *in vivo* conversion are identical to those of the FMN-light model system.

Table 1. *Relative efficiency of ethylene formation from various substrates by apple tissues.* Each plug of apple tissues (1 x 2 cm, 1.3 g) was vacuum infiltrated with 80 µl of 2% KCl solution containing 5 to 100 mµmoles of radioactive substrates. Radioactive ethylene was determined by gas radiochromatography after incubation for 3 hr. Conversion efficiency was calculated assuming that ethylene was derived from the ethylene moiety of the substrates. Relative conversion efficiency for each substrate was then calculated with reference to the conversion efficiency of methionine which was conducted under the identical experiment conditions. The sources of the radioactive substrates were described in the text.

Substrate	Relative rate of conversion
$CH_3-S-CH_2-CH_2-CH(NH_2)-COOH$	100
$CH_3-S-CH_2-CH_2-CH(NH_2)-COOCH_3$	0
$CH_3-S-CH_2-CH_2-CH_2NH_2$	0
$CH_3-S-CH_2-CH_2-CH(NHCOCH_3)-COOH$	3
$CH_3-SO-CH_2-CH_2-CH(NH_2)-COOH$	14
$C_2H_5-S-CH_2-CH_2-CH(NH_2)-COOH$	0
$CH_3-CH_2-CH(NH_2)-COOH$	0
$CH_3-S-CH_2-CH_2-CO-COOH$	65
$CH_3-S-CH_2-CH_2-CHO$	5
$HO-CH_2-CH_2-CH(NH_2)-COOH$	29

In order to gain more insight into the fate of the methylmercapto group, we have fed L-methionine-methyl-^{14}C to apple tissues and analyzed the gas phase of the incubation flasks by gas radiochromatography. The only radioactive compound which was detected was $^{14}CO_2$. When the temperature of the incubation flask was raised, some radioactive dimethylsulfide was also detected. The evolution of dimethylsulfide suggests that it may be derived from S-methylmethionine, $(CH_3)_2-\overset{+}{S}-CH_2-CH_2-CH(NH_2)COOH$. Indeed ^{14}C-S-methylmethionine was identified from the apple extracts, based on the following criteria: (a) affirmative results from paper chromatography and from paper electrophoresis, and (b) upon heating in alkaline solution radioactive CH_3-S-CH_3 was released. Since the dimethylsulfonium group in S-methylmethionine is a better leaving group than the methylmercapto group in methionine, we thought that the direct precursor of ethylene could be S-methylmethionine, rather than methionine. In order to test this possibility we prepared S-methyl-L-methionine-U-^{14}C and fed it to apple tissue. Although S-methylmethionine was converted to ethylene in apple tissue, the efficiency of conversion was not as great as that of methionine. Furthermore in the presence of unlabeled methionine, the conversion of labeled S-methylmethionine was greatly reduced, while the conversion of labeled methionine into ethylene was not significantly reduced by the inclusion of unlabeled S-methylmethionine. These data suggest that S-methylmethionine was a metabolite of methionine unrelated to the biosynthesis of ethylene, and dimethylsulfide was a degradation product of S-methylmethionine.

Physiologically, it is very important that no volatile sulfur compounds are evolved during the conversion of methionine to ethylene. Since all the carbons of methionine are derived from glucose, which is very abundant in apple, the carbon source does not become limiting even if all the carbons of methionine are volatized. However, if during the biosynthesis of ethylene the sulfur were lost as a volatile then there would be a shortage of sulfur. We have analyzed apple tissue and have found that the level of free methionine plus protein methionine is about 60 mµmole/g, while the ethylene production rate is about 4 mµmoles/g-hr. Therefore, the sulfur atom of methionine must be conserved and utilized during the continuous synthesis of ethylene. There are two possibilities which may explain the failure to detect sulfur as methylmercaptan or methyldisulfide in the biosynthesis of ethylene. One explanation is that neither methylmercaptan nor methyldisulfide is a degradation product of methionine. The other explanation is that methylmercaptan or methyldisulfide are primary degradation products, but that these compounds are so efficiently converted to other non-volatile compounds that the volatile compounds cannot be detected. We tested the latter possibility by introducing ^{14}C or ^{35}S methylmercaptan into sealed flasks in which apple plugs were incubated. In both cases more than 95% of the non-volatile counts recovered were incorporated into S-methylcysteine. S-Methylcysteine was identified according to the following criteria: (a) paper chromatography and paper electrophoresis, (b) oxidation with H_2O_2 to S-methylcysteine sulfoxide followed by reduction with mercaptoethanol back to S-methylcysteine. Again in experiments with either ^{14}C or ^{35}S, about 0.5 to 1% of the total non-volatile radioactivity was identified as methionine. These data indicate that both the carbon and the sulfur of methylmercaptan were incorporated as a unit into S-methylcysteine and methionine. The incorporation into methionine was less than 1% of that into S-methylcysteine. In this respect, it is pertinent to note the report of Giovanelli and Mudd (1968) who demonstrated the enzymic formation of S-methylcysteine and methionine through direct methylthiolation of O-acetylserine and O-acetylhomoserine with methylmercaptan as illustrated by the following reactions:

O-Acetylserine   + $CH_3SH \longrightarrow$ S-methylcysteine + acetate

O-Acetylhomoserine   + $CH_3SH \longrightarrow$ methionine + acetate

If methylmercaptan were released from methionine during the conversion to ethylene *in vivo*, and later incorporated into S-methylcysteine, then we should be able to detect labeled S-methylcysteine when methionine-methyl-^{14}C and methionine-^{35}S are fed to apple tissue. Indeed, we have found that this is the case. Since significant formation of radioactive S-methylcysteine was observed from both $^{14}CH_3$- and ^{35}S-labeled methionine, we have concluded that the methylmercapto group of methionine was transferred to S-methylcysteine as a unit. Although S-methylcysteine and its sulfoxide are normal constituents of a number of plants (Thompson, 1967), its metabolism in higher plants is obscure.

The present data suggest that methylmercaptan may be released from methionine during its conversion to ethylene, and that the sulfur is conserved by incorporation into S-methylcysteine.

Since the conversion of methionine into ethylene by fruit tissues is greatly inhibited by hydroxylamine (Baur and Yang, unpublished) and by rhizobitoxine (Owens *et al.*, 1970) which are inhibitors of pyridoxal phosphate-mediated reactions and considering the present evidence that CH_3SH may be the primary degradation product of methionine, a reaction mechanism which accounts for the production of ethylene from methionine as mediated by pyridoxal phosphate is proposed (Fig. 1). A role for oxygen in this conversion is included. It should be noted that pyridoxal phosphate-mediated γ-elimination reactions have been recognized in biological systems (Flavin, 1963).

Is methionine a major precursor of ethylene in vivo? There is little doubt that methionine is a precursor of ethylene *in vivo*. It is now pertinent to ask to what extent the ethylene produced *in vivo* is derived from the methionine pathway. In theory, the ratio of the specific radioactivity of ethylene produced to the specific radioactivity of methionine in the metabolic pool represents the fraction of ethylene derived from methionine. To get some insight into this question, we have measured the ratio of the specific radioactivity of ethylene producted to the specific radioactivity of the methionine supplied exogenously at various concentrations (Table II). This ratio was very low at low methionine concentration, because the radioactive exogenous methionine was greatly diluted by endogenous methionine. When methionine was supplied at 0.5 μmoles per 1.3 g of tissues, the ratio became larger than 0.5. Since there must be some dilution by endogenous methionine, the theoretical ratio may be much higher than 0.5. These results suggest that methionine is the major, if not the sole, precursor of ethylene.

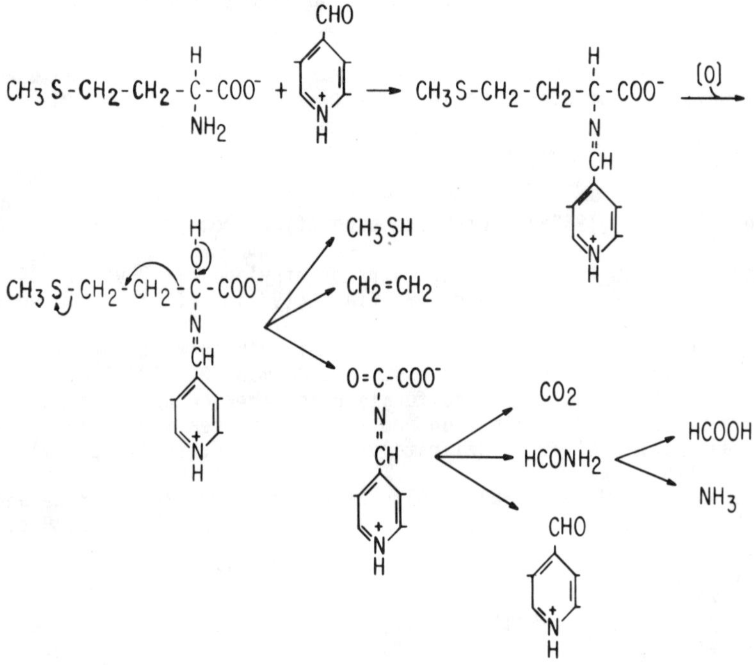

Fig. 1. A postulated mechanism of ethylene production from methionine mediated by pyridoxal phosphate.

TABLE II. *Specific radioactivity of ethylene produced by apple tissues as related to the concentration and specific radioactivity of methionine administered.* Each plug of (1 x 2 cm, 1.3 g) apple tissue was vacuum infiltrated with 0.1 ml of 2% KCl solution containing various amounts of methionine -U-^{14}C. The radioactivity of methionine listed below represents 2/5 of the actual total radioactivity, since only 2 (carbons 3 and 4) out of 5 carbons of methionine are converted into ethylene. Ethylene produced between 140 and 200 min incubation was analyzed.

Methionine			Ethylene			Ratio of B/A
mμmole	mμc	mμc/mμmole (A)	mμmole	mμc	mμc/mμmole (B)	
1	96	96	7.8	3.4	0.44	0.004
10	100	10	5.1	4.9	0.96	0.096
100	174	1.74	5.1	2.6	0.51	0.29
500	268	0.53	7.8	2.2	0.28	0.53

IV. SUMMARY

Evidence available is in accord with the view that methionine is a precursor of ethylene. The data suggest that methionine is the major, if not the sole, precursor of ethylene in apple tissues. Although methylmercaptan could not be found in the apple tissues during the conversion of methionine into ethylene, when supplied exogenously it was efficiently converted to S-methylcysteine. S-methylcysteine was identified as a conversion product of methionine in apple tissues. It is assumed that when methionine is converted to ethylene the methylmercapto group is retained and converted into S-methylcysteine.

Evidence is presented to show that β-alanine, β-hydroxypropionic acid and ethionine pathways are not operative in fruit tissues.

V. REFERENCES

BAUR, A., and S.F. YANG (1969a). Ethylene production from propanal. Pl. Physiol. 44, 189-192.
BAUR, A.H. and S.F. YANG (1969b). Precursors of ethylene. Pl.Physiol. 44, 1347-1349.
BAUR, A.H., S.F. YANG, H.K. PRATT and J.B. BIALE (1971). Ethylene biosynthesis in fruit tissues. Pl. Physiol. 47, 696-699.
BURG, S.P. and C.O. CLAGETT (1967). Conversion of methionine to ethylene in vegetative tissue and fruits. Biochem. Biophys. Res. Commun. 27, 125-130.
FLAVIN, M. (1963). Microbial trans-sulfuration and the mechanism of some pyridoxal phosphate potentiated elimination and replacement reactions, in "Chemical and biological aspects of pyridoxal catalysis" (Ed. E.E. Snell et al.) p.377, MacMillan, New York.
GIOVANELLI, J. and S.H. MUDD (1968). Sulfuration of O-acetylhomoserine and O-acetylserine by two enzyme fractions from spinach. Biochem. Biophys. Res. Commun. 31, 275-280.
JANSEN, E.F. (1965). Ethylene and polyacetylene, in "Plant Biochemistry" (Ed. J. Bonner and J.E. Varner), p. 641. Academic Press, New York.
KU, H.S., S.F. YANG and H.K. PRATT (1969). Ethylene evolution from α-keto-γ-methylthiobutyrate by tomato fruit extracts. Phytochemistry 8, 567-575.
LIEBERMAN, M. and A.T. KUNISHI (1967). Propanal may be a precursor of ethylene in metabolism. Science 158, 938.
LIEBERMAN, M., A. KUNISHI, L.W. MAPSON and D.A. WARDALE (1966). Stimulation of ethylene production in apple tissue slices by methionine. Pl. Physiol. 41, 376-382.
LIEBERMAN, M. and L.W. MAPSON (1964). Genesis and biogenesis of ethylene. Nature 204, 343-345.
MAPSON, L.W., J.F. MARCH, J.J.C. RHODES and L.S.C. WOOLTORTON (1970). A comparative study of the ability of methionine or linolenic acid to act as precursors of ethylene in plant tissues. Biochem. J. 117, 473-479.
MAPSON, L.W., J.F. MARCH and D.A. WARDALE (1969). Biosynthesis of ethylene. 4-Methylmercapto-2-oxobutyric acid: an intermediate in the formation from methionine. Biochem. J. 115, 653-661.
MAPSON, L.W. and A. MEAD (1968). Dual nature of cofactors required for the enzymic production of ethylene from methional. Biochem. J. 108, 575-581.
OWENS, L.D., M. LIEBERMAN and A. KUNISHI (1970). Inhibition of ethylene production by rhizobitoxine. Pl. Physiol. 46, S - 32.
SHIMOKAWA, K. and Z. KASAI (1967). Ethylene formation from ethyl moiety of ethionine. Science 156, 1362-1363.
SIEBERT, K.J. and C.O. CLAGETT (1969). Formic acid from carbon 2 of methionine in ethylene production in apple tissue. Pl. Physiol. S - 30.
STINSON, R.A. and M. SPENCER (1969). β-Alanine as an ethylene precursor. Pl. Physiol. 44, 1217-1226.
THOMPSON, J.E. and M. SPRENCER (1967). Ethylene production from β-alanine by enzyme powder. Canad. J. Biochem. 45, 563-571.
THOMPSON, J.F. (1967). Sulfur metabolism in plants. Ann. Rev. Pl. Physiol. 18, 59-84.

VARNER, J.E. (1961). Biochemistry of senescence. Ann. Rev. Pl. Physiol. 12, 245-264.
YANG, S.F. (1967). Ethylene formation from methional by horseradish peroxidase. Arch. Biochem. Biophys. 122, 481-487.
YANG, S.F., H.S. KU and H.K. PRATT (1967). Photochemical production of ethylene from methionine and its analogues in the presence of flavin mononucleotide. J. Biol. Chem. 242, 5274-5280.
YANG, S.F. (1968). Biosynthesis of ethylene, in "Biochemistry and Physiology of Plant Growth Substances" (Eds. F. Wightman, G. Setterfield). Runge Press, Ottawa, p. 1217-1228.
YANG, S.F. (1969). Further studies on ethylene formation from α-keto-γ-methylthiobutyric acid by peroxidase in the presence of sulfite and oxygen. J. Biol. Chem. 244, 4360-4365.
YANG, S.F. and A.H. BAUR (1969). Pathways of ethylene biosynthesis. Qual. Plant. Mater. Veg. 19, 201-220.
YANG, S.F. (1970). Photosensitizer conversion of ethionine to ethylene by flavin mononucleotide. Photochem. Photobiol. 12, 419-422.

The Measurement of Ethylene from Plant Tissues and Its Relation to Auxin Effect.

R.M. Muir and E.W. Richter

Department of Botany, University of Iowa, Iowa City, Iowa, U.S.A.

I. INTRODUCTION

Treatment of plant tissues with certain concentrations of IAA leads to the formation of measurable amounts of ethylene in closed systems. This observation has been the basis for the idea that the effects of IAA are due to the ethylene which is formed (Burg and Burg, 1966). Since this idea is incompatible with the evidence from structure-activity relationships for a direct effect of IAA through a 2-point reaction with a plant substrate (Muir and Hansch, 1953, 1967), and since the evidence upon which the idea is based is less than adequate (Andreae et al., 1968), the question of the relationship of IAA and ethylene in the control of plant growth has been re-examined.

II. MATERIALS AND METHODS

Seedlings of the pea (*Pisum sativum* L. cv. Alaska) were grown in moist sterile sand in the dark at 24° and 85% relative humidity for a total of 7 days including 24 hr soaking in water with aeration. Under green safe-light segments 7.2 mm long were excised immediately below the hook from the third internode which was 2 to 4 cm long. Root tissue was obtained from plants grown for 48 hr in moist vermiculite after 6 hr of soaking. Apical segments 7.2 mm long were excised from the roots that were 2 to 4 cm in length.

The incubation medium for epicotyl segments consisted of 0.05M phosphate buffer (pH 6.8), 2% sucrose and 5 µM $CoCl_2$ in distilled water (Burg and Burg, 1966). The medium for root segments was 5 mM phosphate buffer (pH 6.2), 0.5% sucrose and 1mM $Ca(NO_3)_2$ in distilled water (Adreae et al., 1968). Usually 20 segments weighing 0.4 to 0.5g were placed in a 50 ml Erlenmeyer flask with 10 ml of the medium containing various concentrations of IAA and sealed with neoprene caps. The flasks were shaken slowly in the dark for 18 hr.

Ethylene was measured with a Varian Aerograph, model 600-D, gas chromatograph equipped with a flame ionization detector and a very sensitive electrometer attached to a Beckman 10-inch variable speed recorder. Other investigators have used activated alumina as the adsorbent for the measurement of ethylene but this material is quite hydrophilic and irreversibly binds carbon dioxide (Burg, 1962). These properties make alumina less than satisfactory as an adsorbent since the reproducibility of measurements will be diminished by the altered adsorption properties. In this investigation a new adsorbent, "Porapak" (Water Associates, Framingham, Mass.), has been used. This material gives excellent separation of ethylene from other gases and it has an extremely low retention volume for polar substances such as water and alcohols. The column was prepared with copper tubing 7 ft long and 1/4 in outside diameter. It was used at a temperature of 80°, a hydrogen flow rate of 64 ml/min and a nitrogen flow rate of 64 ml/min. The proportionality of signal area on the chart and the amount of ethylene was determined with standard mixtures prepared by flushing a liter flask with nitrogen, sealing it with a neoprene cap, injecting a known quantity of ethylene into the flask and mixing with a magnetic stirrer. All measurements were made with injection samples of 5 ml.

III. RESULTS AND DISCUSSION

A reproduction of the actual tracing of the chromatographic signals for an injection of 5 ml of a mixture of air and ethylene in the Porapak column is shown in figure 1. Excellent resolution for oxygen and methane (first and second from the right) as well as for ethylene is demonstrated. The oxygen and methane signals were established by retention times and co-chromatography with known gases. The methane is the common laboratory air pollutant from the natural gas system. For the measurement of small quantities of ethylene by chart area the chart speed is increased from 0.5 in/min to 5 in/min. As small an area as 0.05 in^2 for 20 ppb ethylene can be measured accurately and reproducibly.

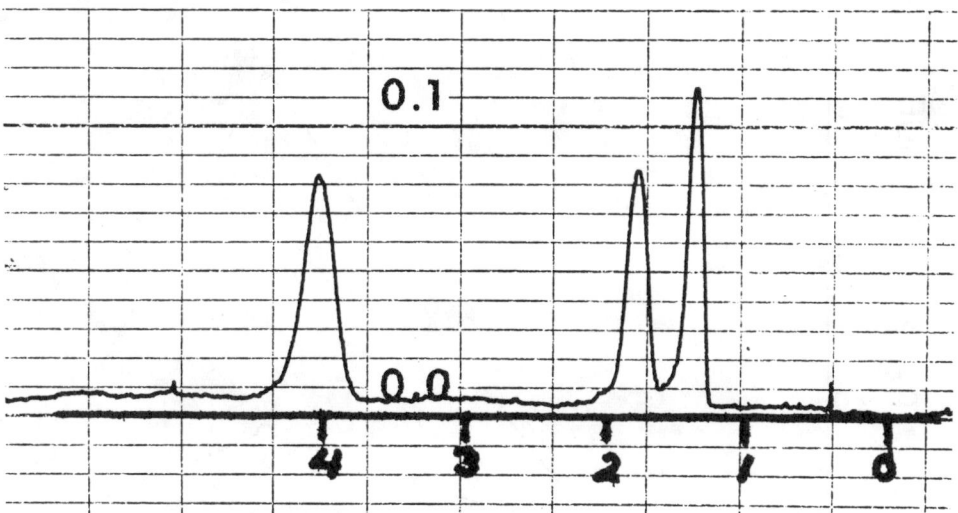

Fig. 1. Chromatographic tracing obtained with a mixture of air and ethylene. Chart speed is 0.5 in/min. Beginning time on the right with signals for oxygen, methane and ethylene in that order.

Measurements of ethylene production during 18 hr by segments of stem and root tissue of the pea in response to treatment with IAA are given in Figure 2. The production of ethylene by the root tissue is much greater than that by the stem tissue. At 10^{-4}M IAA 68 mµl of ethylene were formed per gm fresh weight of root tissue. This value is in agreement with the value of 84 mµl for such tissue by Andreae et al., (1968).

In a report by Chadwick and Burg (1967) the production of ethylene by the root tissue in 10^{-4}M IAA is given as high as 25 mµl/gm/hr while in an adjacent figure the total quantity in 18 hr is only 18 mµl. Since it is unlikely that such differences would be found for the same tissue under identical conditions, it is probable that the discrepancy in measurement is the result of the inadequacy of activated alumina for the gas chromatography of ethylene unless water and CO_2 are removed prior to analysis of the sample. The similarity in IAA concentration-dependency of ethylene production in the root and stem tissue is not found in the IAA concentration-dependency of elongation in these tissues shown in Table 1. While elongation of the root segment is reduced by concentrations of 10^{-7}M IAA and higher, ethylene production is not observed until the concentration of IAA is 10^{-5}M. Elongation of the stem segment is still as great as that of the control at 10^{-5}M IAA yet a large amount of ethylene has been formed. Finally, although maximum inhibition occurs invariably at 10^{-3}M IAA, ethylene production in both the root and the stem tissue is less at this concentration of IAA than at 10^{-4}M IAA. The production of ethylene is not related qualitatively or quantitatively to the effects of IAA on stem or root elongation.

Andreae et al., (1968) realized that a critical test of the dependency of inhibition by IAA on ethylene production by root tissue was simply to determine the reversibility of the inhibition by IAA and the inhibition by ethylene. They found that inhibition of growth of root segments by 100 ppm ethylene was not apparent until 3 to 6 hr after onset of treatment while inhibition by IAA began at zero time, and after 16 hr of treatment the inhibition by ethylene was irreversible while that by IAA was reversible. A recent report by Chadwick and Berg (1970) shows that under a low level of 10 ppm ethylene there is some degree of reversibility of the inhibition in root tissue during the first 8 hr but little at 16 hr. The results of an experiment of similar design with stem segments of the pea are shown in Figures 3 and 4.

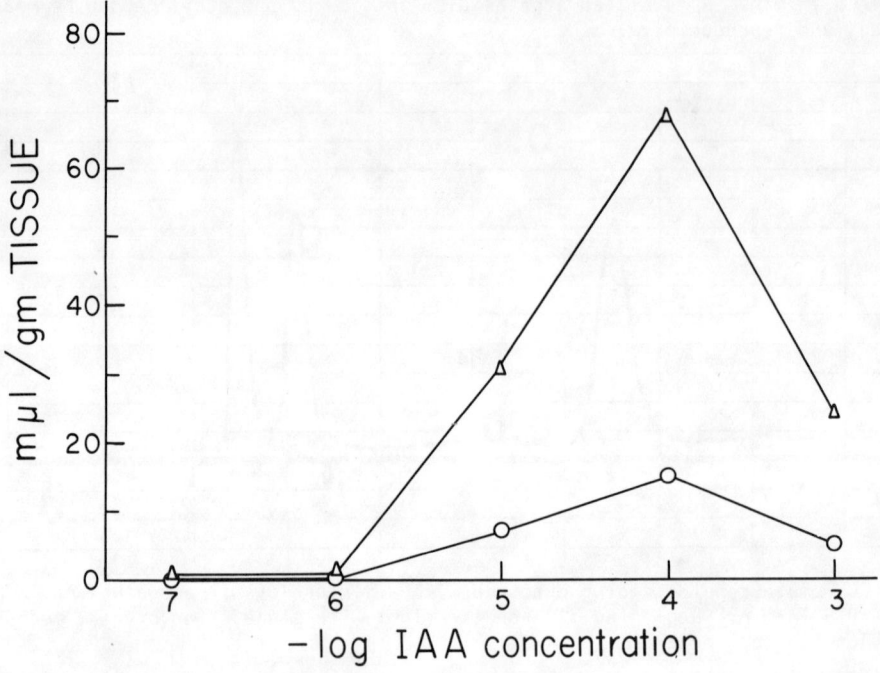

Fig. 2.

Total ethylene produced by root segments (triangles) and stem segments (circles) of the pea after 18 hours treatment with IAA.

The stem tissue was treated with 100 ppm ethylene in the one case and 10^{-3}M IAA in the other. Comparison of elongation or increase in fresh weight with the control shows that inhibition by IAA begins at zero time while that with ethylene does not begin until sometime after 4 hr. By 18 hr inhibition is approximately the same for both treatments. At this time the segments were removed, washed with distilled water and returned either to the basic medium alone or to the basic medium with IAA or ethylene. During the next 6 hr the segments released from the inhibition by IAA underwent elongation and increase in fresh weight at rates approximating the rates during the first 4 hr for the control tissue. The segments released from the inhibition by ethylene showed no change in growth rate. The reversibility of the inhibition by IAA and the irreversibility of the inhibition by ethylene clearly indicate different mechanisms of inhibition.

Another type of experiment which demonstrates the difference in sites of action of IAA and ethylene is shown in Figure 5. In this experiment ethylene at 100 ppm was added to 2 sets of flasks with stem segments and the basic medium. At zero time and at intervals of 3 hr IAA was injected through the caps of one set of flasks to make the medium 10^{-5}M IAA. Three hours after the addition of IAA the segments were measured. The addition of IAA at zero time and at 3 and 6 hr gave the same rates of elongation during the subsequent 3 hr period. Not until the tissue had been exposed

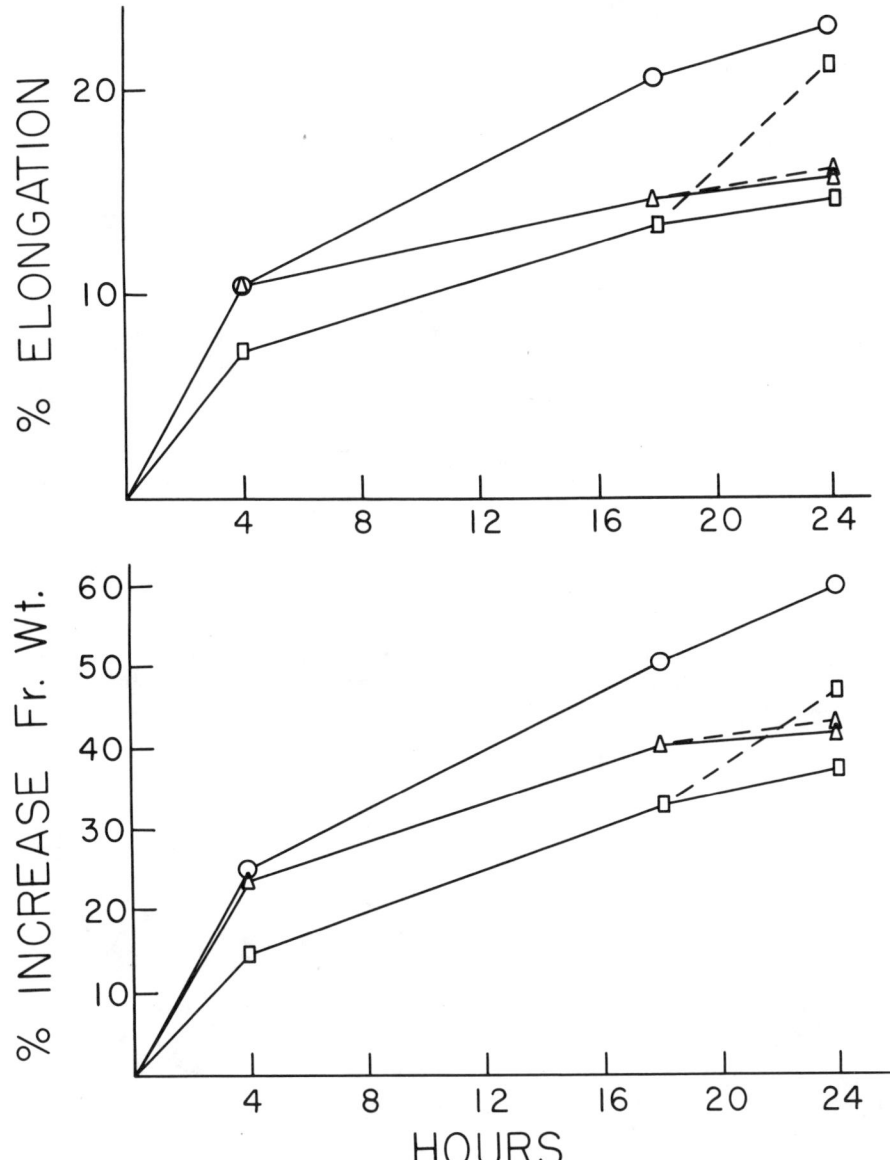

Fig. 3.

Reversibility of inhibition of elongation in stem segments of the pea caused by IAA and ethylene. Control segments (circles); segments in 100 ppm ethylene (triangles); segments in 10^{-3}M IAA (squares); transfer to fresh medium (broken lines).

Fig. 4.

Reversibility of inhibition of fresh weight increase in stem segments of the pea caused by IAA and ethylene. Control segments (circles); segments in 100 ppm ethylene (triangles); segments in 10^{-3}M IAA (squares); transfer to fresh medium (broken lines).

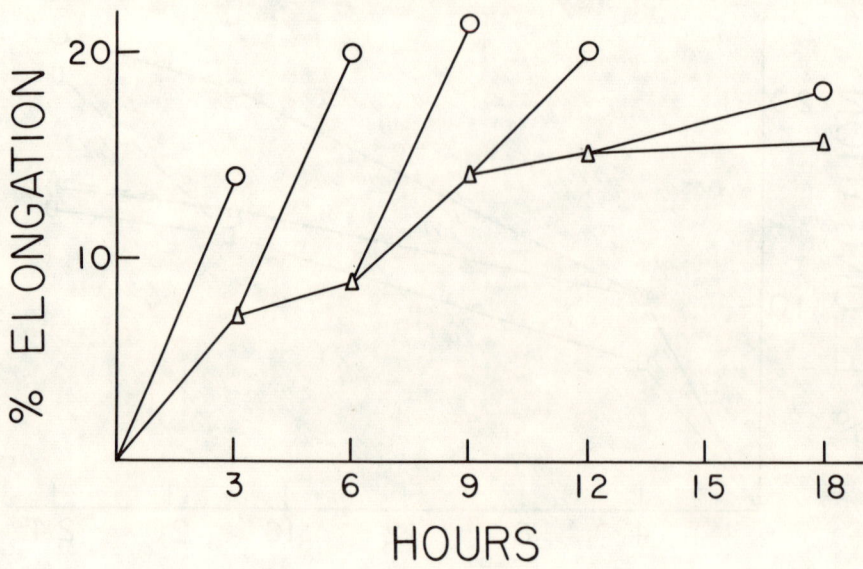

Fig. 5.

Elongation response of stem segments of the pea to additions of IAA during exposure to ethylene. Segments in 100 ppm ethylene continuously (triangles); segments in 100 ppm ethylene and 10^{-5}M IAA for 3 hours prior to measurement (circles).

Table 1. Elongation response of stem and root segments of the pea to IAA after 18 hours.

Molar Concentration of IAA	% Elongation Stem Segments	% Elongation Root Segments
Control	37	60
10^{-8}		68
10^{-7}	50	57
10^{-6}	53	29
10^{-5}	38	8
10^{-4}	33	4
10^{-3}	32	

to the inhibiting effect of ethylene for 9 hr was there any effect on the rate of elongation in response to the addition of IAA. Even though the inhibitory effect of ethylene was complete at 9 hr the tissues could still respond to the addition of IAA at 12 hr. Browning of the tissue after 12 hr treatment with ethylene indicates a general toxicity effect at this time but one which is not related to the effect of IAA.

Since CO_2 delays fruit ripening and ethylene hastens it, Kidd and West (1945) proposed that they are antagonistic. Burg and Burg (1967) have reported that CO_2 reduces the inhibitory effect of both ethylene and IAA on elongation. Their data on the interaction of IAA and CO_2 are not convincing because they are plotted as a percentage of a percentage response. If the inhibition of elongation at high concentrations of IAA is the result of the production of ethylene, and CO_2 is antagonistic to the effect of ethylene, then the presence of 1% CO_2 should reduce significantly the inhibiting effect of IAA on elongation. The data of such an experiment are given in Table 2. The ethylene and/or CO_2 were injected into the sealed flasks and the lengths of the segments were determined after 18 hr. Although the presence of 1% CO_2 does prevent the inhibition of elongation by 10 ppm ethylene, it has no effect either on the promotion of elongation or the inhibition of elongation by IAA. This is additional evidence against the idea of a dependency of the IAA effect on ethylene production.

Table 2. Interaction of carbon dioxide with ethylene and IAA in the elongation response of stem segments of the pea.

Treatment	Without CO_2 % Elongation	With CO_2 % Elongation
Control	19	
10 ppm Ethylene	11	18
10^{-6}M IAA	67	64
10^{-5}M IAA	18	17
10^{-3}M IAA	12	13

Evidence concerning the relationship of IAA and ethylene in the control of growth can be obtained by comparing the effects of another auxin such as 2,4,5-trichlorophenoxyacetic acid (2,4,5-T), which both promotes and inhibits elongation depending upon concentration, with the effects of 2,4,6-trichlorophenoxyacetic acid (2,4,6-T), which only inhibits elongation. The effects of these substances on the growth of stem segments of pea during 18 hr are shown in Table 3, and the production of ethylene by segments in the presence of these substances after 18 hr is shown in Figure 6.

Table 3. Elongation of stem segments of the pea in the presence of 2,4,5-trichlorophenoxyacetic acid (2,4,5-T) or 2,4,6-trichlorophenoxyacetic acid (2,4,6-T).

Treatment	2,4,5-T % Elongation	2,4,6-T % Elongation
Control	33	33
10^{-6}M	43	26
10^{-5}M	30	19
10^{-4}M	24	15
10^{-3}M	16	17

As with IAA, measurable ethylene is found with concentrations of 2,4,5-T and 2,4,6-T of 10^{-5}M and higher but the amount of ethylene at 10^{-3}M is greatest and the amount generally is 2 to 3 times as much as that formed in the presence of IAA. Most significantly, there is little or no difference in the formation of ethylene in the presence of 2,4,5-T or 2,4,6-T while there are extreme differences in the elongation of the tissue. At a concentration of 10^{-6}M 2,4,5-T promotes elongation while 2,4,6-T inhibits elongation and ethylene is not detectable. At 10^{-5}M where ethylene production is approximately the same 2,4,5-T causes little inhibition of elongation while 2,4,6-T causes nearly maximal inhibition. The effects of these substances on elongation cannot be explained in terms of the ethylene they cause to be produced. An explanation in terms of their penetration and reactivity with the plant substrate in growth has existed for a long time.

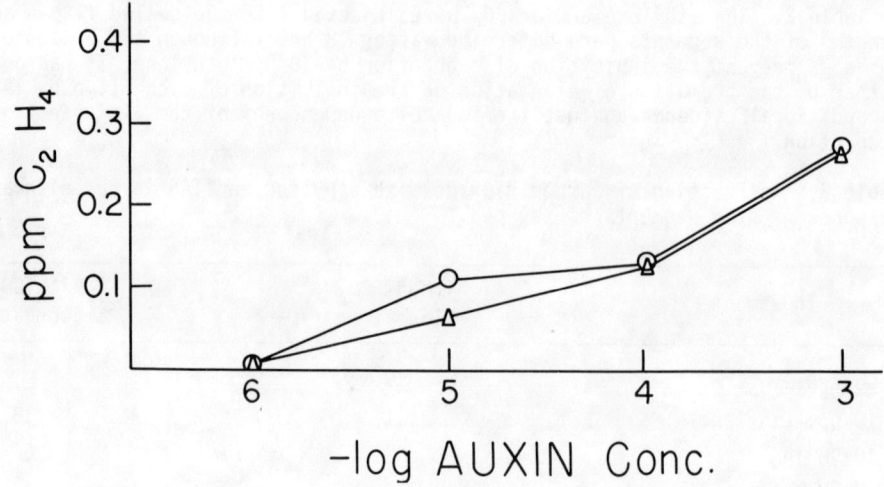

Fig. 6.
Total ethylene produced by stem segments of the pea in the presence of 2,4,5-trichlorophenoxyacetic acid (circles) or 2,4,6-trichlorophenoxyacetic acid (triangles).

That the effects of IAA on growth are not dependent upon the ethylene it causes to be formed is shown by the lack of relationship of elongation to the amount of ethylene produced particularly at a concentration of 10^{-3}M IAA, the irreversibility of ethylene inhibition contrasted with the reversibility of IAA inhibition, the elongation response to IAA when the tissue is under the inhibiting action of ethylene, the interaction of CO_2 in the inhibition by ethylene contrasted with the lack of interaction of CO_2 in the effects of IAA, and the similarity of ethylene production in the presence of 2,4,5-T or 2,4,6-T although their effects on growth are quite different.

The relationship of the production of ethylene in response to treatment with IAA will not be known until we have information on the metabolic pathway whereby ethylene is formed. Surely the diversity of stimuli which induce ethylene production ranging from wound injury (Denny, 1935) to infection by tobacco mosaic virus (Nakagaki, 1969) suggests that its production is an unspecific response to disruption of the normal metabolism and the treatment with IAA may simply represent such a disruption. In this view the ethylene evolved upon treatment with inhibitory concentrations of IAA is the result of inhibition and not the cause.

So far as the effect of ethylene on plant tissue is concerned, the idea of Kurssanov (1941) that ethylene acting as a surfactant may release hydrolytic enzymes from adsorption complexes appears as valid as any other suggestion.

IV. SUMMARY

The measurement of ethylene produced by plant tissue with gas chromatography using a new adsorbent, "Porapak", is described. The production of ethylene by stem and root tissue of etiolated pea seedlings in the presence of IAA is not related either to the promotion or inhibition of elongation in these tissues. Ethylene production is less in the presence of 10^{-3}M IAA than 10^{-4}M while there is more inhibition. The inhibiting effect of 10^{-3}M IAA on elongation and the increase in fresh weight of stem tissue develops immediately upon treatment while the effect of 100 ppm ethylene is not apparent until after 4 hr. The effect of IAA is reversible after 18 hr while that of ethylene is irreversible. Stem tissue in the presence of 100 ppm ethylene responds to the addition of IAA with a rate of elongation comparable to control tissue for as long as 9 hr of treatment with ethylene. Carbon dioxide at 1% prevents 10 ppm ethylene from having an inhibitory effect on elongation but has no effect on the promotion or inhibition of elongation by IAA. Treatment of stem tissue with either 2,4,5-trichlorophenoxyacetic acid or 2,4,6-tricholorophenoxyacetic acid gives rise to the same production of ethylene yet the first compound both promotes and inhibits elongation while the second only inhibits. Ethylene is produced as a result of treatment with IAA but it has no relation to the growth effect of IAA.

V. REFERENCES

ANDREAE, W.A., M.A. VENIS, F.JURSIC and T. DUMAS (1968). Does ethylene mediate root growth inhibition by indole-3-acetic acid? Pl. Physiol. 43, 1375-1379.

BURG, S.P. (1962). Vapor phase chromatography, in "Modern Methods of Plant Analysis" (Ed. H.F. Linskens and M.V. Tracey) Vol. 5, p. 129. Springer-Verlag, Berlin. 1962.

BURG, S.P. and E.A. BURG (1966). The interaction between auxin and ethylene and its role in plant growth. Proc. Nat. Acad. Sci. U.S.A. 55, 262-269.

BURG, S.P. and E.A. BURG (1967). Molecular requirements for the biological activity of ethylene. Pl. Physiol. 42, 144-152.

CHADWICK, A.V. and S.P. BURG (1967). An explanation of the inhibition of root growth caused by indole-3-acetic acid. Pl. Physiol. 42, 415-420.

CHADWICK, A.V. and S.P. BURG (1970). Regulation of root growth by auxin-ethylene interaction. Pl. Physiol. 45, 192-200.

DENNY, F.E. (1935). Testing plant tissue for emanations causing leaf epinasty. Contr. Boyce Thompson Inst. 7, 341-347.

KIDD, F. and C. WEST (1945). Respiratory activity and duration of life of apples gathered at different stages of development and subsequently maintained at a constant temperature. Pl. Physiol. 20, 467-504.

KURSSANOV, A.L. (1941). Untersuchung enzymatischer Prozesse in der lebenden Pflanze. Adv. In Enzymology 1, 329-370.

MUIR, R.M. and C. HANSCH (1953). On the mechanism of action of growth regulators. Pl. Physiol. 28, 218-232.

MUIR, R.M. and C. HANSCH (1967). Structure-activity relationship in the auxin activity of mono-substituted phenylacetic acids. Pl. Physiol. 42, 1519-1526.

NAKAGAKI, Y (1969). Ethylene production by detached leaves infected with tobacco mosaic virus. Virology 70, 1-9.

Auxin and Ethylene in Adventitious Root Formation in *Phaseolus aureus* (Roxb.).

Michael. G. Mullins

C.S.I.R.O. Division of Horticultural Research, G.P.O. Box 350, Adelaide, South Australia

Present address:- University of Sydney, N.S.W. 2006

I. INTRODUCTION

It is 36 years since Thimann and Went (1934) showed that β-Indole acetic acid (IAA) stimulates the formation of adventitious roots in pea stem segments, but knowledge of the mechanisms of rhizogenesis is still fragmentary. It has become clear that auxin is but one of many factors in the regeneration of plants from cuttings. Included are nutritional and environmental factors, plant factors such as age of tissue, juvenility, virus status, degree of lignification (Hartmann and Kester 1959), and numerous endogenous and exogenous compounds (reviewed by Hess, 1969). Of these compounds the "rooting co-factors" (Hess 1964) have received much attention, as have a variety of synthetic and naturally-occurring phenolic compounds (Fernqvist 1966, Bastin 1966, Girouard 1969, Fadl and Hartmann 1967). However, there is at present no clear understanding of the function of these substances in adventitious root formation. The existence of auxin-phenol rhizocalines (Bouillenne and Bouillenne 1955) remains an open question, as does the importance of the IAA-sparing effects of phenolic compounds.

The only working hypothesis to emerge is that root initiation in cuttings is controlled by a complex array of interacting factors (Hess 1969). IAA is a major component of the multi-factor array, but if any one factor is limiting (environmental, nutritional, co-factor etc.) the whole array becomes ineffective as a trigger for root formation. In this paper it will be shown that yet another factor is involved, namely, ethylene.

Some aspects of current plant propagation practice indeed suggest that ethylene is involved in the rooting of cuttings. Treatments which stimulate root formation, such as auxin application or wounding, are known to stimulate ethylene production. In 1933 Zimmermann and Hitchcock showed that unsaturated hydrocarbon gases cause aerial root production in several herbaceous and woody species. With lateral root formation, however, recent work by Radin and Loomis (1970) has shown that ethylene inhibits the formation of adventitious primordia. Finally, there may be significance in the fact that some of the phenolic compounds which stimulate rooting have been implicated in the regulation of ethylene biosynthesis (Yang 1967).

II. PRELIMINARY

The work of Gautheret (1969), on rhizogenesis in explants of *Helianthus tuberosus*, shows that root formation involves sequences of morphogenetic phenomena with differing requirements for nutrients or growth substances. A similar concept may be adopted for adventitious root formation in cuttings.

An analysis was made of root formation in cuttings of mung bean (*Phaseolus aureus* Roxb.). This species was chosen because much of our present knowledge of factors affecting adventitious rooting has been derived from the responses of mung bean cuttings in bioassay systems (Hess 1969). Data was obtained on the sequences of morphological changes between excision and emergence of roots, the timing of these changes under standard environmental conditions, and the responses of cuttings to growth regulators at various times after excision. Measurements were also made of changes in the synthesis of nucleic acids and protein, and changes in the activities of a number of enzymes. A brief account of growth changes will be given here.

Cuttings were made from etiolated or light-grown 5 or 6 day-old seedlings, and they were prepared and grown according to the methods of Hess (1964) and Fernqvist (1966). A special technique was developed for precision time-course experiments on effects of growth substances on root initiation. This involved use of an automatic fraction-collector to dispense growth regulator solutions or fixative. Root primordia were recorded by microscopic examination of fixed, cleared, and stained intact cuttings (modification of the Aniline Blue-Lacto-phenol method of Goodey, 1957).

In dark-grown cuttings (26°C) root initials were observed 20-24h after excision. After 30h the root primordia consisted of deeply-staining conical clumps of tissue, and differentiated root caps were observed after 40-45h. Adventitious roots first protruded through the epidermis after 70h. Root initials were detected after 50h in light-grown cuttings (12h illumination, 2000f.c., 26°C), and the first roots emerged after approximately 100h. Thus, the formation of an adventitious root may be divided into three phases. The first, *induction*, comprises the time between excision of the cutting from the parent plant and the onset of cell divisions in the pericycle which lead to the formation of an adventitious root primordium. Induction is followed by *initiation* and differentiation of primordia, and this is followed by the phase of root *emergence*.

These phases of adventitious root formation are not discrete in mung bean. In etiolated cuttings, for example, new primordia arise continuously during the first 45h. It is not yet clear if all primordia are induced simultaneously but develop at different rates, or if induction is a sequential process (Goldacre 1959).

With γ-Indole butyric acid (IBA), a potent stimulator of adventitious rooting in dark-grown mung bean cuttings, there is an increased rate of primordium production (0-45h) as compared with cuttings grown in water, and root primordia are detectable at an earlier stage. There was a difference of 6h in the time to 10 primordia between IBA-treated cuttings (20mg/1) and controls. The findings of Fernqvist (1966), that the root-promoting effect of IBA declines with increasing delay in the time of application were confirmed, and similar responses were found when dark-grown cuttings were supplied with full nutrient solution.

In the work of Skoog and Miller (1957) on organogenesis in explants, it was shown that root formation is determined by the auxin:cytokinin ratio. Root initiation was favoured when the levels of auxin was high relative to that of cytokinin. With cuttings, however, cytokinin has been found to inhibit the formation of adventitious root primordia. (e.g. Fernqvist 1966). In mung bean, time-course experiments with the "fraction-collector" technique have shown that effects of kinetin on root primordium formation are related to time of application relative to the concentration of auxin supplied at excision. Kinetin (1mg/1) inhibited the formation of primordia when supplied to newly-made cuttings, but later applications had progressively less effect. Cuttings treated with IBA were less susceptible to inhibition by kinetin at all times of application after 24h. When kinetin was applied at the 43rd hour to cuttings which received IBA (25mg/1) at zero hour there was an increase in the total numbers of primordia produced as compared with cuttings which received IBA alone.

These results suggest that auxin:cytokinin ratio is of importance in the rooting of cuttings as well as in undifferentiated tissue. Moreover, it is clear that root formation in cuttings involves sequences of phenomena with differing requirements for growth substances.

III. AUXIN:ETHYLENE RELATIONSHIPS IN ADVENTITIOUS ROOTING

1. Effects of exogenous ethylene

Cuttings of etiolated mung bean seedlings were incubated with ethylene gas for 24 h. Inhibitory effects of ethylene on formation of adventitious root primordia were evident at a concentration of 1.0ppm (Table 1), and similar responses to ethylene were found when wicks impregnated with potassium hydroxide were enclosed with the cuttings (see Radin and Loomis 1969). In further experiments ethylene gas (100ppm) was supplied to cuttings with pulse applications (0.5-9h duration) during the first 40h of rooting. With all treatments there was a reduction in root primordium production as compared with controls.

Ethylene is reported to cause aerial root formation in seedlings of *Tagetes erecta* (Zimmermann and Hitchcock 1933), and this effect was re-examined. Five-week-old marigold seedlings were enclosed with ethylene (1000ppm) for 48h, and then transferred to the glasshouse. After 3-4 weeks numerous aerial roots grew from the hypocotyls and first internodes. However, examination of the hypocotyl and first internode of 5-week-old marigold seedlings revealed an abundance of preformed root initials. Here, as with certain hardwood cuttings (Zimmermann and Hitchcock 1933), ethylene promoted the emergence of pre-formed roots. In mung bean cuttings, where there are no pre-formed root initials, ethylene inhibited the formation of adventitious primordia.

Table 1. Effect of Ethylene (0-24h) on formation of adventitious root primordia in cuttings (40mm) of etiolated Mung bean seedlings.

Ethylene Concentration (ppm)	Mean No. Root Primordia per Cutting	Log Transform
0	23.7	1.3655
1	18.5	1.2581
10	14.6	1.1390
100	12.7	1.0796
1000	9.5	0.0951
	LSD(P=0.05)	0.0830

2. Ethylene production by mung bean cuttings

The pattern of ethylene production by mung bean cuttings was similar in all experiments. The maximum rate of production occurred 10-12h after excision. There was a steady decline in ethylene production during the next 24h, and thereafter a constant low rate of evolution.

Investigations were made of effects on ethylene production of a number of phenolic compounds (guiacol, resorcinol, quinol, caffeic acid, chlorogenic acid, coumaric acid), and kinetin. When used at concentrations which affect rooting of mung bean cuttings, none of these compounds greatly affected ethylene production.

To assess effects of auxins on ethylene production, and possible relationships to adventitious rooting, a comparison was made of the ethylene-inducing properties of IAA and IBA. IAA alone has little effect on the rooting of etiolated or light-grown mung bean cuttings (Fernqvist 1966, Hess 1969), but IBA is a potent stimulator of adventitious rooting. The superiority of IBA in this respect has been attributed to its greater resistance to degradation. However, IAA has a much greater effect on ethylene production than IBA (Fig. 1). During 40h from excision ethylene production by cuttings treated with 1ml of IAA (5×10^{-5}M) was three times that of cuttings treated with 1ml IBA at the same molarity.

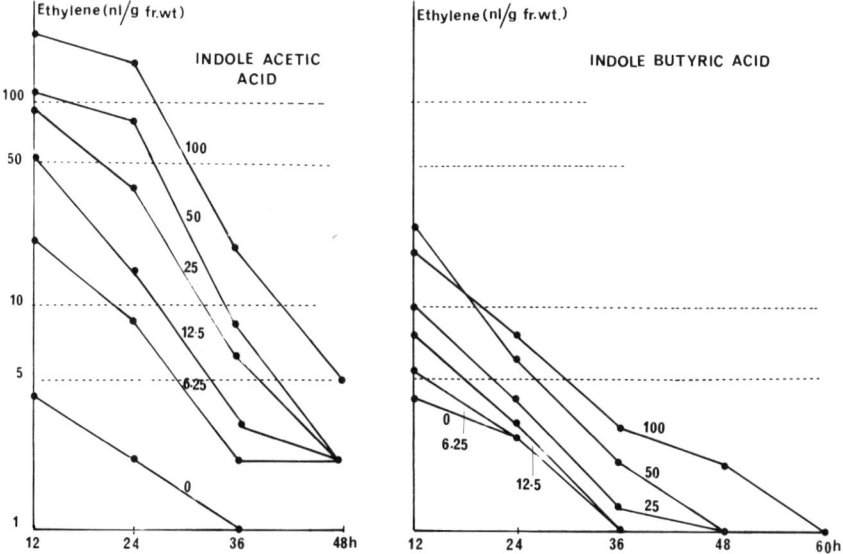

Fig. 1.

Effects of β-Indole acetic acid and γ-Indole butyric acid on ethylene production by cuttings (40mm) of etiolated mung bean seedlings. Cuttings were incubated in darkness (26°C). Estimates of ethylene production were derived from gas chromatographic assay of 1ml aliquots of the headspace atmosphere in the storage vials.

Now ethylene inhibits the formation of adventitious primordia (Table 1), but there was no direct relationship between total ethylene production and the root-inducing properties of IAA and IBA. IBA (50mg/l) and IAA (6mg/l) caused the production of similar amounts of ethylene, but IBA (50mg/l) stimulated root initiation and IAA (6mg/l) was inactive in the mung bean test. There is, however, a relationship between ethylene production relative to auxin concentration and root initiation. Formation of adventitious roots was promoted when high concentrations of exogenous auxin were associated with relatively low rates of ethylene production.

These results suggest that auxin and ethylene are antagonists in root initiation in mung bean cuttings. Auxins promote rooting, but rooting is opposed by the ethylene generated as a consequence of the auxin application. In this feedback system, formation of primordia is promoted when production of ethylene is low relative to the concentration of auxin (IBA). With potent inducers of ethylene (IAA), promotive effects of auxin on root initiation are outweighed by inhibitory effects of ethylene.

To test this theory a first approach was to see if the response of mung bean cuttings to IAA was affected by growing cuttings in ethylene-scrubbed atmospheres (mercuric perchlorate or alkaline potassium permanganate). Results of these experiments were inconclusive, possibly because effects of ethylene were manifested before escape of the gas from the tissues.

Indirect evidence of auxin:ethylene feedback in adventitious rooting was found in experiments with IAA and its synergist in the mung bean test, catechol.

IAA is rapidly metabolised by mung bean tissues, but addition of catechol (3×10^{-4}M) slows the conjugation of IAA with aspartic acid (Hess 1965). Indoleacetyl-aspartic acid is an inactive compound with respect both to rooting and to ethylene production in mung bean cuttings, as are the following IAA-metabolites: indole-3-aldehyde, indole-3-acetaldehyde, indoleacetonitrile (Mullins, unpublished results). Inhibitory effects of catechol (3×10^{-4}M) on IAA-oxidase activity in mung bean cuttings were confirmed in experiments on the decarboxylation of ^{14}C-1-IAA (Fig. 2). Catechol has a conserving effect on applied IAA in mung bean tissues, and there may be effects of the phenol on synthesis of native IAA (Gordon and Paleg 1961).

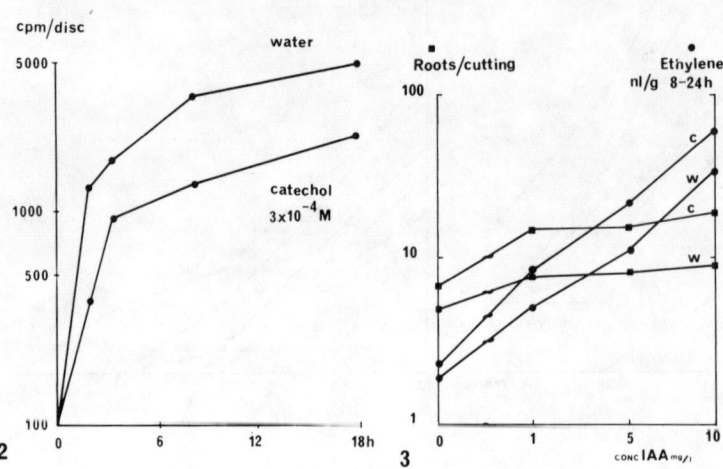

Fig. 2.

Effect of catechol on decarboxylation of ^{14}C-1-IAA. 1.0μl of isotope (48.5mCi/mM) was applied to the bases of light-grown cuttings. After uptake of the droplet (15 mins) groups of 10 cuttings were transferred to vials containing either 1.0ml of water or the same volume of catechol (3×10^{-4}M). Suspended from the closure of each vial was a glass fibre disc impregnated with 2-phenylethylamine. At appropriate intervals the discs were removed and assayed for radioactivity by liquid scintillation methods.

Fig. 3.

Effect of β-Indole acetic acid (0-10mg/1) and catechol (3×10^{-4}M) on root initiation (■) and ethylene production (●) in light-grown mung bean cuttings. Cuttings were grown either in aqueous solution (W) or with catechol (C).

Consistent with these findings is the fact that catechol-(3×10^{-4}M)-stimulated ethylene production by IAA (Fig. 3.). Addition of IAA (1mg/1) to light-grown mung bean cuttings caused a small (40 percent) increase in the numbers of adventitious roots as compared with controls, but higher concentrations of hormone gave no additional stimulation. With catechol there was increased root initiation as compared with cuttings which received IAA alone (1mg/1). Again, there was no additional stimulation of root formation when IAA was supplied at concentrations in excess of 1mg/1. At this low concentration IAA had relatively little effect on ethylene production (Fig. 3). It appears that the rapid inactivation of IAA by mung bean tissues cannot be offset by addition of hormone at high concentration because such treatments induce production of ethylene which is excessive for root initiation. Similarly, it can be argued that effects of catechol on auxin-induced root initiation were attenuated when IAA was applied at concentrations in excess of 1mg/1 because ethylene production was raised to a level which could no longer be accommodated by the system (Fig. 3).

In horticultural practice it is known that IAA promotes rooting in many species when supplied to cuttings at high concentration (Pearse 1948), e.g. 50-1000ppm. It seems certain that such treatments induce large amounts of ethylene. In the nursery, however, growth regulators are applied to cuttings with a pulse application. In the mung bean bioassay the method of growth regulator application involves a long term soak (24h), a high volume of solution (4-10ml), and relatively low concentrations of the active material.

Auxins have greatest effect on root initiation in mung bean when applied immediately after excision (as stated above) but the maximum rate of auxin-induced ethylene production occurs after 10-12h. Thus, induction of the first primordia and high rates of ethylene production are separated in time. IAA is rapidly metabolised by mung bean tissues (Fig. 2). When IAA-2-C14 was applied to mung bean cuttings as a microlitre pulse only 10 percent of the radioactivity recovered after 5h could be attributed to IAA (unpublished results). Chadwick and Burg (1970) have shown that high volume applications of IAA at a given concentration lead to greater persistence of IAA within tissues than low volume applications. In addition they suggest that ethylene production is dependent upon the internal concentration of IAA, and this conclusion is supported by results of experiments with IAA and catechol (Figs. 2 and 3).

It is now clear that failure to demonstrate marked root-promoting effects of IAA in mung bean cuttings has been due to use of high volume applications of hormone. For root promotion IAA must be applied at high concentration but low volume so that (i) initially, sufficient hormone is applied to elicit a rooting response, (ii) subsequently, the internal concentration of IAA falls rapidly so as to minimise ethylene production. A compromise must be reached between the requirement for induction of root primordia, the rate of IAA degradation, and IAA-induced ethylene production.

Table 2. Effect of mode of application of IAA on formation of adventitious root primordia in cuttings (40mm) of etiolated mung bean seedlings

Treatment	Mean No. Root Primordia/cutting	Log Transform
Basal Soak (0-24h, 4ml)		
IAA (20mg/1)	17.2	1.2338
IBA (10mg/1)	26.7	1.4176
Water (control)	14.1	1.1398
LSD (P=0.05)	-	0.1204
2 µl IAA (176 mg/1)		
Apical Application	32.6	1.4917
Basal Application	27.9	1.4309
Water (Control)	16.5	1.2013
LSD (P=0.05) Apical v Basal	-	NS
" " Apical v Control	-	0.0906
" " Basal v Control	-	0.0852

It was found that a 2 µl droplet of IAA (10^{-3}M), applied either to the apical or basal ends of dark-grown cuttings greatly stimulated root initiation (Table 2). Ethylene production was more variable with pulse than with high volume applications of IAA. However, 2µl IAA (10^{-3}M) caused evolution of ethylene (12-24h) at rates of 40-80nl/g fresh wt. Similar rates of ethylene production were associated with high volume applications of IAA at concentrations of 25-50mg/1, i.e. treatments which had little effect on adventitious root formation. In these experiments, as in earlier work, formation of adventitious roots was promoted when high concentrations of exogenous auxin were associated with relatively low rates of ethylene production.

IV. SUMMARY

Adventitious root formation in mung bean cuttings involves sequences of phenomena (induction of initials, initiation and differentiation of primordia, emergence of roots) with differing requirements for growth substances. The time-course of these changes under standard environmental conditions has been determined, as have responses to auxins, cytokinin, and ethylene at various times after excision. Root-promoting effects of auxins decline when application is delayed. Response of cuttings to kinetin is dependent upon time of application relative to auxin concentration. Ethylene inhibits formation of adventitious root primordia in cuttings, but it promotes emergence of roots in stems with pre-formed primordia. Relationships between effects of β-Indole-acetic acid (IAA) and γ-Indole-butyric acid on ethylene production and root initiation have been examined, as have effects of catechol in relation to IAA metabolism, IAA-induced ethylene production, and rooting of cuttings. It is proposed that promotive effects of auxins on induction of root primordia are opposed by inhibitory effects of auxin-induced ethylene (feedback). Adventitious rooting is promoted when rates of ethylene production are low relative to auxin concentration.

V. REFERENCES

BASTIN, M. (1966). Root initiation, auxin level, and biosynthesis of phenolic compounds. Photochem. Photobiol., 5, 423-429.

BOUILLENNE, R. and M.W. BOUILLENNE (1955). Auxines et bouturage Proc. 14th. Int. Hort. Congr., Scheveningen 1, 231.

CHADWICK, A.V. and S.P. BURG (1970). Regulation of root growth by auxin:ethylene interaction. Pl. Physiol. Lancaster. 45, 192-200.

FADL, M.S. and H.T. HARTMANN (1967). Isolation, purification, and characterisation of an endogenous root- promoting factor from basal sections of pear hardwood cuttings. Ibid. 42, 541-549.

FERNQVIST, I. (1966). Studies on factors in adventitious root formation. Lantbruk-shögsk. Ann., 32, 109-204.

GAUTHERET, R.J. (1969). Investigations on root formation in the tissues of Helianthus tuberosus cultivated in vitro. Amer. J. Bot., 56, 702-717.

GIROUARD, R.M. (1969). Physiological and biochemical studies in adventitious root formation: Extractable rooting cofactors from Hedera helix. Can. J. Botany, 47, 687-699.

GOLDACRE, P.L. (1959). Potentiation of lateral root induction by root initials in isolated flax roots. Aust. J. biol. Sci., 12, 388-394.

GOODEY, J.B. (1957). Laboratory methods for work with plant and soil nematodes. Tech. Bull. 2. Ministry of Agriculture, Fisheries, and Food. H.M.S.O. London.

GORDON, S.A. and L.G. PALEG (1961). Formation of auxin from trytophan through action of polyphenols. Pl. Physiol. Lancaster., 36, 838-845.

HARTMANN, H.T. and D.E. KESTER (1959). "Plant propagation; principles and practices". 2nd Edn. Prentice-Hall, N.J.

HESS, C.E. (1964). Naturally-occurring substances which stimulate root initiation. In "Régulateurs naturels de la croissance végétale". Coll. Int. C.N.R.S. 123, 517-527.

HESS. C.E. (1965). Phenolic compounds as stimulators of root initiation. Pl. Physiol. Lancaster. 40, (supp.) XLV.

HESS, C.E. (1969). Internal and external factors regulating root initiation. In "Root Growth". ed W.J. Whittington. pp. 42-53. Butterworths. London.

PEARSE, H.L. (1948). Growth substances and their practical importance in horticulture. Tech.Comm. 20, Commonwealth Bureau of Horticulture and Plantation Crops. C.A.B. London.

RADIN J.W. and R.S. LOOMIS (1969). Ethylene and carbon dioxide in the growth and development of cultured radish roots. Pl. Physiol. Lancaster. 44, 1584-1589.

SKOOG, F. and C.O. MILLER (1957). Chemical regulation of growth and organ formation in plant tissues cultured in vitro. Symp. Soc. Exp. Biol. 11, 118-131.
THIMANN, K.V. and F.W. WENT (1934). On the chemical nature of the root-forming hormone. Proc. Kon. Nederl. Akad. Wetensch., Amsterdam, 37, 456-459.
YANG, S.F. (1967). Biosynthesis of ethylene. Ethylene formation from methional by horseradish peroxidase. Arch. Biochem. Biophys., 122, 481-487.
ZIMMERMANN, P.W. and A.E. HITCHCOCK (1933). Initiation and stimulation of adventitious roots by unsaturated hydrocarbon gases. Contrib. Boyce Thompson Inst. 5, 351-369.

Ethylene and the Growth of Plant Cells:
Role of Peroxidase and Hydroxyproline-rich Proteins

Daphne J. Osborne, Irene Ridge and J.A. Sargent

Agricultural Research Council Unit of Developmental Botany, University of Cambridge, England

I. INTRODUCTION

Of recent years interest and experimentation have escalated in the study of ethylene as a natural regulator of plant growth, though many of the physiological effects of the gas have been known for longer than those of any other plant hormone.

One of the most obvious responses of a plant exposed to ethylene is a reduction in the rate of stem elongation and a lateral swelling of the normally extending parts. This response can be seen in green plants and in plants grown in darkness. Crocker (1948) noted that pea plants are remarkably sensitive to low concentrations of the gas and we have used 6-7 day old etiolated seedlings of *Pisum sativum* var. Alaska in all the studies that I am about to describe.

II. OBSERVATIONS ON CELL GROWTH

The reduction in growth of a marked apical segment of pea plants grown in high concentrations of ethylene is very considerable (Table 1(a)).

Table 1(a). Average growth of a marked 5.5 mm apical segment after an exposure of intact peas for 24 hr to air or ethylene 500 ppm.

	0 hr	24 hr AIR	24 hr C_2H_4
F.wt mg	10.4	51.6	18.2
Length mm	5.5	21.4	6.0

Table 1(b). Average volume of mature cortical cells from control plants or plants exposed for 5 days to ethylene 500 ppm. Measurements in Projectina units.

	AIR	C_2H_4
Area (TS)	108	316
Length (LS)	57.6	23.0
Volume	6220 ± 599	7273 ± 695

However, although the rate of growth is reduced, the final volume achieved by the cortical cells (which constitute the major part of the swollen tissue) remains essentially unaltered (Table 1(b)). Earlier work by Fan and Maclachlan (1966) showed that swelling of etiolated pea stems and cell enlargement occurred when indole-3-acetic acid (IAA) was applied in lanolin to the cut tops of decapitated pea shoots. Associated with this was a large increase in a soluble cellulase enzyme which degraded carboxymethylcellulose. Determination of cellulase activity in soluble cytoplasmic extracts and in extracts of ionically or covalently bound wall protein from the swollen parts of ethylene treated pea plants show that compared with controls there is no increase of this enzyme in any of the fractions (Table 2) although auxin-treated decapitated peas showed the large increases reported by Fan and Maclachlan.

Table 2. Cellulase activity in marked 5 mm segments from a) control peas, b) the swollen parts of peas grown in ethylene 100 ppm, c) decapitated peas treated with 0.5 per cent IAA in lanolin (48 hr). Expressed as per cent decrease in viscosity of carboxymethyl cellulose. For assay methods see Ridge and Osborne (1969). For extraction of ionic and covalent wall fractions see legend to Fig. 1.

	CYTOPLASMIC*	WALL BOUND**	
		IONIC	COVALENT
a) AIR	12.2 ± 0.2	0.9	1.0
b) C_2H_4	10.8 ± 0.9	0	0.4
c) IAA	30.1 ± 1.7	-	-

* Assay at $25°C$/24 hr. Values per unit protein.
** Assay at $40°C$/24 hr. Values per unit wall weight, and corrected to cytoplasmic equivalent.

Examination of the swollen cells resulting from auxin applications showed them to be quite unlike those that occur in ethylene; the volumes of the cortical cells at maturity were some three times as large as those of control or ethylene-treated pea stems (Ridge and Osborne, 1969). Clearly, auxin-induced cell enlargement is different from that induced by ethylene. It would seem however that the expansion induced by ethylene is dependent on the presence of some auxin in the tissue, for decapitated peas do not develop lateral swelling in ethylene unless auxin is simultaneously applied to the cut stump (results not presented). But ethylene above 0.2 ppm will reduce the extension growth of intact peas and of decapitated peas, particularly if the latter are induced to extend by added auxin (Table 3).

Table 3. Length of the marked 5 mm segment below the decapitated tip of 6 day old pea plants following application of IAA in lanolin to the cut stump and exposure to air or ethylene 100 ppm for 48 hr.

	% IAA in lanolin (1 µl drop)				
	0	0.5	0.05	0.005	0.0005
AIR	6.42	8.01	9.47	9.74	9.32
C_2H_4	5.87	7.69	8.39	7.60	7.05

One could conclude from experiments of this type that if auxin is required for lateral expansion, and if the cortical cells from ethylene-treated plants have the same total volume as controls, then wall loosening (plasticity) may not be directly affected by ethylene. Rather, ethylene may direct the orientation in which cell growth takes place, following plasticity changes induced by auxin. Fan and Maclachlan (1967) have associated increased cellulase activity in the presence of auxin with

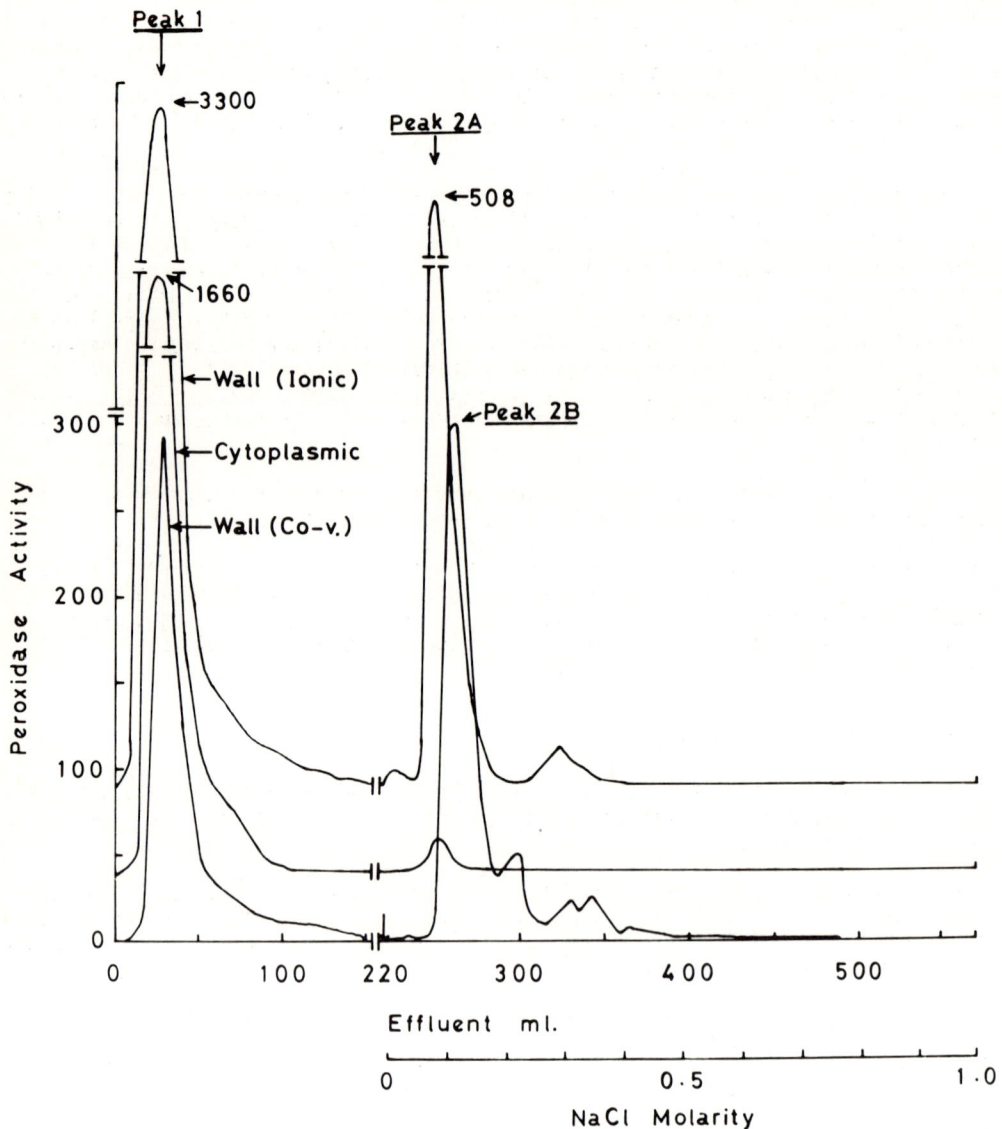

Fig. 1. DEAE cellulose separations of cytoplasmic and wall bound peroxidases. For the cytoplasmic protein, fraction tissue was homogenised in 0.05 M phosphate pH 6.0 and centrifuged for 10 min at 2000 g. For the wall bound protein fractions the remaining pellet was washed to remove contaminating cytoplasm by suspension and centrifugation (45 secs at 800 x g) in 1 per cent 'Triton-X 100' (x 2) and water (x 20 or until washings were free of peroxidase activity). The cell wall pellet was then washed x 3 with M NaCl and these washings constituted the ionically-bound cell wall fraction. The residual wall pellet was washed with 0.5 M $NaHCO_3$ (x 2), with M NaCl until washings were free of peroxidase activity (x 4 - 10), and finally with water (x 3). The pellet then contained only proteins which were presumed to be covalently bound to the cell walls. After lyophilisation and weighing, the cell walls were incubated for 20 hr at $25^{\circ}C$ with 0.5 per cent cellulase in 0.01 M sodium acetate pH 5.5. The incubation was repeated with fresh cellulase solution and the walls washed with water twice. Extracts and washings were pooled, filtered, volumes reduced in a rotary evaporator and dialysed against 0.01 M phosphate pH 8.0. Extracts were separated on DEAE cellulose columns (Whatman 23), 1.5 x 30 cm equilibrated with 0.01 M phosphate pH 8.0.

increased wall plasticity and a potential for cell expansion. The absence of any marked effect of ethylene on total cellulase levels in intact peas would accord with the idea that ethylene may have relatively little effect on total plasticity but instead may dictate the orientation of cell enlargement. Such a system implies a close regulatory balance between auxin and ethylene in the control of shape and size of plant cells.

Preston (1961) and later Lamport (1965) developed the concept that the proteins closely associated with the cellulose in plant cell walls might determine the cohesion and extensibility of the walls and perhaps also regulate their final shape. The proteins of plant cell walls are remarkably rich in the imino acid, hydroxyproline and the content of hydroxyproline rises sharply as cells reach full expansion (Cleland and Karlsnes, 1967). The precise nature of this covalently bound hydroxy-proline-rich wall protein is still largely unknown, but in tomato callus the hydroxyproline residues are linked through O-glycosidic linkages to arabinose (Lamport, 1967). As much as 90 per cent of the hydroxyproline may be glycosylated (Lamport, 1969). The level of hydroxyproline then reflects the extent of the cross linking between protein and wall polysaccharide.

A number of enzymes are tightly bound to cell walls either ionically or covalently, but with one exception, none has been shown to contain hydroxyproline. The exception is peroxidase. Shannon, Kay and Lew (1966) isolated three isoenzymes from soluble horseradish peroxidase which contained small amounts of hydroxyproline and arabinose. Peroxidase activity is commonly found in the cell walls of plants, particularly in mature tissue (Ridge and Osborne, 1970), so both peroxidase and hydroxyproline appear to be associated with tissue maturation, the cessation of cell growth and the loss of wall extensibility.

III. BIOCHEMICAL AND ULTRASTRUCTURAL STUDIES

We now present the result of our studies to date on the biochemical and ultrastructural changes that occur in the extending cell walls of pea stems when seedlings are exposed to ethylene.

The increases in peroxidase content in both cytoplasmic and covalently-bound wall proteins during normal maturation in air and after 5 days exposure of the plants to ethylene are shown in Table 4. Values for the levels of hydroxyproline are included, and it is seen that peroxidase activity and hydroxyproline levels follow a parallel course; both increase with age and in response to ethylene. The content of hydroxyproline in the cytoplasm is relatively low compared with the walls and by far the greatest increase in both peroxidase and hydroxyproline occurs in the wall-bound proteins of the expanding tissue. The localization of the increase to the covalently bound fraction of the wall protein is illustrated clearly in Table 5.

Table 4. Peroxidase and hydroxyproline levels in cytoplasmic and wall bound proteins in the internodes of peas before and after 5 days in ethylene 500 ppm. Extraction methods as in Ridge and Osborne (1970).

INTERNODE	TREATMENT	PEROXIDASE		HYDROXYPROLINE $\mu g/100~\mu g$ protein	
		Cytoplasmic	Wall bound	Cytoplasmic	Wall bound
3rd (apical)	AIR	44.8	3.1	0.20	5.01
	C_2H_4	285.7	21.2	0.23	16.49
2nd	AIR	65.2	9.7	0.25	10.92
	C_2H_4	458.8	22.5	0.39	28.01
1st	AIR	136.8	13.6	0.31	16.42
	C_2H_4	729.5	21.9	0.38	22.05

Variations between replicates do not exceed 4 per cent. Peroxidases estimated by the method of Ponting and Joslyn (1948). Hydroxyproline by the method of Leach (1960).

Table 5. Effect of Ethylene 500 ppm for 5 days on peroxidase activity and hydroxyproline levels in cytoplasmic and cell wall proteins. Methods as in Ridge and Osborne (1970).

	FRACTION	HYDROXYPROLINE µg/100 µg protein		PEROXIDASE ACTIVITY /mg protein		PROTEIN *mg/gm F Wt **mg/100 mg wall Wt.		
		Air	C_2H_4	Air	C_2H_4		Air	C_2H_4
CYTOPLASMIC	(1)	0.20	0.23	44.8	285.7	*	16.2	6.0
	IONICALLY-BOUND (2)	0.01	0.01	18.2	231.6	**	6.3	6.0
CELL WALL	COVALENTLY-BOUND (3)	5.01	16.50	3.1	21.2	**	8.9	5.8

    Proteins from the cytoplasmic and wall-bound fractions from apical pea tissue have been separated on columns of DEAE cellulose in an elution gradient of 0-1.0 M NaCl in 0.05 M phosphate pH 8.0, as described by Ridge and Osborne (1970). Eluent fractions were monitored for peroxidase activity and the active peaks and the intervening troughs were subsequently assayed for their hydroxyproline content and protein levels. Two major peaks of peroxidase activity were found in each sample, Fig. 1,
— Peak 1, which is not absorbed to the column, and a second peak which elutes early in the gradient. Peroxidases of Peak 1 are practically coincident in the three samples and whereas the Peak 2 samples of cytoplasmic and ionically bound wall peroxidases coincide (Peak 2A), Peak 2 of the covalently bound peroxidase is retarded (Peak 2B) and elutes at a higher molarity.

    Further separation of the isoenzymes of these peroxidase peaks by cellulose acetate membrane electrophoresis followed by determinations of the pH optimum for enzyme activity have shown that the cytoplasmic and ionically bound wall peroxidases contain similar isoenzymes with similar pH optima (Fig. 2). The covalently bound wall peroxidases from both Peak 1 and 2B are distinct both in their isoenzymes and pH optima.

    Determinations of the hydroxyproline contents were made on the covalently bound wall proteins separated on DEAE columns from both air-grown and ethylene-treated peas. Fig. 3 shows that the regions of hydroxyproline-rich protein coincide with the regions of high peroxidase activity, rather than with the pattern for total protein. The possibility is therefore considered that wall-bound peroxidase could be one of the hydroxyproline-rich proteins of the cell wall.

    Electron micrographs (Fig. 4) of the cortical cells of the apical segment from control peas and peas exposed to ethylene 100 ppm for 5 days show that the thickness of the cell wall is increased as much as two-fold by ethylene treatment, with a relatively greater thickening occurring on the longitudinal walls. The micrographs indicate a deposition of secondary wall material within the cell and since the total protein content per unit wall weight is lower in ethylene treated peas (Table 5) it seems likely that the deposition is of polysaccharide material.

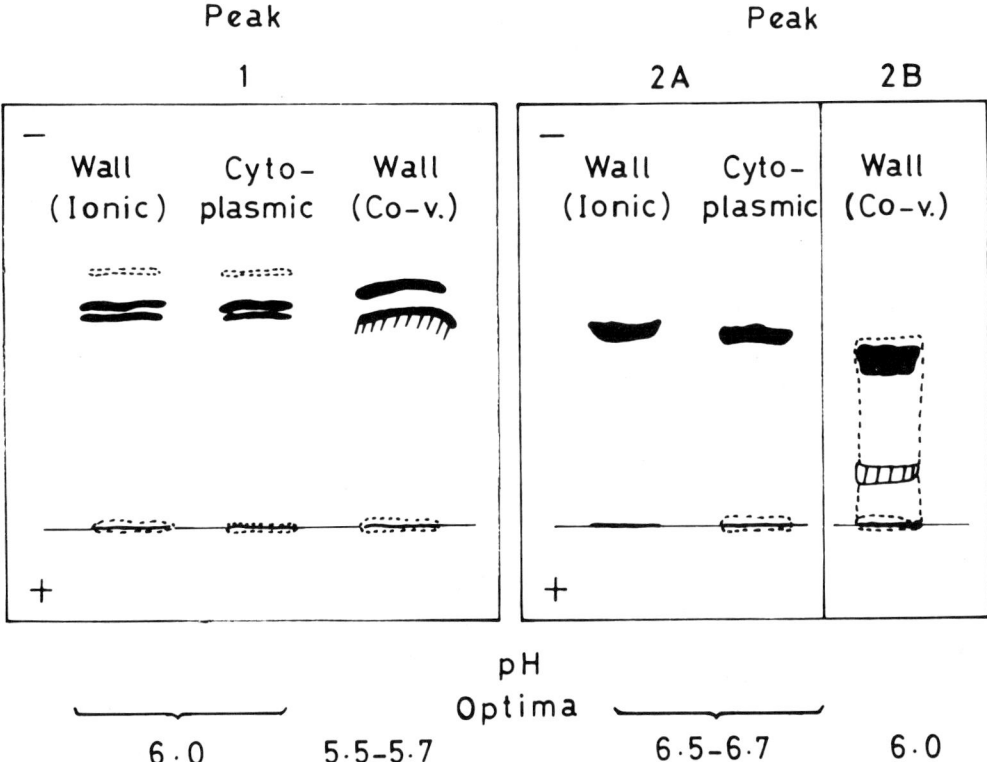

Fig. 2. Electrophoretic separations of peroxidases from the major DEAE cellulose peaks after electrophoresis on cellulose acetate membranes in a Shandon Universal Tank by the method of Kohn and Feinberg (1965).

IV. SUMMARY

The enzymic and ultrastructural changes that take place in the walls of extending cells of pea shoots in response to ethylene are associated with an alteration in both the rate and the orientation of cell expansion. It is proposed that the enhanced level of hydroxyproline and the apparently greater extent of wall deposition could lead to an increased crosslinking by glycosidic bonds of hydroxyproline-rich proteins with polysaccharide in the wall material. Such a change in wall structure could contribute to the decrease in wall plasticity and extensibility and since the distribution of thickening is greatest on the longitudinal walls, could account for the changed orientation of cell growth.

V. ACKNOWLEDGEMENTS

We would like to acknowledge the encouragement of Professor G.E. Blackman of the Department of Agricultural Science, Oxford.

VI. REFERENCES

CLELAND, R. and A.M. KARLSNES (1967). A possible role of hydroxyproline - containing protein in the cessation of cell elongation. Plant Physiol., 42, 669-671.
CROCKER, W. (1948). "Growth of Plants." Reinhold, New York.

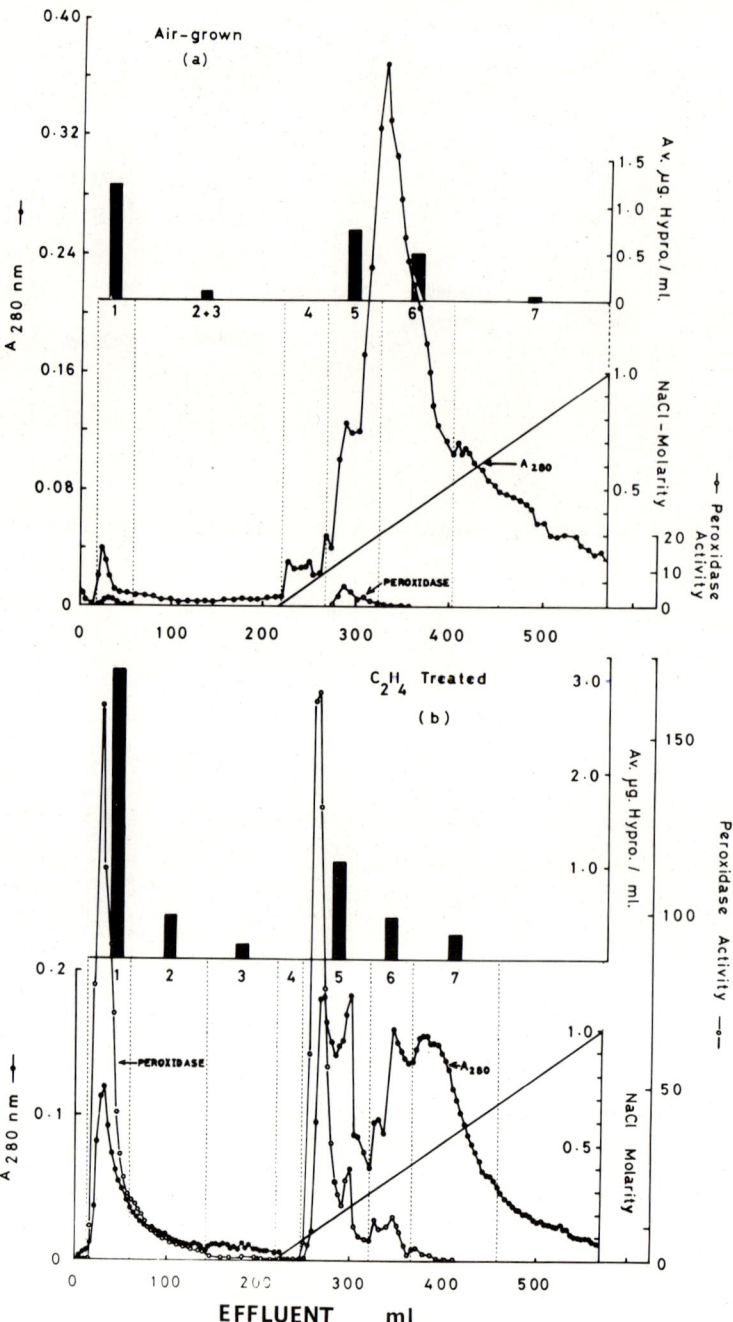

Fig. 3. Elution profiles from DEAE cellulose of cellulase extracts of cell walls from control peas grown in air (a), or from peas exposed to ethylene 500 ppm for 5 days (b). Absorbance at 280 nm was taken as a measure of total protein. Appropriate groups of fractions were pooled, dialyzed and reduced in volume in a rotary evaporator, before lyophilization and hydrolysis in 6N HCl. Hydroxyproline in the hydrolysates was assayed by the method of Prockop and Udenfriend (1960). Separation of cellulase enzyme alone on DEAE cellulose revealed no peroxidase or hydroxyproline on elution up to M NaCl.

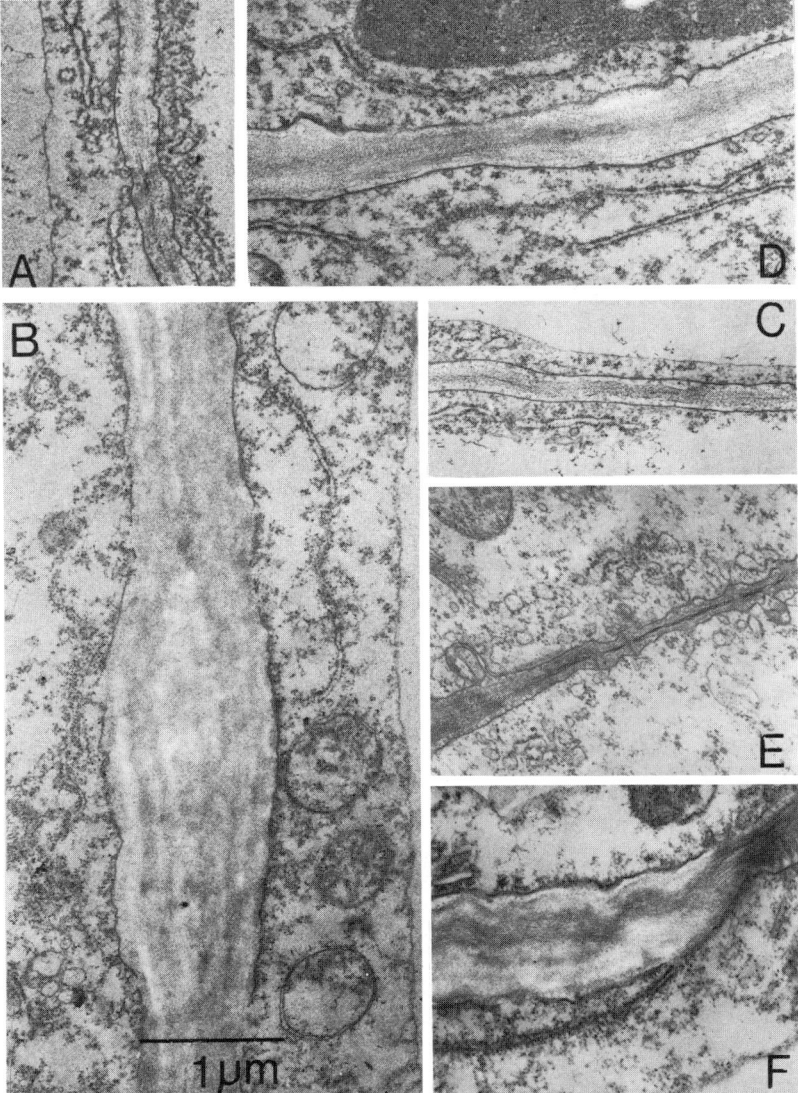

Fig. 4. Electron micrographs of cortical cells from apical segments of pea plants grown for 10 days in air or for the sixth to tenth day in ethylene 100 ppm. Material fixed in glutaraldehyde and osmium tetroxide. Stained with lead citrate.

A. L.S. Longitudinal wall. Control.

B. L.S. Longitudinal wall. Ethylene.

C. L.S. Transverse wall. Control.

D. L.S. Transverse wall. Ethylene.

E. T.S. Control.

F. T.S. Ethylene.

FAN, D.F. and G.A. MACLACHLAN (1966). Control of cellulase activity by indoleacetic acid. Canad. J. Botany, 44, 1025-1034.
FAN, D.F. and G.A. MACLACHLAN (1967). Studies on the regulation of cellulase activity and growth in excised pea epicotyl sections. Canad. J. Botany, 45, 1837-1844.
KOHN, J. and FEINBERG, J.G. Shandon Instrument applications No. 11. Shandon Scientific Co. Ltd. Willesdon, London. 1965.
LAMPORT, D.T.A. (1965). The protein component of primary cell walls. Adv. Bot. Res., 2, 151-218.
LAMPORT, D.T.A. (1967). Hydroxyproline-O-glycosidic linkage of the plant cell wall glycoprotein, extensin. Nature, 210, 1322-1324.
LAMPORT, D.T.A. (1969). The isolation and partial characterization of hydroxyproline-rich glycopeptides obtained by enzymic degradation of primary cell walls. Biochemistry, 8, 1155-1163.
LEACH, A.A. (1960). Notes on a modification of Neuman and Logan's method for the determination of hydroxyproline. Biochem. J., 74, 70-71.
PONTING, J.D. and M.A. JOSLYN (1948). Ascorbic acid oxidation and browning in apple tissue extracts. Arch. Biochem., 19, 47.
PRESTON, R.D.(1961). Cellulose-protein complexes in plant cell walls. In "Macromolecular complexes" (Ed. M.V. Edds), pp 229-253. The Ronald Press, New York. 1961.
PROCKOP, D.J. and UDENFRIEND, S. (1960). A simple method for the analysis of hydroxyproline in tissues and urine. Anal. Biochem. 1, 228-239.
RIDGE, I. and D.J. OSBORNE (1969). Cell growth and cellulases: regulation by ethylene and indole-3-acetic acid in shoots of *Pisum sativum*. Nature, 223, 318-319.
RIDGE, I. and D.J. OSBORNE (1970). Hydroxyproline and peroxidases in cell walls of *Pisum sativum*: regulation by ethylene. J. Exptl. Bot. 21, 843-856.
SHANNON, L.M., E. KAY and J.Y. LEW (1966). Peroxidase isoenzymes from horseradish roots. 1. Isolation and physical properties. J. biol. Chem., 241, 2166.

PLANT GROWTH SUBSTANCES, 1970

Trauma-induced Ethylene Production by Citrus Flowers, Fruit, and Wood

W.C. Cooper

Investigations Leader, Plant Science Research Division, Agricultural Research Service, U.S. Department of Agriculture, Orlando, Florida, 32803

I. INTRODUCTION

Ethylene is usually produced in those parts of the plant which have the highest auxin content, presumably because the production of the gas is stimulated by auxin (Burg and Burg, 1966). In immature citrus leaves and fruits, high rates of ethylene production are associated with high auxin levels, and in mature citrus leaves and fruits, low rates of ethylene production are associated with low auxin levels (Cooper, et al., 1969). However, injury or infection or chemical treatment causes any plant tissue to increase its production of ethylene. For instance, virus-infected rose leaves and shredded rose leaves (Williamson, 1950), freeze-stressed citrus leaves (Cooper, et al., 1969), and citrus fruit treated with beta (2-(3,5-dimethyl-2-oxo-cyclohexyl-2-hydroxyethyl) glutarimide [cycloheximide] (Cooper and Henry, 1970), all produce more ethylene than healthy or uninjured or untreated tissues.

With flower fading, cycloheximide-induced fruit drop, and mechanical wounding, destruction of cell walls probably takes place. In this paper we show that large increases in ethylene production in leaves, flowers, fruit, and wood of citrus trees are associated with trauma-induced phenomena. Although wound ethylene is produced endogenously, it may not be directly associated with the classical auxin-mediated ethylene production exhibited by young developing shoots.

II. METHODS AND MATERIALS

The citrus cultivars used for this work include 'Valencia' oranges (*Citrus sinensis* [L.] Osb.), 'Redblush' grapefruit (*C. paradisi* Macf.), and 'Robinson' tangerine (*C. reticulata* Blanco X [*C. paradisi* X *C. reticulata*]).

An earlier paper from this laboratory described the treatment of citrus trees with cycloheximide and other abscission chemicals, the measurement of ethylene by gas chromatography, and measurement of attachment force by a pull tester (Cooper, et al., 1968). Detached fruit with about 4 inches of stem were used to compare the effects of applying chemicals directly to the abscission zone, in contrast to spraying them on the outer surface of the rind as in field practice. Chemicals were applied through the stem or to the outer surface of the rind and the treated fruit were sealed in plastic containers, with or without $KMnO_4$ to absorb ethylene.

III. RESULTS AND DISCUSSION

Endogenous wounding

I found a large production of ethylene by the pollen, stigmas and petals of freshly opened Valencia orange flowers (Table I), and this is presumably induced by auxin in these tissues. The additional large increase in ethylene production associated with the faded (injured) petals may well be from the gas stimulating its own formation as a consequence of ethylene injury to the petals, as was found by Burg and Dijkman (1967) for fading orchid blossoms.

Table I. Ethylene production by mature leaves, flower parts and small immature fruitlets of Valencia orange[a]

Plant part analyzed	C_2H_4 Production (nl/g/hr)
Mature green leaves	0.02
Pollen from freshly opened flowers	18.00
Stigmas of freshly opened flowers	34.80
Petals of freshly opened flowers	9.40
Petals faded just prior to abscission	23.70

[a] The leaves and flower parts were collected and analyzed on April 4, 1970. Each individual sample consisted of 10 plant parts. Data represent means of 3 sets of samples from each of 3 trees, making a total of 9 samples. The differences between petals freshly opened and petals faded were significant at the .05 level.

Fruit splitting is another example of tissue-wounding due to internal causes. Seedless immature Robinson tangerine fruit often split while attached to the tree. I found such fruit to be producing ethylene at a rate of 9.00 nl/g/hr, as compared to 0.02 nl/g/hr for sound fruit on the same tree. This high rate of ethylene production causes the split fruit to degreen and abscise prematurely.

Mechanical wounding

Almost any mechanical injury to fruit causes ethylene production. Loosening the rind at the stem end of the Robinson tangerine by gently pulling the stem stimulates ethylene production (Table II). Removing the peel from citrus fruits increases ethylene production in both the detached peel and the endocarp. Increasing the injury to the endocarp by cutting into the juice sacs increases ethylene production still further. Extracting the juice from the endocarp doubles the ethylene production by the residual pulp. The extracted juice, however, does not produce ethylene. Thus, it is apparent that the juice was only diluting the ethylene concentration in the endocarp. Likewise, although the peel stimulates ethylene production, homogenizing it into a puree does not. Homogenization apparently destroys the ethylene-producing system.

Mechanical wounding of stems also increases ethylene production of the wounded tissue. When the bark was peeled from the wood of Robinson tangerine terminal shoots, 5 to 10 times as much ethylene was produced in the peeled bark and wood as in the intact stems (Table III). Likewise, slitting the stems into two pieces, or macerating the stems by beating them with a hammer increased the rate of ethylene production. In experiments with Redblush grapefruit (data not presented), I measured ethylene production by the woody portions of branches 2 and 3 inches in diameter and found ethylene production rates equal to those shown for Robinson tangerine shoots. In most instances, I used branches from which the bark slipped readily. Cleaning off the moist cambial cell fragments on the woody stem pieces did not decrease the ethylene-production activity of the stems. The rate of ethylene production was actually increased by macerating these large woody stem sections with a hammer. Thus, ethylene-production activity is not confined to the injured cambial cells on the outer periphery of the woody cylinder, but may possibly result from injured parenchyma cells in the xylem.

Table II. Ethylene production by fruit of Robinson tangerine subjected to various injury treatments[a]

Treatment	C_2H_4 (nl/g/hr)
Whole uninjured fruit (control)	0.020 a
Rind loosened at stem end by pulling stem	0.185 b
Detached peel	1.140 e
Homogenized peel	0.0
Peeled endocarp	0.415 c
Peeled endocarp with cuts into juice sacs	0.550 d
Extracted juice	0.0
Extracted pulp	0.840 e

[a] Data represent means of 25 individual fruits for each treatment. Values followed by the same letter are not significantly different.

Table III. Ethylene production response to wounding of mature stems of Robinson tangerine[a]

Wounding treatment	C_2H_4 production (nl/g/hr)
Whole stems	0.290 a
Stems slit into 2 pieces	3.662 c
Bark from peeled stems	1.820 b
Wood from peeled stems	4.370 c
Whole stems macerated	4.015 c

[a] Data represent means of 10 stem pieces for each treatment. Values with the same letter are not statistically significant.

Cycloheximide-induced rind injury

When mature fruit on a citrus tree is sprayed with solutions containing cycloheximide, small amounts of the chemical are deposited on the rind and on the calyx lobes near the abscission zone in the button. Concentrations of 20 ppm induce rind injury, and loss in attachment force of the fruit develops within 3 days (Cooper, et al., 1971), but there is generally no evidence of injury to the calyx lobes. The rind injury usually consists of barely perceptible necrotic spots at the blossom end of the fruit associated with spray deposits of the chemical.

Ethylene accumulates inside the fruit concomitantly with the appearance of rind injury and the reduction in attachment force. Ethylene accumulates both inside and outside of the fruit if the fruit is detached from the tree and incubated in a sealed chamber (Table IV). If the fruit is placed in the test chamber at the time of treatment and KMnO4 pellets are added to the chamber, the KMnO4 absorbs the ethylene produced by the fruit (presumably by the injured rind), and no ethylene accumulates either inside or outside of the fruit (Table IV). Rind injury occurs but abscission is not accelerated.

Table IV. Relation of site of application of 20 ppm cycloheximide (CHI) to ethylene accumulation and reduction in attachment force of Valencia orange fruits[a/]

CHI application methods with and without $KMnO_4$ in sealed container	C_2H_4 content of air		Initial attachment force of fruit (%)
	Inside fruit (ppm)	Outside fruit (ppm)	
Applied on surface of rind:			
With $KMnO_4$	0.0	0.0	100 b
Without $KMnO_4$	0.495 a	0.120 a	42 a
Absorbed through stem:			
With $KMnO_4$	0.655 a	0.0	110 c
Without $KMnO_4$	1.325 b	0.220 a	110 c

[a/] Data represent means of 10 fruit per treatment. Individual fruit with fruit stems attached were sealed for 48 hr in 1.37 liter chambers. The initial attachment force of the fruit was 7.7 kg. Values in the same column followed by the same letter are not significantly different at the .05 level.

If the cycloheximide is applied directly to the abscission zone (by stem uptake), the same surge of ethylene production occurs (apparently from injury of the cells of the abscission zone), and it accumulates both inside and outside of the fruit (Table IV). When $KMnO_4$ is added to the test chamber, under these conditions, it greatly reduces the accumulation of ethylene inside of the fruit but does not completely eliminate it. Thus, ethylene moves out of the fruit, possibly through the rind, although it could conceivably move out through the stem. When cycloheximide is applied by stem absorption, abscission is inhibited. Thus, cycloheximide, an effective inhibitor of cellulase activity (Horton and Osborne, 1967), inhibits citrus fruit abscission, even in the presence of a large accumulation of ethylene; but inhibition occurs only when cycloheximide is applied directly to the abscission zone. When the chemical is sprayed on the tree, it apparently does not penetrate into the abscission zone in the button. If it did, abscission would be inhibited.

These data substantiate the conclusion that cycloheximide accelerates abscission by a wound reaction. The wounded cells of the citrus rind generate ethylene, and the ethylene accumulates in all parts of the internal atmosphere of the fruit, including the air near the abscission zone in the button. The wound ethylene acts as a triggering agent of the biochemical process involved in cell separation.

Cycloheximide-induced wood injury

When 20 cm cycloheximide was sprayed on Robinson tangerine trees, ethylene production by the leaves and stems is stimulated (Table V). This stimulation of ethylene production occurred before tissue damage could be seen. Possibly some cell wall damage had occurred but was not visible. I did not examine the tissues microscopically.

Four weeks after the cycloheximide was applied to the Robinson tangerine trees, some twigs began to droop, turn brown, form gum, and die. In this experiment, none of these symptoms occurred on any of the control Robinson tangerine trees. Thus, the twig dieback and gumming are associated with the application of cycloheximide, which also stimulated twigs to produce ethylene.

Table V. Effect of cycloheximide (CHI) on ethylene production of leaves and stems of Robinson tangerine after 4 days a/

	Plant part		
Treatment (ppm)	Leaf lamina (nl/g/hr)	Leaf petiole (nl/g/hr)	Intact stem (nl/g/hr)
20 CHI	0.11	2.86	0.85
Control	0.22	0.97	0.29

a/ Data represent means of three sets of 5 leaves and stems from each of 3 trees. The difference between treatments was statistically significant at the .05 level except for the leaf lamina.

Likewise, when Robinson tangerine trees were sprayed with 2-chloro-ethylphosphonic acid or with ascorbic acid (data not presented), the leaves and twigs were stimulated to produce ethylene, and the twigs died back as with the cycloheximide treatment. These chemicals, applied to Valencia orange and Redblush grapefruit trees, also stimulated the tissues of the twigs to produce ethylene, but none of these trees developed the twig dieback symptoms observed for Robinson tangerine trees. These results suggest that some factor in addition to ethylene is involved in dieback.

IV. GENERAL DISCUSSION

A wound ethylene-producing system occurs naturally in flowers, fruit, and stems of citrus trees. Fruit, when wounded either by natural, mechanical, or chemical means, produces ethylene in excess of 0.5 nl/g/hr. In fruit and flowers, the wound ethylene acts as a triggering agent of the biochemical processes involved in abscission.

Woody stems, when wounded mechanically, produce ethylene at rates of 4 nl/g/hr. Because ethylene production is so intimately associated with damaged cells, there is a tendency to think of ethylene as causing tissue damage, rather than as a product of damaged cells. For instance, twigs of Robinson tangerine trees with dieback symptoms from natural causes also produce copious quantities of ethylene, but there is no clear-cut evidence for the role of ethylene in the development of twig dieback.

The significance of wound ethylene as a healing substance should not be overlooked. Ethylene is known to produce intumescences in apple twigs (Wallace, 1926) and callusing and root formation in stem cuttings of tomato (Zimmerman and Wilcoxon, 1936). Thus, ethylene fulfills the requirements of Haberlandt's (1913) definition of a wound hormone: it is a product of damaged cells, diffuses to neighboring cells, and induces cell division leading to healing of the wound.

V. SUMMARY

A wound ethylene-producing system occurs naturally in tissues of the fruit, flowers, and stems of citrus. The rind of fruit and woody stem tissues produce large quantities of ethylene when injured. Cycloheximide-induced abscission is a wound-ethylene reaction. Cycloheximide wounds the cells of the rind, wound-ethylene is produced; and the ethylene, in turn, participates in the abscission process.

Robinson tangerine, Valencia orange, and Redblush grapefruit twigs, when wounded by application of cycloheximide, produce ethylene. Dieback occurs in Robinson tangerine, but not in the branches of Valencia orange and Redblush grapefruit trees. Twigs of Robinson tangerines with dieback symptoms from natural causes also produce copious amounts of ethylene, but there is no clear-cut evidence for the role of ethylene in the development of twig dieback.

VI. REFERENCES

BURG, S.P. and M.J. DIJKMAN (1967). Ethylene and auxin participation in pollen-induced fading of *Vanda* orchid blossoms. Pl. Physiol. 42, 1648-1650.

BURG, S.P. and E.A. BURG (1966). The interaction between auxin and ethylene and its role in plant growth. Proc. Natl. Acad. Sci. 55, 262-266.

COOPER, W.C. and W.H. HENRY (1971). Abscission chemicals in relation to citrus fruit harvest. Agr. Food Chem. 19, 559-563.

COOPER, W.C, G.K. RASMUSSEN, and E.S. WALDON (1969). Ethylene production in freeze-injured citrus trees. Jour. Rio Grande Valley Hort. Soc. 23, 29-37.

COOPER, W.C, G.K. RASMUSSEN, B.J. ROGERS, P.C. REECE and W.H. HENRY (1968). Control of abscission in agricultural crops and its physiological basis. Pl. Physiol. 43, 1560-1576.

HORTON, R.F. and O.J. OSBORNE (1967). Senescence, abscission and cellulase activity of *Coleus* leaves. Am. J. Bot. 45, 673-675.

WALLACE, R.H. (1926). The production of intumescences upon apple twigs by ethylene gas. Bull. Torrey Bot. Club 53, 385-401.

WILLIAMSON, C.E. (1950). Ethylene, a metabolic product of diseased or injured plants. Phytopath 40, 205-208.

ZIMMERMAN, P.W. and F. WILCOXON (1936). Several chemical growth substances which cause initiation of roots and other responses in plants. Contrib. Boyce Thomp. Inst. 7, 209-229.

PLANT GROWTH SUBSTANCES, 1970

THOUGHTS ON THE ROLE OF ETHYLENE IN PLANT GROWTH AND DEVELOPMENT

Morris Lieberman and A.T. Kunishi

Pioneering Research Laboratory for Post Harvest Physiology, Market Quality Research Division, ARS United States Department of Agriculture, Plant Industry Station, Beltsville, Maryland 20705 U.S.A.

I. INTRODUCTION

Of all the regulator substances which influence physiological development of plants ethylene stands apart in its wide-ranging effects on most aspects of growth and development. Table I is a partial list of some responses caused by ethylene. These responses vary from stimulation of seed germination (Vacha and Harvey, 1927), retardation of seedling growth (Knight and Crocker, 1913), induction of flower formation (Traub et al, 1940), to acceleration of senescence (Biale et al, 1954). The marked effect of ethylene is therefore not only observed in fruit ripening, its traditional role, but also in vegetative and reproductive metabolism. Ethylene appears to exert an influence on all aspects of metabolism, throughout the life-cycle of the plant.

Table 1. Some Effects of Ethylene on Plant Tissues.

Effect	Reference
Epinasty	Denny, (1935)
Abscission	Doubt, (1917)
Retardation of Growth	Knight and Crocker, (1913)
Induction of Flower Formation	Traub et al, (1940)
Induction and Acceleration of Ripening	Biale et al, (1954)
Inhibition or Stimulation of Bud Growth	Denny (1926), Burg, (1968a)
Destruction of Chlorophyll	Zimmerman, (1931)
Induction of Stem Swelling	Knight and Crocker, (1913)
Induction of Seed Germination	Vacha and Harvey, (1927)
Stimulation of Root Hair Development	Zimmerman and Hitchcock (1933)
Modification of Sex of Flowers	Robinson et al, (1968)
Removal of Apical Dominance	Elmer, (1932)

Ethylene has been categorized as a plant hormone (Burg, 1962) because it is a normal product of plant metabolism, apparently does not enter into metabolism as substrate or enzyme, and can produce its manifold effects at very low concentrations (fractions of a part per million). However, even after more than 50 years of recognition the role of ethylene in plant metabolism remains vague and confusing. Whereas the functions of auxins, cytokinins, and gibberellins have more or less been defined as growth-promoting agents, the function of ethylene remains an enigma. We wish in this report to suggest a function for ethylene as a growth hormone modulator and support this proposal with examples of interaction of growth hormones with ethylene.

We propose to show that ethylene is an antagonist or modulator of auxins, gibberellins, and cytokinins. Ethylene is presumed to act by modulating these growth-promoting hormones and thus preventing overgrowth or overdevelopment of tissues and organs of plants. Such a hormone would be expected to show a wide range of activities, especially in the very young active plant, the seedling, and also in aging or senescent tissue such as the mature fruit. In all these cases the actions of ethylene are the culminations of the interaction or interplay of auxins, gibberellins, and cytokinins with ethylene. Other plant hormones such as abscisic acid will be briefly mentioned but we shall not consider phytochrome, or other regulators which may also contribute in some way to the effects observed with ethylene.

II. ACTION OF ETHYLENE IN SEEDLINGS

Links between ethylene and growth hormones

The interplay between ethylene and plant hormones can be surmised by observing the startling effect of ethylene on growth of etiolated pea seedlings. Twenty-four hours after applying about 4ppm ethylene to a 4-day old seedling, growth in length of the epicotyl is severly inhibited, there is a marked increase in stem diameter, and a bulbous swelling appears in the subapical region of the epicotyl. A disorientation with respect to gravity also occurs in these seedlings. This disturbance of the growth pattern of the seedlings may well involve the growth hormones. A more direct link of growth hormones to ethylene, particularly IAA and cytokinins, is observed after treating seedlings with high levels of IAA or kinetin (Fig. 1).

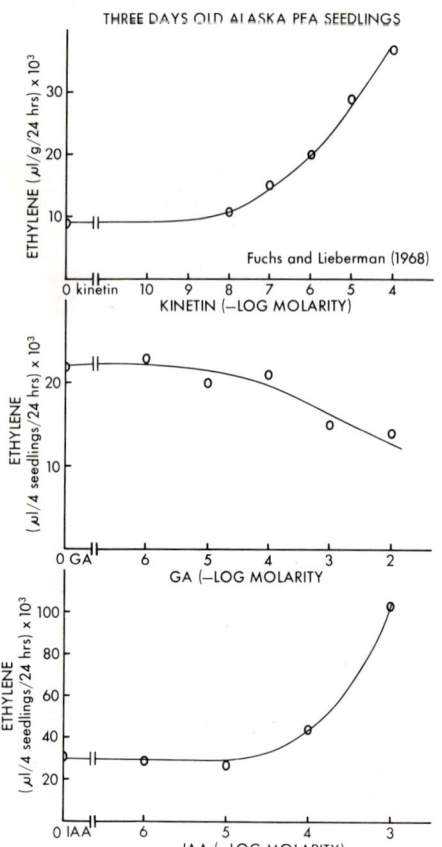

Fig. 1.

Ethylene production by etiolated pea seedlings germinated in various concentrations of IAA, Kinetin, and GA. Four pea seeds were germinated in a 50 ml sealed flask as described by Fuchs and Lieberman (1968).

Concentrations of IAA or kinetin above 10^{-6}M stimulate ethylene production in pea seedlings. At concentrations of 10^{-4}M IAA or kinetin, ethylene production may be doubled or tripled. In contrast GA in concentrations of 10^{-6} to 10^{-2}M does not stimulate ethylene production in pea seedlings. Combinations of IAA and kinetin caused even greater stimulation of ethylene production by pea seedlings (Fig. 2). This increased production of ethylene due to a combination of the 2 hormones was more than an additive effect. However, GA produced little additional stimulation of ethylene either in combination with IAA or kinetin alone, or as a third additive to IAA plus kinetin.

Stimulation of ethylene production by IAA or kinetin, or a combination of these, is related to the age of the tissue. Kinetin (10^{-4}M) stimulates ethylene production maximally in 3-day old seedlings but not at all in 5-day old seedlings (Fig. 2).

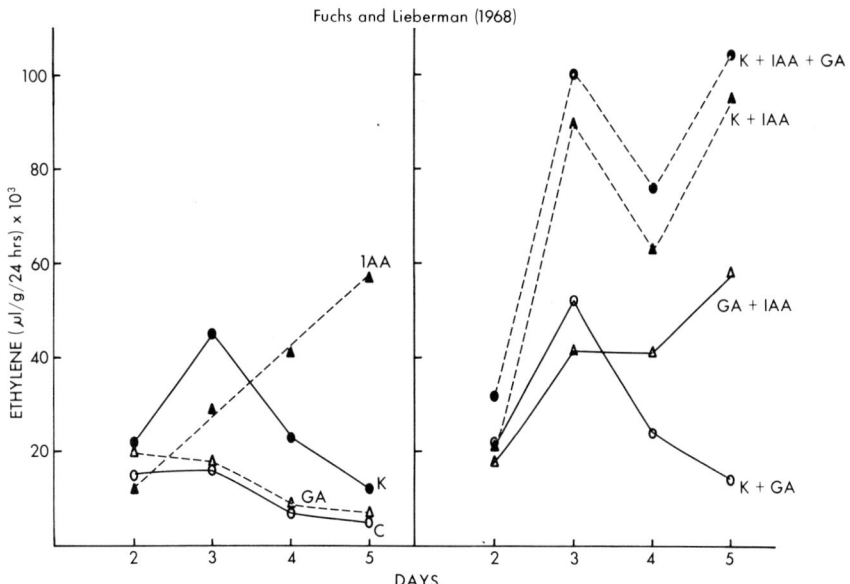

Fig. 2.

Effect of 10^{-4}M concentrations of IAA, Kinetin and GA, singly or in combinations, on ethylene production of etiolated pea seedlings during six days of germination in a sealed flask, as described in Fuchs and Lieberman (1968).

On the other hand IAA (10^{-4}M) continues increasingly to stimulate ethylene production linearly during the 5-day period of the experiment. Although ethylene production in 5-day old seedlings is not stimulated by kinetin alone, it is stimulated about 70 per cent above that obtained with IAA alone, by a combination of IAA with kinetin (Fig. 2). This is an example of hormonal interaction influencing ethylene production.

Rationale for stimulation of ethylene by IAA and kinetin in seedlings

Natural levels of growth hormones (IAA and kinetin) are in the order of 10^{-7} to 10^{-9}M. We believe that seedlings respond to high (10^{-4}M) doses of IAA or kinetin by markedly increased ethylene production. During growth and development, as the levels of IAA, kinetin, and GA rise or as they interact, growth may be accelerated beyond the normal rate and can result in distorted forms or unequal development of tissues and organs. During such conditions of excessive growth rates, ethylene synthesis is stimulated by high levels of IAA or cytokinins. The increased level of ethylene acts as an anti-growth substance or growth modulator, and restores growth rate to a normal balance. Thus there is a feedback relationship within the tissues between IAA and cytokinins on the one hand and ethylene on the other. As the growth hormones, singly or in combination, rise above the threshold point for balanced growth, ethylene is brought into play to restore the balance.

If this feedback relationship exists between IAA and kinetin on the one hand and ethylene on the other, then it should be possible to reduce levels of IAA or cytokinins by applying high levels of ethylene to seedlings. As far as we know there are no data concerning the effect of ethylene on levels of cytokinins in seedlings or other tissue. However, a number of experiments report a considerable decrease in diffusible auxin in ethylene-treated seedlings (Laan, 1934; Michener, 1938; Valdovinos, 1967). Valdovinos (1967), also reported that ethylene-treated plants showed reduced auxin synthesis. These data are summarized in Table II.

This evidence points to a reduction in diffusible auxin levels and also to a reduction in IAA synthesis, in ethylene-treated plants. Additionally ethylene-treated plants are known to show inhibited lateral auxin movement (Burg and Burg 1966), an inhibited capacity for polar auxin transport (Burg and Burg, 1967; Morgan et al 1967), and an increased IAA oxidase activity (Hall and Morgan 1964). Any of these effects of ethylene would be sufficient to reduce the growth-promoting activity of auxin, if it is in excess, and thus restore normal growth in the tissue. At the same time the lowered levels of auxin could also tend to cause a drop of ethylene synthesis, which would in turn tend to raise the auxin to normal levels. Of course the levels of ethylene and IAA, operating in such a hypothetical feedback system, would be much lower (in the order of $10^{-7}M$) than the levels used to demonstrate stimulation of ethylene production by IAA ($10^{-4}M$).

Table II. Effect of Ethylene on Auxin Content of Plant Tissues

Tissue	Treatment	Diffusible Auxin (degree curvature)	Auxin Formation IAA Equiv. x 10^2 n moles/g/FW	Reference
Pea Seedling Stems				
Tip	Control	16.5	---	Michener (1938)
	Ethylene	2.0	---	
Middle	Control	7.5	---	
	Ethylene	0.6	---	
Base	Control	1.4	---	
	Ethylene	0.5	---	
Coleus Stem Apical Bud				
	Control	14.8	858	Valdovinos et al (1967)
	Ethylene	11.4	594	
Pea Seedlings				
	Control	22.6	---	
	Ethylene	8.7	---	

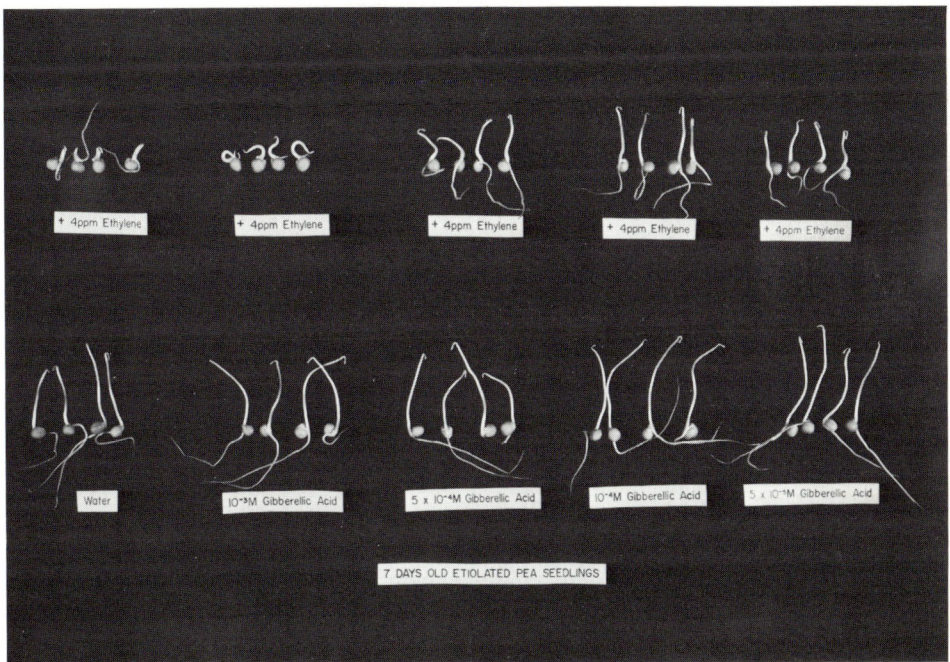

Fig. 3.

Effect of GA in ethylene-treated etiolated pea seedlings. Top row: seedlings treated with 4 ppm ethylene with no GA, and with concentrations of 10^{-3}M, 5×10^{-4}M, 10^{-4}M, and 10^{-5}M GA, from left to right. Bottom row: control and seedlings treated with GA in concentrations of 10^{-3}M, 5×10^{-4}M, 10^{-4}M, and 5×10^{-5}M, from left to right. Seedlings were treated in closed systems on the 4th day of germination and examined on the 7th day.

Ethylene and gibberellins

If ethylene is a universal modulator of growth hormone activity it should also influence activity of gibberellins. As we have already seen, GA does not stimulate the production of ethylene (Figs. 1 and 2). However, GA and ethylene are mutually antagonistic to the growth of etiolated pea seedlings. This is shown in Fig. 3, wherein we can see that GA partially overcomes the inhibition of growth caused by ethylene in etiolated pea seedlings. Some antagonism between the action of ethylene and GA is also observed in Fig. 4. Four day old etiolated seedlings treated with GA followed by ethylene treatment on the 5th day, show some growth, in contrast to ethylene-treated etiolated seedlings with no prior GA treatment, which show no growth at all. Thus GA does not stimulate ethylene production, but it does tend to antagonize the action of ethylene.

Reversal of ethylene-induced growth inhibition by growth hormones

Although there appears to be mutual antagonism between the hormones of growth and development and ethylene, it has not been possible to reverse the growth-inhibiting effect of ethylene on etiolated peas by addition of IAA, kinetin, or GA, either alone or in various combinations in concentrations shown in Fig. 5. Addition of 4 ppm ethylene to 4-day old etiolated pea seedlings completely stops elongation of the epicotyl. Addition of 4 ppm ethylene to seedlings germinating in 10^{-3}M GA decreases epicotyl growth to about 55 per cent of the control. Further additions of IAA or IAA and kinetin to the ethylene-inhibited system restores less growth than addition of GA alone. This probably results from additional ethylene induced by the IAA and kinetin additions. It also explains the lesser growth obtained in the ethylene-GA system to which IAA and kinetin have been added, as compared to adding only IAA (Fig. 5). In the absence of ethylene, a combination of GA, IAA, and kinetin have virtually no effect on the epicotyl growth of etiolated pea seedlings. This suggests that the induced ethylene due to IAA and kinetin, which should cause growth inhibition, is counterbalanced by GA. Conversely, the extension (usually about 25 per cent) of epicotyl growth observed after GA addition, is antagonized by the ethylene increase induced by IAA and kinetin.

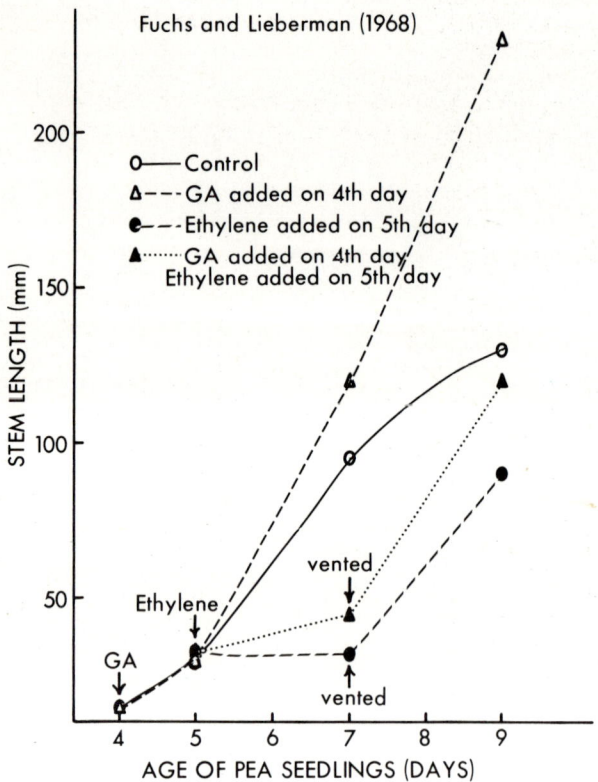

Fig. 4.

Antagonism between GA and ethylene on growth of etiolated pea seedlings.

Taken together these experiments reveal that reversal of ethylene-inhibited growth of the epicotyl, in etiolated pea seedlings, can only be partially achieved by growth hormones. Aside from the possibility that proper ratios of hormones must be restored and are difficult to obtain by addition of large doses of growth hormones, there may be other factors involved in restoring growth in the presence of ethylene. Perhaps levels of IAA-destroying peroxidases are high in its presence. There may also be a continual destruction or obstruction of growth hormones by ethylene at the site of their action. It is also possible that light (through phytochrome) may be important in overcoming growth inhibition by ethylene. Burg (1968b) reports that applied auxin prevents ethylene from inhibiting growth in light-grown pea seedlings. These experiments suggest the presence of a rather complex system in ethylene-mediated growth inhibition which cannot be overcome or restored by simple additions of GA, IAA, or kinetin.

Abscisic acid and growth hormones

ABA is now also recognized as a growth inhibitor. At a concentration of 5×10^{-4}M, ABA was as effective as 4 ppm ethylene in inhibiting epicotyl growth of etiolated pea seedlings. A combination of ethylene (4 ppm) and ABA caused a further 50 per cent decrease in growth of the epicotyl (Fig. 5). This observation suggests that ethylene and ABA may inhibit growth by different mechanisms. Epicotyls inhibited by ABA do not show the bulbous sub-apical growth typically associated with ethylene inhibition, and in general, except for dwarfing, do not exhibit characteristics of ethylene-inhibited epicotyls.

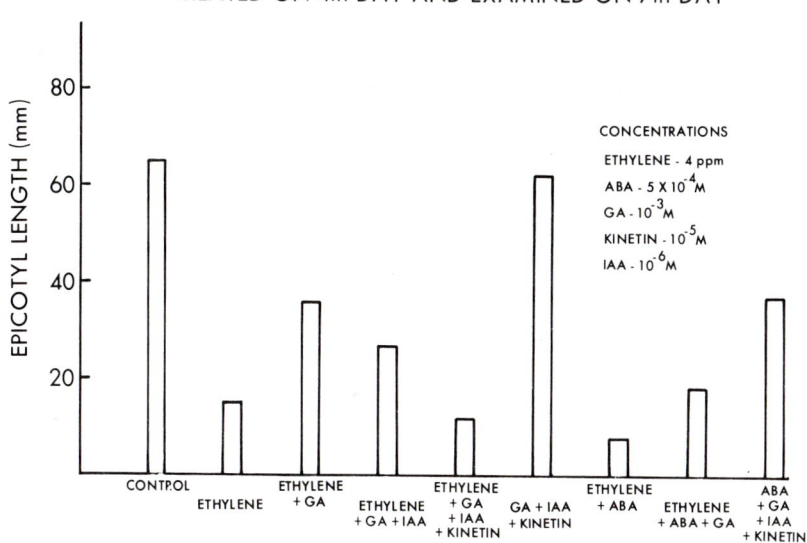

Fig. 5.

Interaction of ethylene, ABA, and growth hormones on growth of etiolated pea seedlings. Seedlings were treated on 4th day of germination and examined on 7th day.

Application of GA (10^{-3}M), ethylene, and ABA together to seedlings, resulted in approximately a doubling in length of the epicotyl, relative to that of seedlings treated with both ethylene and ABA (Fig. 5). Additions of GA, IAA, kinetin and ABA, to etiolated pea seedlings, in the absence of ethylene, produced epicotyls whose length was about 50 per cent of the control. Comparison of the effect of combinations of GA, kinetin, and IAA, in ABA and in ethylene-treated seedlings (Fig. 5), suggests that the additional ethylene induced by IAA plus kinetin further suppresses epicotyl growth in the ethylene-treated seedlings, even in the presence of GA. However, in the ABA-treated seedlings IAA plus kinetin stimulates growth of the epicotyl in the presence of GA. This suggests again that ABA and ethylene act in different ways to cause growth inhibition, since they respond differently to IAA and kinetin added to growth-inhibited seedlings.

ABA in concentrations varying from 10^{-6} to 5×10^{-4}M does not induce ethylene production in etiolated pea seedlings. Actually the stimulation of ethylene induced in seedlings by IAA plus kinetin is suppressed by ABA (Fig. 6). These results indicate that the mechanism of growth inhibition by ABA does not operate through the ethylene inhibition system. The ABA system may be an alternative growth suppression system, with its own interactions with the growth hormones, and independent of the ethylene growth-inhibiting system.

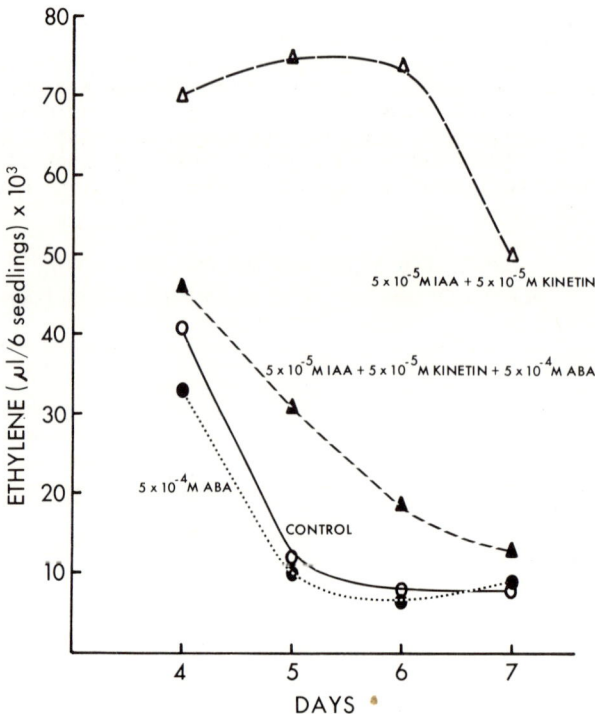

Fig. 6.

Effect of ABA on ethylene production in etiolated pea seedlings induced by IAA plus Kinetin.

III. ACTION OF ETHYLENE IN MATURE TISSUES

Ethylene and Abscission

Abscission is a phenomenon which occurs during senescence and therefore is preceded by a dramatic change in patterns of growth hormone levels. At this stage of the life cycle, the type of growth and development of earlier periods ceases, and hormone levels may drop. We believe that ethylene accelerates abscission due to its anti-growth and anti-growth hormone effects which bring into play the metabolic pathways associated with senescence. An apparent paradox in abscission concerns the action of IAA and kinetin, which tend to retard it even though they stimulate ethylene production, which in turn should accelerate abscission (Rubinstein and Leopold 1963). On the other hand GA promotes abscission (Chatterjee and Leopold 1964).

How can these actions of growth hormones on abscission be explained? The abscission-retarding effects of IAA and kinetin may readily be understood as a growth-promoting action or an anti-ethylene effect (antagonism to senescence), which in the presence of ABA overrides the ethylene-stimulating effect of these growth hormones, at least in the early stages of senescence. Acceleration of abscission by GA is more difficult to explain, since it is known not to stimulate ethylene production in etiolated pea seedlings. However, GA does stimulate ethylene production in citrus explants (R.H. Biggs, 1970) and may do so in other tissues where abscission is accelerated. It is also possible that the growth hormones may antagonize ethylene at its sites of action and thereby retard senescence, in addition to stimulating ethylene production. Both of these possibilities are in accord with the concept of antagonistic effects, between ethylene as an antigrowth hormone and growth hormones, whether they be IAA, kinetin, or GA.

Ethylene and fruit development and maturation

Ethylene is classically considered a ripening hormone because it accelerates ripening in mature fruit and is formed in fruit just prior to ripening (Burg, 1962). We consider ethylene a modulator or antagonist of growth hormones, for reasons already outlined in an analysis of its action on pea seedlings. We will now attempt to further elaborate the role of ethylene as a modulator or anti-growth hormone in fruit ripening.

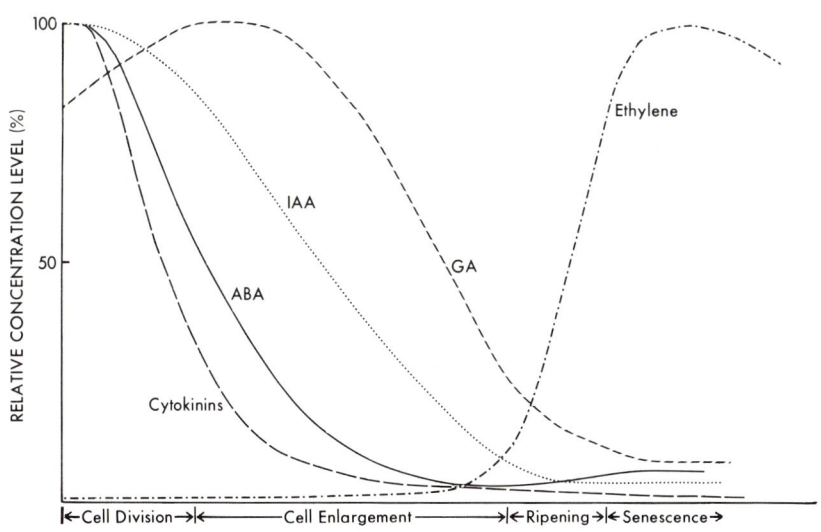

Fig. 7.

Hypothetical scheme of time course of hormonal changes during various phases of life cycle of a fruit.

A theoretical time-course for hormone levels during growth, development, and ripening of fruit is shown in Fig. 7. In the cell division period of growth, in the very early development of the fruit, cytokinins, IAA, and perhaps GA are considered to be relatively high, stimulating the reactions leading to cell division, cell expansion and cell elongation, as well as differentiation. In this time-period ABA was also found to be quite high in the fruit (Faust, 1970; Dorffling, 1970), but ethylene is known to be very low or perhaps absent (Lyons et al, 1962). During the period of cell enlargement, levels of cytokinins, IAA, and ABA are believed to decline much faster than GA which remains fairly high in concentration and regulates enlargement of the developing fruit. During this phase in the life of the fruit IAA probably declines less rapidly than cytokinins or ABA, and ethylene is present in only trace amounts. Towards the end of the phase of cell enlargement the growth hormones and ABA are assumed to be very low, and ethylene starts to rise very sharply and continues to rise to a peak, during the final phases of ripening.

Our interpretation of these hypothetical curves, which are based on the functional attributes of growth hormones and data on ABA and ethylene levels during the life of the fruit, considers both ABA and ethylene as growth modulators or antagonists to growth hormones. The high levels of ABA during the cell division phase may serve to modulate or control cell division and enlargement of the tissue in order to shape the tiny fruit as it grows in size. After the cell enlargement phase ethylene assumes the role of growth antagonist or growth modulator. The cells no longer can enlarge and no further growth ensues. A new course of metabolism is initiated by the growth-arresting action of ethylene. This new course of development brings on the ripening process with its multitude of new enzymes which carry on the metabolism of senescence.

We can account for the sudden rise in ethylene production, at the end of the period of cell enlargement, by assuming that the extremely low levels of growth hormones somehow trigger a rise in ethylene production. This is in contrast to the observations on seedlings, wherein relatively high levels of IAA and kinetin stimulate ethylene production.

Table III. Effect of Ethrel, Alone and in Combinations with Alar and Auxins, on Abscission, firmness and colour of "McIntosh" Apples. (From Edgerton and Blanpied, (1970).)

Treatment[1]	Harvest Date	Drop %	Firm (lb)	Red Color %
Alar	Sept. 24, 1968	2	16.3	56
Alar + Ethrel	"	77	15.5	67
Alar + Ethrel + TP	"	3	14.7	91
Control	"	29	14.5	51

[1] Alar applied at 2000 ppm on Aug. 9, 1968.
Ethrel applied at 250 ppm on Sept. 15, 1968.
TP applied at 20 ppm on Sept. 15, 1968.

Apparently either very high or very low levels of IAA or cytokinins are correlated with a stimulated production of the anti-growth hormone, ethylene. The evidence for assuming that low levels of growth hormones stimulate ethylene production, comes from the observation that ethylene production is always associated with cessation of growth or onset of senescence, which is related to very low levels of IAA, cytokinin and GA. Conversely, the addition of growth hormones to mature tissues, such as leaves or fruit, delays senescence or ripening and usually suppresses ethylene production (Richmond and Lang, 1957; Dostal and Leopold, 1967; Vendrell 1969, 1970). However, IAA can accelerate (Palmer and McGlasson, 1969) or delay ripening in bananas (Vendrell, 1969). This is in keeping with the dual role of auxins as stimulators of ethylene at high concentration levels on the one hand, which would tend to accelerate ripening, and antagonists of ethylene on the other (in abscission) which tends to delay ripening.

An example of the interplay of reactions between growth hormones, ethylene, and a quasi anti-ripening agent, Alar (succinic acid 2,2,dimethyl hydrazide), is shown in Table III. Alar which inhibits ripening and retards ethylene formation in apples (Looney (1968)), also prevents abscission and softening of the fruit. These are anti-ethylene effects which may be related to the anti-senescent effects of GA, auxins and kinetin in retarding abscission, aging of leaves, or fruit ripening. However, Alar also causes an increase in anthocyanin synthesis in apple fruit, a characteristic associated with ripening and senescence (Table III) (Edgerton and Blanpied (1970)). Alar, therefore, appears to act as both a ripening and an anti-ripening agent in development of pome fruit. Addition of Ethrel (which forms ethylene) to Alar-treated fruit accentuates ripening and senescence, causing increased abscission, softening and anthocyanin synthesis. This effect can be modified by also adding TP (2,4,5-

trichlorophenoxy propionic acid), a synthetic auxin, to the fruit treated with Alar and Ethrel. Addition of TP restores ability to prevent abscission but enhances anthocyanin synthesis and softening of the fruit. The additional ripening effect caused by TP may be related to an additonal production of ethylene. The anti-ripening effects of TP, which cause reduction of abscission, must be related to the anti-abscission action of auxins. The interplay of these synthetic regulators representing auxins (TP), ethylene (Ethrel), and Alar, a substance that acts as an auxin in its anti-ripening effects, and also as an ethylene substitute, in its stimulation of red color in apples, exemplifies the antagonistic relationships and interactions which may exist between these hormones.

IV. SUMMARY

An attempt is made to reconcile the conflicting data on the action of ethylene in seedlings, vegetative tissues in general, and in fruit ripening. A common link between differences in the action of ethylene in these different tissues is believed to be associated with its antagonism to the growth hormones IAA, cytokinins, and GA. The function of ethylene may be to modulate the action of growth hormones, keep it within normal bounds and thus prevent excessive growth which might otherwise occur. Ethylene may act by suppressing the synthesis of IAA or cytokinins, by influencing the destruction of IAA or cytokinins, or by suppressing the effect of IAA, GA or cytokinins at their sites of action. Additions of IAA, kinetin, or GA tend to override some effects of ethylene especially those on abscission. ABA may similarly be considered to be an antagonist of the growth hormones, serving to modulate the effects of auxins, cytokinins and GA on growth and development. Although the actions of ethylene and ABA are essentially similar, in that they antagonize or modulate growth, they appear to act independently in different pathways and may act at different stages in the life cycle of the plant.

V. LIST OF REFERENCES

BIALE, J.B., R.E. YOUNG and A.J. OLMSTEAD. (1954). Fruit respiration and ethylene production. Pl. Physiol. 29, 168-174.
BIGGS, R.H. (1970). Personal communication.
BURG, S.P. and E.A. BURG. (1962). The role of ethylene in fruit ripening. Pl. Physiol. 37, 179-189.
BURG, S.P. and E.A. BURG (1966). The interaction between auxin and ethylene and its role in plant growth. Proc. Natl. Acad. Sci. (USA) 55, 262-266.
BURG, S.P. and E.A. BURG (1967). Inhibition of polar auxin transport by ethylene. Pl. Physiol. 42, 1224-1248.
BURG, S.P. (1968a). Ethylene formation in pea seedlings; its relation to inhibition of bud growth caused by indole-3-acetic acid. Pl. Physiol. 43, 1069-1074.
BURG, S.P. (1968b). Ethylene, plant senescence and abscission. Pl. Physiol. 43, 1503-1511.
CHATTERJEE, S.K. and A.C. LEOPOLD. (1964). Kinetin and gibberellin actions on abscission processes. Pl. Physiol. 39, 334-337.
DENNY, F.E. (1926). Hastening sprouting of dormant potato tubers. Am. J. Bot. 13, 118-125.
DENNY, F.E. (1935). Testing plant tissue for emanations using leaf epinasty. Contrib. Boyce Thompson Instit. 7, 341-347.
DORFFLING, K. (1970). Changes in abscisic acid content during fruit development in *Solanum lycopersicum*. Planta 93, 233-242.
DOSTAL, H.C. and A.C. LEOPOLD. (1967). Gibberellins delay ripening of tomatoes. Science 158, 1579-1580.
DOUBT, S.L. (1917). The response of plants to illuminating gas. Botan. Gaz. 63, 209-224.
EDGERTON, L.J. and G.D. BLANPIED. (1970). Interaction of succinic acid 2,2-dimethyl hydrazide, 2-chloroethylphosphonic acid, and auxins on maturity, quality, and abscission of apples Jour. Amer. Soc. Hort. Sci. 95, 664-666.

ELMER, O.H. (1932). Growth inhibition of potato sprouts by volatile products of apples. Science 75, 193.
FAUST, M. (1970). Personal communication.
FUCHS, Y and M. LIEBERMAN. (1968). Effects of kinetin, IAA, and gibberellin on ethylene production and their interaction in growth of seedlings. Pl. Physiol. 43, 2029-2036.
HALL, W.C. and P.W. MORGAN. (1964). Auxin-ethylene interrelationships. Régulateurs naturels de la croissance végétale. Coll. Int. C.N.R.S. 123. 727-745.
KNIGHT, L.I. and W. CROCKER. (1913). Toxicity of smoke. Botan. Gaz. 55. 337-371.
LOAN, P.A. VAN DER. (1934). Der Einfluss von Aethylen and die Wuchstoffbildung bei *Avena* und *Vicia*. Rec. Trav. Botan. Néerl. 31, 691-742.
LOONEY, N.E. (1968). Inhibition of apple ripening by succinic acid 2,2-dimethyl hydrazide and its reversal by ethylene. Pl. Physiol. 43, 1133-1137.
LYONS, J.M., W.B. McGLASSON, and H.K. PRATT. (1962). Ethylene production, respiration, and internal gas concentrations in canteloupe fruits at various stages of maturity. Pl. Physiol. 37, 31-36.
MICHENER, H.D. (1938). The action of ethylene on plant growth. Amer. J. Botan. 25, 711-720.
MORGAN, P.W., E. BEYER, and H.W. GAUSMAN. (1968). Ethylene effects on auxin physiology, *in* "Biochem. and Physiol. of Plant Growth Substances" (Wightman, F., and G. Setterfield, Eds.) Runge Press, Ottawa, p. 1255-1273.
PALMER, J.K. and W.B. McGLASSON. (1969). Respiration and ripening of banana fruit slices. Aust. J. Biol. Sci. 22, 87-99.
RICHMOND, A.E. and A. LANG. (1957). Effect of kinetin on protein and survival of detached *Xanthium* leaves. Science 125, 650.
ROBINSON, R.W., S. SHANNON, and M.D. DE LA GUARDIA. (1968). Regulation of sex expression in cucumber. Bio Science 19, 141-142.
RUBINSTEIN, B. and A.C. LEOPOLD. (1963). Analysis of auxin control of bean leaf abscission. Pl. Physiol. 38, 262-267.
TRAUB, H.P., W.C. COOPER, and P.C. REECE. (1939). Inducing flowering in the pineapple, *Ananas sativus*. Proc. Amer. Soc. Hort. Sci. 37, 521-525.
VACHA, G.A. and R.B. HARVEY (1927). The use of ethylene, propylene and similar compounds in breaking the rest period of tubers, bulbs and seeds. Pl. Physiol. 2, 187-192.
VALDOVINOS, J.G., L.C. ERNEST, and E.W. HENRY. (1967). Effect of ethylene and gibberellic acid on auxin synthesis in plant tissues. Pl. Physiol. 42, 1803-1806.
VENDRELL, M. (1969). Reversion of senescence: effect of 2,4-D and IAA on respiration, ethylene production, and ripening of banana fruit. Aust. J. Biol. Sci. 22, 601-610.
VENDRELL, M. (1970). Acceleration and delay of ripening in banana fruit tissue by gibberellic acid. Aust. J. Biol. Sci. 23, 553-559.
ZIMMERMAN, P.W., A.E. HITCHCOCK, and W. CROCKER. (1931). The effect of ethylene and illuminating gas on roses. Contrib. Boyce Thompson Instit. 3, 459-481.
ZIMMERMAN, P.W. and A.E. HITCHCOCK. (1933). Initiation and stimulation of adventitious roots caused by unsaturated hydrocarbon gases. Contrib. Boyce Thompson Instit. 5, 351-359.

HORMONES AND SENESCENCE

ON THE NATURE OF SENESCENCE IN OAT LEAVES

Kenneth V. Thimann, Hiroh Shibaoka[1] and Colin Martin[2],
Crown College, University of California, Santa Cruz

[1]Present address: Department of Botany, Faculty of Science, University of Tokyo, Hongo, Tokyo

[2]Present address: Department of Botany, University of Witwatersrand, Johannesburg, S. Africa

I. INTRODUCTION

This research came about almost by accident. Three sets of observations of different types came to a focus;

(1) cytokinins are known to release the inhibition of lateral buds exerted by auxin or by their own apices, whether or not roots are present (Wickson and Thimann, 1958; Sorokin and Thimann, 1964);

(2) the presence of roots, however, helps the development and elongation of such lateral buds (Went, 1939) (although sometimes roots may contribute to the inhibition (Libbert, 1955)).

(3) cytokinin activity is present in bleeding sap and in root extracts (Kende, 1965; Weiss and Vaadia, 1965; Nitsch and Nitsch, 1965; Carr and Burrows, 1966; Skene and Kerridge, 1967).

Since we had used peas for the bud inhibition experiments, it was natural to look for the presence of cytokinins in pea roots. To this end, 4 mm root tips cut from 48-hour old pea seedlings were extracted with ice-cold methanol, chromatographed, and the fractions tested for cytokinin activity (Shibaoka and Thimann, 1970).

II. THE TEST METHOD

The test used was a minor modification of one used earlier (Thimann and Sachs, 1966), which in turn was derived from the work of Gunning and Barkley (1963). The first leaves of 7-day old oat seedlings (*Avena sativa* cv. Victory) grown in fluorescent light at 23^0 are cut off and the apical 5 cm placed in petri dishes over moist filter paper. A 10 µl droplet of the test solution is placed in the center of each leaf and the leaves are held for 3 (or occasionally 4) days in the dark. The amount of cytokinin in the test solution is then determined either by the amount of chlorophyll extracted into acetone or by the length of the green zone. Both give equally satisfactory calibration curves. The test clearly detects 1 nanogram of kinetin = 5×10^{-12} moles. If instead of kinetin the droplet contains ^{14}C-leucine the leaf can be extracted with alcohol or 5% TCA, and, after thorough washing, the soluble and insoluble counts determined. In the first 24 hours the entry of leucine proceeds with linear proportionality to its concentration in the droplet, and the fixation of radioactivity into the insoluble fractions shows the same linearity: from 50 to 60% of the ^{14}C is converted to so-called "protein"* (Fig. 1).

---

* For convenience, here and in all incorporation experiments, the word "protein" means "the washed and pigment-free material insoluble in 80% ice-cold ethanol." In chemical experiments the term is used without quotes to mean, as usual, the material containing 16.2% nitrogen which is insoluble in 80% alcohol after boiling 10 minutes.

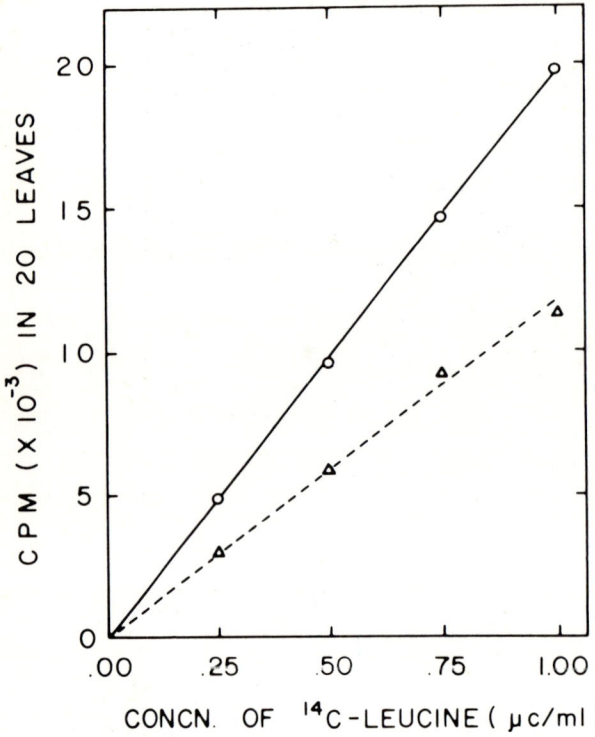

Fig. 1. Proportionality between the amount of ^{14}C-L-leucine in the droplet and its total uptake (solid line) and incorporation into the alcohol-insoluble fraction (dashed line). From Shibaoka and Thimann, 1970.

If kinetin is present in the droplet along with the leucine, the uptake is faster but the proportion fixed as protein is only slightly higher; however, the uptake and fixation continue for longer in presence of kinetin, since in the controls they come almost to a stop after 36 hours (Fig. 2). If kinetin is applied only after the leaf has been detached for 24 hours, its effects are observed in about 6 hours. Thus in the first 24 hours after detachment the leaf cells must undergo a change which makes them more sensitive to kinetin.

Instead of determining radioactivity the droplet can be washed off, the leaves extracted with boiling 80% alcohol, and the chlorophyll, free amino acids and protein determined by standard biochemical methods. Results obtained in this way will be discussed below. Thus the test system is simple, rapid, and well adapted to quantitative techniques of different sorts. Its only disadvantage is that because the droplet is small and much of its contents do not enter, the concentrations needed in it are rather high.

III. RESULTS

To our surprise none of the chromatographed fractions showed much cytokinin activity, although Carr & Burrows (1966) reported such activity in field pea sap and Short and Torrey (1970) have subsequently reported activity in three fractions from these roots. Apparently pea roots contain, at this age, much less activity than the *Helianthus* roots used by Weiss and Vaadia (1965). But instead, the extracts actually had the opposite effect to cytokinins, *i.e.*, they promoted yellowing. Carr & Burrows (1966) noticed a similar opposite effect in extracts of *Impatiens* roots.

The most active fraction in this promotion of senescence was soluble in butanol and was accompanied by a strong ninhydrin reaction; it was thus apparently an amino acid. All known amino acids were therefore tested, with the result that L-cysteine and L-alanine were found to show the yellowing effect weakly, while L-serine

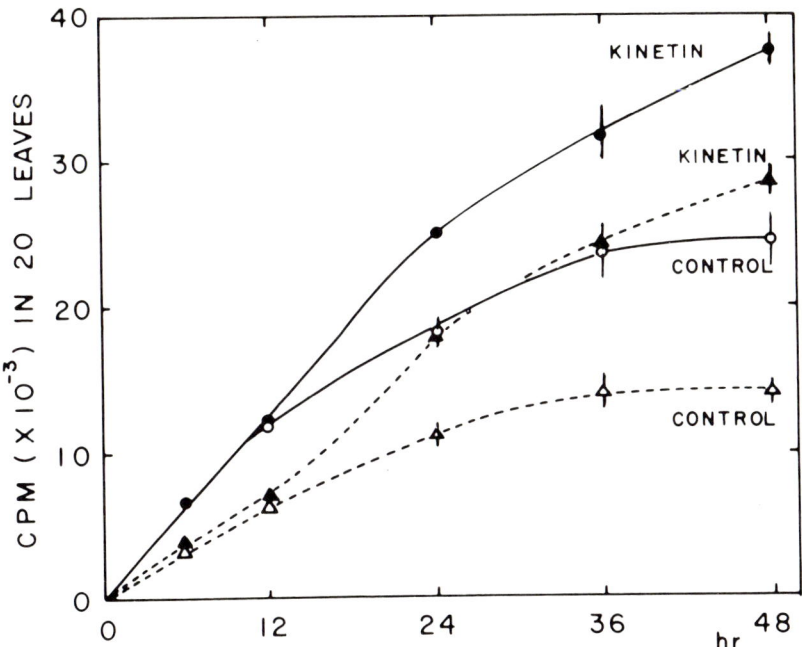

Fig. 2. Effect of kinetin (3 p.p.m.) in increasing and prolonging the uptake of ^{14}C-L-leucine and its incorporation into "protein", from 10 µl droplets applied to oat leaves. Solid lines: total uptake; dashed lines: incorporation. From Shibaoka and Thimann, 1970.

showed it strongly. The effect of L-serine was much more notable in presence of a little kinetin (see below). The related forms, D-alanine, β-alanine, D-serine and DL-homoserine were inactive, *i.e.*, only the amino acids of proteins were involved.

The effect of L-serine was not simply that of an antagonist to cytokinin. To show this, use was made of the known fact that both IAA and adenine at sufficiently high concentrations also cause the retention of chlorophyll. IAA was used at 300 times as high a concentration as kinetin, and adenine at 30,000 times the kinetin concentration. In both cases L-serine 0.03 M had the same antagonistic effect (Fig. 3). Serine therefore acts on the chlorophyll breakdown process itself and not on the hormonal control of it.

When L-serine is applied together with the ^{14}C-L-leucine the results are characteristic (Fig 4; Shibaoka and Thimann, 1970). In the controls, as noted above, incorporation into protein slows down after 24 hours, but if the leucine droplet is removed at 24 hours, the previously incorporated C^{14} is released again. Evidently rapid proteolysis is occurring. When serine is present the incorporation is somewhat reduced, and again on removal of the leucine there is rapid proteolysis. But the presence of serine makes an important difference in that even though the leucine is still present, net proteolysis now begins at 48 hours (Fig 4). Thus serine promotes the proteolysis. Nevertheless, when ^{14}C-serine is applied to the leaf it is readily taken up and, at 1 µc/ml, from 20-40% of the ^{14}C taken up is incorporated into protein after 24 hours. It is important that, although the incorporation is somewhat less than that with leucine, serine is indeed converted to protein.

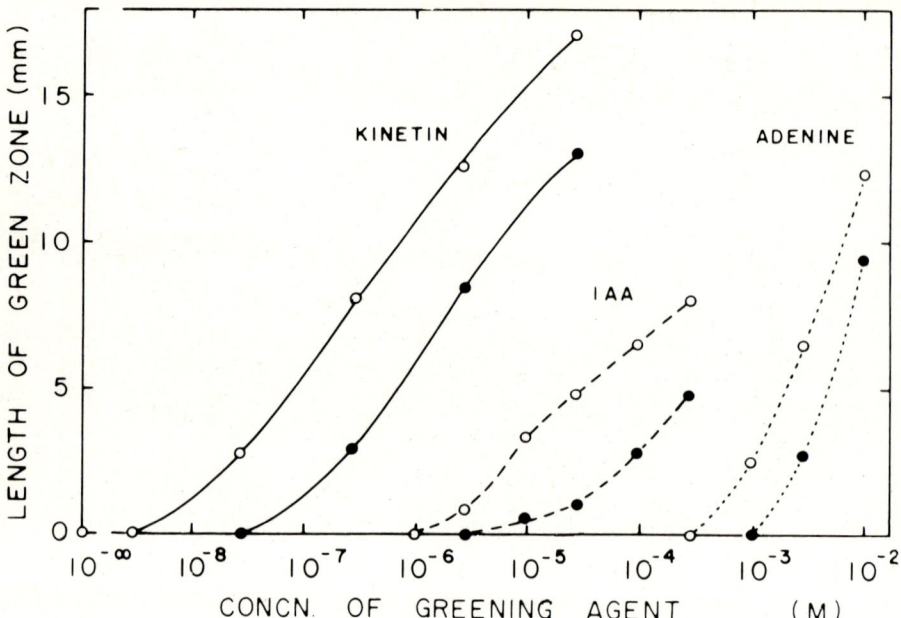

Fig. 3. Effect of L-serine on length of green zones caused by kinetin, IAA or adenine after 3 days in darkness. Open circles: without serine; closed circles: with serine (3×10^{-2}M). From Shibaoka and Thimann, 1970.

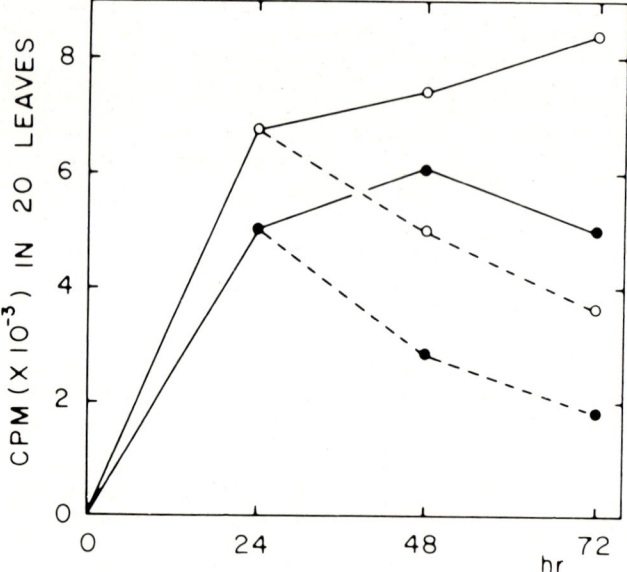

Fig. 4. The time course of incorporation of ^{14}C-L-leucine into "protein" from a 10 µl droplet left on (solid lines) or removed after 24 hours (dashed lines). Open circles: leucine alone; closed circles: plus ^{12}C-L-serine (1.5×10^{-2} M). From Shibaoka and Thimann, 1970.

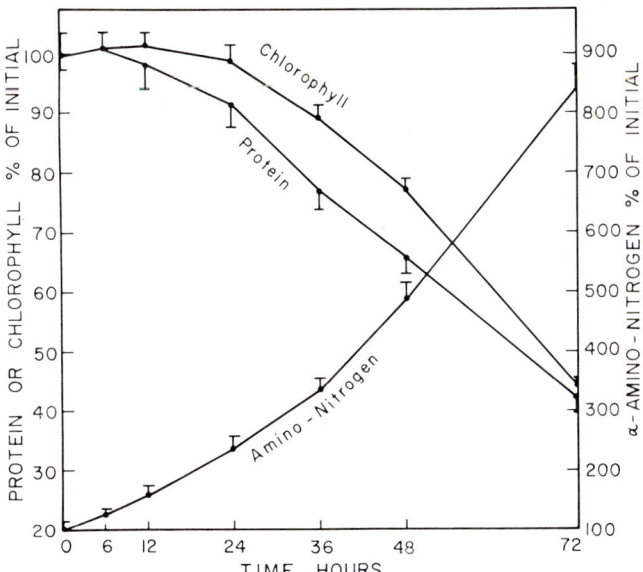

Fig. 5. The time course of chlorophyll and protein breakdown, and of amino acid accumulation, in oat leaves detached and kept 3 days in darkness. The vertical bars represent one half the 95% confidence limits.

In Fig. 5 the reactions in the detached leaf have been followed during the 3 days in the dark, and it is seen that protein breakdown (by the method of Lowry et al., 1961) and amino acid increase (by the ninhydrin color) have both begun by 6 hours, while chlorophyll breakdown is not detectable until about 24 hours; thereafter both proteolysis and yellowing continue in parallel. A tentative deduction from the initial discrepancy in timing is that proteolysis begins in the cytoplasm, and only after some hours are the plastids attacked. This time course is highly repeatable and the figure represents the mean of 3 complete runs. For simplicity as an assay, the yellowing and the proteolysis can be determined after a fixed time as a function of any added substance. In the latter way the effects of L-serine were studied in presence of different amounts of kinetin; Fig 6a shows that serine has only a modest effect on the chlorophyll in controls but acts drastically in presence of kinetin; at 0.1 M, serine overcomes the entire greening effect of even 10 ppm kinetin. The protein-preserving effect of kinetin is antagonized in exactly the same way (Fig. 6b); serine is evidently a powerful promoter of proteolysis and can eliminate the effect of kinetin.

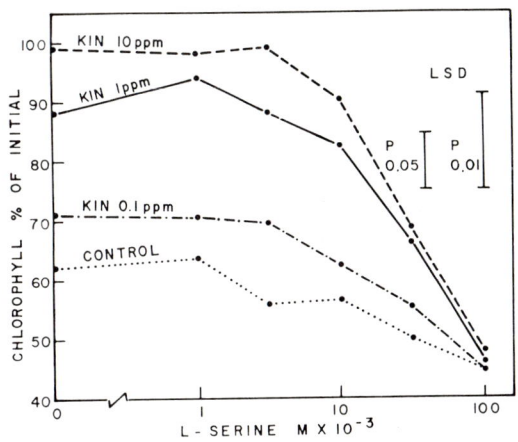

Fig. 6a. Interaction between L-serine and kinetin in the loss of chlorophyll.

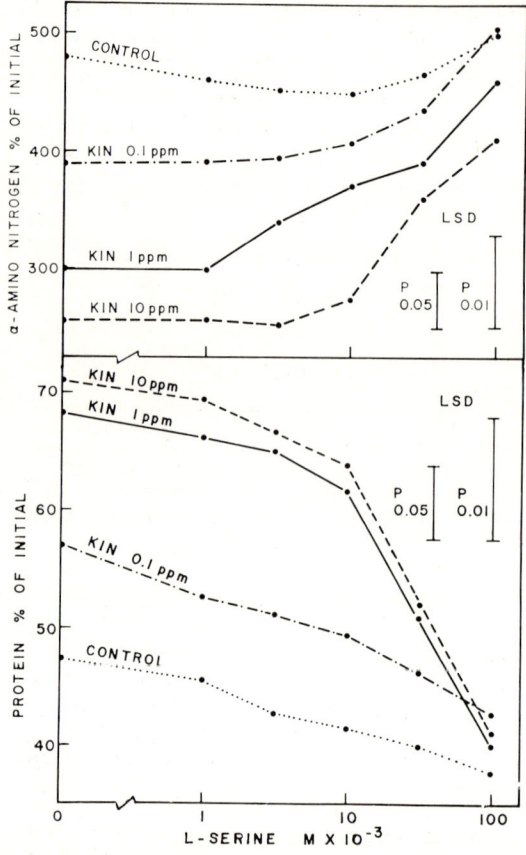

Fig. 6b. Same interaction in the hydrolysis of protein. Both after 3 days of darkness.

From the above groups of facts it was deduced: (1) that L-serine acts by being incorporated into a protein, (2) that this results in increased proteolysis, and hence in accelerated chlorophyll breakdown. Since the L-serine residue is known to provide the active center of many proteinases, it was a further logical deduction that serine forms the active center of a newly produced proteinase.

That proteolysis plays an active role in leaf senescence has been known since the classical work of Yemm (1937), Chibnall (1948) and others. We decided to look for evidence as to whether: (1) the proteinase of these leaves is of the serine type, and (2) if so, whether enzymes have to be newly formed to bring about senescence.

A characteristic of the serine-type proteinases is that they are inactivated by di-isopropyl-fluoro-phosphate (DFP) and phenylmethylsulfonyl fluoride (PMS). Since these compounds act on the hydroxyl of serine, they are subject to hydrolysis in water; they were, therefore, applied in 100% ethanol. Ethanol interferes somewhat with the test, but the results are nevertheless clear; both these compounds inhibit the breakdown of chlorophyll and of protein (Table 1).

Table 1. Effects of two inhibitors of serine-type proteinases on the breakdown of chlorophyll and protein after 3 days in darkness. DFP: di-isopropylfluorophosphate; PMS: phenylmethylsulfonyl fluoride.

Treatment (10 µl droplet)	Percent of initial: Chlorophyll	Protein
Control	61.1	40.5
DFP 10^{-2} M	82.9	44.9
DFP 10^{-1} M	92.5	65.0
PMS 10^{-1} M	85.0	65.0
DFP + PMS (both 10^{-1} M)	91.0	55.7

The effect on protein is less complete than on chlorophyll, suggesting that there is some additional source of proteinase, or that, as in the time course curve, some proteolysis can go on without affecting the chloroplasts. In general, PMS appears to act in essentially the same way as kinetin, but its action both on chlorophyll and on the liberation of amino acids can be more complete than 10 ppm of kinetin - a striking result.

The proteinase active in senescence evidently is of the serine type.

The next question is - what starts the proteolysis? In 1966 H.P. Balz suggested that proteinases can become active by being liberated from lysosomes. If so, then detergents which disrupt membranes should promote senescence. But Triton X-100 and Sodium lauryl sulfate, which are disruptive to membranes, as well as Pluronic F-68 which is not, were found to exert no effect on either the rates or the extent of senescence, even up to concentrations of 20%.

In the absence, then, of any evidence supporting its liberation from an organelle, we deduce that the proteinase must be newly formed. If so, inhibitors of protein

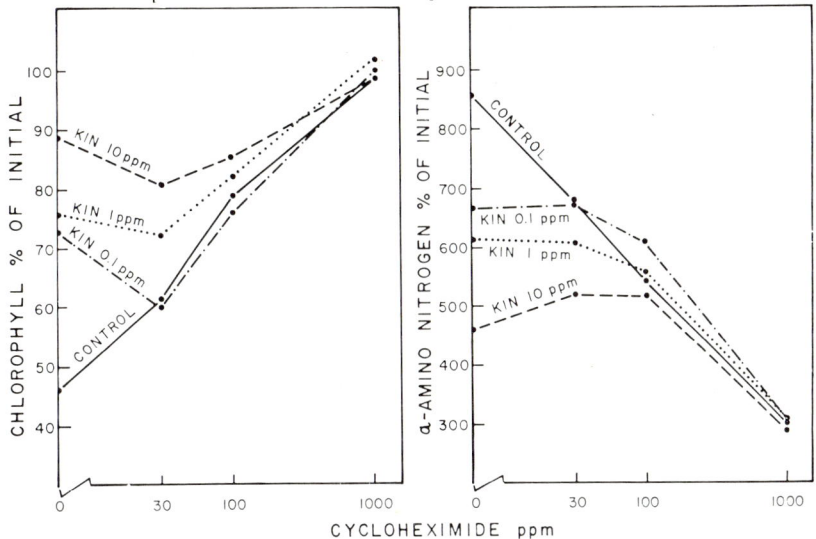

Fig. 7a, left: Effect of cycloheximide (CYC) in preventing loss of chlorophyll, and its interaction with kinetin (0.1, 1.0 and 10 p.p.m.), after both were present for 3 days in darkness.

7b, right: Protein hydrolysis as shown by amino acid accumulation as function of CYC and kinetin. Data corresponding exactly to those of Fig. 7a.

synthesis should, paradoxically, actually inhibit senescence and maintain the chlorophyll. They do. Fig. 7a shows that cycloheximide (CYC) at all concentrations prevents yellowing and at maximum concentration actually maintains the chlorophyll at 100% of its initial value. The interaction of cycloheximide (CYC) with kinetin is complex, and low concentrations have an antagonistic effect which will be discussed below. Just as with serine, the effects on greening are fully paralleled on protein breakdown as shown by amino acid liberation (Fig. 7b).

Table 2. Action of cycloheximide (CYC) on incorporation of ^{14}C-L-leucine into "protein" after 3 days in darkness (c.p.m. per 5 leaves).

Droplet	No addition	Plus kinetin 1 ppm
Control	4600	8400
CYC 45 ppm	1050	1550
CYC 150 ppm	1100	1050

In order to be sure that CYC was exerting its expected effect on protein synthesis in this system, 10 μl droplets containing ^{14}C-L-leucine, with and without kinetin, were applied to the leaves as done previously (Shibaoka and Thimann, 1970) and the incorporation of counts into the alcohol-insoluble fraction ("protein") determined. Table 2 shows that CYC is highly effective against protein synthesis, and indeed can completely wipe out the effect of kinetin. Other experiments have shown that CYC can also prevent the senescence-accelerating action of L-serine. Of course, since CYC prevents protein synthesis it would be expected to prevent L-serine from becoming built into a proteinase. Thus all these tests of the above theory are in agreement.

Neo-formation of protein should be preceded by formation of RNA. Inhibitors of RNA synthesis and action should, therefore, also prevent senescence. Of these, 6-methyl-purine is the most active we have found so far (Fig. 8). The concentrations needed are high, but the droplet volume is, of course, small and much of it does not enter the leaf.

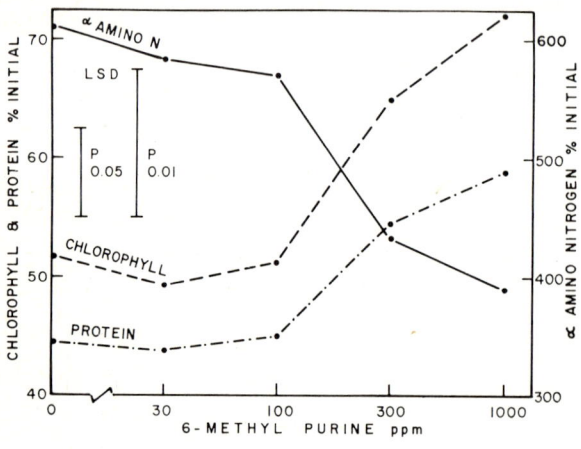

Fig. 8. Effect of 6-methyl-purine in preventing the loss of chlorophyll and protein, and the accumulation of amino acids.

Placing the leaves in 100% nitrogen prevents yellowing, which indicates that an oxidation is required, probably to provide the energy for protein synthesis.

Since proteolysis seems so closely bound to yellowing, a few experiments have been begun on the proteolysis of isolated chloroplasts. The results are tentative, but Table 3 shows that: (a) control (washed) plastids lose no color in 3 days in the dark, *i.e.*, a cytoplasmic constituent is needed to start yellowing; (b) crystalline trypsin caused definite yellowing; (c) oxygen appears not to be needed for this trypsin effect; (d) kinetin does not act against trypsin in these isolated plastids, which is in line with Anderson and Rowan's (1966) finding that kinetin does not react with an isolated leaf endopeptidase.

Table 3. Action of crystalline trypsin (at pH 7.4) on the chlorophyll content of isolated washed oat leaf chloroplasts.

Chloroplasts from <u>Avena</u> washed and suspended in buffer pH 7.4. Treatment for 72 hours in darkness. Chlorophyll content as percent of control.

	Expt. 1	Expt. 2
Initial	100	100
Control	106	96
Trypsin in air	86	81
Trypsin in nitrogen		83
Trypsin in air plus kinetin	86	
Kinetin alone	103	

IV. THE ROLE OF KINETIN

The role of kinetin has not been clarified in the above. From the evidence thus far, its inhibition of senescence is apparently by preventing proteolysis, yet there is no evidence that it directly inhibits proteinases (cf. Anderson and Rowan 1966). Tentatively we suggest that its action may be to promote the synthesis, not of proteins in general, but of something which specifically inhibits proteinase formation. If this "something" were a protein, then the sequence in senescence would be:

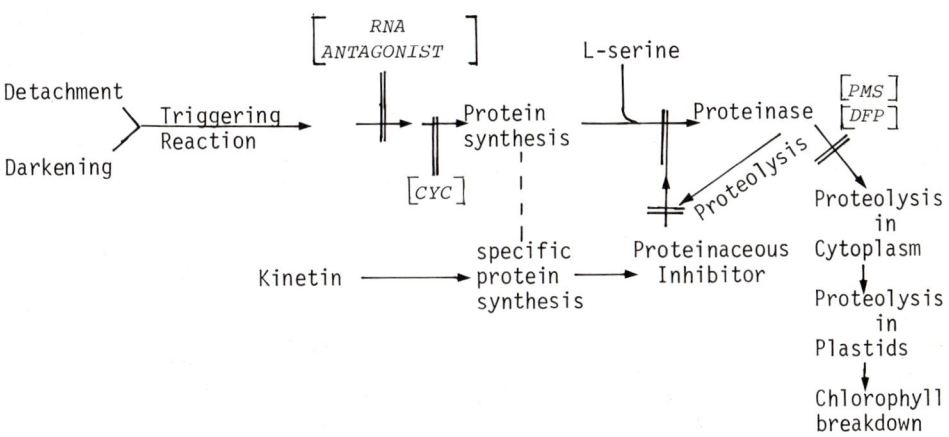

Serine, by increasing the formation of proteinase, would help to hydrolyze (*inter alia*) the proteinaceous inhibitor, and this would explain how its effect would be greater in the presence of kinetin than in its absence. Further, the curious 2-phase interaction of CYC with kinetin could find a ready explanation by the proposed proteinaceous nature of the inhibitor, which makes it sensitive to the proteinase. It is not excluded that a simpler scheme may explain the facts, and work is continuing.

V. REFERENCES

ANDERSON, J.W. and ROWAN, K.S. (1966). The effect of 6-furfuryl-aminopurine on senescence in tobacco leaf tissue after harvest. Biochem. J. 98, 401-404.
BALZ, H.P. (1966). Intrazellulare Lokalisation und Funktion von hydrolytischen Enzymen bei Tabak. Planta 70, 207-236.
CARR, D.J. and BURROWS, W.J. (1966). Evidence of the presence in xylem sap of substances with kinetin-like activity. Life Sci. 5, 2061-2077.
CHIBNALL, A.C. (1948). "Protein Metabolism in the Plant." New Haven, Yale University Press, and lit. there cited.
GUNNING, B.E.S. and BARKLEY, W.K. (1963). Some effects of kinetin on detached oat leaves. Nature (Lond.) 199, 262-265.
KENDE, H. (1965). Kinetin-like factors in the root exudate of sunflower. Proc. Nat. Acad. Sci. U.S.A. 53, 1302-1307.
LIBBERT, E. (1955). Der Einfluss von Blatt und Wurzel auf die Auxin-induzierte korrelative Knospenhemmung. Flora 142, 619-628.
LOWRY, O.H., ROSEBROUGH, N.H., FARR, A.L. and RANDALL, L.J. (1951). Protein measurement with the Folin phenol reagent. J. Biol. Chem. 193, 265-275.
NITSCH, J.P. and NITSCH, C. (1965). Présence de phytokinines et autres substances de croissance dans la sève d'*Acer saccharum* et de *Vitis vinifera*. Bull. soc. bot. fr. 112, 11-18.
SHIBAOKA, H. and THIMANN, K.V. (1970). Antagonisms between kinetin and amino acids: Experiments on the mode of action of Cytokinins. Plant Physiol. 46, 212-220.
SHORT, K.C. and TORREY, J.G. (1970). Natural occurrence of cytokinins in roots of pea seedling. Plant Physiol. 46, suppl., abstr. 47.
SKENE, K.G.M. and KERRIDGE, G.H. (1967). Effect of root temperature on cytokinin activity of root exudate of *Vitis vinifera* L. Plant Physiol., 42, 1131-1139.
SOROKIN, H. and THIMANN K.V. (1964). The histological basis for inhibition of axillary buds in *Pisum sativum* and the effects of Auxins and Kinetin on xylem development. Protoplasma 59, 326-350.
THIMANN, K.V. and SACHS, T. (1966). The role of cytokinins in the "Fasciation" disease caused by *Corynebacterium fascians*. Amer. J. Bot. 53, 731-739.
WEISS, C. and VAADIA, Y. (1965). Kinetin-like activity in root apices of sunflower plants. Life Sci. 4, 1323-1326.
WENT, F.W. (1939). Transplantation experiments with peas. Amer J. Bot. 25, 44-45 (esp. Table 3).
WICKSON, M. and THIMANN, K.V. (1958). The antagonism of auxin and kinetin in apical dominance. Physiol. Plantarum 11, 62-73.
YEMM, E.W. (1937). Respiration of barley plants. III. Protein catabolism in starving leaves. Proc. Roy. Soc. B. 123, 243-273.

HORMONAL REGULATION OF LEAF SENESCENCE
IN INTACT PLANTS

R.A. Fletcher and N.O. Adedipe[1]

Department of Botany, University of Guelph, Guelph, Canada

[1]Present address: Department of Horticultural Science, University of Guelph, Guelph, Canada

I. INTRODUCTION

Decreases in the levels of chlorophyll, protein and RNA are the prominent symptoms of leaf senescence. In detached leaves or leaf discs these symptoms can be delayed by the application of growth substances, including cytokinins (Richmond and Lang, 1957; Letham, 1967), auxins (Osborne and Hallaway, 1964), or gibberellins (Fletcher and Osborne, 1966; Beevers, 1966). The effectiveness of any one or combinations of these compounds is dependent on species and age of the leaf. Senescence of *Taraxacum* (Fletcher and Osborne, 1966) and nasturtium (Beevers, 1966) leaves can be delayed with application of GA or cytokinins and it has been shown that the senescence of these leaves is associated with an endogenous deficiency of these hormones (Fletcher et al., 1969; Chin and Beevers, 1970).

When the senescence of detached leaves is delayed by the application of hormones, the levels of chlorophyll, protein and RNA are maintained, and the incorporation of radioactive precursors into protein and RNA is enhanced (Osborne, 1962; Wollgiehn, 1967). It has been suggested by some (Fletcher and Osborne, 1966) that the hormonal action is mediated through a regulation of DNA-dependent RNA synthesis, while others have suggested that it is by prevention of protein degradation (Tavares and Kende, 1970). There are numerous other reports suggesting that localized retardation of senescence in detached leaves is associated with mobilization of metabolites from untreated zones (Mothes and Engelbrecht, 1961; Gunning and Barkley, 1963; Muller and Leopold, 1966).

Changes associated with senescence in intact plants have been described by Carr and Pate (1967) and Wareing and Seth (1967). In the present report, hormonal regulation of leaf senescence on the intact plant is emphasized.

Cytokinin-like substances produced by the root may regulate senescence of leaves on intact plants (Sitton et al., 1967; Wareing and Seth, 1967). Although the application of kinetin had little or no effect on leaf senescence on intact tobacco plants (Engelbrecht, 1964), Fletcher (1969) has demonstrated that N^6-benzyladenine (BA) could retard leaf senescence on intact bean plants under normal growth conditions. This has been confirmed by Jacoby and Dagan (1970). Retardation of senescence in these plants is associated with numerous metabolic changes. The present report discusses some of these effects and compares them to those reported for detached leaves.

II. RETARDATION OF LEAF SENESCENCE IN INTACT PLANTS

Application of BA (30 mg/l) to the primary leaves of 2-week old intact bean plants grown at 21-25° and a 16 hr photoperiod, retards senescence of the treated leaves (Fig. 1A), and to a lesser extent that of the whole plant (Fletcher, 1969).

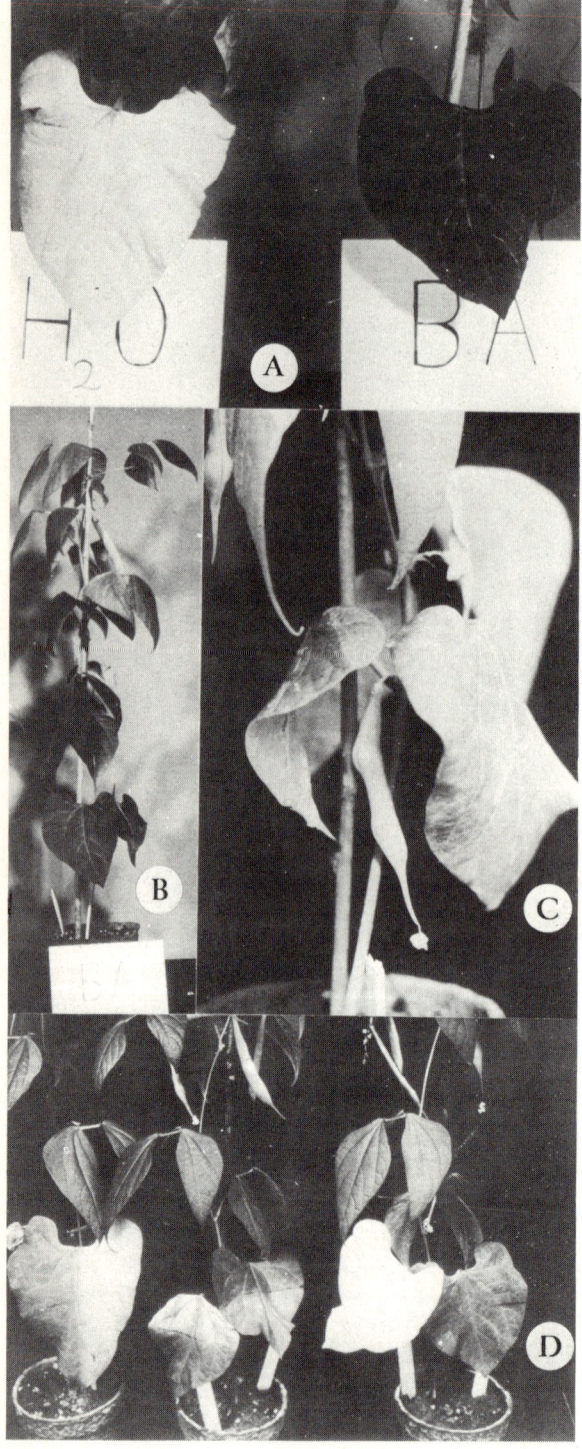

Fig. 1. A. Primary leaves of 7-week old bean plants treated at weekly intervals with water or BA.

B. Senescence of the apical trifoliate leaves of a 7-week old bean plant where the primary leaves were treated with BA at weekly intervals.

C. Pod formation in the axils of the BA-treated primary leaves of 16-week old plant. BA was applied at weekly intervals from the 2nd to the 7th week.

D. Five-week old bean plants in which portions of the primary leaves were treated with water or BA: (from left to right) the left half, the distal half, the proximal half, and the entire leaf on the right was treated with BA. Reproduced by permission of the National Research Council of Canada from the Canadian Journal of Botany 49 (1971).

With the onset of flowering and fruit set, the primary leaves of the control plants turn yellow and show a decline in the levels of chlorophyll, protein and RNA. In the BA-treated plants, although there is no effect on the time of flowering and pod set, the manifestations of senescence are prevented. A single application of the growth regulator in both 2- and 5-week old plants is sufficient to retard senescence. Repeated applications are however more effective.

The bean plant does not exhibit a strictly sequential pattern of senescence. The primary leaves senesce first, followed by the basal trifoliate leaves, and then the whole plant. In 16-week old BA-treated plants (Fig. 1B), the primary leaves remain green while the upper trifoliate leaves start to turn yellow. The senescence pattern is therefore apparently reversed, with the youngest trifoliate leaves senescing first. After pods have matured, shrivelled and dried a second set of flowers and pods emerges at the lower nodes, particularly in the axils of the primary leaves and cotyledons (Fig. 1C). This could occur as a result of a reversal of the gradient of growth substances allowing production of flowers by the basal tissue. An alternative explanation for this observation is that, following senescence of the apical parts, the lower buds are released from apical dominance. It should be pointed out that the BA-treated plants have lived for as long as 21 weeks.

III. LACK OF BENZYLADENINE-DIRECTED MOBILIZATION

In short-term experiments ^{14}C-sucrose was fed through the terminal leaflet of the second trifoliate leaf at weekly intervals, and the distribution of radioactivity determined after 24 hrs (Adedipe and Fletcher, 1970a). In long-term experiments 2-week old plants were fed for 24 hrs and the distribution pattern determined at weekly intervals. In both cases, less than 6% of the total radioactivity was in the BA-treated leaves, with no increase over the water-treated control leaves. Resnik and Montaldi (1968) made similar observations with ^{32}P in bean plants.

Application of BA to one half of a bean leaf, either along or across the midvein, or to one of the 2 primary leaves on the intact plant, retarded senescence of the treated tissue (Fig. 1D). When ^{14}Carbon-labelled CO_2, IAA, leucine or sucrose was fed to the untreated part of the same leaf, or to the untreated opposite leaf, there was less than 2% of the total radioactivity in the BA-treated part after 24 hrs (Adedipe and Fletcher, 1971). It was concluded that localized BA treatment of leaves on the intact plant does not result in mobilization of metabolites from the untreated portions of the leaf. This is contrary to earlier reports of cytokinin-directed transport in detached leaves. We suggest that these reports for detached leaves might be due to the absence of pods and roots which are the "potent sinks" of the plant (Adedipe and Fletcher, 1971). Osborne (1962) and McHale and Dove (1968) have indicated that even in detached leaves, cytokinins operate directly on the treated areas and do not depend on accumulation of metabolites from untreated tissue.

Mobilization of ^{14}C metabolites in detached leaves, as reported by some workers, could occur if their experiments were carried out in a closed system where the cytokinin-treated tissue would preferentially refix respired $^{14}CO_2$. This possibility was investigated by conducting one set of experiments in the open, and another set under a belljar. Attached and detached 5-week old bean leaves were fed $^{14}CO_2$ as indicated in Fig. 2 and the distribution of radioactivity determined after 24 hrs. The leaves were pretreated at weekly intervals with water or BA, starting from the 2nd week. In detached leaves there was no apparent difference in the mobilization pattern between the open and closed system (Fig. 2C,D,E). It therefore appears that refixation of respired $^{14}CO_2$ is not a satisfactory explanation to account for mobilization reported by others. However, in attached leaves (Fig. 2A,B) the BA-treated leaves accumulated more radioactivity in the closed than in the open system. This is understandable on the basis that attached leaves are more efficient in photosynthesis and respiration. In spite of this, the BA-treated attached leaves accumulated about 2 and 5% of the total radioactivity in the open and closed systems respectively. These values are about half of the radioactivity accumulated laterally (Fig. 2C) and basipetally

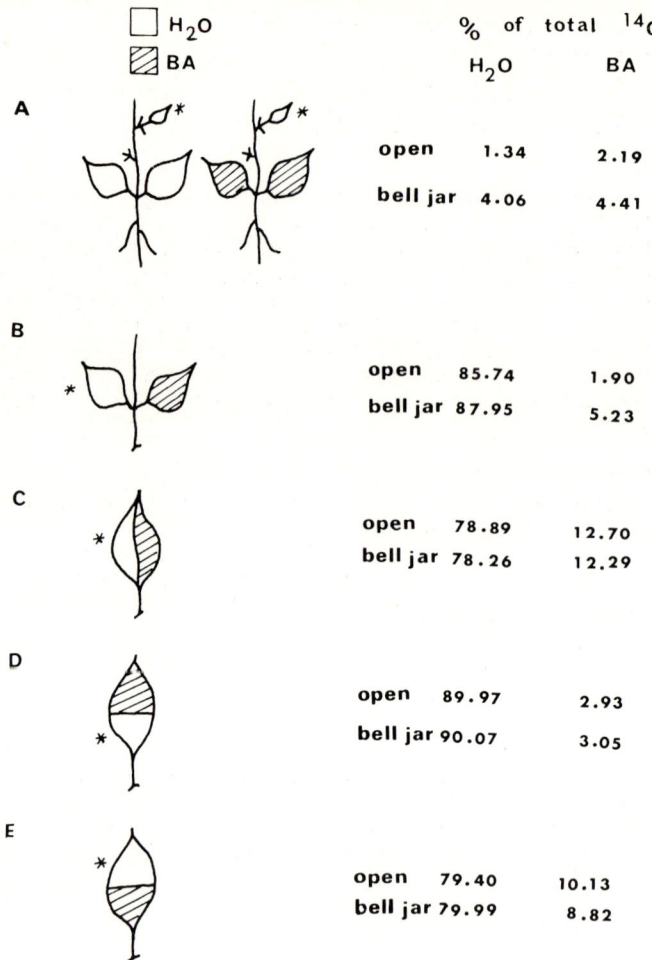

Fig. 2. Distribution of radioactivity in the water- and BA-treated primary leaves or leaf halves of 5-week old plants. $^{14}CO_2$ was fed to the portions marked with asterisks. Prior to feeding: A, the plants were decapitated and defoliated, leaving the primary leaves and the terminal leaflet of the 2nd trifoliate; B, the plants were decapitated, defoliated and derooted, leaving the primary leaves; C, D, E, the primary leaves were detached. The roots of A were intact in soil. The cut stem of B and the petioles of C, D and E were kept in tap water.

(Fig. 2E) by the BA-treated side of detached leaves. From attached leaves the stem accumulated 11% of total radioactivity (Fig. 2B). When the roots were present (Fig. 2A) there was 11% in the stem and an additional 11% in the roots. The greater accumulation of radioactivity in the stem and the roots compared with BA-treated leaves supports a previous report that in the intact plant the "potent sinks" are the roots and pods (Adedipe and Fletcher, 1971).

The lack of mobilization by the BA-treated leaf or parts thereof in these experiments is not due to leaf age since similar results were obtained with 3-week old leaves (Adedipe and Fletcher, 1970a). Neither is it due to lack of functional phloem in the senescing, water-treated leaves because such leaves are capable of exporting as much as 50% of their photosynthate (Fletcher et al., 1970). It is concluded, therefore, that retardation of bean leaf senescence by BA does not depend on mobilization of metabolites from untreated parts, but is accomplished by metabolic self-sustenance of the BA-treated part.

IV. ENHANCED PHOTOSYNTHESIS AND RETENTION OF ASSIMILATES

When the primary leaves were painted with water or BA at weekly intervals commencing from the 2nd week, photosynthetic rates declined in the water-treated leaves, while in the BA-treated leaves the photosynthetic rate was high (Adedipe et al., 1971). The percentage increases over water-treated leaves were 60, 500 and 190 at weeks 4, 5 and 6 respectively. These increased photosynthetic rates were accompanied by decreased mesophyll resistance, but no consistent changes in stomatal resistance. The percentage decreases in mesophyll resistance were 36, 93 and 82 at the 3 respective periods. Meidner (1967) also reported enhanced photosynthesis by kinetin application to detached primary leaves of barley, but such increases were accompanied by decreases in stomatal resistance.

When the leaves from 2-week old plants were fed $^{14}CO_2$, 24 hrs after the first treatment the BA-treated leaves retained about 70% of their assimilates and maintained this high level (70-85%) for the duration of the experiment (Fletcher et al., 1970). In contrast, the primary leaves of control plants retained about 30% at week 2, 80% between weeks 4 and 5, and with senescence at week 6, 50%. Retardation of senescence by BA is therefore not only accompanied by increased rates of photosynthesis but also by increased retention of photosynthetic assimilates.

V. PHOSPHATE AND CARBOHYDRATE METABOLISM

The levels of sugar phosphates and adenosine phosphates were determined in the water- and BA-treated leaves using ion exchange chromatography (Khym and Cohn, 1953). Although the levels of hexose- and adenosine-phosphates were lower in the BA-treated leaves, the incorporation of ^{32}P into these compounds by 3- and 6-week old plants was higher than in the controls (Adedipe and Fletcher, 1970b). This indicates that the retardation of leaf senescence by BA is associated with a rapid turnover and a greater utilization of high energy compounds, rather than a conservation of ATP as suggested by MacLean and Dedolph (1964).

Primary bean leaves were fed $^{14}CO_2$ for 5 min and the incorporation of radioactivity into carbohydrates was determined after 0.25, 2.50 and 25.0 hrs. The sugars were separated by ion exchange chromatography (Khym and Zill, 1952), starch isolated by the method of McCready et al. (1950), and the carbohydrates analyzed by the method of Scott et al. (1967). Incorporation of $^{14}CO_2$ into ribose, glucose and fructose increased with assimilation time, while that into sucrose decreased (Fig. 3). In the ethanol-soluble fraction of both the water- and BA-treated leaves, most of the radioactivity was found in sucrose, irrespective of assimilation time or leaf age. After 0.25 hrs of assimilation, the incorporation of $^{14}CO_2$ into sucrose in the water-treated leaves was 40, 50 and 20% of the total activity, at weeks 2, 4 and 6 respectively. There was a marked decrease in the amount of radioactivity in sucrose 25.0 hrs after feeding, indicating that most of the sucrose synthesized 0.25 hrs after feeding was exported out of the primary leaf. At week 4, just prior to senescence of the water-treated leaves, 50% of the total radioactivity was in sucrose. This value is similar to that reported for Cucurbita leaves (Webb, 1970), but it is recognized that species differ greatly as to the amount of carbon that is assimilated into sucrose (Hofstra and Nelson, 1969).

In the BA-treated leaves, incorporation of $^{14}CO_2$ into ribose and sucrose was lower than for water-treated leaves regardless of assimilation time or leaf age. Most of the carbohydrate exported from leaves is translocated as sucrose. The decreased incorporation of $^{14}CO_2$ into sucrose by BA-treated leaves supports our previous finding that only a small percentage of their photosynthate is exported (Fletcher et al., 1970). Retardation of senescence by cytokinins is associated with a maintenance of

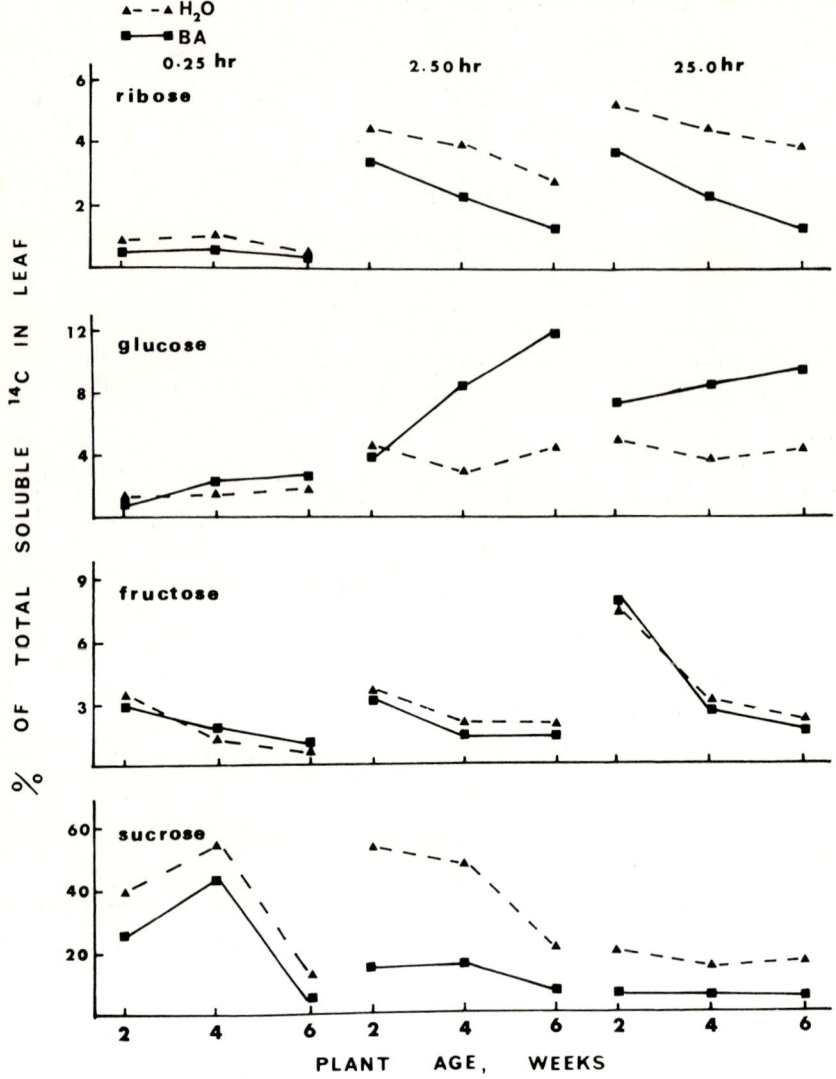

Fig. 3. Incorporation of $^{14}CO_2$ into sugars in H_2O- and BA-treated primary leaves of bean plants. The leaves were exposed to $^{14}CO_2$ for 5 min and the incorporation studied after 0.25, 2.5 and 25 hr.

RNA synthesis (Osborne, 1962; Fletcher, 1969; Skoog and Armstrong, 1970) which would require increased utilization of ribose. This would explain the decrease in radioactivity in ribose. BA treatment increased the incorporation of $^{14}CO_2$ into glucose while it had no effect on incorporation into fructose. Increased assimilation into glucose may be a reflection of the central role of glucose as a primary product of photosynthesis.

The amount of starch in BA-treated leaves was 60 and 100% higher than in the water control leaves at weeks 4 and 6 respectively (Table 1). BA treatment also caused enhanced assimilation of $^{14}CO_2$ into starch. This effect was apparent when the plants were 2 weeks old and had been treated with BA only once. The enhancement was maintained up to week 6 but the magnitude depended on assimilation time and leaf age. Enhanced synthesis of starch has also been reported for kinetin-treated tobacco leaves (Mothes, 1963).

Hormonal Regulation of Senescence in Plants

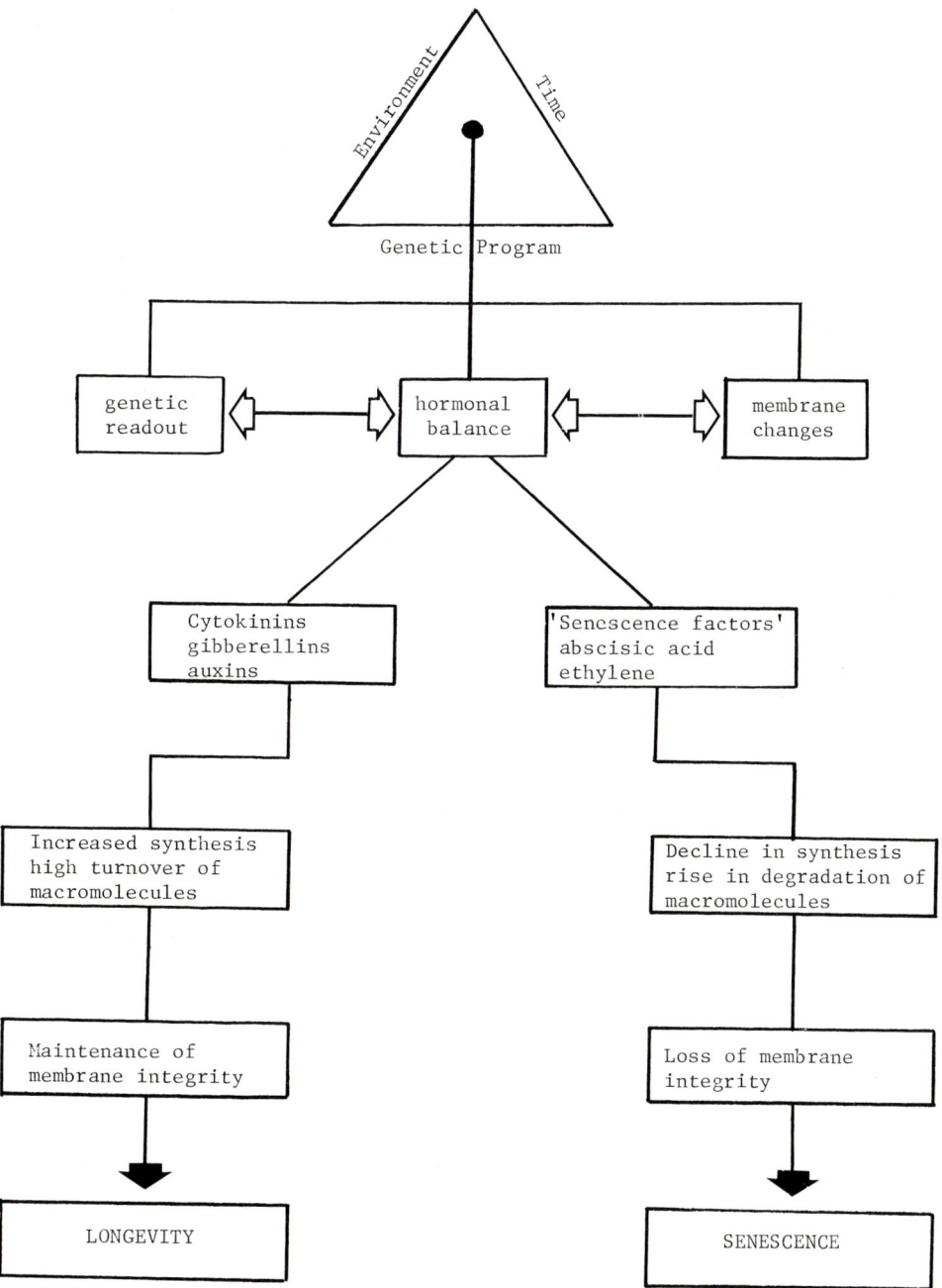

Fig. 4. A model for hormonal regulation of senescence.

Table 1. Incorporation of $^{14}CO_2$ into starch by H_2O and BA treated primary bean leaves

Plant age (weeks)	$^{14}CO_2$ assimilation time (hr)	Starch level (mg/g fresh leaf)		Specific Activity (cpm x 10^{-3}/mg starch)	
		H_2O	BA	H_2O	BA
2	0.25	1.10	0.96	19.55	22.82
	2.50	1.07	1.02	34.16	52.88
	25.0	1.13	0.97	91.71	158.12
4	0.25	1.41	2.28	12.90	18.56
	2.50	1.32	2.19	40.40	45.53
	25.0	1.51	2.33	30.19	36.01
6	0.25	0.67	1.30	6.23	8.94
	2.50	0.68	1.25	22.10	25.30
	25.0	0.63	1.41	10.11	17.89

VI. ROLE OF HORMONES IN LEAF SENESCENCE

The interplay of environment, time and genetic potential determines the levels of endogenous growth regulators (Fig. 4). These regulators may be classified into 2 broad groups: (a) senescence retardants which include cytokinins, gibberellins and auxins, and (b) senescence accelerators which include ethylene, abscisic acid and unidentified "senescence factors". The effectiveness of any one of the senescence retardants in delaying senescence is dependent on the species and plant age. A decline in the levels of senescence retardants and/or a rise in those of senescence accelerators would lead to senescence. The role of senescence retardants is to maintain the synthesis of macromolecules such as chlorophyll, protein and nucleic acids. Such a maintenance confers structural integrity and proper compartmentation of metabolites in cells. This is associated with a high photosynthetic rate, and an increased turnover of metabolites accompanied by an orderly metabolic coordination between energy production and utilization for biosynthetic and growth processes.

On the other hand, a rise in the levels of senescence accelerators and/or a decline in senescence retardants would result in a decrease in the synthesis or an increase in the degradation of macromolecules. The decrease in synthesis is well documented, but an increase in degradation needs to be further substantiated. These changes, accompanied by a loss of membrane integrity, could lead to senescence, and ultimately, death. The numerous metabolic changes associated with senescence could be delayed by the exogenous application of hormones.

VII. SUMMARY

It has been shown that N^6-benzyladenine (BA) can retard leaf senescence in the intact plant. This retardation is accompanied by maintenance of chlorophyll, protein and RNA levels, similar to that reported for detached leaves. There are numerous reports that retardation of senescence in detached leaves by cytokinins is associated with mobilization of metabolites from untreated zones. In our studies, there is no mobilization of ^{14}C-labelled CO_2, IAA, leucine, sucrose or ^{32}P into BA-treated leaves when these metabolites are fed to other parts of the plant. On the other hand, we find that retardation of senescence is accompanied by enhanced photosynthesis and a high retention of photosynthate. Although the levels of hexose phosphates and adenosine phosphates in the BA-treated leaves are lower than in the controls, the incorporation of ^{32}P into these compounds is higher, indicating a more rapid turnover.

It is concluded that retardation of leaf senescence by BA treatment on the intact plant is associated with a high retention of photosynthate, high starch levels, maintenance of protein and RNA synthesis, and a rapid turnover of metabolites. Therefore, these effects of BA in retarding senescence are brought about by an orderly metabolic coordination between the production and utilization of energy for biosynthetic and growth processes.

VIII. ACKNOWLEDGEMENTS

This work was supported by grants from the National Research Council of Canada. The skillful technical assistance of Miss Dianne McCullagh is gratefully acknowledged.

IX. REFERENCES

ADEDIPE, N.O. and R.A. FLETCHER (1970a). Benzyladenine-directed transport of Carbon-14 and Phosphorus-32 in senescing bean plants. J. exp. Bot. 21, 968-974.
ADEDIPE, N.O. and R.A. FLETCHER (1970b). Retardation of bean leaf senescence by benzyladenine and its influence on phosphate metabolism. Plant Physiol. 46, 614-617.
ADEDIPE, N.O. and R.A. FLETCHER (1971). Retardation of leaf senescence by benzyladenine in bean plants is not dependent on mobilization. Can. J. Bot. 49, 59-61.
ADEDIPE, N.O., L.A. HUNT and R.A. FLETCHER (1971). Effects of benzyladenine on photosynthesis and growth in relation to retardation of bean leaf senescence. Physiol. Plant 25, 151-153.
BEEVERS, L. (1966). Effects of gibberellic acid on the senescence of leaf discs of Nasturtium (*Tropaeolum majus*). Plant Physiol. 41, 1074-1076.
CARR, D.J. and J.S. PATE (1967). Ageing in the whole plant, in "Aspects of the Biology of Ageing" (ed. H.W. Woolhouse), pp. 559-599. Cambridge University Press.
CHIN, T.-Y. and L. BEEVERS (1970). Changes in endogenous growth regulators in Nasturtium leaves during senescence. Planta (Berl.) 92, 178-188.
ENGELBRECHT, L. (1964). Über Kinetinwirkungen bei intakten Blättern von *Nicotiana rustica*. Flora (Jena) 154, 57-69.
FLETCHER, R.A. (1969). Retardation of leaf senescence by benzyladenine in intact bean plants. Planta (Berl.) 89, 1-8.
FLETCHER, R.A., G. HOFSTRA and N.O. ADEDIPE (1970). Effects of benzyladenine on bean leaf senescence and the translocation of ^{14}C-assimilates. Physiol. Plant. 23, 1144-1148.
FLETCHER, R.A., T. OEGEMA and R.F. HORTON (1969). Endogenous gibberellin levels and senescence in *Taraxacum officinale*. Planta (Berl.) 86, 98-102.
FLETCHER, R.A. and D.J. OSBORNE (1966). Gibberellin, as a regulator of protein ribonucleic acid synthesis during senescence in leaf cells of *Taraxacum officinale*. Can. J. Bot. 44, 739-745.
GUNNING, B.E.S. and W.K. BARKLEY (1963). Kinetin-induced directed transport and senescence in detached oat leaves. Nature 199, 262-265.
HOFSTRA, G. and C.D. NELSON (1969). A comparative study of translocation of assimilated ^{14}C from leaves of different species. Planta (Berl.) 88, 103-112.
JACOBY, B. and J. DAGAN (1970). Effects of 6N-benzyladenine on primary leaves of intact bean plants and on their sodium absorption capacity. Physiol. Plant. 23, 397-403.
KHYM, J.X. and W.E. COHN (1953). The separation of sugar phosphates by ion exchange with the use of the borate complex. J. Amer. Chem. Soc. 75, 1153-1156.
KHYM, J.X. and L.P. ZILL (1952). The separation of sugars by ion exchange. J. Amer. Chem. Soc. 74, 2090-2094.
LETHAM, D.S. (1967). Chemistry and physiology of kinetin-like compounds. Annu. Rev. Plant Physiol. 18, 349-364.
MACLEAN, D.C. and R.R. DEDOLPH (1964). Phytokinins and senescence in Broccoli. Amer. J. Bot. 51, 618-621.
McCREADY, R.M., J. GUGGOLZ, V. SILVIERA and H.S. OWENS (1950). Determination of starch and amylase in vegetables. Analyt. Chem. 22, 1156-1158.

McHALE, J.S. and L.D. DOVE (1968). Mobilization independent effects of a cytokinin on senescing tomato leaves. Naturwiss. 55, 1-2.

MEIDNER, H. (1967). The effect of kinetin on stomatal opening and the rate of intake of carbon dioxide in mature primary leaves of barley. J. exp. Bot. 18, 556-561.

MOTHES, K. (1963). The role of kinetin in plant regulation, in "Régulateurs Naturels de la Croissance Végétale", 5th Intnl. Conf. on Plant Growth Substances, Gif-sur-Yvette, C.N.R.S., Paris.

MOTHES, K. and L. ENGELBRECHT (1961). Kinetin-induced directed transport of substances in excised leaves in the dark. Phytochem. 1, 58-62.

MULLER, K. and A.C. LEOPOLD (1966). Correlative ageing and transport of ^{32}P in corn leaves under the influence of kinetin. Planta (Berl.) 68, 167-185.

OSBORNE, D.J. (1962). Effect of kinetin on protein and nucleic acid metabolism in *Xanthium* leaves during senescence. Plant Physiol. 37, 595-602.

OSBORNE, DJ. and H.M. HALLAWAY (1964). The auxin, 2,4-dichlorophenoxyacetic acid as a regulator of protein synthesis and senescence in detached leaves of *Prunus*. New Phytol. 63, 334-337.

RESNIK, M.E. and E.R. MONTALDI (1968). Kinetin effects on phosphorus uptake and translocation in plants:*Phaseolus vulgaris* cv. Bountiful. Biol. Prod. Veg. 5, 99-111.

RICHMOND, A.E. and A. LANG (1957). Effect of kinetin on protein content and survival of detached *Xanthium* leaves. Sci. 125, 650-651.

SCOTT, R.W., W.E. MOORE, M.J. EFFLAND and M.A. MILLET (1967). Ultraviolet spectrophotometric determination of hexoses, pentoses and uronic acids after their reactions with concentrated sulfuric acid. Analyt. Biochem. 21, 68-80.

SETH, A.K. and P.F. WAREING (1967). Hormone-directed transport of metabolites and its role in plant senescence. J. exp. Bot. 18, 65-77.

SITTON, D., C. ITAI and H. KENDE (1967). Decreased cytokinin production in the roots as a factor in shoot senescence. Planta (Berl.) 73, 296-300.

SKOOG, F. and D.J. ARMSTRONG (1970). Cytokinins. Annu. Rev. Plant Physiol. 21, 359-384.

TAVARES, J. and H. KENDE (1970). The effect of 6-benzylaminopurine on protein metabolism in senescing corn leaves. Phytochem. 9, 1763-1770.

WAREING, P.F. and A.K. SETH (1967). Ageing and senescence in the whole plant, in "Aspects of the Biology of Ageing" (ed. H.W. Woolhouse) pp. 543-558. Cambridge University Press.

WEBB, J.A. (1970). The translocation of sugars in *Cucurbita melopepo*. V. The effect of leaf blade temperature on assimilation and transport. Can. J. Bot. 48, 935-942.

WOLLGIEHN, R. (1967). Nucleic acid and protein metabolism of excised leaves, in "Aspects of the Biology of Ageing" (ed. H.W. Woolhouse) pp. 231-246. Cambridge University Press.

WOOLHOUSE, H.W. (1967). The nature of senescence in plants, in "Aspects of the Biology of Ageing" (ed. H.W. Woolhouse) pp. 179-213. Cambridge University Press.

Further Stuides of Hormone-regulated Senescence in *Rumex* Leaf Tissue

Jonathan Goldthwaite

The Biological Laboratories, Harvard University, 16 Divinity Ave., Cambridge, Massachusetts, 02138

I. INTRODUCTION

Senescence of leaf tissue and attached leaves can be inhibited by all of the three major groups of higher plant growth hormones. Leaf senescence can also be promoted by ethylene and abscisic acid. Which hormones are active in a given case seems to be primarily determined by the species used, although some evidence indicates this may be an oversimplification (e.g. Aspinall, *et al.*, 1967; Goldthwaite and Laetsch, 1968). Previous work (Whyte and Luckwill, 1966; Goldthwaite and Laetsch, 1968) showed that net breakdown of chlorophyll and protein in excised leaf tissue of *Rumex spp.*, is strongly inhibited by very low concentrations of gibberellic acid (GA_3). Also described were several aspects of the behavior of the *Rumex* system including time courses of chlorophyll and protein loss, reversibility of the GA_3 effect and activity of the cytokinin 6-benzylaminopurine (BAP). This report describes experiments carried out to further our understanding of hormone-controlled senescence in this plant. Included are studies of the effect of GA_3 and BAP on senescence induced in attached leaves and the effect of other growth-regulators and combined applications of GA_3 and cytokinins to excised leaf discs.

II. MATERIALS AND METHODS

Leaf tissue for senescence experiments was obtained from recently fully-expanded leaves of *Rumex crispus* plants grown in a greenhouse as previously described (Goldthwaite and Bogorad, 1971). Samples of leaf discs 6 mm in diameter were punched from the lamina, randomized and floated on test solutions in plastic Petri dishes. Incubation was carried out in darkness at 30°C for the indicated number of days. Senescence of attached leaves was induced by loosely wrapping the entire lamina in aluminum foil. Only one leaf per plant was used for treatment. These plants were returned to the greenhouse under ambient conditions for the number of days indicated. Hormones in aqueous solution containing 0.05% Tween 80 were sprayed in a fine mist over both leaf surfaces until runoff. The plants were kept in darkness at 22-23°C overnight, before wrapping the treated leaves in order to thoroughly dry the leaf surface.

Chlorophylls were extracted from samples of 10 leaf discs in warm 100% ethanol. Total chlorophyll and protein determinations were made as described previously (Goldthwaite and Laetsch, 1968). Chlorophyll content is normally expressed as per cent of the amount initially present in an equivalent tissue sample. In Table I the figures for inhibition of chlorophyll loss are calculated as the chlorophyll content of the treated discs minus the control discs, with chlorophyll units again being expressed as percent of that present initially. *In vivo* absorption spectra were obtained using a Biospect 61 scanning spectrophotometer (courtesy of Dr.W.R. Briggs). The baseline was approximated with layers of paper tissue (Kimwipes, Kimberly-Clark). The baseline absorption decreased only 0.066 O.D. across the wavelength region scanned (550 to 750 nm).

Endogenous senescence-regulating compounds were extracted from fresh *Rumex* leaf tissue in an aqueous buffer containing 0.2M Tris-SO_4 pH8.0, 10 nM $MgSO_4$, 20 mM mercaptoethanol and 2% soluble polyvinylpyrrolidone or in 66% methanol. Methanol was removed *in vacuo* at 40°C. The aqueous extracts at pH 8.0-8.5 were partitioned against petroleum ether to remove pigments. The extracts were adjusted to pH 2.5 and partitioned against ethyl acetate 3 times. The pooled acidic ethyl acetate fraction was dried over anhydrous sodium sulfate, reduced to a small volume *in vacuo* and applied to 20 x 20 cm thin-layer chromatographic plates bearing a 0.4mm thick layer of silical gel H. The chromatograms were developed in a mixture of chloroform, ethyl acetate, and acetic acid (60:40:5 volume ratio). After drying the plate in a vacuum oven, R_f zones were scraped from the plate and growth substances were eluted in distilled water. The eluates were tested for biological activity on chlorophyll degradation in *Rumex* leaf discs and on the extension of the leaf sheaths of d_5 dwarf maize. Standard dose-response curves for the bioassays were prepared with solutions of GA_3. Activity is reported as GA_3 equivalents.

Growth-regulating chemicals and their sources were as follows: GA_3, 80% potassium salt (Calbiochem): BAP, (lot 54347, Calbiochem); GA_4 plus GA_7 mixture, GA_7 46.4%, GA_4 53.6%, (Calbiochem); zeatin, trans-6-(4-hydroxy-3-methyl-but-2-enyl)-aminopurine, (lot 901263), Calbiochem); cyclic AMP, adenosine 3'-5' cyclic monophosphoric acid (Calbiochem and Sigma).

III. RESULTS

The inhibition of chlorophyll breakdown by increasing concentrations of GA_3, BAP, and zeatin is shown in figure 1. All 3 hormones are effective and the magnitude of the senescence-inhibiting effect is similar when saturating levels of each are used. The system is very sensitive to GA_3 with a threshold, half-saturation, and saturation of the effect at about 0.5, 3, and 20 nM respectively. Analogous figures for the BAP effect are 20, 300, and 10,000 nM; and for the zeatin effect are about 50, 800, and 30,000 nM. The response to cytokinins is accordingly 100-250 times less sensitive than to GA_3 and requires a wider concentration range to achieve saturation.

A mixture of GA_4 plus GA_7 is also active in the chlorophyll loss response. Preliminary results showed this effect to be saturated at about 300 nM (in GA_7) indicating that GA_{4+7} is about 15 times less effective than GA_3. Cyclic AMP in aqueous solution was applied to *Rumex* leaf discs in the usual senescence assay. The pH of solutions was adjusted to 6.0-7.5 before assay and the solutions were renewed after 3 days. No inhibitory or promotive effect on yellowing was observed over a range of concentrations from 10^{-7} to 3×10^{-3}M.

The yellowing of attached leaves which are darkened appears to follow a time-course roughly comparable to that found in excised discs at 30° (see Goldthwaite and Laetsch, 1968): a lag period for about 2 days followed by rapid chlorophyll loss, leading to complete yellowing on about the 6th day. The yellowing and protein degradation in attached leaves is also markedly inhibited by GA_3. Figure 2 shows that this effect is half-saturated at 10, and saturated at about 80 nM GA_3. Accordingly the sprayed attached leaves are only about 3-4 times less sensitive to GA_3 than leaf discs floating on GA_3 solutions - a result which could easily be due to the sprayed leaves taking up a smaller amount of the hormone at a given concentration. Application of BAP to attached leaves showed a considerable inhibition of chlorophyll loss at 10^{-6} and 10^{-5} M. Some evidence of a supra-optimal effect was observed at 10^{-4} M, where yellow and green areas were scattered over the leaf surface.

Because both gibberellins and cytokinins are active inhibitors to *Rumex* leaf senescence it was of interest to see whether combined application of hormones of the two classes resulted in any interaction. Simultaneous treatment with GA_3 and BAP at concentrations ranging from about 60% saturating up to supersaturating levels (GA_3 10^{-7}M + BAP 5×10^{-4} M) in all cases gave a maximum inhibition of chlorophyll loss, indicating these hormones do not antagonize each other at high concentrations. Simultaneous application of GA_3 plus BAP or zeatin at concentrations less than half-saturating consistently gave a "synergistic" effect - defined as an inhibition of chlorophyll breakdown greater than that obtained by adding the effects of the gibberellin and cytokinin applied individually at the same concentration. Examples of the gibberellin-cytokinin synergism observed over a range of concentrations are shown in table 1. The synergistic effect is to increase the inhibition of chlorophyll loss from 1.2 to 2.0 times that expected from a simple additive effect. The magnitude of this synergism generally increased as the concentration of the hormones decreased to the threshold level. Preliminary results suggest that BAP and zeatin at (10^{-7} and 10^{-6}M) do not act synergistically. If this is true, it would suggest that the gibberellin-cytokinin synergism is somehow due to the presence of different types of hormones, rather than to some non-specific property of the biological response which would always give synergism when any two active compounds are present at less than half-saturating concentration.

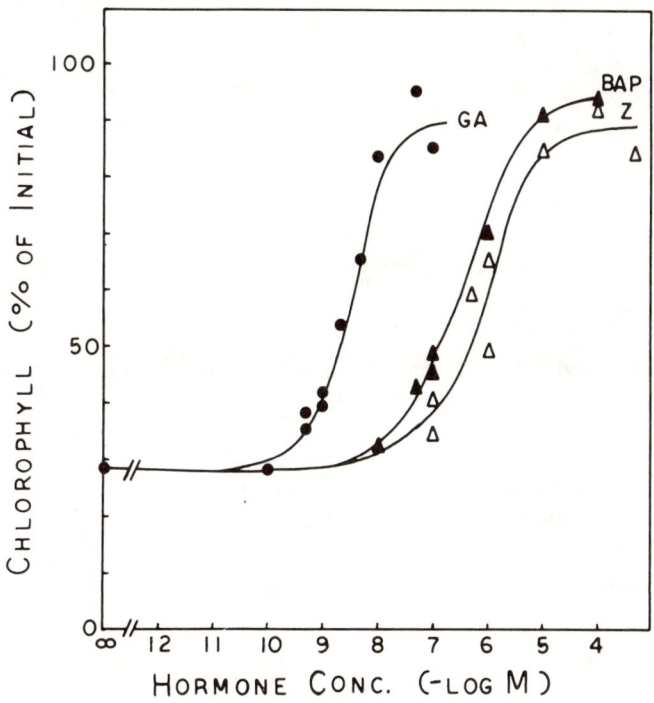

Fig. 1.

Dose-response curves for the inhibition of chlorophyll breakdown in *Rumex* leaf discs by increasing concentrations of gibberellic acid (GA), 6-benzylaminopurine (BAP) or zeatin (Z). Samples of leaf discs were incubated on water or hormone solutions from day zero for 6 days.

Extractable senescence-retarding activity was obtained from lamina tissue of mature leaves (fully-expanded, but not old) of *Rumex*. In two preliminary trials such activity was obtained in both the buffer and methanol extracts and amounted to an *in vivo* concentration of 0.04 and 0.29 μg GA_3 equivalents/kg fresh weight. This is equivalent to 1.2 and 8.3 x 10^{-10} M GA_3 equivalents, assuming the leaf tissue is of unit density. These concentrations are compared in figure 3 with the senescence response curves of comparable leaf tissue to externally supplied concentrations of GA_3. The GA_3-equivalent levels obtained were so low as to be at the threshold of detectibility in the dwarf corn assay.

The endogenous senescence-retarding activity in the *Rumex* extracts was found in R_f zones 0.1-0.3 with a small additional amount possibly present at R_f 0.5. A potent "lethal-factor" was also present in the extracts. This material caused rapid loss of turgor and apparent death without yellowing when tested on *Rumex* leaf disc senescence. The substance is not well separated from the senescence-retarding zones on the chromatograms. Therefore these results must be considered tentative.

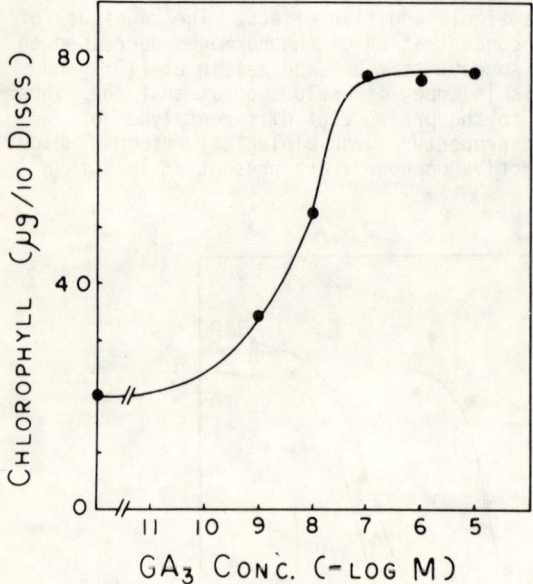

Fig 2.

Retardation of chlorophyll degradation in darkened attached leaves of *Rumex* by increasing concentrations of (GA_3). Disc samples were taken for chlorophyll determination after 6 days. Each point is the average of 3 leaves.

Table 1. Effect of combined treatment with various concentrations of GA, BAP and zeatin (Z) on chlorophyll loss in *Rumex* leaf discs.

Hormone and conc.(nM)	Chlorophyll content (% of Initial)	Inhibition of Chlorophyll loss (% of Initial)	Inhibition expected from additive effect (% of Initial)*
GA .5+BAP 50	78.0	42.9	22.6
" +BAP 1000	89.7	61.4	49.1
" + Z 100	58.4	23.3	20.3
" + Z 1000	92.6	57.5	43.1
GA 1 +BAP 10	67.4	32.3	16.4
" +BAP 50	78.3	43.2	25.8
" + Z 100	72.9	37.8	23.5
GA 2 +BAP 10	78.0	42.9	26.9
" +BAP 50	85.8	50.7	36.3
" + Z 100	80.4	45.3	34.0

* Data for effect of hormones applied individually are those graphed in figure 1.

Chlorophyll breakdown in *Rumex* leaf discs and bean (Goldthwaite, 1968) is not accomplished *via* conversion of any large fraction of the chlorophyll into a stable pheophytin pool. Solvent partitioning of *Rumex* ethanol extracts against diethyl ether indicates that a major conversion of chlorophylls into chlorophyllides by dephytylation probably also does not take place. A more subtle change in the environment of the bulk chlorophyll during senescence without chemical conversion might be detectable using *in vivo* spectrophotometry. This method has revealed such changes in the chlorophyll absorption maximum during greening of etiolated leaves (Kirk and

Tilney-Bassett, 1967). A series of such spectra from yellowing *Rumex* leaf discs is shown in figure 4. No change in the chlorophyll absorption maximum (about 676 nm) in control discs was observed during the entire period of chlorophyll loss. A shift to about 671 nm was seen on day 6, but this is probably a post-mortem change and in any case occurs after the period of chlorophyll breakdown. No change was seen in the absorption maximum in GA-treated discs through the 6th day of incubation.

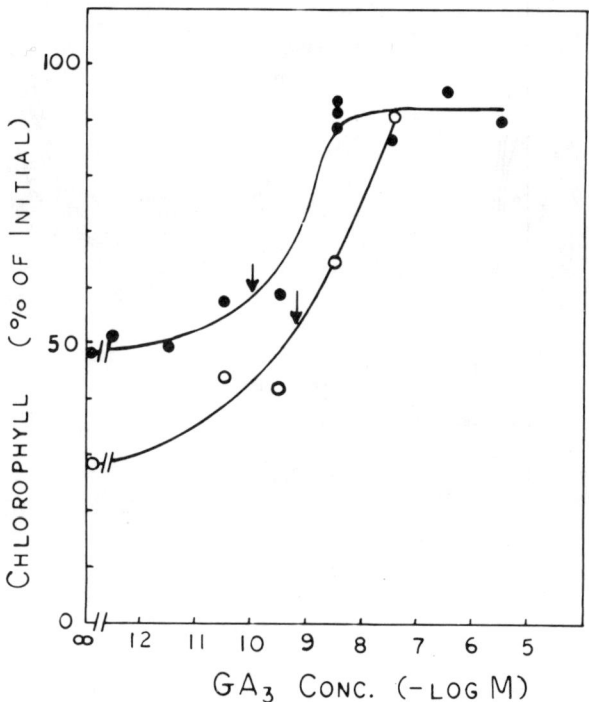

Fig. 3.

Standard dose-response curves for estimation of endogenous senescence-retarding substances in *Rumex* leaf extracts. Incubation periods for the two leaf disc senescence assays were 4.8 and 7.0 days. Arrows indicate total levels of such substances in 2 experiments expressed as the molar concentration of GA_3-equivalents in the tissue extracted.

IV. DISCUSSION

Retardation of senescence in *Rumex* leaf discs is not restricted to the gibberellins. The cytokinins BAP, zeatin, and kinetin are also effective. The sensitivity to cytokinins was not lost when leaf tissue was preaged as might be inferred from the results of Whyte and Luckwill (1966) and Fletcher and Osborne (1966). The threshold sensitivity of *R. crispus* tissue reported here is comparable to that found by Whyte and Luckwill (1966) in *R. obtusifolius*. Apparent saturation of the GA_3 effect was, however, found over a narrower range of concentration than that in either Whyte and Luckwill's or in Fletcher and Osborne's (1966) report on *Taraxacum*. This sensitive dependence on hormone concentration makes the response in *R. crispus* leaf tissue a bioassay capable of determining gibberellin concentrations in test solutions with better precision than most GA bioassays. In using this response as a bioassay, purification steps should be employed to overcome the lack of specificity (Reynolds, 1969). Confirmation of activity by a different biological or nonbiological assay is advisable. Spurious positive responses can be caused by compounds which poison or kill the cells. Such effects can often be detected visually by turgor loss and water-logging of the

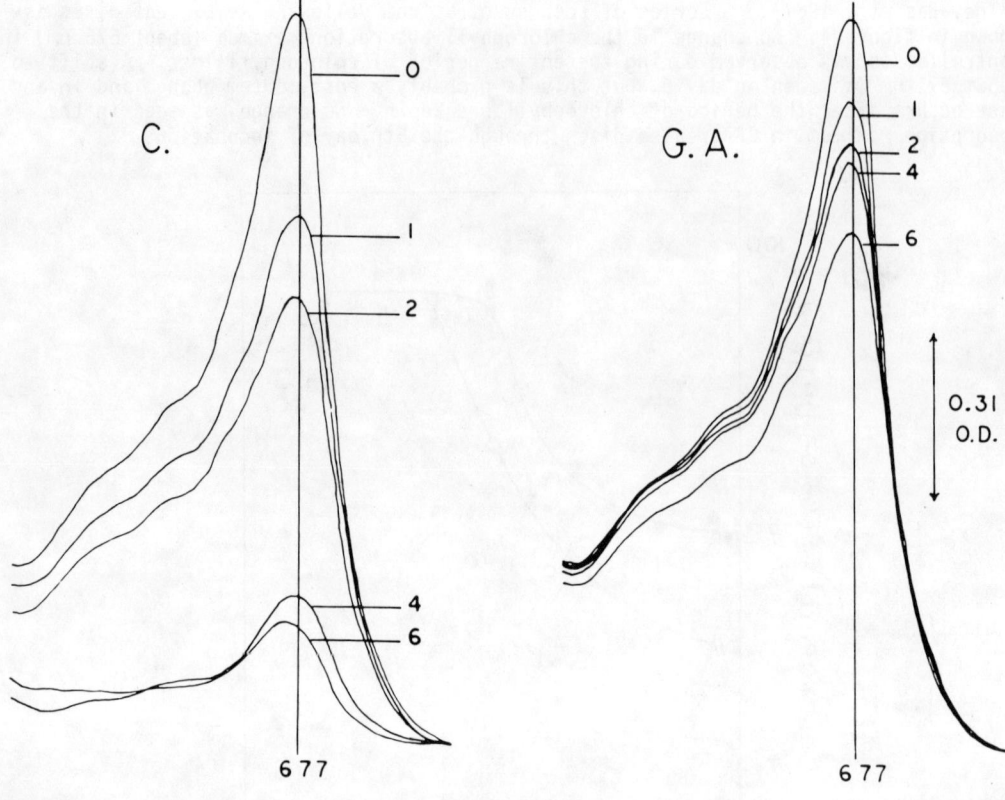

Fig. 4.

In vivo absorption spectra of individual *Rumex* leaf discs incubated for the indicated number of days on water (C) or 10^{-7} M GA_3 (GA). Samples were scanned from 550 to 750 nm.

tissue and by testing for normal completion of senescence after removal of the test material.

This report shows that application of GA_3 together with BAP or zeatin gives an inhibition of cumulative chlorophyll breakdown which is greater than that calculated by adding the separate effects of the 2 hormones (Table 1). This suggests, but does not prove, that GA_3 and cytokinins interact synergistically at low concentrations in *R. crispus*. Back and Richmond (1969) concluded that kinetin and GA_3 interacted negatively at high concentrations in *Taraxacum megallorrhizon* and *Tropaeolum majus*. They later reported that in these plants and *R. pulcher* cytokinins showed an interaction with low concentrations of abscisic acid (ABA) while GA_3 did not interact with ABA (Back and Richmond, 1971). Beevers had previously found no obvious distinction between the antagonism of GA_3 and kinetin effects by ABA in *Tropaeolum*. The results in Table 1 and in most of the work just cited come from measurements of chlorophyll remaining in a sample of leaf discs after a fixed incubation period. Kinetic analysis of such data is not possible because rates of the process are not obtained except in the case where the breakdown rate is constant from the beginning of incubation. This condition is not true in freshly-cut leaf material of *R. crispus* (Goldthwaite and Laetsch, 1968; Goldthwaite, unpublished) and apparently *Tropaeolum majus* (Beevers, 1968). In this material a lag or slow period of chlorophyll breakdown is initially present which can be markedly extended by the senescence-retarding hormones. Another point apparently overlooked is the necessity of comparing the "chlorophyll remaining" data with the amount initially present in order to quantitate the process (breakdown) being measured. This author is presently developing experiments with *Rumex* using the rate of chlorophyll breakdown to study more rigorously the interactions between the senescence-retarding hormones and between them and ABA.

The role of endogenous leaf senescence regulators is a subject of continuing investigation. In the gibberellin-responding species *Taraxacum officinale*, Fletcher, et al (1969) reported a seasonal correlation of levels of endogenous GA-like substances with leaf senescence and sensitivity to exogenous GAs. Chin and Beevers (1970) showed that GA-like substances decreased while ABA-like components increased during senescence of detached and attached *Tropaeolum* leaves. It is not yet clear whether such changes are adequate in amount or timing to account for leaf senescence regulation.

In an attempt to evaluate whether the amounts of endogenous GA-like materials in mature *Rumex* leaves are high enough to account for the stability of that tissue when left attached to normal light-grown plants, the calculated *in vivo* concentrations were compared with the dose-response curve to exogenous GA_3 (figure 3). In this case and in the case of the reported *Taraxacum* results (Fletcher, et al, 1969; cf. Fletcher and Osborne, 1966), the estimated endogenous concentration of GA_3-equivalents falls at a value less than half-saturating, ie. where rapid senescence is taking place in the leaf discs. Thus it appears that the internal levels of GA-like senescence inhibitors in these 2 species may not be sufficient *per se* to account for the stability of the mature green leaf. This conclusion requires the assumption that the intracellular GA_3 concentration becomes greater than or equal to the concentration of exogenous GA_3 supplied continuously in solution to senescing leaf discs. Only if the leaf discs were relatively impermeable to GA_3 or if the exogenous GA was isolated from its site(s) of action or otherwise inactivated in the tissue would this assumption appear to be invalid.

The results reported here, showing the rapid senescence and maintained hormone response of attached darkened leaves as well as previous demonstrations of light-inhibited senescence in excised tissue of other species (e.g. Goldthwaite and Laetsch, 1967) and *Rumex* (Goldthwaite, unpublished), lead to the conclusion that light alone or in combination with endogenous senescence regulators could maintain the mature green leaf. A diminished capacity for light utilization in old leaves such as has been demonstrated in *Perilla* (Hardwick, et al, 1968) could initiate the senescence of older leaves in intact plants.

These light effects may prove to be mediated by changes in the levels of endogenous growth hormones. However some further experiments are needed to fully evaluate the importance of endogenous senescence-regulators and light in leaf maturity and senescence. Among these are studies which quantitate, over time, changes in the levels of such compounds in both attached and detached leaves exposed to light and dark, and correlate these changes with the speed of senescence of similarly treated tissue. It should also be determined whether rates of leaf senescence correlate well with the different endogenous concentrations of these compounds which can occur in plants grown under various conditions or treated with selective hormone antagonists.

V. SUMMARY

The net breakdown of chlorophyll in *Rumex* leaf discs is inhibited by the gibberellins GA_3 and GA_{4+7} and by the cytokinins 6-benzylaminopurine (BAP), zeatin, and kinetin. Cyclic AMP from 10^{-7} to 3×10^{-3} M had no effect on the loss of chlorophyll. Dose-response curves show that this response is about half-saturated at the following hormone concentrations: GA_3 - 3nM, BAP - 300nM, and zeatin - 800nM. The yellowing of attached darkened *Rumex* leaves is also strongly inhibited by GA_3 and BAP. This response is about half-saturated by 10nM GA_3. When threshold to half-saturating concentrations of GA_3 together with BAP or zeatin are added to leaf discs, an effect is obtained which is more than additive. Preliminary results show that extractable endogenous senescence-retarding substances are present in mature green *Rumex* leaves at an average *in vivo* concentration of 0.1 to 0.8 nM GA_3-equivalents. The possibility that this level may not be high enough *per se* to stabilize the mature leaf is discussed. No obvious change in the shape of the *in vivo* chlorophyll absorption spectrum was observed during the entire course of chlorophyll breakdown in control leaf discs.

VI. ACKNOWLEDGEMENTS

This research was supported in part by a grant from the Wm. F. Milton Fund of Harvard University. The technical assistance of David Bateman is greatly appreciated.

VII. REFERENCES

ASPINALL, D., L.G. PALEG and F.T. ADDICOTT (1967). Abscisin II and some hormone-regulated plant responses. Aust. J. Biol. Sci. 20, 869-882.

BACK, A. and A.E. RICHMOND (1969). An interaction between the effects of kinetin and gibberellin in retarding leaf senescence. Physiol. Plant. 22, 1207-16.

BACK, A. and A.E. RICHMOND (1971). Interrelations between gibberellic acid, cytokinins and abscisic acid in retarding leaf senescence. Physiol. Plant. 24, 76-9.

BEEVERS, L. (1968). Growth regulator control of senescence in leaf discs of nasturtium (*Tropaeolum majus*), in "Biochemistry and Physiology of Plant Growth Substances" (Eds. F. Wightman and G. Setterfield). The Runge Press Ltd., Ottawa. 1960.

CHIN, T. and L. BEEVERS (1970). Changes in endogenous growth regulators in nasturtium leaves during senescence. Planta 92, 178-88.

EL-ANTABLY, H.M.M., P.F. WAREING and J. HILLMAN (1967). Some physiological responses to D,L-Abscisin (Dormin). Planta 73, 74-90.

FLETCHER, R.A., T. OEGEMA, and R.F. HORTON (1969). Endogenous gibberellin levels and senescence in *Taraxacum officinale*. Planta 86, 98-102.

FLETCHER, R.A. and D.J. OSBORNE (1966). A simple bioassay for gibberellic acid. Nature 211, 743-4.

GOLDTHWAITE, J.J. (1968). "Experimental investigations of leaf tissue senescence". Thesis, University of California, Berkeley.

GOLDTHWAITE, J.J. and L. BOGORAD (1971). A one-step method for the isolation and determination of leaf ribulose-1,5-diphosphate carboxylase. Anal. Biochem. 41, 57-66.

GOLDTHWAITE, J.J. and W.M. LAETSCH (1967). Regulation of senescence in bean leaf discs by light and chemical growth regulators. Plant Physiol. 42, 1757-62.

GOLDTHWAITE, J.J. and W.M. LAETSCH (1968). Control of senescence in *Rumex* leaf discs by gibberellic acid. Plant Physiol. 43, 1855-8.

HARDWICK, K., M. WOOD, and H.W. WOOLHOUSE (1968). Photosynthesis and respiration in relation to leaf age in *Perilla frutescens*. (L.) Britt. New Phytol. 67, 79-86.

KIRK, J.T.O. and R.A.E. TILNEY-BASSETT (1967). "The Plastids." W.H. Freeman and Company, London.

REYNOLDS, T. (1969). Senescence retardation in dock leaves by soluble peptone. Nature 223, 505-6.

WHYTE, P. and L.C. LUCKWILL (1966). A sensitive bioassay for gibberellins based on retardation of leaf senescence in *Rumex obtusifolius*. Nature 210, 1360.

KINETIN TREATMENT AND PROTEIN SYNTHESIS
IN DETACHED WHEAT LEAVES

Heng Fong Tung and C.J. Brady

Plant Physiology Unit, C.S.I.R.O. Division of Food Preservation, Ryde, and School of Biological Sciences, University of Sydney, N.S.W., 2006

I. INTRODUCTION

In the leaves of *Perilla* (Hardwick and Woolhouse, 1967), and in wheat leaves (Brady et al., 1971), the incorporation of amino acid precursors into protein changes with age such that in the older leaf there is relatively little incorporation into the plastid-located fraction 1 protein (Wildman and Bonner, 1947). However, incorporation into other proteins of the chloroplast continues in the older leaves (Brady et al., 1970) so there is some turnover of the chloroplast proteins. Turnover of chloroplast ribosomal RNA in senescent *Lemna* fronds was demonstrated by Trewavas, 1970. Age-related shifts in enzyme content may be indicative of other changes in protein synthesis within leaves.

Treatment with cytokinins preserves the protein content of senescing leaves, and several authors (Kuraishi, 1968, Mizraha et al., 1970, Tavares and Kende, 1970, Shibaoka and Thimann, 1970) have concluded that the cytokinins regulate protein catabolism in the leaves. These authors suggest that demonstrations of greater incorporation into protein in leaf pieces treated with cytokinins reflect differences in the size of the endogenous amino acid pools rather than greater rates of protein synthesis.

In this paper we examine the effect of applying cytokinins to detached wheat leaves on the content of fraction 1 protein and other soluble proteins, and on the progress of the shifts in protein synthesis which normally occur. We have sought to determine if there is a relationship between the effectiveness with which kinetin maintains proteins, and the apparent turnover rates of the proteins. We were also interested in determining if kinetin treatment would retard the changes in the relative synthetic rates which are a feature of the senescence of the wheat leaf. We felt that by measuring the effect of kinetin treatment on incorporation into one protein relative to others, differences contributed by differential uptake and endogenous dilution could largely be avoided.

II. MATERIALS AND METHODS

Experimental material

We used the second leaf of seedling wheat (*Triticum aestivum* L. var. Stewart). The detached leaves were held in the light (600 foot candles) at 23°C, in a stream of humid air which maintained them turgid and prevented ethylene accumulation. Kinetin (10^{-4} M) was supplied to the leaves in the transpiration stream. To measure incorporation into protein we fed radioactive valine *via* the transpiration stream for one or two hours while the leaves were in the light (1800 foot candles).

In these detached young leaves, the content of free valine increases from about 0.06 μmoles/mg nitrogen when detached to about 0.22 μmoles/mg nitrogen when they have been detached for two days. If kinetin is fed to the detached leaves, the free valine content after two days is about 0.13 μmoles/mg nitrogen. This difference in pool size appears relevant to incorporation studies, for if valine of a range of specific activities is given to the leaves, incorporation is a function of the specific activity of the radioactive amino acid diluted by the endogenous amino acid and not of the amino acid as fed. This is true whether or not the leaves are treated with kinetin. It is true of both freshly-detached leaves and leaves aged for some days after harvest. In each case, endogenous dilution is rapid relative to incorporation.

Protein fractionation

We have considered incorporation into three protein fractions, fraction 1 protein, fraction 2 proteins (Wildman and Bonner, 1947), and the chloroplast lamellae proteins. To measure the soluble proteins, leaves were extracted by grinding in Tris-borate-EDTA buffer pH 8.5 (Raymond, 1962); the homogenate was centrifuged (30,000 x g, 2^oC, 60 min) and the supernatant passed through a Millipore filter (1.2µ pore). Proteins in a portion of this were precipitated by trichloroacetic acid (TCA) 5 per cent ($^W/_v$), and freed of non-protein material by the method of Mans and Novelli (1961). Portions were applied to 6 per cent or 7.5 per cent acrylamide gels and subjected to electrophoresis for 16 hr using Tris-borate-EDTA buffer and a current of 1.5 mamps/tube. The gel cylinders were washed briefly in water and scanned at 280 nm using the scanning attachment of a Shimadzu 50 MPS spectrometer. The content of fraction 1 protein was measured from the scan (Watkin and Miller, 1970) using the extinction co-efficient for fraction 1 protein found by Trown (1965). After scanning, gels were stained with amido black 10B in 7 per cent ($^W/_v$) acetic acid and destained by diffusion and transverse electrophoresis. Fraction 2 proteins were measured as the difference between fraction 1 and total soluble protein.

When the fraction 2 proteins were to be separated from each other, electrophoresis was for about 2½ hr through gels of 7.5 per cent acrylamide using a current of 2.5 mamps per tube. These gels were then stained and destained. To recover radioactivity, gel slices were hydrolyzed (6N HCl, 24 hr, 105^oC) and the hydrolyzate in 1.0 ml of water counted in a scintillant mixture containing Triton x-100 (Patterson and Greene, 1965).

Lamellae proteins were prepared from chloroplasts isolated by the method of Nobel (1967). Chloroplasts were washed twice in the isolation medium, twice in hypotonic buffer, and twice in water. They were then dispersed in 1% ($^W/_v$) Triton x-100 in a medium containing Tris (0.005 M), Mg acetate (5mM) and mercaptoethanol (0.1mM) (Tewari and Wildman, 1969). After 1 hr at 0^oC, the sample was centrifuged (30,000 x g, 60 min.) and the supernatant concentrated by lyophilization. Pigments and detergent were removed in acetone (-15^oC), and the protein was dissolved in 0.25 N sodium hydroxide. The protein was reprecipitated in TCA, and purified (Mans and Novelli, 1961).

III. PROTEINS IN SENESCENT LEAVES

Age-related changes in amino acid incorporation into protein

If the leaf lamina is detached from the wheat plant when it has just reached full expansion and with the ligule emerged, incorporation into both the fraction 1 and fraction 2 soluble proteins, and into the proteins of the chloroplast lamellae proceeds at comparable rates (Table 1). From this time of full expansion, we took leaves each few days and measured incorporation into leaf proteins. Incorporation into fraction 1 protein and into the lamellae proteins was expressed relative to incorporation into the total soluble protein and these ratios are plotted against leaf age in Fig. 1. The ratio for fraction 1 protein declines rapidly as the leaves age and soon reaches a constant low value. In contrast the ratio for the proteins of the chloroplast lamellae remains high for at least two weeks after full expansion, and by this time the fifth leaf is well emerged.

Table 1. Rates of protein synthesis in newly-expanded wheat leaves. Leaves were detached as soon as the ligule was clearly emerged, and were fed L-^{14}C-valine (270 mC mmole, 1.0μC/1ml) *via* the transpiration stream for 60 min in the light. Proteins were prepared from 3 groups each of 12 leaves. Results are means with standard errors.

Protein Fraction	Specific Activity (dpm x 10^{-3}/mg)
Fraction 1	4.72 ± 0.02
Fraction 2	5.72 ± 0.03
Lamellae	5.77 ± 0.09

There is an obvious contrast between these two groups of chloroplast proteins. The result is not dependent on the use of valine as the amino acid precursor. It is observed if other amino acids, or carbon dioxide, contribute ^{14}C to the proteins. Nor is the decline in the ratio for fraction 1 protein much influenced by a change in the rate of incorporation into fraction 2 proteins, for we find no increment with leaf age in the rate of passage of valine into protein (cf. Atkin and Srivastava, 1970).

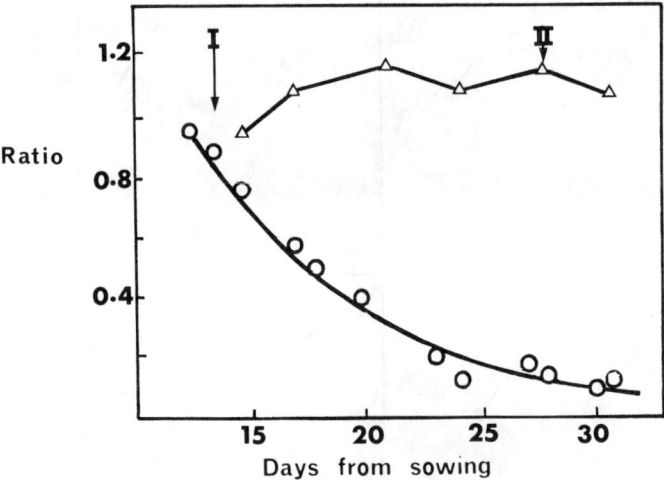

Fig. 1. Synthesis of chloroplast proteins in expanded wheat leaves. Each few days, after the second leaf reached full expansion, groups of these leaves were detached and fed L-^{14}C-valine for 60 min in the light. The specific activity of Fraction 1 protein, the total soluble protein and the lamellae proteins at the end of the feeding period was determined. The specific activity of Fraction 1 protein (O) and of the lamellae proteins (Δ) is expressed as a ratio to the specific activity of the total soluble proteins. Arrows indicate harvests corresponding to stage I and stage II leaves.

The declining ratio of incorporation into fraction 1 to fraction 2 proteins with leaf age and the near equal specific activities of these proteins in the newly expanded leaves, (Tables 1, 2) suggests that the pattern of incorporation is dependent on the development of the whole leaf rather than the individual cells. If this were so age-related differences between the proteins along the length of the leaf would not occur. Experiment confirms this and establishes that into each protein fraction, including fraction 1 protein, there is appreciable incorporation in fully expanded cells when these cells form part of a newly-expanded leaf. The important point appears to be the age of the leaf and not that of the cell. Such circumstances suggest hormonal control.

The protein content of detached leaves

The facts that proteins are hydrolyzed in detached leaves, and that kinetin treatment maintains the protein content of wheat leaves are well known. The points we would like to emphasise are, that fraction 1 protein declines as rapidly in older leaves (stage II, Fig. 1) as in younger leaves (stage I, Fig. I), and that fraction 2 proteins are lost about as rapidly as fraction 1 proteins when older (stage II) leaves are detached (Fig. 2). There is no correlation between the incorporation rate at harvest, and protein loss in the post-harvest period. This fact makes it most unlikely that protein loss results from a declining rate of synthesis coupled with a constant rate of catabolism. It is also clear (Fig. 2a, 2c) that kinetin treatment maintains the content of fraction 1 protein whether or not incorporation into this protein is rapid when the leaves are detached.

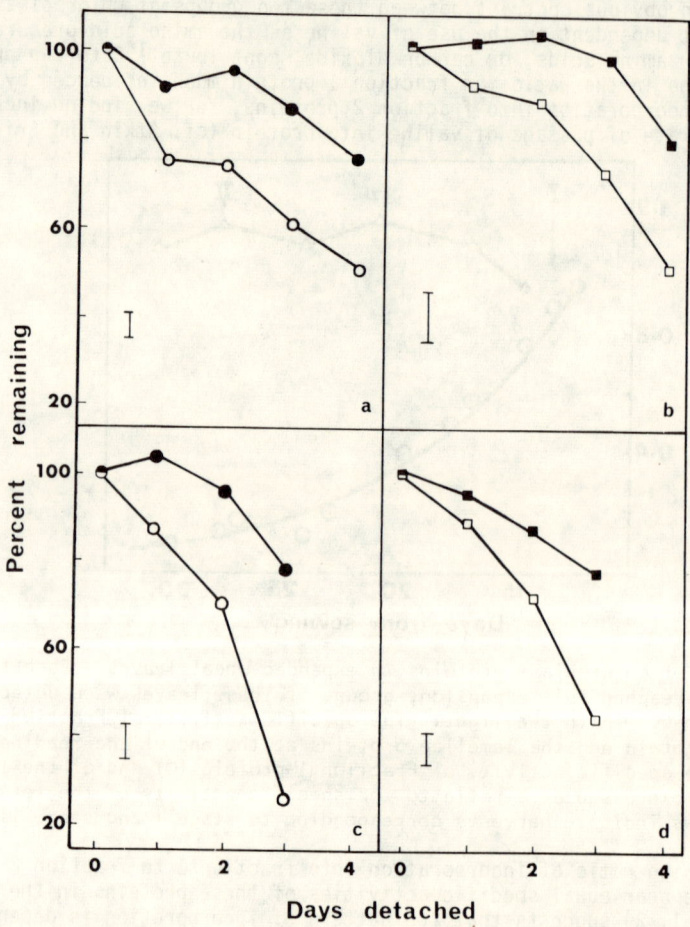

Fig. 2.

The effect of kinetin treatment on the content of Fraction I protein and soluble protein in detached stage I and stage II leaves. Leaves held in water (open symbols) or in 10^{-4} M kinetin solution (closed symbols) were analyzed at daily intervals for their content of Fraction I protein, total buffer-extractable protein and total nitrogen. Fraction I (Fig. 2a, 2c) and soluble protein (Fig. 2b, 2d) were calculated relative to total nitrogen, and their content on each day expressed as a percentage of that on day 0. Fig. 2a and 2b are stage I leaves, and Fig. 2c and 2d, stage II leaves. Bars show L.S.D. (P = 0.05).

Table 2. The synthesis of fraction 1 protein, and fraction 2 proteins in detached stage I leaves. Immediately after detachment, and then at daily intervals, leaves were fed L-^{14}C-valine (270 µC/µmole) for 1 hr, and the specific activities of fraction 1 protein and total soluble protein determined. The specific activity of fraction 2 proteins was calculated from these specific activities and the measured portion of fraction 1 protein in the total soluble protein. The L-^{14}C-valine fed to 8 leaves was 5µC on day 0, 10µC on days 1 and 2, 15µC on day 3 and 20µC on day 4.

Days Detached	Specific Activity (dpm x 10^{-3}/mg)		Fraction 1 / Fraction 2
	Fraction 1	Fraction 2	
0	38.4	35.0	1.1
1	26.6	47.0	0.6
2	9.4	27.1	0.3
3	3.9	73.4	0.05
4	0.0	60.7	0.0

Table 3. The effect of kinetin treatment in the relative synthesis of fraction 1 and fraction 2 proteins in stage I leaves. Leaves were fed L-^{14}C-valine (270 mC/mmole, 1.0 µC/leaf on day 0, and 1.5 µC/leaf on day 2) Errors are standard errors.

Days detached	Kinetin	Specific Activity (dpm x 10^{-3}/mg)	
		Fraction 1	Fraction 2
0	-	114.2 ± 4.0	114.0 ± 6.0
2	-	8.4 ± 0.3	43.1 ± 1.6
2	+	56.3 ± 1.6	160.0 ± 4.8

Table 4. The effect of kinetin treatment on the relative synthesis of fraction 1, fraction 2 and lamellae proteins in detached stage II leaves. Leaves were fed DL-^{3}H-valine (G) (5.5 mC/mmole, 7µC/leaf) for 2 hr. Three batches of 12 leaves each were analyzed for each treatment. Results are means with their standard errors.

Days detached	Kinetin	Specific Activity (dpm x 10^{-3}/mg)		
		Fraction 1	Fraction 2	Lamellae
0	-	0.33 ± 0.04	6.30 ± 0.23	3.30 ± 0.18
2	-	0.08 ± 0.02	5.80 ± 0.38	1.19 ± 0.23
2	+	0.12 ± 0.02	6.82 ± 0.39	2.21 ± 0.16

Kinetin and amino acid incorporation in detached leaves

When we measured valine incorporation into proteins in detached stage I leaves (Table 2), we found that incorporation into fraction 1 protein declined, relative to that into fraction 2 proteins. Again the same pattern emerged if carbon dioxide was used as precursor, and there was no evidence of an increase in the overall incorporation rate with age. We concluded that the synthesis of fraction 1 protein declined rapidly as the leaves senesced.

If the detached leaves were treated with kinetin, the decline in incorporation into fraction 1 protein occurred, but it was much less than it was in leaves given no kinetin (Table 3). Again, feeding $^{14}CO_2$ yielded a similar result. We accept this as evidence that kinetin treatment maintains the synthesis of fraction 1 protein, and limits the normal age-related change in the pattern of protein synthesis in the leaves.

When older leaves (stage II, Fig. 1) are detached, the already low incorporation into fraction 1 protein declines further. The larger effect, however, is the incorporation into the chloroplast lamellae proteins declines, relative to incorporation into the soluble proteins. We do not find any effect of kinetin treatment on incorporation into fraction 1 protein, but it does serve to limit the decline with the lamellae proteins (Table 4).

These experiments show that kinetin treatment of leaves tends to maintain the pattern of protein synthesis in the leaves in the form occurring when treatment was commenced. There is no evidence that a previous decline can be reversed, for fraction 1 protein synthesis cannot be increased at any stage by treating leaves with kinetin. It also seems clear that kinetin does not maintain the content of the protein by an effect on synthesis, for it maintains fraction 1 protein in stage I leaves in which it does maintain the synthesis of protein, and it also maintains fraction 1 protein in stage II leaves in which there is no effect on synthesis.

The changing pattern of synthesis of fraction 2 proteins with age:

The results we have given distinguish the pattern of synthesis of fraction 1 protein from that of the other soluble proteins taken collectively. To investigate differences within the soluble proteins we have combined the techniques of dual-labelling of proteins (^3H and ^{14}C) and protein fractionation by gel electrophoresis. Details of a number of experiments are in the legend to Fig. 3.

Relating incorporation in freshly-detached stage I leaves (^3H) to that in these leaves detached for 2 days (^{14}C) (Fig. 3b), we find that while many of the fraction 2 proteins have comparable $^3H/^{14}C$ ratios (normalized ratio about 1.0) several of the bands have distinctly low ratios, while fraction 1 protein has a very large ratio. The latter reflects the relatively large decline in the synthesis of fraction 1 protein in the detached leaf, and the figure shows that none of the fraction 2 proteins have comparable ratios. It would appear that among the more prominent soluble proteins this pattern of synthesis with age is peculiar to fraction 1 protein.

Fig. 3b shows that a number of the fraction 2 protein bands have unusually low ratios. Relative to the normal pattern, synthesis in these areas is particularly well maintained in the detached leaves. It may well be that there are proteins in these regions which are synthesized at an increasing rate during senescence. In Fig. 3c, normalized ratios comparing synthesis in newly-detached stage I leaves (^3H), and detached leaves kept two days in kinetin solution (^{14}C) are presented. Kinetin treatment has not prevented a decline in incorporation into fraction 1 protein (cf. Table 3), but it has prevented the development of very low ratios among the fraction 2 proteins. One conclusion might be that the synthesis of fraction 1 protein is less well maintained by this kinetin treatment than the synthesis of fraction 2 proteins (this would be so if the low ratios in Fig. 3b result from a decline in incorporation in the normal band, but not in some others). Another interpretation would be that the decline in incorporation into fraction 1 protein is an early stage of senescence, which has occurred to some degree even in the presence of kinetin; the differences between fraction 2 proteins occur at a later stage and have not developed in these leaves. The latter interpretation is indicated by Fig. 3d which shows that the decrease in fraction 1 protein occurs between stages I and II (Fig. 1) unaccompanied by the large differences between fraction 2 proteins which are seen in Fig. 3b.

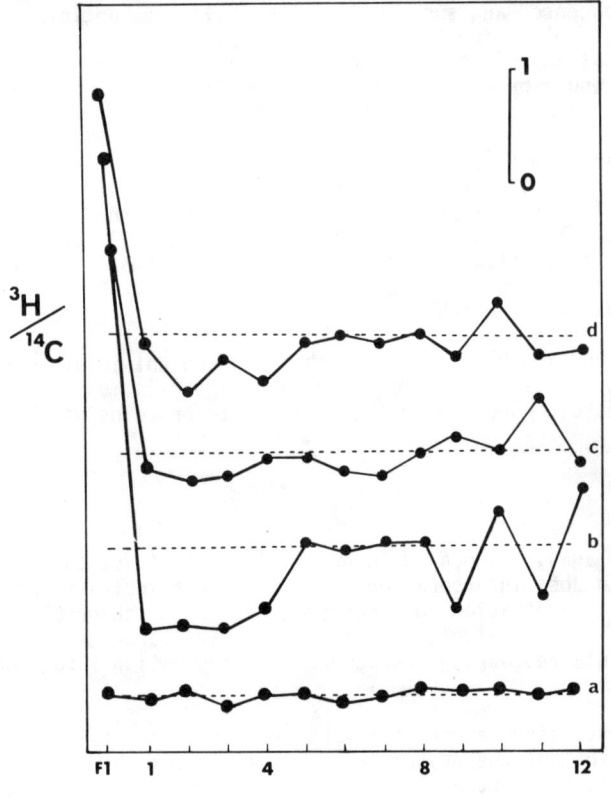

Fig. 3. **Gel slice**

The relative rates of synthesis of Fraction 1(F1), and certain Fraction 2 proteins (1-12) during senescence. The figure summarizes 4 experiments in which DL-^{3}H-valine was fed to one set of leaves, and L-^{14}C-valine to a second set. After the valine had been fed for 2 hr the two sets of leaves were combined, and their soluble proteins separated by disc electrophoresis (Davis 1964). Proteins were stained with amido black 10b in 7% (V/V) acetic acid and radioactivity not in protein removed by electrophoresis and diffusion. Proteins bands were excised and hydrolyzed (6N HCl, 105°, 24 hr) and ^{3}H and ^{14}C measured in the hydrolyzates. The ratio of ^{3}H/^{14}C in each band was normalized to the mean of all bands in the control gel (a), and to band 8 in all other cases.

For (a) ^{3}H-valine or ^{14}C-valine was fed to two similar groups of stage I leaves. In (b) ^{3}H-valine was fed to freshly-detached stage I leaves, and ^{14}C-valine to stage I leaves detached and held transpiring water for 2 days. In (c) ^{3}H-valine was fed to freshly-detached stage I leaves, and ^{14}C-valine to stage I leaves which had transpired 10^{-4}M kinetin for 2 days. For both (b) and (c) sowings made 2 days apart were used to give appropriate leaves. In (d) ^{3}H valine was fed to freshly-detached stage I leaves and ^{14}C-valine to freshly detached stage II leaves.

IV. CONCLUSION

There is now ample evidence that a regulation of protein breakdown is involved in the maintenance of the protein content of leaves by growth factor treatment. There is, however, little evidence as to how this regulation might be exercised (Shibaoka and Thimann, 1970). We have evidence that there are shifts in the direction of protein synthesis during senescence, and such shifts might have been anticipated from the change in ultra-structure and function which occur during senescence. It remains to be demonstrated that any of these shifts are essential in a positive sense to the processes of macromolecule breakdown and product mobilization which characterize leaf senescence.

Kinetin treatment limits or delays these changes in protein synthesis, and in this it appears to be more effective in regulating fraction 2 than fraction 1 protein synthesis. Fraction 1 protein appears to be synthesized on the 70S ribosomes of the chloroplast stroma (Smillie et al., 1967, Chen and Wildman, 1970), and a poorer binding of cytokinins to plastid than to cytoplasmic ribosomes (Berridge et al., 1970) may correlate with our results. Direct evidence relating cytokinin binding to ribosome function is, however, lacking.

If cytokinins function by binding to ribosomes and not in other ways, the delay in senescence-related shifts in fraction 2 protein synthesis must be exercised through translation level control. Such control may extend to proteins which regulate peptidase function.

V. SUMMARY

As a wheat leaf ages, amino acid incorporation into fraction 1 protein declines much more rapidly than does incorporation into other protein fractions. At a later stage of senescence, incorporation into the proteins of the chloroplast lamellae declines. Among the smaller soluble proteins, there are large differences in how well synthesis is maintained during senescence. Incorporation into some proteins may increase with leaf age.

There is no correlation between the apparent turnover rate of a protein at harvest, and the net loss of the protein when a leaf is detached. Kinetin maintains proteins in detached leaves whether or not they were subject to rapid turnover when the leaves were harvested. Kinetin treatment appears to maintain the protein content of detached leaves by limiting protein breakdown.

However, kinetin treatment does limit the shifts in the relative incorporation rates which characterize leaf senescence. In this, it more effectively prevents shifts amongst the smaller proteins than it does the decline in incorporation into fraction 1 protein. It appears to influence synthesis on both chloroplast and cytoplasmic ribosomes, but may be more effective in regulating the latter.

VI. REFERENCES

ATKIN, R.K. and B.I.S. SRIVASTAVA (1970). Studies on protein synthesis by senescing and kinetin-treated barley leaves. Physiologia Pl. 23, 304-315.
BERRIDGE, M.V., R.K. RALPH and D.S. LETHAM (1970). The binding of kinetin to plant ribosomes. Biochem. J. 119, 75-84.
BRADY, C.J., B.D. PATTERSON, HENG FONG TUNG and R.M. SMILLIE. (1971). Protein and RNA synthesis during ageing of chloroplasts in wheat leaves in "Autonomy and Biogenesis of Mitochondria and Chloroplasts", North-Holland Publishing Co. Amsterdam, 453-465.
CHEN, J.L. and S.G. WILDMAN (1970). Free and membrane-bound ribosomes, and nature of products formed by isolated tobacco chloroplasts incubated for protein synthesis. Biochim. biophys. Acta. 209, 207-219.
DAVIS, B.J. (1964). Disc electrophoresis. II Method and application to human serum proteins. Ann. N.Y. Acad. Sci. 121, 404-427.
HARDWICK, K. and H.W. WOOLHOUSE (1967). Foliar senescence in Perilla frutescens (L) Britt. New Phytol. 66, 545-552.
KURAISHI, S. (1968). The effect of kinetin on protein level of Brassica leaf disks. Physiologia Pl. 21, 78-83.

LOWRY, O.H., N.J. ROSEBROUGH, A.L. FARR and R.J. RANDALL, (1951). Protein measurement with the Folin phenol reagent. J. biol. Chem. 193, 265-275.

MANS, R.J. and G.D. NOVELLI (1961). Measurement of the incorporation of radioactive amino acids into protein by a filter-paper disk method. Archs. Biochem. Biophys. 94, 48-53.

MIZRAHI, Y-, J. AMIR and A.E. RICHMOND (1970). The mode of action of kinetin in maintaining the protein content of detached *Tropaeolum majus* leaves. New Phytol. 69, 355-361.

NOBEL, P.S. (1967). A rapid technique for isolating chloroplasts with high rates of endogenous photophosphorylation. Pl. Physiol. Lancaster, 42, 1389-1394.

PATTERSON, M.S. and R.C. GREENE (1965). Measurement of low energy beta-emitters in aqueous solution by liquid scintillation counting of emulsions. Analyt. Chem. 37, 854-857.

RAYMOND, S. (1962). A convenient apparatus for vertical gel electrophoresis. Clin. Chem. 8, 455-470.

SHIBAOKA, H. and K.V. THIMANN, (1970). Antagonisms between kinetin and amino acids. Experiments on the mode of action of cytokinins. Pl. Physiol. Lancaster, 46, 212-220.

SMILLIE, R.M., D. GRAHAM, M.R. DWYER, A. GRIEVE and N.F. TOBIN (1967). Evidence for the synthesis *in vivo* of proteins of the Calvin cycle and of the photosynthetic electron-transfer pathway on chloroplast ribosomes. Biochem. Biophys. Res. Commun. 28, 604-610.

TAVARES, J.E. and H. KENDE (1970). The effect of 6-benzylamino-purine on protein metabolism in senescing corn leaves. Phytochemistry 9, 1763-1770.

TEWARI, K.K. and S.G. WILDMAN (1969). Function of chloroplast DNA. II Studies on DNA-dependent RNA polymerase activity of tobacco chloroplasts. Biochim. biophys. Acta 186, 358-372.

TREWAVAS, A. (1970). The turnover of nucleic acids in *Lemna minor*. Pl. Physiol. Lancaster 45, 742-751.

TROWN, P.W. (1965). An improved method for the isolation of carboxydismutase. Probable identity with Fraction 1 protein and the protein moiety of protochlorophyll holochrome. Biochemistry, N.Y., 4, 908-918.

WATKIN, J.E. and R.A. MILLER (1970). Quantitative measurement of protein in disc electrophoresis by direct ultraviolet absorbance. Analyt. Biochem. 34, 424-435.

WILDMAN, S.G. and J. BONNER (1947). The proteins of green leaves. I Isolation, enzymatic properties and auxin content of spinach cytoplasmic protein I. Archs. Biochem. Biophys. 14, 381-413.

EFFECTS OF SENESCENCE AND HORMONE TREATMENT ON THE β-1,3-GLUCAN HYDROLASE IN *Nicotiana glutinosa* LEAVES

A.E. Moore and B.A. Stone

Russell Grimwade School of Biochemistry, University of Melbourne, Parkville, Victoria, Australia, 3052

I. INTRODUCTION

It has been known for some time that enzymes which hydrolyse β-1,3-glucans are widely distributed in higher plants (Clarke and Stone, 1962; Mandels and Reese, 1963; Barras, Moore and Stone, 1969). During a survey of a number of plant tissues, it was found that leaves of *Nicotiana glutinosa* posses a high level of β-1,3-glucan hydrolase (Clarke and Stone, 1962) and this enzyme has now been purified and its physical and biochemical characteristics examined (Moore and Stone, 1972 a,b). Some of its properties are summarised in Table 1.

Table 1. Properties of *N. glutinosa* β-1,3-Glucan Hydrolase

MOLECULAR WEIGHT	45,000
pH OPTIMUM	5.0
ISOELECTRIC POINT	4.87
pH STABILITY	Stable between 4 and 8.5
TEMPERATURE STABILITY	Inactivated in 10 min at 65°
SUBSTRATE SPECIFICITY	Hydrolyses β-glucans with runs of 1,3-linkages. Laminaritetraose is the smallest substrate.
K_m (CM-pachyman)	0.6% w/v, (0.13 mM)
ACTIVATORS	None demonstrated. B.S.A. stabilizes the purified enzyme.
INHIBITORS	Cu^{2+}, Hg^{2+}, phenylmercurinitrate, carbodiimide, 2-OH-5-NO_2-benzyl bromide, N-acetyl imidazole, N-bromo succinimide, iodine, leucoanthocyanidin from *Lespedeza cuneata*.

II. RESULTS

The *Nicotiana* enzyme is highly specific for β-1,3-glucosidic linkages in glucans, which are attacked with an endo-action pattern. In this way a series of oligosaccharides of D.P. 2-7 is formed from linear β-1,3-glucans such as pachyman, paramylon and laminarin. In contrast to the group of β-1,3-glucan endo-hydrolases of the *Rhizopus arrhizus* type (Barras, Moore and Stone, 1969), the *Nicotiana* hydrolase does not produce oligosaccharides from the β-1,4;1,3-glucans of oat and barley endosperm cell walls. However, it has been found that the *Nicotiana* enzyme does cause a very limited hydrolysis of these mixed-linked glucans (Moore and Stone, 1972 b), probably due to the presence of a few runs of 1,3-linked glucose residues of sufficient length to satisfy the specificity requirements of the enzyme.

Location of the enzyme in the plant and in sub-cellular fractions from leaves

The enzyme is found in all parts of the *Nicotiana* plant, the activities in roots and in leaf lamina being the highest on a fresh weight basis, whilst activity in the petioles is highest on a soluble protein basis (Moore and Stone, 1972c).

The location of the hydrolase within the cells of the leaf has also been investigated, as outlined below, to determine whether it is present in the cytoplasm, in a sub-cellular particle or associated with the cell walls.

The cell wall fraction from leaf homogenates was separated from chloroplasts and other sub-cellular components by repeated density gradient centrifugation using a modification of the method of Bean and Ordin (1961). No more than 4% of the total activity was found in the cell wall fraction, whether the walls were prepared from young green, mature green or yellow leaves.

When a post-chloroplast fraction from a leaf homogenate, prepared by sand-grinding in an osmotically controlled medium, was fractionated into particulate ("mitochondrial") and soluble fractions by the method of Semadeni (1967), virtually all the hydrolase was recovered in the soluble fraction. In the same experiment variable but significant proportions of an acid protease, which is known to be present in the spherosomes (Matile *et al*, 1965; Balz, 1966), were found in the particulate fraction.

These findings indicate that the hydrolase is neither firmly bound to the cell walls nor contained in spherosomal particles, but the possibility remains that it may be present either in the cytoplasm or in larger fragile organelles such as the vacuoles and provacuoles. In maize root tip cells these organelles have been demonstrated to contain acid hydrolases and it has been suggested that these enzymes may also be present in the larger vacuoles of mature parenchymatous tissue (Matile, 1968).

Changes in the level of hydrolase during senescence

The level of the enzyme in the leaves has been followed during their maturation and senescence and the activities compared on the basis of fresh weight of leaf (Moore and Stone, 1972c). The level of the enzyme was lowest in the young top leaves and mature green leaves, rose 14-fold to a maximum in yellow-green leaves and was again low in very yellow leaves. In contrast, the β-1,4-glucan hydrolase level which was very low in young green leaves decreased progressively during maturation and senescence. The pattern of increase in β-1,3-glucan hydrolase level observed in leaves *in situ*, also occurred in detached leaves and in leaf discs floated on water in the dark. In each case the maximum level was reached in about four days.

It is possible that the rise in hydrolase could be due either to an increase in amount or activity of the enzyme present in young leaves or to the production of a different hydrolase. To distinguish between these possibilities some characteristics of the enzymes from green and yellow leaves were compared by electrophoresis in polyacrylamide gel and also with respect to their action patterns. No physical or enzymological differences were apparent.

Experiments have also been performed to decide whether the increase in activity of the hydrolase during yellowing is due to the removal of an enzyme inhibitor present in green leaves or to the presence of an enzyme activator in yellow leaves. The activities of extract from yellow and green leaves were compared with the activity of mixed extracts or of extracts prepared from mixtures of green and yellow leaves, but no evidence of interactions involving inhibition or activation was found (Moore and Stone, 1972c).

Changes in enzyme level following hormone treatment of leaf discs

Using leaf discs the influence of the hormones GA, IAA, kinetin and ABA on the patterns of change in chlorophyll, soluble protein and β-1,3-glucan hydrolase was examined (Moore and Stone, 1972c). Neither GA (50µM) nor IAA (10µM) caused significant alterations in the patterns of change of these components. On the other hand, in discs treated with kinetin (50µM) chlorophyll was maintained at a level significantly higher than that of the controls, whilst the protein, although decreasing, did not drop to the very low value of the control. On treatment with kinetin there was a much slower rise in hydrolase which at four days had reached a value only one third that of the control. ABA (190µM) also altered the pattern of change of chlorophyll and protein, bringing about a more rapid decrease in their concentrations in the leaf discs. However these changes were not accompanied by a rapid rise in hydrolase activity and at four days the level was only one third of the control.

Changes in hydrolase level in virus infection

It has been found that a very substantial increase in activity of the hydrolase accompanies the development of symptoms during infection of *N. glutinosa* by TMV, tomato spotted wilt virus and broadbean wilt virus (Moore and Stone, 1968, 1972d). In leaves infected by TMV and tomato spotted wilt virus, which cause local lesions, there is a high concentration of the enzyme in and around the lesions and a slight increase in enzyme level in the remainder of the leaf.

III. DISCUSSION

Senescence in leaves is marked by a mobilization of protein, nucleic acids and polymeric carbohydrates (Leopold, 1964). The agents responsible for these catabolic changes are the hydrolases specific for these polymers and increases in their activities have been detected during leaf senescence in several species (McHale and Dove, 1968; Atkin and Srivastava, 1969; Uvardy *et al*, 1969; Srivastava and Ware, 1965; Anderson and Rowan, 1966). The β-1,3-glucan hydrolase of *Nicotiana glutinosa* is a further example of such a specific hydrolase which increases in activity during senescence.

The increase in activity of these enzymes may arise in a number of ways including synthesis of new enzyme protein, the liberation of a bound inactive form of the enzyme, the activation of an inactive pro-enzyme, a decrease in the rate of enzyme degradation, the relief of a direct inhibition or the activation of pre-existing enzymes. Experiments have shown that the last two possibilities are not responsible for the rise in activity of the *Nicotiana* hydrolase (Moore and Stone, 1972c). In other systems the effects of protein synthesis inhibitors have suggested that *de novo* enzyme synthesis is responsible for the increase in level of certain leaf enzymes (Atkin and Srivastava, 1970; McHale and Dove, 1968) and it is of interest that Abeles and Forrence (1970) have produced some evidence in favour of *de novo* synthesis of a similar hydrolase during abscission of *P. vulgaris* petioles.

The progressive metabolic changes associated with senescence are controlled by a balance of hormones which may act in various ways, steps concerned with nucleic acid and protein synthesis being particularly important. It is well established that exogenously added plant growth regulators can also influence such metabolic events (Glasziou, 1969). In *N.glutinosa* leaf discs, added IAA and GA did not change the course of senescence under the conditions of our experiments, although a retardation effect occurs in some other species (Harada, 1966). On the other hand, in *N.glutinosa* leaf discs senescence was retarded by kinetin and was accelerated by ABA, these findings being similar to those observed in other plants (Aspinall *et al*., 1967; El-Antably *et al*., 1967; Sankhla and Sankhla, 1968). In *N. glutinosa* the increase in hydrolase level in leaf discs treated with IAA, GA or kinetin showed a correlation with the rate of onset of senescence, but no such correlation was found on treatment with ABA which both promoted senescence and suppressed hydrolase formation. It has been noted that ABA may bring about either an increase or a decrease in the levels of particular enzymes in other systems (Addicott and Lyon, 1969) and it is possible that such a specific action may be responsible for the suppression of the increase in hydrolase which normally occurs in senescence. The role of ethylene in the production of the β-1,3-glucan hydrolase has not been investigated but Abeles and Forrence (1970) have reported that a similar hydrolase in *Phaseolus vulgaris* leaves rises rapidly on ethylene treatment.

In all treatments of *N. glutinosa* leaf discs, with the exception of those involving ABA, a relationship has been shown to exist between the onset of senescence and the rise in hydrolase. If the hydrolase is of functional significance in senescence several roles may be considered. One would be the removal of the β-1,3-glucan, callose, from the pores in the sieve plates of the phloem and from lateral pit fields. Callose has been detected in young *Elodea* leaves in the epidermal pits and sieve tubes (Currier and Shih, 1968) and in the primary leaves of *Phaseolus vulgaris* in mesophyll pits, sieve plates and obliterated protophloem (McNairn and Currier, 1968). In these situations callose is commonly believed to have a blocking function (but see Eschrich, 1970) and its removal could facilitate the translocation of mobilized leaf components during senescence. However, a systematic study of the occurrence and distribution of callose in leaves during senescence is needed before such a functional relationship can be established. Abeles and Forrence (1970) have reported that the amount of callose in the sieve tubes of the phloem in the pulvinus region of *Phaseolus vulgaris* petioles correlates with the level of the β-1,3-glucan hydrolase during abscission which is consistent with the idea that the hydrolase may regulate the amount of callose on the sieve plates.

It has been noted earlier that the *Nicotiana* hydrolase, in addition to extensively hydrolysing β-1,3-glucan substrates, may effectively lower the degree of polymerization of the mixed-linked β-glucans which are present in some plant cell walls. Such depolymerization of cell wall glucans could cause an alteration in wall structure, which could increase wall permeability or increase the accessibility of other constituents of the wall to enzymic attack. There is no information on chemical changes in leaf cell walls during the mobilization process of senescence, but in electron micrographs of wheat leaf cell sections (Mittelheuser and van Steveninck, 1972) cell walls were still apparent at the later stages of senescence although greatly reduced in thickness, suggesting that matrix components may have been removed. Comparable changes in endosperm cell wall structure are known to occur during the mobilization of endosperm cell materials during germination (Preece and Hoggan, 1957), although in this case there is a general increase in polysaccharide hydrolases including the β-1,4-glucan hydrolase, in contrast to the finding in senescent leaves.

In any discussion of the role of the *Nicotiana* hydrolase its subcellular location is relevant. The experimental results have indicated that the hydrolase is located either in the cytoplasm or in an easily disrupted, membrane-bound organelle such as a vacuole. When enzymes are enclosed by a vacuolar membrance they can be effective only after release of the enzyme by membrane destruction or after introduction of the substrate into the vacuole. Shaw and Manocha (1965) have shown that the membranes of the vacuole vesiculate during the later stages of senescence in wheat leaves.

Finally it should be made clear that it is still uncertain whether all leaf cell types contain the hydrolase and whether the increase is general throughout the leaf or occurs only in certain cell types. Such information is needed before a description of the functional significance of the hydrolase can be made.

IV. SUMMARY

The general properties and specificity of a β-1,3-glucan hydrolase from the leaves of *Nicotiana glutinosa* are described. The enzyme is found in all parts of the plant with highest activity in the leaves and roots. The hydrolase is not associated with cell walls or with a particulate fraction containing spherosomes. The level of hydrolase in young leaves is low but rises 14-fold to a maximum in yellow-green leaves. The enzyme present in senescent leaves appears to be the same as that found in young green leaves. No evidence has been obtained to suggest that the rise in activity is due to the removal of an inhibitor as the leaves senesce or to the presence of an activator in the yellow-green leaves. In floating leaf discs, exogenously applied IAA

(10 µM) or GA (50µM) does not alter the patterns of loss of chlorophyll and soluble protein or the increase in hydrolase found in control discs. Kinetin (50µM) retarded both senescence and the increase in hydrolase activity, whilst ABA (190µM) accelerated senescence and also prevented the increase in hydrolase. Possible roles for the enzyme in senescent leaves are discussed.

V. REFERENCES

ABELES, F.B. and L.E. FORRENCE (1970). Temporal and hormonal control of β-1,3-Glucanase in *Phaseolus vulgaris* L. Pl. Physiol., Lancaster 45, 395-400.
ADDICOTT, F.T. and J.L. LYON (1969). Physiology of Abscisic Acid and related substances. Ann. Rev. Pl. Physiol. 20, 139-164.
ANDERSON, J.W. and K.S. ROWAN (1966). The effect of 6-Furfuryl-aminopurine on senescence in tobacco-leaf tissue after harvest. Biochem. J. 98, 401-404.
ASPINALL, D., L.G. PALEG and F.T. ADDICOTT. (1967). Abscisin and some hormone-regulated plant responses. Aust. J. Biol. Sci. 20, 869-882.
ATKIN, R.K. and B.I.S. SRIVASTAVA (1969). The changes in soluble protein of excised barley leaves during senescence and Kinetin treatment. Physiologia Pl. 22, 742-750.
ATKIN, R.K. and B.I.S. SRIVASTAVA (1970). Studies on protein synthesis in senescing and kinetin-treated barley leaves. Physiologia Pl. 23, 304-315.
BALZ, H.P. (1966). Intracellulare Lokalisation und Funktion von hydrolytischen Enzymen bei Tabak. Planta 70, 207-236.
BARRAS, D.R., A.E. MOORE and B.A. STONE (1969). Enzyme — substrate relationships among β-Glucan hydrolases, *in* "Cellulases and their Applications" (Ed. R.F. Gould) pp. 105-138. American Chemical Society, Washington D.C. 1969.
BEAN, R.C. and L. ORDIN (1961). A study of procedures for isolation and fractionation of plant cell walls. Analyt. Biochem. 2, 544-557.
CLARKE, A.E. and B.A. STONE (1962). β-1,3-Glucan hydrolases from the grape vine (*Vitis vinifera*) and other plants. Phytochemistry 1, 175-188.
CURRIER, H.B. and C.Y. SHIH (1968). Sieve tubes and callose in *Elodea* leaves. Am. J. Bot. 55, 145-152.
EL-ANTABLY, H.M.M., P.F. WAREING and J. HILLMAN (1967). Some physiological responses to D, L-Abscisin (Dormin). Planta 73, 74-90.
ESCHRICH, W. (1970). Biochemistry and fine structure of phloem in relation to transport. Ann. Rev. Pl. Physiol. 21, 193-214.
ESCHRICH, W. (1956). Kallose. Protoplasma 47, 487-530.
GLASZIOU, K.T. (1969). Control of enzyme formation and inactivation in plants. Ann. Rev. Pl. Physiol. 20, 63-88.
HARADA, H. (1966). Retardation of the senescence of *Rumex obtusifolius* leaves by growth retardants. Pl. Cell Physiol., Tokyo, 7, 701-703.
LEOPOLD, A.C. (1964). "Plant Growth and Development". McGraw-Hill, New York, 1964.
McHALE, J.S. and L.D. DOVE (1968). Ribonuclease activity in tomato leaves as related to development and senescence. New Phytol. 67, 505-515.
McNAIRN, R.B. and H.B. CURRIER (1968). The influence of boron on callose formation in primary leaves of *Phaseolus vulgaris*. Phyton. B. Aires. 22, 153-158.
MANDELS, M. and E.T. REESE (1963). Enzymatic hydrolysis of β-Glucans, *in* "Advances in Enzymic Hydrolysis of Cellulose and Related Materials" (Ed. E.T. Reese) pp. 197-234 Pergamon Press, Oxford, 1963.
MATILE, P. (1968). Lysosomes of root tip cells in corn seedlings. Planta 79, 181-196.
MATILE, P., J.P. BALZ, E. SEMADENI and M. JOST (1965). Isolation of spherosomes with lysosome characteristics from seedlings. Z. Naturf. 20B, 693-698.
MITTELHEUSER, C.J. and R.F.M. VAN STEVENINCK (1972). Ultrastructural changes in naturally senescing wheat leaves compared with changes induced by (RS)-Abscisic Acid and Kinetin, *in* "Plant Growth Substances, 1970" (this volume).
MOORE, A.E. and B.A. STONE (1968). The occurrence of a β-1,3-Glucan hydrolase in plants of *Nicotiana glutinosa* in normal and pathological states. Proc. Federation European Biochemical Societies, Prague p. 182.
MOORE, A.E. and B.A. STONE (1971a). β-1,3-Glucan hydrolase from *Nicotiana glutinosa* leaves. 1. Extraction, purification and physical properties. Biochim. biophys. Acta (in Press).

MOORE, A.E. and B.A. STONE (1972b). β-1,3-Glucan hydrolase from *Nicotiana glutinosa* Leaves. II. Specificity, action pattern and inhibitors. Biochim. biophys. Acta 258, 248-264.
MOORE, A.E. and B.A. STONE (1972c). Physiological studies on the β-1,3-Glucan hydrolase from *Nicotiana glutinosa* leaves. Planta (in Press).
MOORE, A.E. and B.A. STONE (1972d). Effect of infection with TMV and other viruses on the level of β-1,3-Glucan hydrolase in leaves of *Nicotiana glutinosa* L. Virology (Submitted for publication).
PREECE, I.A. and J. HOGGAN (1957). Carbohydrate modifications during malting. Proc. Euro. Brew. Conv. Copenhagen pp. 72-83.
SANKHLA, N. and D. SANKHLA (1968). Abscisin II - Kinetin interaction in leaf senescence. Experientia 24, 294-295.
SEMADENI, E.G. (1967). Enzymatic characterization of lysosome equivalents (Spherosomes) in corn seedlings. Planta 72, 91-118.
SHAW, M. and M.S. MANOCHA (1965). Fine structure in detached, senescing wheat leaves. Can. J. Bot. 43, 747-755.
SRIVASTAVA, B.I.S. and G. WARE (1965). The effect of Kinetin on nucleic acids and nucleases of excised barley leaves. Pl. Physiol. Lancaster, 40, 62-64.
UVARDY, J., G.L. FARKAS and E. MARRÉ (1969). On RNase and other hydrolytic enzymes in excised *Avena* leaf tissue. Pl. Cell Physiol. Tokyo 10, 375-386.

Ethylene Production and Biochemical Changes in Detached Leaves
of *Nymphoides indica*

D.C. Goldney and R.F.M. Van Steveninck

University of Queensland St. Lucia. Qld. 4067

I. INTRODUCTION

The importance of ethylene in growth regulation, including its ability to promote senescence, has been amply demonstrated (Pratt and Goeschl, 1969). Although the leaves of many plants have been shown to produce this gas (Hall *et al.*, 1957; Morgan and Hall, 1964; Maxie and Crane, 1967; Hallaway and Osborne, 1969) its role in leaf senescence is largely unknown.

The use of leaf discs and detached leaves in current studies on leaf senescence may well initiate increased ethylene production due to wounding (Lipetz 1970). Wound ethylene commonly causes biochemical changes in tissue in which it is produced (Imaseki *et al.*, 1968; Galliard *et al.*, 1968; Riov *et al.*, 1969). Added hormones and growth regulators may also modify leaf senescence through an increased rate of ethylene production (Fuchs and Liebermann, 1968; Maxie and Crane, 1967; Abeles and Holm, 1966; Chadwick and Burg, 1970).

The present study examines the importance of ethylene production in detached, fully expanded leaves of the water plant *Nymphoides indica* ageing in darkness for ten days. The following measurements were made.
1. The rate of production of ethylene.
2. Changes in total chlorophyll, soluble protein and in the activities of the following enzymes: Peroxidase (E.C. 1.11.1.7) Donor: H_2O_2 oxidoreductase; "Malic" enzyme (E.C. 1.1.1.40) L-malate: NADP oxidoreductase (decarboxylating); 6 phosphogluconate dehydrogenase (E.C. 1.1.1.43) 6 phospho-D-gluconate: NADP oxidoreductase, hereafter called 6 PG dehydrogenase; and RUDP carboxylase (E.C. 4.1.1.39) 3-phospho-D-glycerate carboxylyase (dimerizing).
3. The effect of 25 ppm added ethylene on the parameters listed under 2.
4. Changes in the respiratory rate after detachment.

II. MATERIALS AND METHODS

Plant Material

The tropical water plant *Nymphoides indica* (family Menyanthaceae) originating from a single clone, was grown in a glasshouse in plastic bins containing a 6 inch layer of sandy loam mixed with a slowly released complete fertilizer. The crown of the plant was submerged under 6 inches of water. Leaves were used when fully expanded, approximately 9 days after unfolding at the water surface. Leaves were cut in early morning light leaving a short petiolar section attached, and immediately transferred in water to the laboratory where they were placed in darkness at 28°C in an incubator flushed continuously with humidified air, scrubbed free of ethylene by passage through a purafil column.

Added ethylene at 25 ppm was monitored using gas chromatography. Three or four leaves were assayed for each reading, and experiments have been repeated at least twice.

Ethylene Determination and Identification

Ethylene was identified by mass spectrometry, by absorption to and release from mercuric perchlorate, and by comparing the retention times of pure ethylene with unknowns (Poapst et al., 1963) under a range of gas flow conditions using a Pye Series 101 GC single flame ionization unit.

The rate of ethylene production from whole detached leaves was determined in containers sealed with silicone rubber ports. Blanks were run to detect possible release of ethylene from non-biological sources. Containers were flushed with ethylene-free medical air at the end of each build-up period, generally in the 15-30 minute range.

Preparation of Enzyme Extracts

Leaves were washed, halved, blotted dry, weighed, wrapped lightly in silver foil and chilled for 10 minutes. All extraction operations were carried out at 0-4°C. One half of each leaf was used for peroxidase determination and the other for determination of the remaining enzymes, total chlorophyll and soluble protein. For peroxidase determinations leaf samples were finely chopped and ground in a blender at high speed for 1 minute with 0.1M tris buffer (pH 8.3), 1 mM EDTA, and 0.5% polyvinylpyrrolidone (PVP). For the other determinations PVP was replaced with 10 mM mercaptoethanol and 5 mM $MgCl_2$. The homogenates were filtered through "Chux" (non-woven plastic) cloth (Johnson & Johnson, Sydney), 1 ml removed for chlorophyll determination, and the remainder centrifuged at 25,000 g for 15 minutes. The supernatants were used as the "crude" enzyme mixture or 1 ml of the supernatant passed through a Sephadex G-25 column (12 x 0.7) cm, equilibrated with the appropriate buffer mixture and the protein collected in 1.3 ml.

Estimation of Enzyme Activity

Crude and chromatographed extracts were tested for the presence of high and low molecular weight inhibitors by serial dilution techniques and by boiling concentrated crude extracts and adding these to the assay mixtures. Initial rates of enzyme activities were determined at 30°C under optimal conditions, unless otherwise stated. Peroxidase was estimated at pH 7.0 following the oxidation of the dye p-phenylenediamine (Luck, 1963). RUDP carboxylase was assayed measuring the acid stable C^{14} incorporated in the reaction mixture as in the method of Slack and Hatch (1967). Six phosphogluconate dehydrogenase and malic enzyme were assayed at pH 7.7 and pH 7.4 respectively, by following the rate of formation of NADPH at 340 mμ as in the methods of Kornberg and Horecker, (1955) and Ochoa (1955).

Other determinations

Chlorophyll was determined by the method of Arnon (1949). Soluble protein was precipitated with 10% trichloracetic acid redissolved in alkali and determined by the method of Lowry (1951). Respiratory changes were measured using an infra-red gas analyser.

III. RESULTS

Ethylene Production

On excision, ethylene production increases from a basal rate of 3-5 mμl/g FW/hr to a maximum of 20-30 mμl/g FW/hr 3-6 hours later (Fig. 1), gradually declining to a rate of 1.5 mμl/g FW/hr over 10 days. The increased ethylene production is not initiated by bruising associated with detachment.

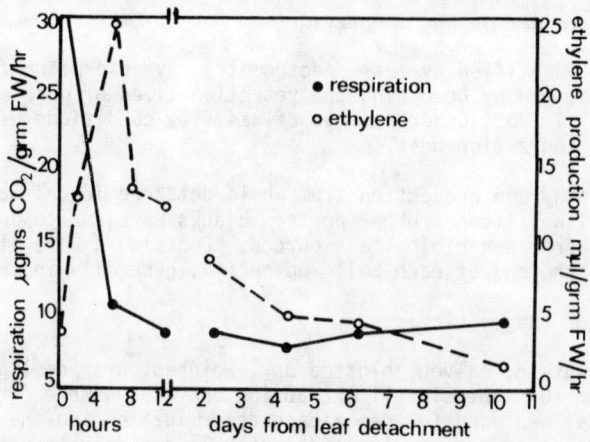

Fig. 1. The rate of respiration and ethylene production in detached leaves of *Nymphoides indica* floating on water in darkness at 28°C.

Respiratory Changes

The respiratory rate falls from 30 μg CO_2/g FW/hr to 10 μg CO_2/g FW/hr in 6 hours and continues to fall slightly after the peak of ethylene production has been reached (Fig. 1).

Changes in total chlorophyll and soluble protein

Total chlorophyll increases by 17% in 26 hours after detachment and then declines to below zero time level after 4½ days on water (Fig. 2). Soluble protein increases by 9% in 18 hours and falls below zero time levels after 2½ days (Fig. 2B).

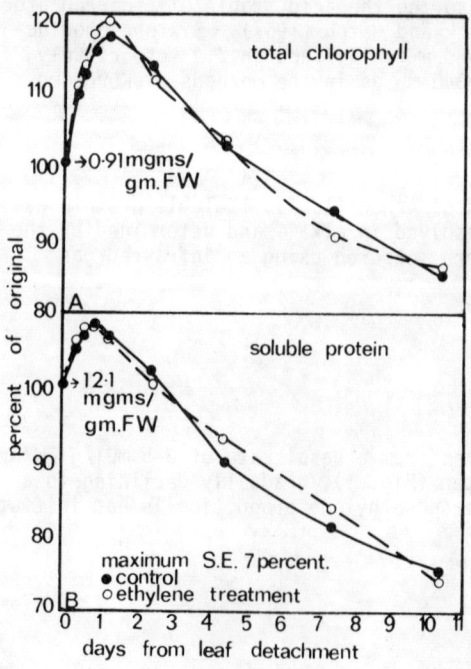

Fig.2. The effect of 25 ppm applied ethylene on total chlorophyll and soluble protein in detached leaves of *Nymphoides indica* in darkness at 28°C.

Changes in enzyme activities

Following detachment, the activities of both malic enzyme and 6 PG dehydrogenase increase rapidly (Fig. 3A, 3C). The increase of malic enzyme and 6 PG dehydrogenase activity after 54 hours ageing on water represents an increase of 315% and 144% respectively. The activities of both enzymes then decline, returning to the initial level after 10 days.

The activity of RUDP carboxylase (Fig. 3B) drops abruptly by 50% in 2 hours following detachment and then climbs to 75% of the initial activity, 18 hours after excision, remaining constant for 4-5 days and then gradually decreasing over the remainder of the experimental period.

Peroxidase activity (Fig. 3D) declines by 80% in 60 hours following excision, climbs to 40% of initial activity after 4½ days and then declines slowly during the remainder of the experimental period.

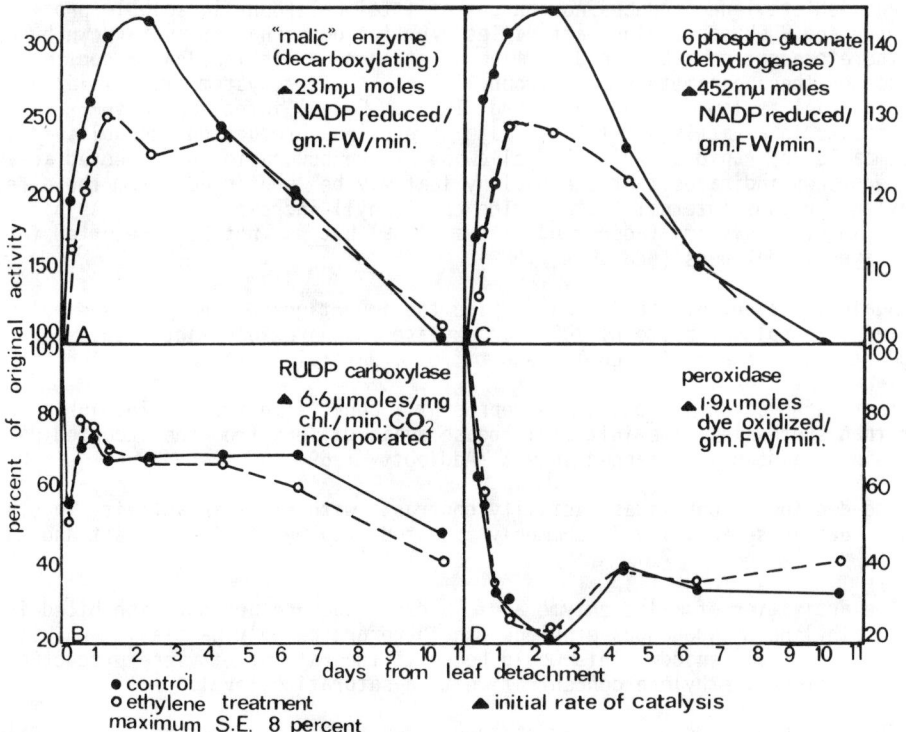

Fig. 3. The effect of 25 ppm applied ethylene on the activities of "malic" enzyme, 6 phosphogluconate dehydrogenase, RUDP carboxylase and peroxidase in detached leaves of *Nymphoides indica* in darkness at 28°C.

The effect of 25 ppm exogenous ethylene on detached leaves

Total chlorophyll, soluble protein, RUDP carboxylase and peroxidase activities are relatively unaffected by exogenous ethylene during 10 days of treatment (Fig. 2; Fig. 3B, 3D), compared to water controls. However, added ethylene inhibits both malic enzyme and 6 PG dehydrogenase during the initial 48-60 hour period (Figs. 3A, 3C). After 6 hours treatment malic enzyme activity is depressed by 18% compared to control levels and 6PG dehydrogenase by 7%. The maximum inhibition of 29% (malic enzyme) and 14% (6PG dehydrogenase) is attained 60 hours after excision. From the 5th day of treatment and throughout the remainder of the experimental period, no difference exists between enzyme activities in ethylene and water treatments.

IV. DISCUSSION

The measured rates of ethylene evolution are relatively high. Even the lowest rate, at excision, of 3-5 mμl/g FW/hr would be sufficient to produce ripening in some fruits. However, the high surface area/volume ratio of the leaf is likely to facilitate diffusion of ethylene. Both theoretical (Baur and Morgan, 1969) and practical values of the internal ethylene concentration were determined. They vary from 0.1-0.3 ppm at excision, 1-2 ppm 6 hours after excision and 0.4-0.8 ppm after 48 hours. The levels following excision appear to be above the threshold at which many ethylene effects are expected to occur and would generally produce at least a half-maximum effect (Burg, 1968).

The decrease in respiratory rate on detachment during the period of increased ethylene production differs markedly from other systems where detachment and bruising commonly cause an "induced" respiration (Lipetz, 1970). The stimulatory effect of ethylene on respiration is well known (Pratt and Goeschl, 1969).

Applied ethylene produces no changes of total chlorophyll, soluble protein, peroxidase and RUDP carboxylase activities over ten days that are not shown by controls. It is therefore not possible to determine if endogenous ethylene has a controlling influence on these parameters. Chlorophyll loss in other systems is initiated by a threshold level as low as 0.02 ppm (Burg, 1968). Fungal infection in *Nymphoides indica* which results in a vastly accelerated rate of ethylene production (unpublished data) is accompanied by rapid and visible yellowing at the boundaries of infected areas. This phenomenon indicates that the healthy leaf may be "protected" from the effects of ethylene by some internal factor. The chlorophyll increase in darkness has been observed in other systems (Popov and Dilova, 1969) but has not been reported for detached leaves in darkness (Addicott, 1969).

Again it is not possible to determine the importance of endogenous ethylene to the changes in soluble protein, RUDP carboxylase and peroxidase activity. The decreased RUDP carboxylase activity could result from a partial light inactivation.

Increasing soluble protein may represent a conversion from an insoluble form rather than synthesis. The initial increase again differs from the accepted patterns of protein breakdown in detached leaves (Addicott, 1969).

The decline in peroxidase activity contrasts with rises in activity in other plants. Peroxidase activity is commonly affected, too by ethylene (Pratt and Goeschl, 1969).

The activities of malic enzyme and 6 PG dehydrogenase are both inhibited initially by added ethylene. Endogenous ethylene can therefore be only partially inhibitory during this initial period. This would be so if the rate of ethylene production never raised the internal ethylene concentration to a saturating level.

Comparison of chromatographed and non-chromatographed extracts, serial dilution techniques and the addition of boiled concentrated crude extracts to assay mixtures, indicate that low and high molecular weight inhibitors and activators are unimportant or play a minor role in change in enzyme activity following detachment.

The depressed respiratory rate of approximately 4 mμ moles CO_2/g FW/min is operative when malic enzyme and 6PG dehydrogenase give *in vitro* values of approximately 730 and 610 mμ moles CO_2/g FW/min respectively, suggesting that respiratory CO_2 may be reassimilated. While PEP carboxylase is generally considered to be primarily responsible for dark fixation, both RUDP carboxylase and malic enzyme could contribute to overall dark fixation. The increased activity of the pentose phosphate shunt enzyme, 6PG dehydrogenase, could provide a source of reduced NADP for the reverse malic enzyme reaction.

The events described in this paper refer to an ageing rather than a senescing system. The decline in respiration and in peroxidase activity appear important in understanding the ability of *Nymphoides indica* to resist senescence. The low respiratory rate ensures a slow utilization of available energy sources. Increased peroxidase activity is often associated with senescing tissue (Osborne, 1968). Peroxidase is commonly regarded as an IAA oxidase, and therefore its lowered activity may result in a maintenance of auxin.

Ethylene may have some controlling influence but generally it does not appear to play an important role in the senescence of *Nymphoides indica*.

V. SUMMARY

The following biochemical changes were observed in detached leaves of the water plant *Nymphoides indica* aged in darkness on water.
1. A progressive increase in the rate of ethylene production which reached a maximum of 20-30 mμl/g fresh weight/hr within 3-6 hours.
2. The respiratory rate fell rapidly during the period of increasing ethylene production and this decline continued for some hours after the maximum rate of ethylene production was reached.
3. A significant increase in total chlorophyll and soluble protein occurred, reaching a maximum within 12-24 hours.
4. The activities of "malic enzyme" and 6 phosphogluconate dehydrogenase increased over a period of 48-60 hours while RUDP carboxylase and peroxidase activities declined.

Treatment with 25 ppm exogenous ethylene caused a decrease in the activities of "malic enzyme" and 6 phosphogluconate dehydrogenase in the initial 48-60 hour period, while total chlorophyll and soluble protein, and the activities of peroxidase and RUDP carboxylase, remained unaffected for ten days when compared to controls.

The significance of these results is discussed.

VI. REFERENCES

ABELES, F.B. and R.E. HOLM (1966). Evidence for an hormonal role for ethylene in abscission. Pl. Physiol. (Sup.), Wash. 41, liii.
ADDICOTT, F.T. (1969). Ageing, senescence and abscission in plants. Hort. Sci. 4, 114-116.
ARNON, D.I. (1949). Copper enzymes in isolated chloroplasts. Pl. Physiol. Wash. 24, 1-15.
BAUR, J.R. and P.W. MORGAN (1969). Effects of picloram and ethylene on leaf movement in Huisache and Mesquite seedlings. Pl. Physiol. 44, 831-838.
BURG, S.P. (1968). Ethylene, plant senescence and abscission. Pl. Physiol. 43, 1503-1511.
CHADWICK, A.V. and S.P. BURG (1970). Regulation of root growth by auxin-ethylene interaction. Pl. Physiol. 45, 192-200.
FUCHS, Y. and M. LIEBERMANN (1968). Effect of kinetin, IAA and gibberellin on ethylene production and their interactions in growth of seedlings. Pl. Physiol. Wash. 43, 2029-2036.
GALLIARD, T., M.J.C. RHODES, L.S.C. WOOLTORTON and A.C. HULME (1968). The development of ethylene biosynthesis during the ageing of disks of apple peel. Phytochem. 7: 1465-1470.
HALL, W.C., G.B. TRUCHELET, C.L. LEINWEBER and F.A. HERRERO (1957). Ethylene production by the cotton plant and its effect upon experimental and field conditions. Physiol. Plant. 10, 306-317.
HALLAWAY, M. and D.J. OSBORNE (1969). Ethylene. A factor in defoliation induced by auxin. Science. N.Y. 163, 1067-1068.
IMASEKI, H., UCHIYAMA and I. URITANI (1968). Effect of ethylene on the inductive increase in metabolic activities in sliced sweet potato roots. Agr. Biol. Chem. 32: 387-389.

LIPETZ, J. (1970). Wound healing in higher plants. Rev. of Cytol. $\underline{27}$, 1-25. Academic Press N.Y.
LOWRY, O.H., N.J. ROSEBROUGH, A.L. FARR and R.J. RANDALL (1951). Protein measurement with the Folin phenol reagent. J. Biol. Chem. $\underline{193}$, 265-275.
LUCK, H. (1963). Peroxidase, *in* "Methods of Enzymatic Analysis". (Ed. H. Bergmeyer) p. 895-897. Academic Press. N.Y.
MAXIE, E.C. and J.C. CRANE (1967). 2,4,5, - Trichlorophenoxyacetic acid: Effect on ethylene production by fruits and leaves of fig tree. Science N.Y. $\underline{155}$, 1548-1550.
MORGAN, P.W. and W.C. HALL. Accelerated release of ethylene by cotton following application of indolyl-3-acetic acid. Nature, $\underline{201}$: 99.
OSBORNE, D.J. (1968). Ethylene as a plant hormone. S.C.I. Monographs No.$\underline{31}$, 236-250.
POAPST, P.A., A.B. DURKEE, W.A. McGUGAN and F.B. JOHNSTON (1968). Identification of ethylene in GA treated potatoes. J. Sci. Fd. Agric. $\underline{19}$, 325-327.
POPOV, K. and S. DILOVA (1969). On the dark synthesis and stabilization of chlorophyll, *in* "Progress of photosynthesis research" (Ed. H. Metzner) Vol. II. 606-610.
PRATT, H.K. and J.D. GOESCHL (1969). Physiological role of ethylene in plants. Ann. Rev. Pl. Physiol, $\underline{20}$. 541-584.
RIOV, J., S.P. MONSELISE and R.S. KAHAN (1969). Ethylene-controlled induction of phenylalanine ammonia-lyase in citrus fruit peel. Pl. Physiol. Wash. $\underline{44}$, 631-635.
SLACK, C.R. and M.D. HATCH (1967). Comparative studies on the activity of carboxylases and other enzymes in relation to the new pathway of photosynthetic carbon dioxide fixation in tropical grasses. Biochem. J. $\underline{103}$, 660-665.

Increase in ABA-like Growth Inhibitors and Decrease in Gibberellin-like Substances During Ripening and Senescence of Citrus Fruits

E.E. Goldschmidt, S.K. Eilati and R. Goren

Department of Citriculture, Hebrew University, Rehovot, Israel

I. INTRODUCTION

Citrus fruits undergoing colour changes during ripening have been used recently for the study of senescence (Eilati *et al.*, 1969; Goldschmidt & Eilati, 1970; Goldschmidt *et al.*, 1970). Senescence in this system involves mainly biochemical and structural events associated with transformation of the chloroplasts into chromoplasts and these phenomena are quite distinct from deterioration and death of the tissue as a whole, which occur much later (Thomson, 1966; Lewis *et al.*, 1967). Use of whole, uninjured fruits for the study of senescence offers advantages, as compared with the widely used floating leaf-disc systems.

Exogenous cytokinins and gibberellins delay the senescent colour changes while ethylene enhances them. Little is known up to now about the endogenous hormonal balance of citrus fruit tissues during ripening and senescence. The present study deals with quantitative changes in ABA-like growth inhibitors and gibberellin-like substances whose presence and amount seem to be closely related to the onset of senescence.

II. MATERIALS AND METHODS

'Shamouti' orange (*Citrus sinensis* L.) fruits were used throughout. Mature, but still green, fruits were harvested and stored in the dark at $25^{\circ}C$. Treatments with growth regulators consisted of dipping the fruits, immediately after harvest, into solutions of 50% ethanol + 0.02% Tween-20 with or without the desired growth regulator. N_6 benzyladenine (BA) was applied at 100 mg/l. 2-chloroethylphosphonic acid (CEP or Ethrel) (Amchem Products ACP 66-329) was used at a concentration of 2500 mg/l. Fruit colour was determined with a Hunter Color & Color Difference Meter (Gardner Laboratories Inc.), using the a/b ratio (a - measure of redness; b - measure of yellowness) for quantitative estimation of colour development (Ayers and Tomes, 1966; Eilati *et al.*, 1969).

The outer, coloured peel layer (the flavedo) was peeled with a carrot peeler, frozen in liquid air and stored at $-20^{\circ}C$ for bioassay determinations. The processes leading to bioassay, described in detail elsewhere (Goren & Goldschmidt, 1970), are here outlined briefly.

Extraction of plant material was carried out overnight at $5^{\circ}C$, in a 5-fold amount (V/W) of 80% methanol. After evaporation of most of the 80% methanol extract *in vacuo* at $45^{\circ}C$, the remaining liquid was washed x 5 with diisopropyl ether at pH 6.0, washed again x5 with diethyl ether at pH 6.0 and finally with diethyl ether at pH 3.0. Paper (Whatman 3MM) strips were run in an ascending system, using isopropanol: ammonia : water (80 : 0.1 : 19.9) as the developing solvent.

The wheat coleoptile section bioassay (Nitsch & Nitsch, 1956), was used for the evaluation of growth inhibitors, with no exogenous auxin.

The barley endosperm sugar release bioassay (Coombe et al., 1967) was used for the detection of gibberellin-like substances. The same barley endosperm system, which is sensitive also to ABA-like growth inhibitors, was used to estimate growth inhibitors, in the presence of 1 ml of 10^{-7}M gibberellin A_3 as described previously (Goldschmidt & Monselise, 1968). The released reducing sugars were determined according to Noelting & Bernfeld (1948) by absorbance at 550 nm.

III. RESULTS

Colour changes

Fig. 1 shows the change in colour of green-harvested oranges during 10 days' storage, as influenced by the cytokinin BA and by the ethylene-releasing regulator CEP (Cooke & Randall, 1968). Slight effects of CEP become visible as soon as 2 days after harvest but the difference between BA and control takes much longer to appear. Fig. 1 outlines the background for the changes in endogenous hormonal balance which will be described in the following.

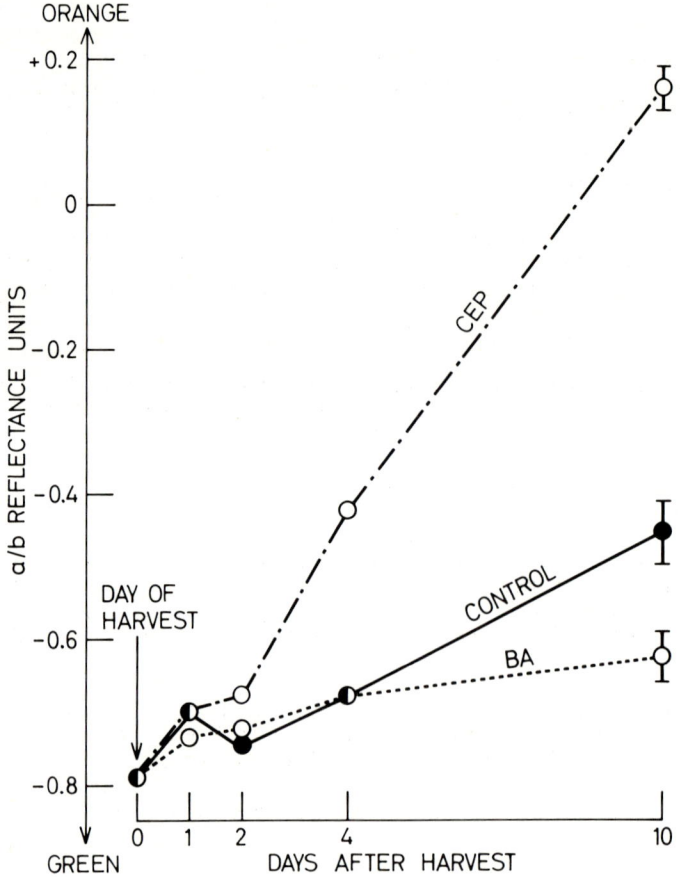

Fig. 1. Change in colour of green harvested oranges during 10 days' storage, as influenced by BA (N_6 benzyladenine, 100 mg/l) and CEP (2-chloroethylphosphonic acid, 2500 mg/l). Fruit colour determined with a 'Hunter' Color Difference Meter as explained in Methods. Each point is an average of data from 6 fruits.

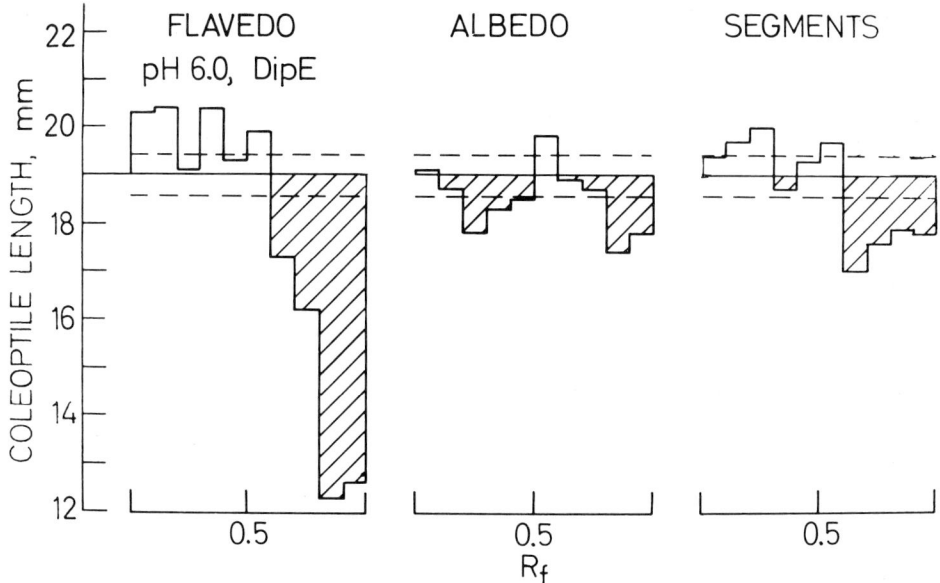

Fig. 2. Comparison of biological activity (wheat coleoptile section bioassay) of the diisopropyl ether fraction (DipE) from flavedo, albedo and segments of citrus fruits. After Goren & Goldschmidt (1970).

ABA-like growth inhibitors

ABA was identified previously in the 'β-inhibitor' zone of citrus fruit extracts (Milborrow, 1967). However, in fruits approaching ripening, considerable amounts of the 'β-inhibitor' remain in the water phase after solvent partition (Goren & Goldschmidt, 1970; see their fig. 3), indicating perhaps the presence of one of the glucose esters of ABA (Koshimizu et al., 1968; Milborrow, 1970).

Comparing different fruit tissues, the pigmented flavedo is by far the richest in ABA-like growth inhibitors (Fig. 2), suggesting perhaps some link between growth inhibitors and carotenoids (Taylor, 1968).

In the present study we attempted to study the overall change in ABA-like growth inhibitors, without going into detailed analysis of components. We preferred, therefore, to bioassay paper chromatograms of the 80% methanol extract, without further purification. Fig. 3 shows that the amount of inhibitors in the flavedo increased considerably during storage, as compared with the concentration on the day of harvest. The different rates of senescence obtained with the control, BA and CEP treatments (Fig. 1) are reflected in the differences in content of endogenous ABA-like growth inhibitors.

The inhibitory activity at R_f 0.5 - 1.0 shown in the histograms increases by 180, 130 and 213 percent for the control, BA and CEP treatments, respectively, as compared with the day of harvest. The absolute differences are in fact even larger, since the height of the histogram columns is more proportional to the log. of the concentration, as indicated also by the ABA standards in Fig. 3.

The existence of quantitative differences in inhibitor content is further demonstrated over a wide range of concentrations in the dilution experiment presented in Fig. 4. The inhibition was measured in this case using the barley endosperm bioassay system which is sensitive to ABA-like inhibitors. The concentration of inhibitors is obviously much higher in control and CEP treatments than in the 'day of harvest' and BA.

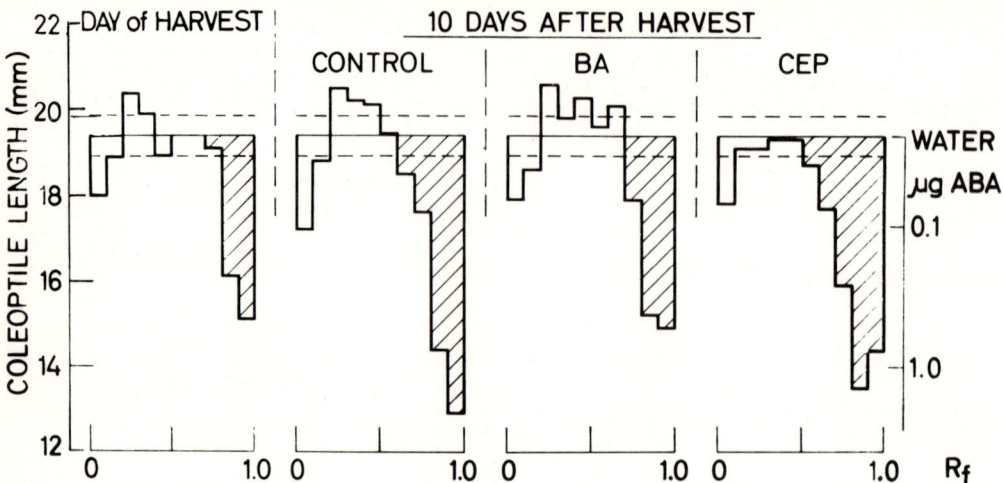

Fig. 3. Histograms of biological activity obtained using the wheat coleoptile section bioassay from chromatograms spotted with 80% methanol extracts of 100 mg fresh material. Each histogram represents the average of 4 replicate samples extracted separately. Statistical treatment as in Goren & Goldschmidt (1970).

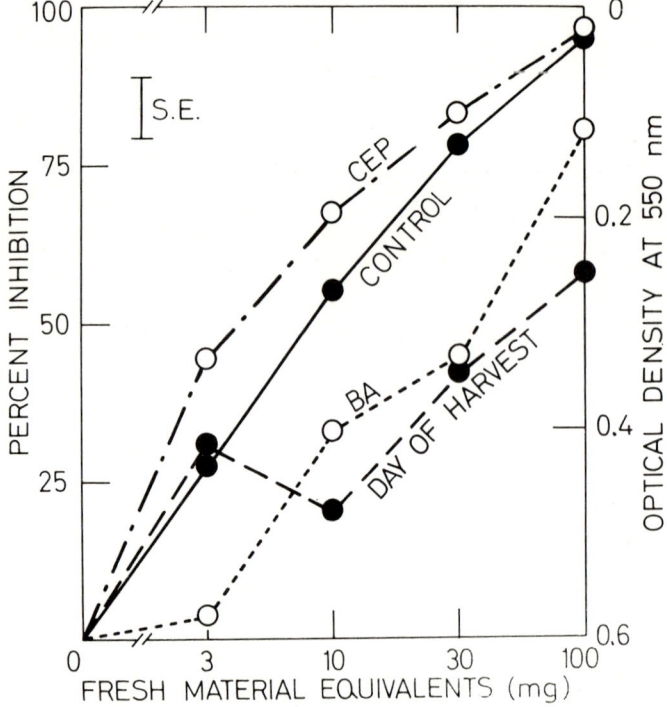

Fig. 4. Dilution experiment, showing the inhibition obtained by increasing amounts of plant material from all treatments of figs. 1 and 3 as determined by the barley endosperm bioassay for gibberellin inhibitors in the presence of 1 ml 10^{-7}M GA_3. Extracts (80% methanol) of 5 gm of plant material were chromatographed (see Methods) on 46 x 40 cm Whatman 3MM sheets. R_f zone 0.8 - 1.0 was eluted with methanol : ether 1:1. Amounts equivalent to 3, 10, 30 and 100 mg fresh material were spotted on paper sections and bioassayed. Each point is the average of 5 bioassay determinations.

It was found that most of the inhibitor accumulation takes place within the first 48 hrs after harvest and perhaps even before that (Fig. 5). This means that the increase in inhibitor occurs rather early in the sequence of senescence processes, before the changes in fruit colour become evident, and the quantitative differences between the three treatments prevail from the 2nd day on.

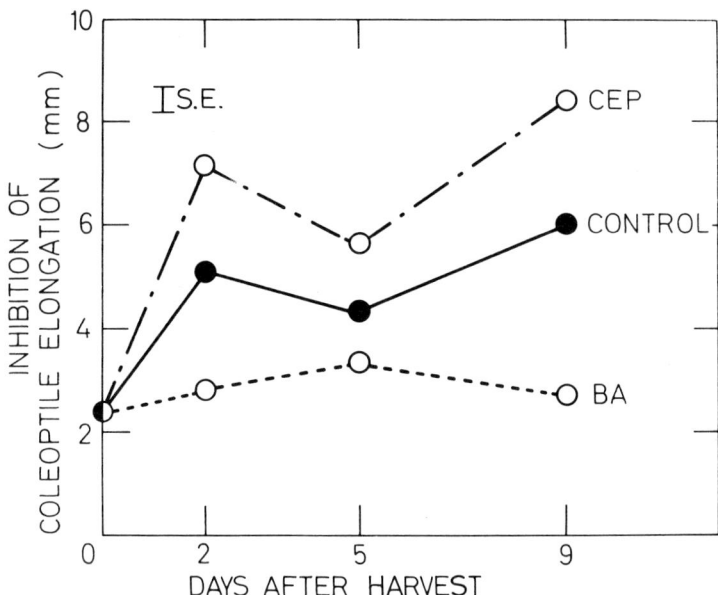

Fig. 5. Changes in the amount of growth inhibitors during storage of control fruits and of BA- and CEP-treated fruits, as measured by the wheat coleoptile section bioassay. Each chromatogram was spotted with extracts of 40 mg plant material and the 3 R_f sections 0.7 - 1.0 were bioassayed. Each point represents the average of three separate samples of plant material.

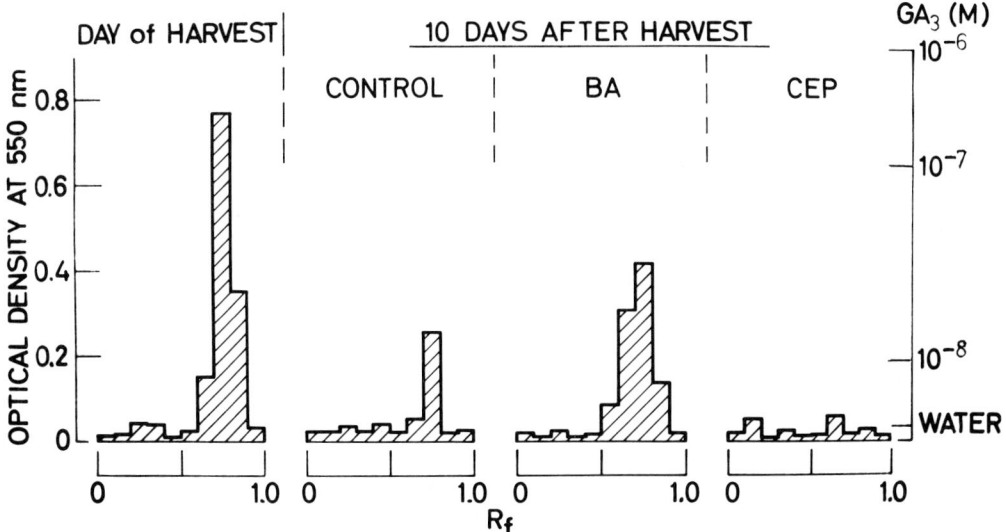

Fig. 6. Histograms of gibberellin-like activity obtained with the barley endosperm bioassay from chromatograms spotted with 250 mg fresh material equivalents of the pH 3.0 ether fractions. Each histogram is the average of 3 replicate chromatograms.

Gibberellin-like substances

Gibberellins are very potent retardants of senescence in citrus fruits and we looked for some changes in gibberellin-like substances in our system. Preliminary assays of all fractions, including the water residue, indicated that such activity was strictly confined to the pH 3.0 diethyl ether fraction, hence only this fraction was bioassayed regularly (Fig. 6).

Flavedo of mature green fruit (day of harvest) contains a considerable concentration of gibberellin-like substances. Ten days after harvest the same tissue had apparently lost most of its gibberellin-like activity, except for the BA treated fruits, in which the level of gibberellin-like substances approached that of the day of harvest. Again there seems to be a correlation, this time an inverse one, between senescence and the concentration of gibberellin-like substances.

IV. DISCUSSION

Senescence of citrus fruit involves an increase in ABA-like growth inhibitors which occurs within the first 48 hrs, before other senescence phenomena become visible. Moreover, the intensity of senescence, which can be controlled by exogenous growth regulators, seems to be correlated with the extent of inhibitor accumulation.

Several recent reports indicate that plants exposed to stress conditions respond by rapid accumulation of ABA-like growth inhibitors (Wright & Hiron, 1969; Mizrachi et al., 1970; Steadman & Sequeira, 1970). Senescence phenomena resemble stress conditions in many instances and the increase in ABA-like growth inhibitors during senescence (reported recently also by Rudnicki et al. (1968) and by Chin & Beevers (1970)) fits into this framework very well. The observed decrease in gibberellin-like substances (Fig. 6) was like that reported recently for other senescent tissues (Fletcher et al., 1969; Chin & Beevers, 1970). It is not clear how closely these changes are related to the onset of senescence and very little is known concerning the inactivation of gibberellins in general.

Our results might be interpreted as suggesting that exogenous growth regulators exert their control on senescence through influences on the endogenous hormonal balance of the senescent tissue. Ethylene might then operate by increasing the level of ABA-like substances in plant tissues. Verification of such hypotheses requires, of course, further studies and careful experimentation.

V. SUMMARY

Citrus fruits have been proposed as a system advantageous for the study of hormonally regulated senescence. Senescent colour changes can be delayed by cytokinins and gibberellins or enhanced by ethylene.

The concentration of ABA-like growth inhibitors in the flavedo (the outer, coloured layer of citrus fruits) increases markedly within the first 48 hrs after harvest. Treatment with N_6-benzyladenine, which delays senescence, largely prevents the increase in ABA-like growth inhibitors. On the other hand, the ethylene-releasing regulator, 2-chloroethylphosphonic acid, which accelerates senescence, causes the largest increase in concentration of ABA-like growth inhibitors.

Gibberellin-like substances decrease and almost disappear in the flavedo of senescent fruits. This change can also be avoided by treating fruits with N_6-benzyladenine or induced more strongly by application of 2-chloroethylphosphonic acid.

VI. ACKNOWLEDGEMENTS

The expert technical assistance of Mr. D. Galili is gratefully acknowledged.

The chemical regulator 2-chloroethylphosphonic acid (ACP 66-329) was kindly supplied by Mr. I. Tobolski of AGAN Chemicals, Tel Aviv.

VII. LITERATURE CITED

AYERS, J.E. and M.I. TOMES (1966). The effect of two uniform ripening genes on chlorophyll and carotenoid contents of tomato fruit. Proc. Am. Soc. Hort. Sci. 88, 550-556.

CHIN, T.-Y. and L. BEEVERS (1970). Changes in endogenous growth regulators in nasturtium leaves during senescence. Planta (Berl.) 92, 178-188.

COOKE, A.R. and D.I. RANDALL (1968). 2-haloethanophosphonic acids as ethylene-releasing agents for the induction of flowering in pineapples. Nature 218, 96-97.

COOMBE, B.G., D. COHEN and L.G. PALEG (1967). Barley endosperm bioassay for gibberellins. I. Parameters of the response system. Plant Physiol. 42, 105-112.

EILATI, S.K., E.E. GOLDSCHMIDT and S.P. MONSELISE (1969). Hormonal control of colour changes in orange peel. Experientia 25, 209-210.

FLETCHER, R.A., T. OEGEMA and R.F. HORTON (1969). Endogenous gibberellin levels and senescence in *Taraxacum officinale*. Plant (Berl.) 86, 98-102.

GOLDSCHMIDT, E.E. and S.K. EILATI (1970). Gibberellin-treated Shamouti oranges: Effects on coloration and translocation within peel of fruits attached to or detached from the tree. Bot. Gaz. 131, 116-122.

GOLDSCHMIDT, E.E., S.K. EILATI, J. RIOV and B. BRAVDO (1970). Acceleration of senescence in young orange fruitlets by gibberellins and cytokinins. Plant Physiol. 46, (Suppl.) p. 10.

GOLDSCHMIDT, E.E. and S.P. MONSELISE (1968). Native growth inhibitors from citrus shoots: Partition, bioassay and characterization. Plant Physiol. 43, 113-116.

GOREN, R. and E.E. GOLDSCHMIDT (1970). Regulative systems in the developing citrus fruit. I. The hormonal balance in orange fruit tissues. Physiol. Plant. 23, 937-947.

KOSHIMIZU, K., M. INUI, H. FUKUI and T. MITSUI (1968). Isolation of (+)-abscisyl-β-D-glucopyranoside from immature fruit of *Lupinus luteus*. Agric. and Biol. Chem. 32, 788-791.

LEWIS, L.N., C.W. COGGINS, C.K. LABANAUSKAS and W.M. DUGGER (1967). Biochemical changes associated with natural and gibberellin A_3-delayed senescence in the navel orange rind. Plant and Cell Physiol. 8, 151-160.

MILBORROW, B.V. (1967). The identification of (+)Abscisin [(+)-Dormin] in plants and measurement of its concentration. Planta 76, 93-113.

MILBORROW, B.V. (1970). The metabolism of abscisic acid. Jour. Exptl. Bot. 21, 17-29.

MIZRACHI, Y., A. BLUMENFELD and A.E. RICHMOND (1970). Abscisic acid and transpiration in leaves in relation to osmotic root stress. Plant Physiol. 46, 169-171.

NITSCH, J.P. and C. NITSCH (1956). Studies on the growth of coleoptile and first internode sections. A new, sensitive, straight-growth test for auxins. Plant Physiol. 31, 94-111.

NOELTING, G. and P. BERNFELD (1948). Sur les enzymes amylolytiques. III. La β-amylase: dosage d'activité et contrôle de l'absence d' α-amylase. Helv. Chim. Acta 31, 286-290.

STEADMAN, J.R. and L. SEQUEIRA (1970). Abscisic acid in tobacco plants. Tentative identification and its relation to stunting induced by *Pseudomonas solanacearum*. Plant Physiol. 45, 691-697.

TAYLOR, H.F. (1968). Carotenoids as possible precursors of abscisic acid in plants. In: "Plant Growth Regulators", S.C.I. Monograph No. 31, pp. 22-35.

THOMSON, W.W. (1966). Ultrastructural development of chromoplasts in Valencia oranges. Bot. Gaz. 127, 133-139.

WRIGHT, S.T.C. and R.W.P. HIRON (1969). (+)Abscisic acid, the growth inhibitor induced in detached leaves by a period of wilting. Nature 224, 719-720.

PLANT GROWTH SUBSTANCES, 1970

Effects of ABA and Kinetin on Ultrastructure of Senescing Wheat Leaves

Cathryn J. Mittelheuser and R.F.M. Van Steveninck

Department of Botany, University of Queensland, St. Lucia, Queensland, 4067

I. INTRODUCTION

The aim of this investigation was to examine changes occurring in naturally senescing primary wheat leaves and in detached leaves induced to senesce either in the light or in darkness in order to establish the pattern of senescence in each case and to detect differences, if any, in organelle structure and rates of senescence. The effects of ABA and kinetin on the ultrastructure of detached leaves were examined to determine whether either plant growth substance induced any specific alteration in ultrastructure or any change in the rate of organelle degeneration. There had been no previous studies of the ultrastructure of the 3 types of senescence in a single plant species and, although ABA has been known to accelerate leaf senescence (El-Antably et al., 1967) there have been no reports of its effect on leaf ultrastructure. Shaw and Manocha (1965) reported that kinetin increased the number of cytoplasmic ribosomes in senescing wheat leaves incubated under low light intensity.

II. MATERIALS AND METHODS

Wheat (*Triticum aestivum* cv Mendos) was grown in vermiculite and nutrient solution under a 16 hr photoperiod (850 lumens/ft^2, 26°± 2°C day, 18°± 2°C night). For experiments involving detached leaves, primary leaves were detached 10 days after sowing and incubated under the above conditions or in the dark at 24° ± 1°C in individual vials containing either 3.8 x 10^{-6}M ABA, 5 x 10^{-5}M kinetin or deionized water. The solutions were changed daily and were protected from light. As wheat leaves senesce basipetally, sections from a number of regions of the leaves were examined.

Leaf sections were fixed in ice cold 4% glutaraldehyde in 0.05 M cacodylate - HCl buffer pH 7.2 for 3-4 hr, and post-fixed in 2% osmium tetroxide in the above buffer at room temperature for 1.5 hr.

After dehydration through a graded ethanol series, the leaf tissue was treated with 1% uranyl acetate in ethanol for 0.5 hr, followed by propylene oxide and embedded in Araldite. Sections were cut on an LKB Ultrotome, post-stained with lead citrate (Reynolds, 1963) for 1.5 min and examined in a Siemens Elmiskop 1 or 1A electron microscope operated at 60 KV.

III. RESULTS

Naturally Senescing Leaves

The mesophyll cells of mature wheat leaves possess large central vacuoles. The nucleus, situated either centrally or peripherally within the cytoplasm have a regular outline within which, with glutaraldehyde-osmium tetroxide fixation, numerous discrete regions of chromatin are revealed. Microbodies are usually situated close to the chloroplasts and possess a finely granular matrix. Cytoplasmic ribosomes (250-260 Å x 200-220 Å) are found free in the cytoplasm, in polysome arrangements or attached to the endoplasmic reticulum and, rarely, attached to the outer nuclear membrane.

Chloroplasts, somewhat ellipsoid in shape, range in length from 3 to 6 μ, the average being about 4.5μ. Two or 3 starch grains can be found within the chloroplasts but usually there was only 1. They vary in size up to 1.6μ. The chloroplasts have well developed grana irregularly situated and joined by inter-grana lamellae. Chloroplast ribosomes (210-220Å x 160-180Å) are found in polysome arrangements, free in the stroma, or attached to the outer lamellae of grana.

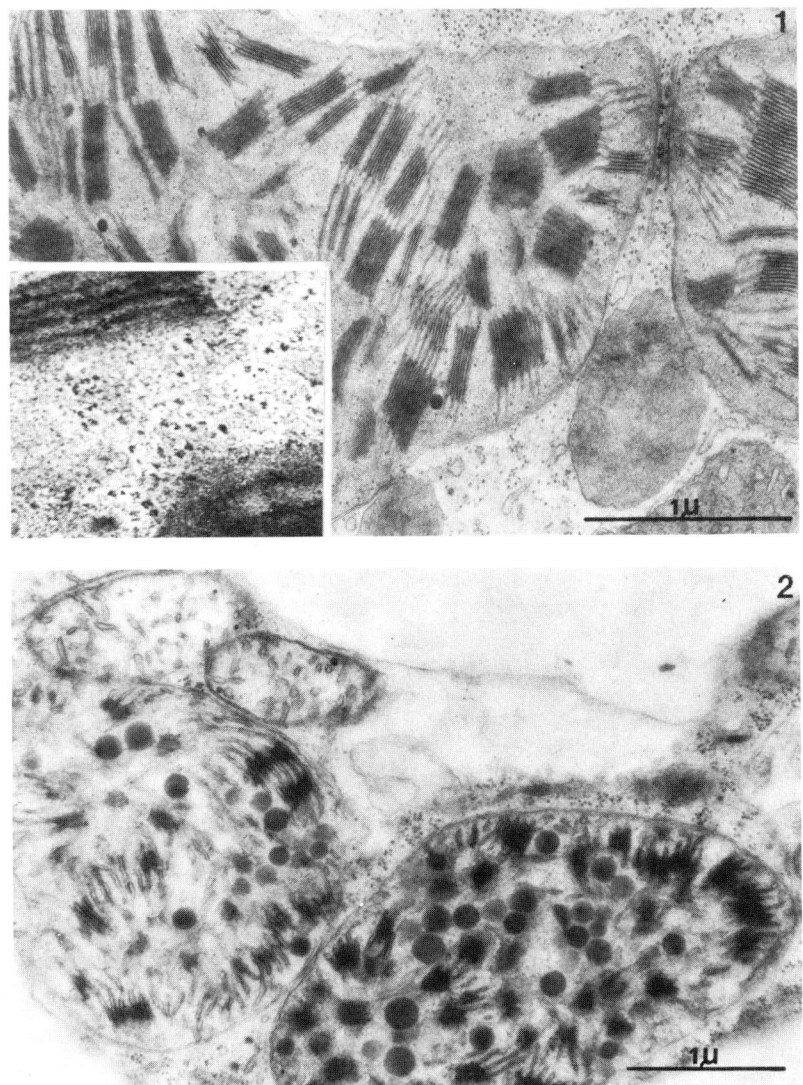

Fig. 1. Chloroplast ribosomes and cytoplasmic ribosomes in a 12 days old wheat leaf. Inset shows detail of chloroplast ribosomes. (X 96,000).

Fig. 2. Cytoplasmic ribosomes in tissue taken from the tip of a detached wheat leaf which had been incubated in water in the dark for 3 days. Chloroplast ribosomes are absent.

Another feature of the chloroplasts of wheat leaves is the presence of wavy "membranes" which were found in the stroma in groups of 5, 6 or 7 and did not appear to be connected to grana or intergrana lamellae. Each membrane had a diameter of 60Å and the spacing between "membranes" was 120Å. The usual length was 0.3 to 0.4µ though some were up to 1.2µ.

As natural senescence progressed, three changes became evident. First was the development of plastoglobuli. These osmiophilic globules were detected first at about 12 days when they were 0.10 to 0.15µ in diameter. The second change detected early in senescence was the development of lipid bodies in the cytoplasm. These usually appeared at about 14 days although occasionally they were detected at 11 days. These lipid bodies possessed a wide size variation some being only about 0.15µ in diameter while others were up to 4.5µ in diameter. Small lipid bodies appeared to coalesce to form large lipid bodies. Light microscopy reveals that they stain an intense red with Sudan IV indicating that they are composed largely, if not entirely, of lipid. Also they are resistant to treatment with pronase. The lipid bodies were roughly circular in outline, devoid of internal structure, and did not appear to be membrane-bound. Usually they distended the tonoplast and, in the late stages of senescence were found within the vacuole. Occasionally they were detected between a chloroplast and the plasmalemma.

The third change detected early in natural leaf senescence was a decrease in the number of chloroplast ribosomes though no change was observed in the cytoplasmic ribosome population. Very few chloroplast ribosomes could be detected in tissue taken from the base of 16 day old wheat leaves and none was detected at 20 days.

Starch grains were still present in the chloroplasts at a stage when signs of increasing senescence, such as degeneration of lamellae and increased size of plastoglobuli were evident. Other structural changes possibly related to increasing senescence were the development of osmiophilic deposits in the microbodies, the formation of vesicles between the plasmalemma and the cell wall, swelling of the endoplasmic reticulum, and a lessening of cytoplasmic density, but mitochondria, nuclei and the tonoplast remained structurally intact until quite late in senescence.

Detached Leaves Incubated under a 16hr Photoperiod

Little change was apparent in water-treated leaves 24hr after detachment and incubation under a 16hr photoperiod. A few very small plastoglobuli had formed. The chloroplasts of ABA-treated leaves also contained small plastoglobuli. Lipid bodies, some of which were up to 3µ in diameter, had developed in the cytoplasm. The starch grains of ABA-treated leaves were smaller and fewer in number than in the water-treated leaves. In contrast, examination of kinetin-treated leaves revealed a considerable development both of the size and number of starch grains, some of the starch grains being up to 2.6µ in length.

After 2 days there was an increase in both the size and number of starch grains in the water-treated leaves also, some being up to 2.2µ in length. On the other hand, the ABA-treated leaves contained very few starch grains at this stage. Osmiophilic deposits were present in the microbodies and there was considerable development of lipid bodies. With kinetin-treated leaves still further development of starch grains was apparent, the increase in starch content distending the chloroplasts so that some were up to 7.5µ in length.

By 3 days the starch content of the water-treated leaves had started to decline. Plastoglobuli and wavy "membranes" were detected in the chloroplasts. Whilst lipid bodies were present in the cytoplasm and osmiophilic deposits in the microbodies, no change was apparent in mitochondrial structure or in nuclear structure. Tissue taken from the base of ABA-treated leaves at this stage revealed very few starch grains, and none was over 0.6µ in length. Although tissue at the base of these ABA-treated leaves was generally in good condition, tissue at the tip of the same leaves showed considerable organelle degeneration. No starch grains or ribosomes were present in the chloroplasts which contained plastoglobuli up to 0.25 µ in diameter and broken-down grana and inter-grana lamellae. However, cytoplasmic ribosomes were still evident, some being in polysome arrangements, although the endoplasmic reticulum was somewhat distended. Little change could be detected in the mitochondria.

No change was detected in either chloroplast or cytoplasmic ribosomes, mitochondria or nuclei following kinetin treatment for 3 days. The chloroplasts contained starch grains up to 3 µ in length. Osmiophilic deposits were beginning to develop in the microbodies and lipid bodies also had formed.

Detached Leaves Incubated in the Dark

The first change detected in detached leaves incubated in water in the dark was the loss of starch grains. After 24 hr only a few small starch grains were observed. At the same time small plastoglobuli formed as well as a few small lipid bodies and some osmiophilic deposits in the microbodies. Starch grains were completely absent in ABA treated leaves at this stage but, in other respects, the ultrastructure was similar to that of the water-treated leaves.

In contrast, tissue taken from kinetin-treated leaves was unchanged in ultrastructural appearance from freshly detached tissue, starch grains still being prominent in the chloroplasts. After 2 days the only change apparent in any region of kinetin-treated leaves was the development of very small plastoglobuli.

At this stage, however, tissue at the tip of water-treated leaves showed ultrastructural signs of advancing senescence. Chloroplast ribosomes had disappeared though there was no change in the cytoplasmic ribosome population. Grana and inter-grana lamellae were degenerating and plastoglobuli up to 0.3 µ in diameter had formed. Also vesicles were beginning to appear between the plasmalemma and the cell wall. Lipid bodies up to 3 µ in diameter distended the tonoplast.

Following dark incubation for 2 days, tissue from the tip of ABA-treated leaves was almost completely degenerate. The chloroplasts were largely filled with plastoglobuli which were up to 0.5 µ in diameter and were much less osmiophilic than in the earlier stages of senescence. A few small grana could be detected but practically no inter-grana lamellae. However, wavy "membranes" were still apparent. Decreased numbers of ribosomes were found both free in the cytoplasm and in polysome arrangements and also attached to considerably swollen endoplasmic reticulum. By contrast, tissue at the base of these ABA-treated leaves was in relatively good structural condition with membrane-bound chloroplast ribosomes still detectable.

By 3 days tissue at the base of water-treated leaves was in fairly good condition but the leaf tip was showing signs of advancing senescence, as the plasmalemma was very fragile and easily disrupted from the cell wall. There was considerable development of plastoglobuli as well as degeneration of inter-grana lamellae and some granal disruption though not to the extent which occurred in the tips of ABA-treated leaves after dark incubation for 2 days. Wavy "membranes" were present in the chloroplasts. Although chloroplast degeneration was relatively advanced, the nucleus and the cytoplasmic ribosomes still showed no signs of deterioration and the majority of mitochondria was structurally intact (cf. Figs 1 & 2.).

At the same time, the entire length of ABA-treated leaves was in a relatively advanced state of senescence with vesicles and coiled membranes lying between the plasmalemma and the cell walls. The chloroplasts possessed little lamellar structure and large numbers of plastoglobuli but there was still ample evidence of both free and endoplasmic reticulum-attached cytoplasmic ribosomes, and mitochondria were well preserved. However, the tissue rapidly degenerated in the following 24 hr.

All organelles of water-treated leaves showed degenerative changes after incubation for 5 days in the dark. The cell contents tended to aggregate at one end of the cell. Little or no cytoplasm was present, a mass of globules had replaced the chloroplast structure and few cristae could be detected in the mitochondria.

Kinetin-treated leaves still possessed remnants of starch grains at 6 days, although lipid bodies were apparent in the cytoplasm and osmiophilic deposits were developing in the microbodies. By 11 days, the chloroplasts of kinetin-treated leaves were somewhat swollen, the stroma appeared less dense and plastoglobuli had formed. However, grana and inter-grana lamellae were intact and both chloroplast and cytoplasmic ribosomes were present although the endoplasmic reticulum was considerably swollen. Mitochondria were structurally intact but the microbodies possessed intense osmiophilic deposits.

IV. DISCUSSION

These investigations have revealed the presence of two structures not previously reported in mature or senescent leaf tissue. These are the chloroplast wavy "membranes" and the lipid bodies. The wavy "membranes" occur both in mature chloroplasts and in chloroplasts in an advanced state of senescence. The function of these structures is unknown but they obviously possess considerable stability.

The development of large lipid bodies in the cytoplasm of leaves during early senescence and their presence in the vacuole with advancing senescence suggests that lipids congregate in a depository.

Another feature of wheat leaf senescence is the loss of chloroplast ribosomes long before the loss of cytoplasmic ribosomes. Also cytoplasmic ribosomes and mitochondria have been found to be of greater structural stability than the majority of other organelles. Brady et al. (1970), in studying RNA metabolism in senescing wheat leaves, found that uracil incorporation into chloroplast rRNA ceased when growth stopped but that incorporation into cytoplasmic rRNA continued at a significant rate. It was suggested that this was the result either of an absence of uracil in the chloroplast rRNA precursor pool or an absence of turnover of chloroplast rRNA with the cessation of growth. It could possibly also result from the absence of chloroplast ribosomes as demonstrated above.

The ultrastructural effects of ABA and kinetin during induced senescence of wheat leaves are of particular interest. With dark-induced senescing leaves ABA did not appear to have any specific effect on ultrastructure but it accelerated the normal pattern of organelle degeneration. However, in detached leaves incubated under a 16hr photoperiod, ABA prevented the increase in starch which occurred in the water-treated control leaves. This is of particular interest in view of ABA-induced stomatal closure (Mittelheuser and Van Steveninck, 1969) and ABA-induced inhibition of the rate of photosynthesis (Mittelheuser and Van Steveninck, 1971) and it is suggested that the paucity of starch grains results from the effects of ABA in enhancing stomatal resistance and reducing the rate of photosynthesis.

A previous report of kinetin-induced increase in the amount of endoplasmic reticulum and cytoplasmic ribosomes (Shaw and Manocha, 1965) could not be confirmed. For the first few days of kinetin treatment in the dark-incubated wheat leaves, metabolic processes appeared to be "suspended" in that, other than for the development of small plastoglobuli and lipid bodies, no changes in ultrastructure could be detected. The presence of chloroplast ribosomes after prolonged treatment with kinetin and at a stage when other organelles were showing signs of advancing senescence was of particular interest. It is possible that the action of kinetin in maintaining the synthesis of fraction 1 protein in detached wheat leaves (Brady et al, 1971) may be due to its effect in maintaining the chloroplast ribosome population.

The increased starch content in kinetin-treated leaves and in the first 48 hr in water-treated leaves may be due initially to the removal of the normal export channels with leaf detachment. However, the greater amount of starch apparent in kinetin-treated leaves than in water-treated leaves suggests stimulation of starch syntheses as a result of kinetin treatment.

V. SUMMARY

A study has been made of the effects of kinetin and ABA on the ultrastructure of detached wheat leaves incubated in the dark or under a 16hr photoperiod and the results obtained were compared with ultrastructural changes occurring in naturally senescing wheat leaves.

It was found that one of the first signs of senescence was the disappearance of chloroplast ribosomes. Cytoplasmic ribosomes either free, in polysome arrangements or attached to the endoplasmic reticulum appeared to be particularly stable. Treatment of detached leaves incubated in the dark with kinetin resulted in the maintenance of the chloroplast ribosome population for up to 11 days. Also, starch grains remained for 5-6 days, whereas in water-treated leaves few could be detected after dark incubation for 24 hr. A massive increase in both the size and the number of starch grains was observed in detached leaves treated with kinetin and incubated under a 16hr photoperiod.

Treatment with ABA prevented the initial increase in starch grains in light-incubated water-treated leaves.

VI. ACKNOWLEDGEMENTS

This research was supported by the Wheat Industry Research Council.

VII. REFERENCES

BRADY, C.J., PATTERSON, B.D., HENG FONG TUNG and SMILLIE, R.M. (1971). Protein and RNA synthesis during ageing of chloroplasts in wheat leaves, *in* "Autonomy and biogenesis of mitochondria and chloroplasts". Eds. N.K. Boardman, A.W. Linnane and R.M. Smillie. North-Holland, pp. 453-465.

EL-ANTABLY, H.M.M., WAREING, P.F. and HILLMAN, J. (1967). Some physiological responses to D, L-abscisin (dormin). Planta, Berl. $\underline{73}$, 74-90.

MITTELHEUSER, CATHRYN J. and VAN STEVENINCK, R.F.M. (1969). Stomatal closure and inhibition of transpiration induced by (RS)-abscisic acid. Nature, Lond. $\underline{221}$, 281-282.

MITTELHEUSER, CATHRYN J. and VAN STEVENINCK, R.F.M. (1971). Rapid action of abscisic acid on photosynthesis and stomatal resistance. Planta, Berl. $\underline{97}$, 83-86.

REYNOLDS, E.S. (1963). The use of lead citrate at high pH as an electron-opaque stain in electron microscopy. J. Cell Biol., $\underline{17}$, 208-212.

SHAW, M. and MANOCHA, M.S. (1965). Fine structure in detached, senescing wheat leaves. Can J. Bot., $\underline{43}$, 747-755.

GROWTH AND MORPHOGENESIS

The Role of Basal and Apical Factors in the Coordination of Growth in the Stems of White Clover (*Trifolium repens L.*)

Roderick G. Thomas

Department of Botany and Zoology, Massey University, Palmerston North, New Zealand

I. INTRODUCTION

The cause of the inability of most isolated stem apices to grow and the role of roots as stimulators of their growth are problems which remain unresolved. They are closely related to other problems of plant growth coordination. During the first forty years of this century some notable investigations were made into these problems (e.g. Child and Bellamy 1919, Snow 1925, Went 1938, 1943), the earlier of which were made in ignorance of plant growth factors and the later of which attempted to determine the role of auxin in coordination. A sound understanding of the basic problems of coordination derived from these studies.

More recently, attention has been paid largely to the mode of action of known growth factors at the cellular level and apart from continuing studies of apical dominance (Phillips 1969) the general problems of coordination have been rather neglected. This paper describes an attempt to elucidate further the mechanism of plant growth coordination in white clover by combining current knowledge of the nature and distribution of plant growth factors with the earlier basic approach to the problem. The former allows us to test the role of specific factors within a general framework; the latter forces us to consider the possibility that the growth-promoting influence of one plant organ on another could be just as easily the result of inhibitor removal as of supply of growth promoter. Thus while on one hand we can consider on the basis of their distribution the possibility that cytokinins and gibberellins act as caulocaline (Phillips and Jones 1964, Carr and Burrows 1966, Went 1938), on the other the possibility must be borne in mind that roots could just as easily promote stem growth by removing inhibitors from them.

II. MATERIALS AND METHODS

Plants used consisted of ramets of a clone of New Zealand Government Stock white clover selected some years previously as non-flowering at temperatures above 15°C and designated clone B (Thomas 1962). These were raised from cuttings and grown in pots in a greenhouse in natural light regimes at 15 to 30°C until required for experimentation. One to two weeks before the start of an experiment plants were transferred to a growth room operating at 23±2°C where they grew in continuous light from a combination of 95% fluorescent (Phillips TLA 80W/55) and 5% incandescent light giving an intensity of about 14 W. m^{-2}.

The basic morphology of the apical region of this clone has been outlined previously (Thomas 1962). For the purposes of the present set of investigations a system of numbering leaves and nodes was used as shown in Fig. 1. At the start of an experiment the youngest leaf with unfolded leaflets was referred to as leaf A, and internode 2 was between 2 and 5 mm long. Four leaf primordia, numbers 3,4,5 and 6, were present between leaf 2 and the apical meristematic dome. A comparison of cell lengths and numbers in internodes A, 1 and 2 (Table 1) shows that at the start of each experiment cell division in the tissues measured was ceasing in internode 1, but that cell elongation in that internode was continuing.

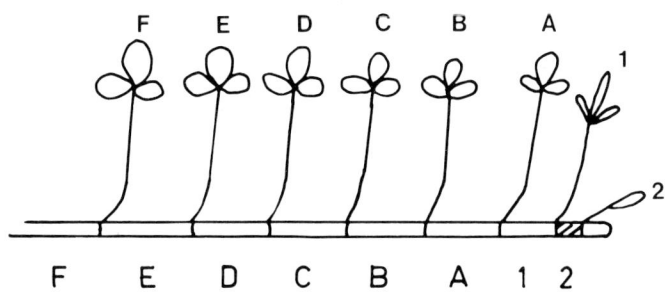

Fig. 1.

Diagram of stem to show leaf and internode numbering. Leaf A = youngest leaf with unfolded leaflets; internode 1 is almost fully elongated and has just commenced secondary thickening.

Table 1. Cell length and cell number in internodes A, 1 and 2. (Lengths are averages of 150 cells/tissue/internode).

	Internode length (mm)	Cell no.			Cell length (μ)		
		Epidermis	Cortex	Pith	Epidermis	Cortex	Pith
Internode A	19.4	306	191	211	65	106	97
Internode 1	14.6	296	203	228	49	72	69
Internode 2	2.1	198	109	112	11	19	19

The effects of experimental treatments on elongation were recorded by measuring lengths of internodes, petioles, leaf laminae, etc. usually at daily intervals for periods of up to a week. In some cases stems were dissected 7 or 10 days after the start of experiments to determine the extent of leaf primordium initiation and growth.

Transverse sections were cut of internode 1 and stained with phloroglucin/HCl or zinc-chlor-iodide to observe fibre development, cambium development and lignification at the time of destructive dissection. The anatomical features of white clover stems have been described by Erith (1924). The arrangement of bundles in the stem is very characteristic, three leaf trace bundles always being readily distinguishable. Throughout this study measurements were confined to these bundles.

The first stages of fascicular cambium development occur in internode 1. Interfascicular cambium, the ring of lignified parenchyma, and the thickening of xylem and phloem fibres are largely not apparent in internode 1 and do not develop strongly until internode A.

Where required, records of fibre development were made by estimating the percentage of cells having less than 25%, between 25 and 50%, and above 50% of their lumen filled by thickening which stains a 'plum' colour in zinc-chlor-iodide. The degree of cambial activity was estimated by counting and averaging the number of secondary cells (secondary xylem, cambium and phloem) in the three longest rows of secondary cells in each of the leaf trace bundles. Their existence in rows in itself renders the identification of secondary cells fairly simple. Cells of interfascicular cambium and lignified parenchyma were counted from the midway point between the leaf trace bundle and its nearest neighbour on one side to the equivalent point on its other side.

III. EXPERIMENTAL AND RESULTS

Effect of leaf removal on apical growth

Four groups of 6 stems each were set up so that groups 1 and 2 remained attached to their parent plants and groups 3 and 4 were excised at internode E. The basal ends of the latter were placed in test tubes containing water.

All leaves and primordia, except for the three youngest, at nodes 4,5 and 6, were removed from groups 2 and 4. Stems were dissected and growth measured 7 days after leaf removal (Tables II & III). The effect of excision was far greater than that of defoliation, internode growth being depressed markedly in excised stems. Defoliation over the period of 7 days had little effect; in both attached and excised stems, leaf removal slightly depressed elongation of internodes 1 and 2 (Table III) and stimulated growth of young leaves at nodes 4 to 6 (Table II). One must conclude that young leaves provide some stimulus to elongation of internodes basal to them but that they do not play a major regulatory role: on attached stems, for instance, internode 2 grew from an initial 2 mm to 19 mm after defoliation while the presence of leaves led to a further increase of 6.5 mm. At the same time defoliation stimulated compensatory growth at the apex leading to greater growth of both internodes and leaves and presumably a transitory higher rate of leaf initiation.

Table II. Petiole lengths (mm) on foliated and defoliated stems attached to and excised from parent plants.

Node no.		A	1	2	3	4	5	6	7
Stems attached	+ Leaves	111	114	104	44	13	4.4	1.6	.
	- Leaves	-	-	-	-	15	5.5	2.3	.
Stems excised	+ Leaves	123	59	23.9	9.7	4.1	1.3	0.4	.
	- Leaves	-	-	-	-	12.9	7.7	2.8	0.6

Table III. Internode lengths (mm) of foliated and defoliated stems attached to and excised from parent plants.

Internode no.		A	1	2	3	4
Stems attached	+ Leaves	18	22.5	25.5	15	3
	- Leaves	23	20.5	19	12	3
Stems excised	+ Leaves	29	24.2	8.5	1.2	.
	- Leaves	30	19.7	6.7	1.6	0.2

Table IV. Total percentage increase in length of internode 2 following various treatments.

	Time (days)	2	4	6
1	Intact	123	292	375
2	Decapitated	19	57	73
3	Excised (Short)	56	56	60
4	Excised (Long)	23	56	113

Effect of excision and decapitation on growth

The results of the previous experiment showed that growth was depressed when the younger, more apical parts of a stem were excised from the older parts and the main root system. A further experiment was set up to compare the effects of removal of the apical region with removal of various amounts of basal tissue on the elongation of internode 2. Four groups of 10 fully foliated stems were set up as below,

1. Intact, attached to parent plant
2. Decapitated above internode 2 to remove the apical tissues
3. Excised immediately below internode 2
4. Excised immediately below internode 8.

Stems in groups 3 and 4 were lain horizontally in water in petri dishes for the duration of the experiment (6 days). Measurements of internode 2 length (initially 2.6 mm) were made at 2-daily intervals (Table IV).

Both decapitation and excision strongly depressed internode elongation but the patterns of response to these treatments differed. Of the treated stems, growth of the short excised pieces was the most rapid over the first 2 days but it had virtually ceased by the end of that period; the longer excised pieces continued to grow over the 6-day period.

The results thus show that both the apical region and the basal region are required for continued growth. Comparison of the growth of the long excised stems and that of stems on intact plants indicates the probable necessity of a root system for maximum growth but comparison of long and short excised stems shows that basal stem tissue itself can play a stimulating role.

Substitution of apical tissues by growth substances

Auxins and gibberellins have both been suggested as growth substances of apical origin which might influence stem elongation in the intact plant. The possible role of both these was tested by comparing the growth of intact stems with that of stems decapitated above internode 2 and variously treated with 10^{-4}M IAA and/or 1 p.p.m. GA_3. The growth substances were incorporated into agar and applied to the cut tips as small blocks. The IAA concentration had previously been ascertained to be optimal for promoting elongation of clone B internode sections, while 1 p.p.m. GA_3 was the lowest concentration found to produce marked stimulatory effects when applied to intact clone B plants. Fresh blocks were applied daily. Treatments were as follows, each consisting of 5 stems replicated twice:

1. Intact. Untreated control.
2. Decapitated. Water applied in blocks.
3. Decapitated. 10^{-4}M IAA applied in blocks.

4. Decapitated. 1 p.p.m. GA applied in blocks.
5. Decapitated. 10^{-4}M IAA plus 1 p.p.m. GA applied in blocks.

Table V. Total percentage increase in length of internode 2 in response to substitution of apical region by IAA and GA.

	Time (days)	1	2	3	4	5
1	Intact	37	95	164	217	243
2	Decap. + water	12	30	50	56	64
3	Decap. + IAA	34	70	89	102	106
4	Decap. + GA	27	66	103	136	150
5	Decap. + IAA/GA	30	71	99	111	117

Daily measurements of internode 2 (Table V) show that IAA effectively replaced the apex over the first 24 hours but that GA was more effective on a long term basis. Neither, however, maintained elongation at the level found in the intact plant over the whole 5-day period. It seems that a major apical factor besides auxin or gibberellin is probably necessary for the long-term maintenance of stem elongation. During the first 24 hours following decapitation (or excision of stem sections) this additional factor may not be limiting, thus allowing almost complete replacement of the apex by an auxin.

The effect of stem ringing on apical growth

Possible roles of basal tissues shown to play a vital role in the growth of apical tissues include (a) the supply of promoters via the xylem, (b) the supply of promoters via the phloem and/or ground tissue and (c) the removal of apical inhibitors via the phloem and/or ground tissue. To test the role of factors supplied by the xylem, stems were treated in three ways:

1. Control attached to parent plant. Unringed.

2. Attached to parent plant but ringed on internode E to kill all tissue but leave xylem functional

3. Attached to parent plant and ringed as in 2. but on internode A.

Ringing was performed using the hot wax technique described by Rabideau and Burr (1945). Transverse sections and tetrazolium tests made of the ringed area at the end of the experiment indicated that no live cells remained.

Measurements made after 2 days (Table VI) show that ringing on internode A strongly depressed elongation of young internodes and petioles, but a similar ring at internode E had a very much smaller effect.

Table VI. Average increment in length (mm) after 2 days in response to 'ringing'.

		Petiole 1	Internode 1	Internode 2
1.	Control	25.9	6.6	8.6
2.	Ringed at E	20.4	5.6	6.3
3.	Ringed at A	12.5	0.6	2.0

Essentially similar results were obtained in a further experiment in which the treatments given were:

1. Control attached to parent plant

2. Excised at base of internode A and placed with base in a Hoagland's mineral nutrient solution. Leaf A was removed

3. Attached to parent plant but ringed on internode A.

The immediate effect of ringing in this experiment (Table VII) was a greater decrease of elongation than in equivalent excised stems. This could have been the result of slight water stress; the day 4 measurements show little differences between extension growth of excised and ringed stems. Apparently xylem does not supply the necessary growth factors. Dissections made after 7 days further showed no significant difference between leaf initiation in excised and ringed stems during the week (Table VIII). Results in Table VIII represent the difference between the initial and the final number of leaves and primordia apical to leaf A.

Table VII. Average increment in length (mm) after 1 and 4 days in response to 'ringing' and excision.

	Day 1			Day 4		
	Petiole 1	Internode 1	Internode 2	Petiole 1	Internode 1	Internode 2
1. Control	9.0	5.4	1.9	39.6	12.3	14.4
2. Excised	5.8	2.7	0.3	16.1	3.2	1.4
3. Ringed at A	3.7	0.6	0.2	15.3	1.1	1.5

Table VIII. Number of leaf primordia initiated following ringing and excision.

	Leaves initiated		S.E.
Control	2.6	±	0.1
Excised	0.8	±	0.1
Ringed at A	1.1	±	0.1

The conclusion that the xylem provides no growth factors influencing elongation and leaf initiation does not hold for all aspects of growth. Anatomical investigations showed that secondary thickening and fibre development are influenced by the xylem connexion with the basal tissues in ringed stems.

After 7 days, transverse sections were cut of internode 1 and the variations in anatomical development recorded (Table IX). There can be little doubt that, whereas severing xylem contact with the stem base (as in excised stems) markedly decreases secondary tissue development and fibre thickening, the development of such secondary tissues and fibres in ringed stems is comparable with that of intact plants. In fact secondary vascular development is somewhat enhanced by ringing, although fibre thickening is not as great as in intact plants.

The effect of gibberellic acid and benzyladenine as substitutes for basal tissue

In view of the evidence that gibberellins and cytokinins are synthesized in roots it seemed possible that the xylem-transported factor leading to secondary and fibre development might be one or both of these growth factors. An experiment was therefore designed to test the effect of such substances on growth and development.

Apical stem tissues were excised at the basal end of internode A and set up in groups of 10, leaf A having been removed, with their bases in a Hoagland's mineral nutrient solution to which was added various concentrations of benzyladenine (BA) and/or GA_3. Untreated excised controls were set up similarly and all were compared with control stems remaining attached to the parent plant. Stems were dissected after 10 days to determine the number of leaf primordia initiated during the experiment and transverse sections were cut of internode 1.

Table IX. Comparative effect of ringing and excision on secondary development and fibre thickening.

	No. of secondary vascular cells per file	No. of interfascicular cambium cells/stem	Fibre thickening score	
			xylem	phloem
Control	4.4	5.6	5.0	5.0
Excised	3.7	1.4	0.6	0.4
Ringed at A	5.0	5.8	2.2	2.8

Table X. Influence of BA and GA on development of internode 1 and leaf initiation on excised stems. Cell numbers represent the average numbers in transverse sections and were recorded as described in Materials and Methods.

Treatment	Percent phloem fibres thickened	No. of thickened xylem fibres/bundle	No. of lignified parenchyma cells/bundle	No. of secondary vascular cells/file	No. of interfascicular cambium cells/bundle	No. of leaf primordia initiated
Attached control	82	4.0	40	5.0	5.5	4.5
Excised control	47	0.3	3	4.4	3.3	1.8
0.00001 p.p.m. BA	56	0.4	5	4.7	4.0	1.5
0.001 p.p.m. BA	69	0.3	4	4.0	3.0	1.9
0.1 p.p.m. BA	17	0.0	0	5.3	5.0	2.4
1.0 p.p.m. BA	9	0.0	0	6.4	6.2	2.2
1.0 p.p.m. GA	91	1.6	30	4.5	1.9	1.7
100 p.p.m. GA	64	0.7	30	5.3	3.1	1.4
0.00001/1.0 p.p.m. BA/GA	80	1.6	33	4.5	1.9	1.5
0.1/1.0 p.p.m. BA/GA	20	0.9	7	5.0	3.5	2.5

The results show (Table X) that BA and GA were able to replace the basal parts of the plant as far as development of internode 1 was concerned; GA increased the fibre thickening and lignification of parenchyma in excised stems to the level occurring in the intact control; BA at high concentrations raised cambial activity in excised stems to the control level. Similar concentrations of BA depressed fibre thickening and lignification, and increased leaf initiation significantly.

IV. DISCUSSION

Experimental evidence favours the hypothesis that growth in the apical region of white clover stems is promoted by basal and apical factors. Elucidation of the nature of these factors is rendered difficult because of their interactions but this investigation has led to the hypothesis illustrated by Fig. 2.

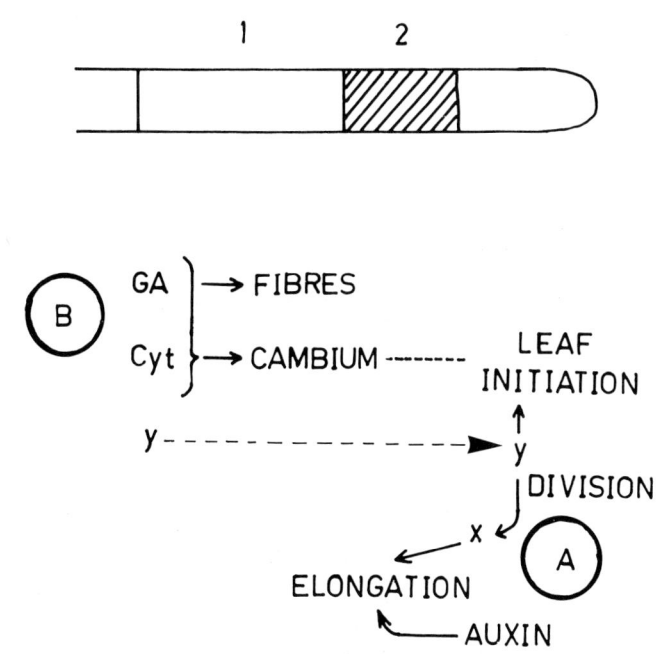

Fig. 2.

Hypothetical scheme showing the relationship between basal factors (B) and apical factors (A). Cyt = cytokinin.

It is proposed that the basal factor (B) consists of at least three components: cytokinin, gibberellin, and an unknown 'y'. The cytokinin and gibberellin stimulate secondary development in internode 1, the former also possibly being required for cell division (leaf initiation) in the apical meristem itself. Cytokinin and gibberellin travel from the plant base to the apex in the xylem and possibly originate in the roots; 'y' on the other hand, is not active when the base is not in communication with the apex via living cells (phloem and/or parenchyma), and it could therefore be envisaged as the removal of an inhibitory condition from the apex just as readily as a directly promoting substance.

Apical factor (A) likewise consists of at least two components: auxin replaces the apex effectively on a short term basis as a stimulus to internode 2 elongation in decapitated stems; longer term, however, a further component,'x,' is required to maintain such elongation; 'x' is dependent on B for its synthesis or action and similarly could involve removal of an inhibitory condition by B.

Interaction of A and B are clearly seen: accumulation of A (perhaps the auxin component) stimulates secondary cell formation to a level above normal in internode 1 on ringed stems; internode 2 elongation cannot respond to x in the absence of B. While the picture obtained as yet is extremely incomplete - x and y remain to be identified and probably will be found themselves to consist of several major components - it is apparent that continued investigation of the coordination of plant growth in the light of current knowledge of plant growth regulators can contribute considerably to our understanding of the plant as a whole.

V. SUMMARY

The roles of basal tissues (roots and older regions of shoots) and apical tissues (leaf primordia and apical meristem) as regulators of the growth of the apical regions of stems (young leaves and internodes) were studied. Both contribute to the elongation of internodes but secondary thickening of internodes is mainly controlled by factors transported to them from the basal tissues via the xylem.

The apical role in internode elongation is replaced by substituted IAA over the first 24 hr. but a further unknown factor (x) is required for continued elongation. Benzyladenine and GA together have the same effect as xylem-transported factors on secondary thickening. The effect of basal tissues on division and elongation of younger apical tissues, however, is not mediated by the xylem. An additional basal factor (y), ineffective in the absence of a continuum of live cells between apex and base, is postulated to be necessary for apical activity in addition to auxin. It may well be that y is necessary for the production of x.

It is not yet possible to decide whether x or y are direct promoters of growth or whether they act by removal of an inhibition.

VI. REFERENCES

CARR, D.J. and BURROWS, W.J. (1966) Evidence of the presence in xylem sap of substances with kinetin-like activity. Life Sci. 5, 2061-77.
CHILD, C.M. and BELLAMY, A.W. (1919). Physiological isolation by low temperature in *Bryophyllum* and other plants. Science 50, 362-5.
ERITH, A.G. (1924). "White clover (*Trifolium repens* L.). A monograph." Duckworth & Co., London.
PHILLIPS, I.D.J. (1969). Apical dominance, in "The physiology of plant growth and development". (Ed. M.B. Wilkins). McGraw-Hill, London. pp. 165-202.
PHILLIPS, I.D.J., and JONES, R.L. (1964). Gibberellin-like activity in bleeding sap of root systems of *Helianthus annuus* detected by a new dwarf pea epicotyl assay and other methods. Planta 63, 269-78.
RABIDEAU, G.S. and BURR, G.O. (1945). The use of the C^{13} isotope as a tracer for transport studies in plants. Amer. Jour. Bot. 32, 349-56.
SNOW, R. (1925). The correlative inhibition of the growth of axillary buds. Ann. Bot. Lond. 39, 841-59.
THOMAS, R.G. (1962). The initiation and growth of axillary bud primordia in relation to flowering in *Trifolium repens* L. Ann. Bot. N.S. 26, 329-44.
WENT, F.W. (1938). Specific factors other than auxin affecting growth and root formation. Pl. Physiol. Lancaster 13, 55-80.
WENT, F.W. (1943). Effect of the root system on stemgrowth. Pl. Physiol. Lancaster 18, 51-6.

PLANT GROWTH SUBSTANCES, 1970

Studies on Leaf Unrolling in Barley

D.J. Carr*, Clements, J.B.**, R. Menhenett*

* Research School of Biological Sciences, Australian National University, Canberra.
** Botany Department, Queen's University, Belfast, N. Ireland

I. INTRODUCTION

The first leaf of a grass seedling is normally folded in the form of a scroll inside the coleoptile (Duval-Jouve, 1875). In light-grown plants the leaf unrolls as it emerges from the tip of the coleoptile. When grass seedlings are grown in the dark unrolling does not occur. Unrolling of the leaf is due to differential cell expansion largely within the mesophyll (Burström, 1942). Dark-grown leaves are roughly circular in cross-section with the margins considerably overlapping each other. When unrolling occurs, the diameter of the cross-section increases and it bears a weakly sigmoidal relationship to the length of the inner arc of the leaf (Virgin, 1962). However, this lack of linearity does not seriously impair quantitative estimates of the degree of unrolling based on measurements of the projected leaf width. Virgin (1962) determined the action spectrum for light-induced unrolling of the first leaf of wheat seedlings and found it to be a process promoted by red light (c. 660 nm), the promoting effect being nullified by far-red illumination (c.730 nm). This evidence, which we and others (e.g. Wagné, 1964, 1965) have extended to barley seedlings, identifies the pigment system involved as phytochrome. A considerable number of morphogenetic phenomena (Mohr, 1969) including the unbending of the plumular hook of dark-grown legume seedlings, a process analogous in some ways to grass leaf unrolling, are known to be mediated by light absorbed in phytochrome. Most of these responses are developmental events involving irreversible steps in differentiation. There is considerable and growing evidence in the literature to support the view that these steps involve gene action (Mohr, 1966, 1969). Inhibitors of protein synthesis are often potent inhibitors of these phytochrome-mediated responses. Carr and Reid (1966) have shown for instance that the unrolling of barley leaves and plumular hook unbending are inhibited by actinomycin D. Evidently protein synthesis, possibly specific to the unrolling response, is set in train by absorption of red light in phytochrome.

Evidence has also accumulated that growth substances, especially gibberellins, may be involved in photoresponses mediated by phytochrome. Indeed Brian had already in 1958 suggested that gibberellin-like hormones might be synthesized in leaves following irradiation with red light. This suggestion was based on an assessment of the available evidence which showed that GA often had effects similar to those observed after exposure of plants or plant parts to red light. It is therefore of considerable interest that exogenous application of GA_3 to dark-grown peas and beans can cause unbending of the plumular hook, replacing the need for red light. Burström (1942) decided that the mesophyll cell expansion responsible for unrolling of etiolated grass leaves must be accompanied by a transient increase in wall extensibility. However, contrary to his expectations auxin, even at concentrations as low as 10^{-7} Molar, only inhibited unrolling. These considerations lead to the possibility of an examination of a phytochrome-mediated response - such as grass leaf unfolding - in terms of regulation by growth substances.

II. MATERIALS AND METHODS

Barley seed (most of our work has been done with var. Pallas) is sown about 1 cm deep in washed, autoclaved river sand saturated with distilled water. No further water is applied, but excess water is allowed to drain away. The trays are consigned to a darkroom at 25±1°C and the first leaves harvested for experiments after 6-8 days. Harvesting is done with the aid of a dim green safelight. The ANU safelight consists of a green fluorescent tube (Philips TL 40w/17) with a filter of 4 sheets of green Cinemoid No.39 and one of primary blue Cinemoid No.20. The energy distribution is asymmetrical about 510 nm at which wavelength the energy at bench height is about $10^{-4}uW/cm^2/sec/nm$. There is a slight unrolling effect of this safelight if leaf segments are left under it for more than 90 minutes after cutting. The red source consists of 2 Philips TL 20w/15 61cm red fluorescent tubes with a 3mm thick red plexiglass filter Cornelius 501. The red source in the Belfast experiments was very much weaker; it consisted of the red plexiglass filter and two 4' red Atlas fluorescent tubes, the output of which was very much lower at the effective wavelength (660 nm) than that of the Philips tubes. This was placed 80 cm above the plant material. The far-red source in the Belfast experiments was a simple one consisting of 3-60w tungsten fluorescent lamps and one layer each of primary red, primary blue and orange Cinemoid. This again was also much weaker than the Canberra far-red source which consists of radiation from a Philips 250 w infra-red lamp with internal reflector, filtered through 10 cm of water, and 3 mm black plexiglas generally with the addition of a sheet each of orange and primary deep blue Cinemoid. This complex filter cuts off at 725 nm.

After experimental treatment leaves are left in the darkroom in darkness for usually 18-24 hours and leaf widths measured by projection at a standard magnification in an enlarger. Chemicals are applied using mild and standardised vacuum infiltration techniques. All treatments are carried out in the darkroom using the safelight as little as possible, usually less than 30 minutes. Further details of methods are given in Reid (1967), Reid, Clements and Carr (1968) and Crossley (1970).

Methods of extraction and bioassay of growth substances from the leaves are those in common use in most laboratories. Growth substances are usually extracted from whole leaves, but a segment 1 cm long is cut 1 cm from the tip of each leaf for use in experiments on effects of irradiation or of chemical substances. Shorter segments are subject to cutting injuries which induce some unrolling. Large numbers of segments are cut, randomised and assigned in batches to the treatments. During handling the segments are kept quite fully turgid by floating them on water in petri dishes or laying them on moist filter paper.

III. RESULTS AND DISCUSSION

If briefly illuminated with red light and then put back into darkness the leaf segments unroll in about 18-24 hours. The response is reversible by far-red but far-red alone produces some unrolling as expected on a hypothesis of phytochrome recycling. The starting point of our interest in this system was the discovery that it is strongly inhibited by actinomycin D (Carr and Reid, 1966), applied at the time of unrolling (Table 1). Also very effective is C.I.P.C. (isopropyl N-(3 chlorophenyl carbamate) (Mann *et al*, 1967) at 100 mg.l. Chloramphenicol at 100 mg.l and puromycin at 50 mg.l are quite ineffective. Cycloheximide is very effective in inhibiting unrolling, and remains effective when applied quite late in the process (Fig. 1). One difficulty with these protein synthesis inhibitors is that they may inhibit processes far removed from those associated directly with the immediate effects of light, which are of course those of greatest interest. Nevertheless, their effects indicated that syntheses followed illumination and were involved in the photoresponse.

It was therefore of great interest to discover that unrolling can be achieved without the aid of light, by applying growth substances to leaf segments. The time course seems not to be different whether unrolling is induced by growth substances or by light. In the Belfast experiments (carried out by Reid, Carr and honours students M.K. Garrett (1967) and J.B. Clements (1968) at Queen's University Botany Department) gibberellic acid at 10 mg l. produced a full unrolling response equivalent to a saturating dose of red light. There are two disturbing features of gibberellin-induced unrolling which still await explanation. The first is that it is less inhibited by actinomycin D than is unrolling induced by red light. The second is that far-red produced a far smaller reversal of gibberellin-induced unrolling than it did of red light-induced unrolling. In effect, far-red increased the variability of the response to GA3. Analysis showed that this was due to the existence at the termination of the experiment of two populations of leaf segments - those which had unrolled more or less completely and those which had unrolled very little. This analysis suggested in turn that perhaps the time interval between application of GA_3 and illumination with far-red might be of importance in determining the degree of reversibility of the GA-induced unrolling. This proved, indeed, the case (Fig. 2).

Table 1. Effects of RNA & protein synthesis inhibitors on red light-induced unrolling.

Applied before red light	Applied after red light	Conc.
ACT.D 4%	ACT.D 0%	90 mg/l
6-AZAURACIL 86%	6-AZAURACIL 86%	100 mg/l
CYCLOHEXIMIDE 0%	CYCLOHEXIMIDE 0%	22.5 mg/l
CIPC 0%	CIPC 7%	100 mg/l
CAP 145%	CAP 129%	100 mg/l
ABSCISIC ACID 0%	ABSCISIC ACID 0%	22.5 mg/l

UNROLLING GIVEN AS % OF RED LIGHT-INDUCED UNROLLING

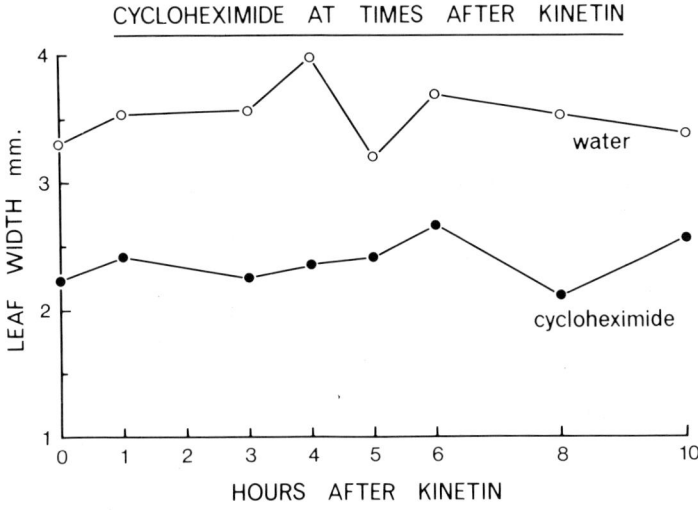

Fig. 1. Effectiveness of cycloheximide (22.5 mg. l) inhibition of leaf unrolling when applied at successive times during leaf unrolling. Unrolling continued for 20 hours after application of kinetin; leaves kept in darkness at all times.

Nevertheless, a majority of leaves unrolled even if irradiated with far-red at the same time as given GA3. The morphactin, IT 3233 (N-butyl-9-hydroxy fluorene-9-carboxylate, Merck, Darmstadt) at 25 mg l. fully eliminated the unrolling due to GA_3, and also that due to red light.

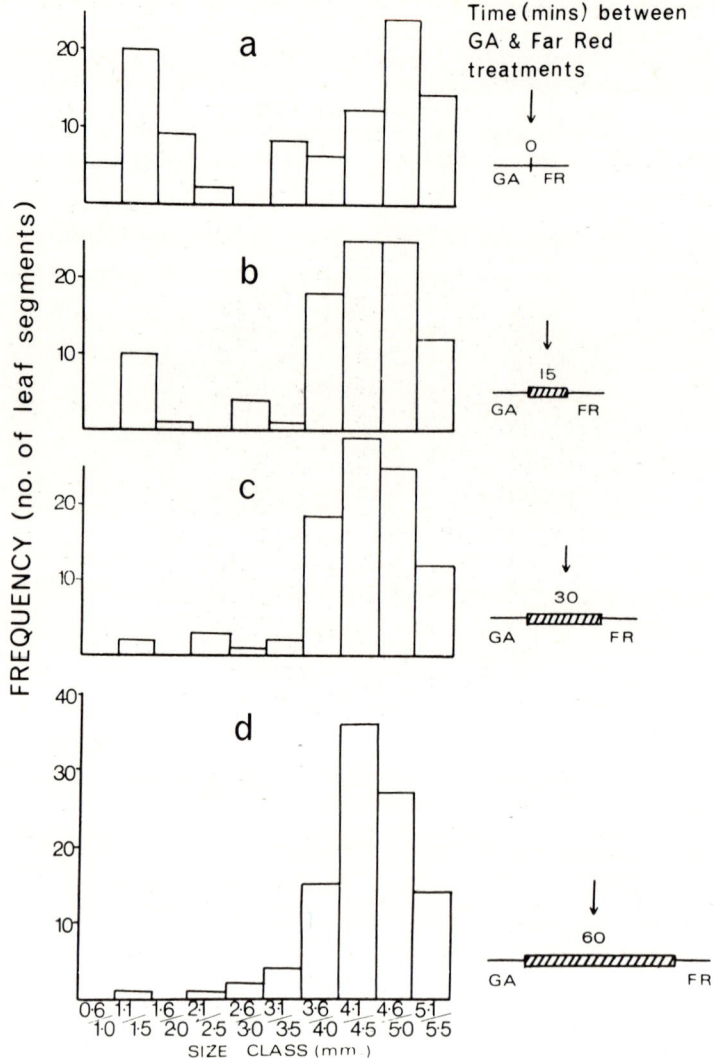

Fig. 2. Following treatment of leaf segments with GA$_3$, the segments were irradiated with far-red either immediately or after intervals of time. The histograms (frequency classes of degrees of unrolling) illustrate the rapidly developed irreversibility of the response to GA by far red.

The fact that gibberellic acid could cause unrolling led to an attempt to detect the appearance of gibberellins in the leaves after red irradiation, as predicted by Brian (1958). This was done initially on a small scale using a method in which barley aleurone layers were mildly homogenized on a glass plate and the cellular homogenate spread over a starch agar plate. Drops of test solutions were applied and the plate was incubated and later developed with iodine. Half an hour of red light (Fig. 3) caused the appearance of gibberellin in the extracts from the leaves, as compared with extracts from leaves kept in darkness.

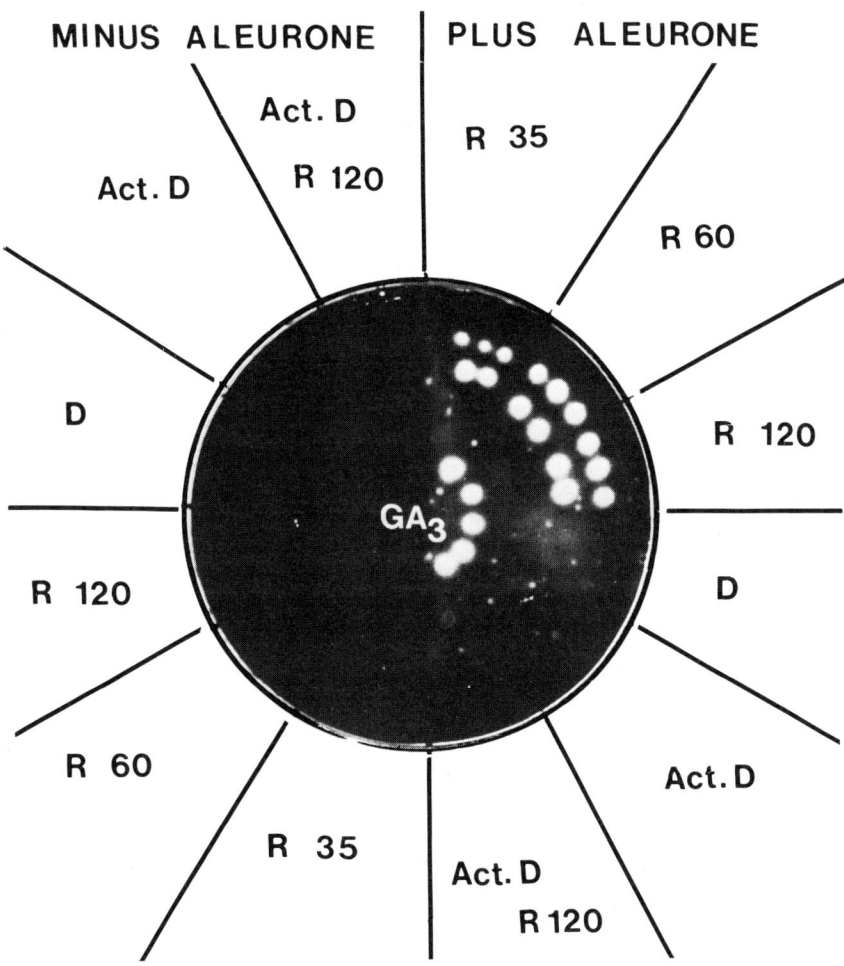

Fig. 3. Bioassay of gibberellin-like substances in barley leaves. The petri dish contained starch-agar: on one half (+ aleurone) was spread an homogenate containing whole cells of barley aleurone (var. Pallas) layers. The other half (- aleurone) had no aleurone cells. Drops (2 μl) of the test solutions (5 drops per sector) were applied and the dish incubated at 25° for 3 hours. The dish was then flooded with iodine solution. The clear plaques at the centre correspond to drops of GA_3 (0.01 mg.l.). R-35, R-60, R-120 correspond to extracts from leaves made 35, 60, and 120 minutes after exposure to 35 min of red light, followed by darkness. Segments were also infiltrated with actinomycin-D in the dark (ActD) and prior to red light Act.D-R). Extracts of these segments were made at 120 min after the light treatment. D refers to extracts of leaf segments kept in darkness. Gibberellin-like substances are apparent in R-35 and even more in R-60. Actinomycin-D inhibits their formation.

Later, the time course was more extensively examined using larger scale extractions and the standard barley endosperm bioassay (Nicholls and Paleg, 1963) and the dwarf-1-maize bioassay (Fig. 4), which appears to be particularly sensitive to the gibberellins in barley leaves.

The period of illumination (red light) given in these experiments was 30 minutes and already at the end of this time there was a much greater amount of extractable gibberellin in the leaves than in the dark controls (Fig. 5).

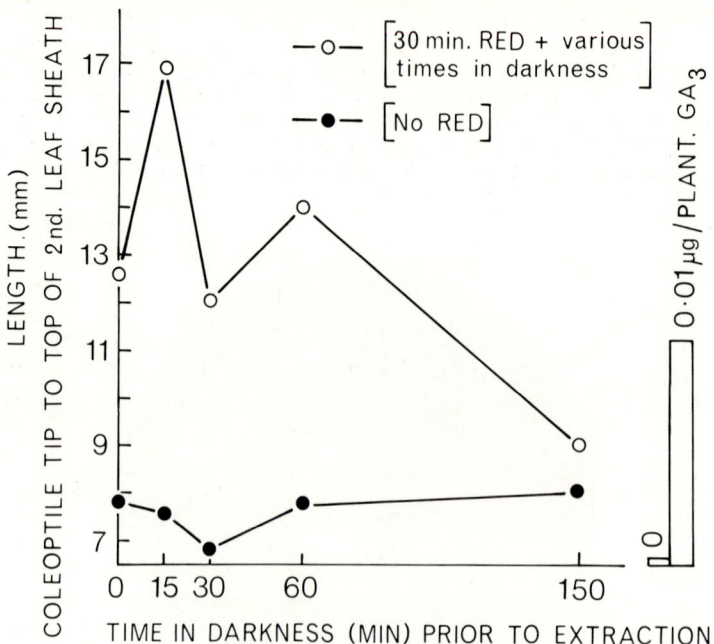

Fig. 4. Dwarf-1 maize bioassay of extracts from illuminated barley leaves.

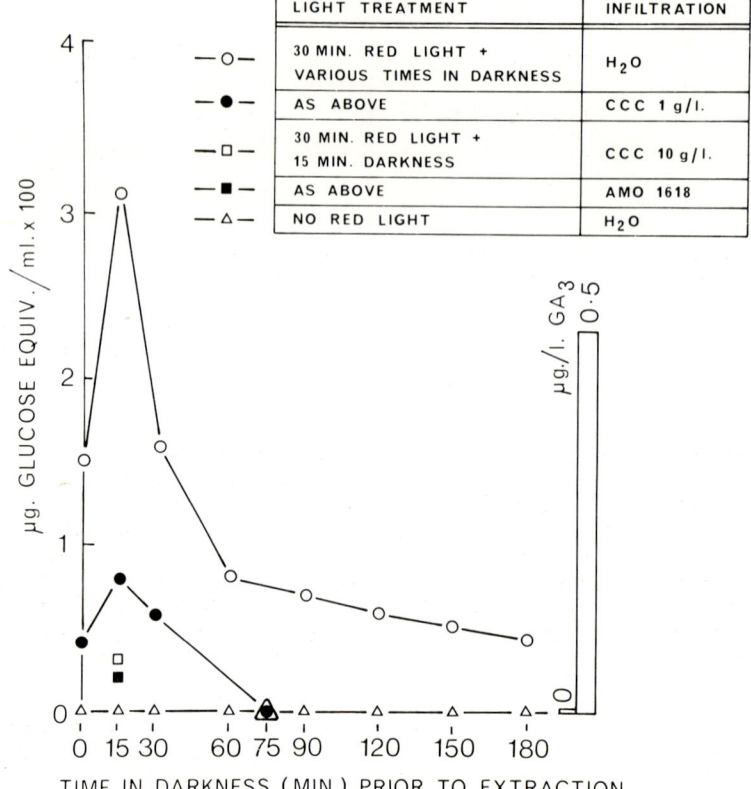

Fig. 5. Barley endosperm assay of extracts from variously treated illuminated barley leaves.

The content of gibberellin reached a peak about 15 minutes after the end of a 30 min illumination period and then slowly declined over a period of an hour. A very short period (<10 min) of illumination with our weak (Belfast) red light source produced no gibberellin. A continuous illumination gave the same rise in level of activity followed by a decline as was given by a 30-minute illumination; thus the decline is not an effect of returning the leaves to darkness. However, after 6 hours of red illumination, there is a steady rise in gibberellin activity which continues until at least 14 hours (Fig. 6).

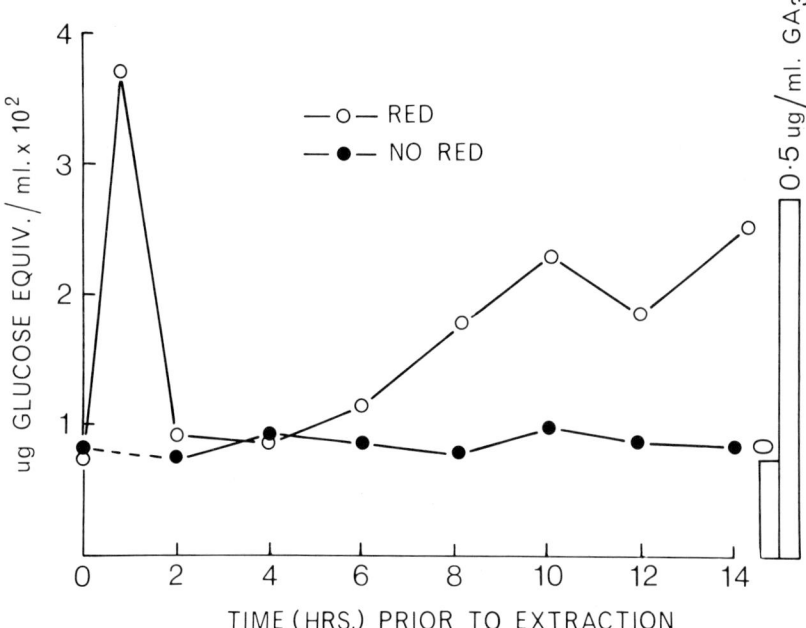

Fig. 6. The effect of longer periods of illumination on gibberellin activity of extracts from barley leaf segments.

As it could be argued that the decline in gibberellin activity after a short illumination could be due to depletion of some precursor stored within dark-grown leaves, experiments were carried out in which intact leaves (or leaf segments) were given pulses each of 30 minutes of red light separated by 90 minutes of darkness. In all, three pulses of light were given. The leaves or leaf segments were then extracted and gibberellin activity bioassayed using the barley endosperm technique (Fig. 7).

There was no difference between whole intact leaves and leaf segments, indicating that gibberellin production does not depend on mobilisation of some material into the leaf during the time of the experiment. Each of the first two pulses of red light produced a burst of gibberellin production, followed by a decline, but the third red light pulse did not. Evidently some precursor is consumed in the induction of gibberellin formation by light. Effects of far-red on gibberellin production were tested by irradiating for 15 minutes (the minimal time) with red and then for 30 minutes with far-red. <u>Far-red irradiation had no effect on the appearance of gibberellin</u> in the leaves, nor on the subsequent decline in gibberellin levels in the leaves in the dark (Fig. 8).

Inhibitors of gibberellin biosynthesis (AMO 1618 and CCC) were very effective in inhibiting gibberellin production following red light (Reid et al., 1968); AMO 1618 was the more effective inhibitor (Fig. 5). IT 3233 had no effect on gibberellin synthesis. Actinomycin D and chloramphenicol almost completely inhibited red-light-induced gibberellin synthesis (Reid and Clements, 1968); cycloheximide and CIPC were much less effective. The two effective inhibitors also inhibited the steady rise due to prolonged illumination for periods of 12 hours or over.

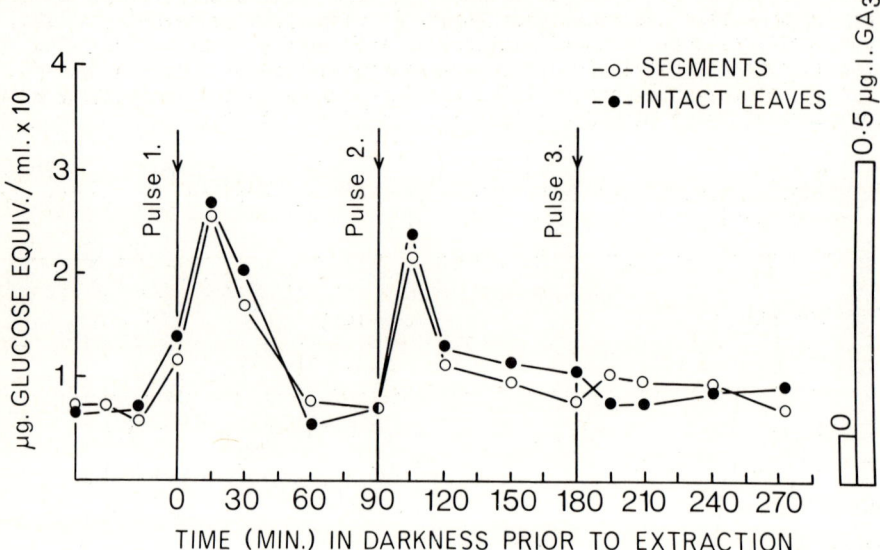

Fig. 7. Gibberellin activity of barley leaf segments v. leaves intact on barley plants, following pulses of red light, each of 30 minutes duration. Note the close similarity between extracts from segments and from intact leaves.

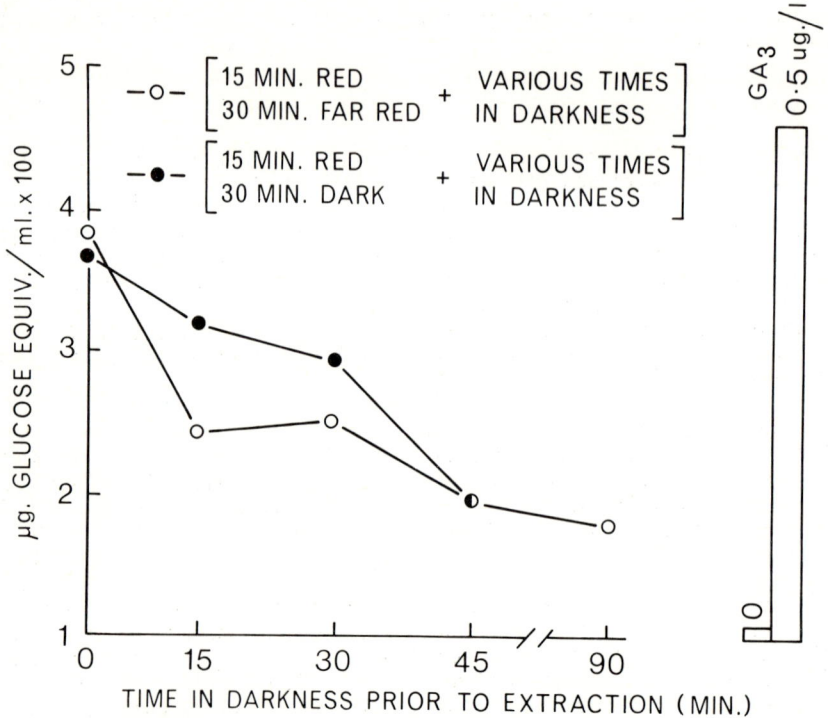

Fig. 8. Lack of effect of far-red on the red light-induced synthesis of gibberellin in barley leaf segments.

One immediate question is: will the gibberellins produced in response to red light cause leaf unrolling? Batches of etiolated leaf segments exposed to 40 minutes of red light or kept in darkness throughout were macerated and extracted in phosphate buffer pH 7.0 acidified and extracted repeatedly with ethyl acetate. The ethyl acetate and aqueous extracts were further fractionated into acidic, neutral and basic fractions which were assayed separately for effectiveness in causing unrolling as well as by conventional gibberellin bioassays. The undiluted extracts tended to be somewhat inhibitory. Results from a X10 dilution are shown in Table 2.

Table 2. Effectiveness of extracts of barley leaves on unrolling of segments of dark grown leaves of barley.

Fraction assayed	Extract of leaves kept in:-	
	Red light	Darkness
	Mean leaf widths, mm.	
Acidic	1.5±0.3**	1.5±0.3
Acidic Butanol*	1.2±0.5	1.2±0.6
Neutral	3.2±0.4	1.4±0.3
Basic	1.5±0.5	1.5±0.5
GA_3 (10 mg.l) + darkness	3.7±0.4	
Red light	3.9±0.5	
Darkness	1.3±0.4	

* Aqueous extract after separation of the acidic fraction was re-extracted with n-Butanol.

** Standard error.

Extracts made from leaves exposed to red light for 2 hours showed the expected overall decline in total gibberellin activity as well as a loss of unrolling activity in the neutral fraction. Thus the transient rise in gibberellin activity in the leaves is accompanied by a transient rise in an extractable gibberellin-like substance active in causing unrolling. In view of the inhibitory action of auxin on unrolling mentioned by Burström it was of interest to see whether the dark grown leaves contained considerable quantities of auxin, and whether leaves after illumination had less. Extracts (cold absolute ethanol) of leaves illuminated for 0,4,8 and 12 hours were chromatographed and assayed using the oat coleoptile straight growth assay (Fig. 9) (Nitsch and Nitsch, 1956).

Significant auxin activity was detected in all the extracts but there was no apparent change in the amount due to irradiation with red light. Extracts were also made for auxin bioassays from segments treated with gibberellic acid. There was a marked fall in auxin activity in these leaf segments with no auxin detectable 6,8 or 10 hours after the commencement of gibberellin treatment. The reasons for this decline in auxin activity are obscure.

At this point in the elucidation of the hormonal regulation the work was taken up in Canberra necessarily with some changes in the conditions of growth of the plant material, the source of seed, lighting conditions, etc. It was surprising and as yet unexplicable that the leaf segments (which in the Belfast experiments did not respond at all or not markedly to kinetin) now gave a full unrolling response to that compound and had a much smaller response to gibberellin. Segments floated for 22 hours on 2.5 mg l. kinetin unrolled completely. With vacuum infiltration techniques (brief

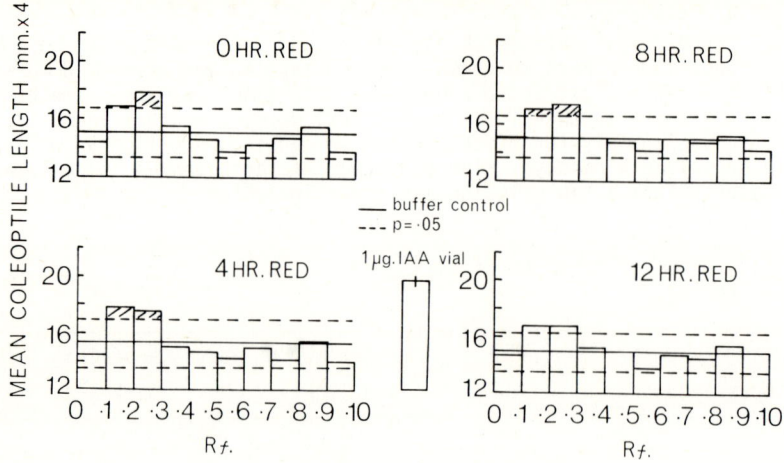

Fig. 9. Results of bioassay of auxin extracts from leaf segments illuminated for different periods. 10 gm of tissue was macerated and extracted by cold absolute ethanol. The auxins were taken up in water after 24 hr in the cold, acidified, extracted with diethyl ether (3x) and the ether washed with 5% NaHCO$_3$ (3x). The aqueous extract was acidified and extracted with ether and the ether fraction chromatographed on Whatman No.1 paper, descending, with water. The bioassay is as referred to in the text.

application) a higher concentration was required. The relative lack of sensitivity to GA$_3$ did not appear to a function of leaf age and other gibberellins (GA$_5$ and GA$_7$) also failed to elicit the former large unrolling response. Cytokinins other than kinetin are effective: zeatin at a given concentration has about one tenth the activity of kinetin. These experiences have naturally led to the hypothesis that cytokinin is synthesized or released in the leaves following irradiation with red light. One of us (Menhenett, 1972) has in fact extracted and bioassayed a cytokinin from dark grown barley leaves. The amounts are small and bioassay has been difficult. We have as yet no evidence of a change in the amount of cytokinin following irradiation.

One of the difficulties experienced in handling extracts from the barley leaves is the presence of considerable amounts of inhibitory materials. The possibility that these might indeed be regulatory has recently been investigated. In fact, there is an extractable inhibitor of unrolling. (Fig. 10). We are not yet certain of its nature but it is certainly not abscisic acid. An important fact about this inhibitor is that it declines in amount during the 6 hours after irradiation.

This is about the length of the lag period before unrolling commences and is also about the duration of the escape time of the phytochrome controlled part of the process of unrolling.

During the course of this work it had become apparent that abscisic acid strongly inhibits the unrolling response to both red light and gibberellic acid. Actually ABA, like cycloheximide, is inhibitory at virtually any stage of the unrolling response and it therefore cannot be determined whether it has any specific effect on the reactions immediately following irradiation. It is known however (Beevers et al, 1970) to suppress the appearance of gibberellin in the irradiated leaves.

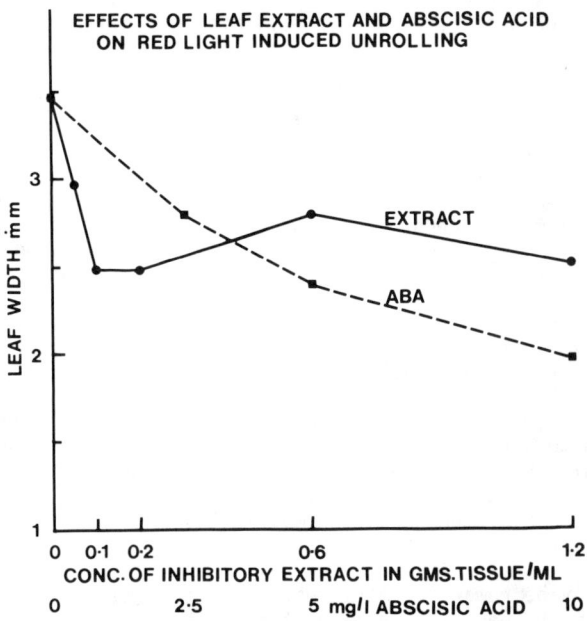

Fig. 10. Bioassay of an unknown inhibitor from dark grown barley leaves - its effects on unrolling of illuminated barley leaves compared with that of ABA, at different concentrations.

IV. DISCUSSION

The apparently relatively simple morphogenetic response here investigated has proved to be very complex in its physiology. We do not yet know how phytochrome acts in this system. The relatively long escape time (6 hours in the system set up in Canberra, Menhenett, 1972) permits the possibility of gene expression but we are still far from demonstrating that specific gene products are involved. As in some other slow phytochrome-mediated responses, inhibitors of protein synthesis inhibit leaf unrolling, some most effectively when applied before, coincident with or soon after irradiation (Carr and Reid, 1966). Actinomycin D is very effective when applied at the time of irradiation; its effectiveness declines after that almost to zero by about 6-8 hours. This suggests a release of information for unrolling, completed by the end of the lag phase (about 8 hours). The possibility of phytochrome operating through hormonal regulation is still open since gibberellins and cytokinins can replace the need for light. In principle, actinomycin D and CIPC applied at time zero could act by inhibiting synthesis of gibberellin in the irradiated leaf and so blocking the unrolling response. However, cycloheximide is less effective in blocking gibberellin synthesis but very effective at any time in blocking unrolling, whereas chloramphenicol has just the opposite effects - inhibiting GA synthesis effectively but having little or no effect on unrolling. Far-red irradiation also fails to suppress red-induced GA synthesis (Fig. 8), contrary to the unsupported assertion of Beevers et al, 1970, although it may exert a small effect on unrolling caused by exogenous GA. Indeed, there has been no unequivocal demonstration (pace Beevers et al, loc. cit.) that the appearance of gibberellins following irradiation of dark-grown wheat or barley leaves is a phytochrome-mediated response. The repeated production of GA following a sojourn in darkness and a further irradiation (Fig. 7) is strongly reminiscent of protochlorophyll synthesis in the dark and its conversion in the light. The contrasting effects of cycloheximide and chloramphenicol, held to be specific inhibitors of protein synthesis on cytoplasmic and chloroplast ribosomes, respectively, on unrolling and GA formation also suggest that the latter process is probably a chloroplast function. Proof of this has yet to be given, but some evidence in its favour was obtained by Reid (1967). From such considerations it would appear that GA synthesis following irradiation may represent an effect of light unconnected, in a causal sense, with leaf unrolling. In a variety of barley (cv. Cape) examined in Canberra, a red light-induced increase in gibberellins occurs (P.B. Goodwin, pers. comm.) but only cytokinins are able to substitute for light in unrolling (Crossley, 1970). The light-induced *synthesis* (it is suppressed by AMO 1618 and CCC) of gibberellins has the nature of a trigger reaction. A certain minimum period of red light (in the Belfast experiments 10 minutes of rather weak red light) was necessary for any gibberellin to appear. After that, the amount of gibberellin produced seems independent of the amount of light given (Fig. 6). This synthesis itself is an intriguing response to light, worthy of study on its own account irrespective of its true role in hormonal regulation of leaf unrolling.

In some unrolling systems exogenously applied gibberellins can substitute for light. Alternatively other hormones may be able to bring about the same action as, or a more effective action than, gibberellin. Cytokinin is present in the leaves of Pallas barley and in the presence of a falling background of inhibitor the unrolling effect of the steady although small level of cytokinin may find expression. It would clearly be rewarding to try to rediscover and investigate the nature of the transient "unrolling factor" discovered by J.B.C. in the neutral fraction. One of the most surprising features is the apparent lack of a role for auxin. Since cell expansion is indicated as the basis of the unrolling movement the inhibitory action of auxin, first discovered by Burström, remains a paradox.

It would clearly be rewarding to investigate the nature of the transient "unrolling factor" in the neutral fraction of solvent extractions.

V. SUMMARY

Unrolling of the first seedling leaf of dark-grown barley seedlings takes place in response to irradiation by red light and is a phytochrome-mediated response. Gibberellic acid and/or cytokinins can replace the need for red light. Following brief but supra-threshold irradiation with red light, gibberellin-like substances rapidly appear in the intact leaf or in leaf segments. The level of gibberellins increases dramatically to a peak shortly after the end of irradiation and then declines over a period of an hour or so. Another smaller burst of gibberellin production can be obtained by a subsequent period of irradiation but not by a third. Steady illumination over a period of 12-14 hours results in a slow rise in gibberellin level, after the initial burst. Inhibitors of gibberellin biosynthesis suppress the production of gibberellins following light. Certain inhibitors of protein synthesis also suppress it. Far-red irradiation does not elicit gibberellin synthesis nor suppress it following red irradiation. Actinomycin D and cycloheximide effectively inhibit unrolling but the latter antibiotic has little effect on gibberellin synthesis. Chloramphenicol on the other hand strongly suppresses gibberellin synthesis but has no, or even stimulatory, effects on unrolling. Extracts from light-treated leaves contain a substance which stimulates non-irradiated segments to unroll. The presence of this factor in the leaf is as transient as that of the gibberellins, following light.

Using the same variety of barley, the system set up in Canberra is more responsive to cytokinins than to gibberellins. However, attempts to demonstrate a higher level of cytokinins in leaves after irradiation have not been successful. Even dark grown leaves contain a low but consistently demonstrable level of cytokinins. On the other hand levels of an extractable inhibitor of leaf unrolling fall following red irradiation. The chemical nature of this inhibitor, as well as that of the transient unrolling factor, is still under investigation. Auxin does not cause unrolling and may be inhibitory even at low concentrations. Auxin levels in the leaves remain unaffected by light treatment but fall after treatment with GA_3.

Barley leaf unrolling does not involve cell division and depends solely on cell expansion. Phytochrome control can be mimicked by hormones, suggesting internal hormonal regulation. The nature of this regulation is still obscure.

VI. ACKNOWLEDGEMENTS

We thank Professor D.M. Reid, Dr. M.K. Garrett and Mr. D.J. Crossley for permission to use information from their theses.

VII. REFERENCES

BEEVERS, L., LOVEYS, B., PEARSON, J.A. and WAREING, P.F. (1970). Phytochrome and hormonal control of expansion and greening of etiolated wheat leaves. Planta, 90, 286-294.

BRIAN, P.W. (1958). Role of gibberellin-like hormones in regulation of plant growth and flowering. Nature, 181, 1122-1123.

BURSTRÖM, H. (1942). Über Entfaltung und Einrollen eines mesophilen Grasblattes. Bot.Notiser, 7, 351-362.

CARR, D.J. and REID, D.M. (1966). Actinomycin-D inhibition of phytochrome-mediated responses. Planta, 69, 70-78.

CLEMENTS, J.B. (1968). Effects of light upon the level of hormones in leaves of barley (B.Sc.Hons.Thesis, Queen's University, Belfast).

CROSSLEY, D.J. (1970). Phytochrome and barley leaf unrolling. (M.Sc.Thesis, A.N.U. Canberra).

DUVAL-JOUVE, J. (1875). Histotaxie des feuilles des Graminées. Ann.Sci.Nat.Bot. Ser. IV, I.

GARRETT, M.K. (1967). Mode of action of phytochrome in leaf unrolling of barley. (B.Sc.Hons.Thesis. Queen's University, Belfast).

MANN, J.D., HAID, H., JORDAN, L.S. and DAY, B.E. (1967). Inhibition of the action of phytochrome by the herbicide CIPC. Nature, 214, 420-421.

MENHENETT, R. (1972). Hormonal and phytochrome-induced unrolling of barley leaves. Ph.D thesis A.N.U., Canberra.

MOHR, H. (1966). Differential gene activation as a mode of action of phytochrome 730. Photochem.Photobiol. 5, 469-483.

MOHR, H. (1969). Photomorphogenesis. In "The Physiology of plant growth and development". M.B. Wilkins, Ed., McGraw Hill, London, 507-556.

NICHOLLS, P.B. and PALEG, L.G. (1963). A barley endosperm assay for gibberellins. Nature, 199, 823-824.

NITSCH, J.P. and NITSCH, C. (1956). Studies on the growth of coleoptile and first internode sections. A new, sensitive straight growth test for auxins. Plant Physiol. 31, 94-

REID, D.M. (1967). Sites of synthesis and mode of action of gibberellins in higher plants. Ph.D. Thesis, Belfast.

REID, D.M. and CLEMENTS, J.B. (1968). RNA and protein synthesis: prerequisites of red-light-induced gibberellin synthesis. Nature, 219, 607-609.

REID, D.M., CLEMENTS, J.B. and CARR, D.J. (1968). Red light induction of gibberellin synthesis in leaves. Nature, 217, 580-582.

VIRGIN, H.I. (1962). Light-induced unfolding of the grass leaf. Physiol.Plant. 15, 380-389.

WAGNÉ, C. (1964). The distribution of the light-effect on partly irradiated grass leaves. Physiol.Plant. 17, 751-756.

WAGNÉ, C. (1965). The distribution of the light effect from irradiated to non-irradiated parts of grass leaves. Physiol.Plant. 18, 1001-1006.

PLANT GROWTH SUBSTANCES, 1970

The Effects of Growth Regulators on RNA Metabolism during the Unrolling of Barley Leaf Segments

Rozanne Poulson and Leonard Beevers*

R.S.B.S., Australian National University, Canberra, A.C.T. 2601

* Dept. of Horticulture, University of Illinois, Urbana - Champaign, Illinois, 61801

I. INTRODUCTION

Photomorphogenesis in the barley leaf is characterised by an increased synthesis of chlorophyll and an associated increase in leaf width. These changes are accompanied by an increased capacity for RNA and protein synthesis. We have previously demonstrated that photoinduced unrolling can be enhanced by GA and, furthermore, this growth regulator can induce unrolling of segments maintained in the dark. ABA prevents photoinduced unrolling (Poulson and Beevers 1970). In view of the light-induced changes in RNA metabolism of barley leaf segments (Poulson and Beevers 1970) it was of interest to determine whether the hormonal treatments which influenced unrolling affected the pattern of RNA synthesis of the segments.

II. MATERIALS AND METHODS

Sections, 7mm in length, were prepared from the primary leaf of etiolated barley seedlings and incubated as described previously (Poulson and Beevers 1970). Chlorophyll content of the leaf segments was determined on an 80% (v/v) ethanol extract. Protein and RNA contents of the ethanol insoluble residue were determined according to the method of Osborne (1962).

In investigations of RNA synthesis segments were incubated for 6hr in the presence of 200 μc carrier-free ^{32}P-orthophosphate prior to extraction of RNA (Poulson and Beevers 1970). The extracted radioactive RNA was separated by gel electrophoresis (Loening and Ingle 1967) and the distribution of radioactivity in the gels determined (Poulson and Beevers 1970).

RNA polymerase was determined as described previously (Poulson and Beevers 1970) according to a modification of the procedure published by Stout and Mans (1968).

Ribosomal preparations were extracted as described previously (Poulson and Beevers 1970), and the distribution of monosomes and polysomes was determined following sucrose density gradient centrifugation. The amino acid-incorporating activity of the ribosomal preparation was determined according to the method of Mans and Novelli (1961, 1964) with a 150,000 x g supernatant from germinating pea seeds serving as a source of activating enzymes and soluble RNA.

III. RESULTS

Segments treated with $1.5 \times 10^{-5}M$ GA in either the light or dark have a greater leaf width, protein and RNA content in comparison to control segments (Table 1). Leaf segments treated with GA in the dark did not unroll as extensively as illuminated segments, indicating that the effect of light on unrolling involves more than a changed gibberellin status of the segments. Photoinduced leaf unrolling is inhibited by 4×10^{-5}M ABA and ABA-treated segments have a reduced chlorophyll, protein and RNA content in comparison to control segments.

Table 1. The influence of illumination, 1.5×10^{-5}M GA and 4×10^{-5}M ABA on leaf width, chlorophyll content, protein content and RNA content of barley leaf segments. Measurements made after a 20 hr incubation.

	Illuminated			Non-Illuminated		
	Control	GA	ABA	Control	GA	ABA
Leaf Width (1.01mm)*	4.45	5.05	1.72	2.02	3.01	1.27
Chlorophyll (0.014 O.D./segment)	0.145	0.143	0.077	0.035	0.040	0.077
Protein (225µg/segment)	387	437	302	275	315	235
RNA (16.9µg/segment)	18.7	19.0	17.7	17.6	18.1	16.8

* Figures refer to initial measurements.

An indication of a requirement for protein and RNA synthesis during the unrolling of leaf segments induced by light or GA was obtained using inhibitors of protein and RNA synthesis. Chloramphenicol and cycloheximide inhibited the unrolling of illuminated and non-illuminated leaf segments (Table 2). Cycloheximide was the most effective inhibitor. The inhibitors of protein synthesis also prevented the GA-induced unrolling of non-illuminated segments. Unrolling of dark incubated segments induced either by light or by GA was only slightly restricted by 5-fluorouracil treatment. Addition of actinomycin D to leaf segments prevented unrolling stimulated by either light or GA. It appears, therefore, that leaf unrolling is at least partially dependent upon the synthesis of a restricted complement of RNA species. Foreseeably the effects of GA and ABA treatments on unrolling could be related to their effects on the synthesis of specific RNA fractions. Studies were therefore made to determine the influence of hormonal treatments on the synthesis of RNA.

Table 2. The effect of inhibitors of protein and RNA synthesis on unrolling of barley leaf segments incubated in the light, in the dark or in the dark in the presence of 1.5×10^{-5}M GA for 20 hr.

Inhibitor	Concentration (µM)	Leaf Width (mm)		
		Illuminated	Non-illuminated	
			- GA	+ GA
Initial		1.01	1.01	1.01
Control		4.10	2.04	3.14
Cycloheximide	50	1.12	1.17	1.65
	500	1.12	1.12	-
Chloramphenicol	100	3.38	1.95	-
	1000	2.68	1.42	1.91
Actinomycin D	10	2.77	1.52	-
	50	1.91	1.33	1.61
5-Fluorouracil	100	3.79	2.40	3.05
	1000	3.25	2.05	3.03

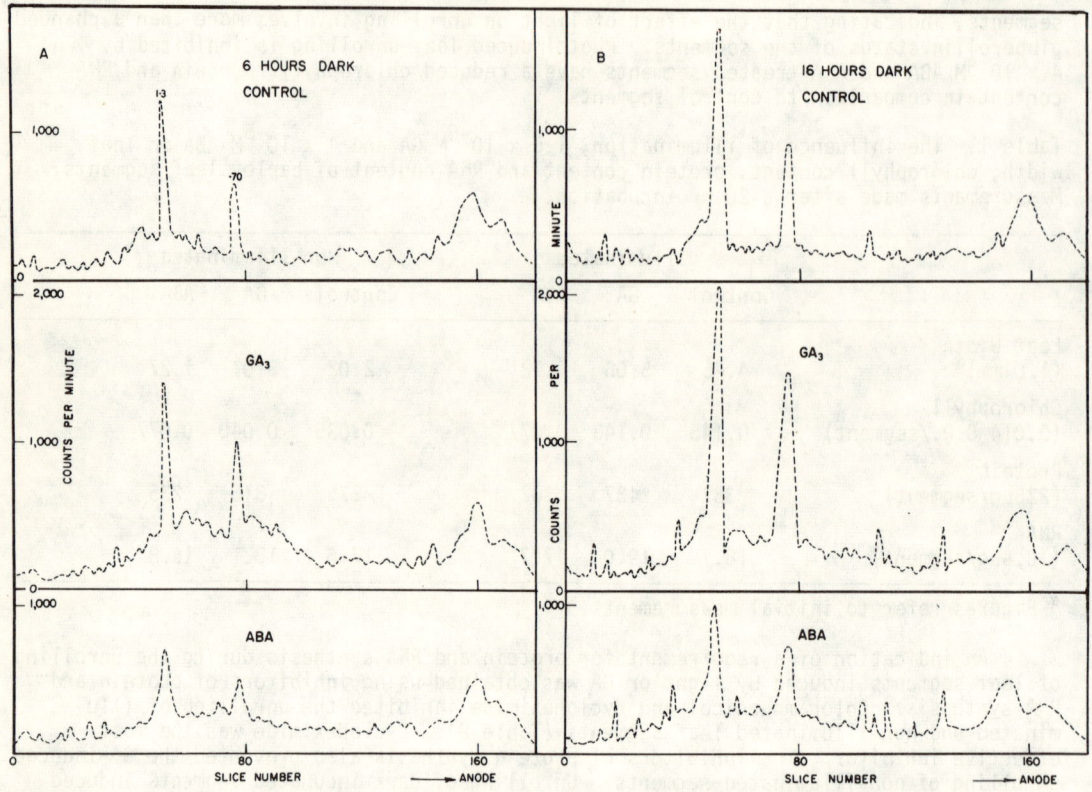

Fig. 1. The distribution of radioactivity following gel electrophoresis of RNA extracted from barley leaf segments incubated with ^{32}P-orthophosphate in the presence of $1.5 \times 10^{-5}M$ GA or $4 \times 10^{-5}M$ ABA.

A. RNA isolated after a 6hr incubation

B. RNA isolated after a 16hr incubation during which isotope was included for the terminal 6hr.

Gel electrophoresis of RNA isolated from segments incubated in the dark for 6hr in the presence of ^{32}P-orthophosphate showed that radioactivity was associated with the soluble RNA and with the regions of the gel which have been previously characterised as having molecular weights of 1.3×10^6 and 0.7×10^6 daltons, corresponding to the heavy and light cytoplasmic rRNAs (Loening and Ingle 1967; Poulson and Beevers 1970). More radioactivity was associated with RNA prepared from leaf segments treated for 6hr with GA than from control segments (Figure 1A). The increased radioactivity was apparent in the cytoplasmic ribosomal regions and there appeared to be a greater amount of radioactivity heterogeneously dispersed in the gel of RNA extracted from the GA-treated segments. The RNA extracted from segments which had been treated for 16hr with GA with ^{32}P present during the terminal 6hr contained more radioactivity than the RNA extracted from control segments. GA treatment enhanced the incorporation of ^{32}P into all the detectable components (Figure 1B). ABA treatment reduced the incorporation of ^{32}P into all the RNA components detected by gel electrophoresis.

The effects of the growth regulators on the incorporation of ^{32}P into RNA were equally apparent in illuminated segments (Figure 2). RNA extracted from segments receiving extended illumination in the presence or absence of GA showed that appreciable radioactivity was associated with RNA species with molecular weights of 1.1×10^6, 0.56×10^6 and 0.40×10^6 daltons, indicating that following prolonged illumination there was synthesis of chloroplast rRNAs (Figure 2B). Thus, following short periods of illumination the bulk of the RNA synthesized is cytoplasmic and chloroplast rRNA synthesis occurs only after prolonged illumination suggesting that chloroplast rRNA synthesis is not required during the initial stages of photomorphogenesis in barley leaf segments. GA treatment enhanced the incorporation of ^{32}P into the detectable RNA components but did not stimulate their synthesis differentially. Segments treated with ABA during a 16hr illumination period showed no evidence of incorporation of radioactivity into the chloroplast rRNA regions and the incorporation of radioactivity into the cytoplasmic rRNA regions was much less than in the control sample (Figure 2B).

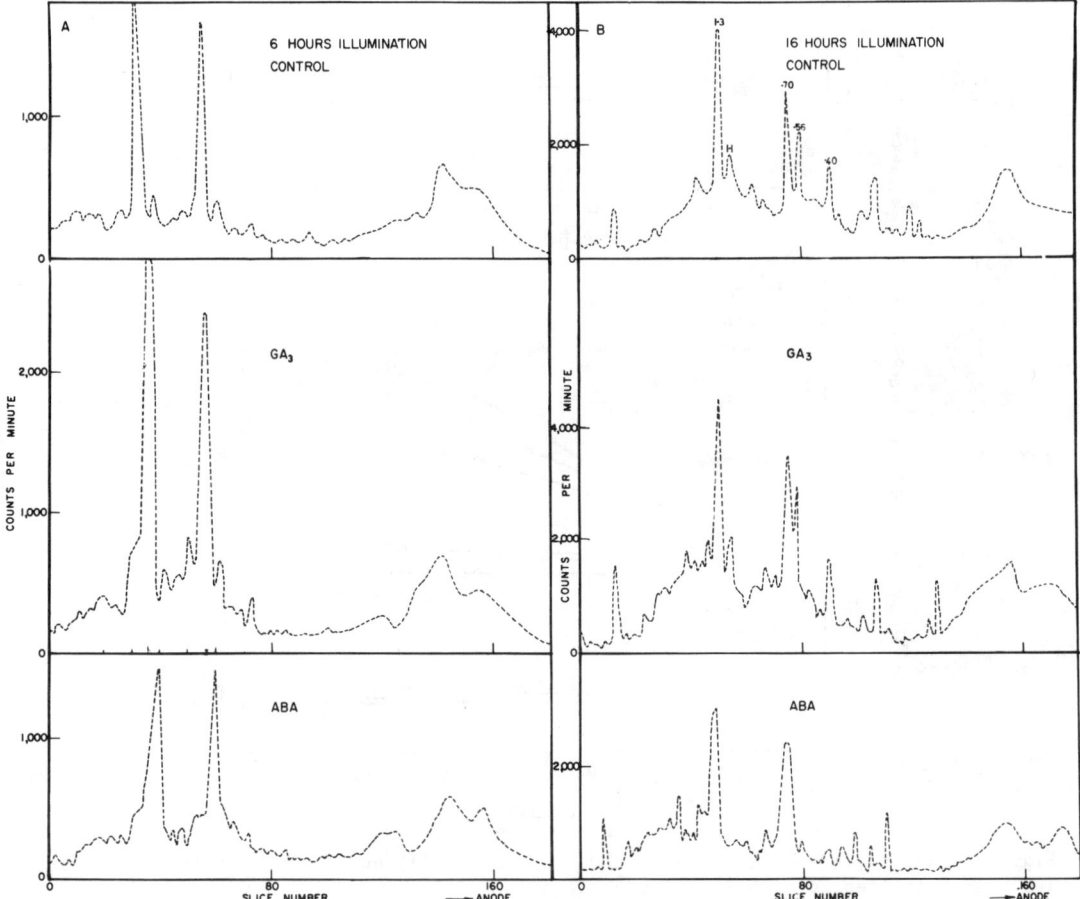

Fig. 2. Gel electrophoresis of ^{32}P-labelled RNA extracted from illuminated barley leaf segments which had been treated with $1.5 \times 10^{-5}M$ GA or $4 \times 10^{-5}M$ ABA

A. RNA isolated after a 6hr illumination period in the presence of isotope and growth regulators.

B. RNA isolated after a 16hr illumination period in the presence of growth regulators. Isotope was included during the terminal 6hr.

The possibility that the increased synthesis of RNA during leaf unrolling may be partly associated with an increase in RNA polymerase level was investigated. In crude preparations the RNA polymerase activity from ABA-treated segments was consistently lower than from control or GA-treated segments (Figure 3A). The failure to detect an increase in RNA polymerase activity in crude extracts from GA-treated segments may be due to the presence of RNase which would lead to an underestimation of polymerase activity.

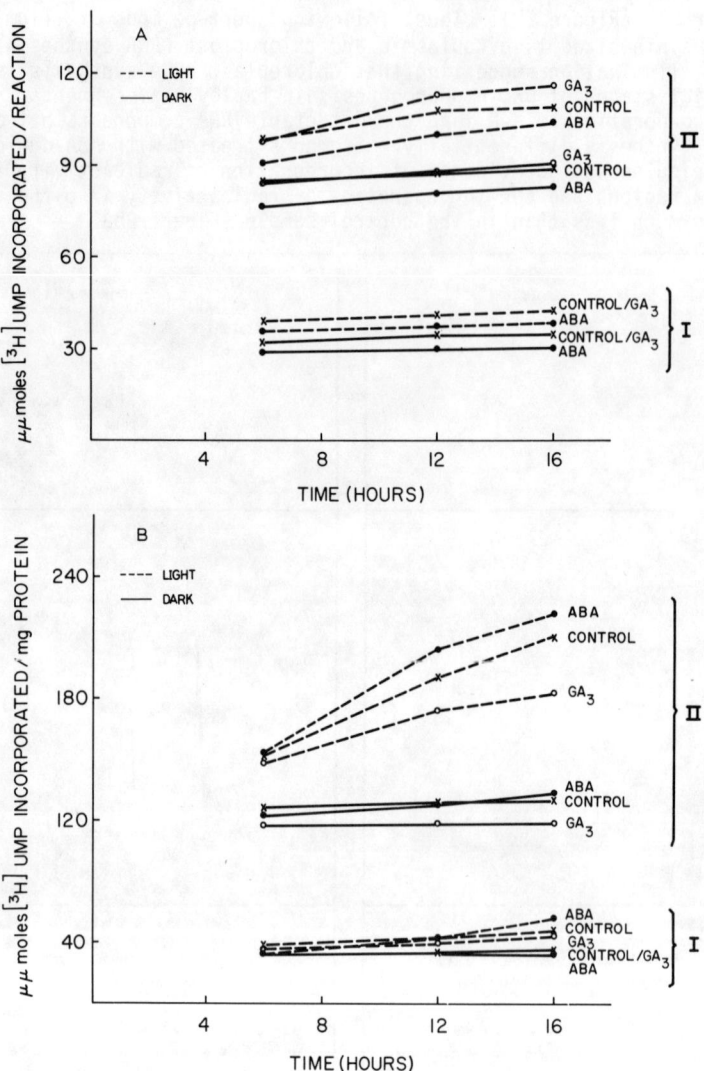

Fig. 3. RNA polymerase activity in crude and partially purified extracts from barley leaf segments treated with 1.5×10^{-5}M GA or 4×10^{-5}M ABA. I - crude preparation; II - partially purified preparation.

A. Results expressed on activity per 0.4ml extract

B. Polymerase activity on a specific activity basis.

Since illumination and growth regulator treatment altered the protein content of segments it was necessary to determine whether the observed changes in polymerase activity represented specific treatment effects or merely reflected overall changes in protein synthesis. ABA-treated segments had a higher polymerase activity per unit protein than GA-treated segments (Figure 3B). It seems, therefore, that the altered polymerase activity of extracts from hormone treated segments is associated with an overall change in protein level.

Light and hormonal treatments influenced the capacity of the segments to incorporate ^{32}P into all the detectable RNA components. However, apart from the heterogeneously dispersed label present in RNA from GA-treated segments the techniques of gel electrophoresis failed to demonstrate the synthesis of specific RNA species which might be required for unrolling. A further assessment of RNA synthesis (specifically mRNA) is provided by analysis of the ribosomal fraction.

It was found that illumination or growth regulators had a marked influence on the relative distribution of polysomes and monosomes. In ribosome preparations from etiolated segments maintained in the dark 45% of the 254 nm absorbing material is associated with the polysome region of the sucrose density gradient (Figure 4). Treatment of the segments with ABA produced ribosomal preparations in which the percentage of polysomes was reduced to 36% whereas GA treatment enhanced polysome formation so that polysomes constituted 66% of the ribosomal preparation.

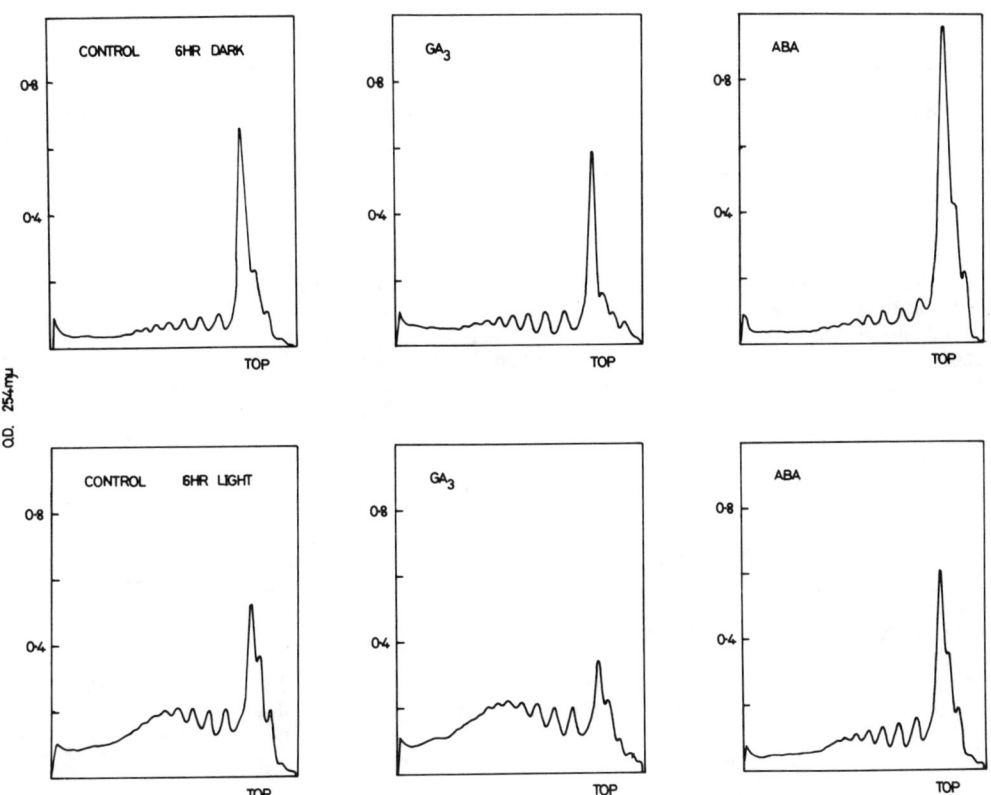

Fig. 4. Sucrose density gradient profiles of ribosome preparations from barley leaf segments incubated for 6hr in the light or dark in the presence of 1.5×10^{-5}M GA or 4×10^{-5}M ABA.

Illumination of the segments increased the population of ribosomes occurring as polysomes to 75%, GA treatment during illumination of the segments resulted in a further enhancement of polysome formation (Figure 4). ABA treatment prevented the photoinduced formation of polysomes.

In addition to an increased polysome content it was found that ribosomal preparations from GA-treated illuminated or etiolated segments had an enhanced capacity to incorporate ^{14}C-leucine into TCA insoluble material (Table 3). Ribosomal preparations from illuminated segments had an increased capacity for amino acid incorporation in comparison to similar preparations from non-illuminated segments. Treatment of the segments with ABA during illumination prevented the increase in amino acid-incorporating capacity of the preparations.

Table 3. In vitro amino acid incorporating capacity of ribosomal preparations extracted from segments treated with 1.5×10^{-5}M GA or 4×10^{-5}M ABA.

Treatment	^{14}C - L - leucine incorporated (cpm/mg rRNA)	
	6hr.	16hr.
Non-illuminated:		
Control	7,013	7,769
GA	10,220	10,126
ABA	7,528	6,113
Illuminated:		
Control	14,523	14,768
GA	17,436	15,783
ABA	8,201	8,236

IV. DISCUSSION

The photoinduced and GA-stimulated unrolling of barley leaf segments is associated with an increased protein and RNA synthesis. In contrast, ABA prevents photoinduced unrolling and eliminates the increase in protein and RNA. Inhibitors of protein and RNA synthesis also prevent unrolling stimulated by GA or light, suggesting that macromolecular synthesis is a necessary prerequisite in the unrolling process. The observation that actinomycin D prevented unrolling whereas 5-fluorouracil did not suggests that the dependence for RNA synthesis during unrolling is restricted to a specific complement of RNA species. It is considered that ABA and GA could differentially affect unrolling by their influence on the synthesis of RNA species required for the unrolling process.

Gel electrophoresis of RNA extracted from segments incubated in the presence of the growth regulators in the dark or during illumination indicated that the unrolling response was associated with quantitative changes in the capacity for incorporation of ^{32}P into RNA components. However, there were no detectable qualitative changes in RNA metabolism (^{32}P incorporation) which could be associated with unrolling. The increased capacity for RNA synthesis in illuminated or GA-treated segments is associated with an increased RNA polymerase in comparison to ABA-treated segments. However, some of the changes in enzyme level are associated with changes in net protein level and thus they do not necessarily indicate specific effects of the growth regulators on the synthesis of RNA polymerase.

The observations that unrolling occurred in the presence of 5-fluorouracil which we have shown to inhibit RNA synthesis in the barley leaf segments suggests that the observed influences of illumination, GA and ABA on the synthesis of the principal RNA species detectable by gel electrophoresis are probably not related to the effects of these treatments on the unrolling process. However it has been demonstrated by Key (1966) that the synthesis of DRNA is insensitive to inhibition by 5-fluorouracil. Therefore the influence of growth regulators and illumination on unrolling may be attributable to changes in the synthesis of the DRNA species.

This concept is supported by the observed changes in polysome levels following growth regulator treatment or illumination of the leaf segments. Insofar as polysome level is an index of mRNA (DRNA) availability, it appears unrolling of barley leaf segments is dependent upon an increased mRNA level. This increased mRNA level can be induced by illumination or GA treatment, in contrast ABA treatment appears to repress the synthesis of mRNA. If this is the case, GA could enhance the availability of mRNA by increasing the amount of DNA available for transcription. A similar role for GA has been suggested by Jarvis, Frankland and Cherry (1968) in studies of seed dormancy. The decrease in unrolling and polysome level following ABA treatment suggests a decreased availability of mRNA via a repression of transcription.

Although the experimental data are consistent with the above conclusions it must be recognized that the control of polysome formation can be exerted at the cytoplasmic level (Baliga et al, 1968) and thus growth regulators or illumination may be influencing the binding of ribosomes to existing mRNA. In this case the regulation of unrolling could be exerted at the translational as opposed to the transcriptional level. A detailed analysis of the mRNA associated with the ribosomes following the various treatments is necessary to resolve these possibilities.

V. SUMMARY

Illumination and/or GA treatment of etiolated segments stimulates unrolling and results in increased levels of RNA and protein. ABA treated segments do not unroll and have a lower content of RNA and protein. Studies with inhibitors of protein synthesis demonstrated a requirement for protein synthesis concomitant with leaf unrolling. Actinomycin D effectively inhibits RNA synthesis and leaf unrolling whereas 5-fluorouracil inhibits the bulk of RNA synthesis but has little effect on leaf unrolling, thus implying that leaf unrolling is dependent on the synthesis of a restricted complement of RNA species.

GA treatment enhanced the capacity of the segments to incorporate radioactivity from ^{32}P-orthophosphate into all the fractions detected by gel electrophoresis; ABA greatly retarded the incorporation of precursors into all RNA components. Associated with a changed capacity for RNA synthesis, ABA-treated segments had a lower level of soluble DNA - dependent RNA polymerase in comparison to illuminated or GA-treated segments. Ribosomal preparations from GA treated segments had a greater percentage of polysomes and a greater capacity for in vitro amino acid incorporation than similar preparations from ABA treated segments. The data suggest that unrolling is regulated by the availability of mRNA and the growth regulators may influence unrolling by controlling the mRNA level.

VI. REFERENCES

BALIGA, B.S., A.W.PRONCZUK and H.N. MUNRO (1968). Regulation of polysome aggregation in a cell-free system through amino acid supply. J. Mol. Biol. 34, 192-218.
JARVIS, B.C., B. FRANKLAND and J.H. CHERRY (1968). Increased DNA template and RNA polymerase associated with the breaking of seed dormancy. Pl. Physiol. 43, 1734-1736.
KEY, J.L. (1966). Effects of purine and pyrimidine anologues on the growth and RNA metabolism in the soybean hypocotyl: the selective action of 5-fluorouracil. Pl. Physiol. 41, 1257-1264.
LOENING,U.E.and J. INGLE (1967). Diversity of RNA components in green plant tissues. Nature 215, 363-367.
MANS, R.J. and G.D. NOVELLI (1961). Measurement of the incorporation of radioactive amino acids into protein by a filter paper disk method. Arch. Biochem. Biophys. 94, 48-53.
MANS, R.J. and G.D. NOVELLI (1964). Stabilisation of the maize seedling incorporating system. Biochim. Biophys. Acta 80, 127-136.
OSBORNE, D.J. (1962). Effect of kinetin on protein and nucleic acid metabolism of Xanthium leaves during senescence. Pl. Physiol. 37, 595-602.
POULSON, R. and L. BEEVERS (1970). Effects of light and growth regulators on leaf unrolling in barley. Pl. Physiol. 46, 509-514.
POULSON, R. and L. BEEVERS (1970). Nucleic acid metabolism during greening and unrolling of barley leaf segments. Pl. Physiol. 46, 315-319.
STOUT, E.R. and R.J. MANS (1968). Partial purification and properties of RNA polymerase from maize. Biochim. Biophys. Acta 134, 327-336.

AUXIN-INDUCED GROWTH OF TUBER TISSUE OF JERUSALEM ARTICHOKE.

VII. EFFECT OF CYCLIC 3',5'-ADENOSINE MONOPHOSPHATE ON THE AUXIN-INDUCED CELL EXPANSION GROWTH*

Seiichiro Kamisaka

Department of Biology, Faculty of Science, Osaka City University, Sumiyoshi-ku, Osaka, Japan

I. INTRODUCTION

Cyclic 3',5'-adenosine monophosphate (cyclic AMP) is well known to be a mediator of the action of some animal hormones (Robinson, et al, 1968). So far this compound has been shown to occur in various types of organisms ranging from microorganisms (Tao and Lipman, 1969) to higher organisms (Robinson, et al, 1968). Recent studies have made it clear that cyclic AMP is involved in various kinds of cellular regulation (Robinson, et al, 1968; Palman and Paston, 1968; Bonner, 1970).

In the field of plant physiology, Galsky and Lippincott (1969) have claimed that cyclic AMP stimulates the synthesis of α-amylase in barley endosperm, like gibberellic acid (GA_3). Cyclic AMP was also found to stimulate the auxin-induced expansion growth of slices excised from cold-stored tubers of Jerusalem artichoke (Kamisaka and Masuda, 1970). These findings suggest that cyclic AMP could be involved in the action of plant hormones, as it is in the case of some animal hormones (Robinson, et al, 1968). However, very little is known about the physiological effect of cyclic AMP on higher plants.

In the present experiments, the physiological effect of cyclic AMP on the auxin-induced expansion growth of tuber slices was studied, paying special attention to the relationship between the action of GA and cyclic AMP.

II. MATERIALS AND METHODS

Tubers of Jerusalem artichoke had been stored at ca. 4°C for 1-4 months after harvest before experimental use. Slices were excised from tubers by the method reported previously (Masuda, 1965).

Tuber slices were transferred to petri dishes containing filter paper held on the surface of growth solution which contained 1 mg/l 2,4-dichlorophenoxyacetic acid (2,4-D). During incubation at $25° ± 1°C$, the increase in wet weight of slices was measured at appropriate intervals. In some experiments, slices were pretreated by soaking them in distilled water at $25° ± 1°C$ for 20 hours. They were then transferred to a growth solution containing 2,4-D. Effects of cyclic AMP on cell expansion growth were tested by adding it to growth solutions with and without 1 mg/l 2,4-D.

---

* Some experiments in the present paper were carried out in School of Biological Sciences, Macquarie University, N.S.W. Australia.

III. RESULTS AND DISCUSSION

Freshly cut slices were incubated in growth solutions, containing cyclic AMP at concentrations ranging from 10^{-7} to 10^{-4}M with and without 1 mg/l 2,4-D. At these concentrations cyclic AMP showed no significant effect on cell expansion growth when applied alone, but, as shown in Fig. 1, it stimulated growth when given with 2,4-D. However, as shown in Fig. 2, the stimulating effect of cyclic AMP on auxin-induced cell expansion growth decreased as the incubation time became longer.

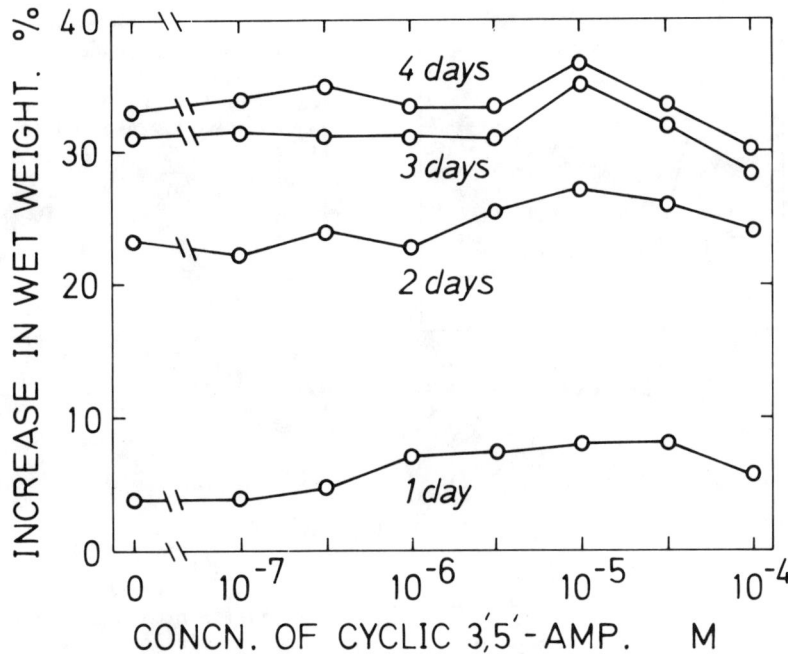

Fig. 1. Effect of cyclic AMP at various concentrations on expansion growth of tissue treated with 1 mg/l 2,4-D.

It was supposed that this decrease in the stimulating effect of cyclic AMP may be caused by the destruction of cyclic AMP during the process of incubation. This possibility was tested by changing daily the growth solutions containing 1 mg/l 2,4-D and 10^{-5}M cyclic AMP. Freshly cut slices were incubated for 4 days in growth solutions which were either changed daily or not changed. As shown in Fig. 3, there was no significant difference in cell expansion growth between the two incubation conditions. This result suggests that the decrease in the stimulating effect of cyclic AMP may be caused by some factor other than the destruction of cyclic AMP during incubation.

The growth response of tissue slices to plant hormones is known to be different between freshly-cut slices and slices cut and then washed for about 1 day in distilled water. For example, freshly-cut slices are more sensitive to the stimulating effect of GA on the auxin-induced cell expansion growth than soaked slices (Setterfield, 1963; Masuda, 1965). Hence, the next experiment was carried out to find out whether or not the response of freshly-cut slices to cyclic AMP is different from that of pre-treated slices.

Slices pretreated for 20 hours in water, and freshly-cut slices were both incubated for 26 hours in 2,4-D solutions with and without 10^{-5}M cyclic AMP. As shown in Table 1, cyclic AMP stimulated the auxin-induced cell expansion growth of freshly cut slices, but not that of pretreated slices. This result indicates that the response of

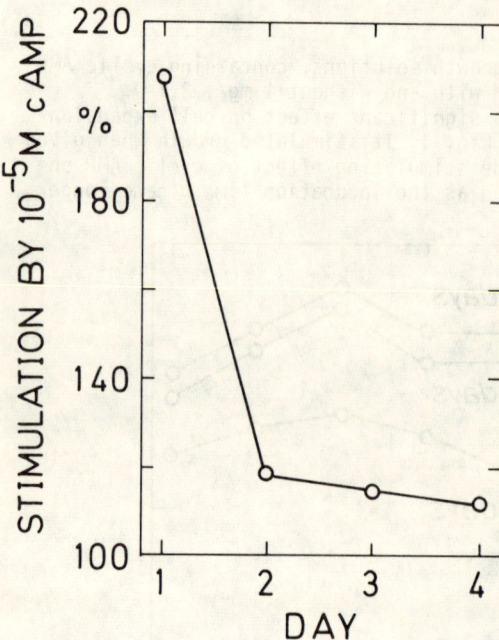

Fig. 2. Change in the stimulating effect of cyclic AMP during incubation. The ratio of % increase in wet weight caused by 1 mg/l 2,4-D and 10^{-5}M cyclic AMP to that caused by 1 mg/l 2,4-D alone.

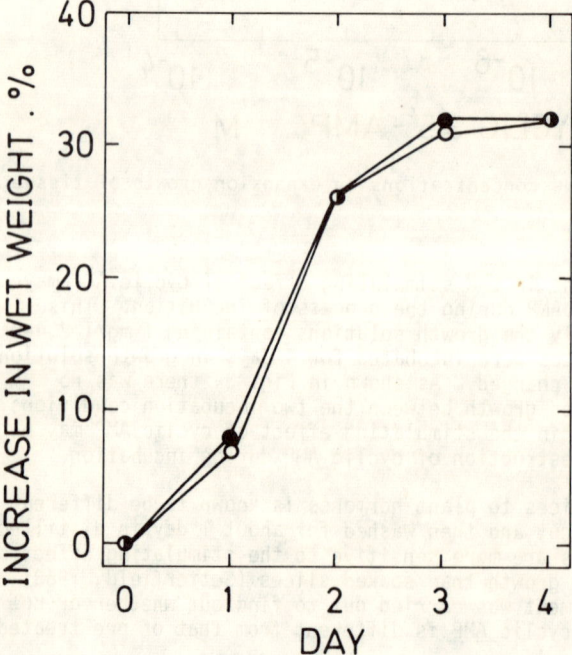

Fig. 3. Time course of cell expansion growth of slices in growth solutions which were changed daily (closed circles) or not changed (open circles). Growth solution contained 1 mg/l 2,4-D and 10^{-5}M cyclic AMP.

slices to cyclic AMP is similar to that of GA. This similarity was investigated further by including CCC, an inhibitor of gibberellin biosynthesis (Kende et al, 1963), in the treatment solutions.

Table 1. Effect of cyclic 3',5'-adenosine monophosphate on the auxin-induced expansion growth of freshly cut slices and slices pretreated in water after slicing.

Treatments		% increase in wet weight in 26 hours	Stimulation by cyclic AMP
Freshly cut slices	2,4-D	4.5%	100
	2,4-D + cyclic AMP	9.6%	213
Slices soaked for 20 hours in water	2,4-D	7.8%	100
	2,4-D + cyclic AMP	7.7%	98.6

2,4-D, 1 mg/l; cyclic AMP, 10^{-5}M.

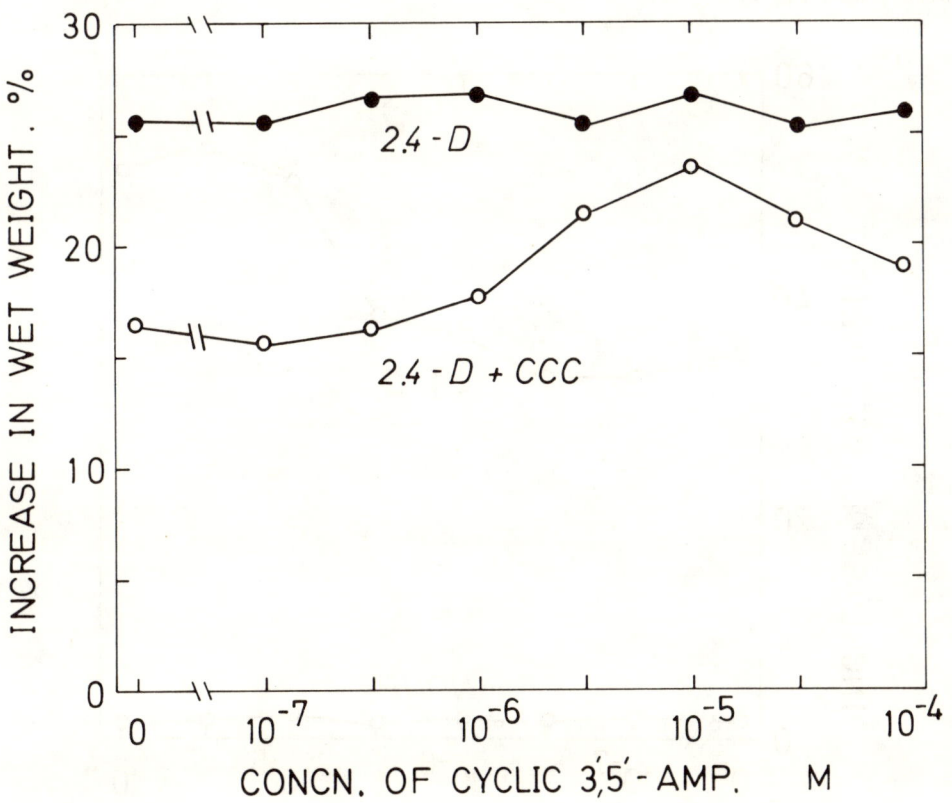

Fig. 4. Effect of cyclic AMP on the inhibition by CCC of the auxin-induced cell expansion growth. Freshly cut slices were incubated for 2 days in growth solutions, to which various concentrations of cyclic AMP were given with 1 mg/l 2,4-D and/or 400 mg/l CCC.

CCC inhibits the auxin-induced expansion growth of tuber slices (Kamisaka and Masuda, 1968), and the inhibitory effect of CCC is eliminated by simultaneous addition of GA (Kamisaka and Masuda, 1968). Hence, the possibility was examined that cyclic AMP might eliminate the inhibitory effect of CCC on the auxin-induced cell expansion growth, as does GA. Freshly cut slices were incubated for 2 days in growth solutions, to which various concentrations of cyclic AMP were given with and without 1 mg/l 2,4-D and/or 400 mg/l CCC. As shown in Fig. 4, CCC inhibited the auxin-induced cell expansion growth. On the other hand, inhibition due to CCC was substantially reversed by cyclic AMP. This experimental result together with that of Table 1 suggests that cyclic AMP, like GA (Setterfield, 1963; Masuda, 1965), may be involved in the control of metabolic changes during the so-called aging process of slices that have been pretreated in water.

As mentioned above, cyclic AMP stimulated the auxin-induced cell expansion growth, but its stimulatory effect was much smaller than that of GA. The effect of sodium dibutyryl cyclic AMP was tested, because the derivative is known from work with animal tissue to be in some cases more active than cyclic AMP. For this experiment, freshly-cut slices were allowed to grow for 4 days in solutions, to which sodium dibutyryl cyclic AMP at the concentration range from 10^{-7} to 10^{-4}M was given with and without 1 mg/l 2,4-D.

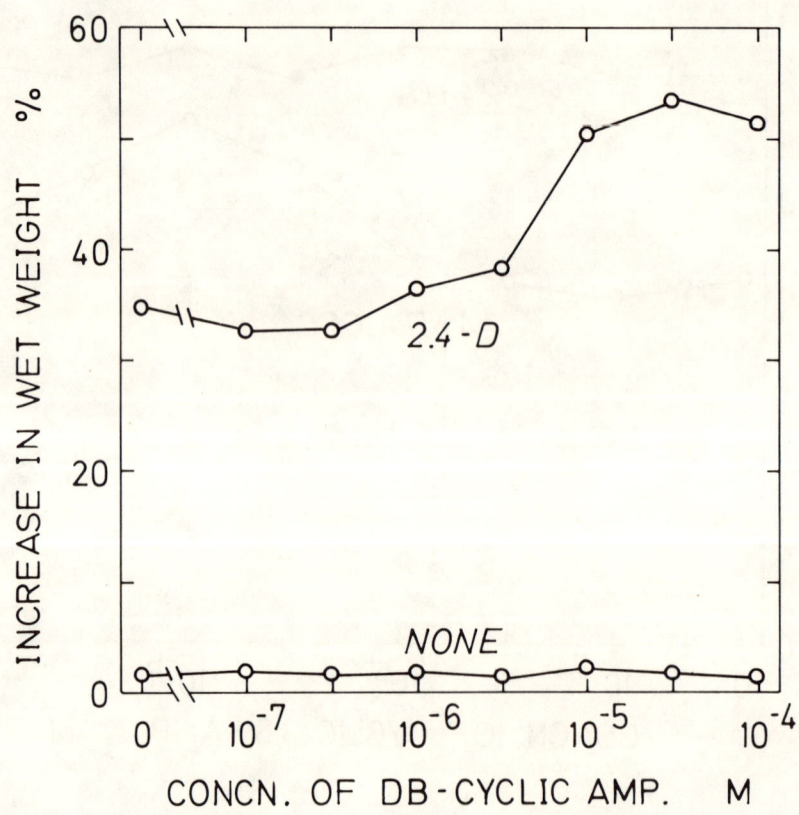

Fig. 5. Effect of sodium dibutyryl cyclic AMP on expansion growth. Freshly-cut slices were incubated for 4 days in growth solutions, to which sodium dibutyryl cyclic AMP was given with and without 1 mg/l 2,4-D.

Fig. 5 shows the percentage increase in wet weight of slices in 4 days, and Fig. 6 shows the time course of expansion growth caused by 2,4-D with or without 3×10^{-4}M sodium dibutyryl cyclic AMP. As indicated in Figs. 5 and 6, the stimulation of auxin-induced expansion growth by sodium dibutyryl cyclic AMP was greater than that caused by cyclic AMP. On the other hand, sodium dibutyryl cyclic AMP showed no significant effect on cell expansion growth when given alone. Sodium dibutyryl cyclic AMP, like cyclic AMP (see Table 1), did not stimulate the cell expansion growth caused by auxin, when applied to slices pretreated for 20 hours in water after slicing.

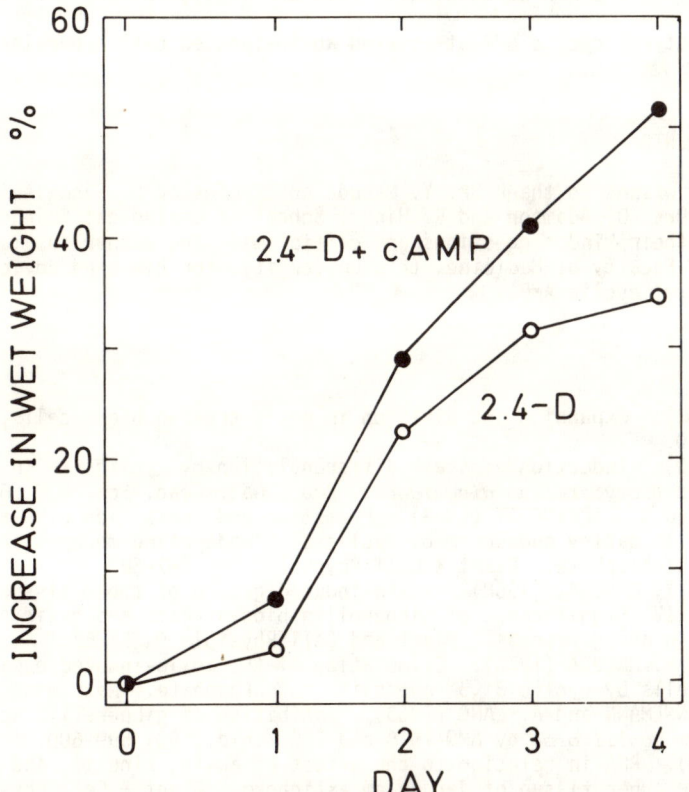

Fig. 6. Time course of expansion growth caused by 1 mg/l 2,4-D and by 1 mg/l 2,4-D with 3×10^{-4}M sodium dibutyryl cyclic AMP.

Cyclic AMP has in part a chemical structure similar to a cytokinin. In tuber slices, cytokinins sometimes stimulate synergistically the auxin-induced cell expansion growth (Adamson, 1952). Judging from these facts, the stimulating effect of cyclic AMP on tuber slices could be thought of as a cytokinin-like activity, rather than a GA-like activity. On the other hand, substances similar in part in their chemical structure to cytokinins, such as adenine, adenosine, AMP, ADP and ATP do not show such a promoting effect on the auxin-induced cell expansion growth, as does cyclic AMP (Kamisaka and Masuda, 1970). Further, unlike GA, cyclic AMP and sodium dibutyryl cyclic AMP, cytokinin stimulates the auxin-induced cell expansion growth of freshly cut slices as well as slices pretreated in water after slicing (Masuda, 1965). These experimental results suggest at present that the action of cyclic AMP is similar to that of GA, rather than that of cytokinin.

IV. SUMMARY

The physiological effect of cyclic 3',5'-adenosine monophosphate (cyclic AMP) on cell expansion growth was studied, using slices excised from cold-stored tubers of Jerusalem artichoke.

Cyclic AMP stimulated auxin-induced cell expansion growth. On the other hand, it showed no significant effect on cell expansion growth in the absence of auxin.

Cyclic AMP enhanced the promoting effect of auxin on cell expansion growth when given to freshly cut slices, but not when given to slices soaked for 20 hours in distilled water prior to auxin application.

CCC, an inhibitor of gibberellin biosynthesis, inhibited cell expansion growth caused by auxin. Cyclic AMP substantially reversed the inhibitory effect of CCC.

Sodium dibutyryl cyclic AMP stimulated auxin-induced cell expansion growth much more than cyclic AMP.

V. ACKNOWLEDGEMENTS

The author wishes to thank Dr. Y. Masuda and Professor N. Yanagishima, this laboratory, and Drs. D. Adamson and R. Hinde, School of Biological Sciences, Macquarie University, for their kind suggestions and criticisms. The author is also indebted to Dr. S. Genba, Faculty of Medicine, this University, for his kind advice and gift of sodium dibutyryl cyclic AMP.

VI. REFERENCES

ADAMSON, D. (1962). Expansion and division in auxin-treated plant cells. Can. J. Bot. $\underline{40}$, 719-740.
BONNER, J.T. (1970). Induction of stalk differentiation by cyclic AMP in the cellular slime mould *Dictyostelium discoideum*. Proc. Nat. Acad. Sci. U.S. $\underline{65}$, 110-113.
GALSKY, A.G. and J.A. LIPPINCOTT (1969). Promotion and inhibition of α-amylase production in barley endosperm by cyclic 3',5'-adenosine monophosphate and adenosine diphosphate. Plant & Cell Physiol., $\underline{10}$, 607-620.
KAMISAKA, S. and Y. MASUDA, (1968). Auxin-induced growth of tuber tissue of Jerusalem artichoke. IV. Significance of gibberellin biosynthesis and basic proteins in chromatin in aging process. Plant and Cell Physiol. $\underline{9}$, 61-67.
KAMISAKA, S. and Y. MASUDA (1970). Stimulation of the auxin-induced expansion growth in plant cells by cyclic 3',5'-adenosine monophosphate. Naturwiss. $\underline{57}$, 546.
KENDE, H., H. NINNEMANN and A. LANG (1963). Inhibition of gibberellic acid biosynthesis in *Fusarium moniliforme* by AMO-1618 and CCC. ibid., $\underline{50}$, 599-600.
MASUDA, Y. (1965). RNA in relation to the effect of auxin, kinetin, and gibberellic acid on the tuber tissue of Jerusalem artichoke. Plant & Cell Physiol., $\underline{18}$ 15-23.
PALMAN, R.L., and I. PASTAN (1968). Regulation of β-galactosidase synthesis in *Escherichia coli* by cyclic 3',5'-adenosine monophosphate. J. Biol. Chem. $\underline{243}$, 5420-5427.
ROBINSON, G.A., R.W. BUTCHER, and E.W. SUTHERLAND. (1968). Cyclic AMP. Ann. Rev. Biochem. $\underline{37}$, 149-174.
SETTERFIELD, G. (1963). Growth regulation in excised slices of Jerusalem artichoke tuber tissue. Symp. Soc. Exptl. Biol., $\underline{17}$, 98-126.
TAO, M., and F. LIPMAN (1969). Isolation of adenyl cyclase from *Escherichia coli*. Proc. N.A.S. (U.S.A.), $\underline{63}$, 86-92.

PLANT GROWTH SUBSTANCES, 1970

Evidence for the Presence and Biological Activity of a Chorionic Gonadotropin-like Plant Growth Substance - (Phytotropin)

Y. Leshem, A. Shomer-Ilan and R.R. Avtalion

Department of Biology, Bar Ilan University, Ramat Gan, Israel

I. INTRODUCTION

In a series of experiments Leshem (1967), Leshem and Lunenfeld (1968) and Leshem et al. (1969) have shown that in several plant species certain parameters of morphogenetic growth, especially adventitious root production, were markedly promoted by exogenously applied human chorionic gonadotropin (HCG). Gonadotropins in animal systems are glycoprotein sex hormones usually produced by the anterior pituitary of the brain; they act on the target gonads, inducing steroidogenesis. HCG differs in source, being found in the chorionic fluid enclosing the foetus during pregnancy and is excreted in the urine of pregnant and post-climacteric females. The standard Ascheim-Zondek 'frog test' for pregnancy is based on spermatogenesis induced in young male frogs treated with urine containing HCG.

Plant species tested included broccoli (*Brassica oleracea* var. *cymosa*) curd cuttings, and stem cuttings of *Begonia semperflorens*, *Vitis vinifera* and *Chrysanthemum morifolium*. Table 1 and Fig. 1 summarize results.

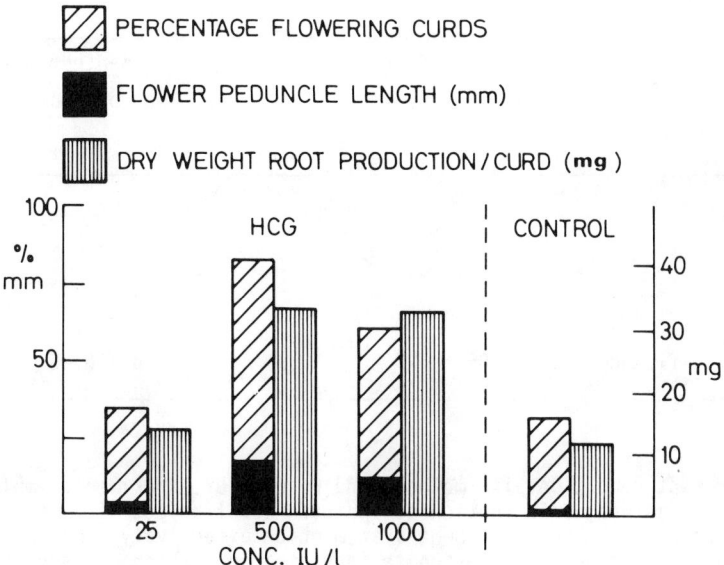

Fig. 1. Effect of HCG on several parameters of growth of broccoli curd cuttings. According to Leshem 1967.

Table 1. No. of roots produced on cuttings of *Begonia semperflorens* and *Vitis vinifera* immersed in solutions with and without HCG. Experimental duration 6 weeks. According to data of Leshem and Lunenfeld (1968).

Treatment	No. of roots/cutting	
	Begonia semperflorens*	Vitis vinifera**
Control	9.0	28.9
HCG 500* - 750**I.U./l	14.0	32.5
HCG 1500 I.U./l	14.5	42.6
HCG 3000 I.U./l	18.1	44.6
level of significance	$p < 0.05$	$p < 0.05$

I.U. = international units

That an endogenous gonadotropin-like plant growth substance exists was suggested by the fact that anti-serum to HCG significantly checked root development (Table 2). An immuno-adsorbent column of insolubilized HCG antibodies was prepared; in such a column any endogenous plant gonadotropin immunologically similar to HCG should be 'trapped'. Upon passing a crude protein extract through such a column, the effluent (theoretically depleted of HCG-like material) had markedly less activity (Fig. 2).

Table 2. Effect of HCG antiserum, serum and HCG on root development in *Begonia semperflorens* and *Chrysanthemum morifolium* growing in nutrient solutions. Leshem et al. 1969.

Treatment	Begonia semperflorens	Chrysanthemum morifolium	
	mg dry wt roots	no. of roots	total root length/cutting - cm.
HCG antiserum*	1.0	0.5	5.0
normal serum	8.3	2.0	27.0
nutrient solution	9.1	---	---
HCG 750 I.U./l	56.0	3.9	47.0
level of significance	$p < 0.05$	$p < 0.05$	$p < 0.05$

The present report details methods and results of experiments employing other techniques such as haemagglutination to indicate HCG-like activity of plant extracts. We also report on an immuno-adsorbent system which effectively binds HCG-like plant material (which we tentatively designate 'phytotropin') which can subsequently be released by appropriate buffer and tested immunologically and biologically. If these tests are positive this would imply the existence of an endogenous plant growth promoter resembling the mammalian gonadotropic sex hormones.

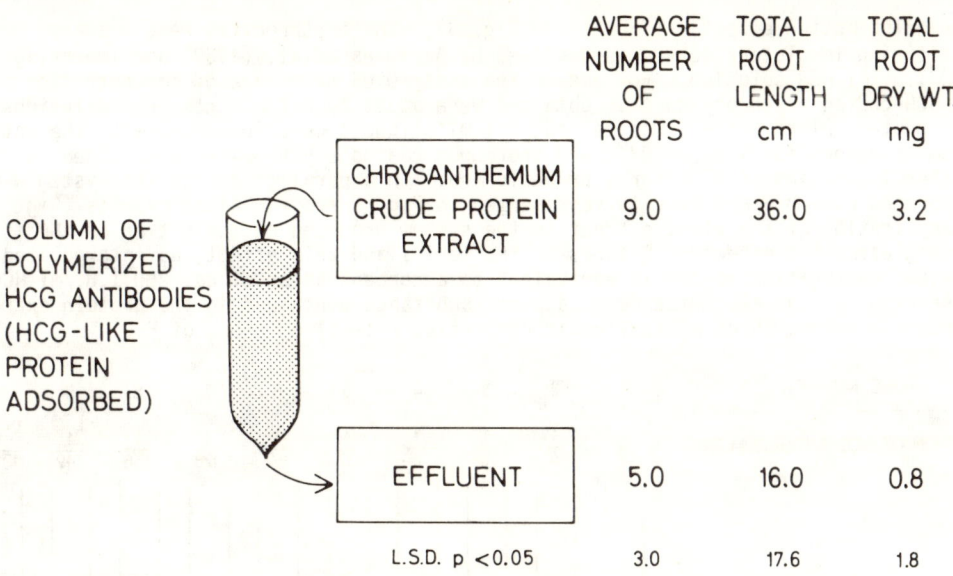

Fig. 2. Effect of immuno-absorption through an HCG antibody column, on rooting evoked by *Chrysanthemum* protein extract. Leshem et al., 1969.

II. METHODS AND RESULTS

Haemagglutination Inhibition

    This technique, using the method of Stavitsky (1964) as modified by Avramaes et al. (1969), employs sheep erythrocytes coated with the antigen HCG. To a series of test tubes containing decreasing dilutions of anti-HCG serum of known antibody titre, a given amount of HCG-coated erythrocytes are added. From a certain concentration of anti-HCG serum and concentrations above it a complete antigen-antibody reaction is obtained. This is a haemagglutination of the erythrocytes discernible as a spread on the bottom of the tube of the complex-HCG-coated erythrocytes and the antibodies and differs markedly from the clustering obtained when the reaction does not occur.

    HCG or plant extract containing HCG-like material added to the anti-HCG serum before addition of the coated erythrocytes, would react with part of the antibodies contained in the serum and cause inhibition of the haemagglutination which is expressed by decrease of antibody titre.

    HCG antiserum was prepared in rabbits by intramuscular injection of 1 ml 5000 I.U. HCG in an emulsion of Freund's complete adjuvant followed at 10 day intervals by three booster injections each of 2500 I.U. Standard medical procedure was used to obtain serum. Plant protein was extracted from 100 gr fresh weight *Begonia semperflorens* cv. Indian Maid leaves plucked from identical phyllotaxial positions about midway along the stem. These were homogenized in the cold in an equal volume of buffer, tris 0.2M pH 9, containing 10^{-3}M EDTA and 10^{-3}M ascorbic acid. The homogenate was passed through a triple layer of gauze and thereupon centrifuged in the cold at 12000xg for 20 minutes and the supernatant collected. Protein was salted out by saturation with $(NH_4)_2SO_4$. The sediment was dissolved in tris buffer 0.2M pH7.6 and exhaustively dialyzed with the buffer system employed in serum extraction.

    The haemagglutination-inhibition assay was carried out by preparing a series of tubes containing a serial 1:2 dilution of HCG antibody serum in normal serum. Dilution is terminated when a standard amount of HCG-coated erythrocytes no longer produces an haemagglutination effect   (Tube 10 in Fig. 3). The erythrocytes were prepared by sensitizing in glutaraldehyde as outlined by Avramaes et al. (1969) and immersing in 1500 I.U./ml HCG solution. For use in the assay 0.05 ml of a 2.5% concentration of the HCG-coated erythrocytes thus obtained were added to 0.5 ml antiserum dilutions. When protein plant extract was tested, 0.1 ml aliquots were incubated with the anti HCG serum dilutions for 1 hr at 37°C and thereupon coated erythrocytes were added and a further incubation at 37°C for 1 hr performed. In a parallel series the system was calibrated with HCG and results represented in Fig. 3 above. Final results (Fig. 3) of application of the plant extract to the system produced an inhibition of the haemagglutination effect of 5 tubes in the series and this effect, as interpolated from the calibration series is equivalent to a concentration of ca. 200 I.U./ml HCG. These results indicate presence of a plant substance contained in the protein extract of *Begonia* leaves which has antigenic properties resembling those of HCG.

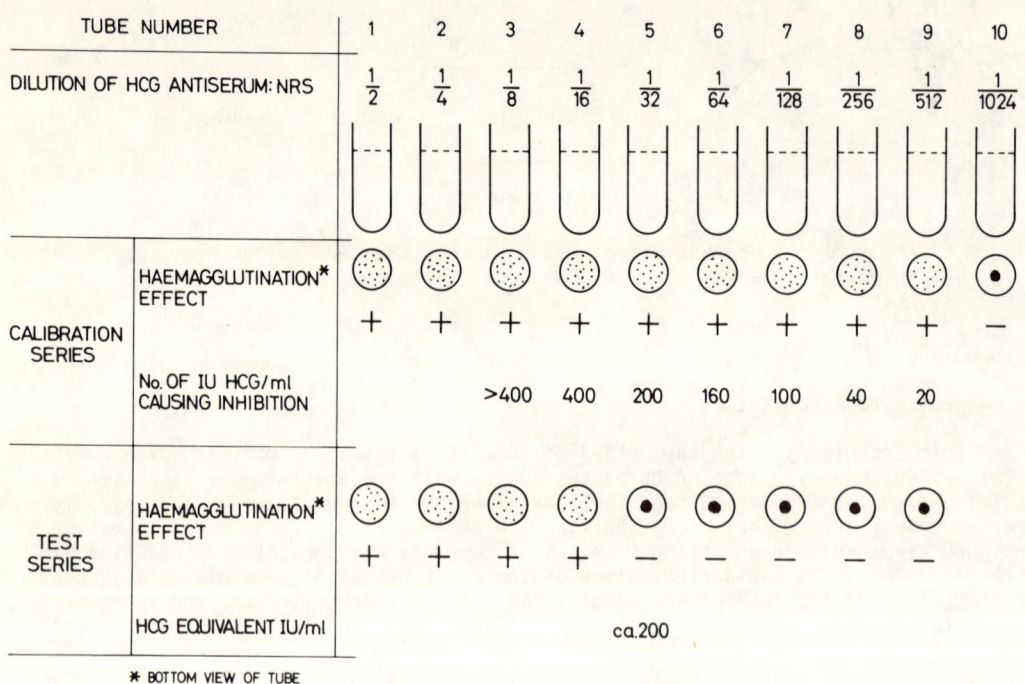

Fig. 3. Haemagglutination inhibition of HCG-coated erythrocytes produced by proteinaceous *Begonia semperflorens* extract in a series of antiHCG serum - normal rabbit serum (NRS) dilutions. See text for details (4 replicate means).

Isolation of 'Phytotropin' and Assay for Hormonal Activity

In a preliminary attempt to isolate the active principle Leshem et al. (1969) used the approach that a column of polymerized and unsolubilized HCG antibodies prepared by the procedure of Avramaes and Ternynck (1967) would trap any HCG-like material present in extracts (see also Fig. 2). However it was later found that in the present system the degree of binding in such columns is low and difficulties were encountered in releasing the bound material. After experimentation on several types of other immunosorbent systems the procedure of Donini and Donini (1969) was adopted. One ml of the antiHCG serum prepared as in the previous experiment was added to 1 ml of acetate buffer 0.4M pH 5.0. While gently stirring, 0.15 ml ethylchloroformate was added and with the aid of dilute acetic acid or NaOH, the pH was kept between 4.6-4.9. After 15 mins a gel is formed and allowed to set at room temp for 1 hr. The gel was then mixed with 0.1M saline phosphate buffer pH 7.2, homogenized and repeatedly washed with buffer to get rid of surplus non-polymerized protein. The final product was lyophilized and the polymer stored in the cold. For final use, 1 gr of the lyophilized polymer is added to 50 ml of the above saline phosphate buffer.

Plant extract was obtained by homogenizing 600 g *Begonia* leaf blades with a Virtis homogenizer in 200 ml of buffer, tris 1M pH 9. After centrifuging down coarse material the supernatant was reduced to half volume by ultracentrifugation in a Diaflo chamber with a molecular exclusion membrane of 20,000 M.W. and the final solution added to 500 ml of the polymer mixture prepared as above, and gently agitated in the cold for 48 hours. The hypothetical polymer-"phytotropin" complex formed was then separated by centrifugation for 15 mins at 8000 g, washed with saline-phosphate buffer and redispersed.

The bound phytotropin was freed from the polymer by adding 6 ml guanidine-HCl buffer, 6M pH 2.4. This buffer had previously been found effective in freeing bound HCG without impairing its immunological activity. After addition of buffer the mixture was gently agitated at room temperature for 1 hour. The mixture was again dispersed and placed in the cold for 24 hours. This buffer presumably released 'phytotropin' from the complex and the released substance was separated from it by centrifugation for 15 mins at 8000 x g. The supernatant thus obtained was exhaustively dialysed against saline phosphate buffer 0.1M pH 7.2 until no traces of the guanidine buffer remained and the pH restored to 7.2.

The final extract was tested by haemagglutination. In a dilution series identical to that described in Fig. 3 the extract produced inhibition between tubes nos. 5 and 6. According to calibration with exogenous HCG, as also seen in Fig. 3, this corresponds to a value of ca. 180 I.U. units of HCG per ml of solution. These results, similar to those of the previous experiment where a value of ca. 200 I.U./ml was obtained, again indicate the presence of a plant substance antigenically resembling HCG.

The method described above was utilized to obtain further quantities of the 'phytotropin' for biological assay.

III. BIOLOGICAL ASSAY IN PLANTS

Since the plant growth parameter mentioned in most previous experiments was adventitious rooting, the extract here obtained was tested on the same response. As it was feared that biological activity might be lost rather rapidly, instead of *Begonia*, *Vitis* or *Chrysanthemum* cuttings which require at least 5-6 weeks to produce discernible results, stem cuttings of 2 month old tomato - (*Lycopersicon esculentum* cv. Marmande) which produce roots within 5-7 days were used. Methods of culture, media and environmental conditions were as described elsewhere (5, 6).

Experiment 1:

Effect of exogenous HCG and indole-butyric acid (IBA) on adventitious rooting of tomatoes. Treatment with IBA was included since this chemical is a known rooting promotor and some indication of HCG promotion of rooting in tomatoes, if any, was considered of interest. This information was necessary before application of 'phytotropin' as isolated above.

Experiment 2:

Effects of immersing ends of stem cuttings in HCG, IBA or a mixture of both. In this variation of the trial, the cut surfaces were immersed in high concentrations of HCG and IBA or mixture of both for 12 hours, and the cuttings subsequently removed and grown in standard rooting media.

Experiment 3:

Effect of 'Phytotropin'. Phytotropin extracted as above was applied to rooting media to provide a final concentration equivalent 200 I.U./l. The medium for control cuttings received an identical amount of buffer.

Table 3. Rooting effects produced on tomato stem cuttings growing in rooting medium. Cuttings treated with HCG, IBA, HCG + IBA or 'phytotropin'. Experimental duration 6 days. Relative values expressed as means of 4 replicates of 10 plants each. Different lower case letters in a given vertical column indicate statistical significance at $p < 0.05$ as determined by Analysis of Variance.

	Treatment	Relative No. of roots per cutting	Weight of roots/cutting Relative values	
			fresh	dry
*Experiment 1	Control	100^a	100^a	100^a
	HCG 100 I.U./l	157^a	861^b	156^a
	IBA 1 p.p.m.	496^b	209^c	134^a
**Experiment 2	HCG 5000 I.U./l	100^a	100^a	100^a
	IBA 100 p.p.m.	475^b	166^b	170^b
	HCG 5000 I.U./l + IBA 100 p.p.m.	585^b	275^c	188^b
***Experiment 3	Control	100^a	100^a	100^a
	Phytotropin (equivalent to HCG 200 I.U./l)	102^a	169^b	176^b

* Cuttings in rooting medium containing no hormone or the stated concentration of hormone.

** Cutting ends dipped in hormone concentrates for 12 hours and subsequently grown in rooting medium containing no hormone.

*** Cuttings grown in rooting medium contain 'phytotropin' extracted from foliage of *Begonia semperflorens*.

Results of these experiments (Table 3) indicate that HCG has a significant promotive effect on fresh weight of roots and this in experiment 1 exceeds that produced by IBA. As compared to either separately, a combination of HCG and IBA, as applied in experiment 2, produces higher figures of all parameters measured and of fresh weight in particular. The phytotropin applied in experiment 3 had significantly promoted the root fresh and dry weight while the number of roots produced was affected only slightly.

IV. DISCUSSION

Results presented in Fig. 3 indicate that leaf extracts of *Begonia semperflorens* contain a substance which antigenically resembles the human sex hormone, HCG. The latter when applied to plants has promotive growth effects (Table 1 and Fig. 1), a reproducible phenomenon (experiments 1 and 2 in Table 3). When applied with IBA, HCG seems to have a synergistic effect, HCG alone promoting root fresh weight, IBA primarily affecting the number of roots.

Antigenic resemblance does not necessarily imply like biological activity since active sites of the latter on the protein molecule may differ from immunological sites. However, preliminary results obtained by application of the hypothetical 'phytotropin' show marked increase of root fresh and dry weight (experiment 3, Table 3) as compared to control. Both the immunological evidence derived from the haemagglutination inhibition and that from biological tests of a specific antigenic protein released from the antiHCG-polymer complex, indicate the presence of 'phytotropin' in *Begonia* leaf extract. This is in accordance with observations of Leshem *et al*. (1969) who reported that antiHCG sera inhibit rooting of both *Begonia* and *Chrysanthemum* and that a protein extract of foliage has a greater promotive effect than the same extract after passage through a column of polymerized HCG antibodies which presumably adsorb gonadotropin-like material.

Experiments are being conducted to test 'phytotropin' in mammalian systems as expressed by the growth of the uterus and ovary in immature female mice. Plant extracts other than from *Begonia* foliage are being assayed as sources for biologically active material.

V. SUMMARY

As tested by haemagglutination inhibition, proteinaceous extracts of leaves of *Begonia semperflorens* cv. Indian Maid showed antigenic activity resembling that of the proteinaceous human sex hormone - chorionic gonadotropin (HCG). An immunosorbent complex containing HCG antibodies was incubated with foliage extract and by antigen-antibody reaction, any HCG-like substance present in the plant extract presumably trapped. The bound material was subsequently released by guanidine-HCl buffer and pH restored to 7.2. Upon testing this freed antigenic substance, HCG-like immunological properties were revealed and marked promotion of rooting of tomato stem cuttings in rooting medium obtained. The latter effect was also evoked by HCG. These observations together with previously reported findings discussed in the text indicate the presence of a proteinaceous plant growth substance, tentatively designated 'phytotropin', which is akin to the mammalian chorionic gonadotropin.

VI. ACKNOWLEDGEMENTS

The authors wish to thank Mr. Kalman Langenthal, Miss Rina Slezak and Miss Ruth Cohen for their technical assistance. This research was financed in part by the Bar Ilan Research Council.

VII. REFERENCES

AVRAMAES, S., TAUDOU, B., and CHAILON, S. (1969). Glutaraldehyde, cyanuric chloride and tetraazotized O-dianisidine as coupling reagents in the passive hemagglutination test. Immunochem. 6: 67-76.

AVRAMAES, S. and T. TERNYNCK (1967). Biologically active water-insoluble protein polymers. J. Biol. Chem. 242: 1651-9.

DONINI, S. and P. DONINI. (1969). Radioimmunoassay employing polymerized antisera. Karolinska Symp. on Res. Meth. in Reprod. Endocrin. Stockholm. 257-78.

LESHEM, Y. (1967). Physiological effects of animal steroid and gonadotropic hormones on curd cuttings of *Brassica oleracea* L. var. *cymosa*. Phyton. 24: 25-9.

LESHEM, Y., R.R. AVTALION, M. SCHWARZ and S. KAHANA (1969). Presence and possible mode of action of a proteinaceous gonadotropic growth regulating factor in plant systems. Pl. Phys. 44: 75-7.

LESHEM, Y., and B. LUNENFELD (1968). Gonadotropin promotion of adventitious root production on cuttings of *Begonia semperflorens* and *Vitis vinifera*. Pl. Phys. 43: 313-7.

STAVITSKY, A.B. (1964). 'Micromethods for the study of proteins and antibodies. I. Procedure and general application of hemagglutination and hemagglutination-inhibition reactions with tannic acid and protein-treated red blood cells'. J. Immunol. 72: 312-360.

PARTIAL AND COMPLETE GROWTH PROMOTING SYSTEMS FOR CULTURED CARROT EXPLANTS:
SYNERGISTIC AND INHIBITORY INTERACTIONS

F.C. Steward and E.F. Bleichert

Laboratory for Cell Physiology, Growth and Development, Cornell University, Ithaca, New York, 14850

I. INTRODUCTION

This paper recognizes that mature living cells throughout the carrot plant remain capable, with appropriate stimuli, of growth even though this would not occur *in situ* (Steward and Caplin. 1954; Steward *et al.*, 1964). This renewed growth is due to substances which act singly, synergistically or sequentially and meet the exogenous requirements that satisfy endogenous limitations upon the growth of the existing cells.

Any system to test growth promoting substances which consists of explanted tissue or an excised organ has inherent characteristics which reflect its past growth and its further potential. The oat coleoptile, used when cell divisions have ceased and growth is confined to cell extension, represents an obvious example. The carrot phloem assay system utilizes small tissue explants cut at a standard distance from the cambium along a gradient which reflects their need for exogenous stimuli at different distances from the cambium (Steward and Degani, 1969, cf. Fig. 10). The explants cut from a single carrot root constitute a clone and the magnitude of the response induced under otherwise standard conditions varies from clone to clone (Degani and Steward, 1969). These clonal differences obviously reflect limitations upon the growth of the quiescent cells as they developed *in situ*. Invariably any clone gives good growth in a basal medium, such as that of White (B_W), supplemented with the liquid endosperm of the coconut (CM) and a source of reduced nitrogen such as casein hydrolysate (CH).

The first idea was that the coconut milk, or its analogues in extracts from immature corn grains (Zea) or fruits of horsechestnut (*Aesculus woerlitzensis*) might contain a single growth stimulatory substance (other than auxin) that could cause the entire response. Progressively it has become apparent that such fluids represent well balanced systems in which many substances take part. A step in this direction was the recognition that the rich hexitol content of these fluids constituted a so-called neutral fraction (NF) which interacted synergistically with an active fraction (AF) to induce growth.

Further fractionation of the AF yielded components which were mediated by either IAA or *myo*-inositol (Shantz and Steward, 1964). At the 6th Growth Substances Conference, evidence was presented for the activity of an isolate from the vesicular fluid of *Aesculus* that interacted with *myo*-inositol; this isolate consisted of rhamnose, glucose and IAA combined in equimolar proportions (Shantz and Steward, 1968). Thus, the concept emerged (Steward and Degani, 1969) that carrot root explants respond in culture to two distinct growth promoting systems (designated Systems I and II, respectively). In System I the active moieties (AF_1) interact with *myo*-inositol and may be represented by the IAA-glycoside from the *Aesculus* active fraction (AF_{aesc}); in System II the active moieties (AF_2) interact with IAA and may be represented by many naturally-occurring or synthetic adenyl compounds, of which zeatin is a convenient prototype. In these distinctive growth-inducing systems there was a synergism between their component parts, and the resultant growth was due to multiple interactions between many substances in the ambient medium. In fact, without iron, no growth occurs (Neumann and Steward, 1968) and, when other requirements are met, casein hydrolysate furnishes an added stimulus. The best conditions for the growth (i.e., the use of B_W + CM + CH) can be still further modulated by light or temperature (Steward, 1970). It is the present knowledge of these partial systems and their component parts, their existence in well balanced combinations (like coconut milk) which supply all exogenous requirements to satisfy all endogenous limitations and their interactions with inhibitors which is now to be considered.

II. THE COMPONENT PARTS OF EXOGENOUS GROWTH PROMOTING SYSTEMS

The active fraction of System I (AF_1)

This class of substances is recognized by interaction with *myo*-inositol. Since the cells in an otherwise complete basal medium grow more with *both* exogenous *myo*-inositol and the AF_1 the presumption is that they are not fully autotrophic for these active principles. The carrot strains which are most useful to demonstrate the activity of substances of this nature are those that are endogenously limited by them and this may often be ascertained by prior tests; moreover, different carrot varieties and stocks exhibit a range of responses to the component parts of System I, separately and in combination.

An isolate with the properties now attributable to AF_1 came from the central cavity of immature *Aesculus woerlitzensis* fruits and it contained rhamnose, glucose and IAA in equimolar proportions (Shantz and Steward, 1968). A complete reisolation of this material was carried out, using fluid collected in 1968 from the same tree that supplied the earlier collections; this was done to build up stocks of the material in question and to be sure that it was not due to the vagaries of season, stage of development, and the minor details of an isolation procedure. Suffice it to say, that the later isolate, when pure, reproduced entirely the properties of the earlier one. This was shown by its identical hydrolytic properties, by its ultraviolet absorption spectrum and its interaction with *myo*-inositol in the induction of growth in carrot explants (Table 1). The contrasts between the total active fraction (AF_{aesc}) and the isolate merely reinforce the point that the isolated glycoside, despite its demonstrable activity, is only partially responsible for the total *myo*-inositol mediated activity of the *Aesculus* fluid or its active fraction. Like coconut milk, the combination of *myo*-inositol and AF_{aesc}, causes cells to multiply, permits them to enlarge in the later stages of growth and it also promotes greening in the light.

Table 1. The Growth Promoting Activity of an Isolate from *Aesculus*, Singly and in Various Synergistic Combinations, on the Growth of Carrot Explants (Clone 939-A)

Treatments	Fresh weight (mg/explant)	Cell number (thousands/explant)	Cell size (mμg/cell)[1]
B + CH	10.8	67	146
B + CH + AF	23.6	429	53
B + CH + INOS + AF	106.7	1829	62
B + CH + IAA + Z	30.8	837	46
B + CH + CM	293.7	2479	137
B + CH + A	20.9	244	75
B + CH + INOS + A	30.8	351	81
B + CH + INOS + Z + A	141.8	1908	75

Key: B = Basal medium modified from White CH = Casein hydrolyzate (200 mg/l)
 CM = Coconut milk (10% v/v) INOS = *myo*-Inositol (25 mg/l)
 AF = *Aesculus* active fraction (2 mg/l) IAA = Indoleacetic acid (0.5 mg/l)
 Z = Zeatin (0.1 mg/l) A = *Aesculus* isolate (2 mg/l)
 [1]millimicrograms (nanograms) per cell = micrograms per 1,000 cells.

The Active fraction of System II (AF_2)

The first published evidence (Caplin and Steward, 1948) was that the activity of coconut milk on carrot explants exceeded any effect attributable solely to IAA; it did not, however, exclude a possible interaction between IAA and elements of the coconut milk system. In retrospect, the partial autotrophy of the tissue for IAA and/ or its presence in the fractions being isolated permitted this interaction to pass unrecognized until sufficiently pure materials were being processed over columns, or via Craig Post liquid-liquid separations, so that "IAA-requiring" as well as "*myo*-inositol-requiring" fractions became recognizable (Shantz and Steward, 1964). With the recognition of zeatin (Letham, 1963) and its availability as a synthetic product (Shaw and Wilson, 1964), it became possible to use it in lieu of AF_2 and to demonstrate its activity, along with IAA, in the growth and other responses of carrot explants. Thus, different clones respond in varying degrees to IAA and to zeatin as the sole growth factors in the medium and, when they do so, the response they elicit (attributable to System II) may be very different from that due to System I (Table 2). While IAA and zeatin may account in very variable degrees for the maximum response attributable to coconut milk, they characteristically promote cell division more than the subsequent cell enlargement; hence the cells tend to remain small, the cultured tissue is also much less prone to turn green.

Table 2. Responses of Explants From Two Carrot Clones to Growth-Promoting Systems I and II in the Presence of Casein Hydrolyzate

(Clone 921-A was more responsive to System II, whereas Clone 921-B was more responsive to System I. Fresh weight in mg/explant; number of cells in thousands/explant; and cell size in mμg/cell.)

		System I B+CH+INOS+AF	System II B+CH+IAA+Z	Complete System B+CH+CM
Clone 921-A	Fresh weight	27.1	43.6	129.4
	Cell number	772	1662	1662
	Cell size	34	25	77
Clone 921-B	Fresh weight	76.4	36.2	170.4
	Cell number	2036	987	2279
	Cell size	34	32	78

Key: B = Basal medium modified from White
 CH = Casein hydrolyzate (200 mg/l)
 CM = Coconut milk (10% v/v)
 Z = Zeatin
 INOS = *myo*-Inositol
 AF = *Aesculus* active fraction; i.e. the ethyl acetate extract of the vesicular fluid of *Aesculus woerlitzensis*.
System I: INOS at 25 mg/l; AF at 2 mg/l.
System II: IAA at 0.5 mg/l; Z at 0.1 mg/l.

Many zeatin-like adenyl compounds have been synthesized. Therefore, a number of these have been tested (at 0.1 mg/l) on carrot explants in a basal medium with IAA (at 0.5 mg/l), *myo*-inositol and casein hydrolysate. Not unexpectedly, effects due to their chemical configuration modulated the activity of these "adenyl cytokinins." In a series of 10 n-alkylaminopurines, the length of the side chain affected activity so that the observed maximum was at 4 to 6 carbon atoms. This activity occurred with different System II-sensitive clones and was diagnosed both by the effect on fresh weight and cell multiplication during 18 days of culture (Figure 1). The cell division activity attributable to zeatin was somewhat less than that due to an unsubstituted side chain (where n = 5), but it concurrently permitted more cell enlargement. In fact, the balance between cell multiplication and cell enlargement promoted or permitted by these substances seemed to be sensitively related to their chemical configuration (Shaw et al., 1971).

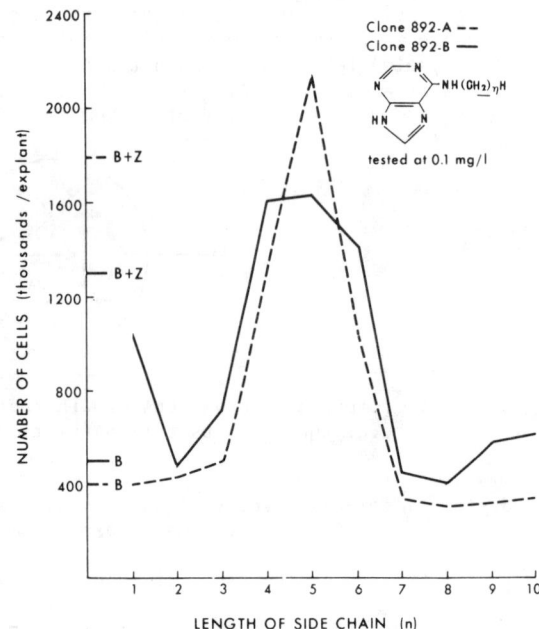

Fig. 1. The effect of a series of n-alkylaminopurines on cell multiplication in cultured carrot explants.

Systems I and II and the maximum growth response

This subject cannot be resolved in a single, simple, overall statement, for it depends upon (a) the endogenous characteristics of the carrot clone and (b) whether growth is assessed by fresh weight, by cell multiplication or by cell enlargement.

Significantly, the partial systems (singly or combined) never produce *greater growth* than that attributable to whole coconut milk (Table 2). However, the maximum effect attributable to the IAA + zeatin system, in relation to whole coconut milk (CM), is to be seen if the response of the explants is measured in terms of thousands of cells per explant or per unit fresh weight of cultured tissue; the maximum effect attributable to *myo*-inositol + the *Aesculus* active fraction (AF_{aesc}) appears if the growth is expressed in terms of total fresh weight or average cell size (i.e., mμg/cell).

The balanced growth stimulus represented by the whole coconut milk affects not only the amount and kind of growth in terms of number and size of cells, but also the course of metabolism. Whereas the nitrogen metabolism under the stimulus of the coconut milk is commonly canalized toward alanine (free or combined in proteins), the stimulus of the partial growth-factor systems has been observed to direct the soluble nitrogen of the same clone toward glutamine (Steward and Rao, 1970). The roles of the respective component parts of the two systems, as they interact with each other and with the trace elements Fe, Mn and Mo, may also be complementary (Steward, 1970).

At this point one should recognize that the sites of action in the cells of the component parts of the partial Systems I and II or of the more complete and balanced natural ones, as in coconut milk, corn or *Aesculus* fluid, are not known. One can, however, visualize (Figure 2) that the known systems may act independently within the cells as exogenous coconut milk supplies the endogenous deficit for any or all of their component parts. When furnished separately, the component parts will only seem to be effective inasmuch as the clone in question is endogenously limited by a particular component; or it will become effective if it is also supplied with other endogenously limiting substances with which its action is linked. Finally, the evidence is (Steward and Degani, 1969) that a source of reduced nitrogen (e.g. casein hydrolysate) tends to "broaden the base of effectiveness of Systems I and II" as though, *via myo*-inositol, it were able to link the activity due to System I to some latent endogenous activity of System II. These ideas are all summarized in Fig. 2.

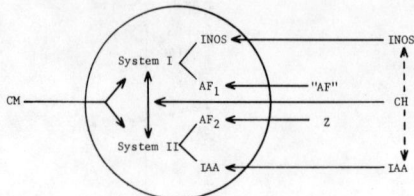

<u>System I</u> comprises the various growth factors (AF_1) mediated by *myo*-inositol (INOS); e.g. the <u>Aesculus</u> active fraction ("AF") which contains the glycoside of IAA.

<u>System II</u> comprises the various natural and synthetic growth factors (AF_2) mediated by IAA; e.g. the many adenyl compounds of which zeatin is an example.

CH extends the range of exogenous Systems I and II and links them together.

CM represents the complete or balanced system which achieves results over and above anything that can be attributed to exogenous Systems I and II in the presence of CH.

Fig. 2. Systems which induce growth in carrot explants: Their component parts and their interactions.

Also, there is room in the superiority of CM and CH, over all other known combinations, for types of growth factors not yet accounted for. The gibberellins have been repeatedly sought as part of the primary growth stimulus of the natural fluids like CM, corn extract and *Aesculus* fluid and highly active fractions isolated from them have been tested, unsuccessfully, for activity in gibberellin assays (Steward *et al.*, 1964).

In the light of recent knowledge of the two distinctive cell division systems, the interaction of gibberellic acid (GA_3) with their component parts was investigated. Table 3 shows that GA_3 interacted to accentuate the growth-promoting activity otherwise due to Systems I or II. This interaction was mediated via the cofactors (endogenous or exogenous). In fact, the GA_3 may furnish a requirement for maximum growth of the carrot explants and may even supplement the action of coconut milk which may be deficient in this respect.

III. EFFECTS DUE TO INHIBITORS: INTERACTIONS WITH GIBBERELLINS

Since the behavior of cultured carrot explants responds to a network of interacting growth factors which, singly and in combination, comprise systems that stimulate growth, several other lines of evidence may be integrated with this view.

Table 3. Interactions of Gibberellic Acid with Coconut Milk and the Component Parts of Systems I and II on the Growth of Carrot Explants (Clone 915-B).

Treatment	Fresh weight (mg/explant)	Cell number (thousands/explant)	Cell size (mµg/cell)
B + CH	15.4	146	105
B + CH + GA_3	15.0	158	95
B + CH + INOS	15.9	146	109
B + CH + INOS + GA_3	16.0	188	85
B + CH + IAA	17.3	279	62
B + CH + IAA + GA_3	24.6	418	59
B + CH + INOS + IAA	53.4	795	73
B + CH + INOS + IAA + GA_3	79.2	1177	61
B + CH + CM	241.4	2189	110
B + CH + CM + GA_3	314.5	2795	112

Key: B   = Basal medium modified from White   CH   = Casein hydrolysate (200 mg/l)
      CM = Coconut milk (10% v/v)   INOS = myo-Inositol (25 mg/l)
      GA_3 = Gibberellic acid (0.5 mg/l)   IAA = Indoleacetic acid (0.5 mg/l)

A regulatory control of growth in carrot cells could flow from means to inhibit growth or reversibly to remove that inhibition. The growth of cultured tissue tends to slow down, even on otherwise competent media, as though the tissue builds up its own resistance to further growth which is usually eliminated when subcultures are made. By contrast, very actively growing, freshly explanted tissues convey to fresh medium properties that may make it peculiarly effective as "conditioned medium" for the growth of less active cells or cultures. These observations lead to interpretations based on the ability of freely growing cells to produce and excrete their own stimulatory substances whereas older cultures may build up inhibitors which at the outset were overcome.

Naturally-occurring inhibitors were demonstrated (Steward and Caplin, 1952) in extracts of storage organs by the responses obtained when they were added in the carrot-coconut milk assay system. A particular class of inhibitors, which behaved like "anti-CM factors" had a simple explanation, for they (L-hydroxyproline and some related compounds) acted as competitive proline antagonists (Steward et al., 1958). But the long-standing knowledge that dormant maple buds yield an inhibitory extract competitive with CM prompted later tests on abscisic acid (supplied by Drs. J.W. Cornforth and B.V. Milborrow) as a possible antagonist of CM and the Systems I and II. Meanwhile, heliangine (supplied by Dr. T. Yamaki), as an inhibitor of cell elongation in shoots, the formation of which is attributable to light, and a glycoside of solanidine (supplied by Dr. T. Tagawa), an inhibitor from potato tubers, also became available. The full study of these substances cannot be given here, but it can be shown that their effects integrate with the concepts outlined above.

The effects of abscisic acid on carrot explants

Figure 3 shows the inhibitory effects of abscisic acid in the range 0.02 to 10 mg/l; complete inhibition of the action of the CM on carrot explants occurred at 10 mg/l. Thus, this substance could have been responsible, at least in part, for the growth inhibition earlier ascribed to extracts of dormant buds of Acer. But, following the known ability of gibberellic acid to counteract the effects of abscisic acid in other systems (Wareing et al., 1968), it was also shown that this substance could overcome the growth inhibition otherwise due to abscisic acid. Thus, when quiescent carrot cells were subject to an overall stimulus conveyed by coconut milk, they could be extensively antagonized by abscisic acid, and this in turn could be overcome by gibberellic acid (Figure 4). Thus, abscisic acid [1] and gibberellic acid may intervene to modulate further the overall system represented by Fig. 2.

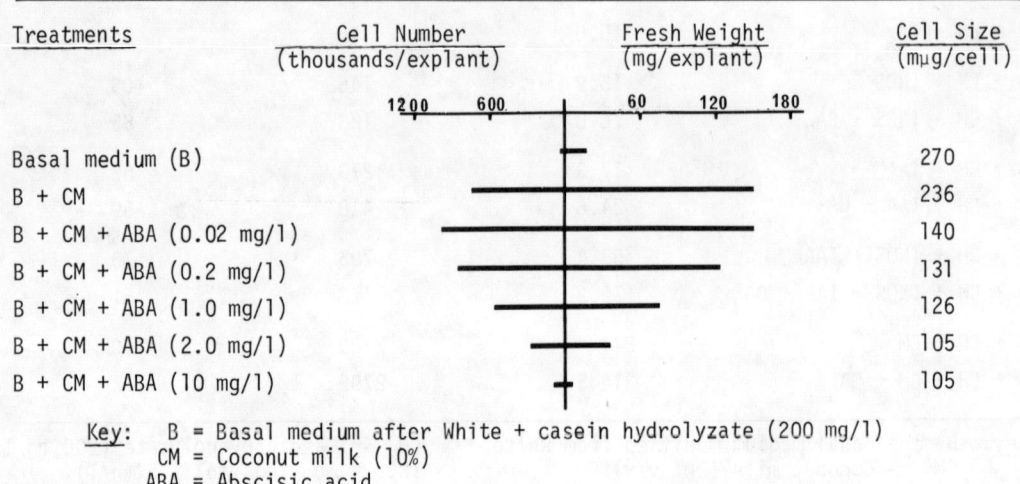

Fig. 3. Effects of abscisic acid concentration on the growth of cultured carrot explants.

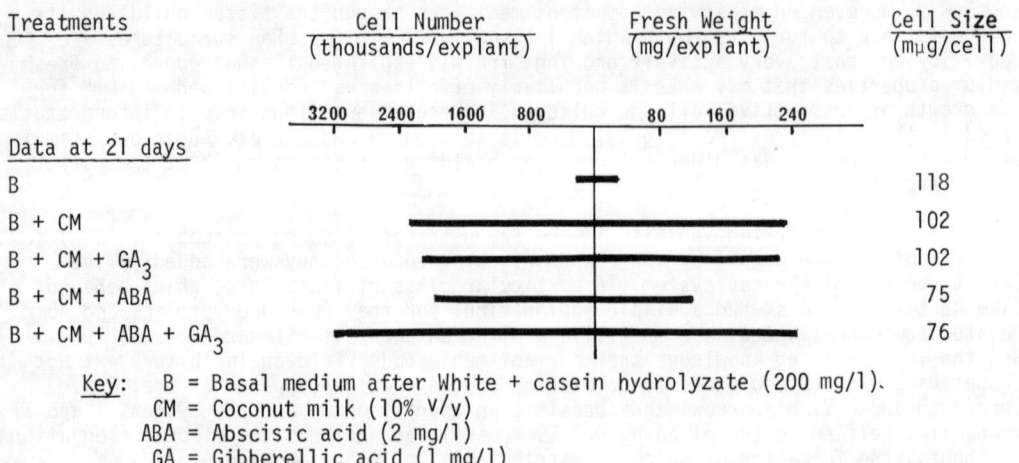

Fig. 4. Contrasted effects of abscisic acid and gibberellic acid on the growth of cultured carrot explants.

---

[1] Although abscisic acid has been found in coconut milk it appears to occur there in concentrations below these at which growth of carrot cultures is limited by its action. (Private communication from Dr. B.V. Milborrow).

Briefly, experiments have been made to test whether the abscisic acid acted selectively upon either System I or II or on reactions mediated specifically by their component parts. The short answer is that abscisic acid interacts more with System I than with System II and this is attributable to its effects on the cell enlargement fostered by System I.

Determinations suggest that abscisic acid may inhibit growth (in fresh weight and by cell enlargement) even without retarding the synthesis of protein or the incorporation and hydroxylation of proline. In fact, slower growing cells inhibited by abscisic acid had a higher combined hydroxyproline content than the faster growing or uninhibited cells (Table 4). Thus, while the coconut milk stimulus reactivates the cells, inducing them to divide, abscisic acid intervenes to retard cell division and to arrest cell enlargement (Fig. 3).

Table 4. Effects of Abscisic Acid (ABA) on the Content of Combined Proline (Pro) and Hydroxyproline (HO-Pro) in Cultured Carrot Explants at 18 Days (Clone 815-B).

Treatment	Fresh weight (mg/explant)	Hydroxyproline (μM/gfwt)	Proline (μM/gfwt)	Ratio (HO-Pro/Pro)
B + CH	10.2	9.4	0.7	1.4
B + CH + CM	141.4	6.4	1.2	1.1
B + CH + CM + ABA (0.2 mg/l)	133.0	5.7	0.9	1.3
B + CH + CM + ABA (1 mg/l)	114.3	6.3	0.8	1.5
B + CH + CM + ABA (2 mg/l)	69.3	6.4	0.8	1.6
B + CH + CM + ABA (10 mg/l)	15.2	13.1	0.4	7.5

Key: B = Basal medium modified from White
CH = Casein hydrolyzate (200 mg/l)
CM = Coconut milk (10% v/v)

In the outcome the inhibited cultures had a lower fresh weight but a higher content of solutes (amino acids, salts, sugars and even protein). The primary effect is, therefore, to retard the uptake of water. The clue is that abscisic acid retards the hydration of cells, whereas a first effect of the CM stimulus, as it reactivates quiescent cells, is to increase their "succulence." In this respect, therefore, abscisic acid acts upon the cultures like a "hardening-off" procedure in horticultural practice. This being so, gibberellic acid has the reverse effect. Thus, one should regard inhibitors like abscisic acid and their antagonists like gibberellic acid along with other such general environmental factors as light and temperature (Steward, 1970), as able to modulate the activity of the whole system through the physiological role of water.

IV. THE INVOLVEMENT OF ETHYLENE IN THE INDUCTION OF GROWTH

This investigation, done in collaboration with Dr. Stanley Burg, can only be summarized here.

Some have claimed that ethylene is a virtually universal agent, a common denominator in fact, which frequently mediates the effect of plant growth regulating substances, whether these belong to one named class or another. The induction of growth in quiescent carrot explants under the impact of coconut milk and the separate Systems I and II and their component parts offered a unique means to test these ideas. Briefly, the carrot explants, appropriately drawn from clones with somewhat different sensitivity toward Systems I and II, were cultured aseptically and, periodically sealed to accumulate ethylene which was sampled and analyzed in Dr. Burg's laboratory.

The effects of various growth promoting substances, separately and in combination were first ascertained after the tissue had passed through the period of growth induction. The first conclusions were that:

(i) There were initial "bursts" of ethylene, which gave their highest values on the first sampling (after 5 days of growth) with the obvious implication that an earlier maximum could have been passed.

(ii) The metabolism in the basal medium (and also in the presence of the cofactors IAA or *myo*-inositol separately) produced virtually no ethylene.

(iii) The interactions of IAA and zeatin, on the one hand, and *myo*-inositol and AF_{aesc}, on the other, produced substantial amounts of ethylene: more in fact than the more complete growth promoting system represented by coconut milk.

(iv) After the initial "burst" of C_2H_4, when the tissue had responded to the growth promoting systems, the output of ethylene fell, even though the tissue grew (in fact very substantially so in the coconut milk medium).

Thus, the tissue explants could be stimulated to grow rapidly in a balanced growth factor system with very little output of ethylene. Furthermore, the conspicuous ethylene production occurred when the growth stimulus was partial or unbalanced, and even this was most evident in the period of transition from one state to another. Therefore, other experiments were made to test these ideas, taking advantage of clones with different sensitivity to IAA and zeatin on the one hand and to *myo*-inositol and AF_{aesc} on the other and paying closer attention to the early periods of contact with the growth substances. Data of this kind obtained on two clones (922-A and 922-B) are to be seen in Fig. 5.

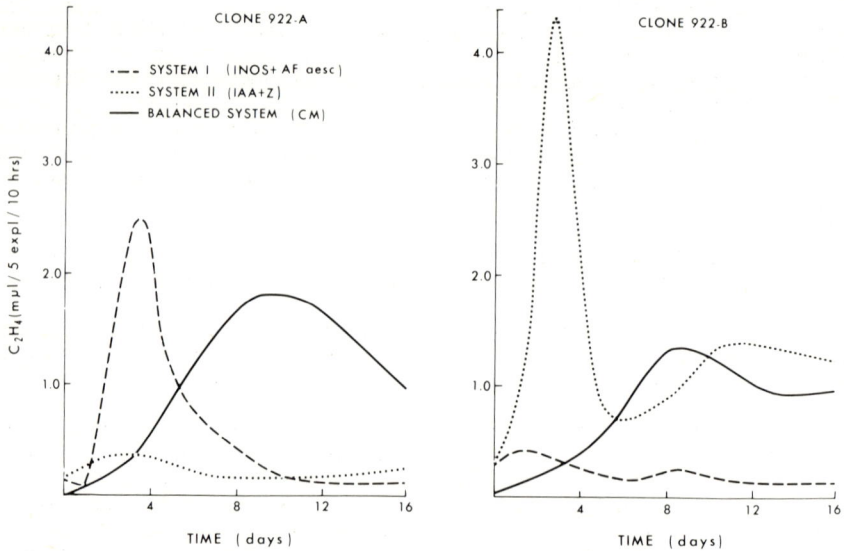

Fig. 5. The time course of ethylene (C_2H_4) output of cultured carrot explants in relation to growth induction.

Despite the normal growth stimulus of coconut milk, the explants so treated failed to produce a large initial burst of ethylene and, despite a large growth during the period of observation, the ethylene output relative to the weight of the explants was low. Neither clone produced ethylene in quantity under the stimulus of the basal medium plus IAA or *myo*-inositol. They differed, however, in that one clone (922-B) produced a large and transient burst of ethylene due to IAA + zeatin, whereas the other (clone 922-A) produced a burst of ethylene in response to *myo*-inositol + AF_{aesc}.

The short conclusion is as follows. Even when the carrot explants are so drastically activated as by the *balanced* coconut milk system, in which both cell enlargement and cell divisions occur, their output of C_2H_4 is unexpectedly small, especially when the explants adjust to a steady growth rate. On the other hand, when the quiescent tissue was activated by the partial systems and was in transition to a more active state, then brief, but very active "bursts" of C_2H_4 occurred. On this view, therefore, C_2H_4 becomes more a symptom than a primary cause of the growth induced by the regulators in question. Moreover, it is significant that symmetrical, balanced, stimuli which satisfy all exogenous requirements to meet the endogenous limitations of cells can induce and maintain their growth without the apparent release of C_2H_4. The supposed involvement of ethylene in other growth situations has led to current frenetic ideas that it is the principal, or even the sole, mediator of growth promoting effects in plants which are otherwise attributable to growth hormones. Later experiments (not quoted) have shown that actual additions of C_2H_4 to carrot cultures in the logarithmic phase of their coconut milk-stimulated growth may mildly inhibit cell division, although the ethylene does allow the cells to enlarge somewhat more than they otherwise would, and thus may even lead to a slight increase in fresh weight per explant.

V. SUMMARY

Using relatively simple measurements of the onset of growth in otherwise mature, small tissue explants of carrot, the controls of cell division, enlargement and maturation have been found to involve many different kinds of molecules which comprise different systems and which act synergistically and sequentially to activate or inhibit the events in question. With due regard to the multiplicity of events in a typical cell cycle (Kihlman, 1966), and to the highly heterogenous system of relatively autonomous organelles to be activated or controlled, a multiplicity of controlling factors and mechanisms with interlocking effects seems much more reasonable than a simplistic concept in which each regulatory substance acts directly and independently of all others (Steward and Krikorian, 1971). In fact, the versatility of different growth factors appears when they are tested in the most totipotent systems and through the network of their interactions (synergistic, sequential, inhibitory), for their often complementary effects ramify throughout the vital machinery. The problem may seem to be one of trying to understand how growth substances assist nuclear genes to tell cells what to do; but at the time the growth factors act on angiosperm cells they may have long passed the stage at which all the cytoplasmic organelles (with their own complements of RNA and DNA) need any longer "ask the nucleus" for permission to do what is already within their competence. Therefore, the importance of the plant growth regulators lies, not in their resemblance to animal hormones, but in that they seem to be very different and behave in accordance with the distinctive way of life that higher plants have evolved. In this way, each presumptively totipotent cell, responds to its immediate environment and uses as stimuli an array, or network or matrix of many interacting molecules.

Individual substances in this array could, under different circumstances, be variously classified in such currently accepted categories as auxins, cytokinins, gibberellins, etc. But neither a given growth factor (cofactor or promotor of cell division as in Systems I and II) nor an individual substance (like ethylene) can have any exclusive claim to be the universal growth regulator. Nor does any specific metabolic response hold the sole key to their mode of action. In fact, such reversible effects as those due to coconut milk (and to the partial systems), to abscisic acid, and to gibberellic acid seem to focus upon the physiological roles of water as they affect the "succulence" of the cells.

VI. ACKNOWLEDGEMENTS

The support of Research Grant No. GM 09609 from the National Institute of General Medical Science is gratefully acknowledged.

VII. REFERENCES

CAPLIN, S.M. and F.C. STEWARD (1948). Effect of coconut milk on the growth of explants from carrot root. Science 108: 655-657.

DEGANI, N. and F.C. STEWARD (1969). The effect of various media on the growth responses of different clones of carrot explants. Ann. Bot. 33: 483-504.

KIHLMAN, B.A. (1966). "Actions of Chemicals on Dividing Cells". xi + 260 pp. Englewood Cliffs: Prentice-Hall, Inc.

LETHAM, D. (1963). Zeatin, a factor inducing cell division isolated from Zea mays. Life Sci. 2: 569-573.

NEUMANN, K.H. and F.C. STEWARD (1968). Investigations on the growth and metabolism of cultured explants of Daucus carota. I. Effects of iron, molybdenum and manganese on growth. Planta 81: 333-350.

SHANTZ, E.M. and F.C. STEWARD (1964). Growth promoting substances from the environment of the embryo. II. The growth stimulating complexes of coconut milk, corn and Aesculus. Coll. Intern. C. N. R. S. (Paris) 123: 59-75.

SHANTZ, E.M. and F.C. STEWARD (1968). A growth substance from the vesicular embryo sac of Aesculus. In: "Biochemistry and Physiology of Plant Growth Substances", pp. 893-909. Ed. by F. Wightman and G. Setterfield, Runge Press, Ottawa.

SHAW, G., B.M. SMALLWOOD and F.C. STEWARD (1971). The structure and physiological activity of some N^6-substituted adenines and related compounds. Phytochemistry. 10: 2329-2336.

SHAW, G. and D.V. WILSON (1964). A synthesis of zeatin. Proc. Chem. Soc. (Lond.) p. 231.

STEWARD, F.C. (1970). The Croonian Lecture, 1969. From cultured cells to whole plants: The induction and control of their growth and morphogenesis. Proc. Roy. Soc. Lond. 175: 1-30.

STEWARD, F.C. and S.M. CAPLIN (1952). Investigations on the growth and metabolism of plant cells. III. Evidence for growth inhibitors in certain mature tissues. Ann. Bot. 26: 477-489.

STEWARD, F.C. and S.M. CAPLIN (1954). The growth of carrot tissue explants and its relation to the growth factors present in coconut milk. I(A). The development of the quantitative method and the factors affecting the growth of carrot tissue explants. Année Biol. 30: 386-394.

STEWARD, F.C. and N. DEGANI (1969). Endogenous characteristics of different clones of carrot explants and their exogenous requirements for growth. Ann. Bot. 33: 615-646.

STEWARD, F.C. and A.D. KRIKORIAN (1971). "Plants, Chemicals and Growth." Academic Press, pp. 232.

STEWARD, F.C., K.H. NEUMANN and K.V.N. RAO (1968). Investigations on the growth and metabolism of cultured explants of Daucus carota. II. Effects of iron, molybdenum and manganese on metabolism. Planta 81: 351-371.

STEWARD, F.C., J.K. POLLARD, B. WITKOP and A.A. PATCHETT (1958). The effects of selected nitrogen compounds on the growth of plant tissue cultures. Biochem. Biophys. Acta 28: 309-317.

STEWARD, F.C. and K.V.N. RAO (1970). Investigations on the growth and metabolism of cultured explants of Daucus carota. III. The range of responses induced in carrot explants by exogenous growth factors and by trace elements. Planta 91: 129-145.

STEWARD, F.C., E.M. SHANTZ, M.O. MAPES, A.E. KENT and R.D. HOLSTEN (1964). The growth promoting substances from the environment of the embryo. I. The criteria and measurement of growth-promoting activity and the responses induced. Coll. Intern. C.N.R.S. (Paris) 123: 45-58.

WAREING, P.F., J. GOOD, H. POTTER and A. PEARSON (1968). Preliminary studies on the mode of action of abscisic acid. In: Plant Growth Regulators, S. C. I. Monograph No. 31, pp. 191-207.

PLANT GROWTH SUBSTANCES, 1970

Multiple Interactions between Media, Growth Factors and the Environment of Carrot Cultures: Effects on Growth and Morphogenesis

F.C. Steward and H.W. Israel

Laboratory for Cell Physiology, Growth and Development, Cornell University, Ithaca, New York 14850

I. INTRODUCTION

Carrot explants appropriately nourished and stimulated proliferate and yield cells which produce plantlets in great abundance; the cells so cultivated are, in fact, totipotent. In this respect, and as they give rise to embryoids, the cultured cells behave like zygotes. Significantly, certain stimuli that induce the growth of initially quiescent carrot cells are drawn from the environment of immature embryos, i.e., the liquid endosperm of coconut (*Cocos*), an extract of immature grains of corn (*Zea*) or the fluid from the vesiculate embryo sac of fruits of horsechestnut (*Aesculus*). It is from such sources that evidence of balanced and partial growth promoting systems and of interactions between their component parts has been drawn. The superiority of the naturally balanced fluids over their defined, but still incomplete, experimental replacements has been stressed (Steward and Degani, 1969). In fact, in a variety of ways, the effectiveness of the environment of the ovule, which may bring one cell that can grow to maturity as an embryo plantlet, has to be recognized. This effectiveness is attributable to a combination of circumstances, of which the interactions here described are but a part. Since carrot cells may give rise to small proembryonic cell clusters that undergo embryogenesis, evidence accumulated that the course of their development could be modulated by controllable properties of the environment (Steward et al., 1970). Such variables as the total concentration of the medium, whether due to nutrient salts or organic solutes; the nature and level of nitrogen in the medium, whether nitrate, ammonium or casein hydrolysate; the balance, or imbalance, of exogenous growth factors such as coconut milk, IAA, NAA or 2,4-D; and light v. darkness, have all produced effects which have been described (Ammirato and Steward, 1971). (These effects range from malformations of growth, exaggerated growth in particular sites or organs, to premature "germination" which bypasses the formation of cotyledons.) This communication, therefore, utilizes known effects attributable to component parts of growth-promoting systems, and it initiates what should ultimately become an interpretation of features of ovules and embryo sacs, which are conducive to the growth and development of zygotes *in situ*. To this end, however, this paper is but a progress report.

II. SITES OF ACTION IN CELLS

It is a current trend to involve every control over growth and development in mechanisms which operate at the level of genes and which take effect as genes are turned "on or off"; when genes are activated or de-repressed, the appropriate mRNA's may transmit their effects through selective synthesis of proteins and enzymes and the metabolism they incite. This scheme, based on bacterial models, permits one to visualize how selected complex molecules are made, within the range of those that are genetically feasible. It is another matter, however, to know how externally applied substances control the working of the whole organization involved in cell growth and morphogenesis of higher plants. One might imagine that cells subjected to such drastic change as in the events of growth induction and which pass from quiescence to active cell division and growth should show, concomitantly, abundant evidence of mRNA's in

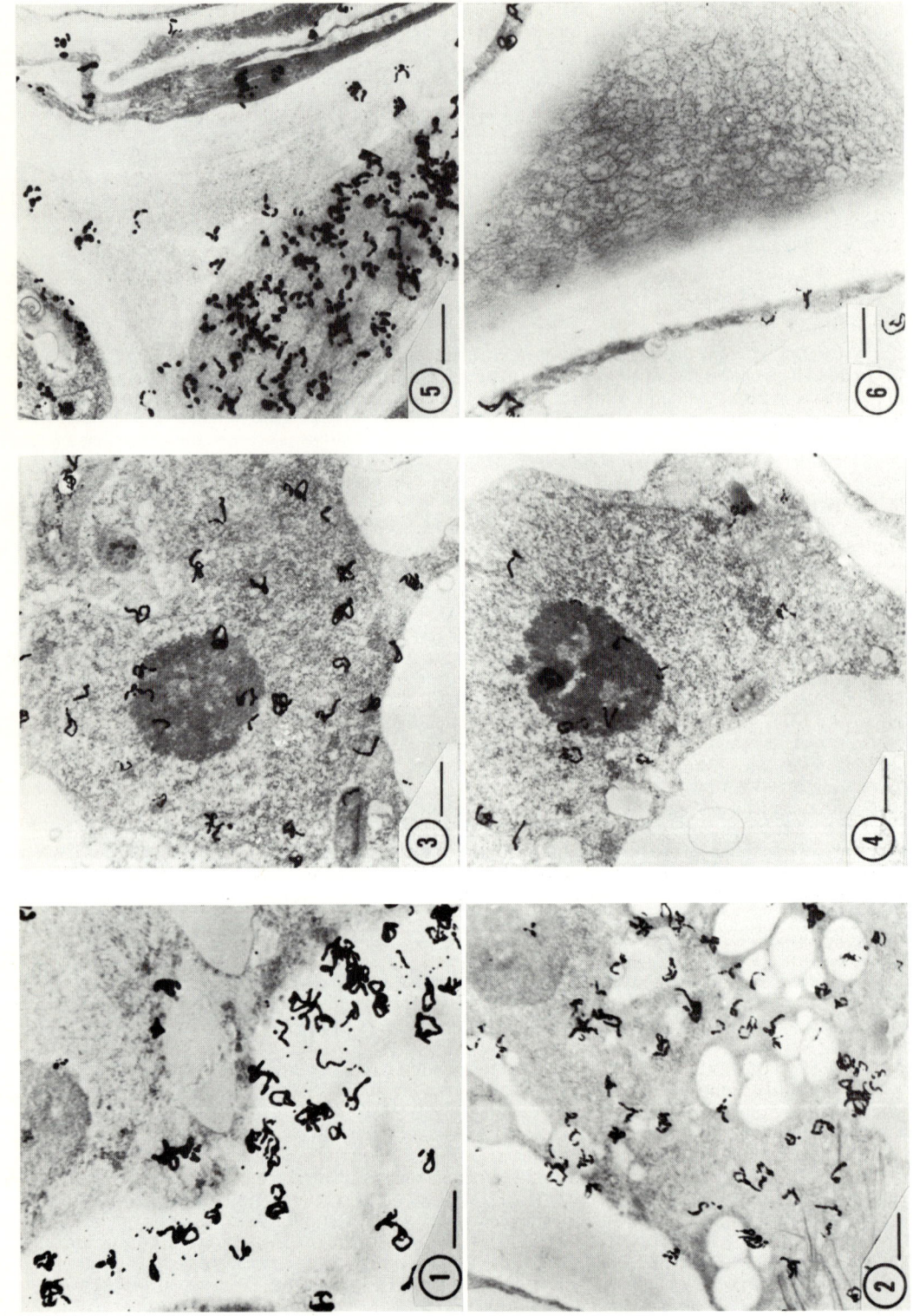

Figs. 1-6. The distribution of component parts of growth factor systems as revealed by high resolution radioautography (calibrations: 1μ).

Fig. 1. H^3-inositol localized in well-formed wall, avoiding the nucleus, of carrot cells cultured in a medium supplemented with coconut milk.

Fig. 2. H^3-inositol as distributed in protoplasm, avoiding wall and nucleus, in carrot cells cultured in a basal medium also containing IAA and zeatin but no coconut milk.

Fig. 3. H^3-IAA in the protoplasm, including the nucleus and nucleolus, of carrot cells cultured in a basal medium also containing inositol and zeatin but no coconut milk.

Fig. 4. H^3-zeatin in the protoplasm, emphasized in the nucleus and nucleolus, but avoiding the wall of carrot cells cultured in a basal medium also containing IAA and inositol but no coconut milk.

Fig. 5. H^3-IAA heavily localized at the juncture of mature walls of expanding cells cultured in a basal medium containing inositol and zeatin but no coconut milk.

Fig. 6. H^3-zeatin avoiding the areas heavily labelled in (5) but present in parietal protoplasmic layers of cells cultured in a basal medium containing IAA and inositol but no coconut milk.

their cytoplasms. So far, no such convincing evidence has emerged (Steward, 1970). Nevertheless, the actions of what may be called the "adenyl-cytokinins", like zeatin, have been attributed (Skoog and Armstrong, 1970) to their presence as "odd-bases" in tRNA's which could thereby influence the synthesis of a protein peculiarly associated with cell division - perhaps a DNA polymerase.

A first step is to localize the component parts of the various growth promoting systems in the cells which they affect. Before speculating upon how the exogenous growth regulators act, one should know where, among the highly compartmented autonomous organelles, the absorbed substances are to be found. To this end generally H^3-labelled IAA, inositol and zeatin have been supplied to carrot cells. If all the labeling, as revealed by high resolution radioautography, were merely random no valid conclusions could be drawn. Since this is not so, it is interesting to note the following trends from the studies made to date.

The method was as follows. Parallel carrot cultures were grown vigorously in a basal medium which also contained a full complement of growth factors in the form of coconut milk and casein hydrolysate and were supplemented with H^3-inositol. Other parallel cultures on the basal medium were supplemented by IAA, inositol and zeatin at low concentration, but each of these supplements was in turn generally labelled with tritium. The complementary labeling patterns then observed with the H^3-IAA, H^3-inositol and H^3-zeatin were, therefore, obtained in cultures subjected to the same total complement of growth factors, only one of which was H^3-labelled while the others were present in the unlabelled form.

Figs. 1-6 only show typical examples from the very large numbers of fields examined for each treatment mentioned above, but they anticipate the later complete quantitation of these distributions when data become available from the use of a tritiated component part of the active complex from *Aesculus* (AF_1).

Figs. 1-6 and their legends tell their own story. The labelled component parts of the growth promoting systems were not randomly distributed within the cells they affect. Inositol or IAA, the cofactors of Systems I and II respectively, may, under the influence of some treatments, become heavily located in well-formed cell walls or at the junctions of enlarging cells where they "pull apart" to form air spaces. Nevertheless, under other treatments, the same factors may avoid the walls of the cells which they enter and be located in the protoplasm, especially in the nuclei and nucleoli. On the other hand, zeatin, the active cell division substance of System II never appeared substantially in cell walls. Nevertheless, zeatin did appear in nuclei and nucleoli of cells which also accumulated IAA in these organelles (cf. Figs. 3 and 4) and in the parietal protoplasm of fully expanded cells with large vacuoles (Fig. 6). Thus, it would be imprudent to limit the mode of action of these substances to any single reaction for they may do different things where and when they are in the ontogeny of the cells they affect.

III. FREE CELLS AND MORPHOGENESIS

The above concerns the role of growth promoting systems in proliferative growth of cells from tissue explants. But totipotent cell clusters grow in an organized embryonic way. This implies that, locally, within a group of attached cells the growth and development is controlled to achieve a predestined pattern. And during differentiation cells which are essentially totipotent must receive in sequence a series of signals, or instructions, to tell them in time and place what to do. Within the range of interlocking growth promoting systems which can act on cells from without, there is scope for changes in the balanced network of exogenous controls to bring this about. The present evidence along these lines and from this laboratory may be summarized as follows:

(a) The first essential is to make cells grow as rapidly as possible in small colonies.

(b) The balanced stimuli from the vicinity of the zygotes, as in coconut milk, as extract of young corn grains, or the fluid from immature *Aesculus* fruits often do this effectively, but, where they are ineffective alone, their synergistic combination with other substances such as 2,4-D or NAA may be more effective, as in the case of potato tuber tissue.

(c) Sequentially applied synergistic combinations (such as B + CM + NAA or B + CM + 2,4-D ⟶ B + CM ⟶ B, etc.) have successfully induced morphogenesis in some otherwise recalcitrant cell cultures.

(d) The form of adventive or somatic embryos which develop in large numbers in cultures of suspended cells may be variable, but it is controllable (Ammirato and Steward, 1971) at appropriate stages by such factors as:

(i) exposure to solutions of high osmotic value (liquid endosperms from immature fruits are commonly at or about 8-10 or more atmospheres). Such higher osmotic pressures affect the embryos and their cells to keep them small.

(ii) Exposure to light or darkness.

(iii) Control of the nitrogen supply, especially when cotyledonary primordia have formed, for they are very responsive to high nitrogen and to ammonia.

(e) The new observations here recorded, as in Figs. 7 to 10, concern:

(i) The controls which may be exercised over the multiplication and uniformity of proembryonic units suspended in liquid; the unexpected discovery here was the contrasted effect of light and darkness for continuous light, which permits proliferation of explants, was markedly deleterious to the multiplication of morphogenetically viable units (cf. Fig. 7 a and b).

(ii) To obtain "poised" cultures of proembryonic cell clusters, in a condition awaiting a stimulus for their subsequent development is a desirable objective, and this has been achieved by suitable control of the osmotic and metabolic values of the medium allocating this between sorbitol (a prime constituent of coconut milk), on the one hand, and sucrose on the other. But again the dramatic evidence is that the sequential transition from a sorbitol-rich to a sucrose-rich medium in the light is also affected, strangely, by a persistent after-effect of a prior exposure to long continued darkness (cf. Fig. 8).

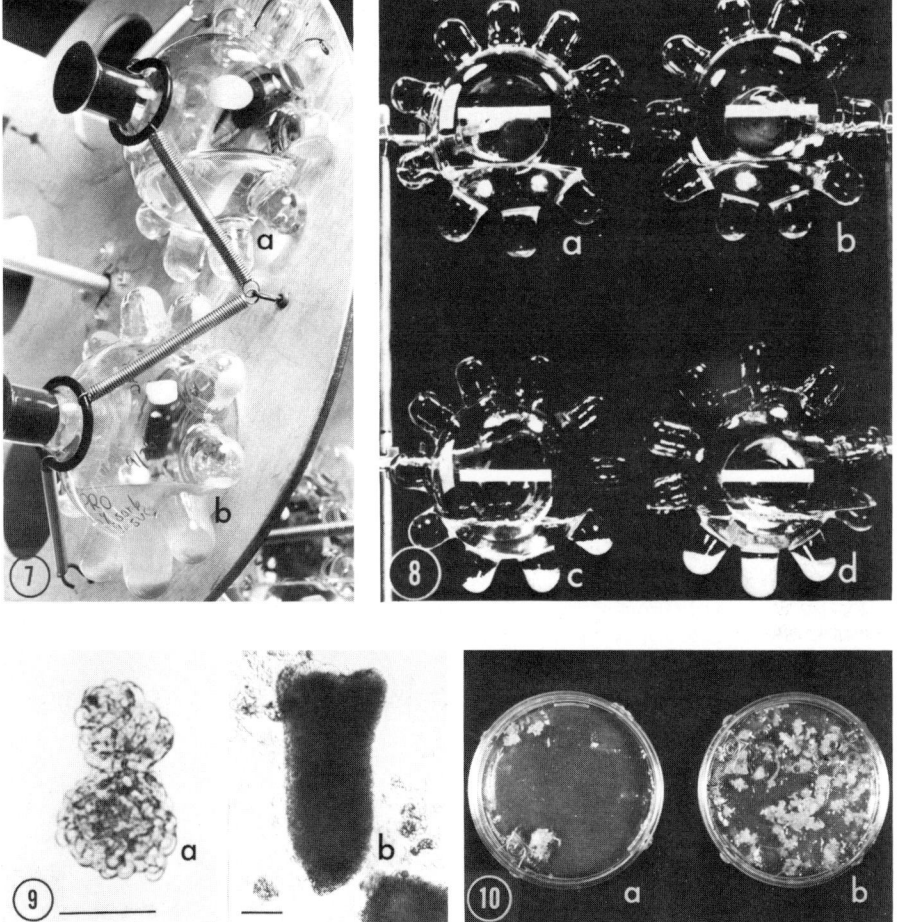

Figs. 7-10. Factors which affect proliferation and development of proembryonic units in carrot cultures.

Fig. 7. Multiplication of uniformly graded proembryonic units in a medium containing NAA, contrasting the effects of continuous light (a) and darkness (b).

Fig. 8. Proembryonic units poised by 3% sorbitol (a and b) and subsequently activated by the addition of 3% sucrose (c and d). Cultures a and c were in continuous light; b and d were in continuous darkness.

Fig. 9 (a). Unorganized globular proembryonic clusters arrested by 3% sorbitol;
(b). Similar units at the torpedo stage after activation by the addition of sucrose (calibrations: 100µ).

Fig. 10. The stimulus of prior darkness to the growth of globular proembryonic units on a conditioned medium (b) in contrast to similar cultures in continuous light (a).

(iii) Lastly, it is still evident - as indeed it was many years earlier - that carrot explants cultured on a medium containing coconut milk can so enrich or modify that medium that it becomes unusually effective as "conditioned medium" in promoting the unorganized growth of cells and cell clusters (cf. Blakely and Steward, 1964, Fig. 5).

(f) Thus unorganized, proembryonic suspension cultures can now be grown, graded for uniformity in size (cf. Fig. 7), held in abeyance by controlling the composition of the medium (cf. Fig. 8 a and b) in readiness for the "signals" that set them off, uniformly and with virtually 100% efficiency, upon their course of growth (cf. Fig. 8 c and d). Thus, the "poised" proembryonic cultures can now be maintained in continuous light in media in which sugar is replaced by sorbitol.

(g) The transfer of such poised cultures to media (on agar or in liquid) which now contain sugar starts off their growth (cf. Figs. 8-10). Procedures (cf. Fig. 10) which foster early and regular morphogenesis are:

(i) The use of a conditioned medium (i.e., a basal medium containing, as one of its constituents, filter-sterilized media on which somatic embryos were grown for 10-12 days).

(ii) The above coupled with culture in continued darkness for periods up to two weeks, followed, if necessary, by subsequent growth in the light.

Thus, the lesson to be learned is that growth and morphogenesis are most feasible when the cells are freed from the environment of the tissue of their origin, when they are caused to grow and to multiply freely, but also when they are subjected to empirically determined sequences of nutrients, environments and stimuli to growth.

These effects may be conceived as perturbations in an otherwise balanced array or network, a matrix, of exogenous controls, of which the partial and complete growth factor systems described for carrot explants are but a part.

IV. SUMMARY

The results in this paper extend the evidence on partial and complete growth factor systems by showing that three component parts may assume different and often complementary distributions in the cells they affect. Moreover, the distribution of one substance (inositol) may be variously affected during the growth of tissue subjected to the more balanced growth promoting stimulus due to coconut milk or, by contrast, to the imbalanced exogenously limited growth due to a basal medium containing only certain of the synergistically active growth factors (e.g. IAA and zeatin). The growth induction stimulus should not, therefore, be circumscribed. The stimulus in question is not to be thought of as a biochemically prescribed single function; there is room, in the regions and organelles of the cells which receive the component parts of such complex growth promoting systems, for many and various effects that may be causally related to the growth which ensues.

But the events of morphogenesis, which range from the arrested development of cells in proembryonic clusters in some media, and their replication and development in others, obviously suggest means by which the behavior of cultured cells in proembryonic clusters may interpret the role played by the content of, and the environment furnished to, zygotes in ovules. The sequential effects of sorbitol followed by sucrose, the importance of prior darkness and subsequent light and the superiority of a "conditioned" medium all add to our knowledge of the causal factors in somatic embryogenesis. In fact, these factors act over and above the balanced and partial growth promoting systems that induce the first growth of the otherwise quiescent cells as they exist in, and are explanted from, the mature plant body.

V. ACKNOWLEDGEMENTS

The support of Research Grant No. GM 09609 from the National Institute of General Medical Science is gratefully acknowledged.

VI. REFERENCES

AMMIRATO, P.V. and F.C. STEWARD (1971). Some effects of environment on the development of embryos from cultured free cells. Bot. Gaz. 132: 149-158.

BLAKELY, L.M. and F.C. STEWARD (1964). Growth and organized development of cultured cells. V. The growth of colonies from free cells on nutrient agar. Am. J. Bot. 51, 780-791.

DEGANI, N. and F.C. STEWARD (1969). The effect of various media on the growth responses of different clones of carrot explants. Ann. Bot. 33, 483-504.

SKOOG, F. and D.J. ARMSTRONG (1970). Cytokinins. Ann. Rev. Pl. Physiol. 21, 359-384.

STEWARD, F.C. (1970). From cultured cells to whole plants: The induction and control of their growth and morphogenesis. Proc. Roy. Soc. Lond., B, 175, 1-30.

STEWARD, F.C., P.V. AMMIRATO and M.O. MAPES (1970). Growth and development of totipotent cells: Some problems, procedures and perspectives. Ann. Bot. 34, 761-788.

Control of Morphogenesis in Plant Tissue Cultures by Hormones and Nitrogen Compounds

J. Reinert

Pflanzenphysiologisches Institut, der Freien Universität Berlin, 1 Berlin 33, Königin-Luise-Str. 12-16a

I. INTRODUCTION

Twelve years ago the first papers of my laboratory on the formation of embryos by somatic cells of tissue cultures from carrots were published. The same work also showed that the nitrogen and the auxin of the nutrient played an important role for this type of morphogenesis (Reinert 1958, 1959). Later on we were able to demonstrate that the production of embryos in the cells of carrots and other species of Umbelliferae could be controlled by changes of the auxin or the nitrogen level of the medium (Reinert, 1963; Reinert et al. 1966). We also had some indication that it was the ratio between these two factors which determined the ability of a cell growing in vitro to form an embryo or not (Reinert and Tazawa 1968). We have now extended these results by further experiments on the interactions of the two factors, and we have investigated the question whether they affect the formation of roots in a similar way to embryogenesis in the carrot cultures.

II. MATERIALS AND METHODS

The materials and the techniques used have already been described (Reinert, 1968). Therefore this part can be restricted to the main facts. In each experiment explants (0.3 x 0.3 cm, 1 cm thick) were cut from the surface-sterilized upper third of five carrots of a commercial strain ("Rote Riesen"). The explants were pooled and distributed in equal parts for transfer to the different media in order to reduce variations which could be caused by the properties of a single carrot. The initial freshweight of subcultures was always 200-400 mg. To estimate embryo and root formation the number of tissues producing embryos was expressed as a percentage of the total number of cultures. All cultures were grown in the dark at a temperature of $28° \pm 1°$ C on synthetic agar (0.8%) media designated as M_S and M_W. M_W contained White's (1954) nutrient, 2 per cent sucrose and 5×10^{-8} g/ml of 2,4-D. M_S differed from M_W only in the composition of the major and minor elements; instead of White's salts and trace elements it contained those of the formula developed by Murashige and Skoog (1962).

III. RESULTS

One of the preconditions for our experiments was a regularly reacting system which had to be developed first. Under the experimental conditions used, the average value for the increase in freshweight of cultures growing on M_S for one transfer period was approximately 300-400%. These cultures on M_S formed only very rarely a root but produced embryos regularly, beginning shortly after the first transfer, that is 4-6 weeks after isolation, and reaching a maximum after 16-20 weeks. Later on the capacity for embryo formation was either lost or declined after 30-40 weeks of cultivation (Fig. 1 and 2).

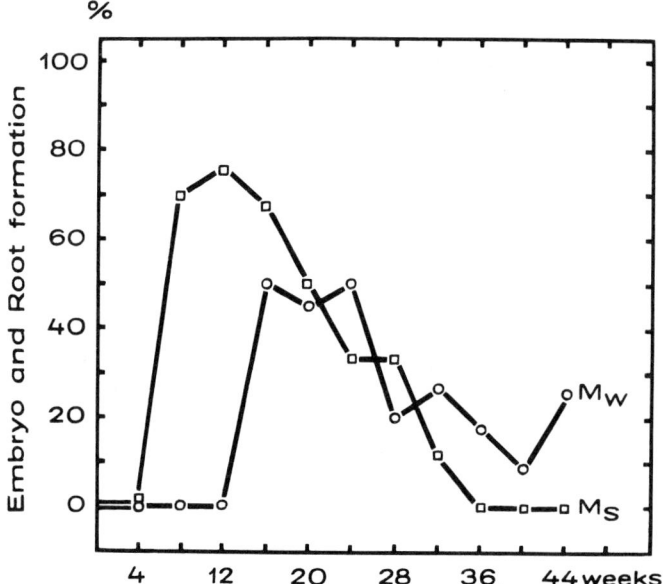

Fig. 1. Embryo (□———□) and root (o———o) formation in carrot cultures on M_S and M_W. Cultures on M_S produced no roots and on M_W no embryos. Abscissa: weeks after isolation of the cultures.

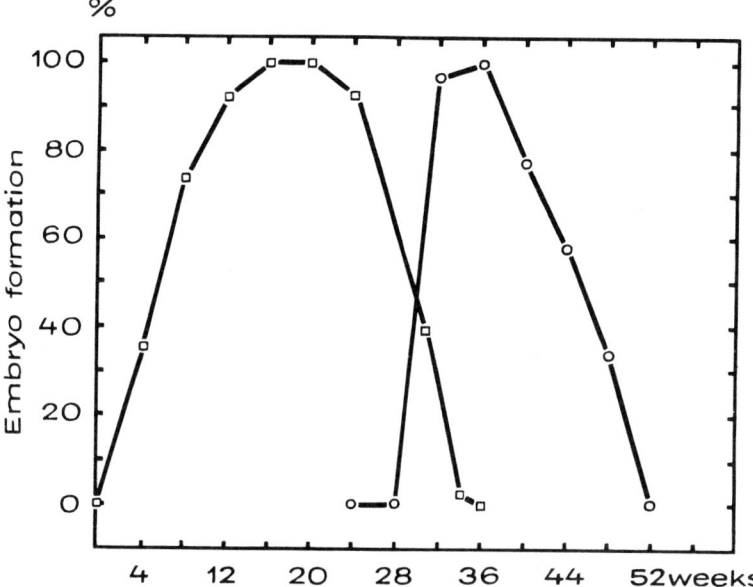

Fig. 2. Normal pattern of embryo formation in carrot cultures isolated and subcultured on M_S (□———□), isolated on M_W and transferred to M_S (o———o). The control was M_W with no embryo formation. Abscissa: weeks after isolation of the cultures.

This pattern was the rule and only in exceptional cases was embryo formation observed over longer periods (70-80 weeks) on M_S. The tissues reacted differently on M_W. In this case the increase in fresh weight was lower (ca. 200%). Moreover, we did not observe embryos on M_W but roots were frequently produced. Transfer of cultures from M_W to M_S resulted in embryo formation and in the cessation of root production (Fig. 2). According to these results it could be assumed that embryo and root formation in the carrot cultures depend upon processes which are mutually exclusive. That this is not the case was shown by varying the nitrogen content of the two media. For instance an increase of the nitrogen level of M_W from 3.2 mM to 10, 20, 40 and 60 mM by addition of KNO_3 resulted in embryo formation which was below 50% at the lower levels and was up to 100% when M_W contained 40 and 60 mM of nitrogen (Fig. 3). In addition to embryos, these cultures formed roots, but root formation was not so strictly determined by the nitrogen level. In one experiment it was almost zero at 60 mM and attained values between 60 and 80% in the presence of 20 and 40 mM of KNO_3, but in other experiments we observed almost no differences for all three nitrogen levels.

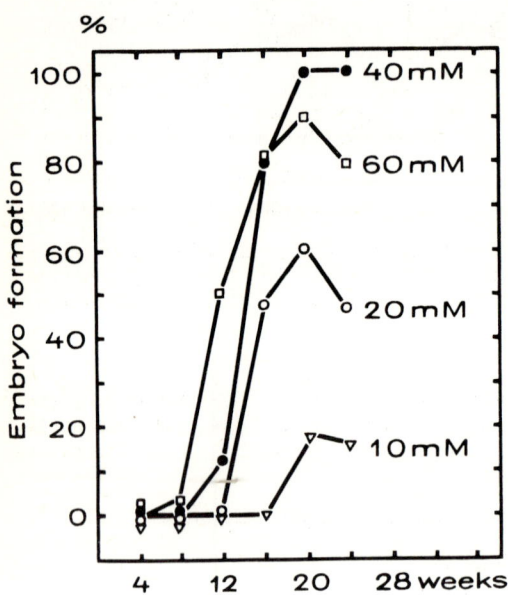

Fig. 3. Effect of increases of the nitrogen level of M_W by addition of various amounts of KNO_3 on embryo formation. The values against the curves correspond to mM of nitrogen in the medium. Abscissa: weeks after isolation of the cultures.

We made similar observations with cultures growing on M_S, which contained 60 mM of nitrogen in the form of NH_4NO_3 (20.6 mM) and of KNO_3 (18.8 mM). A stepwise decrease of the nitrogen of this medium down to 1/16 caused a roughly proportional decrease in embryo formation from more than 90% on M_S to 7% in the presence of only 3.8 mM of nitrogen (Fig. 4a). Root formation in these experiments was zero on M_S and went up to values of 60% and more at the reduced levels of 3.8 and 7.5 mM of nitrogen (Fig. 4b).

In all these experiments the changes in the nitrogen level concerned, at least partially, KNO_3 and therefore also the potassium level of the two media. Some experiments with reduced amounts of nitrogen but with the original potassium level of M_S proved that it was the nitrogen and not K^+-ion which caused the observed changes in morphogenesis.

The second factor which had to be considered was the auxin concentration of the nutrients. It was known that low auxin levels could enhance embryo as well as root production in carrot cultures on media with a relatively high nitrogen content (Reinert, 1959; Halperin and Wetherell, 1964).

In the present experiments we produced even more drastic effects by simply shifting the auxin/nitrogen ratio of M_W in favour of the latter. On this medium only

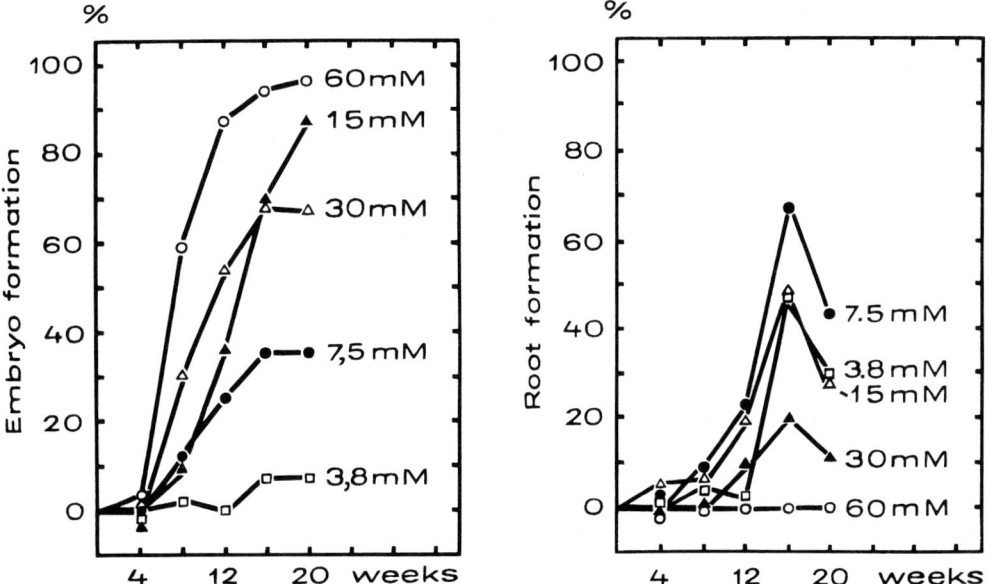

Fig. 4a and 4b. Effect of stepwise decreases of the nitrogen content of M$_S$ on embryo formation (a) and root formation (b) by the carrot cultures. The values against the curves correspond to mM of nitrogen in the medium. Abscissa: weeks after isolation of the cultures.

one part (ca. 40%) of the carrot cultures form roots. Transfer of 70 week old cultures from M$_W$, with only 3.2 mM of nitrogen, to M$_W$-2,4-D caused not only a high percentage of embryo formation but induced root formation or enhanced it when it occurred *per se* (Fig. 5).

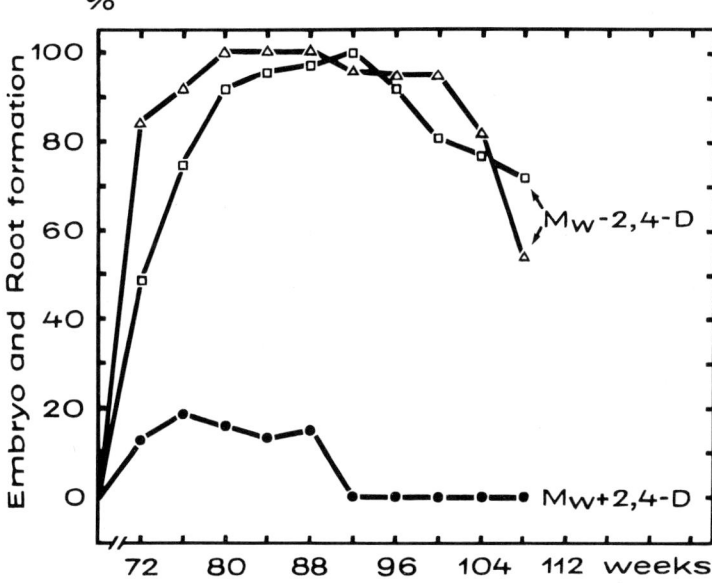

Fig. 5. Root formation in "older" carrot cultures (cultivation time 72 weeks) on M$_W$ (●——●), on M$_W$ - 2,4-D (△——△) and embryo formation on M$_W$ - 2,4-D (□——□). No embryos were produced on M$_W$. Abscissa: weeks after isolation of the cultures.

This reaction to the auxin-free medium was less marked with cultures which were only 4-20 weeks old. But also with these materials we observed on M_W-2,4-D an increase in embryo formation from zero to 60-70% (Fig. 6a) and - contrary to a preliminary result - nearly the same increase for root production (Fig. 6b).

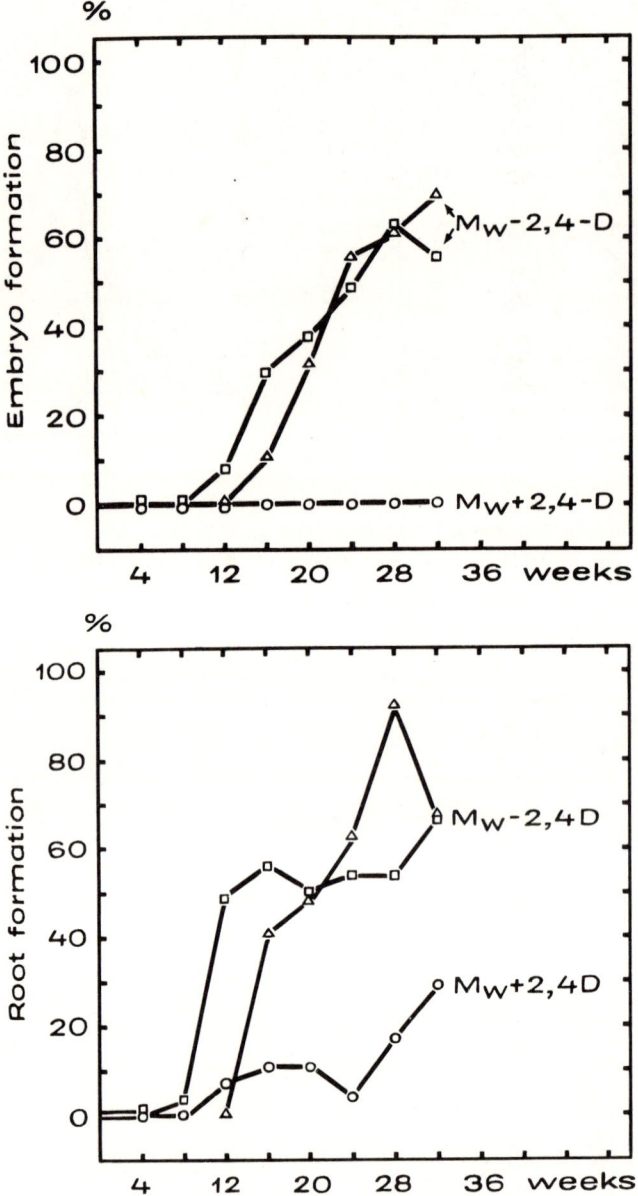

Fig. 6a and 6b. Embryo (a) formation and root (b) formation in "young" carrot cultures (cultivation time 36 weeks) on M_W (o———o) and on M_W-2,4-D. Transfer to the auxin-free medium 4 weeks (□———□) and 12 weeks (△———△) after the isolation of the cultures. Abscissa: weeks after isolation of the cultures.

The main difference between younger cultures, that is 4-30 week old cultures on M_S and on M_S-2,4-D were the development of roots in 38% of them and a prolonged period for embryo formation on the auxin-free nutrient (Fig. 7a, 7b). Furthermore, tissues which had apparently lost their ability to form embryos after a prolonged cultivation time of 80 weeks on M_S regained it when transferred to M_S-2,4-D.

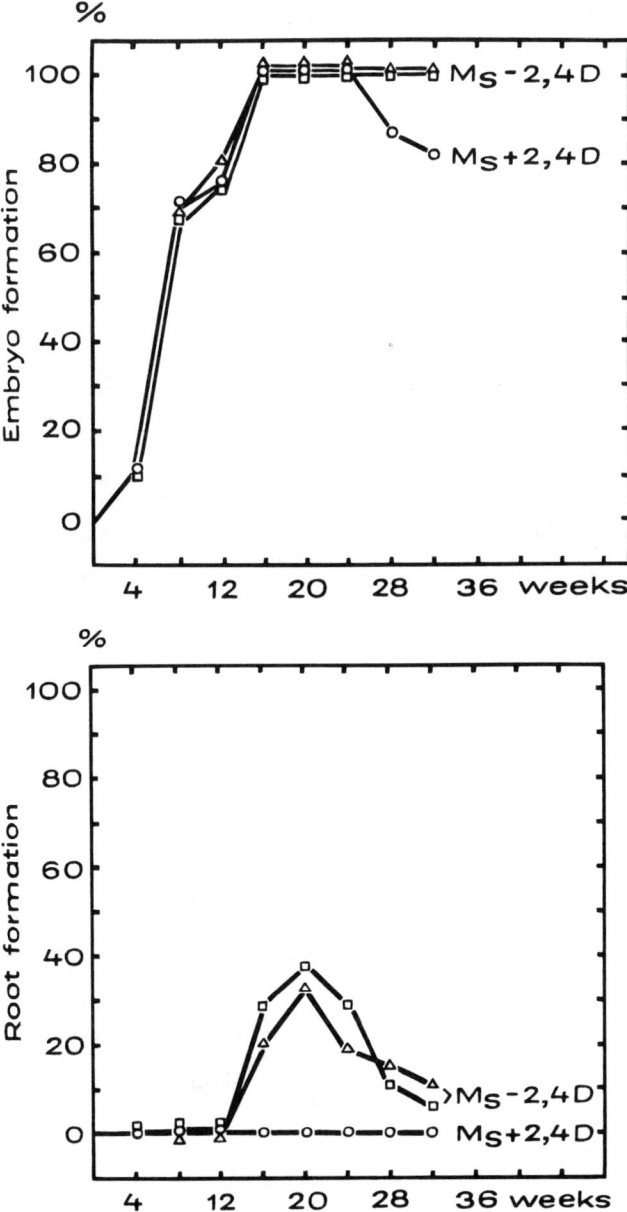

Fig. 7a and 7b. Embryo (a) and root (b) formation in "young" carrot cultures on M_S and on M_S-2,4-D. Transfer to the auxin-free media 4 weeks (□——□) and 8 weeks (△——△) after the isolation of the cultures. Abscissa: weeks after isolation of the cultures.

It is of interest in this connection that the same level of auxin which inhibited morphogenesis was optimal for growth. With both media 2,4-D in the range between

10^{-8} and 10^{-7} g/ml caused a great increase in freshweight (on M_S up to 500%), while 10^{-9} resp. 10^{-6} and 10^{-5} g/ml were only half as or less effective (Fig. 8).

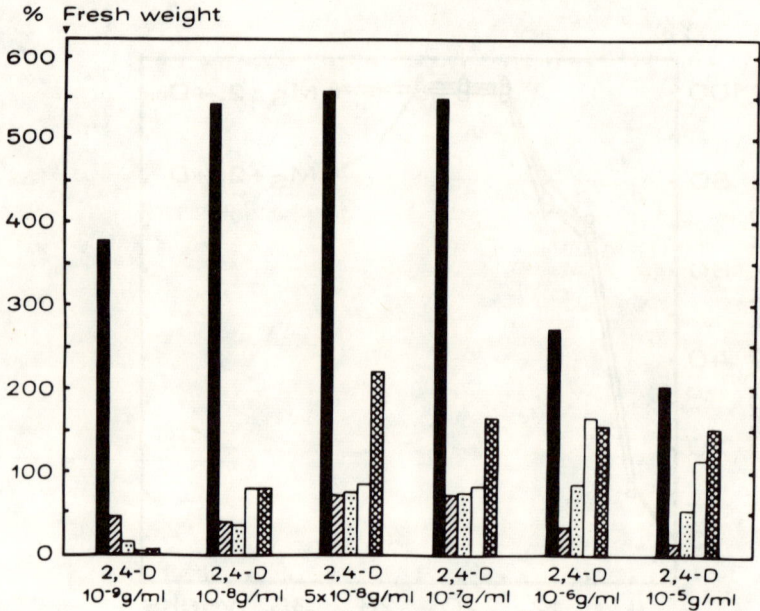

Fig. 8. Effect of various 2,4-D concentrations on the growth of carrot cultures on M_S during a cultivation time of 20 weeks. Each column represents the increase in freshweight for one transfer period. Abscissa: weeks after isolation of the cultures.

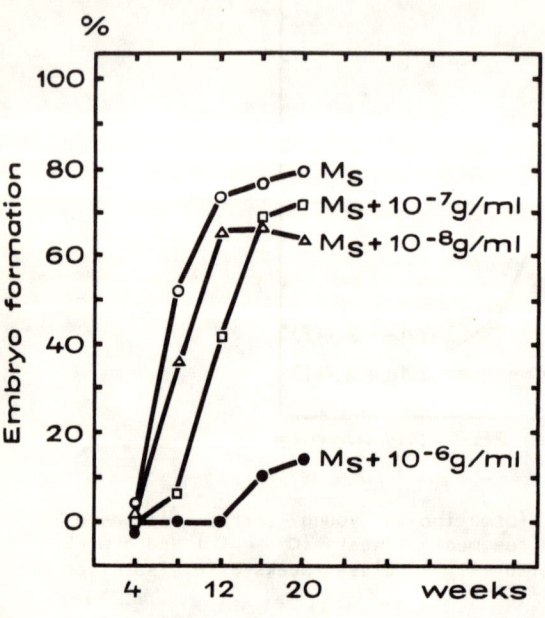

Fig. 9. Inhibition of embryo formation in carrot cultures on M_S by kinetin. The values against the curves correspond to the concentration of kinetin in the medium. Abscissa: weeks after isolation of the cultures.

Another point which deserves attention was the effect of kinetin. Addition of 10^{-6} g/ml and of lower concentrations of the cytokinin to M_S did not significantly change the growth rate of the cultures, while 10^{-5} g/ml inhibited strongly. But concentrations which did not influence growth almost suppressed (10^{-6} g/ml) or inhibited (10^{-7} and 10^{-8} g/ml) embryo production (Fig. 9). We made similar observations with M_W, i.e. a marked inhibition of growth only at 10^{-5} g/ml and of root formation - provided it did occur - already at 10^{-8} g/ml. Finally, although we tried a broad range of concentrations we were not able to initiate embryo formation on M_W by the addition of kinetin.

IV. DISCUSSION

The results of these investigations have a bearing upon several questions. They confirmed previous observations on the decisive role of high nitrogen and low auxin concentrations for the development of embryos in carrot cells (Reinert, 1968). In addition it turned out that stepwise changes in the nitrogen level of both media caused roughly proportional changes in the percentage of embryo formation. The fact that omission of the auxin was sufficient to convert M_W with only 3.2 mM of nitrogen from a "non-inductive" to an "inductive" form proves that it is not the absolute amounts of either but rather the ratio of nitrogen to auxin which is important. The prolongation of the period for maximal embryo formation by young cultures after transfer from M_S to M_S-2,4-D and the reinduced ability for embryo formation by older calli points in the same direction, that is to say, to a control mechanism similar to that for shoot and root formation in tobacco cells by kinetin and auxin (Skoog and Miller, 1957). On the other hand, it is difficult to fit the observed differences in the reaction of cultures of different age into this concept. Although M_S contained sufficiently high amounts of nitrogen for embryo formation in freshly isolated cultures, this process could be reinduced in older cultures only by the omission of auxin. There are also difficulties as far as the control of root development is concerned. In this case we had contradictory results and no clear-cut evidence that high nitrogen, for instance, suppressed root formation regularly and no evidence for a control of this process by the ratio of nitrogen to auxin.

Another question which is of interest concerns the effects of the auxin and the cytokinin. Kinetin could not replace high nitrogen concentrations in the induction of embryos and it inhibited root as well as embryo formation at concentrations which had no influence on growth. Auxin caused similar inhibition but even at concentrations which were optimal for growth. This pertains for 2,4-D (5×10^{-8} g/ml) and also for IAA (10^{-6} g/ml).

These results can be explained theoretically by a changing sensitivity to growth substances during different stages of development (Reinert, 1959; Gautheret, 1966; Torrey, 1966) or by postulating that embryo and root formation, in contrast to growth of the carrot cells, are only inhibited by auxin and cytokinin and may be controlled by other factors. A less ambiguous explanation will only be possible when we have more experimental data on the role of auxin and cytokinin during the early phases of morphogenesis, especially during the induction of embryos and roots.

V. SUMMARY

A study has been made on the role of growth substances and of nitrogen compounds in initiating embryo or root formation in tissue cultures of *Daucus carota* grown on agar. An essential factor for embryo and root formation in these tissues, which grow on chemically defined media (M_S, M_W) and require auxin for growth, is the ratio of the auxin to the nitrogen of the nutrient. High nitrogen content (M_S, 60 mM) favours embryo formation and in the presence of auxin a stepwise decrease of the nitrogen level results in a roughly proportional decrease in embryo production which ceases at concentrations around 3.5 mM. Root production follows a different course, i.e. the reaction to changes of the ratio of nitrogen to auxin is less regular. The same mechanism can be demonstrated on media with a low nitrogen level (M_W, 3.2 mM). In this case, omitting the auxin initiates embryo and root formation; in cultures which form *per se* roots on M_W this process is enhanced by the omission of auxin.

The effect of the changes in the auxin-nitrogen ratio cannot be observed experimentally by similar manipulations of auxins and cytokinins. These substances inhibit rather strongly both embryo and root formation in concentrations which enhance (2,4-D 5×10^{-8} g/ml) or are without influence on the growth of the carrot cultures (Kinetin 10^{-7} g/ml).

These results are discussed in relation to the control of morphogenesis in plant tissue cultures by exogenous and endogenous factors.

VI. REFERENCES

GAUTHERET, R.J. Factors effecting differentiation of plant tissues grown *in vitro*. In: "Cell differentiation and morphogenesis." North Holland Publ. Comp. Amsterdam 1966, pp. 55-95.

HALPERIN, W. and D.F. WETHERELL. Adventive embryony in tissue cultures of the wild carrot, *Daucus carota*. Amer.J.Bot. 51, 274-283 (1964).

MURASHIGE, T. and F. SKOOG. A revised medium for rapid growth and bioassays with tobacco tissue cultures. Physiol. Plantarum 15, 473-497 (1962).

REINERT, J. Untersuchungen über die Morphogenese an Gewebekulturen. Ber.Dtsch.Bot. Ges. 71, 15 (Sonderheft) 1958.

REINERT, J. Über die Kontrolle der Morphogenese und die Induktion von Adventivembryonen an Gewebekulturen aus Karotten. Planta 53, 318-333 (1959).

REINERT, J. Experimental modification of organogenesis in plant tissue cultures. In:"Plant tissue culture - Symposium." Internatl. Soc. of Plant Morphologists (P. Maheshwari and N.S. Rangaswamy, Ed.) pp. 168-177, Delhi 1963.

REINERT, J. Factors of embryo formation in plant tissues cultivated in vitro. In: "Sur les cultures des Tissus de Plantes". Coll. Natl. du C.N.R.S., pp. 33-40, Paris 1968.

REINERT, J., D. BACKS und M. KROSING. Faktoren der Embryogenese in Gewebekulturen aus Kulturformen von Umbelliferen. Planta 68, 375-376 (1966).

REINERT, J., M. TAZAWA and S. SEMENOFF. Nitrogen compounds as factors of embryogenesis in vitro. Nature 216, 1215-1216 (1967).

SKOOG, F. and C.O. MILLER. Chemical regulation growth and organ formation in plant tissues cultured in vitro. Symp. Soc. Expt. Biol. 11 118-130 (1957).

TORREY, J.G. The initiation of organized development in plants. Advances in Morphogenesis 5, 39-91 (1966).

WHITE, P.R. "The cultivation of animal and plant cells." Ronald Press, New York 1954.

TRANSPORT AND TROPISMS

EXPERIMENTS ON THE MECHANISM OF HORMONE-DIRECTED TRANSPORT

J.W. Patrick and P.F. Wareing

Botany Department, University College of Wales, Aberystwyth, U.K.

I. INTRODUCTION

It has been known for many years that when β-indole-acetic acid (IAA) is applied to the stems of bean plants, metabolites tend to accumulate at the point of hormone application so that there is an increase in the content of protein and starch at that point (Mitchell & Martin, 1937; Stuart, 1939). Since, in these early experiments, the hormone was allowed to act for several days, during which time growth was stimulated, it was concluded that the effect of the hormone was an indirect one, resulting from the establishment of a "sink" at the point of application. More recent experiments have shown that when IAA is applied to decapitated pea stems, labelled metabolites and nutrients, such as ^{14}C-sucrose and ^{32}P-orthophosphate, applied to a lower part of the stem, accumulate at the point of hormone application within 6-8 hr., at which time there is no visible sign of growth at the cut surface (Davies & Wareing, 1965). It was concluded, on various grounds, that IAA probably acts by a rather specific and direct effect upon the transport process itself, rather than by the establishment of a "sink". However, it is still possible that the application of the hormone causes an increased rate of metabolism and biosynthesis in the neighbouring tissues, before there are any visible signs of growth, and thereby establishes a "metabolic sink". This paper reports further experiments to investigate this problem.

II. MATERIALS AND METHODS

Seedlings of dwarf bean, *Phaseolus vulgaris* cv. Canadian Wonder were raised in the glasshouse under conditions of low light intensity and a 18 hr. photoperiod. When the second internode had ceased elongating, the plants were transferred to a growth cabinet maintained at $20°C$ with continuous illumination.

The bean plants were decapitated at the second internode, 6-8 cm above the primary node. The cut surface of the stump was covered with a gelatin cap containing either plant hydrous lanolin or 1000 ppm IAA dispersed in lanolin.

Nine hours after decapitation, the upper 1 cm segment of the internode was harvested, cut into slices 1 mm thick and washed in several changes of water at $0°C$ to remove cell debris. Two replicates of each treatment (200-300 mg fr wt) were shaken for 2 hours at $25°C$ in 5 ml of an incubate containing 5 mM sucrose labelled with 0.5 μc of ^{14}C-sucrose and 50 mM KCl adjusted to pH 5.0 with 5 mM di-potassium hydrogen orthophosphate-citric acid buffer. Following incubation, the free-space label was removed by rinsing in tap water. The slices were then extracted with 80% v/v ethanol at $80°C$ for 48 hours and the radioactivity of the soluble fraction was determined by liquid scintillation counting.

III. EXPERIMENTAL

Preliminary experiments confirmed the earlier reports that application of IAA to the decapitated second internode of bean causes it to accumulate ^{14}C-sucrose supplied to the internode 6 cm below the point of decapitation (Table 1).

Table 1. The effect of IAA on the long-distance transport of ^{14}C-sucrose.

Treatment	Radioactivity Accumulated (cpm)
Lanolin	797 ± 265
IAA	5044 ± 651

If IAA affects the mobilization of sucrose by stimulating metabolism and creating a 'sink', then it would be expected that such effects would be reflected in the rates of respiration and protein synthesis. The following experiments were carried out to test this hypothesis.

Measurements of respiration rates

Decapitated bean internodes were pretreated for 9 hours with either IAA or plain lanolin and the rates of dark respiration (measured as rate of carbon dioxide evolution from the upper 2 cm of internode enclosed in cuvettes) were determined in an open gas-flow system by infra-red gas analysis. The results shown in Table 2 indicate that pretreatment with IAA did not significantly increase the respiration rate of the internodal tissue. Confirmatory results were obtained for tissue slices from pretreated internodes using conventional Warburg manometric techniques.

Table 2. Rates of dark respiration ($\mu l CO_2$/g frwt/hr) of decapitated internodes *in situ*.

Treatment	Respiration Rate
Lanolin	224.0 ± 358
IAA	272.8 ± 468

Observations on protein synthesis

The effect of IAA on protein levels and net incorporation of ^{14}C-leucine by internode tissue was determined over the 12 hours following decapitation. 10 µl of ^{14}C-leucine (0.5 µC/plant) was applied to the cut surface and then gelatin caps containing either IAA in lanolin or plain lanolin were placed over the stumps. Upon harvesting the upper 1 cm segment, the free space ^{14}C was removed by several rinses in large volumes of tap water. Protein extraction and separation followed the method of Key (1964). The protein pellet was dissolved in 0.2N NaOH and aliquots were taken for counting.

Table 3. The effect of IAA on protein level and incorporation of ^{14}C-leucine.

Treatment	Protein (µg equivalents/ mg fr wt)	Percentage incorporation of ^{14}C-leucine into protein
Intact	2.3 ± 0.5	-
Lanolin	0.7 ± 0.1	5.6 ± 1.8
IAA	2.3 ± 0.3	27.9 ± 2.7

The results given in Table 3 show that after 12 hours the protein level in the lanolin controls had declined to 30-50% of the initial value, whereas pretreatment with IAA maintained the initial level. Net incorporation of ^{14}C-leucine was 5 times greater in the IAA-treated tissue than in the lanolin controls. These data suggest that IAA maintains the initial protein level by regulating the rate of protein turnover. Whether this is by increased synthesis or decreased degradation is not clear.

Metabolism of ^{14}C-sucrose

Despite the dramatic loss of total protein in the lanolin treated controls, it was seen above that the rate of respiration was not affected by pretreatment with IAA. This observation would suggest that the overall enzyme activity involved in oxidative metabolism was unaffected by the pretreatments. However, it is possible that qualitative changes occurred in biochemical pathways and these could influence the expression of 'sink' activity. For instance, IAA may influence the size of the sucrose pool by increasing the conversion to other compounds. This possibility was investigated in the following experiments.

(a) ^{14}C-sucrose (0.5 µc/plant) was applied to the cut surface of decapitated internodes following the procedure described for the application of ^{14}C-leucine. The radioactivity incorporated into the 80% ethanol-insoluble residue (subdivided into cell wall and remaining fractions) and that remaining in the 80% ethanol-soluble fraction was determined. The results given in Table 4 show that IAA had little or no effect on the proportion of radioactivity incorporated into the cell wall and other insoluble fractions. The 80% ethanol-soluble fraction was partitioned by paper chromatography using ethyl acetate (8): pyridine (2): water (1) as a solvent, and the radioactivity was determined for the various regions of the chromatogram. It was found that the proportion of radioactivity still present as ^{14}C-sucrose was unaffected by pretreatment with IAA.

Table 4. The effect of IAA on ^{14}C-sucrose metabolism.[1]

Fraction	Treatment	
	Lanolin	IAA
80% Alc. sol.	57.0 ± 3.3	50.0 ± 2.3
Sucrose	44.5 ± 0.3	45.4 ± 1.1
80% Alc. insol.	43.0 ± 3.3	50.0 ± 2.3
Cell Wall	26.6 ± 2.9	28.4 ± 2.4
Remainder	16.4 ± 1.3	21.6 ± 0.5

[1]Radioactivity in the various fractions is expressed as a percentage of the total activity incorporated by the tissue.

(b) The effect of IAA on the metabolism of ^{14}C-sucrose was also investigated using isolated stem segments, excised from the top of decapitated internodes which had been pretreated with IAA or plain lanolin. The segments were placed in 3 ml of 5mM ^{14}C-sucrose contained in Warburg flasks. After incubation for 2 hours, the radioactivity in the fractions soluble and insoluble in 80% ethanol, and in the CO_2 respired fraction, were determined (Table 5). As already observed above with the intact internodes there was little difference between the treatments and most of the label remained in the 80% ethanol-soluble fraction, of which 50-60% of the activity still remained as sucrose.

Table 5. Metabolism of accumulated sucrose by pretreated tissue slices.

Fraction	Treatment	
	Lanolin	IAA
80% Alc. sol.	73.3%	77.3%
80% Alc. insol.	6.8%	8.0%
Respired $^{14}CO_2$	19.9%	14.8%

This result suggests that sucrose uptake is probably not governed by a metabolic depletion of endogenous sucrose pools, but rather by an energy-driven reaction at the cell membrane. Evidence that sucrose mobilization is an energy-driven process was provided by the finding that uptake of ^{14}C-sucrose by isolated segments is strongly inhibited by 2,4 dinitro-phenol (Table 6). However, it would appear IAA does not directly affect the process of sucrose uptake.

Table 6. Effect of DNP on sucrose uptake (cpm/g fr wt/hr.) by pretreated tissue slices.

Incubate	Treatment	
	Lanolin	IAA
- DNP	9975 ± 1065	7660 ± 375
+ DNP	168 ± 42	76 ± 11

Negative effects of IAA on "unloading" process

The preceding experiments suggest that the "unloading" of sucrose from the phloem is not directly determined by metabolic demand, and that IAA does not affect the unloading process. Further evidence in support of this conclusion is provided by the following experiment.

In order to distinguish between the effects of hormones on (1) unloading at the 'sink', and (2) the transport processes in the phloem, any effect on these later processes can be eliminated by applying ^{14}C-sucrose to intact internodes at the point of hormone application. After a 9 hr. auxin pretreatment, ^{14}C-sucrose was introduced into the central lacunae of attached, but decapitated internodes. Within 10 and 30 minutes of application, uptake of ^{14}C-sucrose by the IAA-pretreated internodes was lower than by the lanolin controls, but within 60 minutes of feeding there was no difference between the two treatments, possibly because physical diffusion in the sucrose solution (which was unstirred) became limiting as uptake proceeded over the longer period (Fig. 1). Thus, when the possible effects of IAA on the transport process were eliminated, no stimulatory effect of IAA on ^{14}C-sucrose uptake by attached internodes could be demonstrated.

These results are in agreement with those obtained with isolated segments from pretreated internodes (Table 6).

IV. DISCUSSION

The series of experiments reported above were designed to determine whether IAA stimulates the accumulation of ^{14}C-sucrose by creating a metabolic 'sink'. Several

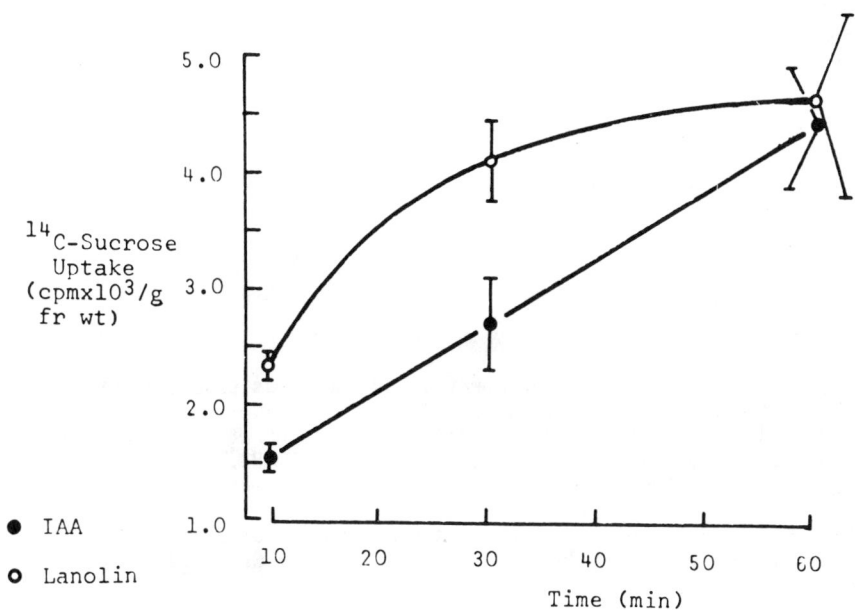

Fig. 1. The effect of IAA on the rate of accumulation of ^{14}C-sucrose from the internodal lacuna.

pieces of evidence militate against this hypothesis, viz. (1) there appeared to be no differences in overall respiration rates between IAA-treated and the control internode segments; (2) the greater part of the radioactivity at the point of hormone application was present as unchanged ^{14}C-sucrose, and (3) the size of the ^{14}C-sucrose pool was unaffected by hormone treatment. Thus, it would appear that the accumulation of ^{14}C-sucrose in the internodes is not controlled primarily by its depletion in metabolism, but is probably regulated by active uptake, presumably at the cell membrane, and that IAA does not affect this latter process. The absence of any effect of IAA when ^{14}C-sucrose is supplied at the point of hormone application is in strong contrast with the marked effects of IAA on the movement of ^{14}C-sucrose applied at a distant point. *Thus, an effect of IAA is only obtained when phloem transport over a distance is involved.* This conclusion suggests that IAA affects the process of phloem translocation rather directly. However, it was seen that the protein content of decapitated internodes declines rather rapidly and that IAA prevents this decline. It is possible, therefore, that the difference in movement of ^{14}C-sucrose in the IAA-treated and control internodes is due to the fact that IAA prevents senescence of the transport tissue and that it plays an essential role in maintaining this tissue in a functional condition, rather than in actively stimulating the transport processes, as suggested by Mullins (1970). It is difficult to determine which of these two alternative interpretations is correct on the present evidence and the precise mode of action of IAA on phloem transport is likely to remain unsolved so long as the mechanism of phloem transport itself remains unknown.

V. SUMMARY

The mechanism of hormone-directed transport remains largely unresolved. Some workers claim that the effect of hormones on metabolite transport is indirect; that is, the hormone maintains or stimulates metabolism and biosynthesis at its point of application thus creating a demand for nutrients. This hypothesis was examined and the effects of IAA on the rates of several parameters of metabolism and sucrose accumulation were determined using the decapitated second internode of *Phaseolus vulgaris* as the plant system. The results presented demonstrate that IAA had no effect on the parameters measured (i.e. sink activity), and this strongly contrasts with the marked effect IAA had on the long-distance transport of metabolites. It is concluded that the stimulation by IAA of metabolite transport is effected in some manner other than by regulating the activity of the sink at its point of application.

VI. ACKNOWLEDGEMENTS

One of us, J.W. Patrick, gratefully acknowledges the support of a C.S.I.R.O. Postdoctoral Studentship.

VI. LITERATURE CITED

DAVIES, C.R. and P.F. WAREING (1965). Auxin induced transport of radio-phosphorus in stems. Planta, 65, 139-156.
KEY, J.L. (1964). Ribonucleic acid and protein synthesis as essential processes for cell elongation. Plant Physiol., 39, 365-370.
MITCHELL, J.W. and W.E. MARTIN (1937). Effect of IAA on growth and chemical composition of etiolated bean plants. Bot. Gaz., 99, 171-183.
MULLINS, M.G. (1970). Hormone-directed transport of assimilates in decapitated internodes of *Phaseolus vulgaris* L. Ann. Bot., 34, 897-909.
STUART, N.W. (1939). Nitrogen and carbohydrate metabolism of kidney bean cuttings as affected by treatment with indole-acetic acid. Bot. Gaz., 100, 298-311.

PLANT GROWTH SUBSTANCES, 1970

The Movement of Plant Hormones:
Auxins, Gibberellins, and Cytokinins

William P. Jacobs

Biology Department, Princeton University, Princeton, New Jersey, U.S.A.

I. HOW GENERAL IS POLAR MOVEMENT OF HORMONES?

Auxins are well known to move with polarity in sections cut from shoots. At first, physiologists thought that non-endogenous auxins like 2,4-D did not move with IAA-like polarity; however, by testing 2,4-D and IAA under identical conditions, McCready and I found polar transport of both (1963). When I proposed, at the Izmir conference on hormone transport in 1967, that polar movement was too valuable to be restricted to the auxins only and that therefore it was likely that *other* hormones would also move with polarity through plants, there was great resistance to the idea - so much resistance, in fact, that I decided to collect more evidence before urging the hypothesis on the transport-physiologists.

We have now collected such evidence for GA_3. This paper will describe some recent work from my lab group on problems of polarity, transport, and ageing, most of which is now being prepared for publication. (Fuller references and details of methods will be found in these papers, which we will submit to the regular journals.)

Gibberellins

Members of this second major class of plant hormone have generally been considered to move *without* basipetal polarity, and in this respect the gibberellins have been considered to contrast strongly with the auxins. The reasons for thinking this are discussed in McComb (1964) and Jacobs and Kaldewey (1970). A critical evaluation of the past papers led us to this research plan: 1) only a small amount of GA_3 as the dose applied in the donor blocks (we used 1 or 0.1 μg of GA_3 - well within the physiological range); 2) organs that we knew responded to GA_3 (our main material was young petioles of *Coleus*, which we had shown responded to GA_3 with both greater elongation and faster abscission (Jacobs and Kirk, 1966)); 3) the same experimental setup that had been used for auxin transport studies (so that the results would be quite exactly comparable to those for the other class of hormone); 4) the most direct bioassay known at the time we started this research in 1965 (the barley endosperm test for the production of reducing sugars (Coombe *et al.*, 1967)).

Our first year of experiments were run on internode sections from *Coleus*, tissues in which we had previously found a 3:1 ratio of a basipetal to acropetal movement of IAA. Although this first material gave results that indicated a similar basipetally polar ratio, the variability was too great to meet our requirements for either reproducibility or level of significance in statistical tests (Greenblatt and Jacobs, 1966). We surmised that part of our difficulty may have been due to the ethyl acetate residues from the extraction of the receiver blocks. Such residues have since that time been found to increase the level of reducing sugars in the barley endosperm bioassay (Briggs, 1966).

Accordingly, we modified the test so that we tested the agar blocks without extraction and switched to the young petioles that were much more polar in their IAA-movement than were the young internodes (Werblin and Jacobs, 1967), as well as being responsive to GA_3.

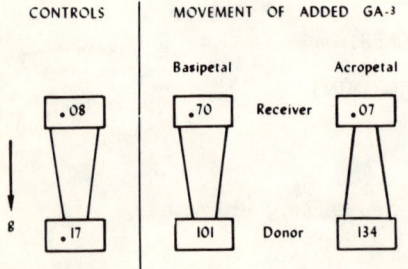

Fig. 1. The amount of gibberellin activity (calculated as ng GA_3) found in each donor or receiver agar cylinder 3 hr after 100 ng GA_3 was added in each donor to sections cut from young *Coleus* petioles. The values for control receivers have been subtracted from the values to the right. Data are averages of all 11 experiments (data from Jacobs and Kaldewey, 1970).

The pooled results of the 11 experiments using 3-hour transport through sections cut from the young *Coleus* petiole #3 are shown in Figure 1 (Jacobs and Kaldewey, 1970). The control values at the left have been subtracted from the transport values that are at the right. There is strong basipetal polarity - 10 times as much "GA activity" moves basipetally as moves acropetally. The GA_3 shows the same direction of polarity as does IAA in the same system. The difference between the apical and basal receivers is statistically significant. Is the "GA activity", found in the receivers, really due to the added GA_3, or to some other material which comes out of the sections after GA treatment? The material in the receivers was GA_3, as judged by the R_f in thin-layer chromatograms of extracts of receivers: the fluorescence was at the R_f zone, of the color, of the intensity, and appeared at the time, expected for GA_3 as contrasted to other known gibberellins. Also, the GA_3 zone on the thin-layer chromatogram was the only zone of the ethyl acetate fraction that contained bioassay activity. Hence, this provides evidence that the gibberellins are a second class of hormone to move with polarity in shoot-sections and that GA_3 apparently comes out into the receiver unchanged as do the auxins.

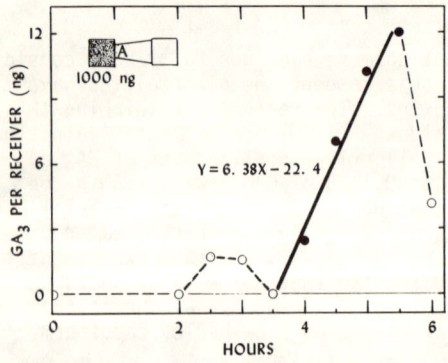

Fig. 2. The time course of basipetal GA_3 movement through sections cut from young *Coleus* petioles. The equation is that for the linear regression (solid line) fitted to the solid points.

Using these same basic techniques, we have recently been studying the time course of movement in these young *Coleus* petioles (Jacobs and Pruett, 1970, and in prep.). Typical results from one of the three experiments (Figure 2) show that gibberellin activity accumulates in the basal receivers following a linear time course, amazingly like the movement of IAA in the many tissues in which it has been studied, including *Coleus* petiole #3. The solid line is the best fitting straight line for the solid points; and by extrapolating the short distance to the X-axis, we can calculate the velocity of movement of GA_3 as being 1.4 mm/hr. This is of the same order of magnitude as the velocity of IAA through the same tissue, although somewhat slower. Note that there is a secondary decline in counts in the receivers at 6 hrs. This secondary decline also is a parallel with both the endogenous auxin, IAA (McCready and Jacobs, 1963; Naqvi, 1964; Keitt and Baker, 1966) and the endogenous cytokinin, adenine (Veen and Jacobs, 1969b). About 1% of the GA added to the original donors appeared as gibberellin activity in the basal receivers during the 6 hr examined: this is the same percentage as that found at 5 hr with IAA movement through *Coleus* petioles of this same age (Veen and Jacobs, 1969a). (Confirmation of this general picture came also from 5 experiments in which the receivers were replaced every hour or two.)

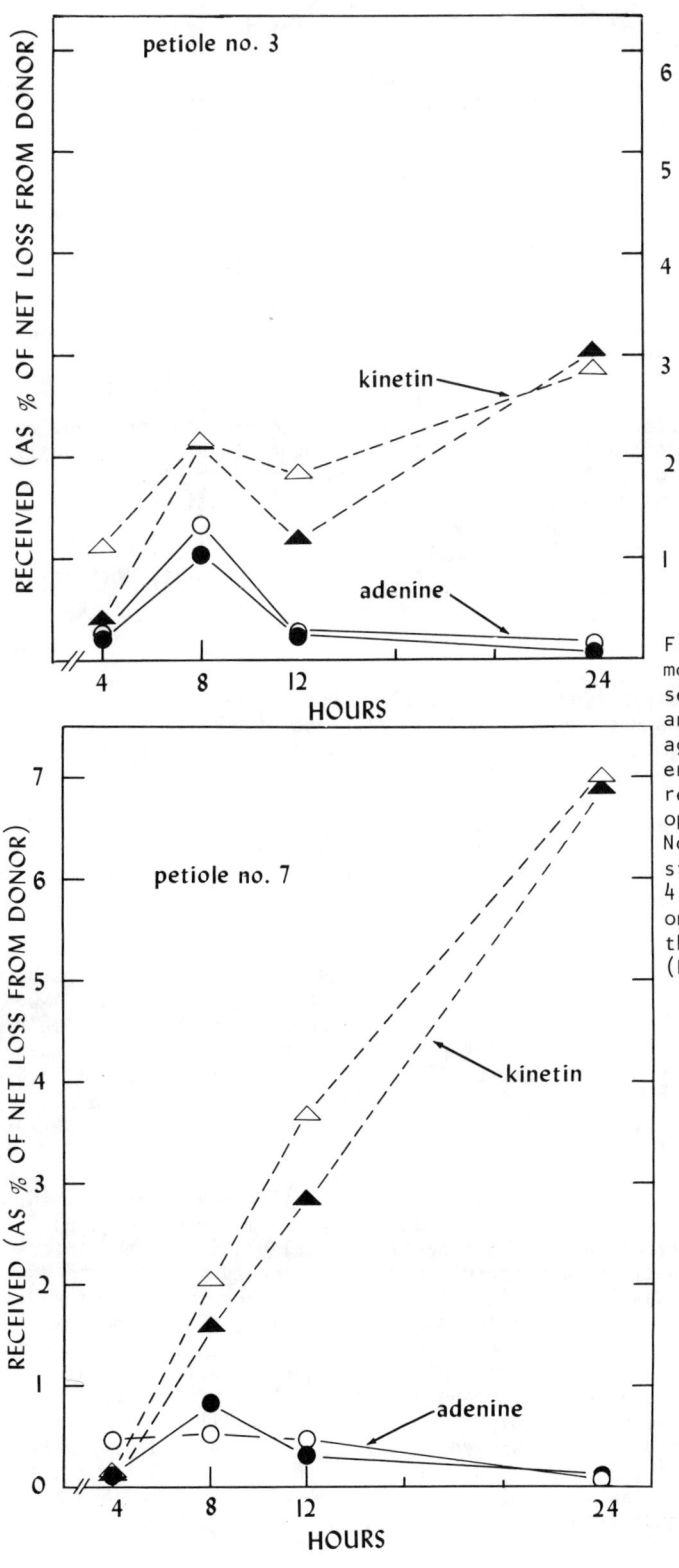

Fig. 3. The time course of movement through *Coleus* petiolar sections of ^{14}C from adenine and kinetin, supplied in donor agar cylinders at apical or basal end of section. Solid points represent basipetal movement; open points, acropetal. Petiole No. 3 is a young developmental stage, petiole No. 7 is roughly 4 weeks older and is usually one node above the oldest leaves that have not yet abscised. (From Veen and Jacobs, 1969b).

Because the time-course experiments revealed that our earlier studies were not run at the hour of maximal basipetal movement, we checked the polarity at 5 hrs: it was still strongly polar in a basipetal direction. The very small amount of GA moving out of the petiolar sections in basipetally polar direction was GA_3 judging by the slight fluorescence at the R_f of GA_3 in thin-layer chromatograms of extracts of basal control receivers.

GA-cofactor From Plant Tissues

One striking difference between GA movement and IAA movement can be seen from Fig. 1: The GA_3 donors, instead of losing activity as the material moves out of them through the sections and into the receiving blocks at the other end of the plant tissue, are seen to gain it (Jacobs and Kaldewey, 1970). We have been investigating the nature of this unexpected activity in the donor blocks (Jacobs and Pruett, 1970, and in prep.). Control receiver blocks, which have been applied to both ends of a petiolar section are without significant GA activity when tested directly. But if we add 10 ng of GA_3 to those blocks once they have been removed from the section, we find that the GA_3 activity recoverable from that block is larger than 10 ng - usually the equivalent of 20 to 30 ng GA_3. Apparently material comes from the tissue and is either converted to GA in the presence of GA_3 or increases the activity of GA_3. There was no clear-cut polarity with regard to the collection of this cofactor from the two ends of the petiolar section. We are actively pursuing its identification and so far can say no more than that it is stable to autoclaving, and its activity disappears upon ashing.

Cytokinins

The third major class of plant hormones currently known are the cytokinins. These include native cytokinins such as zeatin (with high activity per mole) or adenine (with very weak activity per mole) and synthetic ones such as kinetin or benzyladenine. Veen and I (1969b) selected adenine and kinetin for our investigation because they paralleled our earlier studies on IAA and 2,4-D in being endogenous and exogenous respectively, and in showing what seemed to be a similar big difference in activity. We used both young and old leaves from our clonal stock of *Coleus* as the source of the petiolar sections - the old leaves were of particular interest because of the many reports that kinetin was more active in old leaves. Typical results from our time-course studies are shown in Fig. 3.

The radioactivity from both substances can be seen to have moved through both young and old petioles with no polarity whatsoever. Like the exogenous auxin 2,4-D, kinetin showed a steady increase in the receivers up till 24 hr., the longest time examined. There was no secondary decline. The percent of kinetin collected in receivers of *old* petioles in 24 hr. was very similar to that found for basipetal 2,4-D movement through young petioles of bean (7 to 8% of the net loss from the donor blocks). It is striking that much more kinetin moved through the older *Coleus* petioles - the reverse of the situation with IAA (Fig. 4).

Linear regressions were fitted to the data showing the time-course of kinetin accumulation in the receivers, for both basipetal and acropetal movement (Fig. 5). The movement was clearly identical in the two directions. Calculation of the intercepts, from the regression equations, and therefrom of the velocity revealed a velocity of 1.1 mm/hr., which was the same in both directions. The movement of radioactivity added in the form of adenine-^{14}C also showed no polarity; but in contrast to kinetin, the peak of accumulation in the receivers was at 6 to 8 hr. with a decline thereafter until 24 hours. This secondary decline is remarkably similar to that which we have observed for IAA and GA_3 - also endogenous hormones. Thin-layer chromatograms of extracts of the receiver blocks followed by scintillation counting of the various zones of the chromatograms showed that for kinetin the radioactivity was still all with the cytokinin, judging by R_f. For adenine this was mostly so, for extracts made near the hour of peak movement, although there was one other zone of radioactivity on the chromatogram (Fig. 4 and 5 of Veen and Jacobs, 1969b).

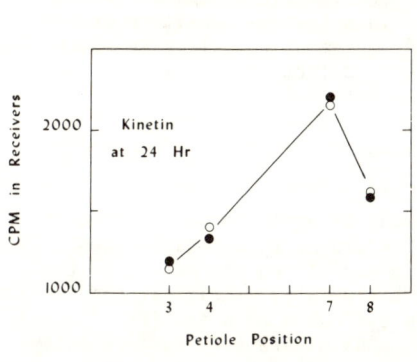 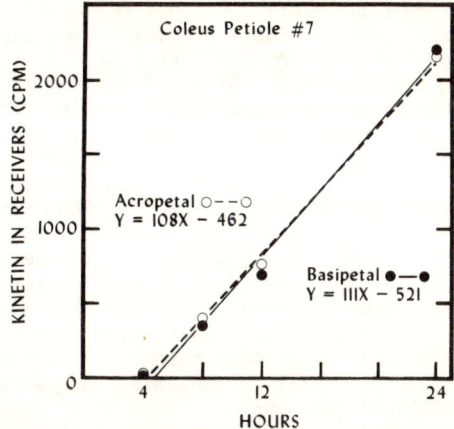

Fig. 4. The amount of kinetin moved basipetally and acropetally in 24 hr through sections cut from *Coleus* petioles of different ages. (Data of Veen and Jacobs, unpublished.)

Fig. 5. The time course of kinetin movement through sections from *Coleus* petioles No. 7, expressed as corrected cpm in receiver cylinders. The equations and lines are for the straight lines that give the best fit to the data. (The raw data are from Table 1 of Veen and Jacobs, 1969b).

II. HORMONE MOVEMENT IN ROOTS

IAA Movement

We presented at the 1967 Ottawa Conference the first report that IAA moved with strong *acropetal* polarity through root sections into receiver blocks (Kirk and Jacobs, 1968). Both the genera investigated showed this. It had been missed by earlier investigators, apparently because so little IAA is moved acropetally (only one-tenth of that moved through shoot sections under comparable conditions). During the year after we reported this, Scott and Wilkins (1968) checked IAA movement in still more genera and confirmed our report, and recently Morris *et al.* (1969) have given evidence that, in the intact plant also, such movement of IAA can take place from the shoot-tip down the stem and then acropetally into the root. This new evidence, in addition to recent papers on the crucial role of the root-cap in controlling the geotropism of roots, necessitates a drastic change in that portion of the Cholodny-Went theory of tropisms that tries to explain root geotropism. It supports the hypothesis that IAA moves acropetally in the root to the tip, where it is redistributed directly or indirectly by the root-cap.

GA Movement

We have been extending our GA_3 studies to roots, using exactly the same methods we had used for petioles (Jacobs and Pruett, in prep.). *Zea* roots were selected as experimental material because Wilkins and Scott have learned so much about IAA movement through sections cut from near *Zea* root tips (1968). When we added 1 μg GA_3 in each donor, and ran time-course studies for 0 to 6 hr., we found no detectable GA activity in either the apical or the basal receivers at any time tested. Control root sections, both ends of which had plain agar receiver blocks, were like the corresponding control petiole sections of *Coleus* mentioned above in giving off so little endogenous GA that the receivers were not significantly higher than values from the plain agar controls of the barley-endosperm bioassay.

Note, however, that if the parallel with IAA movement holds, as we have hypothesized it would, then we would expect only one-tenth as much GA_3 to move through roots as through shoots. By this reasoning the technique we followed in our first 3 transport tests in roots of bioassaying each root receiver block separately might be expected to show no significant GA activity, even in a test as sensitive as the barley-endosperm assay. Also, if the velocity of GA_3 movement were less in roots than in petioles, then we would need a longer time-course to detect it. We are currently checking both these possibilities.

III. "DIFFUSION" OF IAA vs. "ACTIVE TRANSPORT"

For more than forty years, the basipetal movement of auxin through living shoot tissue has been considered to be much faster than it would be by diffusion (Went, 1928, p. 57; and later authors). Much effort has gone into explaining the nature of the presumed "pumping" process that moves so much auxin basipetally. Critical consideration of the evidence led me to wonder if IAA diffusion in cylinders of water (or of agar) were the correct model for movement through tissues. Auxins are very surface-active molecules (Veldstra, 1953). There is an unbelievable number of surfaces in a 5 mm transport section; and I suspected that by ignoring these thousands of surfaces, we all might be using a model too abstract to be pertinent.

Accordingly, a year ago we compared movement of IAA-^{14}C through living sections with that through *Coleus* petiolar sections that had been boiled for 5 minutes or frozen in dry ice (Chang and Jacobs, in prep.). Typical counts in receivers after 3 hr. "transport" are in Table 1. The fresh sections show the expected strong basipetal polarity. The boiled sections show the expected lack of polarity but also show a most surprising feature: much more movement occurred through boiled than through fresh sections. The time-course of such movement is shown in Fig. 6. Prof. Keitt, during a brief stay in our labs in spring 1970, found similar relations in *Phaseolus* petiole (Keitt and Jacobs, in prep.).

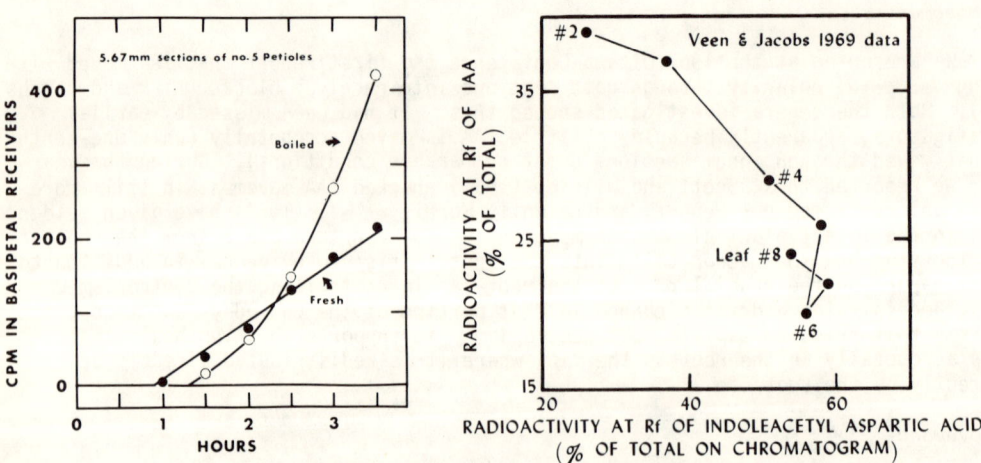

Fig. 6. Time course of movement of ^{14}C-IAA into receiver cylinders through *Coleus* petiolar sections that were used either "fresh" (used as soon as cut) or after 5 min boiling.

Fig. 7. The relations between increasing age of *Coleus* petioles (No. 8 being the oldest), decreasing amount of free IAA, and increasing conjugation of IAA with aspartate - the last two as evidenced by percentage of the extracted ^{14}C that was found at the Rf expected for IAA and IAA-aspartate, respectively, after ^{14}C-IAA had been transported basipetally through the sections. (Data from Table II of Veen and Jacobs, 1969a).

Table 1. Movement in 3 hr. of IAA-^{14}C through boiled, as contrasted to fresh, sections cut from *Coleus* petiole #4. The original donors averaged 52,508 cpm.

Treatment of sections	Radioactivity in Receivers (cpm)		Net Loss from Donors	
	Basipetal	Acropetal	Apical	Basal
Fresh	309 ± 28[a]	3 ± 1[b]	5681	2128
Boiled	670 ± 30[c]	655 ± 29[c]	7458	7865

a, b, c refer to values significantly different at the 5% level by the Duncan multiple-range test. Values are means ± Standard Error (n = 8 in each sample).

These results, it seems to me, require a reversal of our customary line of thought. They suggest that we need to focus, not on what provides the energy for "pumping all that auxin basipetally" in shoot sections, but rather on what *prevents* that acropetal movement.

IV. WHAT CAUSES THE DECREASE IN IAA-MOVEMENT IN AGEING PETIOLES?

IAA movement through sections cut from progressively older *Coleus* petioles shows a progressive decrease in the amount of IAA moving basipetally (Werblin & Jacobs, 1967; Veen and Jacobs, 1969a). This decline in IAA movement with increasing leaf age stands in meaningful relation with phenomena of leaf abscission (Jacobs, 1968). Veen has presented evidence and has hypothesized that conjugation of NAA regulates the levels of free NAA available for transport (1966, 1967). He has extended this interpretation to the movement of the endogenous auxin IAA in *Coleus* petioles. The results support the view that much of the decline in IAA movement is due to increasing conjugation to form indoleacetylaspartate (Fig. 7, data from Veen & Jacobs, 1969a). From young petiole #2 (on the fastest growing leaf) to old petiole #5 there is a progressive increase in IAAaspartate (judging by the R_f) and a concomitant decrease in free IAA.

However, these results leave unexplained the still further decline in IAA movement that is observed in still older petioles #6-8 (Fig. 7). Yaw Pan Chang has been working on the other end of the "ageing series" by investigating the nature and action of the abscission-speeding "Senescence Factor", which is found in detectable amounts only in already senescent #8 *Coleus* leaves (Jacobs et al., 1962). This endogenous material, which has several properties of abscisic acid, decreases the basipetal movement of IAA in transport sections (Chang & Jacobs, 1970; Chang, 1971). It seems likely that Senescence Factor speeds abscission in nature by giving the final *coup de grâce* to the declining trickle of IAA that is able to move down the old petiole to inhibit abscission.

V. SUMMARY

1.   Basipetally polar movement of GA$_3$ through young *Coleus* petioles has been demonstrated. When tested under the same conditions as IAA, GA$_3$ moves with a polarity, a time-course, and a velocity very similar to those of IAA through the same system. Like auxins added in the donor blocks, GA$_3$ is recovered apparently unchanged in the receiver blocks. A new GA-cofactor also comes out of the transport-sections. Applying the same methods to sections cut from near the root-tip showed no detectable movement of the added GA$_3$ in either direction: either the amounts moved through roots are much less (as we would expect from our work on IAA movement in roots), or the velocity in roots is so slow that we need to sample longer time periods, or both.

2.   The long-held assumption that living tissue moves much more auxin basipetally

than would move by merely physical processes has been proven incorrect. In sections from both *Coleus* petioles and *Phaseolus* epicotyls, boiled sections move much more IAA in both directions than the highly polar living tissue-sections do. (The boiled sections, as expected, show no polarity.)

3. A substantial percentage of the decline with petiole age in the basipetal movement of IAA in *Coleus* can be explained by the increasing conjugation of IAA with aspartic acid. Petioles so old that they are actually senescing produce, in addition, Osborne's "Senescence Factor" - a substance with several properties of abscisic acid - which decreases basipetal IAA movement still more.

VI. REFERENCES

BRIGGS, D.E. (1966). Residues from organic solvents showing gibberellin-like biological activity. Nature, London. 210, 419-421.
CHANG, Y.-P. and W.P. JACOBS (1970). Senescence factor decreases IAA transport. Am. J. Bot., Baltimore, Md. 57, 761-762.
CHANG, Y.-P. (1971). "The movement of indoleacetic acid in *Coleus* petioles as affected by abscisic acid, Senescence factor and kinetin." Ph.D. thesis, Princeton University, Princeton, N.J.
COOMBE, B.G., D. COHEN and L.G. PALEG (1967). Barley endosperm bioassay for gibberellins. I.Parameters of the response system. Pl. Physiol., Wash. 42, 105-112.
GREENBLATT, G.A. and W.P.JACOBS (1966). Polar transport of gibberellic acid (GA_3) through isolated internode sections of *Coleus*. Pl. Physiol., Wash. 41, xxxiii.
JACOBS, W.P. (1968). Hormonal regulation of leaf abscission. Pl. Physiol., Wash. 43, 1480-1495.
JACOBS, W.P. and H. KALDEWEY (1970). Polar movement of gibberellic acid through young *Coleus* petioles. Pl. Physiol., Wash. 45, 539-541.
JACOBS, W.P. and S.C. KIRK (1966). Effect of gibberellic acid on elongation and longevity of *Coleus* petioles. Pl. Physiol., Wash. 41, 487-490.
JACOBS, W.P. and P. PRUETT (1970). Polar movement of gibberellic acid-3. Pl. Physiol., Wash. 46 (Supplement), 19.
JACOBS, W.P., J.A. SHIELD, Jr. and D.J. OSBORNE (1962). Senescence factor and abscission of *Coleus* leaves. Pl. Physiol., Wash. 37, 104-106.
KEITT, G.W., Jr. and R.A. BAKER (1966). Auxin activity of substituted benzoic acids and their effect on polar auxin transport. Pl. Physiol., Wash. 41, 1561-1569.
KIRK, S.C. and W.P. JACOBS (1968). The movement of 3-indoleacetic acid-^{14}C in roots of *Lens* and *Phaseolus*, in "Biochemistry and Physiology of Plant Growth Substances" (Ed. F. Wightman and G. Setterfield) pp. 1077-1094. Runge Press Ltd, Ottawa.
McCOMB, A.J. (1964). The stability and movement of gibberellic acid in pea seedlings. Ann. Bot., London. 28, 669-687.
McCREADY, C.C.and W.P. JACOBS (1963). Movement of growth regulators in plants. II. Polar transport of radioactivity from indoleacetic acid-(^{14}C) and 2,4-dichlorophenoxyacetic acid-(^{14}C) in petioles of *Phaseolus vulgaris*. New Phytol., Cambridge. 62, 19-34.
MORRIS, D.A., R.E. BRIANT and P.G. THOMSON (1969). The transport and metabolism of ^{14}C-labelled indoleacetic acid in intact pea seedlings. Planta, Berlin. 89, 178-197.
NAQVI, S.M. (1964). Transport studies with C^{14}-indoleacetic acid and C^{14}-2,4-dichlorophenoxyacetic acid in *Coleus* stems. Ph.D. thesis, Princeton University, Princeton, N.J.
SCOTT, T.K. and M.B. WILKINS (1968). Auxin transport in roots. II. Polar flux of IAA in *Zea* roots. Planta, Berlin. 83, 323-334.
VEEN, H. (1966). Transport, immobilization and localization of naphthylacetic acid-1-^{14}C in *Coleus* explants. Acta bot. néerl., Amsterdam. 15, 419-433.
VEEN, H. (1967). On the relation between auxin transport and auxin metabolism in explants of *Coleus*. Planta, Berlin, 73, 281-295.
VEEN, H. and W.P. JACOBS (1969a). Transport and metabolism of indole-3-acetic acid in *Coleus* petiole segments of increasing age. Pl. Physiol., Wash. 44, 1157-1162.
VEEN, H. and W.P.JACOBS (1969b). Movement and metabolism of kinetin-^{14}C and of adenine-^{14}C in *Coleus* petioles of increasing age. Pl. Physiol., Wash. 44, 1277-1284.

VELDSTRA, H. (1953). The relation of chemical structure to biological activity in growth substances. Ann. Rev. Plant. Physiol., Palo Alto, Cal. 4, 151-198.
WENT, F.W. (1928). Wuchsstoff und Wachstum. Recl. Trav. bot. néerl., Nimègue. 25, 1-116.
WERBLIN, T.P. and W.P. JACOBS (1967). Auxin transport and polarity in *Coleus* petioles of increasing age. Wiss. Z. Univ. Rostock, Rostock. 16, 495-497.
WILKINS, M.B. and T.K. SCOTT (1968). Auxin transport in roots. III. Dependence of the polar flux of IAA in *Zea* roots upon metabolism. Planta, Berlin. 83, 335-346.

PLANT GROWTH SUBSTANCES, 1970

Tropic Stimuli and the Kinetics of Basipolar Transport of Auxin

J. Shen-Miller

Division of Biological and Medical Research, Argonne National Laboratory, Argonne, Illinois, 60439, U.S.A.

I. INTRODUCTION

Auxin transport in the shoot is predominantly basipolar and occurs by still uncharacterized transport mechanisms. In constructing a model to define and predict patterns of auxin flow at the cellular level, data on the fine kinetics of auxin transport are desirable. We began this investigation by an examination of the movement of auxin through oat coleoptiles using a 60 sec pulse of radioactive IAA and followed by an unlabeled IAA "chase". Radioactivity in the coleoptile was determined at 6 min intervals. We find the basipolar movement of auxin is not linear with time - it oscillates. Hertel and Flory (1968) also observed an oscillation of auxin transport through excised corn coleoptile segments. Moreover, in selecting a time for the measurement of curvature after geotropic stimulation of oat coleoptiles, we found that the geotropic curvature also follows a damped oscillation (Shen-Miller, 1970). Since geotropism is mediated by a differential distribution of auxin between the upper and lower half of a horizontal shoot, it seems possible that the gravitational stimulus could alter the pattern of periodicity of auxin transport in the upper and lower halves of a horizontal coleoptile, either *via* phase shifts or by changes in amplitude.

II. MATERIALS AND METHODS

Oat (Victory I) and corn (Wisconsin 64A x 22R) coleoptiles were used. A pulse of ^{14}C-IAA (^{14}C-labeled either at the carboxyl or at the methylene position, 10^{-4}M) in a 1.25% agar block was applied to the apex of the coleoptile with 1 mm of the tip removed. The pulse time was 60 sec. The pulse block was replaced by a chase containing nonradioactive IAA (10^{-5}M). Coleoptiles were divided into 2-mm segments at 6 min intervals; radioactivity of the segments was then determined by scintillation counting. The total transport period was 90 min for oats and 60 min for corn.

In determining the effect of geotropic stimulation on the movement of auxin, an entire coleoptile of corn was cut longitudinally into halves. Each half was placed with its vascular bundle in the *up* or *down* position on a glass slide, resembling the upper and lower halves of a horizontal coleoptile. As controls, we also placed halved coleoptiles in the vertical position, with the vascular bundle facing toward or away from the glass slide. The side away from the glass slide was covered with a thin glass coverslip. The whole assembly was kept in a humid crystallizing dish during the transport period. An IAA pulse was applied to the apex and replaced by a chase as described above. The coleoptile was divided into 2-mm segments at 6 min intervals, and radioactivity measured.

III. RESULTS AND DISCUSSION

The oat coleoptile retains a large amount of auxin in the apex. Figure 1 shows the distribution of IAA in the apical 2-mm segment. The rate of basipolar movement is most rapid for the first 18 min, and is followed by a backflow, or apipolar movement. In corn, this rapid rate of auxin flow lasts for 24 min and is also followed by a backflow. The amount of IAA moved upward is slight. We would have regarded it as an experimental fluctuation, but this phenomenon was observed in 6 of the 7 experiments with oats and in both of the 2 experiments run with corn; the 7th experiment with oats showed a plateau rather than an upsurge. The apipolar surge was observed regardless of the position of the ^{14}C-labeling in the IAA molecule, or whether the chase block was in contact with the apex throughout the transport period. The upsurge of auxin in the apex acts as a check to the subsequent flow of auxin through the subapical segments.

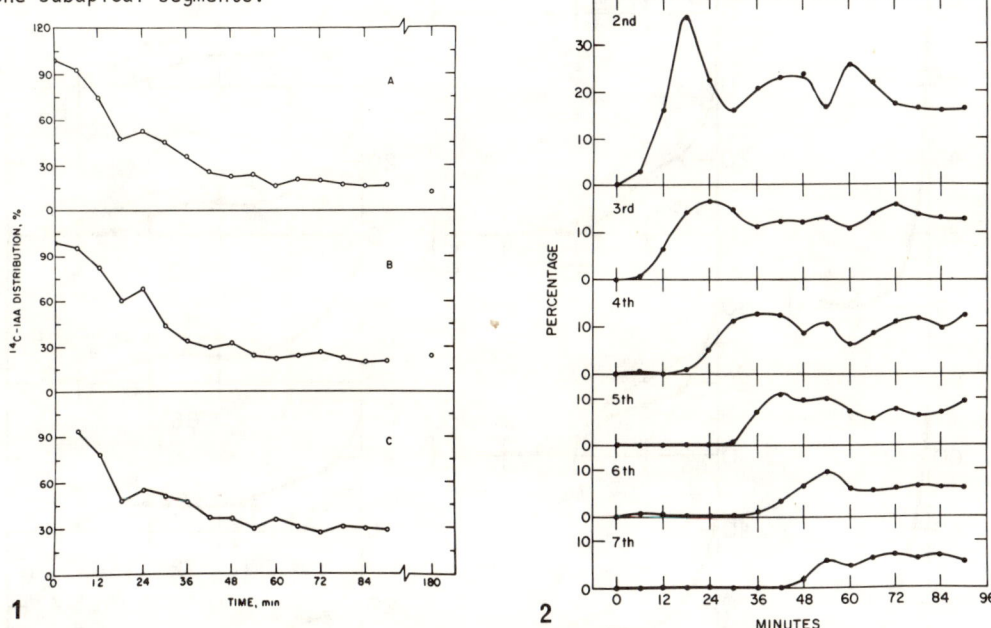

Fig. 1. IAA distribution in the apical 2-mm segment of the oat coleoptile at various times. A. Pulse, IAA-1-^{14}C; chase time, 18 min. B. Pulse, IAA-2-^{14}C; chase time, 18 min. C. Pulse, IAA-2-^{14}C, chase block remained on the coleoptile apex until time of harvest.

Fig. 2. IAA-2-^{14}C distribution in successive subapical oat coleoptile segments (2 mm) at various times.

Figure 2 shows the distribution of IAA in the subapical 2-mm segments. The distribution follows a wave motion and then dampens. The movement of IAA through a vertical coleoptile follows an oscillatory mode. Will an exogenous stimulus, such as gravity, alter this mode of transport in the upper and lower halves of the coleoptile? If it does, would it be in the form of a phase shift or a change in amplitude?

The small size, and the retention of IAA by the tip, of the oat coleoptile made us change our experimental material to corn. We, as well as others, find that auxin is transported more readily out of the apical tissues of corn than out of those of the oat coleoptile (Fig. 3).

Cane and Wilkins (1969) have described a technique of slitting open a coleoptile on one side and flattening the organ into a sheet. They find that if they place the sheet with the outer epidermal side *up* or *down*, resembling the upper and lower half of a horizontal coleoptile, more basipolar movement of auxin occurs in the *down* position. We were unsuccessful in attempting to split coleoptiles in this manner. Therefore, we divided the coleoptile into halves, as described in methods.

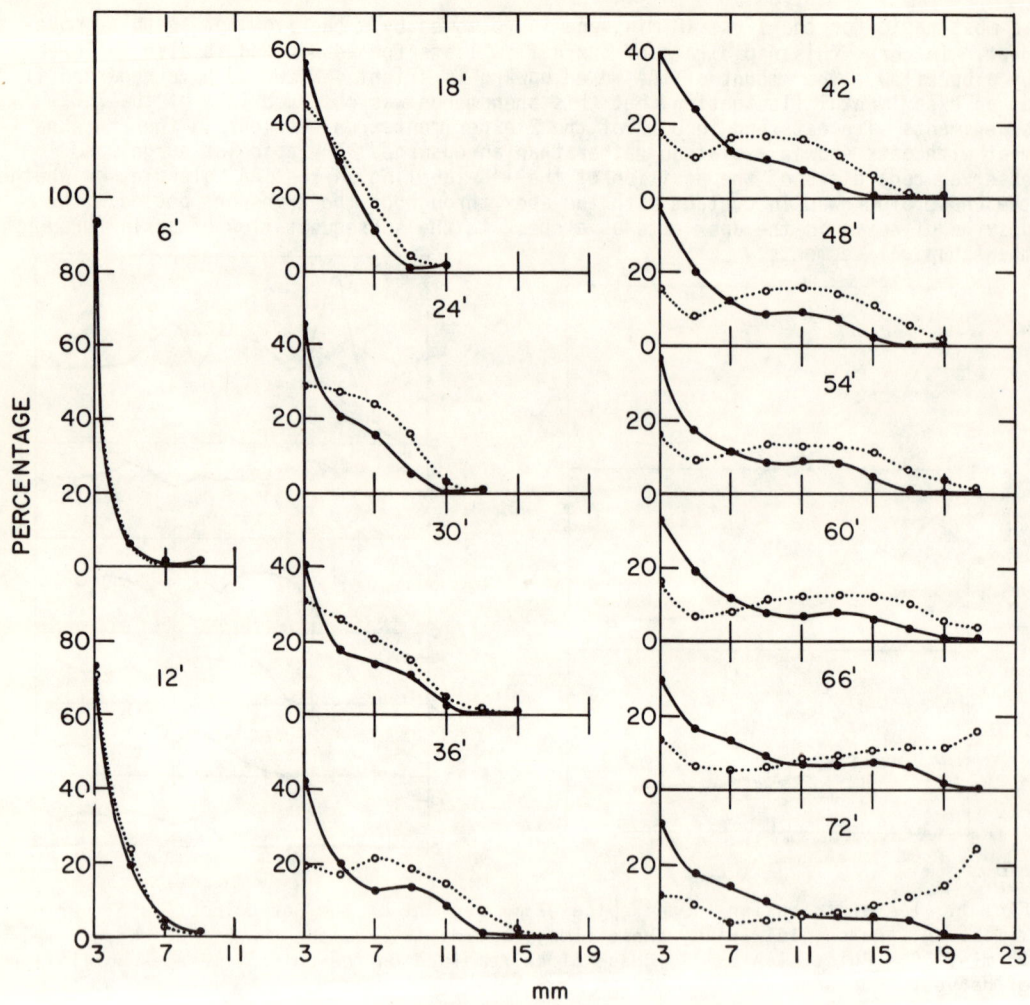

Fig. 3. A comparison of the basipolar movement of ^{14}C-IAA through intact corn (dotted curves) and oat (solid curves) coleoptiles. Abscissa: distance in mm from the apex. Ordinate: percentage distribution of IAA.

Figure 4 shows the movement of IAA through the coleoptile assemblies placed with the bundle in the *up* or *down* position. In the geotropically stimulated coleoptiles (H), during the first 18 min, the apex (2-3 mm) retains more IAA in the *down* position. The rate of auxin transport from the apex is fast for the first 36 min of stimulation. The segment immediately below the apex (4-5 mm) in the *up* position retains more IAA throughout the entire 60 min of experiment. Beginning at 18 min, the basal segments (from 6-8 mm downward) of the *down* tissue receive and transport more IAA in the basipolar direction, supporting the observations of Cane and Wilkins (1969). In the control vertical assemblies, the rate of transport is also fast for the first 36 min through the apex. There is no consistent trend of transport difference between the *up* and *down* positions in the apex until after 36 min of stimulation. There are some differences in the successive segments, but the differences are smaller and less consistent than those in the H-assemblies (see Fig. 4.)

The following is a schematic time sequence of the appearance of auxin in the successive coleoptile segments of the vertical control "*up*" and "*down*" oriented tissues (data from Fig. 4):

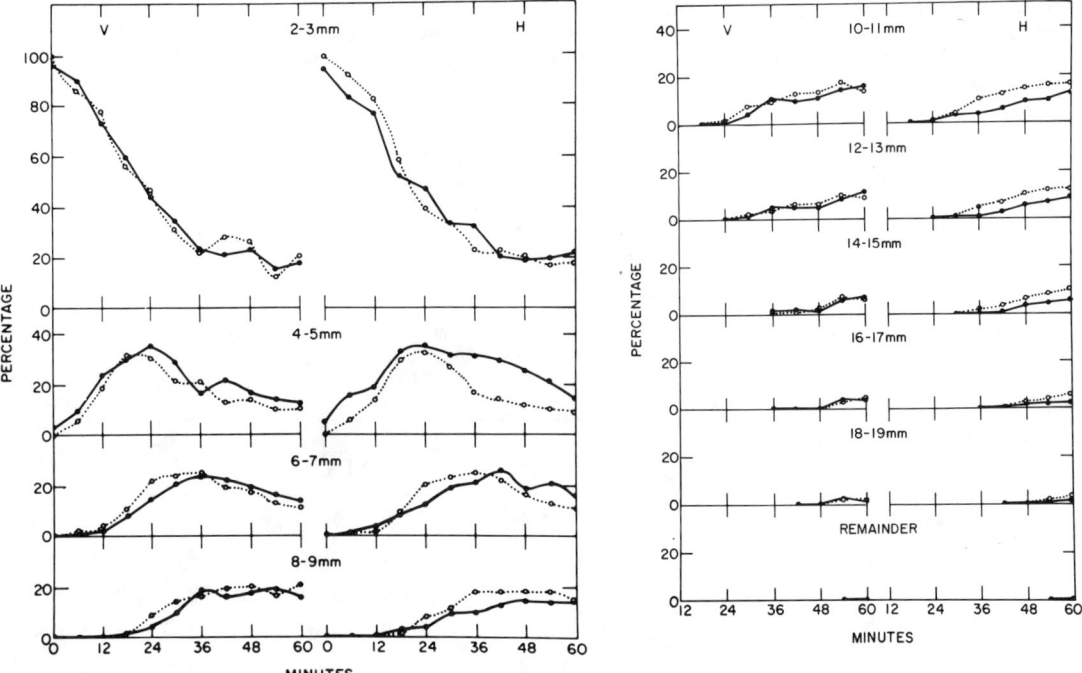

Fig. 4. Distribution of ^{14}C-IAA in the various coleoptile segments from the apex (2-3 mm) to the basal region. V-vertical control. H-geotropically stimulated. Solid curves: upper half-coleoptile; dotted curves: lower half-coleoptile.

"UP"   Donor $\xrightarrow{1'}$ Apex (2-3 mm) $\xrightarrow{1'}$ 4-5 mm $\xrightarrow{14'}$ 6-7 mm $\longrightarrow$
"DOWN" $1'$ $2.5'$ $12'$
(control)

$\xrightarrow{6'}$ 8-9 mm $\xrightarrow{6'}$ 10-11 mm $\xrightarrow{6'}$ 12-13 mm $\xrightarrow{18'}$ 14-15 mm $\longrightarrow$
$6'$ $6'$ $6'$ $18'$

$\xrightarrow{"0"}$ 16-17 mm $\xrightarrow{6'}$ basal
$"0"$ $6'$

The time required for the movement of auxin in the geostimulated tissues is essentially the same as that of the controls, except in the *down* position the transport time between 12-13 mm and 14-15 mm is 12 min, and 14-15 mm and 16-17 mm is 6 min. The tissues in the "*up*" position have a faster rate of auxin movement through the apical two segments (2-3 and 4-5 mm). This could be due to the orientation of this tissue causing a seepage of auxin down the cut surface. Regardless of seepage the movement of auxin through the apex is exceedlingly fast. In the "*down*" position where seepage is nil, it takes only 2.5 min to move the auxin from the apex to the next lower segment. In the 4-5 mm segment however, the retention time of auxin is longer than any of the subapical segments except the 12-13 mm. The segment (14-15 mm) subjacent to the slow auxin-moving segment, 12-13 mm, on the other hand, has a very fast transport rate.

This above data show that different portions of the coleoptile differ in their rates of auxin transport. Dolk (1930) noted that in geotropism, the curvature response began in the first 10 mm of the oat coleoptile at 20 min after stimulation began. The basal portion of the coleoptile required a longer time for the curvature to appear. In this study, we find that the rate of auxin transport is nearly constant, 6 min/2mm, for the subapical region of the coleoptile (6-13 mm) until the rate is checked at the 12-13 mm segment.

In corn the oscillatory characteristics of auxin flow are not as distinct as those observed for the oat. It should be pointed out, however, that a small difference in auxin content between the two sides could result in a substantial curvature difference (Shen-Miller and Gordon, 1967).

Table 1 shows the intervals between peaks of IAA accumulation within individual coleoptile segments. The mean for the vertical coleoptile is about 19.0 min. In the geostimulated segments, the upper tissues (U, *up* position) show a shortening and the lower (D, *down* position) a lengthening of the interval between peaks. The mean of the 2 half-tissues is about 20 min. In the individual segments, the apex has a significantly longer interval between peaks than the rest of the basal segments.

Table 1. Intervals (min) between peaks of IAA accumulation in individual coleoptile segments during the basipolar transport of ^{14}C-IAA. V-vertical control, H-geostimulation, U-upper and L-lower tissues.

Segment (mm)	V-Oat Whole	V-Corn Whole	V-Corn Control U	V-Corn Control L	H-Corn U	H-Corn L
2-3 (apex)	26.7 ± 1.2	30	28.0 ± 8.9	21.0 ± 9.0	20.0 ± 8.0	38.0 ± 4.0
4-5	19.0 ± 2.4	12	18.0 ± 4.6	15.0 ± 0	19.0 ± 4.4	25.5 ± 6.2
6-7	19.0 ± 2.9	12	13.0 ± 1.0	18.0 ± 6.0	14.0 ± 2.0	20.0 ± 4.0
8-9	20.4 ± 3.1	18	18.0 ± 0	15.0 ± 3.0	14.0 ± 2.0	22.5 ± 4.5
10-11	18.0 ± 3.3	--	21.0 ± 1.7	24	18.0 ± 3.5	17.0 ± 1.0
12-13	15.5 ± 2.0	--	21.0 ± 3.0	18.0 ± 0	14.0 ± 2.0	17.0 ± 3.6
14-15	15.0 ± 1.3	--	--	--	--	--
16-17	14.5 ± 1.2	--	--	--	--	--
Grand Mean	18.5	18.0	19.8	18.5	16.5	23.3
Subapical Mean	17.3	14.0	18.2	18.0	15.8	20.4

In determining the growth kinetics of 2-mm corn segments, we find that their elongation follows a stepwise increase (Fig. 5) with a period of 18 min between steps. Further, studies of the kinetics of geotropic response also show an 18 min period of curvature oscillation (Fig. 6). We suggest that the periodicity of 18 min in the development of geotropic curvature, in the growth of the coleoptile segments and the changes in Golgi apparatus (Shen-Miller) could be causally related to that of auxin transport. The question as to whether geo-stimulation alters the transport pattern in the upper and lower half via a change in amplitude or a phase shift is still not resolved. The results so far indicate that both may be affected.

IV. SUMMARY

In the apical region of a coleoptile the rate of auxin transport is nearly constant during the first 18 and 24 min for the vertical oat and corn coleoptile, respectively; there then follows an apipolar movement. In the apex of a geotropically-stimulated corn coleoptile, the lower half retains more IAA during the first 18 min. Subsequently, however, the rate of auxin entry and exit from the subapical coleoptile segments of the lower half is accelerated. There is an 18 min periodicity of auxin flow through the subapical portions of the oat and corn coleoptile. We also find the same periodicity in geotropism, growth, and changes in the Golgi apparatus. It is suggested that the periodicity in growth, curvature, changes in the Golgi, and auxin transport are causally related.

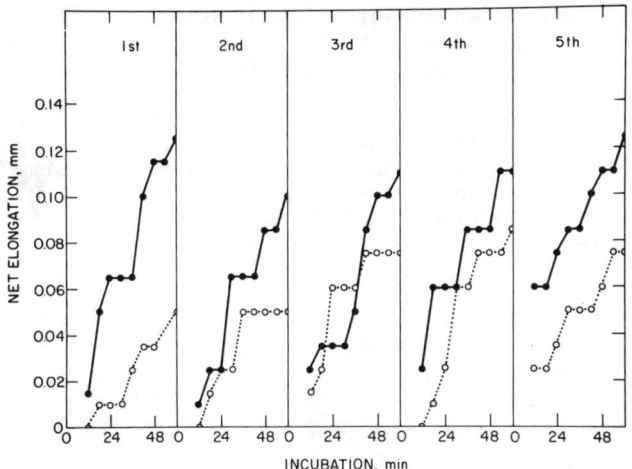

Fig. 5. Representative growth curves of the various 2-mm segments of corn coleoptile beginning from the apex incubated in IAA (10^{-6}M, solid curves) and control (broken curves) solutions.

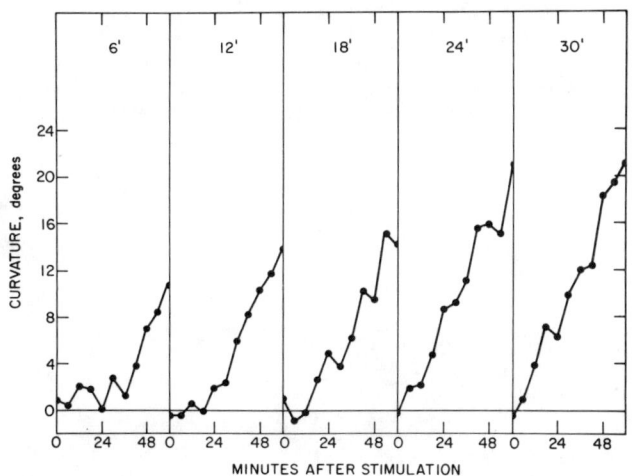

Fig. 6. Geotropic curvature of corn coleoptiles after various periods of stimulation.

V. ACKNOWLEDGEMENTS

I thank E. Gayle Schneider for her technical assistance in part of this study, and S.A. Gordon for his interest and support of this program under the auspices of the U.S. Atomic Energy Commission and the National Aeronautics and Space Administration.

VI. REFERENCES

CANE, A.R. and M.B. WILKINS (1969). Independence of lateral and differential longitudinal movement of indoleacetic acid in geotropically stimulated coleoptiles. of *Zea mays*. Plant Physiol. 44, 1841-1847.

DOLK, H.E. (1930). Geotropie en groeistof. Diss. Utrecht. English trans. in Rec. trav. bot. néerl. 33, 509-585 (1936).
HERTEL, R. and R. FLORY (1968). Auxin movement in corn coleoptiles. Planta (Berl.) 82, 123-144.
SHEN-MILLER, J. and S.A. GORDON (1967). Gravitational compensation and the phototropic response of oat coleoptiles. Plant Physiol. 42, 352-360.
SHEN-MILLER, J. (1970). Reciprocity in the activation of geotropism in oat coleoptiles grown on clinostats. Planta (Berl.) 92, 152-163.
SHEN-MILLER, J. (1972). Participation of the Golgi apparatus in geotropism, *in* "Plant Growth Substances, 1970". Ed. D.J. Carr. pp. 738-744.

PLANT GROWTH SUBSTANCES, 1970

The Source and Transport of Growth Regulators responsible for the Geotropic Response of *Zea mays* Roots

Malcolm B. Wilkins, Griselda S.B. Gibbons and Stanley Shaw

Botany Department, The University, Glasgow, W.2., Scotland. U.K.

I. INTRODUCTION

The geotropic response of coleoptiles depends upon indole-3-acetic acid (IAA) synthesised in the apex of the organ. In a vertical *Zea* coleoptile, the IAA is transported basipetally into the elongating zone and symmetrically distributed, but in a horizontal coleoptile the IAA becomes asymmetrically distributed, the concentration in the lower half being greater than that in the upper half (Gillespie & Thimann, 1963; Goldsmith & Wilkins, 1964). Two apparently independent mechanisms give rise to this asymmetry: (1) a metabolically dependent, lateral, polar transport of IAA from the upper to the lower half of the organ (Goldsmith & Wilkins, 1964; Wilkins & Whyte, 1968; Cane & Wilkins, 1969); and (2) an enhanced basipetal transport of IAA in the lower, as compared with the upper, half of the organ (Naqvi & Gordon, 1966; Cane & Wilkins, 1969).

The mechanism controlling the downward geotropic curvature of a root is less well understood. Re-orientating a primary root from the vertical into the horizontal position causes a significant decrease in the overall growth rate (Sachs, 1874; Bennet-Clark, Younis & Esnault, 1959), and downward curvature is due to a greater reduction in the growth rate of the lower half, than of the upper half, of the organ (Audus & Brownbridge, 1957). The overall depression in the growth rate of the root on geotropic stimulation suggests that there is a net increase in the level of a growth inhibitor in the tissue. The unequal depression in the growth rate of the upper and lower halves of the horizontal root further indicates that a lateral gradient is established either in the effectiveness of the inhibitor, or in its concentration. There is some evidence for the occurrence of an unequal distribution of growth inhibitor in geotropically stimulated root apices (Hawker, 1932).

Several reports suggest that removal of the extreme apex of the root leads to the abolition of geotropic responsiveness. Most recently, removal of just the root cap was shown to abolish the geotropic response of *Zea mays* roots (Juniper, Groves, Landau-Schacher & Audus, 1966; Gibbons & Wilkins, 1970). These findings point to the root cap being the site of (1) the gravi-perception mechanism, (2) the production or release of at least one growth regulator, or (3) both the systems indicated in (1) and (2). Juniper *et al.* (1966) dismissed the possibility that the root cap is the source of growth regulators, but this view is not supported by the recent experiments of Gibbons & Wilkins (1970).

The work reported in this paper shows that the root cap is the source of at least one growth inhibitor upon which the geotropic response of the root depends, and that this inhibitor undergoes lateral transport in the horizontal root. The nature of the inhibitor is discussed briefly in relation to the well established pattern of transport of IAA in the primary roots of *Zea mays* seedlings (Wilkins & Scott, 1968; Scott & Wilkins, 1968; Wilkins & Cane, 1970; Cane & Wilkins, 1970).

II. METHODS AND MATERIALS

Seeds of *Zea mays* L. var. Giant White Horsetooth were grown in darkness for 3 days at $25°C$ by the method of Scott & Wilkins (1968). Immediately before an experiment was begun, the seedlings were brought into, and subsequently kept in, white fluorescent light (5.8×10^{-5} J. cm^{-2} sec^{-1}), since the roots do not exhibit a positive geotropic response in darkness (Scott & Wilkins, 1969). The experiments involved micro-surgery of the root apex, or the insertion into the side of the root of a small mica, polythene or metal foil barrier; details are given with each experiment. The nature of the barrier did not affect the result of a particular treatment. All experiments were repeated on a number of occasions, and the data in the figures show the mean responses of at least 40 individual roots.

III. RESULTS AND DISCUSSION

The root cap as a source of growth inhibitor

In the first experiment the curvatures developed by decapped and non-decapped roots were compared. When the cap is removed the roots show no geotropic response even after 24 hours in the horizontal position, whereas those with an intact cap had developed a curvature of about 40 degrees after only 4 hours in the horizontal plane (Fig. 1A). Despite the absence of a geotropic response in the decapped roots, they continued to elongate (Gibbons & Wilkins, unpublished). The absence of curvature cannot therefore be attributed to a cessation of growth.

This kind of experimentation was extended by removing with fine forceps one half of the root cap and then placing the roots in various orientations with respect to gravity, as shown by the diagrams inset in Figs. 1 B and C. Each root developed a large curvature towards the side upon which the remaining half-cap was located, regardless of whether it was vertical or horizontal (Figs. 1 B and C). Furthermore, the downward curvature of a horizontal root having the remaining half-cap on the lower side of the apex (Fig. 1B, curve a) was nearly twice as great as the upward curvature developed by a horizontal root having the remaining half-cap on the upper side (Fig. 1B, curve b).

It was difficult to dissect away *one half* of the cap of a large number of roots with any degree of precision. An accurate removal of one half of the root cap could be achieved by making two cuts with a razor blade into the apex of the root. Inevitably with this procedure a small portion of the root apex was removed with the half-cap in most of the roots. It was of importance, therefore, to assess whether the root apex itself was the source of growth regulators, since if this were the case, its partial removal with the half-cap would render the experiments difficult, if not impossible to interpret.

To assess whether or not the root apex was the source of growth regulators, the caps were removed from a number of roots. Then either one half of the apex of the root was removed to a distance of 0.5-1 mm behind the apical extremity, or the apex was left intact and a small barrier of mica or metal foil inserted to prevent longitudinal movement of substances from the apex to the growing zone on one side of the root. The barriers were placed 0.5-1 mm behind the apex. Samples of decapped roots treated in these two ways were placed in the vertical and horizontal plane together with the appropriate non-decapped control roots. The resulting curvatures are shown in Fig. 2 in which the various experimental treatments are diagrammatically represented.

When the root cap is present, insertion of a barrier, or the removal of half the cap and a little of the root apex on the same side of the root, resulted in the development of large curvatures regardless of whether the roots were vertical or horizontal. The direction of curvature was towards the remaining half-cap in the case of roots from which one half of the cap had been removed, and away from the side of the root into which the barrier had been inserted. In contrast, when the whole of the cap had been removed, removal of one half of the root apex, or the insertion of a barrier into one side of a root with an intact apex, did not give rise to the development of a significant curvature in either horizontal or vertical roots.

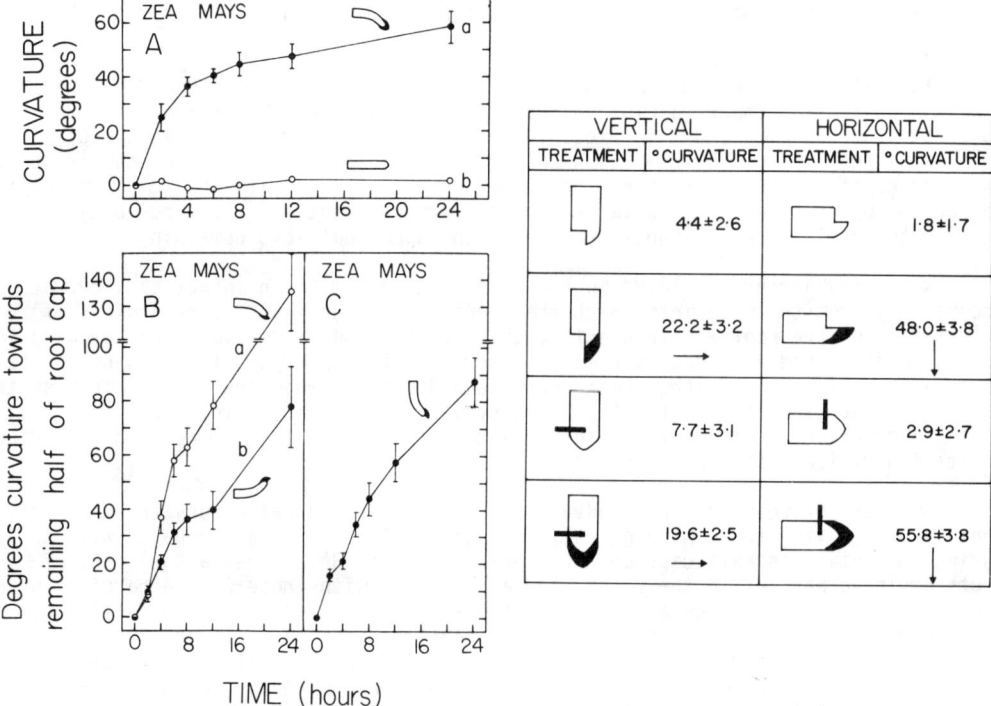

Fig. 1. (left) A. The curvature of horizontal Zea mays roots with time when the root cap is intact (a), and when it is completely removed (b).

B. Time course for curvature of horizontal roots from which either the upper (a) or the lower (b) half of the cap has been removed.

C. Time course for curvature of vertical roots when one half of the cap has been removed.

Vertical lines extending from one side of the points show the standard error of the mean; where they are not shown the standard error is less than $0.5°$. Gravity is acting from the top towards the bottom of figures; the various orientations of the roots are shown by the inset diagrams in which the root cap is shown in black.

Fig. 2. (right). Curvature (with ± S.E. of mean) after 12 hours, of horizontal and vertical roots with and without caps when a barrier is inserted 0.5-1 mm behind apex, or when one half of the apical 0.5-1 mm of the organ is removed. Root caps and barriers are shown in black. Direction of curvature is shown by the arrow. For explanation see text.

These results establish that while the root cap is the source of at least one growth inhibitor, the root apex itself is not the source of any growth regulators. However, this finding does not rule out the possibility that the apex may have a role in regulating root growth, since it might be able to control the transport of the inhibitor from the cap to the growing zone. Since the root apex is not a source of growth regulators, removal of a small part of the apex with the half-cap will not affect the validity of the experiments.

Evidence for the lateral transport of inhibitor

The time courses for curvature developed by roots from which one half of the cap had been removed, and by roots with intact caps but into the side of which a barrier had been inserted, were studied in detail. The roots were placed in several orientations with respect to gravity as shown by the inset diagrams in Figs. 3 A and B.

The data shown in Fig. 3A confirm those shown in Figs. 1 B and C; the roots always bend towards the remaining half-cap, even when they are vertical. There is no doubt from the data shown in Figs. 1B and 3A that the downward curvature of a horizontal root having the remaining half-cap on the lower side of the apex is about twice as great as the upward curvature of a horizontal root with the remaining half-cap on the upper side. The curvature of the vertical root (Fig. 3A, curve c) is similar to that of the horizontal root with the upper half-cap remaining.

When the barrier is placed in the side of a root with an intact cap, curvature occurs away from the side into which the barrier has been inserted regardless of the orientation of the root with respect to gravity (Fig. 3B). However, the downward curvature developed when the barrier is inserted into the upper side of a horizontal root was much larger than the upward curvature developed when the barrier is inserted into the lower side of a root (Fig. 3B, curves a & b). The vertical root (Fig. 3B, curve c) developed a slightly greater curvature than a horizontal one with a barrier in its lower side (Fig. 3B, curve b).

A number of reasons can be advanced to explain the greater curvature of horizontal roots having the remaining half-caps on the lower side as compared with those having the remaining half-caps on the upper side (Fig. 3A, curves a & b), and of roots having a barrier in the upper side as compared with those with a barrier in the lower side (Fig. 3B, curves a & b). The two simplest explanations are: (1) that the lower half-cap produces more inhibitor than the upper half-cap, and (2) that the production of inhibitor by the upper and lower half-cap is similar, but that a downward lateral transport of inhibitor occurs in the cap and root. In the latter case, half-decapped roots would have a more marked concentration gradient of inhibitor between the upper and lower halves of the elongation zone when the remaining half-cap is on the lower side than when it is on the upper side of the apex. In roots with an intact cap and a barrier in the upper side, the inhibitor passing into the growing zone of the lower half of the root would have come from both the upper and lower halves of the cap, and would thus be greater in amount than that passing into the growing zone on the upper side of a root with a barrier in the lower side.

It is possible to argue that downward lateral transport of a growth inhibitor must be taking place in horizontal roots on the basis of the kind of data shown in Figs. 3 A and B. This argument is based on the findings that (1) a half-decapped root with the remaining half-cap on the lower side of the apex (Fig. 3A, curve a) develops a *smaller* curvature than a root with an intact cap but having a barrier inserted in the upper side (Fig. 3B, curve a), and (2) that a half-decapped root with the remaining half-cap on the upper side of the apex (Fig. 3A, curve b) develops a *greater* curvature than a root with an intact cap but with a barrier in the lower side (Fig. 3B, curve b). Since the data in Figs. 3 A and B were not obtained in the same experiment, further experiments were conducted in which the comparisons outlined under (1) and (2) above were made simultaneously. The results, shown in Figs. 4 A and B confirm the conclusions drawn from Figs. 3 A and B.

The greater curvature shown by the root with the barrier inserted in the upper side in Fig. 4A (curve a) can only be attributed to an increased amount of inhibitor passing back in the lower half of the organ, as compared with the half-decapitated root (Fig. 4A, curve b). The only place from which this additional inhibitor can come is the upper half of the root cap by means of a downward lateral movement. Since both roots are horizontal there is no reason to suppose that the sensitivity to the inhibitor of the cells in the growing zone of the lower half of the two roots shown

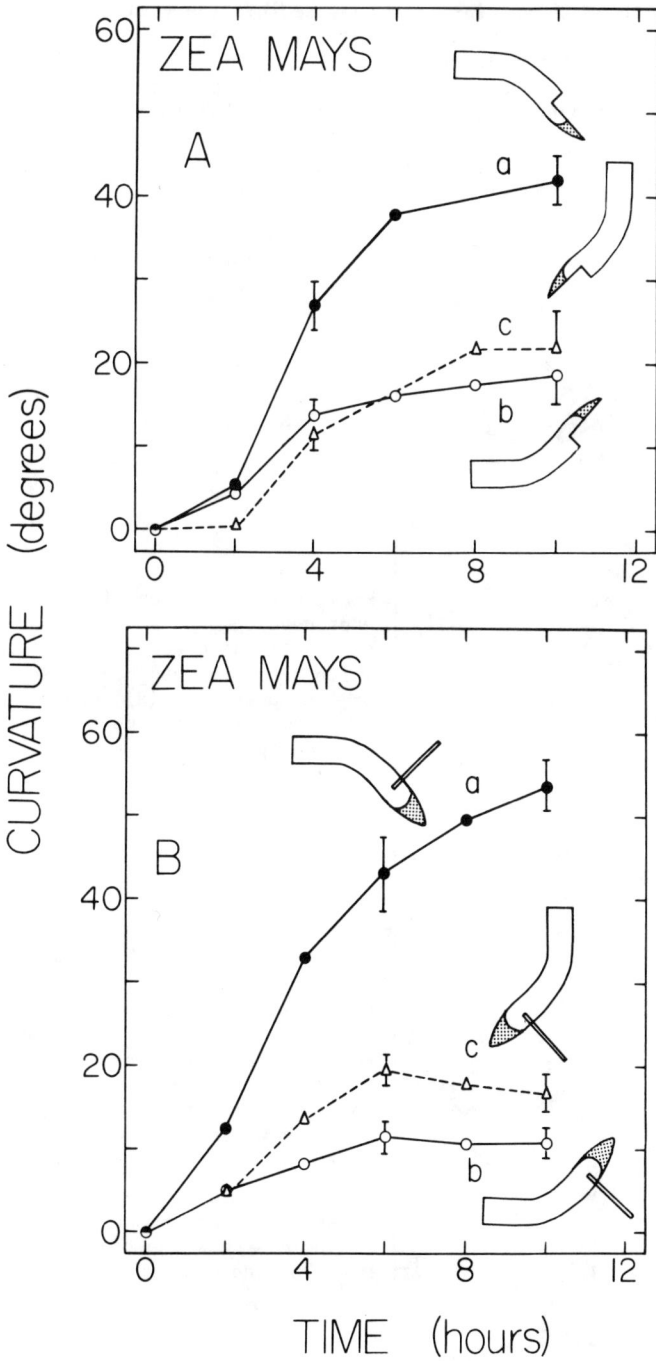

Fig. 3A. Curvature of roots, from which one half of the cap has been removed, and orientated so that the remaining half cap is on the lower (a) or upper (b) half of a horizontal root, or the apex of a vertical root (c).

Fig. 3B. Curvature of roots with an intact cap, but into which a barrier has been inserted on the upper (a) and lower (b) side of horizontal roots, or on the side of a vertical root (c). Root caps are shaded. Vertical lines extending from one side of the points show the standard error of the mean.

in Fig. 4A should be different, nor is there reason to suppose that the basipetal movement of inhibitor in the lower half of the two roots should be different. If downward lateral transport of an inhibitor is occurring, then a half-decapped root with the remaining half-cap on the upper side of the apex must develop a greater curvature than the intact root with a barrier in the lower side. That this is so is shown by the data in Fig. 4B. Lateral transport of inhibitor would be severely reduced in the half-decapped root because of the removal of the receptor lower half (Fig. 4B, curve a). As a result the inhibitor would move basipetally into the growing zone of the upper half of the root. In the intact root, downward lateral movement into the lower half would occur but the inhibitor would then be trapped on the apical side of the barrier (Fig. 4B, curve b). This lateral transport of inhibitor would decrease the amount passing basipetally into the growing zone of the upper half of the root

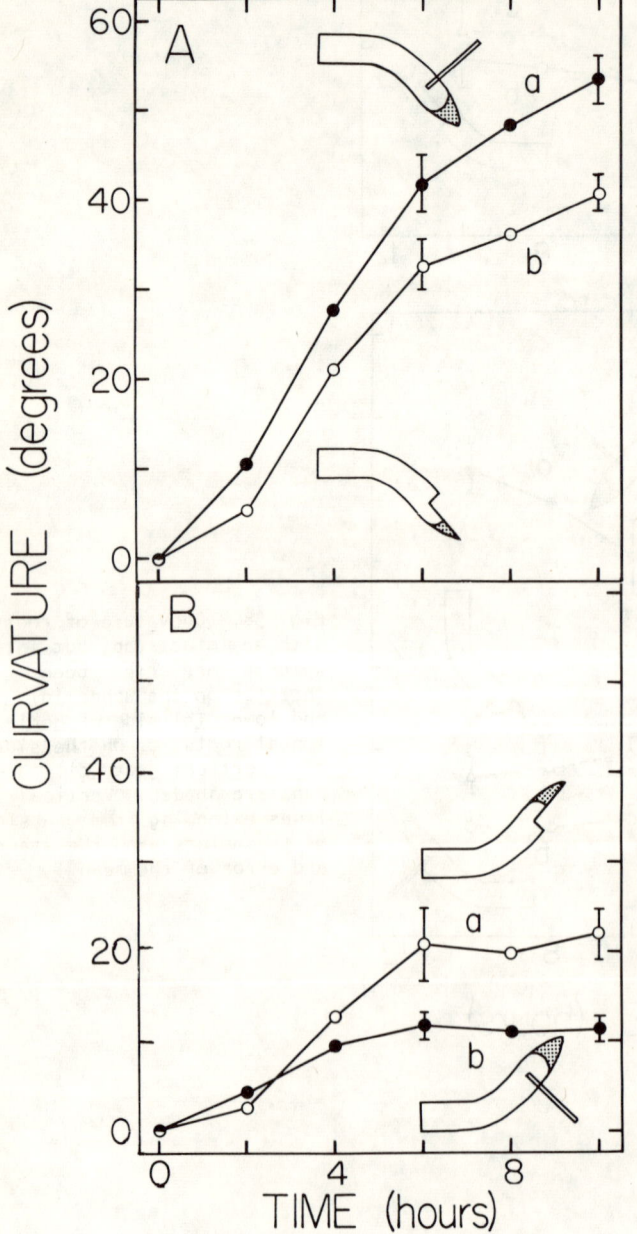

Fig. 4A. Comparison of curvatures developed by a horizontal root with an intact cap, but into which a barrier had been inserted on the upper side (a), and by one from which the upper half of the cap had been removed (b).

Fig. 4B. Comparison of curvatures developed by a horizontal root with an intact cap but into which a barrier had been inserted on the lower side (b), and by one from which the lower half of the cap had been removed (a).

Root caps shaded. Vertical lines extending on one side of the points show the standard error of the mean.

(Fig. 4B, curve b), as compared with the root where the lower half of the root cap had been removed (Fig. 4B, curve a).

The data shown in Figs. 4 A and B therefore indicate that a downward lateral transport of at least one growth inhibiting substance takes place in the root cap, and perhaps also to a small extent in the root apex. Further experimental evidence for the occurrence of lateral transport of a growth regulator has been obtained by inserting horizontal or vertical barriers into the apices of horizontal roots having an intact root cap. These experimental procedures, and the results obtained, are shown in Fig. 5. When the barrier is horizontal so that it blocks the movement of substances from the upper to the lower half of the cap (Fig. 5, curve b), the curvature developed by the root is less than half of that attained when the barrier is vertical and downward lateral movement of inhibitor can occur (Fig. 5, curve a).

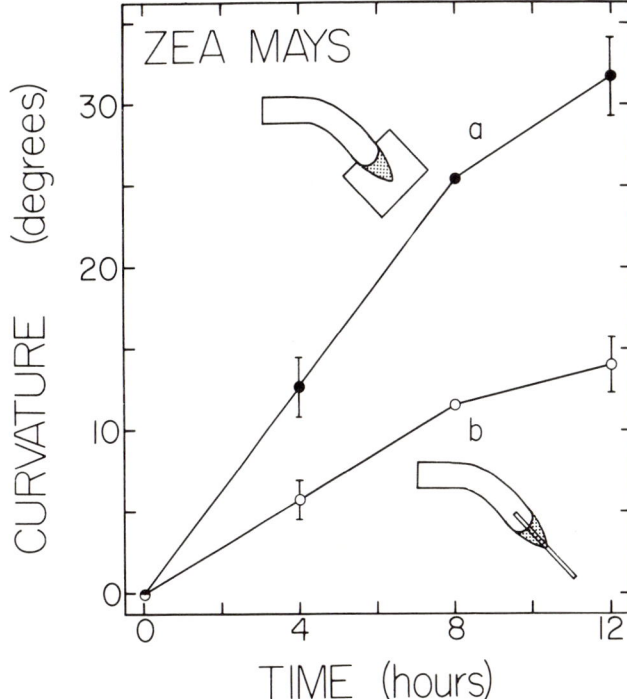

Fig. 5. Comparison of the curvatures developed by horizontal roots with an intact cap but into the apex of which a barrier had been inserted either vertically (a) or horizontally (b). Root caps shaded. Vertical lines on one side of the points show standard error of mean.

The findings presented in this paper show that in light, the geotropic response of the primary root of *Zea mays* depends upon at least one growth inhibiting substance present in the root cap moving basipetally through the root apex and into the extending zone of the root. The downward curvature results from the growth of the lower half of the root being depressed to a greater extent than that of the upper half. These growth rate differences are at least in part attributable to a downward lateral transport of the inhibitor from the upper to the lower half of the cap. Nevertheless, the fact that an intact root with a horizontal barrier (Fig. 5, curve b) developed a slight downward curvature leaves open the possibility that either some lateral transport of inhibitor occurred in the extending zone of the root behind the barrier, or the lower half of the root cap in fact releases a slightly greater amount of inhibitor into the lower half of the root than the upper half-cap releases into the upper half. There may also be slight differences in the sensitivity to inhibitor of the extending cells in the upper and lower halves of the root, or differences in the basipetal movement of the inhibitor in the upper and lower halves of the root. These possibilities, and the problem of identifying the inhibitor, are now being investigated. It is of interest to note, however, that on evidence presently available the possibility of the inhibitor being IAA is not very great, since IAA is transported in a strongly acropetal, polar manner through the growing zone and into the extreme apical 1 mm of these roots (Scott & Wilkins, 1968).

The mechanism whereby the different rates of growth of the upper and lower halves of a horizontal root are controlled during the geotropic response is thus, in part, somewhat similar to that which operates in shoots. However, whereas in coleoptiles it is IAA which is laterally transported into the lower half of the organ, the identity of the substance or substances laterally transported in the root caps is not yet known.

IV. SUMMARY

The findings presented in this paper show that in light, the positive geotropic response of the primary root of *Zea mays* depends upon at least one growth inhibiting substance present in the root cap moving basipetally through the apex and into the extending zone of the root. The downward curvature results from the growth of the lower half of the root being depressed to a greater extent than that of the upper half. These growth rate differences are at least in part attributable to a downward lateral transport of the inhibitor from the upper to the lower half of the cap. The root apex itself is not the source of growth regulating substances.

V. ACKNOWLEDGEMENTS

G.S.B. Gibbons and S. Shaw are supported by Science Research Council Post-Graduate Studentships.

VI. REFERENCES

AUDUS, L.J. and M.E. BROWNBRIDGE (1957). Studies on the geotropism of roots. I. Growth rate distribution during response and the effects of applied auxins. J. Exp. Bot. 8, 105-124.
BENNET-CLARK, T.A., A.F. YOUNIS and R. ESNAULT (1959). Geotropic behaviour of roots. J. Exp. Bot. 10, 69-86.
CANE, A.R. and M.B. WILKINS (1969). Independence of lateral and longitudinal movement of indole acetic acid in geotropically stimulated coleoptiles of *Zea mays*. Pl. Physiol., 44, 1481-1487.
CANE, A.R. and M.B. WILKINS (1970). Auxin transport in roots. VI. Movement of IAA through different zones of *Zea* roots. J. Exp. Bot. 21, 212-218.
GIBBONS, G.S.B. and M.B. WILKINS (1970). Growth inhibitor production by root caps in relation to geotropic responses. Nature, 226, 558-559.
GILLESPIE, B. and K.V. THIMANN (1963). Transport and distribution of auxin during tropistic response. I. The lateral migration of auxin in geotropism. Pl. Physiol. 38, 214-225.
GOLDSMITH, M.G.M. and M.B. WILKINS (1964). Movement of auxin in coleoptiles of *Zea mays* L. during geotropic stimulation. Pl. Physiol. 39, 151-162.
HAWKER, L.E. (1932). Experiments on the perception of gravity by roots. New Phytol. 31, 321-328.
JUNIPER, B.E., S. GROVES, B. LANDAU-SCHACHAR and L.J. AUDUS (1966). Root cap and the perception of gravity. Nature, 209, 93-94.
NAQVI, S.M. and S. GORDON (1966). Auxin transport in *Zea mays* L. coleoptiles. I. Influence of gravity on the transport of indole-acetic acid-2-^{14}C. Pl. Physiol. 41, 1113-1118.
SACHS, J. VON (1874). Über das Wachstum der Haupt- und Neben-wurzeln. Arb. Bot. Inst. Würzburg, 1, 385-474.
SCOTT, T.K. and M.B. WILKINS (1968). Auxin transport in roots. II. Polar flux of IAA in *Zea* roots. Planta, 83, 323-334.
SCOTT, T.K. and M.B. WILKINS (1969). Auxin transport in roots. IV. Effect of light on IAA transport and geotropic responsiveness in *Zea* roots. Planta, 87, 249-258.
WILKINS, M.B. and T.K. SCOTT (1968). Auxin transport in roots. Nature, Lond., 219, 1388-1389.
WILKINS, M.B. and P. WHYTE (1968). Relationship between metabolism and the lateral transport of IAA in corn coleoptiles. Pl. Physiol. 43, 1435-1442.
WILKINS, M.B. and A.R. CANE (1970). Auxin transport in roots. V. Effects of temperature on the movement of IAA in *Zea* roots. J. Exp. Bot. 21, 195-211.

PLANT GROWTH SUBSTANCES, 1970

Asymmetric "Acid Growth" Response Following Gravistimulus

Leonora Reinhold and Dvora Ganot

Botany Department, Hebrew University of Jerusalem, Israel

I. INTRODUCTION

A number of workers have noted that treatment with CO_2, or with acid buffers at approximately the pH of CO_2-saturated water, can bring about a remarkable promotion of extension growth or water uptake (see Evans, Ray and Reinhold 1971 for refs.). Some years ago Zvi Glinka and I (1966) were rather awed to observe that only extremely brief exposure was necessary to obtain the effect - promotion of water uptake by sunflower hypocotyl segments was easily observable after 10 secs. treatment with CO_2. More recently Michael Evans, Peter Ray and I (1971) made a detailed study of CO_2-induced growth and found, among other things, that within 1 min of the start of treatment the rate of elongation of coleoptiles rises 8-16 fold.

In another study (Reinhold 1967) we noted that CO_2 can substitute for the thigmotropic stimulus - it induces coiling of tendrils within a very few minutes - and we were curious as to how it might affect other tropisms. In this paper we report on the effect of CO_2 and acid buffers on geotropism.

II. MATERIALS AND METHODS

We used as experimental material sections of etiolated sunflower hypocotyl (excised 1 cm below the hook) floating in various media; and we followed curvature by measuring the radius of curvature (r) of the inner surface. Curvature is expressed in the Tables as d/r where d = diameter of segment. This ratio reflects the differential behaviour of the upper and lower sides of the segment, since

$$\frac{l_o + \Delta l_1}{l_o + \Delta l_2} = \frac{\frac{\alpha}{360} \cdot 2\pi(r + d)}{\frac{\alpha}{360} \cdot 2\pi r}$$

$$= 1 + \frac{d}{r}$$

where l_o is the initial length of the segment, Δl_1 and Δl_2 the change in length of the lower and upper sides respectively, and α is the angle subtending the arc of curvature. d was assumed to be constant. Δl_1 and Δl_2 were always positive in our experiments.

III. RESULTS

Table 1 shows the effect of CO_2 on degree of geotropic response. In this experiment the segments were transferred every 30 min from water to a saturated air chamber where they were gassed either with CO_2 or with air for 3 min, after which they were returned to water. Care was taken during these transferences not to change their orientation with respect to gravity. This method of treatment was chosen in preference to continuous CO_2, because Glinka and I (1966) had found that a continuous supply of CO_2 is not necessary for the achievement of the full stimulatory effect and one may thus avoid complications due to anaerobiosis and membrane damage. Rather to our surprise - because an auxin shift is believed to be the central feature of the geotropic reaction and auxin-induced growth and acid-induced growth are, as you will hear, two very different phenomena - CO_2 appeared to promote geotropic curvature. The effect, however, though reproducible, depended on careful timing of the first CO_2 treatment with reference to the start of geotropic stimulus. We attributed this to the fact that the effect of CO_2 on straight growth, though spectacular, is short-lived (see Evans et al. 1971) and the response to a second and third treatment is always considerably weaker than that to the first. Acid buffer-induced elongation has much in common with that induced by CO_2, and though the stimulation is less pronounced it is far longer lasting. For further analysis of the effect on geotropism we therefore turned to acid buffers.

Table 1 (Expts. II and III) shows that various acid buffers at the pH of CO_2-saturated water do in fact have a strongly promotive action on geotropic curvature. Effects of pH on geotropism have been noted before, but in relation to exogenous auxin supply - Anker (1956) observed that lower pH shifted the optimum IAA concentration for curvature of Avena coleoptiles. Table I shows the effect of low pH in the absence of added auxin. That the effect of the buffers was in fact attributable to hydrogen ion concentration and not to anion concentration was checked in experiments in which the anion concentration was varied.

Table 1. The effect of CO_2 and of low pH on geotropic curvature of sunflower hypocotyl segments. Curvature was measured after 2 hrs' exposure to gravistimulus and is expressed as $\frac{d}{r} \times 10^3$.

Expt. No.	Medium	pH	
		5.8	3.8
I	H_2O	56 ± 30	
	CO_2*		172 ± 13
II	Glycine-HCl	20 ± 9	153 ± 20
	Acetic-Na acetate	42 ± 13	112 ± 11
III	Citric-Na phosphate	20 ± 15	92 ± 16

* 3 min. CO_2, 27 min. H_2O in repeated cycles.

The response of segments held beneath the surface of aerated solutions was similar to that of floating segments (see Ganot and Reinhold 1970).

The effect of low pH shown in Table 1 would be perfectly consistent with current views on the mechanism of the geotropic reaction if low pH stimulated growth by re-enforcing or enhancing in some way the action of auxin (see Bonner 1934). Our measure of curvature gives the ratio of the lengths of the two sides, $\frac{l_0 + \Delta l_1}{l_0 + \Delta l_2}$.

The growth increments Δl_1 and Δl_2 are unequal, and current theory attributes this to the now well-established fact (see Gillespie and Thimann 1963) that gravitational stimulus causes a displacement of auxin towards the lower side of the treated organ.

If low pH magnified the growth increment due to auxin, i.e. multiplied both Δl_1 and Δl_2 by some factor, the curvature ratio would increase. But the findings of Evans, Ray and Reinhold (1971) have made it very difficult to entertain the view that low pH enhances auxin action. We observed that CO_2-induced elongation, once started, cannot be stopped by a wide variety of metabolic inhibitors which promptly stop auxin-induced growth. Moreover we also found that CO_2 cannot be prevented from producing its effect by pre-treatment with these inhibitors. It was therefore important in the present investigation to test whether acid-promoted geotropic curvature was similarly stable to metabolic inhibitors. In planning the experiment a difficulty presented itself: supply of inhibitors simultaneously with the auxin or acid buffer would lead to an invalid comparison, since the entry or action of the inhibitors might be affected by pH. We therefore gave the segments a 30 min gravistimulus in either auxin or acid buffer and only subsequently transferred them to the inhibitor solutions. Table 2 shows that development of curvature in the case of auxin-treated segments was very severely, almost totally, depressed by arsenite and $HgCl_2$. It was also depressed by cycloheximide but to a lesser extent - probably because this inhibitor penetrates more slowly and the design of the experiment made pre-treatment with inhibitor impossible. The cycloheximide inhibition of curvature, however, is quite clear and is statistically significant. In striking contrast development of curvature after geotropic stimulation in acid buffer was totally unaffected by cycloheximide or by arsenite. Though the figures show a slight effect of $HgCl_2$, the depression is not significant. These results thus strongly suggest that acid-induced geotropic curvature is not mediated via auxin.

Table 2. The effect of metabolic inhibitors on development of curvature in hypocotyl segments after exposure to gravistimulus in IAA or in acid buffer. The segments were exposed to gravistimulus for 40 mins. after which they were rinsed and allowed to react while shaking in H_2O or inhibitor solutions at the following concentrations:

$KAsO_2$ 10^{-4}M; $HgCl_2$ 2×10^{-4}M; cycloheximide 5 mg/l   (From Ganot and Reinhold 1970)

Medium during stimulus	Medium during reaction period	Final curvature ($\frac{d}{r} \times 10^3$)	Increase in curvature during reaction period
IAA 0.05 mg/l Expt. I	H_2O	51 ± 5	32
	$KAsO_2$	27 ± 6	3
	$HgCl_2$	26 ± 5	6
IAA 0.05 mg/l Expt. II	H_2O	67 ± 8	38
	Cycloheximide	43 ± 7	16
Glycine buffer pH 3.8 Expt. I	H_2O	66 ± 15	33
	$KAsO_2$	75 ± 17	34
	$HgCl_2$	52 ± 11	22
Glycine buffer pH 3.8 Expt. II	H_2O	49 ± 7	28
	Cycloheximide	58 ± 7	33

In the experiment just described the concentration of auxin chosen was one near the optimal concentration for curvature in our tissue as we had determined it. The optimal concentration varied between 0.01 and 0.1 mg/l, which is very much below the optimal concentration for straight growth in this tissue. The latter is more than 100 times as great. As this point has relevance in connection with our next experiment, Figure 1 presents the relation we have found between degree of geotropic curvature and external IAA concentration for this tissue.

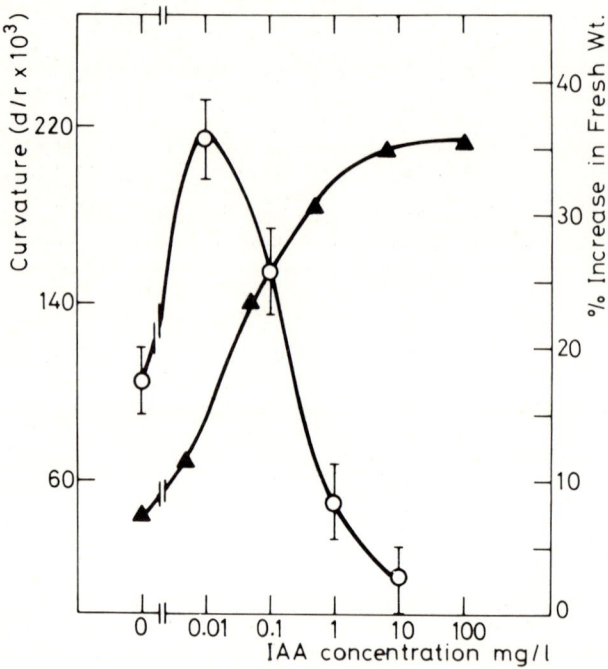

Fig. 1. The effect of external IAA concentration on degree of geotropic curvature (○) and on increase in fresh weight (▲). Curvature was measured after 2 hrs. gravi-stimulus. Gain in fresh weight was determined on segments continuously shaken in IAA solution for 4 hrs.

The optimum is very sharp - again in marked contrast to the plateau commonly observed for straight growth. Anker (1956) has obtained a similar curve for whole coleoptiles and we agree with his suggestion that the curve falls sharply in the post-optimal region because cell elongation becomes less sensitive to small differences in IAA concentration - geotropic curvature depends on the response to the difference between the auxin concentrations on the upper and lower side.

In the last experiment to be presented here we approached the problem of the relation between auxin action and the observed acid effect from a different direction. Following the procedure adopted by Brauner and Hager (1958) we attempted to lower drastically the endogenous auxin content of our tissue. Seedlings were decapitated and left in a saturated atmosphere in the dark for 4 days before the experiment. Segments were then excised as usual, and were given geotropic stimulus either in acid buffer or in 3 concentrations of auxin. As in Brauner and Hager's experiments, prolonged stimulation was required. Table 3 shows that segments taken from starved but light grown plants were still capable of geotropic curvature in auxin solution, although not at a concentration usually optimal for geotropic response. An IAA concentration at least 10 times higher was required. On the other hand, the segments reacted strongly in acid buffer. Segments from etiolated seedlings were totally unreactive in auxin solution even at a concentration 100 times the normal optimum - yet they were able to curve in acid buffer.

Table 3. The effect of low pH or IAA on geotropic curvature of hypocotyl segments excised from auxin-starved seedlings. The seedlings were decapitated and kept for 4 days in the dark in a saturated atmosphere. Segments were subsequently excised 5 mm below the cut surface. The figures in the table give curvature ($d_{/r} \cdot 10^3$) measured after the segments had been floating in the media for 20 hr (green plants) or 32 hr (etiolated plants). (From Ganot and Reinhold 1970).

Pigment status	Medium				
	Water	Glycine buffer pH 3.8	IAA		
			0.1 mg/l	1.0 mg/l	10 mg/l
Etiolated	9 ± 1	64 ± 18	10 ± 5	15 ± 9	13 ± 7
Green	4 ± 3	107 ± 29	12 ± 4	61 ± 16	73 ± 11

IV. DISCUSSION

To consider now the significance of these results: Promotion of geotropic curvature such as we have reported here might conceivably result from an effect of pH on the perception of the gravistimulus, on the response to it, or from a combination of both factors. The response to gravistimulus as we understand it today involves a lateral transport of auxin with the result that subsequent growth is asymmetric. The fact that acid-promoted curvature differed so strongly from auxin-promoted curvature in sensitivity to inhibitors clearly indicates an effect of acid buffer on response reactions. Whether or not pH in addition affected perception could not be determined with the techniques used in this investigation.

Apart from the striking insensitivity of acid-induced curvature to cycloheximide and the other inhibitors tested, two other findings seem to weigh very heavily against the possibility that low pH is giving sharper expression to the asymmetric auxin distribution which is believed to lie at the heart of the geotropic reaction. Firstly, Evans, Ray and Reinhold (1971) observed that the CO_2 effect is independent of the auxin concentration in coleoptiles - the same rate of CO_2-induced elongation was observed whether or not external IAA was supplied. Secondly, in the present investigation IAA at 100 times the normal optimal concentration was incapable of potentiating geotropic curvature in etiolated and starved hypocotyls, but they curved in acid buffer.

These facts seem to suggest strongly that the buffer effect is not mediated via auxin, but is independent of it. If so, and if the acid growth effect is equal on the upper and lower sides, it should *decrease* geotropic curvature, not increase it. The radius of curvature, as I have pointed out, gives the ratio of the lengths of the lower and upper sides, $\frac{l_0 + \Delta l_1}{l_0 + \Delta l_2}$, where Δl_1 and Δl_2 are the unequal increments in growth due to the asymmetric distribution of auxin. If now another increment, x, is added, i.e. acid growth, equal on both sides, it is obvious that

$$\frac{l_0 + \Delta l_1 + x}{l_0 + \Delta l_2 + x} \quad < \quad \frac{l_0 + \Delta l_1}{l_0 + \Delta l_2}$$

The conclusion pointed to is that x is *not* equal on the two sides. There is an asymmetry in the response to acid buffer.

Two possible explanations can be considered: first, that the asymmetry is physiological, the upper and lower side of the hypocotyl responding differently to equal hydrogen ion concentrations; or second, that the hydrogen ion concentration is not uniform through the tissue. Gravitational stimulus might bring about a lateral pH gradient through the hypocotyl. In this connection it is very interesting to recall that both Gundel (1933) and Metzner (1934) reported that the lower half of geotropically-stimulated organs becomes more acid than the upper.

The former suggestion, on the other hand - that the asymmetry is physiological - recalls proposals (Brauner 1966; Wilkins 1966) that gravistimulus causes an increase in sensitivity to auxin of the cells on the lower side. As a possible explanation for this difference in sensitivity Brauner (1966) postulated lateral displacement of a co-factor for auxin. In the case of roots it has been suggested (Audus and Brownbridge 1957; Bennet-Clark et al 1959) that an inhibitor becomes asymmetrically distributed. Since our experiments have produced evidence for an asymmetrical reaction not mediated via auxin, we wish to propose that the polarisation following gravistimulus is more general, and is not restricted to auxin and its cofactors or inhibitors.

V. SUMMARY

Treatment with CO_2 strongly promoted the curvature of etiolated sunflower hypocotyl segments exposed to gravitational stimulus. Similar effects were produced by various buffers at the pH of CO_2-saturated water. IAA also promoted curvature, the optimal concentration for this effect being less than one hundredth of the optimal IAA concentration for promotion of straight growth. The development of curvature by segments stimulated in acid solution did not appear to be mediated via auxin. It was completely stable to inhibitors which severely depressed auxin-induced curvature. Moreover, segments taken from seedlings "starved" by decapitation did not respond to gravistimulus in auxin solution even at 100 times the normal optimal concentration, yet responded in acid buffer. The results point to an asymmetrical "acid growth" response, and suggest that gravistimulus produces a polarisation which is not restricted to auxin and auxin co-factors.

VI. REFERENCES

ANKER, L. (1956). The auxin concentration rule for the geotropism of Avena coleoptiles. Acta bot. néerl. 5, 335-341.
AUDUS, L.J. and M.E. BROWNBRIDGE (1957). Studies on the geotropism of roots. I. Growth-rate distribution during response and the effects of applied auxins, J. Exp. Bot., 8, 105-124.
BENNET-CLARK, T.A., A.F. YOUNIS and R. ESNAULT (1959). Geotropic behaviour of roots, J. Exp. Bot. 10, 69-86.
BONNER, J. (1934). The relation of hydrogen ions to the growth rate of the Avena coleoptile, Protoplasma 21, 406-423.
BRAUNER, L. (1966). Versuche zur Analyse der geotropischen Perzeption. V., Planta (Berl.) 69, 299-318.
BRAUNER, L. and A. HAGER (1958). Versuche zur Analyse der geotropischen Perzeption. I., Planta (Berl.) 51, 115-147.
EVANS, M.L., P.M. RAY and L. REINHOLD (1971). Induction of coleoptile elongation by carbon dioxide. Plant Physiol. 47: 335-341.
GANOT, D. and L. REINHOLD (1970). The "acid growth effect" and geotropism. Planta (Berl.) 95, 62-71.
GILLESPIE, B. and K.V. THIMANN (1963). Transport and distribution of auxin during tropistic response. Plant Physiol. 38, 214-225.
GUNDEL, W. (1933). Chemische und physikalische Vorgänge bei geischer Induktion, Jb. wiss. Bot. 78, 623-664.
MATZNER, P. (1934). Zur Kenntnis der Stoffwechseländerungen bei geotropisch gereizten Keimpflanzen, Ber. dtsch. Bot. Ges. 52, 506-522.
REINHOLD, L. (1967). Induction of coiling in tendrils by auxin and carbon dioxide. Science, 158, 791-793.
REINHOLD, L. and Z. GLINKA (1966). Reduction in turgor pressure as a result of extremely brief exposure to carbon dioxide, Plant Physiol. 41, 39-44.
WILKINS, M.B. (1966). Geotropism, Ann. Rev. Plant Physiol. 17, 379-408.

PLANT GROWTH SUBSTANCES, 1970

The Role of Auxin in Thigmotropism

Leonora Reinhold, Tsvi Sachs and Lea Vislovska

Botany Department, The Hebrew University of Jerusalem, Israel

I. INTRODUCTION

In thigmotropism we are confronted with a series of challenging problems. The first concerns the detection of the stimulus - by what means does a plant organ sense contact? Secondly, what is the nature of the events which intervene between perception of stimulus and the expression of the visible response? Thirdly, in cases where tissues remote from the point of contact respond - when, for instance, a tendril coils not only at the point where it grasps the support but along its entire length - how is the stimulus transmitted? The fourth problem lies in the asymmetry of the reaction - what causes the two longitudinal halves of a tendril to respond unequally?

The nature of at least two of these problems indicates a possible rôle for hormones. The involvement of hormones, however, in the response of plant organs to tactile stimulus does not seem to have been established until very recently. In fact Bünning's article (1959) on thigmotropism for the Encyclopaedia of Plant Physiology makes no mention of hormones at all. Earlier, however, Boresh (see Borgström 1939) had proposed that a lateral displacement of IAA occurred as a result of contact stimulus, analogous to the transverse transport of auxin observed after phototropic and geotropic stimulation.

Three years ago we showed (Reinhold 1967) that symmetric treatment with IAA induced strong coiling of tendrils. IAA treatment could thus *substitute* for the contact stimulus.

Further, when only a few mm of the tip was submerged in IAA solution strong coiling resulted throughout the length of the tendril, indicating that the coiling stimulus was travelling basipetally down the tendril.

These results led us to propose that one of the events in the chain following tactile stimulus was a sharp increase in free auxin concentration in the tissue; that it was this increase in auxin concentration which was responsible for the observed coiling responses; and, further, that the coiling that normally occurs along the entire length of the tendril, bringing the stem closer to the support, involves a supply of auxin translocated basipetally from the point of contact.

II. RESULTS AND DISCUSSION

Experiments on Parthenocissus

We now wish to present further evidence for release of IAA, or accelerated IAA synthesis, following a contact stimulus. Such a stimulus is known to bring about responses other than the coiling of a tendril round its support. A striking example is found in the genus *Parthenocissus* where contact leads to expansion, at the tips of thin thread-like tendrils, of pads which attach firmly to the support. Darwin (1884) described the rapid development of these discs following stimulus as "one of the most remarkable peculiarities possessed by any tendrils". We have found that here, too, treatment with IAA can *replace* the tactile stimulus normally needed to bring about the development of the pads. Figure 1 shows the branched tendrils of *P. tricuspidata*.

In this species the pads are preformed structures present as small swellings at the tips of branches, which expand greatly in size on contact stimulation. The right hand branches of the tendril in Fig. 1 were treated on the plant with IAA-lanolin (0.05%). The photograph, which was taken 48 h later, shows that the tips have expanded as though after contact stimulus (Fig. 2). That this response was in fact elicited by the IAA and not by contact with the lanolin is shown by the fact that the left-hand branches in Fig. 1, which were treated with plain lanolin, retain their typical pre-stimulated form. The time taken for development of the pads is approximately the same as that observed following contact stimulus.

It should be pointed out that these IAA-induced pads do not resemble the amorphous swellings commonly observed when IAA is applied to an organ such as a young part of the stem and which we also observed in the present investigation when we applied IAA to the stem of *Parthenocissus*. In the latter case the swelling has no clear boundary, but merges gradually with the stem. In the case shown in Fig. 1, however, the boundary of the expanded pad is sharply distinct from the tendril and the latter is not swollen at all.

We have also achieved the formation of pads in the absence of contact stimulus in a second species of *Parthenocissus*, *P. quinquefolia*. In this species there are no preformed swellings (Fig. 3). After stimulus the narrow tip of the tendril broadens and flattens. Figures 3, 4 and 5 respectively show tendril tips treated on the plant with lanolin, with IAA-in-lanolin (0.05%), and, for comparison, with a contact stimulus. The photographs, taken 60 hrs after treatment, show the strong resemblance between contact-stimulated and IAA-treated tips.

Experiments on Passiflora

We also wish to present further evidence for our suggestion that IAA, moving basipetally down the tendril from the point of contact, is responsible for the transmission of the stimulus. It is known from a variety of systems that auxin induces the formation of a lignified xylem (Camus 1949; Jacobs 1952; Wareing 1958; Wetmore and Rier 1963; Wangermann 1967). It is also well-known (see Bünning 1959) that tendrils which have made contact with a support become lignified. We therefore made an anatomical comparison of tendrils of *Passiflora caerulea* treated with plain lanolin, with IAA-lanolin (0.5%), or reacting to a tactile stimulus (See Figs. 7, 8 and 9). Auxin-lanolin, but not plain lanolin, induced the formation of lignified xylem. Contact stimulus also caused lignification of the xylem, and this was more pronounced than in the IAA-treated tendrils. Furthermore, contact prevented the disintegration of the pith and brought about the lignification of the phloem fibres. The more pronounced lignification of the xylem and phloem fibres after contact was rather consistently observed, possibly indicating that IAA is not the only signal moving down the tendril.

Auxin thus imitates contact in bringing about both coiling and lignification of tendrils. Lignification, however, becomes apparent some time after coiling, and therefore might be the result of a stimulus produced during coiling, rather than a direct effect of the auxin itself. To test this point we treated *Passiflora* tendrils after the removal of 3/4 of their length. The basal quarter of the tendril does not coil in the absence of the tip, either upon contact or after IAA treatment. Examination of these tendrils after IAA-lanolin or lanolin had been applied to their cut surfaces showed that auxin induced lignification even in the absence of coiling (compare Figs. 7 and 8). Lignification therefore appears to be the direct consequence of auxin treatment. The lignification observed after contact is thus further evidence for the suggestion that contact causes the release of auxin and that auxin is at least part of the signal which transmits the contact stimulus basipetally along the tendril.

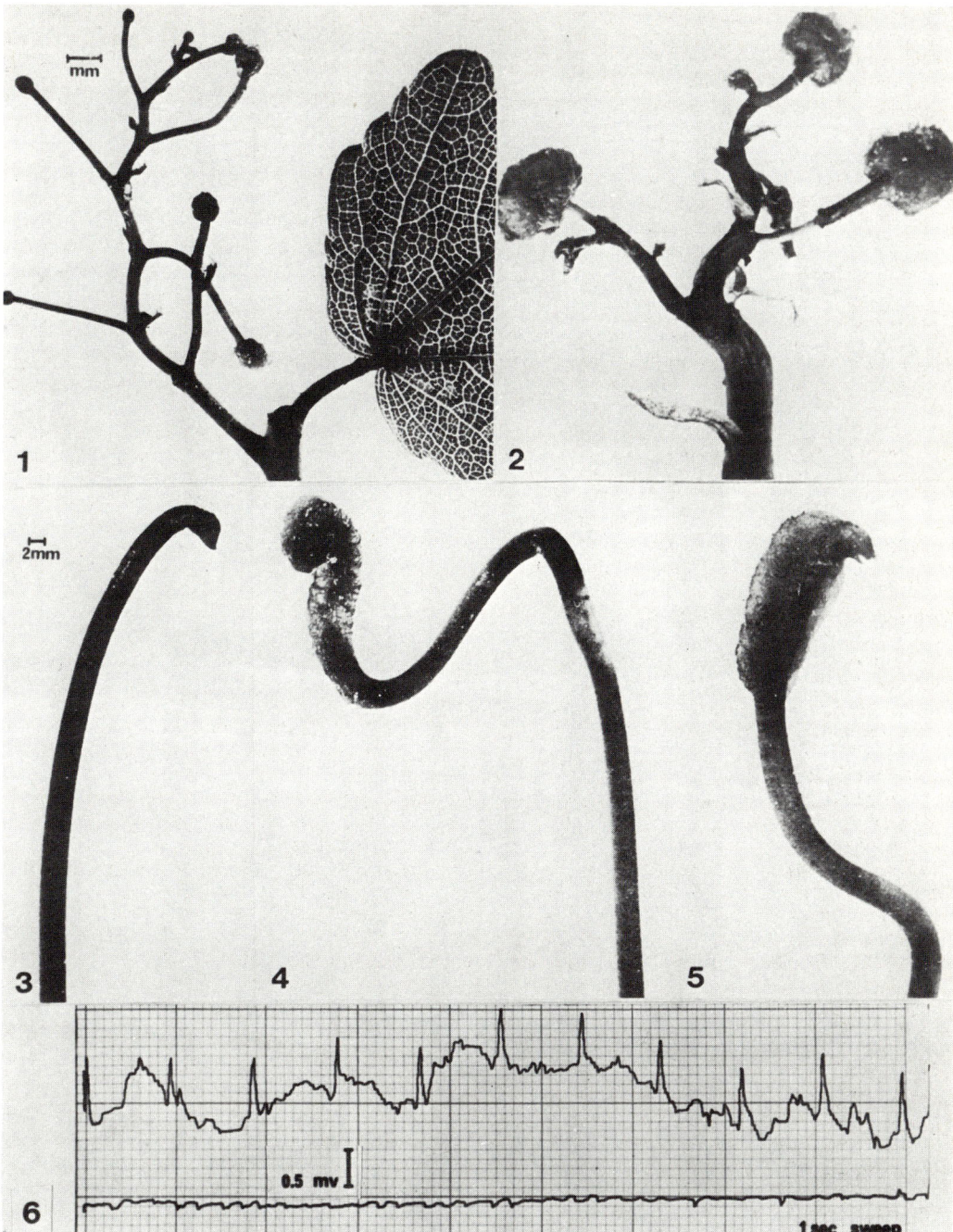

Figs. 1 and 2. Tendrils of *Parthenocissus tricuspidata*.
   Fig. 1. The right hand branches were treated with lanolin and the left-hand branches with IAA-lanolin (0.05%). Photographed 2 days later.
   Fig. 2. Tendrils stimulated by contact with a wall.

Figs. 3 - 5. Tendrils of *P. quinquefolia* treated with lanolin (Fig. 3), IAA-lanolin 0.05% (Fig. 4) and contact stimulus (Fig. 5).

Fig. 6. Fluctuations in electric potential in tendrils of *Cucumis sativus*. "Active" period (above) and quiet period (below) as displayed on the oscilloscope.

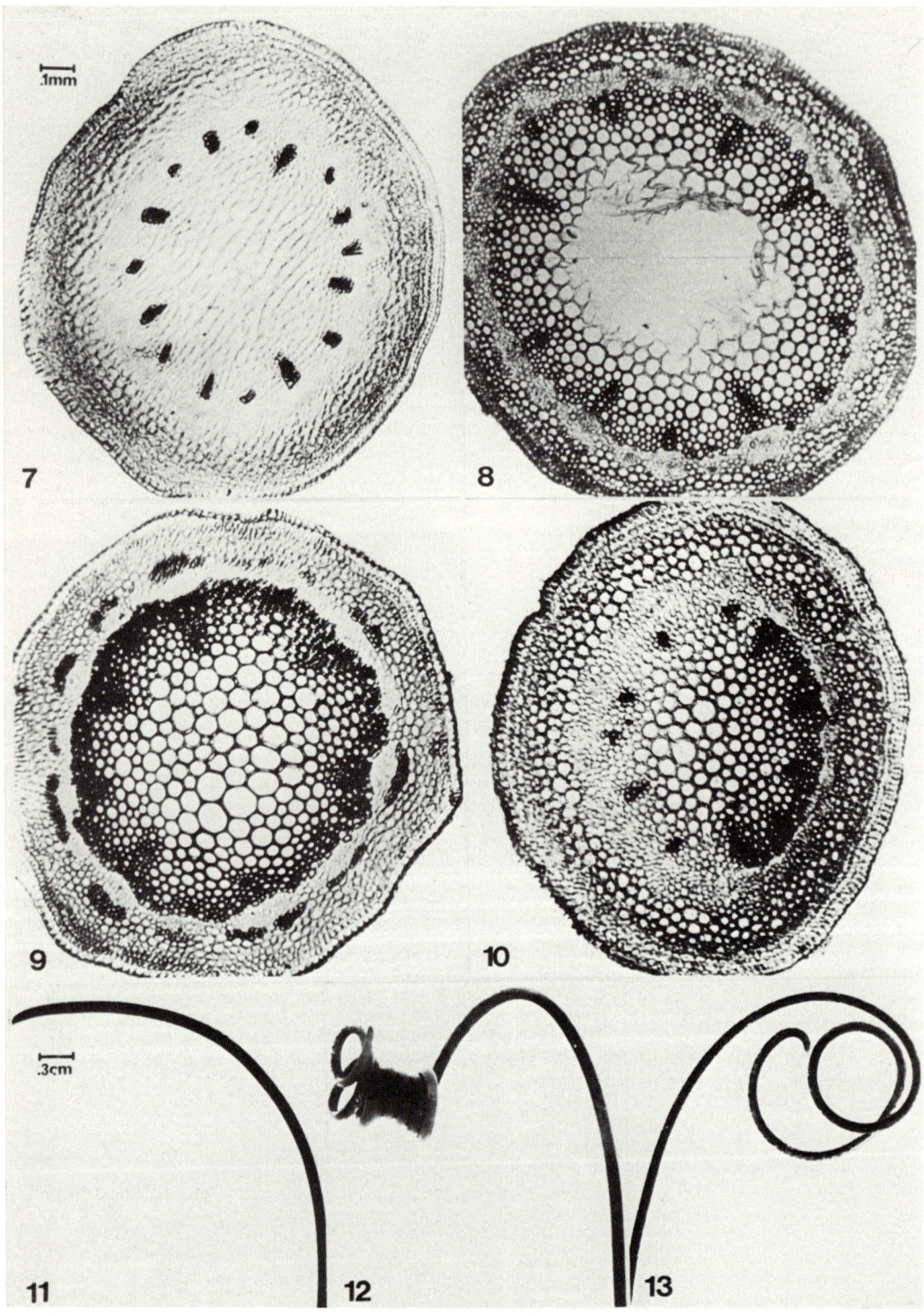

Figs. 7 - 9. Cross sections of *Passiflora caerulea* tendrils, cleared with lactic acid and stained for lignin with phloroglucinol (darker areas). The sections were cut about 1 cm from the base of the tendril 5 days after treatment. Figs. 7 and 8: Three quarters of the tendril were removed and the cut surface was covered with lanolin (Fig. 7) or IAA-lanolin 0.5% (Fig. 8). Fig. 9: An intact tendril stimulated by contact.

Fig. 10. As in Figs. 7 - 9, but cut through the coiled region of the middle of the tendril in an early stage of lignification.

Figs. 11 - 13. Tendrils of *P. caerulea* a day after treatment with plain lanolin on both sides (Fig. 11), IAA-lanolin on concave side, lanolin on convex (Fig. 12) and IAA-lanolin on convex side, lanolin on concave (Fig. 13).

It is thus clear that hormones have an important rôle in thigmotropism. The question arises, is only IAA involved or does ethylene play a part? We reported earlier (Reinhold 1967), and Jaffe (private communication) has also observed, that treatment with ethylene brings about coiling. In the present investigation we have obtained very strong coiling of *Passiflora* tendrils treated symmetrically with ethylene gas in a closed container. Two broad possibilities suggest themselves:- Firstly, ethylene treatment, like contact, may cause release of IAA. Or, secondly, ethylene may be produced in the tissues as a result of the exogenous IAA application and may control the response. Even in the latter case, however, it must be assumed that the basipetal transmission of the response is due to auxin. Choice between these alternatives must await further research.

Asymmetry of the response

Next comes the question of the asymmetric nature of the reaction. Coiling has been shown by a number of careful workers (see Bünning 1959) to be the result of unequal growth on the two sides of the tendril. Furthermore, Jaffe and Galston (1966; 1968) have suggested that contraction of the concave side occurs during the initial stage, mediated by a contractile protein. Our anatomical observations have shown up another asymmetric aspect of the response. This asymmetry is in the appearance of lignin, which occurs sooner on the concave side in tendrils stimulated by either contact or auxin (Fig. 10).

The unequal response of the tendril does not depend on the activity of the tip alone (which in the tendrils we have worked with is manifestly asymmetrical). If the distal half of a tendril is removed and IAA-lanolin (0.5%) is applied homogeneously to the cut surface, the base responds by coiling and unequal lignification. The question thus raised is, does some factor become unequally distributed (cf. Borgström 1939) or is there a "built-in" asymmetrical reaction of the tissues to equal amounts of auxin (cf. Thimann and Schneider 1938)? The following findings favour the second explanation (though not necessarily to the exclusion of the first). Symmetrical treatment with IAA in solution causes coiling. *Passiflora* tendrils also coil regardless of whether IAA-lanolin is applied to their convex or concave surface (See Figs. 11-13). Symmetrical treatment with the gases CO_2 or ethylene also produces rapid coiling. Tendrils dropped into liquid air coil somewhat. This latter reaction clearly cannot involve lateral transport of a stimulating factor. (It might reflect an asymmetric distribution of a contractile protein, cf. Jaffe and Galston, 1966).

Electrical potentials from tendrils

Our observations with regard to coiling, lignification and development of pads on *Parthenocissus* tendrils have provided strong support for a rise in auxin concentration as part of the thigmotropic response. A question of prime importance is, how can a touch on the external surface of an organ give rise to an increase in auxin concentration within the tissue? Our earlier suggestion (Reinhold 1967), based on known properties of mechanoreceptors in the animal kingdom, was that contact stimulus gave rise in the first instance to an action potential. Umrath (1934) has in fact reported the development of such a potential in tendrils after contact, but his work, carried out over 30 years ago, does not seem to have been confirmed. We have been reinvestigating the question. We measure the potential difference between 2 stainless steel microelectrodes sharpened by electrolysis and coated except for their tips with insulating material. The electrodes are inserted into the tendril about 1 cm apart and are connected by shielded wires to a Tetronix 502A oscilloscope (D.C.). We have been very interested to observe the occurrence in these organs of bursts of spontaneous fluctuations in potential. Umrath (1934) also noted briefly that there were spontaneous potentials in tendrils, referring to them as a hindrance in his work. We have observed that there is a rhythm in their production - a burst during which sharp peaks are produced at a rate of approximately 12 per sec. (Fig. 6) is followed by a period of quiet (of variable duration but of the order of 5-10 min) which in its turn is followed by a further burst of potentials. The average size of these potentials is about 0.5 mv, though on occasions they have been nearly 10 times as great. Isolated peaks of 2-3 mv have also occasionally been observed in the "quiet" periods. We are continuing our investigation of the spontaneous fluctuations. We have observed similar bursts of potentials when we inserted electrodes about 1 cm from the tip of tendril-bearing stems (*Cucumis sativus, Passiflora edulis*). We have not observed them in stems of non-climbing plants, or in other plant tissues we have tested. The rhythmic nature of this phenomenon is intriguing and we should like to hazard a preliminary speculation - that the potentials are connected with the rhythmic movements of circumnutation.

We have not so far been able to confirm Umrath's (1934) observation of a potential produced as the result of a contact stimulus. The question of the first reaction to touch thus remains open.

III. SUMMARY

Treatment with IAA can substitute for the contact stimulus not only in inducing coiling of tendrils but in bringing about the development of adhesive pads at the tips of tendrils of *Parthenocissus* spp. Further, auxin imitates contact in inducing lignification of the xylem in tendrils of *Passiflora caerulea*. This lignification was shown to be a direct consequence of auxin treatment and does not require coiling. These results point to an increase in endogenous auxin concentration as a response to contact. Symmetrical treatment with ethylene gas or with CO_2 also induces coiling of tendrils.

The asymmetry of the tendril response has been shown to include lignification, which occurs sooner on the concave side in tendrils stimulated by either contact or auxin. The unequal response does not depend on the presence of the tip. Evidence is adduced for an innate difference between the two sides of the tendril which react differently to the same stimulus.

Bursts of spontaneous fluctuations in electric potential have been observed in tendrils of *Cucumis* and *Passiflora*, alternating with "quiet" periods. During the bursts the sharp peaks (each approx. 10 millisecs in duration) are produced at the rate of approx 12 per sec. It is tentatively suggested that these fluctuations are related to circumnutation.

IV. ACKNOWLEDGEMENTS

We should like to thank Baruch Minka and Aryeh Gilay for help with the potential measurements.

V. LIST OF REFERENCES

BORGSTRÖM, G. (1939). "The transverse reactions of plants." Williams and Norgate, London.

BÜNNING, E. (1959). Die thigmonastischen und thigmotropischen Reaktionen. Encyclopedia Pl. Physiol., W. Ruhland, Ed. vol 17/1 p. 254.

CAMUS, G. (1949). Recherches sur le rôle des bourgeons dans les phénomènes de morphogénèse. Revue Cytol. Biol. vég. 11, 1-199.

DARWIN, C. (1884). "The movements and habits of climbing plants". D. Appleton and Co., New York.

JACOBS, W.P. (1952). The role of auxin in the differentiation of xylem round a wound. Am. J. Bot. 39, 301-309.

JAFFE, M.J. and A.W. GALSTON (1966). Physiological studies of pea tendrils. II. The role of ATP and light in contact coiling. Plant Physiol. 41, 1152-1158.

JAFFE, M.J. and A.W. GALSTON (1966). The physiology of tendrils. Ann. Rev. Pl. Physiol. 19, 417-434.

REINHOLD, L. (1967). Induction of coiling in tendrils by auxin and carbon dioxide. Science 158, 791-793.

THIMANN, K.V. and C.L. SCHNEIDER (1938). Differential growth in plant tissues. Amer. J. Bot. 25, 627-641.

UMRATH, K. (1934). Über die elektrischen Erscheinungen bei thigmischer Reizung der Ranken von *Cucumis pepo*. Planta, Berlin, 23, 47-50.

WANGERMANN, E. (1967). The effect of the leaf on the differentiation of primary xylem in the internode of *Coleus blumei* Benth. New Phytol. 66, 747-754.

WAREING, P.F. (1958). Interaction between IAA and GA in cambial activity. Nature (London) 181, 1744-1745.

WETMORE, R.H. and J.P. RIER (1963). Experimental induction of vascular tissues in callus of angiosperms. Am. J. Bot. 50, 418-430.

PLANT GROWTH SUBSTANCES, 1970

Participation of the Golgi Apparatus in Geotropism

J. Shen-Miller

Division of Biological and Medical Research, Argonne National Laboratory, Argonne, Illinois, 60439, U.S.A.

I. INTRODUCTION

Geotropism results from differences in cell elongation, differences in cell wall extension and/or accretion, in organs stimulated when they have been lying across the direction of an accelerative force. Organelles that may be associated with the growth of the cell wall are the endoplasmic reticulum, the Golgi apparatus or dictyosomes and their associated vesicles.

Work of Pickett-Heaps (1967), Barton (1968), Brown (1969, 1969), Spink (1968), and Ray (1969) showed that material is added to the certain types of cell wall from vesicles that, in some cells, appear to be derived directly from the Golgi apparatus. The rate of Golgi vesicle production (3 vesicles/min) and the time (2.5 min) required for a vesicle to traverse the cytoplasm to the cell membrane and wall (Mollenhauer & Morré, 1966) are compatible with the appearance of visible curvature in geotropism.

What role could the Golgi apparatus play in plant geotropism? It could have the role of a sensor. Griffiths and Audus (1965) and later Sievers (1967) noted a displacement of the Golgi apparatus upon geostimulation of the roots of *Vicia faba* and the rhizoids of *Chara foetida*, respectively. Griffiths and Audus found the displacement of the Golgi to be small but significant, but they consider it as a passive movement resulting from starch sedimentation. We will present data on this point, data that make the Golgi apparatus a contender for the role of a geosensor.

Alternatively, the function of the Golgi apparatus could be in the physiological implementation of geotropism. Geotropic stimulation results in a change of auxin distribution within an organ. More auxin is found in the lower half of an organ placed horizontally. The increased auxin concentration in that tissue could bring about an activation of the Golgi apparatus. Such an activation is supported by the work of Siegesmund (1960) with tobacco cells in culture; more Golgi vesicles were formed in those cells when auxin was added to the medium.

Activation of the Golgi is manifest as an increased production of vesicles. The increased vesicle production could, in turn, enhance the rate of auxin transport. Recent work by Hertel and Flory (1968), Rayle, *et al.* (1969) showed that pretreatment of tissues with auxin increases the subsequent transport of auxin. Thus a positive feedback interaction between Golgi-vesicle production and the rate of auxin transport does not seem unreasonable.

II. MATERIALS AND METHODS

With the above considerations in mind, we undertook to examine the changes in the Golgi apparatus in oat coleoptiles after various periods of gravity stimulation. The maximum length of stimulation was 60 min; samples were collected every 6 min for the tip mm and the 5th mm of the coleoptile. This was done to check if both Golgi distribution and activation can be correlated kinetically with the geotropic response.

The segments were further divided longitudinally into upper and lower halves (with respect to gravity). The halves were fixed either in KMnO4 or in glutaraldehyde solution. (We thank Dr. H.H. Mollenhauer of Kettering Institute, Yellow Springs, Ohio, and Dr. E. Vigil of the University of Chicago for their respective fixation schedules.) The tissues were blind-coded to minimize bias. We scored the number of Golgi apparatus in the top and bottom (with respect to gravity) half-cells in the upper and lower tissues of a coleoptile segment. The number of Golgi expressed in the figures are based on means per cell section, not on means per entire cell. The activity of the Golgi, as indexed by their vesicle production and their size, was arbitrarily classified into active, intermediate and inactive forms (Fig. 1). There are a few vesicles produced by the inactive, some by the intermediate, and many by the active. An average of 25 and 30 cell-sections at each sampling time was scored for glutaraldehyde and KMnO4-fixed tissues, respectively. The work presented here represents data of 2000 cells.

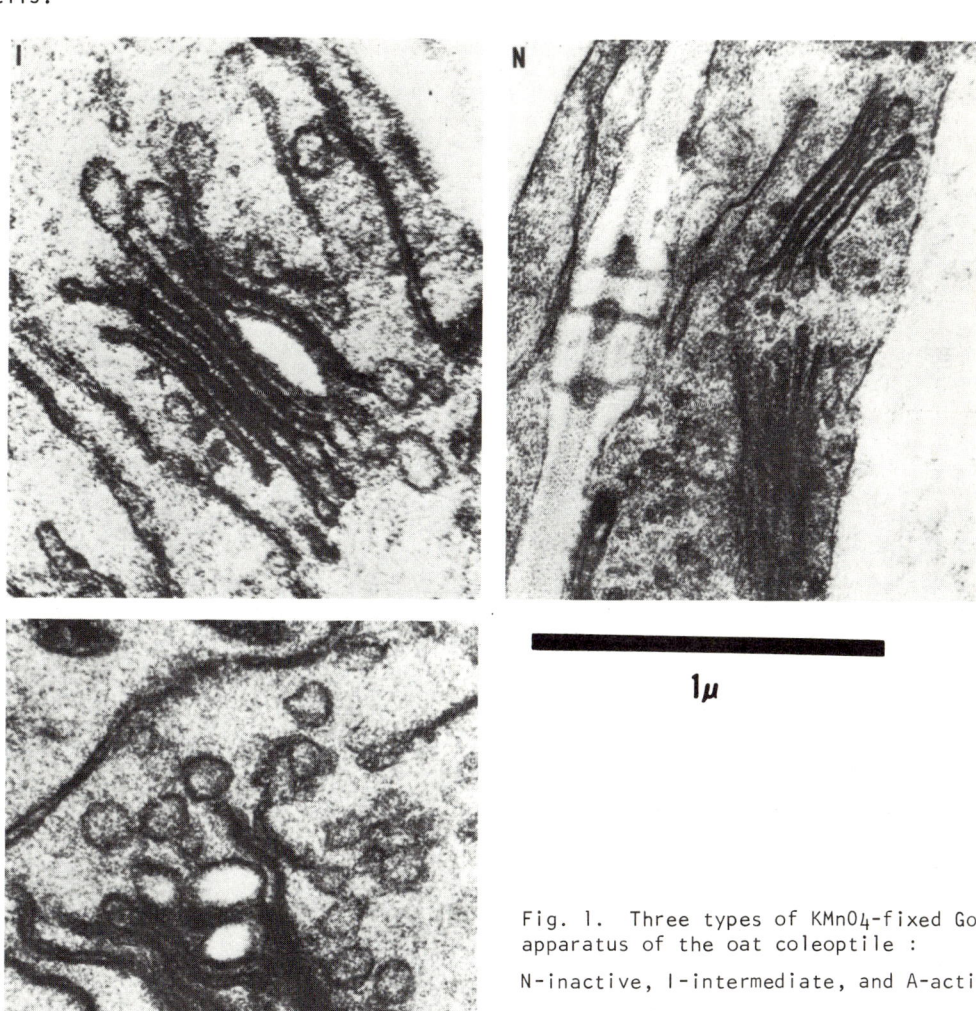

Fig. 1. Three types of KMnO4-fixed Golgi apparatus of the oat coleoptile:

N-inactive, I-intermediate, and A-active.

III. RESULTS AND DISCUSSION

The activity and the localization of activity of the Golgi apparatus are correlated with the direction of the gravitational field.

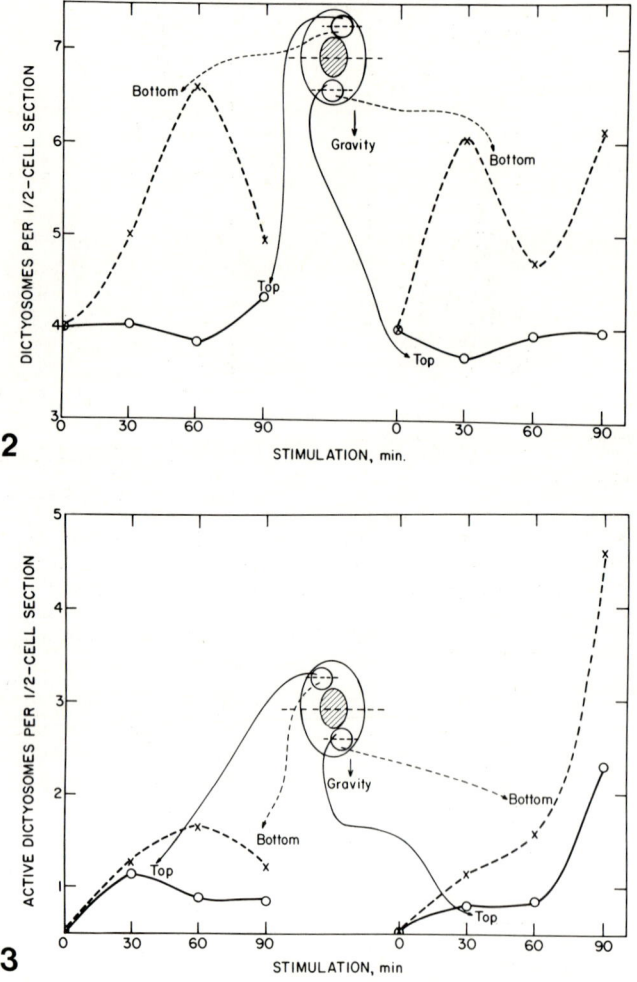

Fig. 2. The effect of gravity stimulation on the distribution of dictyosomes in the top and bottom half-cells in the upper and lower tissues of oat coleoptile tips.

Fig. 3. The effect of gravity stimulation on the distribution of active dictyosomes in the top and bottom half-cells in the upper and lower tissues of oat coleoptile tips.

Figure 2 shows that the total number of Golgi is greater in the bottom half of a cell. For the active form of Golgi, this difference is more striking in the lower tissues than in the upper (Fig. 3). Figure 4 shows the time course of total Golgi distribution in the tip and the 5th-mm segments of the $KMnO_4$ and glutaraldehyde fixed tissues. We were not able to account for as many Golgi in glutaraldehyde fixed material as in the $KMnO_4$-fixed cells. In the 5th-mm segment (Fig. 4A), the difference in the total number of Golgi between the top (T) and bottom (B) half-cell, in both the upper (U) and lower (L) organ halves is significantly different at the 1% level. Figure 5 shows a similar trend for the active form of Golgi. Again in the 5th-mm segment

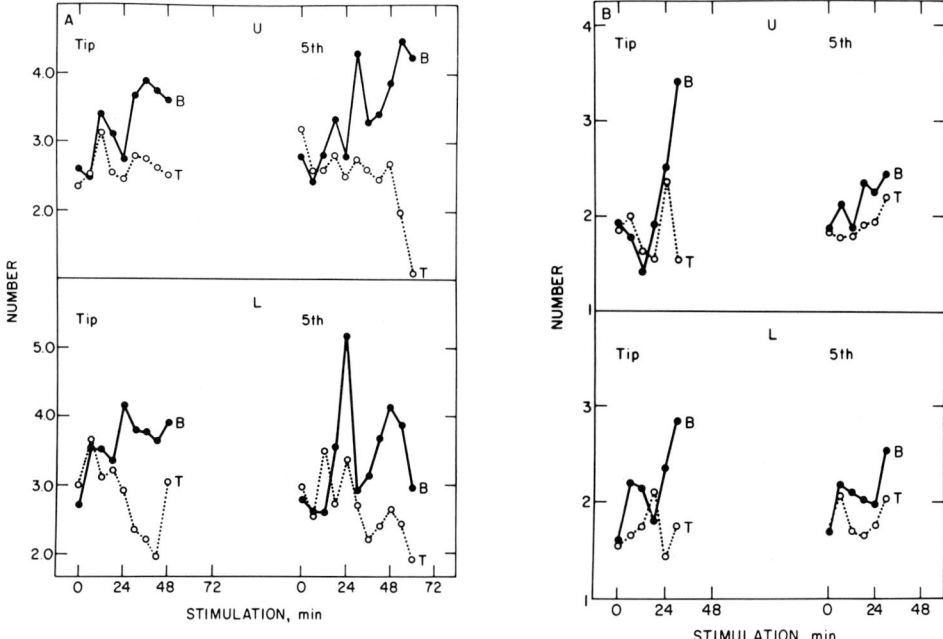

Fig. 4. Kinetics of Golgi distribution in the top (T) and bottom (B) half-cells in the upper (U) and lower (L) tissues of geotropically stimulated oat coleoptile tips and the 5th-mm segments; A.KMnO4-fixed tissues; B.Glutaraldehyde-fixed tissues.

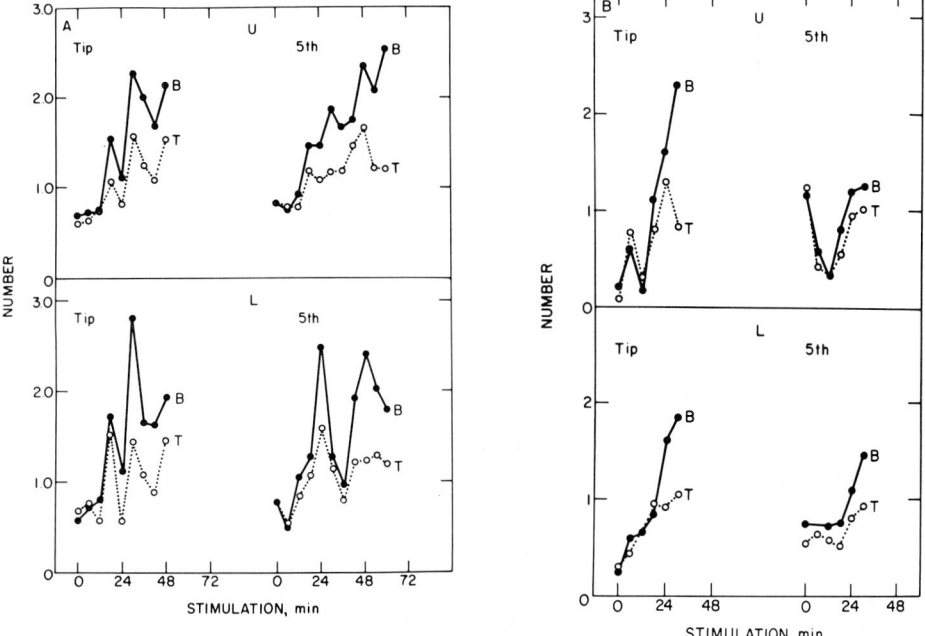

Fig. 5. Kinetics of "active" (intermediate plus active) Golgi distribution in the top (T) and bottom (B) half-cells in the upper (U) and lower (L) tissues of geotropically stimulated oat coleoptile tips and 5th-mm segments; A.KMnO4-fixed tissues; B.Glutaraldehyde-fixed tissues.

(Fig. 5A) the bottom half-cell has significantly (at the 1% level) more active Golgi than the top half-cell. The data for the glutaraldehyde-fixed tissues and for the tips of KMnO4-fixed tissues are as yet not complete; therefore they were not analyzed. Regardless of the fixative used, the number of active Golgi in the bottom half-cell is consistently higher than in the top half. If Golgi apparatus is involved in tropic curvature, its activity change should precede the occurrence of actual curvature. The difference between the two halves, both for the total and active form of Golgi, begins at about 12 min from the initiation of stimulation. In geotropic curvature of corn, we find the appearance of visible geotropism occurring at about 30 min from the start of stimulation (unpublished data).

The significant difference in the total Golgi between the top and bottom half-cells indicates that Golgi apparatus could be the sensor for gravity in the classical sense. The increase in Golgi number in the bottom half-cells could be a result of sedimentation or an increased production in that region of the cell. The present data do not allow us to discriminate between these two possibilities. However, my present opinion is against the likelihood that the Golgi apparatus is acting at the sensor level in geotropic perception. The changes in the Golgi activity and Golgi number do not occur immediately upon stimulation, though they are found within 12 min after the beginning of stimulation. Differences in rates of auxin transport between the upper and lower tissues appear more rapidly than the changes in the Golgi number of activity (Shen-Miller, 1972).

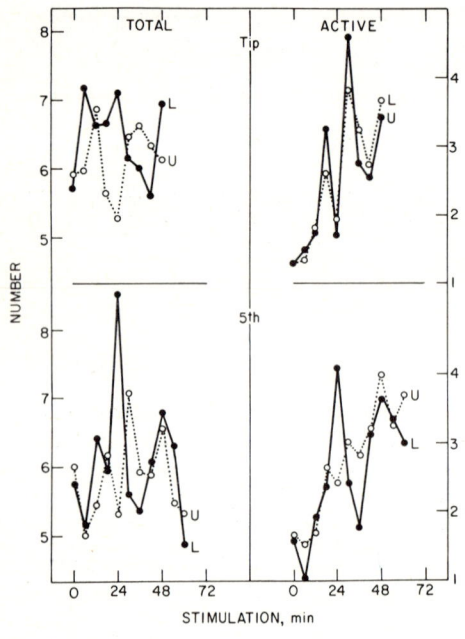

Fig. 6. Distribution of Golgi apparatus in the cells of the upper (U) and lower (L) tissues (KMnO4-fixed) of geotropically stimulated oat coleoptile tips and 5th-mm segments.

On the whole cell basis, the total number of Golgi and the active form of Golgi change significantly (at the 5% level) during geotropic stimulation (Fig. 6). During the 60 min stimulation, there are 2 peaks of increase in Golgi number. In the lower tissues of the 5th-mm, these peaks occur at 24 and 48 min of stimulation. In the tip, the peaks precede those in the 5th mm, and the interval between peaks in the tip is shorter (12-18 min) than that in the 5th mm (Fig. 6). This indicates that change in the Golgi number and Golgi activity occur first in the tip and then in the 5th-mm segment of the coleoptile. It implies also that either geosensing is solely in the tip, or that the tip cells are more sensitive than the cells in the 5th-mm in geotropic perception; or if Golgi activation is a consequence of hormonal changes, the change in hormone occurs first in the tip and then in the 5th-mm segment.

The peak occurrences of Golgi in the upper and lower organ halves are not in phase, as shown by the data on the total Golgi (Fig. 6). This and the periodic increase in Golgi number and activity could be correlates of the oscillatory response of the coleoptile in geotropism. The 24 min periodicity in peak occurrence of the Golgi apparatus in oat coleoptiles is compatible with the 18 min periodicity of auxin transport, geotropism, and growth in the corn coleoptile (Shen-Miller, 1972).

IV. SUMMARY

In the geotropically-stimulated oat coleoptile, in both the upper and lower halves, the lower half of a cell has not only a greater total number of Golgi apparatus but also a greater number of Golgi in the active form. The morphological changes of this organelle from the inactive to the more active form are compatible with the time constants for the occurrence of tropic curvature. However, the lag before these changes occur after stimulation appears to be too long to enable assignment of a *geosensor* role to the Golgi.

The total number and activity of Golgi apparatus oscillate after geotropic stimulation. In the 5th-mm coleoptile segment, there are 2 peaks in activity and in number: at 24 and again at 48 min from the start of stimulation. The interval between peaks is shorter in the tip tissue. A periodicity of 24 min in the changes of Golgi apparatus in oat coleoptile is analogous to the 18 min periodicity of geotropism, growth and auxin transport in the corn coleoptile.

V. ACKNOWLEDGEMENTS

I thank Carol Miller and Ray Hinchman for their technical assistance, and S.A. Gordon for the support of this program under the auspices of the U.S. Atomic Energy Commission and the National Aeronautics and Space Administration.

VI. REFERENCES

BARTON, R. (1968). Autoradiographic studies on wall formation in *Chara*. Planta (Berl) 82, 302-306.
BROWN, R.J., Jr. (1968). Observations on the relationship of the Golgi apparatus to wall formation in the marine chrysophycean algae, *Pleurochrysis scherffelii* Pringsheim. J. Cell Biol. 41, 109-123.
BROWN, R.J., Jr., W.W. FRANKE, H. KLEINIG, H. FALK, and P. SITTE (1969). Cellulosic wall component produced by the Golgi apparatus of *Pleurochrysis scherfellii*. Science 166, 894-896.
GRIFFITHS, H.J., and L.J. AUDUS (1964). Organelle distributions in the statocyte cells of the root-tip of *Vicia faba* in relation to geotropic stimulation. New Phytol. 63, 319-333.
HERTEL, R. and R. FLORY (1968). Auxin movement in corn coleoptiles. Planta (Berl.) 82, 123-144.
MOLLENHAUER, H.H. & D.J. MORRÉ (1966). Golgi apparatus and plant secretion. Ann. Rev. Plant Physiol., 17, 27-46.
PICKETT-HEAPS, J.D. (1967). The use of radioautography for investigating wall secretion in plant cells. Protoplasma 64, 4-66.
RAY, P., T.L. SHININGER and M.M. RAY (1969). Isolation of β-glucan synthetase particles from plant cells and identification with Golgi membranes. Proc. Nat. Acad. Sci., U.S. 64, 605-612.
RAYLE, D.L., R. OUITRAKUL and R. HERTEL (1969). Effect of auxins on the auxin transport system in coleoptiles. Planta (Berl) 87, 49-53.
SHEN-MILLER, J. (1972). Tropic stimuli and the kinetics of basipolar transport of auxin. in "Plant Growth Substances, 1970" Ed. D.J. Carr, Springer Verlag. pp. 710-716.

SIEVERS, A. (1967). Elektronenmikroskopische Untersuchungen zur geotropischen Reaktion. III. Die transversale Polarisierung der Rhizoidspitze von *Chara foetida* nach 5 bis 10 Minuten Horizontallage. Z. Pflanzenphysiologie 57, 462-473.
SIGESMUND, K.A. (1960). Studies of the effects of indoleacetic acid on the fine structure of cultured tobacco parenchyma cells. Ph.D thesis, University of Wisconsin.
SPINK, G.C. (1968). Observations of the Golgi apparatus in *Pisum* primary roots. Proc. Elect. Microscopy Soc. Amer. pp. 14-15.

HORMONES AND FLOWERING

Hormonal Regulation of Plant Flowering in Different Photoperiodic Groups

M.Kh. Chailakhyan

The K.A. Timiryazev Institute of Plant Physiology of the Academy of Sciences of the U.S.S.R., Moscow

I. INTRODUCTION

Hormonal Factors of Flowering

It has been proposed that two groups of substances produced in the leaves enter into the composition of flowering hormones, i.e. the florigen complex: (1) GA, essential for the initiation and growth of flower stems, and (2) substances essential for the flower initiation and provisionally named anthesins (Fig. 1). The absence of flowering of long-day species under short days was found to be due to the lack of GA. Conversely, the retardation of flowering of short-day species under long days results from a deficit of anthesins. (Chailakhyan, 1958).

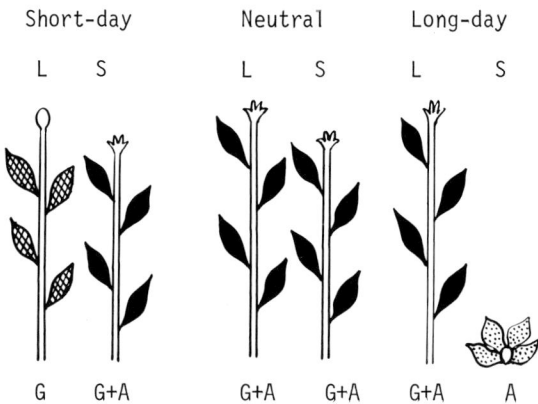

1958. HORMONAL FACTORS OF FLOWERING

▩▩▩ Gibberellins-hormones, necessary for stem formation
▨▨▨ Anthesins - hormones, necessary for flower formation
▬▬▬ Florigen - gibberellins + anthesins, necessary for flowering

Fig. 1. Scheme for the production of flowering hormones in different plant species.

Hormonal regulation of flowering in long-day species is now an incontrovertible fact: extracts of plant leaves as well as GA obtained from *Gibberella* cultures have been shown to initiate stems, flowers and fruits in treated plants. Hormonal regulation of short-day species is still obscure, since anthesins in these plants can still only be detected by biological tests such as (1) the flowering of a short-day species, *Perilla nankinensis*, initiated by metabolites donated from grafted leaves previously detached and induced in the isolated state (Chailakhyan, 1957); (2) coincidence of transmission of the flower inducing principle with the transfer of labelled assimilates from short-day leaves of the same plant to shoot apices (Chailakhyan and Butenko, 1957). The effect of florigenic acid, isolated from flowering cocklebur, causing inflorescence production when supplied to leaves of vegetative plants of the same species (Lincoln, Cunningham and Hamner, 1964; Hodson and Hamner, 1970) is not in itself sufficient from our point of view to consider it as a *hormone* initiating flowers.

Gibberellins and Anthesins of Long-Short Day Species

Our concept of the hormonal regulation of plant flowering is based on the two-component system of hormones in which synthesis of either component proceeds in specific daylengths. This concept is sustained by the behaviour of long-short-day and short-long-day species which have a two-step or two-component photoperiodic response.

In this connection we decided to study the processes of photoperiodism and flowering in long-short-day species of *Bryophyllum daigremontianum* and *Bryophyllum tubiflorum* which flower only after successive exposure to long and then to short days. Our experiments have shown that spraying the plants with a solution of GA completely substituted for exposure to long days and flowering was then initiated under short days. This applied not only to old plants as was shown earlier (Bünsow, and Harder, 1956; Penner, 1960) but also to young plants (Chailakhyan, Yanina and Frolova, 1968). One spray treatment with a solution of GA was found to be sufficient for old plants under short days. A similar treatment with GA substituted also for long days in plants of short-long-day species - *Scabiosa succisa* and *Coreopsis grandiflora* (Chouard, 1957; Kettelapper and Barbaro, 1966). All this suggests that the special metabolites produced in long days, and essential for flowering of long-short-day and short-long-day species, are gibberellins.

Evidence on the nature of the special metabolites produced in short days and essential for flowering of long-short-day species, is not yet available, just as we lack data on the nature of the special metabolites also produced in short days which are essential to the flowering of short-day species. In both cases they are here regarded as hypothetical hormonal substances -anthesins. However, experimental evidence defining the conditions for production and transfer of these metabolites in the plant, is available. In some of our experiments these metabolites have been shown to be synthesized in leaves and to be transported into stem buds inducing in them changes in metabolism and the initiation of flowers. Other experiments have revealed that these metabolites are quite certainly not synthesized in roots, since plants deprived of roots and exposed first to long and then to short days, come into flower at the same rate as rooted plants similarly treated (Chailakhyan, Yanina and Frolova, 1970a). Ultimately, sufficient of these metabolites for full induction of flowering was found to be synthesized in leaves during 10-15 short days (Chailakhyan, Yanina and Frolova, 1970b).

Specific Metabolites of Flowering Under Long and Short Days

According to our concept gibberellins, as long day metabolites, and anthesins, as short day metabolites, are independent groups of substances and not precursors of each other in long-short-day species. This viewpoint is shared by some other authors (Penner, 1960). However, the opposite point of view has also been expressed: that is, that the formation of the specific metabolites of short days is dependent on the previous formation of GA under long days; moreover, according to some authors, gibberellins are supposed to be direct precursors of florigen (Zeevaart and Lang, 1962; Zeevaart, 1969). This idea adequately explains the responses of long-short-day species, where long day metabolites are formed first, followed by those of short days, but is inadequate for short-long-day species, with a necessary inverse order of metabolite formation.

In order to elucidate this problem we carried out grafting experiments with *Bryophyllum daigremontianum* in the greenhouse of the Institute of Plant Physiology in 1969 and 1970.

II. EXPERIMENTAL

In the grafting experiments of 1969 and 1970 each component of the graft continued to be treated to the daylength under which it had been raised. Thus neither component came from an induced plant or was subjected to an inductive photoperiodic regime. Apices from long-day scions were grafted onto short-day stocks (L/S) or *vice versa* (S/L). Control grafts were also made: long-day scion-apices on long-day stocks (L/L) and short-day scion-apices on short-day stocks (S/S); corresponding non-grafted plants were also kept as controls. On the stocks, 4-6 large leaves were left and all new shoots were removed; the scions were deprived of all but some small leaves, surrounding the apical bud, but later on they formed and expanded new leaves. In (S/L) grafted plants the ratio between leaf areas of scions and stocks was regulated by removing leaves during the course of the experiment; in (L/S) grafted plants the ratio between leaves changed in such a way that scion leaves grew fast and reached large sizes, while stocks leaves aged earlier, turned yellow and fell. Therefore unequivocal experimental evidence could be obtained only with grafted plants S/L and control plants L/L and S/S.

Three types of treatment were applied to groups of grafted plants. In the first group, no decapitation of the scion was carried out after grafting; scions exhibited fast growth, producing new leaves, and the apical bud grew rapidly away from the site of grafting; some leaves on the scions were removed but this affected scion growth only slightly. The result of this treatment was that the scion-apices of grafted plants S/L remained in the same vegetative state as scions of control plants L/L and S/S.

In the second group, the scions were decapitated two months after grafting, resulting in the outgrowth of side shoots. These shoots were usually located close to the site of grafting and were therefore influenced by metabolites from both the long-day treated leaves of the stocks and the short-day treated leaves of the scions. As a result, on S/L plants, the side shoots of scions kept under short-days produced flower buds 4.5 months after grafting, and open flowers one month later.

In the third group, that with most variation, apices of short-day scions were grafted on long-day stocks, as in the first and second groups. The decapitation of the scions was followed by the growth of side shoots on both the short-day scions and, later on, on the long-day stocks. The scion shoots were removed while the shoots on the stocks were left. As a result all the grafted plants had stocks with leaves maintained under long days, scions with leaves kept under short days and shoots, between these sets of leaves, kept under long days. These shoots were thus influenced by both short day metabolites and long day metabolites and produced buds 5 months after grafting and open flowers one month later (Fig. 2).

The same results were obtained in experiments with grafted leaves. If plants had 2 shoots and 4 leaves and were exposed entirely to either long or short days, no flowering was recorded. If plants had two shoots and two lower leaves kept under long days, and two grafted leaves situated higher than the shoots kept under short days, then flower bud formation and flowering were observed (Fig. 3).

These experiments unambiguously support the view that substances made specifically under long days and substances made specifically under short days are equally essential for the flowering of a long-short-day plant. They confirm the concept of a separate contribution by both GA and anthesins to the formation of a complex of flowering hormones in the different photoperiodic groups.

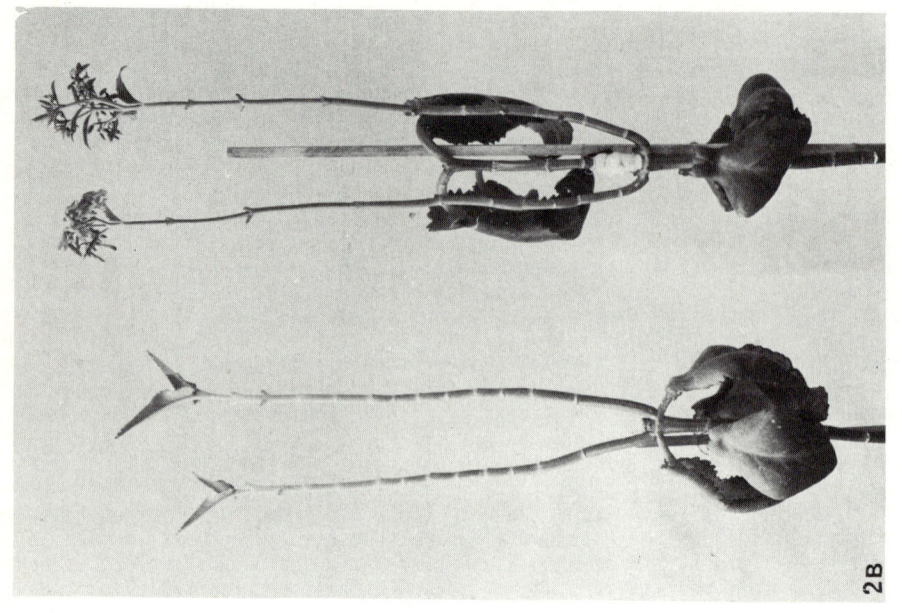
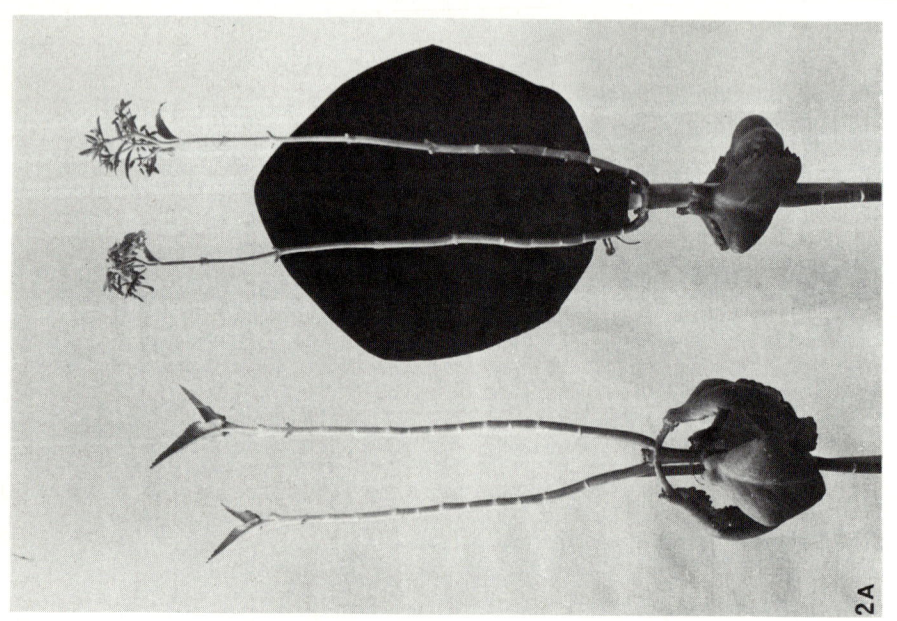

Fig. 2. The effect of a scion, exposed throughout to short days, on the flowering of *Bryophyllum* shoots growing on the stock, kept constantly under long days. Control plant (left) without a short day scion. Grafted plant (right). Fig. 2A. Scion is covered with a cloth bag to provide short days. Fig. 2B. Short day cover removed from scion.

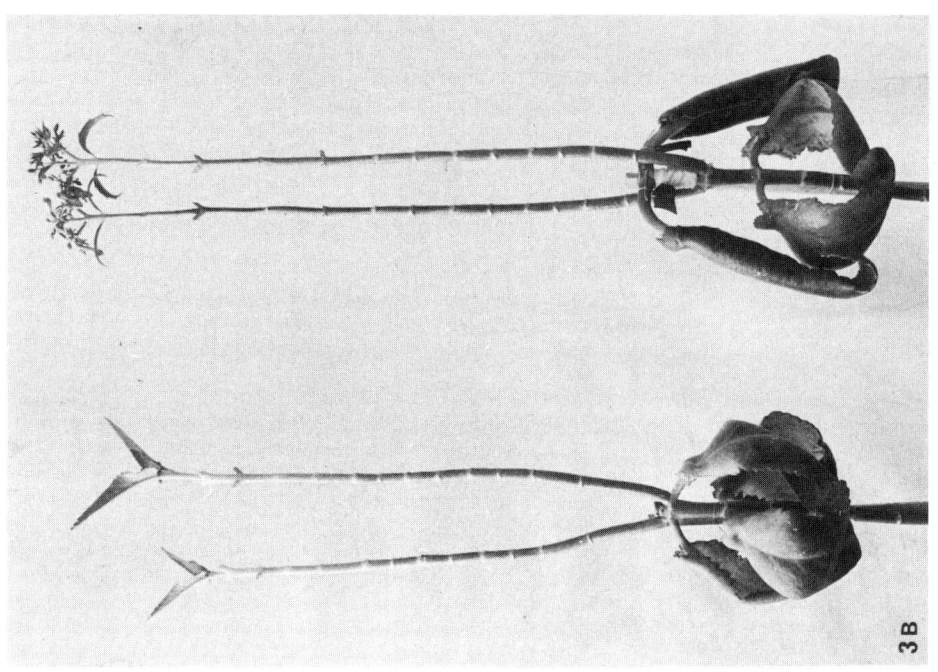

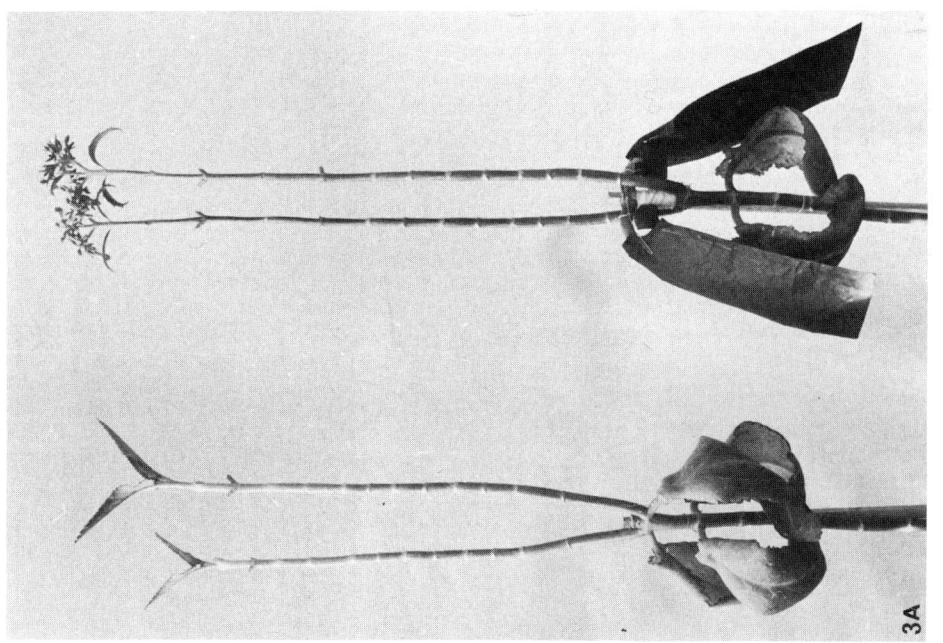

Fig. 3. The effect of grafted leaves, constantly exposed to short days, on the flowering of *Bryophyllum* shoots growing on a stock kept constantly under long days. Control plant (left) (no short day leaves). Grafted plant (right). Fig. 3A. Leaves are covered with short day covers. Fig. 3B. Leaves without covers.

III. INHIBITORS OF FLOWERING OF LONG-DAY AND SHORT-DAY SPECIES

In our early work on the hormonal control of flowering we established that leaves exposed to favourable daylength conditions, formed hormones which stimulated flowering, whereas leaves under unfavourable daylengths retarded flowering. Now, flowering hormones in long-day species are represented by GA and in short-day species by anthesins; therefore inhibitors of flowering should also discriminate between long-day species and short-day species. *A priori*, they should be substances antagonistic to the corresponding hormones.

This view is completely confirmed by experimental evidence. The growth of stems and the formation of flowers in long-day species are inhibited by some natural products, e.g. abscisic acid (ABA), as well as by synthetic compounds: chlorcholine chloride (CCC), retardant B-9, morphactin, maleic acid hydrazide and tri-iodobenzoic acid. These substances are presumed to be flowering inhibitors for long-day plant species irrespective of whether they are specific anti-gibberellins or antiauxins.

Substances known to inhibit flowering or retard floral initiation in short-day species are antimetabolites of nucleic acid metabolism: 2-thiouracil, 5-fluorouracil, 8-azaguanine etc., or inhibitors of steroid metabolism. These compounds are not specific anti-anthesins as may be some as yet unidentified compounds which are formed in leaves in long days, to which we have recently again turned our attention.

Therefore, the regulation of flowering should be regarded in long-day plant species as a process in which there is a balance between the effects of GA and those of antagonistic endogenous inhibitors, and similarly, in short-day species between anthesin-like substances and their endogenous antagonistic inhibitors.

IV. DISCUSSION

Photoperiodic and Autonomous Regulation of Flowering by Hormonal Factors

A study of flowering phases in plants belonging to different photoperiodic groups has shown that the processes involved in flowering as a main criterion of photoperiodism are entirely different for long-day and short-day species. In long-day species, flowering is reflected in the formation of flower-bearing stems, a process which is related to the production and action of GA; in short-day species flowering is reflected by flower initiation itself which is related to the synthesis of substances of the anthesin type.

Genetic information determining flowering in photoperiodically neutral species is realised directly through an internal hereditary program which proceeds regardless of daylength. The hereditary control of flowering, independent of daylength, can be called autonomous regulation. Genetic information determining the process of flowering in species adapted to daylength is also expressed through an internal hereditary program but its direct realization is put under the control of the daylength. This genetic control of flowering related to daylength can be called photoperiodic regulation (Chailakhyan, 1970).

Thus, the ability to synthesize flowering hormones, whether controlled by daylength or not, is not the same in plants belonging to different photoperiodic groups. If the production of GA and anthesins by day-neutral plants is entirely under autonomous control, then the regulation of the synthesis of both types of hormone by long- and short-day species is more strictly under photoperiodic control. Ultimately, the production of flowering hormones by long-short-day and short-long-day species is completely under photoperiodic regulation (Fig. 4).

1970. FLOWERING HORMONES - PHOTOPERIODIC AND AUTONOMIC REGULATION

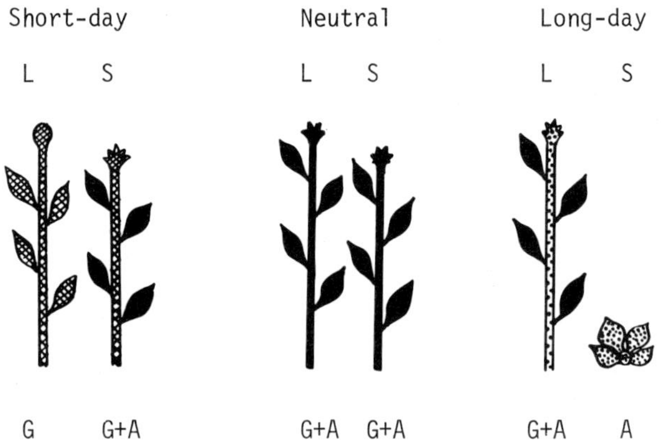

```
    Short-day        Neutral         Long-day
    L      S        L      S        L      S




    G     G+A      G+A    G+A      G+A     A
```

▓▓ Gibberellins-hormones, necessary for stem formation

░░ Anthesins - hormones, necessary for flower formation

■■ Florigen - gibberellins + anthesins, necessary for flowering

Fig. 4. Scheme illustrating autonomous and photoperiodic regulation of flowering in plants belonging to different photoperiodic groups.

V. CONCLUSION

During the last few years it has become evident that there is a plurality of factors affecting plant flowering. These conclusions make it necessary to abandon the simple concepts of the initial florigen hypothesis and to formulate new and more complex hypotheses to meet the challenges arising from the rapid expansion of work in this field (Carr, 1967; Evans, 1969).

We believe that the new experimental evidence and theoretical considerations presented in this paper constitute further progress in the direction of a tenable hypothesis of the hormonal regulation of flowering.

VI. SUMMARY

Subsequent to the discovery of GA we were forced to revise our concept of the nature of florigen, and put forward a new hypothesis according to which florigen is a complex derived from two groups of substances synthesized in leaves: GA and anthesins (1958).

Experiments show that in long-short-day species, the compounds formed in long days correspond to GA. The compounds formed in short days are hypothetical hormones, anthesins, the existence of which can be shown by biological tests. Grafting experiments with *Bryophyllum daigremontianum* show that a supply of both specific long day metabolites and specific short day metabolites is essential for flowering of this long-short-day plant.

Flowering in long-day plant species seems to be regulated by a balance between the effects of GA and of antagonistic inhibitors, and in short-day species between the effects of anthesins and their antagonistic inhibitors.

Photoperiodic regulation is accomplished through control of the production and balance of flowering hormones in leaves as receptor organs. Autonomous control is realized through daylength-independent production of, and changes in, these hormones in stems, buds and leaves.

VII. REFERENCES

BÜNSOW, R. und HARDER, R. (1956). Blütenbildung von *Bryophyllum* durch Gibberellin. Naturwiss. 43, 479-480.

CARR, D.J. (1967). The relationship between florigen and the flower hormones. Annals New York Academy of Sciences 144, 305-312.

CHAILAKHYAN M.Kh. (1957). Photoperiodic sensitivity of excised plant leaves. Dokl. Akad. Nauk SSSR, 118, 197-200.

CHAILAKHYAN M.Kh. (1958). Hormonale Faktoren des Pflanzenblühens. Biol. Zbl. 77, 641-662; Fiziol. Rast., 5, 541-560.

CHAILAKHYAN M.Kh. (1970). Flowering and Photoperiodism of Plants. Usp. Sovrem. Biol., 69, 306-318; Plant Science Bulletin, 16, 1-7.

CHAILAKHYAN M.Kh. and BUTENKO, R.G. (1957). Movement of assimilates of leaves to shoots under differential photoperiodic conditions of leaves. Fiziol. Rast. 4, 450-462.

CHAILAKHYAN M.Kh., YANINA, L.I., FROLOVA, I.A. (1968). Influence of length of day and gibberellin on flowering in *Bryophyllum* of different ages. Dokl. Akad. Nauk. SSSR. 183, 230.

CHAILAKHYAN M.Kh., YANINA, L.I., FROLOVA, I.A. (1970-a). Photoperiodic and chemical regulation of flowering of long-short day species. Fiziol. Rast. 17, 358-370.

CHAILAKHYAN M.Kh., YANINA, L.I., FROLOVA, I.A. (1970-b). Flowering of *Bryophyllum* plants lacking roots. Fiziol. Rast. 17, 709-711.

CHOUARD, P. (1957). La journée courte ou l'acide gibbérellique comme succédanes du froid pour la vernalization d'une plante vivace en rosette, le *Scabiosa succisa*. Compt. rend. Acad. Sci., Paris, 245, 5520-5522.

EVANS, L.T. (1969). The Nature of Flower Induction. In: "The Induction of Flowering: some case histories" (Ed. L.T. Evans) Cornell University Press. Ithaca, New York, 457-480.

HODSON, H.K. and HAMNER, K.C. (1970). Floral Inducing Extract from *Xanthium*. Science. 167, 384-385.

KETTELAPPER, H.J. and BARBARO, A. (1966). The role of photoperiod, vernalization and gibberellic acid in floral induction in *Coreopsis grandiflora* Nutt. Phyton, 23, (1) 33-41.

LANG, A. (1965). Physiology of flower initiation. Encyclopedia of Plant Physiol. 15/1, 1380-1536.

LINCOLN, R.G., CUNNINGHAM and HAMNER, K.C. (1964). Evidence for a florigenic acid. Nature, 202, 559-561.

PENNER, J. (1960). Über den Einfluss von Gibberellin auf die photoperiodisch bedingten Blühvorgänge bei *Bryophyllum*. Planta 55, 542.

ZEEVAART, J.A.D. (1969). *Bryophyllum*. In "The Induction of Flowering: Some Case Histories" (Ed. L.T. Evans) 435-456.

ZEEVAART, J.A.D. and LANG, A. (1962). The relationship between Gibberellin and floral stimulus in *Bryophyllum daigremontianum*. Planta, 58, 531.

PLANT GROWTH SUBSTANCES, 1970

THE USE OF APHIDS IN THE SEARCH FOR THE
HORMONAL FACTORS CONTROLLING FLOWERING

Charles F. Cleland

The Biological Laboratories, Harvard University, Cambridge, Massachusetts

I. INTRODUCTION

The chemical basis for the hormonal control of flowering has been the subject of numerous investigations in which a wide variety of extraction procedures has been used in efforts to isolate a chemical substance (or substances) that could induce flowering in test plants kept under non-photoinductive conditions (Bonner and Bonner, 1948; Hamner and Bonner, 1938; Lincoln et al., 1964). However, except for a recent encouraging report (Hodson and Hamner, 1970) these studies have generally met with little or no success. Consequently, a different approach has been employed in the present study.

Available evidence indicates that the flowering stimulus moves from the photo-induced leaf to the receiving buds in the phloem (Lang, 1965). Aphids feed on materials moving in the phloem by inserting their stylet into the plant tissue until it terminates in an active sieve element (Zimmerman, 1961). Furthermore, actively feeding aphids excrete fairly large quantities of honeydew which qualitatively is fairly similar to phloem sap but quantitatively shows a reduction in certain substances, primarily amino-acids (Mittler, 1958). Recent studies have shown that ABA, GA and IAA could pass through aphids and appear in the honeydew without any apparent loss in biological activity (Bowen and Hoad, 1968; Hoad and Bowen, 1968; Maxwell and Painter, 1962). By analogy it was hoped that the chemical substances responsible for the control of flowering would behave in a similar manner and thus make it possible to isolate them from aphid honeydew.

II. MATERIALS AND METHODS

The Chicago strain of the short-day plant *Xanthium strumarium* L. has been used in this study along with a natural population of the aphid *Dactynotus ambrosiae* (Thomas) found on *Xanthium sp*. The aphids are allowed to feed either on vegetative plants kept on continuous light for the production of vegetative honeydew or on flowering plants kept on a 13L:11D regime for the production of flowering honeydew. The honeydew which they release is collected on glass plates placed around the base of the plants. After one to several weeks the honeydew is removed from the plates, dried in a desiccator and stored at $-20^{\circ}C$.

For extraction the dried honeydew was dissolved in 0.5 M phosphate buffer, pH 8.2. In preliminary experiments the aqueous phase was partitioned against an equal volume of ethyl acetate 4x at pH 8.2 to give a basic fraction (B) and then 4x at pH 2.5 to give an acidic fraction (A). In later experiments it was found that all of the activity could be obtained by partitioning at pH 2.5 and thus the aqueous phase was lowered directly to pH 2.5 and only an acidic fraction was obtained. The basic fraction was evaporated to dryness without further purification. The acidic fraction was further purified by TLC chromatography on silica gel H in the solvent system of chloroform-ethyl acetate-acetic acid (60:40:5, v/v). The TLC plates were allowed to develop for fifteen cm and five 3-cm zones (A-1 to A-5) were scraped off the plate, eluted 3x with 3 ml of water-saturated ethyl acetate and evaporated to dryness.

The long-day plant *Lemna gibba* L., strain G3 or the short-day plant *Lemna perpusilla* Torr., strain 6746 were used as a bioassay. *Lemna gibba* G3 was grown on 20 ml of E medium in 125 ml Erlenmeyer flasks (Cleland and Briggs, 1967), while *L. perpusilla* 6746 was grown on 20 ml of 1/2 Hutner's medium in 25 x 150 mm test tubes (Posner, 1967). The honeydew fractions were dissolved in 3.3 ml of distilled water, gently shaken for 30 min. and then the pH adjusted to between 5 and 7. For each fraction 1 ml was added by sterile filtration to each of 3 flasks of autoclaved medium. For *L. gibba* G3 each flask was inoculated with a single 4-frond colony and grown under previously described conditions (Cleland and Briggs, 1967) for 7 days on the experimental photoperiod followed by 4 days of continuous light. The cultures were then examined for flowering and the flowering per cent (FL%) was determined (Cleland and Briggs, 1967). For *L. perpusilla* 6746 each test tube was inoculated with a single 3-frond colony and grown under the same conditions used for *L. gibba* G3 for 5 days on the experimental photoperiod followed by 2 short days and then counted for flowering.

III. RESULTS

In initial experiments as much as 16 grams dry weight of flowering honeydew was extracted and each of the various fractions applied to 2 vegetative *Xanthium* plants either via the leaves or by use of the stem flap method. However, in 3 separate experiments the results were completely negative. Preliminary attempts to induce flowering in the short-day plant *Pharbitis nil* by applying extracts of flowering honeydew via the roots also gave negative results.

By contrast, when extracts of as little as 2.0 grams dry weight of either flowering or vegetative honeydew were applied to *L. gibba* G3 a striking promotion of flowering was obtained by fractions B and A-4 and in some experiments also by fraction A-1 (Table 1). Most experiments are performed on daylengths slightly longer than the critical daylength of *L. gibba* G3 (about 10 hr.) where the controls show a low flowering percentage because preliminary experiments indicated that the plants are more sensitive under these conditions. However, flowering has also been obtained under a strict short-day regime of 9L:15D indicating that the effect is on flower induction. In addition to the flower-inducing activity there appeared to be some flower-inhibitory activity in fractions A-2 and A-3.

Table I. Influence of honeydew extracts on flowering of *Lemna gibba* G3

Expt.	Photo-period	gram DW starting material	Flowering per cent						
			cont	B	A-1	A-2	A-3	A-4	A-5
		Vegetative honeydew							
1	9L:15D	1.5	0	--	0	0	0	3	0
2	10L:14D	2.0	8	34	14	0	0	40	4
		Flowering honeydew							
1	9L:15D	5.0	0	20	--	0	0	18	3
2	11L:13D	2.0	8	--	52	0	1	55	0
3	12L:12D	5.0	52	74	--	0	0	68	1

The flower-inducing and flower-inhibitory activities in the honeydew extracts are clearly due to substances in the honeydew since when buffer alone was extracted all fractions were only slightly inhibitory to flowering as compared to the controls. Furthermore, in preliminary experiments, extracts of honeydew collected from aphids feeding on a chemically defined synthetic diet (Dadd et al., 1967) were without activity. Therefore, it seems clear that the flower-inducing and flower-inhibitory activities are coming from the Xanthium plants and are not being produced by the aphids themselves.

Table II. Promotion of flowering in Lemna perpusilla 6746 by extracts of flowering honeydew

Expt.	Photo-period	gram DW starting material	Flowering per cent					
			cont	A-1	A-2	A-3	A-4	A-5
1	12L:12D	1.5	11	48	39	19	24	5
2	13L:11D	1.2	1	35	6	2	4	0

The honeydew extracts have also been tested on L. perpusilla 6746 and some promotion of flowering was obtained (Table II). However, the main effect was with fractions A-1 and A-2, the other fractions having little or no effect. The basis for the different response of L. gibba G3 and L. perpusilla 6746 to fractions A-2 and A-4 remains to be resolved.

Attempts have been made to determine whether there are consistent differences in the level of flower-inducing and flower-inhibitory activities in vegetative and flowering honeydew. The results of one such experiment are given in Table III and although the flower-inducing activity in fraction A-4 is slightly higher in the flowering honeydew and the flower-inhibitory activity in fraction A-2 is slightly higher in the vegetative honeydew, the differences in fractions A-1 and A-3 are just the reverse. In other experiments the results have occasionally been encouraging but more often have been ambiguous or negative. Therefore, it would appear that if consistent differences in the level of flower-inducing and flower-inhibitory activities in vegetative and flowering honeydew exist they must be small and hard to resolve.

Table III. Comparison of flower-inducing and flower-inhibitory activities of flowering and vegetative honeydew in Lemna gibba G3

Treatment	Flowering per cent		
	control	Veg H	Flower H
11L:13D control	22		
A-1		32	17
A-1/10		15	17
A-4		50	50
A-4/10		27	41
A-4/30		14	40
12L:12D control	52		
A-2		0	2
A-2/10		1	22
A-2/100		11	40
A-3		6	0
A-3/10		42	2
A-5		49	46
A-5/10		46	56

The flower-inducing activity in fraction A-4 is restricted to a narrow zone in the R_f region 0.65 to 0.70 and rechromatographs to the same R_f region. In UV light of 254 nm a fairly narrow, bright blue fluorescent band is visible in this region, and it can be demonstrated that the flower-inducing activity is restricted to the region of the fluorescent band.

The flower-inducing activity in fraction A-4 appears to be reasonably heat stable since autoclaving for 10 minutes at 15 psi resulted in only a slight reduction in activity. This finding together with the fact that the activity partitions into ethyl acetate only at low pH suggests a low molecular weight substance and possibly a weak acid. Two possibilities might be ABA or GA. No attempt has been made to isolate ABA from honeydew, but it seems extremely unlikely that ABA is involved since concentrations as low as 0.001 ug/ml inhibit flowering in *L. gibba* G3 on continuous light.

GA activity can be demonstrated in both vegetative and flowering honeydew. However, GA involvement can be eliminated because the GA activity is restricted to the R_f region 0.1 to 0.5. Furthermore, it has previously been shown that addition of GA_3 to *L. gibba* G3 does not induce flowering on short days and inhibits flowering on long days (Cleland and Briggs, 1969). Therefore, the chemical nature of the flower-inducing activity in fraction A-4 remains unknown.

IV. DISCUSSION

The long-day plant *L. gibba* G3 exhibits a qualitative flowering response and flowering has never been observed on short days (9L:15D). Furthermore, there are no reports of successful flower induction in *L. gibba* G3 on short days by any known chemical substance. The cytokinins kinetin and zeatin do cause a slight promotion of flowering on marginal daylengths such as 11L:13D but do not induce flowering on strict short days (Cleland, unpublished). Therefore, the results of the present study provide the first evidence for substantial flower induction in *L. gibba* G3 by some means other than long-day treatment.

The lack of consistent differences in the flower-inducing and flower-inhibitory activities between vegetative and flowering honeydew can be explained in at least 2 different ways. First, it may be that the active factors isolated from the aphid honeydew have nothing to do with the control of flowering in *Xanthium*. Possibly the primary effect of fraction A-4 on *L. gibba* G3 is to inhibit growth and by so doing to indirectly induce flowering. However, it should be noted that inhibition of growth by any other means that have been tried either has no effect on flowering or inhibits flowering and certainly does not stimulate flowering. Therefore, if fraction A-4 induces flowering as a result of an inhibition of growth, it would seem that this must be a quite specific and very interesting type of growth inhibition.

Second, flowering may not be controlled simply by the presence or absence of a single flower-inducing substance but rather by the interaction of one or more flower-inducing and flower-inhibiting substances. Presumably photoinduction would lead to a change in the relative concentration of these factors. However, the changes need not be qualitative but could simply be quantitative. Therefore, it is conceivable that quantitative changes do occur upon photoinduction but that the extent of these changes is small enough that it has not yet been possible to resolve them with the approach used in this study. Only further work can help decide between these different possibilities and give an answer to the basic question of what significance, if any, the flower-inducing and flower-inhibitory activities isolated from aphid honeydew have for the hormonal control of flowering in *Xanthium*.

V. SUMMARY

Aphids are allowed to feed either on vegetative or flowering *Xanthium* and the honeydew which they produce is dissolved in aqueous buffer. The acidic ethyl acetate fraction is purified on TLC and the different R_f zones tested for their effect on flowering using the long-day plant *Lemna gibba* G3 for the bioassay. Flower-inducing activity can consistently be obtained from a specific R_f region from either vegetative or flowering honeydew. Most experiments are run on marginal daylengths, but the activity can also clearly be demonstrated on strict short days. In addition to flower-inducing activity, there appears to be at least 2 zones of flower-inhibitory activity from the same TLC plates. Flower induction has also been obtained using the short-day plant *Lemna perpusilla* 6746, but initial attempts to induce flowering in *Xanthium* or *Pharbitis* were negative. Attempts to demonstrate consistent differences in the levels of flower-inducing and flower-inhibitory activities between honeydew from vegetative and flowering *Xanthium* plants have so far been unsuccessful, and therefore the significance of these active factors for the control of flowering in *Xanthium* is unresolved.

VI. REFERENCES

BONNER, J. and D. BONNER (1948). Note on induction of flowering in *Xanthium*. Botan. Gaz. 110, 154-156.

BOWEN, M.R. and G.V. HOAD (1968). Inhibitor content of phloem and xylem sap obtained from willow (*Salix viminalis* L.) entering dormancy. Planta 81, 64-70.

CLELAND, C.F. and W.R. BRIGGS (1967). Flowering responses of the long-day plant *Lemna gibba* G3. Plant Physiol. Wash. 42, 1553-1561.

CLELAND, C.F. and W.R. BRIGGS (1969). Gibberellin and CCC effects on flowering and growth in the long-day plant *Lemna gibba* G3. Plant Physiol., Wash. 44, 503-507.

DADD, R.H., D.L. KRIEGER and T.E. MITTLER (1967). Studies on the artificial feeding of the aphid *Myzus persicae* (Sulzer). IV. Requirement for water-soluble vitamins and ascorbic acid. J. Insect Physiol. 13, 249-272.

HAMNER, K.C. and J. BONNER (1938). Photoperiodism in relation to hormones as factors in floral initiation and development. Botan. Gaz. 100, 388-431.

HOAD, G.V. and M.R. BOWEN (1968). Evidence for gibberellin-like substances in phloem exudate of higher plants. Planta 82, 22-32.

HODSON, H.K. and K.C. HAMNER (1970). Floral inducing extract from *Xanthium*. Science 167, 384-385.

LANG, A. (1965). Physiology of flower initiation, in "Handbuch der Pflanzenphysiologie" (Ed. W. Ruhland). Vol. XV/1, p. 1380. Springer-Verlag, Berlin.

LINCOLN, R.G., A. CUNNINGHAM and K.C. HAMNER (1964). Evidence for a florigenic acid. Nature 202, 559-561.

MAXWELL, F.G. and R.H. PAINTER (1962). Auxin in honeydew of *Toxoptera graminum*, *Therioaphis maculata* and *Macrosiphum pisi* and their relation to degree of tolerance in host plants. Ann. Entomol. Soc. Amer. 55, 229-233.

MITTLER, T.E. (1958). Studies on the feeding and nutrition of *Tuberolachnus salignus* (Gmelin). II. The nitrogen and sugar composition of ingested phloem sap and excreted honeydew. J. Exptl. Biol. 35, 74-84.

POSNER, H.B. (1967). Aquatic vascular plants, in "Methods in Developmental Biology" (Eds. F. Wilt and N. Wessels) p. 301. Thomas Y. Crowell Co., N. Y.

ZIMMERMAN, M.H. (1961). Movement of organic substances in trees. Science 133, 73-79.

PLANT GROWTH SUBSTANCES, 1970

Hormonal Control of Flowering in Citrus and Some Other Woody Perennials

E.E. Goldschmidt and S.P. Monselise

Department of Citriculture, Hebrew University, Rehovot, Israel

I. INTRODUCTION

Citrus trees, like other polycarpic plants, maintain a balance between vegetative and reproductive growth by transforming each year only a certain percentage of their meristems into flowers, while the rest continues to produce vegetative growth which ensures the plant's future.

In terms of control mechanisms, flowering of polycarpic plants might be dependent not so much upon the positive, inducing stimulus from the leaves (which might be present in sufficient amounts) but on a 'negative' control by an inhibitor which holds the meristems in vegetative condition. Only buds which 'escape' this inhibitory influence develop into flowers (Lang, 1965).

Following previous reports (Ayalon & Monselise, 1960; Monselise & Halevy, 1964; Monselise, Goren & Halevy, 1966; Monselise & Goren, 1968) the present study attempts to give a more complete picture of the flowering process and its hormonal control in citrus trees, and shed some light also on the flowering behavior of few other woody perennials.

II. VEGETATIVE AND GENERATIVE SHOOT TYPES

Morphological analysis of the population of lateral shoots emerging as a "flush" of new growth on a citrus tree, revealed the existence of three principal shoot types:

Vegetative shoots, bearing only leaves.

Mixed-type shoots, bearing both leaves and flowers.

Generative shoots, bearing only flowers.

Fig. 1 shows examples of the three shoot types. In fact, the class of mixed-type shoots contains many sub-types, ranging from an almost vegetative shoot with a single apical flower, through shoots which contain equal numbers of leaves and flowers, to an almost generative type with many flowers and a single leaf. This is shown in Fig. 2, which contains besides the vegetative and generative types 3 examples of the mixed-type, which form a bridge between the two extremes. Another feature which can be seen in Figs 1 and 2 is the gradual decrease in shoot length, from the vegetative, through the mixed-type, to the generative type. The relationship between flowering and stem elongation is further illustrated in Fig. 3.

In this case a large sample of lemon shoots was inspected and shoot length is plotted against the ratio between the number of leaves and the number of flowers on each shoot, on a logarithmic scale. Obviously, the ratio for the vegetative type is ∞ and for the generative type 0, while the mixed-type occupies values in between. The multitude of mixed-type combinations forms almost a continuum between the vegetative

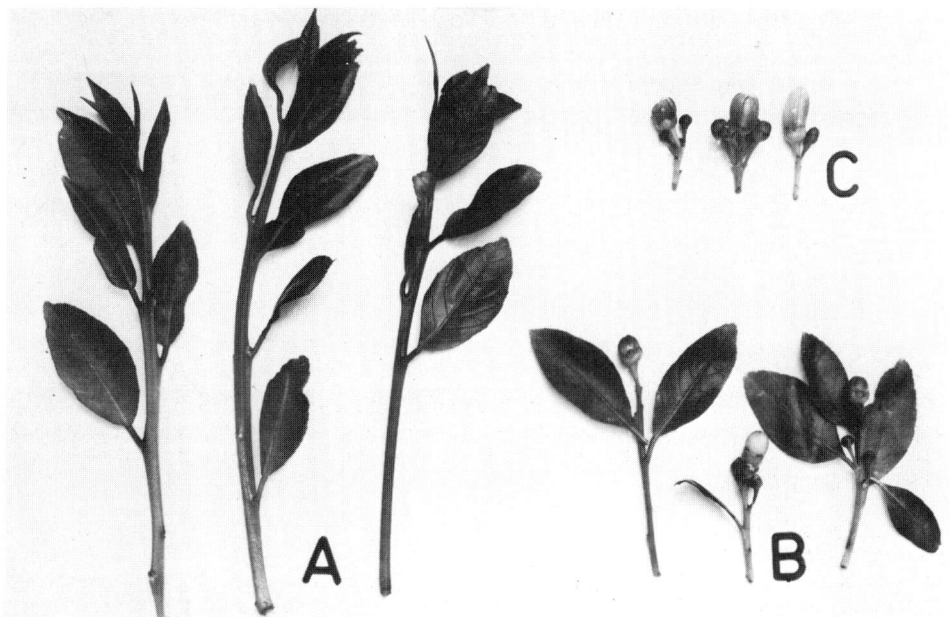

Fig. 1. The 3 shoot types: vegetative (A), mixed-type (B) and purely generative (C). Collected from Eureka lemon trees.

Fig. 2. From right to left: A vegetative shoot, 3 mixed-type shoots and 2 purely generative shoots. Collected from Shamouti orange trees.

and the generative ends. It can be seen in Fig. 3 that stem elongation is inversely correlated with the intensity of flowering, and this correlation exists even among the mixed-type shoots themselves.

The classification of shoot types was found to be a useful criterion for estimating the flowering behaviour of citrus trees. When young trees, which did not yet attain high productivity, were compared with mature trees it was found (Table I) that the younger trees had a higher percentage of the vegetative type and bore most of their flowers on mixed-type shoots. The mature trees, on the other hand, had a smaller reserve of vegetative shoots and a high percentage of purely generative shoots.

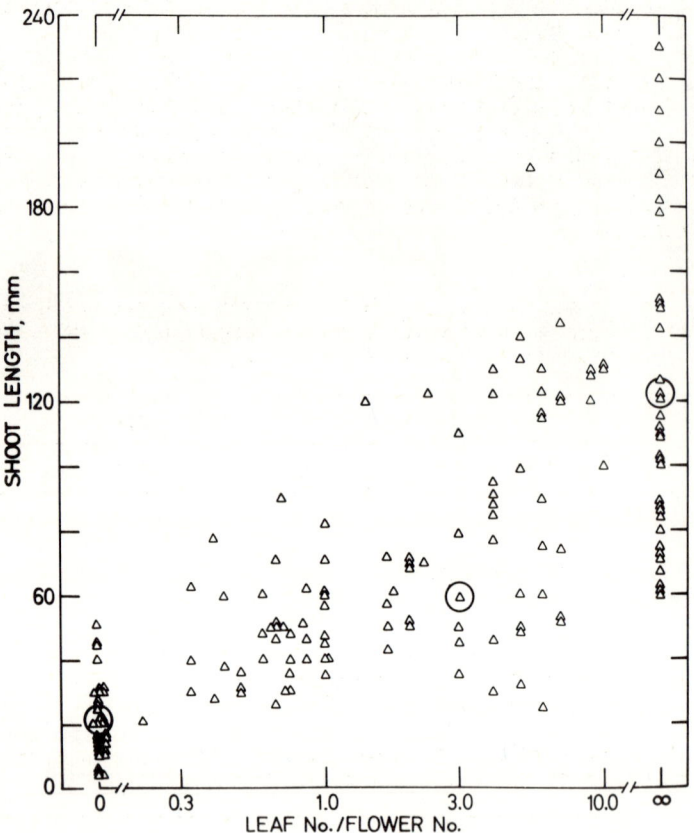

Fig. 3. Morphological analysis of a sample of 161 shoots collected from Eureka lemon trees. Circles mark the mean values for the vegetative, mixed-type and purely generative shoots. See text.

Table I. Percentage of each shoot type in the shoot population of young (5 years old) and mature (10 years old) Shamouti orange trees. Data from young shoots appearing on 10 woody branchlets x 6 trees

	Shoot type		
	Vegetative	Mixed-type	Generative
Young trees	29.3	55.5	15.2
Mature trees	18.5	42.6	38.9

Effects of GA on flower formation

The spring "flush" of Shamouti orange trees (*Citrus sinensis* L.), which includes both vegetative and generative types, emerges as lateral shoots, on thin woody branchlets of the former year. Flower bud differentiation of Shamouti orange trees occurs in the subtropical climate of Israel during the beginning of the cool winter (second half of December, January) and microscopically detectable differentiation is evident by the end of January. The spring "flush" bursts about mid-February and the ensuing uninterrupted development leads to anthesis by the end of March (Ayalon & Monselise, 1960; Monselise & Goren, 1968).

In the present study we examined and compared the effects of GA applications before and after flower bud differentiation in Shamouti orange trees. Ten-μl droplets were applied directly to buds, all the buds along a thin, woody branchlet (usually about 10) receiving the same treatment. This technique, which avoids spraying of whole trees, enabled us to express amounts of applied growth regulators on a "per bud" basis. GA solutions were prepared in 50% ethanol + 0.05% Tween 20 and controls received the same solution without GA.

Table II. Percentage of each shoot type in the shoot population of Shamouti orange trees, as affected by GA_3. Data from young shoots appearing on 4 woody branchlets x 10 trees. Treatments given on 28.11, 10.12 and 19.12,1968.

	Shoot type		
	Vegetative	Mixed-type	Generative
Control	6.6	23.9	69.5
GA_3 (0.075 μg GA_3/bud)*	43.7	28.6	27.7

* Total amount applied in 3 repeated applications

Table III. Effects of GA applications during December on length of vegetative shoots emerging in the following spring flush. Same experiment as Fig. 4.

	Control	GA_3 (μg/bud)			GA_{4+7} (μg/bud)		
		0.03	0.30	3.00	0.03	0.30	3.00
Average length of vegetative shoot, mm	39.2	45.8	42.9	38.0	39.8	56.6	38.9

Applications of GA prior to flower bud differentiation (first half of December) have an inhibitory effect on flowering and cause a strong shift towards vegetative growth (Table II). Comparing different concentrations of GA it was found (Fig. 4) that amounts as low as 0.03 μg GA_3 or GA_{4+7} per bud caused a 75% reduction in the average number of flowers per shoot and higher concentrations caused complete inhibition. Nevertheless, these GA treatments did not affect the length of vegetative shoots which emerged from the treated buds 3 months later (Table III), nor was there any other promotion of vegetative growth following the GA treatments.

Applications of GA during March while shoots were already emerging, had a pronounced promotive effect on shoot elongation, as could be anticipated, but surprisingly enough the number of flowers per shoot was also strongly reduced. When different gibberellins were compared it was found that gibberellins which were most effective in increasing shoot length were also the most potent in reducing flower number. In Fig. 5, shoot length of the vegetative type is plotted against the average number of

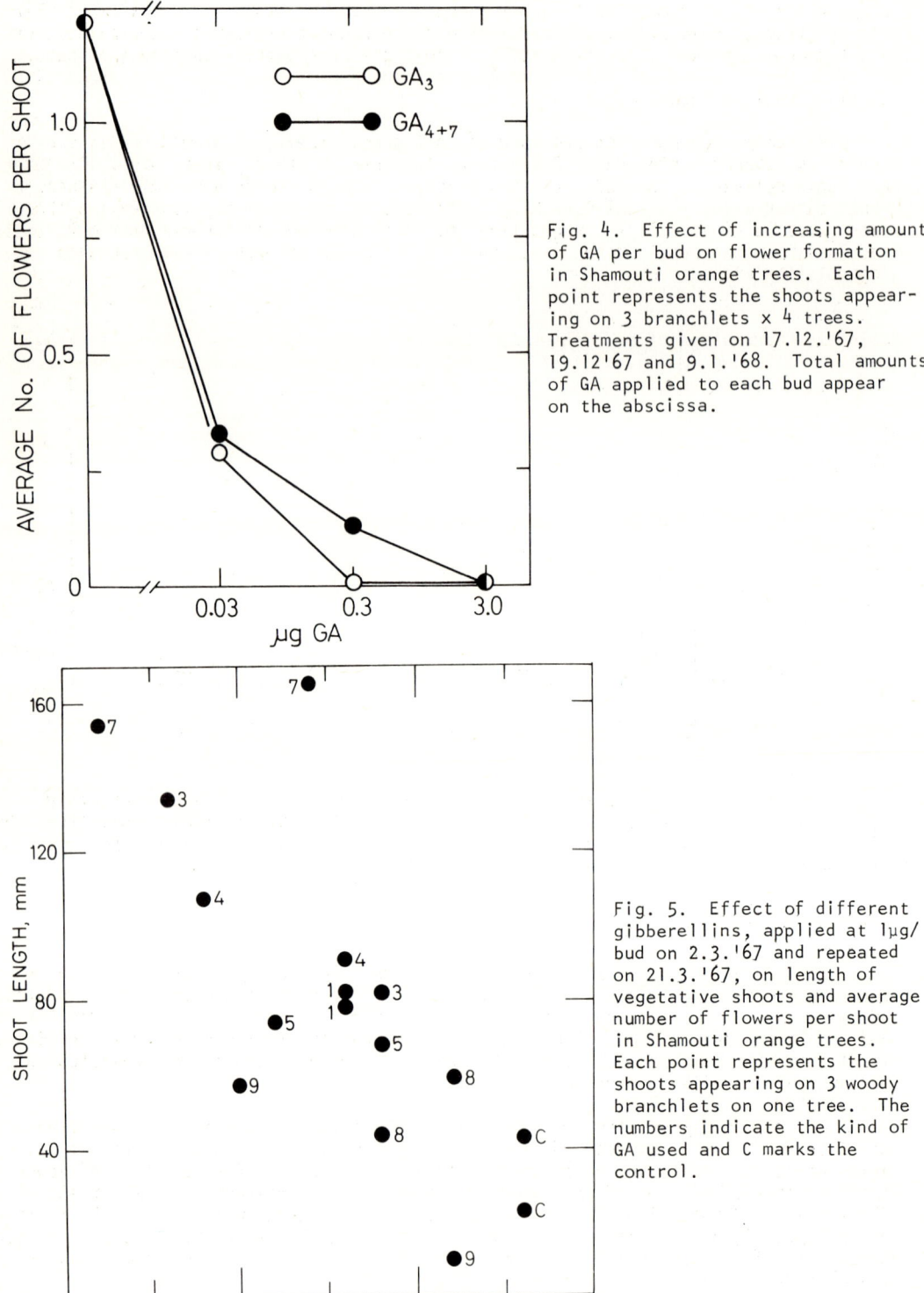

Fig. 4. Effect of increasing amounts of GA per bud on flower formation in Shamouti orange trees. Each point represents the shoots appearing on 3 branchlets x 4 trees. Treatments given on 17.12.'67, 19.12'67 and 9.1.'68. Total amounts of GA applied to each bud appear on the abscissa.

Fig. 5. Effect of different gibberellins, applied at 1μg/bud on 2.3.'67 and repeated on 21.3.'67, on length of vegetative shoots and average number of flowers per shoot in Shamouti orange trees. Each point represents the shoots appearing on 3 woody branchlets on one tree. The numbers indicate the kind of GA used and C marks the control.

flowers per shoot, for controls and 7 gibberellins. The points almost arrange themselves along a regression line, GA7 having the strongest effect on both shoot length and flower number and other gibberellins showing weaker effects in both instances.

Gibberellin-like substances in shoot types

Actively growing lemon shoots were collected, leaves and flowers removed and the stems taken for bioassay. Extraction, solvent partition, paper chromatography and sugar release barley endosperm bioassays were performed as described elsewhere (Goldschmidt & Monselise, 1968). Fig. 6 shows that the level of gibberellin-like substances is highest in the vegetative shoots (A), medium in the mixed-type (B), and lowest in the purely generative type (C), being in accordance with the morphological appearance of shoot types (Fig. 1).

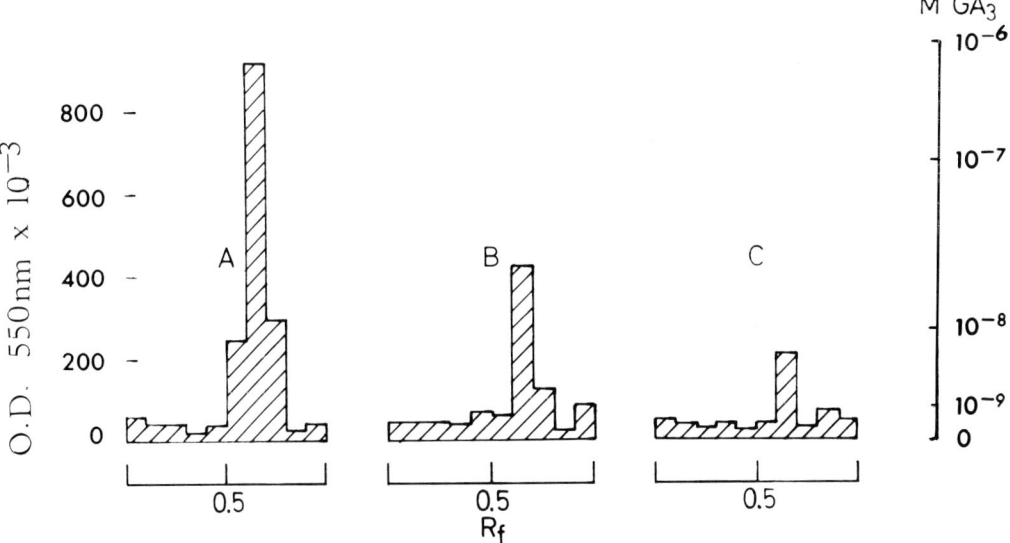

Fig. 6. Reducing sugars released in the barley endosperm bioassay, representing gibberellin like activity, in the pH 3.0 ether fraction (Goldschmidt & Monselise, 1968) of vegetative (A), mixed-type (B) and generative (C) Eureka lemon shoot extracts (250 mg fresh material equivalents per chromatogram).

III. DISCUSSION

The present research work contains a morphological analysis of vegetative and generative growth in citrus, which provides a basis for the study of hormonal control of flowering. The existence of a series of intermediary steps between the vegetative and purely generative types (Figs 2,3) suggests that the flowering response in citrus is of a quantitative nature, rather than an all-or-none response, and indicates the presence of quantitative control factors.

Endogenous gibberellins might fit very well into the role of the postulated controlling factors. Citrus buds are extremely sensitive to minute amounts of GA, which reduce flowering and cause a shift towards production of mixed-type and vegetative shoots (Table II and Fig. 4), while somewhat higher amounts of GA arrest flowering altogether (Fig. 4). The association of flowering with reduced shoot elongation (Figs 2, 3) and the low concentration of gibberellin-like substances in flowering types (Fig.6) are additional clues to the involvement of endogenous gibberellins in the control of flowering in citrus.

In few other cases it has been argued that inhibition of flowering by GA might be an indirect effect due to promotion of vegetative growth (Sachs, Kofranek & Shyr,

1967; Sachs, 1969; Guttridge, 1968). This possibility is almost excluded in our case, since application of GA prior to flower bud differentiation inhibited flowering without having any effect on vegetative growth (Fig. 4 and Table III).

As to the inhibitory effect of GA on flowering when applied at later stages of flower development (Fig. 5), it is difficult to tell whether we have a true reversal of the apical meristem into a vegetative condition (Bernier, 1969) or whether the effect is a result of hormonal competition between an apical meristem which is developing into a flower and a subapical meristem which tends to elongate in response to GA and to suppress flower formation. Anyway it seems that gibberellin is antagonistic to flowering in citrus at all stages from predifferentiation up to late in flower development. We suggest, therefore, that slight variations in concentration of endogenous gibberellins among buds of the same tree might be the predominant factor in determining the intensity of the flowering response and the shoot type to be formed, thereby maintaining the delicate balance between vegetative and generative growth which is so important for the survival of the tree. Similar differences in endogenous gibberellins might account for the behaviour of young and mature trees (Table I) and for other phenomena associated with juvenility in citrus trees.

Table IV. A list of woody perennials in which GA inhibited flowering

Prunus persica (Peach)	Hull & Lewis (1959)
Prunus armeniaca (Apricot)	Bradley & Crane (1960)
Prunus avium (Cherry)	" " "
Prunus amygdalus (Almond)	" " "
Prunus domestica (Plum)	" " "
Malus sylvestris (Apple)	Guttridge (1962)
Pyrus communis (Pear)	Griggs & Iwakiri (1961)
Citrus sinensis (Orange)	Monselise & Halevy (1964)
Weigela sp.	Davidson & Bukovac (1959)
Fuchsia hybrida	Sachs, Kofranek & Shyr (1967)
Cestrum nocturnum	Sachs (1969)
Euphorbia pulcherrima (Poinsettia)	Guttridge (1963)
Ribes nigrum (Black currant)	Tinklin, Wilkinson & Schwabe (1970)
Pittosporum tobira	Goldschmidt & Monselise (Unpublished data)

Our results gain broader significance when considered together with the list of woody perennials in which GA inhibits flowering (Table IV). Many of these plants behave similarly to citrus also in that their flowers are borne on very short shoots. In these plants stem elongation and gibberellin are somehow linked with the vegetative phase, while flowering requires reduced levels of gibberellin. This situation is just the reverse of that found with long-day and/or cold requiring plants in which flowering is usually accompanied by stem elongation and both seem to be associated with high levels of gibberellin.

IV. SUMMARY

The population of young lateral shoots appearing in the spring flush of citrus trees includes a range of shoot types showing degrees of transition from vegetative to generative growth. Shoot types range from a vegetative type which produces only leaves, through intermediate types which produce both leaves and flowers to a completely generative type which produces only flowers. The formation of flowers is accompanied by reduction in shoot length. The purely generative shoot type is the shortest.

Application of GA_3 or GA_{4+7} at 0.03 μg per bud prior to flower bud differentiation results in 75% reduction in number of flowers per bud and somewhat higher concentrations of GA arrest flowering altogether, without having any effect on vegetative growth. Application of GA to buds after flower bud differentiation had occurred results in a reduction of flower number as well as promotion of stem elongation. Endogenous gibberellin-like substances are highest in vegetative shoots and lowest in purely generative types.

Flowering of citrus seems to require reduced levels of endogenous gibberellins. Slight quantitative variations may control the appearance of shoot types and maintain the balance between vegetative and reproductive growth. Comparable situations seem to exist in many other woody perennials.

V. ACKNOWLEDGEMENTS

We are grateful to Mr. S. Milchan and Mr. D. Galili for their expert technical assistance.

VI. REFERENCES

AYALON, S. and S.P. MONSELISE (1960). Flower bud induction and differentiation in the Shamouti orange. Proc. Amer. Soc. Hort. Sci. 75, 216-221.
BERNIER, G. (1969). Sinapis alba L., in "The Induction of Flowering - Some Case Histories"(Ed. L.T. Evans) pp. 305-327. Cornell University Press, Ithaca. 1969.
BRADLEY, M.V. and J.C. CRANE (1960). Gibberellin induced inhibition of bud development in some species of plums. Science 131, 825-826.
DAVIDSON, H. & M.J. BUKOVAC (1959). Amer. Soc. Hort. Sci., 56th Ann. Meeting, Abstr. No. 361.
GOLDSCHMIDT, E.E. & S.P. MONSELISE (1968). Native growth inhibitors from citrus shoots - partition, bioassay and characterization. Plant Physiol. 43, 113-116.
GRIGGS, W.H. & B.T. IWAKIRI (1961). Effects of gibberellin and 2,4,5-trichlorophenoxy-propionic acid sprays on Bartlett pear trees. Proc. Amer. Soc. Hort. Sci. 77, 73-89.
GUTTRIDGE, C.G. (1962). Inhibition of fruit bud formation in apple with gibberellic acid. Nature 196, 1008.
GUTTRIDGE, C.G. (1963). Inhibition of flowering in Poinsettia by gibberellic acid. Nature 197, 920-921.
GUTTRIDGE, C.G. (1968). Hormone physiology of growth regulation in strawberry. in "Plant Growth Regulators" - S.C.I. Monograph 31, pp. 157-169. London. 1968.
HULL, J. Jr. & L.N. LEWIS (1959). Response of one year old cherry and mature bearing cherry, peach and apple trees to gibberellin. Proc. Amer. Soc. Hort. Sci. 74, 93-100.
LANG, A. (1965). Physiology of flower initiation. in "Encycl. Plant Physiol." (Ed. W. Ruhland) Vol. 15/1, pp. 1380-1536.
MONSELISE, S.P. & R. GOREN (1968). Flowering and fruiting-interactions of exogenous and internal factors. In "Proc. 1st int. Citrus Symp." (Ed. H.D. Chapman) Vol. 3, pp. 1105-1112. University of California Press, Riverside. 1968.
MONSELISE, S.P., R. GOREN & A.H. HALEVY (1966). Effects of B-nine, cycocel and benzothiazole oxyacetate on flower bud induction of Lemon trees. Proc. Amer. Soc. Hort. Sci. 89, 195-200.
MONSELISE, S.P. & A.H. HALEVY (1964). Chemical inhibition and promotion of citrus flower bud induction. Proc. Amer. Soc. Hort. Sci. 84, 141-146.
RUDNICKI, R., J. MACHNIK and J. PIENIAZEK (1968). Accumulation of abscisic acid during ripening of pears (Clapp's Favourite) in various storage conditions. Bull. De L'academie Polonaise des Sciences 16, 509-512.

SACHS, R.M. (1969). *Cestrum nocturnum* L. in: "The Induction of Flowering - Some Case Histories" (Ed. L.T. Evans) pp. 424-434. Cornell University Press, Ithaca. 1969.

SACHS, R.M., A.M. KOFRANEK & S.Y. SHYR (1967). Gibberellin-induced inhibition of floral initiation in *Fuchsia*. Am. J. Bot. 54, 921-929.

TINKLIN, I.G., E.H. WILKINSON & W.W. SCHWABE (1970). Factors affecting flower initiation in the black currant, *Ribes nigrum* (L.). J. Hort. Sci. 45, 275-282.

Some Growth Substances associated with Bud Failure of Peach

H.D.R. Malcolm

Biological & Chemical Research Institute, N.S.W. Department of Agriculture, P.M.B. 10, Rydalmere. N.S.W. 2116, Australia

I. INTRODUCTION

Since the discovery and isolation of plant growth substances many workers have attempted to explain dormancy in such terms. In this paper an attempt is made to explain bud failure of peach in similar terms.

In the Murrumbidgee Irrigation Areas (M.I.A.) of New South Wales and also in the Goulburn Valley fruit growing area of Victoria, bud failure of peach can account for as much as 90 per cent loss of crops. Lenz (1963) has shown that 30 to 50 per cent bud failure of stonefruits is common to some cultivars on the M.I.A. Bud failure in its most severe form, however, is accompanied by a characteristic bare lateral condition, the symptoms indicating necrosis of vegetative as well as failure of reproductive buds (Fig. 1a).

There have been reports that bud failure in *Prunus* can be caused by virus (Anon, 1951) and genetic (Saikia et al, 1967) disorders. In this study it was found that some cultivars grown in Victoria, and previously subjected to heat therapy, have shown the bare lateral condition characteristic of severe bud failure but from which none of the currently known viruses associated with the disorder could be isolated. Histological studies have failed to detect the symptoms characteristic of genetic causes. A nutritional imbalance was therefore suspected which may account for the changes of some endogenous growth substances, as proposed by Birch (1969).

It could well be that bud dormancy and failure are closely associated, through either lack of essential growth factors or the presence of active growth inhibitors. The main period of bud abscission on the peach cultivars studied in New South Wales occurs in the late winter, about the time when bud dormancy should terminate, and generally parallels the differentiation of the ovule. Hemberg (1947), for example, first put forward the view that bud dormancy may be due to specific growth inhibiting substances. Wareing and Villiers (1961) later reported that there is no doubt of an annual variation in inhibitor level which is correlated with the state of dormancy of buds.

II. MATERIALS AND METHODS

Extraction of growth substances

The freeze-dried buds (normally 10 g) were macerated in a high speed blendor, following which one litre of 80 per cent methanol was added and left to stand overnight. The liquid was than decanted, filtered, and evaporated under vacuum to about 200 ml; acidified with HCl to pH 3.5, and partitioned three times into ethyl acetate (each aliquot 300 ml). These three aliquots of ethyl acetate combined were designated the acidic fraction of the extract. The remaining solution was then adjusted to pH 9.0, with NaOH, and similarly partitioned into ethyl acetate to yield the basic fraction of the extract. The remaining aqueous solution, readjusted to pH 7.0 with HCl, was designated the neutral fraction. All three fractions were then evaporated to dryness under vacuum, following which the acidic and basic fractions were redissolved in minimal volumes (usually 0.5 ml) of ethyl acetate, and the neutral fraction in 80 per cent methanol.

Fig. 1.   a) "Bare-lateral" of peach typical of severe bud failure.

b) The abnormal abscission layer, associated with severe bud failure, formed in the cortex of the shoot below reproductive and vegetative buds.

c) Early growth on a branch of peach in September, injected two weeks previously with GA.

d) The stimulated extension growth on the same branch (c) in November, when other vegetative growth on the tree had only just begun extending.   (A possible means of rejuvenating the bare lateral condition).

Thin-layer chromatography (TLC)

Thin layer plates (20 x 20 cm) were spread with Keiselgel H at 10 μ thickness. One-fifth, or 0.1 ml, of each extract was spotted onto the thin-layer plates and separation was carried out in two dimensions;   first in isopropanol:ammonia:water (10:1:1) and second in ethanol:water (4:1).   The plates were then divided into 100 equal squares (i.e. 10 x 10);   each square was scraped into a 2 x 1 cm glass vial to which was added 1.25 ml of a pH 5.5 glucose/phosphate/citrate buffer and assayed by a wheat coleoptile test similar to the *Avena* coleoptile test described by Bentley and

Housley (1954). The areas on the plates which caused growth of coleoptile sections significantly different from the background mean were determined by computer using an extension of the analysis of variance method reported by Walker et al. (1958). The Rf's of these areas were compared with the Rf's of standard auxins, naringenin, and abscisic, phenolic and gibberellic acids, previously determined by bioassay of similar plates run in the same solvent systems.

Gas-liquid chromatography (GLC)

The remainder of each fraction of the extracts (0.4 ml) was again evaporated under vacuum. The acidic and basic fractions were each silylated with 0.25 ml BSA, and the neutral fraction with 0.5 ml Tri-Sil. Aliquots (2 or 5 μl) of these silylated fractions were then injected into a Varian Aerograph 1520 Series gas chromatograph, and analysed over a programmed temperature range from 75 to 350°C at 10 and 5°C/min. The columns used were 5ft x 1/8in stainless steel containing 15% SE-30 silicone oil on DMCS acid-washed chromosorb W, 60/80 mesh. The injector temperature was normally set at 300°C and the flame ionization detector at 325°C, with a H_2/N_2/Air gas flow rate of 24 mm^3/min. Emergence temperatures and heights of the peaks found in the extracts were later compared with the data for various concentrations of the synthetic standard growth substances similarly analysed by GLC. To further determine the possible identities of the growth substances found in the extracts significant squares from replicate thin-layer plates were scraped into vials, silylated, and analysed by GLC. Silylated synthetic standard growth substances were also added to the silylated bud extracts to check for parity of emergence temperatures and increased heights of the peaks.

III. RESULTS AND DISCUSSION

Growth substances in viable and failing buds

Although all three fractions of each extract were analysed by both TLC-bioassay and GLC, only the acidic fractions of each could be used to calculate concentrations of the substances in the buds relative to the standard equivalents. Most of the growth substances appeared in the acidic fractions, but probably chemical variations of some of them also appeared in the basic and neutral fractions. Although some of these had similar Rf's in TLC and peak emergence temperatures in GLC, their concentrations in the buds could not be calculated because of their unknown compositions, molecular weights, etc.

(i) Buds from mature peach trees (cv. Elberta) growing in the Macquarie University orchard, were sampled during the winter of 1968 when some bud failure due to inadequate chilling became evident. They were analysed using the methods described, and comparisons made within five groups of growth substances equivalent to synthetic standards (Table I).

Table I. Qualitative comparison within five growth substance equivalents found in viable and failing buds.

Bud Type	Phenolic acid	Auxin	ABA	Naringenin	GA
Viable *	slight increase	constant	Nil	Nil	constant
Failing buds**	greater increase	increasing	increasing	increasing	constant

* Apparently healthy buds, firmly attached at their abscission layers.
** Easily detached at their abscission layers, and revealing a browning of the tissues around the periphery of the exposed abscission zone.

(ii) Analyses of buds from trees (cv. Golden Queen) growing in two orchards on the M.I.A. in 1969, and sampled in April, May, June and August, revealed the trends shown in Figure 2. In one orchard the trees were apparently healthy and had a record of consistently heavy cropping (termed viable buds), whilst in the other, the crops were poor, buds were shed, and symptoms of bare lateral were apparent (termed failure-prone buds).

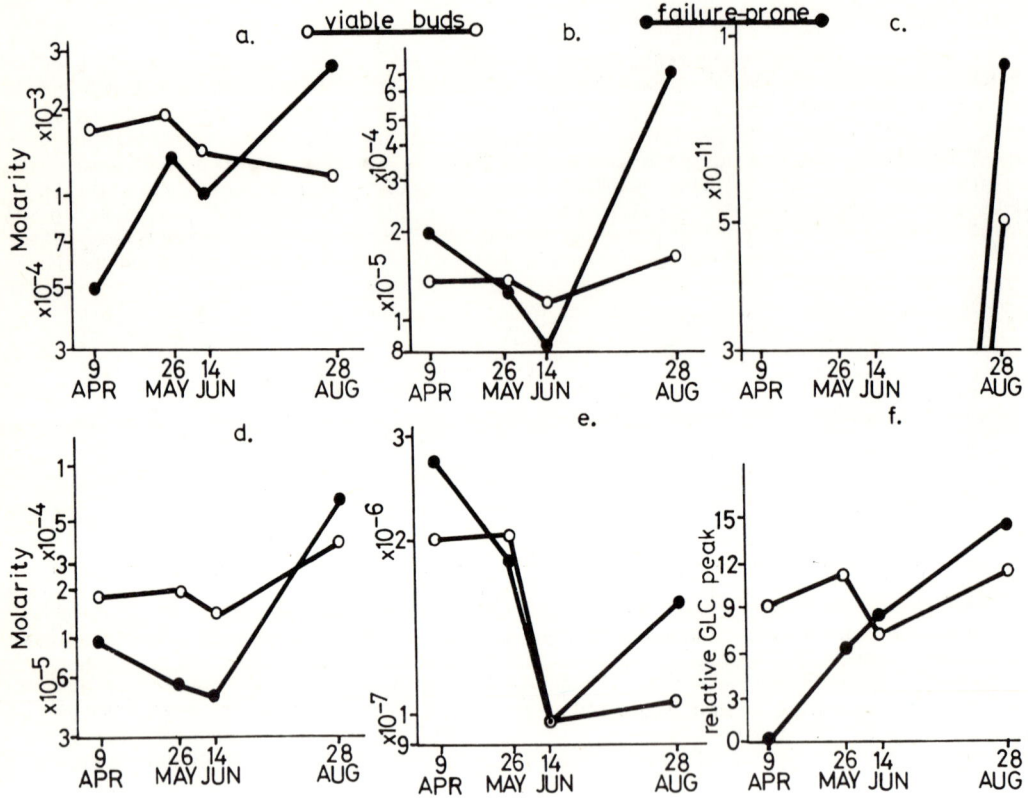

Fig. 2. Relative estimates of six growth substance equivalents found in the buds of Golden Queen peach.

a) Phenolic acids
b) Naringenin
c) Abscisic acid
d) Auxin
e) Gibberellin
f) Hydrolysis and/or reaction products (by GLC only).

The results shown in Figure 2 clearly indicate that the increase in level of growth substance in the buds was closely associated with onset of bud failure, and possibly also failure of the buds to be released from dormancy. Auxin was found in the buds throughout the winter, contrary to the report of Bennett and Skoog (1938) that it only appears towards the latter part of winter. From these data there appeared to be some correlation between an inhibitor:promoter ratio and bud failure (Figure 3). This could implicate the high inhibitor:promoter ratio proposed by Black (1955) to explain prolonged dormancy or the low inhibitor:gibberellin ratio proposed by Thomas et al (1965) to explain failure to terminate dormancy.

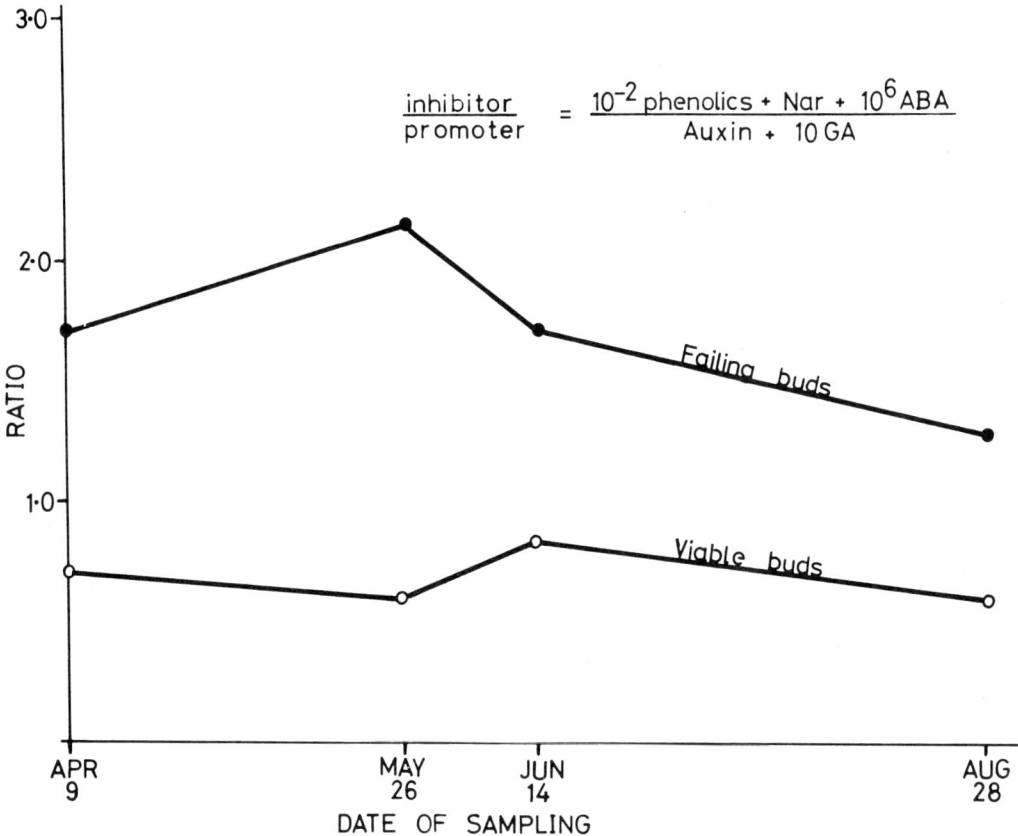

Fig. 3. Relative growth inhibitor:promoter ratios in the buds (calculated from the data of Figure 2).

Effect of shoot injection on bud failure

Because some spectacular empirical effects have been achieved by applying growth substances to plants, it seemed worthwhile to explore direct responses by injecting the shoots, in a manner described by Roach (1938). Substances representing the five equivalents (Figure 2) phenolic acids, ABA, auxin, naringenin, and GA, were injected together with three enzymes thought to be possibly implicated in protein hydrolysis and amino acid metabolism, and their effects on bud failure were assessed. The treatments and results were as follows:- (All growth substance injections were 5 ml of 1mg/ml solutions, and enzymes and other substances 5 ml of 0.1 mg/ml solutions.)

Experiment 1:- Methyl glyoxal, a substance reported to inhibit protein synthesis and cell division (Szent-Gyorgyi, 1968), and α-amylase, an enzyme reported to be implicated in the breakdown of carbohydrates in plant storage organs (Van Overbeek, 1968), were injected into the shoots of three-year-old potted trees (cv. Blackburn) both before and after chilling.

The methyl glyoxal caused abscission of buds on the unchilled trees and almost immediate flowering on the chilled trees. The α-amylase caused necrosis of buds on both chilled and unchilled trees.

Experiment 2:- The enzymes arginase, asparaginase, and glutaminase, thought to be possibly implicated in the metabolism of the buds, were injected into the shoots of "healthy" trees (cv. Blackburn) some four months before bud burst.

All three treatments caused some necrosis of buds but only the glutaminase treatment was significant. The remaining buds on the glutaminase-treated shoots flowered about one week earlier than normal. The arginase treatment caused defoliation to be delayed by about two weeks, and later about one month before flowering, the scales of the remaining viable buds opened and abscissed.

Experiment 3:- Dimethylsulfoxide (DMSO - 0.1% v/v), reported to inhibit protein synthesis and used by Pine (1967) in an attempt to treat virus, and methyl glyoxal (0.1% w/v) were injected into the shoots (cv. Blackburn) one month before bud burst.

The DMSO caused 100 per cent bud necrosis, whilst the methyl glyoxal caused 90 per cent bud abscission.

Experiment 4:- Naringenin, ABA, IAA, IPyA, GA_3, arginase, asparaginase, and glutaminase were injected, about four months before bud burst, into the shoots of trees which had a history of bud failure, (cv. Golden Queen) growing on the M.I.A.

The results of these treatments are shown in Table II.

Table II. Bud failure following shoot injection treatment.

Substance injected	Percent bud failure	Substance injected	Percent bud failure	Substance injected	Percent bud failure
Naringenin =	71*	Control =	54	GA =	33* †
ABA =	63*	Arginase =	40	Indolepyruvic acid =	28*
				Indoleacetic acid =	22*
				Glutaminase =	23*
				Asparaginase =	20*

* Significant at P0.01
† Count corrected for abscission of early flowers.

The growth inhibitors caused increased bud failure whilst the growth promoters and two of the enzymes reduced the bud failure. These results support the view that a high inhibitor:promoter ratio is associated with bud failure, despite the adequate chilling received by the time the buds abscissed just prior to bud burst.

Experiment 5:- Buds from "healthy" trees, with low but differing incidences of bud failure, were analysed for equivalent content of five synthetic growth substances. These results are shown in Table III.

The ABA, Nar, and GA, increased with the incidence of bud failure but no significant increase in the inhibitor:promoter ratio appeared to be implied.

IV. CONCLUSIONS

The distribution of stimulated flower and vegetative growth, and reddish chlorotic colour of the early shoots low down on the branches injected with GA (figure 1C)

Table III. Bud failure and associated concentration of equivalent synthetic growth substance.

Percent bud failure	Equivalent standard growth substance				
	Phenolic acid	Auxin	ABA	Naringenin	GA
22	9.5×10^{-4} M	8.0×10^{-4} M	4.4×10^{-10} M	1.3×10^{-4} M	3.2×10^{-6} M
15	7.2 "	7.4 "	2.7 "	0.7 "	1.7 "
10	6.9 "	2.4 "	1.8 "	0.4 "	1.6 "
5	6.1 "	7.3 "	Nil	0.4 "	2.2 "

was the most marked gross effect on the shoot growth in any of the treatments and suggests that the GA was readily transported within the tree. No such gross effects on growth were observed after injection of the other substances, apart from the high incidences of bud failure caused by methyl glyoxal, DMSO, and α-amylase. A low level of naringenin found in buds and flowers stimulated by GA treatments could perhaps explain the excessive cell division in the flower peduncles, as suggested by Van Overbeek (1968). Growth inhibitors do appear to be closely associated with bud failure, although a further factor such as the virus proposed by Highkin (1955) may well be implicated. The growth substance ratios proposed to explain bud dormancy (Black, 1955; Thomas et al, 1965) could also be related to the failure, but the data on abscising buds from healthy trees (Table III) raise some doubt whether these ratios are valid. If failure to release from dormancy is not associated with bud failure in these terms, a keto-aldehyde, such as methyl glyoxal, which caused abscission of buds early and flowering when applied late in the winter, may well be implicated. In both viable and failure-prone buds there was a rise in the level of GA-equivalent towards the end of winter (Figure 2e), which according to Jackson (1968) could indicate the release of buds from dormancy. Higher levels of GA may also be associated with bud failure. Certainly, a high level of GA, due to the injection treatments, caused abscission of flowers and thus failure of fruit set (Figure 1c and d).

There was a greater effect on increase or decrease of bud failure when the growth substances were injected into the shoots of trees prone to bud failure than when similar injections were made to healthy trees. Methyl glyoxal caused a different form of bud failure to DMSO, and whether or not this occurred *in lieu* of flowering appeared to be dependent upon the amount of chilling received at the time of treatment. Because of its effect on flowering following the chilling requirement, the action of methyl glyoxal appeared to be associated with the release of buds from dormancy. Both DMSO and α-amylase, however, probably caused necrosis of the buds by nutritional depletion in the shoots. Arginase, glutaminase, and asparaginase, might be expected to effect some similar depletion in the shoots, but if so, the mechanism of this action is not known, because the role of enzymes in degradation of nitrogenous components within the tissues of peach is little understood (Taylor, 1967).

Some of the results presented here are purely empirical, making use of a little knowledge of the metabolism of the shoot-bud system in an attempt to find chemical treatments to both increase and decrease bud failure. Once this failure, similar in occurrence to natural conditions, can be duplicated by chemical treatment then its cause may be more fully understood and some means devised for its control.

V. SUMMARY

Growth substances having characteristics in TLC bioassay and GLC similar to those of naringenin, indole auxins, gibberellic and abscisic acids, were found predominantly in the acidic fractions of methanolic extracts of peach flower buds.

Synthetic standards representing these substances, four enzymes, arginase, glutaminase, asparaginase and α-amylase; and dimethylsulfoxide and methyl glyoxal known to affect protein depletion and cell division, were applied to the buds and injected into the shoots. The growth inhibitors caused increased bud abscission, and the growth promoters and glutaminase and asparaginase reduced bud abscission. Dimethylsulfoxide and α-amylase caused necrosis of the buds which did not abscise. Methyl glyoxal caused 90 per cent abscission of buds on partly chilled trees and early flowering on fully chilled trees.

The concentrations of the growth inhibitors in the buds of healthy trees appeared to increase with the incidence of bud failure. There appeared to be no difference in the ratios of growth inhibitors and promoters in the buds of viable and failure-prone trees.

VI.. REFERENCES

ANON (1951). Virus diseases and other diseases with virus-like symptoms of stone fruits in North America. U.S. Govt. Printing Office Washington D.C. 276 pp. (U.S. Dept. Agric., Handbook 10).
BENNETT, J.P., and SKOOG, F. (1938). Preliminary experiments on the relation of growth-promoting substances to the rest period in fruit trees. Plant Physiol. 13: 219-225.
BENTLEY, J.A., and HOUSLEY, S. (1954). Bio-assay of plant growth hormones. Physiol. Plant 7: 405-419.
BIRCH, P.D.W. (1969). The role of plant physiology in advisory work. N.A.A.S. Quarterly Review 84: 169-177.
BLACK, M.W. (1955). The problem of prolonged rest in deciduous fruit trees. Rept. 13th Intl. Hort. Congr. London. 1122-1131.
HEMBERG, T. (1949). Growth-inhibiting substances in terminal buds of *Fraxinus*. Physiol. Plant 2: 37-44.
HIGHKIN, H.R. (1955). Flower-promoting activity of pea seed diffusates. Plant Physiol. 30 (4): 390.
PINE, T.S. (1967). Reactions of peach trees and peach tree viruses to treatment with dimethysulfoxide and other chemicals. Phytopathology 57 (7): 671-673.
ROACH, W.A. (1938). Plant injection for diagnostic and curative purposes. Imp. Bur. Hort. Plantation Crops. Tech. Comm. No. 10., 10 pp.
SAIKIA, B.N., KESTER, D.E., and BRADLEY, M.V. (1967). Dormant vegetative buds in normal and bud-failure forms of almond (*Prunus amygdalus* Batsch.). Proc. Amer. Soc. Hort. Sci. 89: 150-156.
SZENT-GYORGYI, A., EGYUD, L.G., and McLAUGHLIN, J.A. (1967). Keto-aldehydes and cell division. Glyoxal derivatives may be regulators of cell division and open a new approach to cancer. Science 155: 539-541.
TAYLOR, B.K. (1967). Storage and mobilization of nitrogen in peach trees: A review. J. Aust. Inst. Agric. Sci. 33 (1): 23-29.
THOMAS, T.H., WAREING, P.F., and ROBINSON, P.M. (1965). Action of the sycamore dormin as a gibberellin antagonist. Nature, London. 205: 1270-1272.
VAN OVERBEEK, J. (1968). The control of plant growth. Scientific American 219 (1): 75-81.
WALKER, D.R., HENDERSHOTT, C.H., and SNEDECOR, G.W. (1958). A statistical evaluation of a growth substance bioassay method using extracts of dormant peach buds. Plant Physiol. 33 (3): 162-166.
WAREING, P.F., and VILLIERS, T.A. (1961). Growth substances and inhibitor changes in buds and seeds in response to chilling. Proc. 4th Intl. Conf. Plant Growth Regulation. Iowa State Univ. Press Iowa. U.S.A. pp. 95-107.

PLANT GROWTH SUBSTANCES, 1970

The Flowering Process - A New Theory

Paul Baxter

228 Stephensons Road, Mt. Waverley, Victoria

I. INTRODUCTION

The initiation and development of flowers is an event of outstanding significance in the life of a plant as well as for the life of man, since much of agriculture depends on fruit or seed production.

Some plants flower profusely once a certain stage of development is reached but others only flower within a narrow range of environmental restraints such as light (duration, intensity) and temperature. A catalogue of these factors will be found in the book by Salisbury (1963).

But in perennial plants there are two further complications. Flower formation may not occur until the plant is 5-40 years old - the problem of juvenility - and it may not occur following a year of abundant fruit - the problem of biennial bearing.

These two aspects of flowering have traditionally been investigated by two different groups of plant physiologists. The horticultural scientists were concerned with the effect of various tree and soil management methods (rootstocks, cincturing, fertilizer) on the flower and fruit production of fruit or forest trees while the more fundamental work on flower bud initiation (FBI) was carried out by research workers trying to explain the action of light (photoperiodism) and of winter cold (vernalisation) on the flowering process. Two different theories of flowering correspond approximately to these two different approaches. The early horticultural scientists sought the explanation of FBI in changes in plant nutrition in its widest sense; this led from the general theories of Klebs and Poenicke to the particular theory of the carbohydrate:nitrogen ratio of Kraus and Kraybill. The second group postulated a specific flowering hormone, produced by the leaves and acting on the floral apex (Chailakhyan, 1968). It is this second approach, arising mainly from the work on photoperiodism, that today dominates our thinking, strengthened by the knowledge that a variety of plant growth substances (auxins, gibberellins, growth retarding substances) have profound effects on FBI and flower development.

The Swiss physiologist Kobel, in his textbook on the physiological basis of fruit growing, reviewed both theories and concluded that our present emphasis of the hormonal theory to the exclusion of nutritional factors was unwarranted and unwise (Kobel, 1954). Kobel predicted that phosphate should have a large effect on flowering of fruit trees, but the results of his experimental work (Fritzsche et al. 1964) could not be duplicated by Reinken (1963) using apple trees in sand culture.

Recent work at the Scoresby Horticultural Research Station (Victoria, Australia) on the influence of mineral nutrition on flowering has shown that nitrate N decreases flowering of young apple trees compared to ammonium nitrogen (Grasmanis and Leeper 1967). In field trials in our orchards we have shown the flower promoting effect of high concentrations of phosphatic fertilizer on a wide range of fruit trees ranging from plums to lemons.

These results have stimulated us to attempt and combine the knowledge gained from photoperiodic experiments in the glasshouse and that gained from experiences with tree management and fertilizer trials in the orchard. For it seems reasonable to expect that the basic physiology of flower initiation should be the same both in the smallest of flowering plants (*Lemna*) as in the tallest tree. This hypothesis arose out of these endeavours.

II. THE PROBLEM STATED

Evans, in his recent book (1968) describes it as a mystery wrapped in an enigma. The mystery: How does one and the same compound - phytochrome - influence the photoperiodic expression of both short and long day plants; the enigma: How is the floral stimulus transferred from the basal leaves to the floral apex. Even if a floral hormone were found and isolated, this would raise still more questions. How does it originate, how does it act?

But the fruitgrower also wants to know why his young, vigorous apple trees will flower after he has briefly interrupted the phloem flow two months prior to the period of FBI, i.e. 12 months prior to anthesis; why his pear trees develop more flowers when their limbs are bent; why leaf sprays of "Alar" (amino-succinamic acid) induce his young trees to flower and why his pear trees burst into a "second bloom" following a frost which injured the first blossom and aborted the seed of the young pear fruit. Why are young, vigorous trees often devoid of flowers yet stunted trees often flower profusely, particularly those attacked by wood-rotting fungi. Why does nitrogen fertilizer sometimes reduce flowering but sometimes increase it? His concern is understandable as his living depends on adequate, regular flowering to ensure adequate evenly-spaced crops of fruit.

The belief that flowering is always and causally inversely related to growth is not in accordance with observed facts. Recent work on the relationship of seed content of apples and pears to biennial bearing (Chan and Cain 1967 ; Griggs, Martin and Iwakiri 1970) suggests the action of a seed-produced floral inhibitor. The carbohydrate:nitrogen theory appears quite inadequate to explain FBI in fruit trees, as biochemical investigations of buds and wood have shown no causal relationship of either carbohydrates or nitrogen fractions to FBI. Wood is rich in carbohydrates and trees store large quantities of amino and amide nitrogen; it is difficult to conceive that either would ever be insufficient for floral initiation unless the tree were grossly N deficient.

III. THE HYPOTHESIS

A - Flowering as programmed development:

1 - Flowering is the normal expression of the plant's genetic programme. In primitive forms the reproductive ability was found in every cell; with increasing complexity came increasing specialisation. Most plants flower and flower profusely, non-flowering is the exception. Sometimes even trees and shrubs will develop flowers in the seedbed (von Denffer, 1950). Fulford, (1966) who investigated the morphogenesis of apple buds also concluded that "the meristem will always form a flower unless it is prevented from doing so."

11 - The duration of light - or rather the duration of darkness triggers the programme. A light-sensitive pigment - presumably phytochrome - is involved in this timing mechanism which appears to work like an alarm clock, i.e. the effects occur only over a previously programmed period (see for instance the effects of phytochrome on germination (Leopold, 1964)).

111 - Any of the intermediates in the flower initiation sequence could be called a "flowering" hormone, i.e. if no inhibitors are present and sufficient metabolites are available it can induce flowering. It would be graft transmissible.

B - The Control of flowering is achieved through an inhibitor:

1 - The idea that flowering is controlled by inhibitors was fully discussed by von Denffer (1950) who used a number of logical arguments against the still most commonly held view that there is a specific floral stimulus "whose existence is obvious and whose identification is long overdue" (Zeevaart, quoted by Evans 1968).

11 - I suggest that the floral inhibitor achieves its effect by decreasing the permeability of intra-cellular membranes. This serves to decrease the flux of phosphate ion into the cytoplasm and/or into organelles. It is known from the work of Evans (Evans and Rijven, 1967) that P accumulates soon after the floral stimulus is perceived by the apex. Feucht (1967) and Feucht and Arancibia (1968) have shown that the buds of a number of fruit trees accumulate P, particularly organic P, prior to and during the period of FBI. This would explain the finding of a number of workers, apart from our own experiments, that flowering is related to the level of P supplied when P uptake is relatively low (i.e.from soil) (Fukuda and Kondo 1957 ; Anthony and Clarke, 1932). Since excess P often accumulates as inorganic P (Hogue *et al*, 1970) I suggest that it is its transport into the cytoplasm which is here the critical factor, presumably across the plasmalemma. (See also the work by Ullrich-Eberius and Simonis (1970) on algae).

111 - Plant growth substances which affect flowering may do so by changing the permeability and the electric potential of cell membranes. Thus amino-succinamic acid (Alar) will increase membrane permeability (Undurriaga and Ryugo, 1970). Weigl (1969) has also demonstrated that many plant growth substances develop a strong physical association with lecithin and points to the influence of auxins on cell membrane function.

1V - The calcium ion is presumed to be the key ion in regulating energized permeability, i.e. the transport of ions against a concentration gradient. This follows from its general role as a "de-energiser" (Green and Baum, 1968). Auxins are known to be affected greatly by Ca nutrition (Al-Omary, 1968) and excess Ca decreases permeability (Heilbrunn, 1952). Fruit trees which do not initiate flower buds accumulate Ca, especially in and around the developing bud (Feucht, 1969). Conversely, excess Ca should decrease FBI.

C - The Floral Stimulus is transmitted through a combination of electric charges and "hormonal" messengers:

1 - Phytochrome is assumed to be located on membranes and its primary mode of action is thought to be that of changing membrane permeability and electric charges (Kandeler, 1969a, Smith, 1970). The flower-controlling action of phytochromes may be related to photosynthetic phosphorylation,for if ATP production is uncoupled or ATP diverted then *Lemna gibba* will flower much less though its growth is not affected (Kandeler, 1969b). This inhibition can be reversed if ATP is added; it thus acts as a co-factor for phytochrome. Kandeler pointed out that ATP diversion (such as in excessive growth) could account for the well known (but by no means universal) inverse relation of growth to flowering. It also relates well with the fact that a temporary stoppage of the phloem flow to the roots (cincturing) will often increase flowering markedly. Phloem sap is particularly rich in ATP (Kluge *et al.*1970).

11 - I suggest that the transmission of stimuli in plants has a similar biochemical basis to the transmission of stimuli in animals. Thus in the animal's nervous system, changes in the nerve cells permeability induce changes in electric charges, an ionic flux (electric charge) travels through the cell and releases a hormone which bridges the junction (synapse) between cells, inducing a permeability change in the neighbouring cell. In the higher animals the stimuli are numerous, the nerve cells are specialized structures and the transmitter acetyl choline. In plants the stimulus is mainly light (touch-sensitive plants excepted) and the transmitting agent may be a plant hormone or ATP; the latter also forms the transmitting substance in a newly discovered, primitive autonomous nerve system of mammals (Burnstock *et al.*1970).

D - High light energies may overcome the action of the floral inhibitor:

1 - It is known that high light intensity will overcome the phytochrome-mediated inhibition of flowering in short day strains of *Lemna* (Schuster and Kandeler 1971). This appears to be the result of increased photosynthetic production of sugars and of reduction energy. Sunlight and temperature just prior to the period of FBI are critical factors influencing the proportion of flowers initiated in grapevines (Baldwin, 1964) and oranges (Moss, 1971). This may also be one reason for the universal preference of apple growers for a sunny, high altitude location for their orchards. This theory is put forward as a scientific hypothesis, that is, as a tentative explanation of observed facts and as an aid to future investigations. Many of its consequences are experimentally verifiable but it is speculative and intended solely as a model, to be discarded when new found facts no longer fit or a better model becomes available. Experimental results showing the relation between phosphate supply and flower initiation in young fruit trees are in process of publication elsewhere.

Horticultural science owes a great deal to the plant physiologists and biochemists for their discovery of many useful plant growth substances. It is therefore perhaps not out of place that the experiments and experiences of an applied horticultural scientist should provide the stimulation for a more general theory of flowering leading to a synthesis of the nutritional and hormonal approach.

IV. SUMMARY

Experiments with a wide range of young fruit trees have shown that flower bud initiation can be greatly influenced by the level of available phosphate and, in apple trees, also by the form of nitrogen, supplied during the time flowers are initiated. These results, together with a review of the literature on the flowering process form the basis of a hypothesis which has the following postulates:

1 - Flowering is the normal expression of the plant's genetic programme.

11 - The control, i.e. delay or absence of flowering is caused by an inhibitor, which acts by decreasing the permeability of cell membranes, particularly the plasmalemma This reduces the influx of P required in higher amounts for cell differentiation.

111 - Plant growth substances which affect flowering may do so by changing membrane permeability directly or indirectly by directing Ca movement.

1V - The floral stimulus may be transmitted through a combination of electric charges and transmitting substances akin to the action of the animal nervous system.

V - High light energies can decrease the action of the floral inhibitor.

V. ACKNOWLEDGEMENTS

These ideas have emerged from many useful discussions with Dr. R. N. Rowe (Scoresby Horticultural Research Station) and from correspondence with Prof. R. Kandeler (Univ. of Wurzburg). I am particularly grateful to Prof. Sir R.N. Robertson (A.N.U.) and Prof. G. Buchloh (Univ. of Hohenheim) for their continued encouragement.

VI. REFERENCES

ANTHONY, R.D. and W.S. CLARKE (1932). Growth record of fertilized apple trees grown in metal cylinders. J. Agric. Res. 44, 245-266.
BALDWIN, J. (1964). The relationship between weather and fruitfulness of the sultana vine. Austr. J. Agric. Res. 15, 920-928.
BURNSTOCK, G. *et al.* (1970). Evidence that ATP or a related nucleotide is the transmitter substance released by non-adrenergic inhibitory nerves in the gut. Br. J.Pharmac. 40, 90-121.

CHAILAKHYAN, M.K. (1968). Internal factors of plant flowering. Ann.Rev. Pl. Physiol., 19, 1-36.
CHAN, B.G. and J.C. CAIN (1967). The effect of seed formation on subsequent flowering in apple. Proc.Amer.Soc.Hort.Sci. 91, 63-68.
VON DENFFER, D. (1950). Blühormon oder Blühhemmung? Naturwiss. 37, 296-301.
EVANS, L.T. (1968). The Induction of Flowering. MacMillan & Co. Melbourne.
EVANS, L.T. and RIJVEN, A.H.G.C. (1967). Inflorescence initiation in Lolium temulentum L. Part XI. Austr.J.Biol.Sci. 20, 1033-1042.
FEUCHT, W. (1967). Daten zum Stoffwechsel des P,Ca, und N in den Endknospen des Apfelbaumes. Arch. f. Gartenb. 15, 175-182.
FEUCHT, W. and M. ARANCIBIA (1968). La induccion floral en naranjos y su relacion con los minerales. Agrochimica 12, 89-99.
FEUCHT, W. and M. ARANCIBIA (1969). Zur lokalisierten Ca-Oxalatsausscheidung im Knospengebilde des Apfelbaumes während der floralen Induktion. Arch.f.Gartenb. 17, 565-574.
FRITZSCHE, R. et al. (1964). Dungungsversuche mit Äpfel-und Kirsch-bäumen. Schweiz. Z.Obst. und Weinb. 73, 531-560.
FUKUDA, A. and G. KONDO (1957). Studies on the nutrition of the peach tree. Part IV. Studies of the Inst.Hortic.Kyoto Univ. 8, 16-23.
FULFORD, R.M. (1966). The morphogenesis of apple buds. Part III. Ann.Bot. 30, 207-219.
GRASMANIS, V. and G.W. LEEPER (1967). Ammonium nutrition and flowering of apple trees. Austr.J.biol. Sci. 20, 761-767.
GREEN, D.E. and H. BAUM (1969). "Energy and the Mitochondrion." Academic Press. New York.
GRIGGS, W.H., G.C. MARTIN and B.T. IWAKIRI (1970). The effect of seedless versus seeded fruit development on flower bud formation in pear. J.Amer.Soc.hort.Sci. 95, 243-248.
HEILBRUNN, L.V. (1952). "An Outline of General Physiology." W.B. Saunders Company, Philadelphia.
HOGUE, et al. (1970). Effect of soil P levels on phosphate fractions in tomato leaves. Jour.Amer.Soc. hort. Sci. 95, 174-176.
KANDELER, R. (1969a). Phytochrom. Fortschr. der Botanik 31, 152-162.
KANDELER, R.(1969b). Förderung der Blütenbildung von Lemna gibba durch Ammonium. Planta, 84, 279-291.
KOBEL, F. (1954). "Lehrbuch des Obstbaus". Springer Verlag. Berlin.
KLUGE, M., D. BECKER and ZIEGLER, H.(1970). Untersuchungen über ATP im Siebröhrensaft von Yucca und Salix. Planta 91, 68-79.
LEOPOLD, A.C. (1964). "Plant Growth and Development" McGraw-Hill. New York.
MOSS, G.I. (1970). Pers. communic.
REINKEN, G. (1963). Wachstum, Assimilation und Transpiration von Äpfelbäumen und ihre Beeinflüssung durch Phosphor. Phosphorsäure 23, 91-108.
SAIB A AL-OMARY (1968). A preliminary study of the relationship of mineral nutrition to the auxin content of Zebrina pendula. Ann.Physiol.végét. (Bruxelles) 13, 109-135.
SALISBURY, F.B. (1963). "The Flowering Process". Pergamon Press.
SCHUSTER, Maria and R. KANDELER (1971). Die Bedeutung der Photosynthese fur die Langtagblüte der Kurztagpflanze, Lemna perpusilla. Planta (in press).
SMITH, H. (1970). Phytochrome and photomorphogenesis in plants. Nature 227, 665-668.
ULLRICH-EBERIUS, C.I. and W. SIMONIS (1970). Der Einfluss von Na und K Ionen auf die Phosphataufnahme bei Ankistrodesmus braunii. Planta 93, 214-226.
UNDURRIAGA, N.J. and K. RYUGO (1970). The effect of Alar on permeability: a possible explanation of its mode of translocation. J.Amer.Soc.hort.Sci. 95, 348-354.
WEIGL, J. (1969). Wechselwirkungen pflanzlicher Wachstumshormone mit Membranen. Zeitschr. f. Naturf. 24b, 1046-1049.

PLANT GROWTH SUBSTANCES, 1970

The Promotion of Ripening in Cereal Crops by Peduncle-Injection of Aqueous Solutions of Nucleotides

Toyoo Tomita

National Institute of Agricultural Sciences, Konosu, Saitama, Japan

I. INTRODUCTION

In the central part of Japan,"Syurin"(frequent rains in autumn) is an inevitable environmental factors in the season of rice cultivation. Most rice varieties have to pass through this unfavourable period during the important growth stages of heading, flowering and ripening. The lack of sunlight during the ripening period decreases the yield a great deal, producing many unripe and poorly ripened grains.

On the other hand it has been claimed by Japanese agricultural research workers that the use of organophosphorus pesticides seemed to be correlated with a good yield of rice, although there is no experimental proof of this.

The author intended to search for nontoxic chemicals which would promote the ripening of cereal crops such as rice and wheat, and so increase the quality and yield of these crops even under the unfavourable conditions mentioned above.

II. MATERIALS AND METHODS

Preliminary experiment (1)

The rice plants (*var.* Manryo), which were transplanted to the paddy field from the nursery bed in the beginning of July, had almost completed their fertilization around 5th of September, 1968. On the 7th day of September each hill of rice was transplanted again to a Wagner pot (1 a/5000) together with the paddy soil, and all the stems were trimmed off leaving 10 uniform panicles. Then 10 grams of compound fertilizer (16-10-14) was applied to each pot. Fifteen potted plants were transported to a shed where some weak diffused light penetrated through the skylights. The light intensity inside the shed was approximately 1/20 of the outside light. Aqueous solutions (200 p.p.m.) of AMP, GMP, CMP and UMP were sprayed on the plants before this heavy shading treatment. Three plants were used for each nucleotide, and the plants were randomised several times during the shading treatment. On 25th October all the panicles in each plant were harvested, dried in the air for three days, and were then threshed by hand to remove every grain from the rachis branch. The grain was poured into salt water of 1.06 specific gravity, and stirred with a glass rod so that the good grains and the poor grains could be separated. Then, after rinsing and drying, the percentage of ripe grain was calculated. (This is the standard way in Japan for determining the percentage of ripe grain).

Proof of translocation of chemicals injected into the peduncular air space (2)

Plants of both rice and wheat were transplanted into Wagner pots after fertilization, and were placed in the Radioisotope Greenhouse of the Central Agricultural Experiment Station at Konosu. Prior to the peduncle-injection of ^{32}P, pin-holes were made at the top of the peduncle with a needle so that the air could come out when the isotopic liquid was injected into the peduncular air space. Approximately

0.3 ml of ^{32}P was injected through the tissues of the flag leaf sheath and peduncle, as shown in Fig. 1, using a medical syringe. The panicles were harvested 7 days after the injection together with the stem and the upper three leaves. The harvested plant organs were laid out on a board and fixed with pieces of Scotch adhesive tape, and were wrapped with a thin plastic film. A sheet of X-ray film (25.4 x 30.5 cm) was then contacted with the sample, and was exposed for 2 days in a dark room to obtain autoradiograms.

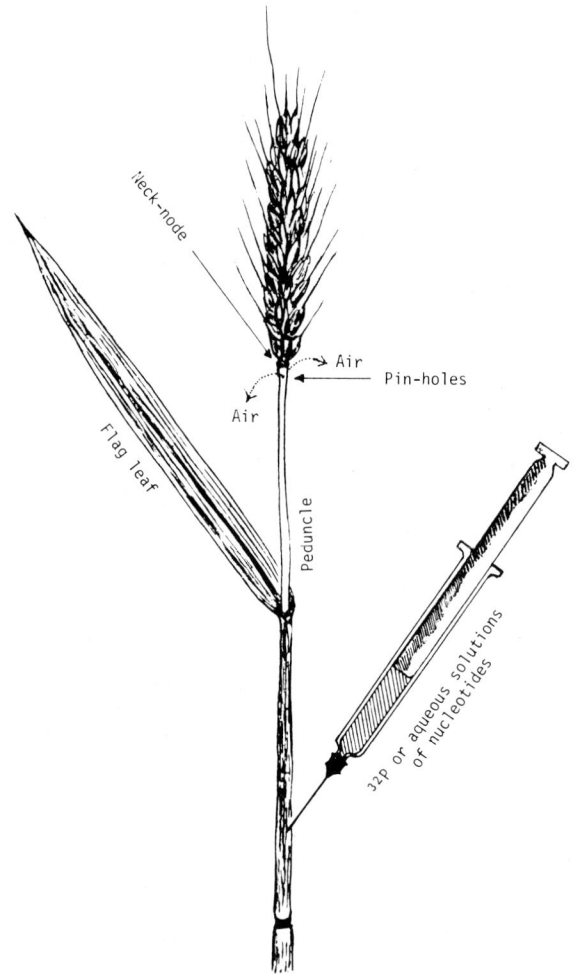

Fig. 1. Injection of ^{32}P (or aqueous solution of nucleotides) into the peduncular air space of a wheat plant.

Application of nucleotides to the crops in the field (3)

Aqueous solutions of nucleotides, such as AMP, GMP, CMP and UMP were also injected into the peduncles of both rice and wheat under field conditions.

Rice (3a) The rice plants (*var.* MANRYO, transplanted 21st May, 1970, at a spacing of 22.2 hills/m^2) were headed evenly around 10th of August, and the peduncle-injection was done on that day. Ten panicles were used per treatment, and they were tagged after injection. The treatments were as follows:

AMP-20 (p.p.m.), AMP-200, AMP-2000;   GMP-20, GMP-200, GMP-2000;   CMP-20, CMP-200, CMP-2000;   UMP-20, UMP-200, UMP-2000;   CONTROL.

Fig. 2. The multilayer shaking screen used in the separation of wheat grains.

Two sheets of reed screen (2 x 3 m/sheet, ca. 10% permeability to sunlight) were used to make a heavy shade after the injection, and were kept above the plants for 50 days until the treated panicles were harvested. The harvested panicles were dried in the air for 3 days, and the percentage of ripe grain was determined as before.

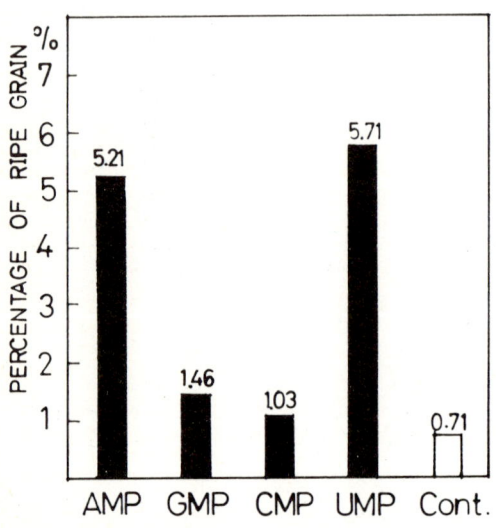

Fig. 3. Comparison of percentage of ripe grain grown in the weak light conditions of a shed after foliar application of aqueous nucleotide solutions. (Numbers above the histogram bars are each averages of three plants).

Wheat (3b) Injection of nucleotides into wheat peduncles was tried on 20th of May, 1970 in the field at the institute mentioned above. The seeds of wheat (*var.* KANTO No. 77) were sown 1st of November, 1969. Aqueous solutions of AMP, GMP, CMP and UMP were applied to the wheat peduncle, all at 1000 p.p.m. Since the peduncle of this variety is rather thick, about 3 ml of liquid could be injected. No shading treatment was done during the ripening period.

Fig. 4. Autoradiograms of ^{32}P injected into the peduncles of rice (a) and wheat (b) Black arrow indicates the injected point. fl: flag leaf, 2: 2nd internode, 3: 3rd internode.

Fig. 5. Autoradiograms of ^{32}P in a wheat ear and its parts (p = whole ear, k = grains, pl = paleas, lm = lemmas, r = rachis).

The treated panicles were harvested on 10th of June, 1970, and the grains were removed from the rachis by hand. A multi-layer shaking screen (OHYA Co. Ltd., Nagoya, Japan, see Fig. 2) was used to sort the grains by their size. Five flat boxes (25 x 20 x 3.5 cm) attached different screen-mesh at the bottom were piled up in the order of 1.8, 2.0, 2.2, 2.4 and 2.6 mm from the bottom. The wheat grains were put into the top box and a horizontal shaking was given for 5 minutes. Then the number of grains and the weight of grains left on the individual screens were measured. The results were expressed as percentages of the total grain number and weight respectively.

III. RESULTS

(1) As shown in Fig. 3, the percentage of ripe grain of rice plants grown inside a shed was affected by treatment with nucleotides. Aqueous solutions of AMP and UMP both seemed to promote ripening, although the percentage increases were rather low.

(2) Clear evidence of translocation of P^{32} from the peduncle to the panicle was obtained, although experiments with labelled nucleotide were not carried out. Fig. 4 shows the translocation of ^{32}P to the panicles of both rice and wheat. Fig. 5 shows incorporation of ^{32}P into the wheat kernel. There was little evidence of movement towards the lower internodes, but ^{32}P was actively incorporated into the kernels together with photosynthetic metabolites.

(3-(a)) As shown in Table 1 and Fig. 6, the ripening of rice grains appears to be influenced by nucleotides injected into the peduncular air space after heading. UMP-200 was the most effective treatment, followed by AMP-200. The effects of GMP and CMP on ripening were lower as in the preliminary experiment mentioned above. The concentration of any of these nucleotides giving the highest percentage of ripe grain is 200 p.p.m.

Table 1. No. of grains separated by salt water (specific gravity 1.06) and the percentage of ripe grain treated with nucleotides aqueous solutions by mean of peduncle injection.

Treatment	No. of Good Grains	No. of Poor Grains	Total No. of Grains	% of Ripe Grains	No. of Collected Panicles
AMP-20	279	695	974	28.64	7 (3 were missed)
AMP-200	397	911	1308	30.35	9 (1 was missed)
AMP-2000	371	1029	1400	26-50	10
GMP-20	241	833	1074	22.44	8 (2 were missed)
GMP-200	396	1035	1431	27.67	10
GMP-2000	245	790	1035	23.67	7 (3 were missed)
CMP-20	246	865	1111	22.14	10
CMP-200	403	1027	1430	28.18	10
CMP-2000	255	1145	1400	18.21	10
UMP-20	226	819	1045	21.63	8 (2 were missed)
UMP-200	456	798	1254	36.36	10
UMP-2000	332	1112	1444	22.99	10
CONTROL	245	1101	1346	18.20	10

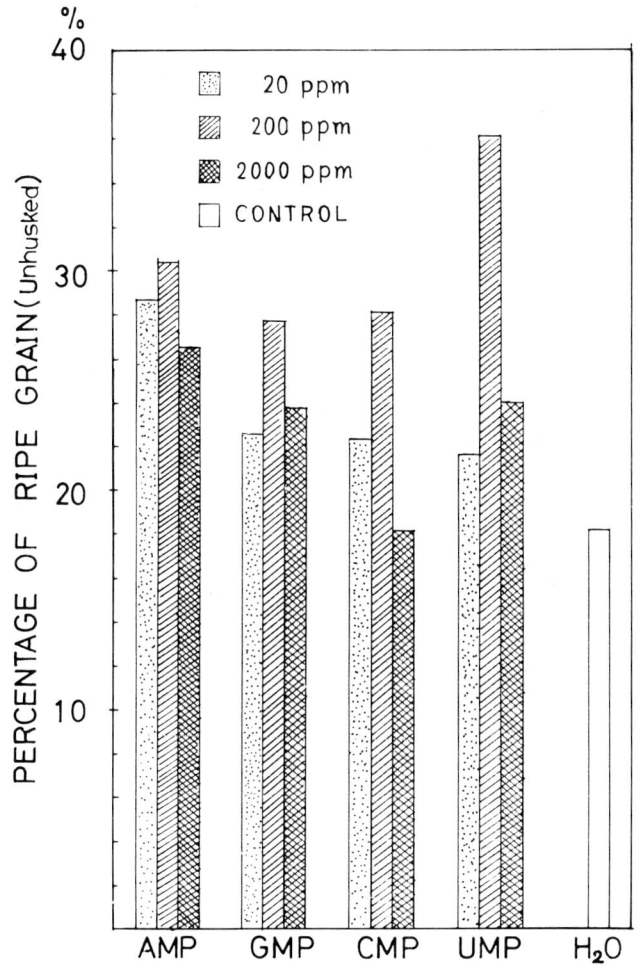

Fig. 6. Percentage of ripe grain grown under a heavily shaded condition in the paddy field after the peduncle-injection of aqueous nucleotides solutions.

(3-(b)) Table 2 and Fig. 7 are the results of the experiment with wheat grown in the field. According to these data, the percentages of both number and weight of wheat grains left on the largest screen-mesh were highest for UMP. The percentages for GMP were the lowest on the larger screen-mesh and the highest on the smaller screen-mesh.

IV. DISCUSSION

A single spray of aqueous nucleotide solution to the leaves of rice plants might not be sufficient to cause promotion of ripening, but the author did observe a tendency towards ripening promotion by AMP and UMP in the preliminary experiment. The heavier the shading treatment, the clearer the effects of AMP and UMP.

The clear demonstration of the incorporation of ^{32}P into the panicles of both rice and wheat leads the author to assume that the injected nucleotides also were incorporated and might contribute to the ripening of spikelets. It is of interest to note that the main incorporation of phosphorus is at the centre of growth (the ripening panicle in this case), as far as the peduncle-injection is concerned. Lis and Antoszewski (1970) studied the translocation of IAA-^{14}C in the strawberry peduncle, and reported that the phloem stream is normally occupied in moving assimilates acropetally because the developing fruit actively attracts many metabolites. Grochowska

Table 2. The number and the weight of wheat grains (var. KANTO No. 77) left on individual screen after 5 minutes' horizontal shaking, and the percentages in number and weight.

PLOT	ITEMS	CLASSES OF GRAIN SIZE (mm)				
		2.6	2.4	2.2	2.0	1.8
AMP	No. of Grains	282	288	77	7	4
	% in Number	42.9	43.8	11.7	1.1	0.5
	Wt. of Grains (gr)	11.1	9.8	2.0	0.2	0.1
	% in Weight	48.1	42.6	8.5	0.7	0.1
GMP	No. of Grains	195	236	103	32	6
	% in Number	34.1	41.3	18.0	5.6	1.0
	Weight of Grains	7.6	8.0	2.8	0.7	0.1
	% in Weight	39.7	41.4	14.9	3.4	0.6
CMP	No. of Grains	241	257	77	9	4
	% in Number	41.0	43.7	13.1	1.5	0.2
	Weight of Grains	9.4	8.7	2.4	0.2	0.1
	% in Weight	45.4	42.1	11.5	1.0	0.1
UMP	No. of Grains	301	264	76	10	3
	% in Number	46.0	40.4	11.6	1.5	0.5
	Weight of Grains	11.8	8.8	2.1	0.3	0.1
	% in weight	51.4	38.5	9.1	0.9	0.1
CONT.	No. of Grains	236	236	50	6	4
	% in Number	44.4	44.4	9.4	1.1	0.7
	Wt. of Grains	9.1	8.0	1.3	0.1	0.1
	% in Weight	49.3	43.1	7.1	0.3	0.2

(1970) injected labelled IAA into growing apple seeds, and found radioactivity in the peduncle, adjacent leaf, and fruitlets within 24 hrs after injection. The author has also found a similar phenomenon in this experiment, and in an earlier experiment, in which he injected uracil-2-^{14}C into the endospermic cavity of a young wheat seedling (Tomita, 1968). As Lis and Antoszewski have pointed out, the peduncles of cereal crops as well as those of apple and strawberry seem to have an active physiological function as reservoirs for accumulated assimilates. Gifford et al. (1963) applied ^{3}H-thymidine and adenine-8-^{14}C to the shoot tips of *Gossypium hirsutum*, *Tropaeolum majus* and *Vitis rupestris*. Applications to the bud and internode resulted in the incorporation of these labelled substances into the shoot tips of these plants. Adenine was readily incorporated into the growing point of grapevine, though both radioactive substances were translocated to the shoot tip within 24 - 96 hrs after the application. The author previously experimented with the injection of nucleosides and nucleotides into the peduncle of rice (Tomita, 1967), and he noticed that rice grains which incorporated a mixture of AMP and GMP produced albino seedlings in high percentage (17 - 20%) when they were germinated after the breaking of dormancy.

From these and other results mentioned above, the author has confirmed that not only phosphorus but also other substances are actively incorporated into the fruits and contribute to development and ripening.

3-(a). Since there are so many factors in the practical field situation, it is rather hard to obtain clear results with growth regulators. Moreover the rice plant is rather insensitive to growth regulators except GA. The heavy shading experiment with a reed screen, however, increased the effect of nucleotides on the ripening of rice in the

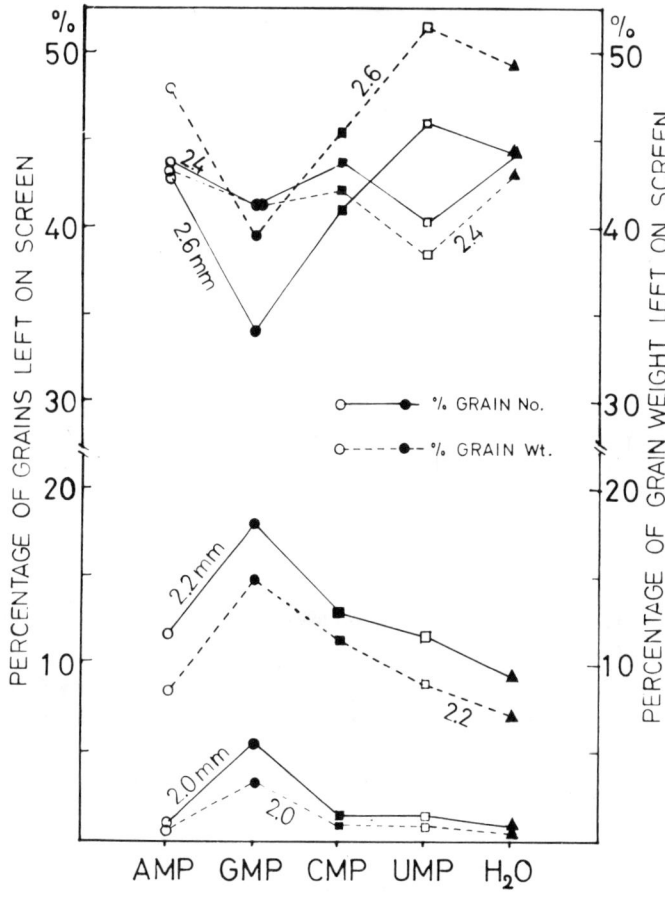

Fig. 7. Percentages of grain number and grain weight of wheat left on the individual screen after 5 minutes' horizontal shaking.

field. Of course, peduncle-injection of nucleotides is not a practical technique, but the result in this report may give some hope of promoting ripening in cereal crops by nucleotides. According to Murakami (1970), nucleotides such as AMP, UMP, ADPG, ATP, UDP are highly localized in rice grains during the ripening period; when supplied to detached ears, the content of AMP, UMP, ADPG and ATP is increased (see Fig. 8). The data of Murakami support my own conclusions. Akazawa et al. (1964) affirm that UDPG and ADPG play an important role in starch synthesis in the rice plant. The ripening of cereal crops may be enhanced by the exogenous application of nucleotides leading to an increase in the supply of these co-factors.

3-(b). Although further work needs to be done on wheat peduncle-injection, almost identical effects of both AMP and UMP on ripening were observed as for rice plants. Since the glumes are open in the case of wheat, differing from the closed glumes of rice, salt water treatment cannot be used in grain classification. Murakami (1970) has reported that AMP, ADP, ADPG, ATP, UMP, UDP, UDPG, UTP, CMP and guanosine derivatives are present in ripening wheat grains, and she concluded that ADP- and UDP-sugars were the principal derivatives of adenosine and uridine in the wheat grains.

A question, "What about the effect of inorganic phosphorus on ripening?", may be asked. Kadowaki and Fujiwara (1966) applied both $H_3{}^{32}PO_4$ and ^{32}P-RNA to rice seedlings, and they found that the incorporation of ^{32}P-RNA into leaves was much higher than that of $H_3{}^{32}PO_4$, three days after application. Organic P linked with nucleosides (=nucleotides) may thus be able to contribute more directly to ripening than inorganic P.

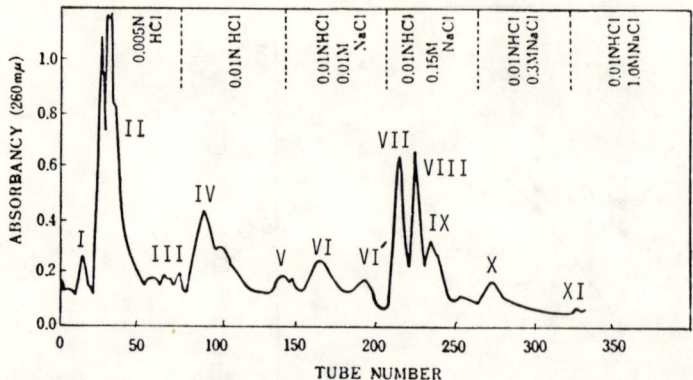

Fig. 8. Ion exchange chromatogram of acid-soluble nucleotides extracted from rice grains (T. Murakami, 1970, with permission from the author and the editor of Proceedings of the Crop Science Society of Japan).

I: unknown. II: AMP. III: CMP. IV: UMP. V: unknown. VI: ADPG. VII: UDPG.
VIII: ATP. IX: UDP. X: UTP. XI: unknown.

Interactions of nucleotide mixtures on ripening were not investigated in these experiments. It is possible that further promotion may result from the application of suitable concentrations and mixtures of nucleotides.

It may be asking a lot to expect agronomic application of some effective nucleotides to cereal crops to secure better ripening and higher yields. However, the mononucleotides mentioned above are not commercially expensive chemicals any longer. These chemicals may well be used in agriculture and horticulture as fruit-developing agents in the near future.

V. SUMMARY

Experiments were carried out with the aim of promoting the ripening of cereal crops such as rice and wheat especially under unfavourable conditions of frequent rainy weather during the ripening period.

Organic phosphorus compounds, nucleotides, were injected into the peduncular air space of rice and wheat after fertilization. Aqueous solutions of AMP and UMP seemed to promote the ripening of both crops, and especially UMP at 200 ppm showed a clear effect in rice grown in the paddy field under heavy shade. GMP and CMP were less effective.

Translocation of injected nucleotides toward the panicle from the peduncle was inferred from injection of ^{32}P, and autoradiography. Incorporation of ^{32}P into the kernels was more active than into the glumes.

The ripening of cereal crops such as rice and wheat may be greatly enhanced by the application of nucleotides under inadequate lighting during the ripening period.

VI. REFERENCES

AKAZAWA, T., T. MINAMIKAWA and T. MURATA (1964). Enzymic mechanism of starch synthesis in ripening rice grains. Pl. Physiol., Wash. 39, 371-378.

GIFFORD, E.M., S. KUPILA and S. YAMAGUCHI (1963). Experiments in the application of H^3-thymidine and adenine-8-C^{14} to shoot tips. Phytomorphol., Delhi, 13, 14-22.

GROCHOWSKA, M.J. (1970). The problems of metabolism and transport of plant growth regulators. BIOLOGIA, Toruń, XIII, 171-174.

KADOWAKI, M. and A. FUJIWARA (1966). The incorporation of $H_3P^{32}O_4$ and P^{32} labelled RNA into various cellular fractions of rice seedlings. Tohoku Jour. Agri. Res., Sendai, 17, 81-88.

LIS, E.K. and R. ANTOSZEWSKI (1970). Translocation of IAA-^{14}COOH and some mineral ions in the strawberry peduncle (in vivo and in vitro studies by mean of double labelling technique). BIOLOGIA, Toruń, XIII, 175-179.

MURAKAMI, T. (1970). Acid soluble nucleotides in rice plants at ripening stage and effect of removal of ears on the nucleotide pool. Proc. Crop Sci. Soc. Japan. 39, 287-294.

MURAKAMI, T. (1970). Acid soluble nucleotides in wheat plants at ripening stage and effect of removal of ears on the nucleotide pool. Proc. Crop Sci. Soc. Japan, 39, 409-417.

TOMITA, T. (1966). The effect of nucleic acid substances injected into the peduncle of rice plant on the growth of next generation (preliminary report). Bull. Tohoku Branch, Crop Sci. Soc., Sendai, 8, 15-16.

TOMITA, T. (1968). Flowering activities of natural and chemical substances injected into the cavity of empty endospermic space of young winter wheat seedlings, in "Biochemistry and Physiology of Plant Growth Substances" (Ed. Wightman F. and G. Setterfield), 1399-1413. Runge Press Ltd., Ottawa, 1968.

APPLICATION OF GLC-MS TO HORMONE STUDIES

A System for the Characterisation of Plant Growth Substances
Based Upon the Direct Coupling of a Gas Chromatogram, a Mass
Spectrometer, and a Small Computer - Recent Examples of its
Application

J. MacMillan

School of Chemistry, The University, Bristol, England

I. INTRODUCTION

At the 6th International Conference on Plant Growth Substances the initial application of combined gas chromatography-mass spectrometry (GC-MS) for the detection and identification of plant hormones was described (MacMillan, 1967). The potential of the method was indicated by some preliminary results on the identification of gibberellins in unripe seed of *Phaseolus coccineus**. In the intervening three years the power of the technique has been amply confirmed. The present paper describes a further development of the technique using a small computer and presents a selection of recent applications which illustrate further the value of the method.

II. COMPUTERISED GC-MS SYSTEM

The limiting factor in the combined operation of a gas chromatogram directly linked to a low resolution mass spectrometer is the processing of the vast amount of spectral data. These data are usually obtained as non-permanent traces on light sensitive paper and their conversion into convenient, permanent records such as normalised line diagrams or a list of m/e values and their intensities require considerable manual effort. In our laboratories, a computer-based system using a DEC Linc 8 has been developed (Binks et al., 1971) which processes on-line the low-resolution data from GC-MS and provides a library system for the storage, retrieval, and comparison of spectra. With this system, a normalised spectrum is obtained in 20 seconds from the end of the magnetic scan of a GC-peak. This spectrum can be displayed on an oscilloscope and photographed by an automatic camera. Alternatively it can be plotted on a slave flat-bed recorder or it can be listed by teletype.

The library system offers 8 options including the following two:

(a) Search Index

This option allows the search of reference spectra, contained on magnetic tape, for spectra which satisfy particular criteria. The spectra are located and listed *via* the teletype. At present two criteria are used - the largest m/e value and the base (most intense) peak - but provision has been made for the future addition of further search criteria to the programme. Those spectra that have the specified largest m/e value ± 2 units or have the specified base peak are listed on the teletype. Alternatively only those spectra which satisfy both criteria can be listed.

---

* The use of the alternative description, *P. multiflorus*, for scarlet runner bean shall now be discontinued in publications from the laboratory.

(b) Compare Two Spectra

To assist the identification of GC-peaks by GC-MS, the normalised spectrum of the peak can be compared with spectra already on reference file. The spectrum of the unknown can be displayed on the oscilloscope together with any selected reference spectrum. A difference spectrum, obtained by subtracting the intensities of the reference spectrum from those of the unknown spectrum, can also be obtained and displayed on the oscilloscope.

III. SELECTED RECENT EXAMPLES

With the kind co-operation of many colleagues we have used GC-MS to investigate the growth substances in extracts from a wide variety of plant sources. A selection of recent applications is presented below.

A. *Phaseolus coccineus*

A comprehensive investigation of the growth substances in immature seed of *P. coccineus* by GC-MS has been made by MacMillan and Pryce (1968). Their results are summarised in Table 1. The study of growth substance in *P. coccineus* is currently being extended to young seedlings in collaboration with Dr. A. Crozier, University of Calgary. Extracts of dark-grown seedlings of *P. coccineus* were purified essentially as described by Crozier *et al.* (1969). In the final step 25 fractions were obtained from the silicic acid partition column (Powell and Tautvydas, 1967). These fractions which were bio-assayed in 12 tests, were further examined by Mr. D.H. Bowen at Bristol the following way. Each fraction was methylated, then examined by GC on both 2% QF-1 and 2% SE-33 before and after trimethylsilylation. This preliminary GC study indicated the presence of $GA_{4/7}$, GA_{19}, and GA_1 in fractions 4, 8, and 12 respectively. The methylated fraction 4 was further purified by TLC on silica gel with 35% v/v acetone in light petroleum (60-80°). The TLC plate was divided into 4 equal sections and the ethyl acetate eluants of these bands were re-examined by GC. One of these bands from fraction 4 was examined by GC-MS before and after trimethylsilylation. In both cases, the mass spectra of the main GC-peak were identical to those of the authentic GA_4 derivatives. The TLC trace of the methylated and silylated fraction is shown in Fig. 1; peaks 1 and 3 have not been identified. It should be noted that GA_4 could have been mis-identified as GA_7 on the basis of the GC data and the biological activity.

Table 1. Growth Substances in *P. coccineus*

Compound	Conc.(mg/kg f.wt.) by GC
A_1	8.0
A_5	1.5
A_6	2.5
A_8	30.0
A_{17}	2.0
A_{19}	0.5
A_{20}	0.5
C-α	0.35
C-β	0.15
Phaseic acid	0.3

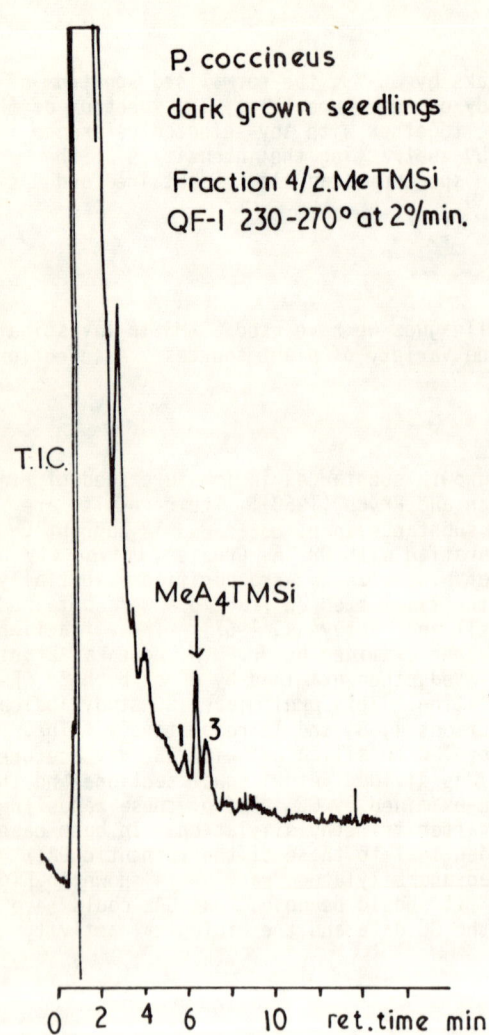

Fig. 1. *P. coccineus* dark grown seedlings; TLC trace of methylated and trimethysilylated fraction 4/2 on 2% QF-1 programmed from $230°$ to $270°$ at $2°$/min.

This identification of GA_4 as the major gibberellin in dark-grown seedlings of *P. coccineus var.* Prizewinner requires comment because this gibberellin was not detected in immature seed of the same variety. From this result it could have been concluded that the accumulation of gibberellins in developing seed has no direct bearing on the subsequent germination process. For it is highly improbable that the 13-hydroxylated gibberellins in the developing seed are converted during germination into GA_4 which lacks this hydroxyl group. However the situation is not so straightforward. Schneider and Sembdner (personal communication) have detected GA_4 in seed of the same variety by TLC and IR and we have confirmed by GC-MS the presence of GA_4 in a fraction provided by them.

B. *Liquid endosperm of Echinocystis macrocarpa*

Information on the gibberellins present in liquid endosperm of *E. macrocarpa* is of obvious interest in connection with the important biosynthetic studies of West and his group. Using TLC, bio-assay, and fluorimetry Elson et al. (1964) detected GA_1, GA_3, GA_4 and GA_7 in very low concentration compared to the very high biological activity, originally reported for the liquid endosperm by Corcoran and Phinney (1962).

With the co-operation of Professor C.A. West who has generously supplied extracts containing the ethyl acetate-soluble acids from liquid endosperm we have re-investigated the growth substances in these extracts. Mr. P. Gaskin, in our laboratory, has examined the crude acids in three separate extracts; all showed low biological activity and gave essentially the same results by GC-MS.

Direct GC-MS of the crude acid fraction after methylation and after methylation and trimethylsilylation, showed the presence of GA_4 and GA_7, together with three new compounds E_1, E_2 and E_3 which appear to be gibberellins from their mass spectra. Compounds E_1 and E_2 are isomers of GA_{13} and GA_{17} and compound E_3 is probably 2,3-dihydroxy-GA_{25} (Fig. 2). Trace amounts of ABA and its *trans*-isomer were also detected after purification by TLC. A major component (M^+ 460 with base peak m/e 199) is not a gibberellin.

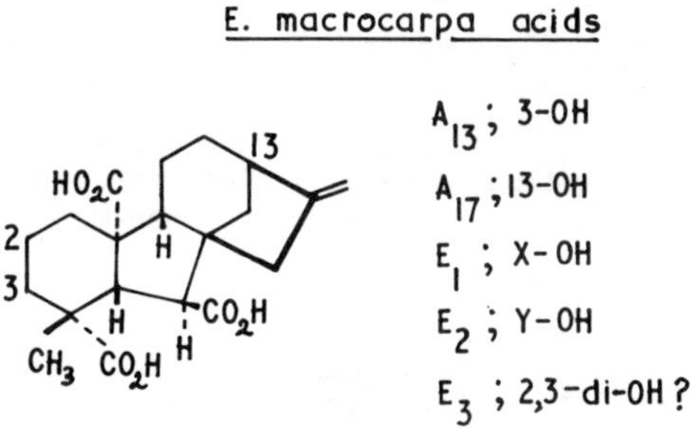

E. macrocarpa acids

A_{13} ; 3-OH
A_{17} ; 13-OH
E_1 ; X-OH
E_2 ; Y-OH
E_3 ; 2,3-di-OH ?

Fig. 2. Gibberellins E_1, E_2, and E_3 from liquid endosperm of *E. macrocarpa*.

The amounts of gibberellins, detected in these extracts are very low ($\leqslant$ 2.0 µg/ml liquid endosperm). Nevertheless, the extracts were obtained from endosperm used by Professor West in his biosynthetic studies. The low level of gibberellins in this endosperm may point to a low capability for gibberellin biosynthesis and may explain the inability, observed to date, of the cell-free systems derived from the endosperm to convert 7β-hydroxykaurenoic acid into gibberellins.

C. *Conversion of gibberellins by in vitro hydroxylating enzyme systems*

The oxidation of gibberellins A_4, A_5 and A_7 by hydroxylating enzyme systems are being studied in collaboration with Drs. J.R. Stoddart and Russell Jones. Mr. C. Cloke, in our group, has been investigating the nature of the conversion product of GA_7 (I; Fig. 3) by peroxidase/dihydroxy fumarate and by tyrosinase/ascorbate systems. Both systems gave in about 5% yield a fluorescent, bio-active product which looked like GA_3 from TLC, GLC and bio-assay.

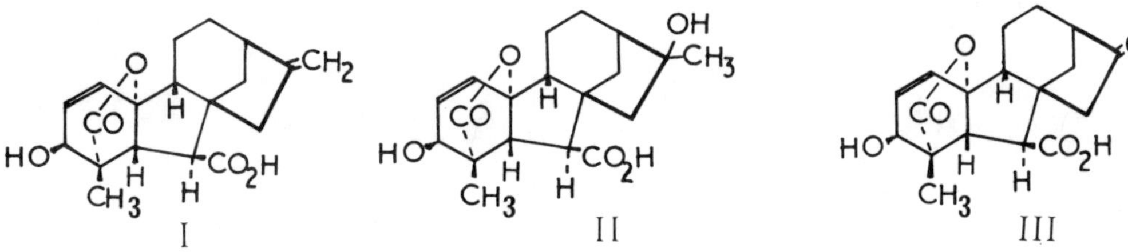

Fig. 3. Gibberellin A_7 and derivatives.

This was an exciting possibility as an analogy for the last step in the fungal biosynthesis of GA_3 (Verbiscar et al. 1967; Spector and Phinney, 1968). However careful GC of the product and the NMR spectrum, obtained on the methyl ester of the partially purified product, excluded this possibility, and suggested the hydrated-GA_7 structure (II; Fig. 3) which has recently been the subject of a patent claim (Sumiki et al. 1969). Comparison of the NMR spectrum, kindly provided by Dr. Kagawa for compound (II) with that of the conversion product of GA_7, showed that the two compounds were not the same. Unfortunately this information did not exclude the structure (II) for the conversion product because the spectroscopic data indicated that the compound described by Kagawa et al. may not have the structure (II) - an observation previously made by Professor N. Takahashi (personal communication). The hydrated structure (II) for the conversion product of GA_7 was finally excluded by GC-MS of the methyl ester and its trimethylsilyl ether. The GC-MS data indicated that the conversion product was gibberellin-A_7 norketone (III; Fig. 3) and this structure was confirmed by comparison of the GC, IR and GC-MS with those of an authentic specimen.

The significance of this unusual and surprising conversion will be discussed elsewhere. The pertinent point here is that without GC-MS, a mistaken identification of the conversion product might easily have been made.

D. *Studies with Gibberella fujikuroi*

GC-MS has proved of great value in studying the production of metabolites by *G. fujikuroi*. Several new metabolites have been detected and their isolation and characterisation are in hand. One example, illustrating the use of the method in time-course studies, is selected from a collaborative programme with Professor B.O. Phinney, Miss M. Fukuyama and Mr. J. Bearder on the biosynthesis of gibberellins by recombinant and mutant strains from wild-types of Japanese origin.

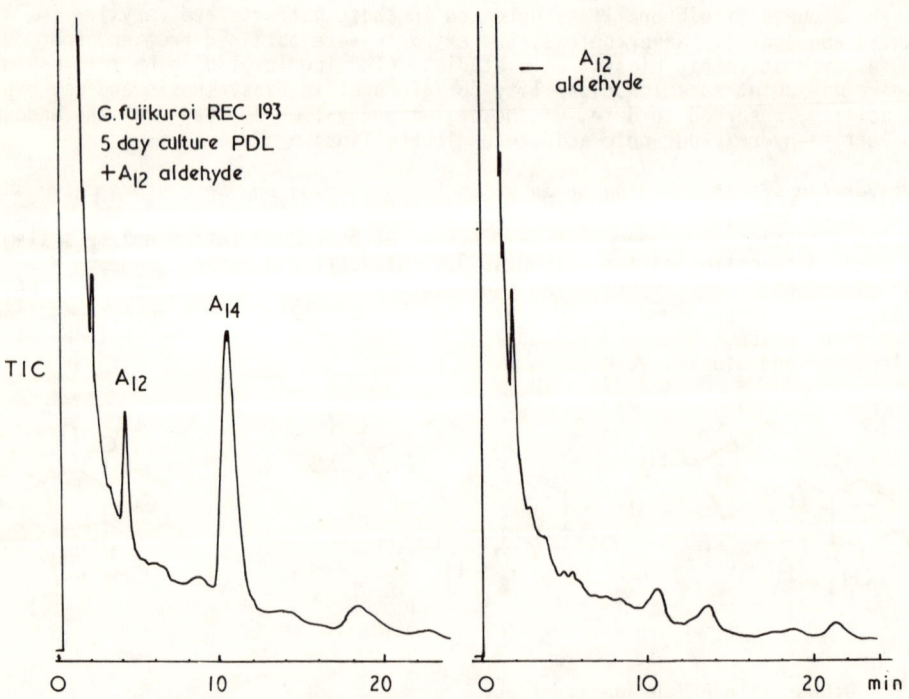

Fig. 4. *G. fujikuroi* REC193 (BOP); TLC trace of methylated acids after 5 days' growth on PDL: (a) plus GA_{12}-aldehyde (2 mg./100 ml.) and (b) control.

A recombinant strain (REC 193) is a good producer of gibberellins and the metabolites detected by GC-MS after 11 days' culture on potato-dextrose liquid medium (PDL) are listed in Table 2. After 5 days' growth on this medium REC 193 produces little or no gibberellins (Fig. 4b). However with gibberellin A_{12}-aldehyde added to the medium (2 mg. per 100 ml) gibberellin A_{12} and gibberellin A_{14} can be identified by GC-MS (Fig. 4a) under the same conditions of culture. Although the conversion of GA_{12}-aldehyde to GA_{12} and GA_{14} must be confirmed by tracer studies, this preliminary result provides direct evidence (cf. Cross et al. 1968) that the first step in the conversion of GA_{12}-aldehyde (or GA_{12}) into gibberellins is hydroxylation at position-3 (diterpene numbering).

Table 2. Metabolites of G. fujikuroi (REC193 BOP) by GC-MS (11-day culture on PDL)

Gibberellins		Non-gibberellins
A_1	A_9	p-Hydroxyphenyl acetic acid
A_3	A_{12}	Di-acid ex Fujenal
A_4	A_{13}	Tri-acid ex Fujenal
A_7	A_{14}	Fusaric acid.

E. *GC-MS Identification of abscisic acid*

Recently abscisic acid (ABA) has been identified by GC-MS in young carob fruit and in carob syrup from a Cyprus source (Most et al. 1970). It is the principal, if not the only inhibitor, in fractions B_1 and C as defined by Corcoran (1966). It should be noted that fractions B_1 and C, kindly supplied by Dr. M. Corcoran from a Californian source, have been found by Mr. P. Gaskin in our laboratories to contain ABA in trace amounts, far below the level to account for the observed biological activity of these fractions.

Recently ABA has been identified by GC-MS in an extract of coffee buds (Browning et al. 1970). The TLC trace (Fig. 5) revealed the presence of palmitic and stearic acids in the fraction together with ABA and an unidentified compound (M^+ 264).

IV. SCOPE AND LIMITATIONS

The main requirement for GC-MS is that the growth substance, or suitable derivatives, can be gas chromatographed. This is a severe limitation and precludes the direct investigation of many conjugates, such as glycosides, although they can be examined indirectly after hydrolysis. Nevertheless, polyhydroxy gibberellins like GA_{32}, recently isolated from seed of *Prunus persica* by Yamaguchi et al. (1970) can readily be gas chromatographed as the MeTMSi-ether or trimethylsilyl ester TMSi ether, and we have shown by GC-MS that GA_{32} and apricot gibberellin (B.G. Coombe and M.E.Tate, this volume) are identical. Another hazard is the possibility of decomposition during gas chromatography.

A further limitation is the amount of material required. Although much less than 1 μg can be detected, a minimum of 5 μg is usually required in practice for derivatisation. This minumum quantity should be in a minimal concentration of 5% in the total fraction. This last requirement is not easily met in cases where the endogenous levels are low and considerable purification may be necessary. The use of pure solvents for extraction and purification steps is essential to avoid contamination with interfering plasticisers such as phthalates and acetyl tri-n-butyl citrate (Binks et al. 1970). Also the use of Parafilm should be avoided (Gaskin et al. 1971).

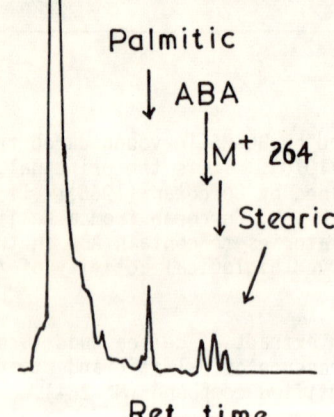

Fig. 5. TLC trace of methylated fraction from *Coffea arabica* on 2% SE-33 programmed from $210°$ to $230°$ at $2°$/min.

Despite these limitations, the value of GC-MS is such that, apart from isolation, it is the only available method for the conclusive identification of known plant growth substances. GC-MS can also assist in the detection of new growth substances and may, in particular cases, allow their structures to be deduced with a fair degree of certainty. However the eventual isolation of new compounds is necessary to establish their structures and evaluate their biological activity unless the deduced structures can be synthesised.

V. SUMMARY

A system for on-line processing of GC-MS data by a small computer is described briefly. The system provides a library system for storage and retrieval of reference GC-MS spectra. Several examples of the use of GC-MS are presented. Gibberellin A_4 has been identified in dark grown seedlings of *Phaseolus coccineus*, but not in seed of the same variety. Gibberellins A_4 and A_7 have been identified in extracts of the liquid endosperm of *Echinocystis macrocarpa* together with three new gibberellins and abscisic acid (ABA); the amounts detected are surprisingly low. The oxidation product of GA_7, obtained *in vitro* with peroxidase/dihydroxyfumarate, is GA_7 norketone. The use of GC-MS in the study of gibberellin production by *Gibberella fujikuroi* is exemplified by the identification of GA_{14} and GA_{12} after the addition of GA_{12}-aldehyde under culture conditions which do not produce these two gibberellins. The GC-MS identification of ABA in extracts of carob and coffee buds is noted.

The scope and limitations of GC-MS are discussed.

VI. ACKNOWLEDGEMENTS

GC-MS spectra were obtained by the skill and industry of Messrs. P. Gaskin and D.H. Bowen on either (a) an LKB9000 instrument through the courtesy of Dr. R.L.S. Patterson, Meat Research Institute, Langford, or (b) a Varian MAT CH7 instrument (NERC Grant GR/3/655 to Dr. G. Eglinton). The Linc 8 computer was purchased on an SRC Grant (B/SR/4959) to Dr. J. MacMillan.

VII. REFERENCES

BINKS, R., R.L. CLEAVER, J.S. LITTLER, and J. MacMILLAN (1970). A small computer on-line to a low resolution Mass Spectrometer. Chemistry in Britain, 7, 8-12.
BINKS, R., R.J. GOODFELLOW, J. MacMILLAN, and R.J. PRYCE (1970). Acetyl tri-n-butyl citrate, a common laboratory contaminant. Chem. and Ind. (London), 565-566.
BROWNING, G., G.V. HOAD, and P. GASKIN (1970). Identification of Abscisic Acid in flower buds of *Coffea arabica*. Planta, 94, 213-219.
CORCORAN, M.R. (1966). Reduction of α-amylase by an inhibitor from Carob. Plant Physiol., 41, 1265-1267.
CORCORAN, M.R. and B.O. PHINNEY (1962). Changes in amounts of Gibberellin-like substances in developing seed of *Echinocystis*, *Lupinus* and *Phaseolus*. Physiol. Plantarum, 15, 252-262.
CROSS, B.E., K. NORTON, and J.C. STEWART (1968). The biosynthesis of Gibberellins Part III. J. Chem. Soc. (London), 1054-1063.
CROZIER, A., H. AOKI, and R.P. PHARIS (1969). Efficiency of countercurrent distribution, Sephadex G-10, and Silicic Acid partition chromatography in the purification and separation of Gibberellin-like substances from plant tissue. J. Exptl. Bot., 20, 786-795.
ELSON, G.W., D.F. JONES, J. MacMILLAN, and P.J. SUTER (1964). Plant Hormones IV. Identification of the Gibberellins of *Echinocystis macrocarpa* Greene by thin-layer chromatography. Phytochemistry, 3, 93-101.
GASKIN, P., J. MacMILLAN, R.D. FIRN, and R.J. PRYCE (1970). "Parafilm" - A Convenient Source of n-Alkane standards for the determination of gas chromatographic retention indices. Phytochemistry, 10, 1155-1157.
MacMILLAN, J. (1967). In "Biochemistry and Physiology of Plant Growth Substances", 101-107 (Wightman, F., Setterfield, G. Ed. Runge, Ottawa).
MacMILLAN, J. and R.J. PRYCE (1968). Recent studies of endogenous plant growth substances using combined gas chromatography-mass spectrometry. Soc. Chem. Ind. Monograph 31, 36-50.
MOST, B.H., P. GASKIN, and J. MacMILLAN (1970). The Occurrence of Abscisic Acid in inhibitors B_1 and C from immature fruit of *Ceratonia siliqua* L. Planta, 92, 41-49.
POWELL, L.E. and K.H. TAUTVYDAS (1967). Chromatography of Gibberellins on silica gel partition columns. Nature (London), 213, 292-293.
SPECTOR, C. and B.O. PHINNEY (1968). Gibberellin Biosynthesis: Genetic studies in *Gibberella fujikuroi*. Physiol. Plantarum, 21, 127-136.
SUMIKI, Y., T. KAGAWA, and T. FUKUMIBARA (1969). Gibberellin A_7 derivatives. Chem. Abs., 70, 37947k (Jap. Pat. 68 10, 149).
VERBISCAR, A.J., G. CRAGG, T.A. GEISSMAN, and B.O. PHINNEY (1967). Studies on the biosynthesis of Gibberellins-II. The biosynthesis of gibberellins from (-)-Kaurenol, and the conversion of Gibberellins ^{14}C-GA-4 and ^{14}C-GA-7 into ^{14}C-GA-3 by *Gibberella fujikuroi*. Phytochemistry, 6, 807-814.
YAMAGUCHI, I., T. YOKOTA, N. MUROFUSHI, Y. OGAWA, and N. TAKAHASHI (1970). Isolation and structure of a new Gibberellin from immature seed of *Prunus persica*. Agric. Biol. Chem. (Japan), 34, 1439-1441.

IDENTIFICATION OF CYTOKININS BY GAS-LIQUID CHROMATOGRAPHY
AND GAS-LIQUID CHROMATOGRAPHY-MASS SPECTROMETRY

C.D. Upper, John P. Helgeson and C.J. Schmidt

Pioneering Research Laboratory, Crops Research Div., Agricultural Research Service, U.S. Dept. of Agriculture and Dept. of Plant Pathology, University of Wisconsin, Madison, Wis., 53706

I. INTRODUCTION

Gas-liquid chromatography (GLC) of cytokinins was demonstrated by Most et al. (1968) for known compounds and by Upper et al. (1970) for isolation of cytokinins from hydrolysates of t-RNA or extracts of bacterial culture filtrates. Babcock & Morris (1970) demonstrated the feasibility of quantitative as well as qualitative measurements of cytokinin ribosides from t-RNA hydrolysates. The following is a report on our progress in improving separation of known cytokinins, in extending our measurements to include free cytokinins (particularly those from higher plants), and using coupled gas-liquid chromatography-mass spectrometry (GLC-MS). These techniques may provide a simple, sensitive and accurate procedure for identifying and measuring levels of free and t-RNA cytokinins *in vivo*.

II. RESULTS AND DISCUSSION

GLC of cytokinins

Most cytokinins are easily separated by GLC of their trimethylsilyl (TMS) derivatives on either DC-11 (Babcock and Morris, 1970) or QF-1 (Upper et al., 1970). The usefulness of these liquid phases is limited by poor separation of the TMS derivatives of 2-methylthio-6-(3-methyl-2-butenylamino)-β,D-ribofuranosylpurine (2MS-2iPA) and *trans*-zeatin riboside. Separation of these two compounds has been obtained on SE-33 or OV-17 (Table 1). In both systems, separation of TMS_4 *cis*-zeatin riboside, TMS_3-2MS-2iPA and TMS_4 *trans*-zeatin riboside are adequate. Unfortunately, the poor resolution of SE-33 limits the usefulness of this column. On the other hand, good resolution and reasonably good separation of all of the cytokinins we have tried (free bases and ribosides) was achieved using temperature programmed operation of an OV-17 column. However, chromatography of TMS-ribosides derived from enzymic hydrolysis of t-RNA did not separate cytokinins from some other, thus far unidentified ethyl acetate-extractable RNA components.

Babcock & Morris (1970) have pointed out that it is possible to recognize cytokinin ribosides from RNA hydrolysates by isothermal chromatography on a DC-11 column. Table 2 shows the relative retention times of TMS derivatives of several cytokinins (both ribosides and free bases) on DC-11, using temperature programmed operation. Although TMS_4 *trans*-zeatin riboside and TMS_3-2MS-2iPA do not separate well in this system, the other cytokinins are relatively well separated.

GLC-MS. Preliminary experiments, using an LKB Model 9000 instrument[1], demonstrated the feasibility of GLC-MS of the cytokinins. However, the apparent GLC resolution of

---

[1] Mention of a trademark name or a proprietary product does not constitute a guarantee or warranty of the product by the USDA, and does not imply its approval to the exclusion of other products that may also be suitable.

Table 1. Relative retention times[1] of TMS derivatives of cytokinin ribosides, guanosine and adenosine.

	3% OV-17[2]	2% SE-33[4]
adenosine	0.773	0.366
	---	0.436
guanosine	0.868	0.621
	0.894	0.671
2iPA	1.00[3]	1.00[5]
zeatin riboside (*cis*)	1.11	2.01
(*trans*)	1.15	2.37
2MS-2iPA	1.19	2.24

[1] Retention time of 2iPA = 1.00
[2] Conditions: 3% OV-17 on gas chrom Q in a 1.8 mm i.d. x 1.5 m Pyrex column, 28 ml N_2/min (outlet), 3 min at 165°, then programmed at 6°/min for 21 min, then isothermal 6 min. Detection by flame ionization.
[3] Retention time = 22.1 min
[4] Conditions same as 2, except temperature isothermal @ 245°.
[5] Retention time = 4.9 min

Table 2. Relative retention times of some TMS derivatives of cytokinins on DC-11[1].

Cytokinin	Relative retention time[2]
2iP	0.29
zeatin	0.61
2iPA	1.00
zeatin riboside (*cis*)	1.13
(*trans*)	1.18
2MS-2iPA	1.15

[1] Conditions: 2% DC-11 on gas chrom Q in a 2 mm i.d. x 1.5 m Pyrex column, 28ml N_2/min (outlet), 3 min at 165°, then programmed at 6°/min for 21 min.
[2] 2iPA - 1.00, retention time = 19.4 min.

the cytokinins which we obtained on a QF-1 column by monitoring total ion current in the LKB was considerably poorer than we expected from analytical runs in our own gas chromatograph, using flame ionization detection. Thus although we obtained good spectra of TMS-cytokinin ribosides, we were unable to resolve individual chromatographic peaks from partially purified biological extracts.

We were unable to obtain any spectra of TMS-cytokinins on an Hitachi RMU-6 fitted with a Biemann-Watson separator. This may have been due to either the separator or to stainless steel tubing downstream from the separator.

Finally, we tried a Finnigan Model 1015 instrument fitted with an all-glass Goelke separator. When properly tuned, this instrument appears satisfactory for GLC-MS of the TMS derivatives of the cytokinins in spite of the relatively low resolution and lack of metastable ions inherent in the quadrapole-type mass spectrometer. This indicated to us that the sensitivity of the mass spectrometer and quality of the separator seem to be critical factors for measurements of this type; the resolution of the mass spectrometer does not (providing that it can handle resolutions greater than about 1:700) seem to be critical.

The fragmentation pattern of the purine moiety of the TMS-derivatives of the cytokinins is essentially the same as that of the nonderivatized compounds. Thus, the mass spectra of the TMS cytokinins are fairly simple and can be interpreted easily by comparisons with spectra of the appropriate nonderivatized cytokinins. The fragments can generally be divided into three classes:

(1) Those predicted by the spectrum of the nonderivatized material but increased in mass due to the expected number of TMS groups.

(2) Those due to the purine moiety which do not contain TMS moieties and are common to the mass spectra of both the derivatized and nonderivatized compounds.

(3) Those attributable to the TMS_3-ribosyl moiety and common to all TMS ribosides (McClosky et al., 1968).

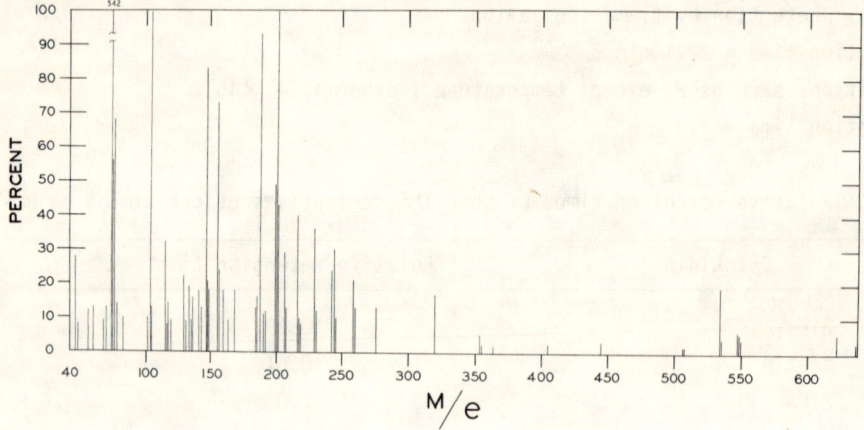

Fig. 1. The mass spectrum of TMS_4 trans-zeatin riboside. Inlet system: GLC with DC-11 column. Mass spectrometer: Finnigan Model 1015.

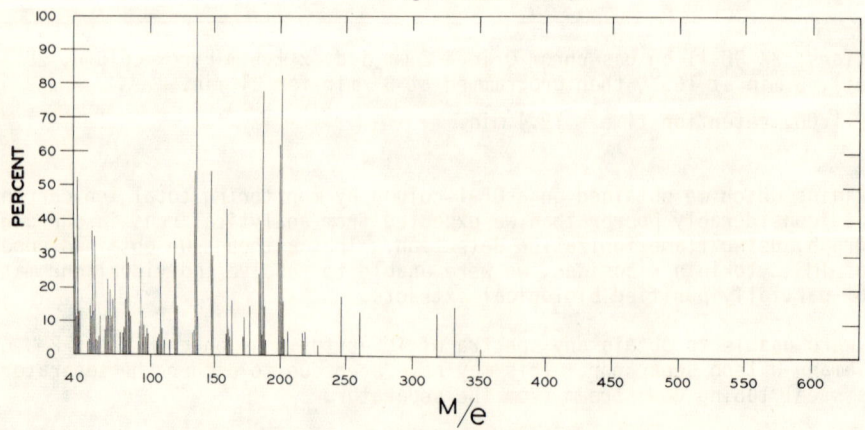

Fig. 2. The mass spectrum of TMS_4 zeatin riboside. Inlet system: solid inlet, $150°$ probe temp. Mass spectrometer: Finnigan Model 1015.

The mass spectrum of TMS4 zeatin riboside as obtained from GLC-MS with the Finnigan instrument is shown in Fig. 1. For comparative purposes, the mass spectrum of nonderivatized zeatin riboside, as obtained with the direct inlet with the same instrument on the same day, is given in Fig. 2.

This spectrum of the nonderivatized material is very similar to that reported by Hall et al. (1967) for cis-zeatin riboside. Comparing this spectrum to that of the TMS4 derivative one finds the molecular ions at M/e 351 and 639 (351 + 4x72) for nonderivatized and TMS4-zeatin riboside, respectively. The ion at M-15 or M/e 624 represents loss of a methyl group, a frequent fragment from TMS derivatives. Class 1 fragments with their corresponding ions from nonderivatized material are detected at M/e 334 (loss of -OH) and 550 and 549 (loss of OTMS, OTMS and H), M/e 320 and 536 (loss of CH_2OH and CH_2OTMS), M/e 292 and 508 (loss of $C(CH_3)CH_2OH$, not visible in Fig. 2 and loss of $C(CH_3)CH_2OTMS$). Class 2 fragments, arising by fragmentation of the zeatin moiety itself, include peaks at M/e 202 (loss of OH and OTMS), 201 (loss of H_2O or TMSOH), 188 (loss of CH_2OH or CH_2O TMS) as well as peaks at M/e 160, 148, 135, 119 and 108.

Class 3 fragments common to all TMS ribosides, include a small TMS_3 ribosyl fragment (with or without loss of a proton, M/e 349 or 348) and fragments at M/e 258, 245, 243, 230, 169 and 103.

Peaks representing TMS transfer (McClosky et al., 1968) are frequently present in mass spectra of the cytokinin ribosides e.g. the peak at M/e 260 in Fig. 1. This, in addition to a tendency toward hydrogen migrations, the many isomers of silicon, and the "sticky" nature of TMS derivatives make these particular derivatives quite undesirable for mass spectroscopy. Nonetheless, they are currently the derivatives of necessity since they can be separated by gas chromatography.

The spectra of TMS_2 zeatin and the TMS_4 zeatin ribosides are the most complex spectra of the derivatized cytokinins which we have examined by GLC-MS. The fragmentation patterns of all of the other TMS derivatives of cytokinins which we have examined are even less complex. For example, the fragmentation pattern of TMS_3 6-(3-methyl-2-butenylamino)-β,D-ribofuranosylpurine (2iPA) obtained on the LKB Model 9000 GLC-MS is shown in Fig. 3.

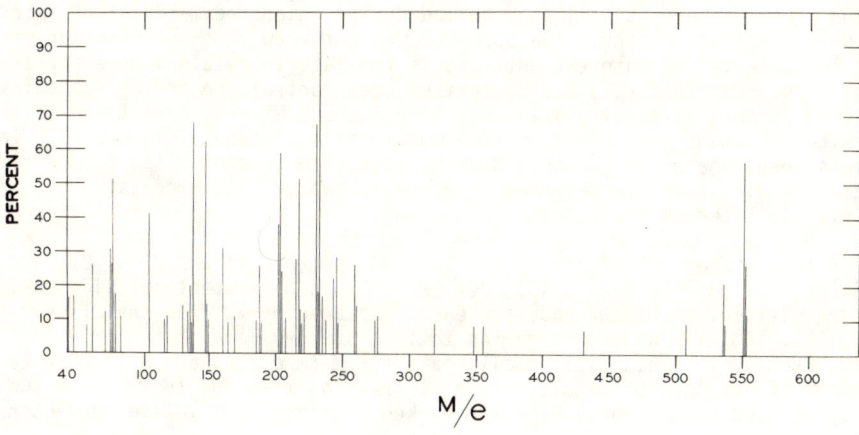

Fig. 3. The mass spectrum of TMS_3-2iPA. Inlet system: GLC with QF-1 column. Mass spectrometer: LKB Model 9000.

Class 1 fragments are the molecular ion (M/e 551), M-15 (M/e 536, loss of CH_3) and M-43 (M/e 508, loss of $CH(CH_3)_2$). Class 2 fragments common to both the spectra of derivatized and nonderivatized 2iPA (Biemann et al., 1966; Hall et al., 1966) include those at M/e 203, 188, 160, 148, 135, 119 and 108. Class 3 fragments include those at M/e 349, 348, 258, 245, 243, 230, 169 and 103. Ribosyl fragments still containing the purine moiety include M/e 318, and part or all of the peak at M/e 232. (This peak can also arise by silyl transfer, as do those at M/e 260 and 275.)

Burrows et al. (1968) showed that the mass spectrum of 2MS-2iPA was very similar to that of 2iPA, but with the appropriate ions all 46 amu larger due to the methylthio group. As can be seen from the spectrum obtained on the LKB 9000, this relationship also holds for TMS_3-2MS-2iPA (Fig. 4) and TMS_3-2iPA (Fig. 3).

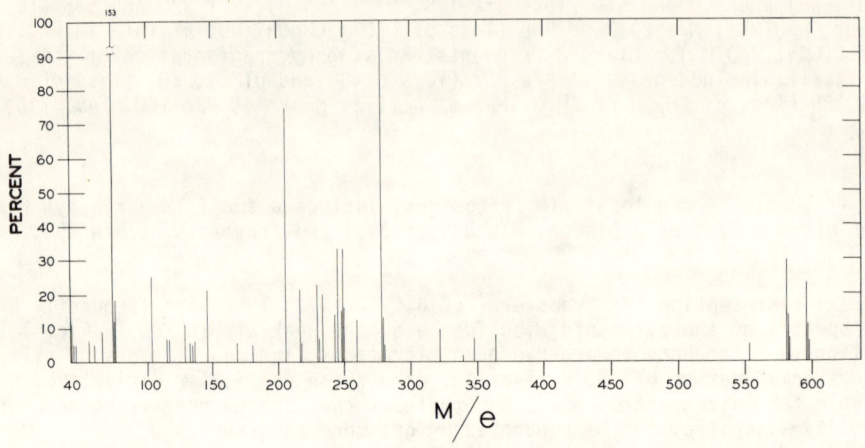

Fig. 4. The mass spectrum of TMS_3-2MS-2iPA. Inlet system: GLC with QF-1 column. Mass spectrometer: LKB Model 9000.

Computer-controlled GLC-MS

During GLC-MS, there is a limited period during which the material of interest is actually in the spectrometer. The operator may not know that a particular peak represents the material of interest until it is too late to obtain a mass spectrum during that peak. For this reason computerized scan control, recording, calculation and processing of data is particularly desirable for GLC-MS. We have tried this technique with a Systems 150 output control module on a Finnigan 1015 GLC-MS. In this mode the mass spectrometer scans continuously, scan rate is controlled by the computer, and all data for all scans are recorded on magnetic tape. Data processing and printout are performed later at the operator's command.

Fig. 5 is a photograph of a computer-plotted spectrum of TMS-2iP. Class 1 peaks at M/e 275 (M), 260 (M-15), 232 (M-43), 207 (M-68, loss of isopentenyl side chain also a peak in the OV-1 column background) are each displaced by 72 from the class 2 peaks at M/e 203, 188, 160, 135 which are common to this spectrum and that of 2iP (Helgeson and Leonard, 1966). The relatively small peaks for M (M/e 275) and M-15 (M/e 260) are an example of the loss of sensitivity at increasing mass due to improper tuning of a quadrapole type instrument. Note the marked increase in relative ion intensity with decreasing ion size.

Zeatin riboside, obtained from Calbiochem, contains both the *trans*-(predominant) and the *cis*-isomers which can be separated on DC-11 (Babcock & Morris, 1970). When scans were examined across these two peaks, only small quantitative differences in relative peak heights were observed in the mass spectra. Thus, unlike the nonderivatized compounds (Hall et al, 1967) the TMS4 derivatives of cis- and trans-zeatin riboside are probably indistinguishable by mass spectroscopy.

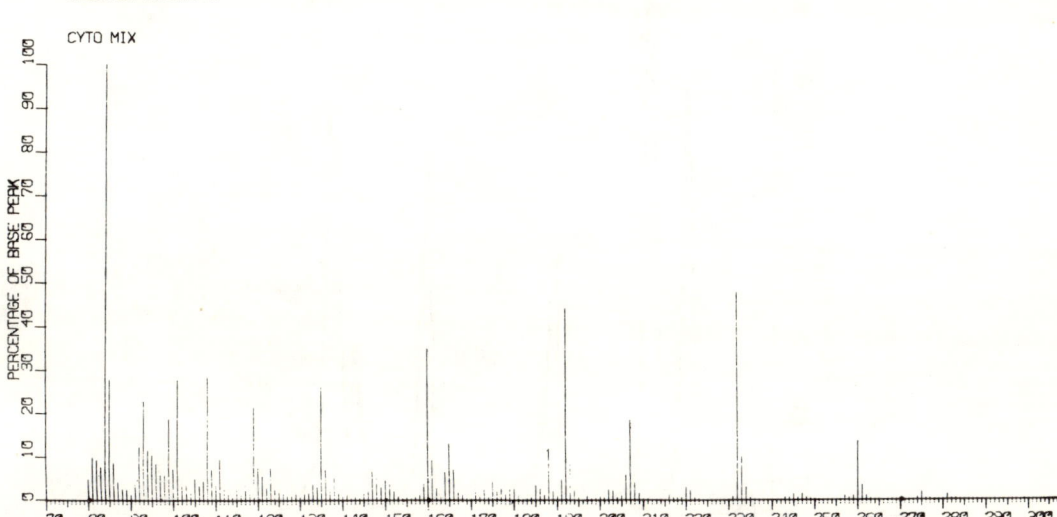

Fig. 5. Computer plotted mass spectrum of TMS-2iP. Inlet system: GLC with OV-1 column. Mass spectrometer: Finnigan Model 1015 with Systems Model 150 control module.

The computerized GLC-MS can also be used to identify components from poorly resolved mixtures. As mentioned above, all of the cytokinins except trans-zeatin riboside and 2MS-2iPA are easily separable as the TMS derivatives. Fig. 6 shows a computer reconstructed gas chromatogram. In this experiment the recording system and the $10°$/min temperature program were started three minutes after injection into an OV-1 column at $195°$. During the experiment the mass spectrometer scanned continuously, and all data were recorded on magnetic tape. In Fig. 6, the sum of all ions received in each scan is normalized on the largest peak and plotted on the ordinate. The unit of the abscissa is scan number. Each scan was approximately ten seconds in duration.

The compounds injected were (in order of elution) TMS derivatives of 2iP, kinetin, 2MS-2iP, 2iPA, cis-zeatin riboside, 2MS-2iPA, and trans-zeatin riboside. The plot of total ions received indicates that all compounds except trans-zeatin riboside and 2MS-2iPA were well separated. However, when individual sequential scans were examined, separation of even these two components became apparent (Table 3). In this experiment M/e 205 was selected as an approximate indication of background, M/e 201 as an indication of both cis- and trans-zeatin ribosides, and M/e 206 as an indication of the presence of 2MS-2iPA. The relative abundance of these ions in each scan indicates that TMS_4 cis-zeatin riboside peaked during scans 62 and 63, that TMS_3-2MS-2iPA peaked about 20 seconds later, during scans 64 and 65, and that TMS_4 trans-zeatin riboside peaked 10 to 15 seconds later, in scan 66. Thus, the computer GLC-MS detected a separation which was not evident from either flame ionization or total ion measurement. Furthermore, this detailed examination of the data would not have been possible had not all data from every scan been stored on magnetic tape.

III. MEASUREMENT OF CYTOKININS IN HIGHER PLANTS

Detection of cytokinins in extracts of higher plants by a GLC procedure is complicated by the large number and quantity of interfering materials. We have used a combination of DEAE-cellulose column chromatography and ethyl acetate extraction to remove or reduce the quantities of many of the interfering compounds prior to GLC

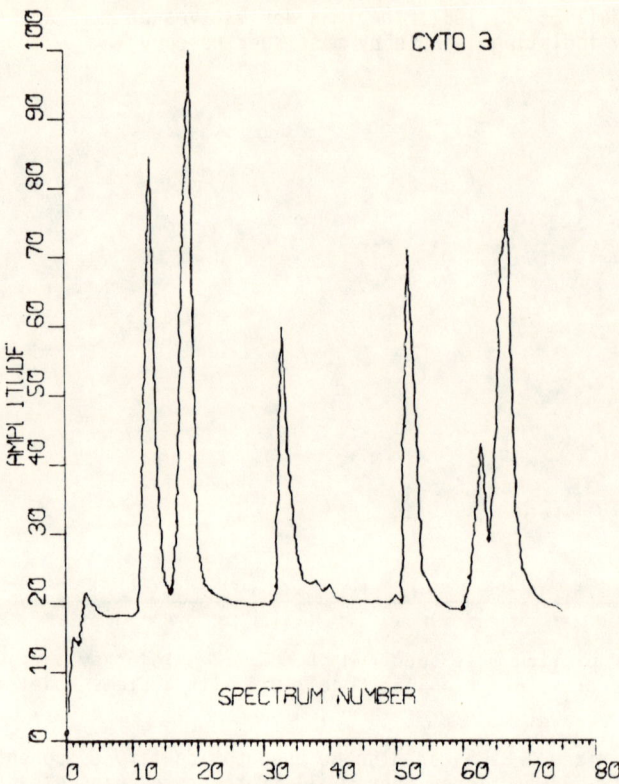

Fig. 6. Computer reconstructed gas chromatogram of a mixture of TMS derivatives of cytokinins. See text for description of experiment.

Table 3. Relative intensities of selected ions in sequential mass spectra from a portion of the gas chromatogram shown in Figure 6.

Spectrum number	Height of peak at M/e		
	201	205	206
	% base peak		
61	2.80	2.80	3.73
62	11.78	2.02	4.37
63	12.56	2.01	3.51
64	4.22	2.34	12.67
65	2.39	1.62	13.90
66	23.83	0.96	3.41
67	12.73	2.54	4.45

analysis. By this procedure, all of the cytokinins which we have tested are carried on to the GLC step unseparated and in at least semi-quantitative yields. (The exceptions are the zeatin ribosides, 60 to 70% of which remain in the aqueous phase during ethyl acetate extraction.) In one experiment, the top (15 g) of a tobacco plant (grown in a growth chamber until just before flower bud opening) was removed, and homogenized in 25 ml of 0.03 M potassium phosphate buffer, pH 7.5 containing 0.02 M

KCl. The debris was removed by centrifugation. The supernatant fluid was passed through a 2.5 x 7.5 cm column of DEAE-cellulose and the column eluted with the same buffer. The fraction emerging between 20 and 154 ml was collected and extracted 4 times with 1/2 volumes of ethyl acetate.

One tenth of the combined ethyl acetate extracts (corresponding to 1.5 g tissue) was concentrated to dryness, dried by azeotroping from ethanol, silylated (10 µl dimethylformamide and 10 µl N,O-bis(trimethylsilyl)trifluoroacetamide, 65°, approximately 45 minutes). Subsequent GLC of 0.7 µl of the silylation mixture (equivalent to approximately 50 mg fresh weight) on 2% DC-11 is shown in Fig. 7.

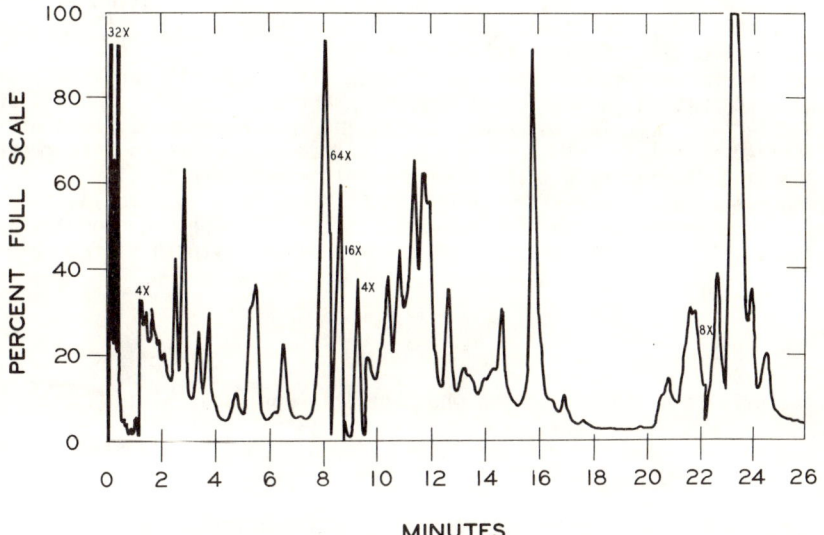

Fig. 7. GLC of a tobacco extract (see text for preparation). Chromatographic conditions were: 2% DC-11 on Gas Chrom Q, N_2 flow rate = 28 ml/min (outlet), flame ionization detection (full scale deflection represents 3×10^{-11} amp at an attenuation of 1). Temperature program: 3 minutes isothermal at 165°, then increased by 6°/min for 21 min, then isothermal for 6 min.

A peak at 4.6 min (2iP) and a shoulder at 11.5 min (zeatin) suggested the presence of these substances in this extract. No peaks corresponding to 2MS-2iP or 2iPA are visible (these would be at approximately 12.1 and 18.5 min, respectively). Several peaks emerged in that region of the chromatogram in which 2MS-2iPA and *cis*- and *trans*-zeatin ribosides normally appear. Coinjection of TMS-derivatives of 2iP and zeatin resulted in increased height of the appropriate peaks; coinjection with TMS_4-zeatin riboside suggested that a shoulder at 22.8 min might contain the *cis*-compound.

The remaining 3/4 of the solution (equivalent to about 1.2 g fresh weight) was injected, 5 µl at a time, into an identical column with an all-glass stream splitter at the outlet. Fifteen sixteenths of the column effluent at retention times corresponding to 2iP, zeatin, 2MS-2iP, 2iPA, *cis*-zeatin riboside and combined 2MS-2iPA and *trans*-zeatin riboside were collected. The "2iP" and "zeatin" fractions contained biological activity in the tobacco bioassay equivalent to 0.5 and 0.8 n moles per gram tissue, respectively. Although this does not constitute a positive identification of 2iP and zeatin in extracts of tobacco, the GLC would have separated them from any other cytokinins which we have examined. Positive identification of these fractions awaits coupled GLC-MS. The bioassays of the areas representing 2MS 2iP and 2iPA were negative, as expected. Bioassays for 2MS 2iPA and the zeatin ribosides were also negative.

The identification and measurement of cytokinins by GLC or GLC-MS is well on the way to becoming a usable technique. Several columns are known which will separate most or all of these compounds as their TMS derivatives. Coupled GLC-MS of the TMS derivatives of the cytokinins yields relatively simple mass spectra. The simple nature of these mass spectra, particularly above M/e 210, will simplify both qualitative and quantitative measurements of cytokinins even under conditions where they are not completely separated from other components.

IV. SUMMARY

Progress in developing a GLC and GLC-MS system for identification and measurement of levels of cytokinins is described. Several column systems that will separate TMS derivatives of most or all of the cytokinins are known. Introduction of the TMS derivatives of cytokinins into a mass spectrometer via a GLC inlet provides both chromatographic separation and efficient sample insertion. The mass spectra so obtained are relatively simple and can be easily interpreted by comparison with spectra of non-derivatized materials. The use of computerized scan control, data recording and data processing can increase the apparent resolution beyond the limits of total ion measurements or flame ionization. Progress has been made in removing interfering materials by column chromatography and solvent extraction. These procedures will simplify both qualitative and quantitative measurements of cytokinins from complex mixtures.

V. ACKNOWLEDGEMENTS

We gratefully thank the following who made this work possible:

F.T. Keeney and the Department of Dairy Science, Pennsylvania State University, for the use of the LKB-9000 GC-MS.

The Finnigan Instrument Company, Palo Alto, Calif., for the use of a Finnigan 1015 GC-MS with computerized control, data recording and data processing.

F. Skoog and N.J. Leonard for samples of cytokinins.

VI. REFERENCES

BABCOCK, D.F. and R.O. MORRIS (1970). Quantitative measurement of iso-prenoid nucleosides in transfer ribonucleic acid. Biochemistry 9, 3701-3705.
BIEMANN, K., S. TSUNAKAWA, J. SONNENBLICHLER, J. FELDMAN, D. DUTTING, and H.G. ZACHAU (1966). Struktur eines ungewöhnlichen Nucleosids aus Serin-spezifischer Transfer-Ribonucleinsäure. Angew. Chem. 78, 600-601.
BURROWS, W.J., D.J. ARMSTRONG, F. SKOOG, S.M. HECHT, J.T.A. BOYLE, N.J. LEONARD, and J. OCCOLOWITZ (1968). Cytokinin from soluble RNA of *Escherichia coli*: 6-(3-methyl-2-butenylamino)-2-methylthio-9-β-D-ribofuranosylpurine. Science 161, 691-693.
HALL, R.H., L. CSONKA, H. DAVID, and B.M. McLENNAN (1967). Cytokinins in the soluble RNA of plant tissue. Science 156, 69-71.
HALL, R.H., M.J. ROBINS, L. STASIUK, and R. THEDFORD (1966). Isolation of N^6-(γ,γ-dimethylallyl)adenosine from soluble ribonucleic acid. J. Amer. Chem. Soc. 88, 2614-2615.
HELGESON, J.P. and N.J. LEONARD (1966). Cytokinins: Identification of compounds isolated from *Corynebacterium fascians*. Proc. Natl. Acad. Sci. U.S. 56, 60-63.
McCLOSKEY, J.A., A.M. LAWSON, K. TSCUBOYAMA, P.M. KRUEGER and R.N. STILLWELL (1968). Mass spectrometry of nucleic acid components. Trimethylsilyl derivatives of nucleotides, nucleosides, and bases. J. Am. Chem. Soc. 90, 4182-4184.

MOST, B.H., J.C. WILLIAMS, and K.J. PARKER (1968). Gas chromatography of cytokinins. J. Chromatog. 38, 136-138.

UPPER, C.D., J.P. HELGESON, J.D. KEMP, and C.J. SCHMIDT (1970). Gas-liquid chromatographic isolation of cytokinins from natural sources. Plant Physiol. 45, 543-547.

INDEX TO AUTHORS

Abdul-Baki, A.A. 81
Abeles, F.B. 239, 278, 500, 501, 508, 602, 609.
Abrams, M. 214, 247, 255.
Ackers, G.K. 239.
Acs, G. 377.
Adamson, D.A. 227, *299*, 305, *428*, 433, 434, 440, 660.
Adamson, Heather. 433, *435*.
Addicott, F.T. 165, 214, 221, *272*, 278, 279, 334, 342, 588, 602, 609.
Addink, C.J. 68.
Adedipe, N.O. *571*, 579.
Agatep, A.D. 426.
Ahmad, A. 125.
Alam, S.N. 474.
Albritton, W.L. 377.
Alexander, D. McE. 483.
Al-Omary, Saib A. 779.
Amir, J. 597.
Ammirato, P.V. 685.
Amos, H. 394.
Andersen, A. Skytt. 101.
Anderson, J.D. 426.
Anderson, J.W. 570, 602.
Andersen, L. 214, 279.
Andersen, M. 290.
Anderson, M.B. *181*, 189, 465.
Andreae, W.A. 61, 100, 500, 501, 525.
Anfinsen, C.B. 247.
Anker, L. 730.
Anojulu, C.C. 214, 279.
Anon, 774.
Anthony, R.D. 778.
Antozewski, R. 789.
Aoki, H. 418, 797.
Aoyama, T. 174.
Apgar, J. 189.
Arancibia, M. 779.
Arch, P.D. 280.
Arglebe, C. 255.
Armstrong, D.J. 189, 254, 255, 466, 580, 685, 806.
Arnon, D.I. 609.
Ashri, A. 447.
Askonas, B.A. 355.
Aspinall, D. 214, 278, 588, 602.
Atkin, R.K. 596, 602.
Atkinson, D.E. 271.
Atkinson, M.R. *459*, 465, 466.
Atsmon, D. *23*, 29, *222*, 227.
Audus, L.J. 81, 278, 427, 724, 730, 743.
Aurich, O. 149, 150.
Avramaes, S. 667.
Avtalion, R.R. *661*, 667.
Awdeh, Z.L. 355.

Ayalon, S. 765.
Ayers, J.E. 617.
Axen, R. 247.
Azakawa, T. 789.

Babcock, D.F. 806.
Back, A. 588.
Backs, D. 694.
Bagi, G. 394.
Bailey, R.W. 52, 61.
Bailey, W.K. 343, 508, 509.
Baker, N. 61.
Baker, R.A. 708.
Baldev, B. 426.
Baldwin, J. 778.
Baliga, B.S. 653.
Balls, A.K. 342, 508.
Balz, H.P. 570, 602.
Bara, M. 233.
Barbaro, A. 752.
Barendse, G.W.M. 149.
Barkley, G.M. 61, 67.
Barkley, W.K. 570, 579.
Barnes, J.E. 234.
Barnes, M.F. 141, 426.
Barras, D.R. 602.
Barrett, F.C. 394.
Barrs, H.D. 29.
Bartlett, G.R. 42.
Barton, R. 743.
Bastin, M. 532.
Baur, A.H. *510*, 516, 517.
Baur, J.R. 609.
Bauer, S. 61.
Baxter, P. *775*.
Beakbane, A. Beryl. 483.
Bean, R.C. 602.
Beath, O.A. 125.
Becker, D. 779.
Beevers, L. 279, 427, 579, 588, 617, 645, *646*, 653.
Bellamy, A.R. 8, 189, 363.
Baum, H. 779.
Bellamy, A.W. 632.
Benn, M.H. 125.
Bennet, J.P. 774.
Bennet-Clark, T.A. 278, 724, 730.
Ben-Tal, Y. 447.
Bentley, J.A. 67, 774.
Bergmann, F. 125.
Bernfeld, P. 617.
Bernier, G. 765.
Berridge, M.V. *248*, 254, 596.
Berry, D.R. 427.
Beye, F. 323.
Beyer, Jr. E.M. 500, *502*, 508, 509, 560.

Bhattacharyya, J.   254.
Biale, J.B.   559.
Bickle, A.S.   67.
Bieleski, R.L.   29, 394.
Biemann, K.   806.
Biggs, R.H.   559.
Bigot, C.   465.
Binks, R.   797.
Birch, A.J.   141.
Birch, P.D.W.   774.
Birecka, H.   233.
Biswas, D.K.   264.
Bjorndal, A.   239.
Black, M.W.   774.
Blackman, G.E.   67, 75.
Blakely, L.M.   685.
Blanpied, G.D.   559.
Bleichert, E.F.   278, *668*.
Blew, Daphne.   205.
Blumenfeld, A.   617.
Blundell, J.B.   278.
Boardman, N.K.   254.
Bock, R.M.   189, 254.
Bogorad, L.   588.
Bollard, E.G.   483.
Bonner, D.   757.
Bonner, J.   8, 22, 42, 50, 95, 214, 247, 597, 730, 757.
Bonner, J.T.   660.
Booth, A.   447.
Bopp, M.   227.
Borek, E.   370.
Borgstrom, G.   736.
Bornman, C.H.   500.
Borthwick, H.A.   100.
Bottger, M.   278, 447.
Bottomley, W.   310.
Bouillenne, M.W.   532.
Bouillenne, R.   532.
Bowen, M.R.   757.
Boyd, E.S.   125.
Boyle, J.T.A.   189, 806.
Bradley, M.V.   765, 774.
Brady, C.J.   *589*, 596, 623.
Bragt, J. van.   427.
Branscombe, E.W.   67.
Brauner, L.   730.
Bray, G.A.   125, 264, 370.
Brian, R.C.   42.
Brian, P.W.   149, 173, 413, 645.
Briant, R.E.   75, 501, 708.
Briggs, D.E.   387, 394, 708.
Briggs, W.R.   100, 757.
Bristow, J.M.   427.
Brock, B.L.W.   271.
Broder, I.   271.
Brooks, G.C.   271.
Brooks, H.J.   447.
Broughton, W.J.   *407*, 413.
Brown, B.T.   *318*.
Brown, C.L.   447.
Brown Jr., R.J.   743.
Brownbridge, M.E.   724, 730.
Bruchovsky, N.   264.
Buchenauer, H.   323.
Buckwalter, B.   458.
Buggy, J.   16.
Bukovac, M.J.   765.
Bullock, M.N.   474.
Bunning, E.   737.
Bunsow, R.   752.
Burden, R.S.   290, 334, 335.
Burdett, A.N.   195.
Burg, E.A.   501, 525, 548, 559.
Burg, S.P.   501, 508, 516, 525, 532, 548, 559, 609.
Burgess, R.R.   247.
Burr, G.O.   632.
Burns, R.G.   363.
Burnstock, G.   778.
Burrows, W.J.   189, 254, 570, 632, 806.
Burstrom, H.   645.
Burton, K.   214, 363, 433.
Butcher, R.W.   376, 660.
Butenko, R.G.   752.
Butler, G.W.   125.
Buttrose, M.S.   395, 405.

Cain, J.C.   779.
Campbell, D.H.   247.
Camus, G.   737.
Cane, A.R.   715, 724.
Caplin, S.M.   678.
Capra, J.D.   370.
Carns, H.R.   278.
Carr, D.J.   *371*, 377, *378*, 387, 419, 427, 570, 579, 632, *633*, 645, 752.
Carter Jr., J.R.   394.
Case, D.B.   448.
Catalano, M.   447.
Ceriotti, G.   317.
Chader, G.J.   247.
Chadwick, A.V.   525, 532, 609.
Chailakyan, M-Kh.   *745*, 752, 779.
Chailon, S.   667.
Chakravorty, A.K.   221.
Chalkley, G.R.   264.
Chalmers, A.H.   465.
Champagnat, P.   447.
Chan, B.G.   779.
Chandra, G. Ram.   363, *365*, 370, 376, 377, 394, 395, 406.
Chang, Y-P.   708.
Changeux, J.P.   264, 271.
Chapman, D.   42.
Chastain, B.H.   370.
Chatterjee, S.K.   559.
Chavez, M.   61.
Chen, C.M.   189, 458, 465.
Chen, D.   394.

Chen, J.L.   596.
Cherayil, J.D.   189.
Cherry, J.H.   *181*, 189, 206, 221, 247, 264, 278, 370, 465, 653.
Chibnall, A.C.   570.
Child, C.M.   632.
Chin, T-Y.   579, 588, 617.
Chisholm, M.D.   126.
Chouard, P.   752.
Chrispeels, M.J.   61, 278, 343, 351, 355, 363, 370, 376, 377, 387, 394, 395, 406.
Church, R.B.   8.
Clagett, C.O.   516.
Clarke, A.E.   602.
Clarke, W.S.   778.
Cleaver, R.L.   797.
Cleland, C.F.   447, *753*, 757.
Cleland, R.   *1*, 8, 16, 22, 29, *44*, 50, 51, 61, 67, 427, 539.
Clements, J.B.   *633*, 645.
Click, R.E.   8, 413.
Cocking, E.C.   254.
Coggins, C.W.   617.
Cohen, D.   *23*, 90, 95, 164, 617, 708.
Cohen, L.A.   131.
Cohn, W.E.   579.
Cole, C.V.   394.
Collier, J.R.   377.
Collins, G.G.   *388*, 405.
Conn, E.E.   125.
Cook, A.R.   617.
Coombe, B.G.   *158*, 164, 165, 395, 405, 617, 708.
Cooper, W.C.   *543*, 548, 560.
Corcoran, Margaret R.   797.
Cousineau, G.H.   221.
Cracker, L.E.   278.
Cragg, G.   797.
Crane, J.C.   610, 765.
Crocker, W.   539, 560.
Cross, B.E.   141, 156, 173, 427, 797.
Crossley, D.J.   645.
Crow, W.D.   *324*.
Crowden, R.K.   413.
Crozier, A.   149, *414*, 418, 419, *420*, 427, 797.
Cruickshank, I.A.M.   310.
Csonka, L.   189, 806.
Cuatrecasas, P.   247.
Cunningham, A.   752, 757.
Currier, H.B.   602.

Dadd, R.H.   757.
Dagan, J.   579.
Dahm, K.H.   36.
Dainty, J.   29.
Dale, J.E.   427.
Dalton, L.K.   *318*.
Daly, J.W.   131.

Damman, L.G.   254.
Danielli, J.F.   335.
Darwin, C.   737.
Datta, A.   254, 264.
David, H.   189, 806.
Davidson, E.H.   221.
Davidson, H.   765.
Davies, B.J.   596.
Davies, C.R.   700.
Davies, E.   206, 254.
Davies, J.W.   254.
Davies, P.J.   278, 465.
Davis, L.A.   165.
Davson, H.   334.
Day, B.E.   645.
Dedolph, R.R.   579.
Deenan, L.L.M. van.   42.
Degani, N.   678, 685.
Degani, Y.   29, 227.
Dem-l, R.A.   42.
De.ffer, D. von.   779.
Dennis, D.T.   141, 156, 173, 427.
Dennis Jr., F.G.   447.
Denny, F.E.   525, 559.
Dijkman, M.J.   548.
Dilova, S.   610.
Dingman, C.W.   363.
Dixon, G.H.   221.
Dolk, H.E.   716.
Donaldson, G.   465.
Donini, P.   667.
Donini, S.   667.
Doree, M.   465, *469*, 474, 475.
Dorffling, K.   447, 559.
Dostal, H.C.   559.
Dotts, M.A.   36.
Doubt, S.L.   559.
Dove, L.D.   579, 602.
Dowben, R.M.   61.
Downs, R.J.   100.
Drowne, M.J.   508.
Drury, R.E.   343.
Duda, C.D.   413.
Duerst, W.   95.
Duffus, J.H.   355, 405.
Dugger, W.M.   617.
Dumas, T.   525.
Dunmore, P.   271.
Durand, H.   62.
Dure III, L.S.   *216*, 221, 278, 395.
Durkee, A.B.   610.
Durley, R.C.   149, 418.
Dutting, D.   189, 806.
Duval-Jouve, J.   645.
Duve, C. de.   405.
Duynstee, E.E.   370, 394.
Dwyer, M.R.   597.
Dyson, W.D.   *449*, 458.

Eagles, C.F.   280, 335.
Eb, A.A. van der.   351.
Edelman, I.S.   264.
Edelman, J.   433.
Eder, F.   323.
Edgerton, L.J.   559.
Eelnurme, I.   125.
Effland, M.J.   580.
Egawa, H.   132.
Egyud, L.G.   774.
Ehrenberg, L.   363.
Eidt, D.C.   298.
Eilati, S.K.   *611*, 617.
El-Antably, H.M.M.   580, 602, 623.
Elliott, Daphne C.   *459*, 465, 466.
Elliott, M.C.   125.
Elmer, O.H.   560.
Elson, G.W.   149, 797.
Embry, J.L.   278.
Engelbrecht, L.   579, 580.
Engelsma, G.   75.
Erdman, D.   323.
Erickson, R.O.   16.
Eriksson, K.E.   239.
Erith, A.G.   632.
Ernback, S.   247.
Ernest, L.C.   *493*, 501, 560.
Esashi, Y.   278, 343, 508.
Eschrich, W.   602.
Esnault, R.   724, 730.
Ettlinger, M.C.   125.
Euyama, A.   132.
Evans, A.C.   447.
Evans, L.T.   752, 779.
Evans, M.L.   8, 16, 50, 61, 67, 75, 81, 730.
Evans, P.K.   434.
Everett, G.A.   189.

Fadl, M.S.   532.
Falk, H.   743.
Fall, R.R.   133.
Fambrough, D.M.   247.
Fan, D.F.   206, 542.
Fanestil, D.D.   264.
Farkaṣ, G.L.   363, 394, 603.
Farkaṣ, V.   61.
Farr, A.L.   95, 214, 247, 370, 570, 597, 609.
Faust, M.   560.
Fawcett, C.H.   335.
Fedorcsak, I.   363.
Feigelson, M.   377.
Feigelson, P.   377.
Feinberg, J.G.   542.
Feldman, J.   189, 806.
Felippe, G.M.   427.
Fernquist, I.   532.
Ferrari, T.E.   278.
Feucht, W.   779.

Filner, P.   278, 351, 363, 370, 394, 405.
Finkelstein, A.   395.
Firn, R.D.   334, 797.
Fischer, R.A.   298.
Fittler, F.   189, 254, 466.
Flavin, M.   516.
Fleissner, E.   370.
Fletcher, R.A.   271, *571*, 579, 588, 617.
Flory, R.   716, 743.
Foard, D.E.   227, 278.
Focke, I.   *143*, 149.
Forest, J.C.   89.
Forrence, L.E.   602.
Fox, J.E.   189, *449*, 458, 465, 474.
Francki, R.I.B.   254.
Franke, W.W.   743.
Frankel-Conrat, H.   239.
Frankland, B.   214, 653.
Freebairn, H.T.   509.
Friedman, L.   36.
Fritzsche, R.   779.
Frolova, I.A.   752.
Frydenberg, O.   355.
Fuchs, Y.   560, 609.
Fuente, R.K. de la.   67, 501.
Fuganti, C.   317.
Fujiwara, A.   789.
Fukuda, A.   779.
Fukui, H.   290, 617.
Fukumibara, T.   797.
Fukushima, E.   174.
Fulford, R.M.   779.
Furlong, N.B.   466.

Gaillard, B.D.E.   61.
Gaines, T.P.   116.
Galbraith, M.N.   *299*, 305, 434.
Galliard, T.   609.
Galsky, A.G.   42, 660.
Galson, E.C.   215.
Galston, A.W.   75, 95, 116, 205, *228*, 233, 247, *256*, 264, 278, 355, 465, 737.
Ganot, Dvora.   *725*, 730.
Garrett, M.K.   645.
Garrod, J.G.   447.
Gaskin, P.   797.
Gaspar, T.   214, 279.
Gausman, H.W.   501, 509, 560.
Gautheret, R.J.   532, 694.
Gawer, M.   475.
Gayler, K.R.   278.
Gefter, M.L.   189, 254, 370, 465.
Geissman, T.A.   797.
Gerhart, J.C.   264.
Gibbons, Griselda S.B.   *717*, 724.
Gibson, R.A.   *82*, 89, 90.
Gifford, E.M.   789.
Gilbert, W.   264.
Giles, K.W.   247, 394.
Giles, W.G.   465.

Gillam, I.   205.
Gillbank, L.R.   76.
Gillespie, Barbara.   724, 730.
Giorgio, N.A.   247.
Giovanelli, J.   516.
Girouard, R.M.   532.
Glasziou, K.T.   278, 602.
Glenn, J.L.   *441*.
Glew, R.H.   278.
Glinka, Z.   730.
Gmelin, R.   125, 131.
Goad, L.J.   406.
Goeschl, J.D.   610.
Goldacre, P.L.   532.
Goldberg, I.E.   377.
Goldney, D.C.   *604*.
Goldschmidt, E.E.   611.
Goldschmidt, E.E.   617, *758*, 765.
Goldsmith, Mary G.M.   724.
Goldthwaite, J.   *581*, 588.
Good, J.   280, 678.
Good, N.E.   100, 500, 501.
Goodey, P.L.   532.
Goodfellow, R.J.
Goodwin, P.B.   *371*, 377, *378*, 387.
Gordon, S.A.   89, 95, 532, 716, 724.
Goren, R.   *611*, 617, 765.
Graebe, J.E.   151.
Graham, D.   597.
Graham, J.S.D.   405.
Grasmanis, V.   779.
Gray, M.W.   205.
Gray, R.A.   125.
Green, D.E.   779.
Green, P.B.   *9*, 16.
Greenblatt, G.A.   708.
Greene, R.C.   597.
Greenman, W.D.   377.
Greenwood, C.T.   387.
Grieve, A.   597.
Griffiths, Hillary J.   743.
Griggs, W.H.   765, 779.
Grochowska, M.J.   789.
Gross, P.R.   221.
Grossman, F.   323.
Grove, J.F.   173.
Groves, S.   724.
Guardia, M.D. de la.   560.
Guern, J.   465, *469*, 474, 475.
Guernsey, F.S.   427.
Guggolz, J.   405, 579.
Guha, S.   264.
Guinn, G.   363.
Gulyas, A.   363.
Gundel, W.   730.
Gunning, B.E.S.   570, 579.
Gupta, S.   264.
Guttenberg, H. von.   501.
Guttridge, C.G.   765.

Haber, A.   227, 278.
Haber, E.S.   508.
Haberlach, G.T.   *484*, 492.
Hacker, B.   370.
Hackett, D.P.   8, 413.
Hacskaylo, J.   278.
Hagan, R.M.   298.
Hager, A.   75, 730.
Haglund, H.   239.
Haid, H.   645.
Hale, W.S.   342, 508.
Halevy, A.H.   427, 447, 448, 765.
Hall, M.A.   433.
Hall, R.H.   189, 254, 255, 458, 465, 466, *467*, 474, 806.
Hall, W.C.   508, 560, 609, 610.
Hallaway, H.M.   580, 609.
Halperin, W.   501, 694.
Hamner, K.C.   752, 757.
Hampel, A.C.   189.
Hand, J.J.   474.
Hansch, C.   525.
Hansen, H.J.   465.
Hanson, J.R.   141, 142, 206, 214.
Harada, H.   149, 427, 602.
Harada, N.   317.
Harder, R.   752.
Hardin, J.W.   247.
Hardman, J.G.   376.
Hardwick, K.   588, 596.
Harris, G.D.   447.
Harrison, A.   50.
Hartmann, H.T.   532.
Harvey, R.B.   343, 560.
Hassid, W.Z.   405.
Hatch, M.D.   610.
Haughton, P.F.   *1*, 50, 51.
Hawkins, A.R.   90.
Hayashi, Y.   305.
Hawker, William E.   724.
Heacock, R.A.   323.
Hecht, S.M.   189, 254, 806.
Heftman, E.   448.
Heilbrunn, L.V.   779.
Heinz, D.E.   165.
Heit, C.E.   214, 279.
Helgeson, J.P.   *484*, 492, *798*, 806, 807.
Hellmuth, E.O.   413.
Hellyer, R.O.   329.
Hemberg, T.   214, 278, 774.
Hemming, H.E.   149, 413.
Hendershott, C.H.   774.
Hendricks, S.B.   100.
Heng, Fong Tung.   *589*, 596, 623.
Henry, E.W.   501, 560.
Henry, W.H.   548.
Henson, W.   427.
Henrick, C.A.   173, 174.
Herrero, F.A.   609.
Hertel, R.   8, 67, 81, 447, 716, 743.

Heslop-Harrison, J.   351.
Hess, C.E.   329, 532.
Heuvel, F.A. van den.   36.
Heyes, J.K.   61.
Heyn, A.N.J.   8, 22.
Hill, T.A.   447.
Hillman, J.   602, 623.
Hillman, W.S.   75, 95, 100.
Hinde, R.   *428*.
Hinman, R.L.   116.
Hiron, R.W.P.   290, *291*, 298, 617.
Hitchcock, A.E.   533, 560.
Hoad, G.V.   757.
Hodson, H.K.   752, 757.
Hofstra, G.   579.
Hoggan, J.   603.
Hogue, .   779.
Holley, R.W.   189.
Holm, R.E.   227, 609.
Honda, O.   305.
Honma, H.   305.
Hopkins, W.G.   100.
Hokanson, R.   61, 67.
Holm, R.E.   501.
Holmes, W.L.   142.
Holsten, R.D.   678.
Horn, D.H.S.   *299*, 305, 434.
Horton, R.F.   239, 501, 548, 579, 617.
Housley, S.   67, 774.
Hsiao, T.C.   298.
Huang, R.C.   214.
Hudson, V.W.   36.
Hull Jr., J.   765.
Hulme, A.C.   609.
Hunt, L.A.   579.
Hurlbert, R.B.   466.
Hutzinger, O.   247.
Hyde, B.   351.

Ihle, J.N.   *216*, 221, 278.
Iinuma, H.   406.
Ikegami, S.   *306*.
Ilan, I.   67, 75.
Imamura, H.   305.
Imaseki, H.   609.
Ingle, J.   205, 233, 363, 653.
Inui, M.   290, 617.
Iriuchijima, S.   174.
Isherwood, F.A.   394.
Ishii, S.   132, 206.
Isogai, Y.   305, 310, *311*, 317.
Israel, H.W.   679.
Ito, S.   305.
Iwakiri, B.T.   765, 779.

Jacks, T.J.   351.
Jackson, D.I.   165.
Jackson, G.A.D.   278.
Jackson, M.B.   501, 508.
Jackson, P.   116.

Jacob, F.   271.
Jacobs, W.P.   448, *701*, 708, 709, 737.
Jacobsen, J.V.   279, *336*, *344*, 351, *356*, 377, 387.
Jacoby, B.   579.
Jaffe, M.J.   737.
Jansen, E.F.   516.
Jarvis, B.C.   206, 214, 653.
Jeffcoat, B.   447.
Jefferies, P.R.   173, 174.
Jenner, C.F.   *388*, 394, 405.
Jenner, D.V.   67.
Jensen, T.E.   *493*, 501.
Jesensky, C.   370.
Johnson, K.D.   95.
Johnston, F.B.   610.
Johnston, C.S.   405.
Johri, M.M.   215, 227, 247, 264, 395.
Jones, D.F.   797.
Jones, O.P.   483.
Jones, R.   *435*.
Jones, R.J.   298.
Jones, R.L.   343, 351, 419, 632.
Jong, H.G. Bungenburg de.   42.
Jordan, L.S.   645.
Joslyn, M.A.   542.
Jost, M.   602.
Jung, J.   149.
Juniper, B.E.   724.
Jursic, F.   525.

Kadowaki, M.   789.
Kagawa, T.   797.
Kahan, R.S.   610.
Kahana, S.   667.
Kaiser, R.   95.
Kaldeway, H.   448, 708.
Kaminek, M.   254.
Kamisaka, S.   67, 395, *428*, 440, *654*, 660.
Kandeler, R.   779.
Kapoor, M.   466.
Karlsnes, A.M.   539.
Kasai, Z.   516.
Kasamo, K.   205.
Katchalski, E.   394.
Kato, J.   174, 180, 227, *352*, 355.
Katsumi, M.   29, 173, 174, 227.
Katsura, K.   132.
Kau-Lawbney, R.   116.
Kaur-Sawhney, R.   *37*, *228*, 233, 264, 406.
Kavanagh, J.A.   131.
Kay, E.   233, 542.
Kefeli, V.I.   90, 100, 125, 132.
Kefford, N.P.   116, 264, 279.
Keitt Jr., G.W.   708.
Kelly, R.B.   195.
Kemp. J.P.   807.
Kende, H.   142, 149, 255, 458, 466, 483, 570, 580, 597, 660.

Kennedy, E.P.   395.
Kenny, F.T.   377.
Kent, A.E.   678.
Kepes, A.   67.
Kerkof, P.R.   395.
Kerridge, G.H.   483, 570.
Kessler, B.   227, 427.
Kester, D.E.   532, 774.
Ketring, D.L.   343, *502*, 508, 509.
Kettelapper, H.J.   752.
Key, J.L.   8, 67, 195, 205, 227, 264, 370, 458, 653, 700.
Khan, A.A.   *207*, 214, 279.
Khym, J.X.   579.
Kidd, F.   525.
Kiger, Jr., J.A.   195.
Kihlman, B.A.   678.
Kindl, H.   90, 125.
Kinsky, C.B.   42.
Kinsky, S.C.   42.
Kipling, J.J.   264.
Kirk, J.T.O.   588.
Kirk, S.C.   708.
Kissman, H.M.   466.
Klambt, D.   190.
Kleinig, H.   743.
Kline, L.K.   189, 466.
Kluge, M.   779.
Klyne, W.   149.
Knight, B.E.A.   298, 335.
Knight, L.I.   560.
Knight, P.R.   465.
Knox, R.B.   *344*, 351.
Kny, H.   131.
Kobayashi, Ko.   196.
Kobel, F.   779.
Kodama, M.   305.
Kofranek, A.M.   766.
Kohler, D.   61, 310.
Kohn, J.
Komoda, Y.   310.
Komoto, N.   *306*.
Kondo, G.   779.
Konig, K.H.   149.
Kong, Huei-Kuen.   189.
Kormanik, P.P.   447.
Korn, M.   394.
Kornberg, H.L.   221.
Koshimizu, K.   290, 617.
Koshland, D.E.   271.
Kosuge, T.   247.
Kramer, P.J.   298.
Krieger, D.L.   757.
Krikorian, A.D.   678.
Krishnakumaran, A.   36.
Krohn, K.   317.
Krosing, M.   695.
Krueger, P.M.   806.
Ku, H.S.   516, 517.
Kulkarni, V.G.   36.

Kumar, D.   448.
Kumar, S.A.   125.
Kunert, R.   95.
Kunishi, A.T.   516, *549*.
Kuo, Chung-Chi.   149, 418, *441*.
Kuraishi, S.   205, 255, 474, 596.
Kurssanov, A.L.   525.
Kutacek, M.   90, 100, 125, 132.

Laan, P.A. van der.   501.
Labanauskas, C.K.   617.
Laboureur, P.   149.
Lacey, H.J.   483.
Ladbrooke, B.D.   42.
Laetsch, W.M.   588.
Lamport, D.T.A.   50, 51, 542.
Landau-Schachar, B.   724.
Lane, B.G.   205.
Lang, A.   141, 142, 149, 157, 174, 227, 310, 426, 427, 448, 560, 580, 660, 752, 757, 765.
Lang, J.   116.
Lantican, B.P.   100, 501.
Larsen, P.   *102*, 109.
Larson, J.E.   377.
Lavee, S.   205, 233, 355.
Lawson, A.M.   806.
Leach, A.A.   542.
Leavenworth, C.S.   405.
Leeper, G.W.   779.
Leggett, P.A.   280.
Legler, G.   149.
Leguay, J.J.   474.
Leinweber, C.L.   609.
Leo, P. de.   279.
Leonard, N.J.   189, 254, 806.
Leopold, A.C.   67, 278, 343, 447, 501, 508, 559, 560, 580, 602, 779.
Lerman, L.S.   247.
Leshem, Y.   *228*, *661*, 667.
Letham, D.S.   *248*, 254, 255, 458, 466, 475, 483, 579, 596, 678.
Lew, F.J.   234, 427.
Lew, J.Y.   233, 542.
Lewis, L.N.   *234*, 239, 279, 617, 765.
Libbert, E.   95, 570.
Lieberman, M.   516, *549*, 560, 609.
Light, E.N.   141, 426.
Lincoln, R.G.   752, 757.
Link, C.B.   335.
Linsmaier, E.M.   195, 492.
Lipe, J.A.   *502*, 509.
Lipetz, J.   609.
Lipmann, F.   255, 660.
Lippincott, J.A.   42, 660.
Lippold, P.C.   214.
Lis, E.K.   789.
Little, C.H.A.   298.
Little, E.C.S.   75.
Littler, J.S.   797.
Livne, A.   29, 298.

Loan, P.A. van der.   560.
Lockhart, J.A.   16.
Loeffler, J.E.   279, 483.
Loening, V.E.   233, 363, 653.
Loomis, R.S.   532.
Loveys, B.   645.
Lowe, E.   413.
Lowenstein, J.M.   271.
Lowry, O.H.   95, 214, 247, 370, 570, 597, 609.
Lubke, K.   247.
Luck, H.   609.
Luckwill, L.C.   174, 483, 588.
Lucy, J.A.   42.
Luescher, E.   247.
Lundeen, A.J.   125.
Lunenfeld, B.   667.
Lyon, J.L.   221, 278, 279, 342, 602.
Lyons, J.M.   560.

McAfee, J.A.   509.
McAlpine, R.G.   447.
McCarthy, B.J.   8.
McCalla, D.R.   466, 475.
McChesney, J.   *449*, 458, 465.
McCloskey, J.A.   806.
McComb, A.J.   *407*, 413, 708.
McCready, C.C.   708.
McCready, R.M.   405, 579.
McCune, D.C.   233, 355.
McGlasson, W.B.   343, 560.
MacGregor, K.W.   387.
McGugan, W.A.   610.
McHale, J.S.   579, 602.
MacLachlan, G.A.   206, 254, 542.
McLaughlin, J.A.   774.
MacLean, D.C.   579.
McLennan, B.D.   189, 474, 806.
MacLeod, Anna M.   387, 405.
McMeans, J.L.   278.
MacMillan, J.   174, 180, 290, 419, *790*, 797.
McNairn, R.B.   602.

Madison, J.T.   189.
Madison, M.   279.
Mahadevan, S.   *117*, 125.
Maheshwari, S.C.   264.
Malcolm, H.D.R.   *767*.
Mandava, N.   36.
Mandels, M.   602.
Manichi, A.   95.
Mann, J.D.   387, 645.
Manocha, M.S.   255, 623.
Mans, R.J.   597, 653.
Mansfield, T.A.   298.
Manteuffel, R.   95.
Manuel, J.   280.
Mapes, M.O.   678, 685.
Mapson, L.W.   516.
March, J.F.   516.

Marino, M.L.   317.
Marinos, N.G.   67.
Marquisus, N.   189.
Marre, E.
Marshall, D.C.   61.
Marshall, J.S.   247.
Martin, C.   *561*.
Martin, G.C.   779.
Martin, W.E.   700.
Marushige, K.   8.
Mason, H.H.   116.
Mason, M.I.R.   279.
Masuda, Y.   *17*, 22, 67, 75, 195, 206, 227, 395, 440, 660.
Masuko, M.   132.
Matile, Ph.   351, 602.
Matsubara, S.   189.
Matsuo, M.   125.
Matthysse, Anne G.   214, 227, 247, 255, 264.
Matzner, P.   730.
Maurer, H.R.   264.
Maxie, E.C.   610.
Maxwell, F.G.   757.
Mazelis, M.   125.
Mead, A.   516.
Meheriuk, M.   509.
Meidner, H.   580.
Meijer, G.   *68*, 75.
Menhenett, R.   *633*, 645.
Merrill, S.N.   189.
Mertz, D.   427.
Meudt, W.J.   *110*, 116.
Michener, H.P.   501, 560.
Milborrow, B.V.   279, *281*, 290, 617.
Miller, A.S.   387.
Miller, C.O.   458, 466, 483, 533, 694.
Miller, R.A.   597.
Miller, R.L.   255.
Millet, M.A.   580.
Mills, A.K.   413.
Millward, S.   205.
Minamikawa, T.   789.
Mishimura, S.   206.
Mitchell, J.W.   36, 700.
Mitsui, T.   290, 617.
Mittelheuser, Catherine J.   298, 602, *618*, 623.
Mittler, T.E.   757.
Mizrachi, Y.   597, 617.
Mohr, G.   323.
Mohsin, M.   335.
Mohr, H.   645.
Mollenhauer, H.H.   501, 743.
Momotani, Y.   *352*, 355.
Mondelli, R.   317.
Mondon, A.   317.
Monod, J.   264, 271.
Monro, J.   8, *52*.
Monselise, S.P.   610, 617, *758*, 765.
Montaldi, E.R.   448, 580.
Montreuil, J.   61.

Moore, A.E.   *598*, 602, 603.
Moore, T.C.   426.
Moore, W.E.   580.
Morf, W.   448.
Morgan, Page W.   343, 500, 501, *502*, 508, 509, 560, 609, 610.
Morinaga, T.   174.
Morre, D.J.   8, 22, 279, 466, 475, 501, 743.
Morris, D.A.   75, 501, 708.
Morris, R.O.   806.
Morton, R.K.   405.
Mosettig, E.   174.
Moss, G.I.   779.
Most, B.H.   797, 807.
Mothes, K.   580.
Mudd, J.B.   395.
Mudd, S.J.   516.
Muir, R.M.   *96*, 100, 101, 501, *518*, 525.
Mulholland, T.P.C.   173.
Muller, K.   580.
Muller-Hill, B.   264.
Mulliken, J.A.   271.
Mullins, M.G.   279, *526*, 700.
Munck, A.   264.
Munro, H.N.   653.
Murakami, T.   789.
Murakami, Y.   22, *166*, 174, 180, 419, 789.
Murao, K.   206.
Murashige, T.   694.
Murata, T.   789.
Murofushi, N.   150, 165, 174, *175*, 180, 797.
Murphy, P.J.   142, 157, 427.
Murray, A.W.   *459*, 465, 466.
Myers, A.   247, 394.
Myers, P.L.   141.

Nadler, S.B.   465.
Nagao, M.   *127*, 132, 355.
Naik, G.G.   434.
Nakagaki, Y.   525.
Nakamura, T.   206.
Nakanishi, K.   317.
Naqvi, S.M.   708, 724.
Neet, K.E.   271.
Nelsen, N.   239.
Nelson, C.D.   579.
Nelson, H.   67, 75.
Neuberg, C.   125.
Neuman, W.F.   125.
Neumann, K.H.   678.
Newhall, W.F.   142.
Nicholls, P.B.   466, 645.
Nicholls, W.   *324*, 329.
Nielsen, G.   355.
Nieuwdorp, P.J.   351.
Ninneman, H.   142, 660.
Nissl, D.   42, 67, *75*, 271.
Nitsan, J.   227.
Nitsch, Colette.   116, 132, 570, 617, 645.

Nitsch, J.P.   116, 132, 173, 483, 570, 617, 645.
Nobel, P.S.   597.
Noddle, R.C.   290.
Noelting, G.   617.
Noma, M.   *306*.
Nomota, M.   *127*.
Nooden, L.D.   8, 75.
Norton, K.   141, 797.
Novelli, G.D.   370, 597, 653.

Oberg, B.   413.
Obreiter, J.B.   36.
O'Brien, T.J.   206, 216, 247.
Occolowitz, J.   189, 254, 806.
Ockerse, R.   355, 427.
Oegema, T.   579, 617.
Ogawa, Y.   165, 180, 395, 797.
Ohkuma, K.   221.
Okamoto, T.   305, 310, *311*, 317.
Olmstead, A.J.   559.
Olson, A.C.   8, 22.
Ono, H.   305.
Ordin, L.   602.
Oritani, T.   279.
Osborne, Daphne J.   239, 255, 279, 466, 475, 501, 508, *534*, 542, 579, 580, 588, 609, 610, 653, 708.
Osborne, O.J.   548.
Osborne, H.D.   427.
Oster, M.   427.
Ouitrakul, R.   743.
Overbeek, J. van.   22, 466, 483, 774.
Owens, H.S.   405, 579.
Owens, L.D.   516.

Paleg, L.G.   *37*, 42, 165, 214, 278, 351, 355, 363, 387, *388*, 395, *396*, 405, 532, 588, 602, 617, 645, 708.
Painter, R.H.   757.
Palman, R.L.   660.
Palmer, J.K.   560.
Palmer, J.H.   448.
Pan, Foo.   370.
Parker, K.J.   807.
Pastan, I.   377, 660.
Patchett, A.A.   678.
Pate, J.S.   579.
Paton, D.M.   *324*, 329.
Patrick, J.W.   *695*.
Patterson, B.D.   596, 623.
Patterson, M.S.   597.
Peacock, A.C.   363.
Pearse, H.L.   532.
Pearson, A.   280, 678.
Pearson, A.P.   214.
Pearson, J.A.   279.
Pegg, G.F.   90.
Pelton, J.S.   174.
Penman, M.   377.
Penman, S.   377.

Penner, J.   752.
Penny, E.D.   8.
Penny, D.   36, *52*, 61.
Penny, P.   8, *52*, 61.
Penny, P.J.   75.
Pensky, J.   247.
Penswick, J.R.   189.
Perdue, Stella W.   227, 278.
Perley, J.E.   501.
Perlman, R.   377.
Perrin, D.R.   310.
Peterkovsky, A.   370.
Pharis, R.P.   149, 418, *441*, 448, 797.
Phelps, R.   90.
Philipson, L.   413.
Phillips, B.J.   280.
Phillips, C.   227, 247, 264, 448.
Phillips, I.D.J.   419, 448, 632.
Phillips, M.L.   395, *396*.
Phinney, B.O.   29, 173, 174, 227, 797.
Pickett-Heaps, J.D.   743.
Piette, L.   116.
Pilet, P.E.   279.
Pine, T.S.   774.
Piozzi, F.   317.
Plaut, Z.   448.
Plimmer, J.R.   36.
Poapst, P.A.   610.
Pollard, C.J.   43, 387, 395.
Pollard, J.K.   678.
Ponting, J.D.   542.
Pool, R.M.   483.
Popov, K.   610.
Porath, J.   247.
Posner, H.B.   757.
Potter, H.   280, 678.
Poulson, R.   279, *646*, 653.
Poux, N.   351.
Powell, L.E.   797.
Pratt, H.K.   516, 517, 560, 610.
Preece, I.A.   603.
Prescott, D.M.   377.
Preston, W.H.   335.
Preston, R.D.   542.
Prochazka, Z.   125.
Prockop, D.J.   542.
Pronczuk, A.W.   653.
Pruett, P.   708.
Pryce, R.J.   180, 290, 419, 797.
Pryor, L.D.   329.
Pucher, G.W.   405.
Purves, W.K.   29, *91*, 95, 227.
Pustovoitova, T.N.   298.

Rabideau, G.S.   632.
Radin, J.W.   532.
Radley, Margaret.   149.
Raison, J.K.   405.
Rajagopal, R.   95, *102*, 109.
Ralph, R.K.   8, *248*, 254, 255, 363, 596.

Randall, D.I.   617.
Randall, R.J.   95, 214, 247, 370, 570, 597, 609.
Rao, K.V.N.   678.
Rappaport, L.   279.
Rasmussen, G.K.   548.
Rasmussen, H.   43.
Ray, M.M.   743.
Ray, P.M.   8, 16, 50, 51, 61, 67, 75, 81, 116, 730, 743.
Rayle, D.L.   *1*, 8, 29, *44*, 51, 67, 81, 95, 743.
Raymond, S.   597.
Reddy, M.N.   131.
Reed, C.   351.
Reece, P.C.   548, 560.
Reese, E.T.   602.
Reich, E.   377.
Reid, D.M.   *414*, 418, 419, *420*, 427, 645.
Reid, P.B.   234.
Reinert, J.   *686*, 694.
Reinhard, E.   149.
Reinhold, Leonora.   50, 67, 75, *725*, 730, *731*, 737.
Reinken, G.   779.
Resnik, M.E.   580.
Reynolds, E.S.   623.
Reynolds, T.   588.
Rhodes, M.J.C.   609.
Rhodes, J.J.C.   516.
Richmond, A.E.   419, 560, 580, 588, 597, 617.
Richmond, P.A.   16.
Richter, E.W.   *518*.
Rickards, R.W.   141.
Rideal, E.K.   42.
Ridge, I.   *534*, 542.
Riecke, E.   95.
Rier, J.P.   737.
Riov, J.   610.
Riviere, J.   149.
Roach, T.S.   774.
Robertson, J.D.   395.
Robins, M.J.   189, 255, 806.
Robinson, D.R.   142, 149, 290, 427.
Robinson, G.A.   660.
Robinson, P.M.   280, 335, 343, 774.
Robinson, R.W.   560.
Robison, G.A.   376.
Rogers, B.J.   548.
Rogers, W.S.   483.
Roller, H.   36.
Rose, M.S.   255.
Rose, R.J.   227.
Rosebrough, J.N.   95, 214, 247, 370, 570, 597, 609.
Rosenfeld, I.   125.
Ross, C.   394.
Rothfield, L.   395.

Rothwell, K.   279, 335.
Rowan, K.S.   *76*, 81, 570, 602.
Rowsell, E.V.   406.
Roy, S.C.   254.
Roychoudhury, R.   264, 395.
Rubinstein, B.   500, 560.
Ruddat, M.   174, 448.
Ruge, U.   509.
Ruesink, A.W.   8, 16, 51, 67, 81.
Russell, R.L.   189, 254, 370, 465.
Ryback, G.   280, 290.
Ryugo, K.   779.

Saccone, G.T.   *459*.
Sacher, J.A.   43.
Sacher, J.A.   279.
Sacher, R.   149.
Sachs, J. von.   724.
Sachs, R.M.   765, 766.
Sachs, T.   570, *731*.
Sadorge, P.   465, 474.
Sadri, H.A.   150.
Saikia, B.N.   774.
Sakan, T.   305.
Salisbury, F.B.   779.
Saneyoshi, M.   206.
Sankhla, D.   214, 323, 603.
Sankhla, N.   204, 323, 603.
Sargent, J.A.   *534*.
Sarid, S.   394.
Sarkissian, I.V.   *265*, 271.
Sasse, Jenneth M.   *299*, 305, 434.
Saunders, P.F.   67, 279.
Savona, G.   317.
Scandalios, J.S.   351, 377, 387.
Scatchard, G.   264.
Schachman, H.K.   264.
Sharma, O.K.   370.
Schechter, I.   142, 150.
Schilling, G.   149.
Schmalstieg,   271.
Schmialek, P.   36.
Schmidt, C.J.   *798*, 807.
Schmidt, R.   75.
Schneider, C.L.   22, 737.
Schneider, Elnora A.   *82*, 89, 90.
Schneider, G.   *143*, 149, 323.
Schneiderman, H.A.   36.
Schotz, M.C.   61.
Schraudolf, H.   125, 132.
Schreiber, K.   *143*, 149, 150.
Schroder, E.   247.
Schroder, R.   95.
Schuster, Maria.   779.
Schwabe, W.W.   766.
Schwarz, M.   36, 667.
Schweet, R.   255.
Scott, R.W.   580.
Scott, T.K.   448, 708, 709, 724.
Sechet, M.   149.

Sedat, J.W.   195
Segal, S.J.   264.
Sell, H.M.   67, 447.
Selman, I.W.   90.
Semadeni, E.   602.
Sembdner, G.   *143*, 149, 150.
Semenoff, S.   695.
Sen, S.P.   254, 264, 395.
Sequeira, L.   90, 617.
Sessa, G.   43.
Seth, A.K.   448, 580.
Setterfield, G.   95, 280, 377, 395,
   434, 440, 660.
Sexton, R.   280.
Shafiq, A.   377.
Shannon, L.M.   233, 542.
Shannon, S.   560.
Shantz, E.M.   678.
Shaw, G.   458, 678.
Shaw, M.   90, 255, 623.
Shaw, S.   *717*.
Shen-Miller, J.   *710*, 716, *738*, 743.
Sherwin, J.E.   91, 95.
Shibaoka, H.   *561*, 570, 597.
Shield, Jr., J.A.   708.
Shih, C.Y.   602.
Shih, L.M.   *228*.
Shilo, R.   427.
Shimokawa, K.   516.
Shimshe, D.   29.
Shininger, T.L.   743.
Shmueli, E.   448.
Shomer-Ilan, A.   *661*.
Short, K.C.   570.
Shugart, L.S.   370.
Shulka, P.A.   125.
Shyr, S-Y.   766.
Siebert, K.J.   516.
Siegel, B.Z.   205, 233, 355.
Siegel, S.M.   116, 247.
Sievers, A.   744.
Sigesmund, K.A.   744.
Silveira, V.   405, 579.
Simmonds, J.A.   427.
Simonis, W.   779.
Singh, B.N.   387, 395.
Singh, M.   271.
Sinsheimer, R.L.   195.
Sitte, P.   743.
Sitton, D.   419, 483, 580.
Skene, K.G.M.   *476*, 483, 570.
Skoog, F.   189, 195, 254, 255, 466, 492,
   533, 685, 694, 774, 806.
Slack, C.R.   610.
Slama, K.   36.
Smallwood, B.M.   458, 678.
Smillie, R.M.   596, 597, 623.
Smith, H.   141, 427.
Smith, M.S.   335.
Smith, M.V.   36.

Smith, O.E.   150, 221, 278, 279.
Smith, T.A.   335.
Smithies, O.   233.
Snedecor, G.W.   774.
Snir, I.   227.
Snow, R.   632.
Soloff, M.S.   247.
Solymosy, F.   363.
Sondheimer, E.   215, 280.
Sonnenblichler, J.   806.
Sonnet, P.E.   36.
Sood, C.   *449*, 458, 465.
Soofi, G.S.   280.
Sopori, M.L.   264.
Sorokin, H.   570.
Spector, C.   797.
Spencer, D.   255.
Spencer, M.   509, 516.
Spenser, I.D.   125.
Sperca, R.J.   247.
Spiegel, S.   427.
Spik, G.   61.
Spink, G.C.   744.
Spring, A.H.   *76*, 81.
Srere, P.A.   271.
Srinivasan, P.R.   370.
Srivastava, B.I.S.   215, 255, 279, 475, 596, 602, 603.
Stasiuk, L.   255, 806.
Stavitsky, A.B.   667.
Steadman, J.R.   617.
Steinmetz, E.   501.
Sterns, M.   329.
Steveninck, R.F.M. van.   280, 298, 602, *604*, *618*, 623.
Steward, F.C.   278, 280, 458, *668*, 678, *679*, 685.
Stewart, E.C.   509.
Stewart, J.C.   141, 156, 797.
Stillwell, R.N.   806.
Stinson, R.A.   516.
Stoddart, J.L.   141, 156.
Stokstad, E.L.R.   474.
Stone, B.A.   *598*, 602, 603.
Stone, B.P.   370.
Stout, E.R.   653.
Stowe, B.B.   *30*, 36, *117*, 125, 501.
Straus, W.   351.
Strong, F.M.   475.
Stuart, N.W.   700.
Stuart, R.N.   67.
Stulberg, M.P.   370.
Suge, H.   174.
Sumida, S.   395.
Sumiki, Y.   797.
Sunagawa, M.   305.
Sussman, M.   377.
Sutcliffe, J.F.   280.
Suter, P.J.   797.
Sutherland, E.W.   376, 660.

Svoboda, A.   61.
Szego, C.M.   247.
Szent-Gyorgyi, A.   774.

Tabachnik, M.   247.
Taiz, L.   351.
Takahashi, N.   150, 165, 174, *175*, 180, 797.
Takahashi, S.   305.
Takahashi, T.   305.
Talwar, G.P.   264.
Tamura, S.   *127*, *132*, 150, 174, 180, *306*, 355.
Tang, Y.W.   95.
Tanimoto, E.   17, 195, 206.
Tao, M.   660.
Tapper, B.A.   125.
Tata, J.R.   395.
Tate, M.E.   158.
Taudou, B.   667.
Tautvydas, K.J.   *256*, 264, 797.
Tavares, J.   255, 458, 466, 580, 597.
Taylor, B.K.   774.
Taylor, H.F.   290, 298, 334, 335, 617.
Tazawa, M.   695.
Tener, G.M.   205.
Ternynck, T.   667.
Terrine, C.   *469*, 474, 475.
Tewari, K.K.   597.
Thedford, R.   189, 255, 806.
Thimann, K.V.   8, 22, 75, 116, 533, *561*, 570, 597, 724, 730, 737.
Thomas, R.G.   *624*, 632.
Thomas, T.H.   280, 343, 774.
Thompson, D.L.   29.
Thompson, J.E.   516.
Thompson, P.G.   75, 501, 708.
Thompson, W.F.   *1*, 8, 50.
Thomson, W.W.   617.
Tigerstrom, Margaret von.   205.
Tilney-Bassett, R.A.E.   588.
Ting, I.P.   239.
Tinklin, I.G.   766.
Tishel, M.   125.
Tobin, N.F.   597.
Tobolsky, A.W.   8, 22.
Toki, T.   132.
Tomaszewski, M.   448.
Tomes, M.I.
Tomita.   *780*, 789.
Tookey, H.L.   126.
Toole, E.H.   343, 508, 509.
Toole, V.K.   343, 508, 509.
Torii, Y.   305, *311*, 317.
Torrey, J.G.   570, 694.
Trapy, F.   474.
Traub, H.P.   560.
Trewavas, A.   75, 597.
Trown, P.W.   597.
Truchelet, G.B.   609.
Truelsen, T.A.   67.

Tscuboyama, K.   806.
Tsunakawa, S.   806.
Tullio, N.W. di.   142.
Turner, N.A.   29.
Tuttle, A.A.   75.
Tzou, D.   215.

Uchiyama, Y.   609.
Udenfriend, S.   542.
Uhrstrom, I.   67.
Ullrich-Eberius, C.I.   779.
Umrath, K.   737.
Underhill, E.W.   125, 126.
Undurriaga, N.J.   779.
Upper, C.D.   141, 142, 427, *484*, 492, *798*, 807.
Uritani, I.   609.
Uvardy, J.   603.

Vaadia, Y.   298, 419, 570.
Vacha, G.A.   343, 560.
Valdovinos, J.G.   *493*, 501, 560.
Varner, J.E.   215, 227, 239, 247, 264, 278, 279, 343, 351, 355, 363, 364, 370, 376, 377, 387, 394, 395, 405, 406, 419, 517.
Veen, H.   708.
Vegis, A.   343.
Veldstra, H.   43, 709.
Vendrell, M.   560.
Venis, M.A.   *240*, 247, 525.
Verbiscar, A.J.   797.
Vesco, C.   377.
Vickery, H.G.   405.
Vickery, L.W.   91.
Villiers, T.A.   280, 774.
Virgin, H.I.   645.
Virtanen, A.J.   125.
Vislovska, Lea.   *731*.

Wada, S.   22, 195.
Wada, Y.   174.
Wagne, C.   645.
Wagner, J.   125.
Wain, R.L.   279, 323, *330*, 335.
Wakabayashi, N.   36.
Waldon, E.S.   548.
Walker, D.R.   774.
Wallace, R.H.   548.
Walton, D.C.   280.
Wangermann, Elisabeth.   737.
Wardale, D.A.   516.
Ware, G.   603.
Wareing, P.F.   195, 214, 279, 280, 335, 343, 448, 580, 588, 602, 623, 678, *695*, 700, 737, 774.
Watanabe, E.   174.
Waters, E.C.   214.
Waters, L.C.   221, 395.
Watkin, J.E.   597.

Wattiaux, R.   405.
Waygood, E.R.   466.
Weaver, P.   174.
Weaver, R.J.   483.
Webb, J.A.   580.
Weber, H.   132.
Weber, R.P.   95.
Webster, B.D.   501.
Weigl, J.   779.
Weiland, J.   *143*, 149, 150.
Weisner, R.   377.
Weiss, C.   570.
Weiss, M.J.   466.
Weiss, S.B.   395.
Weitzman, P.D.J.   271.
Wells, R.D.   377.
Went, F.W.   533, 570, 632, 709.
Werblin, T.P.   709.
West, C.A.   *133*, 141, 142, 149, 150, 156, 157, 427.
West, C.   525.
Westphal, U.   247.
Wetherell, D.F.   694.
Wetmore, R.H.   737.
Wetter, L.R.   126.
Whaley, W.G.   501.
White, A.F.   141, 142.
White, P.R.   694.
Whitfeld, P.R.   255.
Whitty, C.D.   370.
Whyte, P.   483, 588, 724.
Wichner, E.   95.
Wicks, W.D.   377.
Wickson, M.   570.
Wigglesworth, V.B.   36.
Wightman, F.   *82*, 89, 90, 95, 280, 323, 335, 377, 395.
Wilchek, M.   247.
Wilcoxon, F.   548.
Wildman, S.G.   254, 596, 597.
Wilkins, M.B.   708, 709, 715, *717*, 724, 730.
Wilkinson, E.H.   766.
Williams, J.C.   807.
Williams, M.W.   483.
Williams, P.H.   131.
Williams, R.M.   42.
Williamson, A.R.   355.
Williamson, C.E.   548.
Willing, R.R.   329.
Wilson, D.V.   678.
Wilson, J.D.   264.
Wimmer, E.   205.
Wira, C.   264.
Witkop, B.   131, 678.
Wolff, I.A.   126.
Wollgiehn, R.   580.
Wood, A.   *37*, 406.
Wood, M.   588.
Woodham, R.C.   483.

Woolhouse, H.W.   580, 588, 596.
Wooltorton, L.S.C.   516, 609.
Worley, J.F.   36.
Wray, J.L.   278.
Wright, S.T.C.   280, 290, *291*, 298, 335, 617.
Wu, J.Y.   271.
Wyatt, G.R.   395.
Wyman, J.   264.

Yamaguchi, I.   165, 797.
Yamaki, T.   61, *196*, 205, 206.
Yamamoto, R.   17.
Yamazaki, I.   116.
Yanagashima, N.   395.
Yang, J.T.   271.
Yang, S.F.   *510*, 516, 517, 533.
Yanini, L.I.   752.
Yatsu, L.Y.   351.
Yemm, E.W.   570.
Yeoman, M.M.   434.
Yeung, D.   413.
Yokota, T.   149, 150, 165, *175*, 180, 797.
Yomo, H.   364, 387, 406.
Yoshida, R.   279.
Young, H.   458, 466.
Young, R.E.   559.
Younis, A.F.   724.
Yung, H.K.   387.
Ysselstein, M.H.W. van.   100, 501.

Zachau, H.G.   189, 806.
Zamir, A.   189.
Zaoral, M.   36.
Zeevaart, J.A.D.   142, 427, 447, 752.
Zenk, M.H.   29, 42, *62*, 67, 75, 247, 271.
Ziegler, H.   779.
Zill, L.P.   579.
Zimmerman, M.H.   757.
Zimmermann, P.W.   533, 548, 560.
Zolotov, Z.   427.
Zubay, G.   206, 255.
Zwar, J.A.   *356*.

INDEX TO ORGANISMS

Acacia farnesiana (huisache) : *609*.
Acer pseudoplatanus (sycamore) : 333, 467, *474*, 673, 774.
Acer saccharum (maple) : 570.
Aesculus woerlitzensis (horse chestnut) : 668, *678*, 679.
Agrobacterium tumefaciens (crown gall) : *427*.
Agropyron repens : *448*.
almond : (see *Prunus amydalus*).
Althaea rosea (hollyhock) : 144, *149*.
Amaranthus caudatus : *465*.
Amaranthus tricolor : 459, 460.
Ananas sativus (pineapple) : *560*, *617*.
Ankistrodesmus braunii : *779*.
apple : (see *Pyrus malus*).
apricot : (see *Prunus armeniaca*).
Arabidopsis thaliana : 318, *323*.
Arachis hypogea (peanut) : 216, 343, *447*, 502, *508*, *509*.
Arbacia (sea urchin) : *221*.
Avena sativa (oat) : 1, *8*, 9, *16*, 17, 22, *42*, 44, 49, *50*, 52, *61*, 62, 65, 66, *67*, 75, 76, *81*, 96, 100, 102, *109*, 115, 128, 164, 190, *195*, 203, *206*, 223, 226, 303, 311, *395*, 494, *501*, *560*, 561, *570*, *579*, *603*, *617*, 668, 710, *716*, *730*, 738.
avocado : (see *Persea*).

bamboo : 170, 171, *174*.
Bacillus cereus : *501*.
barley : (see *Hordeum vulgare*).
bean : (see *Phaseolus vulgaris*).
Begonia evansiana : *278*.
────── semperflorens : 661, *667*.
Beta vulgaris (sugar beet) : *274*.
Bos bovis (calf) : 181, 468, *474*.
Brassica oleracea (cabbage) : 99, 100, *101*, 118, *125*, 596.
────── ────── var. *botrytis* (cauliflower) : 265, *395*.
────── ────── var. *cymosa* (broccoli) : *579* , 661, *667*.
────── ────── var. *gemmifera* (brussels sprout) : 294.
────── *pekingensis* (chinese cabbage) : 127, 248, *254*, *255*.
────── *rapa* (rape) : *36*, 127, 132.
Bryophyllum sp. : *632*, *752*.
────── *daigremontianum* : 746, 747, *752*.
────── *tubiflorum* : 746.

Calonyction aculeatum (moonflower) : 175, 180.
Cannabis sativa (hemp) : 306.
carrot (see *Daucus carota*).
Carya illinoensis (pecan) : 509.
castor-bean : (see *Ricinus*).
Ceratonia siliqua (carob) : 795, *797*.
Cestrum nocturnum : 764, *765*.
Chara foetida : 738, *743*, *744*.
cherry : (see *Prunus avium*).
Chrysanthemum morifolium : 661.
Citrus spp. : *548*, 556, *610*, *617*, 758, *765*.
Citrus limon (lemon) : *271*, 758, *765*.
Citrus paradisi Macf. (grapefruit) : 543.
Citrus reticulata Blanco X (*C. paradisi* X *C. reticulata*) (tangerine) : 543.
Citrus sinensis (*L.*) Osb. (orange) : 543, 611, *617*, 760, *764*, *765*, 779.
Cochlearia armoracia (horseradish) : *233*, *517*, 542.
Cocos nucifera (coconut) : *483*, 668, *678*, 679.

Coffea arabica (coffee) : 796, *797*.
Coleus sp. : 701, *708*, *709*.
────── *blumei* : 280, 442, *448*, 493, *500*, *548*, 552, *737*.
────── *rehneltianus* : 273, *278*, *447*.
Convolvulaceae : 175, 179.
Coreopsis grandiflora : 746, *752*.
corn : (see *zea*).
Corylus avellana (hazel) : 211, *214*.
Corynebacterium fascians : 181, *189*, *570*, *806*.
cotton : (see *Gossypium*).
Crambe abyssinica : *126*.
Cruciferae (Brassicaceae) : 85, 90, 99, *125*, *132*.
cucumber : (see *Cucumis sativus*).
Cucumis melo var. *cantalupensis* (canteloupe) : *560*.
Cucumis sativus (cucumber, gherkin) : 23, 24, *29*, 68, *75*, 91, *95*, 145, 223, 225, 227, 560, 736.
Cucurbita Pepo (pumpkin) : 151, 156, *737*.
────── *melopepo* (water melon) : *580*.
Cupressus arizonica : 442, 443, *448*.
Cynodon dactylon (Bermuda grass) : *448*.

Dactynotus ambrosiae (Thomas) : 753.
Daucus carota (carrot) : 181, 274, 277, *501*, 668, *678*, 679, *685*, 686, *694*.
Dianthus caryophyllus (carnation) : 441.
Dictyostelium discoidium : 660.

Echinocystis macrocarpa (wild cucumber) : 136, 138, *142*, 151, *156*, *157*, 792, *797*.
Elodea aquatica : 601, *602*.
Escherichia coli : 67, 139, 181, *189*, *206*, 243, *247*, 248, *271*, 365, *370*, *394*, 660, 806.
Eucalyptus deglupta : 326.
────── *grandis* : 324, *329*.
Euphorbia pulcherrima (poinsettia) : 764, *765*.

Ficus carica (fig) : *610*.
Fragaria vesca (strawberry) : 765, *789*.
Fraxinus (ash) : 774.
French bean : (see *Phaseolus vulgaris*).
frog : 661.
Fuchsia hybrida : 764, *766*.
Fusarium moniliforme (*Gibberella fujikuroi*) : 134, 135, 138, 141, *142*, 143, 144, 148, *149*, *150*, 151, 168, 332, 424, *426*, *427*, *660*, 745, 794, *797*.

Galleria (wax moth) : 33, 34, 35.
Gallus domesticus (chicken) : 181, 468, *474*.
Gibberella fujikuroi : (see *Fusarium*).
Gladiolus sp. : *351*, *427* .
Glycine max (soybean) : 182, *189*, *206*, *214*, 223, 227, 450, 459, *465*, 477, *653*.
Gossypium hirsutum (cotton) : 67, 216, *221*, 274, *278*, 291, *500*, 502, *508*, *509*, *609*, *610*, 786.
grape : (see *Vitis*).

Hedera helix (ivy) : 532.
Helianthus annuus (Sunflower) : *95*, 414, *419*, 427, *483*, 562, *570*, *632*, 725.
────── *tuberosus* (Jerusalem artichoke) : *227*, *305*, 428, *433*, *434*, 435, *440*, 526, 532, *654*, *660*.
Homo sapiens (human) : 181.
Hordeum vulgare (barley) : 37, *42*, *43*, 65, 82, *89*, 90, *214*, *215*, 274, 276, *278*, 298, 330, 336, *343*, 344, *351*, 352, *355*, 356, *363*, *364*, 365, *370*, 371, 378, *387*, 388, *394*, 400, *405*, 420, *570*, *580*, *596*, *603*, 612, 633, 646, *653*, 654, *660*, 706, *708*.

Impatiens Roylei (touch-me-not) : 562.
Ipomoea batatas (sweet potato) : *609*.
Isatis tinctoria (woad) : 117, 118, *125*.

Jerusalem artichoke : (see *Helianthus tuberosus*).
Juniperus : 442.

kidney bean : (see *Phaseolus vulgaris*).

Lactobacillus acidophilus : 181.
Lactuca sativa (lettuce) : 164, *214*, 223, *227*, 273.
Lemna, spp : 274, 276, 277, 589, 776.
────── *gibba* : 754, 777, *779*.
────── *minor* (duckweed) : *597*.
────── *perpusilla* : 754, *779*.
lemon : (see *Citrus limon*).
Lens culinaris (lentil) : 207, 212, 223, *227*, 276.
Lepidium sativum (cress) : 323, 333.
Lespedeza cuneata : 598.
lettuce : (see *Lactuca sativa*).
Linum usitatissimum (flax) : *532*.
Lolium perenne (perennial rye-grass) : 333.
────── *temulentum* : *779*.
Lupinus sp. (lupin) : *797*.
────── *luteus* (yellow lupin) : 280, 290, 335, *617*.
────── *angustifolia* (blue lupin) : 52.
Lycopersicon esculentum (tomato) : 82, *89*, *90*, 91, *95*, 281, 330, *419*, 441, *447*, 493, 501, *516*, 559, 579, *617*, 665, *779*.
Lycoris radiata : *305*, 311, 317.

Macrosiphum pisi : *757*.
maize : (see *Zea*).
Musa sapientum (banana) : *343*, 558, *560*.
mung bean : (see *Phaseolus aureus*).
Mus musculus (mouse) : 377.
Myzus persicae : *757*.

nasturtium : (see *Tropaeolum*).
Nicotiana glutinosa : 598, *602*, *603*.
────── *rustica* (wild tobacco) : *579*.
────── *tabacum* (tobacco) : 181, *189*, 190, 228, *233*, *254*, 303, 316, 449, *458*, 459, *492*, 493, *501*, *525*, 537, *570*, 571, *596*, *617*, 738, *744*, 804.
Nitella : 15, *16*.
Nymphoides indica : 604.

oats : (see *Avena*).
orange : (see *Citrus sinensis*).
Oryctolagus (rabbit) : 463.
Oryza sativa (rice) : 146, 166, 167, 168, 173, *174*, 176, 303, 316, *395*, 415, *419*, 423, 780, *788*, *789*.
Ovis aries (sheep) : 181.

Parthenocissus quinquefolia : 732.
────── *tricuspidata* : 731.
Passiflora caerulea (passionflower) : 732.
peach : (see *Prunus persica*).
peanut : (see *Arachis*).
pear : (see *Pyrus communis*).
Pelargonium : *233*.
Penicillium digitatum : *509*.

Perilla frutescens : 587, *588*, 589, *596*.
────── *nankinensis* : 746.
Persea gratissima (avocado) : 281.
Pharbitis nil (morning glory) : 144, 145, *149*, *150*, 175, 176, *180*, *427*, 754.
Phaseolus aureus (mung bean) : *95*, 196, *205*, 526.
────── *coccineus (Ph. multiflorus)* (scarlet runner-bean) : 144, 148, *149*, *150*, 176, *180*, 290, 414, *418*, 420, *427*, 790.
────── *vulgaris* (French bean, kidney bean) : *89*, 234, *239*, 265, *271*, 274, *280*, 282, 291, 294, 330, *427*, *501*, 560, 571, *579*, *580*, *588*, 601, *602*, 633, 695, *700*, *708*, *797*.
Phoma medicaginis var. *pinodella* : 147, 148.
Picea : 442.
Pinus sp. (pine) : *441*, 442.
Pisum sativum (pea) : 1, 20, 21, 22, 30, *36*, 52, *61*, 75, 95, 96, *116*, 145, *149*, 181, 227, *233*, 240, 241, 243, *247*, 252, *254*, 256, *264*, 291, 299, 306, *310*, 306, 318, 330, *394*, *405*, 407, *413*, 420, *426*, 431, *433*, 441, *500*, 518, 526, 534, *542*, 550, *559*, *570*, 633, *737*, *744*, 774.
Pisum sp. (black-eyed peas) : 216, *221*.
Pittosporum tobira : 764.
Plasmodiophora brassicae : 127, 130, 131, *132*.
Pleurochrysis scherfellii : *743*.
plum : (see *Prunus domesticus*).
Podocarpus : 302, 303, 431.
potato : (see *Solanum tuberosum*).
Prosopis juliflora (mesquite) : *609*.
Prunus amygdalus (almond) : 764, 774.
────── *armeniaca* (apricot) : 158, *165*, *298*, 764.
────── *avium* (cherry) : 764, *779*.
────── *domestica* (plum) : 764, *765*.
────── *persica* (peach) : 163, *165*, 764, 767, *774*, *779*, 795, *797*.
Pseudomonas fluorescens : 147.
────── *solanacearum* : *617*.
Pseudotsuga : 443.
Pyrrocoris : 35.
Pyrus communis (pear) : 207, *214*, 441, *447*, 483, *532*, 764, *765*, *779*.
────── *malus* (apple) : 166, 172, 173, *174*, 479, *483*, 510, *516*, *525*, *542*, 558, *559*, *560*, *609*, 764, *765*, 775, *778*, *779*, 786.

Raphanus raphanistrum (radish) : *195*, 274, 276, *532*.
Rattus (rat) : 181, *247*, *264*, *370*.
Rhizopus arrhizus : *598*.
Rhodnius prolixus : 35, *36*.
Rhoeo discolor : 274.
Ribes nigrum (black currant) : *335*, 764, *766*.
rice : (see *Oryza sativa*).
Ricinus communis (castor bean) : 136, *142*, 274, *278*, *427*.
Rosa sp. : 543, *560*.
────── *sherardi* (wild rose) : 273, *278*.
Rumex spp. (dock) : .64, 581.
────── *crispus* (dock) : 581, *588*.
────── *obtusifolius* (dock) : 585, *588*, *602*.
runner bean : (see *Phaseolus coccineus*).

Saccharomyces cerevisiae (yeast) : 181, *189*, 196, *205*, 248, *271*.
Saccharum officinarum (sugar cane) : 274, 276, *278*.
Salix viminalis (willow) : *757*.
Salvia occidentalis : *75*.
Scabiosa succisa : 746, *752*.
Secale cereale (rye) : 65, 66.
Sequoia gigantea
────── *sempervirens* : 442, *448*.

Silene armeria : 447.
Sinapis alba (white mustard) : 119, 223, *765*.
Solanum tuberosum(potato) : 214, *278*, *343*, 400, *427*, *447*, *448*, *559*, *560*.
soybean : (see *Glycine*).
Spinacia oleracea (spinach) : 181, 516, *597*.
Spirodela oligorrhiza : *394*.
Stereum sanguinolentum : *239*.
sunflower : (see *Helianthus annuus*).
Sus (pig) : *271*, *369*.
sycamore : (see *Acer pseudoplatanus*).

Taraxacum megallorhizon : 586.
─────── *officinale* (dandelion) : 571, *579*, 585, 587, *588*, *617*.
Tenebrio (flour beetle) : 35.
Therioaphis maculata : *757*.
Thuja : 442.
tobacco : (see *Nicotiana tabacum*).
tomato : (see *Lycopersicon*).
Toxoptera graminum : *757*.
Trichoderma viride : *396*.
Trifolium pratense (red clover) : *61*.
─────── *repens* (white clover) : 624, *632*.
─────── *subterraneum* (subterranean clover) : *343*, *508*.
Triticum aestivum (wheat) : 65, 76, 81, 196, *205*, 223, *227*, *254*, 273, 281, *290*, 291,
    298, *305*, 342, *351*, 388, *394*, *395*, 396, *405*, *406*, *433*, *508*, 589, *596*, *602*, *603*,
    618, *623*, *645*, 781, *789*.
Tropaeolum majus (nasturtium) : 118, *125*, 126, *427*, 571, *579*, 586, *588*, *597*, *617*,
    786.
Tsuga : 442.
Tuberolachnus salignus : *757*.

Vanda : *548*.
Verticillium albo-atrum : *90*.
Vicia Faba (broad bean) : *501*, *560*, 738, *743*.
Vitis Longii : 476.
───── *lubrusca* : 476.
───── *riparia* : 476.
───── *rupestris* : 786.
───── *vinifera* (grape, sultana) : 164, 476, *483*, 570, *602*, 661, *667*, *778*.

Weigela sp. : 764.
wheat : (see *Triticum*).

Xanthium sp. : *214*, *560*, *580*, *653*, *757*.
─────── *strumarium* L. : *753*.
Xanthomonas sp. : 147, *255*.

Zea mays (maize, corn) : 62, 65, 76, 145, 164, 166, 168, *173*, *174*, 176, 181, *189*,
    233, *239*, 241, *255*, 352, 442, *466*, *580*, *597*, *603*, 638, *653*, *678*, 705, *708*, 710,
    715, *716*, 717, *724*, *743*.
Zebrina pendula : 779.

INDEX TO SUBJECTS

Abscisic acid: 301, 431, 673.
——— ———, biosynthesis: 281 et seq.
——— ———, and flowering: 750.
——— ———, and growth stimulation: 273.
——— ———, and leaf unrolling: 635.
——— ———, and peach bud failure: 769 et seq.
——— ———, and RNA synthesis: 207 et seq, 219 et seq, 276, 649.
——— ———, and senescence: 581, 611 et seq, 618 et seq.
——— ———, effect on amylase synthesis: 336 et seq.
——— ———, effects on GA biosynthesis: 143, 642.
——— ———, in carob fruits: 795.
——— ———, in ripening fruits: 613.
——— ———, interaction with IAA: 431, 555.
——— ———, interaction with cytokinins: 213, 273, 555.
——— ———, interaction with gibberellins: 336, 555, 586, 634 et seq, 646 et seq.
Abscission: 234, 272 et seq, 493, 502 et seq, 546, 549, 558, 767 et seq.
Acid growth effect: 44, 66, 725 et seq.
Acid phosphatase: 196, 345.
Acrylic acid, as precursor of ethylene: 511.
Actinomycin D: 216 et seq, 228, 257, 358, 371 et seq, 388, 404, 408, 633, 647 et seq.
Adenine: 251, 462, 471, 563, 702, 786.
——— phosphoribosyl transferase: 460.
Adenosine: 251, 799.
——— aminohydrolase: 467 et seq.
——— diphosphate (ADP):   390, 392.
——— monophosphate (AMP):   780 et seq.
——— monophosphate, cyclic: (see CYCLIC AMP).
——— triphosphate (ATP):   390, 777.
ATP, and gibberellin biosynthesis: 152.
Affinity chromatography: 240.
Age, effects on senescence of leaves: 576, 589, 591 et seq, 608.
———, effects on leaf unrolling: 642.
β-alanine, and ethylene formation: 511.
Alanyl-tRNA: 181.
Alar (= B9 = N,N-dimethylsuccinamic acid):   139, 441, 558, 750, 776.
Aldehyde oxidase: 102.
Aleurone cells: 336-406.
——— grains: 346, 350.
n-alkylaminopurines, effects on cell division: 671.
Allosteric proteins: 262, 265 et seq.
Amiben (2,5-dichloro-3-aminobenzoic acid):   240.
AMO 1618:   138, 145, 420 et seq, 441, 639.
α-amylase: 223, 274, 336, 344, 352, 371, 378, 396, 369, 771.
——— synthesis, inhibition by O-phenanthroline: 383, 384.
———, isozymes: 352 et seq, 378.
β-amylase: 274, 350, 372, 378, 407.
Anthesin: 745 et seq.
Anthrone: 400.
Antibodies: 344 et seq, 662.
Anticodon: 186, 204, 248, 449.
Anti-gibberellin: 306 et seq, 750.
Antitumor, properties of inhibitor: 315.
Apex, shoot (see also Dominance, apical):   624 et seq.
Aphids: 753.
Arabinose: 537.
Arginase: 772.
Argon: 6.
Ascorbigen: 99.
Asparaginase: 772.

Asymmetry, of response to tropistic stimulation: 729, 730, 735 et seq.
Auxanometer: 52, 62, 68.
8-azaguanine: 750.
6-azauracil: 635.
Auxin, and ABA interaction: 273.
——, as allosteric effector: 265 et seq.
——, and cell extension: 1 et seq, 30, 53, 62 et seq, 68 et seq, 76 et seq, 628, 654.
——, and ethylene production: 520, 530, 531, 551.
——, and flowering: 772.
——, and lignification: 732 et seq.
——, and root formation: 526, 530.
——, and tendril coiling: 731 et seq.
——, and thigmotropism: 731 et seq.
——, binding to proteins: 110, 240, 256, 265.
——, binding to RNA: 205.
——, cell wall effects: 1, 5, 15, 19, 23.
——, concentration and embryo formation: 688 et seq.
——, conjugation: 96, 494, 707.
——, extraction: 494.
——, growth limiting proteins: 3.
——, (IAA) oxidase: 91, 110, 494.
——, inhibition of leaf unrolling: 633, 642.
——, interaction with blue light: 70.
——, interaction with cyclic AMP: 655 et seq.
——, interaction with ethylene: 537, 549 et seq.
——, NMR spectra: 128, 129.
——, rapid action: 44-81, 190 et seq.
——, respiration: 76.
——, RNA synthesis: 2, 73, 190 et seq, 196 et seq.
——, transport: 26, 64, 494, 502, 552, 701-716, 738, 785.
Azathioprine: 459.
8-azauracil: 371.
Azide: 11, 13.

Benzimidazole: 463.
Benzoic acid: 453.
3-benzyladenine: 251.
6-benzyladenine (6-BA):   181, 250, 251, 450, 467 et seq, 571, 582, 612, 629.
6-BA riboside: 450.
9-benzyladenine: 251.
6-benzyladenosine: 461.
Betacyanin: 459, 462.
Biennial bearing: 776.
Bioassay, for inhibitors of rooting: 324.
————, method for growth regulators: 299, 428, 768.
Bioelectric potential: 733 et seq.
Biosynthesis, abscisic acid: 281 et seq, 330 et seq.
—————, auxins: 82-132, 494.
—————, cytokinins: 449 et seq.
—————, ethylene: 510 et seq, 543 et seq.
—————, gibberellins: 133-180, 417, 425, 794 et seq.
Bud failure, in peach: 767.
Bulbs: 311.
N-butyl-9-hydroxy fluorene-9-carboxylate (MORPHACTIN: IT 3233):   635.

Calcium, in relation to flowering: 775 et seq.
——, requirement in amylase synthesis: 376, 385, 404.
Callose, and senescence: 601.
Callus tissues: 190 et seq, 277, 459, 468, 484 et seq, 668-694, 738.
Cambium, growth: 443, 615.

Carbohydrate metabolism, auxins and: 697.
——————— ———————, cytokinins and: 575.
——————— ———————, gibberellins and: 407 et seq.
Carbon dioxide, effect on growth: 44, 522, 523, 725, 735.
Carboxymethyluridine: 196.
Carvadan: 139.
Casein hydrolysate: 668.
Catechol: 530.
Caulocaline: 624.
Cell, division: 224 et seq, 408, 428, 435, 484, 669 et seq, 771.
—— extension: 1-81, 113, 223, 412.
——, expansion: 52, 431, 634 et seq, 654 et seq, 669 et seq.
Cell-free system, protein synthesis by: 354.
Cellulase: 196, 234 et seq, 274, 275, 410, 500, 535.
Cell wall, extensibility: 1 et seq, 7, 8, 17 et seq, 44 et seq, 49, 58, 62 et seq, 535.
—— ——, metabolism: 499, 538, 598 et seq.
—— ——, polysaccharides: 52 et seq, 537 et seq, 598 et seq.
Cell, water relations; water potential: 9, 11, 23, 25, 28, 59, 76.
Chemical creep, and cell wall extension: 7, 58.
Chloramphenicol: 226, 251, 356, 429, 634, 647.
Chlor-Choline Chloride (CCC; 'Cycocel'):   138, 143, 420 et seq, 435 et seq, 477, 480, 639, 657, 750.
2-chloroethyl phosphonic acid: (see Ethrel).
p-chloromercuribenzoate: 236.
Chlorophyll, breakdown: 549, 564, 565, 573, 581 et seq, 600.
Chlorophyll, increase in darkness: 608.
Chloroplast: 226, 248, 260, 568, 589, 596, 599, 619, 643.
——————————, lamellae proteins: 590.
——————————, ribosomes: 620 et seq, 643, 649.
——————————, synthesis of gibberellins: 133, 643.
Choline: 139.
—————— chloride: 420.
Cholodny-west theory of tropisms: 705.
Chromatin: 207 et seq, 222, 240, 242, 263, 408, 618.
Citrate synthase: 265 et seq.
Clones, differences between: 668.
Club roots, auxins in: 127 et seq.
Coconut milk, growth regulators in: 668, 679.
Colchicine: 223.
Computer, application: 56, 496, 790 et seq, 798 et seq, 802 et seq.
Conditioned medium: 683, 684.
Condurit B epoxide: 146.
Conjugation of hormones: 96, 144.
Copalyl pyrophosphate: 136.
Cortisone: 376.
CTP: 390, 392.
Cyanide, (KCN):   77, 112.
Cyclic AMP (3',5'-adenosine monophosphate):   376, 582, 654 et seq.
Cyclofarnesol: 287.
Cycloheximide: 251, 543, 567, 634, 647, 727.
——————————, and cell extension: 3, 20, 21, 47, 53, 68, 76, 216 et seq, 727.
——————————, inhibition of ethylene production: 546.
——————————, inhibition of leaf unrolling: 633-653.
Cysteinyl-tRNA: 181.
Cystine, incorporation into glucobrassicin: 120.
Cytokinin, and ABA, interaction: 273.
——————, and apical dominance: 561, 624 et seq.
——————, and leaf unrolling: 641 et seq.
——————, and RNA synthesis: 181 et seq, 449 et seq.
——————, binding to ribosomes: 248, 596.

Cytokinin, bioassay: 181, 300, 477, 561.
Cytokinin-directed transport: 572, 573.
Cytokinin, in bleeding sap: 476 et seq, 561 et seq, 571, 629.
—————, inhibition of senescence: 561-597.
—————, in RNA hydrolysate: 798.
—————, metabolism of: 449 et seq, 467 et seq.
—————, mode of action: 186, 248 et seq, 449 et seq, 459 et seq, 484 et seq.
—————, transport: 701 et seq.
Cytokinins, non-purine: 454.
—————, protein synthesis: 248 et seq.
—————, 9-substituted: 454, 459 et seq.
—————, rapid action: 248, 253.
Cytosine monophosphate (CMP):   780 et seq.

Day-neutral plants: 745 et seq.
Deaminases: 467.
Dehiscence, ethylene effects on: 502.
Deoxyglucose, inhibitor of polysaccharide synthesis: 59.
Desthioglucobrassicin: 120, 121.
Dieback, of citrus: 548.
Diethyl pyrocarbonate: 357, 407.
Di-isopropyl fluro-phosphate (DFP):   566.
N,N-Dimethyl succinamic acid: 139, (see also ALAR).
DNA, content: 223 et seq, 428 et seq.
DNA -RNA, hybridisation: 2 et seq.
———, synthesis: 222, 277.
2,4-dichlorophenoxyacetic acid (2,4-D):   240, 330, 430, 435, 655, 686 et seq, 704.
2,4-D-lysine: 240.
3,5-dichlorophenoxyacetic acid (3,5-D):   330.
Dimethyl sulphoxide: 772.
2,4-dinitrophenol (DNP):   698.
Diphenylurea: 187, 450.
Diterpenes: 134, 152, 426, 431.
Dithionite, inhibition of AAld oxidase: 106.
Dominance, apical: 441 et seq, 549, 561.
Dormancy: 272 et seq, 342, 479, 506, 507, 767 et seq.
Dwarfing compounds, interaction with ethylene: 558.
Dwarf plants: 98, 144 et seq, 166 et seq, 407, 414, 637.
Dwarfing compounds: 138, 139, 143 et seq, 330 et seq, 420 et seq, 435 et seq, 477, 558, 639, 657, 750.

Ehrlich carcinoma: 315.
Electrofocusing, gel: 352.
Embryoids: 679.
Embryo, production from cultured cells: 679, 686 et seq.
Embryos, RNA metabolism in: 207 et seq, 216 et seq.
Endoplasmic reticulum: 344, 393, 497, 622, 738.
Endosperm: 133, 151 et seq, 792 et seq.
Enzymes, IAA biosynthesis: 88, 91 et seq.
—————, inhibition by ABA: 274.
Epinasty, induced by BuBA: 464.
—————, induced by ethylene: 549.
Esterase: 350.
Estradiol: 369.
Estrogen: 369.
Ethionine, as ethylene precursor: 511.
Ethrel (2-chloroethyl phosphonic acid, CEP):   558, 611.
Ethylene, ABA antagonism: 342.
—————, and auxin: 493 et seq, 502, 518, 526, 549 et seq, 676.
—————, and growth: 300 et seq, 493, 518, 526, 534 et seq, 549 et seq, 675 et seq.
—————, and hormone transport: 552.

Ethylene, and root formation: 526, 528.
————, and senescence: 581, 604 et seq.
————, and tendril coiling: 735.
————, effects on abscission: 502 et seq, 556 et seq.
————, effect on amylase synthesis: 336 et seq.
————, mode of action: 493, 549 et seq.
————, production by callus tissues: 675 et seq.
Ethionine, effects on barley aleurone system: 365, 369.
Ethylene diamine tetraacetic acid (EDTA):   385.
Ethyl-3-indoleacetate: 112.
n-ethyl maleimide: 236.

Fatty acid synthetase, ABA and: 275.
Florigen: 745 et seq, 775.
Florigenic acid: 746.
Flowers, sex affected by ethylene: 549.
Flowering, hormones and: 320, 549, 745-779.
————, woody species: 758-779.
Fluorenols, (morphactins):   318 et seq.
5-fluorodeoxyuridine: 223.
p-fluorophenylalanine: 355.
5-fluorouracil: 647, 750.
Fraction-1 protein: 589 et seq, 622.
Fraction-2 proteins: 589 et seq.
Freeze-sectioning: 344.
β-fructofuranosidase: 409, 410.
Fruits: 145, 273, 281 et seq, 502 et seq, 543, 557, 611 et seq, 668.
————, ethylene production by: 545, 557.
Fruit, production, effects of rootstocks: 482.
———— ripening: 549, 558, 611 et seq, 780 et seq.
Fujenal: 135.
Fungal infection, and ethylene production: 608.

Gamma-irradiated seedlings: 223.
Gas chromatography of hormones: 162, 445, 518 et seq, 605, 767 et seq, 790-807.
——————————, mass spectrometry: 135, 163, 790-807.
Geotropism: 441, 705, 710-730, 738 et seq.
Geranyl-geranyl pyrophosphate (GGPP):   425, 426.
Germination: 207, 216 et seq, 272 et seq, 388, 502, 549.
Gibberellic acid, effects on amylase synthesis: see "Aleurone".
—————— glucoside: 147.
——————————, interaction with coconut milk: 673.
Gibberellin, as Florigen precursor: 746.
————————, bioassay: 167, 176, 414, 420, 612, 637, 763.
————————, cell expansion: 25, 302.
————————, decrease in fruit ripening: 611.
————————, effects on auxin content: 99, 100, 641, 642.
————————, effects on DNA synthesis: 222 et seq.
————————, effects on flowering: 746 et seq, 760 et seq.
————————, effects on metabolism: 407.
————————, effects on shoot growth: 98, 223, 306, 407, 441 et seq, 627 et seq.
———————— glucosides: 144, 176.
————————, inhibition of flowering: 764.
————————, inhibition of senescence: 582.
————————, in seeds: 791 et seq.
————————, interaction with ABA: 674.
————————, interaction with auxin: 441.
————————, interaction with cytokinin: 441.
————————, interaction with ethylene: 336 et seq, 550.
————————, interaction with light: 98.
————————, microbial degradation: 147.

Gibberellin, oxidation in vitro: 793 et seq.
——————, replacement of red light by: 633 et seq.
——————, transport: 701 et seq.
—————— action, inhibitor of: 306.
β-1,3-glucan hydrolase: 196, 598 et seq.
Glucobrassicin: 99, 117, 130.
α-Glucosidase: 350.
β-Glucosidase, inhibition by Condurit B epoxide: 146.
Glucosinolates: 117, 118.
Glucotropaeolin: 118.
Glutaminase, and peach flowering: 772.
Glutamine: 671.
Golgi apparatus: 497, 738 et seq.
—————————, as geosensor: 738.
Gonadotropin, chorionic: 661 et seq.
Grafting, experiments and flowering: 747.
Growth, acid effect: 44.
——————, cell division: 408, 428 et seq, 435, 484, 625, 669 et seq.
——————, co-ordination: 441 et seq, 624 et seq.
——————, cell expansion: 1, 6, 23, 534 et seq, 300 et seq, 408, 624, 625, 669 et seq.
——————, and flowering: 758-779.
——————, kinetics: 9, 52, 56, 62, 68.
——————, rate: 9 et seq, 53, 68 et seq.
——————, stimulated by CCC: 435 et seq.
——————, stored: 47.
——————, temperature coefficient: 47, 49, 63.
Guanosine: 799.
—————— monophosphate (GMP):   780 et seq.
—————— triphosphate (GTP):   390, 392.

Haemagglutination: 661 et seq.
Heliangine: 673.
Helminthosporol: 352.
Hemicellulase: 196.
Hemicelluloses: 59.
Hexitol: 668.
Honeydew, aphid, hormones in: 753.
Hormone binding: 240-271.
Hydrocortisone: 376.
Hydrogen ion, effect on growth: (see acid growth effect).
β-hydroxykaurenoic acid: 793.
L-hydroxyproline, as inhibitor of growth: 673.
Hydroxyproline, in cell walls: 50, 537.
β-hydroxypropionic acid, and ethylene formation: 511.
8-hydroxyquinoline sulphate: 383.

Immuno-absorption: 662 et seq.
Immunofluorescence technique: 345.
Indol: 112.
β-indoleacetic acid (IAA): see auxin.
Indoleacetic acid glycoside: 668, 669.
—————————— lysine (IAA-lysine):   240.
—————————— oxidase: 91, 110 et seq.
Indoleacetaldehyde: 82, 91, 102, 530.
3-Indoleacetaldoxime (IAOX):   117.
Indoleacetamide: 130.
Indoleacetonitrile (IAN):   117 et seq, 127, 530.
Indoleacetyl aspartate: 96, 499, 530, 707.
Indole butyric acid (IBA):   527.
Indole-3-carboxylaldehyde: 112.
Indole-3-ethanol: 91.

Indolelactic acid: 82, 84, 91.
Indolenine: 115.
Indolepyruvic acid: 82, 91, 772.
Inhibition, of growth by light: 72.
—————, of lateral bud outgrowth: 441 et seq.
Inhibitor "β":   272, 291, 613.
Inhibitors, DNA synthesis: 222 et seq.
——————, growth: 272, 281, 291, 299, 306, 311, 318, 324, 330, 521, 523, 534, 553, 615 et seq, 639, 657, 673, 717 et seq, 767 et seq, 769.
——————, metabolic: 11, 13, 59, 76, 146, 260, 268 et seq, 698.
——————, nuclease: 188.
——————, of flowering: 750, 756, 758, 776 et seq.
——————, of proteases: 566.
——————, protein synthesis: 3, 20, 21, 47, 53, 68, 73, 76, 216 et seq, 226, 228 et seq, 251, 252, 355, 358, 371 et seq, 429, 543, 633 et seq, 647, 727, 750, 771.
——————, transport: 302, 717 et seq.
Inosine monophosphate (IMP):   474.
Inositol: 668 et seq, 681.
INSTRON, technique: 1, 2, 17, 44, 52.
Inumakilactone: 303, 433.
Invertase: 274, 429.
IPA (γ,γ-dimethylallyladenosine):   181.
Iron, requirement for amylase synthesis: 380.
———, requirement for callus tissue growth: 668.
Isatan B: 117.
Isocitratase: 216.
Isopentenyladenine (IPA):   181 et seq, 455.
Isopentenyladenosine: 469.
Isopentenyl, pyrophosphate: 181, 459.
Isopropyl N-(3-chlorophenyl carbamate) (CIPC):   634.
Isoenzymes: 228 et seq, 234 et seq, 352 et seq, 378, 534.
Isotherm, binding: 257.

Kaurene: 134 et seq, 151 et seq, 168, 424.
————— synthetase: 136 et seq.
Kaurenal: 152 et seq.
Kaurenoic acid: 134, 135, 151, 168, 418.
Kinetin, (Furfuryladenine):   251, 430, 435, 460, 468, 550, 561, 589, 622, 641.
—————, binding to ribosomes: 250.
—————, and RNA: 187, 207 et seq, 212, 467.
—————, and root initiation: 527.
—————, inhibition of embryo formation: 692.
—————, stimulation of starch synthesis: 622.

Laminarin: 598.
Leaf, and production of flower hormones: 746 et seq, 753 et seq.
——, effect of removal on apical growth: 626.
——, ethylene production in: 543.
——, inhibitors in: 291, 642 et seq.
——, senescence in: 561-610, 618-623.
——, unrolling: 633-653.
Leptospermone: 325.
Light, and ethylene production: 554.
——, and floral inhibitor: 778.
—— and gibberellin production: 633.
——, effects on IAA synthesis: 97, 554.
——, effects on growth: 68, 96 et seq, 633-653.
Limonene derivatives: 139.
Lineweaver-Burk analysis: 336 et seq.
Lipids, association with AAld oxidase: 106.

Lipids, and membranes: 30-43.
Long-day plants: 745 et seq, 754 et seq.
Long-short day plants: 745 et seq.
Lycoricidine: 311.
Lycoricidinol: 303, 311, 431.
Lysosomes: 404.

Maleic acid hydrazide: 750.
Malic enzyme: 604.
Mannitol: 6, 76.
Mass spectrography of hormones: 128, 326, 605, 790-807.
Membranes: 30 et seq, 37 et seq, 332, 394, 405, 620, 777 et seq.
—————, association of enzymes with: 108.
6-mercaptopurine: 459.
p-Mercuribenzoate: 268, 269.
Meristem: 223, 224.
Methionine, ethylene precursor: 510.
Methoxyglucobrassicin: 130.
Methoxy indoleacetonitrile: 128.
Methyl abscisate: 282.
Methylase, tRNA: 365.
6-(3-methyl-2-butenylamino)-purine (2iP):   484, 798.
Methyl glyoxal: 771.
Methylmercaptan: 514.
S-methylmethionine: 513.
6-methylpurine: 371.
Methylthionyl-IPA: 181.
Mevalonate: 152, 168, 181, 187, 286, 459.
Microbodies: 619 et seq.
Microtubules: 223.
Mitochondria: 248, 265 et seq.
Morphactin: 302, 318, 635, 750.
6-Morpholinopurine: 251.
Myo-inositol: 668.
Myrosinase: 117 et seq.

NADP oxidoreductase: 604.
Nagilactone: 302, 433.
Naphthalene acetic acid (NAA):   443.
Naringenin: 769.
Neoglucobrassicin: 117, 130.
Neoxanthin: 333.
Nitrate reductase: 274.
Nitrilase: 124.
Nitrogen nutrition, effects on flowering: 775 et seq.
—————, nutritional effects on growth: 443, 667 et seq.
NMR spectra: 128, 163, 178, 284, 313, 326, 460.
Nuclease, IPA as inhibitor of: 188.
Nucleotides: 388 et seq, 780 et seq.
Nucleus, binding of auxin to: 256.

Optical rotatory dispersion: 285, 291, 314, 334.
Ovule, Abscisic acid in: 219.
2-oxindole: 112.

Pachyman: 598.
Paramylon: 598.
Parthenocarpy, ABA and: 273.
—————, induced by GA, NAA: 443, 445.
Pectinase: 274, 410.
PEP, and gibberellin biosynthesis: 152.

Peroxidase: 196, 228 et seq, 274, 345, 352, 534, 537, 604 et seq.
Petals, ethylene produced by: 543.
6-PG dehydrogenase: 604, 606.
PH, optimum for amylase synthesis: 380 et seq, 404.
Phaseic acid: 282.
O-phenanthroline: 376, 383.
Phenazine methosulphate: 102.
Phenolic acids, in peach: 769.
Phenolic compounds, and root initiation: 528, 530.
Phenylalanine ammonia lyase (PAL): 274, 275.
Phenylalanyl-tRNA: 185.
Phenylmethylsulfonyl fluoride (PMS): 566.
Phloem, effects of auxin on: 695 et seq.
Phosphate metabolism, cytokinin and: 575.
────── nutrition, in relation to flowering: 775 et seq.
3-phospho D glycerate carboxylase: 604.
Phospholipid synthesis: 393.
Phosphon D: 138, 331.
────── S: 139.
Photoperiod, effects on geotropism: 443.
Photoperiodism: 745-757.
Photosynthesis, enhanced by cytokinins: 575.
Phthalates, interference in GC-MS: 795.
Phytoalexin: 307.
Phytochrome, and morphogenesis: 633, 776.
Phytotropin: 661 et seq.
Pisatin, inhibitor of GA action: 307.
Plagiotropism: 441-443.
Plastoglubuli: 620 et seq.
Podolactones: 299 et seq, 431.
Polarity, of transport: 701.
Pollen, ethylene produced by: 543, 544.
Polysomes: 651.
Polycarpic plants: 758.
Polyteny: 224.
Porapak: 518.
Proembryonic cultures: 682 et seq.
Proline: 673, 675.
6-propylaminopurine: 468.
Protease: 216, 274, 350, 376, 563 et seq.
Protein, growth limiting: 2 et seq, 47, 55.
────── synthesis: 66, 73, 190, 219, 222, 228, 248 et seq, 276, 354, 561 et seq, 589 et seq, 600, 696.
Proteolysis: 563, 571.
Purafil: 495.
Puromycin: 252, 634.

Quaternary ammonium compounds: 331.

Reducing sugars, and callus growth: 485 et seq.
Respiration: 76, 375, 506, 573, 608, 696.
Retardants, see Dwarfing compounds.
Rhamnose: 669.
Ribonuclease (RNA-ase): 274, 350, 355, 376, 393.
Ribosomes: 248 et seq, 276, 596, 619, 643, 646.
Ringing, effect on apical growth: 628.
RNA, messenger: 1 et seq, 66, 196, 216 et seq, 362, 388, 652.
──, ribosomal: 194, 276, 409, 648.
──, soluble: 196 et seq, 276, 409, 648.
──, transfer: 181 et seq, 365 et seq, 248, 276, 449, 459, 798.
── polymerase, effects of growth substances: 190, 207 et seq, 277, 408, 646, 649.

RNA, synthesis: 1. 2, 181, 190 et seq, 207 et seq, 216 et seq, 242, 256 et seq, 276, 277, 356 et seq, 371 et seq, 378, 408, 467, 646 et seq.
Root apex, as source of growth substances: 414 et seq, 719.
Root-cap: 705, 717 et seq.
Root formation, callus cultures: 686 et seq.
—— growth: 315, 318 et seq, 522.
Rooting, (adventitious root production):   324, 526 et seq, 547, 661 et seq.
Roots, and gibberellin metabolism: 414, 479, 624, 629.
——, and cytokinin biosynthesis: 476 et seq, 561 et seq, 571, 624, 629 et seq.
——, auxin synthesis in: 127 et seq.
——, ethylene production by: 520.
——, geotropism in: 717 et seq.
——, hormone transport in: 705.
——, in relation to flower hormones: 746.
——, RNA synthesis in: 212, 369.
Rootstocks, influence on growth of scions: 476.
RUDP carboxylase: 604, 607.

Sap, xylem: 476, 479, 561.
Schardakov method, water potential by: 23, 59.
Seeds: 144, 216 et seq, 336-396, 502 et seq, 133, 151, 155, 158, 166, 175 et seq, 182, 207, 216 et seq, 223, 273, 282, 306, 505, 790 et seq.
Senescence: 561-618, 707.
————— factor: 707.
L-serine, anti-cytokinin action: 562 et seq.
Seryl-tRNA: 181, 185.
Shoot types, in woody plants: 758 et seq.
Short-day plants: 745 et seq, 754 et seq.
Short-long day plants: 745 et seq.
Skatol: 112.
Solanidine: 673.
Sorbitol: 682.
Spherosomes: 599.
Starch phosphorylase: 410.
Steroid synthesis inhibitors: 139.
Steviol: 168.
Stigma, ethylene produced by: 544.
Stomata: 295.
Sulphoglucobrassicin: 117.

Temperature, and amylase synthesis: 378.
————— coefficient: 47, 49, 63.
—————, optimum for amylase: 378.
—————, root, and cytokinin production: 477.
————— stepdown incubation: 380.
Tendrils: 731 et seq.
Thigmotropism: 731 et seq.
2-thiouracil: 750.
Thymidine: 786.
Thyroxine: 240.
Tin, enzyme inhibition by: 105.
Tissue cultures: 484 et seq, 668-694.
Transpiration: 446.
—————, effect of ABA: 297.
Transport, hormone: 23 et seq, 330, 441, 499, 502, 552, 701-724.
—————, hormone directed: 441, 571, 695 et seq.
—————, lateral, of inhibitor in roots: 720.
Tree form: 442.
Trees, flowering of: 767 et seq, 775 et seq.
2,4,5-trichlorophenoxyacetic acid (2,4,5-T):   523, 558.
2,4,6-trichlorophenoxyacetic acid (2,4,6-T):   523.

2,4,5-trichlorophenoxypropionic acid (TP):   559.
Tri-iodobenzoic acid: 302, 750.
Tryptamine: 82, 100.
Tryptophan: 82, 100, 124, 494.
Tryptophol: 82.
Tyrosyl-tRNA: 181, 185.
Turgor pressure: 9, 11 et seq.

UDP-glucose: 390.
Ultrastructure: 344 et seq, 401-403, 493, 538 et seq, 601, 618 et seq.
Uracil: 786.
Uridine monophosphate (UMP):   390, 780 et seq.
Uptake, auxin: 64.
———, cytokinin: 467.
———, auxin effects on: 697 et seq.
UTP: 390, 392.

Vacuoles: 599, 601, 618.
Valyl-tRNA: 185, 186, 196.
Violaxanthin: 288, 333.
Virus infection: 543, 600, 767.

Waterlogging: 291.
Wilting: 288, 291.
Wounding, and ethylene production: 524, 543 et seq, 604.
———, and root formation: 526.
———, effects on tissue sugars: 486.
———, effects on ribonuclease: 393.

Xanthoxin:'   333.

Zeatin: 252, 300, 477, 582, 641, 668, 681, 704.
Zeatin nucleoside: 477.
——— nucleotide: 477.
——— riboside: 469, 798.